ENCYCLOPEDIA OF
INFANT AND EARLY CHILDHOOD DEVELOPMENT

ENCYCLOPEDIA OF INFANT AND EARLY CHILDHOOD DEVELOPMENT

EDITORS-IN-CHIEF

MARSHALL M. HAITH
and
JANETTE B. BENSON
Department of Psychology, University of Denver,
Denver, Colorado, USA

ELSEVIER

AMSTERDAM • BOSTON • HEIDELBERG • LONDON • NEW YORK • OXFORD
PARIS • SAN DIEGO • SAN FRANCISCO • SINGAPORE • SYDNEY • TOKYO
Academic Press is an imprint of Elsevier

ACADEMIC
PRESS

Academic Press is an imprint of Elsevier
The Boulevard, Langford Lane, Kidlington, Oxford OX5 1GB, UK
525 B Street, Suite 1900, San Diego, CA 92101-4495, USA

First edition 2008

British Library Cataloguing in Publication Data
A catalogue record for this book is available from the British Library

Library of Congress Catalog Number: 2007930619

ISBN: 978-0-12-370460-3

For information on all Elsevier publications
visit our website at books.elsevier.com

PRINTED AND BOUND IN CANADA
07 08 09 10 11 10 9 8 7 6 5 4 3 2 1

EDITORS-IN-CHIEF

Marshall M. Haith received his M.A. and Ph.D. degrees from U.C.L.A. and then carried out postdoctoral work at Yale University from 1964–1966. He served as Assistant Professor and Lecturer at Harvard University from 1966–1972 and then moved to the University of Denver as Professor of Psychology, where he has conducted research on infant and children's perception and cognition, funded by NIH, NIMH, NSF, The MacArthur Foundation, The March of Dimes, and The Grant Foundation. He has been Head of the Developmental Area, Chair of Psychology, and Director of University Research at the University of Denver and is currently John Evans Professor Emeritus of Psychology and Clinical Professor of Psychiatry at the University of Colorado Health Sciences Center.

Dr. Haith has served as consultant for Children's Television Workshop (Sesame Street), Bilingual Children's Television, Time-Life, and several other organizations. He has received several personal awards, including University Lecturer and the John Evans Professor Award from the University of Denver, a Guggenheim Fellowship for serving as Visiting Professor at the University of Paris and University of Geneva, a NSF fellowship at the Center for Advanced Study in the Behavioral Sciences (Stanford), the G. Stanley Hall Award from the American Psychological Association, a Research Scientist Award from NIH (17 years), and the Distinguished Scientific Contribution Award from the Society for Research in Child Development.

Janette B. Benson earned graduate degrees at Clark University in Worcester, MA in 1980 and 1983. She came to the University of Denver in 1983 as an institutional postdoctoral fellow and then was awarded an individual NRSA postdoctoral fellowship. She has received research funding form federal (NICHD; NSF) and private (March of Dimes, MacArthur Foundation) grants, leading initially to a research Assistant Professor position and then an Assistant Professorship in Psychology at the University of Denver in 1987, where she remains today as Associate Professor of Psychology and as Director of the undergraduate Psychology program and Area Head of the Developmental Ph.D. program and Director of University Assessment. Dr. Benson has received various awards for her scholarship and teaching, including the 1993 United Methodist Church University Teacher Scholar of the Year and in 2000 the CASE Colorado Professor of the Year. Dr. Benson was selected by the American Psychological Association as the 1995–1996 Esther Katz Rosen endowed Child Policy Fellow and AAAS Congressional Science Fellow, spending a year in the United States Senate working on Child and Education Policy. In 1999, Dr. Benson was selected as a Carnegie Scholar and attended two summer institutes sponsored by the Carnegie Foundation program for the Advancement for the Scholarship of Teaching and Learning in Palo Alto, CA. In 2001, Dr. Benson was awarded a Susan and Donald Sturm Professorship for Excellence in Teaching. Dr. Benson has authored and co-authored numerous chapters and research articles on infant and early childhood development in addition to co-editing two books.

EDITORIAL BOARD

FOREWORD

This is an impressive collection of what we have learned about infant and child behavior by the researchers who have contributed to this knowledge. Research on infant development has dramatically changed our perceptions of the infant and young child. This wonderful resource brings together like a mosaic all that we have learned about the infant and child's behavior. In the 1950s, it was believed that newborn babies couldn't see or hear. Infants were seen as lumps of clay that were molded by their experience with parents, and as a result, parents took all the credit or blame for how their offspring turned out. Now we know differently.

The infant contributes to the process of attaching to his/her parents, toward shaping their image of him, toward shaping the family as a system, and toward shaping the culture around him. Even before birth, the fetus is influenced by the intrauterine environment as well as genetics. His behavior at birth shapes the parent's nurturing to him, from which nature and nurture interact in complex ways to shape the child.

Geneticists are now challenged to couch their findings in ways that acknowledge the complexity of the interrelation between nature and nurture. The cognitivists, inheritors of Piaget, must now recognize that cognitive development is encased in emotional development, and fueled by passionately attached parents. As we move into the era of brain research, the map of infant and child behavior laid out in these volumes will challenge researchers to better understand the brain, as the basis for the complex behaviors documented here. No more a lump of clay, we now recognize the child as a major contributor to his own brain's development.

This wonderful reference will be a valuable resource for all of those interested in child development, be they students, researchers, clinicians, or passionate parents.

T. Berry Brazelton, M.D.
Professor of Pediatrics, Emeritus Harvard Medical School
Creator, Neonatal Behavioral Assessment Scale (NBAS)
Founder, Brazelton Touchpoints Center

PREFACE

Encyclopedias are wonderful resources. Where else can you find, in one place, coverage of such a broad range of topics, each pursued in depth, for a particular field such as human development in the first three years of life? Textbooks have their place but only whet one's appetite for particular topics for the serious reader. Journal articles are the lifeblood of science, but are aimed only to researchers in specialized fields and often only address one aspect of an issue. Encyclopedias fill the gap.

In this encyclopedia readers will find overviews and summaries of current knowledge about early human development from almost every perspective imaginable. For much of human history, interest in early development was the province of pedagogy, medicine, and philosophy. Times have changed. Our culling of potential topics for inclusion in this work from textbooks, journals, specialty books, and other sources brought home the realization that early human development is now of central interest for a broad array of the social and biological sciences, medicine, and even the humanities. Although the 'center of gravity' of these volumes is psychology and its disciplines (sensation, perception, action, cognition, language, personality, social, clinical), the fields of embryology, immunology, genetics, psychiatry, anthropology, kinesiology, pediatrics, nutrition, education, neuroscience, toxicology and health science also have their say as well as the disciplines of parenting, art, music, philosophy, public policy, and more.

Quality was a key focus for us and the publisher in our attempts to bring forth the authoritative work in the field. We started with an Editorial Advisory Board consisting of major contributors to the field of human development – editors of major journals, presidents of our professional societies, authors of highly visible books and journal articles. The Board nominated experts in topic areas, many of them pioneers and leaders in their fields, whom we were successful in recruiting partly as a consequence of Board members' reputations for leadership and excellence. The result is articles of exceptional quality, written to be accessible to a broad readership, that are current, imaginative and highly readable.

Interest in and opinion about early human development is woven through human history. One can find pronouncements about the import of breast feeding (usually made by men), for example, at least as far back as the Greek and Roman eras, repeated through the ages to the current day. Even earlier, the Bible provided advice about nutrition during pregnancy and rearing practices. But the science of human development can be traced back little more than 100 years, and one can not help but be impressed by the methodologies and technology that are documented in these volumes for learning about infants and toddlers – including methods for studying the role of genetics, the growth of the brain, what infants know about their world, and much more. Scientific advances lean heavily on methods and technology, and few areas have matched the growth of knowledge about human development over the last few decades. The reader will be introduced not only to current knowledge in this field but also to how that knowledge is acquired and the promise of these methods and technology for future discoveries.

CONTENTS

Several strands run through this work. Of course, the nature-nurture debate is one, but no one seriously stands at one or the other end of this controversy any more. Although advances in genetics and behavior genetics have been breathtaking, even the genetics work has documented the role of environment in development and, as Brazelton notes in his foreword, researchers acknowledge that experience can change the wiring of the brain as well as how actively the genes are expressed. There is increasing appreciation that the child develops in a transactional context, with the child's effect on the parents and others playing no small role in his or her own development.

There has been increasing interest in brain development, partly fostered by the decade of the Brain in the 1990s, as we have learned more about the role of early experience in shaping the brain and consequently, personality, emotion, and intelligence. The 'brainy baby' movement has rightly aroused interest in infants' surprising capabilities, but the full picture of how abilities develop is being fleshed out as researchers learn as much about what infants can not do, as they learn about what infants can do. Parents wait for verifiable information about how advances may promote effective parenting.

An increasing appreciation that development begins in the womb rather than at birth has taken place both in the fields of psychology and medicine. Prenatal and newborn screening tools are now available that identify infants at genetic or developmental risk. In some cases remedial steps can be taken to foster optimal development; in others ethical issues may be involved when it is discovered that a fetus will face life challenges if brought to term. These advances raise issues that currently divide much of public opinion. Technological progress in the field of human development, as in other domains, sometimes makes options available that create as much dilemma as opportunity.

As globalization increases and with more access to electronic communication, we become ever more aware of circumstances around the world that affect early human development and the fate of parents. We encouraged authors to include international information wherever possible. Discussion of international trends in such areas as infant mortality, disease, nutrition, obesity, and health care are no less than riveting and often heartbreaking. There is so much more to do.

The central focus of the articles is on typical development. However, considerable attention is also paid to psychological and medical pathology in our attempt to provide readers with a complete picture of the state of knowledge about the field. We also asked authors to tell a complete story in their articles, assuming that readers will come to this work with a particular topic in mind, rather than reading the Encyclopedia whole or many articles at one time. As a result, there is some overlap between articles at the edges; one can think of partly overlapping circles of content, which was a design principle inasmuch as nature does not neatly carve topics in human development into discrete slices for our convenience. At the end of each article, readers will find suggestions for further readings that will permit them to take off in one neighboring direction or another, as well as web sites where they can garner additional information of interest.

AUDIENCE

Articles have been prepared for a broad readership, including advanced undergraduates, graduate students, professionals in allied fields, parents, and even researchers for their own disciplines. We plan to use several of these articles as readings for our own seminars.

A project of this scale involves many actors. We are very appreciative for the advice and review efforts of members of the Editorial Advisory Board as well as the efforts of our authors to abide by the guidelines that we set out for them. Nikki Levy, the publisher at Elsevier for this work, has been a constant source of wise advice, consolation and balance. Her vision and encouragement made this project possible. Barbara Makinster, also from Elsevier, provided many valuable suggestions for us. Finally, the Production team in England played a central role in communicating with authors and helping to keep the records straight. It is difficult to communicate all the complexities of a project this vast; let us just say that we are thankful for the resource base that Elsevier provided. Finally, we thank our families and colleagues for their patience over the past few years, and we promise to ban the words "encyclopedia project" from our vocabulary, for at least a while.

Marshall M. Haith

and

Janette B. Benson
Department of Psychology, University of Denver
Denver, Colorado, USA

HOW TO USE THE ENCYCLOPEDIA

The Encyclopedia of Infant and Early Childhood Development is intended for use by students, research professionals, and interested others. Articles have been chosen to reflect major disciplines in the study of infant and early child development, common topics of research by academics in this domain, and areas of public interest and concern. Each article serves as a comprehensive overview of a given area, providing both breadth of coverage for students, and depth of coverage for research professionals. We have designed the encyclopedia with the following features for maximum accessibility for all readers.

Articles in the encyclopedia are arranged alphabetically by subject in the Contents list. The index is located in volume 3. Some topics are covered in a multitude of articles from differing perspectives, while other topics may have only one entry. We encourage use of the index for access to a subject area, rather than use of the Contents list alone, so that a reader has a full notion of the coverage of that topic. The influence of the family on an infant, for example, may be covered under separate articles on Family Dynamics, Birth Order, Siblings and Sibling Rivalry, Family Influences, and Parenting Styles and Their Effects. A reader searching under F for family in the Contents list would easily find one of these articles but would miss the others.

Each article contains a glossary, cross-references to other related encyclopedia articles, and suggested readings where applicable, and relevant websites for additional information. The glossary contains terms that may be unfamiliar to the reader, with each term defined *in the context of its use in that article*. Thus, a term may appear in the glossary for another article defined in a slightly different manner or with a subtle nuance specific to that article. For clarity, we have allowed these differences in definition to remain so that each article is fully understandable on its own.

Each article has been cross-referenced to other related articles in the encyclopedia at the close of each article. We encourage readers to use the cross-references to locate other encyclopedia articles that will provide more detailed information about a subject.

The suggested readings include recent secondary sources to aid the reader in locating more detailed or technical information. Review articles and research articles that are considered of primary importance to the understanding of a given subject area are also listed. These suggested readings are not intended to provide a full reference listing of all material covered in the context of a given article, but are provided as next steps for a reader looking for additional information.

CONTRIBUTORS

B Ackerson
University of Illinois at Urbana–Champaign, Urbana, IL, USA

D Adams
National Institutes of Health, Bethesda, MD, USA

K E Adolph
New York University, New York City, NY, USA

A Ahuja
National Jewish Hospital, Denver, CO, USA

N Akhtar
University of California, Santa Cruz, Santa Cruz, CA, USA

A Almas
University of Toronto, Toronto, ON, Canada

S Al'Otaiba
Florida State University, Tallahassee, FL, USA

H Als
Harvard Medical School, Boston, MA, USA

K M Andrews
McGill University, Montreal, QC, Canada

M E Arterberry
Colby College, Waterville, ME, USA

J B Asendorpf
Humboldt-Universität zu Berlin, Berlin, Germany

D H Ashmead
Vanderbilt University Medical Center, Nashville, TN, USA

R N Aslin
University of Rochester, Rochester, NY, USA

J W Astington
University of Toronto, Toronto, ON, Canada

J Atkinson
University College London, London, UK

L E Bahrick
Florida International University, Miami, FL, USA

D B Bailey
RTI International, Research Triangle Park, NC, USA

L A Baker
University of Southern California, Los Angeles, CA, USA

A Balasubramanian
University of California, Riverside, Riverside, CA, USA

R Barr
University of British Columbia, Vancouver, BC, Canada

P J Bauer
Emory University, Atlanta, GA, USA

A Baxter
University of South Alabama, Mobile, AL, USA

A Belden
Washington University School of Medicine, St. Louis, MO, USA

D Benoit
University of Toronto, Toronto, ON, Canada;
The Hospital for Sick Children, Toronto, ON, Canada

D Bergen
Miami University, Oxford, OH, USA

K Bernard
University of Delaware, Newark, DE, USA

B I Bertenthal
Indiana University, Bloomington, IN, USA

J Bhagwat
Cornell University, Ithaca, NY, USA

N Bhullar
Widener University, Chester, PA, USA

A E Bigelow
St. Francis Xavier University, Antigonish, NS, Canada

M M Black
University of Maryland, Baltimore, MD, USA

E Blass
University of Massachusetts, Amherst, MA, USA

N J Blum
University of Pennsylvania School of Medicine, Philadelphia, PA, USA

C A Boeving
Yale University School of Medicine, New Haven, CT, USA

C F Bolling
Cincinnati Children's Hospital Medical Center, Cincinnati, OH, USA

M H Bornstein
National Institutes of Health, Bethesda, MD, USA

L Bosch
Universitat de Barcelona, Barcelona, Spain

O Braddick
University of Oxford, Oxford, UK

J Brooks-Gunn
Columbia University, New York, NY, USA

R T Brouillette
McGill University, Montreal, QC, Canada

A W Burks
Duke University Medical Center, Durham, NC, USA

S C Butler
Harvard Medical School, Boston, MA, USA

J P Byrnes
Temple University, Philadelphia, PA, USA

M L Campbell
Kennedy Krieger Institute, Baltimore, MD, USA

R L Canfield
Cornell University, Ithaca, NY, USA

S M Carlson
Institute of Child Development, Minneapolis, MN, USA

A S Carter
University of Massachusetts Boston, Boston, MA, USA

M Casasola
Cornell University, Ithaca, NY, USA

L M Casper
University of Southern California, Los Angeles, CA, USA

I Chatoor
Children's National Medical Center, Washington, DC, USA

A I Chin
University of California, Los Angeles, Los Angeles, CA, USA

R Clark
University of Wisconsin, Madison, WI, USA

A Clarke-Stewart
University of California, Irvine, Irvine, CA, USA

C M Connor
Florida State University, Tallahassee, FL, USA

J Coolbear
University of Toronto, Toronto, ON, Canada;
The Hospital for Sick Children, Toronto, ON, Canada

M L Courage
Memorial University, St. John's, NL, Canada

M J Cox
University of North Carolina, Chapel, NC, USA

A Crawford
University of Toronto, Toronto, ON, Canada;
Mount Sinai Hospital, Toronto, ON, Canada

E M Cummings
University of Notre Dame, Notre Dame, IN, USA

L A Dack
University of Toronto, Toronto, ON, Canada

M W Daehler
University of Massachusetts, Amherst, MA, USA

S R Daniels
University of Colorado Health Sciences Center, Denver, CO, USA

R B David
St. Mary's Hospital, Richmond, VA, USA

G Dawson
University of Washington, Seattle, WA, USA

L F DiLalla
Southern Illinois University School of Medicine, Carbondale, IL, USA

J A DiPietro
Johns Hopkins University, Baltimore, MD, USA

B M D'Onofrio
Indiana University, Bloomington, IN, USA

R L Doty
University of Pennsylvania School of Medicine, Philadelphia, PA, USA

M Dozier
University of Delaware, Newark, DE, USA

W O Eaton
University of Manitoba, Winnipeg, MB, Canada

C Edwards
University of Nebraska–Lincoln, Lincoln, NE, USA

K K Elam
Southern Illinois University School of Medicine, Carbondale, IL, USA

R R Espinal
University of Chicago, Chicago, IL, USA

R S Everhart
Syracuse University, Syracuse, NY, USA

D Fair
Washington University School of Medicine, St. Louis, MO, USA

F Farzin
University of California, Davis, Davis, CA, USA

D J Fidler
Colorado State University, Fort Collins, CO, USA

T Field
University of Miami School of Medicine, Miami, FL, USA

B H Fiese
Syracuse University, Syracuse, NY, USA

K W Fischer
Harvard Graduate School of Education, Cambridge, MA, USA

H E Fitzgerald
Michigan State University, East Lansing, MI, USA

D R Fleisher
University of Missouri School of Medicine, Columbia, MO, USA

M J Flory
New York State Institute for Basic Research, Staten Island, NY, USA

B Forsyth
Yale University School of Medicine, New Haven, CT, USA

S Fowler
University of Kansas, Lawrence, KS, USA

R C Fretts
Harvard Vanguard Medical Associates, Wellesley, MA, USA

J J Gallagher
University of North Carolina at Chapel Hill, Chapel Hill, NC, USA

J M Gardner
New York State Institute for Basic Research, Staten Island, NY, USA

J-L Gariépy
The University of North Carolina at Chapel Hill, Chapel Hill, NC, USA

M A Gartstein
Washington State University, Pullman, WA, USA

M Gauvain
University of California at Riverside, Riverside, CA, USA

D R Gemmill
California Sudden Infant Death Syndrome Advisory Council, Escondido, CA, USA

I R Gizer
Emory University, Atlanta, GA, USA

M M Gleason
Tulane University Health Sciences Center, New Orleans, LA, USA

M M Gleason
Brown University School of Medicine, Providence, RI, USA

R L Gómez
The University of Arizona, Tucson, AZ, USA

L Godoy
University of Massachusetts Boston, Boston, MA, USA

W A Goldberg
University of California, Irvine, CA, USA

E C Goldfield
Harvard University, Boston, MA, USA

S Goldin-Meadow
University of Chicago, Chicago, IL, USA

H H Goldsmith
University of Wisconsin–Madison, Madison, WI, USA

C Golomb
University of Massachusetts Boston, Boston, MA, USA

E L Grigorenko
Yale University, New Haven, CT, USA

J E Grusec
University of Toronto, Toronto, ON, Canada

L M Gutman
University of London, London, UK

M de Haan
University College London Institute of Child Health, London, UK

J W Hagen
University of Michigan, Ann Arbor, MI, USA

R J Hagerman
University of California, Davis, Medical Center, Sacramento, CA, USA

N Halfon
University of California, Los Angeles, Los Angeles, CA, USA

G S Halford
Griffith University, Brisbane, QLD, Australia

J Harel
University of Haifa, Haifa, Israel

K M Harrington
Emory University, Atlanta, GA, USA

H Hayne
University of Otago, Dunedin, New Zealand

L J Heffner
Boston Medical Center, Boston, MA, USA

N Heilbron
University of North Carolina, Chapel, NC, USA

R W Hendershot
University of Colorado at Denver Health Sciences Center, Denver, CO, USA

M Hernandez-Reif
University of Miami School of Medicine, Miami, FL, USA

K Herold
University of California, Santa Cruz, Santa Cruz, CA, USA

A H Hindman
University of Michigan, Ann Arbor, MI, USA

J A Hofheimer
University of North Carolina at Chapel Hill, Chapel Hill, NC, USA

C von Hofsten
Uppsala University, Uppsala, Sweden

G Hollich
Purdue University, West Lafayette, IN, USA

J R Hollister
University of Colorado at Denver and Health Sciences Center, Denver, CO, USA

A H Hoon
Kennedy Krieger Institute, Baltimore, MD, USA

N Howe
Concordia University, Montréal, QC, Canada

C Howes
University of California, Los Angeles, Los Angeles, CA, USA

A Hupbach
University of Arizona, Tucson, AZ, USA

J S Hyde
University of Wisconsin, Madison, WI, USA

J Isen
University of Southern California, Los Angeles, CA, USA

J S Jameson
Coloradoes State University, Fort Collins, CO, USA

T Jirikowic
University of Washington, Seattle, WA, USA

R Jochem
University of California, Davis, Davis, CA, USA

M H Johnson
University of London, London, UK

S P Johnson
New York University, New York, NY, USA

T R B Johnson
The University of Michigan, Ann Arbor, MI, USA

M V Johnston
Kennedy Krieger Institute, Baltimore, MD, USA

J Jones-Branch
University of Nebraska–Lincoln, Lincoln, NE, USA

T A Jusko
University of Washington, Seattle, WA, USA

J Kagan
Harvard University, Cambridge, MA, USA

J Kapala
McGill University Health Centre, Montréal, QC, Canada

B Z Karmel
New York State Institute for Basic Research, Staten Island, NY, USA

T G Keens
Keck School of Medicine of the University of Southern California, Los Angeles, CA, USA

D J Kelly
The University of Sheffield, Sheffield, UK

S King
University of Washington, Seattle, WA, USA

P Kitchen
University of Southern California, Los Angeles, CA, USA

J Koontz
University of California, Riverside, Riverside, CA, USA

C B Kopp
Los Angeles, CA, USA

C D Kouros
University of Notre Dame, Notre Dame, IN, USA

L L LaGasse
Warren Alpert Medical School of Brown University, Providence, RI, USA

F G Lamb-Parker
Columbia University, New York, NY, USA

M Lampl
Emory University, Atlanta, GA, USA

A L Lathrop
University of Rochester, Rochester, NY, USA

R A Lawrence
University of Rochester School of Medicine and Dentistry, Rochester, NY, USA

M S Leidy
University of California, Riverside, CA, USA

E M Lennon
New York State Institute for Basic Research, Staten Island, NY, USA

T A Lenzi
Vanderbilt University Medical Center, Nashville, TN, USA

S E Lerman
University of California, Los Angeles, Los Angeles, CA, USA

B M Lester
Warren Alpert Medical School of Brown University, Providence, RI, USA

H Liang
King's College London, London, UK

K Libertus
Duke University, Durham, NC, USA

E Lieven
Max Planck Institute for Evolutionary Anthropology, Leipzig, Germany

B Lozoff
University of Michigan, Ann Arbor, MI, USA

J S Lu
David Geffen School of Medicine at UCLA, Los Angeles, CA, USA

M C Lu
David Geffen School of Medicine at UCLA, Los Angeles, CA, USA

J Luby
Washington University School of Medicine, St. Louis, MO, USA

R Lucas-Thompson
University of California, Irvine, Irvine, CA, USA

L E Lurye
New York University, New York, NY, USA

M Macaoay
Children's National Medical Center, Washington, DC, USA

S C Mangelsdorf
University of Illinois at Urbana–Champaign, Champaign, IL, USA

M Martinos
University College London Institute of Child Health, London, UK

D Matthews
University of Manchester, Manchester, UK

K McCrink
Yale University, New Haven, CT, USA

M McIlreavy
University of Georgia, Athens, GA, USA

L M McKelvey
University of Arkansas for Medical Sciences, Little Rock, AR, USA

G W McRoberts
Haskins Laboratories, New Haven, CT, USA

A N Meltzoff
University of Washington, Seattle, WA, USA

D Messinger
University of Miami, Coral Gables, FL, USA

S Meyer
University of California Davis, Davis, CA, USA

L J Miller
Sensory Processing Disorder Foundation, Greenwood Village, CO, USA

M A Miller
University of California, Riverside, CA, USA

W R Mills-Koonce
University of North Carolina, Chapel, NC, USA

K Minde
Montreal Children's Hospital, Montreal, QC, Canada

J L Miner
University of California, Irvine, Irvine, CA, USA

K L Morris
University of California, Riverside, Riverside, CA, USA

F J Morrison
University of Michigan, Ann Arbor, MI, USA

M C Moulson
Massachusetts Institute of Technology, Cambridge, MA, USA

M E Msall
University of Chicago, Chicago, IL, USA

M Muenke
National Institutes of Health, Bethesda, MD, USA

P Y Mullineaux
Southern Illinois University School of Medicine, Carbondale, IL, USA

J P Murray
Kansas State University, Manhattan, KS, USA

A D Murray
Kansas State University, Manhattan, KS, USA

L Nadel
University of Arizona, Tucson, AZ, USA

A Needham
Duke University, Durham, NC, USA

C A Nelson
Harvard Medical School, Boston, MA, USA

L M Oakes
University of California, Davis, Davis, CA, USA

H Carmichael Olson
University of Washington, Seattle, WA, USA

M Y Ono
University of California, Davis Medical Center, Sacramento, CA, USA

C W Oppenheimer
University of North Carolina at Chapel Hill, Chapel Hill, NC, USA

T Ostler
University of Illinois at Urbana–Champaign, Urbana, IL, USA

K P Palmer
University of Arkansas for Medical Sciences, Little Rock, AR, USA

R Panneton
Virginia Tech, Blacksburg, VA, USA

R D Parke
University of California, Riverside, Riverside, CA, USA

O Pascalis
The University of Sheffield, Sheffield, UK

D L Paulhus
University of British Columbia, Vancouver, BC, Canada

F S Pedroso
Universidade Federal de Santa Maria, Santa Maria, Brazil

J L Petersen
University of Wisconsin, Madison, WI, USA

S L Pillsbury
Richmond, VA, USA

J Pinkston
University of Kansas, Lawrence, KS, USA

F Pons
Universitat de Barcelona, Barcelona, Spain

G Posada
Purdue University, West Lafayette, IN, USA

A Pressel
University of North Carolina, Chapel, NC, USA

H H Raikes
University of Nebraska–Lincoln, Lincoln, NE, USA

J T Rapp
St. Cloud State University, St. Cloud, MN, USA

H E Recchia
Concordia University, Montréal, QC, Canada

M Regalado
University of California, Los Angeles, Los Angeles, CA, USA

J M Retrouvey
McGill University Health Centre, Montréal, QC, Canada

C A Reynolds
University of California, Riverside, Riverside, CA, USA

J E Richards
University of South Carolina, Columbia, SC, USA

J Richmond
Harvard University, Boston, MA, USA

J Robinson
University of Connecticut–Storrs, Storrs, CT, USA

M K Rothbart
University of Oregon, Eugene, OR, USA

D N Ruble
New York University, New York, NY, USA

J A Rudolph
Children's Hospital Medical Center, Cincinnati, OH, USA

P A Rufo
Children's Hospital Boston, Boston, MA, USA

S Russ
University of California, Los Angeles, Los Angeles, CA, USA

A Sadeh
Tel Aviv University, Tel Aviv, Israel

R C Schaaf
Thomas Jefferson University, Philadelphia, PA, USA

A Scher
University of Haifa, Haifa, Israel

B L Schlaggar
Washington University School of Medicine, St. Louis, MO, USA

T J Schofield
University of California, Riverside, Riverside, CA, USA

E K Scholnick
University of Maryland, College Park, MD, USA

S Schwartz
McGill University Health Centre, Montréal, QC, Canada

D C Schwebel
University of Alabama at Birmingham, Birmingham, AL, USA

N Sebastián-Gallés
Universitat de Barcelona, Barcelona, Spain

R Seifer
Brown University, Providence, RI, USA

M Shah
University of Pennsylvania School of Medicine, Philadelphia, PA, USA

E Simonoff
King's College London, London, UK

D P Sladen
Vanderbilt University Medical Center, Nashville, TN, USA

D L Smith
The Children's Hospital, Denver, CO, USA

K A Snyder
University of Denver, Denver, CO, USA

J B Soep
University of Colorado at Denver and Health Sciences Center, Denver, CO, USA

K C Soska
New York University, New York, NY, USA

M M Stalets
Washington University School of Medicine, St. Louis, MO, USA

L Sterling
University of Washington, Seattle, WA, USA

M Sumaroka
National Institutes of Health, Bethesda, MD, USA

H N Switzky
Northern Illinois University, DeKalb, IL, USA

D E Szwedo
University of Virginia, Charlottesville, VA, USA

A Taddio
The Hospital for Sick Children, Toronto, ON, Canada and University of Toronto, ON, Canada

B Taubman
University of Pennsylvania School of Medicine, Philadelphia, PA, USA

D M Teti
The Pennsylvania State University, University Park, PA, USA

A M Tharpe
Vanderbilt University Medical Center, Nashville, TN, USA

C R Thomann
University of Massachusetts, Boston, MA, USA

R A Thompson
University of California Davis, Davis, CA, USA

N Tolani
Columbia University, New York, NY, USA

M Tomasello
Max Planck Institute for Evolutionary Anthropology, Leipzig, Germany

C M Torrence
University of Denver, Denver, CO, USA

N Towe-Goodman
The Pennsylvania State University, University Park, PA, USA

S E Trehub
University of Toronto at Mississauga, Mississauga, ON, Canada

R E Tremblay
University of Montréal, Montreal, QC, Canada

E Tronick
University of Massachusetts, Boston, Boston, MA, USA

A Tullos
The University of Texas, Austin, TX, USA

C D Vallotton
Harvard Graduate School of Education, Cambridge, MA, USA

I D Waldman
Emory University, Atlanta, GA, USA

J S Wallerstein
The Judith Wallerstein Center for the Family in Transition, Corte Madera, CA, USA

S E Watamura
University of Denver, Denver, CO, USA

N Wentworth
Lake Forest College, Lake Forest, IL, USA

R A Williamson
University of Washington, Seattle, WA, USA

K Willoughby
University of Toronto, Toronto, ON, Canada

M A Winter
Syracuse University, Syracuse, NY, USA

M S Wong
University of Illinois at Urbana–Champaign, Champaign, IL, USA

J D Woolley
The University of Texas, Austin, TX, USA

K Wynn
Yale University, New Haven, CT, USA

P D Zeanah
Tulane University Health Sciences Center, New Orleans, LA, USA

C H Zeanah
Tulane University Health Sciences Center, New Orleans, LA, USA

P D Zelazo
Institute of Child Development, Minneapolis, MN, USA

P Zelkowitz
McGill University, Montreal, QC, Canada

C Zera
Brigham and Women's Hospital, Boston, MA, USA

D Zlotnik
National Institutes of Health, Bethesda, MD, USA

K M Zosuls
New York University, New York, NY, USA

CONTENTS

VOLUME 1

A

B

VOLUME 2

G

H

I

L

VOLUME 3

R

S

T

V

Abuse, Neglect, and Maltreatment of Infants

D Benoit and J Coolbear, University of Toronto, Toronto, ON, Canada; The Hospital for Sick Children, Toronto, ON, Canada
A Crawford, University of Toronto, Toronto, ON, Canada; Mount Sinai Hospital, Toronto, ON, Canada

Glossary

Adrenocorticotropin-releasing hormone (ACTH) – Hormone released from the pituitary gland through the action of corticotropin-releasing hormone (CRH) as part of the hormonal cascade triggered by stress. ACTH then acts on the adrenal glands to stimulate the release of cortisol.

Corticotropin-releasing hormone (CRH) system – In response to stress, a hormonal cascade is triggered by the release of CRH from the hypothalamus. Release is influenced by stress, by blood levels of cortisol, and by the sleep/wake cycle. CRH activates the release of ACTH, which in turn stimulates the release of cortisol from the adrenal glands.

Cortisol – Stress hormone that mediates the body's alarm response to stressful situations. It is produced by the adrenal glands as a result of stimulation by ACTH. Cortisol, secreted into the blood circulation, affects many tissues in the body, including the brain.

Hypothalamic–pituitary–adrenal (HPA) axis – The HPA axis is one of the two stress response systems of the body (the other is the sympathetic–adrenal–medullary system), which consists of the hypothalamus, the pituitary gland, and the adrenal glands. The HPA axis activates and coordinates the stress response, through the action of hormones, by receiving and interpreting information from other areas of the brain (amygdala and hippocampus) and from the autonomic nervous system.

Reported case of maltreatment – A case where physical, sexual, and emotional abuse, neglect, or exposure to interpersonal violence is suspected and reported to a child protection agency. In many jurisdictions, the reporting of cases of suspected child maltreatment is required by law.

Substantiated case of maltreatment – A case where child maltreatment is confirmed following an investigation.

Introduction

> The history of childhood is a nightmare from which we have only recently begun to awake. The further back in history one goes, the lower the level of child care and the more likely children are to be killed, abandoned, beaten, terrorized and abused.
>
> Lloyd De Mause, *The History of Childhood*

Infant maltreatment has existed across all cultures, all socioeconomic strata, and in all historical epochs. In fact, there is evidence of infanticide from antiquity. The increasing recognition that children have the right to protection, and that they are not the property of their caregivers, led to the modern child protection movement. In 1874, the advocacy of the Society for the Prevention of Cruelty to Animals in the case of Mary Ellen, a young girl who was severely abused by her stepmother, led to an unprecedented judicial intervention and protection. Shortly afterward, the New York Society for the Prevention of Cruelty to Children was established, which gave rise to the founding of similar societies. Since then the complex social and familial dynamics of child maltreatment have been increasingly recognized. It was not until 1962, however, following a medical symposium the previous year, that several physicians, headed by Denver physician C. Henry Kempe, published the landmark the 'battered child syndrome' in the *Journal of the American*

Medical Association. The battered child syndrome described a pattern of child abuse that included both physical and psychological aspects and established it as an area of academic and clinical focus. In the early twenty-first century, the enormous social burden of child maltreatment remains timely, unresolved, and an important public health and policy issue. Every day, clinicians and investigators continue to attend to individual infants and children who are maltreated and make their way through the complexities of healthcare and judicial systems. The impact of maltreatment on infants and children, particularly early and repeated abuse, is one of the most significant emotional and psychological traumas that a child can endure. Unlike other traumatic events in which the infant or child may be soothed by the ameliorating comforting of their caregiver, child maltreatment is most often committed by a caregiver or attachment figure. This double rupture, the lost sense of the safety and predictability of the world, and the loss of caregiver protection and security, make maltreatment a breach of profound magnitude for many infants.

Incidence and Prevalence

The incidence and prevalence rates of maltreatment in infancy (i.e., ages 0–3 years) are difficult to ascertain, in part because of the lack of universally accepted definitions of various types of maltreatment across countries. Further, there is consensus that much maltreatment goes unreported and that each year infants die as a result of their caregivers maltreating them. In the US, ~3 million reports of child abuse or neglect are made each year and at least 1.5 million are substantiated. In Canada, recent data indicate that, in 2003, over 38 child abuse investigations per 1000 children were conducted and nearly half of the cases were substantiated. Estimates from various European and Eastern European countries reveal that between 3 and 360/1000 of children are maltreated. The wide range of incidence and prevalence rates reflect the varying definitions of maltreatment used in various jurisdictions around the world and the inconsistent reporting, investigation, and recording practices. In every country where relevant data have been collected, neglect occurs up to three times as often as abuse and incidence rates of maltreatment are highest for infants from birth to age 3 years.

Definitions

There are no universally accepted definitions of infant or child maltreatment. Definitions also vary depending on the professional discipline involved (e.g., child protection, law enforcement, judiciary, clinical). This inconsistency hinders the collection of reliable vital statistics and interferes with scientific research on infant maltreatment. The lack of universally accepted definitions of maltreatment may also contribute to delays in protecting maltreated infants and in providing them and their families with adequate assessment and intervention. **Table 1** lists various definitions of child maltreatment.

Risk Factors for Maltreatment

Infant maltreatment occurs in complex social and interpersonal circumstances. There is no single factor that predicts risk to an infant, and the absence of identifiable risk factors does not confer immunity from maltreatment. Rather, a profile of risk indicators must be considered within the individual, familial, economic, and social contexts of each infant. Most of the data on risk indicators for child maltreatment come from the study of child physical and sexual abuse. Data regarding risk indicators for emotional abuse and neglect are limited. Risk indicators may be broadly separated into child and household or caregiver characteristics. Further, there is support for the position that environmental factors beyond the child's immediate family or household – such as factors within the local community – may also play a role in creating high-risk caregiving situations. This perspective on the human ecology of child maltreatment posits that social impoverishment, such as low socioeconomic neighborhoods, poor community social support networks, observable criminal behavior within the community, poor housing conditions, and poor access to social services and programs, are environmental correlates of child maltreatment, and that rates of child maltreatment may be responsive to social change. Most information about risk factors related to child maltreatment comes from research on children older than age 3 years and this is reflected in the information provided in the following.

Child Factors

1. *Age.* American epidemiologic data indicate that incidence rates for child maltreatment are highest in infants, up to age 3 years.
2. *Gender.* In the 0–3 age group, based on Canadian data, rates of substantiated maltreatment for males and females are similar overall (51% vs. 49%, respectively). More females are physically abused (57%) sexually abused (53%), and emotionally maltreated (56%) in this age group, while more males are neglected (58%).
3. *Child psychological and developmental functioning.* Problems in the areas of psychological and developmental functioning and disability in children who are maltreated are likely under-reported, as not all children receive professional assessment. A large-scale Canadian study

Table 1   Definition of child maltreatment

1. *Emotional maltreatment*
   a. Emotional abuse (child has suffered or is at substantial risk of suffering from mental, emotional, or developmental problems caused by overly hostile, punitive treatment, or habitual or extreme verbal abuse such as threatening, belittling, etc.)
   b. Nonorganic failure to thrive
   c. Emotional neglect (child has suffered or is at substantial risk of suffering from mental, emotional, or developmental problems caused by inadequate nurturance/affection)
   d. Exposure to nonintimate violence (between adults other than caregivers) – e.g., child's father and an acquaintance
2. *Exposure to domestic violence*
   a. Child directly witnesses the violence
   b. Child indirectly witnesses the violence (e.g., sees the physical injuries on caregiver the next day or overhears the violence)
3. *Neglect*
   a. Failure to supervise – physical harm (including situations where child was harmed or endangered as a result of caregiver's actions, e.g., drunk driving with a child, or engaging in dangerous criminal activity with child)
   b. Failure to supervise – sexual abuse (caregiver knew or should have known of risk and failed to protect)
   c. Physical neglect (e.g., inadequate nutrition, clothing, unhygienic or dangerous living conditions)
   d. Medical neglect (caregiver does not provide, refuses, or is unavailable/unable to consent to treatment, including dental services)
   e. Failure to provide psychological/psychiatric treatment (also includes failing to provide treatment for school-related problems such as learning or behavior problems, infant development problems)
   f. Permitting criminal behavior (caregiver permits or fails/unable to supervise enough)
   g. Abandonment (caregiver died or unable to exercise custodial rights and no provisions made for care of child)
   h. Educational neglect (knowingly allows chronic truancy ($\geq$5 days/month), fails to enroll child, repeatedly keeps child at home)
4. *Physical abuse*
   a. Shake, push, grab, or throw (including pulling, dragging, shaking)
   b. Hit with hand (e.g., slapping and spanking)
   c. Punch, kick, or bite (also hitting with other parts of the body – e.g., elbow, head)
   d. Hit with object (e.g., stick, belt; throwing an object at a child)
   e. Other physical abuse (e.g., choking, stabbing, strangling, shooting, poisoning, abusive use of restraints)
5. *Sexual abuse*
   a. Penetration (penile, digital, or object penetration of vagina or anus)
   b. Attempted penetration
   c. Oral sex
   d. Fondling
   e. Sex talk (proposition, encouragement, or suggestion of a sexual nature; face to face, telephone, written, internet, exposing child to pornographic material)
   f. Voyeurism (perpetrator observes child for own sexual gratification)
   g. Exhibitionism (perpetrator exhibited self for own sexual gratification)
   h. Exploitation (e.g., pornography, prostitution)

Adapted from Trocmé N, Fallon B, MacLaurin B, *et al*. (2005) Canadian incidence study of reported child abuse and neglect – 2003: Major findings. Minister of Public Works and Government Services Canada. http://www.phac-aspc.gc.ca/ncfv-cnivf/familyviolence/index.html (accessed on May 2007).

that relied on reports by child protection workers, found that child functioning, in the areas of physical, cognitive, behavioral, and/or emotional health, is estimated to be impaired in ~50% of cases where child maltreatment has been substantiated. In about one-third of cases at least one problem related to physical health and emotional and/or cognitive functioning is documented, with the most common concerns being depression or anxiety, followed by learning disability. Ten per cent of maltreated children have a developmental delay. In 40% of cases where child maltreatment is investigated, behavioral concerns are identified. It is important to remember that these child-functioning characteristics are not necessarily causal in the maltreatment, and may be sequelae of the maltreatment. An American study reported that, in 34 states surveyed,

6.5% of all victims of child maltreatment had a disability, defined as mental retardation, emotional disturbance, visual impairment, learning disability, physical disability, behavioral problem, or medical problem.

Household and Caregiver Factors

1. *Family structure.* Estimates suggest that 43% of maltreated children live in single-parent families. Nearly one-third of cases involve children living with both biological parents. Approximately 16% of maltreated children live in blended families with a step-parent as caregiver. In cases of sexual abuse, the absence of a biological parent in the household or the presence of a stepfather are particular risk indicators, whereas

single-parent status is a risk indicator for physical abuse and neglect.

2. *Age of primary caregiver.* Overall, both male (80%) and female (64%) caregivers who maltreat children tend to be over 30 years of age. The proportion of females under 30 years of age is somewhat increased for neglect and emotional maltreatment.

3. *Gender of perpetrator.* The majority of nonmentally ill caregivers who cause child maltreatment fatalities are male; however, the younger the maltreated child is, the more likely the perpetrator is to be the child's mother. Men and women both appear to be equally culpable of nonaccidental injury. Men are overwhelmingly more often the perpetrators in the sexual abuse of both girls and boys (95% and 80% of the time, respectively). Children are twice as likely to be neglected by women than by men, reflecting the fact that women are more often primary caregivers of young children than men.

4. *Number of siblings in the household.* In ~65% of cases the maltreated child has at least one other sibling who is living in the household and is also investigated for allegations of child maltreatment.

5. *Socioeconomic status.* The primary income in families where there is child maltreatment is from full-time employment in the majority of cases (57%); 24% of the time, income is from benefits and/or social assistance; and 12% of the time from part-time or seasonal work. In cases of neglect, a higher proportion of families obtain their income from benefits or part-time employment.

6. *Housing.* The majority of children who are maltreated live in rental accommodations (56%), while 32% live in purchased homes, and 1% live in hostels or shelters.

7. *Mental illness.* American data demonstrate that of caregivers convicted of criminal offenses pertaining to child maltreatment, more than 50% had received psychiatric treatment, and almost one-third have been admitted to hospital for psychiatric treatment. Forty two percent of these mothers were suffering from either major depression or schizophrenia. Another study estimated that 27% of female caregivers and 18% of male caregivers were identified as having a mental health impairment.

8. *Substance abuse.* Approximately 18% of female caregivers and 30% of male caregivers abuse alcohol in cases of substantiated child maltreatment. Retrospective data show that rates of physical and sexual abuse are doubled in cases where caregivers are also reported to have a history of alcohol abuse, with rates markedly increased when both caregivers are substance abusers.

9. *Caregiver history of maltreatment as a child.* There is controversy and conflicting research evidence as to whether a childhood history of maltreatment in the caregiver increases the risk for abusive or neglectful behavior as a caregiver. In retrospective studies documenting a link between a history of childhood abuse or neglect and abuse or neglect of one's children, the link is weak. For example, one study indicated that 25% of abusive female caregivers and 18% of abusive male caregivers were maltreated as children; these rates were higher in cases of child neglect and emotional maltreatment. In general, ~20% of caregivers who were abused as children go on to abuse their own children, whereas 75% of perpetrators of child sexual abuse report having been sexually abused as children.

10. *Prior history of criminality.* Men who injure their children more commonly have a history of prior criminality and antisocial personality traits. One study estimated that 16% were involved in criminal activity. Women in these partnerships often have a psychiatric history, and may be incapable of providing protection to the child.

11. *Domestic violence.* Approximately 50% of female caregivers who maltreat their children have themselves been victims of domestic violence, including physical, sexual, or verbal assault, in the 6 months prior to the child maltreatment.

Impact of Maltreatment

During infancy, abuse, neglect, or exposure to interpersonal violence are stressful experiences that can be devastating and may result in pervasive psychological, behavioral, cognitive, and biological deficits. An infant or young child may witness interpersonal violence by being present; or hearing the violence from another room; or seeing bruises, black eyes, broken bones on the caregiver; or by having an incapacitated or unavailable caregiver. Infants and toddlers are more negatively affected when they witness their primary caregiver being threatened or harmed (e.g., being exposed to interpersonal violence) than when they are injured themselves. During infancy, most maltreatment is perpetrated by a caregiver or attachment figure rather than a stranger, and this may have a particularly deleterious impact on the infant. The infant who is maltreated, or is not protected from harm by a caregiver or attachment figure, comes to view the world as unsafe and dangerous; adults as untrustworthy; and the self as unworthy of love, affection, and protection. Such an infant is likely to develop an attachment relationship with his or her primary caregiver that is insecure-disorganized. In turn, insecure-disorganized infant–caregiver attachment is linked to the most negative socioemotional outcomes and the most severe forms of psychopathology (e.g., aggression, social incompetence, dissociation, difficulty regulating and expressing negative emotions, low self-esteem, and poor school achievement). There is growing evidence to suggest that emotional abuse and neglect, including exposure to interpersonal violence, can create even more harmful consequences for the child's functioning and outcome than physical and sexual abuse. Chronic childhood trauma interferes with the capacity to integrate and process sensory, cognitive, and emotional information and sets the stage for unfocused and maladaptive responses to subsequent stress. Long-term maltreatment has more pervasive effects than single-incident traumas.

Impact on Brain and Development

There is considerable evidence to indicate that maltreatment experiences in the early years have a profound effect on the developing brain, affecting both acute and long-term development of neuroendocrine, cognitive, and behavioral systems. Alterations in the central neurobiological systems that occur in response to adverse early-life stress lead to increased and abnormal responsiveness to stress, increase the risk of psychopathology in both childhood and adulthood, and can lead to lifelong psychiatric sequelae such as mood disorders and anxiety disorders (e.g., generalized anxiety disorder, post-traumatic stress disorder (PTSD), and panic disorder). The association between childhood trauma and the development of mood and anxiety disorders may be mediated by changes in the same neurotransmitter and endocrine systems that modulate the stress response and are implicated in adult mood and anxiety disorders (**Figure 1**). The impact of early adversity may differentially

affect individuals; some people with a history of severe maltreatment are well adjusted, while others manifest more profound developmental and psychiatric consequences. This likely has to do with complex gene–environment interactions which are only beginning to be delineated. One theory underlying the relation between genetic predisposition to major psychiatric disorders and the impact of early traumatic experiences during critical phases of development is that persistent changes occur in specific neurobiological systems in response to early stress, which later mediate adaptation to subsequent stressful life events and mood and anxiety symptoms. Specifically, stress has a major impact on the hypothalamic–pituitary–adrenal (HPA) axis, which is one of the two stress response systems of the body and consists of the hypothalamus, the pituitary gland, and the adrenal glands (**Figure 1**). The HPA axis activates and coordinates the stress response by receiving and interpreting information from other areas of the brain (amygdala and hippocampus) and from the autonomic nervous system. In response to acute situations of stress, a hormonal cascade is triggered with the release of corticotropin-releasing hormone (CRH) from the hypothalamus, which stimulates the release of adrenocorticotropin-releasing hormone (ACTH) from the pituitary gland. ACTH then triggers the production of cortisol within the adrenal cortex which is secreted into the blood circulation. Cortisol then provides negative feedback at the level of the hypothalamus, the pituitary, and the hippocampus, thereby shutting off the stress response. This sequence of hormonal responses and negative feedback allows humans to deal with experiences of stress in ways that allow them to recover from stressful events.

There is empirical evidence to suggest that following early-life stress, the set point of HPA-axis activity in response to stress is permanently altered so that subsequent adaptation to stressful situations throughout the lifespan may be affected. In other words, infants who are maltreated and traumatized might later react with overwhelming stress to innocuous or mildly stressful events. There is also evidence to suggest that early-life stress is related to persistent sensitization of pituitary–adrenal and autonomic stress responses, most likely caused by CRH hypersecretion, and may increase risk for psychopathology during adulthood. For example, research shows the implication of the CRH system in adult mood and anxiety disorders. This is because the HPA axis is involved not only in the stress response but also in the development of mood and anxiety disorders. Dysregulation of the CRH and the other downstream hormones (ACTH and cortisol; **Figure 1**) may explain the symptoms of increased vigilance and enhanced startle response observed in patients with anxiety disorders, such as PTSD, and may in part explain the high incidence of comorbid anxiety and mood disorders. It is important to note that most clinical studies evaluating the impact of childhood trauma on the brain

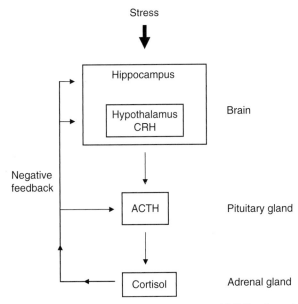

Figure 1   The hypothalamic–pituitary–adrenal (HPA) axis.

have been conducted in adults or children who have a history of physical or sexual abuse. However, different results in these various studies suggest that the effects of early-life stress may be variable and influenced by numerous factors.

When the HPA axis is overactivated over long periods (e.g., when an infant is repeatedly stressed by experiences of maltreatment), it becomes dysregulated and creates the production of stress hormones at levels that can be harmful, particularly to a developing brain. Some structural brain changes have been documented in individuals who are victims of child maltreatment, specifically in the hippocampus, prefrontal cortex, and amygdala. Recent data suggest that CRH hypersecretion itself (leading to high levels of cortisol) may be one causative factor in these structural alterations. The stress hormone cortisol prepares us to withstand threatening or stressful events. However, too much cortisol for too long is detrimental to the brain and linked to marked changes in brain activity and structures. Multiple brain regions may be affected by chronic and frequent high levels of cortisol. Specific areas of the brain that are negatively affected by sustained elevations in cortisol over time include:

1. The hippocampus, the brain structure involved in learning and explicit memory (remembering where one left one's keys is an example of explicit memory); a shrinkage of the hippocampus has been documented in adults who experienced PTSD and presumably produced high levels of cortisol at the time of trauma.
2. The anterior cingulate gyrus, the brain structure involved in selective attention; disruption in this may lead to difficulty focusing attention and inhibiting inappropriate actions.
3. The amygdala, the brain structure involved in the processing of frightening and negative events; the affected individual becomes more sensitive to negative emotions and is more likely to produce a hormonal stress reaction in situations of perceived threat.
4. The prefrontal cortex is the brain structure that is sensitive to information about the social environment and social partners; affected individuals may find it difficult to act appropriately in social situations (especially for children; however, this area is also developing until late adolescence and early adulthood).
5. The cerebral cortex and corpus callosum. Studies have shown lower intracranial volumes in individuals with PTSD compared to carefully matched controls, in addition to smaller volumes of the corpus callosum (and hippocampus). More global effects include intelligence, which was negatively correlated with duration of maltreatment, and intracranial volume which was correlated with age of onset of maltreatment (**Figure 2**).

Recent data suggest that effects of exposure to increased levels of maternal cortisol, in cases where pregnant women have PTSD, can be observed very early in the life of the offspring and underscore the relevance of *in utero* contributors to putative biological risk for PTSD. Taken together, these findings strongly suggest that early trauma can be toxic to the developing brain.

Neuroimaging studies have documented significant neurobiological changes in three specific areas of the brain of individuals with PTSD compared to individuals without PTSD: the hippocampus (responsible for some aspects of memory), the amygdala (responsible for the emotional and somatic contents of memories), and the

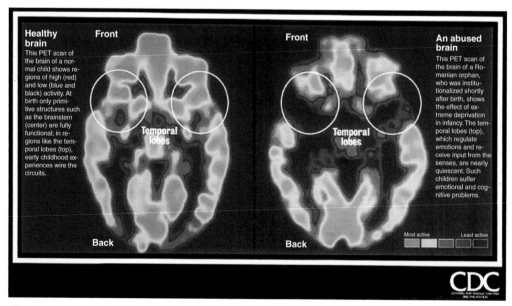

Figure 2   Effects of maltreatment on brain structures. Reproduced from the CDC website.

medial frontal cortex (responsible for the modulation of the cognitive control of the anxiety response and is probably essential for habituation in normative stress reactions). A current hypothesis attributes the hallmark symptoms of PTSD, exaggerated startle response and flashbacks, to the failure of the hippocampus and medial frontal cortex to dampen the exaggerated symptoms of arousal and distress that are mediated through the amygdala, in response to reminders of the traumatic event.

Impact on Behavior

The internal neuroendocrine and neurobiological changes associated with early exposure to maltreatment are often 'translated' into observable behavioral symptoms. For example, a subgroup of maltreated infants and young children can suffer from PTSD (**Table 2** lists symptoms of PTSD in infants). PTSD is important to recognize in infants exposed to violence and maltreatment as its symptoms are not likely to resolve spontaneously and the associated risk for long-term adverse outcomes if left untreated is high. However, it is important to recognize that not all infants exposed to a traumatic event will develop PTSD and that some infants who develop PTSD will resolve their PTSD symptoms – for example, with appropriate intervention, without long-term consequences.

While PTSD is a serious sequela of early exposure to violence and maltreatment that requires treatment, clinicians must be aware that a group of infants exposed to traumatic events, especially infants who are chronically traumatized by their attachment figures' abusive and/or neglectful caregiving, may not display prominent symptoms of PTSD. Instead, infants and toddlers who have endured repeated maltreatment, complex trauma, exposure to violence, and other chronic forms of maltreatment often do not meet criteria for PTSD but experience developmental delays across a broad spectrum, including physical, cognitive, affective, language, motor, and socialization skills. As a result of their multiple developmental delays, they tend to display complex disturbances with a variety of often fluctuating presentations that are qualitatively different from the clinical presentation of an infant with PTSD. The lack of capacity for emotional self-regulation is probably the most striking feature of infants who have experienced chronic and complex trauma and may contribute to the various associated symptoms which can be grouped into five major categories:

1. Intrapersonal thoughts/self-concept, such as lack of a continuous, predictable sense of self, a poor sense of separatedness, disturbances of body image, low self-esteem (and related behaviors), shame and guilt, and negative life view.

Table 2   Diagnostic criteria for post-traumatic stress disorder in infants and young children

1. The child has been exposed to a traumatic event – i.e., an event involving actual or threatened death or serious injury or threat to the physical or psychological integrity of the child or another person
2. A re-experiencing of the traumatic event(s) as evidenced by at least one of the following:
   a. Post-traumatic play
   b. Recurrent and intrusive recollections of the traumatic event outside play
   c. Repeated nightmares
   d. Psychological distress, expressed in language or behavior, at exposure to reminders of the trauma
   e. Recurrent episodes of flashback or dissociation
3. A numbing of responsiveness or interference with developmental momentum, appearing or being intensified after the trauma and revealed by at least one of the following:
   a. Increased social withdrawal
   b. Restricted range of affect
   c. Markedly diminished interest or participation in significant activities
   d. Efforts to avoid activities, places, or people that arouse recollection of the trauma
4. Symptoms of increased arousal that appear after a traumatic event, as revealed by at least two of the following:
   a. Difficulty going to sleep, evidenced by strong bedtime protest, difficulty falling asleep, or repeated night waking unrelated to nightmares
   b. Difficulty concentrating
   c. Hypervigilance
   d. Exaggerated startle response
   e. Increased irritability, outbursts of anger or extreme fussiness, or temper tantrums
5. This pattern of symptoms persists for at least 1 month.

   Associated features include a temporary loss of previously acquired developmental skills; aggression toward peers, adults, or animals; fears not present before the trauma (e.g., separation anxiety, fear of toileting alone, fear of the dark); and sexual and aggressive behaviors inappropriate for a child's age.

Adapted from The DC:0–3R Revision Task Force (2005) *DC:0–3R – Diagnostic Classification of Mental Health and Developmental Disorders of Infancy and Early Childhood*, Rev. edn. Arlington, VA: Zero to Three Press.

2. Emotional health, such as dissociative experiences (e.g., distinct alterations in states of consciousness, amnesia, depersonalization and derealization, impaired memory for state-based events); problems with affect regulation (e.g., difficulty with emotional self-regulation, difficulty labeling and expressing feelings, problems knowing and describing internal states, and difficulty communicating wishes and needs); impaired behavioral control (e.g., poor modulation of impulses, self-destructive behavior, aggression toward others, pathological self-soothing behavior, sleep and eating disturbances, substance abuse, excessive compliance, oppositional behavior/ difficulty understanding and complying with rules, re-enactment of trauma in behavior or play with sexual, aggressive themes); anxiety disorders (e.g., separation anxiety disorder, PTSD); mood disorders; suicidal thoughts (e.g., children exposed to domestic violence have a six times higher likelihood of attempting suicide compared to children who did not grow up in violent homes); personality disorder (e.g., borderline, narcissistic, paranoid, obsessive–compulsive).

3. Interpersonal relationships (e.g., disorganized infant–caregiver attachment; problems with boundaries; distrust and suspiciousness; social isolation), interpersonal difficulties (low social competency, difficulty attuning to other people's emotional states, decreased capacity for empathy/sympathy for others, difficulty with perspective taking); noncompliance; oppositional defiant disorder; disruptive or antisocial behaviors; delinquency/criminality (74% greater chance of committing crimes against a person); sexual maladjustment (abuse toward dating partner; 24% greater chance of committing sexual assault crimes; sexual dysfunctions in women); dependency.

4. Learning/cognition (e.g., difficulties with object constancy, attention regulation, focusing on and completing tasks, executive functioning, planning and anticipating, processing novel information, understanding responsibility; lack of sustained curiosity); learning difficulties or low academic achievement; problems with language development and orientation in time and space; impaired moral reasoning.

5. Physical health/biology (e.g., increased medical problems or complaints across the lifespan such as failure to thrive, asthma, skin problems, pseudoseizures, somatization, pelvic pain, autoimmune disorders; high mortality; sensorimotor developmental problems; analgesia; problems with coordination, balance, muscle tone).

Assessment

Maltreated infants represent a heterogeneous population. Maltreatment refers to a range of abusive/neglectful caregiver behavior that varies along a number of different dimensions (e.g., severity, duration) and, as a result, the outcomes for these infants are not uniform or universal. Some infants may be asymptomatic, while others present as being significantly impacted by their adverse experiences. A comprehensive clinical assessment helps to determine the unique impact of maltreatment on the individual infant. Because of potential police, child protection, and court involvement, assessments need to be forensically sound. Various published guidelines summarize the domains to be addressed when assessing the impact of child maltreatment and determining the most appropriate treatment recommendations. The American Academy of Child and Adolescent Psychiatry has published several separate assessment guidelines depending on the age of the child, the presenting problem, and the focus of the assessment. For example, the following assessment guidelines would be relevant when assessing concerns related to child maltreatment: the assessment of infants and toddlers, the forensic evaluation for children and adolescents who may have been sexually abused, the assessment of PTSD, the assessment of sexually abusive children, and the assessment of reactive attachment disorder. The American Professional Society on the Abuse of Children has also published guidelines, including guidelines for the assessment of suspected psychological treatment in children and adolescents. Finally, the Zero to Three/National Center for Clinical Infant Programs also provides guidelines for the assessment of very young children.

These various guidelines generally recommend a multidimensional approach to gathering information, including obtaining information from multiple sources (e.g., caregivers, child, daycare or school, child protection workers, police) and using a variety of assessment methods (e.g., clinical interview, structured and semistructured diagnostic interviews, questionnaires, observation). Evaluation of the young child's strengths and vulnerabilities within the various overlapping domains of development (e.g., biological, social, emotional, behavioral, cognitive) is essential. This information must then be placed within the child's environmental context (e.g., caregiver–child relationship, family systems and beliefs, socioeconomic circumstances).

Interviews with the child's caregivers allow the assessor to gather information about the developmental history of the child to determine the child's overall level of functioning before and after the child's experiences of maltreatment. It also allows the assessor to gather information about the child's caregivers (including trauma history, mental health history, substance abuse history, and environmental stressors such as poverty, exposure to domestic or community violence) in order to determine the caregivers' strengths and vulnerabilities and their ability to support the child and participate in recommended interventions.

A direct interview with the very young child may not be possible due to language limitations and cognitive immaturity. Even a young child, however, may be able to provide valuable information about his or her experiences. Information may be gathered from a younger child during a play-based interaction with the assessor using materials appropriate for this age group (e.g., age appropriate toys representing aspects of daily life), and/or direct observation of the child interacting with significant others (e.g., caregivers, teachers, peers).

Collateral information provides the assessor with information about the nature and history of the child and family's involvement with other services and agencies (e.g., mental health, child protective services, education). It is important to gather information about previous child welfare involvement to determine the extent of previously reported child maltreatment. This provides information about the chronic nature of the maltreatment, and the child and family's response to previous intervention. Interviews with the child's siblings and other family members (e.g., grandparents) may yield additional information. The main goals of gathering this information are to determine the child's level of functioning before and after the incident(s) of maltreatment, to determine the presence of any specific psychiatric disorder (e.g., PTSD; **Table 2**), and to develop an appropriate treatment plan for the child and family.

During the first 3 years of life, the quality of the caregiver–child relationship is of primary importance, and therefore is often the central focus of both assessment and intervention. Components of the caregiver–child relationship to be assessed include both the observable interactions between child and caregiver during various structured and unstructured activities (e.g., play, feeding, limit setting) and the caregivers' perceptions and subjective experience of the child and their relationship with the child (e.g., attributions about the child's behavior, importance of their role as caregivers). In addition, an assessment of the quality of the child's attachment relationships with his or her caregivers should be completed. Structured protocols should be used to assess the internal and external aspects of the caregiver–child relationship. Structured protocols can provide valuable information about areas of strength and vulnerability in the caregiver–child relationship which can be targeted during treatment.

The assessment should focus on both the child's general functioning and any maltreatment-specific issues. The assessment of the child's general functioning is informed by the various overlapping domains of development and the salient developmental tasks and challenges for a child at a particular age and stage of development. The various domains of functioning include:

1. Neurophysiological regulation (e.g., eating, sleeping, and capacity to self-soothe).

2. Affect regulation (e.g., accurate identification of internal emotional states, differentiation, interpretation, and application of appropriate emotional labels; safe emotional expression; and ability to modulate/regulate internal experiences). When children have an impaired capacity to self-regulate and self-soothe, they may present as emotionally labile, often in response to minor stressors.
3. Social skills and relational difficulties.
4. Emotional – including anxiety, mood, and attachment (separation anxiety, establishing a secure attachment relationship); self-esteem, self-efficacy.
5. Behavioral regulation – undercontrolled (e.g., aggressive, controlling, oppositional) or overcontrolled (e.g., compulsive compliance) behavioral patterns.
6. Cognitive/language development (e.g., expressive/receptive language, problem-solving, attention, abstract reasoning, executive function skills).
7. Temperament and constitutional characteristics.

The assessment of maltreatment-specific issues involves gathering details about each incident of maltreatment that the child has experienced. Relevant information includes the frequency, severity, and chronic nature of all incidents of maltreatment; the nature of the relationship between the child and the individual(s) who is/are maltreating the child; and the family/situational context in which the abuse has occurred. Gaining an understanding of the relationship between each of these factors assists in determining an appropriate intervention.

The response of the nonoffending caregiver(s) to the child's disclosure of maltreatment is one of the strongest predictors of outcome for young children. The level of caregiver support has a significant impact on the child's level of functioning, and therefore is an important aspect of assessment, and a target for intervention. The presence of a supportive primary caregiver, or a supportive relationship with another important adult, is associated with decreased levels of distress and lower levels of behavior problems. The assessment of the caregiver's support involves determining the caregiver's level of belief in and validation of the child's experience, the caregiver's emotional availability for the child (e.g., caregiver's ability to experience a range of emotions, to label the child's emotional experiences accurately, to tolerate the child's distress), the caregiver's own level of distress, and how the caregiver is managing his or her own emotional response.

Treatment

Young children who have been maltreated and their families represent a heterogeneous population. Therefore, they require individualized treatment approaches that address the unique needs of the child and family. Some treatments target specific individuals (e.g., child, caregiver, family, caregiver–child dyad), specific issues (e.g., anger management, caregiving or parenting skills, addressing mental health concerns, child behavior management), or vary according to treatment modality (e.g., individual, family, group). When children are very young, however, caregivers play a particularly significant role in the child's assessment, treatment, and recovery. Although interventions vary according to the unique needs of the child and family, and may specifically target the child, caregivers, family, or environment, or various targets simultaneously, all forms of treatment for maltreated infants and their families have three essential, basic components in common, including:

1. Establishing a sense of safety by providing reassurance to the child, and in some situations actually creating a safe environment by removing the child from an unsafe situation, or removing the individuals who are creating an unsafe and/or high-risk situation for the child. The treatment process is hindered if the child experiences repeated exposure to unsafe and stressful situations (e.g., remaining in a home where there is ongoing exposure to domestic violence).
2. Addressing issues of engagement/motivation, as many caregivers involved with the child protection system are obligated to attend treatment rather than seeking treatment voluntarily.
3. Addressing practical issues that may create obstacles to attending treatment (e.g., child-care, transportation, provision of snacks, financial assistance).

Other components of interventions may then focus specifically on helping the child and/or the caregiver in the following ways:

1. Helping the 'child' to:
   - Reduce the intensity of affect (e.g., fear, anger) and to regulate their affect, as experiencing maltreatment is often associated with affective dysregulation.
   - Develop a coherent narrative (the complexity of the narrative will vary depending on the age of the child) of their negative experiences, and to integrate these experiences at a level appropriate to the child's developmental stage. An aspect of this process may also involve the therapist challenging distorted cognitions associated with the negative experiences (e.g., guilt, responsibility) with children who are old enough.
2. Helping 'nonmaltreating caregivers' to:
   - Be emotionally available and able to respond empathically to the needs of the child. This may include psychoeducation about outcomes associated with different types of maltreatment and helping

caregivers link specific symptoms to the child's adverse experiences, helping caregivers manage the child's symptoms within the home environment and develop effective behavior management strategies, and assisting caregivers to negotiate the child welfare and legal systems. This may also involve referring the caregiver for individual treatment as many nonoffending caregivers may also have experienced trauma or violence within the home.
3. Helping the 'child and caregiver(s)' to:
   - Deal with the negative sequelae of the maltreatment (e.g., manage the child's behavioral disturbance, developmental delay, adjusting to a change in residence, separation from caregivers, financial hardship). Referral for specialized assessment may be necessary (e.g., occupational therapy, speech and language pathology).
   - Address both abuse-specific (e.g., PTSD) and general psychopathology (e.g., depression, disrupted behavior) in the child and/or caregivers.

In recent years there has been an increase in the research exploring the efficacy of a number of different interventions that target maltreated children and their families and incorporate the aforementioned components of intervention. In 2003, the National Crime Victims Research and Treatment Center published a report summarizing the review of several different interventions that have some level of empirical support. Several of these interventions are now considered 'best practice' when working with maltreated children and their families. However, these interventions have not been validated for use in children under 3 years of age.

The intervention that has received the highest rating and the most empirical support is trauma-focused cognitive behavioral therapy (TF-CBT). This intervention is designed for children as young as 3 years who have experienced sexual abuse, and who are displaying symptoms of PTSD and associated mental health problems (e.g., anxiety, depression, inappropriate sexual behaviors). The treatment model can be adapted to the developmental level of the child. TF-CBT is based on learning and cognitive theories, and is designed to reduce children's negative behavior and emotional responses, and to identify and correct maladaptive attributions and beliefs related to the sexual abuse. This intervention also involves providing support and teaching skills to the nonoffending caregiver(s) to enhance their coping and their ability to respond to the child's needs. No comparable intervention has been validated for use with children under 3 years of age.

Based on both learning theory and behavioral principles, abuse-focused cognitive behavioral therapy (AB-CBT) focuses on child, caregiver, and family characteristics related to physical abuse. This intervention addresses both the risk factors associated with physical abuse and the common sequelae for children who have experienced physical abuse (e.g., aggression, poor social competence and relationship skills, trauma-related symptoms). The intervention is comprised of primary caregiver, child, and family systems components and is appropriate for maltreated infants and their families.

The third intervention that received a high rating is parent–child interaction therapy (PCIT). This intervention is used with physically abusive caregivers who have children as young as 4 years. PCIT is a caregiver–child relationship intervention that focuses on several goals including improving parenting skills, decreasing child behavior problems, and improving the quality of the caregiver–child relationship. Specifically, the intervention addresses the coercive relationship that has developed between the caregiver and child and pattern of parent response to the child (e.g., high rates of negative interaction, low rates of positive interaction, ineffective parenting strategies, over-reliance on punishment). It also addresses the child's behavioral difficulties (e.g., aggression, defiance, noncompliance, and resistance in response to caregivers' requests). Although there are no published reports of its efficacy in treating infants, there is clinical evidence that PCIT may be appropriate for maltreated infants and their families.

Lieberman and Van Horn's (2000) child–parent psychotherapy for young children who have been exposed to family violence is a relationship-based treatment model that has several basic premises. These include the premise that the child–caregiver attachment relationship is of paramount importance as the main organizer of children's responses to danger and safety within the first 5 years of life, that emotional and behavioral problems in young children need to be addressed within the context of the child's primary attachment relationships, that risk factors during the first 5 years of life operate within the context of transactions between the child and the child's ecological environment (e.g., family, neighborhood, community), and that interpersonal violence is a traumatic stressor that has specific adverse effects on those who witness and/or experience it. Although this intervention has not yet received the empirical support of the previously described interventions, it is based on sound theory, and is an accepted clinical approach used by experts in the field. Research exploring the efficacy of this intervention would provide additional support for its use.

Conclusion

Maltreatment during infancy, a formative period of both physical and psychological growth, presents serious challenges to development. Such disruptions continue to impact many maltreated infants and produce deleterious short- and long-term effects on the infant's brain and behavior. Maltreated infants require early identification along with appropriate assessment and interventions. The aim and ongoing task, at both a policy and clinical practice level, involves the prevention of serious, negative long-term sequelae of maltreatment.

See also: Attachment; Brain Development; Emotion Regulation; Endocrine System; Mental Health, Infant; Mortality, Infant; Nutrition and Diet; Risk and Resilience; Safety and Childproofing; Stress and Coping; Temperament.

Suggested Readings

Glaser D (2000) Child abuse and neglect and the brain – a review. *Journal of Child Psychology and Psychiatry* 41: 97–116.
Kaplan SJ, Pelcovitz D, and Labruna V (1999) Child and adolescent abuse and neglect research: A review of the past 10 years. Part 1: Physical and emotional abuse and neglect. *Journal of the American Academy of Child and Adolescent Psychiatry* 38: 1214–1222.
Larrieu JA and Zeanah CH (2004) Treating parent–infant relationships in the context of maltreatment: An integrated systems approach. In: Sameroff AJ McDonough SC, and Rosenblum KL (eds.) *Treating Parent–Infant Relationship Problems – Strategies for Interventions*, pp. 243–267. New York: Guiford Press.
Lieberman AF and Van Horn P (2000) *Don't Hit My Mommy! A Manual for Child–Parent Psychotherapy with Young Witnesses of Family Violence.* Washington DC: Zero to Three Press.
Nemeroff CB (2004) Neurobiological consequences of childhood trauma. *Journal of Clinical Psychiatry* 65(supplement 1): 18–28.
Osofsky J (ed.) (2004) *Young Children and Trauma: Intervention and Treatment.* New York: Guilford Press.
Perry BD (2004) *Maltreatment and the Developing Child: How Early Childhood Experience Shapes Child and Culture.* The Margaret McCain Lecture Series. http://www.lfcc.on.ca (accessed May 2007).
Scheeringa MS and Gaensbauer TJ (2000) Posttraumatic stress disorder. In: Zeanah CH (ed.) *Handbook of Infant Mental Health* pp. 369–381. New York: Guilford Press.
The DC:0–3R Revision Task Force (2005) *DC:0–3R – Diagnostic Classification of Mental Health and Developmental Disorders of Infancy and Early Childhood,* Rev. edn. Arlington, VA: Zero to Three Press.
Trocmé N, Fallon B, MacLaurin B, et al. (2005) Canadian incidence study of reported child abuse and neglect – 2003: Major findings. Minister of Public Works and Government Services Canada. http://www.phac-aspc.gc.ca/ncfv-cnivf/familyviolence/index.html (accessed on May 2007).

Relevant Websites

http://www.nctsn.org – National Child Traumatic Stress Network.
http://www.musc.edu/ncvc – National Crime Victims Research and Treatment Center – *Child Physical and Sexual Abuse: Guidelines for Treatment (Revised Report: April 25, 2004).*
http://www.apsac.org – Practice guidelines from the American Professional Society on the Abuse of Children.
http://www.aacap.org – Practice parameters from the American Academy of Child and Adolescent Psychiatry pertaining to the psychiatric assessment of infants and toddlers (0–36 months).

ADHD: Genetic Influences

I R Gizer, K M Harrington, and I D Waldman, Emory University, Atlanta, GA, USA

Glossary

Allele – One of the alternate forms of a DNA marker.

Association – A nonrandom difference in the frequency of alternate forms of a DNA marker between individuals with and without some diagnosis or across levels of a trait.

Candidate gene study – A study that conducts a targeted test of the association of one or more DNA markers in a specific gene with a disorder or trait.

Endophenotype – Constructs posited to underlie psychiatric disorders or psychopathological traits, and to be more directly influenced by the genes relevant to disorder than are manifest symptoms.

Exon – The nucleotide sequences of a gene responsible for the coding of proteins that comprise the gene product.

Genome scan – An exploratory search across the whole genome for genes related to a disorder or trait.

Haplotype – A particular configuration of alleles at multiple DNA markers in close contiguity within a chromosomal region.

Insertion/deletion – An insertion (deletion) occurs when one or more nucleotides are added to (removed from) the genetic sequence. It can be difficult to discern whether a given polymorphism is the result of an insertion or a deletion, and thus, such polymorphisms are often referred to as insertion/deletions.

Intron – The nucleotide sequences of a gene that lie between the exons and are not involved in the coding of proteins that comprise the gene product.

'Knockout' gene studies – Studies in model organisms, such as mice, in which one or both copies of a gene are deactivated and the effects on behavior and/or cognition are examined.

Linkage – The correlation of a disorder and DNA markers within families, typically tested by examining the co-segregation of the presence or absence of the disorder with sharing particular allele(s) of a DNA marker.

Polymorphism – A DNA marker that varies among individuals in the population.

Population stratification – An association between a DNA marker and a disorder or trait that is not due to the causal effects of the gene, but is instead due to the mixture of subsamples (e.g., ethnic groups) that differ in both allele frequencies and symptom levels or diagnostic rates.

Promoter – A DNA sequence involved in the initiation of transcription of the associated gene.

Repeat sequences (STR and VNTR) – DNA markers that consist of a number of base pairs that are repeated a varying number of times across individuals in the population. The length of the repeat can vary, with repeats of just 2 or 3 base pairs (bp) (i.e., dinucleotide repeats or short-tandem repeats (STRs)) to repeats of between 10 and 60 bp (i.e., variable number of tandem repeats (VNTRs)).

SNP – Single-nucleotide polymorphism: a single nucleotide base that varies among individuals in the population.

Transmission disequlibrium test (TDT) – A within-family test of association and linkage that is robust to the potentially biasing effects of population stratification, the TDT contrasts the transmitted and nontransmitted alleles from heterozygous parents only (i.e., parents with two different alleles) to their children diagnosed with the target disorder.

UTR – An untranslated region of the gene, meaning a part of the gene that is not involved directly in the coding of proteins, but which may contain regulatory elements that are involved in gene expression.

3′ and 5′ – The nucleic acid sequences of genes are written from left to right with the 5′ end lying to the left of the genetic sequence and the 3′ end lying to the right.

Introduction

Since the mid-1980s, considerable progress has been made in understanding the etiology of childhood 'attention deficit hyperactivity disorder' (ADHD), largely due to the publication of numerous twin studies of ADHD symptoms conducted in both clinically referred and large, nonreferred, population-based samples. Findings from these studies are consistent in suggesting substantial genetic influences (i.e., heritabilities ranging from 60% to 90%), nonshared environmental influences that are small to moderate in magnitude (i.e., ranging from 10% to 40%), and little-to-no shared environmental influences. Following from the findings of these quantitative genetic studies, numerous molecular genetic studies of association and linkage between ADHD and a variety of candidate genes have been conducted since the mid-1990s. While the majority of the candidate genes studied underlie various facets of the dopamine neurotransmitter system, researchers also have examined the etiological role of candidate genes in other neurotransmitter systems (e.g., norepinephrine, serotonin), as well as those with functions outside of neurotransmitter systems (e.g., involved in various aspects of brain and nervous system development).

The current review describes recent findings from the behavior genetic and candidate gene literatures of childhood ADHD. It begins with an introduction to the key features of ADHD. This is followed by a brief review of quantitative behavior genetic studies that have attempted to estimate the genetic and environmental influences underlying ADHD. This leads to a review of the extant molecular genetic literature on ADHD, first summarizing genome scan studies and then summarizing candidate gene studies of childhood ADHD. Finally, the review concludes with a consideration of some of the emergent themes that will be important in future studies of the genetics of ADHD.

Background of ADHD

ADHD is a childhood disorder characterized by inattention, hyperactivity, and impulsivity. The prevalence of ADHD has been estimated as 3–7% in school-age children, with male-to-female ratios ranging from 2:1 to 9:1. The definition of ADHD has evolved over time and has been known previously as hyperkinetic reaction of childhood, hyperkinetic syndrome, hyperactive child syndrome, minimal brain damage, minimal brain dysfunction, minimal cerebral dysfunction, minor cerebral dysfunction, and attention deficit disorder with or without hyperactivity.

Currently, ADHD is defined by two distinct, but correlated symptom dimensions, namely an inattentive and a hyperactive–impulsive symptom dimension, each consisting of nine symptoms. The inattentive symptoms consist of behaviors such as 'often has difficulty sustaining attention in tasks' and 'often has difficulty organizing tasks and activities'. The hyperactive–impulsive symptoms consist of behaviors such as 'often fidgets with hands or feet' and 'often has difficulty waiting turn' (see **Table 1** for a complete list of symptoms). Because an individual can present with just inattentive symptoms, with just hyperactive–impulsive symptoms, or with both inattentive and hyperactive–impulsive symptoms, three subtypes of ADHD corresponding to these patterns of presentation have been defined: the predominantly inattentive type, the predominantly hyperactive–impulsive type, and the combined type, respectively.

Table 1 The symptoms of ADHD

Inattentive symptoms
1. Often does not give close attention to details or makes careless mistakes in schoolwork, work, or other activities.
2. Often has trouble keeping attention on tasks or play activities.
3. Often does not seem to listen when spoken to directly.
4. Often does not follow instructions and fails to finish schoolwork, chores, or duties in the workplace (not due to oppositional behavior or failure to understand instructions).
5. Often has trouble organizing activities.
6. Often avoids, dislikes, or does not want to do things that take a lot of mental effort for a long period of time (such as schoolwork or homework).
7. Often loses things needed for tasks and activities (e.g., toys, school assignments, pencils, books, or tools).
8. Is often easily distracted.
9. Is often forgetful in daily activities.

Hyperactive symptoms
1. Often fidgets with hands or feet or squirms in seat.
2. Often gets up from seat when remaining in seat is expected.
3. Often runs about or climbs when and where it is not appropriate (adolescents or adults may feel very restless).
4. Often has trouble playing or enjoying leisure activities quietly.
5. Is often 'on the go' or often acts as if 'driven by a motor'.
6. Often talks excessively.

Impulsive symptoms
1. Often blurts out answers before questions have been finished.
2. Often has trouble waiting one's turn.
3. Often interrupts or intrudes on others (e.g., butts into conversations or games).

Theoretical accounts of ADHD have long focused on deficits in sustained attention, and more recently, executive functions deficits have been hypothesized as another possible core feature of the disorder. The term 'executive functions' refers to a list of 'higher-order' cognitive processes required for goal-directed behavior, which includes inhibitory control, working memory, strategy generation and implementation, shifting between subordinate tasks, and monitoring. Common assessment measures hypothesized to assess executive functioning include the 'Wisconsin card sorting task', 'go/no-go tasks', and the 'Stroop color/word task'. The presence of executive functions deficits in ADHD has been well documented in recent reviews, which provide strong support suggesting that both children and adults diagnosed with ADHD show impaired performance on these tasks relative to control subjects. Though the term 'executive functions' has long been synonymous with the frontal lobes, more recent accounts of the neurobiology of executive functions have begun to take seriously the reciprocal connections between the prefrontal cortex and subcortical brain areas such as the basal ganglia, and as a result, these brain regions have been implicated in the pathophysiology of ADHD.

The most common treatments for ADHD consist of psychostimulant medications such as methylphenidate and psychosocial treatments focusing on behavior management. Treatment outcome studies have tended to suggest that the gains achieved with medication are greater than those achieved by psychosocial treatments, though there are beneficial aspects to both approaches. Nonetheless, psychostimulant medications have proven extremely effective with studies demonstrating that between 75% and 92% of children diagnosed with ADHD will show improvement in symptoms following treatment. These medications have been shown to act on the dopamine, norepinephrine, ans serotonin neurotransmitter systems, which allow for communication between neurons throughout the brain including the frontal lobes and basal ganglia. Importantly, studies focusing on the specific mechanisms by which psychostimulant medications influence these neurotransmitter systems have been highly informative for molecular genetic studies of ADHD, as will be reviewed.

Behavioral Genetic Studies of ADHD

Research designs for investigating genetic and environmental influences include family studies, adoption studies, and twin studies all of which have suggested that ADHD is transmitted within families from parents to their offspring. Twin study designs have certain advantages over both family and adoption studies, however, in that they are more generalizable, more powerful, and better able to provide accurate estimates of the magnitude of genetic and environmental influences. Twin studies examine the etiology of a trait by taking advantage of the fact that MZ twin pairs share 100% of their genes identical by descent, whereas DZ twin pairs share 50% of their genes on average. By using this information and comparing the correlations of the trait or disorder in MZ and DZ twin pairs, the magnitude of genetic and environmental influences acting on a trait or disorder can be estimated.

More than 20 twin studies have now been published that have attempted to disentangle the genetic and environmental influences underlying ADHD, and though these studies have differed in many ways including how attention/hyperactivity problems are operationalized, the source of participants, the age range of the subjects, and the statistical methods used, several general conclusions about the etiology of ADHD can be drawn. Most importantly, both ADHD symptoms in the general population and extreme levels of ADHD in selected populations appear to be highly heritable (with most h^2 estimates ranging from 0.6 to 0.9), and demonstrate little evidence of shared environmental influences. Further, researchers who have conducted behavior genetic studies examining the etiology of inattention and hyperactivity–impulsivity as two separate dimensions rather than as a single disorder have reported similarly high heritability estimates for each symptom dimension.

Molecular Genetic Studies of ADHD

Before proceeding with the review, a brief introduction to some key concepts commonly used in molecular genetic studies is necessary. With the discovery of the double-helix structure of DNA, it was determined how paired nucleotide bases form the basic building blocks of life. These bases, defined by the letters A, C, G, and T, for adenine, cytosine, guanine, and thymine, respectively, make up the basic language of DNA. Each base on one strand of DNA forms a pair with its complement on the second strand to form the double helix structure with adenine and thymine always pairing together and cytosine and guanine always pairing together. The human genome has been shown to be made up of approximately three billion such base pairs (bp). Importantly, these three billion bp do not occur on a single length of DNA, but are divided into 23 pairs of chromosomes, with one set of chromosomes inherited from the mother and one set inherited from the father. The structure of a chromosome consists of a centromere at the center and two arms, a short arm and a long arm, that project from the centromere.

Population geneticists have estimated that 99.9% of the human genome is identical across individuals, which means that 1 in every 1000 bp represents a point of variation across individuals. These points of variation or polymorphisms are the source of genetic variation that contribute to differences between individuals, and thus, are the focus of attention for molecular genetic studies. There are several types of polymorphisms, though two commonly studied types are repeat sequences and single-nucleotide polymorphisms (SNPs). Repeat sequences consist of a set of bp that can be short in length (i.e., 2–4 bp) or quite long (i.e., 10–60 bp), and the different variants of the polymorphism, or alleles, are defined as how many times the sequence is repeated (e.g., two-repeat vs. four-repeat vs. seven-repeat). A SNP consists of a change in a single bp, however; thus, the alleles at a SNP are defined by the observed bp (e.g., A vs. C).

Polymorphisms throughout the genome are of interest to molecular geneticists, but those that lie within or near actual genes are of particular interest. The human genome is estimated to contain around 20 000 genes, each of which is responsible for the production of a specific protein(s). The structure of a gene consists of a promoter region that is involved in the initiation of transcription of the gene, a process that ultimately leads to the production of the gene product, and the gene sequence itself. The gene sequence consists of exons, which are elements of the gene sequence responsible for the coding of proteins,

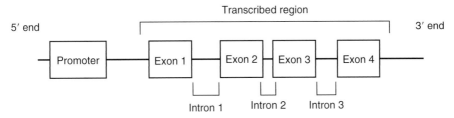

Figure 1 Diagram of a gene.

and introns, which are elements of the gene sequence not involved in the coding of proteins (see **Figure 1** for an illustration). As a result, polymorphisms that lie within the exons are the most likely to result in functional changes in the gene product, though recent research suggests that polymorphisms in the promoter region and introns may also result in functional changes in the gene product and differences in levels of gene expression. Ultimately, the aim of molecular genetic research is to identify polymorphisms that result in these types of functional changes that are related to disorders of interest.

Genome Scans for ADHD

Given the strong evidence suggesting that genetic influences are substantively involved in the etiology of ADHD, researchers have begun conducting molecular genetic studies that attempt to identify the specific genes or genomic regions related to ADHD. Broadly speaking, such studies use one of two general strategies to accomplish this. The first is a genome scan, in which linkage or association is examined between a disorder and evenly spaced DNA markers (approximately 10 000 bp apart, though this spacing continues to decrease as genotyping technologies continue to advance) distributed across the entire genome. Evidence for linkage or association between any of these DNA markers and the trait or disorder of interest implicates a broad segment of the genome that may contain hundreds of genes. Thus, genome scans may be thought of as exploratory searches for putative genes that contribute to the etiology of a disorder.

Four independent genome scans for ADHD have been published to date. Across these studies, 22 different genetic loci have provided evidence that was either significant or at least suggestive of linkage, and although many of these linkage regions were unique to a particular study, several loci demonstrated replicable evidence of linkage with ADHD in multiple studies. The most robust finding is a linkage region on the short arm of chromosome 5 with each of the published genome scans reporting evidence that was suggestive of linkage for this region. Interestingly, the dopamine transporter gene (*DAT1*), which will be discussed in detail, is found near this region, though further studies are needed to determine whether

the linkage peak can be attributed to this gene. Nonetheless, the consistent evidence of linkage across the four genome scans provides strong support for a gene or genes in this region to be involved in the pathophysiology of ADHD.

In addition to the short arm of chromosome 5, three loci have been independently identified in three of the four genome scans, which include the long arms of chromosomes 9 and 11 and the short arm of chromosome 17. Further, two loci have been independently identified in two of the four genome scans, which include the short arm of chromosome 8 and the long arm of chromosome 20. Thus, there are now six promising regions of the genome that have been identified for future studies attempting to identify the actual genes in these regions involved in the etiology of ADHD.

Although the initial findings from these genome scans are encouraging, the 16 novel loci identified that are unique to each study also highlight some of the difficulties inherent in drawing inferences regarding linkage from a few studies with relatively small samples, in which it is likely that the genomic regions suggestive of linkage will differ appreciably across studies for statistical reasons alone; that is, although there may be other reasons for the discrepant findings across these samples, such as differences in the populations sampled or in the assessment or diagnostic methods used, the stochastic fluctuations associated with few studies of small sample size are sufficient to cause such discrepancies. Thus, while these findings provide promising directions for future research, they also highlight the necessity for future studies conducted with larger samples and for meta-analytic reviews of the results of genome scans, as have appeared for schizophrenia and bipolar disorder.

Candidate Genes for ADHD

The second strategy for finding genes that contribute to the etiology of a disorder is the candidate gene approach. In many ways, candidate gene studies are polar opposites of genome scans. In contrast to the exploratory nature of genome scans, well-conducted candidate gene studies represent a targeted test of the role of specific genes in the etiology of a disorder as the location, function, and etiological relevance of candidate genes is most often known or strongly hypothesized *a priori*. With respect to ADHD, genes underlying the various aspects of the dopaminergic, and to a lesser extent the noradrenergic and serotonergic, neurotransmitter pathways have been widely studied based on several lines of converging evidence suggesting a role for these neurotransmitter systems in the etiology and pathophysiology of ADHD. For example, stimulant medications, the most common and effective treatment for ADHD, appear to act primarily by regulating dopamine levels in the brain, and also affect noradrenergic and serotonergic function. In addition, 'knockout' gene studies in mice have further demonstrated the potential relevance of genes within these neurotransmitter systems. Such studies breed genetically engineered mice lacking one or more specifically targeted genes. These mice are then studied, and if they display behaviors similar to those that characterize the disorder of interest, it can be inferred that the gene that has been 'knocked out' may be causally related to the disorder. Results of such studies have markedly strengthened the consideration of genes within the dopaminergic system, such as the dopamine transporter gene and the dopamine D1 and D4 receptor genes, as well as genes within the serotonergic system, such as the serotonin 1β receptor gene, as candidate genes for ADHD.

In the following section, studies of association and linkage between ADHD and candidate genes within the dopaminergic and other prominent neurotransmitter pathways, including the noradrenergic and serotonergic pathways, are reviewed. These studies are being published at a rapid rate, and the number of candidate genes that have been explored in relation to ADHD is continually increasing. Further, many genes that have been examined have led to largely negative results (e.g., the dopamine D2 receptor gene (*DRD2*), the dopamine D3 receptor gene (*DRD3*), and the tyrosine hydroxylase gene (*TH*)) that will not be discussed in the current review. Thus, the following is meant to be a representative though not exhaustive review and should provide the reader with a sense of current findings from studies of association and linkage between ADHD and several prominent candidate genes.

Dopamine Transporter

The dopamine transporter is involved in regulating dopamine neurotransmitter levels in the brain. Neurons transmit impulses from one neuron to the next across small junctions called synapses. This is accomplished when a nerve impulse causes the first, or presynaptic neuron, to release a neurotransmitter into the synapse, which then triggers the postsynaptic neuron. Once this is accomplished, any excess neurotransmitter is cleared from the synapse to allow for effective transmission of future nerve impulses. Transporter proteins help to accomplish this by binding to the neurotransmitter and transporting it

back to the presynaptic neuron. The dopamine transporter is an example of such a protein. It is densely distributed in the striatum and nucleus accumbens, which are areas in the brain involved in motor control and reward pathways, respectively, and represents the primary mechanism of dopamine regulation in these brain regions.

The gene that codes for the dopamine transporter, *DAT1*, has generated interest as a candidate gene for ADHD based on several lines of converging evidence. For example, stimulant medications (e.g., methylphenidate), which are among the most effective treatments available for ADHD symptoms, act by inhibiting the function of the dopamine transporter and thereby increasing the levels of available dopamine in the synapse. Further, a study of *DAT1* 'knockout' mice demonstrated that mice lacking both copies of the gene, and thus lacking any dopamine transporter, exhibit behaviors analogous to ADHD, such as greater motor activity, compared to mice with intact copies of the gene. This suggests these nice experience a downregulation of the dopamine system as a compensatory mechanism for the lack of dopamine transporter, and this downregulation results in a hypoactive dopamine system. In addition, studies using single photon emission computed tomography (SPECT), which can measure levels of targeted proteins in the brain, have suggested that adult participants with ADHD show differences in dopamine transporter availability that is related to a specific polymorphism in *DAT1*.

Each of the lines of research described above suggests involvement of the dopamine transporter in the etiology and pathophysiology of ADHD. Thus, *DAT1* has been one of the most widely researched genes in relation to ADHD. These studies have focused almost exclusively on a repeat polymorphism at the $3'$ end of the gene in an untranslated region (UTR) of *DAT1* that consists of a variable number of tandem repeats (VNTR) in the genetic sequence. This repeat sequence is 40 bp in length and the most common alleles are the 10 (480 bp) (71.9%) and 9 (440 bp) (23.4%) repeats. By the end of 2005, approximately 20 published studies had evaluated this relation in clinic-referred samples, and of these studies, approximately half reported positive evidence suggesting that the 10-repeat allele was associated with increased risk for developing ADHD. Given that a large number of studies failed to detect a significant relation between *DAT1* and ADHD, it is not surprising that recently published meta-analyses of these studies suggest that there is not a significant relation between *DAT1* and ADHD across studies. Nonetheless, these meta-analyses have also reported that there is greater heterogeneity in the effect sizes across studies than would be expected by chance with odds ratios ranging from 0.81 to 2.90. An odds ratio represents the ratio of having a risk factor to not having the risk factor, and thus, values of 1 indicate no increased risk, values less than one indicate reduced risk, and values greater than indicate 1 increased risk. As stated, the odds ratios for studies testing for association between *DAT1* and ADHD ranged from 0.81 to 2.9, which suggests there may be important moderating variables related to the sample characteristics of each study that influence the strength of the relation. Thus, meta-analyses evaluating specific variables that quantify specific sample characteristics (e.g., use of a clinic-referred sample vs. community-based sample, ethnicity of the sample, proportion of ADHD subtypes in each sample, etc.) as moderators of the relation between *DAT1* and ADHD are needed to elucidate what role, if any, *DAT1* plays in the pathophysiology of ADHD.

Further, as stated, the studies described thus far that have tested for association and linkage between *DAT1* and ADHD have focused almost exclusively on a single polymorphism, the VNTR in the 3′ UTR of the gene. Although the 10-repeat allele of the VNTR has been shown to be associated with increased *DAT1* transcription, it is not currently known whether the VNTR itself is a functional polymorphism that contributes directly to susceptibility for ADHD, or whether the VNTR simply is in close linkage disequilibrium with a functional polymorphism that represents the actual susceptibility allele. Linkage disequilibrium (LD) refers to the nonrandom association of alleles at multiple DNA markers that results from their close proximity to one another within a chromosome and co-inheritance. Researchers have begun to examine multiple markers in candidate genes, including *DAT1*, and to create haplotypes, which summarize the genetic information across a set of identified markers in close proximity to one another into a single descriptor. In doing so, these haplotypes capture a greater degree of the genetic variation in that region than a single marker and, thus, provide a more powerful method to test for association and linkage. These studies have suggested a relation between *DAT1* and ADHD, and importantly, the results from these studies have tended to yield stronger and more consistent results than studies that include only tests of individual markers. Thus, studies that test for association and linkage between ADHD and multiple markers that lie within or near *DAT1* have the potential to further our understanding of the potential involvement of *DAT1* in the pathophysiology of ADHD.

Dopamine D4 Receptor

As described, neurotransmitters convey nerve impulses from one neuron to the next across small junctions called synapses. When these neurotransmitters successfully cross the synapse, they bind to specific receptor on the postsynaptic neuron which then trigger that postsynaptic neuron to give. Abnormalities in the dopamine neurotransmitter system have been hypothesized to underlie ADHD, and thus, the five genes that code for the five different types of dopamine receptors have been identified as candidate loci for ADHD. The dopamine D4 receptor gene (*DRD4*) has been the most widely studied of the dopamine receptor genes in relation to ADHD primarily due to association

studies that initially linked the gene to the personality trait of novelty seeking, which has been compared to the high levels of impulsivity and excitability often seen in ADHD. It is also highly expressed in the frontal lobes, which are significantly involved in executive functioning. As a result, the deficits in executive functioning associated with ADHD also suggest a possible relation between *DRD4* and ADHD. Further interest has been generated from studies of *DRD4* knockout mice. For example, one study compared the behavior of *DRD4* knockout mice and 'wild-type' controls following administration of cocaine and methamphetamine, which belong to the same family of drugs as methylphenidate that is commonly used to treat ADHD. The investigators noted that the knockout mice showed a heightened response to cocaine and methamphetamine injection relative to controls, as measured by increases in locomotor behavior. In addition, it has been suggested that the seven-repeat of a 48-bp VNTR in exon 3 of the gene differs, albeit slightly, from the two- and four-repeats in secondary messenger (i.e., cAMP) activity and also possibly in response to the antipsychotic medication, clozapine.

Following from this suggested involvement of *DRD4* in the pathophysiology of ADHD, several studies have investigated the relation between the exon 3 VNTR of *DRD4* and ADHD, the findings and methods of which have been described in a number of previous reviews. The findings of association between ADHD and *DRD4* were replicated in some studies but not in others, similar to the pattern of findings reported for *DAT1*. Thus, it is noteworthy that meta-analytic reviews of these studies have repeatedly suggested a significant *DRD4*–ADHD association with odds ratios of approximately 1.4. Further, some studies have also examined whether the strength of the association between *DRD4* and ADHD might differ by subtype, and though these studies are few in number, they tend to suggest that *DRD4* is more strongly associated with the inattentive than with the combined subtype of ADHD.

More recently, studies testing for association and linkage between *DRD4* and ADHD have examined other polymorphisms in addition to the exon 3 VNTR. The most frequently studied marker after the exon 3 VNTR has been a 120-bp VNTR in the 5′ UTR of the gene. These studies have typically created haplotypes using multiple markers within *DRD4* to test for association and linkage with ADHD. Overall, this has tended to strengthen the relation between *DRD4* and ADHD, but such studies still yielded both significant and nonsignificant results, again demonstrating the necessity for meta-analytic reviews before drawing substantive conclusions from the existing literature regarding the relation between *DRD4* and ADHD.

Catechol-*O*-Methyl-Transferase

Catechol-*O*-methyl-transferase (COMT) is an enzyme responsible for the degradation of catecholamines, such as dopamine and norepinephrine. COMT is highly expressed in the frontal lobes and plays an important role in regulating synaptic dopamine levels in this region because the dopamine transporter is not significantly expressed in the frontal lobes. Thus, because frontal lobe dysfunction has been hypothesized as a possible causal factor in ADHD several studies have recently tested for association and linkage between this gene and ADHD. These studies have focused on a functional SNP in exon 4 that leads to an amino acid substitution (valine → methionine), and has been shown to substantially affect COMT enzyme activity such that homozygosity for valine shows 3–4 times greater activity than homozygosity for methionine. Given that the higher activity of the valine allele leads to less synaptic availability of dopamine than does the methionine allele, it is reasonable to consider the valine allele as the high-risk allele for ADHD.

Despite such evidence suggesting that the *COMT* gene would represent a strong candidate gene for ADHD, the results from studies testing for association and linkage between *COMT* and ADHD have been largely negative. The initial study to test this relation yielded positive evidence for association, suggesting that the valine allele was associated with increased risk for ADHD. Nonetheless eight studies that have attempted to replicate this association have failed to support this relation with one exception. A single study examined the relation between *COMT* and ADHD and examined subtype and gender differences as moderators of genetic association. They found that the evidence for association and linkage was strengthened when analyses were restricted to male subjects with the inattentive ADHD subtype, showing significant preferential transmission of the methionine allele (rather than the valine allele) to boys with ADHD. Furthermore, there was significant evidence for association between *COMT* and ADHD among girls with the valine allele being over-represented, consistent with the original association reported. Thus, these findings suggest an important sex difference in the relation of *COMT* to ADHD. Importantly, these results are consistent with the findings from a study of *COMT* knock-out mice, which found similar gender differences. As a result, additional studies of association and linkage between *COMT* and ADHD are needed that focus on identifying moderating variables such as children's sex and ADHD subtypes or symptom dimensions.

Dopamine D5 Receptor

The dopamine D5 receptor belongs to a class of dopamine receptors distinct from the dopamine D4 receptors and is expressed in different areas of the brain, most predominantly in the hippocampus which is involved in spatial mapping and memory. Studies that have tested for association and linkage between ADHD and the dopamine D5 receptor gene (*DRD5*) have almost exclusively focused on a highly polymorphic dinucleotide repeat 18.5 kb 5′ of the

gene. Initial studies reported at least suggestive evidence for association and linkage between ADHD and *DRD5*, but an interpretation of their results was not straightforward with respect to allelic association, given that some of the studies' findings differed as to which allele was being preferentially transmitted.

In an attempt to clarify the nature of the relation between *DRD5* and ADHD, a combined analysis of the data from 18 independent samples was performed that examined the evidence for association and linkage between ADHD and the 148-bp allele of the *DRD5* dinucleotide repeat. Importantly, the authors of this combined analysis did not detect significant heterogeneity among samples, and thus were able to conduct their analyses on the combined samples. The combined samples showed clear evidence for the preferential transmission of the 148-bp allele ($p = 0.00005$, odds ratio = 1.24) providing strong support for association and linkage between *DRD5* and ADHD.

Dopamine D1 Receptor

The dopamine D1 receptor gene (*DRD1*) gained attention as a candidate gene for ADHD due to several converging lines of evidence suggesting its involvement in the development of ADHD symptoms. First, dopamine D1 receptors are present in the prefrontal cortex and striatum, two brain regions widely believed to be involved in ADHD. Second, dopamine D1 receptors have been shown to influence working memory processes localized in the prefrontal cortex, which appear to be impaired in ADHD. Third, *DRD1* knockout mice have displayed hyperactive locomotive behavior, and thus provide a promising animal model of ADHD.

Based on these converging lines of evidence, two studies of association and linkage between ADHD and *DRD1* have been conducted. The first used four previously identified nonfunctional, biallelic polymorphisms including one marker in the 3′ UTR, two in the 5′ UTR, and one that lies upstream of the promoter region. Tests of association at each marker yielded statistical trends toward association for the two markers in the 5′ UTR and the marker in the 3′ UTR. There was less evidence for association and linkage between ADHD and the marker upstream of the promoter region. The authors then constructed haplotypes from the four markers, and found three that were common in their sample, one of which was preferentially transmitted to ADHD children. Further, it was demonstrated that this haplotype appeared to be more strongly associated and linked with inattentive than hyperactive–impulsive symptoms. These findings were partially replicated in an independent sample that tested for association and linkage between ADHD and *DRD1* using the two identified SNPs in the 5′ UTR. Thus, the studies conducted to 2007 provide promising evidence suggesting a relation between *DRD1* and ADHD.

Dopamine Beta Hydroxylase

Norepinephrine is a widely distributed neurotransmitter in the brain hypothesized to be involved in processes of behavioral arousal and learning and memory. Dopamine beta hydroxylase converts dopamine to norepinephrine and thus represents an interesting candidate gene for ADHD given the suggestion that the underlying pathophysiology of ADHD involves norepinephrine as well as dopamine. Further, a functional polymorphism within the *DβH* gene has been shown to strongly influence dopamine beta hydroxylase levels in plasma and cerebrospinal fluid, providing strong evidence for *DβH* involvement in noradrenergic regulation in the brain. Of direct relevance to ADHD, *DβH* knockout mice display hypersensitive responses to amphetamine treatment, such that they exhibit increased locomotive behavior relative to wild-type, control mice.

Five research groups have published studies of association and linkage between ADHD and *DβH*, with most of these studies focusing on a *Taq*I polymorphism in intron 5 of the gene. Of the studies that have focused on this marker, each one reported evidence that was significant or suggestive of association with ADHD, but importantly the studies differed with regard to which allele was associated with increased risk for developing the disorder. More specifically, four studies suggested that the A2 allele was related to ADHD, whereas one study reported that the A1 allele was associated with increased risk. Nonetheless, it is noteworthy that those studies that examined additional markers found no evidence for association between any polymorphisms other than the *Taq*I polymorphism and ADHD. Further, of those studies that conducted haplotype analyses, the authors reported that the evidence for association with these haplotypes were no stronger than those for the *Taq*I polymorphism by itself. Thus, given the potential role of both norepinephrine and dopamine in ADHD, as well as the positive association reported in several studies, *DβH* represents an interesting candidate gene for ADHD that warrants further study.

Norepinephrine Transporter

Like the dopamine transporter, the norepinephrine transporter is a protein responsible for the reuptake of neurotransmitters, in this case norepinephrine, from the synaptic cleft back to the presynaptic neuron. Unlike the dopamine transporter, however, it is highly expressed in the frontal lobes, and thus represents an important mechanism for the regulation of norephinephrine activity in the prefrontal cortex. Given the hypothesis that noradrenergic dysregulation might be an underlying cause of ADHD, researchers have begun to examine the potential role of the norepinephrine transporter in ADHD. Much of this attention has come from pharmacological studies demonstrating that stimulant medications lead to reductions in ADHD symptoms through increases in dopamine and norephinephrine activity, as well as from studies showing that tricyclic antidepressant medications also lead to reductions in ADHD symptoms, via blocking activity of the norepinephrine transporter. Most recently, treatment outcome research has shown that a drug that specifically blocks the reuptake of norephinephrine (i.e., atomoxetine) leads to significant improvements in ADHD-related symptoms. Thus, the gene that codes for the norepinephrine transporter (*NET1*) has recently received attention as a candidate gene for ADHD.

Four studies have been published examining the relation between *NET1* and ADHD, which have yielded largely negative results. An initial study examined three polymorphisms within the gene, located in exon 9, intron 9, and intron 13, and a second study examined a SNP in intron 7 and the same intron 9 SNP genotyped in the first study. Although both studies failed to detect evidence of association between *NET1* and ADHD, it is important to note that the markers selected in both studies were located at the 3′ end of the gene and were in strong LD with each other. As a result, these studies might have failed to detect an association between *NET1* and ADHD if the susceptibility locus was found to be at the opposite end of the gene (i.e., the 5′ end). To evaluate this possibility, a more recent study examined the relation between *NET1* and ADHD using 21 SNPs that were spaced across the length of the gene to provide a more comprehensive test of association. Nonetheless, this study also failed to detect a significant relation between *NET1* and ADHD. Despite these negative findings, another study that examined just two SNPs within *NET1* did report significant evidence of association and linkage between these SNPs and ADHD. Nonetheless, this study tested for association and linkage between ADHD and 11 other genes, in addition to *NET1*, without correcting for multiple testing. Thus, it is possible that this result represents a false positive. As a result, there is little current evidence to support a relation between *NET1* and ADHD, though this gene is likely to receive further interest as a candidate for ADHD given the research literature suggesting that noradrenergic dysregulation may represent an underlying cause of ADHD.

Adrenergic 2A Receptor Gene

The noradrenergic and adrenergic neurotransmitter systems are hypothesized to influence attentional processes and certain aspects of executive control. More specifically, it has been suggested that adrenergic neurons influence attention and executive processes through the inhibition of noradrenergic neurons and that abnormalities in this regulatory system might contribute to a specific subtype of ADHD. Thus, genes involved in the adrenergic neurotransmitter system represent interesting candidate genes for ADHD. As specific genes in the noradrenergic system have already been discussed (i.e., *DβH* and *NET1*) the following section focuses on published studies that have examined evidence for association and linkage between the adrenergic 2A receptor gene (*ADRA2A*) and ADHD.

The *ADRA2A* gene has been widely studied and there are now seven published studies that have examined the relation of this gene with ADHD. Each of these studies has focused on a *Msp*I restriction site polymorphism in the promoter region of the gene, though some studies have also genotyped additional polymorphisms. The first association that was reported between *ADRA2A* and ADHD was detected in a sample that was initially selected for the presence of Tourette's syndrome and was subsequently diagnosed with ADHD. The authors reported that the G allele of the *Msp*I polymorphism, which indicates the presence of the restriction site, was positively associated with ADHD. Given that the sample was originally selected for Tourette's syndrome, several additional research groups tested this relation in samples of children selected for ADHD, without the presence of comorbid Tourette's syndrome, to determine if the original reported association would generalize to the wider ADHD population. Of the six studies that followed up this initial report, four have yielded significant evidence for association between the G allele of the *Msp*I polymorphism and ADHD, one yielded evidence suggesting a trend for such an association, and one study failed to detect any evidence of such an association. It is also noteworthy that two of these studies yielded evidence suggesting that *ADRA2A* is strongly associated with both the hyperactive–impulsive and inattentive ADHD symptom dimensions. Thus, the results are fairly consistent across studies providing support for the involvement of *ADRA2A* in the pathophysiology of ADHD.

Serotonin Transporter

Like the dopamine and norepinephrine transporters, the serotonin transporter is a solute carrier protein responsible for the reuptake of neurotransmitters, in this case serotonin, from the synaptic cleft back to the presynaptic neuron. Serotonin dysregulation has been related to impulsive and aggressive behavior in children and thus has been hypothesized as a causal factor in ADHD. Involvement of the 5-HT transporter gene (*5-HTT*) in ADHD is suggested by studies that have demonstrated that the binding affinity of the platelet serotonin transporter shows a positive relation with impulsive behavior, such that increases in binding affinity, which corresponds to lower levels of available serotonin, are associated with increases in impulsive behavior in children with ADHD. In addition, pharmacological studies have demonstrated that the serotonin-selective reuptake inhibitors used to

treat depression by blocking activity of the serotonin transporter, thereby increasing levels of available serotonin, also lead to reductions in ADHD symptoms. In light of this evidence, *5-HTT* has been widely studied as a candidate gene for ADHD.

Seven studies have been published examining the relation between *5-HTT* and ADHD, and all of these have focused on a 44-bp insertion/deletion in the promoter region leading to long and short alleles that are believed to have functional consequences. More specifically, the long variant appears to be associated with more rapid serotonin reuptake, and thus, lower levels of active serotonin, whereas the short variant appears to be associated with reduced serotonin reuptake. Of the seven studies, five have reported evidence suggesting that the long allele is associated with ADHD providing fairly strong evidence for a relation between *5-HTT* and ADHD. In addition, one of these studies also found that the evidence for association was stronger among the ADHD combined subtype than the inattentive subtype, and, while this finding clearly requires replication, such studies have the potential to further our understanding of the relation between *5-HTT* and ADHD.

Serotonin 1B Receptor Gene

As described, serotonin dysregulation has been hypothesized to underlie the impulsive symptoms of ADHD. In addition to the serotonin transporter, the serotonin 1B receptor gene (*HTR1B*) has received attention as a candidate gene for ADHD. Specific evidence supporting *HTR1B* involvement comes from a study of knockout mice lacking this gene suggesting that these mice show increased aggression and impulsive behavior and fail to show the normal hyperlocomotion associated with amphetamine administration.

Five studies have been conducted examining *HTR1B* as a candidate gene for ADHD, with all of the studies focusing on the G861C polymorphism. Four of these studies utilized clinic-referred samples and the fifth study utilized a community-based sample. Importantly, each of the studies utilizing a clinic-referred sample reported evidence that the 861G allele was associated with increased risk for ADHD, whereas the single study utilizing a community-based sample failed to detect a relation between the G861C polymorphism and ADHD. This difference in findings might suggest that the association between *HTR1B* and ADHD may not generalize beyond clinic-referred samples, but additional studies utilizing community-based samples are needed before such a conclusion can be made. Nonetheless, several of the studies utilizing clinic-referred samples conducted important follow-up analyses in an attempt to explain the relation between *HTR1B* and ADHD further. For example, two studies found that the evidence for association and linkage between *HTR1B* and ADHD was stronger for the inattentive subtype than the combined subtype. Taken together, the evidence suggesting a relation between *HTR1B* and ADHD is fairly consistent, providing strong support for the involvement of this gene in the pathophysiology of ADHD.

Tryptophan Hydroxylase and Tryptophan Hydroxylase 2

Tryptophan hydroxylase (TPH) is an enzyme crucial to the synthesis of the neurotransmitter serotonin. The *TPH* gene was originally thought to be solely responsible for TPH production, but more recently, a second gene, *TPH2*, was identified that is highly involved in TPH production. Researchers have since focused on both genes as candidates for behavioral disorders characterized by impulsivity and aggressiveness, which have been related to serotonin dysregulation. Nonetheless, results from studies testing for an association between *TPH* and ADHD have been largely negative, and thus, are not reviewed.

The two studies that have tested for association between *TPH2* and ADHD, however, have yielded positive evidence suggesting that this gene may be involved in the etiology of ADHD. The authors of the first study genotyped eight SNPs located in introns 4, 5, 7, 8, and 9 of *TPH2*, and they reported significant evidence for association between a SNP in intron 5 and ADHD that was strengthened when a haplotype was created using this SNP as well as a second SNP in intron 5. The second study genotyped three different SNPs, two of which were located in the regulatory region of the gene at the 5′ end of the gene and a third that was located in intron 2. The authors reported significant evidence for association between the two SNPs in the regulatory region of *TPH2* and ADHD. In addition, the evidence for association was strengthened when a haplotype constructed from the two regulatory region SNPs was tested in relation to ADHD. Thus, despite including SNPs from different regions of *TPH2*, both studies were suggestive of an association between *TPH2* and ADHD.

Monoamine Oxidase Genes

The monoamine oxidase genes (*MAOA* and *MAOB*) are located in close proximity to one another on the X chromosome and encode enzymes involved in the metabolism of dopamine, serotonin, and norepinephrine. Treatment studies have suggested that monoamine oxidase inhibitors (MAOIs) can reduce ADHD symptom levels. Given that each of these neurotransmitters are thought to be involved in the etiology of ADHD, the two monoamine oxidase genes, *MAOA* and *MAOB*, represent interesting candidate genes for ADHD. More specific support for *MAOA* comes from a linkage study conducted in a large Dutch family, demonstrating a relation between *MAOA* and impulsive,

aggressive behavior. In addition, *MAOA* knockout mice have been shown to display increased levels of aggressive behavior associated with increased levels of monoaminergic neurotransmitter levels.

Five published studies have examined a possible relation between *MAOA* and ADHD, largely focusing on a dinucleotide repeat in intron 2 of the gene, a 30-bp VNTR in the promoter region of the gene, and a SNP in exon 8. The VNTR has received particular interest due to studies suggesting an association between this polymorphism and impulsive, aggressive behavior. The VNTR consists of alleles containing 2, 3, 3.5, 4, and 5 copies of the repeat sequence. The two- and three-repeat alleles have been shown to be less efficiently transcribed and have been associated with impulsivity and aggression in previous studies. Thus, the two- and three-repeat alleles have been designated 'low-activity' alleles, while the remaining alleles have been designated as 'high-activity' alleles.

Studies testing for association and linkage between *MAOA* and ADHD have all reported significant evidence suggesting such a relation. Nonetheless, the reported findings differed across studies, both with regard to which polymorphism yielded significant evidence of association and which allele within a polymorphism was the risk-inducing allele. As a result, there is consistent evidence implicating *MAOA* in the pathophysiology of the disorder, but the differences across reports make it difficult to offer substantive conclusions regarding the nature of this association. In contrast, findings from two published studies that have tested for association between *MAOB* and ADHD have been more consistent. These studies focused on a dinucleotide repeat in intron 2 of the gene and both studies failed to detect evidence for association, suggesting that *MAOB* is not involved in the pathophysiology of ADHD.

Synaptosomal-Associated Protein 25 Gene

Researchers have also examined association and linkage of ADHD with candidate genes outside of the major neurotransmitter systems. Synaptosomal-associated protein 25 gene (*SNAP-25*) is an example of such a gene, as it codes for a protein involved in the docking and fusion of synaptic vesicles in presynaptic neurons necessary for the regulation of neurotransmitter release. The *coloboma* mouse strain, which has been bred lacking one copy of the *SNAP-25* gene following a radiation-induced deletion of a segment of DNA on one chromosome, displays hyperactive behavior and provides a potential animal model of ADHD. Thus, several studies have tested for linkage and association between *SNAP-25* and ADHD.

The two most commonly studied markers in the *SNAP-25* gene are SNPs at positions 1065 and 1069. Three initial studies tested for association between *SNAP-25* and ADHD using these two markers, and each study reported evidence suggestive of such a relation. The first published study also reported that a haplotype constructed from these markers showed significant evidence of association and linkage with ADHD. The second study reported a significant association between the polymorphism at position 1069 and ADHD, but the result conflicted with the association reported in the initial study as to which allele within the polymorphism was the risk-inducing allele. Nonetheless, the third study reported results that were consistent with the initial published report. The authors of this study reported a trend for biased transmission of the same haplotype implicated in the first study, and importantly, follow-up analyses revealed a parent-of-origin effect for the transmission of this haplotype. They found that the evidence for association became significant when paternal transmission of the haplotype was examined but not when maternal transmission was examined suggesting that genomic imprinting may be involved. Imprinting refers to specific regions of the genome where only the maternal or paternal copy of a gene is expressed. Thus, the expressed copy of an imprinted gene is either paternally or maternally inherited, and as result, if a disorder is associated with an imprinted gene it will follow the same inheritance pattern.

Evidence suggesting association and linkage between *SNAP-25* and ADHD has also been detected in studies that have examined polymorphisms other than the two SNPs described. For example, association between *SNAP-25* and ADHD has been reported in one study that identified a tetranucleotide repeat polymorphism that lies in the first intron of *SNAP-25*. Further, the authors of this study expanded their analyses to include seven additional polymorphisms, including the SNPs at positions 1065 and 1069. They reported significant evidence for association between three individual polymorphisms and ADHD, namely a SNP in the promoter region of the gene, the tetranucleotide repeat, and a SNP in exon 7. In addition, they reported that several haplotypes showed significant evidence for association that was stronger than the evidence obtained from the individual markers. Finally, follow-up analyses suggested that the findings for the individual markers were stronger when only paternal transmissions of the putative 'high-risk' alleles were included in the analyses. Thus, this study not only provides additional evidence supporting the involvement of *SNAP-25* in the etiology of ADHD, but it also provides additional evidence suggesting that genomic imprinting may be involved in the transmission of genetic risk for ADHD at the *SNAP-25* gene.

Future Directions

This review concludes with a consideration of some of the more important themes emerging from molecular genetic studies of ADHD and related psychopathology that will inform future research in this area. These include the replicability and consistency of findings of association and linkage between a candidate gene and a disorder, the transition from testing single to multiple markers in candidate genes, the specificity of association and/or linkage findings to particular diagnostic subtypes or symptom dimensions, the heterogeneity of association and/or linkage with a particular disorder due to characteristics of individuals such as age, sex, or age of onset, or due to aspects of the environment (i.e., gene–environment interactions). The last theme involves the use of endophenotypes in molecular genetic studies of psychopathology, namely examining association and/or linkage with some underlying biological or psychological mechanism that is thought to reflect the gene's action more directly than does the disorder of interest.

It should be clear from the preceding section that for each candidate gene studied, there is a mixed picture of positive and negative findings. This is true not only for candidate gene studies of ADHD, but also for those of all other psychiatric and complex medical disorders. Such mixed findings tend to appear as studies of a particular candidate gene accumulate and the effect size typically diminishes from that in the original published study. This phenomenon is well illustrated by the studies of *DAT1*, *DRD4*, and *DRD5* reviewed above. Fortunately, meta-analytic procedures are becoming more common as a framework for systematically evaluating the consistency and replicability of findings of association and linkage of candidate genes with disorders across multiple studies. Such analyses can also test whether there is significant heterogeneity of the effect sizes across studies and are capable of mapping such heterogeneity on to substantively meaningful or methodologically important differences across studies. Meta-analytic methods have recently been used to good effect in reviewing the findings of association and linkage between ADHD and *DRD4* and *DRD5*, as they demonstrated consistent, significant association across studies, even in the presence of mixed findings.

A second theme in the research literature on association of candidate genes with ADHD is the transition from studying a single marker to studying multiple markers in candidate genes. In the literature reviewed above, most of the studies examined a single polymorphism in a particular candidate gene for its association with ADHD. This is problematic for at least two reasons. First, negative findings for the association between a single polymorphism in a candidate gene and a disorder are ambiguous because they would appear to indicate that the gene is not involved in the disorder's etiology. It may be the case, however, that a studied marker in an etiologically relevant candidate gene may not be associated with a disorder simply because it is not in strong enough LD with (e.g., not close enough to) the functional, etiologically relevant polymorphism(s) in the gene. Second, and somewhat paradoxically, positive findings for the association between a single polymorphism in a candidate gene and a disorder also are ambiguous because one may not know whether the studied polymorphism is functional, and thus the risk-inducing polymorphism. Further, it is possible that certain genes contain multiple functional, etiologically relevant polymorphisms. Thus, even if the studied marker is known to be functional, it may not be the only functional marker in the gene. The difficulty this poses is that even if one finds a significant association between a disorder and a single marker in a candidate gene, one cannot estimate accurately the magnitude of the gene's role in the etiology of the disorder because one is limited to inferring this from only one of the possible functional, etiologically relevant markers in the gene.

A third theme for future studies of candidate genes and ADHD involves the specificity of association and/or linkage findings to particular diagnostic subtypes or symptom dimensions. It is highly unlikely that whatever genes confer risk for ADHD work at the level of the overall diagnosis, and that nature so closely resembles the current version of the Diagnostic and Statistical Manual (DSM). Thus, it is possible that whatever genes contribute to risk for ADHD do so by conferring risk for specific diagnostic subtypes or symptom dimensions. Although this area of molecular genetic research is only in its infancy, there have been a few examples of such findings in this research domain. For example, some studies have suggested that *DRD4* is more strongly associated and linked with the inattentive than the combined ADHD subtype, and appears to be related more strongly to inattentive than to hyperactive–impulsive symptoms. Although other researchers have focused on examining genetic influences on higher-order diagnostic constructs, such as an externalizing symptom dimension, and have advocated the utility of studying the genetics of broad diagnostic constructs that span several DSM-IV diagnoses, the results cited above suggest that examining association and linkage with more specific diagnostic subtypes or symptom dimensions also will be a fruitful approach. Pursuing both of these possibilities simultaneously in a two-pronged approach is ideal, given the primitive stage of our knowledge of the association between specific genes and disorders, and the likelihood that some genes will be risk factors for several related disorders whereas others will only confer risk on narrower disorder phenotypes.

A fourth theme that is important for future studies of candidate genes and ADHD involves the heterogeneity of association and/or linkage with a particular disorder due to characteristics of individuals such as age, sex, or age of onset, or due to aspects of the environment (i.e., gene–environment interactions). Few molecular genetic studies of ADHD have examined such sources of heterogeneity, and, given that additional analyses such as these will increase the rate of false-positive results due to multiple statistical tests, some caution is warranted when conducting such analyses. Nonetheless, prudently selected characteristics, particularly those shown to be biologically relevant to the disorder of interest and/or the candidate gene being studied, have the potential to inform future candidate gene studies. For example, age and sex represent potential sources of heterogeneity given research showing that several candidate genes show important sex or age differences in expression. Within the dopamine system, for instance, levels of the dopamine transporter have been shown to be higher in males than females and to decline appreciably with age. Despite these findings, none of the published studies of *DAT1* and ADHD have examined sex or age differences in association. As reviewed above, studies of association and linkage between *COMT* and ADHD have yielded mixed findings. In the most recent of these studies, however, analyses were conducted separately by sex and suggested sexually dimorphic findings, with the low-activity methionine allele being associated with ADHD in boys but the high-activity valine allele being associated with ADHD in girls. Although these results are preliminary, confined to one study, and need to be replicated, they embody the type of heterogeneity analyses that may be useful for elaborating the nature of the relations between candidate genes and ADHD.

Although developmental psychopathology researchers have long been excited by the prospect of gene–environment interaction, and many have contended that one cannot understand the development of psychopathology without the consideration of such processes, initial studies identifying specific gene–environment interactions for psychopathology have only recently been published. For example, one such study found that risk for adolescent antisocial behavior and violence was in part determined by an interaction between the presence of abuse during early childhood and alleles at a functional polymorphism in the *MAOA* gene. Importantly, it remains an empirical question whether such gene–environment interactions are present in the etiology of ADHD. As described above, twin studies suggest substantial genetic influences, small-to-moderate nonshared environmental influences, and little-to-no shared environmental influences in the etiology of ADHD. Thus, such gene–environment interactions may not be as relevant to molecular genetic studies of ADHD as for other conditions. Nonetheless, twin studies typically assume that genetic and environmental influences combine in an additive rather than multiplicative fashion and cannot be used either to support or refute the presence of gene–environment interactions. Thus, studies of gene–environment interactions in ADHD should be pursued, but few such interactions have been posited and/or studied in the ADHD literature.

Gene–environment interactions that have been studied in relation to ADHD include interactions between

DAT1 genotype and maternal smoking and maternal alcohol consumption during pregnancy, two environmental risk factors that have been related to ADHD. Unfortunately, the results from these studies have been mixed in their support of such interactions, and thus are not described in detail here. Nonetheless, the relation of such environmental risk factors, as well as other factors such as pre- or perinatal complications and early child abuse, to genetic influences underlying ADHD represents an important line of research that is likely to gain further consideration.

The final theme of this review involves the use of endophenotypes in molecular genetic studies of ADHD. Clearly there is a large gap between candidate genes and the manifest symptoms of disorders such as ADHD as typically assessed by interviews or rating scales. It is desirable from both a conceptual and empirical perspective to find valid and meaningful mediational or intervening constructs that may help to bridge this gap. The term 'endophenotype' is often used to describe such constructs and the variables that are used to measure them. More generally, endophenotypes refer to constructs that are thought to underlie psychiatric disorders. Thus, they are hypothesized to lie closer to the immediate products of such genes (i.e., the proteins they code for) and are thought to be more strongly influenced by the genes that underlie them than the manifest symptoms that they in turn undergird. Endophenotypes also are thought to be 'genetically simpler' than the manifest disorders or their symptom dimensions such that there are fewer individual genes (or sets thereof) that contribute to their etiology suggesting that they may be more straightforward to study.

The use and evaluation of putative endophenotypes in molecular genetic studies of ADHD is in its infancy, such that only a few studies have examined association and linkage between candidate genes and plausible measures of endophenotypes for ADHD. These studies have focused almost exclusively on measures of sustained attention and executive functions as endophenotypes for ADHD, given their posited relation to ADHD. Researchers proposing that sustained attention and executive functions might serve as useful endophenotypes for ADHD cite empirical studies demonstrating that children with ADHD perform poorly on these tasks relative to control children. Importantly, such studies provide the basis for recent theoretical accounts of ADHD that focus on deficits in executive functioning as the core mechanisms underlying the disorder.

Results from early studies that have included measures of sustained attention and executive functions as endophenotypes for ADHD, however, have yielded a mixture of positive and negative findings that have proven to be as complex as those reported for the ADHD diagnosis itself. As such, these studies have yet to provide results that are more informative or more consistent in explaining the relation between specific candidate genes and ADHD than studies that have focused solely on ADHD as the phenotype. Nonetheless, there are several possible explanations as to why sustained attention and executive function measures have thus far proved of limited utility in molecular genetic studies of ADHD. For example, a prerequisite for the validity and utility of putative endophenotype measures is that they represent heritable traits and demonstrate shared genetic influences with the disorder of interest. Nonetheless, large-scale, quantitative genetic studies with sufficient statistical power to estimate the heritability of such measures and the etiology of their overlap with ADHD symptoms have yet to be conducted. Thus, some measures may prove to be inappropriate as endophenotypes. Therefore, while putative endophenotypic measures hold much promise for identifying susceptibility genes and explaining their relation to psychiatric disorders, such issues must be addressed before any findings of association between candidate genes and endophenotypes can be fully interpreted.

Summary

This review has attempted to summarize current studies and some of the most exciting recent developments in molecular genetic research on ADHD. In addition to reviewing extant findings for the association and linkage of ADHD with candidate genes, the review also focused on several emerging themes in this literature that should guide future research. These themes include the 'replicability and consistency' of findings of association and linkage between candidate genes and ADHD, the transition from the use of single to multiple polymorphisms to characterize variation in candidate genes, the 'specificity' of association and/or linkage findings to particular ADHD diagnostic subtypes or symptom dimensions, the 'heterogeneity' of association and/or linkage between candidate genes and ADHD due to characteristics of individuals or to aspects of their environments (i.e., gene–environment interactions), and the use of 'endophenotypes' (i.e., underlying biological or psychological mechanisms thought to reflect more directly the gene's action) in molecular genetic studies of ADHD. It is hoped that these themes not only provide a glimpse of extant molecular genetic research on ADHD and its development, but will help to set the research agendas for future studies.

Acknowledgments

Preparation of this article was supported in part by NIMH grants F31-MH072083 to I R Gizer and K01-MH01818 to I D Waldman.

See also: Behavior Genetics; Developmental Disabilities: Cognitive; Fetal Alcohol Spectrum Disorders; Fragile X Syndrome; Genetic Disorders: Sex Linked; Genetic Disorders: Single Gene; Genetics and Inheritance; Learning Disabilities; Mental Health, Infant; Nature vs. Nurture; Sensory Processing Disorder; Sleep; Special Education; Television: Uses and Effects.

Suggested Readings

Doyle AE, Faraone SV, Seidman LJ, *et al.* (2005) Are endophenotypes based on measures of executive functions useful for molecular genetic studies of ADHD? *Journal of Child Psychology and Psychiatry and Allied Disciplines* 46: 774–803.

Faraone SV, Perlis RH, Doyle AE, *et al.* (2005) Molecular genetics of attention-deficit/hyperactivity disorder. *Biological Psychiatry* 57: 1313–1323.

Heiser P, Friedel S, Dempfle A, *et al.* (2004) Molecular genetic aspects of attention-deficit/hyperactivity disorder. *Neuroscience and Biobehavioral Reviews* 28: 625–641.

Thapar A, O'Donovan M, and Owen MJ (2005) The genetics of attention deficit hyperactivity disorder. *Human Molecular Genetics* 14: R275–R282.

Waldman ID (2005) Statistical approaches to complex phenotypes: Evaluating neuropsychological endophenotypes for attention-deficit/hyperactivity disorder. *Biological Psychiatry* 57: 1347–1356.

Waldman ID and Gizer I (2006) The genetics of attention deficit hyperactivity disorder. *Clinical Psychology Review* 26: 396–432.

Adoption and Foster Placement

K Bernard and M Dozier, University of Delaware, Newark, DE, USA

Glossary

Adoption – The permanent placement of a child in surrogate care involving the legal transfer of parental rights from the biological parents to the adoptive caregiver(s).

Attachment – The affectional tie from a child to his or her caregiver which is further characterized by a child's use of that figure as a safe haven for comfort and a secure base for exploration.

Foster care – The temporary placement of a child in surrogate care through the public child welfare system.fm>/glossary-def>

Institutional care – A common pre-placement experience of internationally adopted children involving group care in a residential facility, such as an orphanage.

Kinship care – The formal or informal foster care placement with biological relatives rather than unrelated foster caregivers.

Maltreatment – Acts of physical abuse, sexual abuse, emotional abuse, and/or neglect against a child.

Open adoption – Type of adoption involving continued contact among biological parents, adoptive parents, and children.

Surrogate care – The general term for a caregiving arrangement in which someone other than the biological parent is caring for the child; different types of surrogate care vary in duration and degree of permanency.

Introduction

Children's early relationships have important effects on physical, emotional, and social development. Needs of nutrition, affection, and stimulation are all met within the immediate context of caregiving and the broader context of family environment. Adverse prenatal conditions (e.g., malnutrition, drug exposure) along with postnatal adversities (e.g., poverty, maltreatment, neglect) threaten the well-being of a child and may result in removal from biological parents and placement in alternative care. Whereas these disruptions in care may be necessary for the safety of the child, any changes may have considerable effects on development.

Foster care and adoption are two types of surrogate care with inherent developmental risk factors. The foster care system serves to protect children from adverse living and family environments by placing them in out of home care. Placement in foster care may result from child neglect, abuse, homelessness, abandonment, or parental problems (e.g., incarceration, substance abuse, illness). Adoption is the permanent placement of a child with substitute caregivers involving the legal transfer of parental rights. Adoptions can be classified as domestic or international. Domestic adoption (i.e., adoption of children from within the US) often takes place through the public child welfare system. Private domestic adoptions can be arranged directly between birth parents and adoptive parents with the help of an intermediary or through private state-licensed agencies. Although nearly twice as many children are adopted domestically as internationally, the number adopted internationally has increased dramatically since the mid-1990s. Children adopted internationally have often spent a considerable amount of time in institutional care, many experiencing inadequate nutrition, poor medical care, and lack of social interaction during that time.

While we will primarily focus on foster care and adoption, there are other types of care that are worth mentioning. Kinship care refers to arrangements where relatives care for children when biological parents are unable to do so. In some instances, kinship care allows for children to continue contact with family members. Children in kinship care, however, often remain in problematic environments. Foster children are sometimes placed with relatives through the child welfare system, but informal arrangements, both temporary and long-term, are often made as well.

Adopted and foster children have a range of experiences before and during care which account for individual differences in later adjustment. Infants adopted at birth experience continuous care and show positive outcomes as a group. These children look comparable to children raised continuously by a biological parent. Thus, we will mainly focus on children who have not experienced continuity in care, specifically children placed in foster care or adoptive homes following experiences with a previous caregiver (e.g., biological parent, institutional caregiver). Pre-placement experiences of children who are not placed at birth may involve multiple stressors. Furthermore, changing caregivers represents a major disruption in a child's life. Whether or not significant problems result depends considerably on the quality of subsequent care.

Attachment

According to attachment theory, as proposed by John Bowlby, there is an evolutionary benefit of forming a close relationship with a primary caregiver. Attachment behaviors (e.g., crying, reaching, crawling) serve to increase proximity between an infant and his or her caregiver. The attachment is the tie from a child to a specific attachment figure characterized by the use of that figure as a secure base for comfort and exploration. The attachment behavioral system is activated when an infant perceives a threat. An infant seeks his or her primary caregiver upon becoming frightened, hurt, or distressed, but engages in exploration of the environment when threat is minimal.

Typically, a pattern of attachment develops within the first year of life. By 12 months of age, most infants will have expectations of attachment figures that are based on repeated interactions. Infants form coping strategies, or organized behavioral responses, that reflect these expectations. Given that a key developmental task

for infants and young children is forming and maintaining attachments to primary caregivers, it is not surprising that the conditions associated with foster and adoptive care are often challenging for children.

History of Care

Substitute care was a necessary social convention long before formal legal policies were established. Orphaned children were often cared for by relatives or placed in group care facilities. Early foster care and adoption practices generally served the needs of the caregiver over the needs of the child. Children were placed into homes or adopted into families in order to provide indentured service or labor. In the 1800s, there was an increase in the number of orphaned children in urban areas due to the Industrial Revolution and massive immigration. These dependent children were often sent west by way of 'orphan trains' to homes of farm families who provided free care in exchange for the children working for them. In the early 1900s, local foster families were proposed as an alternative to previously accepted solutions for orphans. Formal agencies were established to supervise this practice.

In 1851, the first legal adoption policies were established in Massachusetts which outlined the nature and requirements of transferring care. By 1929, all states had developed legislation for adoption practice. Infant adoption became popular during the early 1900s due to decreased birth rates. Following World War II, international adoption became prevalent to aid in the care of children orphaned as a result of the war. International adoption persisted as the number of adoptable infants in the US was fewer than the number of couples wishing to adopt.

Recent US legislation has focused on policy regarding the domestic adoption of children in foster care. In 1980, the Adoption Assistance and Child Welfare Act established the goal of permanent placement of children in foster care through either timely return to biological parents or planning for adoption. Although there was a reduction in the number of children in foster care immediately following this legislation, it did not last and length of time spent in foster care remained high. The Adoption and Safe Families Act (ASFA) of 1997 reiterated goals of serving the best interests of the child. The ASFA stressed that children's safety was of primary concern when planning for reunification or adoption. The legislation further ordered that attempts at reunification with biological parents should not continue after 15 months of foster care placement; that is, after a child has been in foster care for 15 months of a 22-month period, a petition should be filed for the termination of parental rights. Thus, a primary goal of the ASFA is shorter timeframes for permanent placements. To facilitate permanency planning, the ASFA provides guidelines for adoption policies and increased funding to support adoption planning.

Overview of Foster Care

The US Department of Health and Human Services estimated that there were 518 000 children under the age of 20 years in foster care in 2004. This figure represents a significance increase since 1980. The mean age of children in care is 10 years old, but recent trends show increasing numbers of infants and younger children in the system. Of the 305 000 children who entered foster care in 2004, it was estimated that one-third were between birth and 3 years old. In attempts for prevention and early intervention, child welfare agencies have increasingly focused on identifying infants and toddlers who have experienced abuse or neglect. For example, the Child Abuse Prevention and Treatment Act (CAPTA) amendments of 2003 addressed the needs of infants born affected by illegal substance abuse by requiring the notification of child protective services and the development of a plan of safe care. As a result, the identification of cases of prenatal drug exposure may account for the growing number of infants and toddlers entering care.

Neglect, the failure to provide adequate care for a child, is a common reason for foster care placement. Abandonment and failure to provide healthcare are considered acts of physical neglect, whereas emotional neglect includes inattention to needs for affection, failure to provide psychological care, and domestic violence. Neglect is associated with numerous stressors including parental substance abuse, poverty, and homelessness. Other reasons for entry into foster care include child abuse, parental illness, parental incarceration, and parental death.

Decisions about where to place a child involve multiple factors, including availability or willingness of relatives to provide care, proximity of caregivers to birth parents, special needs of the child, and goals of permanence. Placement in nonrelative foster family homes accounted for 46% of the foster care settings in 2004. Other placements included kinship care (24%), institutions (10%), group homes (9%), pre-adoptive homes (4%), trial home visits (4%), runaways (2%), and supervised independent living (1%) as reported by the US Department of Health and Human Services.

For the most part, case goals reflect the ASFA of 1997 in supporting reunification with parents or adoption in a relatively short timeframe. Other goals include long-term foster care, emancipation, living with other relatives, and guardianship. Estimates from the US Department of Health and Human Services report that 283 000 children exited foster care in 2004, of whom 54% were reunited with parents or primary caregivers, and 18% were adopted.

Overview of Adoption

Domestic Adoption

Adoption is ideal for foster care children for whom reunification with biological parents is not an option because it establishes a stable and permanent home and family environment. According to the US Department of Health and Human Services, there were approximately 52 000 children adopted from foster care in 2004, which represents a recent increase likely due to the ASFA. Of these children, 59% were adopted by foster parents, 16% by other nonrelatives, and 24% by relatives. About one-third years were between the ages of 0 and 3.

Private domestic adoptions involve the adoption of infants within the US. Independent adoption refers to the selection and placement of an infant directly between birth parents and adoptive parents, possibly involving a third party for legal assistance. Private adoptions can also be arranged through a profit or nonprofit agency. Stepparent adoptions are another common subcategory of private domestic adoptions, but they typically are not associated with a change in primary caregiver. The number of private domestic adoptions is not easily measured because states are not required to collect or report this information.

International Adoption

Approximately 6000 children were adopted into the US from overseas following World War II. According to the US Department of State, that number has grown to approximately 23 000 children who were adopted from other countries in 2005. Asia and Eastern Europe have generally been the major sources of internationally adopted children. In 2005, the top countries of origin for adopted children included China (35%), Russia (20%), Guatemala (17%), and Korea (7%). Many of the children adopted from outside of the US have spent 8 months or more in an institution.

Challenges to Children in Surrogate Care

Many children who are adopted or placed in foster care face multiple challenges that put them at risk for maladjustment. Some of these risks relate to the circumstances that they encounter prior to placement (e.g., drug exposure, maltreatment, institutional care) and others relate to the nature of surrogate care itself (e.g., changing caregivers, instability of placement). In considering how these children develop as compared to a normal sample, it is important to keep these factors in mind.

Prenatal Substance Exposure

According to the National Institute of Drug Abuse, 5.5% of mothers report using illicit drugs while pregnant. Prenatal exposure to harmful substances (e.g., cocaine, tobacco, alcohol) is common among children who are removed from the home. Testing positive for substance use at the time of delivery is the primary reason for foster placements at infancy. Findings concerning the immediate and long-term effects of prenatal drug exposure are inconsistent, but a number of studies do suggest an increased risk for developmental problems. Challenges of studying children with prenatal drug exposure arise due to the confounding effects of other prenatal adversities, such as poor maternal nutrition and poor prenatal care during pregnancy.

In substance-exposed infants, there is an increased tendency for physical deficiencies, specifically low head circumference, low birth weight, and growth retardation. Prenatal substance exposure also has subtle developmental effects on the quality of motor responses and regulatory behavior displayed at 1 month of age. Prenatal substance exposure introduces a general susceptibility to significant developmental problems; the environment plays an important role in mediating its effects.

Maltreatment

Children who are placed into foster or adoptive care have often experienced maltreatment. Maltreatment poses a serious problem, especially when it occurs early in life, at a time when children depend on their parents for almost everything. Adverse experiences during these first few years threaten the optimal development of attachment relationships, neurobiological regulation systems, and emotional stability. When needs are not met (i.e., cases of neglect) or interactions are frightening (i.e., cases of abuse), children are unable to depend on their caregivers for support. Although infants form attachments to maltreating caregivers, these attachments are often disorganized, leaving children without a strategy for interacting with parents when distressed.

Experiences of maltreatment can be overwhelmingly stressful to a child. Facing trauma is especially difficult for infants because they are dependent upon caregivers for help with regulating behavior and physiology. Evidence at the neurobiological level supports the disrupting effects of early adversity. For example, the hypothalamic–pituitary–adrenocortical (HPA) system serves as a regulator of daily functioning and also as a stress response system. Children who have experienced maltreatment often show disruptions to diurnal patterns of hormone (i.e., cortisol) production as well as abnormal neuroendocrine reactions to stressful situations.

Institutional Care

Many internationally adopted children are in institutional care prior to placement. Early research has been critical in illuminating the debilitating effects of institutional rearing and driving policy change worldwide. In the 1940s and 1950s several researchers, including Rene Spitz and John Bowlby, observed the conditions of institutions and described the devastating effects of minimal stimulation and social isolation. They suggested that sterile caregiving led to significant and sometimes irreparable delays in cognitive and socioemotional development. Researchers continue to study the effects of institutional care through longitudinal studies using comparison samples. The Bucharest Early Intervention Project, for example, studies children raised in Romanian institutions, previously institutionalized Romanian children raised in foster care, and Romanian children raised continuously by their birth parents. Ongoing research initiatives are beneficial in exposing the nature of present-day institutions and in informing policy decisions for children in out-of-home care.

Although there are differences in levels of privation between institutions and even within institutions, there are multiple factors that potentially put children at risk. For one, there are often problems with providing physical care and healthcare for children in institutions. Due to the nature of institutional care as a public facility serving many children at the same time, these basic needs may go unmet if funding is poor and number of staff members is low. Delays in physical growth result from inadequate nutrition and medical care, but many adopted children do catch up to the normal range after leaving institutions.

The environment of an institution also inhibits development in multiple ways. Limited resources, both interpersonal (e.g., staff) and environmental (e.g., toys), lead to inadequate stimulation. Infants may be kept in cribs without opportunities to explore their environment. Another major issue with institutional care is the changes in caregivers. Due to frequent changes in staff and high staff-to-child ratios, children rarely have one primary caregiver. Interactions are often minimal and unaffectionate. The formation of an attachment relationship is difficult when interactions are infrequent and inconsistent. Immediately following institutional care, children also show developmental delays in motor, cognitive, and language abilities as a result of suboptimal levels of stimulation. Recovery of functioning is seen in some domains following adoption, but there is often limited catch-up in areas such as developing discriminating attachments.

Changing Caregivers

With the exception of children placed into foster or adoptive care at birth, all children in surrogate care have experienced a transition to a new caregiver at least once. Children in foster care often face multiple placements before permanency is established. The experience of changing caregivers has important implications for a child's representation of self as effective and others as reliable. Older children may reflect on the experience of foster placement or adoption as a form of rejection or abandonment. Infants and younger children, who are unable to conceptualize this experience consciously, are still affected by separations from caregivers. Instead of verbally expressing feelings of rejection, they show difficulty in adjusting to new attachment relationships and difficulty in self-regulation.

Issues in Providing Care

Adoptive and foster parents have a unique role in providing care to a child who is not biologically their own. The decision to take on this role is made for different reasons, such as infertility, or a desire to help children in need. Regardless of the reason, providing surrogate care can be a rewarding yet challenging experience.

The Caregiving System

John Bowlby suggested that there is a behavioral caregiving system that involves a set of parental behaviors (e.g., picking up, carrying) that serve to protect a child. Evolutionarily, the caregiving system functions to ensure reproductive fitness through the survival of one's child. The development of this set of caregiving behaviors occurs across the lifespan. Thus early experiences with a caregiver have implications for later experiences as a caregiver. Also contributing to a caregiver's behaviors are the specific experiences and history with his or her child. A child's set of characteristics and behaviors affects how that child's parent will provide care (parenting style); similarly a parent's set of characteristics and behaviors affects how that parent's child will seek and accept care (attachment quality).

Commitment

Whereas in a biologically linked dyad there is the assumption of a stable lasting relationship, this is not always the case with foster care dyads. In foster care, the level of emotional investment from the caregiver is challenged by the nature of foster care as a temporary situation and the lack of biological relatedness. Mary Dozier and colleagues have found that the degree to which foster parents are committed to their children varies with past experience as a foster parent and age of child placement. Specifically, caregivers who have had higher numbers of children in the past reported lower levels of commitment to children presently in their care. Caregivers reported higher levels of commitment to children who were placed at younger ages compared to children placed at older ages. Further, commitment is an important determinant of whether a placement disrupts or endures.

Quality of Attachment

Attachment quality refers to a child's expectations of his or her caregiver's availability and responsiveness. Mary Ainsworth developed the Strange Situation procedure to measure attachment quality. From observations of infants' behaviors in response to multiple stressful stimuli (e.g., an unfamiliar room, an unfamiliar person, brief separations and reunions from a primary caregiver), Ainsworth generated three primary classifications: secure, avoidant, and resistant. An infant with a secure attachment generally has a caregiver who is nurturing and sensitive to his or her needs. This infant seeks out the caregiver directly when distressed for reassurance. An infant with an avoidant attachment to a particular caregiver typically ignores or turns away from that caregiver in times of stress. Rejecting and unresponsive caregivers generally have infants with avoidant classifications as these infants learn that their caregivers are not available in times of need. An infant with a resistant attachment tends to have inhibited exploration and a mixed strategy in using the caregiver as a secure base characterized by both proximity seeking and angry resistance. A resistant pattern of behavior is the result of inconsistent responding by a caregiver to an infant's needs.

A fourth category of attachment quality was identified by Main and Solomon in 1990 to account for infants who did not clearly fit into the established organized patterns of attachment behavior. The disorganized/disoriented category reflects a breakdown in an infant's strategy. Behaviors displayed by infants in this category may include contradictory behavior (e.g., approaching the parent with sharply averted head), apprehensive behavior (e.g., jerking away from the parent with a fearful expression), or confused behavior (e.g., greeting the stranger upon the return of the parent). Disorganized attachment appears to be at least partially the result of caregiving experiences that are frightening, such as abuse. Although infants need their attachment figure as a secure base, they simultaneously fear that figure.

Within intact mother–infant dyads, attachment formation is a gradual development over the first year. Because foster children are often placed at developmental points when they would have already developed attachments, the process by which new attachments develop can be observed at an accelerated rate. When young children older than about 1 year of age are first placed with new caregivers, they often show avoidant or resistant behaviors when distressed. These behaviors elicit non-nurturing behaviors from caregivers. Thus, these young children in foster or adoptive care seem to be 'leading the dance' with their parents initially. Nonetheless, after several months, children develop attachments to parents based on parent characteristics rather than their own. Unfortunately, these children are prone to develop disorganized attachments unless parents behave in nurturing ways.

Some infants in foster care and institutional care display behaviors toward strangers that are extremely disordered, including indiscriminate friendliness and responses of terror. Indiscriminate friendliness describes attempts by infants to use all adults as potential attachment figures. Terror of strangers refers to infants' responses to all new adults as threatening. Both patterns of response place infants at significant risk, as seeking of any available adult is dangerous and failing to form new relationships is equally detrimental. These anomalous behaviors are captured in the Diagnostic and Statistic Manual (4th edition) criteria for reactive attachment disorder (RAD).

Adjustment Outcomes

Although it is difficult to disentangle the effects of surrogate care from the effects of pre-placement experiences and disruptions in care, numerous studies report a heightened risk for maladjustment among these children. In considering how surrogate care affects children's abilities to regulate their behavior, it is important to look at later outcomes. Due to differences between types of care, we will consider adjustment for each group separately.

Infants adopted at birth consistently show favorable outcomes, whereas later-placed children are at increased risk for adjustment problems. Adopted children are at risk for developing problems across multiple domains, including problems in school (e.g., poor concentration, restlessness, 'attention deficit hyperactivity disorder') and in peer relationships (e.g., oppositional behavior, aggression). Externalizing behaviors (e.g., delinquency, substance use) are more common for adopted children than internalizing problems (e.g., depression, anxiety). Adjustment problems greatly diminish by young adulthood.

Compared to adopted children, children in foster care are at a higher risk for behavioral and psychological problems. Foster children are more frequently diagnosed with internalizing and externalizing disorders than comparison peers. Children who have experienced foster care are at significant risk for high rates of problems in academic adjustment, social functioning (e.g., antisocial behavior), and emotional competence (e.g., low self-esteem, negative emotionality). The differences in adjustment between foster care children and adopted children may be the result of variations in several factors, such as number of disruptions in care, caregiver characteristics, and pre-placement experiences.

Factors Affecting Children's Adjustment

Resiliency of children in adoptive and foster care is significantly affected by experiences in subsequent care. Characteristics within the new environment contribute to child functioning, including aspects of the family (e.g., number of children in care, level of income), aspects of the home (e.g., availability of a stimulating and safe environment), and aspects of the community (e.g., school district, support resources). Positive adjustment is associated with authoritative parenting styles, parental acceptance, realistic parental expectations, and flexibility, whereas poorer adjustment is associated with parental annoyance, unrealistic expectations, excessive physical punishment, and inflexibility in parenting. Parental state of mind (autonomous, dismissing, preoccupied, or unresolved), as measured by the 'Adult Attachment Interview', reflects how responsive a caregiver is to his or her child's attachment needs. Autonomous parents, who are consistently sensitive to their infants' needs, tend to have securely attached infants. Security of attachment is also associated with children's social and emotional competency.

Children's perceptions of experiences in surrogate care can further contribute to their adjustment. If adopted children represent placement experiences as rejection by biological parents, they may develop negative self-concepts. Furthermore, these children may have difficulties with identity formation because they do not have access to information from the biological family (e.g., culture, race, history). Open adoption permits the continued connection among all units of the adoptive triad: birth parents, children, and adoptive parents. This practice is becoming more common. Potential benefits include the availability of a child's medical and preadoption history to the adoptive parent, ability of a birth parent to select an adoptive family, and fewer feelings of abandonment experienced by the child.

Interventions

Adoption and foster care are interventions in and of themselves. Despite the positive intentions of these practices, changes in caregiving pose significant challenges to children. Further intervention programs have been developed to target the needs of children in surrogate care. Research concerning the effectiveness of many of these programs is ongoing.

Several interventions for foster children target the need for permanent care. Such programs either focus on achieving timely adoption or preserving the birth mother as a primary caregiver. Shared family foster care is one example of the latter, in which foster parents care for both a biological mother and her child. Thus, caregiving is continuously provided by the biological mother. She is supported and mentored in developing appropriate parenting techniques. Though the models of these programs are empirically based, evidence for their effectiveness is limited at this time.

Other intervention programs serve to enhance attachment to a caregiver. Mary Dozier and colleagues developed the Attachment and Biobehavioral Catch-up program which focuses on fostering attachment quality and self-regulation. This 10-session intervention has three primary aims. First, it teaches foster parents how to reinterpret signals from an infant who may appear not to want support. Second, it helps foster parents overcome their own difficulties in providing sensitive care. Third, it helps foster parents learn to provide a very responsive interpersonal world to improve children's biobehavioral regulation. Thus, parents are helped to change the way in which they respond to their infants' needs (e.g., behavioral cues, need for contact). Preliminary results from this program support the possibility of helping develop secure behaviors and better regulatory capabilities.

Interventions beyond infancy are generally behaviorally based. Philip Fisher developed the Early Intervention Foster Care (EIFC) program for preschool-age children. Through parent training and family therapy, it promotes the development of behavioral control abilities in the child. The EIFC targets several domains including case management, child needs, and the caregiver–child relationship. Parents are taught to respond to their children's needs, support positive child behavior, set limits, and maintain close supervision. The EIFC program also aims to affect neuroendocrine regulation by decreasing child behavior problems and supporting positive parenting processes. Behavioral interventions in middle childhood also teach parents skills in behavior management. The strategies of these programs reflect the changing nature of parent–child relationships later in life. By focusing on parenting strategies, they continue to address any problems as occurring within the dyad rather than within the child.

Conclusion

Children in surrogate care face many challenges that put them at risk for maladjustment. Postplacement experiences have a significant effect on the development of problems later on. Research on interventions that can increase the protective effects of subsequent care can inform policy regarding adoption and foster care practices.

See also: Abuse, Neglect, and Maltreatment of Infants; Attachment; Behavior Genetics; Depression; Emotion Regulation; Endocrine System; Family Influences; Parenting Styles and their Effects; Self-Regulatory Processes.

Suggested Readings

Brodzinsky DM and Palacios JP (eds.) (2005) *Psychological Issues in Adoption.* Westport, CT: Praeger.
Brodzinsky DM, Smith DW, and Brodzinsky AB (1998) *Children's Adjustment to Adoption: Developmental and Clinical Issues.* London: Sage Publications.
Dozier M, Albus K, Fisher PA, and Sepulveda S (2002) Interventions for foster parents: Implications for developmental theory. *Development and Psychopathology* 14: 843–860.
Gunnar MR, Bruce J, and Grotevant HD (2000) International adoption of institutionally reared children: Research and policy. *Development and Psychopathology* 12: 677–693.
Lawrence CR, Carlson EA, and Egeland B (2006) The impact of foster care on development. *Development and Psychopathology* 18: 57–76.
Stovall KC and Dozier M (1998) Infants in foster care: An attachment theory perspective. *Adoption Quarterly* 2: 55–88.

Relevant Websites

http://www.adoptioninstitute.org – Evan B. Donaldson Adoption Institute.
http://www.acf.hhs.gov – US Department of Health and Human Services, Administration for Children & Families.
http://travel.state.gov – US Department of State: Children & Family.

AIDS and HIV

C A Boeving and B Forsyth, Yale University School of Medicine, New Haven, CT, USA

Glossary

Adherence – Routine maintenance of illness management regimen, typically referring to successful compliance with the medication schedule.
Health-related quality of life (HRQOL) – Inclusion of the impact of a disease and its treatment in the assessment of a person's functioning and life satisfaction; domains include physiological, social, educational, emotional, and cognitive functioning.
Highly active antiretroviral therapy (HAART) – Approved in 1998 for use with children, this medication regimen includes a combination of at least three medicines from different classes of medications. The aim of treatment is to suppress

viral replication and reduce the emergence of
resistant viral variants.

Mother-to-child transmission (MTCT) – Route of
HIV infection by which the infant acquires HIV either
prenatally, during the birth process, or postnatally
(typically through breast milk).

Opportunistic infection – Illnesses incurred as a
result of suppression of the immune system. HIV
diminishes the ability of the individual's immune
system to respond to infections that would be
innocuous in immunocompetent individuals.
Opportunistic infections are considered
AIDS-defining illnesses.

Introduction

Since the first description of AIDS in the 1980s, the
epidemic has exploded to impact millions of adults and
children throughout the world. As the epidemic has
expanded, children have become increasingly affected,
both by infection with the virus as well as by parental
illness and death. With more recent advances in prevention and treatment science, the current impact of
the epidemic is glaringly disproportionate between the
western world and resource-poor settings. In developed
countries, the infection rates of children living with the
virus have dropped dramatically, as have rates of orphan-
ing since those individuals who are infected are remaining healthier longer due to availability of life-saving
medications. A very different picture has emerged in the
developing world where the epidemic continues to expand. The implementation of prevention programs and
availability of treatment in these parts of the globe have
grown slowly, and healthcare systems in these settings
need substantial revision and bolstering to handle the
care of millions of people living with the virus.

For those children who have access to treatment, pediatric HIV infection is now considered a chronic, although
life-threatening, illness. This is in contrast to earlier in
the epidemic when it was almost certainly viewed as a
terminal illness. The advent of life-prolonging treatment
facilitates infected children's development into adolescents
and young adults living with HIV, only recently creating
the opportunity to research the impact of HIV/AIDS
throughout the developmental trajectory. This article presents a synthesis of the current state of knowledge regarding infant and young child development in the context of
pediatric HIV/AIDS. Great strides have been made in
prevention and treatment, and there is now a considerable body of work examining the neurodevelopmental
sequelae of HIV in children. However, in many respects,

developmental research in the context of HIV is in its
nascency. The majority of pediatric HIV research has
been conducted in developed nations, which, for certain
aspects of child development, may have limited trans-
latability to low-resource settings. Child development is
inextricably linked with the environment, thus a child's
physiological, psychological, and social development is
closely tied to the child's community and familial
resources. Further research on the impact of the epidemic
upon the development of children in low-resource settings
(e.g., in sub-Saharan Africa) is essential to combat the
effects of the epidemic and to increase our understanding
of the developmental impact of pediatric HIV/AIDS upon
the world's children.

Epidemiology of Pediatric HIV/AIDS: Global Epidemic

The epidemiology of HIV/AIDS is an ever changing
landscape. Not only does the absolute number of infected
people continue to grow, but the demographic of people
living with HIV has broadened. Whereas AIDS was first
identified in homosexual men and intravenous drug users
in the 1980s, currently, heterosexual transmission drives
the epidemic in most parts of the world. Likewise, the
reach of HIV/AIDS now extends beyond adults to children across the globe. HIV/AIDS exerts a harrowing
impact on the world's children, reflected in the World
Health Organization (WHO) estimates that a child is
newly infected with HIV every minute. Global estimates
indicate that approximately 2.3 million children were living with the virus in 2006, which is quite likely to be an
underestimate (higher estimates reach 3.5 million).

The epidemic has followed a very different course
in the developed world as compared to low-resource
settings. In the West, findings from a landmark study
were released in 1994 that identified a protocol to successfully prevent transmission of the virus from a mother
to her child, and now, in the US, fewer than 200 babies are
born infected with the virus each year. The advent of
highly active antiretroviral therapy (HAART) followed
just a few years later, prolonging life and greatly improving the functioning of infected individuals.

These improvements in prevention and treatment
have not been widely implemented in other parts of
the world, resulting in a differential impact of the HIV/
AIDS epidemic upon the children living in developing
nations. According to the United Nations Children's Fund
(UNICEF), approximately 1% of women worldwide are
infected with HIV; however, the vast majority (95%) of
these infected women live in developing countries. The
impact of the epidemic upon children is not only due to
infection with the virus; parental loss due to AIDS has

created 15 million AIDS orphans, the vast majority of whom live in the developing world (**Figure 1**).

Specifically, children living in sub-Saharan Africa have born the brunt of the pediatric epidemic. According to the UNAIDS 2006 Epidemic Update, a preponderance (87%) of children living with HIV/AIDS live in sub-Saharan Africa. Further, estimates released from WHO and UNAIDS indicate that 1400 children die every day from causes attributable to AIDS and about 90% of these deaths occur in sub-Saharan Africa.

The context of HIV disease is also important when considering how children are affected by the epidemic. It is a disease that is commonly associated with poverty and, in the developed world, may also be associated with intravenous drug use. In developing countries, poverty is aligned with nutritional concerns and an increased prevalence of other illnesses such as tuberculosis. Advances in decreasing high child mortality have been reversed. In all societies, AIDS is a stigmatizing disease which contributes to the isolation of those affected by the disease, both those who are infected and also their family members. Millions of children have been orphaned by AIDS. Others are experiencing the illness of parents, sometimes taking on roles of caregiver for their ill parents or assuming the parenting role for younger siblings. Aging grandparents take an expanding number of children into their homes and, with fewer adults generating incomes, food is dispersed more thinly among members of the household.

Transmission of HIV/AIDS to Infants

The growth of the global pediatric HIV epidemic is primarily due to transmission of the virus from HIV-positive mothers to their children. Before the routine use of preventive medications in Europe and the US, approximately 15–25% of children born to HIV-infected mothers were infected with the virus. Rates of transmission vary, however, between developing and developed nations. The virus can also be transmitted through breastfeeding, so in parts of the world where breastfeeding is prevalent, or even essential, rates of transmission may reach as high as 45%. Aside from availability of preventive measures, transmission rates may be influenced by rates of Cesarian section (a protective factor), breastfeeding patterns, prematurity (influenced by availability of obstetric care and the woman's access to proper nutrition), and the extent to which the mother's disease has advanced.

There are three primary mechanisms through which a woman's HIV infection may be transmitted to her infant: (1) *in utero* (prenatally), (2) during the birth process (intrapartum), and (3) via breast milk (postnatally). Although less common than other modes of transmission, the fetus can become infected with HIV *in utero* during pregnancy, particularly if exposed to the mother's blood through bleeding in the placental lining. Invasive procedures, high maternal viral load, vaginal delivery, and prolonged rupture of the amniotic sac during the birth process are all factors associated with increased rates of intrapartum transmission. Postnatally, breastfeeding can pass the virus to the child through colostrum and breast milk. Again, higher maternal viral loads increase the risk of transmission to the infant. Furthermore, likelihood of transmission is enhanced if the infant has sores in his or her mouth or if the mother's nipples are cracked and bleeding. Infants who become infected via breastfeeding typically acquire the virus in the first 6 months although transmission can occur anytime during the period of breastfeeding.

Researchers have striven to identify the transmission mechanisms as well as effective prevention methods. In 1994, results were released from a landmark study

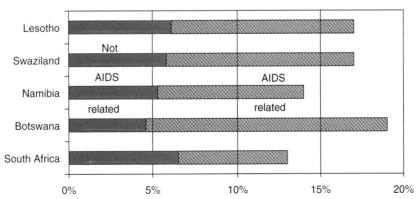

Figure 1   Percentage of children in sub-Saharan Africa who were orphans in 2005. Source UNAIDS/UNICEF, 2006.

conducted by the US and France indicating that mother-to-child transmission (MTCT) can be reduced by two-thirds with the targeted use of antiretroviral medication. Within 2 months of analysis of initial data, the randomized controlled trial was ceased and the protocol (referred to as the AIDS Clinical Trial Group Study, ACTG 076) was deemed the standard of care for HIV-positive pregnant women in countries where it can be afforded. The protocol involves administering the drug to a woman while she is pregnant, giving it intravenously during labor, and then to her newborn for the first 6 weeks of life. Because this regimen is considered too expensive for most resource-poor countries, there has been extensive research to identify less expensive interventions to reduce perinatal transmission. Now, in a number of developing countries, the accepted regimen is the use of two doses of one of the antiretroviral medications, nevirapine. One dose of the drug is given to the mother when she is in labor and one dose to the child shortly after delivery. In total, this regimen costs only about $4.

Along with this protocol, breastfeeding guidelines have been developed to reduce transmission rates. In countries where it is considered affordable, safe, and sustainable, it is suggested that HIV-infected women feed formula to their infants rather than breastfeed. In developing countries (e.g., in sub-Saharan Africa), however, women often do not have consistent access to canned formulas or to clean water to mix with powder formulas; thus, the use of infant formulas can lead to an increase in child mortality due to increased rates of infectious illnesses, malnutrition, and dehydration. In these circumstances, the medical community advises that women exclusively breastfeed, meaning that nothing else, not even water, should be given to the baby. There is now evidence that mixed feeding (i.e., the combination of other liquids, such as water and formula, with breastmilk) results in higher rates of HIV transmission than exclusive breastfeeding, although the mechanism for this is poorly understood.

It is important to note that dissemination of prevention of mother-to-child transmission (PMTCT) programs has been slow in the developing world. Barriers to widespread implementation of these protocols have included low rates of HIV testing of pregnant women, lack of availability of medications, and systems of care that cannot handle the influx of women and newborns in need of care. As a first step, it is of paramount importance that women have access to and utilize HIV testing to determine whether they need to be placed on a PMTCT protocol. Equally essential, there must be a concomitant scale-up of the systems of medical care in these resource-poor settings. More health professionals need to be trained and retained (as many leave to work in developed countries) to provide care for the HIV-affected population.

Social–Ecological Model of Child Adaptation: Application to Pediatric HIV

The information presented in this article is conceptually organized according to a social–ecological framework of child adaptation and development, which has been widely applied in research with chronically ill children. The 'transactional stress and coping model' described by Robert Thompson, Jr. and Kathryn Gustafson in 1997 is an application of social–ecological theory to childhood chronic illness that further explicates the role of environment in the child's development. The transactional model incorporates developmental processes with attention to the family and social environments as influences upon the illness–outcome (medical and psychological) relationship. Chronic illness is conceptualized as a stressor to which the child and his or her family must adapt. Thompson indicates that adaptation to pediatric illness is impacted by family functioning, methods of coping, and cognitive processes. Specifically, cognitive processes include expectations, self-esteem, and a sense of ability to control one's own health (health locus of control). Family environment is conceptualized as one of three types: supportive, conflictive, and controlling. Descriptors of the child's illness, such as severity and treatment demands, as well as demographic indices (e.g., socioeconomic status) are included in the model. All of these elements interact to influence child development and adaptation.

Pediatric HIV particularly fits with a social–ecological model because of the many influential environmental and familial factors relating to this disease. The model also lends well to cross-cultural adaptation. In keeping with the transactional perspective, this article highlights the medical, psychological, and social (including familial) aspects of child development in the context of pediatric HIV/AIDS.

Medical Aspects of Pediatric HIV/AIDS

When considering the impact of HIV/AIDS upon infant and early child development, it is important to understand key aspects of the progression and treatment of the disease.

Medical Impact of Pediatric HIV/AIDS

HIV manifests and progresses differently in children than in adults. The clinical course of the disease in children is much faster than in infected adults. Even without treatment, adults often live for many years with HIV before becoming symptomatic, but this is generally not true for children. The disease tends to follow a bimodal presentation in children, with some children having very rapid progression to AIDS and others experiencing a more indolent course. In western countries before treatment was available, approximately 25% of children progressed to AIDS within 1 year. These 'rapid progressors' usually became seriously ill during their infancy, whereas the majority of children progress more slowly, developing AIDS later in childhood (typically between the ages of 6 and 10 years). In developing countries children fare even more poorly – in an analysis using data from seven prospective studies in Africa, 35% of untreated children died by 1 year of age and over half (52%) of infected children did not reach their second birthday.

Comorbid physical conditions

By definition, AIDS is a condition in which the damage to the immune system by HIV leads to the development of opportunistic infections and other disorders such as cancers. The most common of these infections is *Pneumocystis carinii* pneumonia (PcP). Without appropriate HIV management, this pneumonia often occurs in perinatally infected children from 3 to 6 months of age and has a very high mortality. *Mycobacterium avium* complex disease is another prevalent opportunistic infection, typically presenting with fever, diarrhea, and night sweats. Other opportunistic infections include tuberculosis, chronic herpes viruses, and infections of the central nervous system (CNS) with organisms such as *Cryptococcus* and *Toxoplasma*. Children living with HIV have increased risk of malignancies (most commonly lymphoma). The disease can also directly affect different organ systems such as the heart, kidneys, liver, gastrointestinal tract, and bone marrow, the latter causing hematologic abnormalities such as anemia.

Growth and physical development

Pediatric HIV disease is associated with growth deficiencies in over half of untreated children. These children not only experience difficulties maintaining their weight, but also may have problems with linear growth and depletion of lean muscle. Chronically poor growth, or progressive stunting, is associated with a higher risk of mortality. The causes of growth abnormalities are multiple and include neuroendocrine abnormalities, gastrointestinal dysfunction and malabsorption, vitamin and mineral deficiencies, and low levels of growth hormone; however, study results have been varied and therefore these linkages are not conclusive and sometimes are not considered to be causal.

Impact upon CNS

Unlike adult HIV, which begins in a fully mature and myelinated nervous system, pediatric HIV infects infants and children with developing and vulnerable central nervous and immune systems. Hence, children tend to become symptomatic faster than adults with a higher incidence of CNS disease.

Manifestations of the virus's impact upon a child's developing CNS include encephalopathy, neoplasms, opportunistic infections, disruption of the blood–brain barrier, and vascular changes associated with strokes. HIV-associated progressive encephalopathy presents with three primary clinical symptoms: impaired head growth, loss of developmental milestones or stagnation of developmental progression, and progressive motor dysfunction. The clinical course follows one of three patterns: (1) early presentation with a rapid course, (2) subacute course with periods of stability, and (3) static, slower progression. The early, rapid encephalopathy occurs over 1–2 months and is accompanied by a loss of developmental milestones. The typical postdiagnosis survival rate without treatment is less than 2 years. The children presenting with static encephalopathy do not lose milestones, but rather fail to achieve new milestones at an age-appropriate rate, evidencing increasing developmental delay over time.

Medical Treatment of Pediatric HIV/AIDS

In the 1980s and early 1990s, treatment for pediatric HIV/AIDS consisted of single medication regimens that were far less effective than the cocktail, or combination, approach used today. This includes the use of at least three antiretroviral medications with at least one of the three being from a different class of medication. This approach to treatment, referred to as HAART, was first tested and approved for use with adults and approximately 2 years later became approved for use with children (1998). The present approach is to initiate HAART when there is evidence of progression of disease such as a decrease in CD4 T lymphocyte count, which are part of the immune system and suppressed by the activity of the virus.

HAART is a complex medication regimen that can place significant adherence demands upon a child and family. It is extremely important that infected individuals do not forget to take the medicines; lack of adherence can lead to drug resistance, which can be very problematic, particularly in settings in which the different types of available medications are limited.

There are a number of adverse effects of long-term antiretroviral use that can have both physiological and psychiatric consequences. Lipodystrophy syndrome is a condition that is associated with changes to the body shape, including thinning of the face, arms, and legs accompanied by fattening of the abdomen. This can obviously be very distressing for a child or adolescent. Other potential side effects include metabolic abnormalities such as increasing cholesterol and glucose levels (**Figure 2**).

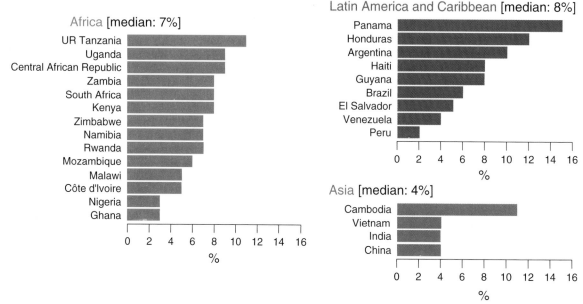

Figure 2 Percentage of people on treatment who are children by country, 2005.

Psychological Aspects of Pediatric HIV/AIDS

Neurodevelopment and Cognitive Functioning

Neurological functioning is very closely tied to a child's cognitive development and brain abnormalities resulting from CNS disease can broadly influence a child's functioning. Deficits that have been associated with pediatric HIV infection include impaired cognitive functioning, attentional difficulties, behavioral and emotional disruption, and problems in academic performance. Children may experience difficulty with expressive language as well as visual–spatial skills and memory tasks. Research has demonstrated that environmental elements (e.g., poverty, correlates of maternal illness) that often accompany pediatric HIV infection can also impact a child's performance on these tasks. At times it can be very difficult for researchers to tease apart the specific contribution of the medical illness and environmental factors to a child's developmental outcomes. In addition, small sample sizes and variability in findings have increased the complexity of interpreting research for clinical application.

In light of these challenges, researchers have striven to identify mechanisms and pathways via which HIV affects a child's cognitive development. Studies have linked structural abnormalities in the brain to cognitive dysfunction. In 1995, Pim Brouwers and colleagues examined brain scans of 87 children with symptomatic HIV who were previously untreated. The scans indicated a higher rate of brain abnormalities in perinatally infected children than in children who acquired the virus from blood

transfusions. The severity of abnormalities (including cortical atrophy, white matter changes, and ventricular enlargement) was found to be predictive of cognitive deficits. However, other studies have reported neuropsychological deficits without the accompanying structural abnormalities, indicating that deficits may also be influenced by other factors.

In an effort to examine the role of CNS disease in the neurodevelopment of HIV-infected children further, Wanda Knight and colleagues undertook a prospective study of children's mental and psychomotor development. At two time points, 20 HIV-infected and 25 noninfected infants who had been born to infected mothers (aged 3–30 months) participated in neurological examinations that included the Bayley Scales of Infant Development (BSID). Specific information regarding the children's antiretroviral treatment was not available to the researchers, but all the children were receiving care at an HIV primary care clinic and consistent treatment guidelines were applied. The results of the study demonstrated that HIV-infected infants scored significantly lower at baseline than noninfected infants on the mental development component of the BSID. At follow-up, the infants no longer differed significantly on mental development, but HIV-infected children scored lower than noninfected children on the motor scale. Interestingly, at both time points, the HIV-infected infants with CNS disease scored significantly lower on both mental and psychomotor indices than all other children. These findings suggest that HIV affects children's mental and psychomotor development via the CNS as a primary mechanism.

There is now evidence that treatment with HAART results in clinical improvement of certain cognitive deficits and adaptive behavior, including communication and daily living skills. Although asymptomatic children living with HIV still tend to fall below the average (or norm) on neuropsychological tests, the results are certainly improved since the advent of HAART. Unfortunately, however, children living with HIV in many parts of the world continue to have limited access to these life-saving treatments.

Psychological Distress

A great deal of research has been conducted on the psychological adjustment of HIV-positive adults and a number of interventions have been described for adults living with the virus. Adolescents have also received attention in investigations of mental health and prevention, and, not surprisingly, results suggest that adolescents living with HIV suffer greater psychological distress than their healthy peers. There is, however, a relative paucity of research on the mental health of infants and young children infected with HIV.

Psychiatric disorders including major depression, anxiety, attentional disorders, and behavioral disruption (conduct and oppositional defiant disorders) have all been associated with pediatric HIV infection. In 2006, Claude Mellins and colleagues published findings based upon clinical interviews with 47 perinatally infected youths (aged 9–16 years) and their primary caregivers and reported that 55% of these youths met criteria for a psychiatric disorder, a rate that is substantially higher than that found in the general population (although, these children were not directly compared to a control group with similar environmental stressors). In a second investigation, Mellins examined rates of behavioral problems among very young perinatally infected children (aged 3 years) and compared these to perinatally exposed, but uninfected, children. Very interestingly, the results of this study failed to demonstrate a relationship between HIV status and behavioral problems; instead, the findings showed that sociodemographic characteristics were the strongest predictors of behavioral symptoms.

In other studies, perinatally infected children have been reported to have higher rates of psychiatric symptoms than children infected via blood transfusion. Again, it is important to consider the social–ecological framework in interpreting these findings; it may be that children who acquired the virus earlier in development are physiologically more prone to psychiatric distress than those who became infected later. Alternatively, perinatally infected children are more likely to experience additional stressors, including maternal illness and correlates of maternal HIV (poverty, poor nutritional status, higher rates of intravenous drug use). Likely, the impact of HIV/AIDS upon children's psychological development is exerted environmentally as well as physiologically. Family stress related to childhood chronic illness can be conceptualized in terms of the overall illness burden in the family (including medication demands, hospital visits, emotional strain); if the child and mother are both infected, the family will experience a very high illness burden. As demonstrated in the model, family stress directly impacts the child's psychological and emotional adjustment, so this constellation of stressors presents significant challenges.

Quality of Life

The quality of a child's life has become an increasingly important consideration as many children with chronic illnesses (including HIV/AIDS) are living longer, but not necessarily more comfortable, lives. The concept of quality of life generally refers to an individual's health, culture, beliefs, values, and life conditions that support the person's wellbeing. In considering an HIV-positive child's health-related quality of life (HRQOL), it is very important to assess the impact of the HIV disease and its treatment upon the child's functioning and life satisfaction.

Quality of life has emerged as a significant marker of how well any particular medical regimen works, as a primary goal of treatment is to not only increase the child's longevity, but also to improve the child's adaptive functioning and ability to enjoy life. In evaluating the impact of pediatric HIV disease and the effectiveness of treatment, it is important to consider child-oriented aspects of quality of life. These include access to education, positive social structure with peers and family, emotional health, physical health, and age-appropriate cognitive and attentional ability. HIV/AIDS can exert a negative influence upon each of these elements. Although HAART has been extremely instrumental in reducing CNS disease and facilitating children's healthy physical functioning, as discussed, this treatment approach involves some trade-offs in adverse effects. Thus, the field still has a great challenge to improve the quality of life of these children.

Coping

Young children living with HIV face a sobering set of challenges. Many of these children must cope with environmental stressors, educational problems, and missing school due to their own or even their parents' illnesses, as well as the physical symptoms and treatment demands associated with their illness.

Children's coping strategies for stressful experiences can render the child either vulnerable or resilient to subsequent stressors. According to the original conceptualization by

Richard Lazarus and Susan Folkman in 1988, coping can be broadly delineated into two categories of behavior: problem-focused and emotion-focused coping. Problem-focused coping efforts directly target the stressor in an attempt to resolve the stressful situation. Emotion-focused coping efforts are the individual's attempts to regulate the negative emotional state that is aroused by the stressor, without directly targeting the stressor itself. Pediatric HIV poses complex challenges to the child and family, requiring an array of coping strategies. Controllable aspects of the illness (such as the taking of medications) call for an approach to coping that is problem focused. However, many aspects of the illness cannot be directly ameliorated, requiring coping strategies that facilitate the child's healthy emotional and social adjustment in the midst of illness-related stress. Coping interventions developed for children must be targeted and developmentally flexible.

Social Context of the HIV Epidemic: Impact upon Child Development

Public health and social consequences of the epidemic, including the escalating numbers of children being orphaned, widespread poverty, unpredictable availability of food and safe shelter, and lack of medical and requisite psychosocial care all contribute to poorer outcomes for children. HIV/AIDS has been referred to as a social disease, as it often affects the marginalized, underserved, or socioeconomically disadvantaged segments of a population. The challenges already present in individuals' lives are compounded by the illness and treatment demands of HIV. The stigma associated with the disease can also have a detrimental impact. Families often suffer an emotional exile because of fears of others knowing about the infection, not disclosing their diagnosis to others who might ordinarily provide support. This may also affect an individual's use of healthcare and taking of medications.

The social context for children's development with HIV/AIDS is significantly impacted by parental illness and loss due to the illness. UNICEF released sobering estimates in 2006 indicating that 15.2 million children worldwide had lost one or both parents to AIDS; 12 million of these children live in sub-Saharan Africa. The impact of parental loss on a child's developmental trajectory can be obviously detrimental in the absence of protective factors, particularly if the child has his or her own illness demands and functional limitations. Additionally, lack of schooling, either due to their own illness or due to the necessity of providing for ill parents or younger siblings, can disadvantage these children even further.

Caregiver and Family Functioning in the Context of Pediatric HIV/AIDS

Globally, families have become increasingly stressed and resources fewer due to the consequences of HIV/AIDS. Many children who are perinatally infected will lose their mothers to AIDS, and grandmothers will assume the caregiving role. Particularly in the developing world, HIV-related trauma is intergenerational; grandmothers who have lost children to AIDS are now caring for a grandchild infected with HIV. Grandmothers are often the primary caregiver for more than one grandchild and the majority of families with an infected child has at least one additional HIV-infected person living in the household. Children may also be cared for by multiple caregivers, usually in an extended family network. This may pose a distinct challenge for the management of the child's HIV, particularly if there is secrecy and lack of disclosure regarding the child's HIV diagnosis.

The research investigating the influence of family functioning upon the child's adjustment to chronic illness has been primarily conducted in the western world. However, given the cultural import regarding familial relationships, it is likely that family functioning is similarly, if not more, influential in children's adaptation to illness in African cultures. Family functioning is consistently a strong predictor of the child's adjustment to chronic illness, and many studies have linked maternal coping and adjustment to child psychosocial functioning and illness adaptation. Family functioning has also been linked to specific disease correlates, such as adherence and disease management over time. Findings across studies suggest that family cohesion and emotional expressiveness, as well as open communication and hopeful attitudes about the child's illness, are characteristics of system functioning that reliably predict desirable child adjustment outcomes. Given these demonstrated relationships, harnessing the strength of families for pediatric HIV intervention is considered critical.

Illness Management of Pediatric HIV

As conceptualized in the social–ecological framework, child and family psychosocial functioning directly impact the management of pediatric illness and adaptation. In the following discussion of disclosure and adherence, the role of the family is illustrated as central to the child's successful adaptation to illness demands.

Disclosure: Beginning the Process of Illness Management

Disclosure of the child's HIV status to the child and within the family system is an important aspect of illness management of pediatric HIV. The American Academy of Pediatrics strongly urges caregivers to discuss the diagnosis with their HIV-infected child, and it is the policy in many pediatric infectious disease clinics to promote disclosure to school-aged children.

Family functioning is closely tied to successful negotiation of the disclosure process. Open communication within the family about the child's HIV status is linked to improved psychological and behavioral adjustment outcomes of the child and greater family expressiveness has been linked to earlier diagnostic disclosure to HIV-infected children. Children who know their HIV status also display fewer symptoms of depression than those children who do not know their diagnosis. In addition, anxiety has been shown to increase when children are not allowed to discuss their fears regarding illness. In research conducted by Des Michaels and colleagues in 2006, 126 caregivers of children on antiretroviral therapy in South Africa were interviewed regarding HIV infection in the family and disclosure to the child. The majority of caregivers indicated a belief that children should learn their HIV status between the ages of 6 and 10 years. However, only a small minority of caregivers interviewed had disclosed to their children. This discrepancy reflects the barriers to disclosure, including fear of familial stigmatization, the caregiver's own difficulty in coping with the child's diagnosis, fear of the child's emotional reaction, and lack of psychosocial support in disclosing to the child.

Clearly, disclosure to the child is crucially important in preparing him or her to cope with and manage the HIV infection, and to develop skills and patterns early for adherence and health-promoting behaviors. However, the timing and method of disclosure is critical with regard to the child's emotional reaction to diagnosis. The child will invariably learn his or her HIV status, and it is clearly best for this process to occur in a controlled, supportive environment. Disclosure of HIV status to the child is best conceptualized as a process, not a one-time conversation with the child, and should be developmentally appropriate. There is significant need for support of the family during the disclosure process. Although research has indicated the benefits of disclosure, empirical investigations of interventions to facilitate disclosure are limited and there is little information and understanding of the importance of the cultural context in issues of disclosure.

Treatment Adherence – Critical to Illness Management

Pediatric HIV is an illness that demands lifelong adherence to a challenging medical regimen. Research on children's medical adherence indicates that adherence is most poor for regimens that are complex, interfere with the child's activities, produce negative side effects, and are long-term and future oriented. By these criteria, pediatric HIV poses clear challenges for children's lifetime adherence. Complicating the demands upon adherence, many children (particularly in the developing world) are diagnosed with comorbid conditions (such as tuberculosis) that require additional medications. Although medication adherence is a challenge in the developed world, when drug resistance does develop because of nonadherence, there are other medications that can sometimes be used. As previously discussed, however, nonadherence in the developing world can have dire consequences given the few pediatric HIV medications available to children. Near-perfect adherence is required for medications to be successful; thus, children's adherence to HAART requires substantial attention and is a crucial target of pediatric intervention.

Children's adherence is positively influenced by disclosure. The child's adaptive functioning and coping also impact adherence, as do the social–ecological processes of family functioning and parenting. Similarly to facilitating disclosure, family environments that are low in conflict and high in support facilitate children's adherence. Communication and consistency among caregivers as well as parent–child communication are critical in promoting adherence to pediatric HIV-treatment regimens.

Future Directions for Intervention Efforts in Pediatric HIV/AIDS

Intervention goals for children living with HIV/AIDS must be targeted and developmentally applicable to the child's particular mental health and illness management needs. As discussed throughout this article, intervention with HIV-affected families requires attention to the social–ecological framework. It is important for interventions to promote coping skills and bolster resiliency within the family for adapting to illness-related stress (**Figure 3**).

Additionally, there is significant need for structured interventions that directly target children's acquisition of illness management skills and engage the caregiver in supporting the child's adaptation to HIV. Given the significance of family functioning, coping, and resiliency to successful illness management, it is of the utmost importance to integrate psychosocial intervention with adherence intervention for children. Further, it is critical to intervene early to bolster the child's skills for illness management and prevent nonadherence and engagement in risky health behaviors, which is often the typical trajectory in adolescence.

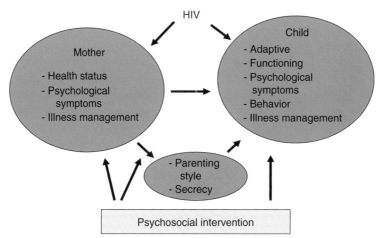

Figure 3   Model of psychosocial intervention with HIV-affected families, attending to the social–ecological processes of the mother and the child.

Along with providing appropriate medical treatment and family-based support, the child's cognitive development should be monitored with routine neuropsychological testing. Consistent evaluation provides a baseline of cognitive functioning and alerts the child's caregivers and medical providers if there is a drop-off in skills. Based upon a child's functioning, she or he may require treatment from a child psychologist to mitigate loss of functioning. Intervention may include cognitive remediation tasks such as memory work and focus on expressive language. Unfortunately, very few of the millions of children living with HIV have access to monitoring or remediation therapy.

Finally, intervention with children and families should be grounded in the cultural context. Little is currently known about the psychosocial impact of pediatric HIV in African children, and there is sobering potential for unfolding psychological effects of HIV-related trauma. As advances in the availability of antiretroviral therapy improve children's survival rates, health researchers have a responsibility to address these children's needs through specific and targeted efforts to alleviate distress and supply resources with which to face their challenges. Pediatric HIV/AIDS is a complex medical illness that is accompanied by an array of familial and societal stressors. It is of paramount importance to the world's children that researchers continue to attend to the amelioration and mitigation of children's suffering attributable to this disease.

See also: Bayley Scales of Infant Development; Birth Complications and Outcomes; Endocrine System; Healthcare; Immune System and Immunodeficiency; Mental Health, Infant; Mortality, Infant; Screening, Newborn and Maternal Well-being; Stress and Coping; Teratology.

Suggested Readings

Brouwers P, DeCarli C, Civitello L, Moss H, Wltrs P, and Pizzo P (1995) Correlation between computed tomographic brain scan abnormalities and neuropsychological function in children with symptomatic human immuno-deficiency virus disease. *Archives of Neurology* 52: 39–44.

Brown LK, Lourie KJ, and Pao M (2000) Children and adolescents living with HIV: A review. *Journal of Child Psychology and Psychiatry* 41(1): 81–96.

Chakraborty R (2005) HIV-1 infection in children: A clinical and immunologic overview. *Current HIV Research* 3: 31–41.

Forsyth BWC (2003) Psychological aspects of HIV infection in children. *Child and Adolescent Psychiatric Clinics* 12: 423–437.

Knight WG, Mellins CA, Levenson RL, Arpadi SM, and Kairam R (2000) Brief report: Effects of pediatric HIV infection on mental and psychomotor development. *Journal of Pediatric Psychology* 25(8): 583–587.

Lazarus R and Folkman S (1988) The relationship between coping and emotion: Implications for theory and research. *Social Science and Medicine* 26: 309–317.

Mellins CA, Brackis-Cott E, Dolezal C, and Abrams EJ (2006) Psychiatric disorders in youth with perinatally acquired human immunodeficiency virus infection. *The Pediatric Infectious Disease Journal* 25(5): 432–437.

Mellins CA, Smith R, and O'Driscoll P (2003) High rates of behavioral problems in perinatally HIV-infected children are not linked to HIV disease. *Pediatrics* 111(2): 384–393.

Michaels D, Eley B, Ndhlovu L, and Rutenberg N (2006) *Horizons Final Report: Exploring Current Practices in Pediatric ARV Rollout and Integration with Early Childhood Programs in South Africa: A Rapid Situation Analysis*. Washington, DC: Population Council.

Thompson RJ and Gustafson KE (1997) *Adaptation to Childhood Chronic Illness*. Washington, DC: American Psychological Association.

Relevant Websites

http://www.nih.gov – National Institutes of Health (NIH), Department of Health and Human Services.

http://www.unaids.org – The Joint United Nations Programme on HIV/AIDS.

http://www.unicef.org – UNICEF, Unite for Children.

Allergies

A W Burks, Duke University Medical Center, Durham, NC, USA
K P Palmer, University of Arkansas for Medical Sciences, Little Rock, AR, USA

Glossary

Allergens – Normally harmless proteins or glycoproteins encountered in the environment which stimulate the production of IgE and result in allergic responses when bound by IgE on the surface of mast cells and basophils.

Atopy – The genetic predisposition to develop IgE-mediated responses to allergens encountered in the environment.

CAP FEIA – CAP Fluorescent Enzyme Immunoassay; newer, more sensitive *in vitro* assay used to detect allergen-specific IgE.

Cross-linkage of IgE – The process of allergen binding to multiple IgE molecules on the surface of mast cells and basophils which leads to signaling through the IgE molecule and the resultant allergic response.

Hypersensitivity – Immune-mediated response directed toward normally harmless substances; may be IgE-mediated but also includes various other immune mechanisms.

IgE – One of the immunoglobulin isotypes secreted by immune plasma cells which binds to IgE receptors on the surface of mast cells and basophils; cross-linking of surface-bound IgE by allergen leads to the signs and symptoms of immediate hypersensitivity.

Immunoglobulin – Protein produced by plasma cells of the immune system that acts to neutralize invading microorganisms and toxins.

Immunotherapy – The repeated administration of specific allergens to an individual that changes the IgE-mediated response so as to reduce symptoms when naturally exposed.

RAST – Radioallergosorbent test; commonly used *in vitro* assay used to detect allergen-specific IgE.

Spirometry – Technique for measuring lung function that is used for the diagnosis and management of asthma.

Introduction

Allergic diseases affect over 20% of the US population, and the prevalence of these conditions is rising. They are the sixth leading cause of chronic disease; an estimated 14.1 million physician office visits occur each year for allergic rhinitis alone. These disorders significantly impact quality of life of affected individuals and account for billions of dollars in direct and indirect costs every year.

The Allergic Response

Immunologically mediated events directed at common, harmless substances characterize the allergic response. Although many parts of the immune system are involved, the principal mediator is immunoglobulin E (IgE). IgE was first discovered in 1967 and, along with IgG, IgA, and IgM, is one of the immunoglobulin isotypes produced and secreted by plasma cells. Production of allergen-specific IgE depends on both the genetic predisposition of an individual to form IgE and the pattern and timing of environmental allergen exposure. Individuals not affected by allergy do not produce allergen-specific IgE and therefore do not respond immunologically to allergens.

The IgE-mediated allergic response, or immediate hypersensitivity reaction, is shown in **Figure 1**. Sensitization occurs during the initial exposure to a specific allergen and is the process whereby an individual initially forms allergen-specific IgE. In order for sensitization to occur, the allergen must first be recognized by specialized cells called antigen-presenting cells that process the allergen into antigenic peptides and present these peptides to T cells. This process triggers the production of cytokines by T cells which directly interact with B cells to stimulate the production of allergen-specific IgE. Secreted IgE then binds to high-affinity IgE receptors on the surface of mast cells and basophils that are located in the skin, mucosal surfaces, and circulation.

Upon re-exposure, allergen binds to and links multiple surface-bound IgE molecules on mast cells and basophils, and thus initiates the early-phase response. Mast cells and basophils contain granules with numerous preformed chemical mediators such as histamine, tryptase, and heparin, which are released upon binding of the allergen to IgE on these cells. Newly synthesized mediators, such as leukotrienes and prostaglandins, are also released. These mediators, of which histamine is paramount, cause increased vascular permeability, mucus secretion, smooth muscle constriction, vasodilatation, and sensory nerve stimulation. When these changes occur in various organs, they result in the clinical symptoms observed in an immediate allergic reaction (**Table 1**). The late-phase response, which usually occurs 2–4 h after allergen exposure, is due to infiltrating inflammatory cells, such as eosinophils and mononuclear cells. These cells release various cytokines, protein mediators that have various inflammatory effects on other types of cells, and this process results in clinical chronic allergic inflammation.

Characteristics of Common Allergens

Allergens are common proteins or glycoproteins found in the environment. They may function as enzymes, structural or regulatory proteins, or ligand-binding proteins, thus

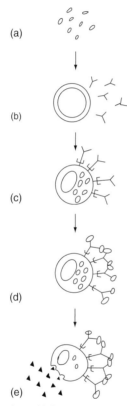

Figure 1 Mechanism of IgE-mediated allergy. (a) Allergen enters the body via inhalation, ingestion, injection, or direct contact. (b) B cells secrete allergen-specific IgE. (c) IgE binds to high-affinity IgE receptors on the surface of mast cells. (d) Upon re-exposure, allergen cross-links allergen-specific IgE on the surface of mast cells. (e) Mast cells degranulate, releasing chemical mediators such as histamine.

representing a variety of biologic activities. Their solubility allows them to penetrate the respiratory mucosa when inhaled, but they can gain entry by other means including ingestion, injection, and direct contact. House-dust mite, animal dander, and cockroach are important indoor allergens, whereas the primary source of outdoor allergens are plants and fungi, although the latter may be present in both settings. Other important allergens are found in foods, latex, drugs or drug metabolites, and insect venom.

Pollens

Allergenic pollen grains typically originate from wind-pollinated trees, grasses, and weeds. They are usually 10–100 µm in diameter, allowing them to reach the upper and lower respiratory tract. Exposure and sensitization to pollens is largely dependent on the geographic distribution of these plants and individual characteristics of the pollen grain such as size, dispersibility, and buoyant density. In the US, trees pollinate in early- to mid-spring, whereas grasses pollinate in late spring to early summer and weeds in late summer to early fall. Although there are numerous species of trees, grasses, and weeds in the US, only a limited number of these are responsible for allergic disease.

Fungi

Fungi are organisms with rigid cell walls that are classified according to their sexual reproductive structures. Allergenic fungi are usually microscopic (mold spores); however, macroscopic fungi like mushrooms may also be allergenic. Fungi are common throughout the US, and outdoor mold spore counts are usually highest in the summer and early fall. Common indoor mold sources include damp indoor spaces, baths, showers, crawlspaces, and basements.

Animal Dander

Animal allergens are proteins found in saliva, urine, and secretions from sebaceous oil glands in the skin. Allergenic proteins from cats and dogs comprise the most common, clinically relevant animal allergens; however, birds, rabbits, and multiple other animals can also produce allergenic proteins. The major allergens of cat and dog are extremely

Table 1   The allergic response in target organs

Eye	Nose	Lung	Gastrointestinal tract	Skin	Heart and blood vessels
Ocular itching, redness, watery eyes	Nasal itching, runny nose, sneezing, congestion	Cough, shortness of breath, wheezing	Abdominal pain, vomiting, diarrhea	Hives, itching, angioedema, flushing	Decreased blood pressure

lightweight and can be carried through the air easily so that they accumulate on furniture, clothing, and carpets. Therefore, their distribution is virtually ubiquitous with measurable amounts of allergen located in homes with and without pets, schools, offices, and public buildings.

House-Dust Mite

The allergenic proteins of house-dust mite are enzymes excreted in mite feces. Two species of dust mite, *Dermatophagoides pteronyssinus* and *D. farinae*, account for over 90% of allergen found in house-dust samples in the US. Dust mites tend to thrive in warm environments with high humidity, and primary reservoirs include mattresses, carpets, upholstered furniture, draperies, and stuffed toys.

Food Allergens

Food allergens are 10–70 kDa proteins or glycoproteins that are relatively resistant to heat, acidity, and digestion. Although any food can cause an IgE-mediated allergic reaction, reactions to milk, soy, egg, wheat, peanuts, tree nuts, fish, and shellfish account for 90% of food allergies. Proteins identified as major allergens include whey and casein in milk, tropomycin in shellfish, ovomucoid in egg white, and the seed storage proteins vicilin and conglutin in peanuts.

Allergic Diseases in Childhood

Allergic Rhinitis

The term rhinitis refers to inflammation of the nasal mucous membranes which may be due to underlying allergic disease, nonallergic disease, or both. Allergic rhinitis is one of most common allergic diseases in the US affecting approximately 40 million individuals. It has a significant impact on health-related quality of life and results in millions of school and work days missed each year. Children with allergic rhinitis may have sleep disturbances, school problems, anxiety, difficulty concentrating, and familial dysfunction. Furthermore, allergic rhinitis and the use of first-generation antihistamines, such as diphenhydramine, have been found to adversely impact learning in children. Characteristic symptoms of allergic rhinitis include repetitive sneezing, nasal itching, congestion, and clear nasal drainage. These symptoms often develop in young children and may persist into adulthood. Furthermore, allergic rhinitis is commonly associated with other conditions such as asthma, allergic conjunctivitis, and atopic dermatitis.

Allergic rhinitis accounts for approximately 50% of all rhinitis and the symptoms are frequently similar to various forms of nonallergic rhinitis, so it is important

to differentiate the two. Infectious rhinitis, often confused with allergic rhinitis, is commonly caused by respiratory viruses and may be accompanied by low-grade fever and purulent nasal secretions predominantly containing neutrophils. Nonallergic eosiniphilic rhinitis is characterized by sneezing, nasal itching, congestion, and nasal discharge, but differs from allergic rhinitis in that there is no evidence of allergen-specific IgE. As the name implies, significant nasal eosinophilia is present on nasal scrapings obtained from these patients. Other forms of nonallergic rhinitis include vasomotor rhinitis, the result of mucosal hyperresponsiveness to various changes in environmental stimuli such as humidity, temperature, strong odors, and chemicals, and hormonally induced rhinitis which may occur during pregnancy, puberty, menses, or in endocrine disorders such as hypothyroidism. Various anatomic abnormalities, such as nasal septal deviation, nasal foreign body, choanal narrowing, and adenoidal hypertrophy can cause rhinitis because normal respiratory mucosal physiology may be altered in these conditions. Finally, medications such as antipsychotic and antihypertensive drugs can induce rhinitis. Topical decongestants, in particular, can lead to significant rebound rhinitis if used for a prolonged time period.

Allergic rhinitis can be classified as either seasonal, caused by allergens such as trees, grasses, and weeds, or perennial, usually caused by allergens such as house-dust mite and animal dander, which do not have seasonal variation. Perennial allergic rhinitis and mixed patterns of perennial rhinitis with seasonal exacerbations account for the majority of cases. In pure seasonal allergic rhinitis, symptoms are episodic and correlate with the implicated seasonal allergen. The profuse watery nasal drainage, repetitive sneezing, and nasal and ocular itching characterize what is commonly referred to as hay fever. Perennial symptoms are similar but are more prominently characterized by postnasal drainage and chronic nasal congestion.

Important risk factors for the development of allergic rhinitis include an immediate family history of allergic disease and a personal history of other allergic diseases such as asthma or atopic dermatitis. The relationship between allergic rhinitis and asthma is particularly important, and the same allergens can be responsible for exacerbations of both diseases. Both the nose and the lungs share a common respiratory mucosa and poorly controlled IgE-mediated responses in the nose can potentially lead to or worsen allergic inflammation of the lower airway.

When evaluating a patient with suspected allergic rhinitis, one must identify the onset and duration of symptoms, severity, relationship of symptoms to seasons, and other identifiable triggers such as dust, animals, or pollen. Examination of the eyes, ears, nose, and throat may reveal characteristic findings that are helpful. Allergic shiners are darkened areas of skin underneath the eyes which

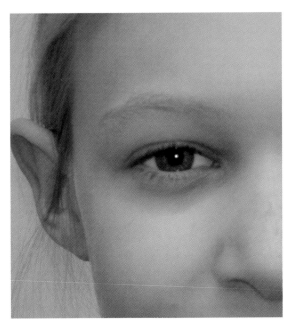

Figure 2 Allergic shiners in a child with perennial allergic rhinitis. Reproduced with permission from the parent.

result from chronic venous congestion and nasal obstruction (**Figure 2**). Persistent nasal obstruction may also lead to mouth breathing, and nasal pruritis can lead to the formation of a horizontal crease of the mid-to-lower nose due to repetitive rubbing and wiping. The nasal turbinates may appear pale and boggy, at times almost completely obstructing the nasal airway.

The first step in the management of allergic rhinitis is allergen avoidance, and recommendations should be based on tests for allergen-specific IgE (discussed later), in combination with the clinical history. Occasionally, allergen avoidance alone is all that is required for symptom control; however, in most cases pharmacologic therapy is needed. Medications used to treat allergic rhinitis include antihistamines, decongestants, intranasal steroids, leukotriene modifiers, mast cell stabilizers, and anticholinergics. In many comparison studies with antihistamines and leukotriene modifiers, intranasal steroids have proved to be most effective in the treatment of all symptoms of allergic rhinitis. In severe or refractory cases, immunotherapy should be considered as it is effective in alleviating symptoms. Ongoing education and follow-up regarding allergen avoidance and compliance with medication and/or immunotherapy is imperative.

Asthma

Asthma is the most common chronic disease of childhood affecting approximately 4.8 million children in the US. Morbidity rates have increased since the mid-1970s despite significant advances in the understanding of pathophysiology and treatment. Asthma can develop at any time, but approximately 80% of individuals who have asthma develop symptoms before 5 years of age. The most important risk factor is atopy, but, a personal or immediate family history of allergy, passive tobacco smoke exposure in childhood, and sensitization to certain inhalant allergens are also important.

Individuals with asthma have recurrent episodes of coughing, shortness of breath, chest tightness, and wheezing. Infants may also demonstrate difficulty feeding, rapid breathing, and grunting. These symptoms are a result of chronic airway hyperresponsiveness, inflammation, and partially reversible airway obstruction. The inflammation of airways in asthma is characterized by mast cell activation, infiltration of inflammatory cells, edema, mucus hypersecretion, and damage to the bronchial epithelium. Over time, these changes lead to airway wall remodeling with collagen deposition underneath the basement membrane, mucus gland hyperplasia, smooth muscle hypertrophy, and vascular proliferation. Once airway remodeling has occurred, it is irreversible. Asthma that is predominantly triggered by allergens may be referred to as extrinsic asthma. In contrast, some asthmatic patients have no identifiable allergies and so are referred to as having nonallergic, or intrinsic, asthma. Triggers such as upper respiratory infections, exercise, cigarette smoke, and cold air can contribute to disease in both types of asthmatic patients.

Objective measurements of airway hyperresponsiveness and pulmonary function are useful for diagnosis and management. Spirometry, measurement of the volume and speed of inhaled and exhaled air, is the most commonly performed method of measurement and is used both for diagnosis and for determining asthma control and response to therapy. Spirometry is used to produce a flow–volume curve which depicts changes seen with inspiration and expiration. The expiratory portion of the flow–volume curve in an asthmatic patient characteristically appears concave. The forced expiratory volume at 1 s (FEV1), measured before and after the administration of a short-acting bronchodilator, can be used to confirm the diagnosis. Reversibility, or response to bronchodilator, is defined as an FEV1 increase of 12% or more. In children who are too young or unable to complete spirometry, a diagnostic trial of inhaled bronchodilators or anti-inflammatory medication can be useful.

Asthmatic patients demonstrate airway hyperresponsiveness to nonspecific parasympathomimetic stimuli such as methacholine, and measurement of this response is commonly known as a methacholine challenge. Histamine can also be used because both these agents act directly on bronchial smooth muscle to cause constriction. Inhaling these pharmacologic agents induces a decrease in lung function – usually a minimum 20% decrease in FEV1. These challenges are particularly useful in clinical trials and in the presence of normal spirometry when the diagnosis of asthma is questionable.

Another useful tool is the peak expiratory flow rate (PEFR), which is a measure of the maximum ability to expel air from the lungs and primarily reflects large airway function. Because it is highly effort-dependent, it is not as sensitive as FEV1 for diagnosing obstruction and should not be used as a diagnostic tool for asthma. Serial measurements are helpful, however, as a monitoring tool that can be performed at home using a peak flow meter. While diurnal variation (difference between measurements obtained in the morning and evening) of PEFR in nonasthmatic individuals is about 5%, variation of more than 20% commonly occurs in individuals with asthma.

The treatment of asthma is aimed at preventing symptoms, maintaining normal pulmonary function, and minimizing exacerbations and need for hospitalization and emergency care. The National Asthma Education and Prevention Program (NAEPP), a multidisciplinary coalition coordinated by the National Heart, Lung, and Blood Institute (NHLBI) of the National Institutes of Health have published guidelines regarding the evaluation and management of asthma. These guidelines use specific criteria to classify asthma severity, and pharmacologic therapy is escalated or decreased in a stepwise fashion according to severity classification (**Table 2**).

There are multiple medications available for the treatment of asthma. Quick relief medications are used for the rescue of acute symptoms, and long-term controller medications treat underlying airway inflammation and prevent the long-term remodeling changes of chronic asthma. Rescue medications, called β-agonists, can be either short acting or long acting. These medications act on airway smooth muscle and cause relaxation resulting in bronchodilation. While all individuals with asthma require short-acting bronchodilators for rescue, those with persistent asthma, in particular those with night-time symptoms,

may benefit from the addition of a long-acting bronchodilator. Asthmatic patients with persistent symptoms require a long-term controller medication, of which the most effective are inhaled corticosteroids. Well-designed studies have shown that treatment with inhaled corticosteroids leads to decreased airway hyperresponsiveness as well as decreased frequency of asthma symptoms, exacerbations, hospitalizations, death from asthma, and improved quality of life. Systemic corticosteroids may be needed for individuals with acute exacerbations or in those with severe, refractory asthma. Leukotriene modifiers such as montelukast can also be used for long-term control in those with mild, persistent asthma and as adjuncts in those with more severe asthma. Finally, allergen immunotherapy is effective in the treatment of allergic asthma, and can be particularly useful in those patients with co-morbid, poorly controlled allergic rhinitis. However, the risk of systemic reactions from allergen vaccines may be increased in those with poorly controlled symptoms and decreased pulmonary function.

Atopic Dermatitis

Atopic dermatitis is a chronic remitting and relapsing inflammatory disease of the skin affecting 10–20% of children. While skin of normal individuals is usually well hydrated and free from redness or irritation, the skin of those affected with atopic dermatitis is chronically dry, extremely itchy, red, and inflamed. It is usually present in infancy and early childhood, with half of all affected individuals developing symptoms during the first year of life. Atopic dermatitis can cause significant morbidity and adversely affects quality of life leading to missing days at school and emotional stress. It may precede the development of allergic rhinitis and asthma, and so is usually the first manifestation of what is commonly referred to as the atopic march.

The pathophysiology of atopic dermatitis is complex and a result of both genetic and environmental factors. Sensitization to aeroallergens and food allergens, atopy, colonization with *Staphylococcus aureus*, and an altered skin barrier all play significant roles in the inflammatory process. Most individuals with atopic dermatitis have positive skin prick tests or *in vitro* assays for allergen-specific IgE. Multiple controlled studies have demonstrated that these specific allergens especially food allergens, house-dust mite, and animal danders, contribute to the severity and course of skin disease. Food allergy plays a significant role in approximately one-third of those with moderate-to-severe atopic dermatitis. The most commonly implicated foods are milk, soy, egg, wheat, peanuts, and fish. Relevant food allergens are identified by dietary history, skin prick testing, *in vitro* assays for food-specific IgE, and double-blind, placebo controlled food challenges. This is discussed in more detail in further sections.

Table 2   Approach to asthma management

Measurements of asthma control
Frequency of daytime symptoms
Frequency of night-time symptoms
FEV1
Peak flow variability
↓
Classification of asthma severity
Mild intermittent
Mild persistent
Moderate persistent
Severe persistent
↓
Pharmacotherapy based on severity classification
Short-acting-β-agonist
Low-, medium-, or high-dose inhaled corticosteroids
Long-acting β-agonist
Leukotriene modifiers
Oral corticosteroids

Recent studies have elucidated an important role of *S. aureus* in the pathophysiology of atopic dermatitis. This organism is found on approximately 90% of skin lesions. *S. aureus* secretes exotoxins, such as enterotoxins A and B and toxic shock syndrome toxin, which act as superantigens that polyclonally stimulate immune cells and contribute to persistent and worsening inflammation. Furthermore, IgE to these toxins has been found in affected individuals.

Significant itching is a hallmark feature of atopic dermatitis and is usually worse at night. Itching leads to repetitive scratching that increases inflammation and leads to additional itching, a pattern commonly called the itch–scratch cycle. The mechanical trauma from repetitive scratching and the immunologic changes in the skin result in an altered skin barrier. This altered barrier allows increased evaporative losses and an increased portal of entry for allergens and chemical irritants that can lead to exacerbations of the disease.

The diagnosis of atopic dermatitis is based on the presence of major and minor clinical criteria. Central to the diagnosis of atopic dermatitis is the presence of intense itching and scratching. Other major criteria include a chronically relapsing course, lichenification of the skin, and a personal or family history of atopy. Skin lesions have variable appearance, but acute lesions are typically red patches of intensely itchy, excoriated, dry skin (**Figure 3**). Vesiculations with thin, clear, or discolored oozing may also be present. Chronic lesions generally are thickened or lichenified. The classic distribution of skin lesions in children is on the face and extensor surfaces of the extremities. In adults, involvement of the flexor surfaces is more common. Affected skin contains infiltration of activated lymphocytes, eosinophils, mast cells, and macrophages.

Treatment of atopic dermatitis is aimed at maintaining hydration and restoring the skin barrier, treating and controlling ongoing inflammation, controlling the itch–scratch

cycle, and using anti-infective therapy when appropriate. Daily soaking or baths in lukewarm water followed by the application of an emollient act to hydrate the skin and seal in moisture. Moisturizers in the form of creams or ointments should be applied several times daily. Antihistamines act by blocking histamine receptors in the skin and are used in an effort to suppress the itch–scratch cycle. There are numerous anti-inflammatory medications available to treat atopic dermatitis, and the most commonly used are topical corticosteroids of varying potencies. Potency of the topical agents is based on severity of inflammation and location of affected areas. Also available are topical immunomodulators such as tacrolimus and pimecrolimus that can be used in place of, or in addition to, topical steroids. Systemic and/or topical antibiotics are useful in the treatment of individuals colonized with *S. aureus*. In cases of severe atopic dermatitis refractory to these therapies, the addition of systemic immunosuppressants, phototherapy, or wet dressings may be of added benefit.

Food Allergy

Adverse food reactions can result from both immune and nonimmune mechanisms, and such reactions are common in children. One study of 480 children under 3 years of age revealed that 28% reported an adverse food reaction, but only 8% of these were actually proven to be food related and even fewer were IgE mediated. Examples of nonimmune adverse food reactions include food toxin-mediated effects, such as in scromboid fish poisoning, and host metabolic or enzymatic alterations, as in lactose intolerance. Adverse food reactions due to immune mechanisms can be further classified into IgE-mediated and non-IgE-mediated reactions. IgE-mediated food allergy occurs in approximately 6–8% of young children and 3–4% of adults. These reactions may be severe and are the leading cause of anaphylaxis in children.

The prevalence of IgE-mediated food allergy is greatest during the first few years of life, and children with atopic disease, particularly atopic dermatitis, are more likely to have coexisting food allergy. Ninety percent of food allergies in children are caused by six foods: milk, egg, peanuts, wheat, soy, and tree nuts. In older children and adults, the majority are due to peanuts, tree nuts, fish, and shellfish. Allergies to milk, eggs, wheat, and soy are more likely to be outgrown than those to peanuts, tree nuts, fish, or shellfish.

Food allergens are heat and acid stable proteins that are absorbed via the gastrointestinal tract, crosslink food-specific IgE on the surface of mast cells, and lead to the signs and symptoms of immediate hypersensitivity. When an allergic food reaction occurs, symptoms occur within minutes to a few hours after ingestion. Symptoms may include hives, swelling of the lips or tongue, flushing, itching, wheezing, shortness of breath, abdominal cramping,

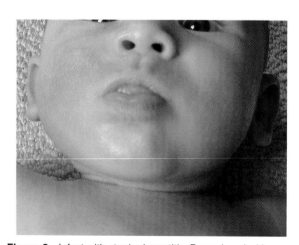

Figure 3   Infant with atopic dermatitis. Reproduced with permission from the parent.

vomiting, decreased blood pressure, or loss of consciousness. A history of a temporal relationship between the ingestion of a particular food and onset of symptoms is important in establishing a diagnosis of food allergy. When skin-prick testing to the suspected allergen is performed properly and quality extracts are used, a negative test essentially excludes food allergy. Conversely, a positive skin-prick test has a positive predictive value of approximately 50%, so it is important that testing be correlated with clinical history. *In-vitro* measurements for food-specific IgE, preferably the CAP fluorescent immunoassay (CAP FEIA), are particularly useful in food allergy diagnosis because predictive values have been established for the major food allergens based on CAP FEIA levels and results of double-blind, placebo-controlled food challenges in children. These challenges are the gold standard for the diagnosis of food allergy, and decisions regarding food challenges are often based on a combination of clinical history, skin-prick testing, and CAP FEIA results.

Current treatment of IgE-mediated food allergy involves strict elimination of the offending allergen, and ongoing patient and family education regarding how to respond to accidental ingestions and reactions. Food allergic individuals with a history of a severe allergic reactions or anaphylaxis should carry and be able to use injectable epinephrine. (epinephrine is a medication with adrenergic activity that reverses the pathophysiologic changes observed in anaphylaxis.)

Non-IgE-mediated food reactions are usually present in the first few months of life and include food protein-induced gastrointestinal disorders. A comparison of these processes with IgE-mediated reactions is shown in **Table 3**. The most common foods implicated are milk and soy, although foods such as rice, wheat, and poultry have also been implicated. Food-induced enterocolitis can cause protracted vomiting and diarrhea leading to failure to thrive or episodic vomiting and diarrhea 2 h or more following ingestion. The latter may lead to severe dehydration and decreased blood pressure and may be confused with fulminant infection, especially in young infants. Food-induced enteropathy is a malabsorption syndrome characterized by

vomiting, failure to thrive, and greasy, foul-smelling stools. In contrast, protocolitis, inflammation of the lower part of the gastrointestinal tract or rectum, is not associated with vomiting or diarrhea but rather is characterized by gross or occult blood in the stools. Although the exact pathophysiologic mechanisms of these processes are not well understood, studies have demonstrated that T cells and cytokines such as tumor necrosis factor (TNF)-α and transforming growth factor (TGF)-β1 play important roles in food protein-induced enterocolitis. Treatment of all of these entities involves strict elimination of the suspected food, and they are usually outgrown within the first several years of life.

Anaphylaxis

Anaphylaxis is a severe, IgE-mediated allergic reaction to a specific allergen to which an individual has been sensitized and results from the systemic, rather than local, release of inflammatory mediators from mast cells and basophils. These reactions are potentially life threatening and are considered medical emergencies. In children, foods are the most common cause of anaphylaxis outside of the hospital setting, but other common causes include medications, latex, and insect stings. Anaphylaxis may also occur with exercise or in rare cases, the cause may be unknown. Risk factors include a prior history of reaction, asthma, food allergy, atopy, sensitivity to multiple antibiotics, and use of certain cardiovascular medications.

Symptoms of anaphylaxis may include hives (urticaria), swelling of the face or other body parts (angioedema), itching, repetitive vomiting, anxiety, shortness of breath, wheezing, throat fullness, decreased blood pressure, and collapse. Tryptase is one of the mediators released upon degranulation of mast cells and may be elevated following an anaphylactic reaction for up to 24 h. Measuring serum tryptase levels can be useful particularly if the diagnosis of anaphylaxis is in question.

The primary treatment of anaphylaxis involves initiating basic cardiopulmonary support measures and administering epinephrine.

Table 3 Comparison of immune-mediated food reactions

	IgE-mediated food allergy	Non-IgE-mediated food allergy
Time course of symptoms	Immediate symptoms – usually within minutes to 2 h following ingestion	Delayed symptoms – usually >2 h following ingestion
Commonly implicated foods	Milk, soy, egg, wheat, peanut, tree nuts, fish, shellfish	Milk and soy
Signs and symptoms	Hives, angioedema, itching, wheezing, cough, vomiting, abdominal pain, throat tightness, shortness of breath	Vomiting, diarrhea, failure to thrive, bloody stools
Examples	Anaphylaxis to peanut	Food protein-induced • enterocolitis • enteropathy • proctocolitis

Additional medications such as antihistamines, steroids, bronchodilators, and intravenous fluids are often required as well. Close monitoring of vital signs, airway patency and breathing, and perfusion is extremely important. Approximately 20% of anaphylactic episodes have a late phase component 2–4 h later during which symptoms may recur.

Latex Allergy

Latex allergy is an IgE-mediated response to the proteins of natural rubber latex which is the fluid obtained from the cultivated rubber tree, *Hevea brasiliensis*. Healthcare workers, individuals who have had multiple surgeries, patients with *spina bifida*, and those who work in the rubber industry are most at risk because sensitization to latex proteins is more likely to occur with repeated exposure. Accordingly, the prevalence of latex allergy has increased since the application of universal precautions in the healthcare industry and the resultant routine use of latex gloves. Atopy is also an important risk factor for the development of latex allergy. Sensitization to latex can occur by wearing latex gloves, inhaling powder from latex gloves, or using medical devices, such as barium enema applicators or urinary catheters, which contain latex.

Allergic reactions to latex present with signs and symptoms similar to other allergic reactions, and some individuals may have anaphylactic reactions. Diagnosis of latex allergy involves a consistent history and positive laboratory testing for latex-specific IgE. Skin-prick testing in latex is not usually performed due to lack of a standardized latex reagent and reports of life-threatening anaphylaxis associated with latex skin testing. *In-vitro* measurements are available and should be correlated with the patient's history. Treatment involves strict avoidance of all latex products and the prescription of injectable epinephrine to be used in case of accidental exposure and reaction.

Insect Hypersensitivity

IgE-mediated systemic reactions can occur to the venoms of several stinging insects. These insects are members of the order *Hymenoptera* and include honeybees, bumblebees, yellow jackets, hornets, and fire ants. While reactions to fire ants are particularly common in the Gulf Coast region, most insect sting reactions in the United States are due to yellow jackets.

Most individuals develop localized swelling, redness, and pain at the site of an insect sting which resolves within several hours. However, some individuals may experience more extensive swelling and redness over a larger but still localized area that may last for several days. Both these types of local reactions can be treated symptomatically and are not indicative of future, more severe reactions. In contrast to local reactions, signs and symptoms of a systemic sting reaction occur at sites distant to the site of sting and may include, generalized urticaria, itching, flushing, angioedema, wheezing, shortness of breath, nausea, vomiting, hypotension, and collapse. These symptoms generally occur within minutes of the sting. Studies have shown that children under 16 years of age with isolated skin symptoms (i.e. hives and angioedema), even if generalized, are not at increased risk for a more severe reaction with subsequent stings.

Individuals with a history consistent with a systemic reaction to an insect sting should undergo skin testing. Venoms of the order *Hymenoptera* are available for immunotherapy and this is the preferred treatment for individuals with a history of systemic reaction to an insect sting confirmed by skin testing or radioallergosorbent assay test (RAST). Affected individuals should continue to carry self-injectable epinephrine in case of accidental sting.

Urticaria and Angioedema

Both urticaria and angioedema can occur as part of an acute, IgE-mediated allergic reaction but may be due to other disease processes as well. Urticaria are red, itchy, blanchable, elevated areas of the skin that are due to mast cell degranulation, venule dilatation, and dermal edema. Lesions can coalesce or be discreet, with individual lesions usually resolving within 24 h. Angioedema is similar but also involves the deep dermis and subcutaneous tissue causing swelling of the face, tongue, genitalia, or extremities.

Urticaria which consistently appear shortly following exposure to a particular allergen are likely IgE-mediated while those that occur on a regular basis with no clear, identifiable triggers may be idiopathic or a sign of other processes such as autoimmune disease, chronic infection, or neoplasm. Viral infections are an important common cause of acute urticaria in children. Likewise, angioedema may be IgE-mediated; idiopathic; or induced by certain physical conditions such as cold, heat, or pressure. Hereditary angioedema is an autosomal dominant condition that results from deficiency of C1 esterase inhibitor and results in recurrent angioedema of the face and extremities as well as repetitive attacks of severe abdominal pain due to bowel wall edema.

Drug Allergy

Adverse reactions to medications include any unintended response elicited by the drug. Those that are immune-mediated are called allergic, or hypersensitivity reactions, and approximately 5–10% of all adverse drug reactions are in this category. Drug reactions can be classified based on underlying immune mechanisms. Immediate hypersensitivity reactions are IgE-mediated and have been extensively described above. In another type of reaction, certain components of the immune system, such as antibodies, interact with drug allergens that associate with cell membranes. This interaction leads to destruction of cells such as platelets, and red and white blood cells. Drug reactions can also be mediated by immune complexes in which the drug acts as antigen and is bound by antibody. These antigen–antibody complexes aggregate in blood vessels and basement membranes and cause significant inflammation. Serum sickness is an example of this type of reaction, occurring when immune complexes enter the circulation, leading to joint pain, enlarged lymph nodes, rash, fever, and hepatitis. Finally, some drug reactions are mediated primarily by T cells. Following drug exposure, these cells become activated and mediate a robust inflammatory response as is seen in allergic contact dermatitis.

The most common medications causing IgE-mediated drug reactions are antibiotics although, theoretically, any drug can be implicated. Many drugs act as haptens which bind to carrier proteins which go on to elicit the immune reponse. In many of these reactions, IgE is specific for metabolites of the drug rather than for the parent drug.

The evaluation for drug allergy is complicated by the lack of standardized skin-testing reagents for drugs other than penicillin. Diagnosis is based on clinical history and the appropriate tests for drug-specific IgE when available. When these are consistent with an IgE-mediated event, the drug should be avoided, and if alternative unrelated medications cannot be used, desensitization can be performed in a controlled medical setting.

Diagnostic Testing for Allergies

Specific testing for allergies is necessary to confirm that certain symptoms are allergic in nature and to guide treatment for allergic diseases. This is especially true when symptoms persist or worsen despite therapy or when immunotherapy is being considered. Currently available tests for allergen-specific IgE include skin testing and *in-vitro* assays for allergen-specific IgE.

Skin-prick tests are performed by introducing a small amount of allergen extract just underneath the top layer of skin and measuring the size of wheal and flare response. These tests are commonly placed on the back but can be performed on the inner surface of the forearm as well. Extracts of a wide variety of allergens can be used and results are obtained within a short time period. The results of skin-prick tests alone are not diagnostic and should always be correlated with the individual's history. Furthermore, the size of a reaction to a skin test is not related to clinical significance and does not predict severity of reaction. There are certain circumstances in which skin tests administered intradermally are indicated. Intradermal tests are generally more sensitive than prick tests and can be useful in the evaluation of medication allergy,

venom hypersensitivity, or when the prick test is negative to an allergen that is strongly suspected by history (but not in food allergy).

The second method for detecting allergen-specific IgE is by *in-vitro* assays such as the RAST and CAP FEIA. These tests detect the presence of allergen-specific IgE antibody in the serum. As with skin testing, these results must be correlated with the individual's clinical history and environmental exposures.

Allergy skin-prick testing is generally the preferred method of testing and is more sensitive than *in-vitro* assays. Furthermore, these tests are less expensive, and the results are available immediately. Under certain circumstances, however, *in-vitro* assays are particularly useful, such as in individuals with severe eczematous rashes or in those who cannot discontinue the use of antihistamines (which suppress skin-prick test results).

Management of Allergic Diseases

Although each of the allergic diseases affecting children must be approached differently, there are several general principles of management that are common to all. These include environmental control of allergen exposure, pharmacologic therapy, allergen immunotherapy, and ongoing education.

Environmental Control of Allergen Exposure (Allergen Avoidance)

The development and pathology of allergy depends on initial and ongoing exposure to allergen. Therefore, the first line in management of allergic disease is identification of clinically relevant sensitivities and education about how to minimize exposure to these allergens effectively. In doing so, it is important to take into consideration the individual's lifestyle, occupation, and hobbies. Because it is not always practical or feasible to remove the individual from the allergen source completely, advice must be tailored to the individual family.

Reservoirs for house-dust mite include bedding, carpet, upholstered furniture, and draperies. Effective techniques for minimizing dust-mite exposure include covering the bed with dust-mite impermeable encasings, laundering bed linens in hot water at least once weekly, and vacuuming frequently or removing carpet. Other suggestions may include minimizing the amount of upholstered furniture pieces and draperies, and reducing the indoor humidity to less than 50%.

Sources of indoor mold include bathrooms without vents or windows, crawl spaces, sites of water damage, and moldy air conditioners, and humidifiers. To minimize mold growth and exposure, indoor humidity should be kept low, water leaks promptly repaired, and crawlspaces ventilated.

Complete avoidance of outdoor mold exposure is virtually impossible, but exposure may be minimized by refraining from walking through wet forests and raking leaves.

When animal dander is identified as a clinically relevant allergen, the animal should be removed from the home. If this is not acceptable for the family, the pet can be washed weekly, kept off upholstered furniture, and out of the child's bedroom. In order to minimize cockroach allergen exposure, food and garbage should be kept in closed containers and disposed of regularly.

Pharmacologic Therapy

In some cases, controlling exposure to the offending allergen is sufficient, particularly when an individual has only a limited number of sensitivities. However, because complete avoidance is often not feasible, additional treatment is needed. Multiple medications are available to treat allergic diseases, and the pharmacologic treatment of each allergic process has been outlined in the preceding sections. The general characteristics of the most commonly used medications will be discussed in more detail here. These medications include antihistamines, decongestants, corticosteroids (oral, inhaled, and intranasal), and leukotriene modifiers.

Antihistamines are generally used for allergic rhinitis, atopic dermatitis, allergic conjunctivitis, urticaria, and in the treatment of acute allergic reactions or anaphylaxis to foods, medications, insect stings, and latex. These medications target the histamine receptor and are classified according to their chemical structure and sedative properties. They affect histamine release, the production of adhesion molecules, and recruitment of inflammatory cells. First generation antihistamines have been available for over 50 years and are effective in treating allergic disease; however, they penetrate the central nervous system and have a variety of side effects such as sedation, changes in appetite, dry mouth, and urinary retention. They also may cause psychomotor impairment, and studies have shown evidence of this effect while driving. However, second-generation antihistamines are now available that are not associated with these side effects.

When nasal congestion is the predominant symptom of allergic rhinitis, patients may benefit from decongestants that decrease respiratory mucosal edema. These are available in oral and topical formulations and are often used in conjunction with antihistamines. It is important to emphasize that prolonged use of topical nasal decongestants can lead to rebound effects with a paradoxical increased nasal congestion.

Corticosteroids are anti-inflammatory agents used to treat a variety of allergic diseases and are available in oral, intranasal, and inhaled formulations. They act by binding to special receptors inside the cell and altering the transcription of genes encoding inflammatory proteins.

They are usually used in combination with other agents and are the mainstay of therapy for persistent asthma as described above. Intranasal and inhaled corticosteroids are the preferred method of delivery; however, systemic corticosteroids may be required in patients with severe, persistent symptoms.

Leukotrienes are newly synthesized mediators produced by mast cells during the allergic response. Medications, called leukotriene modifiers, are available that interfere with the interaction of leukotrienes with their receptor. These medications are administered orally and are used in the treatment of asthma and allergic rhinitis.

Allergen Immunotherapy

Immunotherapy, or allergy vaccines, is the repeated administration of specific allergens to an individual that changes the IgE-mediated response so as to reduce symptoms when naturally exposed to these allergens. Circumstances in which immunotherapy must be considered include: (1) when symptoms are severe or persistent despite maximal pharmacologic and avoidance management and (2) when allergen exposure is unavoidable. Immunotherapy is effective in the treatment of allergic rhinitis, allergic conjunctivitis, allergic asthma, and stinging-insect hypersensitivity. There are no well-controlled studies that support the use of conventional immunotherapy for food allergy or atopic dermatitis. Allergy vaccines are usually administered over a period of 3–5 years, and studies have shown that improvement in symptoms persists for at least 3 years following vaccine discontinuation. Immunotherapy should only be given under the supervision of a specialist in allergy and immunology and in a setting where trained personnel are available to respond to allergic emergencies.

Education

Individuals with allergic diseases should have ongoing education regarding their diagnosis, allergen avoidance, and medication regimen. This is most helpful when tailored to the child and family and reinforced at regular intervals. Furthermore, written emergency action plans containing guidelines for managing exacerbations and/or severe allergic reactions are useful for individuals with food allergy, stinging-insect hypersensitivity, and asthma.

Summary

Allergic disease is one of the most common chronic conditions in childhood. At a cellular level, it results from the interaction of allergen with allergen-specific IgE on the surface of mast cells and basophils, resulting in the release in chemical mediators and the influx of inflammatory cells. The effect of this process in various target organs results in the clinical signs and symptoms of allergy. Asthma, atopic dermatitis, allergic rhinitis, and food allergy are the most common allergic diseases affecting children, and the prevalence of these conditions has risen in recent years. Recent advances in the understanding of allergic pathophysiologic mechanisms are leading to advancement in the prevention and treatment of these diseases.

See also: Asthma; Immune System and Immunodeficiency.

Suggested Readings

Adkinson NF, Yunginger JW, Busse WW, *et al.* (eds.) (2003) *Middleton's Allergy Principles and Practice,* 6th edn., 2 vols. Philadelphia, PA: Mosby.
Bielory L, Bock SA, Busse WW, *et al.* (eds.) (2000) *The Allergy Report,* 3 vols. The American Academy of Allergy, Asthma and Immunology.

Bock SA (1987) Prospective appraisal of complaints of adverse reactions to foods in children during the first 3 years of life. *Pediatrics* 79: 683–688.
Leung DYM, Sampson HA, Geha RS, and Szefler SJ (eds.) (2003) *Pediatric Allergy Principles and Practice.* St. Louis, MO: Mosby.
Meltzer EO (2006) Allergic rhinitis: Managing the pediatric spectrum. *Allergy and Asthma Proceedings* 27(1): 2–8.
National Asthma Education and Prevention Program (1997) Expert Panel Report 2: Guidelines for the diagnosis and management of asthma. NIH publication 4051. Bethesda, MD.
Sicherer SH and Leung DYM (2005) Advances in allergic skin disease, anaphylaxis, and hypersensitivity reactions to foods, drugs and insects. *The Journal of Allergy and Clinical Immunology* 116: 153–163.
Sicherer SH and Sampson HA (2006) Food allergy. *The Journal of Allergy and Clinical Immunology* 117: S470–S475.

Relevant Websites

http://www.aaaai.org – American Academy of Allergy, Asthma & Immunology.
http://www.theallergyreport.com – The Allergy Report.
http://www.foodallergy.org – The Food Allergy & Anaphylaxis Network.

Amnesia, Infantile

P J Bauer, Emory University, Atlanta, GA, USA

Glossary

Autobiographical memory – Constituted of memories of specific events from the past that have relevance to the self; also characterized by a sense of reliving at the time of retrieval from memory.
Autonoetic awareness – Conscious or self-knowing awareness; in the case of autobiographical memory, awareness that the source of one's memory is an event that happened at some point in the past.
Childhood amnesia – Synonymous with infantile amnesia; the relative paucity among adults for memories of events from the first 3 years of life and a gradual increase in memories of events from ages 3–7 years, when an adult-like distribution is reached.
Elaborative style – A narrative style adopted by some parents in conversations with their children, marked by more cues to and details about past events, and invitations to the child to co-construct the story of the event.
Hippocampus – Medial temporal structure implicated in the encoding and consolidation of new memories of events and experiences.

Infantile amnesia – Synonymous with childhood amnesia.
Repetitive style – A narrative style adopted by some parents in conversations with their children, marked by repetition of questions seeking specific information, fewer cues and details, and overall fewer contributions to conversations.
Temporal cortical network – Network of neural structures implicated in memory for past events and experiences; includes cortical and medial temporal structures, including the hippocampus.

Introduction

Infantile or childhood amnesia is the relative paucity among adults for autobiographical memories from early childhood. It is virtually universal yet there are individual and group differences in its offset and density. Explanation of the amnesia requires understanding of the development of autobiographical memory in childhood, the course of which is multiply determined, by factors ranging from brain development to culture. These multiple dynamic sources of variance contribute to differences in the rate of formation of memories and the rate at which they are forgotten. The crossover of these complementary functions marks the offset of infantile amnesia.

Infantile Amnesia

Adults have relatively continuous personal histories from about the age of 7 years onward. That is, adults can recall events that took place from age 7 years or older, place them in spatial and temporal context, and attribute to them some degree of personal relevance or significance. Prior to age 7 years, however, most adults suffer from an amnesic syndrome that has two phases. From the first phase – prior to age 3 years – adults have few if any personal or autobiographical memories. From the second phase – between the ages of 3 and 7 years – adults have a smaller number of autobiographical memories than would be expected based on forgetting alone. In the literature, this two-part phenomenon is known as infantile amnesia or childhood amnesia.

For What Is There Amnesia?

Childhood amnesia is a paucity of a certain type of memory known as autobiographical memory. Autobiographical memories relate to events that happened to one's self; events in which one participated; and about which one had emotions, thoughts, reactions, and reflections. From a theoretical standpoint, childhood amnesia is interesting and important because of its implications for one's sense of self. Although we consider ourselves as continuous in space and time, there is a point in development at which that continuity ends. That moment in time is the boundary of childhood amnesia. Childhood amnesia thus presents itself as apparent evidence of discontinuity in development.

In addition to the defining feature of self-relevance, autobiographical memories have a number of characteristic features. They tend to (1) be of unique events that happened at a specific place, at a specific time; (2) entail a sense of conscious, autonoetic, or self-knowing awareness that one is re-experiencing an event that happened at some point in the past; (3) be expressed verbally; (4) be long-lasting; and (5) be veridical. This family resemblance definition of autobiographical memory (i.e., a concept specified by characteristic, as opposed to defining, features) has important implications for how we conceptualize its developmental course (see section entitled 'Explaining autobiographical memory development').

The Phenomenon of Infantile or Childhood Amnesia

Age of Earliest Memory

Research on the phenomenon that would come to be called childhood amnesia dates back to the 1890s when scholars conducted surveys of adults asking them to think about the earliest experience they could remember, and how old they were at the time. The results were consistent, with respondents dating their earliest memories to age 3 years. The phenomenon received its name in 1905, when Sigmund Freud coined the term to describe the early memories of some of his patients. He noticed that few of the adults he saw in his psychoanalytic practice had memories from their early years and that the memories they did have were sketchy and incomplete.

Subsequent to these early investigations, numerous studies of adults' memories of their childhoods have been conducted. They have yielded one of the most robust findings in the memory literature, namely, that in Western cultures, among adults, the average age of earliest autobiographical memory is 3–3.5 years. This age is obtained whether participants are asked to respond to free recall prompts such as those used by the early researchers of the phenomenon, or whether they are asked to remember a specific event the date of which is clearly known, such as the birth of a younger sibling.

Distribution of Early Memories

The age of earliest memory is only one component of the definition of childhood amnesia. The second component is that from the ages of 3 to 7 years, the number of memories that adults are able to retrieve is smaller than the number expected based on forgetting alone. Normal forgetting is a linear function of the time since experience of an event. Among adults, however, there is an under-representation of memories from ages 3 to 7 years. The strongest evidence of this under-representation is from studies employing the cue-word technique: respondents are asked to provide a memory related to each of a number of cue words (e.g., ice cream), and to estimate how old they were at the time of the event. From these data researchers have created distributions of memories over the first decade of life; a sample such distribution is illustrated in **Figure 1**. Both components of childhood amnesia are clearly apparent in the figure. Respondents report few memories from the period before age 3 years. The number of memories reported increases gradually from 3 to 7 years, at which time a steeper, more adult-like distribution is observed. This pattern has proved to be quite robust. It is observed regardless of the specific method used to elicit the memories and the age of the respondents at the time the memories were cued.

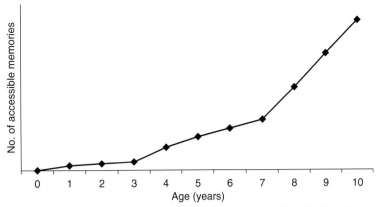

Figure 1 Schematic representation of the distribution of autobiographical memories from the first 10 years of life.

The Universality of Childhood Amnesia

One of the features of childhood amnesia that makes it so compelling is its universality. That is, virtually every adult suffers from it, to one extent or another. That does not mean that its form is identical across individuals, however. There are individual and group differences in the timing of its offset, in its density, and in the content of reports of early memories. The differences are interesting and important in their own right. They also are critical to evaluation of theories as to the sources of childhood amnesia in that an adequate theory must account not only for the normative trend, but for individual and group differences as well.

Individual Differences

Whereas more than a century of research has revealed age 3–3.5 years as the average age of earliest memory, from the beginning of the study of adults' early autobiographical memories, individual differences in the age of earliest memory have been apparent. In virtually all historical and contemporary reports of adults' early autobiographical memories that provide information on variability, there are instances of a small number of memories from the first year of life. Memories from at least some respondents from age 2 years are more the rule than the exception. There are also differences in the latest early memory, with some adults reporting their earliest memory from as old as 6–9 years of age. There are few developmental phenomena for which the age of onset is so variable. In addition to differences in the age of earliest memory, there are individual differences among adults in the density of early memories. That is, some adults are able to recall many memories from their childhood years, whereas others remember only a few.

Group Differences

Table 1 summarizes some of the groups for which differences in the age of earliest memory have been reported, three of which are discussed below.

Table 1 Group differences in the age of earliest memory

Source of variance in age of earliest memory	Groups reporting earlier and later memories	
	Earlier early memories	Later early memories
Gender	Women	Men
Birth order	First-borns	Later-borns
Culture group	Western cultures	Eastern cultures
Family moved before age 4 years	Yes	No
Attended preschool	Yes	No

Gender. Of all the possible sources of group difference in the age of earliest memory one could contemplate, possible gender differences have received the most attention. A consistent finding is that women have memories from earlier in life than do men. In some cases the differences are statistically reliable, whereas in other cases they are not. Regardless of their statistical reliability, differences in the ages of earliest memories for women and men typically are small in magnitude.

There are also some reports of differences in the length of women's and men's reports of early childhood events, in their affective qualities, and in the interpersonal themes represented in them. Specifically (1) women tend to provide longer, more detailed, and more vivid accounts of their early memories, relative to men; (2) women more often refer to anger, shame, and guilt in the earliest memories relative to men; and (3) women's early emotional memories tend to concern attachment issues (i.e., concerns regarding security, approval, separation, and reunion), whereas those of men tend to concern competence issues (i.e., concerns regarding ability, performance, achievement, and identity).

Birth order. Although it has received significantly less research attention, relative to gender, birth order also has been found to be systematically related to age of earliest memory. Children who are first-born have earlier memories than children who are later-born. There also appear to be differences in the age of earliest memory as a function of the number of children in the family, and their spacing. Only-children have earlier autobiographical memories, relative to the oldest children in multichild families. First-borns for whom there is a larger difference in ages between themselves and their oldest siblings report autobiographical memories from earlier in life, relative to first-borns whose oldest siblings are more closely spaced.

Culture. Most of the research on adults' early memories has been conducted with individuals from Western societies, including Canada, the UK, and the US. Studies with individuals from Eastern cultures make clear the limits to generalizability of the findings from Western culture. There are systematic differences in the age of earliest memory with Western samples reporting memories from earlier in life, relative to the Eastern samples. In some cases the differences are pronounced, with American adults reporting memories that are a year or more earlier than adults from Eastern cultures.

There are also cultural differences in the content of adult women's and men's reports of their early experiences. For example, relative to respondents from Eastern cultures, Americans provide longer memory narratives; their memories are more frequently of a single event or specific memory, as opposed to a more general memory; and they more frequently comment on their own experiences and attitudes, including emotions and feeling states.

Explaining Infantile or Childhood Amnesia

Why is it that adults experience amnesia for events from the first 3–3.5 years of life, and have fewer memories from the ages of 3 to 7 years than would be expected based on forgetting alone? There are two major categories of theories to explain childhood amnesia. By one category of accounts, memories for early-life events are formed, but later functionally disappear or become inaccessible. By the other category of accounts, adults lack memories of events from infancy and very early childhood because, in effect, no memories were created. Alternatively, memories were created but lacked an important feature or features that precluded them from being entered into the autobiographical record.

Memories Are Formed but Become Inaccessible

One category of explanation for childhood amnesia suggests that young children and perhaps even infants form memories of the events of their lives, but that over the course of time and development, these early memories become inaccessible.

Freud's explanation in terms of repression and screening. In 1905, Sigmund Freud delivered a lecture in which he gave infantile or childhood amnesia its name. In the lecture he commented on the fact that the first 6–8 years of life are full of accomplishments (e.g., children learn to walk and talk, they accrue a lot of knowledge about the world, and so forth) yet adults remember few of the experiences that led to them. He further commented that the memories that survive seem unreliable. He deemed them unreliable because the early memories that his patients reported were not of the psychic struggles that Freud assumed consumed much of mental life but of bland, unemotional, and often commonplace events and experiences. Freud was also impressed by the observation that his patients often described their memories from the unrealistic third-person perspective. That is, rather than through the eyes of the beholder, the memories were described as from the perspective of a third party. Given that this was an impossible perspective for an autobiographical memory, Freud concluded that these memories were the result of reconstructive processes. Based on these observations, Freud advanced the theory that early memories were blockaded or screened from consciousness. He suggested that the relative paucity of early memories was due to repression of inappropriate or disturbing content of early, often traumatic (due to their sexual nature) experiences. Events that were not repressed were altered to remove the offending content. In effect, he hypothesized that the negative emotion in these memories was screened off, leaving only the bland skeleton of a significant experience.

Freud's explanation for childhood amnesia in terms of repression and affective screening was internally consistent with his larger theoretical framework. External to the theory, however, the explanation has not fared especially well. One issue is that although adults remember fewer early-life events than would be expected based on forgetting alone, they nevertheless have more memories from childhood than would be expected by Freud's model of repression. In addition, contrary to the suggestion that memories of early-life events would be devoid of emotion or overwhelmingly positive, both traumatic and non-traumatic events from childhood are recalled. In some studies, memories of negative episodes actually outnumber positive episodes. A second issue is that although many early memories are from the third-person perspective, there are also many from the first-person perspective. Moreover, many later memories are from the third-person perspective. Some scholars suggest that the perspective adopted has more to do with the event being remembered than age at the time it occurred. Today, Freud's suggestions of repression and screening of early memories generally are not considered adequate explanations for childhood amnesia.

Different cognitive lenses. The second exemplar of an explanation of childhood amnesia in terms of memories that are formed but become inaccessible is actually not itself a unified theory. Instead, it is a category of explanations that has in common the suggestion that there are different cognitive lenses for different times of life. As individuals change lenses over the course of their lives, they lose the ability to access memories created with the old lens type (not unlike the inaccessibility that results from a change in operating systems on a computer). By some accounts, the lenses differ in their reliance on language. Because they lack language, infants and young children encode memories visually or imaginally, but not symbolically. With the advent of language skills, exclusively nonverbal encoding gives way to primarily verbal encoding. As the system becomes more and more verbally saturated, it becomes increasingly difficult to gain access to memories encoded without language.

Other accounts place emphasis not on language but on differences in life periods, each of which has distinct hopes, fears, and challenges, for example. Life periods may correspond to elementary vs. secondary school vs. college, or before vs. after marriage, or before vs. after retirement. Memories from different lifetime periods may differ from those from the current period not only because of the passage of time, but because of the new phase of life, which may herald concomitant changes in thinking or world view.

Models that implicate different cognitive lenses as the explanation for childhood amnesia make two critical predictions, namely, that (1) early memories are not accessible later in life, and (2) memories from within a life period should be more readily accessible than memories across life periods. Though it is negative in nature, there is overwhelming evidence for the first of these predictions. If we allow that infants and children form memories (see section titled 'Explaining autobiographical memory development' for evidence that they do), then the very phenomenon of childhood amnesia is one of later inaccessibility of early memories. Critically, there is no direct evidence that developments in language actually *cause* early memories to become inaccessible. Moreover, although preverbal memories do not readily lend themselves to verbal description, under some circumstances, they can be described with language once it is acquired.

There is evidence that memories from within a life period are more readily accessible than memories across life periods. Some of the most compelling illustrations come from studies in which immigrants are asked to retrieve memories of events that took place before vs. after they emigrated. Memories retrieved from the time before immigration more frequently are in the native language and memories retrieved from the time after immigration more frequently are in the language of the adopted home. Similarly, cue words from the native language elicit memories from before immigration, whereas cue words from the language of the adopted home elicit memories

from after immigration. Thus, it seems that memories from within a life period are more accessible in the language of that period. Other reasons why memories might become differentially accessible over time are discussed in 'Explaining autobiographical memory development'.

Autobiographical Memories of Early Life Events Are Lacking

The second category of explanation of childhood amnesia suggests that adults have few autobiographical memories from infancy and very early childhood because during this period no such memories were formed, due to general or more specific cognitive deficits. Similarly, these accounts explain that the number of memories that adults have from the preschool period is smaller than would be expected based on forgetting alone because during this period autobiographical memory competence is under construction and so, consequently, there are relatively fewer memories from this period.

The suggestion that cognitive deficits explain the relative paucity of memories from early in life has had a number of proponents but the name that is most readily associated with the perspective is Jean Piaget. Although Piaget did not advance a specific theory of childhood amnesia, he nonetheless provided a compelling explanation for it. He maintained that for the first 18–24 months of life, infants and children did not have the capacity for symbolic representation. As a result, they could not mentally represent objects and entities in their absence. They thus had no mechanism for recall of past events.

Piaget further suggested that even once children had constructed the capacity to represent past events, they still were without the cognitive structures that would permit them to organize events along coherent dimensions that would make the events memorable. One of the most significant dimensions that Piaget suggested preschool-age children lacked was an understanding of temporal order. Specifically, he suggested that it was not until children were ~5–7 years of age that they developed the ability to sequence events temporally. Without this fundamental organizational device, children were not able to form coherent memories of the events of their lives. The more contemporary, so-called neo-Piagetian perspectives suggest that limits on cognitive capacity (e.g., working memory capacity) either prevent information from being encoded in an accessible format to begin with, or that limitations on retrieval mechanisms prevent it from being recalled at a later time.

There are also suggestions that specific conceptual changes play a role in the explanation of childhood amnesia. By some accounts, adults have few memories from early in life because, for the first 2 years, there is no 'cognitive self' around which memories can be organized. As a consequence, there is no 'auto' in autobiographical.

By other accounts, for the first 5–7 years of their lives, children lack autonoetic awareness, rendering it impossible for them to create memories that have this characteristic feature. As a consequence, autobiographical memories are not formed and thus are not available to be retrieved by adults.

The suggestions that infants and very young children lack the symbolic capacity to form memories and that the memories of preschool-age children are disorganized are no longer tenable. As will be seen in the next section, even in the first year of life infants encode and later retrieve memories of past events. Nevertheless, there are pronounced changes in the basic processes of memory, which have implications for the reliability, robustness, and temporal extent of memory through infancy and early childhood. Thus, although infants and very young children are no longer seen as total mnemonically incompetent, neither are their memory systems as effective and efficient as those of adults. The differences have implications for the density of representation of autobiographical memories from the early years of life (see section–entitled 'Explaining autobiographical memory development'). Finally, it is increasingly apparent that no single factor, such as development of a self concept or absence of autonoetic awareness, will provide a sufficient explanation for why autobiographical memory seems to begin when it does or why adults lack autobiographical memories from a period of their lives (respectively). Rather, it is recognized that autobiographical memory is a complex, multifaceted capacity, the development and operation of which are influenced by many factors.

Explaining Autobiographical Memory Development

Autobiographical memory involves a number of capacities and skills. It requires that events and experiences be encoded and stored in an accessible manner, and later retrieved. Once retrieved, the memory must be expressed. The most informative expression is via a narrative that provides the listener with information about the who, what, where, when, why, and how of the event. Moreover, autobiographical memories are of events that happened to the self at a specific place and time. They are accurate, long-lasting, and when they are retrieved, there is a sense of awareness that they are based on past experience. Critically, each of these aspects of the ability has its own developmental course. An adequate explanation of the development of autobiographical memory – and thus of achievement of an adult-like distribution of memories – must recognize this complexity as well as account for the individual and group variability that is apparent in autobiographical records. This is best accomplished by undertaking analysis at multiple levels, ranging

from the brain systems that support memory to the cultural influences on verbal expression of memory.

The Neural Substrate of Autobiographical Memory and Its Development

The ability to encode, store, and later retrieve autobiographical memories depends on a multicomponent neural network that includes structures in the medial temporal lobes (including the hippocampus), as well as neocortical structures. The network is schematically represented in **Figure 2**. Specifically, primary, secondary, and association cortices register what we are seeing, smelling, hearing, and so forth, and integrate it all into a coherent experience. For that experience to endure beyond the moment, it must be consolidated into a memory trace. Consolidation depends on neurochemical and neuroanatomical changes that create a physical record of the experience. The processes are carried out by medial temporal structures in general and the hippocampus in particular, in concert with the cortex. Throughout the period of consolidation – which may take weeks to months in the human – memories are vulnerable to disruption and interference. Eventually, however, they become stabilized, and no longer require the participation of the hippocampus for their survival. Rather, they are maintained in the cortices that gave rise to the original experience. Finally, the prefrontal cortex in particular is implicated in retrieval of memory traces from these long-term stores.

Portions of the medial temporal structures mature relatively early. For instance, the cells that make up most of the hippocampus are formed in the first half of gestation and, by the end of the prenatal period, virtually all

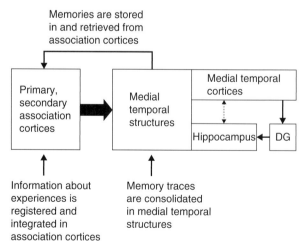

Figure 2 Schematic representation of the flow of information through the temporal cortical network responsible for formation of autobiographical memories. DG, dentate gyrus.

have migrated to their adult locations. In some areas of the hippocampus, synapses are present as early as 15 weeks gestational age. By ~6 postnatal months, the number and density of synapses have reached adult levels, as has glucose utilization in the temporal cortex. In contrast, development of the dentate gyrus of the hippocampus is protracted. At birth, the dentate gyrus includes only ~70% of the adult number of cells and it is not until 12–15 postnatal months that the morphology of the structure appears adult-like. Maximum density of synaptic connections in the dentate gyrus is also reached relatively late. The density of synapses increases dramatically (to well above adult levels) beginning at 8–12 postnatal months and reaches its peak at 16–20 months. After a period of relative stability, excess synapses are pruned until adult levels are reached at ~4–5 years of age.

The association areas also develop slowly. For instance, it is not until the seventh prenatal month that all six cortical layers are apparent. The density of synapses in prefrontal cortex increases dramatically beginning at 8 postnatal months and peaks between 15 and 24 months. Pruning to adult levels is delayed until puberty and beyond. Other maturational changes in the prefrontal cortex, such as myelination, continue into adolescence, and adult levels of some neurotransmitters are not seen until the second and third decades of life.

The full temporal cortical network can be expected to function as an integrated whole only once each of its components, as well as the connections between them, has reached a level of functional maturity. This leads to the prediction of emergence of long-term memory by late in the first year of life, with significant development over the course of the second year, and continued (albeit less dramatic) development for years thereafter. The time-frame is based on increases in the formation of new synapses beginning at ~8 months in both the dentate gyrus and prefrontal cortex, with continued synaptogenesis through 20 and 24 months, respectively. The expectation of developmental changes for months and years thereafter stems from the schedule of protracted selective reduction in synapses both in the dentate gyrus (until 4–5 years) and in the prefrontal cortex (throughout adolescence or early adulthood).

Developments in Basic Memory Processes

The relatively late development of aspects of the temporal cortical network has implications for behavior. Because of their involvement in all phases of the life of a memory, protracted development of cortical structures can be expected to impact the encoding, consolidation, and storage, as well as retrieval, of memories. Late development of the dentate gyrus of the hippocampus is critical because, at least in the adult, it is the major means by which information makes its way from the cortex into the hippocampus where new memory traces are consolidated for long-term storage (see **Figure 2**). The immaturity of these structures and connections between them would present challenges to these processes. As they develop, we would expect to see age-related changes in behavior.

The time course of changes in behavior matches what is known about developments in the temporal cortical memory network. Using nonverbal measures of memory (elicited and deferred imitation in which props are used to produce novel sequences of actions that infants are invited to imitate), researchers have demonstrated that between 9 and 20 months of age, the length of time over which recall is apparent increases dramatically, from 1 to 12 months. Over the same period, the robustness of memory increases such that infants remember more, based on fewer experiences of events. In addition, long-term memory is more reliably observed. Whereas at 9 months of age only ~50% of infants show evidence of long-term recall, by 20 months, individual differences in whether or not infants recall are the exception rather than the rule (though there remain individual differences in how much is remembered).

Because the actions and sequences on which infants are tested are novel to them, their behavior provides evidence that they are able to remember unique events. Moreover, because infants recall sequences in the correct temporal order, there is evidence that they remember when events occurred. Infants also demonstrate that they remember specific features of events, in that they reliably select the correct objects from arrays including objects that are different from, yet perceptually similar to, those used to produce event sequences. These behaviors make clear that by the end of the second year of life, children have many of the memory skills necessary to form an autobiography.

Over the course of the preschool years, memory abilities change in at least two important ways that are relevant to developments in autobiographical memory. First, memory processes sharpen to the point that a single experience of an event is sufficient to ensure retention over the long term. Prior to the preschool years, very long retention is seemingly dependent on multiple experiences of an event. In contrast, by the age of 3–4 years, children remember events experienced only once. This age-related change likely is linked to developments in the temporal cortical network supporting memory. Second, children develop the ability to locate events in a particular time and place. This development is apparent in age-related increases in the use of temporal markers such as yesterday and last summer, for example. Such markers serve as a time line along which records of events can be ordered. These and other changes in basic memory abilities mean that more events can be stored with more of the elements of autobiographical memories: unique events, with distinctive features, accurately located in time and place, that are maintained for long periods of time.

Developments in Nonmnemonic Abilities

Developments in nonmemory abilities also contribute to age-related improvements in autobiographical memory. Two of the most prominent have already been mentioned: changes in self-concept and development of autonoetic awareness.

Self-concept. Because autobiographical memories are about one's self, a self-concept is a necessary ingredient for an autobiographical memory. Children's first references to themselves in past events occur at about the same point in developmental time as they begin to recognize themselves in a mirror. In the second half of the second year of life, children who recognize themselves in the mirror have more robust event memories and they make faster progress in independent autobiographical reports, relative to children who do not yet exhibit self-recognition. Over the preschool years, there are further developments in recognition of continuity of self over time, both in physical features and in psychological characteristics. These developments have implications for autobiographical memory: for a past event to be relevant to present self, the rememberer must realize that the self who is remembering is the same as the self who experienced the event in the past. Also over the preschool years children develop a more subjective perspective on experience. This facilitates inclusion of events in an autobiographical record because experiences are not just objective events that play out, but are events that influence the self in one way or another. The personal significance of the events is conveyed, in particular, by references to the emotional and cognitive states of the one who experiences. Such references indicate the sense of personal ownership and unique perspective that is so characteristic of autobiographical memories. Together, these and other developments in the self concept mean that more aspects of more events have relevance to the self thus providing more opportunity for formation of self-relevant memories.

Autonoetic awareness. Retrieval of autobiographical memories is accompanied by autonoetic awareness: an understanding that the recollected event is one that happened in the past. It is not until children are 4–6 years of age that they reliably identify the sources of their knowledge. This ability aids in location of events in space and time, thereby contributing to the specificity of memories (as discussed earlier in this section). Understanding that the source of a current cognition is a past event can also contribute to better event narratives. Children who have this realization can be expected to provide their listeners with orientation to the circumstances of the past; to advise them of the specifics of the event, such as who was there and where it took place; and to provide their own subjective perspective on the event. Consistent with this suggestion, 3.5–4.5-year-olds who perform better on tasks

that measure their understanding of knowledge also have more sophisticated conversational skills. Autonoetic awareness may foster autobiographical memory development more directly as well. As children come to appreciate that the sources of their cognitions are representations, and that others too have representations, both of which are unique to the individuals, they can begin to construct personal perspectives on events. Over time, the practice of reflecting on one's evaluation of an event would be expected to foster further development of the self concept, in that children have the opportunity to reflect on the continuities (as well as discontinuities) in their own and others' reactions to events and experiences. In a variety of ways then, both indirectly and directly, conscious appreciation that the source of a representation is a past event contributes to increases in autobiographical memories.

Developments in Language and Narrative Expression

Because autobiographical memories are expressed verbally, developments in language and in narrative expression play a role in autobiographical memory development. In the first years of life, children who have larger vocabularies and more sophisticated syntax make more contributions to memory conversations relative to children with less developed language skills. Over the preschool years, children play increasingly active roles in conversations. For example, they provide (1) more of the elements of a complete narrative (i.e., the who, what, where, when, why, and how of events), (2) more descriptive details, and (3) more evaluative information, thereby adding texture to their narratives. A more complete narrative not only makes for a better story for the listener but also provides the storyteller with a structure for organizing memory representations, for differentiating events from one another, and for creating associative links between events. It thus works to facilitate encoding and consolidation of event memories in a way that simultaneously preserves their uniqueness and integrates them with other memories in long-term stores, thereby strengthening their representation. The organizational frame provided by a complete narrative also may aid memory retrieval.

The Social Context of Remembering

The development of autobiographical memory does not occur in a vacuum. From early in their lives, children participate in the activity of sharing their own and others' memories. At first, much of the work of recollecting past experiences falls to more verbally and narratively accomplished partners, typically the children's parents. Parents tell what happened in an event and children participate by affirming the parents' contributions, and by adding a bit of memory content here and there. Through these conversations, children begin to learn what to include in their memory reports and also how to organize their narratives. As noted earlier, as they internalize the narrative form, it comes to serve important mnemonic functions at encoding as well as retrieval. Through conversations about past events, children also learn the social function of talking about the past, which is to share thoughts, feelings, reactions, and experiences with other people. Families or cultures that place a high premium on talking about the past, and on the child's own experience of events, likely will promote more rapid development of structures for organizing autobiographical memories, relative to families and cultures that place less emphasis on these aspects of experience. These variables will interact with characteristics of the individual child, resulting in individual as well as group variation.

Variability among children. There are numerous individual differences that may affect the development of autobiographical memory. For example, there are individual differences in the most basic element of the self concept, namely, self-recognition. In the middle of the second year, ∼50% of children already indicate self-recognition. Whereas another 25% recognize themselves by the end of the second year, the remaining 25% still do not. A similar range is apparent on tasks that assess the temporally extended self. That is, at 3 years of age, 25% of children already show evidence of recognition that the past self and present self are one and the same. A full year later, 75% of children show this evidence but 25% still do not.

Individual differences also are apparent in many of the other domains that relate to developments in autobiographical memory, including (1) the amount that infants and children remember, (2) the accuracy of memories, (3) acquisition of temporal concepts that aid in location of events in time and in relation to one another, and (4) understanding of a variety of cognitive concepts that are hypothesized to relate to autonoetic awareness. Children also differ in their verbal and narrative sophistication and thus in their abilities to express their memories. These sources of variability have direct, indirect, and as described next, even interactive effects on autobiographical memory development.

Variability among families. Individual children bring their individual differences into home environments that themselves are variable. One of the ways that home environments differ is in the narrative style of the parents. For example, some mothers use a large number of evaluative terms in autobiographical memory conversations. Over time, their children come to use a larger number of such terms when they report on events. More broadly, some parents exhibit an elaborative style of talking about the past, providing cues and details about events and

inviting their children to join in on the story. Other parents exhibit a more repetitive style, asking children questions for which they seem to have a particular answer in mind. Children exposed to the elaborative style report more about events both concurrently and over time. It seems that they are internalizing a narrative form that helps them organize, remember, and subsequently retrieve stories of previous life events. Importantly, parental style is, to a certain degree, a misnomer, in that characteristics of the child and even of the dyad influence it. For example, parents are more elaborative with children who are more verbal and with daughters relative to sons. The attachment security of the dyad also is related to parental style: mothers in securely attached dyads are more elaborative. Even family demographics may relate to autobiographical memory development. For instance, on tests of understanding of the representational nature of mind, a concept implicated in autobiographical memory, children who are first-born and thus have no older siblings have low rates of success. In sum, there are numerous differences among families that may affect the course of development of autobiographical memory in any given child.

Variability among cultures. Just as important as differences in the child who is doing the remembering and in the familial environment in which memory is being shaped are differences in the larger cultural milieu of experience. Illustrative examples come from research that involves contrasts between children from Eastern and Western culture groups. Briefly, the early autobiographical memory reports of children from Eastern cultures include fewer references to themselves and fewer personal evaluations, relative to reports from children in Western cultures. In addition, the autobiographical memory reports of children from Eastern cultures tend to feature generic as opposed to specific events, and they are shorter and less detailed, relative to those provided by children in the West. Thus, on at least three critical features – significance to the self, specificity in place and time, and verbal expression – the early narratives of children from Eastern cultures may be viewed as less prototypically autobiographical, relative to the narratives of children from Western cultures.

Linking Autobiographical Memory Development and Infantile or Childhood Amnesia

If, over the course of infancy and the preschool years, memories of events appear more and more autobiographical, why then do adults have so few personal memories from this period? Addressing this question requires consideration of the rate at which autobiographical memories are formed and the complementary rate at which they are forgotten. Although the preschool years are marked by an increasing rate of formation of event memories with autobiographical features, they also are marked by a rate of forgetting that is accelerated, relative to the rate of forgetting in later childhood and adulthood. At some point, the rate at which new, more autobiographical memories are formed overtakes the rate at which they are forgotten. From adults' retrospective perspective, that point is the offset of infantile or childhood amnesia.

The Rate at Which Memories Are Formed

There is incontrovertible evidence that over the course of the preschool years, children form memories that are more and more autobiographical. As early as they are able to use past tense markers, children refer to past events of relevance to themselves. As they gain in narrative sophistication, children's stories become more complete and more coherent. Their narratives about past events also take on more and more elements of drama and they contain an increasing amount of evidence of the significance of the event for the child. Stories are told not only about routine events but about unique experiences that happened at a particular place, at a particular time. Some events – though certainly not all – are remembered for months and even years. Although prior to 4–6 years of age children do not pass tasks that permit researchers to say that they are aware of the sources of their representations, children's narratives certainly contain evidence of vivid recollections of events from the past: they include elements that provide a sense of the intensity of experience, elements of suspense, and information about the internal states of the participants, for example. In sum, over the course of the preschool years, children's stories of their lives bear more and more of the marks of typical autobiographical reports. As a result, children exhibit an increasing number of memories that are recognized as autobiographical.

The Rate at Which Memories Are Forgotten

Even infants and young children remember, but they also forget. The younger the infant or the child, the faster the rate of forgetting. Age-related differential forgetting results from a number of sources. One source is the relative immaturity of the neural structures responsible for formation and maintenance of memories over the long term. Because the temporal cortical network is relatively less developed, encoding and consolidation processes are less effective and efficient in younger infants and children than in older children. As a result, they exhibit a faster rate of forgetting.

Differential rates of forgetting also likely result from several nonmnemonic sources. In effect, the memories that the young child is asking her or his immature brain to consolidate and store contain fewer of the features that typify autobiographical memories: (1) the self to

which they are referenced is not as stable and coherent a construct as it will be later in development; (2) relative to later memories, early memories tend to contain fewer distinctive features and are less specifically located in space and time; (3) younger children likely encode fewer of the elements that make for a good narrative, relative to older children, thereby denying themselves an effective organizational tool; and (4) early-memory representations contain fewer indications of their origin in events from the past, relative to those encoded with a more mature understanding of the representational nature of the human mind. In short, in early childhood we have less than optimal processes operating on less than optimal raw materials. The quality of the resulting output is simply not as high as it is in later childhood and adulthood, when we have more optimal processes operating on more optimal materials. The net result is a faster rate of forgetting. Importantly, this analysis implies that the rate of forgetting is not accelerated as suggested by the second component of the definition of infantile or childhood amnesia. That is, the analysis implies that the number of memories of events from the ages of 3 to 7 years is not smaller than would be expected based on forgetting alone. That characterization holds only when an adult rate of forgetting is applied. Consideration of a more developmentally appropriate rate of forgetting is expected to yield a more normal distribution.

The Crossover of Two Functions

The net effect of this analysis is a model that suggests that, among adults, we see an increase in autobiographical memories dating from around age 4 to 6 years because, as depicted in **Figure 3**, this is the point at which the functions of memory formation (ascending solid line) and memory loss (descending solid line) cross over. Prior to the age of 4 years, the rate at which memories are lost is faster than the rate at which they are gained; after the age of 6 years, the rate at which memories are formed is faster than the rate at which they are lost. Considering the adult phenomenon of childhood amnesia to be a result of the crossover of two functions provides a ready account of individual and group differences in the age of, and distribution of, autobiographical memories from early in life. Two individuals who as children had the same rate of forgetting may nevertheless have vastly different offsets of childhood amnesia as a function of differences in the slopes of change in the remembering function. Children in a family and cultural environment that places a premium on narrative, and that encourages reflection on the meanings of events and their significance for the child, may have autobiographical memories from earlier in the preschool years (dashed line crossing at age 4 years). In contrast, children in a family and cultural environment that uses a less elaborative

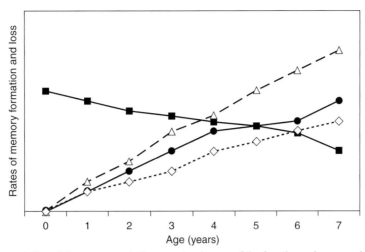

Figure 3 Schematic representation of the crossover in the preschool years of the functions of memory formation (the rate of which increases with age; solid ascending line) and memory loss (the rate of which decreases with age; solid descending line). The period of intersection between ages 4 and 6 years is recognized as the offset of infantile or childhood amnesia. Also illustrated are possible patterns of individual differences in the offset of infantile amnesia as a function of differences in the slope of change of the remembering function. The dashed line represents a steeper increase in the memory function and an earlier crossover with forgetting (at age 4 years). The dotted line represents a more shallow increase in the memory function and a later crossover with forgetting (at age 6 years).

style and which does not encourage reflection on the self may have autobiographical memories from later in the preschool years (dotted line crossing at age 6 years).

Individual differences also could result from differences in the slopes of forgetting functions (not shown in **Figure 3**). That is, individuals whose rates of increase in the formation of autobiographical memories are the same could nevertheless experience a different course of development of autobiographical memory because of differences in the rate at which memories are forgotten. Variability in the forgetting function no doubt is associated with a variety of factors, including different rates of maturational change in the temporal cortical network and associated differences in the basic mnemonic processes of encoding, consolidation, storage, and retrieval. As such, the conceptualization provides a ready account of individual and cultural differences: they result from differences in the quality of the autobiographical memories that are formed during the period and from the likelihood of survival of the memories over time.

Summary and Conclusions

Whereas adults have a wealth of memories from later childhood and early adulthood, there is a virtually universal paucity of memories from the first years of life. For over a century, there have been reports that the average age of earliest memory among Western adults is age 3–3.5 years. Adults report a larger and gradually increasing number of memories from the ages of 3 to 7 years. The number of events adults remember from the age of 7 years

onward is consistent with what would be expected, based on adult rates of forgetting.

From a theoretical standpoint, this distribution of autobiographical memories is interesting and important because of its implications for the self concept. For much of the lifetime there is a continuous time line of events and experiences. The boundary of childhood amnesia represents a break in the otherwise continuous history and thus a challenge to a fully integrated sense of self. Infantile or childhood amnesia also is interesting and important because of its clinical and forensic implications. Theories of personality and psychopathology look to early experiences as an important source of adult attitudes and behaviors. For these analyses, determination of the nature of the trace that early experiences leave behind is crucial. Forensic concerns also compel research on memory for events during the period obscured by childhood amnesia. The veridicality and accessibility of memories of events from early in life are questions the answers to which can have profound consequences for childhood victims of crimes, as well as their alleged perpetrators.

The fact that the phenomenon of childhood amnesia is almost universal does not mean that there are no individual and group differences in early memory. On the contrary, there is wide variation in the age of earliest memory, and there are differences as a function of gender, birth order, and culture group. Theories that hope to explain the relative lack among adults of memories of specific events from early in life must account not only for the age of earliest memory and distribution of early memories, but for these systematic sources of individual and group variability as well.

Theories to explain infantile or childhood amnesia have suggested either that memories for early life events are formed but then later become inaccessible, or that memories are not accessible later in life because they were never formed. Although these explanations may seem incompatible, it is likely that elements of both figure in the development of autobiographical memory and thus the offset of childhood amnesia. Consideration of age-related developments in the neural substrate of autobiographical memory, in the processes that it subserves, and in a number of nonmnemonic concepts and domains (including language), implies that, in infancy and the preschool years, memories are formed but that they are relatively quickly forgotten. Conversely, developments in all of these domains imply that, over the same period of time, the memories that are formed are increasingly autobiographical. The point at which the two functions – rate of forgetting and rate of remembering – cross over one another is the point that we recognize as the offset of infantile or childhood amnesia. The crossover point varies as a function of individual differences in children, in their families, and even in the culture group in which they are raised. This analysis makes infantile or childhood amnesia less enigmatic. At the same time, it makes additional research on the processes and determinants of remembering and forgetting in infancy, early childhood, and beyond, all the more imperative.

See also: Birth Order; Cognitive Development; Gender: Awareness, Identity, and Stereotyping; Hippocampus; Memory.

Suggested Readings

Bauer PJ (2006) Event memory. In: Damon W and Lerner RM (eds.) *Handbook of Child Psychology: Cognition, Perception, and Language,* 6th edn., pp. 373–425. Hoboken, NJ: Wiley.

Bauer PJ (2006) *Remembering the Times of Our Lives: Memory in Infancy and Beyond.* Mahwah, NJ: Erlbaum.

Brewer WF (1996) What is recollective memory? In: Rubin DC (ed.) *Remembering Our Past: Studies in Autobiographical Memory,* pp. 19–66. New York: Cambridge University Press.

Markowitsch HJ (2000) Neuroanatomy of memory. In: Tulving E and Craik FIM (eds.) *The Oxford Handbook of Memory*, pp. 465–484. New York: Oxford University Press.

Mullen MK (1994) Earliest recollections of childhood: A demographic analysis. *Cognition* 52: 55–79.

Nelson CA, Thomas KM, and de Haan M (2006) Neural bases of cognitive development. In: Damon W and Lerner RM (eds.) *Handbook of Child Psychology, Vol. 2: Cognition, Perception, and Language,* 6th edn., pp. 3–57. Hoboken, NJ: Wiley.

Nelson K and Fivush R (2004) The emergence of autobiographical memory: A social cultural developmental theory. *Psychological Review* 111: 486–511.

Rubin DC (2000) The distribution of early childhood memories. *Memory* 8: 265–269.

Wang Q (2003) Infantile amnesia reconsidered: A cross-cultural analysis. *Memory* 11: 65–80.

Wetzler SE and Sweeney JA (1986) Childhood amnesia: An empirical demonstration. In: Rubin DC (ed.) *Autobiographical Memory*, pp. 191–201. New York: Cambridge University Press.

Anger and Aggression

R E Tremblay, University of Montréal, Montreal, QC, Canada

Glossary

Aggression – The initiation of an attack. These attacks can directly or indirectly attempt to hurt the victim. The attacks can use different means: physical (e.g., fists, guns, noise), psychological (e.g., insult, blame, shame), or social (e.g., exclude, reject, prevent access to resources).

Anger – Considered one of the basic emotional states. Anger can vary in intensity from mild irritation and annoyance to intense fury and rage. It is usually well-expressed by facial expression. Anger is sometimes distinguished from rage with reference to the intensity of the emotional reaction and the capacity to maintain control of one's behavior.

Development – Refers to the physical, psychological, and behavioral changes that occur with time from conception to death. The developmental studies of a specific phenomenon (e.g., anger, aggression) describe the changes over time, the phenomena that are associated with the changes, the causes of these changes, and their consequences.

Prevention – Planned activities to avoid the development of a disorder (e.g., chronic anger and aggression). The activities can target an entire population or a specific subpopulation considered at high risk following studies of risk factors (see 'risk factor').

Risk factor – A characteristic of individuals shown to precede and be statistically related to the disorder we want to predict (e.g., chronic anger or aggression). A risk factor may be a dichotomy (e.g., boys/girls), a category (e.g., ethnic groups), a rank order (e.g., birth order of sibling), or a continuum (e.g., age). There are generally many risk factors associated with a mental health problem. Risk factors are generally used to divide the population into a high-risk subpopulation and a low-risk subpopulation for preventive interventions (see 'prevention').

Introduction

To understand present-day ideas concerning human development it is useful to take a historical perspective. Many of our apparently new ideas are in fact old ones that are now packaged with a modern flavor. Much of this article takes this historical perspective to the development of anger and aggression, as it relates to infancy and early childhood.

Anger is considered one of the basic emotional states. Joy, sadness, and fear are among this group of 'primitive' emotions, while pride, guilt, and shame are apparently more complex emotions and appear later in life. Aggression is a behavior which often follows anger, but nonhuman animals and humans also use aggression without being angry. A lion hunting for its prey, like a hunter or a hired killer, will be more effective if it remains calm and collected.

Anger and aggression are often studied separately. Anger has probably most often been studied in the context of coronary heart disease. In these medical studies of adults, anger is an important part of the Type A Behavioral Pattern, considered one of the best predictors of coronary heart disease. Aggression is also part of the Type A Behavioral Pattern, but it has been most often studied in relation to children's antisocial behavior problems, conduct disorders, and adult criminality.

The aims of this article are to summarize knowledge on the early development of anger and aggression, to highlight to what extent these two developments are linked, and to what extent early development of anger and aggression has consequences for development during later childhood, adolescence, and adulthood. An intergenerational approach is also taken to show that early childhood development of anger and aggression is not only linked to future development of the individual, but also to the past development of his parents, and the future development of his own children. Chronic problems of anger and aggression are thus discussed with reference to risk factors, outcomes, and prevention.

The Development of Anger

In the 1972 edition of his psychology textbook, Donald Hebb, a founder of modern neuroscience, wrote: "Neither a human nor a chimpanzee baby needs to learn how to have a temper tantrum." Is this statement really true? Are humans wired for angry feelings and expression of anger? Have we always believed that this could be true?

One of the first extensive discussions on anger was written by Seneca, a Roman lawyer–philosopher–writer born 5 years before Christ. Seneca had been exiled to the isle of Corsica by Emperor Claudius on a charge of adultery. He was obviously angered himself by this exile and decided to reflect on the origin, development, consequences, and prevention of this terrible emotion, of which he said: "No plague has cost the human race more dear. You will see bloodshed and poisoning, the vile counter charges of criminals, the downfall of cities and whole nations given to destruction, princely persons sold at public auction, houses put to the torch, and conflagration that halts not within the city-walls, but makes great stretches of the country glow with hostile flame."

His description of an angry human is an excellent introduction to our topic: "His eyes blaze and sparkle, his whole face is crimson with the blood that surges from the lowest depths of the heart, his lips quiver, his teeth are clenched, his hair bristles and stands on end, his breathing is forced and harsh, his joints crack from writhing, he groans and bellows, bursts out into speech with scarcely intelligible words, strikes his hands together continually, and stamps the ground with his feet; his whole body is excited and performs great angry threats; it is an ugly and horrible picture of distorted and swollen frenzy – you cannot tell whether this vice is more execrable or more hideous."

This description of an angry man by Seneca is a splendid caricature of the Type A man who has been shown to be at high risk for coronary heart disease. Seneca went on to compare this behavior pattern to the rage of boars, lions, snakes, and dogs. However, he concluded that only humans experienced anger, because anger is a loss of reason and animals do not possess reason. Seneca had observed early childhood angry reactions and noted that children will, for example, hit the floor in anger because they fell. He thus emphasized the importance of calm educators for children because he believed that children learn angry behavior by imitating their tutors, especially their parents. From this perspective he had anticipated, by 2000 years, the social learning theory of modern psychology. In this he was following the Greek philosopher Aristotle who wrote, in his book on Politics published 400 years earlier, that children's misbehavior was the result of imitation of adults.

One of the first descriptions of anger in an infant by a modern 'emotion scholar' comes from Charles Darwin (1809–82). This British geologist and naturalist is also a pioneer of modern psychology with his works on evolution and the expression of emotions in humans and animals. He seems to have written his first description of a temper tantrum in a letter to one of his sisters following a visit to the London Zoo, some 30 years before he published his book on emotions humans and animals : "I saw also the Ourang-outang in great perfection: the keeper showed her an apple, but would not give it to her, whereupon she threw herself on her back, kicked & cried, precisely like a naughty child. She then looked very sulky & after two or three fits of passion, the keeper said, 'Jenny if you will stop bawling & be a good girl, I will give you the apple.' She certainly understood every word of this, &, though like a child, she had great work to stop whining, she at last succeeded, & then got the apple, with which she jumped into an arm chair & began eating it with the most contented countenance imaginable." In the following year, Darwin became father to a son and carefully noted his development. He wrote the following description of anger expression at 11 months: "During the last week has got several times in passion with his playthings, especially when the right one has not been given him. When in a passion he beats & pushes away the offending object."

The difference between Darwin's and Seneca's view of anger was the evolutionary perspective that Darwin slowly came to construct. Seneca believed that only humans were endowed with reason and passion, while animals had only instincts. Children behaved like animals until they were old enough to receive the ultimate gift, reason. From then on they could feel and express anger. For Seneca, the tantrums of young children were similar to the tantrums of adults, as the tantrums of adults were similar to the rage of animals, but the dividing line was the gifts of reason and passion that were given only to humans.

Darwin produced a quantum leap in our perception of the place of humans in the nature of things when he identified the mechanisms that could explain how humans were the product of evolution. The gifts of reason and passion were now simply differences that appeared over time, but functioned with the mechanisms shared with our cousins the great apes, rats, birds, and even mollusks. Darwin suggested that anger had evolved as a mechanism to attain a desired goal. Human and nonhuman animals become angry when they do not get what they want, and anger gives them the energy and determination to overcome the blockage. In fact, that was essentially what Aristotle appeared to mean when he wrote: "Anger is necessary; without it we cannot succeed at anything. We must use it, not as a leader, but as a soldier."

Investigators have recently tried to model temper tantrums of young children mathematically. The main behaviors during a tantrum are kicking, throwing, stamping, and crying. The modal tantrum duration is between 30 and 60 seconds. Tantrums are shorter when the child drops to the floor at the beginning of the tantrum. The longer the tantrum lasts the more likely the child will go from anger expression to distress. Kicking and throwing are mostly associated with anger expression, while stamping and crying tend to be more frequent in the last phase of the tantrum when distress dominates.

Modern research with infants has also shown that anger expression can be observed very soon after birth. Infants who express anger on their face when frustrated are more likely to work harder at overcoming the obstacle than children who express sadness. The experiments that reveal these facts are in the true tradition of Darwin's experiments with animals and children. The infants are seated in a reclining seat. They face a projection screen and a speaker. Infants' faces are filmed. Their gestures and the emotional expressions on their faces are eventually categorized with a sophisticated coding system of muscle movements. A string is attached to the wrist of the infant. The movement of his arm activates the projection of infant smiling faces and recording of singing by children. Once children have learned that their arm movement produces the attractive visual display and music, the projector and speakers are turned off. As expected, a substantial increase in frequency of angry facial expressions follows this manipulation. Sad facial expressions also increase, but are almost four times less frequent than anger expressions. At the same time, expressions of joy and surprise almost completely disappear. Investigators have recently shown that infants who show an increase in expressed sadness also have an increase in the cortisol levels assessed from their saliva. Saliva cortisol usually increases when an individual is stressed. Interestingly, no increase in cortisol is observed for infants who become angry. This difference in hormonal reaction gives support to Aristotle and Darwin's hypothesis that anger is a tool for solving problems, clearly a better tool than sadness. Trainers and athletes clearly use this primitive tool for achieving a goal.

The presence of angry reactions following frustration, in the first few months after birth, is a good indication that Seneca was probably wrong when he concluded that children learn from adults to react with anger. However, although infants appear to react with anger spontaneously, they may learn from their environment to gain control over that emotional reaction. Studies of infants' perceptions of anger expression can help in understanding to what extent their expressions of anger are influenced by their environment.

One of the clever observational strategies to test this hypothesis with infants who can crawl (around 8 months of age) is the 'visual cliff': a large table made of thick transparent safety glass, with a design under the glass that gives the crawling child the impression that he is crawling over a shallow cliff that suddenly becomes a deep cliff. The child has to do this crawling in order to get to the other side of the table, where his mother is waiting for him. Mothers are asked through earphones to attract the child by smiling and showing an attractive toy. When the child gets to the middle of the table over the deep end of the cliff, mothers are asked to express on their face one of the following emotions: joy, interest, sadness, anger, or fear. Each mother was previously trained to make the exact facial expression they were meant to use. Results of this experiment clearly show that parent's expression of the basic emotions are well perceived by the infant, and that their behavior changes accordingly. Indeed, 75% of the children crossed the deep end of the cliff when their mothers expressed joy or interest. Results for the other three emotions were 33% for sadness, 11% for anger, and 0% for fear. Similar results are obtained in a different experiment where children can not see their mother's facial expressions, but can hear her express the emotions with her voice. The results are similar even when a stranger expresses anger or disgust.

Thus, soon after birth children express anger, and very soon after, if not at the same time, will react appropriately to angry expressions of parents and strangers. Are children influenced by their environment to the point where they will increase the frequency of angry reactions as they are exposed more and more to angry reactions around them? In the late 1920s, Florence Goodenough, a professor at the University of Minnesota Institute of Child Welfare, did a study that provided interesting information on this question. She asked parents, mostly college graduates, to keep records of their child's anger outbursts during a 1-month period. Two children were not yet 1 year old (7 and 11 months), nine were between 17 and 23 months, 13 were between 26 and 35 months, 10 were between 36 and 46 months, and the 11 others ranged from 53 to 94 months. One thousand eight hundred seventy-eight anger outbursts were recorded, for a mean of approximately one outbursts per 10 h of observation. As shown in **Figure 1(a)**, the number of angry outburst reached a peak between 17 and 23 months of age (1.3 outbursts per 10 h) and decreased afterwards. Comparisons of children with different ages (cross-sectional studies)

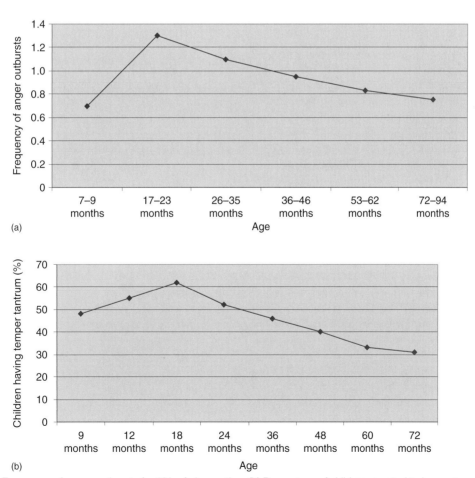

Figure 1 (a) Frequency of anger outbursts for 10 h of observation. (b) Percentage of children reported to have a temper tantrum at least every day. (a) Adapted from Goodenough FL (1931) *Anger in Young Children*. Westport, CT: Greenwood Press. (b) Adapted from Sand EA (1966) *Contribution à l'Étude du Développement de l'Enfant: Aspects Médico-Sociaux et Psychologiques*. Bruxelles: Éditions de l'Institut de sociologie de l'Université libre de Bruxelles.

such as this one can provide an indication of developmental changes with age. However, following the same children as they grow up (longitudinal studies) will provide more adequate estimates of changes in behavior with age. There are very few longitudinal studies of anger development during early childhood. Interestingly, a Belgian longitudinal study of temper tantrums reported by the mothers of a sample of 126 children at the end of the 1950s essentially replicated the results Goodenough obtained with her Midwest US sample 30 years earlier. **Figure 1(b)** shows that the percentage of children who threw at least one temper tantrum per day increased from 50% at 9 months of age to 70% at 18 months, and then steadily decreased to a low of approximately 35% at 72 months of age. A recent cross-sectional study of temper tantrums between 18 and 48 months of age among Midwest US children shows the same type of increasing and decreasing curve, but the peak is between 30 and 36 months with somewhat higher percentages at each age.

The Development of Aggression

Who Becomes a Physically Violent Adolescent?

Temper tantrums of 2-year-olds do not often make the newspaper headlines. But the angry adolescents who shot dead 13 school friends in a Columbine high school in 1999 shocked the whole world. Is there a link between temper tantrums of infants and physical violence by adolescents and young adults?

To answer this question we need to follow children from early childhood to adolescence. But let us start with one of the longest longitudinal studies of criminal behavior ever. This is a recent analysis of official crimes committed by 500 males between 7 and 70 years of age who had been labeled delinquents during their adolescence. The analysis showed clearly what criminologists are calling the 'age–crime curve': the frequency of arrests for physical aggressions increases with age up to early adulthood, and then decreases steadily up to old age. Thus, compared to pre-adolescents and adults, the adolescents and young adults are more likely to be arrested for physical aggression. Much research has been done on adolescents and young adults to understand this phenomenon. Seneca and modern learning theorists believed the explanation came from the fact that young humans learn to physically aggress by imitation of aggressions observed in the home, in the neighborhood, and in the media. This learning would obviously take time, since it peaks during adolescence. Many believe that this is caused by the fact that adolescents are bombarded by media violence and are also victims of bad-peer influence. The older they are, the more they have observed acts of physical aggression, hence are more likely to imitate what they have seen.

Two important recent reviews of the literature of violent behavior concluded that physical violence appears during adolescence. Note that the term 'adolescent violence' is generally used to refer to behaviors ranging from bullying at school to murder. The first is a 2001 report on violence by the Surgeon General of the US. The second is a 2003 World Health Organization report on violence and health. The latter cites the former and concludes: "The majority of young people who become violent are adolescent – limited offenders who, in fact, show little or no evidence of high levels of aggression or other problem behaviors during their childhood."

However, there is a rival hypothesis to the social learning theory of aggression. This one suggests that there is a biological cause to the age–crime curve. The level of a hormone, testosterone, increases exponentially during adolescence, and this increase is much larger in males than in females. After its peak in early adulthood, testosterone level decreases steadily up to old age. Thus, the hypothesis that testosterone is the cause of increase in violent crimes during adolescence is based on the observation that increase and decrease in levels of testosterone with age mirrors the increase and decrease in violent crimes, including the sex difference. There is also evidence that the most aggressive animals and humans tend to have higher levels of testosterone.

The beauty of the age–crime curve is that it appears to be an extremely robust finding. Adolphe Quetelet, a nineteenth century Belgian astronomer–statistician, came to the same conclusion when he analyzed data he collected on criminal statistics from France in the late 1820s. More recently, a Swiss sociologist replicated the age–crime curve with data on homicides in the seventeenth, eighteenth, and nineteenth centuries. Thus, the violent adolescent and young adult is not a result of life in our modern societies and the advent of movies, television, and computer games.

There are, however, at least two major problems with the belief that the age–crime curve describes well the development of human physical aggression: first, it describes the mean trajectory of a sample of highly deviant juvenile males, not the trajectory of the majority of humans; and second, the studies of that curve limited their focus on adolescents and adults. The curve gives the impression that physical aggressions appear with the legal age for criminal responsibility, as if lawmakers had chosen the age for criminal responsibility after detailed studies of child development. Those who decided to study elementary school children to understand the precursors of adolescent delinquency discovered that the children used physical aggression much more than expected. In fact, a North Carolina longitudinal study in the 1980s and 1990s showed that the mean level of physical aggression was decreasing from 10 to 18 years of age in samples of males and females.

One explanation for this observation could be that, although the majority of youth reduce the frequency of physical aggressions as they grow older, a minority are increasing their frequency of physical violence and are being processed through the legal system. This would generate the statistics represented in the age–crime curve. To test this hypothesis, we needed to go beyond a description of the mean developmental trajectories of delinquents, and identify the different types of developmental trajectories that children are following. This was done with a longitudinal sample of more than 1000 boys in schools from low socioeconomic areas in Montréal, Canada. The Canadian boys were followed during the 1980s and 1990s, as were the North Carolina children, but this time from kindergarten to high school. Why use males from low socioeconomic areas to study the development of physical aggression? Because poverty has long been associated with higher levels of physical aggression.

Results from teacher ratings of the Montréal boys confirmed the North Carolina data. The large majority of boys from the poorest inner city areas of Canada were using physical aggression less frequently as they grew older. Only a very small group of boys (4%) did not show the declining trend; these were the boys who had the highest level of physical aggression in kindergarten and remained at that level until adolescence. When interviewed at ages 15 and 17 years, they were the boys who reported the highest frequency of physical violence. They were also the ones most frequently found guilty of infractions before 18 years of age. Thus, the increase in physical aggression observed in the age–crime curve during adolescence appears to be produced by the fact that, during this period, the police and judicial system start arresting and convicting individuals who have been physically aggressing others at least since kindergarten. An international team of investigators replicated these finding using five other longitudinal studies in Canada, New Zealand, and the US.

When Does Physical Aggression Start?

The studies of physical aggression during the school years described earlier show that school children are at their peak frequency during their kindergarten year. If this is the case, when do they start to use physical aggression? There is a long history of case studies of young children being physically aggressive with siblings, peers, and parents. One of the first observation of the 'developmental origins of aggression' was published by Augustine of Thagaste in AD 397. This Roman citizen from North Africa, later known as Saint Augustine, was writing his best-selling book *Confessions* and wanted to reveal to the readers all his sins. Realizing the recall limits of childhood memories, he wondered: "Who will bring to my mind the sins of my infancy?" His solution to the problem was what most modern investigators of aggression development almost never did, being focussed on the learning hypothesis. Since aggression peaked in adolescence, learning of aggression must occur either during adolescence or during elementary school. Augustine reasoned differently. He followed Aristotle in thinking that if the development of something is to be understood, it should be done so from the beginning. He believed that he would learn how he behaved at the beginning if he observed young children, "for in him I now perceive what I do not remember about myself?" So, after having carefully observed very young children he concluded: "My parents, and many other prudent people who would not indulge in my whims . . . I struck at them and tried to hurt them as far as I could because they did not obey orders that would be obeyed only to my harm . . . . Thus it is not the infant's will that is harmless, but the weakness of infant limbs . . . These things are easily put up with, not because they are of little or no account, but because they will disappear with increase in age. This you can prove from the fact that the same things cannot be borne with patience when detected in an older person."

Almost 1600 years later, in the 'rage' section of the *The Expression of the Emotions in Man and Animals*, Darwin wrote: "Every one who has had much to do with young children must have seen how naturally they take to biting, when in a passion. It seems as instinctive in them as in young crocodiles, who snap their little jaws as soon as they emerge from the egg."

Many present-day child-development experts deny that the physical aggressions of young children are 'true' aggressions. Like Seneca's reasoning concerning anger, they argue that to qualify as an aggression, there must be intent, and infants are not endowed with the capacity to 'will'. Augustine was probably responding to Seneca when he wrote: "Thus it is not the infant's will that is harmless, but the weakness of infant limbs". We have seen in the section titled 'The development of anger' that experiments with infants clearly show 'will' and 'anger' within the first few months of life. There is certainly a huge difference between the will of a 16-month-old and the will of a 16-year-old; but the reasoning power of a 16-year-old who is physically aggressive with a rival or a girlfriend in a state of rage is often not that different from the 16-month-old who attacks a peer to take an attractive toy. As Augustine writes, the main difference is the weakness of the infant's limbs. That difference in strength is not in the eyes of the peer who is attacked, but in the eyes of the adult who concludes that there is no real physical aggression in infants because of lack of the ability to will. If we accepted the 'lack of will' argument, we would have to declare that there are no true physical aggressions between animals since they cannot 'will' like human adults can.

Recently, a number of longitudinal epidemiological studies of thousands of children followed from infancy showed that, from the first to the third year after birth, the proportion of children reported by their mothers as using physical aggression increases substantially. This remarkable increase is then followed by a continuous decline in frequency, as seen in **Figure 2**.

The same general developmental picture is drawn whether we use data from different periods, data from different countries, data from different reporting sources (filmed interactions in daycare, parents' detailed records, or parents' recall of behavior in the past months), or data from different methodologies (cross-sectional or longitudinal, calculated as an absolute frequency over a given period of time, a relative percentage of social behaviors, a percentage of individuals using the behavior, or a general estimate of the frequency of the behavior). As we described for anger in the previous section, frequency of physical aggression increases rapidly from the first year after birth to approximately the third, and then decreases. Unfortunately, none of the longitudinal studies tracing the developmental trajectories during early childhood are old enough to report on trajectories of physical aggression during adolescence and adulthood. However, we know from predictive studies that aggression during childhood is the best predictor of aggression during adolescence and adulthood. Thus, although physical aggressions by very

young children appear qualitatively different from physical aggressions by adolescents and adults, the trajectory of the former appears to generally lead to the trajectory of the latter.

An overview of the available data on the development of physical aggression over the lifespan is summarized in **Figure 3**: most humans started using physical aggression before they reached 20 months of age; humans use physical aggression most often between 18 and 42 months after birth; if humans are learning to be physically aggressive through imitation, this learning is happening in the first 2 years after birth, and not by watching television and playing video games in middle childhood or adolescence. Humans clearly learn not to use physical aggression, they learn to use alternative solutions, and this learning starts well before they enter school. It appears important to note that this phenomenon does not appear to be restricted to physical aggression and anger. There is growing evidence that it is true also for hyperactivity, stealing (taking things from others), vandalism (destroying others' belongings), and fraudulent behavior (e.g., lying).

From Physical Aggression to Other Forms of Aggression

The preceding discussion focused on the development of physical aggression for two reasons. It is the first form of

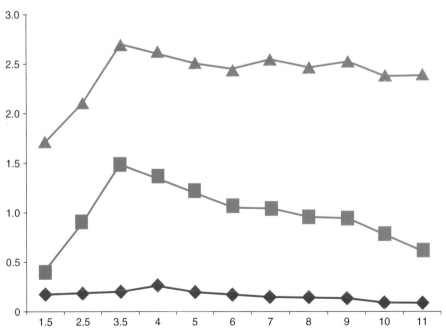

Figure 2 Trajectories of physical aggression from 1.5 to 11 years after birth. Adapted from Tremblay RE, Nagin D, Séguin JR, *et al.* (2004) Physical aggression during early childhood: Trajectories and predictors. *Pediatrics* 114: e43, and Côté S, Vaillancourt T, LeBlanc JC, Nagin DS, and Tremblay RE (2006) The development of physical aggression from troddlerhood to pre-adolescence: A nation wide longitudinal study of Canadian children. *Journal of Abnormal Child Psychology* 34: 71–85.

aggression to appear during development, and it is the form of aggression that has been best studied from a developmental perspective. The second form of aggression to appear is verbal aggression. Children do use their voice to express anger from the start (they yell), but verbal aggression requires ability to use language, like physical aggression requires motor coordination. So, verbal aggression will appear with the development of language during the second and third year after birth. A third form of aggression will appear still later: indirect aggression (sometimes named relational and social aggression). In a physical and verbal aggression the victim is directly confronted; hence the victim knows that he or she is being attacked. An indirect aggression is different, in that it is done behind the victim's back. For example, the aggressor convinces others not to play with the child he dislikes, or he attempts to destroy his reputation. This type of aggression obviously needs more sophisticated cognitive abilities than physical and verbal aggression. Recent studies have shown that it is clearly present before the end of the fourth year after birth, and that it tends to increase in frequency with age. Interestingly, girls start to use indirect aggression before boys do, and tend to use it more often. This is the exact opposite of the sex differences for physical aggression. One obvious reason is that cognitive skills, language, and self-control develop more rapidly in girls than in boys. From an evolutionary perspective, it also makes sense that females learn to use alternatives to physical aggression more rapidly than males, since they are physically less likely to have an advantage in a physical fight. Recent studies have also documented the links between the development of physical aggression and indirect aggression. As physical aggression decreases, indirect aggression increases, but almost all of the children on a trajectory of very frequent physical aggression since infancy go on a high trajectory of indirect aggression.

From Anger to Aggression

We have seen that anger appears in the first few months after birth, and physical aggression increases from the end of the first year after birth to a peak around the third year. How does anger get transformed into physical aggression? To answer this question we must take into account that motor development is important for physical aggression. To push, kick, attack, and throw objects at others, one needs to have the required motor coordination. At 2 months after birth, the angry infant already expresses himself with his muscles. The facial muscles create the angry face, the limb muscles make his arms and legs move in a way that looks erratic, and the muscles of his trunk and legs often succeed in making his whole body arc. By

6 months, the hand and arm movements will be much more coordinated. The child can grasp objects at will and let them drop but throwing objects towards a target will come later. Learning to use his muscles to crawl will then enable the child to approach others and their toys. Aggressive interactions around toy possession can already be seen between two crawling 9-months-olds. Standing up, walking, and running will then be a major motor progress from the end of the first to the end of the second year. The hands are freed, and running towards and from someone gives the liberty needed to threaten, to attack, and to flee. By 24 months, a normally developed child can do most of the physically aggressive acts that an adult can. The coordination will get much better with neural maturation and practice but, as noted by Saint Augustine, the main difference between a child and an an adult is the weakness of the limbs.

There is often no better way to understand development than to observe it on fast-forward. You probably have all seen movies using the fast-forward approach to show the growth of a flower or the movement of an animal. To understand the links between anger in the first year of life and aggression over the next 2.5 years, we can use this approach with notes written by Charles Darwin. In his book *Expression of the Emotions in Man and Animals*, he writes: "With one of my own infants, from his eighth day and for some time afterwards, I often observed that the first sign of a screaming-fit, when it could be observed coming on gradually, was a little frown, owing to the contraction of the corrugators of the brows; the capillaries of the naked head and face becoming at the same time reddened with blood. As soon as the screaming-fit actually began, all the muscles round the eyes were strongly contracted, and the mouth widely opened...; so that at this early period the features assumed the same form as at a more advanced age."

In the following notes he describes the anger and aggressions of his son, William Erasmus (Doddy), from 11 to 38 months. Doddy was born on 27 December 1839 when Darwin was 30 years of age. We start with the entry we cited earlier in this article.

- 1840, December 8 (11 months):
  During the last week has got several times in passion with his playthings, especially when the right one has not been given him. When in a passion he beats & pushes away the offending object.
- 1841, January 26 (13 months):
  Has for some time often gone into passions for smallest offences – for instance with Anne the nurse for trying to take piece of cake from his lips with her fingers, when he wished her to take it with her mouth – out of his mouth–he tried to slap her face, went scarlet, screamed & shooked his head. How has he learned that

slapping gives pain-like the just-born crocodile from egg, learns to snap with its weak jaws, i.e., instinctively.

- 1841, April 15 (16 months):
  Jealous of a doll for last fortnight, jealous of Anny (born 2 March 1841), few days ago jealous of my weighing Baby.
- 1842, February (25 months):
  I have long observed that the horse-shoe lip of misery is the endeavour to keep mouth closed just before it opens wide for a roar.

  I have observed during last two months, how curiously Doddy expresses by a modulation of humph of assent "Yes to be sure": As he formerly used to express by a negative whine "No that I won't" or rather a sort of defiance as much as to say "if you do so–I will do so"–I suspect many expressive modulation of tone, come to children before appropriate expressions for their feelings.
- 1842, March 20 (27 months):
  Doddy is a great adept at throwing things & when choleric he will hurl books or sticks at Emma. About a month since, he was running to give Anny a push with a little candlestick, when I called sharply to him & he wheeled around & instantly sent the candlestick whirling over my head—He then stood resolute in the middle of the room as if ready to oppose the whole world–peremptorily refused to kiss Anny, but in short time, when I said "Doddy wont throw a candlestick at Papa's head" & he said "no wont kiss Papa"–I shall be curious to observe whether our little girls take so kindly to throwing things when so very young. If they do not, I shall believe it is hereditary in male sex, in the same manner as the S. American colts naturally amble from their parents having been trained.
- 1842, March 26 (27 months):
  Doddy was generous enough to give Anny the last mouthful of his gingerbread & today he again put his last crumb on the sofa for Anny to run to & then cried in rather a vainglorious tone 'Ok kind Doddy, kind Doddy'.

- 1842, April 2 (27 months):
  Doddy used bit of a stick as lever to break doll.
- 1842, April 4 (27 months):
  During the last fortnight has shown much suspicion (his characteristic) & could not endure anyone laughing, thinking it directed at him. He has lately been very contradictory; by mistake he one day graciously gave Elizabeth a kiss, but repenting said "Doddy did not kiss Dziver" & when Elizabeth remonstrated by saying "you may be sorry for it, but you did kiss me." He stuck to it, "No Doddy did not."
- 1843, January 20 (37 months):
  Threw some cards at my head for alluding to something he used to say when a baby (Mother writes).
- 1843, February (38 months):
  Willy (Doddy) says "No" in the fiercest possible way possible unlike any other child I ever saw (Mother writes).

Risk Factors for Anger and Aggression Problems

We have seen that all infants experience anger and use physical aggression. However, most learn to deal with angry emotions and aggressive behavior in a way that will help them integrate well in society, live a healthy life, and have a sense of wellbeing. Children who have problems with anger and aggression are those who are on the high-frequency trajectory shown in **Figure 3**. The behavior of these children is extremely hard to accept both by their peers and adults, including parents and teachers. Thus, they rightly feel rejected, and this feeling supports their anger and aggression. They are at high risk for school failure, accidents, delinquency, substance abuse, sexual promiscuity, adult criminality, poverty, spouse abuse, child abuse and neglect, coronary heart disease, suicide, and of having children who will repeat

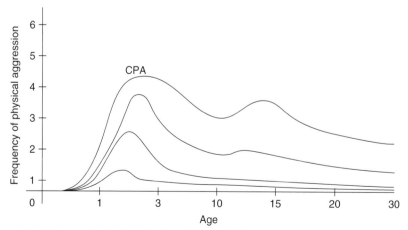

Figure 3   Probable age–physical aggression trajectories from birth to adulthood. Adapted from Tremblay RE (2003) Why socialization fails? The case of chronic physical aggression. In: Lahey BB, Moffit TE, and Caspi A (eds.) *Causes of Conduct Disorder and Juvenile Delinquency*, pp. 182–224. New York: Guilford. With permission from Guilford Press.

this pattern of behavior. Interestingly, there is recent evidence that chimpanzees, who have frequent behavior tantrums, are the most likely to behave in an antisocial way with other members of their group.

There are at least three important reasons to understand the risk factors for anger and aggression problems during early childhood. First, because risk factors give an indication of the causal mechanisms leading to the problems. Second, because they suggest means of preventing these problems during early childhood. Third, because prevention of these problems during early childhood should help in preventing lifelong problems, since early childhood appears to be the time when humans naturally learn to regulate anger and aggression.

Individual and Environmental Risk Factors

As Aristotle suggested a long time ago, if we are to try to understand something, we might as well start at the beginning. Unfortunately, few longitudinal studies have looked at the association between prenatal and perinatal risks of anger and future aggression problems during early childhood. The studies that did address this issue showed that the results were very similar to results from studies with older children, adolescents, and adults. The main risk factors are family characteristics, parent characteristics, and child characteristics.

Family characteristics

Anger and aggression appears to run in families. Genetic causes are clearly involved, and are discussed ahead in the section titled 'Genetically informative studies'. As children with anger and aggression problems are more at risk of failing in school and not finding a stable job, there is also greater possibility of them forming families that will have low income and thus live in poor environments. These angry and aggressive parents are also more likely to have communication problems, to create a dysfunctional family environment, and to divorce. Thus family poverty, dysfunctional family, and single parenthood are predictors of early childhood anger and aggression problems.

Parent characteristics

The intergenerational chain of consequences can especially be seen when we focus on maternal characteristics during pregnancy. Mothers who show more anger during pregnancy are more likely to have infants with early brain activity typical of children with frequent anger expression (greater right-to-left frontal brain activity). Mothers who smoke during pregnancy are more likely to have children on the high physical aggression trajectory. This is true also for pregnant women with low education, pregnant women who report behavior problems during adolescence, pregnant women who had their first child before age 21 years,

and mothers of 5-month-old infants who report angry reactions towards their child. Note that many of these characteristics are found in the same mothers, and each additional negative characteristic increases the risk.

Child characteristics

Sex of the child is one of the most basic (genetic) risk factor for both anger-expression and high frequency of physical aggression. Girls appear to learn more quickly to regulate their emotions and behavior during early childhood. Thus, with time, the difference between females and males increases. This appears to be true for humans and a number of nonhuman species. From a socially adaptive perspective this makes sense, because in most mammal species females are physically less strong than males. It also makes sense from a reproductive perspective since, as we have seen, female control of their behavior is important for the development of their offspring, both during and after pregnancy. Other child characteristics associated with anger and aggression also tend to have important sex differences, for example, temperament and right-to-left frontal brain functioning in infancy, as well as language development and effortful control during toddlerhood.

Genetically Informative Studies

The longitudinal data on the development of anger and physical aggression from infancy onwards, summarized earlier, indicates that anger and physical aggression are not learned like reading or writing, or an illness children 'catch' such as poliomyelitis or smallpox. They are more like crying, smiling, eating, grasping, throwing, and running, which young humans do when the physiological structure is in place. The young human learns to regulate these 'natural' behaviors with age, experience, and brain maturation. The learning-to-control process implies regulating your needs to adjust to those of others, and this process is generally labeled 'socialization'.

It is clear that the evolutionary process would have given humans a genetic program coding for all the basic mechanisms so as to react to care, hunger, and threat. Young children's muscles are activated to cling, yell, push, kick, grab, hit, throw, and run with extreme force when hungry, when angry, or when they are strongly attracted by something. However, stating that humans are genetically programmed to become angry or to physically aggress when needed, is different from stating that the frequency of the anger and physical agressions is genetically programmed. Since all infants who have developed normally show anger and physical agression, but not all do so at the same frequency and with the same vigor, to what extent are these individual differences due to the genetic program they have inherited or to the environment in which they have been growing?

The developmental trajectories of physical aggression shown in **Figures 2** and **3** clearly indicate that these individual differences exist at any given point, starting in early childhood, but the most interesting phenotype is the development over time. There is obviously intraindividual change over time. We see from the reduction in temper tantrums and physical aggression with age that most children learn to reduce the frequency of behaviors, which they apparently did not need to learn. However, relatively stable differences among individuals remain. What are the gene–environment mechanisms that explain the change and stability? They are possibly very similar to the mechanisms, which explain the developmental trajectories of growth in height. Genes code for the growth mechanisms, but there are individual differences in these codings, as well as environmental differences (e.g., access to food) which lead to stable individual differences. Thus, the individual differences in the frequency of temper tantrums and physical aggressions at one point in time, and over time, can be due to a large number of 'causes', for example, individual differences in the genetic coding for neuromodulators (e.g., serotonin and dopamine); hormones (e.g., testosterone); language development, cognitive development, or environmental differences such as mother's tobacco use during pregnancy; birth complications; sibling and peer behavior; and parental care. However, the individual differences that we observe are very likely to be due to interactions between many of these mechanisms, and to what has been labeled epigenetic mechanisms, that is, the changes in gene expression under the influence of the environment.

Knowledge on genetic factors which could explain the early development of anger and physical aggression is perilously close to zero. The first reason is that gene–environment interaction studies are recent. The second reason is that there are very few longitudinal studies that have included repeated assessments of anger and physical aggression from early childhood to later childhood. However, the most important problem is that molecular genetic studies and twin studies have concentrated on a global antisocial behavior phenotype and a global type A behavior phenotype. This is an old problem in both the type A behavior and antisocial behavior literature. Genetic studies have simply followed the main trend which tends to rely on measurement scales constructed by lumping items that are shown to correlate at one point in time.

Twin studies

Twins are used to study the genetics of behavior for the same reason that some twins physically look identical while others do not. The twins who look alike have exactly the same genes, while those who do not look alike share only half of their genes, like normal brothers and sisters. If angry and aggressive behavior is determined by genetic factors, we expect that the twins who share the same genes

(monozygotic (MZ)) will behave in a similar fashion more than the twins who share only half of their genes (dizygotic (DZ)). The genetics of anger was recently studied with a sample of 2500 adult twin pairs living in the Netherlands. Results showed that almost half (49%) of the variation in anger for the males could be explained by genetic effects. For females, genetic effects explained only 36% of the variation in anger.

A strong heritability of the frequency of antisocial behavior was also shown with British, Swedish, and US preadolescent samples. The earliest study of twins to assess gene–environment effects on physical aggression was done with 18-month-old Canadian children. Results showed that 58% of the variation in mother's reports of physical aggression was explained by genetic factors and 42% by environmental factors, suggesting that there are strong genetic effects on frequency of use of physical aggression during early childhood.

Molecular genetic studies

A few molecular genetic studies have shown that levels of anger in males are related to genetic polymorphisms which code for the neurotransmitter, serotonin, often associated with different forms of behavior problems. The lack of association for females may partly explain the lower genetic effects observed for females in twin studies. Recall that in social situations females appear to be more in control of their anger than men. One of the rare molecular genetic studies of infants' behavior showed that children with a specific gene (*L-DRD4*), which codes for the neuromodulator, dopamine, expressed less anger in a frustration test (physical restraint). Interestingly another version of this dopamine-related gene (*DRD4-7*) appears to prevent the association of cognitive problems and behavior problems.

Many molecular genetic studies have attempted to identify polymorphisms related to aggressive behavior, mainly with animal and human adult samples. A large longitudinal study of children born in New Zealand was used to understand specifically to what extent violent behavior during adolescence and adulthood is related to a specific gene (*MAOA*) which influences neurotransmitters (e.g., dopamine and serotonin) linked to behavior problems. The investigators also took into account to what extent the children who had been brought up in an abusive environment , a well-known risk factor for aggression problems, were affected. Results showed that, of all the most maltreated males, the ones who were at higher risk of being convicted of a violent crime before 27 years of age had the short version of the functional polymorphism in the gene coding for *MAOA*. This association was also found for conduct disorder assessed between 10 and 18 years of age, antisocial personality symptoms, and disposition to violence measured at 26 years of age. Individuals with a history of chronic physical aggression may be the driving force in these associations, since they are the most likely to be found in each of the assessed categories.

Conclusion: Prevention of Anger and Aggression Problems

Seneca was convinced that we could prevent people from ixhibiting angry behavior, and that this would have a major impact on society. He wrote: "Let us be rid of it altogether; it can do us no good. Without it, we shall more easily and more justly abolish crimes, punish the wicked, and set them upon the better path. The wise man will accomplish his whole duty without the assistance of anything evil, and he will associate with himself nothing which needs to be controlled with anxious care." We saw that Aristotle's approach was substantially different. He believed that anger and the aggressive disposition that it generates was useful to solve problems when well-controlled. His observation of human development led him to the conclusion that with age and experience, the young child learns to use his cognitive capacity to control and make good use of his impulses: "Anger is necessary; without it we cannot succeed at anything. We must use it, not as a leader, but as a soldier . . . as the body is prior in order of growth to the soul, so the irrational is prior to the rational. The proof is that the anger and will and desire are implanted in a child from their very birth, but reason and understanding develop as they grow older." He added: "Moral virtue is concerned with pleasures and pains. It is pleasure which makes us do what is base, and pain which makes us abstain from doing what is noble. Hence the importance of having a certain training from very early days, as Plato says, so that we may feel pleasure and pain at the right objects; for this is true education. (. . .) So the difference between one and another training in habits in our childhood is not a light matter, but important, or rather, all-important."

We can see that the importance of early childhood development is not a new idea. But have there been any scientific experiments showing that we can prevent the developmental of anger and aggression problems by educational interventions with infants? The majority of anger-control experiments have been done with adults at risk of coronary heart disease, and with school-age children with behavior problems. These experiments target individuals who have failed to learn to control themselves during early childhood. They show some impact on anger control. The experiments to change the trajectories of physical aggression showed long-term impact when they started in the early elementary grades and were very intensive, but only if parents, teachers, and children were all supported for at least 2 years.

Aristotle would argue that interventions during early childhood will have a greater impact. But from our knowledge of genetic and intergenerational effects we can be bolder than Aristotle and suggest that interventions, which would start during the pregnancy of at-risk parents, would have still more of an impact. Young pregnant women with a history of behavior problems could be trained to control their anger and aggressive behavior during pregnancy, and then trained to help their child learn to gain control of these emotions. There have been some long-term experiments where investigators randomly allocated a nurse home-visitation program to young underprivileged pregnant women at high risk of child abuse and neglect. These children were obviously also at high risk of chronic anger problems and physical aggression. The long-term follow-up of the children from the intervention groups have shown that, compared to the control group, they were less frequently abused and neglected, and were also less likely to exhibit delinquent behavior during adolescence. One would expect that the intervention group learned more rapidly to control their anger and regulate physical aggression. Unfortunately the development of anger and physical aggression was not included in the follow-up assessments of these experiments. There is also clear evidence that the same interventions do not have any positive impact if they are done with mothers who are already abusing their children.

The conclusion from our knowledge of the development of anger and aggression, as well as from these prevention experiments, seems clear. It is the same that was reached 2400 years ago by Plato, Aristotle's mentor: that to help humans learn to control their emotions and their behavior, the most appropriate time is early childhood. In fact, it seems that the closer to conception the intervention, the more likely it will have beneficial effects for the parents, their children, and their grandchildren. There is no doubt that if Socrates, Plato, Aristotle, Saint Augustine, and Darwin were to read this conclusion, they would be very surprised that knowledge gained from our sophisticated data collections and statistical analyses are simply concluding what to them appeared obvious. But modern science makes it more probable that they were right. Hopefully modern science will help us find ways of giving the needed cognitive control to those who are most at risk of living a life of misery because they have not acquired the necessary control.

This is again not unlike the British philosopher Thomes Hobbes' 1647 conclusion concerning the importance of early education for becoming a citizen: "It is evident therefore that all men (since all men are born as infants) are born unfit for society; and very many (perhaps the majority) remain so throughout their lives, because of mental illness or lack of training (disciplina) (. . .) Therefore man is made fit for society not by nature, but by training."

See also: Emotion Regulation; Self-Regulatory Processes; Temperament.

Suggested Readings

Cosmides L and Tooby J (2000) Evolutionary psychology and the emotions. In: Lewis M and Haviland-Jones JM (eds.) *Handbook of Emotions,* 2nd edn., pp. 91–115. New York: Guilford.

Côté S, Vaillancount T, LeBlanc JC, Nagin DS, and Tremblay RE (2006) The development of physical aggression from toddlerhood to pre-adolescence: A nation wide longitudinal study of Canadian children. *Journal of Abnormal Child Psychology* 34: 71–85.

Darwin C (1872/1998) In: Ekman P (ed.) *The Expression of the Emotions in Man and Animals,* 3rd edn. London: Harper Collins (US edit.: New York: Oxford Press).

Goodenough FL (1931) *Anger in Young Children.* Westport, CT: Greenwood Press.

Lemerise EA and Dodge KA (2000) The development of anger and hostile interactions. In: Lewis M and Haviland-Jones JM (eds.) *Handbook of Emotions,* 2nd edn., pp. 594–606. New York: Guilford.

Lewis M (2000) The emergence of human emotions. In: Lewis M and Haviland-Jones JM (eds.) *Handbook of Emotions,* 2nd edn., pp. 265–280. New York: Guilford.

Sand EA (1966) *Contribution à l'Étude du Dévelopment de l'Enfant: Aspects Médico-Sociaux et Psychologiques.* Bruxelles: Éditions de l'Institut de sociologie de l'Université libre de Bruxelles.

Seneca LA (1979) On Anger in *Seneca: Moral Essays.* Loeb Classical Library.

Tremblay RE (2003) Why socialization fails? The case of chronic physical aggression. In: Lahey BB, Moffitt TE, and Caspi A (eds.) *Causes of Conduct Disorder and Juvenile Delinquency*, pp. 182–224. New York: Guilford.

Tremblay RE, Hartup WW, and Archer J (eds.) (2005) *Developmental Origins of Aggression.* New York: Guilford.

Tremblay RE, Nagin D, Séguin JR, *et al.* (2004) Physical aggression during early childhood: Trajectories and predictors. *Pediatrics* 114: e43.

Artistic Development

C Golomb, University of Massachusetts, Boston, Boston, MA, USA

Glossary

Divergent perspective – Lines that depict the edges of an object by diverging rather than converging on a vanishing point.

Frontal plane – The plane lying perpendicular to the viewer's line of sight.

Globals – A circle or oblong with facial features that represent the human figure.

Intellectual realism – The child draws what he knows, not what he sees.

Linear perspective – Lines that represent the edges of an object in a scene that converge on a single point, called the vanishing point near the back of the center.

Representation – The ability to evoke mentally the image of an absent object and give it form in drawing or modeling.

Synthetic incapacity – The young preschool child's inability to coordinate the parts of a figure into a coherent drawing.

Tadpole figure – The global human sprouts arms and legs.

Viewpoint – The notional position occupied by a monocular viewer in relation to a scene (Willats).

Introduction

Representational development in drawing and sculpture examines the evolution of forms that can stand for the objects in a scene. Representation is a mental activity of symbol formation concerned with creating forms of equivalence in a given medium. Unlike imitation of reality which aims for a one-to-one correspondence to its referent, artistic representation implies finding structural equivalents for the referent. Its development documents the manner in which young children invent simple, economical forms and the processes of differentiation that lead to a more effective depiction. This development is orderly and rule governed, universal in its early phases that encompass phenotypical diversity on basic structural equivalents.

Children's Drawings

For over 100 years, psychologists and educators have shown a fascination with children's drawings, an interest that dates from the latter part of the nineteenth century and coincides with the beginnings of a systematic psychological study of child development. The first publications of children's drawings stem from this period, suggestive of later stage theories of mental development. The topic of child art subsumes drawing, painting, and modeling, but given the easy access to paper, pencils, and magic markers, drawings have been most widely studied.

Child art, which emerges in the early preschool years, is a symbolic activity that is unique to human beings. Non-human primates such as the great apes are able to recognize photographs and line drawings of familiar objects, but despite extensive training in the use of symbols, they do not create the simple representational forms that most 3-year-old children spontaneously draw and name. It is an amazing accomplishment of the human child to create, without training, these first representational shapes for which there are no readily available models (see **Figures 1(a)** and **1(b)**). These early representations comprised of a large circle with facial features and legs are simple but recognizable representations of a human or an animal.

Psychologists who at the end of the nineteenth century set out to study children's drawings were faced with the peculiarities of child art. Young children's drawings appeared bizarre and they were puzzled by the omission of significant features, their frequent displacement, the lack of proportion and perspective, mixed views, the arbitrary use of colors, the transparencies of features not visible to an observer, and many other faults. Concluding that the drawings of the young were indicative of a conceptually immature mind, they studied the changes in children's drawings as an index of the growth of intelligence. Over the next decades, many extensive cross-sectional and longitudinal studies were undertaken with the intention to chronicle the stages in graphic development and their anticipated progression toward a realistic representation. With Florence Goodenough's Draw-a-Man test published in *Measurement of Intelligence by Drawings* (1926), drawings of the human figure were standardized and scored according to the number of parts depicted and the realism of some of the features that were assumed to correspond to the child's conceptual maturity or intelligence quotient (IQ). This test of intelligence became a widely used instrument that was restandardized by Dale Harris in 1963.

A different conception of child art was held by the modernist artists at the beginning of the twentieth century. Artists and art educators organized the first exhibitions of child art and often displayed their own work along with the drawings and paintings of children. Artists such as Paul Klee, Wassily Kandinsky, Ernst Ludwig Kirchner, Gabriele Münter, and Pablo Picasso appreciated the spontaneity and esthetics of children's drawings and paintings; they considered the language of child art an authentic expression of a creative mind unencumbered by social conventions.

These contrasting approaches to child art and its development have found further elaboration in the writings of Jean Piaget and Rudolf Arnheim both of whom have had a profound impact on research in this field.

Jean Piaget, in his influential book co-authored with Bärbel Inhelder *The Child's Construction of Space* (1956), proposes to view drawing development in distinct stages that correspond to the stages of spatial-mathematical reasoning. The first stage pertains to the preschool years; the child draws closed shapes that differentiate the figure from its background but ignores the true shape and size of the object they represent. These drawings are based on topological relations that distinguish the inside from the outside of a form, and the manner in which they are connected. Only gradually are children able to order the various parts of the figure, and the difficulty in organizing the major parts of a figure into a coherent representation Piaget terms synthetic incapacity. Piaget relates this phase of drawing to the early preoperational period of cognitive development (ages 3–4 years). The next stage sees progress in the differentiation of parts of a figure and the adoption of more varied forms that are also better organized. Following the art historian George Henry Luquet who provided a longitudinal account of his daughter's drawing development, Piaget calls this phase intellectual realism (ages 5–7 years). This term signifies that the drawing child has a better conception of the object he or she is representing although the resemblance to the model remains crude and the perspective of the observer is ignored. This phase is often described as 'the child draws what he knows, not what he sees'. During the concrete operational period of cognitive development (ages 8–11 years), visual realism becomes the dominant form of drawing. Children now consider their viewing point when drawing, and their intuitive understanding of Euclidean concepts of measurement and of projective geometry lead to a more realistic depiction of a scene. New techniques appear for the depiction of depth and volume, experimentation now leads to the use of occlusion of parts hidden from view and to size diminution, and eventually to the use of perspective. By the end of the concrete operational period children are supposed to have made progress toward optical realism in their drawings, a highly valued endpoint in drawing development.

Rudolf Arnheim in his influential book *Art and Visual Perception* published in 1974 provides a new perspective on the psychology of art. On the basis of his extensive reading of the history of art he rejects the notion that realism is a natural endpoint of artistic development. Arnheim points out that in its long history, perspective was invented only once, by the artists of the Renaissance, and that before and since that time there have been multiple forms of artistic expression that do not rely on perspective in their art form. Arnheim contrasts the nature of representation with that of replication. Unlike replication, which aims for a faithful rendition of the elements that comprise an object, representation requires the invention of forms that are structurally or dynamically equivalent to the object. Artists are not motivated to imitate nature, they do not aim for one-to-one correspondence of elements, and all artistic thinking begins with highly abstract and simplified forms. At an early level of development, and that applies

Figure 1   (a) Global humans. (b) Tadpole figure. (c) Open trunk figures. (d) Figures with a graphically differentiated torso. (e) Tadpole animal.

to all beginners, simple generalized forms are the only options available to the inexperienced artist. Thus, inexperience not childhood is the true starting point for representational development in the arts, and the experience with the two-dimensional medium, its possibilities and constraints, becomes a major force leading to the differentiation of forms and their composition. Differentiation does not imply a single developmental trajectory, and

Arnheim emphasizes that there are multiple solutions to representational problems. The perspective inventions of the artists of the Renaissance are only one among many achievements, and they ought not to be seen as the ideal endpoint of ontogenetic or cultural development.

These contrasting approaches have animated numerous studies of children's drawings and the following sections report on general trends that, in the absence of formal training, characterize the evolution of form, space, color, and composition. Most of the research has been conducted on ordinary, normally developing children, but a brief section on talented children and some information on mentally handicapped children will be included. The impact of sociocultural factors will be reviewed, followed by an account of representational development in a three-dimensional medium that highlights the role of the medium.

Form

Representation is intimately linked to the creation of form or shape that resembles the referent, and thus the true beginning of drawing as a representational activity can be seen in the drawing of simple but recognizable forms that can stand for the object. When preschoolers begin to control their earlier scribble actions and recognizable shapes emerge, they tend to identify them as either humans or animals. The first forms are globals, consisting of a large circular or oblong shape that is endowed with facial features (see **Figure 1(a)**). Fairly soon thereafter, the global figure becomes graphically more differentiated, it sprouts arms and legs and thus a universally seen early figure of a human or animal is created, often called the tadpole figure (see **Figure 1(b)**). As the global circle shrinks in size, the lines that stand for legs increase in length and yield a new figure, the open trunk figure with the trunk implied between the two verticals (see **Figure 1(c)**). Children discover several solutions to graphically differentiate the trunk section: drawing a horizontal line that connects the verticals thus closing the open trunk, drawing of a stick figure composed of a single one-dimensional vertical line that joins head and limbs, and/or drawing a separate circle underneath the head (see **Figure 1(d)**). In the case of animal figures, the generic tadpole sprouts four legs, followed by a horizontally drawn body that marks its distinction from the human figure (see **Figure 1(e)**).

In the beginning, while focused on evolving basic forms, children are most concerned with creating a basic likeness to the human or animal figure, and relative size, proportion, orientation, and color are of minor concern. Once the basic forms and their organization have been mastered to the child's satisfaction, some attention is paid toward those other variables. Relative size now enables the drawer to indicate age differences, for example, between adults and children; detail can convey information on gender, endow the figures with individual characteristics (braids, mustache, earrings, eye glasses), portray emotions (tears, anger, happiness); and orientation is useful when action is portrayed. Although the child artist is yet unconcerned with anatomical fidelity or optical realism, attention to these variables creates a more successful narrative. It is worth noting that throughout the childhood years children prefer the frontal orientation of objects, often called the canonical view that provides the most salient information about the referent. Overall, the young artist attends to the essential characteristics of an object, an object centered view, and generally tends to ignore its momentary and changing view. The latter involves varying degrees of distortion of the true shape of an object, for example, in drawing a rectangular table with converging lines or the foreshortening of the human figure. There is a price for deviating from the real shape of an object and focusing on a specific view as seen in tasks in which the handle of a cup has been turned to occlude it from view. Although the handle is invisible, the younger children, invariably, draw a cup with a handle since this is its distinct feature that differentiates it from a glass or a bowl. The older and more experienced children will consider viewpoint as they go beyond depicting some of the invariable characteristics useful for identifying the referent, to less static and more dynamic forms that call for some of the distortions they avoided earlier on.

Along with greater graphic differentiation of the parts of a figure, we also note changes in the use of lines: from one-dimensional lines for limbs to two-dimensional ones, from right-angular relations of the arms to the body to oblique ones useful for the depiction of action and, somewhat later, to interaction among the figures. By the age of 5 or 6 years, children also experiment with a single contour line that encompasses all the major parts of the figure in one comprehensive outline (see **Figure 2(a)**). Further on, children discover that occluding lines of parts that are hidden from view can be useful for the depiction of volume and depth of a figure (see **Figure 2(b)**).

In this process of graphic differentiation, which is not strictly age dependent, the two-dimensional shapes or regions of the early tadpoles and of figures with a separate trunk (see **Figures 1(b)–1(d)**) represent whole volumes, that is, the totality of the object that is not further specified. With practice and the growing ambition to provide more information about the object, regions become more specified and come to represent the faces or surfaces of objects. With a more advanced understanding of the function of lines, one-dimensional lines come to represent edges or contours of an object.

In general, the developmental trends in the untutored drawings that have been outlined tend to reach a plateau in the middle childhood years that still carry some of the typical features of the child art style. Without specific training programs only highly motivated and talented

Girl 5.10

(a)   (b)

Figure 2   (a) Figures drawn with a continuous outline. (b) Occlusion of body parts that are hidden from view.

children are the exception, searching for and discovering new techniques for the depiction of their pictorial world.

Space

The drawing child does not create figures in total isolation, and in addition to mapping out the spatial relations that are internal to a figure, relations among different figures need to be worked out and placed in a common spatial field. This brings us to the question how children organize their figures in the pictorial plane and how they deal with the missing third dimension. Representing a three-dimensional object on a two-dimensional surface presents a formidable challenge for all beginners and the child's earliest organization of pictorial space begins with a principle of proximity, placing items near each other, indicating that they belong together. Soon thereafter, a new directional rule yields side-by-side placements, mostly along the horizontal or vertical axis. The horizontal direction specifies left–right directions, while the vertical represents up–down and near–far dimensions. Only gradually do children come to understand the different

demands made upon the vertical axis, by discovering, for example, that the vertically drawn game of hopscotch competes with the sun and the clouds for the same space. From their different trials they learn that diminishing the size of an item, placing it higher on the page, and partially occluding objects can suggest distance, that color gradients can serve to unify the foreground, middle ground, and background of a scene, and that diagonals can be used to depict the sides of an object. In the case of highly motivated children, one notes experimentation with foreshortening (the proportionate contraction of some of the parts of a figure that suggest its depth and volume) and the converging lines of linear perspective. However, only a few children arrive at these technical skills on their own, without explicit training or working from ready-made models.

Color

The order in which representational skills evolve privileges forms over color. Once the drawing of basic shapes has been accomplished, children use bold and contrasting colors for their sheer enjoyment, at first disregarding their realistic function. Thus, children begin with a subjectively determined choice of color and gradually the drawer imposes some restrictions on their use, such as monochromatic outlines for humans and animals, red for strawberries, green for grapes, brown for tree trunks, yellow for the sun and the moon. Although the general trend is toward a more naturalistic use of color, color remains central in the decorative designs that embellish many drawings and paintings, designs that are unaffected by the rules of realism. By the middle childhood years, color no longer is subservient to form, it becomes a dominant force uniting the diverse elements in a drawing or painting as is the case when a light coloring of the background creates the impression of the outdoors and provides some continuity between the different elements of a scene. Above all, color can become the carrier of mood and feelings with bright colors indicating happy events, dark colors sadness and illness. Color is a major factor in children's attraction to art, in the esthetics of child art, and its ornamental tendencies.

Composition

Drawings are meant to tell a story or to express feelings and this requires the organization of all the elements that comprise a work, that is, the use of line, form, space, and color. Two compositional principles seem to underlie children's drawings: a grid-like alignment of figures along the horizontal or vertical axes and centering strategies that organize items around a pictorial center.

Following a very short-lived phase of forms that appear to be distributed arbitrarily across the page, children begin to organize their figures along one or more horizontal axes. At first, these alignments are imprecise and give the impression that the items are floating in an unspecified space. Gradually, alignments tend to become more organized, with attention to the size of the figures and the distance between them, followed by the introduction of a ground or base line that anchors all elements firmly in the common plane. Depending on the theme, subgroupings appear that indicate that the items belong together or that the actors have a special relationship or a common interest (see **Figure 3(a)**). Such groupings are formed on the basis of similarity of size, color, form, or activity, and convey to the viewer what the picture is about. For example, a family picture can be organized according to the age of its members, from the youngest to the oldest, or grouped in terms of relationships among the members. The compositional strategy of creating meaningful subunits and coordinating the parts into a coherent and balanced configuration can yield a successful and pleasing picture based on thematic unity (see **Figure 3(b)**).

The second compositional principle is expressed in the tendency to center figures on a page and the creation of symmetrical arrangements. Symmetry can be defined as the correspondence in size, shape, and relative position of items that are drawn on opposite sides of a dividing line or distributed around a center. Simple forms of centering and symmetry can already be seen in the earliest drawings of 3- and 4-year-olds, and with more experience the complexity of these arrangements increases. These include equal spacing among figures, similar distance from the edges of the page, systematic variation in the size of figures, pair formation, and the repetition of patterns (see **Figure 3(c)**). Compositions based on complex symmetry maintain spatial order and meaning with more varied items whose organization no longer demands strict one-to-one correspondence of items. The more advanced symmetrical arrangements create a more dynamic balancing of the individual elements that enhance the meaning of the work and its esthetic appeal (see **Figure 3(d)**). Visual narratives that tell a story in a series of frames also represent a dynamic form of symmetry (see **Figure 3(e)**).

Overall, compositional development proceeds from multiple local graphic solutions of isolated descriptions, in which each object is an independent unit, to mixed views of varying degrees of interdependence and coordination. A unitary conception that organizes all the elements into a coherent whole is rarely achieved in childhood or adolescence.

Motivation and the Expression of Feelings

Drawings and paintings are expressive statements about what one knows, feels, and wants to understand. But so far we have looked at child art mostly from a cognitive perspective, as a problem-solving enterprise that highlights the representational abilities that underlie the evolution of form and spatial organization. Drawing children are not only inventors of a pictorial vocabulary, they are motivated to tell a story, to give expression to their experiences, to the joys, sorrows, fears, struggles, victories, and defeats of their daily lives. The motivation to create a pictorial world on a previously blank page draws on deeply felt desires, fantasies, and wishes that endow the maker with a sense of power, to make and to unmake, to create and to obliterate at will. Drawings depict events that affect the child, the family, and, beyond it – the wider community. They depict holidays, birthdays, contests, pregnancy, birth and death, earthquakes, and terrorist attacks. In some cases, illness and the fear of dying can lead, even in the drawings of young children, to surprisingly effective pictorial metaphors.

The theme that the child chooses is the main carrier of the affectively charged message. In many cases the subject matter of a drawing and its simple composition convey the intended meaning without ambiguity as seen in a birthday party that features balloons, a cake, and presents or in the drawing of a crying child, tears running down its cheeks, and a broken toy on the floor. In cases that involve complex emotions such as anger

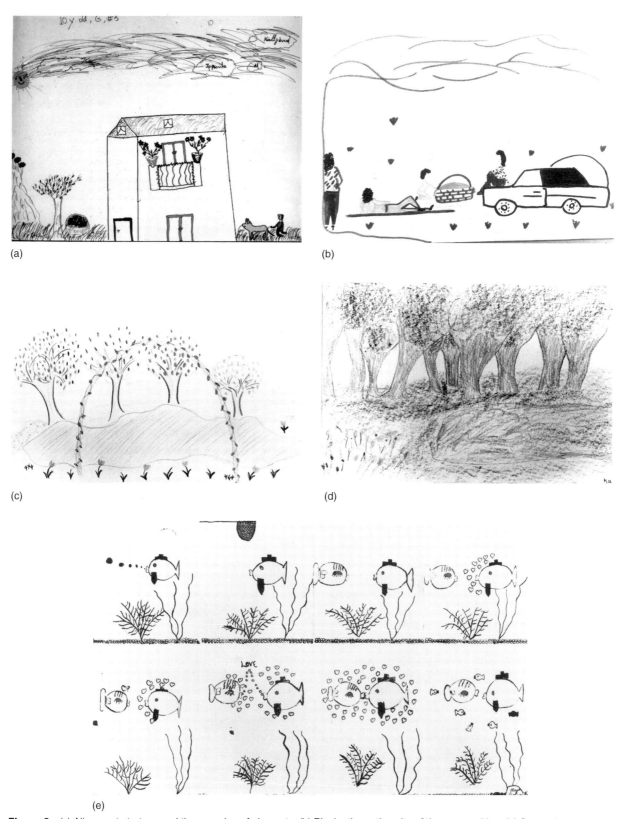

Figure 3   (a) Alignment strategy and the grouping of elements. (b) Picnic: thematic unity of the composition. (c) Symmetry. (d) Complex symmetry. (e) Love: a visual narrative.

at a sibling, feelings of rejection, ambivalence toward a parent, competition, or the desire for recognition, the child may not fully understand the range of emotions that are embedded in a scene he has just created. The following account of the pictorial expression of deeply felt emotions that are not fully understood by the drawing child comes from a retrospective account of an artist who reviewed one of her childhood paintings. Her comments illustrate the complexity of motivational factors that may not reach the level of conscious awareness and verbalization:

> I see myself standing in front of a big white sheet of paper … and I begin to work. I take a wide brush, dip it into paint, and on the white paper arises the large figure of a black woman. The black woman grows, surrounded by decorative ornaments and symbols of great beauty. Below her feet, at the bottom of the page, I add a little grave of the child who died, over which the black woman weeps. How did she get onto my paper? Over whom did I grieve when I was 10 years old? What feelings were stirred up in me that here appear in my painting?
>
> Gering (1998)

While the theme and composition are intimately linked and able to convey message in a general way, the ability to depict the mood of the protagonist graphically emerges only gradually, with the head or the face singled out as the carrier of affective meaning. Thus, happiness, sadness, and anger are portrayed by changes in the drawing of the mouth, with an upwardly curving line depicting a smiling face, a downward curve sadness, while anger may be shown in a set of prominently displayed teeth, a straight or zigzag line for the mouth, and somewhat later also in the diagonal slashes of the eyebrows.

Changes in posture that are congruent with the experience of contrasting emotions occur infrequently in the drawings of younger children, and are used sparingly by the older ones who may avail themselves of verbal commentary that augments a message, such as 'beat it'. With few exceptions, notably the direction of the arms, body posture remains essentially static, that is, undifferentiated. At times exaggerating the form and size of body parts can be an effective means to convey affect, and the use of energetic brush strokes, choice of primary colors, and an emphasis on symmetry can heighten the expressive power of a drawing or painting.

It is important to note that the depiction of feelings in a child's drawing does not tell us who the referent is. The drawing of a sad child need not indicate that the child artist is sad, and we need to be aware that a drawing is not a simple printout of the child's heart and mind. In the presence of an empathic participant observer the highly personal meaning of a drawing might be elucidated. Indeed, a large literature attests to the desirability of using drawings in a therapeutic context. In such a context, drawing may be a useful technique to help an emotionally distressed child to discover and convey feelings he or she harbors about the self and others and to develop the ability to express them in a meaningful way. For some youngsters, drawing is a substitute for verbal communication, for others it is an additional avenue for discovering and communicating important feelings that can be shared with the therapist. Drawing in the presence of the therapist encourages the child to understand his or her own message and to become more active in one's own behalf. Above all, drawing in the presence of an accepting adult encourages the child to face the inner world of demons as such a confrontation is now less threatening. A drawing, however, does not provide solutions magically, and in most therapeutic encounters drawings are seen as aids in the working through of the child's emotional problems.

Talent or Giftedness in the Arts

Children who at an early age perform at the near-adult levels valued in their culture have generally been identified as gifted and at various times have been apprenticed to an artist's studio. In our Western culture, a child who at an early age masters three-dimensional techniques and represents objects in a naturalistic style is most commonly identified as gifted (see **Figures 4(a)** and **4(b)**). Under the impact of modern art and such influential art educators as Franz Cizek and Viktor Lowenfeld the conception of giftedness has been broadened to include compositions of the child art style that show originality and cohesion, the use of vibrant colors, and the ornamental qualities reminiscent of folk art. One notes a division between researchers working in the Piagetian tradition who consider giftedness mostly in terms of the acquisition of skills of optical realism and educators who bring a broader esthetic to bear on the question of giftedness.

Regarding the developmental progression of talented children, most investigators agree that they do not skip stages or phases but move much faster through them, sometimes in the case of hours or days. While there are dramatic differences in the style of artistically gifted children, they are likely to be endowed with a heightened sensitivity for the appearance of objects, develop an early awareness of the function of lines and planes, and are able to extract the rules that underlie graphic depiction.

Noteworthy are the marked individual differences in the style gifted children develop ranging from realists, to colorists, expressionists, and cartoonists. Some children are enamored of the appearance of objects and strive to represent them with utmost fidelity. These are the youngsters who will teach themselves the major projective drawing systems and other three-dimensional techniques that can portray the objects of their interest with vitality and verisimilitude. **Figures 4(a)–4(d)** provide an example of

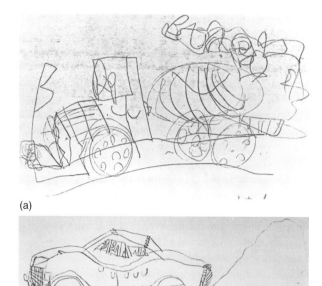

(a)

(b)

(c)

(d)

Figure 4 The pull of realism. Drawings of a precociously talented boy. (a) Cement truck, age 2; (b) Car in divergent perspective, age 3; (c) City of Jerusalem, car drawn in convergent perspective, age 4; (d) Construction scene, age 5.

accelerated development coupled with an unusual graphic talent that captures the excitement and realism of the objects that engaged this child's imagination. Having just turned 2 years, he drew his family as tadpole figures, but within the scope of a few months he mastered, on his own, some of the major projective drawing systems, incorporating different faces into his drawings of cars, trucks, buses, airplanes, tractors, combines, tricycles, and bicycles, and by the age 4 years he began to experiment with convergent perspective (see **Figure 4(d)**). With uncommon persistence, this preschooler developed his own graphic vocabulary and his daily efforts demonstrate the ongoing processes of problem setting and solving, the kind of visual thinking that enabled him to discover or invent useful representational techniques.

Other gifted young artists are colorists, they tend to use dramatic form rather than realistic ones, they relish expressive and decorative attributes of color, texture, and design (see **Figures 5(a)** and **5(b)**). They may also be more attuned to their internal states unlike the realism-oriented child artist whose attention is focused outward, on the appearance of his world. In contrast to the above-mentioned gifted children, cartoonists create worlds of adventure populated with villains and heroes, based on models that inspire their graphic work.

Altogether, children who are highly talented in the arts pursue their own pathways whether they are colorists, expressionists, or realists. They appear to be driven, in the words of Ellen Winner they experience a 'rage to master', and are determined to teach themselves what they need to know. But, of course, all children, even the talented ones, do not grow up in social isolation, they are influenced by the material made available to them (paper, crayon, charcoal, paints, brushes, ink, etc.), by the images that surround them, by peers, teachers, magazines, etc. This is particularly well illustrated in the case of Yani, a Chinese prodigy, who from an early age painted in her father's studio. Although her work with brush, paint, and rice paper (the traditional tools of Chinese painters) is recognizable child art, one is struck by the impact of Chinese artistic traditions in her work.

Drawing of Mentally Handicapped Children

The great majority of children with mental retardation for whom no organic impairment has been established are most commonly classified as familial or cultural retardates and their IQ scores range from 50 to 70. Early investigators assumed that the drawings of these children deviate from the normal course of drawing development. However, later and more carefully controlled studies documented that familial retarded children draw like normally developing children of matched mental age. Findings that identify mental age as a determining factor seem to

(a)

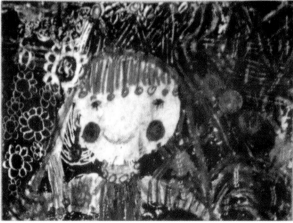

(b)

Figure 5 Color, expression, and love of ornamentation. Drawings of talented children: (a) Warrior, boy, age 6. (b) An imaginary world, girl, age 5.

support the assumptions underlying the Goodenough-Harris drawing of the human figure as an index of conceptual development or IQ. Indeed, studies on Western populations show a good correlation between the Goodenough-Harris Draw-a-Man test and IQ scores for ages 5–10 years.

However, this linkage between drawing of the human figure and IQ scores has been seriously challenged in the drawings of children known as savants, mentally retarded

children with autism who show an unusual artistic ability. The term savant refers to individuals with serious mental handicaps who have a special island of ability that stands in marked contrast to their handicap of mental retardation and autism. The publication of *Nadia: A Case of Extraordinary Drawing Ability in an Autistic Child* by Lorna Selfe (1977) posed a serious theoretical problem given the absence of a correlation between the child's low IQ and her spectacular drawing ability. With the publication of additional cases the question whether savant talent is an indication of a unique conceptual deficit or a mark of unusual talent moved center stage. The results of a series of studies that matched savants with mental retardation to a group of normally developing talented adolescents indicated that, in terms of their graphic talent, the two groups were indistinguishable, which presents a serious challenge to the commonly assumed relationship between IQ (defined as linguistic and mathematical ability) and artistic talent. The artistic accomplishments of several savant artists, one of whom attended an art academy and successfully completed his studies and others whose paintings have been acquired by individual collectors, attest to their talent. From these studies it appears that talent is a major variable that is somewhat independent of IQ scores.

The Sociocultural Milieu

The developmental trends outlined so far, provide an account of children's drawings in Western industrialized settings. To what extent is this representational development characteristic of all children, regardless of time and place? This raises the question of universals in drawing development and the influence of the social milieu on the form children's drawings take.

An early extensive collection of 60 000 drawings from the non-Western world assembled by Paget in 1932 revealed that unschooled preliterate children can, upon request, create drawings of the human figure that, in their diversity, resemble the drawings of children commonly found in Western settings: stick figures, tiny heads, contourless heads and bodies, triangular, squarish, oval, and scribble trunks. Some of these models can be found in any of the large collections assembled in the beginning of the twentieth century, and also resemble the rock carvings and paintings found in such diverse locations as northern Italy, Utah, Guadeloupe, and Hawaii. More recent cross-cultural studies confirm such findings that are especially striking when one observes how naive subjects, never before exposed to paper and pencil, create the same timeless models of animate figures, humans and animals. They attest to a universal factor in the creation of graphic equivalents, at least in early stages of drawing development. There is an underlying logic based on a principle of structural equivalence that generates the diversity of the early models. Local traditions, peers, availability of graphic models, and teachers affect the particular way in which the underlying structure of child art finds expression. In Japan, a culture with a rich artistic heritage, the tadpole figures and their descendents tend to show large eyes, tiny noses, and broad and somewhat flattened heads, clearly influenced by the prevailing comics literature. While it demonstrates the distinct impact of the culture, the similarity to the structural characteristics of early child art is unmistakable.

Divergent Interpretations

The developmental progression outlined so far, especially in regard to the evolution of form and space, finds support among major students of child art. This consensus, however, does not extend to the interpretation of findings. Researchers differ regarding the meaning of the developmental steps, with Piagetian and neo-Piagetian researchers emphasizing the cognitive limitations that underlie the typical childhood drawings, and investigators in the tradition of Arnheim emphasizing the problem-solving intelligence at work and the productivity of visual thinking in a difficult medium. Underlying much of this disagreement is the question of the hypothesized endpoint of drawing development. Researchers in the Piagetian tradition emphasize that the end goal is some degree of optical realism and that the typical childhood drawings are immature and flawed productions that can be attributed to either conceptual deficiencies, production difficulties, and/or limitations on working memory that ought to be overcome by the end of the concrete operational period (ages 10–11 years). Researchers influenced by Arnheim's emphasis on the creation of equivalence of forms rather than imitation as the motivator for art making emphasize the well-established discrepancies between what the child or the inexperienced adult knows and understands about the world he or she wishes to depict, and the specific skills necessary to do so. Many studies indicate significant intra- and interindividual differences depending on the nature of the task and the child's motivation. Above all, the hypothesis of a singular idealized endpoint of optical realism in art whether attained during the concrete or the formal operational period has not been supported. Without training and the high motivation of children talented in the visual arts, there is no evidence of reaching such a state.

The Development of Sculpture

The evolution of children's ability to represent figures in a three-dimensional medium is of interest in its own right while also providing a useful perspective on drawing development. In drawing we observe the difficulties children encounter with the flat two-dimensional medium of paper and their efforts to deal with the missing third dimension. Does the same problem manifest itself in the three-dimensional medium? If so, do children develop their spatial conceptions in analogy to drawing, beginning first with one-dimensional sticks, progressing to two-dimensional flattened slabs arranged in one plane, and only later achieve a three-dimensional conception and production of an object or, alternatively, do they show an early, albeit intuitive, understanding of the three-dimensional nature of representation in this medium? This question has been addressed in a series of studies that closely examine children's modeling strategies on a variety of human and animal tasks. Of particular interest are strategies that might be based on a three-dimensional conception such as modeling multiple faces of an object, emphasizing an upright posture, shaping protrusions and using hollowing out procedures that suggest the inside as well as the outside of a figure, and modeling several distinct layers of an object.

Similar to the earliest scribble actions in drawing, young preschoolers explore clay by squishing, stretching, patting, and rolling it until they discover some likeness in their clay shapes and begin to name them. At first, the three-dimensional medium tends to foster imitative actions, bouncing a piece of clay in imitation of a ball or moving a blob across the table like a train. Between the age of 3 and 4 years children tend to discover the representational possibilities in this medium and make their first attempts to model humans and animals.

Modeling the Human Figure

Early efforts to create a human figure yield three distinct models: a global figure comprised of a sphere with facial features, an erect standing column, and a layout model composed of separately shaped, disconnected parts, mostly facial features and occasionally a tummy and limbs (see **Figures 6(a)–6(c)**). These early models are short lived and soon the one-unit sphere sprouts arms and legs, thus becoming a tadpole figure; the erect standing column undergoes differentiation such that various body parts are either internally represented or the head is modeled separately and placed on top of the erect standing body-column; the layout figure develops into an outline graphic model in clay (see **Figures 7(a)–7(c)**). The columns and their variants are free standing or held upright, while the tadpole and graphic models are placed horizontally on the tabletop. These different models are the simplest structural equivalents for the human figure. From these early and primitive sculpting models one can infer the representational concepts that give rise to them. As in drawing, they are characterized by generality, such that a

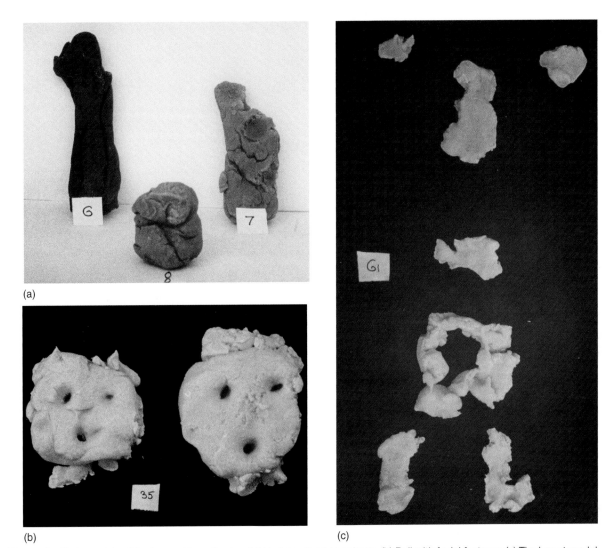

(a)

(b)   (c)

Figure 6   Early models of the human figure in clay. (a) Upright standing column. (b) Ball with facial features. (c) The layout model.

global form comes to stand for another global entity, in this case a person. Verticality, uprightness, and facial features serve as the defining attributes of these early representations of the human figure.

Progress in modeling can be seen when body parts are differentiated: the upright standing figure modeled from separately formed solid parts, the horizontal figure constructed of rounded or flattened parts, and the graphic outline figure enriched with greater detail. With the exception of the graphic model, simplicity and economy of form dominate this art form.

Simplicity of form does not imply ignorance on the part of the young artist. Children comment freely on the perceived flaws in their clay figure, with the younger ones providing corrections verbally or reinterpreting the original intention in line with the perceived outcome. What appears as indifference to the correct proportions of a figure is often a function of the child's working style.

Thus, when the arms are first endowed with fingers, their size is determined by the enthusiastic action of rolling the clay and the need for symmetry rather than by realistic proportions. Most frequently the sun-radial is adopted for the hand, frequently in the form of three fingers that create a balanced structure (see **Figure 8(a)**).

With age and practice, the figure gains in the number of its modeled parts, in attention to detail, proportion, and measurement. The overall trend is toward increasing complexity of the figure and its subdivision into distinctive parts. These include the distinction between the upper and the lower torso, shoulders, neck, clothing in the form of shirts with sleeves, flaring skirts or pants, and accessories.

With increasing complexity of the modeled parts of the human figure, children face the difficulty of maintaining an upright stance when head and body are joined to two spindly legs. This leads to the somewhat paradoxical finding that with increasing age and the sophistication to

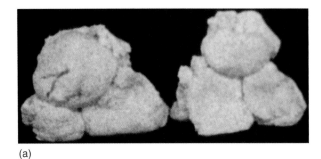

(a)

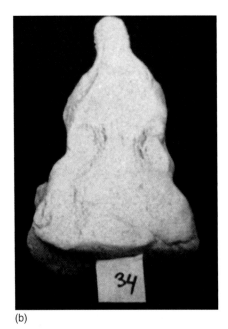

(b)

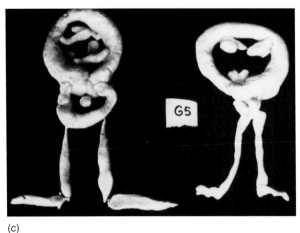

(c)

Figure 7   Beginning differentiation of the modeled human figure: (a) tadpole; (b) column with internal differentiation of parts; and (c) graphic model in clay.

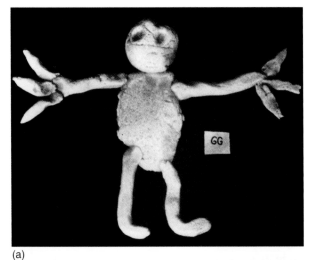

(a)

(b)

Figure 8   (a) Man with large hands. (b) Cow with udders and bell.

model the different parts of the human figure, verticality is often sacrificed out of frustration with the difficulty to maintain the upright posture without the help of an armature.

With age and practice come technical skills and some refinements in the appearance of the sculptures, even though many continue to be modeled quite crudely. Often the technique of some of the older children is not better than that of the younger ones, but their serious reflection and concentrated work distinguish between them. Along with more planning and increased skill come a more critical awareness and often a negative evaluation of the outcome.

In analogy to drawing, the frontal part of the human figure receives the greatest attention since it carries the most identifying characteristics of the human that indicate gender and individuality. However, unlike the distinctly two-dimensionally drawn figures, in modeling attention is paid to three-dimensional aspects, turning the figure and adding or subtracting clay as children model the back and sides, and comment on their work. It is interesting that unlike drawing, the facial features in the clay figure play a less important role and are altogether omitted in 30% of the human figure sculptures, while the trunk section is represented earlier than in the corresponding drawings.

Modeling Animal Figures

Modeling of four-footed animals (dog, cow, turtle) provides much insight into children's three-dimensional conceptions in clay. Given the basic symmetry of the animal structure with the two major sides near duplicates of each other, and the body resting on four legs, children past the stage of the early globals, modeled their animals, with few exceptions, standing upright with attention paid to more than a single side, with some 4-year-olds modeling up to six sides of the figure. Noteworthy is the strategy of turning figures upside down to attach the front and hind legs, sideways to model the tail, frontally to attach the head and differentiate the orientation of the head from the sides of the body. Again, with age and practice, there is a growing interest in size differences, and attempts are made to enliven the figure by action and gesture and the addition of such defining characteristics as spots on the turtle, a collar on the dog, a bell and udders on the cow. (See **Figure 8(b)**). In general, differentiation of form is mostly age related, with older children being more skillful and able to model better balanced sculptures. However, age per se does not guarantee skill, and for most of the children representational intention exceeds their ability to give it adequate expression.

Although facial features are lacking in 40% of the animal figures, the orientation of head and body is most commonly well differentiated, and ears are modeled to indicate the direction of the head. Mixed views so common in drawing, are infrequent in clay modeling. Most interesting is the observation that children adopt a variety of different models on the human and animal figures tasks, clearly demonstrating that there is no single underlying conceptual template that guides production. Different representational models coexist in the same child.

At the beginning of this section the question whether the dimensional progression established for drawing would also be applicable to the domain of sculpture was raised. The results reported so far do not support such a view and instead suggest that from the beginning of children's intention to model a figure they employ some basic, albeit simple, three-dimensional representational concepts as indicated by their attention to multiple sides, volume, and upright posture. Although the untrained efforts of the preschoolers and of many older children yield mostly crude figures, there is early on some basic intention to model in the round.

Concluding Comments

The comparison between drawing and modeling has provided greater insight into the principles that underlie artistic thinking and problem solving and the diverse routes the search for equivalences can take. Studies on drawing and modeling support the conception of an orderly sequence in the acquisition of representational skills. In both domains there is a rule-governed progression in the problem-solving activity that underlies the child's discovery of a graphic and plastic language. The search for forms of equivalence in drawing and modeling is of a universal nature, it manifests itself in ordinary, talented, and mentally handicapped children, and across diverse cultures. Progress in the differentiation of form and space enables young artists to eliminate ambiguity, to express their thoughts and feelings better, and to articulate their sense of esthetics. With age and practice evolve the ability to attend to multiple variables, to plan, monitor, review critically, and to revise the emerging representation. Artistic development, in the absence of training, is a spontaneous and largely self-regulated process that flourishes during the early childhood years and tends to reach a plateau during the middle childhood years.

During the elementary school years, a gradual decline in spontaneous artistic productions is noted, and the work of most children signals the end of a very creative and productive period in child art. There are likely to be diverse reasons and competing interests that lead to this decline of artistic activity. Above all, alternative outlets for self expression can be found in the widening horizons of middle childhood that afford access to sports, dance, music, chess and computer games, and the opportunities for social activities. For some children, the technical problems associated with more advanced pictorial strategies spell the end of their artistic explorations. In some cases, the arts are taken up again in adulthood as a hobby or even a serious endeavor.

See also: Cognitive Development; Imagination and Fantasy; Symbolic Thought.

Suggested Readings

Arnheim R (1968) *Visual Thinking.* Berkeley: University of California Press.

Fineberg J (1997) *The Innocent Eye.* Princeton, NJ: Princeton University Press.

Freeman N (1980) *Strategies of Representation in Young Children.* London: Academic Press.

Hermelin B (2001) *Bright Splinters of the Mind.* London: Jessica Kingsley.

Willats J (2005) *Making Sense of Children's Drawings.* London: Lawrence Erlbaum Associates.

Winner E (1996) *Gifted Children: Myths and Realities.* New York: Basic Books.

Asthma

R W Hendershot, University of Colorado at Denver Health Sciences Center, Denver, CO, USA

Glossary

Airway – The part of the respiratory system through which air is carried, from the mouth, through the trachea, bronchi, and throughout the alveoli of the lung.

Airway or bronchial hyperreactivity – A term used to describe one of the three main features of asthma. It describes how the airways of the lungs in an asthmatic patient are easily triggered to constrict causing an acute asthma attack and airflow obstruction.

Albuterol – A generally inhaled medication, known as a bronchodilator, that relieves narrowing of the airways caused by bronchospasm.

Allergen/antigen – A substance, usually a protein, that can cause an allergic reaction.

Allergy – A clinical condition in which the body has an exaggerated response to an allergen, usually hypersensitivity; the expression of atopy.

Asthma action plan – A written summary of what actions a patient and their family are to take when their child's asthma worsens. In young children, the asthma action plan is based on symptoms.

Atopic march – The progression of allergic disease an individual may undergo as they mature; generally begins with eczema (atopic dermatitis) and food allergies as an infant and progresses through to asthma and eventually allergic rhinitis (hay fever).

Atopy – A genetically determined state of IgE-mediated hypersensitivity to allergens, the likelihood of being clinically allergic.

Basophil – A type of circulating immune cell that plays a role in the allergic response by releasing chemicals such as histamine when exposed to allergens.

β-blocker – A medicine used to treat high blood pressure and heart disease by blocking β-adrenergic receptors. Its use can make asthma worse.

Bronchoscopy – A procedure performed to examine the airways of the lungs (bronchi) visually with a flexible lighted tube.

Bronchospasm – A tightening of the muscles around the airways of the lungs (bronchi) causing the diameter of the airways to constrict.

Endotoxin – A molecule found in the outer membrane of Gram-negative bacteria, exposure to which is hypothesized to protect against the expression of atopy–allergy and asthma.

Eosinophil – A type of immune cell generally used by mammals to fight parasitic infections but primarily responsible for the damaging effects of atopic disease.

FEV1 – The volume of air an individual can exhale in the first second of a forced exhalation.

Gastroesophageal reflux disease (GERD)/acid reflux – A disease in which acid 'refluxes' into the esophagus from the stomach.

IgE – Allergic antibody that triggers allergic reactions when it comes in contact with allergen.

Inflammation – The body's response to injury typically manifested by redness and swelling.

Lymphocyte – A type of circulating immune white blood cell responsible for coordinating the immune response to infection or injury.

Mast cell – A type of tissue immune cell that, like the basophil, is part of the allergic response.

Mendelian genetics – The set of primary tenets that govern the transmission of an organism's physical characteristics to its offspring. Based upon principles originally put forth by Gregor Mendel. Mendel was a nineteenth-century Austrian monk who discovered and published his results based upon plant hybrid experiments. Mendel explained that an organism inherits two copies of each gene, one from each parent. Each gene is either dominant or recessive in its physical expression. A dominant gene expresses itself as a physical characteristic no matter what gene is inherited from the other parent. A recessive gene is only expressed when both inherited genes are the same.

Nebulizer – A machine that aerosolizes medicine for inhalation into the lung.

Obstructive lung disease – A category of lung disease classified as such because it makes getting air out of the lungs more difficult. In its most severe form it traps air in the lungs making air exchange impossible, which can lead to death.

Peak flow meter – A tool used to measure the severity of a patient's asthma by measuring the maximal velocity with which air exits the lungs.

Reactive airway disease (RAD) – A disease that commonly causes wheezing in infants and young children. It differs from asthma in that RAD is common and transient, usually resolving by the time the child is 5 years old. If the child has persistent wheezing the diagnosis is more likely asthma.

Spirometry – A lung test that evaluates how well air moves in and out of a patient's lungs. It requires patient coordination and cooperation to perform the test.
Therapeutic index – A way of comparing the benefits and risks of medical interventions such as a medication or surgery.
Upper respiratory infection (URI) – Most commonly called a 'cold', the technical term for a viral infection of the nose and sinus.
Wheeze – The sound created as air is breathed through constricted airways.

Introduction

The word asthma is derived from the Greek *aazein*, which means to exhale with an open mouth or to pant. It was first used in the Iliad to describe a short-drawn breath. From a clinical standpoint, Hippocrates was the first Western physician to write about it in *Corpus Hippocraticum*. By AD 1, when asthma was described by the clinician Aretaeus of Cappadocia, its place as a clinical entity became well known. Sir William Osler was among the first to describe asthma as inflammation of the bronchi, and by 1909 allergenic sensitization of smooth muscle in animals was demonstrated. Today asthma is defined as a chronic disease of the lung manifest clinically as episodic obstruction of pulmonary airflow. Airflow obstruction is caused by inflammation, mucous plugging, and bronchial hyperreactivity that leads to wheezing, chest tightness, excessive mucous production, shortness of breath, and sensitivity to irritants.

The American Lung Association marks asthma as the seventh-ranked chronic health condition in the US and the leading cause of chronic illness among children. While many patients perceive their disease to be only episodic in nature, 80% of asthmatics have persistent symptoms. There are estimated to be over 20 million Americans who suffer from asthma. Of these, approximately 9 million are under 18 years of age, with 3–5% of adults and 7–10% of children affected. The Centers for Disease Control and Prevention (CDC) estimated that for 2004, asthmatic children missed 12.8 million schooldays, made 7 million outpatient visits, had 750 000 emergency room visits, and 198 000 hospitalizations for asthma. Asthma hospitalizations alone represented 3% of all hospitalizations among children. There were 186 deaths from childhood asthma in 2004 in the US. Asthma continues to be the leading cause of school absenteeism due to a chronic illness, and asthma-related direct and indirect healthcare costs are estimated to exceed $14 billion a year. Even though the prevalence of asthma increases with age, healthcare use is highest among the very young.

The prevalence of asthma has steadily increased since at least 1980. However, since 1997 the frequency of asthma-related hospitalizations and deaths have declined. We assume this is due to the progress made in both, treatment and education, of those affected with the disease. Despite great progress in therapy, there remains a major disparity in both morbidity and mortality among racial and ethnic populations. Specifically African–American inner-city asthmatics have a threefold higher risk of both death and hospital admission than asthmatics in other segments of the population.

It is estimated by the Asthma and Allergy Foundation of America that everyday in the US:

- 40 000 people miss school or work due to asthma,
- 30 000 people have an asthma attack,
- 5000 people visit the emergency room due to asthma,
- 1000 people are admitted to hospital, and
- 14 people die due to asthma.

Asthma is defined as a triad of inflammation, airway hyperreactivity, and mucous plugging. However, patients identify most with episodes of acute episodic airflow obstruction causing wheezing and chest tightness. These episodes are extremely anxiety-inducing and very memorable. While 20% of asthmatics have what is referred to as mild intermittent disease, 80% have persistent disease that requires a daily medication. Many asthmatics, with good control, feel completely well between acute episodes and have no symptoms. Most asthmatics achieve that control through strict adherence to their daily medications. In general, asthma does not mean limiting an individual's activity. It is rarely grounds for labeling a child as 'brittle' or the reason why an individual cannot compete on the athletic field. John Weiler, reporting about the 1996 Summer Olympics, commented that more than 20% of the athletes who participated might be considered asthmatic, and that 10.4% of the athletes stated that they took asthma medications either during the games or on a regular basis. While asthma has no cure and there is no treatment that will end the disease, there are some asthmatics that experience a spontaneous remission or outgrow their childhood disease. For most, asthma is a significant but manageable disease, and appropriate care and attention to detail allows the asthmatic to live a normal life.

Effects of Asthma on the Airways

Asthma targets the airways of the lungs. It interferes with our ability to breathe by impeding the process with which our lungs move air in or out. For this reason asthma is referred to as an obstructive disease. In an acute exacerbation, as an asthmatic inhales, air moves into the lungs because expansion of the chest wall creates a negative pressure, allowing air to flow into the lung. However, during exhalation, outflow is obstructed. As the chest wall collapses and pushes air out, the resultant force compresses the airways of an asthmatic, and air is trapped in the lungs. This process can continue until the lungs are hyperexpanded to the point that no more air can get in and a catastrophic failure occurs.

In the 1950s asthma was thought to be a disease characterized by reversible airway obstruction that either resolved spontaneously or resolved following therapy. In the 1960s asthma was viewed as an episodic disease with airway obstruction caused by airway hyperreactivity. The goal of therapy then became relief of bronchoconstriction. This was and still is accomplished by bronchodilator medications known as relievers. In 1969, Dunhill and co-workers detailed the first histopathological evidence of inflammation in asthma when postmortem examinations were performed on asthmatics who died from the disease. It became obvious that tissue damage from airway edema, infiltration by immune cells (especially eosinophils) and excessive mucus secretion were complications of fatal asthma. However, the extent to which inflammation played a role in patients with more mild disease was still unclear. In the 1980s and 1990s, laboratory research and the use of flexible bronchoscopy as a diagnostic tool, led to the understanding that inflammation plays a part in all asthmatics. The beneficial effect of inhaled corticosteroids (ICSs), introduced in the 1970s and coming into wide acceptance in the 1980s for even mild asthmatics, emphasizes the central role of inflammation in asthma. ICSs are now referred to as controllers because of the long-term effect they have in controlling inflammation.

The asthma inflammatory process begins with an acute insult such as an allergen causing both an early (5–15 min) and late (2–6 h) response. The early response, characterized by immediate bronchial constriction and relieved by bronchodilators, begins as antibodies in the immunoglobulin E (IgE) class bind allergen, causing mast cell activation. IgE is the class of antibody responsible for allergies. Activated mast cells then release inflammatory mediators such as histamine and tryptase. These mediators lead to airway hyperreactivity and bronchoconstriction. The late response occurs because mast cell tryptase induces an influx of neutrophils and eosinophils into the airways. The contents of these immune cells are also released causing further inflammation and airway hyperreactivity. These two phases of asthma, early bronchoconstriction and late inflammation, are what reliever and controller medications, respectively, treat.

The director of the asthma orchestra is the T cell, particularly a subgroup known as T-helper cells (Th cells). T cells are so-called because they mature in the thymus, an organ found in the chests of young children which, if not functioning properly *in utero*, leads to immunodeficiency in the newborn. Two distinct Th-cell populations are described, based on which cytokines they release. Th1 cells preferentially secrete interleukin (IL)-2, interferon-γ (IFN-γ), and tumor necrosis factor (TNF)-β. IFN-γ directly inhibits B cells from producing IgE antibodies. Asthmatics are skewed toward responding to allergen exposure with a Th2 response. Th2 cells preferentially secrete IL 4, IL 5, and IL 13. IL 4 and IL 13 induce B cells to produce IgE antibodies. IL-5 is important for eosinophil survival, maturation, and migration from blood into tissues. Th1 cells and their cytokines are important for fighting viral and bacterial infections. Th2 cells, eosinophils, and IgE are important for fighting parasitic infections. Immunotherapy (allergy shots) works in part by shifting the immune system from a Th2 to a Th1 response.

Eosinophils are the primary immune cells believed to cause most of the damage in the lung that leads to asthma. This damage causes increased mucus production, a thickening of the basement membrane that surrounds the airways of the lung, and bronchial hyperreactivity. These processes reduce the caliber of the airway causing wheezing and the disease we refer to as asthma. If this process goes unchecked, it can lead to airway remodeling and an irreversible loss of lung function.

The amount of inflammation in the lungs of an asthmatic can be quantified several ways. It was first measured by bronchoscopy. Bronchoscopy is a procedure performed that allows a visual examination of the bronchi via a flexible lighted tube and permits the bronchoscopist to obtain microscopic samples of the airways and bronchoalveolar lavage (BAL) fluid washed from the alveoli of the lung. With the tissue, BAL fluid, and a microscope the pathologist can directly measure the amount of inflammation. However, because the risks involved with sedating a child and introducing a foreign object into the lungs is significant, this is only done when essential to help the patient. More commonly, a specialist may measure the inflammation indirectly. This can be accomplished by measuring the percentage of eosinophils in a patient's sputum or the amount of exhaled nitric oxide (NO, a byproduct of eosinophilic inflammation). The long-term treatment of asthma focuses on interruption of the inflammatory cascade via ICSs. The success of that treatment can be evaluated by measuring the response of an asthmatic's inflammation to therapy, as described earlier. This makes these tools very valuable.

In addition to inflammation, asthma is characterized by airway hyperreactivity. 'Reactivity' here refers to the contraction of smooth muscles around the airways resulting in a clinically significant reduction in the internal diameter of the airway, which leads to wheezing. Hyperreactivity denotes an airway that constricts with a minimal amount of stimulus. Stimuli are varied but most commonly include cold air, exercise, allergen, infection, or toxic stimuli such as tobacco smoke and pollution. Of the primary physiological responses in asthma, hyperreactivity is the most widely appreciated because of its immediate clinical effect. Patients understand that if they are exposed to one of their exacerbating stimuli, asthma worsens.

Hyperreactivity is quantifiable. Done by performing an evaluation known as a methacholine, or bronchoprovocational, challenge, it is useful because the more reactive a patient's airways, the more severe their disease. A bronchoprovocational challenge is performed by delivering an agent such as methacholine or histamine in increasing doses until the patient experiences a 20% decrease in their forced expiratory volume in 1 second (FEV1). FEV1 is the volume of air an individual can exhale in the first second of a forced exhalation as measured with a spirometer. A spirometer is a machine used to measure lung function by measuring the volume and flow rate of air during both inhalation and expiration. The PC20 in a bronchoprovocational challenge is the dose of methacholine or histamine at which the patient's FEV1 drops by 20%. There is an inverse relationship between the dose of the agent needed to drop an individual's FEV1 by 20% and the severity of their disease.

Asthma in Young Children

Generally children under 5 years old are incapable of performing the spirometry studies necessary to calculate a PC20. However, some specialists are successful in getting children as young as 4 years old to do it. There is a fair amount of cooperation needed to perform the test. When a child is unable to cooperate to the degree needed, a physician may evaluate a child by performing the provocational challenge and measuring tissue oxygenation and auscultating the chest as the agent is given. Auscultation is the act in which a physician listens to a patient's chest with a stethoscope as they breathe. This allows the physician to assess if a child wheezes or develops other symptoms of concern. While this test is generally unnecessary, it can be helpful if the diagnosis is questionable after the history and physical examination are completed.

Infants are born with small airways compared with adults. Because of their smaller airways and frequent exposure to viral upper respiratory infections (URI), infants and young children are prone to wheezing. If a small child experiences an insult to the airways, they need constrict only minimally before wheezing and increased work of breathing occur. Asthma is the most common cause of wheezing in children 5–18 years old. This is not so in infants and toddlers, whose wheezing is classified as 'reactive airway disease' (RAD). This response is most commonly associated with a viral URI, which causes inflammation and constriction of the infant's airways, leading to wheezing. Most children who experience URI-induced wheezing,

however, will not go on to develop asthma. According to the Tucson Respiratory Children's Study:

- 60% of all children will wheeze transiently in early life ('transient wheezers'),
- 85% of children who transiently wheeze will not have asthma,
- 15% of transient wheezers will go on to develop asthma, and
- the most common cause of transient wheezing is a viral URI like the respiratory syncytial virus (RSV), rhinovirus, or the parainfluenzae virus.

In most asthmatics, allergy (or atopy) plays a major role. Atopy is the hereditary predisposition to develop IgE-mediated hypersensitivity to foods, pollen, and animal dander. Allergy is the clinical manifestation of atopy. The immune processes that are in disarray in asthma are stimulated by the interaction of allergens with IgE. For most patients: good control of allergies, strict adherence to asthma medications, and avoidance of asthma triggers allows them to live normal lives.

Natural History of Allergic Diseases

Allergic diseases are closely related. These include atopic dermatitis (commonly called eczema), food allergy, allergic rhinitis (hay fever), and asthma. Most people who are allergic have more than one of the allergic diseases and having one increases the chances of a child having a combination of them. For example, if a child has severe atopic dermatitis there is a more than 80% chance that the child will eventually have asthma as well. More than 80% of adult asthmatics become so before 6 years of age. Most start down the path toward asthma going along what is referred to as the 'atopic march'. Atopy is a genetically determined state of IgE-mediated hypersensitivity to allergens, the likelihood of being clinically allergic. The atopic march often begins in infancy when the child may have food allergy and atopic dermatitis. Many children grow out of these conditions but others progress toward having asthma and allergic rhinitis. The peak prevalence of atopic dermatitis and food allergy is at 1 year of age. Asthma peaks between 7 and 9 years old, while the peak prevalence of allergic rhinitis occurs in adolescence.

Most infants and young children will wheeze at some time in their life as mentioned earlier. Of course predicting which wheezing children will advance to asthma and which will not is of great interest. A clinical study from the Tucson Children's Respiratory Group in 2000 sought to define an asthma-predictive index for children. In this study it was shown that risk factors for the development of asthma were: a parent with asthma, physician-diagnosed atopic dermatitis, evidence of allergy shown by a positive allergic skin test, wheezing apart from colds, and eosinophilia. (Eosinophilia occurs with increased numbers of circulating eosinophils.) Risk factors for adult-onset asthma include: respiratory infections, tobacco smoke, obesity, occupational exposures, medications, and allergies. A triad of asthmatic disease more common in adults than children is chronic sinusitis associated with nasal polyposis (a condition in which small sac-like growths of inflamed nasal tissue occur in the nose), and a hypersensitivity to nonsteroidal anti-inflammatory medications such as aspirin or ibuprofen.

What Causes Asthma

The etiology of asthma is not straightforward. Asthma is not inherited based on the principles of simple Mendelian genetics. This means that there is no single asthma gene that works in either an autosomal recessive or autosomal dominant manner. Instead, asthma is linked to multiple genes, and the likelihood that an individual will have the disease is multifactorial. Not only are the genetics difficult but asthma, as a disease, is a complex interaction between the environment and the genetic make-up of the individual. It appears that both contribute equally. Research studying the influence genes have on asthma is ongoing and should increase our understanding of, and ability to treat, the disease. Studies attempting to understand the impact the environment has on asthma are ongoing. This research includes everything from allergens to viral and atypical bacterial infections to endotoxin, a component of Gram-negative bacteria. There are several theories about how the interaction between a human's genes and

environment leads to asthma. Some studies lend credence to the theories (**Figure 1**).

Originally it was assumed that air pollution was the primary driver behind the increase in asthma worldwide. However, in 1992 and again in 1994, Erika Von Mutius compared the rates of allergy and asthma in East and West Germany. She hypothesized that in these two genetically identical populations, individuals growing up in the more polluted East would suffer from increased allergies, especially asthma. To her surprise, her studies found the opposite to be true. East Germany had less allergic disease than West Germany.

Extending Von Mutius's findings, D. J. Keeley noted that asthma was more prevalent in children from wealthier, urban environments than in those from poor rural areas. Further studies proposed various causes for the increased prevalence of asthma, including increased tobacco exposure, vaccinations, antibiotic use, obesity, acetaminophen (Tylenol) use during pregnancy, decreased parasite exposure, breastfeeding, or endotoxins. Altered patterns of microflora in the gastrointestinal tract, daycare attendance, and birth order have all been scrutinized. Presently the most popular theory as to the cause of the increasing prevalence of asthma is the 'hygiene hypothesis'.

The 'hygiene hypothesis' states that, from the uterus, the immune system is primed to fight infection. Once the baby is born, if minimal infectious exposures occur, the immune system will expend energy responding to proteins such as animal dander and pollen, and the individual will be allergic. Thus, the good hygiene of contemporary society allows the developing infant to avoid infection, and, as a result, allergies occur and the prevalence of asthma increases.

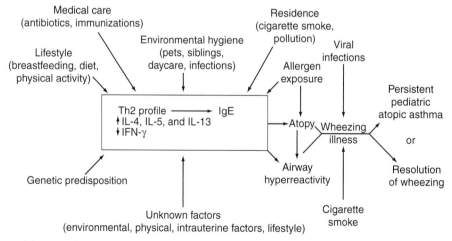

Figure 1   Diagram of the many factors involved in the development of pediatric asthma. The central reaction is the interaction of an individual's genetic composition and their environment that leads to a Th-2 immune response, as indicated by increased release of the mediators of allergy (interleukins 4, 5, and 13) and decreased release of the central mediator of the Th1 response (IFN-γ). The end result is an increase in the key mediator of atopy, IgE, which leads to the symptoms of atopy and bronchial hyperreactivity. This can result in wheezing and potentially persistent atopic asthma. Th, T-helper cell; Ig, immunoglobulin; IL, interleukin; IFN-γ, interferon gamma. Adapted from Johnson CC, Ownby DR, Zoratti EM, et al. (2002) Environmental epidemiology of pediatric asthma and allergy. *Epidemiologic Reviews* 24: 154–175.

Diagnosing Asthma

Asthma can be difficult to diagnose. A child who presents with complaints of a persistent cough, wheezing, shortness of breath, or chest tightness likely has asthma. With these complaints, an alternative diagnosis might include acute viral or bacterial infections, postnasal drip secondary to chronic sinusitis or rhinitis, gastroesophageal reflux disease (GERD), chronic lung disease, cystic fibrosis, aspiration, or a congenital anatomic abnormality. By evaluating a patient's history, doing a physical examination, and performing appropriate testing, a physician is usually able to differentiate between asthma and the alternatives (Table 1).

An asthma evaluation begins with the history and physical examination. Historical elements important in the diagnosis of asthma include when and how the symptoms began, what were the symptoms and their frequency and severity, what triggers symptoms, and how do they respond to therapy like albuterol or prednisone. A history consistent with a diagnosis of asthma includes: onset at an early age with symptoms of cough and wheeze triggered by URIs, tobacco smoke, or allergen exposure such as dogs and cats. Patients may complain of coughing and wheezing that awakens them from sleep or occurs shortly after waking. More severely affected patients will experience symptoms all the time. Infants and young children may even be hypoxic. A positive response to albuterol or prednisone, while helpful, is not definitive.

Once the diagnosis is made, a physician will focus on potential comorbidities. For example, GERD is known to make asthma worse, as will a dog or cat in the home, chronic sinusitis or rhinitis, and tobacco-smoke exposure.

Table 1 Differential diagnosis for asthma in infants and young children

Infectious
Bronchiolitis
Pneumonia
Croup
Bronchiectasis (consider cystic fibrosis, aspiration, and ciliary
  dyskinesia)
Bronchiolitis obliterans
Chronic sinusitis and rhinitis

Anatomic
Gastroesophageal reflux
Cystic fibrosis
Tracheomalacia or bronchomalacia
Congenital heart disease with failure
Tracheoesophageal fistula
Fixed upper airway obstruction (vascular rings or laryngeal webs)
Tumor or enlarged lymph nodes
Aspiration from swallowing disorder

Other
Foreign object in the lung
Bronchopulmonary dysplasia
Allergic bonchopulmonary aspergillosis
Churg–Strauss syndrome

Classification System for Asthma

The National Institutes of Health (NIH) established the National Asthma Education and Prevention Program (NAEPP) in March of 1989 as a task force to reduce asthma-related illness and death, and to improve the quality of life of asthmatics. Today there are 40 major medical associations and health organizations in addition to numerous Federal agencies that comprise the NAEPP Coordinating Committee. The NAEPP published its first set of guidelines for dealing with asthma in 1991. The guidelines were updated in 1997 to include:

- A stepwise approach to asthma putting an emphasis on the early use of inhaled corticosteroids.
- A classification system for asthma severity used to guide therapy.
- A discussion about asthma prevention.
- A discussion about reducing environmental exposures, including tobacco smoke, as a part of asthma therapy.
- Information about identifying an individual's specific triggers.
- Emphasis on education and self-management.
- Recommendations for aggressive detection and treatment of children.
- Tools to help physicians incorporate the guidelines into their practices.

The guidelines were again updated in 2002 with the incorporation of a new class of medication, the leukotriene receptor antagonists (LTRA) as an alternative controller medication in mild persistent asthma. Another update of the guidelines is expected in 2007. Today, once asthma is diagnosed it is classified according to a patient's clinical severity, pulmonary function, and the amount of medicine required to maintain control of symptoms. In infants and young children, where pulmonary function is difficult to obtain, a physician focuses on the clinical symptoms. The goal of therapy is to reduce a patient's symptoms to what would be consistent with mild intermittent asthma.

A mild intermittent asthmatic is defined as an individual with symptoms requiring use of their albuterol medication no more than twice a week, and with nocturnal symptoms no more than twice a month. Patients who have more persistent symptoms need more intensive therapy. The NAEPP guidelines for a stepwise approach to the treatment of asthma are illustrated ahead. The classifications of asthma are: mild intermittent, mild persistent, moderate persistent, and severe persistent (Table 2).

Specifically, the goals of the NAEPP guidelines are: minimal or no chronic symptoms day or night, minimal or no exacerbations, no limitations on activities, no missed work days for parents, maintenance of normal or near-normal pulmonary function, minimal use of albuterol, and minimal or no adverse effects from medications.

Table 2   Classification of asthma severity in infants and young children

Severity	Frequency of symptoms day/night	Most commonly used medications used to establish control in infants and young children
Mild intermittent	≤2 days/week ≤2 nights/month	No daily medication needed
Mild persistent	>2 days/week, <1 time/day <2 nights/month	Recommend: low-dose ICSs, alternatives available
Moderate persistent	Daily symptoms >1 night/week	Medium-dose inhaled corticosteroids, addition of a leukotriene antagonist
Severe persistent	Continous daytime symptoms Frequent nocturnal symptoms	High-dose ICSs, LTRA, and if needed, systemic corticosteroids

Adapted from the National Asthma Education and Prevention Program (NAEPP) guidelines. ICSs, inhaled corticosteroids; LTRA, leukotriene receptor antagonists.

Asthma Therapy

Therapy revolves around the triad of asthma and is the reason all persistent asthmatics need two categories of therapy. Controller therapy is prescribed to treat inflammation and reliever therapy is given to treat bronchoconstriction. Controller therapies are anti-inflammatory in nature and include medications such as ICSs, cromolyn, and LTRAs. Reliever medications are primarily the β-agonists but may also include anticholinergics or theophylline to dilate the airways. Next comes good compliance. No matter how effective a treatment may be, if a medication is not taken as it should be, it will not work. Additionally, triggers that make asthma worse must be avoided. The simplest interventions are often the most effective. Sometimes smoking cessation or removing a cat from the home will work well; and while this seems obvious, this is frequently the most difficult choice for parents or asthmatics to make. Finally, the acute exacerbation must be treated.

The NAEPP guidelines for therapy are based upon expert opinion derived from clinical experience and multicenter, double-blind, placebo-controlled studies. The studies and expert clinicians support the model that the basis of asthma is inflammation of the airway. Consequently, ICSs are the safest and single most effective therapy for asthma reducing both morbidity and mortality. It is because of ICSs that the majority of asthmatics can live relatively normal lives.

There are side effects with all medications. With ICSs the most common side effect is a growth delay of 1 cm in height in the child's first year of use. In long-term studies this delay does not accumulate over the years or affect the ultimate height of the child. Another less common adverse affect is thrush or oral candidiasis (a yeast infection of the mouth), which is usually mitigated by rinsing the mouth or brushing teeth after the ICSs are administered. Less commonly, high-dose ICSs can cause suppression of the adrenal–hypothalamic axis, which is the means with which our bodies make and regulate endogenous cortisol (a corticosteroid). So, while there can be adverse effects from using ICSs, by placing the medicine where it is needed most, the total amount of corticosteroids are reduced. This decreases the likelihood of side effects from the medication and makes the risk–benefit ratio of ICSs favorable.

In mild persistent asthmatics, LTRA are another treatment option, which may limit the amount of ICSs needed. In young children, ICSs and LTRA are the most frequently used and best-studied medications available. Most other medications such as long-acting β-agonists (salmeterol) are unavailable for use in young children because of the difficulty in administering them. There are orally administered medications such as theophylline and albuterol that have a less favorable risk–benefit ratio.

The medications for treating asthma are generally effective. The limiting factor in achieving control is rarely the medicine. Most frequently it is poor compliance, poor technique in administering ICSs, or both. Often it is a combination of poor compliance and a trigger of some sort that leads to an exacerbation. While the appropriate diagnosis and therapeutic plan is essential to a patient's health, ultimately the most important factor is that medication is taken regularly and appropriately. The final piece of controller therapy is avoidance of triggers. Some would rather live with an annoying cough than not have a cigarette or the family cat, however.

Therapy for an acute severe asthma exacerbation has not changed much in recent years. Inhaled short-acting β2-agonists in combination with systemic corticosteroids are the mainstay. Generally, this is effective. However, on occasion an exacerbation is severe enough to require supplemental oxygen or epinephrine (adrenaline). Either way, ambulance transport to the nearest hospital with the resources to care for such a patient is necessary.

Self-Care and Monitoring: Asthma Action Plans

The asthma action plan is a written summary of what actions a patient and his or her family are to take and is considered the standard of care. In young children, the asthma action plan is based on symptoms. An asthma action plan has three zones. The green zone means all is well and to continue routine medications. The yellow zone indicates caution; a patient should begin using their reliever medications, continue routine medications, and if there is no improvement, contact their physician. Sometimes a physician may recommend that a patient double the dose of ICS if in the yellow zone. The red zone means immediate therapy with a reliever medication is needed, followed by contact with a physician. Asthma is a disease that causes a lot of anxiety; the asthma action plan allows patients to take control of their disease.

Taking control is what makes the asthma action plan work. Asthma that is difficult to treat results most often from a lack of 'buy-in' from patients and their families. As with all chronic diseases, if an asthmatic understands that regular use of their medication will improve their life, they will probably take it. A recent study showed that almost 70% of patients with moderate-to-severe disease would be classified as mild if they took their medications appropriately.

Asthma in Special Populations

Infants and young children are a unique challenge to diagnose and treat. There is no blood test to diagnose asthma, and because infants are unable to perform spirometry, evaluating breathing problems is difficult. For similar reasons the treatment of their asthma is difficult. ICSs depend on patient-coordination to work and are much more difficult to administer than a pill or a liquid. In infants and young children, ICSs take time and effort from the parent. When one combines the difficulty of administering the medication with parental concerns about delayed growth, many parents give up, leaving their child at risk for a severe exacerbation.

The inner city is a difficult environment for an asthmatic. The living conditions may include cockroaches, dust mites, pollution, tobacco smoke, and animal dander, all potential triggers of asthma. Additionally, the inner city is often overcrowded. Asthmatics may have challenging social situations as well. There is often a lack of access to healthcare and limited education or family support. Young inner city asthmatics require close monitoring to avoid a catastrophic outcome.

The asthmatic athlete deserves special attention. Most importantly, with appropriate care almost all of them should be able to compete at their highest level. For example, Jerome Bettis, an all-pro National Football League running back, often speaks about how asthma nearly sidelined his career, and that following his physician's advice on how to control his asthma allowed him to perform. Too many athletes try to power their way through

an asthma exacerbation as they do many of the challenges faced in athletics. If an athlete is hindered by asthma, good care of his asthma will generally fix the problem.

Conclusion

Asthma is one of the most common ailments in society today, and its effect is far-reaching in both human and financial terms. The good news is that for most asthmatics, therapy can result in an active normal life.

Some of the key points can be stated as below:

- Asthma is a chronic disease characterized by airway inflammation, bronchial hyperreactivity, and mucous plugging causing airflow obstruction and wheezing.
- While 85% of children with early wheezing do so transiently, the prevalence of asthma has risen for decades and only now seems to be leveling.
- The mainstay of asthma therapy is control, with ICSs as the first-line agents. If a patient is using a short-acting bronchodilator medication more than twice a week for their asthma symptoms, their disease is out of control and their management plan needs revision.
- In general, asthma should pose no limitation to normal activity with appropriate monitoring, therapy, and symptom awareness.
- Compliance with taking medications is the key to asthma control.
- If asthma is limiting a person's lifestyle they need to see their doctor.

See also: Allergies; Immune System and Immunodeficiency.

Suggested Readings

Hannaway PJ (2002) Asthma – An Emerging Epidemic. Marblehead, MA: Lighthouse Press.
Johnson CC, Ownby DR, Zoratti EM, et al. (2002) Environmental Epidemiology of Pediatric Asthma and Allergy. Epidemiologic Reviews 24: 154–175.
Leung DYM, Sampson HA, Geha RS, and Szefler SJ (2003) Pediatric Allergy Principles and Practice 2–3, 32–43: 10–38, 337–472.
Plaut TF and Jones TB (1999) Asthma Guide for People of All Ages. Amherst, MA: Pedipress.

Relevant Websites

http://www.aaaai.org – American Academy of Asthma, Allergy and Immunology.
http://www.aap.org – American Academy of Pediatrics.
http://www.aafa.org – Asthma and Allergy Foundation of America.
www.nhlbi.nih.gov – National Heart Lung and Blood Institute.
http://health.nih.gov – National Institutes of Health (NIH).

Attachment

G Posada, Purdue University, West Lafayette, IN, USA

Glossary

Attachment – An enduring emotional bond that an individual forms toward another, and that ties them together in time and across contexts.

Attachment internal working models – Mental representations about attachment relationships.

Attachment security – Confidence in the attachment figure's availability and responsiveness, and skillful use of an attachment figure as a secure base.

Secure base phenomenon – The apparently purposeful balance between exploring away from an attachment figure and going back to that figure.

Sensitivity – The ability to be aware of and interpret an infant/child's signals correctly, and respond to them promptly and appropriately.

Introduction

John Bowlby and Mary Ainsworth's theory of attachment is concerned with the development of human close relationships. Also, it is concerned with the role that those relationships play in individuals' development. Since its inception in 1958, attachment theory has become an influential theoretical framework to study social, emotional, and relationship development.

The point of Bowlby's departure was a series of observations about children's reaction to separation from, and loss of, important figures in their lives, and the effects that such experiences may have on personality development. Those observations pushed Bowlby to search for an explanation to such an emotional reaction. His initial theoretical account provided the basis for Ainsworth's ground-breaking studies in Uganda and Baltimore. These studies contributed to Bowlby's further elaboration of his developmental model, and to Ainsworth launching the study of individual differences in the development of infant–mother attachment relationships in infancy.

A central feature of the Bowlby–Ainsworth perspective is the role attributed to real-life experiences in relationships. According to Bowlby, it is in the context of interactions that infants develop preference for, organize their behavior around, and become attached to their principal caregivers. Yet, he did not consider the organization of the attachment behavioral system simply as the outcome of learning, but of the interactions between biases in infant abilities and continuous experience in infant–caregiver transactions. Furthermore, such experience provides the raw materials for children's and individuals' representations of attachment relationships later on, and for the maintenance or change of such representations. Interaction experience plays a key role in attachment theory and was a point of departure from traditional psychoanalytic theory at the time Bowlby proposed his account.

Being a clinician, some of Bowlby's core interests were to investigate the impact that attachment relationships have on the notion of the self, emotional security, intimate relationships, and social adaptation, and to explain the role that child–parent relationships play in personality disorders and maladaptation. Indeed, attachment theory has proved both generative, in terms of the number of articles, chapters, and books published, and extensive, in terms of the degree to which attachment is a central construct for several subdisciplines within psychology (e.g., developmental, social, and clinical/health).

This article focuses on attachment relationships during infancy and early childhood. First, it will briefly introduce John Bowlby and Mary Ainsworth and address the definition of central constructs in the theory. Then, it describes the evolutionary rationale Bowlby proposed and attends to normative and individual differences matters in the development and construction of attachment relationships during infancy. Subsequently, it presents and discusses the caregiver sensitivity–child security link. Finally, it covers the same topics during the preschool years. Methodological issues in the study of attachment relationships both in infancy and childhood are also discussed. However, it is important to note that this article does not pretend to be a review of the literature related to those topics, and that most of the literature reviewed is concerned with infant/child–mother attachment. The term 'mother' will be used in many instances instead of caregiver, as most children are still being raised predominantly by their mother. Also, occasionally when mentioning child–mother interactions, we refer to the child as 'he' to avoid confusion.

John Bowlby and Mary Ainsworth

John Bowlby was a British psychoanalyst who emphasized the context provided by relationships when studying individual development. Bowlby introduced attachment theory in 1958. During the following years, he elaborated it and published the now famous trilogy of books *Attachment, Separation,* and *Loss.* The theory he proposed offers (1) an integrative perspective that brings together notions from ethological, evolutionary, control systems, psychoanalytic, and cognitive psychology in an attempt to explain the construction and elaboration of attachment relationships; (2) a clear dialog between research and theory elaboration; and (3) a wide range of implications for central domains of people's lives (e.g., intimate relationships, self, and emotional security, among others).

Mary Ainsworth, born in Ohio and raised in Toronto, became familiar with Bowlby's ideas, when she worked with him in London during the 1950s. Subsequently, she went to Uganda where she conducted her initial study of 28 infants and their mothers, visiting them every 2 weeks during 9 months. Her study was the first to document attachment in the making. Based on her observations, Ainsworth proposed a normative developmental model. Further, she described individual differences as far as infant–mother relationships were concerned. Due to her interest and training on Blatz's security theory, and specifically on the notion that parents provide a secure foundation from which children feel the can safely initiate explorations of their world, Ainsworth construed those differences in terms of security. The Uganda study provided the basis for Ainsworth's landmark Baltimore study in which she aimed at systematically replicating her previous study.

Central Constructs in Attachment Theory and Research

Faced with the problem of accounting for children's reactions to separation from and/or loss of their mother, Bowlby elaborated a theoretical account that provided an initial explanation. In collaboration with James Robertson, he proposed that children's emotional reaction to separation and loss follows a sequence: protest, despair, and, if the separation is long lasting, detachment. This emotional reaction, he argued, is due to the nature of the relationship that children and their mothers develop. Based on studies about child–mother separation and infant–mother interactions, research on different primate species, and on control systems and psychoanalytic theories, Bowlby suggested a normative developmental model (see below) that starts with infant–mother exchanges that interact with an infant's learning biases, grows into an infant's discrimination and preference for those who provide care, and then, consolidates such preference during the second half of the first year, when an infant's behavior becomes clearly organized into a system with the goal of proximity to one or more caregivers. Behavioral organization later develops and expands into a representational

modus operandi that impacts the child–mother relationship and transforms it into a goal-corrected partnership.

Attachment

Bowlby defined attachment as an emotional tie that an infant constructs and elaborates with his principal caregiver(s) in the context of everyday interactions. This bond is specific in that it is directed toward a particular individual, is long lasting, and ties the two individuals together across contexts. Attachment is distinct from attachment behavior; while attachment refers to the emotional bond, and to a strong predisposition to seek proximity to, and contact with, a specific caregiver, attachment behavior is concerned with the conduct manifestations that allow the child to achieve the desired proximity and contact. Importantly, attachment is conceptually different from dependency. While the latter has negative connotations (e.g., clinginess), attachment does not. On the contrary, attachment enables children to explore and learn.

Secure-Base Behavior

Mary Ainsworth contributed the notion of the secure-base phenomenon. This construct was empirically derived from field observations conducted in rural Uganda and it captures well Bowlby's notion that while attachment ties the child to his mother, it also allows him to explore his surroundings. In other words, the secure-base phenomenon refers to the apparent purposeful balance between proximity seeking and exploration away from an attachment figure at different times and across contexts. How a child uses his mother as a secure base, that is, the organization of a child's secure-base behavior is the basis for inferring how secure he feels when exploring his surroundings and retreating to her, and whether a child demonstrates confidence in mother's availability and response.

Internal Working Models of Attachment

John Bowlby recognized preschool children and suggested that infants create a primitive cognitive map, that is, a representation of their relationship with mother. Initially, those maps are about mother moving in a specific spatial and temporal layout. These representations are based on interactions and are expected to be elaborated through relationship experience and cognitive development. Bowlby called those representations internal working models.

Quality of Care

Maternal behavior plays a central role in attachment theory and research. Bowlby suggested that a caregiving system complementary to the attachment behavioral system is necessary for attachment relationships to develop. Further, based on her studies, Ainsworth concluded that it was not the quantity but the quality of care that mattered most in accounting for the different types of infant–mother relationships. Although she observed specific maternal behaviors during interactions with infants, Ainsworth conceptualized four categories of behavior to describe the overall features of maternal care: sensitivity–insensitivity, cooperation–interference, acceptance–rejection, and accessibility–ignoring. Because those categories of maternal behavior turned out to be highly intercorrelated, the overall quality of maternal care was subsumed under the label of sensitivity.

Sensitivity to an infant's signals and communications refers to a mother's ability to see things from the baby's perspective; that is, a mother's ability to perceive her baby's signals, interpret them correctly, and respond to them appropriately and promptly. Cooperation–interference refers to a mother's ability to respect her baby as a separate individual, to intervene in the baby's activities in a skillful and collaborative manner so that the baby does not experience it as interfering. Acceptance–rejection refers to the balance between a mother's positive and negative feelings about her baby, and the extent to which she is able to resolve those negative feelings. Accessibility–ignoring refers to a mother's ability to notice and attend to her baby's signals despite demands from other sources on her attention.

Attachment in Infancy

An Evolutionary Rationale

Bowlby proposed an account of how child–mother attachment relationships develop and are organized during the first 3 years of a child's life. He argued that all infants organize an attachment behavioral control system in the context of interactions with their mother during the course of the first year. That is to say that the system does not emerge automatically and it is not fully functional at birth. Its elements, behaviors such as crying, sucking, smiling, clinging, and following become integrated and coordinated into a system of behavior through learning, practice, and feedback from the mother. The system goal is that of proximity and contact with an attachment figure. The behavioral system becomes functional during the second half of the first year and it is clearly seen in operation during children's implementation of the secure-base game in everyday life circumstances and especially, not exclusively, in emergency situations.

To explain the existence of this particular behavioral system, Bowlby offered a rationale that included the survival advantages of proximity seeking and contact afforded during the course of human evolution. Specifically, infants

who sought and maintained some degree of proximity to, and/or contact with, their caregivers, he suggested, were more likely to survive than those who did not. To be sure, Bowlby proposed that by virtue of our primate heritage, human infants are endowed with learning biases and predispositions that make it possible, even likely, for them to put together an attachment behavioral system, as long as they are exposed to a caring environment. He used available evidence concerning offspring–mother relationships in each of the great apes species, chimpanzees, gorillas, and orangutans to support his reasoning. The biological function of attachment behavior is that of protection and, although it is most obvious during infancy and childhood, it can be observed throughout the lifecycle.

A testable hypothesis derived from such a rationale is that human infants in all cultures have the capacity to construct secure-base relationships in the context of interactions with caregivers. That is, all infants exposed to ordinary parental care will organize their behavior in interactions with caregivers in ways that resemble the secure-base phenomenon. Curiously, despite being at the core of Bowlby and Ainsworth's theoretical foundations, very little research work on the cross-cultural generality of secure-base behavior has been conducted. Empirical tests of this hypothesis are important to help resolve discussions and debates regarding the ubiquity of attachment relationships.

In the only explicit test of the cross-cultural generality of secure-base behavior, Germán Posada and colleagues in the mid-1990s gathered information about secure base behavior at home in infants and children from seven different countries representing a variety of cultures. Specifically, data were collected in China, Colombia, Germany, Israel, Japan, Norway, and the US. Using a methodology that does not presuppose the existence of the secure base phenomenon (i.e., the attachment q-set), descriptions of infants' and children's behavior during interactions with their mother at home indicated that secure base behavior was evident in all countries.

This study, in conjunction with other reports based on samples from one culture, provides support for the idea that secure base behavior characterizes infant and child behavior when interacting with its mother in naturalistic settings. Far from settling the issue, those studies open the door for further exploration of the relationships between hypothesized propensities and cultural and social contexts. In other words, even if secure base behavior turns out to be characteristic of children's behavioral repertoire in diverse cultural contexts, its specific forms and patterning may not be the same.

What has been selected for in evolution is better understood in terms of a propensity to organize a secure base behavioral system within the context of child–mother interactions. The study of specific manifestations or configurations of the secure base phenomenon is not trivial for it will help us determine its flexibility (range) and the impact of context (e.g., cultural) in the construction of attachment relationships. We also need to be aware that existing research has barely included developmental considerations as far as changes over time in secure base behavior are concerned. This issue is key when studying the cross-cultural generality hypothesis for specific cultural influences may be more limited during the first 2 years of life, when most research has taken place. That is, the impact of specific cultural influences on secure base behavior may be obvious once the acquisition of language and representation is in full motion and new channels of influence in child–parent relationships are open. Even if common across cultures, secure base relationships may be used and impacted differently depending on the specific cultural mandates of the group as the child grows older. Little, if anything at all, is known about this.

Normative Issues

All infants form an attachment to their main caregivers. Yet attachment does not appear at once. It takes time to be constructed and elaborated. Indeed, one of Bowlby's key insights was to suggest attachment as a relational outcome tied to interaction experience. To be clear, although Bowlby referred to attachment as the child's emotional bond toward his main caregiver(s), he discussed, at considerable length its dyadic nature, placing child–mother exchanges and the contributions of each member of the dyad at the center of the phenomenon; this is frequently ignored. He proposed four phases in the development of attachment: (1) orientation and signals with limited discrimination of figure; (2) orientation and signals toward one (or more) discriminated figure(s); (3) maintenance of proximity to a discriminated figure by means of locomotion as well as signals; and (4) formation of a goal-corrected partnership.

Orientation and signals with limited discrimination of figure

Although from very early on infants show discrimination of their mother's voice and scent, they do not exhibit a clear preference for any particular caregiver. Infants are not attached to their caregivers at birth. Any caregiver responding to their needs would be as effective; infants tend to respond similarly to any individual who tends to their signals or interacts with them. Yet, from the beginning, babies contribute to their interactions and exchanges with others. Built-in bias to orient toward, look at, and listen to certain stimuli will contribute to paying attention to and eventually developing preference for those who interact with him and provide care on a regular basis. The infant uses characteristic reflexive responses in his behavioral repertoire (e.g., crying, head-turning, reaching, grasping) when interacting with others. These behaviors

typically have as a consequence to increase the time the baby is in proximity with those around him.

Orientation and signals toward one (or more) discriminated figure(s)

In the course of everyday exchanges, the infant experiences patterns of interaction and care from those looking after him, typically but not necessarily the infant's biological parents. These repeated experiences allow the child to learn, through exposure, the perceptual and behavioral features of his caregivers, and discriminate them from other individuals. Repeated exposure to patterns of care leads to familiarity with those figures and their interaction routines. Familiarity leads to preference. The infant continues to be friendly and open, but now distinguishes his caregivers from others and responds differently to them. Thus, for example, differential smiling toward mother and differential crying (i.e., infant cries when held by someone other than mother, and stops crying when mother holds him) were reported by Ainsworth as early as 9–10 weeks of age.

During these first months, experience, that is, practice, establishes the secure base game foundations: the baby signals, mother comes close and joins the child in interaction either cooperating with his behavior and vocalizations, or easing the discomfort the infant is experiencing. The caregiver expands the child's activities in time and space, and/or restructures his behavior and context in ways that the infant is comfortable and/or can re-engage his surroundings. These experiences are likely to provide the behavioral and cognitive substrates for the rapid appearance of the secure base phenomenon soon after the child develops locomotion.

Maintenance of proximity to a discriminated figure by means of locomotion as well as signals

The third phase begins soon after locomotion arises. The infant's way of participating is now more complex. Behavior becomes organized in a goal-corrected basis. Not only does the infant orient and signal toward his main caregiver(s), but in addition, he is active in approaching and maintaining proximity by crawling and/or walking. Now he has new means; his motor behavior is increasingly more integrated and efficient and thus he can use it to achieve his goals for proximity and contact when the situation requires it (e.g., a visitor, a loud noise, a new place). Strangers are treated with caution and attachment figures are clearly preferred. The foundations of the secure base phenomenon are in place and readily observable by 1 year of age in most children. This does not imply that its development is complete. Although established, it needs to be consolidated. In sum, the process of attachment development thus far can be succinctly stated as going from interaction to familiarity and preference to attachment.

Formation of a goal-corrected partnership

This is the last phase in Bowlby's developmental model. By continuously participating in interactions with its mother, and observing maternal behavior and the factors that influence it, the child begins to conceive of the mother as an individual with her own set goals and plans to achieve them. Thus, according to Bowlby, at around a child's third year, a partnership begins where the child will increasingly modify his behavior and expectations based on those of the attachment figure. There the rudiments of a goal-corrected partnership, one that will eventually move the child–mother relationship increasingly and slowly toward a more symmetrical one. Bowlby was general about this phase and did not elaborate much.

Individual Differences in Attachment Relationships

Although all infants who are cared for form attachments to their caregivers, not all attachment relationships are the same. They vary in their quality. Ainsworth pioneered the study of individual differences in the quality of attachment relationships. Her studies established and shaped the field. Her findings have subsequently generated hundreds of studies. Investigating the effects of weaning on Ugandan infants, Ainsworth both sketched the normative development of attachment during the first year or so of life, and set the foundations for her influential Baltimore longitudinal study of 23 infant–mother dyads during the first year of life.

In Baltimore, she conducted careful, frequent (every 3 weeks from week 3 until the infant was 54 weeks of age), and extensive (4 h) observations of child–mother interactions at home. When infants were 12 months old, Ainsworth brought infant–mother dyads to the laboratory and conducted the now well-known strange situation procedure.

This laboratory procedure consists of eight episodes. Each episode lasts a maximum of 3 min except for the first episode that lasts 30 s. In episode 1, infant and mother are brought into a room with toys and two chairs by a research associate. Episode 2 begins when the associate leaves; mother and infant stay in the room and the infant plays with the toys; mother is instructed to be responsive if infant initiates interaction. In episode 3, a stranger walks in; initially, she remains silent for 1 min, then, she talks to mother, and after 1 min, she talks and/or plays with the infant during another minute. In episode 4, the first separation occurs; mother leaves and the infant remains with the stranger; if the infant is too upset and the stranger cannot soothe him, the episode is cut short. In episode 5, the first reunion episode mother comes back in the room and stranger leaves unobtrusively. During episode 6, the second separation episode, mother leaves and the infant stays alone; again if the infant is too upset the episode is cut short. In episode 7, the stranger returns; if the infant

continues to be upset, the episode is cut short and mother comes back into the room. Finally, in episode 8, the second reunion episode, mother returns and stranger leaves.

After careful examination of infants' behavior during the entire strange situation procedure, but especially during the reunion episodes, Ainsworth came up with three major groups of infants based on the configuration or patterning of behavior exhibited. Ainsworth construed the differences among the groups in terms of security when using mother as a secure base. Two of these groups, 'A' and 'C', are considered as consisting of infants in anxious/insecure attachment relationships with their mothers; the other group, 'B', of infants in secure relationships.

Specifically, secure infants (group B) are able to use their mother as a secure base for exploration in the novel room. If distressed during separation, they seek proximity and contact with mother during reunion, and contact is effective in promptly reducing stress. If not openly distressed by separation, the baby responds to mother with active greeting and interaction during reunion episodes. There is little or no tendency to avoid or to resist and be angry with mother upon reunion. As just mentioned, the infant may or may not be distressed during the separation episodes, but when he is, it is clear that he wants his mother, even though he may be somewhat consoled by the stranger. Although secure infants tend to be affiliative with the stranger in mother's presence, they are clearly more interested in contact and interaction with their mother than with the stranger.

Infants in group A are considered to be anxiously avoidant. They exhibit little affective sharing with mother and readily separate to explore toys. They treat the stranger much as they treat their mothers, and are affiliative with the stranger in mother's absence; they show little preference for mother. 'A' infants show active avoidance of proximity to, and contact and interaction with, their mother in reunion episodes. The babies look, turn, or move away, and ignores their mother when she returns. Alternatively, they greets her casually. If there is approach, the infants mix their welcoming with avoidance. If the babies are picked up by mother, there is little or no tendency to cling or resist being put down. During separation episodes, the babies are typically not distressed; but if there is distress, they seems to be due to having been left alone for they tends to be alleviated when the stranger returns; there is little or no stranger avoidance.

Infants in group C are labeled anxious resistant. They exhibit poverty of exploration even in preseparation episodes; they seem wary of novel situations and of the stranger. These infants are likely to be very distressed upon separation and are not easily calmed by the stranger. Upon reunion, babies in group C are not easily calmed by mother's return. They may show proximity seeking and contact mixed with resistance (hitting, squirming, or rejecting toys); alternatively, they may continue to cry and fuss, and show extreme passivity. Babies in this group show no or little tendency to ignore their mother during the reunion episodes. Because upon reunion 'C' infants are likely to seek proximity and contact, but these are not effective in calming them down as shown by their resistance and inability to be soothed, they have also been labeled anxious ambivalent. Ainsworth considered that these infants may show general maladaptive behavior in the strange situation, because they tend to be angrier than infants in the other groups.

Mary Main and Judith Solomon proposed in 1986 a fourth classification group 'D'. Infants in this group often cannot maintain a clear and coherent strategy in the organization of their attachment behavior. Because of this, infants in this group are labeled 'disorganized/disoriented' and are considered to be anxiously attached. This classification is assigned in addition to an alternate best-fitting category of A, B, or C. Infants classified into this group exhibit patterns of behavior that lack a readily observable goal, purpose, or explanation. The most characteristic theme in the list of behaviors is that of disorganization or an observed contradiction in movement pattern. A lack of orientation to the immediate environment is also characteristic of these children. Indices of disorganization and disorientation are: sequential and/or simultaneous display of contradictory behavior patterns, undirected, misdirected, incomplete, and interrupted movements and expressions; stereotypies, asymmetrical movements, mistimed movements, and anomalous postures; and freezing, stilling, and slowed movements and expressions, direct indices of apprehension regarding the parent, and of disorganization and disorientation.

The importance of that classification system and the resulting groups is determined by the associations reported between infants' organization of behavior both at home and in the strange situation. In other words, it is the patterning of behavior, not individual behaviors that were found significantly related in both contexts. The validity of the strange situation attachment classification system rests on the demonstration of the association between patterns in the organization of secure base behavior at home and in the laboratory. Importantly, discrete behaviors were not found to be associated in both contexts. For example, in the case of crying, while securely attached infants were found to cry the least at home, they may or may not have cried in the strange situation. In contrast, anxiously attached babies who cried the least (avoidant) or a lot (resistant) in the strange situation were the infants who cried the most at home during both the first and fourth quarter of their first year, and were not distinguishable from each other. Thus, Ainsworth assigned meaning to an infant's behavior in the strange situation based on her findings about different 'patterns' of interaction she had observed at home.

Individual Differences in Attachment Security and Quality of Care

While studying the development of child–mother relationships, Ainsworth also gathered information about maternal behavior that led her to formulate the hypothesis that the quality of care is important when studying individual differences in infants' attachment security. Findings from her Baltimore study indicated that many aspects of caregiving behavior were significantly related to infant's quality of attachment at 12 months. Ainsworth collected detailed information on specific maternal behavior such as responsiveness to infant crying, behavior relevant to separation and reunion, close bodily contact, face-to-face interactions, feeding, and behavior relevant to child obedience (e.g., frequency of physical interventions). In addition, she rated mothers on the four broader categories of behavior described above. Each category was found to be highly and significantly related to attachment security in the strange situation.

Ainsworth's identification of sensitivity to signals, cooperation with ongoing behavior, acceptance, and accessibility as important dimensions of infant care has provided a valuable framework for empirical research on this issue. Her model of early care has served as the theoretical foundation for empirical studies investigating the factors that account for individual differences in infants' organization of secure base behavior. Most ensuing research on the topic has, however, observed maternal behavior and scored maternal sensitivity in contrived situations, once, and for periods usually lasting under 1 hour. Overall, results indicate that ratings of sensitivity are significantly, if modestly, related to attachment security in middle-class samples.

A meta-analysis of 65 studies conducted by Marianne de Wolff and Marinus van IJzendoorn published in 1997 reported a correlation coefficient of 0.24 for studies that have investigated the relation between sensitivity and security when assessing the constructs in similar ways to Ainsworth's. Those findings are remarkable especially in consideration of the fact that most studies, subsequent to Ainsworth's, have drastically reduced the window of observation time and situations, and thus, perhaps, the representativeness of the phenomena being observed. Low correlations sometimes reflect measurement problems rather than weak effects. Ainsworth's many hours of naturalistic observations throughout the first year afforded a better assessment of maternal sensitivity and child behavior than the less extensive, structured observations, and narrowly focused measures typical of most subsequent studies. Indeed, the results of more recent studies (conducted by David Pederson, Gregory Moran and colleagues, and Germán Posada and colleagues) that involved observations and measures more akin to Ainsworth's have yielded comparable results (i.e., correlation coefficients of 0.40–0.61). Clearly, the issue of effect size in research on maternal care and infant security requires further study with special attention to construct definitions, observational strategies, sampling of behavior, and measurement issues. In the meantime, even small correlations should not be dismissed out of hand in contexts where they can be projected through large numbers of events or interactions to produce important effects.

Assessment of Attachment and Caregiving Behavioral Organization

Information on infant attachment behavior has mainly been collected by means of the strange situation procedure. More recently, Q methodology has been used to describe observations conducted at homes, playgrounds, and hospitals. A third approach consists of observation in naturalistic settings such as Ainsworth did; this alternative, however, has scarcely been employed, presumably because of the extremely high costs of personnel, time, and effort.

The strange situation

Ainsworth's studies have also been foundational because of the methodological innovations offered. The strange situation procedure (see above) was pivotal in launching a generation of studies on individual differences in attachment security, their antecedents, and correlates. The success of the strange situation was such that research with this procedure became synonymous with attachment research up until the mid-1980s. Paradoxically, it could be argued that the success of studying attachment with the strange situation limited the study of infant/child–mother relationships. Thus, observational studies, central to launching the study of attachment relationships, were set aside likely due to their time- and resource-consuming features compared to studies using the strange situation. One of the consequences of this emphasis on the strange situation has been a lack of research that provides a descriptive base to account for the development of attachment relationships after 12–18 months of age. In fact, the most common strategy followed when studying attachment relationships after infancy has been that of using the strange situation to validate assessment tools for older children, with very little empirical work to study the phenomenon in the contexts where it develops (i.e., naturalistic settings), and using secure base behavior in real life circumstances as the key validation criterion.

The attachment Q-set

The attachment Q-set (AQS) was proposed by Everett Waters in the mid-1980s as an economic alternative to Ainsworth's observational methodology, one that allows for efficient descriptions of secure base behavior in infants and children, and affords more analytical alternatives. The AQS allows researchers to describe the functioning of the attachment behavior control system. In brief, it assesses how characteristic secure base behavior is during infant/child–caregiver interactions. The AQS can be used with children aged between 1 and 5 years.

It consists of 90 descriptive items; most of them refer to behavior in context, and tap the secure base phenomenon; there are some filler items. After observing an infant interacting with his mother, researchers use the items to describe child behavior by placing the items into nine piles from most characteristic to least characteristic of the infant/child. To determine whether an infant's behavior is organized in ways similar to the secure base phenomenon, his q-description is correlated with a theoretical description that indicates optimal use of the caregiver as a secure base. The higher the correlation the more an infant's behavior is organized in ways that resemble the secure base phenomenon.

The validity of the AQS has been documented in various studies. More recently in 2004, Marinus van IJzendoorn and colleagues presented a meta-analysis of 65 q-sort studies that supports the AQS validity. The authors indicate that the AQS, as used by trained observers, is a valid tool to assess the organization of attachment behavior. They further suggest caution when using AQS data provided by mothers. Care is need here, however, for the key issue, as with any other methodology, is that of training those who will report on 'what' and 'how' they are to do it. This includes mothers. For example, Douglas Teti in 1996 showed that mothers, like any other observer, can provide valid and useful information when trained properly.

Ainsworth's maternal sensitivity scales

Most research on the associations between caregiving and attachment security has been based on Ainsworth's construct of sensitivity. Of course, the degree of similarity to her assessment approach has varied from study to study with some investigators using a conceptualization close to Ainsworth's and some others using a notion removed from the one she offered (e.g., maternal self-efficacy as defined by her attribution style and mood state). This is so, in part, because Ainsworth's evaluation of maternal care was based upon many hours of naturalistic observation during the first year, and no abbreviated procedure has been devised. The use of the 9-point scales proposed by Ainsworth requires a good deal of insight into a mother's behavior, and this is dependent on a good observational base. In short, appropriate use of the scales is based on familiarization with, and knowledge of, a mother's behavior. Here it is important to note that no study has sampled maternal behavior as extensively as Ainsworth did.

The maternal behavior Q-set

Much like the AQS, the maternal behavior Q-set (MBQS), designed by David Pederson and Gregory Moran, was proposed as an economical descriptive alternative to Ainsworth's caregiving scales. It allows researchers to describe the quality of maternal care and its support of an infant's attachment behavioral system. Although the MBQS has mainly been used with infants, it also has been used with 31-year-olds. It consists of 90 items about maternal behavior based on Ainsworth's conceptualization. One of its great advantages is the operationalization of Ainsworth's sensitivity construct.

After observing a mother interacting with her infant, researchers use the items to describe maternal behavior by placing the items into nine piles from most characteristic to least characteristic of the infant/child. To determine whether a mother's behavior is organized in ways similar to the hypothetically sensitive mother, her q-description is correlated with a theoretical description that indicates optimal sensitive caregiving. The higher the correlation the more a mother's behavior is organized in ways that resemble a sensitive mother. The validity of the MBQS has been built in a series of studies by Pederson, Moran, and colleagues during the 1990s.

Attachment in Early Childhood

A look at the empirical work on child–mother attachment relationships reveals a strong initial emphasis on research with infants and, subsequently, for the past 20 years or so, on adult mental representations with not enough consideration of the development and elaboration of attachment relationships during childhood. Bowlby suggested that the child–mother relationship enters a new phase at around 3 years of age (see above). Yet, he did not expand much on the development of those relationships during the preschool years. Although he conceptualized attachment as a lifespan phenomenon, Bowlby was explicit about the period of time for which he had available data. There was not much information available about what transpires after 2 years of age.

Even today, information about what takes place in the child–mother relationship after infancy is scant. Robert Marvin and colleagues' reports stand as some of the few descriptive data on Bowlby's proposed fourth phase. Marvin's data showed associations between security and cooperation and self-control tasks, when mother is busy, and between security and tolerance of separation. This scarcity is likely to reflect the difficulties associated with assessing the functioning of control systems. Defining and describing patterns of behavior is a complex task. Assessing control systems requires focusing on the efficiency and success with which the system maintains itself within its set goals, while taking into account their developmental transformations. Thus, performance assessment of a control system cannot be equated with the quantity of behavioral output. A child who cries more over separation from his mother than another cannot be said to be more attached to her.

In an attempt to close the existing gap, Everett Waters and colleagues in the early 1990s suggested a normative developmental sequence for attachment relationships after the first year of life that takes into account Bowlby's work and elaborates it. Thus, after the appearance of the secure base phenomenon by age 1 year, a consolidation phase is proposed. That is, the appearance of secure base behavior toward the end of the first year is just the beginning of a longer period that expands early childhood during which the secure base game is put together and firmed up. Its temporal and spatial parameters will get solidified, expanded, and reworked as practice, and new representational abilities and circumstances allow.

Here it is important to remember that the organization of an infant's behavior is closely tied to that of the attachment figure. That is, secure base support is essential in maintaining the organization of infant's secure base behavior. In all likelihood, it takes years for the phenomenon to be consolidated and elaborated; but this has not been studied. There is no reason to expect that a phenomenon, as complex as using mother as a secure base in consideration of information about her, the infant's own state, and contextual characteristics at the moment, would appear fully developed and in final form at 12 months. It is likely that the secure base phenomenon, similar to other developmental phenomena, takes time to be organized and established.

Likely mechanisms involved in the consolidation of the phenomenon include practice (dyadic practice), operant learning, improved locomotion, and the development of the semiotic function (i.e., the increased ability to use representational systems like language and representational thought). Also, children's increased knowledge of diverse environments, and increased familiarity and confidence on their own skills are likely to play a significant role on their ability to use important figures as a secure base. In addition, it is probable that practice in different secure base relationships, each of them unique, allows the child to learn variations on the issue at hand. Nonetheless, input from the attachment figure(s) is key in maintaining behavioral organization.

But what does the empirical research literature on attachment relationships during early childhood indicate? Most studies have focused on the investigation of individual differences in security and used as a point of reference the infant strange situation classification groups. Modifications to the strange situation procedure and/or the coding system have been introduced. Fewer studies have undertaken the examination of individual differences in naturalistic settings and have related them to the quality of maternal care.

Interactions and Continuous Secure Base Support

Attachment relationships are a dyadic phenomenon. Thus, exchanges between child and mother (or father) continue to be important and likely feed the child–mother relationship. Maternal secure base support remains relevant as far as security outcomes and the ability to use the mother as a secure base are concerned. Indeed, various researchers have hypothesized that the child–mother relationship continues to be shaped throughout childhood, and the quality of caregiving has been suggested as central to the maintenance and construction of such relationships. Surprisingly, the topic has remained relatively unexplored.

Empirically demonstrating that the concurrent quality of care (e.g., secure base support) is related to children's secure base behavior and security in early childhood will substantiate claims about attachment relationships as relatively open systems that continue to be elaborated. Also, it would contribute to attributing due weight to interaction experiences after infancy as an important factor in determining attachment relationship outcomes. For years, Alan Sroufe has eloquently argued that although experiences during infancy are important and influential in development, they do not by themselves determine later outcomes.

Few studies have assessed the relations between secure base behavior and caregiving during early childhood. Specifically, maternal interactive behavior observed during relatively short intervals at home and in laboratory settings has been shown to be associated with children's organization of attachment behavior in the expected direction. Overall, these studies provide summary indices (e.g., sensitive responding) of maternal behavior observed in semi-structured situations. These findings suggest that concurrent sensitive caregiving is an important influence on children's attachment security during early childhood.

Further research is needed to broaden our understanding of the association between quality of care (e.g., sensitivity) and security. Is sensitivity during the first year the same as sensitivity at 3–5 years? Is it useful to refer to a comprehensive construct in all instances, or are we to gain by focusing on parental domains directly related to the phenomenon under consideration (e.g., the secure base phenomenon)? The specification of developmentally appropriate caregiving domains beyond the global notion of sensitivity may help us understand the associations hypothesized better and assess the constructs in natural contexts where the mother–child relationship is formed.

In that line, German Posada and colleagues have recently found, in two separate studies, that the overall quality of concurrent maternal caregiving behavior was significantly related to preschool children's organization of secure base behavior in naturalistic settings (home and park). These findings are consistent with Ainsworth's assertion that the underlying characteristic of maternal behavior associated with child security is the ability to establish an atmosphere of harmony and cooperation in interactions with the child. In both studies, the more mothers' behavior contributed to smooth child–mother transactions, the more secure their children were. In other words, attachment security is directly tied to what transpires in child–mother interactions.

Maternal behavior was clearly interlocked with that of the child, and children's contributions to establishing harmonious child–mother interactions (e.g., actively participating in activities with their mothers, exploring away and returning to mother, and following mothers' suggestions and agreements established) were as important. That is, child behavior is significant in facilitating the different caregiving tasks. Longitudinal studies that investigate child–mother interactions in childhood will provide much needed evidence about the increasingly important role played by the child in the construction and maintenance of the relationship. This would validate Bowlby's notion of a child–mother goal-corrected partnership.

Importantly, specific age-relevant aspects of maternal caregiving behavior such as maternal secure base support and supervision were investigated in an effort to understand the sensitivity–security link by identifying important domains of care that may impact a child's secure base behavior organization. Findings indicated that the quality of maternal behavior in those domains was significantly associated with the child's use of the mother as a secure base. Providing a secure base by supporting a child's increasing departures away from the caregiver and by enhancing a child's experiences in his surroundings, as well as providing a haven of safety by facilitating a child's returns, and being readily responsive in stressful situations, are important to foster a child's sense of security. In those daily exchanges, children construct and maintain their trust on their caregivers as a secure base.

Results also indicated that keeping track of the child's whereabouts and activities, and anticipating problematic situations (e.g., being attentive to intervene when necessary), were salient in child–mother interactions. How mothers went about these issues (e.g., whether mothers kept track of their child's whereabouts and were balanced in their role as supervisors and participants in their activities) was significantly related to attachment security.

In sum, these findings point to the need to elaborate and be explicit about our notions of sensitivity and secure base behavior in child–mother relationships. This will serve attachment researchers well and help establish connections with other relevant domains of child development as child–parent relationships are concerned. Also, they highlight the importance of conducting observations of child–mother interactions in different naturalistic contexts and for relatively long periods of time.

Representational Issues

The increasing use of representation, evident after infancy, presents a significant challenge to attachment researchers. Clearly, there exists an empirical gap between the biologically grounded behavioral control system biasing offspring toward proximity to caregivers, and mental representations of relationships serving as sources of security in the caregiver's absence and biasing the child's behaviors and expectations about interactions and relationships with others. Bowlby recognized this and occasionally invoked the well-established mechanisms of learning (e.g., habituation, observational learning with guided participation) and somewhat less well-established notions, at the time, from cognitive psychology such as episodic and autobiographical memory and even the working model concept itself.

Bowlby believed that mental representations of secure base relationships were derived and abstracted from experience; that is, they were learned and co-constructed in the context of dyadic interactions and with the support of attachment figures. He also hypothesized that those representations, although stable, were open to revisions. In this sense, attachment representational models are a product of socialization processes.

The study of the development of children's attachment-related representations and their dyadic co-construction process during early childhood is central to our understanding of child–caregiver secure base relationships transition from a sensory-motor to a representational mode, and of the individual's contributions to his own development. This has been recognized by attachment scholars. However, despite the increasingly frequent use of concepts such as attachment representations and internal working models, after Mary Main and collaborators' pioneering paper in 1985, relatively little is known about what those representations consist of, and what their development is the development of. The issues have remained elusive as far as advancing our understanding about the organization (i.e., structure) of such budding attachment representations.

To be clear, there is an extensive literature examining various aspects of relationship verbal representations of children, adolescents, and adults and the range of correlates these representations may have. The availability of the adult attachment interview (AAI) has been central to this research endeavor in that it has inspired and allowed researchers to assess the current state of mind of attachment related representations via narratives individuals produce. Further, researchers concerned with the role that maternal representations play during child–mother relationships have demonstrated a significant association between maternal security as per the AAI and infant attachment classification in the strange situation.

Relatively recent conceptual and methodological innovations, by Harriet Waters, which integrate cognitive psychological (e.g., scripts) and attachment (e.g., the secure base phenomenon), related concepts have made assessments of the organization of secure base knowledge in adults and children readily accessible and have opened a window to the study of 'how' secure base knowledge is mentally organized and represented. This is an important area of inquiry that may allow researchers to begin specifying the structure of attachment representations and tie those representations to specific aspects of experience.

Based on Inge Bretherton's work and on social-cognitive developmental theory proposed by Katherine Nelson in the 1980s and early 1990s, Waters and associates have suggested that attachment representations could be understood as scripts (temporal–causal sequence) about secure base relationships. The secure base script has an internal order of events: child is engaged in the environment or the attached dyad is interacting in a warm manner, an obstacle or conflict is introduced (such as loss of a desired toy or a minor injury to the attached person), assistance is requested and is offered by the caregiving member of the dyad, the help is successful in resolving the conflict, and the child is able to go back to activity or the dyad is able to return to productive interaction. That order leads to a typical story.

Waters rated preschool children's narratives in response to attachment-related events in terms of secure base scriptedness when children were 37 and 54 months old, and found them significantly related to their attachment security as derived from child behavior at home. The association between the temporal–causal structure of preschoolers' attachment narratives and their secure base behavior organization at home and playgrounds has been replicated recently.

Waters and associates have also shown that this script is present in adult stories from many different sociocultural groups, including North and South America, Europe, the Mediterranean countries, the Middle East, and South Africa. Studies in the US suggest that maternal secure base scriptedness is positively and significantly associated with infants' strange situation classifications and with coherence scores from the adult attachment interview. Further, Brian Vaughn in 2006 reported that script scores derived from narratives provided by mothers in Colombia, Portugal, and the US were associated positively and significantly with AQS security scores for 1–3.5-year-old children.

The development of those procedures for children and adults is important because they allow us to address questions about the basic organization of secure base knowledge; the relations between caregivers' organization of secure base knowledge, the quality of their care, and children's organization of secure base behavior; and the cross-domain relations between secure base behavior and secure base representation at different points in time and contexts. Child–mother co-constructive processes that involve both behavioral as well as cognitive-verbal input are likely to be at play.

Concluding Comments

Attachment theory was proposed by John Bowlby to account for the consequences of maternal separation and loss on an individual's personality development. Based on empirical data available at the time, he suggested a model of normative development. Mary Ainsworth, who was exposed to Bowlby's ideas, set the basis for the study of individual differences in infant–mother attachment relationships by conducting the now famous Uganda and Baltimore studies. The conceptual and methodological foundations that Bowlby and Ainsworth crafted are still essential to understand current research and theoretical explorations.

Bowlby's prospective approach and theoretical elaboration led to an initial emphasis on research about attachment relationships during infancy. Ainsworth's methodological contributions (e.g., the strange situation) opened the gate for a substantial number of studies with infants aged 12–18 months old. More recently, attachment researchers have begun to push the theory beyond infancy into childhood. Explicit attention to both behavioral and representational issues becomes relevant during the preschool years, as does the development of conceptually and age-appropriate methodological tools to expand the study of attachment relationships.

See also: Divorce; Emotion Regulation; Parenting Styles and their Effects; Self-Regulatory Processes; Separation and Stranger Anxiety; Social-Emotional Development Assessment; Social and Emotional Development Theories; Social Interaction; Socialization in Infancy and Childhood; Temperament.

Suggested Readings

Ainsworth MDS, Blehar MC, Waters E, and Wall S (1978) *Patterns of Attachment.* Hillsdale, NJ: Erlbaum.

Bowlby J (1969/1982) *Attachment and Loss, Vol. I: Attachment.* New York: Basic Books.

Bowlby J (1988) *A Secure-Base.* New York: Basic Books.

Bretherton I and Waters E (1985) Growing points of attachment theory and research. *Monographs of the Society for Research in Child Development* 50 (1-2, Serial No. 209).

Cassidy J and Shaver PR (1999) *Handbook of Attachment: Theory, Research, and Clinical Applications.* New York: Guilford Press.

Greenberg MT, Cicchetti D, and Cummings EM (1990) *Attachment in the Preschool Years.* Chicago: University of Chicago Press.

Sroufe LA (2002) From infant attachment to promotion of adolescent autonomy: Prospective, longitudinal data on the role of parents in development. In: Borkowski JG, Ramey SL, and Bristol-Power M (eds.) *Parenting and the Child's World*, pp. 187–202. Mahwah, NJ: Lawrence Erlbaum Associates.

Thompson RA (1998) Early sociopersonality development. In: Damon W and Einsberg N (eds.) *Handbook of Child Psychology Vol. 3: Social, Emotional and Personality Development,* 5th edn., pp. 25–104. New York: Wiley.

Waters E, Kondo-Ikemura K, Posada G, and Richters J (1991) Learning to love: Milestones and mechanisms. In: Gunnar MR and Sroufe LA (eds.) *Self Processes and Development. The Minnesota Symposia on Child Psychology,* vol. 23, pp. 217–255. Hillsdale, NJ: Lawrence Erlbaum Associates.

Waters E, Vaughn BE, Posada G, and Kondo-Ikemura K (1995) Caregiving, cultural, and cognitive perspectives on secure-base behavior and working models: New growing points of attachment theory and research. *Monographs of the Society for Research in Child Development* 60 (2-3, Serial No. 244).

Relevant Websites

http://www.johnbowlby.com – Attachment Theory and Research at Stonybook.

http://www.richardatkins.co.uk – The attachment theory website.

Attention

M L Courage, Memorial University, St. John's, NL, Canada
J E Richards, University of South Carolina, Columbia, SC, USA

Glossary

Attention – The selective enhancement of some behavior at the expense of other behavior. Several different types of attention exist, including stimulus orienting, sustained attention, and executive attention.

Cognitive neuroscience – Cognition is the study of mental processes, such as attention, learning, discrimination, memory, and decision making. Cognitive neuroscience is a field of research that studies brain areas (neuroscience) that control cognitive processes.

Event-related potential – The electroencephalogram (EEG) can be linked to specific experimental or internal (cognitive) events, and EEG responses linked to such events are labeled event-related potentials (ERPs). ERPs may provide a direct and noninvasive measure of brain functioning controlling cognitive processes. One such ERP, the Nc (negative central) has been shown to index orienting of attention to novel stimuli and may be generated by the anterior cingulate of the brain.

Executive attention – An executive function ability to allocate attention in a way that is consistent with self-established goals and plans. Executive attention show the most extended developmental changes, beginning in the late phases of infancy (12 months) and showing changes throughout early childhood and into the first years of adolescence.

Habituation – Habituation refers to a cognitive process in which the initial stimulus orienting to a stimulus is diminished with repeated presentation. This process has been used profitably in the study of infant attention to show which objects elicit stimulus orienting and how infants learn about the objects through repeated presentation.

Heart rate-defined attention phases – Infants heart rate changes during attention. Heart rate can be used to define several types of attention, including stimulus orienting, sustained attention, and inattention.

Neurotransmitters – Chemicals that are used in the brain for transmitting information in neurons and synapses. The noradrenergic and cholinergic neurotransmitters are controlled by an extended network of neural connections and are important in the control of arousal and attention.

Paired-comparison procedure – A procedure in which two visual stimuli are show side by side, and the infant's preference for one or the other stimulus is measured by the amount of time the infant looks at one stimulus. Infants will show a preference to a novel stimulus, so that novelty preference is an index of the familiarity with the nonpreferred familiar stimulus.

Stimulus orienting – A form of attention in which sensory and perceptual systems are excited and often involves moving sensors (ears, eyes) toward environmental events. This type of attention begins development very early in infancy and is fully developed by 6 months of age.

Sustained attention – An extended engagement of a behavior system that enhances social and cognitive processes. This is very similar to arousal and is controlled by brain systems involved in behavioral state and arousal. Sustained attention shows dramatic developmental changes from 3 to 18 months of age.

Introduction

The study of attention has a long history in both cognitive and brain sciences. One of the founders of experimental psychology in the US, William James, wrote in his 1890 text *The Principles of Psychology* that

> Everyone knows what attention is. It is the taking possession by the mind in clear and vivid form, of one out of what seem several simultaneously possible objects or trains of thought. James (1980, p. 403)

This statement reflects a certain commonsense view that attention is a ubiquitous psychological process whereby we can select certain elements of the environment for scrutiny and ignore others. It also implies that the selected elements can be external, such as a particular sound or image, or internal, as when one is 'deep in thought'. James went on to discuss the significance of related processes such as alerting, focus, shifting, and distractibility – all key components of the modern study of attention. Notably absent from James's writing was any reference to the underlying neural mechanisms and processes that might mediate these components of attention. Currently, there is a substantial research literature on these underlying substrates and their role in attention. This research has confirmed that attention is not a single process, but rather a complex and multidimensional one. It performs a variety of functions including the self-regulation of emotion and cognition, is tuned to many sources of information in the environment, and depends on a variety of separate neural systems in the brain. Moreover, attention can be shared, as in joint attention with others or divided as when one talks on a cell telephone while driving a car.

Developmental research on attention has a recent history. In its formative years the study of attention was of secondary interest in research where processes such as perception, discrimination, learning, strategy acquisition, planning, or working memory were of primary concern. The study of attention is currently important in its own right as cognitive developmental research has become increasingly integrated with research in cognitive neuroscience. Significant practical issues have also contributed to the growing interest in the development of attention. For example, individual differences in attention observed during infancy are predictive of achievements in language, cognition, and play later in childhood. Moreover, deficits and anomalies in attention processes contribute to or are symptomatic of the learning and behavior difficulties experienced by children with conditions such as attention-deficit hyperactivity disorder (ADHD), fetal alcohol spectrum disorder (FASD), autism, and early exposure to teratogenic agents and environmental contaminants.

This developmental research indicates that attention changes dramatically over the period of infancy and early childhood. Infants who are less than 3 months of age have very poor vision and attend primarily to salient physical characteristics of their environments. Behavioral state control is limited during this period of time, such that very young infants attend with nonspecific orienting only during their limited periods of alertness. Arousal and visuo-spatial orienting are key emergent aspects of attention in this timeframe. Between about 3 and 18 months of age the development of alert, vigilant, sustained attention occurs. Sustained attention allows the focusing of processing resources and thus during this age infants begin to engage in active processing. This alertness not only affects cognitive processes such as remembering, perceiving, and discrimination, but also allows the infant to engage in social interactions and maintain attentive states for social–emotional drives. Beginning at about 12 months of age, an internal executive system of control begins to operate. This system is used to voluntarily guide the deployment of attention. This executive attention system has an extended developmental time course, showing changes throughout early childhood and into the first adolescent years. Many of the early changes in attention are based largely on age-related changes in the brain, whereas later advances are impacted by the child's experience and specific training activities.

The present article has three objectives. First, behavioral changes in the development of visuo-spatial orienting, sustained attention, and endogenous or executive attention in infancy and early childhood will be described. The major brain structures that are involved in the development of attention will be identified. These brain systems include a general arousal system that affects many cognitive functions as well as specific attention systems that are more limited in their effects on cognition and attention. Second, psychophysiological measures that have been useful in the study of brain–attention relations in infants will be discussed. The use of heart rate as a measure of the general arousal system will be emphasized. The usefulness of these psychophysiological methods in research on attention in infants will be described and illustrated with relevant research on recognition memory. Third, the relationship between developments in visual fixation and visual attention will be discussed with illustrative examples from research on individual differences in attention. It should be noted that this article focuses on the development of visual attention as there is a vast literature on this topic. However, comments about visual attention should generalize to other sensory systems as well.

Attention in Infancy and Early Childhood

Newborn infants orient toward and attend selectively to stimuli during the brief periods when they are alert and awake. This occurs in spite of the immaturity in both structure and function of the human visual system. Robert Fantz in pioneering work on infant visual abilities showed that between birth and 2 months of age infants looked longer at patterned than at unpatterned stimuli and that depending on factors such as familiarity, size, and the amount of contrast, they preferred to look (i.e., looked longer) at some stimuli over others. Some of the infants' early abilities are based on visual reflexes (e.g., saccadic and pursuit movements) that are present at birth at least in rudimentary form. However, more general visuo-motor immaturity restricts infants' ability to scan stimuli extensively or to detect stimuli beyond about 30° in the peripheral visual field. Moreover, when infants visually capture an element in the environment, they seem to have trouble disengaging or looking away from it, a phenomenon that has been referred to as obligatory looking or sticky fixation.

The early immaturities in the young infants' visual system do not last long. There is rapid neurological development in the retina and in the visual pathways to the cortex between 2- and 3-months of age. This neurological development coincides with significant improvement in all aspects of visual functioning (e.g., visual acuity, color vision, smooth pursuit), an expansion of the visual field, moderation of inhibitory mechanisms that restricted eye movements, and the onset of more mature perceptual abilities whereby infants come to recognize objects and to determine their spatial layout. Coincident with this shift toward greater cortical control of vision, infants begin to spend more time awake and in an alert state. They also begin to look about the environment in a way that is less reflexive and appears to be more voluntary in nature. This emerging competence in visual attention reflects developments in the specific structures and pathways within the visual system and the brain.

There are changes in behavioral state that are very important for infant attention development. The physiological and psychological state of being awake and alert is called arousal. Arousal acts to energize primary sensory areas in the cortex and to increase the efficiency of responding in those areas. Its effects are generally nonspecific and affect multiple modalities, cognitive systems, and cognitive processes. Newborn infants spend most of their time in sleeping states and have only very short periods of time when they are in such a state of arousal. Both behavioral control of sleep state (i.e., infants sleeping through the night) and increases in alert states of arousal show dramatic changes from 2 to 6 months. The mature form of arousal invigorates, energizes, and regulates cognitive processes leading to increased processing efficiency, shorter reaction times, better detection, and sustaining of cognitive performance for extended periods of time. The infants' increases in its ability to attain and then maintain a state of alertness or arousal is fundamental to the development of effective information processing.

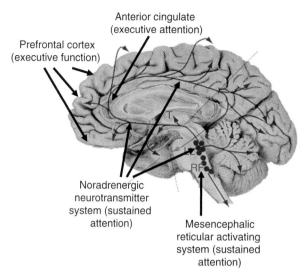

Prefrontal cortex
(executive function)

Anterior cingulate
(executive attention)

Noradrenergic
neurotransmitter
system (sustained
attention)

Mesencephalic
reticular activating
system (sustained
attention)

Figure 1 Some areas of the brain involved in attention. Sustained attention is controlled by the noradrenergic and cholinergic neurotransmitter systems. Executive attention is controlled by areas in the prefrontal cortex, including the anterior cingulate cortex.

The arousal aspect of attention is controlled by several systems in the brain. These can be seen in **Figure 1**. They include the neuroanatomical connections between the mesencephalic reticular activating system in the brainstem, the thalamus, and the cortex. The thalamus is the major sensory connection area between incoming sensory information and the cortex and its reticular nucleus is enhanced bidirectionally, both by the ascending reticular activity and by feedback mechanisms from the primary sensory cortex. The mesencephalic reticular activating system also stimulates extrinsic neurotransmitters that in turn influence the limbic system areas. The noradrenergic and cholinergic transmitter systems are thought to be the neurochemical systems that are most closely involved in cortical arousal as it is related to attention. The dopaminergic system is thought to affect the motivational and energetic aspects of cognitive processing, and the serotonin system is thought to affect the overall control of state. These four neurochemical systems also show changes over the period of infancy and this implies that the arousal controlled by these systems also develops in that time period.

Orienting

Holly Ruff and Mary Rothbart characterized infant's visual behavior in the first postnatal year of life as dominated by an 'orienting/investigative system' of attention. There are two inter-related components in this system. First, a spatial-orienting network (including the posterior parietal cortex with several subcortical systems such as the superior colliculus, pulvinar, and the locus coeruleus in the brainstem), is alerted by peripheral stimuli and

directs attention to potentially important locations (i.e., where) in the environment. This mediates attentional functions such as, engagement, disengagement, shifting, and inhibition of return. Second, an object recognition network (including pathways from the primary visual cortex to the parietal cortex and the inferior temporal cortex) mediates attention to object features and gathers detailed information about form, color, and pattern that enables the identification (i.e., what) of objects. Ruff and Rothbart suggest that there is a marked developmental transition in the structure and function of this system between 3 and 9 months of age. Rapid maturation at all levels of the visual system with increased arousal and alertness enables infants to deploy their attention more flexibly and quickly. They begin to respond to stimulus objects and patterns around them in terms of experiential factors such as the novelty or complexity of objects and events rather than by their salience or intensity alone.

Sustained Attention

A major advance that contributes to this marked developmental transition is the infant's ability to sustain his or her attention. Sustained attention to object features, also called focused attention, is the extended selective engagement of a behavior system that primarily enhances information processing in that system. It is similar to a state of arousal during which cognitive processing is enhanced. Many of the infant's cognitive and social activities occur during episodes of sustained attention. For example, we know that infants prefer to look at relatively novel objects, faces, and sounds. Novel stimuli elicit an initial stimulus orienting, followed by sustained attention; maintenance of sustained attention then leads to the infant being able to learn about and remember aspects of the stimulus as it becomes familiar. Infants as young as 3 months of age will engage in 5–10 s periods of sustained attention. Beginning at about this age and extending into the second year, the duration that an infant can sustain attention increases markedly. As with visuo-spatial orienting, the development of sustained attention during the period of infancy is closely related to the development of brain systems controlling arousal and state. Sustained attention is a manifestation of this global arousal system of the brain that controls responsiveness to events in the environment and affects sensory systems. Thus, sustained attention represents the activation of this arousal system in situations calling for attention.

Executive or Endogenous Attention

At the end of the first year, Ruff and Rothbart suggest that the rudiments of another major attention system, one in which the infant begins to acquire a system of higher level controls over the allocation and deployment of cognitive resources, begins to emerge. This capacity is evident in a wide range of behaviors. For example, infants' look duration to static stimuli and simple objects continues to decline whereas their look duration to complex objects increases. Infants also look more to their caregivers in situations that call for social referencing and joint attention and they begin to show the beginnings of behavioral inhibition on the A-not-B task. Further evidence of emerging intentionality is evident in improvements in deferred imitation, means-end problem solving, and recall memory that occur late in the first year of life. By about 18 months of age, this endogenous control of attention acquires an increasingly executive function. Executive function is a description of psychological activities that control behavior, allocate cognitive resources, evaluate behavior progress, and direct activity with goals and plans. For example, if a toddler decides to stack rings on a dowel he or she must select the relevant rings from a toy box and ignore other irrelevant objects such as blocks that might also be available. Similarly, other activities in the environment that might be of interest to the child must be ignored so as not to disrupt this planful behavior. Executive attention and functioning are closely related to brain activity, in particular to the prefrontal cortex, anterior cingulate, and frontal eye fields. These changes in visual attention and their neural substrates enable (and may be enabled by) coincident changes in language

(e.g., comprehension), cognition (e.g., representation), and self-regulation (e.g., behavioral inhibition) that begin in this timeframe and continue to advance across the preschool years. The timing of these developments in attention from birth to early childhood can be seen in summary form in **Figure 2** that has been compiled by John Colombo.

Psychophysiological Measures of Infant Attention

Psychophysiological measures have been useful in the research on infant attention and brain development. Psychophysiology is the study of psychological processes using physiological measures. The physiological measures are noninvasive and can be used with infant participants. The use of heart rate, electroencephalogram (EEG), and event-related potential (ERP) data as psychophysiological measures of attention will be reviewed to illustrate the type of information that this approach provides.

Heart Rate

The most common measure used by psychophysiologists studying young infants is heart rate. The infant's heart rate can be measured in response to psychological manipulations and used as an index of attention. John

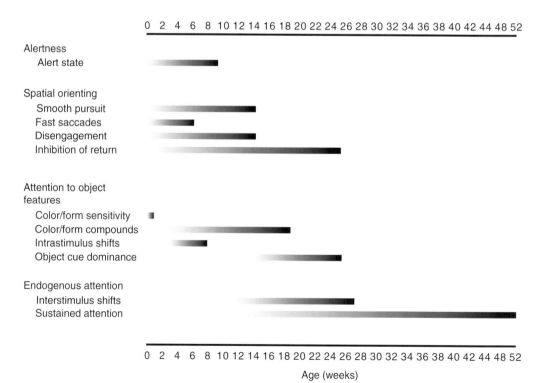

Figure 2 Summary table of the developmental course of visual attentional functions in infancy. The relative darkness of the line indicates the relative degree of maturity at each age. From Colombo J (2001) The development of visual attention in infancy. *Annual Review of Psychology* 52: 337–367.

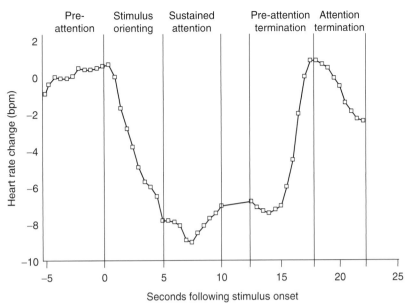

Figure 3 Heart rate changes that define attention. Stimulus orienting is an initial orienting toward a visual pattern or sound. Sustained attention represents engaged arousal. Attention termination represents inattentiveness. bpm, beats per minute. From Richards JE and Casey BJ (1991) Heart rate variability during attention phases in young infants. *Psychophysiology* 28: 43–53.

E. Richards proposed a model in which the phasic changes in infants' heart rates that occur as they look at a stimulus correspond to different levels of attentional engagement. There are four key phases in this model and these are illustrated in **Figure 3**. First, when the stimulus is initially presented there is a brief automatic interrupt phase that reflects the detection of a change in environmental stimulation. A very brief, reflexive, biphasic deceleration–acceleration in heart rate occurs. This automatic interrupt may activate the subsequent attention phases. The second phase is stimulus orienting. Stimulus orienting indicates the beginning of attentional engagement and initiates preliminary processing of the information in the stimulus. A large, rapid deceleration in heart rate from its prestimulus level occurs during this phase. The third phase is sustained attention. Sustained attention reflects the activation of the alertness/arousal system of the brain and involves voluntary, subject-controlled cognitive processing of the stimulus information. The heart rate deceleration that was reached during stimulus orienting is maintained during this phase. Heart rate also shows decreased variability and certain other somatic changes that facilitate attentiveness such as reduced body movement and slower respiration may also occur. The fourth phase is attention termination. During this last phase the infant continues to look at the stimulus but is no longer processing its information (i.e., is inattentive). The heart rate begins to return to its prestimulus levels during this phase.

Many research studies validate this model across infancy and early childhood and show that the phases are elicited by a wide range of stimuli and conditions. These include brief exposures to black and white achromatic patterns and faces, extended exposure to audiovisual material (children's television programs), and the visual and manual examination of toys. Moreover, there is evidence that infants' information-processing activity occurs primarily during the heart-rate phase of sustained attention rather than during the other phases. For example, it is in sustained attention that infants (1) encode information and demonstrate later recognition of it, (2) show attenuated localization of a distracter stimulus located in the peripheral visual field, and (3) are more resistant to distraction from competing stimuli during toy play.

The phases of sustained attention and attention termination (inattention) are markers of the nonspecific arousal system of the brain. The neural control of this heart rate change originates from cardioinhibitory centers in the cortex via the vagus (tenth cranial) nerve. This area has reciprocal connections with the limbic system and through these connections is involved in modulating activity within the mesencephalic reticular formation arousal system and the related neurotransmitter systems. The cardioinhibitory centers act through the parasympathetic nervous system to slow heart rate when the arousal system is engaged. The heart rate changes occurring during sustained attention (sustained heart rate slowing) index the onset and continuing presence of this arousal. The heart rate changes during attention termination (return of heart rate to its prestimulus level) index the lack of activation of this arousal system. These two phases of attention therefore reflect the nonspecific arousal that may affect a number of sensory and brain systems.

EEG and ERPs

EEG activity has been used with both adults and infants as a measure of nonspecific arousal. This measure is important because it is a more direct measure of neural activity than is heart rate. Therefore, it is possible that EEG could be used as a noninvasive measure of the neural activity affected by attention. Scalp-recorded ERPs are derived from the EEG recording. The ERPs are derived from EEG by linking the EEG recording to a specific experimental event and then averaging out the varying EEG activity unrelated to the event. The remaining averaged EEG is thus electrical potential change that is related to the event. If the experimental event is known to be associated with a cognitive process controlled by brain activity, the ERP reflects specific cognitive processes that originate from particular brain areas. Therefore they provide a noninvasive and direct measure of functioning in those areas. For example, specific components of the ERP change in response to familiar and unfamiliar visual stimuli. One such ERP change is closely related to attentional responses. This is a negative-going electrical potential change over the central leads on the scalp. This has been labeled the Nc (negative central). The Nc is thought to represent a relatively automatic alerting response to the presence of a visual stimulus (see **Figure 4**).

Attention and Recognition Memory in Infancy

The usefulness of ERP and heart rate measures can be nicely illustrated from research on the development of recognition memory in infants. Behavioral research on infant recognition memory is typically studied with the paired-comparison procedure. In this procedure infants are shown a single stimulus (familiar stimulus) during a brief familiarization phase. Later, during the recognition memory test phase, the familiar stimulus is paired with a novel stimulus that has not been seen previously. Recognition memory for the familiar stimulus is inferred if the infants show a novelty preference (i.e., look longer at) the novel stimulus than at the familiar stimulus during the paired-comparison test. This method has provided a large database in the conditions under which infants encode, store, and retrieve information. John E. Richards used this procedure in conjunction with the heart rate-defined phases of attention. Infants between the ages of 3 and 6 months of age were first presented with a recording of the 'Sesame Street' television program. Then, when the infants were showing sustained attention to the television program, simple geometric patterns were shown for about 5 s. Infants showed recognition of the stimuli only if

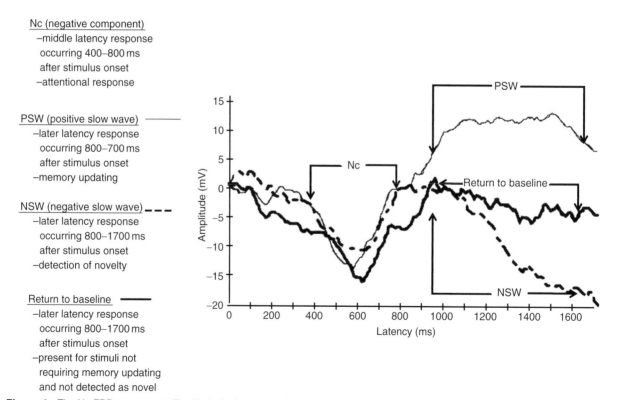

Nc (negative component)
  –middle latency response
   occurring 400–800 ms
   after stimulus onset
  –attentional response

PSW (positive slow wave) ——
  –later latency response
   occurring 800–700 ms
   after stimulus onset
  –memory updating

NSW (negative slow wave) – – –
  –later latency response
   occurring 800–1700 ms
   after stimulus onset
  –detection of novelty

Return to baseline ——
  –later latency response
   occurring 800–1700 ms
   after stimulus onset
  –present for stimuli not
   requiring memory updating
   and not detected as novel

Figure 4 The Nc ERP component. The Nc is the large negative deflection in EEG potential occurring about 600 ms after stimulus onset. It is closely related to attention. Later slow waves in the ERP represent other aspects of visual information processing. From De Haan M and Nelson CA (1997) Recognition of the mother's face: A neurobehavioral study. *Child Development* 68: 187–210.

they were showing sustained attention to the television programs. Interestingly, the demonstration of a novelty preference that is taken as the measure of recognition of the familiar stimulus likely reflects the infant's attempt to acquire new information from the previously unseen stimulus during sustained attention.

Recognition memory has also been examined in a variant of the paired-comparison procedure called the oddball paradigm. Here infants are exposed to repeated presentations of two different stimuli. They are then exposed to one of the familiar stimuli on 60% of the trials (frequent familiar), the other familiar stimulus on 20% of the trials (infrequent familiar), and novel stimulus presentations on the remaining 20% of the trials (infrequent novel). Stimulus presentations are very brief (500 ms) in this procedure. Charles Nelson tested infants in a series of studies using the oddball procedure with ERP and found a large Nc component occurring about 400–800 ms after stimulus onset (see **Figure 4**). He found no differences in the Nc component for any of the stimulus presentation conditions for 4-, 6-, and 8-month-old infants and concluded that the Nc is indicative of a general alerting or orienting response. Other (later) components of the ERP did differ between stimulus presentation conditions for older infants. They demonstrated a negative slow wave following novel stimulus presentations and a positive slow wave following infrequent familiar stimulus presentations. The positive slow wave is likely a response associated with an updating of working memory following presentation of a familiar yet only partially encoded stimulus, whereas the negative slow wave represents the initial processing of new information provided by a novel stimulus. Thus, it is likely that the late slow wave ERP components reflect processes associated with recognition memory, while the Nc component reflects general orienting and attention.

These findings were refined in a series of reports by John E. Richards in which he showed how the ERP measures of brain activity in the oddball paradigm were affected by infants' phases of attention and inattention as defined by their heart rate changes. Infants were 4.5, 6, and 7.5 months of age. The ERP responses to frequent familiar, infrequent familiar, and novel stimuli were recorded and separated into those that occurred when then infant was in sustained attention or was inattentive. He found that there was a larger Nc during sustained attention than during inattention regardless of the familiarity (familiar, novel) or frequency (frequent, infrequent) of the stimulus (**Figure 5**). There were also age changes in the amplitude of the Nc during sustained attention. The close association of the Nc with attention supports the view that this component reflects a general attention process of orienting to the stimulus. Concerning recognition memory, late slow waves were found at about 1000–2000 ms following stimulus onset. During attention, 4.5-month-olds demonstrated a positive late slow wave

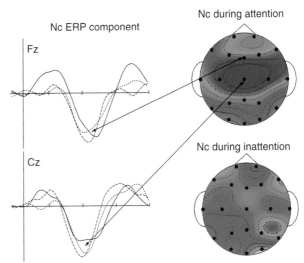

Figure 5 The Nc ERP component. The left graphs show a tracing of the ERP from the onset of a brief visual stimulus. The top left graph is from the Fz electrode and the bottom left graph from the Cz electrode (arrows point to location on scalp). The right graphs are topographical maps reflecting the electrical potential changes across the scalp at the largest negative deflection of the ERP during the Nc. The dark blue on the upper left topographical map shows the enhanced Nc occurring during sustained attention.

that was similar for familiar and novel presentations. The older infants displayed a negative late slow wave following presentations of the novel stimulus, which stood in contrast to the positive slow wave these groups displayed following infrequent familiar stimulus presentations. The age differences suggest that (during sustained attention but not inattention) the older two age groups were sensitive to stimulus novelty and probability, whereas the younger infants were sensitive only to stimulus probability. Finally, although ERPs are voltage changes recorded from the scalp, Greg Reynolds and John E. Richards used high-density EEG recording and specialized analyses to estimate the cortical sources locations of the ERP components. For example, the cortical source of the Nc component appears to be in regions of the prefrontal cortex including the anterior cingulate. The anterior cingulate is part of the cingulate cortex, a region of the brain that shares reciprocal connections with several subcortical, cortical, and limbic regions. Studies have shown that the anterior cingulate is involved in visual target detection and the control or direction of attention.

The general importance of this work is in showing that the arousal component of attention is related to complex infant cognition. Recognition memory involves several brain areas and cognitive functions. It requires the acquisition of stimulus information and memory storage over some period of time as well as demonstration of the existence of the stored memory. It is likely that the arousal aspect of attention invigorates each of these cognitive processes. This enhances familiarization when information acquistion is occurring, may facilitate memory consolidation during the waiting period, and enhances the processes involved in the exhibition of recognition memory. The facilitative effect of attention on infant recognition memory may occur because specific brain areas responsible for information acquisition or recognition are enhanced during attention.

Visual Fixation and Attention: What Does Looking Mean?

The direction and duration of visual fixation (or looking) are core dependent measures in research on infants' attention. Over many years these measures have provided a wealth of information about the development of infants' sensory, perceptual, and cognitive processes. More recently, these measures and the attention processes they imply have become a focus of interest in their own right. In particular, the duration of infants' fixations obtained from research using habituation and selective looking procedures has been informative. A typical and very robust finding is that older infants habituate in less time, at a faster rate, and with shorter look durations than do younger infants. Older infants take less familiarization time than younger infants to show the preferences for novel stimuli indicative of recognition memory. These findings have been attributed to an increase in speed and/or efficiency in information processing with age.

Individual differences in these same look duration measures within age have also been observed. Infants who show brief look durations during visual fixation compared to age mates who show long look durations

also encode information more quickly, show better recognition memory, and disengage fixation from a stimulus more readily. In addition, they show higher performance on certain measures of language, representational play, and cognition in later childhood. Longer and less efficient look duration is ostensibly less mature and is typical of very young infants and has also been observed in those who are at risk for developmental delay (e.g., preterm infants, infants with Down syndrome). Although these findings about look duration and the quality of information processing have been widely replicated, much of it has been gathered from infants who were between 3 and 6 months of age when tested initially. In contrast, research from other paradigms that involved object examination or extended media viewing indicated that when infants' attention is actively engaged with complex and/or interactive stimuli (e.g., toys or video clips), look duration actually increases across age. Much of this research was done with older infants in their second year of life.

John Colombo conducted a meta-analysis of look duration data obtained from infants of various ages across the first year of life in order to reconcile these apparently discrepant sets of findings. The results suggested a model in which infant look duration might follow a triphasic course of development over the first year or so of life. In this model, different patterns of looking reflect the maturational state of the different neural mechanisms and processes underlying infants' attention systems. This model is illustrated in **Figure 6**. In the first phase of the model, the early increase in look duration between birth and 2 months indicates the emergence of the alertness or arousal aspect of attention that is elicited most readily by external or exogenous stimuli and events. Advances in

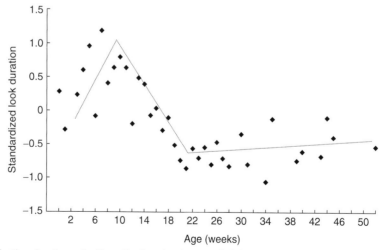

Figure 6   Changes in fixation duration reflecting attention development in infants. The figure shows a systematic decline in fixation from 3 to 6 months to simple stimuli reflecting an increase in processing speed and decline in the amount of time necessary to process simple visual patterns. From Colombo J (2002) Infant attention grows up: The emergence of a developmental cognitive neuroscience perspective. *Current Directions in Psychological Science* 11: 196–200.

alertness over this time likely reflect an increase in sub-cortical (i.e., brainstem reticular system and the associated ascending pathways and neurotransmitters) influence on higher-order (i.e., neocortical) structures. In the second phase, there is a decline in look duration from 3 to about 6 months that primarily reflects improvements in the ability to disengage from a stimulus. Disengagement and shifting of attention have been linked to developments in the posterior orienting network that includes the posterior parietal cortex in conjunction with a number of subcortical connections. The decline may also reflect an increase in sustained attention, leading to more efficient cognitive processing, and a decline in the amount of time necessary to process simple visual patterns. Finally, improvements in visual function (e.g., visual acuity, binocularity), object recognition, and speed of information processing emerge in this timeframe as maturation of the visual and inferior temporal areas of the cortex advance. In the third phase, the plateau or increase in look duration reflects the emergence of an endogenous, apparently volitional directing of attention as a function of the task at hand, in particular the inhibition of the tendency to shift attention away from a task that is interesting or demanding. This endogenous aspect of attention and the control of looking are mediated by developments in the frontal cortical areas that occur late in the first postnatal year, though these are still integrally related to the lower level brainstem systems that direct arousal and alertness.

Mary Courage, Greg Reynolds, and John E. Richards examined this model. They presented infants from 3 to 12 months of age with a variety of visual stimuli. These included simple black and white geometric patterns, faces, and clips from the Sesame Street television program. They measured how long the infant looked at the stimulus and the heart rate changes that indicate the occurrence of attention or inattention. **Figure 7** shows the average look durations to clips from the television program Sesame Street, to faces, and to simple black and white geometric patterns. There was a decrease in look duration for all stimuli during the first 6 months of age. After that there was an increase in look duration from 6 to 12 months for the Sesame Street material and to a lesser degree the faces, but no change in the amount of time spent looking at the geometric patterns. The increase in looking time to complex visual patterns indicates that infants will engage in selective enhanced processing if sufficient complexity exists in the stimulus.

The difference in the development of looking at complex and simple stimuli has been shown to continue into the later parts of childhood. John E. Richards has presented children ranging in age from 6 months to 2 years with the Sesame Street movie, 'Follow that Bird', and with simple computer-generated black-and-white geometric forms. The movie was accompanied by the appropriate soundtrack and the geometric forms with computer-generated music. As with the complex stimuli mentioned above, there was an

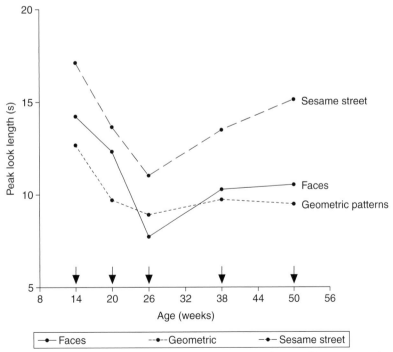

Figure 7 Mean peak look durations for static and dynamic versions on each stimulus type (faces, geometric patterns, Sesame Street) as a function of age. Note the decline over ages in all functions until 6 months, the lack of change after 6 months in look duration for the faces and geometric patterns, and the increase from 6 to 12 months in the looking time toward Sesame Street.

increase in the duration of the look toward the Sesame Street movie from 6 months to 2 years. Alternatively, the 6-month-old infants looked for equal duration at the movie and the computer-generated stimuli. However, there was no change in the amount of time spent looking at the simple forms. In a similar vein, Daniel Anderson has shown that 3–5-year-old children are very sensitive to the comprehensibility of television programs. Young children will show very long and extended looks at programs that are at their comprehension level, but only brief looks at programs that are either too complex or too simple for their cognitive level. These findings indicate that from 6 months to 5 years of age there is an increase in the selective enhanced processing of audio and visual information.

An important implication of these endings is that look duration means different things at different ages. In the early months of life infants who show short look durations appear to be at an advantage on certain cognitive tests later in childhood. Their brief looks indicate that they process stimulus information quickly and efficiently and that they can readily disengage from a stimulus and shift their attention to another object or location in the environment as needed. However, by the end of the first year, longer look durations are likely more adaptive than short looks, especially in situations where the affordances of the environment are complex and engaging and resistance to distraction during information processing is important. John Colombo suggested that the longer look durations seen at older ages reflect the ability to voluntarily sustain attention to an object or to maintain focus on a short-term goal that signals the operation of endogenous attention. By early childhood, attention is directed to objects or audiovisual information that is meaningful to the infant, and information that is too complex or not meaningful is ignored.

Finally, there is an important caveat to the strict interpretation of the duration of infants' looking as an index of visual attention and information processing. This is clearly evident from John E. Richards' model of the heart rate of defined phases of attention and the data that support it. Although many researchers define visual attention in terms of the duration of time that the infant keeps his or her eyes directed toward a stimulus, young infants often will continue to look at a visual pattern even when their heart rate (i.e., sustained attention is not occurring) shows that they are no longer attending to it or presumably processing its information. This may be particularly likely in the first 3 months of life when an infant's looking may be captured by a pattern from which they cannot easily disengage although it is evident later in infancy and early childhood as well. Indeed, even older children and adults can appear to be looking intently at something but are in fact not processing its information at all but have covertly directed their attention elsewhere. An implication of this is that it may be important to provide convergent evidence from

both behavioral measures (looking) and physiological measures (cardiac change) of attention when developmental changes in infants' information processing are of interest.

Individual Differences in Attention

There are marked individual differences in several aspects of attention during infancy and toddlerhood that have implications for attention, learning, and memory in later childhood. The precise meaning of these individual differences as well as their developmental trajectory are not entirely clear and will continue to be explored in future research. However, one might speculate that they will have implications for a range of practical developmental issues such as academic performance, the impact of the extensive exposure to television and other screen media that infants and toddlers experience, and in the design of interventions for infants and children who are at risk for attention disorders and difficulties. A good example in which research on attention processes has informed individual differences in attention is attention deficit hyperactivity disorder (ADHD).

Attention Deficit Hyperactivity Disorder

This is one type of attention difficulty seen in young children, particularly as they enter school. ADHD affects from 5% to 10% of school-aged children. It is characterized by inattentiveness, hyperactivity, poor impulse control, and behavior management problems. These children often come to the attention of healthcare professionals when entering school because they do poorly in situations demanding sustained behavior control. ADHD is usually separated into three subtypes: ADHD-inattentive (ADHD-I), ADHD-hyperactive (ADHD-H), and ADHD-combined type (ADHD-C). The children who are diagnosed as ADHD-I have problems in attention control, sustained attention, and are often inattentive. The children diagnosed as ADHD-H show poor impulse control and exhibit high levels of activity. The ADHD-C children show signs of both inattentiveness and hyperactivity. The treatment for ADHD is generally pharmacological (e.g., methylphenidate (Ritalin)).

The subtypes of ADHD may be related to the distinction between sustained attention and executive attention. A common hypothesis about the cause of ADHD-H is a poorly functioning executive attention system. Supporting this, ADHD-H children have been shown to do poorly on tasks requiring plans, the inhibition of reflexive or automatic behavior, and impulse control. Several studies link deficits in the prefrontal cortex to ADHD-H children. Alternatively, children with the diagnosis of ADHD-I perform nearly as well as normal children on executive function tasks. However, they show deficits on tasks requiring sustained attention, covert shifting of attention, and selective attention. The ADHD subtypes are likely due to differences in the brain regions controlling sustained attention and executive attention. Individual differences in sustained attention in the early part of infancy may be related to ADHD-I outcomes, particularly in infants showing extremely low amounts of sustained attention. Deficits in parts of the brain that allow the sustaining of attention over extended time periods may be impaired leading to consistent poor performance for these infants and children. Alternatively, ADHD-H is not predicted by individual differences in attention observer prior to 2 or 3 years of age. This is likely due to the factor that the areas of the frontal cortex controlling executive attention are not yet sufficiently developed. As early identification is critically important to early intervention and potentially better prognosis, information from developmental and brain sciences could possibly serve to ameliorate attention difficulties and to optimize outcomes for these children.

Conclusion

Attention is the selective enhancement of some behavior at the expense of other behavior. Marked developmental changes occur in stimulus orienting and sustained attention during the first 2 years of life. Stimulus orienting involves the general orientation of sensory systems and receptors to important environmental events and sustained attention involves the enhanced and selective processing of information for specific psychological behaviors. Thus, by the end of infancy these aspects of attention are fully developed. The development of attention in the toddler years and into early and middle childhood involves the development of executive attention, which is the ability to carry out tasks with planfulness, allocate attention to self-established goals and plans, and monitor one's progress in complex tasks. Changes in orienting, sustained attention, and executive attention are closely linked to brain development. One of the most recognized childhood disorders, ADHD, is closely related to problems in the executive attention system and may be caused by a deficit in the brain areas involved in the development of executive function and executive attention.

See also: ADHD: Genetic Influences; Brain Function; Habituation and Novelty; Memory; Perceptual Development; Visual Perception.

Suggested Readings

Colombo J (2001) The development of visual attention in infancy. *Annual Review of Psychology* 52: 337–367.

Colombo J (2002) Infant attention grows up: The emergence of a developmental cognitive neuroscience perspective. *Current Directions in Psychological Science* 11: 196–200.

Courage ML, Reynolds GD, and Richards JE (2006) Infants' visual attention to patterned stimuli: Developmental change and individual differences from 3- to 12-months of age. *Child Development* 77: 680–695.

De Haan M and Nelson CA (1997) Recognition of the mother's face: A neurobehavioral study. *Child Development* 68: 187–210.

James W (1980) *The Principles of Psychology*. New York: Dover Publications.

Reynolds GD and Richards JE (in press) Infant heart rate: A developmental psychophysiological perspective. In: Schmidt LA and Segalowitz SJ (eds.) *Developmental Psychophysiology: Theory, Systems and Applications*. New York: Cambridge Press.

Richards JE (1998) *Cognitive Neuroscience of Attention: A Developmental Perspective*. Mahwa, NJ: Erlbaum.

Richards JE (in press) Attention in young infants: A developmental psychophysiological perspective. In: Nelson CA and Luciana M (eds.) *Handbook of Developmental Cognitive Neuroscience*. Cambridge, MA: MIT Press.

Richards JE and Casey BJ (1991) Heart rate variability during attention phases in young infants. *Psychophysiology* 28: 43–53.

Ruff HA and Rothbart MK (1996) *Attention in Early Development: Themes and Variations*. New York: Oxford University Press.

Auditory Development and Hearing Disorders

D H Ashmead, A M Tharpe, and D P Sladen, Vanderbilt University Medical Center, Nashville, TN, USA

Glossary

Auditory brainstem response (ABR) – Electrical activity in the auditory nerve and brainstem in response to sound.

Cochlear implants – A hearing technology in which environmental sounds are picked up by a microphone and converted to electrical signals that are delivered to the inner ear.

Decibel – A measure of the relative intensities of two sounds, expressed in logarithmic units.

Evoked otoacoustic emissions (OAE) – Faint sounds produced in the inner ear in response to external sounds, used to screen for auditory function in infants.

Hearing aids – A hearing technology in which environmental sounds are picked up by a microphone, amplified, and delivered as sounds to the ear canal.

Hearing level (HL) – Measure of the intensity of a sound relative to the intensity at which young adults with good hearing can hear a sound of that frequency.

Sound pressure level (SPL) – Measure of the intensity of a sound relative to a single reference value of 20 Pa.

Introduction

The auditory system responds to sounds, which are fluctuations in the pressure of a medium such as air or water. Sounds are created by physical events, many of which are important to listeners, such as approaching footsteps, spoken words, or falling water. **Figure 1** shows the overall structure of the outer, middle, and inner parts of the ear.

Overview of Auditory System

Sound is gathered by the pinna and directed along the ear canal to the eardrum, or tympanic membrane, which is the boundary between the outer and middle ear. The shape of the pinna funnels sound toward the ear canal, and

also filters sound differently depending on the direction of the sound source, which provides a cue for localizing sounds. The pressure changes in the arriving sound make the eardrum vibrate. The middle ear is a cavity normally filled with air. The Eustachian tube connects the middle ear to the throat, allowing the air pressure in the middle ear to equalize to the outside air pressure. Movement of the eardrum is picked up by the three bones, also called the ossicles, in the middle ear (malleus, incus, and stapes). The stapes connects to the cochlea of the inner ear at a flexible membrane called the oval window, so that movement of the stapes creates pressure waves in the fluid inside the cochlea. The overall path from the outer ear to the oval window works to amplify sound by a substantial amount. Some of this amplification comes from the funneling effect of the outer ear and the mechanical structure of the middle ear bones, but most comes from the fact that the area of the eardrum is about 17 times larger than that of the oval window, just as the head of a nail is much larger than its tip. The amplification is greater at middle sound frequencies than at low or high frequencies.

The cochlea in mammals is a snail-shaped coil, making about two-and-a-half turns in humans. The cochlea has three fluid-filled spaces running along its length, with a surface called the basilar membrane also running lengthwise. A complex sensory structure called the Organ of Corti rests on the basilar membrane. Among other details, the Organ of Corti has a single row of inner hair cells and three rows of outer hair cells running along its length. The inner hair cells, about 3500 to 4000 per ear, are activated when pressure changes in the cochlear fluids cause motion of the basilar membrane, to which the bases of the hair cells are attached. These hair cells are the sensory

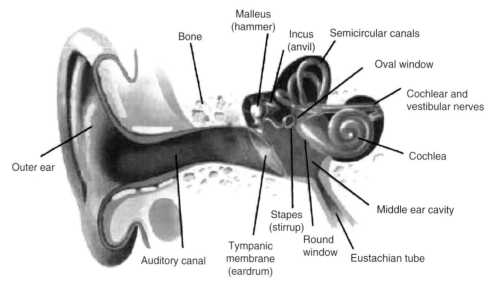

Figure 1   Structure of the auditory system. Reproduced from Widex South Africa, Understanding your Ear.

transducers of the auditory system, converting sound energy into neural signals going to the brain. When an inner hair cell moves, the stereocilia (hairs) at its tip are deflected. This deflection allows potassium ions to enter the cell, which in turn sends neurotransmitter chemicals to the nearby auditory nerve fibers. Action potentials (spikes of electrical energy) arise in those fibers and are carried along the auditory nerve toward the brain. About 40 000 auditory nerve fibers connect to each ear, so each inner hair cell is linked to a number of nerve fibers.

The mechanical properties of the inner ear are elegantly structured so that sounds of high frequency mostly activate inner hair cells closer to the oval window or base of the cochlea, while sounds of lower frequency activate inner hair cells further along toward the apex. This 'place principle' is an important part of how the auditory system distinguishes between sounds of different frequencies. A Nobel Prize was awarded to Georg von Békésy for research begun in the 1920s on cochlear mechanics. The other way that the auditory system keeps track of sound frequency is that the action potentials in the auditory nerve fibers can be timelocked to the sound frequency. This process works for low-to-middle frequency sounds.

The outer hair cells, about 11 500 per ear, are also activated by sounds, but their activity is, for the most part, not relayed to the brain. Instead, the outer hair cells change their length slightly in response to sounds, which has the effect of making that section of the basilar membrane vibrate more strongly, providing a stronger signal to the inner ear cells. This process is especially important when the incoming sound intensity is low to medium. For this reason, the activity of the outer hair cells is referred to as the cochlear amplifier. An important practical point is that this outer hair cell activity generates faint noises that can be recorded by a microphone placed in the ear canal. These 'otoacoustic emissions' provide the basis for testing if young infants can hear. Another key point about both the inner and outer hair cells is that, in mammals, they cannot regenerate after being damaged. The term 'sensorineural hearing impairment' is used to refer to such damage.

Proceeding from the inner ear toward the brain, the auditory nerve of each ear connects to brainstem structures called the cochlear nuclei, one on each side. The ventral part of each cochlear nucleus projects to the superior olivary complex structures in the brainstem, on both the same and opposite sides. The superior olives in turn project to the inferior colliculi. The dorsal part of each cochlear nucleus projects to the same and opposite side inferior colliculi. An important functional point is that all structures in the ascending auditory pathways beyond the cochlear nuclei receive input from both ears. In many listening situations, such as noisy settings, the differences between sounds arriving at the two ears provide a basis for enhanced perception.

The absolute auditory sensitivity of adults varies with sound frequency. Sensitivity is best for the middle range of frequencies from about 500 to 4000 Hz. At frequencies below and above this range, sensitivity worsens substantially. Comparing measures of auditory sensitivity across studies is complicated by the fact that subtle details of the methodology can have a big effect on the results. For this reason, audiologists conduct clinical measures of sensitivity using standardized procedures and instruments.

Another complication is that there are several measurement systems for reporting sensitivity. Each system uses units called decibels (dB). In hearing science the decibel expresses the ratio between the pressures of two sounds, p_1 and p_2:

$$dB = 20 \log_{10}\left(\frac{p_1}{p_2}\right)$$

The term p_1 is the pressure of a sound of interest, such as the softest sound someone can detect in a hearing test, and the term p_2 is a reference value. There are two widely used reference systems, known as the SPL system and the HL system. SPL stands for 'decibels, relative to a sound pressure level of 20 µPa. That reference pressure was selected a century ago as the level at which young adults with excellent hearing could just barely detect a 1000 Hz tone. Any observed sound pressure (p_1, measured in micropascals, µPa) can be related to this reference pressure by the formula

$$dB\ SPL = 20 \log_{10}\left(\frac{p_1}{20}\right)$$

For example, normal conversational speech measures about 65 dB SPL at the listener's position, in which case p_1 is approximately 36 000 µPa. This illustrates the wide dynamic range of the auditory system, allowing us to hear both faint and more intense sounds. We can tolerate brief exposure to very intense sounds, such as machinery sounds above 100 dB SPL. To make matters even more complicated, sound measurements can be made with different filters. A commonly used filter is the 'A-weighted' one, designed to mimic the natural frequency filtering of human hearing. Such a measure might be described in units of dB SPL A-scale. As a semantic note, the terms pressure, intensity, and level are all used to refer to the amount of energy in a sound. Although these terms have distinct physical meanings, there is a tendency to use them interchangeably.

The HL or 'hearing level' sound measurement system is used by audiologists when measuring hearing sensitivity to specific sounds. Instead of using a single reference value, this system uses a different reference value for each sound. For example, an audiologist might measure someone's sensitivity to each of a set of tones ranging from 250 to 8000 Hz. For each tone frequency there is a reference level at which typical young adults can just hear the tone. If someone hears the tone at that reference level, then their sensitivity is 0 dB HL. If the person needs the tone to have a pressure four times the reference level in order to hear it, then their sensitivity is about 12 dB HL. In the HL system, then, 0 dB HL corresponds to typical hearing of young adults, while positive values indicate worse hearing sensitivity.

Development of Auditory Structures

The auditory system begins to develop at a gestational age of 25 days, becomes well differentiated by 9 weeks, and is probably functional by 22 weeks or so. The external, middle, and inner ear parts have distinct embryonic histories and can be associated with different kinds of hearing loss if their development is interrupted. There is currently a strong research focus on molecular and genetic influences on ear development, especially using mouse models in which specific genes are altered.

External Ear

At 6 weeks the outer ears start developing from folds on the front neck area of the embryo. A set of six auricular hillocks or bumps arises for each ear. These hillocks are visible as early features of the adult ear. The ear canal forms at 5 weeks, and at about 12 weeks a plate forms at the inner end of the canal. This plate remains until about 7 months, when it dissolves and the remaining tissue forms the tympanic membrane or eardrum. The outer ear and ear canal continue to grow longer after birth, reaching adult size at about 9 years. Failure of the outer ear to develop at all is called anotia, limited growth is called microtia, while partial or complete closure of the ear canal is called atresia.

Middle Ear

At 5 weeks the middle ear cavity begins as an indentation. Meanwhile the three middle ear bones take shape, and are enveloped by the cavity structure. Part of the cavity also forms the Eustachian tube. Disorders of the middle ear can involve the mechanical connections (either loose connections between the bones or total fusion of the bones) for which surgical adjustment is often possible. There are also hearing aids that bypass the middle ear by transmitting sounds to bones of the skull.

Inner Ear

The inner ear development begins around 3.5 weeks of gestational age as the otic placode, which forms into a pit and over the next 12 days closes to become the otic vesicle. The otic vesicle consists of three parts, each leading to different components of the vestibular (balance) and auditory functions of the inner ear. The cochlea develops as an offshoot of one part of this early structure. The turns of the spiral-shaped cochlea are complete by about 9 weeks' gestational age, and the size of the bone structure of the inner ear is adult-like by 17–19 weeks. The structures of the inner ear, including the hair cells and connected auditory nerve fibers, develop on a timetable so that from a neuroanatomical perspective, it is likely that functional hearing begins around 22 weeks' gestational age. Tests of behavioral or electrophysiological responses to sound by fetuses are complicated by methodological problems, but prematurely born infants show brain activity related to sounds by 25 weeks. Also, full-term infants have been shown to respond preferentially to sounds to which they were exposed *in utero*. Considering that full-term gestation is about 40 weeks, it is clear that during the last 3–4 prenatal months the auditory system is actively processing sounds. Nevertheless, the inner ear continues to develop during this period and for many months following full-term birth.

Approximately half of the cases of hearing loss in infants are from known causes. Of these, about 40% are from confirmed genetic causes and 60% from others. Genetic factors are thought to underlie many of the currently unknown cases of infant hearing loss. One example of a genetically based hearing loss involves a protein called connexin 26, which is critical for the formation of junctions between cells. Development of the inner ear can also be affected by prenatal exposure to viruses, bacteria, and drugs. For example, maternal infection by the rubella virus during pregnancy can cause hearing loss, as documented in numerous epidemics across different countries.

Auditory Pathway

A consistent finding in neuroscience studies of the development of vision and hearing is that sensory brain organization depends on neural activity. This can be understood as an efficient way for the brain to adjust to the fine-grained details of the individual's sensory structures. For example, the cochlea maps sounds from high to low frequency, but the specific relationships between individual inner hair cells and sound frequency must be 'learned' by each developing nervous system. Similarly, language development requires exposure to the specific sound system of one's linguistic environment. For many aspects of auditory development, from basic sensitivity to complex activities such as speech perception, it is agreed that refinement of the auditory pathways above the level of the inner ear has a time course extending for months and even years after birth. One important question in this regard is the age at which to provide hearing aids or cochlear implants for infants with hearing loss, an issue taken up later in this article.

Sensitive Periods and Teratogens

As noted earlier, although many cases of hearing loss evident during infancy are the result of genetic causes, the developing auditory system is susceptible to a variety of teratogenic influences from viruses, bacteria, chemicals, and drugs. There appear to be different sensitive periods, that is, times of greatest risk for these influences. For example, rubella infection during pregnancy can cause damage to a number of organs including the ear, but the prospects are worse during the first 16 weeks of gestation than later on. Effects of cytomegalovirus follow a different and more complex time course. Although most infants exposed to this virus do not experience hearing loss, among those who do, the impairment may exist at birth or may become apparent later. Clearly, the timing of sensitive periods differs for different teratogenic agents. Also, some genetically based hearing losses are progressive in nature or have late onset, so that hearing worsens following birth. An important practical implication is that not all hearing losses are apparent in the newborn period, so that regular ongoing assessment of infant hearing is important.

Age Trends in Hearing Sensitivity

Measurement Considerations

The goal in measuring hearing sensitivity is to find the lowest sound level at which a person can just barely detect a sound, a value known as auditory threshold. In clinical settings this is usually done separately for each ear, and for sounds of different frequencies. A distinction is made between behavioral and electrophysiological measures of infant hearing sensitivity. Several behavioral testing procedures have been devised, somewhat independently, by pediatric audiologists and developmental psychologists. All of these procedures involve presenting infants with sounds and noting whether they make an observable response. Unfortunately, infants do not tend to make reliable, robust responses to soft sounds, especially during the first half year or so after birth. One procedure used with these younger infants is behavioral observation audiometry (BOA), in which the tester looks for any sign of a response to sound, such as eye widening, head turning, or cessation of movement. However, this method is difficult to interpret because of their low responsiveness, so audiologists do not consider it a valid measure of infants' level of hearing sensitivity. Developmental psychologists doing research studies tend to use procedures like BOA, sometimes called observer-based psychoacoustics (OBP). Generally, in research studies by either audiologists or psychologists, measures are taken to keep the observer 'honest' by not knowing whether a weak or strong sound was presented. In clinical settings the observer typically knows the sound level, in order to enhance assessment in a

limited time period. Audiologists use a method called visual reinforced audiometry (VRA) for infants older than about 6 months. Infants are taught to turn their heads when a sound occurs in order to see an attractive toy display or a video event. With this method, the question is whether a head turn was made following a given sound.

For most clinical and some research purposes it is customary to present sounds to one ear at a time using small insert earphones. Otherwise, sounds may be played through loudspeakers, an arrangement known as free field. By presenting sounds with different intensities, the examiner finds a threshold intensity at which the infant can just hear the sound. In clinical settings, it is often necessary to obtain thresholds at several different sound frequencies in order to characterize an infant's hearing. One problem with behavioral tests of infant hearing, noted above, is that they are relatively ineffective at ages below about 6 months. This is because young infants do not reliably make overt responses to sounds, even when we know they can hear the sounds well. Therefore, electrophysiological measures of hearing sensitivity also have an important role.

Several electrophysiological measures are used to estimate infant hearing, but the real workhorse is the auditory brainstem response (ABR). The ABR consists of electrical potentials measured by surface electrodes placed on the head, primarily reflecting neural activity of the auditory nerve and along the brainstem. Usually this response is recorded from sounds presented to one ear at a time through insert earphones, although binaural stimulation is possible. It is not only possible but desirable to do this testing while infants are sleeping or sedated, otherwise body movements interfere with the electrical signal. A threshold is found by noting the sound intensity below which the characteristic waveform of the electrical response is absent. A consistent finding, especially during the first year or so after birth, is that hearing sensitivity estimated by the ABR is more acute than by behavioral measures. One reason is that during behavioral testing infants are easily distracted by nonauditory events. However, it is also likely that the ABR reflects auditory processing at a very peripheral level, whereas behavioral testing reflects not only the entire auditory pathway, but also motor responses linked to hearing. Thus, behavioral testing of infants may reveal immaturities in higher-level auditory processing which are not reflected in the ABR. In fact, audiologists regard the ABR and the otoacoustic emissions test (discussed next) not as measures of 'hearing' but rather as measures of functioning in particular parts of the peripheral auditory system.

Another important means of estimating infant hearing is by measuring the presence or absence of otoacoustic emissions (OAE), a technique widely used in screening programs. As noted earlier, these emissions reflect activity of the outer hair cells. This test is not generally used for measuring hearing thresholds, but rather for determining whether a young infant can hear moderately intense sounds. A probe tip is placed in one ear and a sound is delivered through the probe. A very short time later, a microphone in the probe picks up faint sounds produced in the inner ear as part of the auditory system's response to an incoming sound. Most screening programs for newborn-hearing ability use OAE as the initial test. If these emissions are absent or very weak, then follow-up testing with the ABR is carried out.

Age Trends

Newborn infants are much less sensitive to sounds than adults, but typical newborns can hear sounds in the intensity range of normal conversational speech. It is a cause for concern if a newborn infant does not seem to hear sounds of moderate intensity, and the goal of newborn-hearing screening programs is to identify those infants. We turn now to measures of the absolute sensitivity of newborns to sounds of different frequencies. **Figure 2** shows newborn- and adult-hearing sensitivity measured by ABR at frequencies ranging from 500 to 8000 Hz.

Whereas adults have a bow-shaped curve with better sensitivity at middle frequencies, newborn sensitivity is modest across the entire audible frequency range. In fact, an older child or adult with the hearing sensitivity of a newborn would be considered to have a mild-to-moderate hearing loss. The newborn–adult difference is least at low frequencies, but this is because adults do not hear low-frequency sounds very well. Thus, auditory sensitivity improves during development at all frequencies, but there is more room for improvement with respect to middle- and high-frequency sounds compared to lower frequencies. The modest hearing sensitivity of newborns is

partly attributable to peripheral factors, including the acoustic resonance properties of the ear canal, the mechanical efficiency of the middle ear bones, cochlear function, and coordination of activity across auditory nerve fibers. However, neural organization of the auditory pathways from brainstem to cerebral cortex is also responsible for age changes in auditory sensitivity.

Postnatal development of auditory sensitivity can be characterized by two general trends. First, the pace of development is quite rapid during the first 6–9 months after birth, by which time adult-like sensitivity is approached for middle-to-high frequencies. Second, the rate of development is greater for high-frequency sounds than for low frequencies. In fact, adult sensitivity levels for low-frequency sounds are not achieved until about 4–6 years of age. **Figure 3** shows behaviorally tested sensitivity thresholds across a range of frequencies in 3- and 6-month-olds and adults (9-month-olds were also tested, with results virtually identical to the 6-month-olds).

The general pattern of age-related improvement in auditory sensitivity is paradoxical in that infant–adult differences are initially smallest at low frequencies, yet low-frequency sensitivity reaches adult levels latest during development. This can be understood as a high-to-low-frequency gradient of development, superimposed on the overall human audibility function that has rather poor low-frequency sensitivity.

Most studies of infant hearing sensitivity have been conducted with single age groups, or cross-sectionally with different infants across ages. Longitudinal studies are

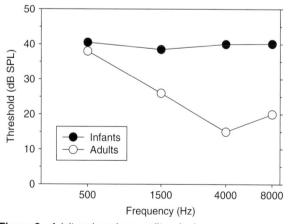

Figure 2 Adult and newborn auditory brainstem response thresholds as a function of sound frequency. SPL, sound pressure level. Adapted from Sininger YS, Abdala C, and Cone-Wesson B (1997) Auditory threshold sensitivity of the human neonate as measured by the auditory brainstem response. *Hearing Research* 104(1–2): 27–38.

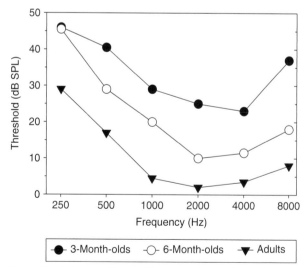

Figure 3 Behavioral thresholds for absolute auditory sensitivity in 3-month-olds, 6-month-olds, and adults. The study also included 12-month-olds, for whom the thresholds were nearly identical to those of the 6-month-olds shown here. SPL, sound pressure level. Adapted from Olsho LW, Koch EG, Carter EA, Halpin CF, and Spetner NB (1988) Pure-tone sensitivity of human infants. *Journal of the Acoustical Society of America* 84(4): 1316–1324.

required to reveal patterns of change within individuals. **Figure 4** shows findings from a study of infants followed during the first year, tested for sensitivity to a noise stimulus filtered to match the frequency composition of speech.

These findings confirm, for individual infants, the pattern of rapid development during the first 6–9 months after birth, followed by continued but very gradual improvement in sensitivity. This developmental profile is an important factor when parents and clinicians make intervention decisions about infants with hearing loss.

The factors underlying both the overall rate and the frequency dependency of the development of auditory sensitivity are not well understood. Some of the changes are the result of physical factors in the external and middle ear. The size and shape of the outer ear and ear canal act as frequency filters, and young infants' ears are tuned to favor somewhat higher frequencies than children or adults. Part of the ear canal itself undergoes a cartilage-to-bone transition during infancy. Another physical factor is the transmission of acoustic energy from the ear canal to the middle ear. Again there is a filter property, with an age-related increase in the amount of energy passed to the middle ear, especially at frequencies above about 1500 Hz. In young infants this may be offset to some extent by higher sound levels in the ear canal at some frequencies. Another likely developmental factor is the transmission of energy through the middle ear to the cochlea. As we know from shouting to underwater swimmers from above

the surface, sound does not travel well from air to water. The fluid-filled cochlea is essentially an 'underwater' device which would strongly limit direct transmission of airborne sound (technically, there is an impedance mismatch). The middle ear overcomes this problem by transforming sound energy into mechanical energy, which drives the cochlear fluid via the oval window. Most of the mechanical advantage of the middle ear is because the tympanic membrane is much larger than the oval window, while some additional power comes from the lever action of the middle ear bones. Although it seems likely that middle ear function changes during development, little systematic research has been reported. However, we do know that there is a change in the angle of the tympanic membrane and that the middle ear bones become sleeker over time, thus contributing to some of the changes in sensitivity noted. Although the overall size of middle ear does not change much after birth, its shape and mechanical properties do change in subtle ways.

Cochlear function is an obvious candidate for explaining age changes in hearing sensitivity, but it is difficult to assess cochlear processes independently. Studies of transient otoacoustic emissions (TEOAE) evoked by click stimuli (which carry a wide frequency spectrum) show that even in newborns the response can be elicited by clicks only slightly stronger than needed in adults. This provides some indication that cochlear function is intact

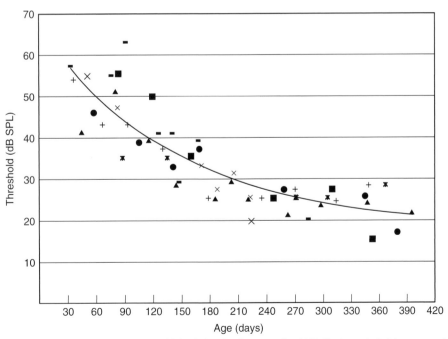

Figure 4 Longitudinal changes in infant auditory sensitivity during the first year after birth. Each symbol style represents a different infant. The curve showing the overall group function was closely matched by individual curves (which are not shown here). SPL, sound pressure level. From Tharpe AM and Ashmead DH (2001) A longitudinal investigation of infant auditory sensitivity. *American Journal of Audiology* 10(2): 104–112.

by early infancy, but the method is not well suited for assessing thresholds or frequency effects. Another variation on this kind of testing is distortion product otoacoustic emissions (DPOAE), in which several tones differing in frequency serve as the stimuli. Good hearing requires frequency selectivity, but there is a tendency for lower frequency components of sounds to mask or suppress higher frequencies. Studies using this method indicate that cochlear frequency selectivity is immature across the age range of newborn to 3 months, and perhaps older. This could reflect both the passive mechanical properties of inner ear structures and the process of active mechanical adjustments via activity of the outer hair cells. Although these findings on frequency selectivity cannot be applied directly to absolute auditory sensitivity, there is a strong implication that cochlear processes change during the months following birth.

Even when factors involving the external, middle, and inner ear are taken into account, the magnitude of infant–adult threshold differences makes it likely that more central processes are also involved in the development of auditory sensitivity. Studies of postnatal development of the auditory cortex in various species indicate, for example, that although there is a basic plan for frequency-specific zones of cortical neurons, the response properties of these cells depend strongly on an animal's auditory experience. Another idea that has been proposed is that there may be more 'internal noise' (e.g., variability in auditory nerve signal transmission) in infants than in adults.

Hearing Disorders

Classification and Incidence of Hearing Disorders

It is estimated that approximately 278 million people worldwide have some form of disabling hearing loss and that two-thirds of those reside in developing countries. Hearing loss is commonly described in terms of severity and type. Severity of hearing loss is broken into the following five categories based on the individual's auditory thresholds: normal range $= 0-20$ dB HL; mild loss $= 20-40$ dB HL; moderate loss $= 40-60$ dB HL; severe loss $= 60-80$ dB HL; and profound loss $= 80$ dB HL or more.

Hearing loss is also classified by type. Specifically, hearing loss is described as conductive, sensorineural, or mixed. Conductive hearing loss refers to a decrease in auditory thresholds resulting from an interruption in the conduction of sound along the ear canal, ear drum, middle ear space, or the middle ear bones (i.e., an impaction of the ear canal, middle ear fluid, or otosclerosis). A purely conductive hearing loss does not involve the cochlea or

remaining auditory system and can often be treated with medical intervention (i.e., surgery or antibiotics). The severity of conductive hearing loss may range from mild to moderate, depending on the cause and severity of the pathology. For example, disarticulation of the ossicles completely interrupts the transmission of energy to the cochlea and can result in hearing thresholds up to 60 dB HL.

A sensorineural hearing loss originates from either the cochlea (sensory) or auditory nerve (neural). This type of hearing loss may occur when there is damage or congenital malformation to the cochlea or the auditory nerve, and is most often permanent. The causes of sensorineural hearing loss in children typically include disease (e.g., meningitis), ototoxic medications (e.g., gentamicin, vancomycin), maternal teratogens (e.g., ethanol ingestion), and genetic factors (e.g., connexin 26). A sensorineural hearing loss in adult years may occur because of excessive noise exposure, the aging process, head trauma, or tumors along the auditory nerve.

It is possible for hearing loss to have both a conductive and sensorineural component. For example, an individual may have an ear infection in addition to an already diagnosed sensorineural hearing loss. In this case, the conductive component is expected to resolve. In other cases, an individual may have a permanent conductive hearing loss in addition to a permanent sensorineural hearing loss.

The prevalence of permanent hearing loss with onset in gestation or infancy in European and North American countries is about 1–6 per 1000, with the lower number referring to more severe degrees of bilateral impairment. Many infants and young children experience transient hearing loss as a result of otitis media, a condition in which there is inflammation and possibly fluid buildup in the middle ear, usually associated with bacterial or viral infection. Young children are at greater risk of this condition because the angle of their Eustachian tube does not favor drainage of infectious organisms. Otitis media is among the leading conditions for pediatric office visits. The degree of hearing loss secondary to otitis media is variable. About half of infants and young children with otitis media experience mild hearing loss and 10% have moderate loss. However, evidence for long-term consequences for hearing and language development does not suggest a strong linkage. Practice guidelines have shifted away from aggressive treatment with antibiotics and surgical intervention toward an emphasis on 'watchful waiting'.

Congenital hearing loss can be syndromic or nonsyndromic. A syndromic hearing loss presents itself with other symptoms. For example, Usher syndrome results in visual impairments as well as auditory deficits. However, most children with genetic hearing loss are nonsyndromic and are born to parents who have no family history of hearing loss.

Screening for Hearing Loss

Efforts to identify infants with hearing loss are based on the reasonable assumption that communication skills are a critical component of infant development, with a goal of providing intervention by 6 months of age. Some known risk factors such as premature birth or family history of genetically based hearing loss can be used to target infants for audiological assessment. However, the application of risk-based screening programs is generally thought to miss substantial numbers of infants with hearing problems. Therefore, the goal of universal screening has become popular as a public health approach. These programs have been directed at screening of full-term newborns or prematurely born infants when they are approaching full-term gestational age. A drawback of the focus on newborns is that these programs miss some infants and children with progressive or late-onset hearing loss, as well as some infants with mild degrees of hearing loss.

Newborn hearing screening

Most newborn screening programs use a two-stage procedure. In the first stage, each ear is evaluated separately using an automated OAE test. These tests are usually administered by nursing staff with little training in audiology. As noted earlier, OAE reflect activity of the outer hair cells, which provides an indirect measure of whether the peripheral auditory system is functioning. If either ear fails the OAE test, then an ABR screen is typically performed on each ear. Infants who fail this second screening are referred for a complete audiological evaluation. Infants born prematurely who have extended hospitalization should be tested as close as possible to their discharge date or their full-term date (whichever comes first), in order to reduce the possibility of failing the screening because of various changes in the ear canal, middle ear, or brainstem.

In general, these screening programs have rather modest epidemiological specificity, meaning that quite a few infants fail the test even though their hearing is normal. This not only has financial costs but also a psychological impact on parents, who are in the position of wondering whether their child really has a hearing loss. The screening programs also have a problem with sensitivity, or failing to identify some infants with hearing problems. By some estimates, about 20% of infants with hearing loss at 1 year of age were missed by newborn screening programs. This may reflect a combination of examiner error, equipment error, failure to screen, and occurrence of progressive or late-onset hearing loss that would not have been evident in the newborn period. However, the lesson is that ongoing assessments of infants after the newborn period are essential.

Screening in older infants and children

Mild, progressive, and late-onset hearing loss may not be detected by newborn hearing screening. Therefore, infants and older children should be screened for hearing loss during regular visits to the pediatrician's office or at least once during the preschool period. Overall communication development can effectively be screened using structured parent-interview scales. It is also possible to screen older infants using OAE, as described earlier for newborn screening. Children over about 2 years of age can be screened using behavioral techniques such as play audiometry. Play audiometry involves instructing a child to respond to sound by performing a motor act such as putting a block in a box. Typically, screening is recommended at 1000, 2000, and 4000 Hz in both ears at 20 dB HL. If a child does not respond at least two out of three times at any frequency in either ear, or if the child cannot be conditioned to the task, a referral to an audiologist with pediatric experience is recommended.

Audiological Management of Children with Hearing Loss

Pediatric audiological assessment

When an infant or young child does not pass a hearing screening or is suspected of having a hearing loss for any reason, an audiological evaluation should be performed by an audiologist who has expertise with infants and young children. The evaluation of infants and young children incorporates a test battery approach that should include tests of middle ear function, hearing sensitivity, and functional use of hearing. The specific procedures used to measure hearing sensitivity depend on the age and developmental abilities of the infant or child, but typically include a combination of the ABR, OAE, immittance audiometry, and one of several available behavioral procedures discussed previously. Sensitivity should be assessed for both ears across a range of sound frequencies including those that predominate in speech (i.e., 0.5, 1.0, 2.0, and 4.0 kHz). Through the use of a test battery, the audiologist can determine the level of hearing sensitivity and, if hearing sensitivity is not normal, the degree, type, and configuration of the hearing loss.

An important component of the evaluation process is engagement of parents, both in assessing the infant's or child's communication needs and abilities, and in planning for rehabilitation services. This can involve a number of caregiver questionnaires used to gain information about the infant's or child's functional use of hearing and language in real-world settings.

Modes of communication

A primary consequence of severe-to-profound hearing loss is the inability to hear spoken language. This problem is compounded by the finding that approximately 90% of all children with congenital hearing loss are born into hearing (speaking) families. Therefore, families of children with severe hearing loss are faced early on with making a decision about which communication modality to use for teaching language to their child. The modalities are distinguished as an oral approach with spoken language, a manual approach with sign language, or a total communication approach using a combination of speech and signs. With current technology, such as hearing aids and cochlear implants, and under ideal rehabilitative circumstances, it is possible for many young children with hearing loss to have access to auditory information and develop spoken language skills that are comparable to normal hearing peers. However, several factors need to be considered when deciding which communication modality to use with a child who has hearing loss. For example, consideration should be given to the child's capacity to develop spoken language, the severity of the child's hearing loss, access to various forms of intervention, and how the family feels about deaf culture'. Ultimately, a family is not tied to any modality and may change among them as the needs of the child and family shift. Regardless of which communication modality a family chooses to use with their child, the goal is to facilitate the development of a fluent communication system by school age.

Hearing technology

Many infants with permanent hearing loss can benefit from hearing aids or cochlear implants. Hearing aids can be effective for mild-to-profound degrees of conductive or sensorineural hearing loss. Most hearing aids work by presenting amplified acoustic signals to the ear canal, with the understanding that inner ear function in that individual is adequate to take advantage of the input. The primary goal of amplification for infants and young children is to make speech input comfortable and audible. Thus, it is crucial, for purposes of spoken language development, that infants be fitted with hearing aids as soon as their hearing loss is identified and that they wear their hearing aids all of their waking hours. Unless there is a contraindication, children are fitted with bilateral hearing aids because two ears are optimal for sound localization and for hearing in the presence of background noise.

Hearing aids can take many different styles including body, behind-the-ear (BTE), in-the-ear (ITE), and in-the-canal (ITC) aids. Style will be dictated by a child's degree of hearing loss, potential for growth of the outer ear, and individual needs. The outer ear may continue to grow until around 9 years of age, thus dictating the BTE style. Otherwise the hearing aid would need to be re-cased on a regular basis. If using a BTE hearing aid, when growth occurs, only the earmold (the piece that

couples the hearing aid to the ear) has to be replaced. Other reasons for fitting young children with BTE aids are that it is more durable (with no circuitry directly exposed to ear wax) than ITE styles, is less likely to produce feedback (the whistling noise produced by amplified sound re-entering the hearing aid microphone), and allows for a variety of features that may be desirable for children (i.e., circuitry for using the aid with a telephone, connections for coupling to television, stereo, or computer inputs).

As described previously, some infants have conductive hearing loss, meaning that sound is not transmitted well by the ear canal or middle ear, with intact inner ear function. In many cases, these problems can be addressed through surgery. If not, and if a traditional air-conduction hearing aid cannot be fitted properly because of a malformation, a bone-conducted hearing aid may be used. Acoustic input is converted into vibrations that are transmitted by the bones of the skull to the cochlea. Two varieties of bone-conducted hearing aids are available, with the stimulator either worn externally on the skull surface, or surgically implanted near the ear ('bone-anchored'). At the time of this writing, the bone-anchored aids are not used with infants and young children in the US.

Another type of hearing technology commonly used with children is a frequency-modulation (FM) system. Because the microphone on a hearing aid will amplify everything in the surrounding area, it is often desirable (especially when there is background noise) to have just an individual's voice amplified. An FM system includes a microphone worn by the speaker, which is typically pinned to the clothes or hung around the neck, to pick up his or her voice, a transmitter to send the signal to the listener, and a receiver worn by the listener. The FM system can be coupled to a child's hearing aid or can be worn alone depending on the style. Regardless of the style used, FM systems are noted for providing a clear acoustic signal to the listener and are often used in educational settings.

Many infants with sensorineural hearing loss that cannot be compensated by hearing aids can benefit from cochlear implants. A cochlear implant is a small ribbon-like array of electrodes that is surgically implanted in the cochlea. The electrodes are wired to a receiver/stimulator, which is implanted in the skull near the ear. An external microphone and acoustic processor transmit signals to the receiver through the skin, causing the stimulator to activate the electrodes in the cochlea. Contemporary cochlear implants have between 12 and 20 electrodes or frequency channels, which are distributed approximately along the natural sound frequency gradient of the cochlea. The match to the natural frequency gradient is not exact, mainly because the implant does not fit all the way to the end of the cochlea. The frequency mismatch requires some adaptation on the part of the cochlear implant user. Thus, when the incoming sound to the microphone has energy at a certain frequency, the stimulator activates a corresponding electrode. Cochlear implants have been widely used with adults since the mid-1980s, and are now approved for use with infants and young children. Regulations about the youngest age at which implants may be used vary by country but implanting children as young as one year of age has become common practice. The question naturally arises as to the optimal age for initiating cochlear implant usage with respect to language development. Research on this topic is complicated by ethical constraints on study designs and by the rapid pace of technical improvement. However, there is general consensus that spoken language development is optimized by use of cochlear implants before the age of 3 years.

It is now common for adults to receive cochlear implants in both ears. Several studies have shown a binaural advantage for adult bilateral cochlear implant users and there is also a growing trend for children to receive bilateral implants as well. There is evidence that implanting bilaterally at a young age will facilitate the development of binaural neurons and allow young children to enjoy the benefits of binaural hearing, although some argue the logic of saving one ear for future technology.

Family support services

It is generally accepted that it is not the lack of hearing but rather the lack of language that can result in psychoeducational and psychosocial difficulties for deaf children. Therefore, regardless of the communication approach decided upon by the family, the early introduction of services that enhance the development of language is crucial. These services should start as soon as an infant is identified with hearing loss, often in the newborn period. Families have repeatedly indicated that they want factual information about the hearing loss and its effects on their child's development. They express interest in learning about causes of hearing loss, communication and educational options, and medical and technological interventions. There are many different models for the provision of services to these families. Services such as these should optimally be delivered via a team approach including audiologists, speech-language pathologists, otologists, deaf educators, and early interventionists. Some service delivery models include home-based services and others include center-based services. There are also numerous online information sources available to families worldwide.

Infant–parent services typically consist mostly of educating the family on the impact of hearing loss and the importance of creating a rich language-learning environment for the child. This includes learning about hearing technology (use and care of hearing aids or cochlear implants), good communication strategies (selecting an effective mode of communication), and creating a good listening environment (reduction of background noise in the home). As the child gets older, more direct intervention with the child occurs. This intervention is likely to include speech-language therapy, preliteracy skill development, and auditory training. The family, including the child when appropriate, should receive information on the services available to them through government and other agencies.

See also: Auditory Perception; Brain Development; Developmental Disabilities: Cognitive; Developmental Disabilities: Physical; Genetic Disorders: Single Gene; Genetics and Inheritance; Language Development: Overview; Language Disorders; Nature vs. Nurture; Neurological Development; Prenatal Development; Preverbal Development and Speech Perception; Screening, Newborn and Maternal Well-being; Special Education; Speech Perception.

Suggested Readings

Abdala C (2004) Distortion product otoacoustic emission (2f1–f2) suppression in 3-month-old infants: Evidence for postnatal maturation of human cochlear function? *Journal of the Acoustical Society of America* 116(6): 3572–3580.

McConkey Robbins A, Koch DB, Osberger MJ, Zimmerman-Phillips S, and Kishon-Rabin L (2004) Effect of age at cochlear implantation on auditory skill development in infants and toddlers. *Archives of Otolaryngology Head and Neck Surgery* 130(5): 570–574.

Olsho LW, Koch EG, Carter EA, Halpin CF, and Spetner NB (1988) Pure-tone sensitivity of human infants. *Journal of the Acoustical Society of America* 84(4): 1316–1324.

Sininger YS, Abdala C, and Cone-Wesson B (1997) Auditory threshold sensitivity of the human neonate as measured by the auditory brainstem response. *Hearing Research* 104(1–2): 27–38.

Tharpe AM and Ashmead DH (2001) A longitudinal investigation of infant auditory sensitivity. *American Journal of Audiology* 10(2): 104–112.

Trehub SE, Schneider BA, and Endman M (1980) Developmental changes in infants' sensitivity to octave-band noises. *Journal of Experimental Child Psychology* 29(2): 282–293.

Werner LA and Marean GC (1996) *Human Auditory Development*. Boulder, CO: Westview Press.

Widex South Africa, Understanding your Ear. http://www.widex.co.za/guide-book/your-ear.htm.

Relevant Websites

http://www.agbell.org – Alexander Graham Bell Association for the Deaf.

http://www.asha.org – American Speech–Language–Hearing Association (ASHA).

http://www.shhh.org – Self-Help for Hard of Hearing.

http://www.nidcd.nih.gov – US National Institutes of Health.

http://www.med.unc.edu – UNC School of Medicine.

Auditory Perception

D H Ashmead, Vanderbilt University Medical Center, Nashville, TN, USA

Glossary

Auditory scene analysis – A collective term referring to the ability to keep track of multiple sound sources in complex listening situations.

Cochlear implant – A device that is surgically placed in the inner ear, to translate sounds arriving at the listener into direct electrical stimulation of the auditory nerve.

Frequency selectivity – The ability to attend to a sound of a given frequency without distraction from competing sounds with nearby frequencies.

Intensity resolution – The ability to detect small increases or decreases in sound intensity.

Loudness – The psychological experience of how strong or weak a sound is, an experience that depends only partly on the actual physical intensity of the sound.

Pitch perception – The psychological experience of how low or high a sound is on the frequency dimension, and of how two or more sounds are related to one another on this dimension.

Sound localization – Perception of the spatial location of a sound source, in terms of horizontal direction, vertical direction, and distance.

Temporal resolution – The ability to resolve very brief changes in a sound, such as the occurrence of a silent gap injected into an ongoing sound.

Introduction

Other articles in this encyclopedia cover topics such as basic structure/function of auditory system, absolute auditory sensitivity, speech perception, music perception, and intersensory perception. This article focuses on certain aspects of listening experiences such as loudness, pitch, and spatial localization. These topics fit under the rubric of perception because they are about how we can use sound to learn about the properties of events going on around us. In a very real way, auditory perception is event driven, because something has to move in order to create a sound. Whether it is recognizing someone by their vocal pitch or attending to the snap of a nearby twig, auditory perception is all about events that make sound.

Basic Auditory Capabilities

The disciplines of audiology and hearing science have traditionally emphasized behavioral measures closely linked with acoustic variables and processing at the more sensory and peripheral levels of the auditory system. This approach has proved valuable both for clinical practice and basic research understanding. Investigators working in these disciplines have translated the measures and concepts into procedures workable with infants and young children. In this article, three basic auditory capabilities are reviewed with respect to early development. Frequency selectivity is the ability to listen within a narrow range of frequencies without interference from nearby frequencies. Temporal resolution is the ability to perceive changes across time in attributes of a sound such as intensity or frequency, and is related to the timespan across which incoming acoustic energy is integrated or averaged. Intensity resolution is the ability to detect small changes in intensity. Taken together, sensitivities to changes in frequency and intensity across time are considered important characterizations of the auditory system's ability to provide meaningful information about events.

Frequency Selectivity

Just as the visual system is organized to keep track of the spatial distribution of stimulation on the retina, so the auditory system preserves information about the frequency composition of sounds. This is useful because the events leading to acoustic stimulation, such as a spoken word or a footstep, have 'signatures' in the frequency spectrum that allow us to identify them. In a setting with multiple sound sources it is important for a listener to focus on activity in one frequency region without interference from nearby frequencies. Classic studies beginning during the 1920s established the concept of the auditory filter, based on the masking paradigm. Detection of a tone or a narrow frequency band of noise can be masked (made more difficult) by other sounds at nearby frequencies, whereas the masking effect from more remote frequencies is far smaller. This led to the idea that much of our listening occurs within fairly narrow frequency regions, sometimes called critical bands. A number of processes promote this frequency filtering, including the fact that frequencies are processed from high-to-low along the extent of the cochlea in the inner ear, local physical enhancement of cochlear activity ('cochlear amplifier'), neural segregation by frequency in most stages of the auditory pathway, and active neural inhibition of inputs from nearby frequencies at central processing sites such as the auditory cortex. Each of these processes is a candidate for developmental change in frequency selectivity during the early postnatal period. For example, the cochlear amplifier is a process whereby stimulation at one place along the cochlea results in a temporary retuning of the mechanical structures there, such that there is even greater sensitivity to sound at the frequency that caused the stimulation.

The most widely used measure of frequency selectivity is the auditory filter width, based on the masking paradigm. The listener's task is to detect a target tone or a narrow frequency band of noise that is embedded in a background masker noise. By presenting masker sounds of different frequencies and intensities, the amount of masking needed to prevent detection of the target can be measured. When the masker closely resembles the target in frequency, even a weak masker blocks target detection, because neural processing units are activated readily by both stimuli. With greater frequency separation between

target and masker, the amount of masker strength needed grows, because the neural units focused on the target frequency are less easily affected by the masker. The function relating masker strength to masker frequency is approximately V-shaped, and the width of the V near its tip is defined as the auditory filter width. An example of such functions is shown in **Figure 1**. It must be noted that if the auditory filter was too wide, there would be a lot of interference across sounds from different sources that ought to be processed separately. Masking studies may be carried out behaviorally, based on observations of infants' or children's responses to sounds; electrophysiologically, using the auditory brainstem response (ABR, recorded by scalp electrodes); or by a method called distortion product otoacoustic emissions (DPOAE), which measures subtle mechanical adjustments of the inner ear in response to sounds.

Behaviorally based masking studies of infants and children beginning in the 1980s suggested a paradox, in that auditory frequency selectivity improved (filter width narrowed) from birth to adult-like levels by 6 months, whereas children around 4–6 years old had poorer frequency selectivity than adults. The paradox was resolved

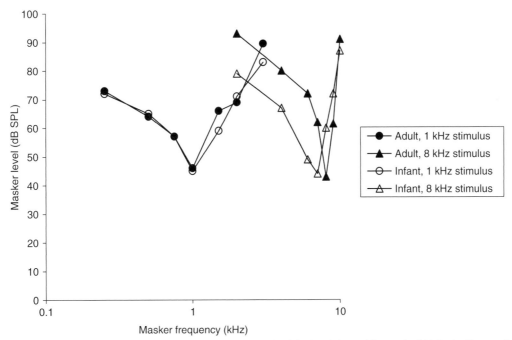

Figure 1 Illustration of masker functions to investigate frequency selectivity in adults and 3-month-old infants. The y-axis shows the magnitude of noise from a given frequency region that is needed to mask, or block, a stimulus from being heard. For example, when the stimulus was a 1 kHz (1000 Hz) tone, a masking noise centered on 1 kHz had to be about 45 dB in magnitude in order to mask the tone. This was true for both infants and adults. When the masking noise was not centered on the stimulus frequency, the masker had to be stronger in order to block perception of the stimulus. Thus, any sound is more easily masked by other sounds from nearby frequencies than by sounds from very different frequencies. The steepness of the masker function is a measure of how selective the auditory system is at keeping track of sounds differing in frequency. At low frequencies (1 kHz stimulus) infants and adults have similar frequency selectivity. At higher frequencies (8 kHz) infants have shallower masker functions than adults, suggesting that infants have less sharply tuned frequency selectivity at high frequencies. Adapted from Folsom RC and Wynne MK (1987) Auditory brainstem responses from human adults and infants: Wave V tuning curves. *Journal of the Acoustical Society of America* 81(2): 412–417.

by showing that the results with children were an artifact of the way they experience loudness, and that children had adult-like filter widths under appropriate test procedures. Current thinking, based on a combination of behavioral and otoacoustic emissions studies, is that the auditory filter width for high-frequency sounds (>5 kHz or so) is rather wide from the newborn period through to about 3 months, probably improving to adult width by about 6 months. For lower-frequency sounds the filter width is adult-like from the newborn period.

The poor-frequency selectivity of very young infants described in the foregoing summary is likely caused by immaturity of the cochlea and its amplifier process. This process is an enhancement of mechanical activity on the basilar membrane, mediated by the outer hair cells. However, frequency selectivity is also affected by more central stages of auditory processing. For example, animal studies show a protracted developmental course for inhibitory processes in the auditory cortex. Cortical neurons are responsive to a range of sound frequencies, and sensitivity to one frequency is improved by inhibiting inputs from nearby frequencies. These inhibitory processes take time to develop after birth. Moreover, active experience with spectrally complex acoustic stimulation is needed for the normal rate of development, although adult animals raised with frequency-restricted sounds quickly acquire inhibitory processes when exposed to a normal sound environment.

This research on cortical inhibitory processes that enhance frequency selectivity raises an interesting question about young children who receive cochlear implants – how does a system with a history of limited or absent input respond to the 'switching on' of sensory stimulation? Current implants have multiple channels to represent a range of frequencies. These channels are programmed with respect to frequency width and sound level. Presumably, the user's frequency selectivity is based on a combination of mapping the programming to the basic structure of the auditory system (e.g., the frequency representation along the cochlea) and the flexibility or plasticity to adapt to the somewhat arbitrary stimulation pattern from the devices. An interesting aspect of this issue is that physical constraints prevent the cochlear implant from being inserted into the full length of the cochlea. Thus, the range of audible frequencies is 'squeezed' into a smaller distance along the cochlea, with the innermost end that normally receives the lowest frequencies not being stimulated. Studies of insertion depth in adult cochlear implant users indicate that speech perception is enhanced when the implant is inserted deeper. This implies that there is limited ability to process very-low-frequency components of sounds outside the 'natural' cochlear region for those frequencies. Whether young children with little or no history of useful auditory experience would show more flexibility at adapting to altered frequency regions along the cochlea is not yet known. However, most young children (aged about 3–5 years or less) who receive cochlear implants acquire useful spoken language ability, in terms of both understanding and production. Thus, there must be a substantial capacity for adapting to the stimulation provided by cochlear implants, despite an early history of deafness.

Temporal Resolution

The auditory system excels at keeping track of rapid changes in sounds. This is important for speech perception, because the differences between many speech sounds are based on timing information. For example, the initial consonant sounds of 'ba' and 'pa' are distinguished by how rapidly the leading release of acoustic energy is followed by the next phase of the sound. Temporal resolution has been studied using a variety of procedures, most of which focus on the ability to detect time-varying changes in the intensity of a sound. One idea underlying these procedures is that the auditory system has a minimal integration time window, such that acoustic events happening within that timeframe are integrated and cannot be distinguished from one another. One such procedure is gap detection, in which there is a silent interval between two sounds. If the silent interval is too short, then a single sound with no gap is heard.

Adults can hear gaps of two- or three-thousandths of a second in a burst of noise, and can detect modulations of noise amplitude at rates up to 1000 times per second. By comparison, the visual system is more sluggish, detecting modulations of light up to only about 60 times per second. This difference shows that the sense of hearing is specialized for perceiving rapid changes, whereas vision is good at registering spatially distributed light patterns over a longer time window. Behavioral studies of gap detection show that in early infancy it is perhaps 10 times worse than in adults, and improvement may continue through about 5 years. However, studies using electrophysiological measures of the brain's response to gaps show nearly adult-like gap detection at 3–6 months. Discrepancies between behavioral and electrophysiological tests are always difficult to interpret, but given the subtlety of the behavioral gap detection task (hearing a very brief silent interval), it seems likely that the time-scale over which sounds are integrated is fairly well established by 6 months or so. Even so, it is possible that there are individual differences in the developmental course of auditory temporal processing. Problems with auditory temporal perception have often been regarded as contributors to developmental problems such as delayed language acquisition, dyslexia, and attention deficits. While the research literatures in these areas are complex and contradictory, the basic idea that poor auditory temporal resolution might lead to communication difficulties is plausible.

Intensity Resolution

The auditory system handles sounds varying from very low to very high levels of physical intensity. Three topics that relate to this intensity range are the absolute threshold for hearing a sound (i.e., the softest audible sound), the ability to hear changes in sound level, and the psychological experience of loudness (covered in the section titled 'Perception of higher-order properties'). Intensity resolution refers to the smallest change in sound level that one can detect. It is relevant to everyday activities like setting volume controls on loudspeakers and adjusting one's vocal output for the social situation. Adults can detect intensity increases or decreases as small as one-half a decibel, which is a difference in physical sound level of about 6%. This works so that, going from barely audible to very strong sounds, we can detect more than 200 gradations in intensity. By about 6 months, and perhaps earlier, infants can detect intensity changes of several decibels, putting them close to adult range. This suggests that infants have adult-like changes in neural activity in the ear and auditory nerve as sound levels change, and that they are behaviorally aware of the change. An interesting twist is that infants are more likely to respond to increases in sound intensity than to decreases, but this may reflect the psychological salience of increases, rather than an inability to hear decreases. It has been speculated that intensity increases may be more salient because in real world settings they specify that the distance between listener and sound source is getting shorter. Several important features of the development of intensity resolution have not been studied, such as how it varies with sound frequency, but on the whole it appears that even young infants can resolve fairly subtle changes in sound level.

Perception of Higher-Order Properties

For purposes of this article, the ability to resolve frequency changes, the temporal integration window, and the ability to notice intensity changes are regarded as basic auditory capabilities. The rationale is that these abilities are largely enabled by peripheral structures and processes at the inner ear and auditory nerve. Some other important auditory capabilities may be regarded as higher order, in that they probably involve more central stages of auditory processing, including listening experiences with a substantial cognitive overlay. The term 'central' does not have an exact specification, but historically it has been used to contrast with 'peripheral' processes at the outer ear, middle ear, inner ear, and auditory nerve. For the higher-order properties described ahead, it is likely that the neural structures involved include the auditory cortex, and even cortical structures outside the traditional auditory areas. Specific topics considered in this context are loudness, pitch, auditory scene analysis, and spatial hearing.

Loudness

The concept of loudness refers to our psychological experience of how the apparent strength of a sound varies with its physical intensity. Although the word 'loud' is often used in everyday language as a description of a sound's physical intensity, the listening experience of loudness varies across individuals and is context dependent. For example, if you raise the volume on a television in order to hear it while a household appliance is running, the television sounds too loud when the appliance goes off, even though the television sound level has not changed. Methods for measuring loudness include magnitude estimation using various rating scales, crossmodal comparisons to dimensions such as visual brightness, and adjusting physical intensity to achieve a loudness match between the target sound and a standard sound (usually a 1000-Hz tone). Loudness increases with physical sound intensity by a power law, such that an equivalent change in loudness requires less physical change at low intensities than at high intensities. Not surprisingly, loudness varies with the frequency composition of sounds in a way that is roughly the flip side of auditory sensitivity, that is, frequencies to which we are more sensitive at threshold are also perceived as louder in the above-threshold range. Despite efforts to quantify loudness, it involves very subjective elements as well, particularly when the notion of a comfortable listening level is involved. For example, people often vary in their impressions of whether a public address system is too loud. No conclusive research has been reported on infants' experience of loudness, and very little has been done with children under 5 years of age. Children older than about 5 years make loudness judgments quite similar to adults, although there tends to be more variability across individuals among children.

Loudness has practical implications for the design and optimal setting of hearing aids and cochlear implants. Many people with hearing disabilities not only have trouble hearing soft sounds, but also experience unusually strong growth of loudness as the intensity of sound (or electrical stimulation by cochlear implants) increases. In audiology this is called atypical recruitment of loudness. When a hearing aid or cochlear implant user is fitted with the device, it is important to make adjustments that keep the range of sound intensity within a zone of comfortable listening. Older children and adults can provide feedback about whether sounds at various frequencies are audible and comfortable, but younger children and especially infants are more limited in this regard. Audiologists have developed various fitting strategies for pediatric clients, but the ability to tailor these to individuals is still limited. One promising research approach is to examine, in older children, the relationship between certain electrophysiological measures of hearing and loudness/comfort judgments. The idea is that the electrophysiological measures could then be extrapolated to infants and young children.

Pitch Perception

'Pitch' refers to the listening experience that varies in a rise-and-fall sense with the frequency of a sound. Pitch is considered the result of constructive listening, rather than being a direct result of which areas along the basilar membrane are stimulated by a sound. A familiar example of this constructive process is that the melodic structure of tunes is preserved across changes in overall frequency, provided that relative pitch relations are maintained. At least three topics can be identified in research on the early development of pitch. One is the salience of pitch changes in infants' organization of acoustic input, the second is the acquisition of adult-like assignment of pitch in harmonic sounds, and the third is the question of whether pitch is perceived in relative or absolute terms. In addition to these topics, pitch has been studied in research on early musical perception. It should also be noted that pitch perception may be involved in the appeal of infant-directed speech ('motherese'), in which the frequency range of the adult's voice widens, especially at higher frequencies.

Perhaps the most basic fact to observe about pitch perception during early development is that changes in the frequency composition of sound are quite salient as early as the newborn period and perhaps even during fetal development. This has been demonstrated, for example, by showing that brain electrical responses to events that differ from an ongoing event are engaged when the dimension along which the events differ is sound frequency. Responsiveness to pitch changes is also suggested by psychophysical studies of the smallest changes in sound frequency to which infants respond. Those studies have used not only discrete jumps in frequency, but also upward or downward glides in frequency. Although studies of infants' responses to changes in sound frequency have focused on a variety of underlying questions, a general implication is that pitch is a very important part of infants' listening experiences. This is consistent with the idea that pitch is a critical dimension of sound for such varied functions as perceiving differences in speech sounds, recognizing people from their voices, and perceiving events like footsteps and animal sounds.

Most natural sounds contain a fundamental frequency as well as harmonics or overtones at integer multiples of the fundamental. This arises from the physics of vibrating objects that produce sound. For example, the middle C key on a typical modern tuned piano produces a fundamental vibration at about 262 Hz, with additional harmonics at 524, 786, etc. This occurs because the string vibrates strongest at a fundamental frequency set by the string's length, at another frequency corresponding to half the length, another for one-third the length, and so on. Our pitch experience of such a sound is dominated by perception of the fundamental frequency, to such an extent that pitch is perceived as corresponding to that frequency even if the sound is modified to physically remove it, leaving only some of the harmonics. This 'missing fundamental' technique provides some of the strongest evidence that pitch is a constructive process in the brain.

By about 7 months of age, infants appear to experience the missing fundamental similar to adults. **Figure 2** illustrates sound waveforms used in some of this research. For example, Clarkson and Clifton trained 7–8-month-olds on a task in which a sound like the middle waveform in **Figure 2** was presented once every second. This consisted of a tone complex with a fundamental frequency of 160 Hz, plus a series of overtones. Occasionally the sound was changed to a tone complex with a different fundamental frequency, 200 Hz (this is not shown in **Figure 2**, but it would resemble the middle waveform there, with the major repeating elements occurring closer together in time). Infants learned that if they turned their heads when the sound changed, they could see an attractive animated toy display. Half of the infants were trained to respond to the change to 200 Hz on a repeating background of 160 Hz, as just described, and half had the opposite frequency rule. After an infant learned to discriminate in this way between the 160 and 200 Hz sounds, different sounds were presented, such as the missing fundamental sound shown in the bottom of **Figure 2**. The full experiment included other variations on the sounds, based on inclusion of different combinations of overtones. Infants learned to respond equivalently to any sound that corresponded to the fundamental frequency to which they had been trained to respond. Other experiments subsequently showed that this pitch-based discrimination is not attributable to subtle differences between waveforms that might be picked up in the auditory periphery. Rather, infants appear to perceive pitch based on the timing of the major elements of signals, as shown in **Figure 2**, disregarding many of the differences between sounds. Thus, pitch perception is a constructive listening process that searches for common elements across sounds, even though the sounds are not identical. It has been difficult to extend this work to younger ages because the conditioned head turn measure does not work well below 7 months of age, but in principle, other behavioral or electrophysiological measures could be used to explore the earlier development of pitch classification using the missing fundamental approach.

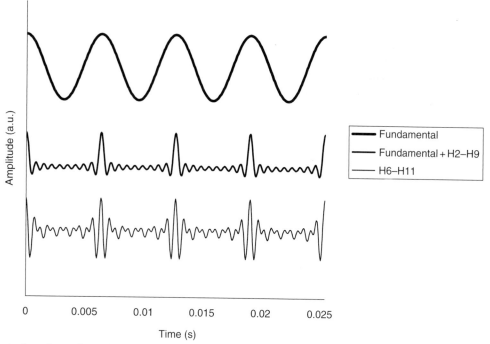

Figure 2 Illustration of sounds used in some studies of pitch perception. Each waveform shows the amplitude of a signal across a short span of time. Amplitude units are arbitrary, since an investigator would scale the amplitude to a comfortable listening level. The three waveforms have been vertically offset in order to show each one individually, but in practice they have a common zero point. The top signal is a pure tone with a frequency of 160 Hz, which is the value of the 'fundamental' frequency in this example. The middle signal is a 160 Hz tone combined with harmonics (overtones) 2 through 9 (320, 480, . . ., 1440 Hz). The bottom signal is a 'missing fundamental' signal, consisting of harmonics 6 through 11 (960, 1120, . . ., 1760 Hz). This graph shows that the major feature of all three signals is the repeating element corresponding to the 160 Hz fundamental frequency. Therefore, listeners will readily categorize these signals as equivalent, even though they do not sound exactly the same. Also, listeners will easily distinguish these signals from others that are based on a different fundamental frequency. Figure prepared by the author, to illustrate stimuli from Clarkson MG and Clifton RK (1985) Infant pitch perception: Evidence for responding to pitch categories and the missing fundamental. *Journal of the Acoustical Society of America* 77(4): 1521–1528.

There has been considerable interest in the ability of some people to have absolute pitch perception, also known as perfect pitch. In practice, this is usually defined as the ability to determine whether a musical sound matches a specific standard such as the note 'A over middle C' (440 Hz) on a musical scale. However, the concept of absolute pitch perception refers more deeply to the listening ability to match sounds to fixed, specific standards. The exact criterion for what constitutes absolute pitch perception is not fully agreed on, and some experts would say that being able to judge whether two sounds are identical, even when separated by a long time such as a week, is an example of absolute pitch. Regardless of definitional issues, a consensus in this research area is that only a small proportion of older children and adults have absolute pitch, and that it requires early and sustained musical exposure. Possessors of absolute pitch excel at being able to categorize sounds according to pitch, but as a group they are no better than others at perceiving fine gradations in sound frequency. Thus, absolute pitch is sometimes described as an achievement of cognition rather than perception. Because of the apparent importance of early experience for absolute pitch perception, the question whether infants perceive pitch in absolute or relative terms has been investigated.

Most of this work has been done with infants between about 6 and 12 months of age. Studies of infants' memory for melodies over the course of several days to weeks indicate that infants are quite responsive to changes in relative pitch (i.e., changes in melodic contour), whereas they do not respond to differences in absolute pitch (transposed versions of the same melody). For example, if infants are familiarized to a tune such as, "Mary had a little lamb," and then tested several days later, they will perk up more if some of the individual notes are changed than if the entire tune is played in a different key. The change in key preserves relative pitch relations between the notes, but the change in individual notes does not.

Another type of study suggests the opposite interpretation, that infants respond more to absolute pitch than relative pitch. In such a study, infants are presented with fairly long sequences of brief tones varying in frequency, organized so that some frequencies are more likely to occur close together in time than other frequencies. This is an example of what has been called statistical learning, in which relationships among perceptual events are learned by repeated associations. The procedure is illustrated in **Figure 3**. Infants were presented with a 3-minute-long sequence of 'tone words', each word consisting of three tones of different frequencies presented in a fixed order. For any given infant, there were four such words, presented repeatedly with the order random across words. The timing of the tones was the same within and between words, so the only basis for learning about the word structure was that tone orders were more predictable within than between words. The overall objective was to give infants the opportunity to learn pitch relations among the tones within words. Following the 3-minute familiarization period, infants were presented with more tone words, but now with each word repeated so that the level of attention to the word could be gauged. Some of the words in this test phase were the same as in the familiarization phase, and some were new. The caption of **Figure 3** describes an example of a new test word. Each new test word was like one of the familiarization words with respect

to relative pitch relations among the tones, but different in terms of absolute pitch. Thus, the test word should seem familiar if infants were focusing on relative pitch, but should seem novel if the focus were on absolute pitch. Infants attended strongly to the new test words, supporting the conclusion that they were using absolute pitch perception. Thus, it seems that although infants can use relative pitch information quite well by late in the first year, they may find absolute pitch relations more salient than typical adults do. This links with the idea that adults who are said to have perfect pitch may have begun their musical exposure and training at an age when absolute pitch relations are a fairly conspicuous part of the listening experience.

Auditory Scene Analysis

In many situations listeners are presented with acoustic input from several sources at the same time, such as when several people in a room are talking at once. Even for a single stream of input, there is often a need to break it down into meaningful units, especially with complex patterns like speech or music. The phrase, auditory scene

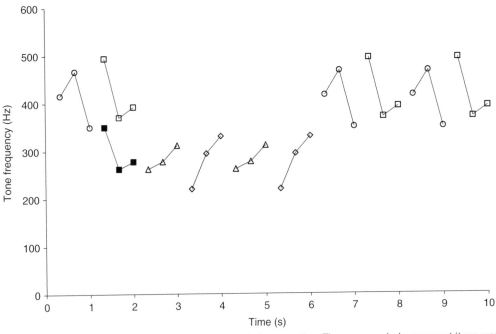

Figure 3 Illustration of tone sequences used in studies of infant pitch perception. The open symbols represent 'tone words' in which each set of three tones is a word. There are four tone words, shown by circles, squares, triangles, and diamonds. Within a word the order of the three tones was always the same, but the order across words was random. Thus, infants could learn the relations among tones within words, but not across words. This figure shows 10 s of the tone sequence presented during familiarization, but the overall sequence continued in like manner for a total of 180 s, that is, 180 tone words. The filled squares show one of the 'test' tone words presented after the familiarization sequence had ended. It is inserted into this figure to compare it to the tone word just above it (open squares). These two tone words are equivalent in terms of relative pitch relations among their tones, but the test tone word is shifted to a lower frequency range. Infants were quite attentive to this test tone word, suggesting that during familiarization they had encoded absolute pitch relations among tones within a word, not relative pitch relations. Based on Saffran JR and Griepentrog GJ (2001) Absolute pitch in infant auditory learning: Evidence for developmental reorganization. *Developmental Psychology* 37(1): 74–85.

analysis, is often used to refer to these aspects of listening. There is a sizeable literature on these topics with adults, but very little direct work on infants or young children. However, certain topics within the domains of music, speech, and crossmodal perceptions are relevant to issues in scene analysis. For example, by about 4–5 months of age, infants distinguish between musical sequences of identical notes with different temporal properties, suggesting that they group certain notes together. In crossmodal perception, temporal synchrony between changes in auditory and visual stimulation is very salient to infants from birth, with other temporal relations such as common duration emerging later. Infants are quite sensitive to rhythmic actions involving events with both visual and auditory components. For example, rhythm perception has been studied with movies of bouncing balls in which each bounce is accompanied by an impact sound. Infants between 4 and 10 months of age were familiarized with a movie in which the rhythm was either two bounces, pause, three bounces, pause, and one bounce (2-3-1), or 2-1-3. Then they were presented alternately with movies of each rhythm. Their attention picked up for the novel rhythm, and this finding held up even when infants had to learn the rhythms with variable overall tempo. Findings like this support the idea that hearing plays a strong role in helping infants to parse the world into meaningful events.

The development of speech perception provides some of the most complex and interesting examples of auditory grouping and segmentation. If you have had the experience of traveling in a country where you have no familiarity with the language, then you have some idea of what speech may sound like to young infants. One question is how infants come to recognize boundaries between words. To some extent there are acoustical markers for the boundaries, such as stress patterns and certain features of acoustical waveforms. Also, speech perception often occurs while the listener is watching the speaker's face, providing the opportunity to watch speaker movements that are linked in time to acoustic events. Thus, grouping and segmentation of spoken units is supported by acoustic and visual information. An interesting recent addition to this topic is a phenomenon called statistical learning, which was mentioned earlier in this article where pitch was discussed. This term has very wide application, but in speech perception it refers to learning that certain sound units belong together, such as the syllables in a word. Infants are presented with long sequences of nonsense speech sounds, presented in a way that removes natural speech tendencies like prosody, stress, and pauses between words. Some subsets of the sounds always occur together in a mini-sequence, and over just a few minutes infants aged 8 months or so learn these linkages. This suggests that infants and young children have a natural tendency to learn about the structure of language by virtue of regularities in the sequences of acoustic events.

Spatial Hearing

The auditory system supports perception of the three-dimensional locations of sound sources, that is, their distance and their direction in the horizontal and vertical dimensions. This is a remarkable feat of perceptual construction. Through analysis of sounds arriving at the ears, we can discern the location of the physical event that produced the sound. Localization is most precise in the horizontal dimension, since sounds from different directions reach the two ears at different times and with different intensities. These interaural time and level differences change with head growth during infancy, reaching nearly adult values around 1 year after birth, which is also when horizontal sounds localization becomes adult-like. Newborn infants can turn their heads in the direction of a sound presented far off to one side, but their localization is probably rather crude. After the newborn period most infants go through several months when they will not turn toward sounds. This is not apparent during everyday interactions because most sounds are accompanied by visible events to which infants respond. At around 4 months of age, infants resume turning toward sounds, but now more briskly and deliberately than in the newborn period. Also, newborns fail to localize sounds when they are presented in a way that mimics echoes, whereas older infants appear to suppress the echo and localize the original sound. For these reasons, it appears that the mature processes for sound localization begin to emerge at about 4 months after birth.

The minimum audible angle is a measure of the smallest change in sound direction someone can perceive, which is about 1 to 2° for adults in the horizontal dimension. It is measured in infants by observing which way they turn their heads in response to a change in location of a sound source. By this measure localization is fairly crude at 4 months, about 20°, improving to around 4° at 1 year. **Figure 4** summarizes the development of infant sound localization. In this figure the minimal audible angle measure has been converted into an equivalent interaural time difference. The figure also shows (square symbols) direct measures of infants' sensitivity to interaural time differences. Curiously, even at 4 months, infants have good sensitivity to the interaural time differences that partly underlie sound localization; in fact, their sensitivity is much better than would be predicted from their minimal audible angles. This is another hint that auditory perception involves integration across multiple aspects of the listening experience, rather than simpler relationships between physical events and perception. As the head grows during infancy, the relationships among various physical cues for sound localization change. These relationships probably stabilize around 1 year of age as the rate of head growth slows, and that is when more adult-like sound localization is reported. There is continued gradual improvement during the preschool years to the adult level of 1 to 2°.

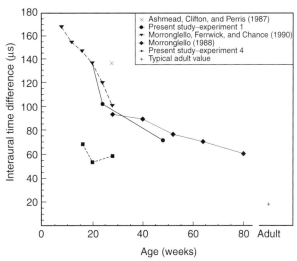

Figure 4   Thresholds for infant interaural time difference discrimination (squares) compared to interaural time differences corresponding to minimum audible angles, as a function of age. Reproduced from Ashmead DH, Davis DL, Whalen T, and Odom RD (1991) Sound localization and sensitivity to interaural time differences in human infants. *Child Development* 62(6): 1211–1226, with permission from Wiley-Blackwell Publishing Ltd.

Localization of sounds in the distance and vertical dimensions has not been studied as much as for the horizontal dimension. In adults, distance perception is strongly influenced by the fact that sound intensity falls off with increasing distance between the source and the listener. Infants around 5–6 months old are similarly influenced, in that they look toward approaching or receding objects depending on whether sound intensity is increasing or decreasing. Studies of reaching by 6-month-old infants show that infants do not attempt to reach for sounding objects in the dark if they are not within reach. Interestingly, this result held even when the natural intensity cue for distance was removed, a situation that left adults uncertain about distance. Vertical sound localization depends mostly on the fact that the outer ears filter the high-frequency components of sound differently, based on the elevation of the source. Perception of source elevation via these spectral cues has been studied in 6–18-month-old infants, by finding how far above or below ear level a sound had to change for the infant to correctly look upward or downward. There was steady improvement to adult-like levels by 18 months of age. Infants, like adults, require high-frequency acoustic energy to localize sounds vertically.

The general trend in the early development of spatial hearing is that mature processes for sound localization seem to begin around 4 months after birth, with substantial improvement up to a year or so, and gradual refinement thereafter. Infants appear to really hear sound sources as external events in their surroundings, as evidenced by turning their eyes and heads toward them, reaching for them, and sometimes crawling to them. The developmental factors that link hearing to these actions on the world are not well understood. Unfortunately, infants who are totally blind show substantial delays in behaviors such as reaching or locomoting toward sound sources, even with caretaking circumstances that attempt to promote such activity. The early linkage between hearing and goal-directed actions may depend strongly on visual support. A substantial amount of research on early visual deficits in animals bears on this issue, but the findings vary substantially by species and discussion is beyond the scope of this article.

See also: Auditory Development and Hearing Disorders; Learning; Music Perception; Perceptual Development; Sensory Processing Disorder; Speech Perception.

Suggested Readings

Abdala C and Sininger YS (1996) The development of cochlear frequency resolution in the human auditory system. *Ear and Hearing* 17(5): 374–385.

Ashmead DH, Davis DL, Whalen T, and Odom RD (1991) Sound localization and sensitivity to interaural time differences in human infants. *Child Development* 62(6): 1211–1226.

Clarkson MG and Clifton RK (1985) Infant pitch perception: Evidence for responding to pitch categories and the missing fundamental. *Journal of the Acoustical Society of America* 77(4): 1521–1528.

Folsom RC and Wynne MK (1987) Auditory brain stem responses from human adults and infants: Wave V tuning curves. *Journal of the Acoustical Society of America* 81(2): 412–417.

Lewkowicz DJ and Marcovitch S (2006) Perception of audiovisual rhythm and its invariance in 4- to 10-month-old infants. *Developmental Psychobiology* 48(4): 288–300.

Morrongiello BA and Rocca PT (1987) Infants' localization of sounds in the median vertical plane: Estimates of minimum audible angle. *Journal of Experimental Child Psychology* 43(2): 181–193.

Saffran JR and Griepentrog GJ (2001) Absolute pitch in infant auditory learning: Evidence for developmental reorganization. *Developmental Psychology* 37(1): 74–85.

Werner LA, Marean GC, Halpin CF, Spetner NB, and Gillenwater JM (1992) Infant auditory temporal acuity: Gap detection. *Child Development* 63(2): 260–272.

Relevant Websites

http://www.psych.mcgill.ca – McGill Department of Psychology.
http://asa.aip.org – The Acoustical Society of America.

Autism Spectrum Disorders

G Dawson and L Sterling, University of Washington, Seattle, WA, USA

Glossary

Autistic regression – A period of typical development for at least 18–24 months followed by a significant decline or loss of language and/or social skills, and the development of autism symptoms.

Broader autism phenotype – The presence of one or more impairments in social functioning, communication, and/or a restricted range of interests/behaviors, but without sufficient severity to meet criteria for a diagnosis of an autism spectrum disorder.

Concordance – The presence of a given diagnosis or impairment in both members of a pair of twins.

Dizygotic twins – Fraternal twins (derived from separately fertilized eggs), sharing approximately 50% of genetic material.

Functional play – The conventional and appropriate use of an object or toy as it is intended, or the association of two or more objects according to their common functions.

Joint attention – The coordination of attention between interactive social partners with respect to objects or events in the environment in order to share awareness of the objects or events.

Monozygotic twins – Identical twins (derived from a single fertilized egg), sharing 100% of genetic material.

Parallel play – Play that occurs independently but alongside another child or group of children.

Social orienting – Volitional visual orienting to naturally occurring social stimuli in one's environment.

Symbolic or pretend play – The engagement in imaginative activities with toys involving substituting one object (the symbol) for another or reference to objects that are not actually present.

Introduction

Autism spectrum disorder (ASD) is the term used to describe the broad range of pervasive developmental disorders, including autistic disorder, asperger syndrome, and pervasive developmental disorder-not otherwise specified (PDD-NOS). As described in the fourth edition of *Diagnostic and Statistical Manual of Mental Disorders* (DSM-IV), individuals with ASD have qualitative impairments in reciprocal social interaction and language and communication, in addition to stereotyped, repetitive, or restrictive behaviors and interests. Of these impairments, deficits in social interaction are considered to be a core aspect of the disorder. The prevalence of ASD is currently estimated to be 1 in 166, which is 3–4 times higher than in the 1970s. Males are affected about 3–4 times more often than females. On average, females diagnosed with ASD are more severely affected (in terms of cognitive functioning and symptomatology) than males.

Impairments in social interaction can include a general lack of interest in others, a lack of affective sharing, and poor peer relationships, among others. Some children with ASD may appear disinterested or disconnected from their peers or their caregivers, whereas others may seek others out for interaction but do so in an odd or awkward manner. Specific signs of social impairment include lack of eye contact with others, restricted range of facial expressions or facial expressions that are not appropriate to the particular situation, lack of seeking to share enjoyment with others, lack of showing and directing attention to things that are of interest (also referred to as 'joint attention'), and difficulty initiating social interactions and establishing peer relationships. In infants, particular social difficulties may also include poor eye contact, failure to orient to their name being called, delayed or absent joint attention, and lack of reciprocal social smiling. Even high-functioning individuals with ASD tend to have difficulty interpreting social cues from others. This can be especially apparent when interacting or conversing, during which the person with ASD may not read subtle changes in someone else's tone of voice, facial expression, or eye gaze.

Communication impairments include delay in language without use of gestures or other nonverbal forms of communication to compensate. Unlike children with autism and those with PDD-NOS, children with Asperger syndrome do not show significant delays in spoken language, although significant impairments in social use of language are present. Although many children with ASD (about 70%) do develop at least some spoken language, the quality of their speech is often atypical. For example, some children with ASD demonstrate stereotyped use of language, such as repeating lines from videos, repeating phrases they have heard others say, or using odd phrases. The rate, rhythm, or volume can also be atypical. For example, a child with an ASD may consistently talk very loudly, too fast, or speak in monotone. Individuals with ASD also typically have difficulties with the pragmatic use of speech, or language used to start, maintain, or end a conversation, making it difficult to establish a reciprocal conversation.

The third category of impairments includes restrictive, repetitive, and stereotyped behaviors and interests. These can be motor stereotypies, such as hand and finger mannerisms and complicated whole-body movements, such as rocking back and forth and spinning in circles. Symptoms in this domain also include preoccupations, a restricted range of interests, and sensory interests. For example, some children with ASD may engage in prolonged visual examination (e.g., of themselves in the mirror or of objects near them), peering at things out of the corner of their eyes, repetitive feeling of textures, touching, sniffing, biting, and sensitivity to sounds or lights. Other children may use objects in a repetitive manner (e.g., lining toys up, spinning the wheels of a toy car), rather than using the objects or toys flexibly or as they are intended to be used. Some individuals with ASD also engage in specific rituals in routines, which can include exact placement of objects or arrangement of items, as well as following a certain sequence of actions that must be performed in a particular order or will result in anxiety (e.g., touching the doorknob every time he or she leaves a room). Finally, many individuals with ASD have intense interests that can take up the majority of their time. For example, a child with ASD may have a particular interest in trains or cameras, and spend a large proportion of time playing with these items, talking about them, and possess a high degree of factual knowledge about them. Such interests can interfere with engaging in other more functional or prosocial behaviors.

Although there is not currently a 'cure' for autism, behavioral intervention can improve a child's level of functioning, quality of life, and prognosis. The early intervention for infants and toddlers has been a recent focus of research. Current studies suggest that children with ASD are benefiting from approaches such as applied behavioral analysis, especially when delivered in a naturalistic context with a focus on social relationships. Researchers anticipate that as children's behavior improves as a result of early intervention, the neural pathways involved in the development of language and social processing will also change. These implications are promising and underscore the benefits of early diagnosis and intervention at a young age.

Diagnosis and Early Recognition

ASD is a developmental disorder, with symptoms present before the age of 3 years. Speech delay, or the loss of previously acquired speech, is often the first developmental milestone alerting parents and professionals to possible difficulties related to ASD. Deviances from typical development can sometimes be apparent by 8–12 months of age; however, a diagnosis of ASD cannot be reliably made until the age of 2 years. Because there is currently no known biological test for ASD (e.g., blood test), a diagnosis is based on developmental history and observed behaviors. Structured diagnostic measures, such as the Autism Diagnostic Interview and the Autism Diagnostic Observation Schedule, are administered by highly trained clinicians, and are considered gold standards for diagnosing ASD. Clinicians use information from these measures as well as information collected through any other observation to help determine whether an individual meets criteria for ASD based on symptoms outlined in the DSM-IV. Because a significant proportion of children presenting with ASD have an identifiable genetic disorder, such as Fragile X syndrome, karyotyping and other biomedical tests are often conducted to define the cause of the symptoms for a given child better.

Much of current research in the field of autism focuses on methods to detect ASD at even earlier ages, with the goal of providing behavioral and psychosocial interventions to children as early as possible. Clinicians and researchers specializing in early development have begun to make diagnoses of ASD at or before the age of 2 years, and screening measures are currently being tested for infants as young as 6–12 months of age. Given the variability in developmental trajectories and the variability of symptom expression in individuals with ASD, diagnoses before the age of 2 years are considered provisional and the stability of the diagnosis should always be reinvestigated at a later age.

There is great heterogeneity among individuals on the autism spectrum. Many individuals with ASD score in the mentally retarded range on IQ tests, often with significant variability in functioning across subtests. The rate of mental retardation also tends to vary depending on the child's specific ASD diagnosis. For example, recent estimates suggest that the rate of mental retardation in children with autistic disorder is about 67%, compared to approximately 12% of children with PDD-NOS and 0% of children with Asperger syndrome. Virtually all have significant challenges in adaptive functioning and self-care. Up to 25% of individuals with ASD are high functioning, have average-to-above-average cognitive skills, and are able to live independently in society and establish successful relationships. Symptom severity also varies. A portion of children never acquire spoken language, while others achieve a level of communicative competence that is near normal, with subtle impairments in the pragmatic aspects of language. In terms of social interaction, some individuals show high motivation to interact with others, although they may execute this in an awkward or inappropriate manner. Play skills and repetitive behavior also vary across individuals. This variability and range in functioning contributes to the uniqueness of each individual with ASD. It also makes diagnosis and treatment of the disorder complicated. Diagnosing children at earlier ages, and growing recognition of individuals with more subtle impairments who have ASD have broadened the definition of autism.

Genetic Influences in ASD

For approximately 5–10% of autism cases, there is an identifiable disorder with a known inheritance pattern, such as Fragile X syndrome, untreated phenylketonuria, (PKU), tuberous sclerosis, and neurofibromatosis. For the remaining cases of autism, there is no known specific inheritance pattern or genetic test for the disorder. Researchers are currently investigating the role of genetics in the development of ASD.

In order to determine whether a disorder has a genetic basis, it is first necessary to evaluate whether it is familial, that is, whether it runs in families. Evidence suggests that there is a higher rate of ASD and related conditions among family members of individuals with ASD than would be expected in the general population. Much of this evidence comes from studies investigating the concordance of ASD among twins. When one twin has ASD, and the other twin also meets criteria for the disorder, the twins are said to be concordant for ASD.

Studies of monozygotic, or identical twins, are especially helpful in investigating the genetic basis of autism. Monozygotic twins share 100% of their genes, whereas dizygotic or fraternal twins share approximately 50% of their genes. Therefore, for the development of a disorder to be completely determined by genes, in theory monozygotic twins should be nearly 100% concordant for the disorder. Studies have reported concordance rates of approximately 70%, with rates of concordance reaching 90% if one considered autism-related symptoms, such as social and language impairment. Thus, some monozygotic pairs are discordant for ASD (i.e., when one twin has ASD, the other does not). This suggests that environmental factors (e.g., infectious agents, toxins, trauma, and pre-, peri-, and postnatal factors) must also play a role in the etiology of autism. Nevertheless, the concordance rates for monozygotic twins are substantially greater than those for dizygotic twins, providing evidence for a strong genetic component in the development of the disorder. The concordance rates for dizygotic twins are similar to those reported for siblings. The reported sibling risk rates for ASD range from about 2.8% to 7.0%, which is still much higher than rates found in the general population, though less than rates reported for identical twins. This would suggest that having more genes in common (monozygotic vs. dizygotic twins) is associated with greater risk

for ASD concordance. The increased concordance rate for siblings in general, as well as the discrepancy in concordance rates between monozygotic and dizygotic twins, indicates that genes play a significant role in the development of ASD.

Even when siblings do not meet full criteria for the disorder, it has been found that about 10–20% show symptoms related to ASD, including language, learning, and communication impairments, as well as social difficulties. Studies of infant and toddler siblings of children with ASD have shown that young non-ASD siblings can show delays in receptive and expressive language, use of gestures, social smiling, adaptive behavior skills, and social communication and social–emotional functioning. Studies have also shown that parents from families that contain at least two children with ASD have elevated risk of particular personality traits (e.g., aloof, rigid, socially anxious), establish fewer closer relationships than typical adults, have communication impairments (including a history of language delay and pragmatic language deficits), and certain cognitive impairments.

The tendency for some siblings and parents to exhibit one or more difficulties in social functioning, communication, and interests/behaviors, without actually meeting criteria for ASD, has been termed the broader autism phenotype, and is often conceptualized as a 'lesser variant' of autism. Because of the variation and spectrum of autism symptoms and severity, researchers often describe autism as a dimensional disorder, with many different components of varying degrees depending on the individual. Rather than one single gene accounting for the entire autism syndrome, it is possible that multiple genes (perhaps 10 or more) act as risk factors for the development of the components making up the autism disorder. A combination of these susceptibility genes may increase risk for the development of autism, with a greater number of genes leading to a greater risk of development of symptoms. This makes identifying the genes responsible for autism very challenging, given that each of these multiple genes, by itself, could have a very small effect size and the fact that symptoms vary significantly across individuals with ASD. Nevertheless, there is great hope that detection of autism susceptibility genes could ultimately lead to better diagnosis and clues to underlying cause and treatment.

Early Brain Development in ASD

Brain imaging and autopsy studies of individuals with ASD suggest that the disorder affects a wide range of brain regions, including the prefrontal cortex, the medial temporal lobe (especially the amygdala), and the cerebellum. The earliest apparent finding is an abnormal head circumference growth trajectory, characterized by unusually rapid head growth during their first year of life. Although children with ASD do not necessarily have larger head circumferences at birth, by 1 year of age, head circumference, on average, is one standard deviation larger than that reported by the national Centers for Disease Control (CDC) norms. Some evidence suggests that, after 12 months of age, the rate of head circumference growth in children with ASD may decelerate such that the rate of head growth does not differ from the normative CDC sample. It has also been reported that first-degree relatives (e.g., siblings) of children with ASD tend to have larger than normal head circumferences. This is especially relevant given the findings from genetic studies of autism, suggesting that behavioral symptoms associated with ASD occur at a higher prevalence in first-degree relatives than would be expected in the normal population. The possible association between head circumference trajectories and behavioral manifestations of autism is of particular relevance. If it can be established that head circumference trajectories are associated with specific autism-related behaviors, this may become a useful screening marker in alerting professionals to the possibility of the development of autism symptoms.

By 2–3 years of age, brain imaging studies using magnetic resonance imaging have documented larger than normal total cerebral volume and an unusually large amygdala, which is a structure in the medial temporal lobe that is associated with emotional functioning. Autopsy studies have revealed abnormal neuron development in the amygdala, as well. Similar studies of the cerebellum have consistently shown cellular abnormalities. The cerebellum is involved in complex motor activities, attention, and language. These recent findings show promise in implicating brain regions that might play a role in the development of autism-related symptoms. However, it is important to interpret the findings with caution, given that brain imaging studies often contain small sample sizes and results may vary depending on the age of the children in the sample and other individual characteristics.

Recent functional brain imaging studies have also shown that ASD is associated with poor functional connectivity between different brain regions. During resting and complex tasks, whereas typically various regions of the brain operate in synchrony with each other, studies have shown reduced long-range connectivity, especially between the frontal cortex and other regions of the brain. This may help explain why individuals with autism often have difficulty on tasks requiring high-order complex reasoning.

Early Behavioral Development in ASD

Joint Attention

Children with ASD have general impairments in the ability to attend to social stimuli in their environment.

Social stimuli can include the sound of a mother's voice, and the movements and features of a human face, particularly eye gaze. Typically developing infants show sensitivity to social stimuli from the first weeks of life. The failure of young children with ASD to attend to these naturally occurring social stimuli in their environment spontaneously has been termed a 'social-orienting impairment'. This impairment is one of the earliest and most basic social deficits in autism and may contribute to social and communicative impairments that emerge later in life.

Young children with ASD also show impairment in joint attention, which refers to the ability to coordinate attention between interactive social partners with respect to objects or events in the environment in order to share awareness of the objects or events. It is a means by which a child can monitor and regulate the attention of another person in relation to objects or events taking place in the outside world. Joint attention behaviors include use of alternating eye gaze, following the attention of someone else by following their eye gaze or point, and directing the attention of someone else through eye gaze or gesture. For example, if a child makes eye contact with an adult, then looks at a toy in the room, and then back to the adult, the child has initiated joint attention by attempting to direct the adult's attention to the toy. Similarly, if a child follows another person's gaze, point, or head turn toward a toy across the room, the child has responded to joint attention.

Typically developing infants tend to demonstrate joint attention abilities by around their first birthday. In children with ASD, this fundamental social-communication impairment is evident by 12–18 months of age and is actually incorporated into the diagnostic criteria for autism. Because joint attention is a discrete observable behavior, it provides a direct measure of social impairment in ASD. Research has shown that joint attention impairments distinguish preschool-age children with ASD from typically developing children and from children who have developmental delay without autism. Degree of joint attention impairment has also been found to be correlated with present and future language ability in children with ASD, and is considered a skill necessary for the acquisition of communicative language.

Play

Typically developing children generally develop symbolic pretend play between 14 and 22 months of age. Pretend play is the engagement in an imaginative activity, and includes using an object to represent another object (e.g., using a block to represent a cup), using absent objects as if they were present (e.g., pretending to feed nonexistent food to a doll), or animating objects and using them as independent agents of action (e.g., making an action figure walk, talk, or interact with other figures). Pretend play is an expression of a child's imagination. Studies have shown that toddlers with ASD produce significantly less pretend play compared with typically developing and developmentally delayed same-aged peers. In addition to the reduced amount of symbolic or pretend play, children with ASD also tend to produce fewer novel play acts, and engage in play that is less elaborate, spontaneous, flexible, and diverse than would be expected given their age. Their pretend play activities tend to be more simplified and rehearsed, as if carried out as part of a ritual, with little variation.

Because pretend play is not typically present until the second year of life, impairment in pretend play is not a distinguishing feature of very young children with ASD; children with other developmental disabilities would also be expected to show deficits in these skills. For example, a child with delayed language development would not be expected to demonstrate elaborate make-believe play skills. However, improving pretend play skills is often a focus of behavioral intervention for children with ASD, as it has been shown to be related to the development of other important skills (e.g., language).

Functional play is defined as the conventional and appropriate use of an object or toy as it is intended to be used, or the use of two or more objects according to their common functions. Examples of functional play include using a spoon to stir in a bowl or playing appropriately with miniatures. In typically developing children, functional play skills tend to emerge during the first year of life. In general, children with ASD engage in less functional play and produce fewer functional play acts. Their functional play also tends to be less diverse and elaborate.

When provided with opportunities to play freely with toys, children with ASD tend to explore toys less, and often play in isolation, without making attempts to involve others in their play. Their play may also be repetitive, often repeating certain activities over and over (e.g., pressing the same button on a pop-up toy). In addition, a child with ASD is more likely to engage in sensory exploration of a toy or play materials. This can include mouthing a toy or banging toys together, as well as more repetitive activities such as spinning the wheels of a toy car or lining up objects.

Both independent and peer play are considered to be social activities, because children's themes and scripts incorporate aspects of their surrounding environment, in addition to serving as a means of reciprocal social interaction. In terms of peer play, children with ASD often play independently and alongside a group of children, rather than joining them in their play. This is referred to as 'parallel play'. This lack of interactive play with peers can exacerbate the social impairments characteristic of children with ASD, making it difficult to develop the foundations necessary to form friendships or nurture social relations with others. It can also contribute to the isolation often experienced by children with ASD, resulting in fewer opportunities to interact with others and greater social impairment.

Motor Imitation

Impairments in motor imitation – both immediate and deferred – are common in ASD. Typically developing infants demonstrate the ability to imitate actions, such as facial expressions, from birth, and by 9 months of age, can actually imitate actions on objects. An infant's ability to imitate is a manifestation of their social connectedness with others, in that it involves attending, listening, and learning from others. Imitation also serves as a means of communicating and sharing experiences with social partners. Children with ASD show significant impairments in object imitation, imitation of facial and body movements, and imitation of actions on objects.

Studies have shown that the ability to imitate body movements (e.g., waving a hand) is more difficult for children with ASD than the imitation of actions with objects. This pattern of imitation skills is also found among typically developing children and children with developmental delay. In fact, research has indicated that children with ASD acquire simple motor imitation skills in a typical sequence, suggesting that the impairments in imitation in young children reflect a delayed, rather than disordered pattern of acquisition.

Infants and toddlers with ASD also show impairments and delays in the development of social imitative play. This includes engaging in social imitative games with others, such as peek-a-boo and pat-a-cake, that involve the tracking and imitation of another person's movements. Additionally, although infants and toddlers with ASD may participate in such activities with a parent, a child with ASD will rarely initiate such a game or take on both roles, reflecting a lack of social initiation and perspective taking. In fact, it has been suggested that failure to engage in such social imitative activities is associated with deficits in social reciprocity, joint attention, play, and language skills in children with ASD. Imitation skills in general have been shown to correlate with early language ability in children with ASD.

Language

Although social impairments are core diagnostic symptoms in children with ASD, parents of children with ASD often first become concerned about their child's development when speech is delayed or when previously acquired speech is lost. There is extreme variability in the developmental outcome of language ability in children with ASD. While some children (about 30%) never learn to talk, others develop speech but continue to show qualitative communication challenges, such as impairments in nonverbal and pragmatic language skills, atypical speech patterns, and repetitive and stereotyped use of language.

In terms of nonverbal communication, children with ASD tend to use less frequent eye contact, pointing, and other gestures. In very young children, there is a delay in the use of gestures that are used for the purpose of sharing interest and directing social interaction. Examples of these gestures include extending one's arms to show a toy or object that he or she is holding, extending arms to be picked up, requesting something by extending one's arm and opening and closing a hand, and waving goodbye when someone leaves. Children with ASD are more likely to use pointing or other gestures for the purpose of requesting something in particular (e.g., a toy or snack) than for indicating and sharing interest in an object or activity. Even when using language to communicate, children with ASD are less likely to coordinate gestures and eye contact with their vocalizations.

Some children with ASD engage in the direct manipulation of an adult's hand in order to request help. For example, a child with ASD may take the finger or hand of a nearby adult and use their finger or hand as a tool, to press a button, turn a knob, or open a door. In the majority of cases, this act is not accompanied by eye contact or vocalization; it is not a social act, but rather a means of acquiring help without actually involving the adult in the child's play or behavior.

The pragmatic use of speech refers to language used to start, maintain, or end a conversation. Although deficits in these skills are often more apparent and applicable as individuals become older, emerging difficulties with reciprocal conversation are also evident in young children. Deficits in pragmatic language can include a failure to respond to questions and comments made by another person. For example, a young child with ASD may not use eye contact, facial expression, or vocalization to acknowledge that another person has directed a statement or question toward them. This failure to respond to others marks a lack of reciprocity on the part of the child. Additional examples of pragmatic language impairments include providing excessive details in conversation or a tendency to monopolize a conversation, often when a child is focused on a particular topic of interest. Although the child is participating in conversation, this also makes it difficult to establish reciprocity, as the conversation is one-sided. Children with ASD may also have difficulty understanding how to interact appropriately with someone else during a conversation, such as knowing how close to stand to someone else or how to participate without interrupting.

Atypical speech patterns in children with ASD include immediate or delayed echolalia, (i.e., a child's immediate or delayed repetition of a word or statement), unusual prosody (e.g., atypical intonation, volume, rhythm, or rate), and semantic difficulties such as pronoun reversal (e.g., "you want a drink" or "he wants a drink" instead of "I want a drink"). These atypical speech patterns can persist into adulthood. Repetitive and stereotyped use of language refers to the repetition of phrases that may or may not be used in combination with functional speech. Often, children have heard these phrases in movies, as part of a game, song, or routine. A child with ASD may repeat the phrase while playing on their own, when attempting to converse with someone else, or even when agitated. In some cases, repetition of these phrases becomes compulsive, such that the child feels compelled to repeat the phrase over and over or even insists on someone else repeating the phrase or responding in a particular way. Another example of stereotyped speech includes the use of words or phrases that are more formal than would be expected given the child's developmental level.

Delays in language development affect both receptive and expressive language. Studies have shown that while typically developing children show early signs of language understanding before the end of the first year of life, such signs may not occur in children with ASD until much later in development. The production of words tends to be significantly delayed in children with ASD; word and phrase comprehension have been shown to be even more delayed. Moreover, even once a child has developed the ability to use speech and demonstrates the ability to respond to the speech of others, he or she may continue to have difficulty actually deciphering the meaning of a sentence. For example, children with ASD may interpret phrases in concrete or literal terms, making metaphors or jokes challenging to understand. Studies have also shown that children with ASD not only show a delayed course of language development; in many cases, the course of development is actually atypical. For example, compared to the normative pattern of development, the production of words may be relatively advanced in comparison to the understanding of words and phrases for children with ASD. It is also important to note that among children with ASD, there is great variability in the development of language, with some children achieving language competence in the typical timeframe. Nonetheless, as a whole, children with ASD tend to have significant delay in the development of language compared to typically developing children.

In addition to a frank delay in language development, children with ASD tend to show a lack of communicative intent, or the motivation to communicate with others. In other words, the language impairment evident in young children with ASD is a result of fewer attempts to comment, engage in conversation, or direct someone else's attention through the use of words, gestures, and eye contact, in addition to the lack of specific language skills.

It has been reported that in approximately 20–47% of cases, children with ASD develop typically for about 18–24 months, at which point they experience a decline or loss of language skills. This decline or loss of skill has been termed a 'regression' in development. The period of regression in ASD can be acute or gradual, during which parents report that their child stops using words or word-like sounds that he or she consistently used in the past. Although additional impairments become evident during regression, such as less frequent use of eye gaze and pointing and failure to respond to their name being called, the decline in previously acquired language skills often first alerts parents of a potential developmental problem.

Conclusion

Children with ASD experience deficits in communication and social skills, and have repetitive and restricted interests and behaviors. Individuals diagnosed with ASD are characterized by heterogeneity in terms of their symptom profiles and severity of impairments. High concordance rates are found among identical twins with ASD, and siblings and family members are also at increased risk for the development of autism-related symptoms and the disorder itself. Evidence for a strong genetic component in the development of autism has led to an increase in current research efforts focusing on the genetic etiology of the disorder. It has also been shown that on average, children with ASD exhibit atypical head circumference trajectories, a reflection of abnormal brain growth patterns. One study found that these trajectories correlate with the development of autism symptoms in the second year of life, and may be a useful marker to alert professionals to autism symptom vulnerability. Infants and toddlers with ASD show specific impairments in early development and cognition, specifically in the areas of play, joint attention, language, and imitation. As the age of diagnosis becomes increasingly younger, children with ASD will have the opportunity to receive effective behavioral and psychosocial intervention in the first years of life, hopefully leading to improvement in functioning and preventative methods for the development of the disorder. Recent studies have shown that early intervention can have a significant impact on outcome in individuals with ASD.

See also: Attention; Behavior Genetics; Brain Development; Developmental Disabilities: Cognitive; Empathy and Prosocial Behavior; Fragile X Syndrome; Genetic Disorders: Sex Linked; Imitation and Modeling; Intellectual Disabilities; Language Development: Overview; Play; Sensory Processing Disorder; Social Interaction; Special Education; Stereotypies; Symbolic Thought.

Suggested Readings

Charman T, Baron-Cohen S, Swettenham J, *et al.* (2003) Predicting language outcome in infants with autism and pervasive developmental disorder. *International Journal of Language and Communication Disorders* 38: 265–285.

Charman T, Swettenham J, Baron-Cohen S, *et al.* (1997) Infants with autism: An investigation of empathy, pretend play, joint attention, and imitation. *Developmental Psychology* 33: 781–789.

Dawson G and Toth K (2006) Autism spectrum disorders. In: Cicchetti D and Cohen D (eds.) *Developmental Psychopathology*, 2nd edn., vol. 3, pp 311–357. *Risk, Disorder, and Adaptation*. New York: Wiley.

Dawson G, Toth K, Abott R, *et al.* (2004) Early social attention impairments in autism: Social orienting, joint attention, and attention to distress. *Developmental Psychology* 40: 271–283.

Dawson G, Webb S, Schellenberg GD, *et al.* (2002) Defining the broader phenotype in autism: Genetic, brain, and behavioral perspectives. *Development and Psychopathology* 14: 581–611.

Mundy P and Crowson M (1997) Joint attention and early social communication: Implications for research on intervention with autism. *Journal of Autism and Developmental Disorders* 27: 653–676.

Stone W, Ousley OY, and Littleford CD (1997) Motor imitation in young children with autism: What's the object? *Journal of Abnormal Child Psychology* 25: 475–485.

Stone W, Ousley OY, Yoder PJ, Hogan KL, and Hepburn SL (1997) Nonverbal communication in two- and three-year-old children with autism. *Journal of Autism and Developmental Disorders* 27: 677–696.

Zwaigenbaum L (2001) Autism spectrum disorders in preschool children. *Canadian Family Physician* 47: 2037–2042.

Relevant Websites

http://www.autism-society.org – Autism Society of America.
http://www.autismspeaks.org – Autism Speaks.
http://www.washington.edu – University of Washington, UW Autism Center.

Bayley Scales of Infant Development

E M Lennon, J M Gardner, B Z Karmel, and M J Flory, New York State Institute for Basic Research, Staten Island, NY, USA

Glossary

Adjusted age – In the assessment of preterm infants, age is usually calculated based on expected date of birth, rather than on actual date of birth, in order to adjust for the preterm infant's neurological immaturity relative to same-age peers born full-term. Standard scores are derived by referring to the norms table appropriate for this adjusted age.

Basal and ceiling rules – In developmental scales covering a wide age range, only a subset of the test items are administered at each age. Basal and ceiling rules determine the specific test items administered. Different tests have different criteria for establishing basal and ceiling levels. For example, on the first edition of the Bayley Scales, the child was required to pass 10 successive items (ordered by degree of difficulty) to establish a basal level and fail 10 successive items to establish a ceiling level on the mental scale before testing was discontinued and a score could be obtained.

Norm-referenced test – A test that has been standardized and tested on a large sample selected to be representative of the larger population, in order to establish a reference group. Standard scores are obtained by comparing an individual's performance to the performance of same-age peers in this reference group.

Standardization – A consistent set of procedures for administering test items and scoring a test, so that all test-takers are assessed under the same standard conditions. Standardized procedures are developed so that children assessed by different examiners and in different locations are all tested under comparable test conditions, and therefore differences in test scores among children are less likely to be due to differences among examiners or test settings. Adhering to standardized procedures is especially important when examiners wish to compare an individual to a particular reference group of others who have taken the same test under standardized conditions.

Introduction

The Bayley Scales of Infant Development (BSID) a set of individually administered developmental scales designed to measure current developmental functioning in the areas of cognition, motor skills, and behavior. Nancy Bayley's unique contribution to the field of infant assessment was to create a standardized test for infants and young children with a flexible administration format. Bayley worked with various research editions of her scales for more than 40 years before the first commercially available edition of the Bayley Scales was released in 1969. The second edition was published in 1993, shortly before Bayley's death in 1994. The third and most recent edition, now titled the Bayley Scales of Infant and Toddler Development, was released in 2006. In the first edition, many test items were borrowed or adapted from items on other developmental scales; others were created by Bayley during her work on the Berkeley Growth Project. In the more recent editions, the age ranges were extended, and many items that were based on infant and child development research from a variety of different theoretical perspectives were added. Thus, all three editions of the Bayley Scales are considered theoretically eclectic, because their content and structure were influenced by a variety of different sources rather than being based on a specific developmental theory. Nevertheless, certain underlying assumptions about the nature of cognitive and motor development guided the construction of the original Bayley Scales, although subsequent editions reflected somewhat different theoretical positions.

It is clear that Bayley had her own theoretical perspective, based on an epigenetic developmental approach that integrated seemingly disparate points of view. This is evident in her interactionist position on the structure of intelligence and the factors contributing to intellectual growth. Bayley felt that cognitive growth occurred as a result of an interaction between a process of maturational unfolding and environmental influences. She believed that later intelligence arose out of earlier simpler functions, such as visual perception, that were subsequently integrated into more complex functions such as visually guided reaching for objects. Eventually, higher-level abilities emerge and become differentiated into separate domains such as memory or problem solving. While development of the earlier simpler functions may follow a maturational timetable, emergence of more complex abilities also depends on environmental influences. Thus, Bayley anticipated contemporary perspectives that consider maturational and environmental influences on infant development to be inseparable and dynamically interactive over time.

The first part of this article will focus on the theoretical and historical background relevant to infant assessment, specifically on the extent to which Bayley's theoretical perspective shaped the construction of her scales. Each of the three editions of the Bayley Scales will be described in turn. The final section will describe contemporary research that has addressed some of the same questions that Bayley anticipated at the beginning of her career.

Historical Background

The study of intellectual development and its measurement historically has involved debate about several theoretical issues, all of which have implications for test design. One issue concerned the structure of intelligence, which was described as either a general global intellectual capacity, complex but unified, or a combination of many separate abilities. The determinants of individual differences in intellectual capacity and rate of development also were the subject of much debate when Bayley first started the construction of her scales. For example, proponents of the view that intelligence is predetermined by either the child's genetic heritage or by a maturational timetable argued that intellectual capacity is fixed and unchanging across the lifespan, while others maintained that environmental influences contributed to changes in performance on intelligence tests over time. Finally, an important theoretical issue as concerned with whether cognitive development can be characterized as qualitative change, involving reorganization and restructuring at each new developmental level, or as a quantitative change involving the steady and incremental accumulation of knowledge with no qualitative change in the nature of intelligence. During the first half of the twentieth century, these issues influenced both the intelligence-testing movement and the movement to catalog the course of human growth in large longitudinal studies, as exemplified by the work of Alfred Binet and Arnold Gesell.

In 1905, Binet developed an intelligence test for young children to determine placement in special remedial classes in the French school system. Binet's test contained items selected to assess complex mental processes such as memory, attention, and comprehension. Its structure was based on the concept of general intelligence, which was widely accepted in the early years of the twentieth century. According to this view, intelligence tests could be constructed of a variety of items measuring performance in different domains combined together into a single scale, rather than divided into separate subscales for different abilities, since each individual item presumably tapped into the same underlying level of general intelligence. Binet test scores for individual children were fairly stable over time, leading some to conclude that this general intellectual capacity was constant throughout the lifespan, although Binet himself was opposed to this idea.

Working with a longitudinal sample in the early 1920s, Gesell developed test items designed to measure age-related changes in infant abilities in five areas: postural, presensory, perceptual, adaptive, and language–social behavior, and created a test divided into five corresponding scales. Gesell believed that responses to test items presented in early infancy were precursors to more mature forms of related behaviors observed at later ages. His five separate scales with no overall composite score implied that Gesell believed abilities in these areas developed independently. Gesell's position was that predetermined biological maturation accounted for the development of these separate abilities, which unfolded in a fixed sequence, and that heredity determined the upper limits of an individual's capacity and rate of development.

Even though Bayley's views differed from many of the theoretical positions of Gesell and Binet, both the content and the structure of the Bayley Scales were strongly influenced by their work. For example, much of the content of her early scales consisted of developmentally ordered items taken directly from Gesell's work, and she employed Gesell's method of administering the same test items at different ages in order to elicit a range of different responses. Bayley shared Gesell's interest in motor development, although she emphasized the inseparability of motor and mental skills in early development. She disagreed with Gesell that items measuring infant abilities could be organized into different subscales, believing that mental abilities were not divided into separate factors, particularly in early infancy. Accordingly, the Bayley Scales were structured like the Binet scales, with items testing performance in different domains grouped together by age levels rather than divided into separate subscales for different kinds of mental abilities. Despite the influence of Binet and Gesell, Bayley did not subscribe to most of the underlying assumptions of the intelligence-testing movement, particularly the idea of a fixed intelligence.

The notion of intellectual capacity as fixed and unchanging across the lifespan, based on the idea that intelligence was genetically predetermined, had two important implications for infant assessment: it should be possible to predict later intelligence from performance on tests given in infancy, and performance on intelligence tests should not be influenced by environmental factors such as level of parental education or the amount of stimulation in the home. Bayley's early research directly addressed these issues and helped to formulate her own theoretical perspective on the structure of intelligence and the nature of intellectual growth. Evidence from her longitudinal studies demonstrating that performance on infant tests did not predict later outcome led her to conclude that intellectual capacity is not immutable. If intelligence is not fixed, but rather changes over time, consideration of the determinants of intellectual growth necessarily included environmental factors. Thus, Bayley's developmental perspective led her to take an inter-actionist position on this issue, and her work supported the idea that maturation and environment had different effects on the course of cognitive development at different points in time. For example, she found that in early infancy, individual differences in performance on the mental scale appeared to be a function of different rates of sensorimotor maturation, while environmental influences on mental scores became more evident in the second year. Her earlier finding that correlations between mental and motor scores decreased with age in typically developing infants supported this conclusion. Bayley took a similar developmental perspective on the question of whether intelligence was best described as a general mental ability or as separate factors, writing that early in development mental abilities consist of the most simple and basic functions, out of which more complex functions gradually emerge and are eventually differentiated into separate abilities in different domains.

Development of the Early Scales

In 1928, Bayley started work on the Berkeley Growth Study, a longitudinal study of 61 children tested at regular and frequent intervals starting at 2 months of age and continuing into adulthood. A primary focus was to examine the normative development of cognitive and motor abilities. To achieve this goal, Bayley developed a series of tests published between 1933 and 1936: The California First-Year Mental Scale, the California Infant Scale of Motor Development, and the California Preschool Mental Scale. These scales were used to test her sample monthly for the first 15 months and every 3 months thereafter until 3 years, after which they were tested at 6-month intervals until 5 years of age.

The original California scales were revised in 1958 and tested on a large nationwide sample of infants from 1 to 15 months. In contrast to the primarily middle-class families in the Berkeley Growth Study, these infants constituted a more diverse sample representative of the US population in terms of level of parental education, which Bayley considered a valid index of socioeconomic status. Bayley found that level of parental education did not predict performance on either the mental or the motor scale at any age between 1 and 15 months, supporting her earlier conclusion that individual differences in performance during the period of rapid development in the first year may be less subject to environmental influences, and more a function of sensorimotor maturation.

This research edition, which also included a version of the Infant Behavior Record, was expanded in 1960 to extend the age range to 30 months. It formed the basis for the first commercially available edition of the Bayley Scales, published in 1969 and considered the most carefully standardized infant test of its time. Bayley used a sample of 1262 children, with between 83 and 95 children at each of 14 age groups between 2 and 30 months, selected to be representative of the 1960 US population by gender, race, level of parental education, and geographic location. Children were excluded if they were born more than 1 month prematurely or had severe emotional or behavioral problems. Additionally, children over 12 months from bilingual homes who had difficulty using English were not included in the final standardization sample.

Bayley Scales of Infant Development

The first edition of the BSID (BSID-I) consisted of three complementary scales designed to assess developmental status of young children between 2 and 30 months of age. The Mental and Motor Scales were composed of tasks directly administered to children in order to elicit a set of behavioral responses. Although it was not constructed as an intelligence test, the Mental Scale was designed to measure cognitive abilities considered to be the foundation for later intelligence. It included tasks measuring responses to visual and auditory stimulation, perceptual discrimination, imitation and social communication, object permanence, memory, eye–hand coordination, problem-solving, classification, vocal ability, and receptive and expressive language skills. The Motor Scale was designed to measure motor control, gross motor coordination, and manipulatory or fine motor skills. The Infant Behavior Record, completed after the administration of the Mental and Motor Scales, allowed the examiner to rate the child on factors such as activity level, attention span, motivation, persistence, and other characteristics displayed during the test session.

Particular care was taken to make the BSID-I as appealing as possible to infants and young children. The original test manual and the manual supplement published in 1984 presented strategies to produce the most reliable and valid test performance in a playful and child-friendly test session. The test materials were items likely to engage the interest of young children, such as rattles, blocks, puzzles, crayons, and picture books. The BSID-I had a very flexible administration format, with order of item administration, as well as the pacing of transitions from one item to the next, based on the child's responsiveness, in order to elicit the child's best performance. The emphasis on flexibility included crediting of incidentally observed behavior and caregiver administration of some items. The re-administration of items initially failed was encouraged, based on the reasoning that the failure may have been the result of inattentiveness, or the child's initial interest in exploring the new item rather than complying with the examiner's instructions. This flexible approach was augmented by specific administration directions to insure a standardized administration of each item.

Each administration of the BSID-I involved a subset of the 163 items on the Mental Scale and the 81 items on the Motor Scale, with the specific items administered dependent on a combination of the child's age and ability level. Although the test items were organized on the record form in order of difficulty, they were not administered in this order. Instead, sequential presentation of items of increasing difficulty using the same materials or testing the same underlying constructs, such as object permanence or language skills, was recommended. This approach allowed the examiner to gain a great deal of information without overtiring the child. In one example cited in the manual it was noted that a total of 13 items, 10 from the Mental Scale and three from the Motor Scale, could be scored from the presentation of three blocks to a young infant.

Items measuring different areas such as language, social skills, problem solving, and eye–hand coordination were interspersed on the Mental Scale record form, arranged according to the age level assigned to each item, which corresponded to the age that it was passed by 50% of the children in the normative sample. Basal and ceiling criteria required passing and failing a specified number of items consecutively ordered on the record form. On the Mental Scale, the recommended basal criterion was 10 consecutive passes, and the ceiling criterion was 10 consecutive failures, while on the Motor Scale the

recommended criterion was six consecutive items passed or failed. There was flexibility even in this, however, and the final decision of when to stop testing was left to the judgment of the examiner. Examiners typically would begin with an item about 1 month below the child's age, ideally an item part of a sequence spanning several age levels. Based on the child's performance the examiner would be able to score several items and determine whether it was necessary to drop back to a younger age level to meet the basal criterion. Following the administration of an adequate number of items in a particular sequence, the examiner would select items from sequences measuring skills in different domains and continue to score for passes and failures until the basal and ceiling criteria were met. Examiners were cautioned that testing should not be discontinued until it was clear that the child would not be able to pass any further items, and emphasized that the goal was to capture the child's full range of successful functioning. The basal and ceiling rules insured that the upper and lower limits of behaviors in all domains would contribute to the final score.

Thus, the specific sequences of items presented were determined by the child's overall level of ability across domains. Children who had particular strengths in one area, such as language, would sometimes be close to meeting the ceiling criteria but then would pass an item in their area of strength, and additional items were administered to obtain the ceiling level. A major advantage of this test design was that the child's final score would be based on a range of items measuring skills in different domains such as problem solving, language, and fine motor skills. Following the administration of the Mental and Motor Scales, the raw scores were tabulated and referenced to norms appropriate for the child's age to obtain the standardized score for each scale, the Mental Developmental Index (MDI) and the Psychomotor Developmental Index (PDI). Performance on the Mental and Motor Scales yielded standard scores with a mean of 100 and a standard deviation of 16.

In contrast to the Mental and Motor Scales, which measured the infant's performance of specific skills, the Infant Behavior Record was a record of the examiner's impression of infant behaviors observed during the test session, such as social responsiveness, activity level and attention, emotional tone, cooperation and persistence, orientation to objects, and muscle tone. Most of these behaviors were measured with rating scales. A few additional items consisted of the examiner's judgment of whether the test performance appeared to be an accurate reflection of the child's optimal behavior, as well as more qualitative descriptions of unusual behaviors observed and an overall summary of the child's performance. Although initially there was a great deal of interest in the inclusion of a measure of behavior in a test of infant development, in practice this scale was not used as frequently.

Much of the early research using the Bayley Scales was concerned with the issue of whether mental scores in infancy predicted performance on childhood measures of intelligence, based in part on the concept of a general intelligence that was fixed and stable over time. In general, studies done with typically developing infants replicated Bayley's earlier findings that there was limited evidence for predictive validity based on mental scores in the first year, although toward the end of the second year mental scores were more strongly related to preschool outcome. Better evidence for prediction to later outcome was found for infants with the lowest scores. When it became clear that global scores on tests given to young, typically developing infants did not reliably predict later outcome, the focus shifted to attempts to identify the specific items or sets of items that might predict more effectively, and some limited ability to predict later outcome in related areas was found. In this endeavor, subsets of items measuring skills related to language development emerged as a relatively stronger predictor of later outcome. Moreover, different subsets of items were better predictors of later performance at different ages – for example, visual–motor tasks given in early infancy, and language items in the second year. These latter results were consistent with Bayley's view that there was a discontinuity in the very nature of intelligence itself, such that intelligence takes different forms at different points in development.

Performance on the Bayley Scales was considered an accurate assessment of current level of functioning, however, and soon emphasis shifted from investigation of its predictive validity to the use of the BSID-I as an outcome measure in studies investigating prediction from early risk factors. Consistent with Bayley's early findings with respect to different influences at different points in development for typically developing children, many studies with high-risk infants indicated a greater influence of biomedical risk factors during the first year while environmental effects became more evident in the second year. While group means indicated an association between increased neonatal risk and subsequent lower BSID-I scores, it was not possible to predict outcome for individual children based on early risk factors. Researchers who investigated patterns of performance over time, rather than relying on scores from one point in development, found that individual developmental trajectories based on scores from three or four different points in time were better predictors of later outcome than either single scores or early risk factors, except in extreme cases.

In the 1984 BSID-I manual supplement it was recommended that examiners calculate age based on expected date of birth for premature infants, and refer to the norms table appropriate for this adjusted age. This was based on research suggesting this may provide a better estimate of the developmental level of premature infants. This issue became more important during the 1980s and 1990s when advances in obstetric and perinatal medicine led to new procedures for treating infertility as well as an improved ability to sustain high-risk pregnancies for longer periods. There was a subsequent increase in deliveries of viable high-risk infants, including preterm and multiple gestation infants. Furthermore, innovative techniques in neonatology led to even higher rates of survival for very premature and other medically fragile infants who were being born in greater numbers and at increased risk for poor neurodevelopmental outcome. As follow-up programs were created to monitor the development of these infants, there was an increased need for objective methodologies to assess them. Consequently, the BSID-I was being used increasingly to assess children with atypical development, not just in follow-up programs but also to test for eligibility for early intervention programs for infants and children with developmental disabilities. Although the normative approach Bayley took in developing her test allowed clinicians to determine the extent to which infants with atypical development deviated from the normative sample, the test was not designed to provide diagnostic information about specific patterns of abnormality. For example, many infants with genetic syndromes or significant central nervous system (CNS) injury due to loss of oxygen at birth have lower mental and motor scores relative to their typically developing peers. Statistical deviation from normative performance does not provide differential diagnostic criteria for abnormality, however, and fails in many cases to establish early abnormality unless it is severe.

The question of whether to adjust for degree of prematurity in calculating scores on developmental assessments is related to theoretical issues about the relative influences of maturation and experience. Premature infants, although they have increased extra-uterine experience compared to full-term infants matched for time from conception, actually show behaviors based on neurological maturity rather than experience, particularly on tasks measuring earlier emerging functions such as visual perception. In general, performance of premature infants on the BSID-I was more consistent with full-term infants with the same time from conception than with full-term infants with the same date of birth. Thus, a 9-month-old infant born 3 months prematurely might be considered delayed compared to a full-term 9-month-old, but would most likely perform comparably to a full-term 6-month-old. If this infant earned a raw score of 74 on the BSID-I Mental Scale, this score would be converted to a scaled score of 100 using the 6-month norms table, indicating age-appropriate cognitive skills. This same child would receive a scaled score of 63 based on the 9-month norms table, indicating significantly delayed cognitive development. Adjustment for degree of prematurity was often discontinued at 2 years of age, based on the idea that premature infants should have caught up to their same-age peers by the age of 2 years, and that continued use of adjusted age might mask underlying deficits or developmental delays. While many premature infants may indeed have obtained scores appropriate for chronological age as they got older, these scores may have reflected proportionally smaller differences between adjusted and chronological ages and a correspondingly smaller difference between scores based on adjusted and chronological age. In some cases, improved test performance for preterm infants at older ages also may have reflected environmental influences at home or in early intervention programs, since test items in the upper age ranges on the Mental Scale measured complex skills influenced by environmental differences.

A method for obtaining age-equivalent scores to determine functional levels for children with atypical or delayed development was described in the manual, although in general this practice was discouraged. Because these children may display very different patterns of abilities from typically developing peers with the same raw scores, examiners were advised to study the test protocol carefully to discover the children's specific areas of strength and weakness. The performance of high-risk infants does tend to be more variable across domains, and this variability would contribute to some of the controversy surrounding the changes made in the second edition of the Bayley Scales.

Bayley Scales of Infant Development-II

The second edition of the BSID (BSID-II) was published in 1993. It maintained the three scales, the Mental Scale and the Motor Scale, which were extensions from the first edition, and the Behavior Rating Scale, which replaced the Infant Behavior Record. One of the goals of the revision was to preserve the basic qualities of the Bayley Scales, thus the BSID-II similarly was designed to assess the infants' current level of functioning through the presentation of a series of items and observation of the infant's behavioral responses. It also combined standardized item administration and scoring procedures with flexibility in the sequence and pacing of the item administration. Some of the changes made in the second edition had the unintended effect of limiting the flexibility that had been a hallmark of the first edition of the Bayley, however.

There were several substantive changes in the second edition. The norms were updated to reflect the US population as of 1988 in terms of race and ethnicity, geographic location, and level of parental education, and the age range was extended to include children between 1 and 42 months. The standardization sample of 1700 consisted of 50 boys and 50 girls at each of 17 age groups. The BSID-II, like its predecessor, was considered to be a carefully standardized infant test. Nevertheless, characteristics of the normative sample may have influenced scores obtained at certain ages or by specific groups of children. For example, children who were born prior to 36 weeks, gestation, had significant medical problems, or were receiving services for mental, physical, or behavioral problems were excluded from the normative sample. This latter criterion for exclusion would have disproportionately excluded older children, as full-term infants without medical problems typically are not referred for services in early infancy unless they exhibit global delays. In particular, referrals for services for language delays often do not occur until late in the second year. Therefore, in the upper age ranges the normative sample consisted of children who were increasingly higher functioning relative to the children in the younger age ranges of the sample as well as to the general population. Moreover, the standardization sample contained very few low-scoring children, and, therefore, their performance did not reflect the full range of behavior found among low-scoring children in the larger population. This is especially relevant in light of the increasing use of the Bayley Scales with atypical populations, as will be discussed further below.

Stimulus materials were updated to make them more colorful and more durable, and new items were developed for the extended age ranges. Even more important for developmentalists, items were added to reflect advances in infant and child development research. In addition to the skills measured by the first edition, the BSID-II Mental Scale contained new items designed to improve measurement of cognitive abilities, including visual preference tasks and visual and auditory habituation in younger infants, and problem-solving and categorization skills in older infants. There also was an increase in the percentage of items measuring language skills, especially by the middle of the second year. Tasks intended to assess early school readiness skills, such as number concepts, categorization, and prewriting skills were added in the upper age range of the test. The Motor Scale measured gross and fine motor control and coordination, and also included some new items designed to measure sensory integration and perceptual–motor development. The percentage of fine motor items was increased at the upper ages, as the first edition had been criticized for being weak in this area.

The Behavior Rating Scale consisted of five-point rating scales designed to assess behaviors observed during the test session that yielded scores for the areas of attention/arousal, orientation/engagement, emotional regulation, and motor quality, although not all of these factors were assessed at each age.

The original Bayley Scales grew out of the movement to study infant and preschool development in order to establish normative data for the developmental changes that rapidly occur during this period, but in practice the scales were frequently used to evaluate high-risk infants in clinical settings. Reflecting this shift in emphasis, the second edition was described as being appropriate for diagnosing developmental delay and planning intervention strategies, and included suggestions to facilitate the clinician's description of the child's strengths and weaknesses. Information on the performance of children from eight high-risk clinical samples was presented in the BSID-II test manual, intended to provide examiners working with these populations with profiles of children in these different groups. The number of children in each clinical sample was small, however, and these data were not necessarily designed to improve the discriminant validity of the test. Large-scale developmental studies investigating performance over time would provide a better picture of the range of functioning of children in these various high-risk groups.

Other changes were made based on critiques of the first edition. Because previous research found that the BSID-I had limited predictive validity, specific items thought to be predictive of later outcome were added, mostly based on the recommendations of specialists in the field of early childhood assessment working with children with delayed or atypical development. Guidelines were not provided for their interpretation or use in planning intervention strategies, however. Since studies that investigated performance on related sets of items suggested that this approach might predict later outcome better than the global mental and motor scores, facet scores were constructed and added to scoring procedures to facilitate interpretation of the child's performance in the areas of cognitive, language, personal/social, and motor skills. In practice, however, several issues limited the usefulness of these facet scores; for example, the lack of equivalent representation across all facets at each age, most notable in the Personal/Social facet that consisted of only five items above the age of 4 months.

The biggest and most controversial change in the second edition was the introduction of age-based item sets, which were developed in an effort to shorten the administration time of the test and to insure that infants of the same age achieved scores that were based on the same items. Initially, infants were administered a predetermined set of items based on their age level. Social, language, memory, visual–motor, and problem-solving items were intermingled on the Mental Scale item sets, while the Motor Scale item sets consisted of a mixture of gross and fine motor items. On the Mental Scale, the basal criterion was passing at least five items within the item set, and the ceiling criterion was failure of at least four items within the item set. For the Motor Scale, basal and ceiling rules required the child to pass at least three items and fail at least two items within the item set. As long as the child met the basal and ceiling criteria, the index score was based strictly on the child's performance on the items within the item set. All items below the item set were credited as if they were passed, and if the child demonstrated proficiency on any items above the item set, credit for these items was not included in determining the final score. Children who did not meet either the basal or ceiling criteria for the item set based on their age were administered items in the items sets above or below their age level until these criteria were satisfied and a score was obtained.

In both the BSID-I and the BSID-II, the test items were arranged on the record form in order of difficulty, and therefore items from a number of different domains were interspersed. The BSID-I had basal and ceiling rules based on a specified number of consecutive items in the order that they appeared on the record form, insuring that testing would be discontinued only when the lower and upper limits of the child's abilities across domains was determined. In contrast, the basal and ceiling rules in the BSID-II did not require the child to pass and fail a number of consecutive items, instead they were just required to pass and fail a specified number of items within the item set based on their age. This meant that children whose development was uneven across domains were likely to meet their basal criteria on tasks testing skills in one domain, while they met their ceiling criteria on tasks testing skills in another domain. In order to illustrate this problem, we present the example of a hypothetical 12-month-old infant whose language and memory skills are somewhat better developed than her visual–motor skills. This child says several words, follows verbal commands, and has good memory and problem-solving abilities. While she can perform many tasks measuring visual–motor skills appropriately for her age, there were several visual–motor tasks that she did not pass at the upper end of her item set. She satisfied the ceiling criteria by failing four items measuring visual–motor skills within the 12-month item set. There are eight additional items measuring language and memory skills above her item set that she would have been able to pass, however, but performance on these items would not be included in her final score because she met the ceiling criterion for her item set. If this child was tested on the BSID-I, she would have failed the same items measuring visual–motor skills but these items were interspersed on the record form with other items measuring language, memory, and problem-solving skills that she would have passed. The ceiling rule based on failure of 10 consecutive items would have insured that testing would have continued until she had demonstrated the limits of her abilities in all domains.

The use of the item sets was a departure from the BSID-I, which had a structure that seemed to provide a more complete picture of the infant's true range of abilities. The BSID-I basal and ceiling rules virtually insured that an approach of testing the child's limits was incorporated into the standardized administration of the test. As a result, test performance on a wide range of items yielded a profile of the child's strengths and weaknesses across the different domains. In contrast, the item sets and the basal and ceiling rules in the BSID-II meant that scores were based on a restricted range of items, not necessarily reflective of the full range of the child's strengths and weaknesses.

Thus, the structure of the first edition, with items testing different domains combined together in one scale but with basal and ceiling rules designed to insure that abilities in all domains are included in the final score, reflected Bayley's theoretical perspective that mental abilities are initially unified but with development they gradually differentiate into separate areas. Bayley also recognized that development is typically uneven across domains, particularly for high-risk infants. The use of the item set approach in the second edition reflects a somewhat different theoretical perspective, specifically that skills develop at the same rate across domains and that performance on a relatively small number of items is an accurate indication of the range of abilities across multiple domains. Much of the early research on the BSID-II was designed to address the item set problem, and demonstrated that these assumptions were not valid. Studies with both typically developing infants and infants with different risk factors for poor developmental outcome found that performance in both groups can be uneven across domains. As a result, infants frequently satisfied basal and ceiling criteria for multiple item sets, resulting in different scores depending on which item set was used, thereby increasing the likelihood of either underestimating or overestimating the child's true ability level.

This issue was even more complex for premature infants, as selection of item sets typically was based on adjusted age, with scaled scores derived from the corresponding norms table. Recommendations for testing premature infants were not very specific, however, and examiners were encouraged to select the item set that was closest to the their estimation of the child's level of functioning, leading to inconsistencies across testing sites. Children could receive different raw scores depending on which item set was administered. If the item set was selected based on chronological age, the scaled score was derived from the norms table appropriate for the child's chronological age. Unless they had global delays, premature infants tended to have higher scaled scores based on the administration of the item set appropriate for their chronological age than they would have if they were tested based on their corrected age. These scores might have been inflated because infants were being given credit for items below the item set that they might not have passed if the items actually were administered, or conversely the scores may have been more accurate than those based on the corrected age item set because the child received credit for items that they were able to pass above the corrected age item set. In contrast, premature infants tested with the BSID-I would have the same raw score regardless of whether adjusted age or chronological

age was used, because the raw score was based on their actual performance rather than on *a priori* assumptions about their level of functioning, as in the BSID-II.

Studies comparing the same child's performance on the first and second editions of the Bayley typically found that scores on the second edition were lower than on the first edition. The BSID-II manual reported a mean difference of 11.8 points on the Mental Scale and 10.1 points on the Motor Scale. Discrepancies between scores on the two tests as large as 30 points have been reported, and it is interesting to note that the largest discrepancies were observed in children who scored in the accelerated range on the BSID-I. Higher scores on the BSID-I compared to the BSID-II have been attributed to the Flynn effect, the finding that performance on psychometrically based scales with outdated norms usually produced inflated scores, whereas scores on subsequent editions of the same test with more recent norms were likely to be lower. An alternative explanation with respect to the BSID-I is that lower scores on the BSID-II may in fact have been due to the item set problem, since the restricted range of items presented during the administration of the BSID-II was less likely to yield scores reflecting the true range of the child's abilities. This argument is supported by a study comparing the Bayley-III to the Bayley-II, which found that in contrast to what would be predicted by the Flynn effect, the more recently normed Bayley-III had mean composite scores that were approximately 7 points higher than BSID-II Mental and Motor Index scores for a group of children administered both versions of the test. The extended range of functioning allowed by the different basal and ceiling rules in both the BSID-I and the Bayley-III may contribute to higher and presumably more accurate scores on both of these scales. The full extent to which the item set problem affected the results of studies using the BSID-II as an outcome measure is unknown, although it is frequently cited as a possible contributing factor to scores that were lower than expected.

Bayley Scales of Infant and Toddler Development

The Bayley-III was developed for the stated purpose of updating the normative data and making the test more appropriate for use in clinical settings, while maintaining the basic qualities of the previous editions. For the first time, the standardization sample included children with various risk conditions, in an attempt to be more representative of the range of functioning in the general population of infants and toddlers. The content and structure of the test were revised, and the item set approach was discontinued. The most significant departure from the previous editions, in direct contrast with Bayley's theoretical positions, was the creation of five distinct scales to measure cognitive, language, motor, social–emotional, and adaptive behavior. The first three scales consist of items that are directly administered to the child, while the latter two scales are caregiver questionnaires. These areas were selected to make the test more appropriate for determining eligibility for early intervention programs, thereby broadening its appeal and increasing its utility to practitioners involved with providing services to the developmentally delayed. Adjustment for prematurity is recommended through 24 months of age.

Many items on the newly constructed Cognitive Scale were taken from the Mental Scale in the previous edition and were chosen to assess abilities such as sensorimotor manipulation and exploration, early memory and problem-solving skills and concept formation, and to differentiate these domains from language skills. Changes in item directions and administration rules were made to reduce the verbal content of the cognitive tasks and to decrease the reliance on motor skills. The newly formed Language Scale is divided into Receptive and Expressive Communication subtests, and the Motor Scale is divided into Gross and Fine Motor subtests. Many items that were part of the Mental Scale in the previous edition now are placed on either the Expressive or Receptive Communication subtests, the Cognitive Scale, or the Fine Motor subtest. New items also were added, increasing the estimated administration time. A Behavior Observation Inventory incorporates examiner observations of behavior during the test session and caregiver ratings of how representative these behaviors were of the child's typical behavior, to assist in the interpretation of the child's scores on these three scales.

The Social–Emotional and Adaptive Behavior Scales consist of questionnaires completed by the child's parent or primary caregiver. The Social–Emotional Scale was designed to measure the acquisition of major social–emotional milestones occurring in infancy and early childhood. The Adaptive Behavior Scale assesses daily functional skills in 10 different areas, and scores from these areas are combined to form a composite score for Adaptive Behavior.

The separate Language and Cognitive Scales on the Bayley-III address the criticism that low scores on the BSID-II Mental Scale might have reflected poor performance on the language items or deficits in fine motor skills or both, but this could not be determined from the global Mental Scale score. A potential disadvantage with the Bayley-III, however, especially for researchers using the Bayley Scales in their longitudinal studies, is that there is no composite mental score, making it difficult to compare performance on the Bayley-III with performance on the previous edition. Based on the reasoning that a global score would not be appropriate for clinicians who use the tests for determination of eligibility for early intervention programs, there are no plans to develop a composite score comparable to the BSID-II MDI. Researchers

desiring to use the third edition in longitudinal research may wish to investigate whether it is feasible to construct a composite score for the purpose of comparison with the previous edition.

The Bayley-III discards the often-criticized item-set approach used in the previous edition. Its five subtests all have the same basal and ceiling rules: the infant must pass three consecutive items and fail five consecutive items before testing is discontinued. Separate scales and subtests insure that acceleration in one area will not obscure deficits in another area. Items with the same content area are generally administered as part of a series of related items, although the administration format of the Bayley-III is somewhat less flexible for items within a subtest. There is flexibility in the order of administration of the different scales and subtests, although it is recommended that the Receptive Communication subtest be administered before the Expressive Communication subtest.

The Bayley-III should provide added information from the separate scales and subtests measuring different areas of development, and the removal of the item sets should lead to a better sense of the true range of the infant's abilities. The controversy over the item sets in the BSID-II led to a series of studies that yielded evidence for variability in performance across multiple domains, and made it clear that the structure of the BSID-II did not allow for a complete picture of the infant's true range of abilities across domains. Bayley recognized that development is typically uneven across domains, especially in high-risk children, although she argued against the use of separate subscales in tests measuring infant development. This raises the question of what Bayley would have thought about the Bayley-III. Evidence from subscales derived from the BSID-I indicated that different sets of skills at different points in infancy were predictive of later outcome, however, and this supported Bayley's theoretical perspective that intellectual development consists of a number of qualitative changes that occur particularly rapidly during infancy. Additionally, the Bayley-III appears to be similar to the original BSID-I in that the structure of the test is designed to test the upper and lower limits of the infant's abilities across domains. Therefore, it is likely that Bayley might have approved of the structure of the Bayley-III, although she might have argued for a composite score, in addition to the scale and subtest scores, or perhaps advocated for some other way to capture the complex developmental relationships among skills in different domains.

Contemporary Research using the Bayley Scales

An important theoretical issue during the time that Bayley constructed her scales was the question of whether cognitive and motor development is a process that involves mostly predetermined sensorimotor maturation or whether it is influenced by the environment. Bayley took an interactionist position, anticipating contemporary views, and maintained that there were different influences on performance at different ages. This view is supported by the results from our longitudinal sample of infants born at varying risk for poor neurodevelopmental outcome. These infants were classified into four CNS risk groups: those with no discernable brain injury, those with abnormal auditory brainstem responses (ABR) but no structural damage (ABR only), those with mild-to-moderate CNS involvement, and those with strong-to-severe CNS injury, as measured by cranial ultrasound. As a part of a series of studies investigating attention and arousal in high-risk infants, the BSID-II was administered every 3 months between the ages of 4 and 25 months. Bayley's description of development in the first year reflecting the rate of sensorimotor maturation and in the second year increasingly reflecting the influence of environmental factors was supported by the performance of the children in our sample who had no discernable CNS injury. For these infants, mental and motor scores were correlated in the first year, but the strength of this relationship decreased throughout the second year, and the relationship between level of maternal education and performance on the Mental Scale increased throughout the second year. Different developmental patterns were displayed by children who had different degrees of perinatal risk.

The results presented here are based on analyses conducted especially for this article using all of the available data from our ongoing longitudinal studies, comprising 11 170 tests on 2132 infants tested at ages 4 to 25 months. **Figure 1** shows scores combined across all ages tested for each of the four CNS risk groups. In our overall sample, degree of CNS injury predicts both BSID-II mental and motor scores, and, as expected, scores are the lowest for children with the most severe CNS injury. The group means presented in this figure obscure developmental patterns, however, and when scores are presented by test age, a more complex picture emerges. The importance of using trajectories of performance over time rather than attempting to predict from one point in development or from scores combined across age cannot be overstated. By modeling cognitive and motor development using risk factors such as degree of CNS injury, level of maternal education and gender, evidence for complex relationships among these variables emerged. Interactions of these risk factors with time were not straightforward, and differed between the MDI and the PDI.

Figure 2 presents the MDI and PDI scores across age for children in each CNS risk group. For all four groups, BSID-II mental and motor scores have an inherent downward trend during the second year. The extent to which

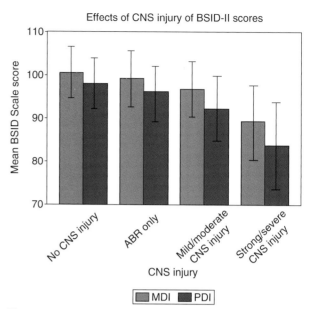

Figure 1 Mean BSID-II Mental Development Index (MDI) and Psychomotor Developmental Index (PDI) scores for four CNS risk groups. ABR, auditory brainstem response; CNS, central nervous system.

characteristics of the BSID-II normative sample or the item-set problem contributed to this downward trend even in our low-risk groups cannot be determined. For the MDI, however, this trend may reflect in part the increasingly verbal content of the BSID-II after the middle of the second year. Indeed, there were significant relationships between MDI scores and a parent report measure that indicated delayed language development in over a quarter of our sample. This association was particularly strong for second-year MDI scores. For both the MDI and the PDI, the decline in scores in the second year is the greatest for children with severe CNS injury.

CNS injury was the strongest predictor of MDI scores until about the middle of the second year, when maternal education becomes a stronger predictor. CNS injury was the strongest predictor of PDI scores at all ages, and there was no relationship between maternal education and the PDI. The effect of maternal education on MDI scores was greatest among infants with mild or moderate CNS injury; high maternal education did little to raise the scores of more severely injured infants. **Figure 3** shows the relationship between maternal education and MDI scores for the four CNS risk groups.

The interactions among these factors differ for male and female infants. In general, girls' MDI and PDI scores are higher than those of boys in our sample, and the effects of CNS injury on scores, especially in the second year, are stronger for boys than for girls. Boys with severe CNS injury show a markedly sharper decline in second-year MDI and PDI scores than do girls with severe CNS injury or boys with less severe CNS injury.

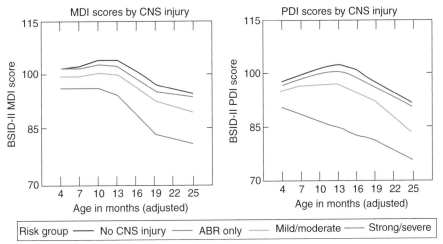

Figure 2   Mean BSID-II Mental Development Index (MDI) and Psychomotor Developmental Index (PDI) scores across age for four central nervous system (CNS) risk groups.

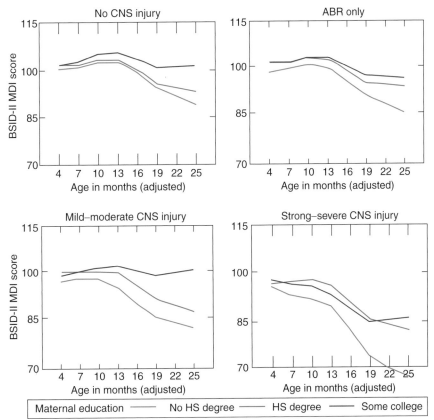

Figure 3   Mean BSID-II Mental Developmental Index (MDI) scores across age by central nervous system (CNS) risk group for three levels of maternal education. HS, high school.

Summary

Nancy Bayley's developmental perspective, derived from her work with a group of infants growing up in Berkeley, CA in the early years of the twentieth century, continues to have relevance today. The most recent editions of the Bayley Scales are routinely used to assess infants who have much smaller birthweights, earlier gestational ages, and more significant neonatal risk factors than the infants who were assessed with the earliest editions of the Bayley Scales, yet contemporary research with typically developing and high-risk infants supports Bayley's theoretical perspective on the nature of cognitive and motor development in the first few years of life.

See also: Developmental Disabilities: Cognitive; Milestones: Cognitive; Milestones: Physical.

Suggested Readings

Aylward GP (2002) Cognitive and neuropsychological outcomes: More than IQ scores. *Mental Retardation and Developmental Disabilities Research Reviews* 8: 234–240.

Bayley N (1969) *Bayley Scales of Infant Development.* San Antonio, TX: The Psychological Corporation.

Bayley N (1970) Development of mental abilities. In: Mussen J (ed.) *Carmichael's Manual of Child Psychology,* 3rd edn., pp. 1163–1209. New York: Wiley.

Bayley N (1993) *Bayley Scales of Infant Development,* 2nd edn. San Antonio, TX: The Psychological Corporation.

Bayley N (2006) *Bayley Scales of Infant and Toddler Development,* 3rd edn. San Antonio, TX: Harcourt Assessment.

Black MM and Matula K (2000) *Essentials of Bayley Scales of Infant Development-II Assessment.* New York: Wiley.

Brooks-Gunn J and Weinraub M (1983) Origins of infant intelligence testing. In: Lewis M (ed.) *Origins of Intelligence: Infancy and Early Childhood,* 2nd edn., pp. 25–66. New York: Plenum Press.

Gardner JM, Karmel BZ, Freedland RL, *et al.* (2006) Arousal, attention and neurobehavioral assessment in the neonatal period: Implications for intervention and policy. *Journal of Policy and Practice in Intellectual Disabilities* 3: 22–32.

Stott LH and Ball RS (1965) Infant and preschool mental tests: Review and evaluation. *Monographs of the Society for Research in Child Development* 30(3, Serial No. 101).

Bedwetting

A I Chin and S E Lerman, University of California, Los Angeles, Los Angeles, CA, USA

Glossary

Atrial natriuretic factor – Hormone produced by the heart involved in homeostatic control of body water and sodium.

Cystometrogram – Test of bladder and urethral sphincter function during the storage and passage of urine, performed by placing a catheter with a sensor into the bladder (catheter cystometry). This procedure is incorporated in urodynamic (see below) testing.

Dysfunctional voiding – Abnormal storage or emptying voiding pattern without neurological or anatomical disease.

Dysuria – Painful or difficult urination, a common voiding symptom.

Encopresis – Repeated involuntary loss of feces in a child of chronologic and mental age of at least 4 years.

Enuresis – Involuntary discharge of urine from the urethra, derived from the Greek word *enourein,* meaning 'to void urine'.

Glomerular filtration rate – Volume of fluid filtered from the kidney often used as a measure of kidney function and typically measured in milliliters per minute.

Hyponatremia – Electrolyte imbalance when the plasma sodium level falls below 135 mmol l^{-1}; severe hyponatremia can cause an osmotic shift of water from the plasma into the brain cells with symptoms including nausea, vomiting, headache, and malaise.

Micturition – Discharge of urine.

Osmolality – Measure of concentration of particles; in humans normal values of serum osmolality is 285–295 mmol kg^{-1}.

Polydipsia – Abnormally large intake of fluids by mouth usually associated with excessive thirst.

Polyuria – Excessive volume of urination; a symptom with multiple etiologies including excessive fluid intake, diabetes, diabetes insipidus, or use of diuretic medications.

Urodynamics – Series of diagnostic tests used to evaluate voiding disorders. These tests may include the measurement of urinary flow, detrusor pressure, sphincter muscle electrical activity, and radiographic imaging.

Voiding cystourethrography – Radiographic study to evaluate anatomic features of bladder and urethra, the ability to empty the bladder, and to diagnose whether urine backs up into the kidneys.

Introduction

The classification of enuresis can be divided into primary enuresis, defined as wetting in patients who have never been dry for extended periods of time, and secondary enuresis, the onset of wetting after a continuous dry period of at least 6 months (**Figure 1**). The degree of incontinence can range from urinary incontinence a few times per week to multiple episodes daily. Enuresis can be further categorized into nocturnal or diurnal enuresis. Nocturnal enuresis is the involuntary wetting that occurs at night or during sleep beyond the age of anticipated bladder control, usually placed at 5 years of age. This condition contrasts to diurnal enuresis, defined as involuntary loss of urine occurring while awake, a problem that is more likely to have an underlying neurologic or physiologic diagnosis. Nocturnal enuresis can be distinguished as monosymptomatic or uncomplicated, defined as nocturnal enuresis associated with normal daytime urination, or polysymptomatic or complicated, defined as nocturnal enuresis associated with other lower tract urinary symptoms. These symptoms include urinary urgency, frequency, poor urinary stream, urinary tract infections, neurological deficit, chronic constipation, or elimination dysfunction.

Epidemiology

Enuresis beholds no prejudice, seen worldwide in all cultures and races, and has been problematic from antiquity to present. The true incidence of enuresis may not be known because of under-reporting and the lack of uniform definitions of what constitutes a wet child. For instance, a family whose siblings were not dry until later years may not seek medical care concerning a child with enuresis at 5 years of age, while conversely a child still wetting at 4 years may be brought to the pediatrician if his or her siblings were dry at that age. Providing this caveat, 15% of 5-year-old children have primary nocturnal enuresis, having never achieved a period of continent nights. Although the incidence is high in young children, the majority spontaneously resolve, an important consideration when discussing treatment options with parents. In a seminal study of over 1000 enuretics, resolution of symptoms occurred at a rate of approximately 15%

each year, such that only 5% of 10-year-olds and 1% of adolescents remain wet. However, historical data suggest that up to 1–2% of adults experience enuresis, based on military recruits during World War II rejected for bedwetting. Monosymptomatic nocturnal enuretics comprise 80–85% of children with enuresis.

The epidemiologic data pose a dilemma on when and if enuretics should be evaluated and treated. Nocturnal enuresis is generally defined as starting at age 5 years, which translates to 15% (or 1 in 6 or 7) 5-year-olds that would require evaluation and treatment for enuresis, which seems impractical. In our experience, most children in this young age group do not consider the bedwetting to be a problem, nor are they sufficiently interested in achieving dry nights through a rigorous treatment program. The economic impact of treating 15% of 5-year-olds is also overwhelming and should not be borne by society. Fortunately, most enuretic children at 5 years of age spontaneously resolve their bedwetting, as anxious parents can be assured that approximately half of children at this stage will improve if treatment is delayed until 8 years of age.

One can then argue that with such a high spontaneous rate of resolution whether or not treatment is necessary at all. Indeed, the 15% per year spontaneous resolution rate of enuresis is the gold standard against which all treatment regimens must be compared. For several reasons, we do not feel these data support a nontreatment approach. First, a lonely, isolated, and painful childhood may evolve while spontaneous cure is awaited. Second, spontaneous resolution does not come to all who wait. The probability that a child at any given age will continue into adulthood with enuresis rises with increasing age. Children who remain enuretic past the age of 8 years have an increasing risk for not resolving their symptoms. Combined with studies concluding that enuresis does not interfere with socialization until 7 years of age, we define enuresis as a clinical condition beginning at age 7–8 years. At this age, most patients are interested in achieving dry nights and their socialization begins to be adversely affected by nocturnal enuresis.

Physiology

Understanding the development of continence provides insight in the etiology and management of enuresis. Bladder storage and emptying requires the activity of sympathetic, parasympathetic, and somatic voluntary nerves coordinated by the spinal cord, brainstem, midbrain, and higher cortical centers (**Figure 2**). The infant bladder relies on the sacral spinal cord micturition center to empty involuntarily and to completion. Bladder distention stimulates the afferent limb to the sacral micturition center, which generates signals

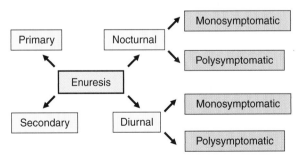

Figure 1 Classification of enuresis. Types of enuresis from primary or secondary to nocturnal or diurnal. Both nocturnal and diurnal enuresis can further by classified into monosymptomatic or polysymptomatic.

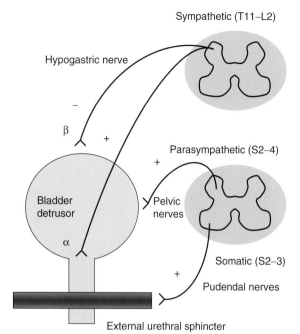

Figure 2   Innervation of the bladder. Sympathetic hypogastric nerves originating from the lateral gray columns of the lumbar spinal cord from T11 to L2 synapse with α-receptors responsible for tonic contraction of the internal sphincter and β-receptors for relaxation of the detrusor muscle to allow bladder filling and storage. Parasympathetic pelvic nerves travel from the ventral gray matter from S2 to S4 to the cholinergic receptors in the detrusor muscle to mediate bladder contraction. Somatic pudendal nerves from the pudendal motor nucleus of S2 to S4 supply the voluntary external urethral sphincter.

through the efferent limb leading to detrusor contraction with coordinated relaxation of the striated muscle of the voluntary external sphincter. Infants empty their bladders approximately 20 times per day, of which roughly 40% occurs during sleep. At 6 months of age, infants decrease the frequency of micturition with a concomitant increase in voided volumes. These changes have been attributed to the development of a cortical inhibition of the voiding reflex and to the proportionately greater increase in bladder capacity relative to increased urine production that occurs with growth. Between 1 and 2 years of age, children begin to achieve conscious sensation of bladder fullness, and the voluntarily ability to void or inhibit voiding until a socially acceptable time is achieved in the second or third year of life, mediated by the frontal lobe and pontine micturition center. Normally inhibited by signals from the frontal lobe, the pontine micturition center coordinates urethral sphincter relaxation and detrusor contraction. When urination is desired, the frontal lobe sends excitatory signals to the pons allowing voiding to occur. By 4 years of age, most children have achieved an adult pattern of micturition. However, like other

developmental milestones, bladder control does not appear at a set time in a child's maturation. The typical sequence of developing bladder and bowel control is first nocturnal bowel control, then daytime bowel control, next daytime bladder control, and finally nocturnal bladder control.

Etiology

The persistence of nocturnal enuresis in children has likely multifactorial contributions. Multiple theories have been proposed, including genetic predisposition, physiologic problems including low functional bladder capacity and bladder instability, increased night-time urine production from behavioral or endocrine factors, and maturation delay. These factors ultimately lead to common mechanisms including increased night-time urine production relative to bladder capacity. The following section reviews the current understandings.

Genetic Inheritance

It has been observed that enuresis runs in families. Statistically, at 5 years of age, approximately 15% of children from nonenuretic parents are bedwetters, while 44% of children with one parent who had enuresis are wet, and 77% of children where both parents had enuresis are afflicted. Other support for genetic predisposition arises from twin studies, with monozygotic twins concordant for enuresis twice as often as dizygotic twins. In one family study, linkage to a region on chromosome 13 has been identified, while other studies have described linkage to regions on chromosome 12 and 22. The practical aspect of this genetic predisposition lies in predicting when a child will likely become dry based on family history, but plays little role in treatment. Currently, no gene for enuresis has been identified and there is no role in gene therapy for treatment of enuresis.

Changes in Bladder Physiology

Two alterations in bladder physiology, detectable by urodynamics, are associated with enuresis, a reduced functional bladder capacity and bladder instability. Prior reports have demonstrated a reduced functional bladder capacity in bedwetters compared with healthy children, with reduction in enuretic children up to 50%. Normal bladder capacity in children can be estimated by the formula age in years plus two equaling bladder capacity in ounces. Enuretic children do not appear to have differences in true or anatomic bladder capacity, which has been shown to be similar to functional bladder capacity when measured in children under general anesthesia. Decreased functional bladder capacity limits night-time urine storage whereby the child either is aroused to void or wets the bed. However, not all studies confirm a generalized decrease in functional bladder capacity among enuretics.

Bladder instability has been widely investigated as an etiology of enuresis, but its contribution is not clear. Initial studies of children with monosymptomatic nocturnal enuresis undergoing awake catheter cystometry showed significant variation in bladder instability, ranging from 16% to 84%. However, subsequent studies found that approximately 16% of children with monosymptomatic nocturnal enuresis experience bladder instability during cystometrograms while awake, a number no different from healthy children. This finding predicts the high failure rate of anticholinergic drugs in the treatment of most enuretics. While performing urodynamic studies it is important not to create false positives by infusing at supraphysiologic flow rates given age and size, as increasing the infusion flow rate can provoke instability. A maximum flow rate at 10% of the patient's normal glomerular filtration rate is recommended.

Endocrine Factors

It has long been observed that less urine is secreted at night than during the day. In 1952, relative nocturnal polyuria was proposed as an etiologic factor for enuresis. Antidiuretic hormone (ADH) or arginine vasopressin (AVP) is secreted in the hypothalamus, stored in the posterior pituitary, and released by factors such as elevated serum osmotic pressure and low fluid volume. It naturally follows a circadian variation with increased secretion at night. Studies demonstrated lack of the normal circadian variation and increased night-time AVP excretion in enuretic children, suggesting that increased urine production resulting from decreased plasma vasopressin led to urine production exceeding diurnal output. Another study confirmed that 25% of children with enuresis had low nocturnal serum AVP levels compared with controls. Sleep-pattern changes are also known to affect AVP expression.

However, not all studies identify differences in day and night urine production or urine osmolality between enuretics and healthy children when diet and fluid intake are controlled, suggesting that only a subset of children have altered AVP levels as a causative factor for enuresis. Other opponents argue the cause and effect relationship between enuresis and AVP levels. Bladder distention has been shown to stimulate AVP secretion. Thus, lower nocturnal AVP levels may be a response to the smaller functional bladder capacity of enuretics and the empty bladder following an enuretic event.

Developmental Delay and Psychological Factors

Achieving nocturnal urinary control is a normal transitional phase of a child's development between the infant and adult control of urinary function. Another theory of enuresis is that it represents a delay in development by both internal and external factors. Accordingly, children with enuresis may have increased fine and gross motor clumsiness, delayed speech, and perceptual dysfunction compared with healthy children. This theory of developmental delay is supported by the high rate of spontaneous cure and improvement in other aspects of motor skills as well. Furthermore, environmental factors can contribute to this delay. The 2–4-year-old age range is a particularly sensitive time for the development of nocturnal bladder control. Stressors and social pressures during this time have been associated with enuresis, with children in families subject to stressful circumstances such as a divorce experiencing a threefold higher risk. These events are more commonly associated with the development of secondary enuresis. Bedwetting is also more common in lower socioeconomic groups. The theory that enuresis is a psychopathologic disorder has been largely refuted. Several studies have confirmed that enuretics do not have significant differences in psychological disorders compared with healthy children.

Behavioral Factors

Behavioral factors leading to night-time polyuria may also influence enuresis. This may occur with fluid intake and patterns between daytime and night-time as well as the role of diuretics. In 1886, a pharmacist named Pemberton changed the US and the rest of the world when he created a carbonated beverage to treat hangovers and headaches by mixing an extract of coca with an extract of the African Kola nut. The diuretic activity of caffeine is well documented. Today, consumption of caffeinated soft drinks grows at a dizzying pace, up more than 500% since 1950. A typical 12 oz of soft drink contains anywhere between 30 and 60 mg of caffeine. With a pharmacological dose of caffeine at approximately $2 \, \text{mg kg}^{-1}$ of body weight, a 30 kg child receives a pharmacologic dose in each 12 oz can. Surveys have shown that people aged 12–24 years consume the most caffeine, followed by people in the 6–11-year age group. A 1986 survey revealed that 25% of 1–3-year-old children consume an average of 7 oz of soft drink per day. Chocolate and cocoa also contain caffeine.

Sleep Factors

The role of sleep disorders in enuresis has long been controversial. A hypothesis suggests that enuresis is a disorder in sleep arousal, with anecdotal evidence reporting that enuretic children are more difficult to arouse.

However, controlled studies have found that children with enuresis do not sleep more soundly than do healthy children, with enuretic episodes occurring in random stages of sleep. An association between nocturnal enuresis and obstructive sleep apnea, however, suggested that apneic episodes increased secretion of atrial natriuretic factor. Elevated levels of atrial natriuretic factor lead to inhibition of renin secretion which leads to decreased aldosterone levels and subsequent increased diuresis and night-time polyuria. Treatments of obstructive sleep apnea have been shown to decrease nocturnal enuresis.

Organic Disease

The majority of monosymptomatic nocturnal enuretics are unlikely to have an organic cause for their bedwetting. However, conditions such as urinary tract infections, spinal cord pathology, posterior urethral valves, ectopic ureter, constipation, diabetes insipidus, and diabetes mellitus all need to be considered and understood in the evaluation of enuretic patients. In a study of over 9000 school-aged girls, the prevalence of significant bacteriuria was 1.5% in healthy girls and 5.6% in bedwetters with treatment of infection alone curing a significant number of enuretics. Urinary infection is also commonly associated with diurnal enuresis. In relapsing enuretics, particularly in girls, irritated bladders with chronically inflamed mucosa termed cystitis cystica can cause enuresis with negative urine cultures, but may respond to long-term antibiotic therapy. Although not a common cause of enuresis in the US, pinworm infection is an important cause of sudden onset of enuresis in young girls. Diagnosis is made by recovering eggs in feces, the perineal skin, and under fingernails, and the infection responds well to antihelmintic therapy.

Evaluation

Timing of Presentation

The timing of parents presenting with a child to the pediatrician is individualized depending on that family's concept of 'normal'. It is important to initially present to parents the natural progression of bladder control and to work with them in choosing when to begin treatment. In the initial discussion, it is crucial that the parents understand that the child is not wetting the bed on purpose and that enuresis is not a volitional event. The child also needs to learn to take responsibility for staying dry. Enuresis may disrupt sleep and be a source of turmoil for the entire family and parents may become frustrated with the child. In fact, up to 20–36% of parents punish their children for bedwetting. Although we do not advocate watchful waiting in children older than 8 years, if a child does not view the bedwetting as a problem and is not interested in treatment, then we do not insist.

History

The evaluation of children with nocturnal enuresis begins with a thorough history and physical examination (**Table 1**). Specifically, the timing and onset of enuresis needs to be established. Is it primary, or has the child had previous dry intervals? How many wet episodes occur each night, and how frequently is the child wet? Specific questions need to be asked to characterize the child's voiding. Does the child or parent complain of urgency, diurnal incontinence, frequency, or slow or intermittent stream? Is there a history of constant wetness suggestive of an ectopic ureter? Have the parents noticed posturing such as leg crossing, squatting, or Vincent's curtsey which may be signs of urgency? Vincent's curtsey refers to adopting a low crouch position and pressing the heel against the perineum in an attempt to prevent urinary leakage from an uncontrolled bladder contraction by increasing sphincter pressure. What about a history of polyuria or polydipsia? Are there any prior urinary tract infections or dysuria? Does the child have a normal bowel habit, associated constipation, or alternatively encopresis, defined as fecal soiling or the involuntary passage of feces? Indeed, dysfunctional voiding is often associated with bowel changes as 15% of children with enuresis have encopresis, and 25% of children with encopresis have enuresis. A complete voiding and stooling diary for several days to record the timing and volume of voids and stools per day can be extremely helpful (**Figure 3**). In addition, a careful diary of the amounts, timing, and types of fluids consumed is vital with particular notation of caffeinated beverages. Milestones to assess for developmental delay should be evaluated as well as neurological problems such as changes in gait. Other medical problems that may contribute to enuresis such as sleep apnea, diabetes, or epilepsy need to be inquired. It is important to consider that shame and abuse are important elements in the lives of enuretic children who may be subject to abuse and wrongful punishment. This anxiety and embarrassment may make them reluctant to discuss the wetting.

Table 1   Common voiding symptoms

Symptoms of voiding dysfunction

- Frequency
- Urgency
  Leg crossing
  Squatting
- Urinary incontinence
  Daytime
  Night-time
- Dysuria
- Urinary tract infections
- Incomplete emptying
- Slow stream
- Constipation
- Stool incontinence

Intake and voiding record

Name _____

Start date _____

	Day 1		Day 2	
	Fluid intake	Void/stool	Fluid intake	Void/stool
12 am	_____	_____	_____	_____
1 am	_____	_____	_____	_____
2 am	_____	_____	_____	_____
3 am	_____	_____	_____	_____
4 am	_____	_____	_____	_____
5 am	_____	_____	_____	_____
6 am	_____	_____	_____	_____
7 am	_____	_____	_____	_____
8 am	_____	_____	_____	_____
9 am	_____	_____	_____	_____
10 am	_____	_____	_____	_____
11 am	_____	_____	_____	_____
12 pm	_____	_____	_____	_____
1 pm	_____	_____	_____	_____
2 pm	_____	_____	_____	_____
3 pm	_____	_____	_____	_____
4 pm	_____	_____	_____	_____
5 pm	_____	_____	_____	_____
6 pm	_____	_____	_____	_____
7 pm	_____	_____	_____	_____
8 pm	_____	_____	_____	_____
9 pm	_____	_____	_____	_____
10 pm	_____	_____	_____	_____
11 pm	_____	_____	_____	_____
12 am	_____	_____	_____	_____

Fluid intake: record intake in ounces, c, caffeinated beverage
Void/stool: v, void; a, accident; b, bowel movement; s, bowel accident

Figure 3 Sample intake and voiding record. An example of a 48-h voiding diary.

Given the genetic predisposition, a family history of enuresis is required. A history of enuresis in either parent or in any siblings should be ascertained, as well as the age in which nocturnal continence was achieved. A review of socioeconomic conditions and stresses at home may help yield information on precipitating causes of the bedwetting, such as an acrimonious divorce coinciding with the onset of secondary enuresis. The parents and child need to be interviewed regarding their attitudes toward bedwetting and how it has affected their familial relationships. Finally, the physician should question the parents and child carefully for a history of punishment and abuse.

Physical Examination

The physical examination should carefully review several specific systems for signs of underlying abnormalities to suggest organic disease and thus complex enuresis, including a thorough neurologic, genitourinary, and abdominal examination. For the neurological examination, first the child's back should be examined carefully for signs of occult sacral dysraphism such as sacral dimpling, hairy patches, or sacral agenesis. Examination of the L5 to S3 motor and sensory nerves should be made with assessment of muscle tone, strength, and sensation. Peripheral reflexes, anal sphincter tone, and the bulbocavernosus reflex should be tested. An abdominal exam should focus on bladder distention following voiding to identify a large postvoid residual and signs of constipation or fecal impaction. The genitalia should be inspected to evaluate for aberrant anatomy that may result in incontinence such as meatal stenosis, epispadias, bifid clitoris or labial adhesions, or evidence of meatal pits which correlate with the incidence of ectopic ureters. The physician should also be aware of signs of sexual abuse including tearing, scars, or trauma. Hypospadias itself is not an etiology for incontinence. Observed voiding is essential for evidence of slow stream or intermittency. Furthermore, we always inspect the child's underwear for evidence of wetness, suggesting diurnal and therefore complex enuresis. Physical examinations in monosymptomatic nocturnal enuresis are almost always completely normal.

Laboratory and Radiographic Examination

Part of the goal for the history and physical examination is to identify the small number of children who need further investigation and treatment. A routine urinalysis and culture identifies children who may have a urinary tract infection and rules out inadequate urine concentration by the kidneys if a specific gravity of 1.022 or greater can be achieved. If dilute urine is identified, it can be confirmed by a morning void. Glucose in the urine may identify children with diabetes and resulting polyuria.

Approximate functional bladder capacity can be ascertained by two described methods: physiological voluntary end fill whereby the child keeps a 48-h diary noting time and voided volume to determine the maximum volume, or the rapid oral fill, where the child ingests a oral water load approximately $20 \, ml \, kg^{-1}$ over 1 h and asking the child to hold his or her urine for as long as possible and then to measure the volume. These values are the estimated functional bladder capacity and can be compared to age-appropriate nomograms as discussed prior. This can be performed in conjunction with an uroflow measurement to identify children with a slow stream, and bladder ultrasounds to measure postvoid residuals.

Monosymptomatic vs. Polysymptomatic Enuresis

At the end of the history, physical examination, and initial laboratory and radiographic imaging, a decision is made whether the child has monosymptomatic or polysymptomatic enuresis. Any child with findings on the review of symptoms and physical examination suggestive of neurologic or anatomic abnormalities, or persistent positive laboratory studies needs to be referred for further urologic evaluation which may include voiding cystourethrography and bladder ultrasonography to evaluate the lower urinary tract and renal ultrasonography to evaluate the upper urinary tract for aberrant anatomy. More recently, magnetic resonance imaging has become a powerful imaging tool to evaluate for genitourinary anomalies. A complete urodynamic test to elucidate abnormal physiology of voiding may be required. The child may require surgical intervention and reconstruction and the discussion is reviewed elsewhere.

Treatment

Once a decision has been made that the bedwetting may not spontaneously resolve, children need to take an active role in treating enuresis. Children should begin treatment only if they are interested and committed. A child unwilling to invest time and energy, regardless of parental concern, is unlikely to respond to therapy. We generally institute treatment at 7–8 years of age. Treatment can be frustrating for parents and children and taxing on physicians. Rates of relapse can be significant and parents need to be made aware and have realistic expectations. Age-appropriate norms should be reviewed. Given the benign nature of the condition, it is also extremely important to implement measures that have virtually no side effects and a risk no greater than that of treatment failure. We also stress the effect of the placebo which shows that treatment is measurably more effective when the patient perceives it as being personally and individually administered by a committed physician rather than through an impersonal protocol. Here we review both behavioral therapy and medical management with pharmacotherapy (**Table 2**). Ultimately, multiple treatment modalities may be employed to achieve a balanced program with the goals of a significant improvement in the number of wet episodes.

Behavioral

Behavioral therapy for enuresis is safe, effective, and often used in conjunction with medical therapy. Traditional therapies include regulating fluid intake, bladder training, and an enuresis alarm. Recently, developments in motivational therapy and hypnotherapy hold promise in the treatment of nocturnal enuresis.

Fluid intake

Although fluid restriction is often recommended, it has not been shown to be effective, and could place the child at risk for dehydration. More effective is a redistribution of fluid intake to decrease nocturnal polyuria. The child's fluid requirements are calculated using the approximation of $100 \, ml \, day^{-1}$ for the first 10 kg body weight, 50 ml day^{-1} for the second 10 kg, and $25 \, ml \, day^{-1}$ for every additional kilogram body weight. The child is instructed to consume approximately 40% of total fluid intake in the morning, 40% in the afternoon, and 20% in the evening. We never ask children to restrict fluid intake. This 'redistribution' helps to promote healthy drinking habits while seeking to decrease urine production at night. Caffeine is eliminated from the diet.

Table 2   The common classes of behavioral and pharmacological treatments for nocturnal enuresis

Behavioral	Pharmacological
Fluid intake regulation	
Bladder training	Desmopressin
Bed-wetting alarm	Imipramine
Motivational therapy	Anticholinergic drugs
Hypnotherapy	Acupuncture

Bladder training

Retention-control training strives to increase the functional bladder capacity of enuretic children with decreased functional capacity. Bladder retention training involves increasing bladder capacity by asking children to hold urine for successively longer intervals after first sensing an urge to void. Voiding diaries are kept to document frequency and voided volumes. The largest volumes are plotted reflecting functional bladder capacity, and compared to calculated age-appropriate bladder capacity. Although results have been variable, mean bladder capacity can be increased over twofold in some studies with cure rates estimated up to 35%. When retention training is combined with other modalities, the results can be impressive. We stress the importance of increasing functional bladder capacity prior to pharmacotherapy with antidiuretic hormone desmopressin acetate (DDAVP).

Alarms

Bladder alarms originate back to 1902, and are arguably the most effective behavioral therapy. Initial models consisted of a detector pad on which the child slept that was activated by urine and awakened the child with an alarm. Current models use transistor technology and consist of a battery-operated device attached to the collar or wrist with an audible alarm or vibrator that is connected to a thin wire attached to the child's underclothes, resembling a small pager in a self-contained unit. A few drops of urine activate the system. The alarm acts as a conditional stimulus that awakens the child when micturition occurs. The mechanism by which the alarm paired with bedwetting is not fully understood. One theory is the alarm conditioned patients to wake up when the bladder is full. However, when using the alarm, many children learned to sleep throughout the night without waking or wetting, with findings of increased functional bladder capacity over the treatment course. This suggested that the alarm may condition the inhibition of the detrusor muscle allowing increased bladder capacity to compensate for bladder fullness.

Randomized, controlled trials showed the enuresis alarm system to be quite effective with a 40–80% response rate, and a relapse rate of only 20–40% and minimized with retreatment. Compared to other behavioral treatments and medical management of nocturnal enuresis, the alarm has the best long-term efficacy. Treatment duration is usually stopped after four dry weeks are achieved with interval usage possibly decreasing the rate of relapse. However, the conditioned response can be a lengthy process and may disrupt the family more than is acceptable with an average of 16 weeks of therapy required. Some families will not tolerate the frequent night waking or the alarm may not wake the child, or may wake other siblings. Success improves with motivation of the child and parents willing to get up with their children at night.

Motivational and hypnotherapy

Motivational therapy promotes behavior modification by making the child responsible for his or her enuresis and crediting for their successes. The therapy requires that the parents and child develop a positive relationship with respect to enuresis and a good rapport between the physician and the family. Although it is difficult to assess the success of motivational therapy, it has been estimated to be as high as 25%. If the child does respond to therapy, a low rate of relapse can be expected.

In hypnotherapy, the patient is first induced to enter a hypnotic state, and then the therapy begins by suggesting that the child will wake up if he or she needs to urinate during the night, that the bladder will be able to hold more urine, and that the child will be able to control his or her urination. A clinical trial has shown that three 30 min sessions of hypnotherapy had similar immediate response rates compared to imipramine, but with improved response rates at 6 months of 68% compared to 24%, respectively. Case series have shown response rates of 60–70%. These two modalities will need to be studied further but reveal the promise of positive suggestions in the treatment of nocturnal enuresis.

Scheduled voiding

In addition to punishment, there are a number of practices that parents should be discouraged from employing. Picking up a child during the night to allow urination without fully awakening the child or waking the child at night to void before wetting has occurred has not been shown to resolve bedwetting. The time of night when a child wets is variable and these practices may teach the child to empty without fully awakening. Also, prolonged use of diapers or pull-ups is inappropriate because it encourages regressive behavior.

Medical Management

Pharmacotherapy for enuresis targets the proposed physiologic alterations contributing to nocturnal enuresis, mainly altered AVP secretion and bladder instability. These medications include DDAVP, imipramine, and oxybutynin. Promising reports on the use of acupuncture will also be discussed in this section.

Desmopressin acetate

The newest medication approved for enuresis appears the most promising. In 1990, the (FDA) Food and Drug Administration approved DDAVP for the treatment of nocturnal enuresis. A synthetic analog of vasopressin with a long half-life, DDAVP decreases urine output by retaining water at the level of the distal tubules. It theoretically functions by reducing the nocturnal urine output below the functional bladder capacity and may be most effective in the approximately 25% of children who do not demonstrate the normal diurnal increases in AVP secretion.

Its efficacy is maximized once children regulate their fluid intake and have normalized their functional bladder capacity. A review of randomized, controlled trials using DDAVP showed that 10–40% of patients became totally dry with a mean of 25%. Unfortunately, relapse rates have been as high as 60%, suggesting a symptomatic control of enuresis.

DDAVP is available as a nasal spray or tablet. The nasal spray is started at 20 µg and titrated up to 40 µg daily, while the oral tablet is started at 0.2 mg daily at bedtime and can be titrated up to 0.6 mg daily. Treatment is continued for 3–6 months with a gradual taper of the medication. Side effects are rare with the most common complaint of nasal irritation for the nasal spray formulation, abdominal discomfort, nausea, and headache. The most severe, yet rare side effect is hyponatremia, with no reports occurring in controlled clinical trials.

Imipramine

The tricyclic antidepressant imipramine was first shown to improve bedwetting in 1960. It is still the most widely prescribed in its class although other family members have been used. Imipramine has multiple effects, with a weak anticholinergic activity and an α-adrenergic activity on the internal urethral sphincter to promote continence. Imipramine has also been shown to stimulate ADH secretion. Results with imipramine are variable with cure rates reported up to 50%. However, high relapse rates (up to 60%) with long-term cure rates of only 25% after discontinuation of the medicine suggest a predominant role in the symptomatic control of enuresis.

We recommend an initial 2-week trial, and if a child responds, continuing a 2-month course with gradual weaning by decreasing dosage and frequency. If the patient relapses, therapy is continued with trial cessation at 3-month intervals. Imipramine is taken once per day, usually before bedtime, at $0.9–1.5 \, \mathrm{mg \, kg^{-1} \, day^{-1}}$. Average doses given for children 5–8 years old are 25 mg with 50 mg given to older children and adolescents. Side effects are generally uncommon and include nervousness, sleep disturbance, and gastrointestinal irritability. However, the largest risk is in children who accidentally ingest the drug in large quantities with toxic overdose leading to cardiac arrhythmias, conduction block, hypotension, convulsions, and death. Parents must secure and control the drug carefully.

Anticholinergic Drugs

Anticholinergic Drugs such as oxybutynin have been found to be largely ineffective in children with monosymptomatic nocturnal enuresis with a partial response rate ranging from 5% to 40% in bedwetting reduction, which was no better than placebo. In children with polysymptomatic nocturnal enuresis with documented bladder instability by urodynamics, oxybutynin may be effective in decreasing symptoms of daytime urgency, frequency, and diurnal incontinence. However, recent studies investigating combination therapy with oxybutynin plus imipramine, and oxybutynin plus desmopressin, suggested a synergistic effect with combinatorial therapy more effective than monotherapy with either agent along.

Acupuncture

Several case series have shown impressive results for the use of acupuncture in the treatment of nocturnal enuresis with 73–98% cure rates, requiring anywhere from one to more than 40 sessions. Although one randomized controlled trial showed similar responses compared to desmopressin, further investigation in the effect and number of treatments needs to be performed for this promising alternative therapy.

Conclusion

Nocturnal enuresis is a common childhood urinary condition that has a high rate of spontaneous cure, but its prolongation can interfere with socialization and raise anxiety for children and their families. Multiple etiologies contribute to nocturnal enuresis including physiological and behavioral conditions. Proper diagnosis is critical and identifying signs of complicated or diurnal enuresis in the initial evaluation warrants further investigation and studies. We recommend seeking therapy when primary monosymptomatic nocturnal enuresis persists at 7–8 years of age. Behavioral therapies including altering fluid intake, bladder training, and motivational therapy are benign and have shown positive results. The conditioned response from enuresis alarms have proved to be the most efficacious treatment, while pharmacological therapy with desmopressin has demonstrated significant responses with minimal side effects. No one treatment modality has a 100% response rate; thus, combinatorial therapy tailored to the individual child and family proves the most expedient solution to resolution of primary nocturnal enuresis.

See also: Endocrine System; Sleep; Toilet Training.

Suggested Readings

Blum NJ (2004) Nocturnal enuresis: Behavioral treatments. *Urologic Clinics of North America* 31: 499–507.
Jalkut MW, Lerman SE, and Churchill BM (2001) Enuresis. *Pediatric Clinics of North America* 48: 1461–1488.
Koff SA and Jayanthi VR (2002) Nocturnal enuresis. In: Walsh PC, Retik AB, Vaughan DE, and Wein AJ (eds.) *Campbell's Urology*, 8th edn., pp. 2273–2283. Philadelphia, PA: W.B. Saunders.
Mammen AA and Ferrer FA (2004) Nocturnal enuresis: Medical management. *Urologic Clinics of North America* 31: 491–498.
Rushton HG (1989) Nocturnal enuresis: Epidemiology, evaluation, and currently available treatment options. *Journal of Pediatrics* 114: 691–696.

Relevant Websites

http://www.bedwettingstore.com – Bedwetting Store.
http://www.pottypager.com – Potty Pager.

Behavior Genetics

H H Goldsmith, University of Wisconsin–Madison, Madison, WI, USA

Glossary

Allele – One of the versions of a gene that occurs at a polymorphic locus.

Centimorgans – A measure of genetic distance derived from the chance of recombination between two genetic markers at meiosis. A 1% chance of recombination corresponds to 1 centimorgan.

Dizygotic – 'Two zygotes'; refers to fraternal twins.

Endophenotype – Relative to a behavioral phenotype, a characteristic, usually biological in nature, which is associated with the same genetic factors but is considered to be 'closer to' (endo-) the genotype than the behavioral phenotype.

Genetic heterogeneity – The condition in which the same ostensible phenotype can be due to different genetic influences.

Heritability – The proportion of the observed (or measured) phenotype that is associated with genetic differences among individuals in a population.

Monozygotic – 'One zygote'; refers to genetically identical twins

Phenotype – A measurable or observable characteristic of an individual; a trait.

Polygenic – Influenced by multiple genes.

Polymorphism – 'Multiple forms', a genetic locus that has more than one version (allele) in the population.

Proband – The individual who is first identified as possessing a particular phenotype in a family. This index case qualifies the family for study.

Quantitative trait loci (QTLs) – Genes in polygenetic systems that individually contribute small amounts to the continuous (quantitative) phenotypic variation.

Single nucleotide polymorphism (SNP) – A polymorphism that is due to change of only one base pair in the DNA sequence.

Introduction

The field of behavioral genetics addresses the multifaceted interplay of genes and experience as they affect behavior. The framework for studying how genes affect behavior continually evolves; it incorporates any method of linking genetic influences to behavior, and entails connections between human and nonhuman (mostly mouse) research strategies. In broad perspective, human behavior–genetic approaches can be divided into quantitative genetic approaches (or genetic epidemiology) and molecular genetic approaches (mainly linkage and association studies). Prospects are brightening for widespread integration of molecular and quantitative genetic approaches and for dealing with complex, nonlinear systems from a genetic perspective.

Designs to Study Genetic Influences

A key principle is that no single method, such as twin studies, stands alone in the behavior–genetic (BG) framework as a basis for inference. Most BG designs are quasiexperimental or correlational, and each has its strengths and shortcomings. Confident inferences depend on replication within and confirmation across the various designs, which are reviewed briefly in this section.

Family Studies

Perhaps the most basic design is the family study, which quantifies the degrees of similarity among various classes of family members (e.g., parent–child, sibling–sibling) and tests whether degree of behavioral similarity aligns with degree of genetic similarity. Ordinary family studies do not distinguish genetic from environmental transmission, but can establish vertical transmission of traits (from parent to offspring). Demonstrating vertical transmission is a key piece of evidence for genetic effects. Data on the risk to siblings of affected individuals also constitute crucial baseline information for other research designs, such as twin studies and genome scans. However, family history of a disorder is not synonymous with genetic risk, and negative family history does not dictate lack of genetic risk. For instance, although genetic influences in the liability to schizophrenia are widely appreciated, 89% of persons with schizophrenia do not have a parent with schizophrenia, and 63% do not have any first- or second-degree relative with schizophrenia. Such a pattern is the expectation for complex disorders with multifactorial inheritance and environmental contributions. When individuals are chosen for study in high-risk designs by virtue of an affected parent, those offspring who eventually do become affected are likely to be at greater genetic risk than other affected persons. Therefore, these individuals will probably be unrepresentative of the full population of affected individuals in terms of etiological factors.

For many areas of early behavioral development, more family studies are needed. For instance, despite ubiquitous family lore, we know very little about parent–offspring similarity for infant characteristics, given the generational span needed to conduct such studies.

Twin and Adoption Studies

These two traditional methods are dominant in early stage human behavior–genetic research. A twin study takes advantage of the contrast in degree of genetic similarity between monozygotic (MZ, identical) twins, which is 100%, vs. dizygotic (DZ, fraternal) twins, which averages 50%. Assuming, and in some cases testing, the propositions that MZ and DZ co-twins share trait-relevant environments to the same degree and are representative of the nontwin population allows researchers to estimate the strength of heritable and nonheritable (environmental) sources of variation on a behavioral trait, or phenotype. The idea of decomposing a phenotype to highlight heritable and nonheritable features involves an interaction between genetic studies and the process of conceptualizing and measuring the phenotype. The longitudinal extension of behavior–genetic designs (e.g., studying twins during infancy and following them up as first graders) allows inferences regarding genetic influences on not only a trait's level, but also on its continuity and change. Controlling for genetic similarity (e.g., by studying twins) allows for the investigation of environmental factors, including their interaction and covariance with genetic factors.

Whereas twin studies capitalize on an experiment of nature, adoption studies take advantage of a cultural practice. Adopted children's behavior can be compared with the behavior of their biological parents, their adoptive parents, their adoptive siblings, and, in rare cases (such as twins reared apart), with their biological siblings reared in different families. The degree of similarity among adoptive siblings can be compared with the degree of similarity among biological siblings raised together. Issues affecting the validity of inferences drawn from the various adoption methods include the representation of biological parents who place children for adoption, representation of the adopted children themselves, and representation of the families who successfully adopt children. These issues are open to empirical investigation as to their degree of violation and whether any violation affects conclusions regarding a specific behavioral pattern under study. Certain inferences from adoption studies require controlling for the developmental level of the persons being compared. For instance, for most phenotypes, the 'environmental similarity' of an adoptive child with a biological child of about the same age in the same family probably has quite different roots than the 'environmental similarity' of an adoptive child and the much older adoptive parents. Adoption studies offer ways to disentangle gene–environment (GE) correlation when that correlation is based on parents typically providing both genes and environment. Although adoption within the US is less frequent than it once was and privacy laws impede some study designs, the frequency of divorce after having children and remarriage followed by more births provides a tremendous potential resource for complex versions of traditional adoption studies.

Linkage and Association Studies

Identifying individual genes involved in etiology is currently feasible although still in early stages for behavioral phenotypes with complex inheritance. The basic question is whether possession of a specific version of a single gene (an allele) is associated a behavioral phenotype. To answer this question, the first requirement is that specific genetic variants be identified. The Human Genome Project and the succeeding HapMap projects have hastened this task of identifying genes. Here, our attention is focused on 'complex inheritance', a term that can encompass single genes of discernible effect operating in a background of both polygenic and environmental effects. Complex inheritance also entails the possibility of genetic heterogeneity and environmentally molded phenotypes that resemble genetically influenced ones (phenocopies).

The methods that are applicable to complex inheritance are varieties of linkage analysis and association studies. For humans, linkage methods include classic pedigree analysis in which statistical transmission models are tested in large families with many affected members. Also relevant to complex inheritance are the newer methods using gene chips and expressed sequence arrays, the latter of which can detect the expression, in mRNA, of hundreds or even thousands of genes simultaneously. Expressed sequence arrays are currently used mainly in nonhuman animals. Ever larger gene chips are used extensively in human studies. Polymorphisms (different genetic variants) in thousands of genes can be identified on a single chip, and various types of polymorphisms can be studied. The most common type is the single nucleotide polymorphism (SNP) which, as its name implies, involves a single change in the DNA base sequence that will lead to a single changed amino acid in a protein if it resides in an expressed portion of a gene.

A class of nonparametric analyses called 'allele-sharing methods' involves studying affected relatives to determine whether a gene or marker is shared by these affected relatives, through inheritance from a common ancestor, more frequently than expected. The most common implementation of allele-sharing methodology is the analysis of affected sibling pairs in 'genome scans', often with a finer-grained scan following an initial scan. Genome scans have been employed to study autism, attention deficit hyperactivity disorder, and other complex phenotypes that characterize young children. In a genome scan, linkage analysis is conducted using a series of anonymous polymorphisms, or genetic probes, spaced at relatively constant intervals over the entire genome – about 350 markers with an average spacing of 10 centimorgans in the late 1990s and early 2000s, but much more closely in recent research. The results of the scan are used to identify candidate chromosomal regions that are shared at rates greater than 50% (the rate expected for randomly selected markers for full siblings). Recently, in addition to using many more genetic markers, usually SNPs, parent–parent–child trios rather than sibling pairs have been studied, and highly discordant, in addition to the usual concordant, pairs of relatives have been employed. This genome scan approach offers the potential of finding genes previously unsuspected of having an influence on the phenotype of interest. Genome scans are complicated by several factors, for example, instead of a single test for linkage, one must conduct multiple tests across the entire genome and genetic heterogeneity and phenotyping mistakes (e.g., misdiagnosis) can seriously distort results.

Allelic association studies compare the frequency of a marker (e.g., a blood group) in two samples (e.g., an affected vs. a control group). A classic association is between the blood group O and duodenal ulcer. Unless the marker locus and the actual gene responsible for the disorder are very tightly linked, results from properly conducted association studies will be negative. Of course, association studies can be used to test candidate genes that might themselves be responsible for a phenotype. A key limitation of association studies is that spurious association can occur due to population stratification, that is, the samples studied might come from a population that contains groups with differing frequencies of the allele under study. If these groups differ for any reason in occurrence of the disorder, then a spurious association of allele and the phenotype will result. Some association studies have been conducted with genetic probes that systematically cover large expanses of the genome, thus allowing a search for association with so-called quantitative trait loci (QTLs). QTLs are genetic regions responsible for small but discernible quantitative effects on a polygenically based phenotype. The smallest QTL effect that can be detected by association with a marker depends on the distance from the marker to the QTL, the size of population, and the heritability of the trait or disorder. In plant genetics, QTLs accounting for as little as 0.3% of the phenotypic variance have been reported.

Different mutations in a single gene can have different effects on a phenotype. That is, if a major gene leading to qualitative dysfunction is identified, other alleles at the same locus could affect more subtle quantitative variation. The allele that leads to qualitative dysfunction could code for an inactive protein whereas a different allelic variant of the same gene might lead to an active but physiologically nonoptimal protein.

Combinations of Methods

The real power of genetic studies of behavior comes from replication of findings using different methods. Although a critic might think that the assumptions of the twin method are dubious for a given application, if the twin study results are consonant with the results from family and adoption studies, then the criticism becomes hard to sustain. Similarly, linkage is often used to confirm association. In contemporary research, association studies are incorporated into twin designs, allowing simultaneous molecular and quantitative genetic approaches to the same phenotype in the same sample. The future of BG research will undoubtedly depend on these integrated approaches.

Behavior–Genetic Concepts

Heritability and Environmental Variance

The relative strength of genetic and environmental factors can be inferred from patterns of co-variation among family members who have varying degrees of genetic overlap and shared environments. The heritability statistic estimates the association between degree of genetic overlap and similarity on behavioral traits in relatives. Thus, heritability is a statistical rather than a biological concept. The genetic component can be divided into additive and nonadditive effects, and the environmental component can be divided into effects shared vs. nonshared by relatives. This basic biometric model can be expanded for longitudinal, multivariate analysis. The demonstration that a trait is heritable is considered a prerequisite for conducting molecular genetic studies to reveal which genes are involved.

The concept of heritability has many interpretive limitations. Heritability applies only to differences among persons within populations, not to the development of single individuals or to differences between populations. Heritability may also change at different points in the lifespan, due to the dynamic nature of gene action, changes in the effects of environmental factors, changes in the nature of the trait being measured, or other constellations of factors. Because heritability is a ratio involving both genetic and environmental effects, a relevant environment that varies widely across a population will reduce heritability whereas more nearly uniform environmental conditions increase it. Another caveat is that similar heritability estimates for two traits, even conceptually related traits, do not imply common genetic underpinnings. In fact, the great majority of reliable heritability estimates for human behavioral traits lie in a 30–70% range. Despite the consensus that most human behavioral traits are moderately heritable, nonheritable, or weakly heritable, exceptions do exist; examples include early positive emotionality, parent–child security of attachment, and romantic love.

The developmental implications of simple heritability estimates are also limited. Classic behavior–genetic inferences are confined to genetic and environmental effects on phenotypic variance, not gene action and the direct action of environmental influences *per se*. Genes exert their effects on behavior via complex pathways that involve feedback loops, rate-limiting processes, and other non-linear mechanisms. Environments undoubtedly also exert many of their effects in inherently nonlinear ways. Nevertheless, analyzing individual differences by linear regression of outcome on sources of variation can be informative, even when the individual differences result from highly contingent, interactive developmental processes operating in the lives of individuals. Such linear models relating predictors and outcomes are used profitably in other fields, such as economics and decision making, where the underlying processes involved are also complex and interactive.

Environmental effects are also subject to BG analysis. Environmental effects can be shared or nonshared by a particular set of family members. Shared environmental effects explain similarity between twins and relatives in addition to that accounted for by common genes. Shared environmental variance also accounts for the similarity of genetically unrelated individuals who are reared together. The nonshared environmental variance is the remainder of the variance not explained by genes or by shared environment. It includes the effects of experiences that are unique to each individual and independent of genetic factors. Nonshared environmental variance can be directly estimated from differences between MZ co-twins. The estimate of nonshared environment is often confounded with measurement error. A common misunderstanding is that environmental effects are 'what is left over' after genetic effects are estimated. Typical BG methods treat genetic and shared environmental effects in an even-handed manner, given the assumptions about how these effects can be partitioned. The more cogent criticism is that the genetic partitioning is based on sound theories of Mendelian inheritance whereas the environmental partitioning is based on familial units that might not be the most important markers of environmental influence. Despite extensive explication of the issue in the BG literature, the meaning of 'shared' and 'nonshared' environments continues to create confusion. The variance component referred to as 'shared environment' can differ from one kinship design to the next. That is, the environment shared by co-twins has a somewhat different quality from the environment shared by ordinary siblings or adopted siblings, and certainly different from the environment shared by parents and offspring. Of course, these differences in the quality of the shared environment might well be irrelevant to the behavior under study. Fortunately, many of these issues can be subjected to empirical tests, which are common in the more technical BG literature.

Parents can exert strong experiential effects on their offspring behavior in the absence of a shared environmental factor in biometric analyses. For example, hypothetically, aggressive fathers might induce inhibited behavior in children of a certain age. Such a hypothetical effect would not emerge in a univariate analysis of parent–offspring data of either aggressiveness or inhibition alone. Part of the solution to problems like this is use of multivariate analysis. A more important part of the solution is integrating theories of how the environment works into these multivariate designs. In summary, the distinctions between shared and nonshared environmental effects and transmitted environmental effects in BG designs do not map well onto some issues concerning the nature and effects of interaction among family members. Testing for environmental risk and protective factors in ways that are also sensitive to genetic factors requires programmatic application of several research strategies.

Endophenotypes

A concept with substantial currency is that of the 'endophenotype'. Endophenotypes underlying behavioral phenotypes can be biochemical, neurological, anatomical, psychophysiological, endocrine, sensorimotor, perceptual, cognitive, or affective in nature. Despite its short history, the endophenotype concept has proven quite useful. Among the desirable properties of a good endophenotype for genetic analysis are (1) an etiologic association with the complex behavioral phenotype, (2) presence in some unaffected relatives, and (3) being amenable to objective measurement.

Complex behavioral phenotypes such as schizophrenia, anxiety, and autism do not yield readily to genetic analysis. The same is true for complex medical disorders such as coronary artery disease and idiopathic generalized epilepsies. For these medical disorders, endophenotypic indicators such as plasma cholesterol and abnormal electroencephalogram (EEG) patterns have facilitated genetic analyses. These and other medical examples indicate that a greater understanding of the disease processes yields endophenotypes, as well as subtypes, that can themselves become the focus of genetic analysis.

GE Covariance and Interaction

Upon first exposure to behavioral genetics, scientists often question how mutual or interactive processes are treated. The concepts of GE covariance and interaction provide a partial answer. These two concepts figure into almost every 'textbook' presentation of the BG paradigm. However, in empirical research, both concepts are difficult to operationalize and are often simply uninvestigated.

GE interaction refers to the differential effects on the phenotype of an environmental factor for persons with different genotypes. In this sense, GE interaction is a statistical rather than a process-oriented interaction. For example, having poor sibling, peer, or adult models for the development of empathy might affect the development of aggressive tendencies primarily for those children with genes associated with, say, low frustration tolerance (assuming that lack of empathy and frustration tolerance are both precursors of aggressive behavior). Scattered early attempts to isolate these interactions in human data were largely unsuccessful, but a recent literature is more promising. An example of GE interaction is the demonstration that a polymorphism in the promoter region of the gene for the serotonin transporter (5-HTT) interacts with maternal reports of social support to predict inhibited behavior with unfamiliar peers in middle childhood. Children who possessed the short 5-HTT allele and whose mothers had low social support showed higher rates of behavioral inhibition in middle childhood.

Genetic analyses also include consideration of GE correlation. Individuals can be differentially exposed to environments that contribute to further development of heritable traits. This gene-conditioned, differential exposure is often viewed as reflecting three types of GE correlation: passive, reactive, and active. If a child's genotype is correlated with the environment provided by parents and siblings and such provision is associated with heritable traits of the parents or siblings, then passive GE correlation results. The classic example is smart parents providing intellectually stimulating environments to their children. 'Reactive', or evocative GE correlation refers to others' reacting to a particular child on the basis of the child's inherited characteristics. For example, children who are inattentive in school may be taught less material in a less effective manner. Thus, the environment becomes correlated with genotypic differences. Lastly, an active GE correlation refers to the situation in which a child seeks an environment conducive to further developing some of his or her heritable tendencies. Thus, aggressive youths may actively choose to become friends with peers who are also easily frustrated and prone to attribute hostile intent to benign actions of others, and these friendships would contribute to further development of the aggressive phenotype. Obviously, naturally occurring situations might reflect combinations of all three types of correlation. Developmental psychologists have postulated a shift from passive to reactive and then to active GE correlation during childhood, as children begin to select more of their own experiences and environments.

Versions of the idea of GE correlation have captured the attention of many behavioral scientists; however, some behavioral geneticists argue that active GE correlation cannot be meaningfully distinguished from 'direct' genetic effects. That is, even direct genetic effects are always instances of genes correlated with environments although the environments might occasionally be entirely nonpsychological in nature. In the case of social environments, suppose that genotypic differences are correlated with, say, antisocial behavioral tendencies. As described earlier, these tendencies might be manifest, in part, by seeking peers who are experienced in antisocial behaviors themselves. The association with peers might be the most proximal influence on antisocial acts. This scenario would usually be characterized as active GE correlation, of the type in which the individual selects an environment on the basis of genetically influenced behavioral predispositions, that is, 'birds of a feather flock together'. But do not all genetic effects on behavior, involve selection of relevant environments, in the sense that genes and their proximal and distal products must be expressed in a facilitative context, where the 'context' may range from the physiological to the social? Perhaps so, but it may still be useful to retain the concept of active GE correlation for scenarios in which the environment is a measurable experience. One empirical example of a GE correlation involved the study of mutuality, the notion that healthy parent–child relationships exhibit emotional reciprocity and have a bidirectional, responsive quality. They used dyadic observational measures such as mutual eye gaze and shared positive emotion with mothers and a pair of their 3–4-year-old children, where the pairs were MZ or DZ twins, full siblings, or genetically unrelated adopted siblings. The similarity of siblings for the mutuality measures with their mothers suggested genetic influence based on child characteristics: MZ, correlation = 0.61; DZ, correlation = 0.26; full siblings, correlation = 0.25; and adopted siblings, correlation = −0.04. Because the child-specific factors are heritable and the parent's contribution to mutuality constitutes an aspect of the child's environment, GE correlation is induced in this situation. Further BG studies of mutuality suggested that it is child-specific and linked to child behavior problems.

Although the idea is intuitively appealing, the importance of GE correlations for emotion-related phenotypes, including those related to effect and psychopathology, largely remain to be demonstrated. GE correlations may be negative in sign, such that caregivers and others provide, for example, inhibited children with extra opportunities to engage their environment without eliciting fear. On the other hand, GE correlations may be positive, such that emotionally labile children experience unsettled and unpredictable environments.

Genetic Heterogeneity

The notion of genetic heterogeneity refers to subsets of a patently single clinical disorder having different genetic roots. At least in the early stages of understanding a disorder, genetic heterogeneity is probably the norm rather than the exception. That is, distinct genes (or distinct alleles at the same gene locus) can lead to the 'same' disorder. Some genetic approaches initially assume that all cases of a disorder have a common cause, but empirical results often reveal unsuspected heterogeneity.

An example of how genetic heterogeneity with regard to depression is investigated involved a large population-based sample of female adult twins. Concordant twin pairs had the same type of depressive syndrome more often than chance, and resemblance was greater in MZ than DZ twin pairs. In this example, a genetic marker that co-segregates with depression in some families but not in others (the best standard of evidence for genetic heterogeneity) has not been found. However, another type of evidence for genetic heterogeneity has been found: certain clinical features of depression 'run true' in families. Another type of evidence for genetic heterogeneity would be a finding that the risk to relatives differs according to the characteristics of the affected individual (e.g., as in early- and late-onset Alzheimer's disease, in which a gene on chromosome 14 affects only the early-onset form).

The general idea of genetic heterogeneity applies not only to disorders but also more generally to traits. That is, different genes might affect traits in the low, medium, and high ranges of trait values. And it is also possible that different combinations of genes result in the same phenotype in different individuals or different subgroups of the population.

Developmental Behavioral Genetics

To this point, the concepts reviewed apply largely to static views of behavior. However, these concepts can be supplemented with notions that help us to understand quantitative genetic aspects of development. First, genetic analyses can be extended from the occurrence or strength of a disorder or trait to developmental (time-related) features of disorders or traits. Thus, we can analyze the genetics of the age of onset and of the clinical course of a disorder. Questions about whether the genetic liability to the disorder itself overlaps with the liability to age of onset can then be addressed. These questions can be framed for polygenic or single-gene effects.

Developmental behavioral genetics received a great boost from the development of multivariate methods, mostly using structural equation modeling. Multivariate methods involve the simultaneous biometric analysis of more than one phenotype, including the case of one or more phenotypes measured on multiple occasions. Even if cross-sectional studies show that a phenotype is heritable at two different ages, and other studies show that the phenotype is stable, we cannot assume that this stability is mediated genetically without longitudinal, genetically informative studies.

It is especially important to consider the potential changing influences of genes and the environment across developmental transitions, when reorganization in one system is often followed by reorganization in another. Where such change is detected, many developmental factors are plausible correlates and perhaps causes, including locomotor and cognitive transitions or reorganizations. It is also important to note that the tendency to equate genetics or biology with 'constraint' and experience with 'possibility', while perhaps holding some validity, should probably be resisted. At best, these equations are oversimplifications.

Non-Mendelizing Genetic Influences

Family, twin, and adoption studies coupled with linkage and association studies are not the whole story of genes and behavior. Certain sources of genetic variability are non-Mendelian in their mechanisms and thus are not captured by the usual BG designs. Three examples of non-Mendelian mechanisms are imprinting (wherein a gene's expression is modulated based on whether it is transmitted from the father or the mother), progressive amplification of repeated DNA sequences, and the degree of homozygosity. Links of these mechanisms to behavioral phenomena during infancy and early childhood largely remain to be demonstrated. However, the mechanism of progressive amplification has received tremendous attention over the past 20 years. Fragile X disorder, a leading cause of mental retardation, exemplifies the pattern of progressive amplification through pedigrees. A CGG trinucleotide in an untranslated segment of the *FMR-1* gene on the X chromosome is normally repeated less than 50 times. When the gene mutates, these unstable CGG repeats are amplified, resulting in hundreds or even thousands of copies. The permutations are expanded to full mutations (termed 'triplet repeats') only when transmitted by the mother. Fathers with the permutation have unaffected daughters who might be at risk to pass on the disorder. Huntington's disease also follows the progressive amplification pattern of inheritance. The extent of the amplified region (a CAG repeat) is associated with age of onset, with earlier onset cases having longer repeated regions. Early-onset cases are more likely to be those with the 'at-risk' allele transmitted from the father, an apparent example of genomic imprinting. Genomic imprinting confers functional differences on specific genes derived from paternal versus maternal genomes during development. Only one of the two alleles is expressed in a particular tissue during a stage of development; the other is silenced. The epigenetic phenomenon of genomic imprinting has been well established in mice and has been hypothesized to be important in mammalian and primate forebrain evolution. Experimental evidence from mice shows that maternally derived genes are expressed preferentially in cortex, striatum, and hippocampus, and paternally derived genes are expressed preferentially in certain hypothalamic regions. Given the different roles of these regions in behavior, such differential expression could affect the behavioral resemblance of relatives if the genes involved are polymorphic. The question of whether imprinting is actually a significant and widespread factor in normal human development and for affective phenomena awaits further study, but studies of human mutations suggest that imprinted genes do affect brain development and behavior in some syndromes, most notably Prader–Willi and Angelman syndromes. These syndromes are due to deletions, rearrangements, or other anomalies in the region of chromosome 15q11-q13, which is a meiotically unstable, imprinted region of the genome. Current research attempts to elucidate the mechanisms of DNA methylation and demethylation that presumably underlie imprinting.

Recent work also suggests the plausibility of variability in degree of homozygosity as a source of genetic variability for psychopathology. A key construct in this line of theorizing and research is 'developmental instability', which can have causes other than homozygosity. Indicators of developmental instability include some minor physical anomalies and fluctuating asymmetries (deviations from symmetry in bilateral physical characteristics). Several scientists have documented associations between developmental instability – likely during the prenatal period – and psychopathology, including childhood disorders. Developmental instability reduces the organism's metabolic buffering and thus renders it susceptible to insult from environmental perturbations.

Summary and Prospects for Integration

Quantitative genetic research (twin, family, and adoption studies) has been undertaken increasingly since the mid-1980s and many projects have been extended longitudinally. The following generalizations can be offered. Heritability of behavioral patterns seldom rises above a figure of 60–70% of the observed variance. In longitudinal analyses, much of the stability – in some cases practically all the stability – of behavioral traits is due to the stability of genetic influences rather than stability of environmental factors. Both genes and environments seem to influence change in many behavioral traits. Adoption and stepfamily evidence still need to be integrated more fully with the more widely available twin results. A continuing need in quantitative genetic studies is for assessment other than self- or parental-report of emotion-related phenotypes. Some laboratory-based assessment of temperament/emotionality has occurred in twin and adoption studies, but much more is needed. Classic twin and family studies need to continue the present trend of becoming highly multivariate, incorporating both dimensional (e.g., temperament trait) and categorical (e.g., diagnosis) variables, and multiple, theoretically relevant occasions of study. Synthetic data-analytic approaches exist for measuring specific

genes within such classic quantitative designs, and these approaches are increasingly being adopted.

Our brief treatment of non-Mendelizing genetic effects caution us not to become paradigm-bound – even to useful paradigms such as the twin design and genome screens – if we wish to discern the entire panorama of genetic influences on behavior.

Another need in this area is integration of assessment of experiential processes into the twin and adoption studies to 'actualize' the environmental component of the observed variation. On the social side of the environment, measures of specific interactional processes need to be refined and incorporated into studies. Environmental assessment needs to extend into the realm of biological measures as well. Identified environmental factors should be quantified (e.g., in terms of increases in relative risk) and compared with other known risk factors to evaluate their importance.

The genetic component of the variation is 'actualized' by detecting specific genes associated with the trait. Molecular genetic studies have recently been initiated in this area, mostly as association studies and mostly for disorders. There is a question of whether investigation will more profitably be gene-centered (wherein the behavioral correlates of identified genes are sought) or behavior-centered (wherein the investigator begins with an emotion-related phenotype and searches for associated genetic markers). Progress in this area has been accelerated by the Human Genome Project, and in particular by mapping of genes that are expressed in the human brain. These efforts promise a new synergism with animal research, where gene function in the neurophysiological sense is more easily investigated.

Perhaps the most general recipe for progress in the field is interdisciplinary research. As in most areas of science, fuller understanding requires the joint perspectives of several methodologies, including epidemiology, neuroscience, and various approaches to investigating the effects of experience. Enthusiasm for reductionistic explanations based on genetics for the origins of the neural underpinnings of behavior should not overshadow the appreciation of developmental plasticity in the manifestation of behavior.

See also: ADHD: Genetic Influences; Depression; Family Influences; Genetic Disorders: Sex Linked; Genetic Disorders: Single Gene; Genetics and Inheritance; Nature vs. Nurture; Temperament; Twins.

Suggested Readings

Goldsmith HH (2003) Genetics of emotional development. In: Davidson RJ, Scherer K, and Goldsmith HH (eds.) *Handbook of Affective Sciences*, pp. 300–319. New York: Oxford University Press.

Gottesman , II and Gould TD (2003) The endophenotype concept in psychiatry: Etymology and strategic intentions. *American Journal of Psychiatry* 160: 636–645.

Lander ES and Schork NJ (1994) Genetic dissection of complex traits. *Science* 265: 2037–2048.

Neale MC and Cardon LR (1992) *Methodology for Genetic Studies of Twins and Families*. Dordrecht, The Netherlands: Kluwer.

Plomin R and Daniels D (1987) Why are children in the same family so different from each other? *Behavioral and Brain Sciences* 10: 1–16.

Reik W and Walter J (2001) Genomic imprinting: Parental influence on the genome. *Nature Review Genetics* 2: 21–32.

Rutter M, Pickles A, Murray R, and Eaves L (2001) Testing hypotheses on specific environmental causal effects on behavior. *Psychological Bulletin* 127: 291–324.

Bilingualism

N Sebastián-Gallés, L Bosch and F Pons, Universitat de Barcelona, Barcelona, Spain

Glossary

Communicative development inventory (CDI) – A tool for gauging the vocabulary of infants and toddlers.

Mutual exclusivity principle – Strategy that children may use for inferring the meaning of a novel noun. It is based on object terms being mutually exclusive, that is, each object should only have one basic level label.

Perceptual reorganization – In the early months of life, infants are able to discriminate both native and non-native contrasts easily. However, by the end of the first year, these perceptual abilities change, involving both decreasing sensitivity to non-native speech contrasts and realignment of initial boundaries.

Rhythmic class – Language rhythms are separated into three major classes reflecting linguistic metrical timing units: stress (English), syllable (Spanish), and a subsyllabic unit termed mora (Japanese). There is also a quantitative analysis of the acoustic characteristics of consonant and vowel units to classify it.

Simultaneous bilingual – When a child becomes bilingual by learning two languages at the same time from birth.

Voice onset time (VOT) – The length of time that passes between when a consonant is released and when voicing, the vibration of the vocal folds, begins.

Introduction

Learning a language is a complex achievement that human beings accomplish with great ease in the first years of life. To learn a language, human infants must discover the fundamental properties of the language of the environment. As discussed elsewhere in this encyclopedia, this is a complex process and to attain it, infants and children must learn multiple properties of the language they hear. If, as adults, we are fascinated by the easiness with which infants learn a language, we are even more astounded when bilingual infants are considered. However, one wonders if such a phenomenal task is out of the reach of young infants; thus, early bilingual exposure might be an undesirable circumstance. It is common that parents and caregivers wonder if infants acquiring two languages at the same time follow a troubled course, marked by delays and confusion.

In this article we describe what is known about very early stages of bilingual language acquisition. We concentrate on infants who are exposed to two languages from the very first day of their lives. These infants are called simultaneous bilinguals, as opposed to successive bilinguals (this type of acquisition has also been referred to as 'bilingual first language acquisition'). For these individuals there is not a first or a second language, in a chronological sense. However, because perfect identical exposure is practically impossible, there is always a language to which infants are more exposed; in this sense, there is a dominant language and a nondominant language in the environment.

As it will be seen, there is no evidence of trouble or confusion; nevertheless, infants learning two languages sometimes follow a developmental path different from that of infants learning just one language. We pay special attention to the perceptual abilities of bilingual infants, in particular, in the preverbal stage. This will lead to focus on the development of phonology and the lexicon, as opposed to the development of morphosyntax, a topic usually addressed in studies analyzing toddlers, and children's production abilities.

Early studies of how young infants and children become bilingual are relatively scarce, in particular when compared with studies with monolingual infants. However, this topic has a long history. The first empirical study on bilingual development dates from 1913. Jules Ronjat, a French linguist, was the first scholar to describe bilingual behavior. Thirty years later, another linguist, Werner Leopold, published a monumental monograph of four volumes describing the acquisition of language in a French–English bilingual child. In fact, still today, most of the studies on early language acquisition rely on observational data of infants' utterances. This type of methodology poses severe limitations if we want to know how language develops from birth. If we exclude babbling, by the time young children start to speak, they have acquired a fairly sophisticated knowledge of the language of the environment. If only studying language production is an important limitation for the study of language development in monolingual infants, as it will be seen below, it is even more important in the case of bilingual children.

Early Differentiation

To be able to learn two languages, infants must realize the existence of two different auditory systems in the speech input. But human beings are already able to make some distinctions at birth. As research on initial language differentiation capacities has shown, newborns can notice the differences between languages belonging to different rhythmic classes, but they cannot if both languages belong to the same rhythmic class. For instance, newborns can detect differences between Japanese and Dutch or French and Russian, but not between English and Dutch or Spanish and Italian. So, even without previous experience with multiple languages, the auditory system is already able to make some language distinctions. Therefore, an infant born in a bilingual family should be able to differentiate between languages if they belong to different rhythmic groups. Two studies carried out with bilingual infants, exposed to English and Tagalog (i.e., to languages belonging to different rhythmic groups) and to Spanish and Catalan (i.e., to languages belonging to the same rhythmic group) have shown converging results in this respect (Spanish and Catalan are, like French, Italian, and Portuguese, Romance languages; Tagalog is one of the major languages spoken in the Philippines and rhythmically very different from English). Two different measures have been obtained to draw this conclusion.

In an auditory preference procedure, infants can control the delivery of sentences by giving high-amplitude sucks (energetic). Every time they make a high-amplitude suck, a sentence is played. It has been observed that when presented with sentences of their maternal language and of a foreign language, newborns show higher sucking rates for the maternal language than for the unknown one.

In the study with newborn infants who were exposed to both English and Tagalog approximately equally throughout gestation, babies were presented with alternated minutes of these two languages. The results showed equivalent sucking rates for English and Tagalog. Importantly, newborn infants who were exposed to only English prenatally showed higher rates for English sentences than for Tagalog ones. Therefore, it is not that English and Tagalog sentences are equally attractive *per se* to newborns, but that bilingual infants are equally attracted to both (unlike English monolinguals).

The other experiment showing equivalent behavior for the two languages of the environment in bilingual infants studied 4.5-month-old infants with a procedure based on a visual orientation latency measure. In this procedure, infants listen to sentences of two languages, presented in a random order. Sentences can be played back from two different loudspeakers located at the right and the left of a central monitor in front of the infants (see **Figure 1**). For every infant, there is no correspondence between a particular loudspeaker and language being heard. At the beginning of each trial, an attractive image is shown in a central monitor. After a short variable time interval, the image disappears and a sentence is played in one of the lateral loudspeakers. Eye movements are recorded and orientation latencies are estimated. It has been observed that infants orient faster to the maternal language, than to unknown ones. When young bilingual infants (born in Spanish and Catalan bilingual families) are confronted with their two first languages, they show equivalent latencies, therefore, indicating that both languages are equally familiar to them (at this age, as described below, Spanish–Catalan bilingual infants when studied in a different paradigm can differentiate both languages).

The results of these experiments indicate that already in the very first months of their life, infants exposed to bilingual environments have been able to extract distinctive acoustic properties of both languages and that they treat them in a special way. However, these results do not answer the question of how early bilingual infants can notice the existence of two systems in their environment. So, do bilingual infants differentiate the languages?

The first evidence of language differentiation in infants exposed to bilingual environments has been obtained with 4.5-month-old Catalan–Spanish bilinguals. In this study, infants were tested using an adaptation of the familiarization-novelty preference procedure. Infants were familiarized with six different sentences of the language of their mother. After 2 min of familiarization, eight new sentences were presented (test sentences). Half of them were in the same language of the familiarization phase (the language of the mother) and the other half in the other language (the language of the father). Infants raised in bilingual environments increased their visual attention times to test sentences in a language different

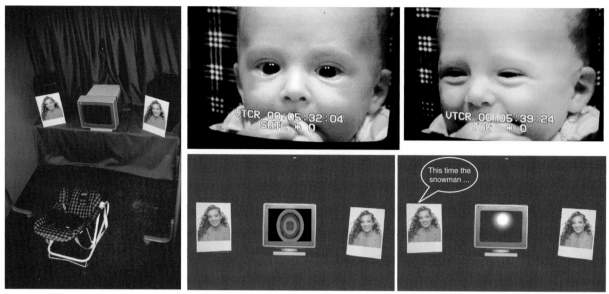

Figure 1 Preference: orientation time procedure.

than that of the familiarization phase, when compared with test sentences in the same language of the familiarization phase. Their results were fully equivalent to those obtained with infants exposed to either Catalan or Spanish monolingual environments.

Taken together, these results indicate that a simultaneous bilingual exposure does not generate specific problems in the processes of language differentiation. Importantly, this precocious differentiation was observed in a perceptually challenging situation, where the two languages of exposure are rhythmically very similar (and cannot be discriminated at birth), as it is the case with Catalan and Spanish.

Nevertheless, bilingual infants do not discriminate languages in the same way as monolingual infants do. Using the visual orientation latency procedure just described, monolingual (Spanish and Catalan) and Catalan–Spanish bilingual infants were also tested presenting the maternal language (Spanish or Catalan) and English as a foreign language. As expected, monolingual infants orient faster to the maternal language than to a foreign one (solid line in **Figure 2**). The opposite pattern is observed with bilingual infants. Although they also show significant differences between the orientation times to the maternal language and the unknown one, they orient slower to the maternal language (the language of the mother in this particular experiment) than to the unknown one (dashed line in **Figure 2**). Importantly, both monolingual and bilingual infants show equivalent orientation times to the unknown language: the differences are specific to the orientation times to the maternal language. This atypical pattern of orientations has been replicated with different language pairs (Spanish–English, Catalan–English, Spanish–Italian, and Catalan–Italian), as well as different ages (4.5 and

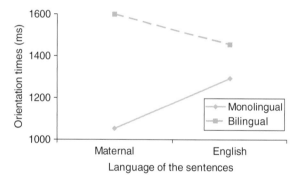

Figure 2 Results of orientation times to familiar and unfamiliar languages. Reproduced from Bosch L and Sebastián-Gallés N (1997) Native-language recognition abilities in four-month-old infants from monolingual and bilingual environments. *Cognition* 65: 33–69, with permission from Elsevier.

6 months). Although at present no satisfactory explanation to this response pattern can be given, it is a clear index of precocious adaptation of the speech-processing mechanisms in the rich and varied environment bilingual infants are exposed to.

Building Up Phonetic Categories

Research with monolingual infants has shown that starting at 6 months of age infants specialize in the phonetic repertoire of their maternal language. Before this age, infants can distinguish phoneme contrasts they have never heard before and that their parents are unable to discriminate. For instance, monolingual Japanese young infants can differentiate between /r/ and /l/ in a fully equivalent way than their English-learning peers. At 6 months, as a consequence of exposure to the maternal language, a perceptual reorganization takes place and infants start to lose this capacity to discriminate sounds in all languages. This reorganization will affect first the perception of vowels and later that of consonants. If language exposure is the main determinant to perceptual reorganization, exposure to two languages must shape the phonetic repertoire of infants in a specific way. So, what happens with bilingual infants?

It is almost impossible that two languages share the very same phoneme repertoire, even for the case of similar languages. For example, English and German share a common origin and are typologically related, but they are quite different in terms of their phoneme repertoires. The same happens with French and Spanish. Furthermore, not only languages have different phonemes, but the precise acoustic realization of the same phonemes varies from language to language. Just consider how phoneme /b/ is pronounced by English and Spanish natives. Or for an even more extreme case, how phoneme /r/ is pronounced in English, French, and Italian. So, if the language of the environment shapes the way perceptual reorganization of phonemes takes place, several possibilities remain open for bilingual individuals.

As just described, we know that language differentiation occurs very early in life; in fact, it occurs before perceptual reorganization takes place. So, one possibility is that infants exposed to bilingual environments acquire two parallel phoneme systems, one for each language. Another possibility is that they melt both systems, only developing a single system where phonemes from both languages are integrated. One way of having a hint of how acquisition will eventually evolve is that of looking into adult data. One prevailing view in adult speech perception models is that bilinguals only possess one phonetic space, integrating and adapting the phonetic repertoires of both languages. However, most of these results have been obtained with successive bilinguals, that is, individuals who were only exposed to one language in the first months (or years) of their lives. The literature with simultaneous bilinguals is scarce and not all the available data fit with this notion of a single system. Finally, an additional issue is that the development of phonetic categories continues to refine well into late childhood-puberty (in particular in the production side). So, it is crucial to look into how bilingual infants and very young children establish their phonetic categories.

Very few studies have addressed the evolution of the initial perceptual capacities in the period when perceptual reorganization takes place. These studies have analyzed the evolution of vowels and consonants in Catalan–Spanish (both vowel and consonant contrasts) and French–English (a consonant contrast) bilingual infants. Let us consider each group of studies in turn.

Catalan–Spanish infants were tested in two phoneme contrasts only existing in one of the languages of exposure and not in the other. In particular, they were tested in the Catalan-specific /e-ɛ/ vowel and /s-z/ consonant contrasts. These contrasts are very difficult to be perceived by adult monolingual Spanish listeners. This is the case because Spanish only has one /e/ vowel and one /s/ consonant (in fact, these are similar situations to those that Japanese listeners face when learning the English-specific /r-l/ contrast, Japanese only has one /l/ sound). The studies of acquisition of phoneme repertoire with bilingual infants have shown a peculiar developmental pattern. Two ages were *a priori* determined to evaluate the perceptual reorganization of infants learning Catalan and Spanish. As mentioned previously, before 6 months, infants show discrimination capacities even for contrasts their parents cannot perceive. Thus, one group of participants were infants before this age (specifically, they were 4.5-month-olds). Three different linguistic backgrounds were tested: monolingual Catalan, monolingual Spanish, and Catalan–Spanish bilingual. As expected infants from all three language environments perceived the contrasts. The other age selected was beyond perceptual reorganization, for the vowel contrast it was chosen 8 months and for the consonant 12 months (perceptual reorganization takes place before for vowels than for consonants). As expected, Catalan and Spanish monolinguals behaved differently: Catalan monolinguals showed a discrimination response, while Spanish monolinguals did not. However, contrary to the hypothesis that exposure should be enough to maintain discrimination, bilinguals did not show discrimination behavior. This result was also surprising because Catalan–Spanish adult bilinguals can actually perceive and produce these contrasts. So, discrimination should be regained at some point. Another group of older bilingual infants were tested (12- and 16-month-olds for the vowel and consonant contrast, respectively), at these ages, they showed discrimination.

To what extent these results are generalizable to other phoneme contrasts and bilingual populations? The studies reviewed thus for refer to a very particular case, that is, to phoneme contrasts only existing in one of the languages and which monolingual adults would find very difficult to perceive. Other studies with infants growing up in bilingual environments have shown that this developmental pattern is widespread. In particular, the very same U-shaped pattern (discrimination, then nondiscrimination, then discrimination) has been obtained with the same population and ages, but tested with a common Spanish–Catalan vowel contrast (/o-u/). However, 8-month olds have shown discrimination behavior when the vowel contrast involved acoustically distinct phonemes, in particular, the /e-u/ contrast.

Converging evidence has been gathered with French–English bilingual infants. Both French and English have the /b-p/ contrast in their phoneme repertoire. However, the specific acoustic parameters are not the same. These consonants differ in a specific acoustic parameter called the voice onset time (VOT). The important feature of French and English /b-p/ contrast is depicted in **Figure 3(a)**. English–French bilinguals were tested in their ability to perceive the French and English contrast. Infants were habituated to the central exemplar [pa], and then they were tested in two test trials, a change to [ba] or a change to [pʰa]. Discrimination of the French boundary would be indicated by increasing looking time to [ba], whereas discrimination of the English boundary would be indicated by an increasing of looking time to [pʰa]. Both bilingual and monolingual young infants (aged 6–8 months) were better able to discriminate the French than the English boundary, but importantly, as expected,

both groups performed similarly. Also, as expected, English monolingual infants did not show discrimination for the French contrast by 10–12 months. Again, in parallel with the Spanish–Catalan bilinguals, French–English bilinguals failed to show discrimination for the French contrast at 10–12 months, when perceptual reorganization is taking place. Older bilingual infants (aged 14–17 months) regained discrimination for this contrast (**Figure 3(b)**: summary of the results of all the experiments).

Both bilingual populations were tested in totally different situations and still the results are fully consistent. What are the mechanisms underlying this peculiar developmental path for bilingual infants? The answer to this question is still under debate. However, a strong candidate is particular frequency distributions of phonemes across languages. As mentioned, Spanish has only one /E/ vowel,

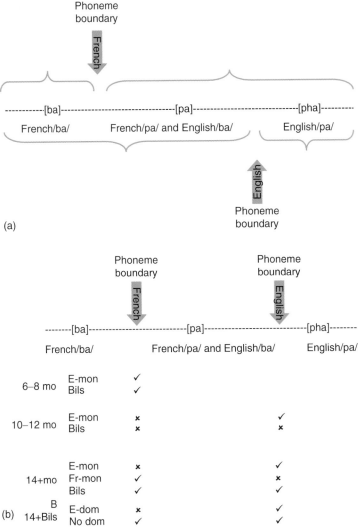

Figure 3 (a) Distribution of phoneme boundaries in the /b-p/ French and English consonant contrasts. (b) Summary of the discrimination results of /b-p/ French and /b-p/ English contrasts by monolingual and bilingual infants at different ages. From Burns TC, Werker JF, and McVie K (2003) Development of phonetic categories in infants raised in bilingual and monolingual environments. In: Beachley B, Brown A, and Conlin F (eds.) *Proceedings of the 27th Annual Boston University Conference on Language Development*, Vol. 1, pp. 173–184. Somerville, MA: Cascadilla Press.

falling roughly in between both Catalan /e/ and /ɛ/ ones. Thus, considering that the Spanish /E/ vowel is very frequent and that Catalan /e/ and /ɛ/ are not so frequent, the frequency distribution of the three vowels in monolingual and bilingual environments would look like something similar to that depicted in **Figure 4**. The distribution of English and French consonants /d-t-tʰ/ in a bilingual environment would also look quite similar, and therefore more frequently heard, in the center (see also **Figure 4**). It has been shown that humans are very sensitive to frequency distributional cues to build up phonetic categories. It has also been shown that frequency distributions of the type that bilingual infants are exposed to for these particular sounds induce the formation of a single phonetic category. Thus, it is quite possible that particular sound distributions in bilingual environments may hamper, in some very specific cases, the creation of the appropriate phonetic categories in bilingual infants. However, this situation does not last for a long period. Both studies show that within few months, bilingual infants regain discrimination behavior. It is an open question what are the underlying mechanisms driving this change: mere frequency accumulation and computation of separate frequency statistics for each language are just two possibilities.

Babbling in Bilingual Infants

The studies on babbling productions in bilingual contexts are scarce. After having explored the early exposure to two languages in the early stages of the language-acquisition processes (onset of canonical babbling, vocal performance, first word production, and multiword combinations), the question whether babbling productions can offer evidence of an early differentiation of the ambient languages still remains. In monolingual infants it has been shown that

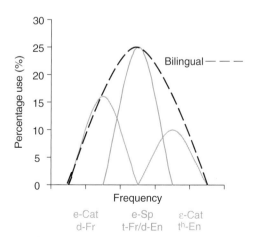

Figure 4 'Bimodal' and 'unimodal' distributions of different phoneme contrasts in Spanish–Catalan and English–French environments.

babbling drifts in the direction of the speech that they hear by 10 months of age, when features of the environmental language are gradually incorporated in these early productions. Although adult listeners may not be able to identify the specific language background after hearing babbling samples from monolingual infants growing up in different linguistic contexts, the analyses of these productions have revealed differences relative to the use of the vowel space, pitch contours, and labial consonants that are characteristic of the ambient language.

By 10–14 months of age, babbling patterns are less universal and become more language-specific. The study of babbling samples from infants growing up in bilingual contexts can thus offer relevant data about early language differentiation and language dominance, two issues that were discussed above. Results from the very few studies available on these issues give preliminary support for the hypothesis that infants in bilingual contexts develop differentiated language systems during the babbling period, although this is a tentative conclusion due to the limited evidence gathered so far: a single case study from an English–Spanish infant and a group study from French–English infants. The early differentiation conclusion derives from the fact that in this single case study the infant produced a different range of sounds depending on the context and the language of his interlocutors. In contrast, the group study showed that infants babbled in a dominant language with no mixing evidence (as far as consonantal sounds are concerned), which is a more indirect evidence for differentiation. It is important to mention that the infant dominant language seems to be assessed by the acoustic saliency of the sounds under analysis instead of the proportion of exposure to each language that the infant receives. A number of variables (amount of exposure in each language, extent of infant-directed speech from the interlocutors when addressing the infant, segmental repertoires in the ambient languages, dialectal differences, etc.) must be taken into account in order to understand and clarify the results that have been obtained better.

Learning Words

Segmenting Words: Phonotactics

Studies with monolingual infants have shown that at a very early age (between 7 and 8 months), infants can segment and recognize words embedded in continuous speech. They can do it, even if a very similar stimulus is presented in the test phase (a minimal pair, that is, familiarized with 'cup' and tested with 'tup'). One particular type of knowledge that infants use is differences in the frequencies of sequences of phonemes occurring within and between words. Infants compute the transitional probabilities of segments within and across word boundaries.

In the second half of the first year of life infants can use this type of information (known as phonotactics) to segment words. Because this source of information is language specific (for instance, no Spanish words can start with three consonants, while English words can), to segment the speech stream and build up appropriate lexicons properly, bilingual infants should compute two different sets of statistics, one for each language. Although there is only one study addressing specifically this issue, the results are consistent with the notion of early language dominance; that is, bilingual infants show sensitivity to the phonotactic information of their dominant language, in an equivalent way as monolingual infants of the same age.

Learning Word Forms

A major achievement in the second and third years of life is that of acquiring a sizable vocabulary. It is in this period when the 'vocabulary spurt' takes place. It has been argued that because of the relatively small vocabulary size in this age, the phonological representation of words is relatively coarse, as compared with adults' for whom fine acoustic–phonetic distinctions are represented. Nevertheless, research with monolingual infants has shown that from the very beginning infants represent words in a very detailed phonological form. For instance, they react to familiar words if slightly mispronounced. In these studies, infants are presented with two pictures of familiar objects (for instance, a dog and a car). After a short familiarization with the materials, while the images are visible, they hear a sentence either presenting a proper pronunciation (i.e., 'Where is the dog?') or a mispronunciation (i.e., 'Where is the tog?'). In these experiments, participants' eye movements are recorded. When presented with a correct pronunciation, infants look faster and longer to the right image, than when presented with the incorrect pronunciation. This difference is interpreted as an indication of infants' sensitivity of phonological lexical representations.

Research with Spanish–Catalan bilingual infants has shown that at this age they are not so sensitive to mispronunciations involving a phoneme contrast existing only in one of their languages. Bilingual infants aged 18–24 months were tested with mispronounced words made by exchanging the /e-ɛ/ vowels (as mentioned, a contrast only existing in Catalan). While infants raised in Catalan monolingual families reacted to this mispronunciation Spanish–Catalan bilinguals did not; that is, they treated the mispronunciation as correct pronunciations. This lack of reaction could be considered as an indication of less accurate lexical representations in bilinguals. However, when bilingual infants were tested with a contrast existing in both languages (/e-i/, acoustically similar to the /e-ɛ/ distinction), they reacted to the mispronunciation in the

same way as monolinguals did. These results show some parallel with those previously observed with phoneme discrimination capacities. It is probable that the particular distribution of phonemes in both languages may be making some mispronunciations more salient than others.

But still, another possibility is that sound distributions are not the sole factor to explain why bilingual infants do not react to some slight mispronunciations. It could be something in the very nature of bilingual language development. Indeed, bilinguals' task is more complex than monolinguals': they must code not only information about the relationship between object/concept and form, but also they must code to which language the word belongs. Infants' attention and memory capacities are clearly limited at this early age, so the task of learning a more complex mapping between form and meaning may be putting a heavy burden on them. Under perceptually difficult situations (either because acoustic similarities, or particular phoneme distributional properties, or both), they may have their processing capacities exhausted, therefore, failing to react in the most difficult perceptual discrimination cases. This possibility is supported by studies carried out with younger infants using a word-learning task.

Although at the end of the first year of life, infants show an excellent capacity to perceive phoneme contrasts existing in their own language, they may fail to use this capacity when learning new words. In the 'switch task' infants are shown with two new objects, paired with two new words. After some trials of exposure, they are shown one object either with the correct label (nonswitch trials) or with the incorrect label (switch trials). If they can succeed in learning the new object–label pairing, infants show surprise by increasing their looking times in switch trials. It has been shown that 14-month-olds cannot learn the pairing if the two-word labels are very similar ('bih-dih'), but they succeed if the differences between both labels are perceptually salient ('nim-leef'). The problem does not lie in the infants' capacity to perceive the contrast in the similar pair at 14 months, but on the computational demands (attention and memory) of the word-learning situation. Thus, between 17 and 20 months, with better processing capacities, infants are able to learn the pairing for the similar labels in this experimental situation.

Infants raised in English–French bilingual environments have been tested with this task. If the hypothesis of a resource limitation interfering with the learning of perceptually difficult minimal pairs is correct, then, bilingual infants should show difficulties for more extended periods. The results confirmed this hypothesis. Bilinguals not only showed difficulties at 14 months, but also at 17 months. So, it seems that because of having to build two different lexicons, it is probable that bilingual infants cannot make use of all phonetic detail when learning new words. With the available data, it is impossible to decide if bilingual infants sometimes do not encode fine detail in their lexical representations or if it is access to this information that takes place later for bilingual than for monolingual infants. However, considering that a few months later bilingual infants show response patterns indicating fine phonological representation, it is unlikely that phonetic detail is not encoded in their mental lexicon.

Vocabulary Size

The fact that infants growing up in bilingual homes may store, or retrieve, word forms in a different way to their monolingual peers does not necessarily entail that their vocabulary size is different from that of a monolingual infant. However, if as mentioned, they may devote some resources to encoding to which language words belong to, it might be the case that their storage capacity is affected. As in other aspects of language development, studies about vocabulary size in infants growing up bilingual must take into consideration that although they are exposed to both languages from birth, this condition, however, is not a homogeneous one, as a consequence of differences in the proportion of exposure to each ambient language and differences in the contexts of exposure that the child may experience (differences in the distribution of languages across speakers, that is, parents/caregivers using mainly one language or speaking both languages to the child). Most of the studies have been done on populations exposed simultaneously to English and Spanish, but work with German–English and Spanish–Catalan is also available (single case studies include a wider variety of language combinations).

When vocabulary size is the measure and vocabulary in both languages is taken into account, bilingual toddlers show a total vocabulary size comparable to the same measure in monolingual children. However, in bilinguals, total vocabulary measures may differ from 'conceptual' vocabulary, when translation equivalents (words with the same meaning, usually present from the early stages of lexical development) are counted just once. This is a highly relevant issue to take into account, especially because bilingual toddlers may easily be misidentified as having smaller vocabularies than monolingual toddlers; if the tool used to measure expressive vocabulary is not well adapted to bilingual contexts a 'communicative development inventory' (CDI) should be administered for each of the ambient languages for a correct comparison with monolingual data.

Other studies have focused on the relationship between vocabulary size in each of the ambient languages and the amount of input received. Results show a positive correlation between these two factors, at least in the early stages, up to the moment when a critical mass of lexical items has been reached. It has to be mentioned, however, that beyond direct exposure provided by parents, media, and community, predominant usage of one of the languages may favor a greater than expected vocabulary growth in one of these languages (this is true for English vocabulary measures in English–Spanish homes for studies run in the US, but the situation can possibly be extrapolated to other bilingual communities in countries other than the US). Studies on expressive vocabulary development in young bilingual children should be undertaken on larger samples and on a wider variety of language backgrounds to get a more complete picture of this process.

On the receptive side, there is a single study concerned with the organization of language-relevant brain activity as a function of vocabulary size in each language. In this study, also carried out with infants being exposed to Spanish and English, electrophysiological measures (ERPs) have been measured. Infants were exposed to lists of words of both languages that the children could know or not (lists were customized for every child). The results showed that the specific brain activity pattern depended on both the number of words known for each language (separate language vocabulary size) and the total conceptual vocabulary. The results suggest that children with smaller total conceptual vocabulary (thus slower in their overall language development) are slower in the processing of lexical items of the nondominant language, but the processing of the dominant language is not affected. Although these results need to be confirmed with future studies, they are in agreement with other research (see the phonotactics section) indicating that if some delay is to be found in bilinguals' early development it is restricted to the nondominant language.

Semantics

The semantic and conceptual knowledge underlying bilingual infants' first words has not been studied extensively. Bilingual infants learn words from both of their languages early in development and it has been demonstrated that they reach language milestones such as the onset of productive language on a similar timeframe as monolinguals: They start producing single-word sentences; they then go on the production of two-word sentences and after producing multiword sentences for a while, they start using complex sentences as well.

However, comparing to monolinguals, infants acquiring two languages face another difficulty while learning words. They must solve the problem of discerning the semantic meanings and related concepts of two lexicons across their two languages. It is well known that one fundamental principle guiding monolingual children's acquisition of new words is the principle of mutual exclusivity, or the assumption that new words tend to refer to new referents or objects. Disambiguation tasks have been the most common way to study mutual exclusivity. In these studies, many objects are presented to the children, one of which does not have a known label, then a novel label is given (that might name one of the objects). The mutual exclusivity effect is evident if the child chooses the previously unlabeled object as the referent of the new word. In a study of disambiguation with 24- and 36-month-olds, bilinguals did not differ from monolinguals in their demonstration of the mutual exclusivity effect in a pointing task. In a recent study, the principle of mutual exclusivity in younger bilingual infants (English–Cantonese 18-month-olds) was examined. The results of this study indicated that both monolinguals and bilinguals have the same pattern of response in the disambiguation task, although bilinguals seemed to respond somewhat slower and less robustly. It has been argued that these results could be either due to lower exposure to and proficiency in English, or due to the dual structure of the bilingual lexicon.

Another interesting issue regards bilingual children's acquisition of translation equivalents. A translation equivalent implies to possess a word for a specific object in both languages, like ball in, for example, Spanish 'pelota' and, at the same time, for example, in English 'ball'). This question has been of interest because this would violate the principle of mutual exclusivity. The reported data have shown that bilingual children acquire many translation equivalents. In a study carried out with Spanish–English bilingual children, it was reported that infant's translation equivalents constituted around 30% of their total vocabulary. Another study of an English–French child reported 50% of translation equivalents (compared to the child's total vocabulary) at the age of 14 months, and 36% at the age of 17 months. This high rate of translation equivalents, a clear violation of mutual exclusivity, suggests that at least from this age on children seem to have two distinct lexical systems. It is possible that the ability to violate mutual exclusivity may be learned through experience of interpreting people's intentions about what words mean.

Another approach to explore conceptual knowledge focuses on the lexical-referent parings and mistakes that infants make. In a study with English–French bilingual babies it was observed that, like monolingual infants, bilingual infants rarely overextended their first words in either of their two languages. With few exceptions, a word used to indicate an object was used only to stand for that object and/or sometimes the class for related objects. It was not used to connote other associative properties of that object (actions, locations, etc.).

Taking into consideration all the infant bilingual studies we have at this point, it seems that early semantic and, thus, conceptual knowledge underlying language acquisition is equivalent in both monolingual and bilingual infants.

Morphosyntax

Research in early capacities of morphosyntactic knowledge in language production faces some specific problems in bilingual populations. Infants start producing single-word utterances, and they increase utterance length. While there is some debate about the optimal utterance length to estimate morphosyntactic knowledge in monolingual infants, this problem is even worse when bilingual infants are considered. The major problem is a consequence of differences between languages, as far as their reliance on constituent order and bound morphology. Even with closely related languages, differences arise: to say 'They did not go' (or 'They didn't go'), Spanish natives say 'no fueron' (no subject, no auxiliary), but French speakers would say 'ils ne sont pas allés' (or 'elles ne sont pas allées' if a group of women did not go; subject, two negative particles, past participle) and Catalan natives would say 'no varen anar' (no subject, past tense auxiliary). So analyzing speech productions of children who can only produce one or two words can easily lead to problems of overestimating or underestimating a child's grammar skills. To explore the morphosyntactic development we take into consideration that solid evidence refers to productions starting around 24 months or later (depending on the child abilities to produce speech).

Children who have been exposed to two languages from birth and who actually speak those languages are not different from children growing up with just one language as far as the general course of morphosyntactic development is concerned.

Bilingual children are able to produce utterances that are clearly relatable to each of their different languages from the very beginning. It is claimed that the morphosyntactic development of one language does not have any important consequence on the morphosyntactic development of the other. However, there is evidence of transfer of specific morphosyntax features from one language to the other, leaving open the possibility of interaction and cross-linguistic influence between the languages. For example, it has been reported that Australian children learning English and German simultaneously used verb–object (VO) word order much more in their German than native monolingual speakers of German (German uses both -VO and -OV word order; English uses -VO order). In another study it has been observed that language dominance could be an important factor in cross-linguistic transfer: Cantonese–English learning children were more likely to incorporate structures from their dominant into their weaker language, than vice versa.

It is important to be aware that there are few studies reporting instances of cross-linguistic transfer. They concern specific aspects of the child's developing grammars and they seem to occur only under particular conditions. Thus, isolated examples of the influence of one language on performance in the other are not sufficient to demonstrate general interdependence. It is also essential to understand properly these examples of cross-linguistic transfer. For example, although it is true that bilingual infants often produce mixed sentences (they insert words of one of the languages into sentences build with the grammar of the other language), this insertion is usually restricted to content words (most of the time nouns). This type of insertion occurs exactly in the same way when bilingual adults are trying to speak one language and they do not know (or cannot find) one word in one of the languages, but they know the word in the other. So, bilinguals (infants or adults) make use of all their resources to communicate.

Summarizing, monolingual and bilingual children acquiring the same language since birth use that language in very similar ways. They produce the same sort of utterances as similar types of errors. There is no systematic evidence of morphosyntactic influence from one language to the other in children who have been exposed to two languages since birth.

Cognitive Development

There is ample evidence indicating that children between 4 and 8 years, growing up in bilingual environments, show improved performance in executive function tasks. The usual result is that bilingual children perform in tasks of cognitive control, at the same level as their monolingual peers 1 year older. The underlying driving force of this enhanced development would be the natural training that being exposed and using two languages involves. Indeed, bilinguals need to monitor attention to two competing and active language systems continuously. In particular, when speaking, they need to select from the appropriate lexicon and grammar system the right words and syntactic structures. Research with bilingual adults has shown that they activate all their languages, even if they are in a monolingual situation; for instance, when asked to name in English the picture of a dog, an English–Spanish bilingual, not only activates the lexical entry for the English word 'dog', but also activates the corresponding Spanish word ('perro'). To be able to say 'dog' and not be confused and say 'perro', bilinguals must monitor the language production system. Different studies support the assumption that the mechanisms used by bilinguals to carry out this monitoring process are the same used in other nonverbal tasks involving executive function. The development of executive function takes place late in life. Frontal lobes, the brain substrate responsible for these functions, are the last cortical areas to mature. The impact of bilingual exposure has been studied with relatively old children (from 4 years on) but some of its effects should be noticeable with younger children. Indeed, as mentioned earlier, infants raised in bilingual environments not only are able to tell apart the two languages, but as production data show, their initial utterances indicate the ability to separate them. The extent to which bilingual exposure affects other cognitive domains very early in life is an open question that needs to be addressed.

Conclusions and Future Directions

One major problem of research with preverbal bilingual infants is that of data scarcity. Most results refer to studies carried out with Spanish–Catalan bilinguals. To what extent are these results generalizable to other language pairs?

For instance, when describing the representation of words in the lexicon, the lack of sensitivity for some phoneme contrasts was explained in terms of frequency of distribution. However, there are other factors that must be taken into consideration and deserve further research. One of them is difference between languages. For instance, Spanish and Catalan are Romance languages; thus, they share a common origin and also many words are from the same root. For instance, the Spanish and Catalan words for 'door' and 'cat' are 'puerta/porta' and 'gato/gat'. Thus, it may be the case that bilingual infants of languages sharing a common origin are used to hear different pronunciations for different objects. An important issue is that of studying infants learning typologically distant languages, so that objects are almost always labelled with very different forms. Furthermore in the study described, only cognate words of Catalan and Spanish (i.e., words with a common origin) were used. It would be interesting to test infants with noncognate words (for instance, the words 'window' and 'dog' are very different in Spanish and Catalan: 'ventana/finestra' and 'perro/gos').

However, as mentioned at the beginning of this article, taken together the results of research on very early language development of infants growing up in bilingual environments indicate that this development does not significantly differ from that of their monolingual peers: major achievements are attained at the equivalent ages. However, the mechanisms by which the two languages are acquired may not be fully equivalent to those of infants only acquiring one language. This aspect relates to studies indicating that children growing up bilingual display speedy development of cognitive control mechanisms.

See also: Auditory Perception; Habituation and Novelty; Language Acquisition Theories; Learning; Semantic Development; Speech Perception.

Suggested Readings

Bosch L and Sebastián-Gallés N (1997) Native-language recognition abilities in four-month-old infants from monolingual and bilingual environments. *Cognition* 65: 33–69.

Burns TC, Werker JF, and McVie K (2003) Development of phonetic categories in infants raised in bilingual and monolingual environments. In: Beachley B, Brown A, and Conlin F (eds.) *Proceedings of the 27th Annual Boston University Conference on Language Development*, Vol. 1, pp. 173–184. Somerville, MA: Cascadilla Press.

Cenoz J and Genesee F (2001) *Trends in Bilingual Acquisition.* Amsterdam: John Benjamins.

Genesee F, Paradis J, and Crago (eds.) (2004) *Dual Language Development and Disorders: A Handbook on Bilingualism and Second Language Learning.* Baltimore, MD: Brookes.

Kroll JF and de Groot AMB (2005) *Handbook of Bilingualism. Psycholinguistic Approaches,* (section I. Acquisition: Chapters 1 and 3, particularly). Oxford: Oxford University Press.

Birth Complications and Outcomes

D L Smith, The Children's Hospital, Denver, CO, USA

Glossary

Bronchopulmonary dysplasia – A chronic lung condition that is caused by tissue damage to the lungs, is marked by inflammation and scarring, and usually occurs in immature infants who have received mechanical ventilation and supplemental oxygen as treatment for respiratory distress syndrome.

Cerebral palsy – A nonprogressive central nervous system disorder characterized by abnormal development of movement and posture that results in impaired motor activity and coordination.

Extremely preterm infant – An infant born at less than 28 completed weeks of gestation.

Hydrocephalus – An abnormal increase in the amount of cerebrospinal fluid within the cranial cavity that is accompanied by expansion of the cerebral ventricles and enlargement of the skull.

Hypoxic ischemic encephalopathy (HIE) – A clinical condition in newborn infants characterized by damage to the central nervous system resulting from an inadequate supply of blood and oxygen to the brain.

Ischemia – Deficient supply of blood to a body part (heart or brain) that is due to obstruction of the inflow of arterial blood.

Necrotizing enterocolitis (NEC) – A gastrointestinal disease that mainly affects premature infants characterized by infection and inflammation of the bowel wall that can lead to irreversible damage and destruction.

Perinatal asphyxia (also known as intrapartum asphyxia) – An interruption in placental blood flow during labor that is significant enough to cause decreased oxygen delivery to the infant.

Preeclampsia – A serious condition developing in late pregnancy that is characterized by a sudden rise in blood pressure, excessive weight gain, generalized edema, proteinuria, severe headache, and visual disturbances.

Preterm infant – An infant born at less than 37 completed weeks of gestation.

Retinopathy of prematurity (ROP) – A disease of premature infants that is caused by abnormal growth and development of retinal blood vessels that can lead to permanent damage to the retina and vision loss.

Very preterm infant – An infant born at less than 32 completed weeks of gestation.

Introduction

There are roughly 4 000 000 infants born in the US each year and the vast majority of these babies do well. However, about 10% of these newborn infants will be sick enough to require care in the neonatal intensive care unit (NICU). Of those infants cared for in the NICU, a small percentage will sustain injury significant to produce long-term morbidities that include neurodevelopmental impairment. There are a number of reasons that infants are admitted to the NICU and a number of conditions that result in long-term impairment. This article focuses on the two most common identifiable causes of neurodevelopmental delay and cerebral palsy (CP): premature birth and perinatal asphyxia.

Premature Birth

Premature birth is a significant problem in the US that has a major impact on the development of infants and children. Despite major advances in the care of pregnant women and premature infants, premature birth remains a leading cause of infant mortality and childhood disability. This section offers an overview of the risk factors and causes of preterm birth. The mortality and morbidity associated with prematurity are then described in detail with a focus on long-term outcome. Specific therapeutic interventions that have significantly altered the outcome of premature infants are also discussed.

Gestational Age and Birth Weight

Preterm birth is defined as delivery of an infant at less than 37 completed weeks of gestation. Premature infants are further classified as 'very preterm' and 'extremely preterm' because the complications associated with preterm birth vary based on the infant's gestational age. Infants can also be classified based on their birth weight. The definition of a low birth weight (LBW) infant is an infant weighing less than 2500 g (5.5 lb) at delivery. Very low birth weight (VLBW) is defined as a weight of less than 1500 g (3.3 lb) and extremely low birth weight (ELBW) is a weight of less than 1000 g (2.2 lb).

In general, there is agreement between gestational age and birth weight. Most infants born at less than 28 weeks, gestation weigh less than 1000 g and are therefore extremely preterm and ELBW. However, only two-thirds of all LBW infants are born preterm. The remaining third of LBW infants are born at term but are small for their gestational age. In addition, premature infants can also be small for their gestational age and these infants are at risk for additional complications.

It is important to recognize the distinction between gestational age and birth weight when reviewing studies about preterm infants. These two classifications are not interchangeable. For example, two infants may be born weighing 1300 g but one infant is an appropriately sized 31 week preterm infant and the other is a small for gestational age 34 week preterm infant. These two infants should not be treated the same, either in clinical practice or in research studies.

Epidemiology of Preterm Birth

The percentage of preterm births in the US in 2004 was 12.5%, which is equal to one in eight infants being born preterm. The incidence of preterm birth has steadily increased since the early 1980s (9.4% in 1981). The percentage of infants born very preterm in 2004 was approximately 2%, which has been relatively stable since about the mid-1990s. The rate of LBW infants was 8.1% in 2004, which is the highest level since 1970. One reason for the increase in preterm and LBW infants is the increase in multiple births in the US. The risk of preterm delivery is six times greater for a multiple birth than for a singleton pregnancy. In addition, approximately half of all multiple births are LBW infants.

The incidence of preterm delivery varies with maternal age and maternal race. Women under 20 and over 40 years of age have the highest rates of preterm delivery. In addition, black women have a significantly higher rate of preterm delivery compared to all other racial and ethnic groups. The disparity in preterm birth between black and white women has been persistent over a number of years and the reasons for the differences are not completely known. This increased risk of preterm birth is not fully explained by differences in socioeconomic status or access to prenatal care. The rate of preterm birth is higher for black women compared to nonblack women for all educational levels.

Etiology of Premature Birth

Table 1 lists the common causes of preterm delivery. Preterm birth is a heterogeneous condition and there are important differences in outcomes based on the cause of delivery and associated complications. The most recent data from the US show that approximately 50% of preterm deliveries occur after the spontaneous onset of preterm labor. The cause of spontaneous pre-term labor remains unclear but there is substantial evidence that infection plays an important role in the etiology of preterm birth. In addition, perinatal infection and inflammation is associated with increased morbidity for the infant and is an important risk factor for later neurodevelopmental impairment. Pregnancies complicated by premature rupture of membranes are also at increased risk of infection and corresponding neurological deficits.

Roughly 40% of preterm infants are the result of medically indicated preterm delivery. Indications for preterm delivery can be either fetal compromise (poor growth or poor placental perfusion) or maternal medical conditions (severe hypertension or intrauterine infection).

Table 1 Causes of pre-term delivery

Spontaneous pre-term labor
Multiple gestation
Cervical incompetence
Uterine malformations
Pre-term premature rupture of membranes
Medically indicated pre-term delivery
    Pregnancy induced hypertension/preeclampsia
    Intrauterine growth retardation
    Antepartum hemorrhage
    Fetal distress

There has been a significant increase in the number of medically indicated preterm deliveries in the US over the 1990s. This increase in medically indicated preterm deliveries was associated with a decrease in perinatal mortality and stillbirths over that same time period. It should be viewed as a success of the advances made in obstetrical and neonatal care. Many infants with significant intrauterine compromise are now delivered prematurely and have an excellent prognosis, especially if delivered after 32 weeks, gestation. However, there is still a need to decrease the incidence of spontaneous preterm delivery as this remains a leading cause of infant mortality and morbidity.

Mortality Associated with Premature Birth

The incidence of both mortality and major morbidity increases with decreasing gestational age and birth weight. Infants born between 23 and 26 weeks, gestation are at the highest risk for death or major disability. **Figure 1** shows a summary of the current survival rates for very preterm infants. Data from large network studies show that the chance of surviving to hospital discharge for an infant born between 23 and 24 weeks, gestation is only 30%. The rate of survival steadily increases after 24 weeks, gestation from 50% survival at 24 weeks, to 80% survival at 26 weeks, completed gestation. Survival is generally good for infants born after 27 weeks, gestation, with 90% being discharged home from the hospital. Mortality is uncommon for infants delivered after 31 weeks, gestation and greater than 95% of those infants survive.

The Centers for Disease Control and Prevention states that preterm birth is a leading cause of infant mortality in

the United States, accounting for over 30% of the infant deaths reported in 2002. Premature birth was second only to birth defects as a cause of death in the first year of life. The majority of premature infants who died in the 2002 study were born at less than 32 weeks, gestation with birth weights less than 1500 g. The number of infant deaths attributed to preterm delivery was highest at 23 weeks, gestation and decreased steadily as gestational age increased. The majority of the infants died soon after birth, with two-thirds dying within the first 24 h of life.

The mortality reported above are population-based statistics that can be used as general guidelines when discussing the likelihood of survival of infants born at a given gestational age. There are a number of additional factors, however, that must be taken into account when discussing the expected survival of a particular infant. Both the race and gender of the infant have been shown to affect survival. These differences are most pronounced at earlier gestational ages. For infants born at less than 28 weeks, gestation, there appears to be a slight survival advantage for black infants over white infants. In addition, female infants have a survival advantage when compared to male infants regardless of race. Another important determinant of the infant's chance of survival is birth weight. Infants who are growth-restricted have an increased mortality compared to appropriately grown infants of the same gestational age. Just like race and gender, the effect of birth weight on mortality is most pronounced for infants delivered at less than 28 weeks.

Care at the Limits of Viability

Infants that are born between 22 and 24 weeks, gestation are often referred to as being at the threshold of viability. There continues to be active debate on whether or not these patients should receive aggressive obstetrical and neonatal care. There are significant gaps in knowledge concerning the care of pregnant women at risk for delivering between 22 and 24 weeks and how to best care for the newborn infant. It is safe to say that most neonatologists consider intensive care therapy at 22 weeks, gestation to be of no benefit to the infant. Many neonatologists even consider therapy at 23 weeks, gestation to be of questionable benefit. These conclusions are based on the very poor survival rate for these infants despite intensive care. In addition, of the infants that do survive, the majority will have severe disabilities in childhood. These disabilities include CP, mental retardation, and blindness.

Given the uncertain benefit of therapy for this population, many obstetricians and neonatologists do not recommend aggressive obstetrical management, such as Cesarean section delivery, or active resuscitation for infants born at less than 24 weeks, gestation. Despite the overall poor prognosis of these infants, some parents request that everything be done on behalf of their baby.

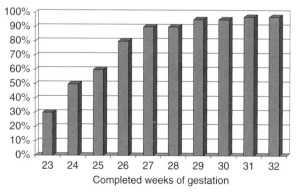

Figure 1 Likelihood of survival based on gestational age. Summary of data from: Bolisetty S, Bajuk B, ME A, *et al.* (2006) Preterm outcome table (POT): A simple tool to aid counseling parents of very preterm infants. *Australian and New Zealand Journal of Obstetrics and Gynecology* 24: 189; Jones HP, Karuri S, Cronin C, *et al.* (2005) Actuarial survival of a large Canadian cohort of preterm infants. *BMC Pediatrics* 5: 40; and Lorenz JM (2001) The outcome of extreme prematurity. *Seminars in Perinatology* 25: 348–359.

In these cases it is appropriate to offer resuscitation and aggressive intensive care. The care of the infant must be decided on a case-by-case basis and cannot be generalized for all patients.

It is imperative that the healthcare team be aware of the most current data on the survival and outcomes of these infants in order to help the parents make an informed decision about how to best care for their child. Whenever possible there should be a joint discussion with the obstetrical and pediatric team. Unfortunately, most of these conversations occur under suboptimal circumstances when a mother presents with an acute complication and the parents are asked to make difficult decisions in a short amount of time. If the parents do not elect to provide aggressive resuscitation for their infant, the baby should receive comfort care and the parents offered ongoing emotional support.

Morbidity Associated with Premature Birth

The majority of complications associated with premature delivery occur in the 1–2% of infants that are born before 32 weeks, gestation, in particular those babies born at 28 weeks or less. The two most significant morbidities associated with premature birth are lung disease and neurodevelopmental delay. Other less common, but potentially devastating, morbidities include retinopathy of prematurity (ROP), necrotizing enterocolitis (NEC), and infection.

Respiratory distress syndrome and chronic lung disease

One of the most common complications experienced by preterm infants is respiratory distress syndrome (RDS). RDS and the resulting respiratory failure is the leading cause of death in premature infants and infants born at less than 28 weeks, gestation are at greatest risk. RDS results from a lack of surfactant in the newborn's lungs. Surfactant is a chemical that is normally produced by the lungs between 34 and 37 weeks, gestation. The main role of surfactant is to reduce the surface tension of the small air sacs (alveoli) in the lung and prevent their collapse. Premature babies born with surfactant deficiency have significant respiratory distress and hypoxia which requires mechanical ventilation. In addition to a lack of surfactant, extremely premature infants are born with structurally very immature lungs that have both a decreased number of alveoli and incomplete development of the alveoli that are present. These infants are at very high risk to develop significant lung disease in the neonatal period which may have long-term implications.

There have been two significant advances since the late 1980s that have altered the incidence and severity of RDS. These therapies are worth discussing in detail because they have allowed for the increased survival of extremely premature infants. The first intervention is the administration of steroid therapy to the mother before delivery. Corticosteroids given to the mother in anticipation of pre-term delivery have been shown to decrease early neonatal mortality, RDS, and the need for mechanical ventilation. Despite the fact that antenatal steroid therapy does reduce the incidence and severity of RDS, the majority of infants born before 28 weeks will still require assisted ventilation at delivery. Some of these infants will then go on to develop chronic lung disease.

In addition to antenatal steroid therapy, infants routinely receive exogenous surfactant after delivery. Surfactant therapy immediately after birth decreases the severity of lung disease. Numerous studies from 1980 on have demonstrated that surfactant therapy rapidly improves oxygenation and leads to decreased ventilator support. There have also been several observational studies that show increased survival of VLBW infants after the introduction of surfactant into routine practice. These studies give further evidence of its beneficial effects. Unfortunately, there are no consistent data that surfactant decreases the incidence of chronic lung disease. This is because the pathophysiology of lung disease in extremely premature infants is more than a lack of surfactant. There is also a component of abnormal alveolar development. These infants' lungs demonstrate injury from oxygen toxicity and trauma that results from the mechanical ventilation of structurally immature alveoli. The remodeling that results from this injury has long-term consequences for pulmonary function.

Chronic lung disease, also known as bronchopulmonary dysplasia, remains a common long-term complication of babies born at less than 28 weeks, gestation. The exact incidence in preterm infants is hard to determine since there is no universal definition used in the literature but most studies report that between 30% and 40% of infants born before 28 weeks, gestation develop chronic lung disease. Chronic lung disease is relatively uncommon in babies born after 30 weeks or with a birth weight greater than 1200 g.

The morbidity associated with chronic lung disease includes a greater likelihood of re-hospitalization in the first 2 years of life with up to 50% of infants being re-admitted to the hospital for respiratory illnesses. Although hospitalizations decrease after the age of 4–5 years, these children continue to have a greater frequency of chronic respiratory symptoms such as recurrent wheezing and cough. In addition, infants with chronic lung disease have a worse developmental outcome compared to premature infants who do not develop chronic lung disease. The association between chronic lung disease and impaired neurodevelopment is complex and not completely understood. It is likely that the same conditions that predispose to developing chronic lung disease, namely inflammation and infection, also predispose to developmental delay.

Neurodevelopmental disability

The second common morbidity associated with prematurity is neurodevelopmental disability. The first two questions a parent of a premature infant asks are whether their baby will survive and be normal. The risk of neurologic impairment for premature infants is well known. The following sections offer an overview of the current information available concerning the neurodevelopmental outcomes of premature infants and describe factors that adversely affect outcome.

Numerous studies have looked at the changes in outcomes of VLBW and ELBW infants over the 20 years between the 1980s and the 1990s. The most consistent finding in all of these studies is that although survival of these infants has increased, rates of neurodevelopmental disability have remained stable. Some studies have even shown an increased rate of disability in this population over time. As obstetrical and neonatal intensive care has improved over the past two decades many of the smallest and youngest infants are able to survive, but they are still at great risk for complications that lead to long-term neurodevelopmental impairment.

When reviewing the literature on neurodevelopmental outcomes in premature infants some common limitations need to be discussed. The first is the length of follow-up. The majority of studies reporting developmental outcomes of preterm infants only follow children up to the age of 2 or 3 years. This is particularly true in studies that are evaluating a particular therapeutic intervention in the nursery. Developmental outcomes in early childhood do not always accurately predict outcomes in later childhood, for example, school performance. In general, a severe disability that is identified in early childhood is predictive of significant disability throughout childhood. However, there are a number of children who have severe disabilities and limitations at school age that are not predicted by an evaluation in early childhood.

Another limitation of these outcome studies is the number of patients who are lost to follow-up and unable to be evaluated. The authors should clearly state whether the patients that were unavailable for follow-up differed in any way from the patients who were evaluated. In this way the reader is aware of any potential bias of the study. One final potential limitation of an outcome study is the reference or control group that is being compared to the premature infant population. Studies can show different rates of neurodevelopmental impairment depending on whether the reference population is contemporary full-term classmates or the mean of a standardized test.

Neurodevelopmental outcome of premature infants

Using the data available from large multicenter studies it is possible to make some general statements about the likelihood of neurodevelopmental impairment in premature infants. Most studies focus on infants born before 28 weeks, gestation or less than 1000 g because this is the most-at-risk population. The two most commonly reported adverse outcomes are CP and developmental delay. CP is a term for a number of related conditions that are characterized by abnormal development of movement and coordination. CP results from central nervous system injury in fetal life or early infancy and does not progress over time. The diagnosis of CP covers a wide spectrum of disorders ranging from impaired coordination to an inability to walk independently.

The prevalence of CP between 18 and 24 months of age is between 10% and 15% in infants born before 28 weeks. This is compared to less than 0.5% in full-term infants. The prevalence of developmental delay in this same group of infants at 18–24 months of age is greater than 20%, compared to less than 3% in full-term infants. Other studies have looked at functional impairment during early childhood as a primary outcome measure for pre-term infants. Using these criteria, roughly half of VLBW infants who survive will have some functional disability and one-quarter of these children will have severe impairments. Keep in mind that even in this group of extremely preterm infants the rate of major disability varies based on gestational age. Infants born at less than 25 weeks have a significantly worse outcome than infants born at 27 weeks.

Studies evaluating these high-risk infants at school age show a high degree of cognitive impairment and need for special education. Large, international, population-based studies have shown that only about half of these infants have IQ scores in the normal range between the ages of 8 and 11 years and that up to one-quarter of children evaluated have an IQ of at less than 70. Up to 50% of children born at less than 1000 g require special education services or have repeated a grade. The special needs of these children are substantial for both the family and the education system. Families should be counseled about these outcomes in early childhood so they have a realistic expectation about the challenges their child is likely to face.

Factors that adversely affect neurodevelopmental outcome

Just as the likelihood of survival is impacted by conditions specific to each individual baby, there are a number of complications that can occur in the neonatal period that significantly increase the risk of poor neurologic outcome. These complications include both central nervous system injury and systemic conditions.

Intraventricular hemorrhage (IVH), also known as germinal matrix hemorrhage, is the most common intracranial hemorrhage in the neonatal period and seen almost exclusively as a consequence of prematurity. It can be a devastating injury and is a major cause of both mortality and long-term disability. Premature infants are at increased risk for IVH because of their inability to regulate cerebral blood flow and cerebral blood pressure. Fluctuations in cerebral blood flow can lead to rupture of the immature and friable blood vessels in the premature infant's brain. This causes bleeding into the fluid-filled spaces in the brain, called ventricles. IVH may also be complicated by hemorrhage in the brain tissue, also known as the parenchyma, surrounding the ventricles. Hemorrhage into the brain parenchyma is associated with permanent brain injury. This type of IVH is often referred to as a hemorrhagic infarct.

IVH is graded from I to IV based on the appearance of the bleed on head ultrasound. An ultrasound is an imaging study that uses sound waves to generate a picture of the infant's brain and the fluid-filled ventricles. Grade I and II IVH are small bleeds that are considered mild. A grade III IVH is often referred to as a moderate hemorrhage and is complicated by enlargement of the ventricles. The ventricles can continue to enlarge over time and lead to an increased accumulation of fluid in the brain, a condition known as hydrocephalus. The most severe IVH is a grade IV bleed which is a hemorrhagic infarct of the brain tissue surrounding the ventricle. The incidence of IVH varies among nurseries, but on average is between 15% and 20% of extremely premature infants born in the US. The incidence of moderate-to-severe IVH is as high as 30% for infants born at 23 weeks, gestation and decreases to 10% for infants born at 26 weeks. Significant IVH is rare in infants born after 30 weeks, gestation with an incidence of only 1%.

The extent of neurologic damage associated with IVH depends on the size of the hemorrhage, whether or not there is involvement of the brain parenchyma, and the development of hydrocephalus. Grade I and II IVH do not lead to serious neurologic impairment but studies have shown that premature infants with mild IVH do have deficits that can be seen on specific cognitive tests when compared to premature infants without IVH. Infants with grade III and grade IV IVH are at increased risk of neurologic impairment with rates of CP and developmental delay that approach 50%.

In addition to IVH the premature infant is also at increased risk for developing periventricular leukomalacia (PVL). PVL is an injury of the white matter adjacent to the lateral ventricles. Like IVH it is seen almost exclusively in premature infants and the injury is thought to occur in the perinatal period. PVL is diagnosed by the presence of cysts in the white matter surrounding the lateral ventricles seen on ultrasound. It is often bilateral and the ultrasound findings appear 4–6 weeks after birth. The most common clinical outcome is CP.

The presence of PVL on ultrasound is currently the best predictor of poor neurodevelopmental outcome in preterm infants. However, the absence of PVL on ultrasound does not ensure a normal long-term outcome.

A number of infants with normal head ultrasounds in the neonatal period will go on to develop neurological deficits in childhood. The reason for this is presumed to be that ultrasound only identifies the most severe form of white matter injury and many premature infants have sustained significant injury that goes undetected. Recently, magnetic resonance imaging (MRI) of the brain has been used to define white matter injury in preterm infants better. MRI studies done at 40 weeks-corrected gestational age have detected both white matter and gray matter abnormalities in very premature infants. These abnormalities have been shown to predict adverse outcomes in early childhood. As more data becomes available on the ability of MRI findings to predict neurodevelopmental outcome it is likely that this will become a valuable tool for identifying those pre-term infants most at risk for poor outcomes before they leave the nursery.

In addition to the clear relationship between IVH and white matter injury with long-term impairment, there has been an increasing awareness of the role of infection as a major contributor to poor neurodevelopmental outcome. The relationship between infection and adverse neurodevelopmental outcome is complex. Intrauterine infection, either clinical or subclinical, is estimated to be involved in nearly half of all preterm deliveries and the lower the gestational age at delivery the greater the likelihood of infection. It is generally accepted that preterm infants born to mothers with intrauterine infection, also known as chorioamnionitis, have an increased risk of PVL and CP. In fact, the timing of delivery for women with premature prolonged rupture of membranes, a condition with a significant risk of intrauterine infection, is often dictated by the desire to avoid infection.

In addition to intrauterine infection, infection in the nursery also increases the infant's risk of neurodevelopmental impairment. A large study performed by the National Institute of Child Health and Human Development Neonatal Research Network demonstrated that two-thirds of ELBW infants had at least one infection documented during their stay in the nursery. Infants who had infections in the neonatal period had a significant increase in the incidence of CP and developmental delay when evaluated at 2 years of age. More research is clearly needed to define further the association between infection and adverse neurodevelopmental outcome, in particular the mechanism underlying brain injury that results from inflammation and infection. The prevention of infection remains a major goal of the obstetrical and neonatal team caring for premature infants.

Retinopathy of prematurity

Another common morbidity seen in premature infants, in addition to chronic lung disease and neurodevelopmental impairment, is retinopathy of prematurity (ROP). ROP is an abnormality of the vascular development of the retina that is seen exclusively in premature infants. It is a major cause of visual impairment and blindness in children. Development of the retinal blood vessels begins at 16 weeks, gestation and is essentially complete by 36 weeks. Preterm delivery causes an arrest of normal vascular development and incomplete vascularization of the retina. This incomplete vascularization leads to hypoxia, which induces new, but abnormal, vessel growth. If this process continues it can result in progressive damage of the retina and ultimately retinal detachment.

The incidence of ROP varies among institutions but it is diagnosed in roughly two-thirds of infants born at less than 1250 g. The rate of severe ROP that threatens vision and requires surgical therapy is between 25% and 35%. Although the overall incidence of ROP has not increased since the mid-1980s, the incidence of severe ROP has increased. This is due to the increased survival of extremely premature infants who have the greatest risk of developing ROP.

Therapy for ROP has focused on prevention, universal screening for at-risk infants, and surgical laser therapy. ROP cannot be completely prevented because it is a natural consequence of premature birth; however, we can try to decrease the incidence and severity of the disease. There is a clear relationship between oxygen toxicity early in life and the subsequent development of ROP. It appears that sudden and wide fluctuations in oxygen saturations are associated with an increased severity of ROP. Many nurseries have developed protocols to maintain the oxygen saturation of premature infants at a constant level to try and reduce the incidence of ROP. Another mainstay of therapy for ROP is universal screening of all infants born at less than 1500 g to identify those high-risk infants likely to need surgery. The current standard for treating severe ROP is laser therapy to prevent retinal detachment. Despite aggressive attempts to prevent and treat ROP, it remains an important morbidity among premature infants.

Necrotizing enterocolitis

The last common complication associated with prematurity that can result in an adverse long-term outcome is necrotizing enterocolitis (NEC). NEC is the most common acquired intra-abdominal emergency in newborns. It affects approximately 10% of all VLBW infants and has significant mortality and morbidity. NEC is characterized by inflammation of the bowel wall that progresses to necrosis (tissue death) that may lead to intestinal perforation in the most severe cases. NEC may involve an isolated segment of bowel or it may involve multiple areas of the intestine. Although full-term infants can develop NEC, 90% of cases are seen in premature infants and infants with birth weights below 1500 g are most at risk. The majority of cases are sporadic but epidemics of NEC are well described in the literature which has suggested a role for infection in the etiology of NEC.

The cause of NEC is not clearly known, despite decades of research. It is thought to be the result of a combination of factors including a susceptible premature intestine which lacks normal integrity and defense mechanisms, ischemia or compromised blood flow, and bacterial overgrowth. The only risk factor clearly associated with NEC is prematurity. Other conditions thought to be associated with NEC are severe growth restriction, significant perinatal hypoxia or asphyxia, congenital heart disease, and certain medications. The relationship between feedings and NEC is the focus of many research studies. Enteral feedings are strongly associated with NEC because greater than 90% of infants that develop NEC have been fed. However, feeding an infant is clearly not the only cause of NEC because the majority of premature infants that are fed do well. It is hypothesized that the presence of milk feedings places additional stress on an already compromised intestine that can lead to local hypoxia and tissue necrosis.

NEC is a serious complication of prematurity with a mortality ranging from 10% to 30% despite aggressive management. The mainstay of therapy for NEC is supportive medical care and antibiotics. Approximately one-third of patients will require abdominal surgery to remove necrotic bowel. In a small number of babies the amount of intestine lost leads to an inability to tolerate feedings and prolonged dependence on intravenous nutrition, a condition known as short bowel syndrome. For those infants who do survive, NEC has been shown to be an independent risk factor for poor neurodevelopmental outcome and, therefore, can have lifelong implications.

Summary

Premature delivery remains a common cause of death and disability in infants in the US despite years of research and advances in perinatal care. The most vulnerable infants are those born before 28 weeks, gestation or at less than 1000 g. These infants have a high mortality and commonly experience complications related to their prematurity. The morbidities associated with preterm birth are numerous and often have lifelong consequences. We have seen an increase in adverse long-term outcomes as more ELBW infants are surviving the newborn period and early childhood. The morbidity associated with preterm delivery exerts a large burden on the child and his or her family as well as the heathcare and public education systems.

Perinatal Asphyxia and Hypoxic Ischemic Encephalopathy

Perinatal asphyxia refers to an interruption in blood flow and oxygen delivery to the infant which occurs during labor and delivery. One possible consequence of perinatal asphyxia is hypoxic ischemic encephalopathy (HIE). HIE is a clinical term used to describe a newborn infant with neurologic abnormalities resulting from an interruption of blood flow to the brain. The combination of perinatal asphyxia and HIE is often referred to as perinatal hypoxic ischemic brain injury.

Perinatal hypoxic ischemic brain injury is one of the most common identifiable causes of long-term neurodevelopmental disability and CP. It occurs in one to two out of every 1000 term infants. This means that a hospital with a moderate size delivery service is likely to have at least one case each year and a large tertiary care NICU will see several cases a year. This is most often an unexpected outcome of an otherwise uncomplicated pregnancy and is devastating for families. Despite advances in obstetrical care, the incidence of long-term disability associated with perinatal hypoxic ischemic injury has not changed since the late 1970s. Up until the early 2000s, the only therapy we had to offer these infants was supportive care. There are now promising new therapies being developed that may actually prevent ongoing injury and therefore improve outcome.

Mechanism of Injury

The brain injury associated with HIE occurs in two phases. The primary injury occurs as a consequence of decreased cerebral blood flow most often caused by an interruption in placental blood flow. This interruption in blood flow is often called perinatal asphyxia. The interruption in placental blood flow can be caused by acute umbilical cord compression, rupture of the uterus, a sudden separation of the placenta from the uterus (abruption), or maternal cardiovascular compromise. The decrease in cerebral blood flow and oxygen delivery results in a series of metabolic changes in the brain that will eventually lead to neuronal cell death. If, however, the infant is delivered and successfully resuscitated in the delivery room, cerebral blood flow is restored. After a brief recovery period a second phase of neuronal injury begins which has been called reperfusion injury.

Reperfusion injury is a term that refers to the damage that occurs after blood supply is restored to a tissue or organ after a period of ischemia. The return of blood flow leads to an influx of oxygen and inflammatory white blood cells that causes local oxidative damage and inflammation. The result is ongoing cell damage and cell death despite an adequate blood supply. This second phase of injury

lasts from 6 to 48 h after the initial insult. Recent therapies for HIE that have been investigated are designed to alter the second phase of injury in an attempt to preserve as much function as possible.

Risk Factors for Asphyxia

Table 2 lists common conditions that increase an infant's risk of suffering from perinatal asphyxia. Asphyxia can be the result of an acute and severe interruption in blood flow, such as compression of the umbilical cord. It can also occur from a less severe compromise in blood flow in a susceptible fetus. For example, severely growth-restricted infants are chronically hypoxic and even the transient decrease in blood flow that can occur during labor can lead to significant asphyxia. Obstetricians are well aware of the potential for asphyxia in these patients and will often choose to deliver at-risk infants by Cesarean section to avoid labor.

Clinical Markers of Asphyxia

One critical aspect in caring for infants with hypoxic ischemic brain injury is to identify those infants who are most at risk for adverse outcomes. This is important for targeting therapy that may be neuroprotective. Individual clinical markers of perinatal asphyxia are not well defined and do not readily correlate with long-term developmental outcome. It is the combination of factors that is most helpful for identifying infants at risk for neurologic injury.

Fetal heart rate tracings are used routinely during labor and delivery to monitor fetal wellbeing and tolerance of labor. However, fetal heart rate tracing abnormalities do not predict neonatal or long-term outcome. This has been demonstrated by the fact that the widespread use of fetal heart rate monitoring during labor has not significantly altered the incidence of CP in term infants since the 1970s. Fetal acidemia is another clinical marker used for asphyxia. At the time of delivery a blood gas can be measured from the umbilical cord to measure fetal acid–base status. If the umbilical cord blood gas shows evidence of severe acidosis this is suggestive of significant hypoxia and increases the risk of adverse neurologic outcome. Fetal acidemia alone, however, does not always predict poor outcome. In fact, the majority of infants do

Table 2 Factors associated with increased risk of perinatal asphyxia

Maternal medical conditions	*Obstetric complications*	*Fetal medical conditions*
Pregnancy-induced hypertension/preeclampsia	Placental abruption	Intrauterine growth retardation
Diabetes mellitus	Umbilical cord compression/prolapse	Prematurity
Collagen vascular disease	Multiple gestation	Infection/septic shock
Substance abuse	Premature rupture of membranes	Congenital anomalies
Hemorrhage	Prolonged labor	
Acute cardiorespiratory collapse	Prolonged pregnancy (>41 weeks)	

well in the newborn period and only a small number develop evidence of significant injury.

Another marker that has been used for predicting adverse outcomes in neonates is the Apgar score. The Apgar score is a universally used assessment tool in the delivery room to evaluate newborn infants. It is based on a scale of 0 to 10 and encompasses five easily measured clinical features. These include heart rate, respiratory effort, muscle tone, color, and response to irritating stimuli. A sore of 7 or above indicates an infant in good or excellent condition. The Apgar score is given at 1 and 5 min after birth and the 5 min score is most predictive of outcome. A low 5 min Apgar score, defined as 3 or less, is associated with increased neonatal mortality and long-term neurologic morbidity. A persistently low Apgar score at 10 and 20 min after delivery indicates a poor response to resuscitation and an even greater increase in mortality and morbidity.

The Apgar score was created to predict the likelihood of survival during the neonatal period and not as a marker for asphyxia. However, many clinicians use the 5 min Apgar score as an indicator of perinatal hypoxia or asphyxia. This is incorrect because there are a number of factors that can cause a low Apgar score, such as congenital anomalies and maternal medications, which are not a consequence of asphyxia. The American Academy of Pediatrics and the American College of Obstetricians and Gynecologists clearly state that "Apgar scores alone should not be used as evidence that neurologic damage was caused by hypoxia..."

The infant's overall clinical condition at delivery is probably the most helpful predictor of the degree of asphyxia and the extent of injury. The need for cardiopulmonary resuscitation in the delivery room increases the risk for adverse neurologic outcome. This relationship is particularly true in the face of significant fetal acidemia. The combination of fetal acidosis and cardiorespiratory failure at birth indicates an infant who has suffered profound hypoperfusion and is likely to have sustained central nervous system injury. Asphyxiated infants can also show evidence of renal failure, liver injury, cardiac dysfunction, or hematologic abnormalities.

In summary, there is no single clinical marker that indicates perinatal asphyxia significant enough to result in lifelong impairment. However, the more risk factors that an infant exhibits, the worse the overall prognosis. A joint statement by the American Academy of Pediatrics and the American College of Obstetricians and Gynecologists summarizes it this way: "A neonate who has had asphyxia proximate to delivery that is severe enough to result in acute neurologic injury should demonstrate all of the following: profound metabolic or mixed acidemia (pH <7.00) on an umbilical cord arterial blood sample, if obtained; an Apgar score of 0 to 3 for longer than 5 minutes; neonatal neurologic manifestations, e.g., seizures, coma, or hypotonia; and

multisystem organ dysfunction, e.g., cardiovascular, gastrointestinal, hematologic, pulmonary, or renal system."

Hypoxic Ischemic Encephalopathy

Neonatal HIE is a clinical description of a combination of neurologic abnormalities that may include altered consciousness, abnormal muscle tone and reflexes, inability to feed, and seizures. It should be clearly stated that not all infants with HIE go on to develop permanent neurologic impairment, but neurologic abnormalities present shortly after birth are one of the best predictors of long-term outcome. The most widely used classification system for neonatal encephalopathy is the Sarnat staging system. This classifies infants into three groups based on level of consciousness, activity, muscle tone, reflexes, and the presence of autonomic dysfunction. Infants with mild encephalopathy do very well and should be considered to have a normal neurologic outcome. Infants presenting with moderate encephalopathy are described as lethargic with decreased muscle tone and poor sucking behavior. Between 20% and 25% of these infants will have an abnormal neurodevelopmental outcome. Infants with severe encephalopathy are comatose and flaccid with no spontaneous activity and commonly have seizures in the first 24 h of life. These infants have a poor outcome, with essentially all patients that survive having significant impairment.

An infant's neurologic examination is likely to evolve over time and most infants who survive will demonstrate some degree of improvement in the nursery. The degree and timing of recovery can be helpful in providing information about the infant's long-term prognosis. For example, infants who recover quickly and are able to leave the nursery on full nipple feeds have a better prognosis than infants who have prolonged hospitalizations and difficulty feeding.

In addition to the neurologic examination, special imaging studies of the brain can be done to better define the prognosis for an infant. MRI provides a detailed picture of the brain and is very helpful in defining the location and extent of injury. Hypoxic ischemic injury has a distinct appearance on MRI that changes over time. Therefore, MRI can also be used to describe the timing of injury. This is important in determining if an injury occurred during delivery or some time remote from the delivery.

In addition to an MRI, an electroencephalogram (EEG) is routinely performed on any infant with significant HIE. An EEG is a noninvasive test that measures the electrical activity of the brain recorded by electrodes placed on the scalp. The EEG can be used to document seizure activity, which can sometimes be difficult to assess clinically in newborns. In addition, the EEG can provide prognostic information on the severity of long-term deficits. A severely and persistently abnormal EEG indicates a very poor prognosis of either death or severe long-term disability. In contrast, a moderately abnormal EEG that normalizes over the first week of life is associated with a more positive prognosis.

Therapy for Hypoxic Ischemic Encephalopathy

Up until the past few years there has been little to offer an infant with HIE except supportive medical care. Immediate intervention includes resuscitation in the delivery room and correction of acidosis as well as cardiovascular support to maintain adequate perfusion. The use of moderate hypothermia has been the first therapy since the 1970s that has shown the potential to improve the outcome of these infants by decreasing the extent of central nervous system injury. Animal studies have shown that hypothermia initiated soon after hypoxic ischemic brain injury decreases neuronal loss and cell injury. The degree of neuroprotection is dependent on the severity of injury and the timing of hypothermia. The more severe the injury and the later hypothermia is initiated, the less effective the therapy. Hypothermia must be initiated within 6 h of injury to demonstrate neuroprotection.

Moderate hypothermia has now been studied in newborn infants and has been demonstrated to be safe without clinically significant adverse effects. Two large multicenter trials have recently been published evaluating whether moderate hypothermia can reduce the incidence of death or neurologic impairment in infants with moderate-to-severe HIE. The two studies used two different methods to produce hypothermia. One study evaluated whole body hypothermia using a cooling blanket to maintain core body temperature at 33.5 °C (92.3 °F). The other study evaluated selective cooling of the head with a cooling cap along with mild whole body hypothermia, keeping the body temperature between 34 °C (93.2 °F) and 35 °C (95.0 °F). These two studies showed promising results in that infants with moderate HIE who underwent 72 h of hypothermia had improved outcomes at 18 months of age. There was no significant benefit, however, for infants who presented with severe encephalopathy.

Despite the initially promising results, hypothermia is not recommended for the routine therapy of infants with perinatal asphyxia and HIE. There are a number of questions that still must be answered to determine which infants would benefit the most from hypothermia and which is the most effective method of hypothermia. In addition, long-term follow-up studies are needed to determine if the neurologic benefits seen at 18 months continue as these children go through childhood.

Summary

Perinatal asphyxia and resulting HIE is one of the most common identifiable causes of CP and poor neurologic outcome in full-term newborns. The clinical assessment of perinatal asphyxia is based on a specific combination of criteria that identify those infants most likely to have suffered acute injury. There are a number of maternal and fetal conditions that can predispose to HIE, but the underlying etiology is decreased cerebral blood flow. The infant's neurologic examination, MRI findings, and EEG results are used to offer a prognosis about long-term neurologic impairment. Modest hypothermia is a promising new therapy that has been shown to be neuroprotective and improve the outcome of infants with moderate HIE.

See also: Birth Defects; Breastfeeding; Cerebral Palsy; Developmental Disabilities: Cognitive; Developmental Disabilities: Physical; Intellectual Disabilities; Maternal Age and Pregnancy; Mortality, Infant; Newborn Behavior; Obesity; Premature Babies; Prenatal Care; Prenatal Development; Reflexes; Screening, Newborn and Maternal Well-being; Screening, Prenatal; Suckling.

Suggested Readings

Bolisetty S, Bajuk B, ME A, *et al.* (2006) Preterm outcome table (POT): A simple tool to aid counseling parents of very preterm infants. *Australian and New Zealand Journal of Obstetrics and Gynecology* 24: 189.

Committee on Fetus and Newborn, American Academy of Pediatrics and Committee on Obstetric Practice, American College of Obstetricians and Gynecologists (1996) Use and abuse of the Apgar score. *Pediatrics* 98: 141–142.

Higgins RD, Delivoria-Papadopoulos M, and Raju NK (2005) Executive summary of the workshop on the border of viability. *Pediatrics* 115: 1392–1396.

Hoyert DL, Mathews TJ, Menacker F, Strobino DM, and Guyer B (2006) Annual summary of vital statistics: 2004. *Pediatrics* 117: 168–183.

Jones HP, Karuri S, Cronin C, *et al.* (2005) Actuarial survival of a large Canadian cohort of preterm infants. *BMC Pediatrics* 5: 40.

Lorenz JM (2001) The outcome of extreme prematurity. *Seminars in Perinatology* 25: 348–359.

Marlow N, Wolke D, Bracewell MA, and Samara M (2005) Neurologic and developmental disability at six years of age after extremely preterm birth. *The New England Journal of Medicine* 352: 9–19.

Msall ME (2006) The panorama of cerebral palsy after very and extremely preterm birth: Evidence and challenges. *Clinics in Perinatology* 33: 269–284.

Recchia FM and Capone A (2004) Contemporary understanding and management of retinopathy of prematurity. *Retina* 24: 283–292.

Saigal S, Ouden L, Wolke D, *et al.* (2003) School-age outcomes in children who were extremely low birth weight from four international population-based cohorts. *Pediatrics* 112: 943–950.

Speer M and Perlman JM (2006) Modest hypothermia as a neuroprotective strategy in high-risk term infants. *Clinics in Perinatology* 33: 169–182.

Vannucci RC (2002) Hypoxia-ischemia: Clinical aspects. In: Fanaroff and Martin RJ (eds.) *Neonatal–Perinatal Medicine,* 7th edn., pp. 867–878 St. Louis: Mosby.

Volpe JJ (2001) Intracranial hemorrhage: Germinal matrix-intraventricular hemorrhage of the premature infant. In: Volpe JJ (ed.) *Neurology of the Newborn,* 4th edn., pp. 428–496. Philadelphia: Saunders.

Relevant Website

http://www.marchofdimes.com – March of Dimes – Saving babies, together.

Birth Defects

D Adams and M Muenke, National Institutes of Health, Bethesda, MD, USA

Glossary

Anterior – The front of the body, for example, the heart, is anterior to the shoulder blades.
Association – A set of medical conditions and/or physical features that occur together more often than would be expected by chance.
Deformation – Unusual forces acting on otherwise normal tissue to cause an alteration in structure.
Disruption – Destruction of otherwise normal tissue to cause an alteration in structure.
Distal – Further from, for example, the hand is more distal to the shoulder than the elbow.
Dorsal – An anatomic term, used for some embryonic structures, meaning the back of the structure. For instance, the dorsal surface of the hand is the back of the hand.
Dysmorphology – The study of variants in anatomic structure, specifically those variants that are associated with significantly impaired function and/or cosmesis.
Dysplasia – The abnormal organization of a tissue.
Embryopathy – A general term for a disease or other process that has an adverse effect on a developing embryo.
Idiopathic – A problem (pathology) of unknown cause, for example, idiopathic pulmonary fibrosis, is pulmonary fibrosis for which the underlying cause is unknown.
Malformation – The abnormal formation of a tissue.
Posterior – The back of the body, for example, the shoulder blades are posterior to the heart.
Proximal – Closer to, for example, the elbow is more proximal to the shoulder than the hand.
Syndrome – A set of medical conditions and/or physical features that occur together more often than would be expected by chance. In addition, the word syndrome implies a common causation or small set of causations.
Teratogenic – A teratogenic event is one in which the fetal environment is changed in, manner that causes an alteration in normal fetal development. The word teratogen is often used when referring to a drug with potential teratogenic effects.
Ventral – An anatomic term, used for some embryonic structures, meaning the front of the structure. For instance, the ventral surface of the hand is the front of the hand.

Introduction

The term birth defect describes a large and complex field of study. For most measurable characteristics, there is a broad range of normally functioning states. The March of Dimes defines a birth defect as, "An abnormality of structure, function or metabolism 'body chemistry' that results in physical or mental disabilities or death." This definition is designed to be inclusive but illustrates some of the difficulties inherent in the term. An example of a newborn with medical pathology who would not be described as having a birth defect is an otherwise healthy newborn emerging from a difficult delivery. He might need medical intervention only during the first hours of life. Abnormalities of metabolism are certainly present, but are transient and will not cause permanent disability. A second newborn may have a DNA change that will predispose her to an adult-onset illness. This is a structural and functional problem in a molecular sense, but does not present an immediate health threat and may not be recognized in the newborn period. The Centers for Disease Control and prevention (CDC) provide a narrower definition. To be named a birth defect, a condition must be:

1. present at birth;
2. the result of a malformation, deformation, or disruption in one or more parts of the body; and
3. characterized by serious, adverse effects on health, development, or functional ability.

This definition may be overly restrictive. First, as we will see, the terms malformation, deformation, and disruption have specific, yet imprecise meanings in the context of congenital (present at birth) abnormalities. A newborn may have features that do not easily fit into those definitions. The truth is likely somewhere in between, including those conditions that generally come to attention in the newborn period, have serious/prolonged health (including social) consequences, and are due to some deviation from the range of normal embryogenesis.

Birth defects are extremely variable. One only need consider the complexity of a process that allows a single cell to transform into a newborn baby to appreciate the scope of things that could go awry. Even 'normal' embryogenesis results in a range of features 'minor variations' that have no functional significance. These can include small skin tags around the ear and variation in the patterning of palm creases. Sometimes, particular patterns of minor variations can be used to identify a more severe underlying condition. 'Major variations' are those that, by themselves, cause an alteration in the functioning of the individual. Examples include some cardiac defects, severe anomalies of the hands and feet, and clefts in the lip and palate. Major variations are detected in 2–3% of pregnancies. Whether or not a variation is 'normal' may also be a function of cultural context. The particular shape, color, or size of a part of a body part may have cultural significance (positive or negative) that is not accompanied by any change in physical function. Other conditions, such as some congenital heart defects, are clearly present at birth, but do not manifest until days, weeks, or years later. The large range of possible birth defects makes their study both interesting and daunting.

Birth Defects: General Principles

As with any major field of study, the initial understanding of a phenomenon is based on simple observation. Ancient statues and paintings depicted conjoined twins and persons with achondroplasia, a specific form of short stature. Religion and mysticism were often convoluted together with birth defects such that an unusual baby was considered to be a consequence or sign of supernatural agency. As the medical profession grew to embrace the principles of scientific investigation, a systematic approach understanding birth defects arose. 'Dysmorphology' (the study of abnormal form) is a medical field of study concerned with the relation between external physical features, for instance, the shape of a nose or a fingerprint pattern, and health. Birth defects are often first recognized by suggestive physical features, thus creating a use for dysmorphology in the diagnosis of birth defects. Dysmorphologists initially relied on the combined clinical experience to ascertain patterns of features that were associated with health implications. As more and more individuals with a given pattern were recognized, summed clinical experience could be used to devise rational approaches to diagnosis, family counseling, and treatment. It was soon recognized that many (although not all) birth defects had a component of heredity. The alliance of dysmorphology with the study of the health implications of inheritance gave birth to a medical specialty, medical genetics.

Early on, the efforts of medical geneticists were complicated by a number of characteristics possessed by inherited traits, both pathogenic and normal. Such characteristics included genetic heterogeneity, defined as an inherited trait that has multiple genetic causes (more than one gene). Variable expressivity is defined as a trait that has different manifestations in different individuals. Penetrance is defined as the likelihood that an individual who has an inherited trait shows any associated characteristics of the trait. Penetrance values less than 1.0 indicate that a

fraction of the population of people who carry the trait gene will show no characteristics. Beyond inheritance, many birth defects arise solely or partly because of nongenetic environmental factors. Given an insufficient number of characteristic physical features, a lack of evidence of a known heritability pattern, or a lack of a known environmental etiology, the root cause of many birth defects remains unknown.

The most important advances in our ability to understand the etiology of birth defects have come in the form of progress in research pertaining to cell biology, molecular (DNA-related) biology, and development. Progress in those fields is yielding an expanding understanding of how disruptions to normal development can lead to birth defects.

Epidemiology and Social Impact

Birth defects are common. The CDC estimates that approximately 120 000 (1 in 33) newborns are born with birth defects each year. Birth defects are the number one cause of infant mortality in the US. Despite rapid advances in our understanding of the causes of birth defects, 70% occur for unknown reasons.

A recent research report released by the CDC attempted to quantify some of the financial impact of birth defects. They noted that birth defects account for 20% of the total infant deaths in the US. They calculated that the associated hospital costs alone exceeded $2.5 billion.

Variability

As discussed in the introduction, the term 'birth defect' covers a wide variety of variation present among newborns. Variability in severity is exemplified by hand malformations. At one end of the spectrum is postaxial polydactyly type B, which manifests as a skin tag on the on the fifth-digit side of the hand. Such skin tags are common and do not cause any functional impairment (**Figure 1**). At the family's request, skin tags can be removed in the newborn nursery using a simple procedure. In contrast, split-hand/split-foot malformation (SHFM) can cause significant functional impairment (**Figure 2**). Multiple, complex surgeries may be required to achieve usable function. SHFM is an example of genetic heterogeneity as it can be cause by several different genes.

Causation

Inherited factors

Birth defects can be caused by genetic factors – generally changes in the affected individual's DNA sequence. Some genetic changes arise for the first time in the affected individual, while others are inherited from either affected or unaffected ('carrier') parents. Some inherited birth

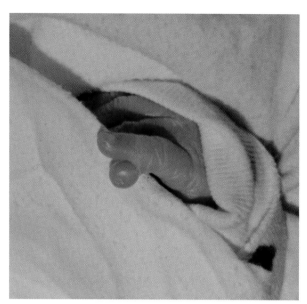

Figure 1 Cutaneous postaxial polydactyly. The fifth finger of the hand of this newborn has an attached extra digit. The digit is partially formed and is not likely to contain any bones. Often such digits are even simpler and do not have a partially formed fingernail as this example does. Such digits are easily removed with a minor surgical procedure.

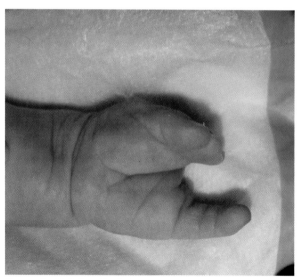

Figure 2 Split-hand/split-foot malformation. The pictured hand of a newborn infant shows anterior/posterior clefting and digit separation. Existing digits are not fully forme.

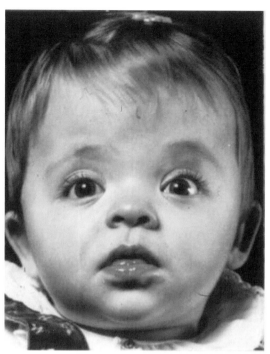

Figure 3 Muenke syndrome. The pictured child has Muenke syndrome with several facial dysmorphisms including facial asymmetry. Muenke syndrome can be considered in patients with craniosynostosis leading to a tall and wide forehead. Because of the wide spectrum of findings, some children with Muenke syndrome may not be diagnosed in the newborn period.

defects are transmitted through families across multiple generations. Some conditions are so severe that affected children do not live long enough to produce offspring. Other birth defects are mild enough that they have little impact on reproduction.

Craniosynostosis, the premature fusion of the sutures of the skull, occurs in 1/2500 live births. Craniosynostosis syndromes, a subset of all craniosynostosis, show varying severity in the newborn period. Apert syndrome (**Figure 3**) is a rare condition that includes abnormal shape of the skull bones and abnormalities of the hands and feet. Apert syndrome is caused by changes in the fibroblast growth factor receptor 2 (*FGFR2*) gene and most (>90%) of new cases arise as new mutations in the affected child. The manifestations of Apert syndrome make the condition universally diagnosable at birth. In contrast, another craniosynostosis syndrome, Crouzon syndrome, also caused by mutations in *FGFR2*, may not be diagnosed in the first affected individual in a family. Some mildly affected individuals are themselves diagnosed when they have an affected child. Newborns with oculocutaneous albinism are often recognizable at birth by skin and hair coloration that is markedly lighter than other family members. Most forms of albinism do not reduce the individual's ability to have children later in life.

Environmental factors

Teratogens

The process of embryonic development is susceptible to disruption by a number of environmental factors including a woman's use of prescribed or nonprescribed drugs during her pregnancy. Certain drugs may be known to affect pregnancies most severely at early or late gestational dates (**Table 1**). Drugs that are approved for medical use are classified by a system that rates their potential averse affects on a fetus (**Table 2**).

One of the most infamous teratogens is the medication thalidomide, marketed widely in the 1950s as a treatment for pregnancy-associated nausea. A dramatic increase in the incidence of severe congenital malformations of the arms and legs (phocomelia) was eventually attributed to the use of thalidomide during pregnancy. The drug was taken off the market for several decades until it was discovered that it was uniquely useful for the treatment of some of the symptoms present in a few uncommon diseases including leprosy. Today, the drug is once again available for use; however, strict monitoring requirements are in place in an attempt to prevent its use during pregnancy. Debate continues as to whether the benefits of the drug outweigh the risk that further cases of phocomelia might occur.

Infections

Certain infectious diseases are known to cause birth defects. One example is German measles (rubella), a viral illness. In healthy children and adults, rubella causes a usually mild, self-limited illness with a rash, swollen lymph nodes, and achy joints. If a pregnant woman becomes infected, there is a risk that the virus will be transmitted to her developing child causing a potentially severe embryopathy. Birth defects that can result from rubella infection include growth retardation, cognitive impairment, cataracts, deafness, and cardiac defects. Immunization for rubella during childhood has decreased the frequency of the disease in the general population, and subsequently, the frequency of rubella embryopathy.

Maternal illness

Some noninfectious maternal illnesses can cause birth defects. Elevated maternal blood sugar during pregnancy, such as that which can occur with diabetes mellitus, is a well-known cause of congenital abnormalities. The risk of a major malformation among all newborns is approximately 2–3%; among children of diabetic mothers, it is 6–9%. Diabetes-related birth defects can cause abnormalities in lower extremity development (caudal regression) in addition to spinal, brain, and cardiac malformation. Careful control of blood sugar levels during pregnancy is critically important for women with diabetes.

Mixed factors

Holoprosencephaly (HPE) is the result of the failure of the developing brain to divide into two separate hemispheres. HPE is associated with structural abnormalities of the midface. HPE demonstrates significant variation in severity and manifestations (variable expressivity), meaning that two people with the same genetic predisposition to HPE can have significantly different physical features. At the mild end of the spectrum is the presence of a single central incisor (one upper front tooth rather than the usual two). At the severe end of the spectrum, newborns present with a single central eye (cyclopia) and related facial abnormalities. Intermediate manifestations often include a cleft lip. **Figure 4** shows examples of the wide variation in the condition. Around 25% to 50% persons with HPE have a microscopically visible chromosomal abnormality, while 20–25% of persons with normal chromsomes have detectable defects in one or more of the genes known to be associated with HPE. In addition to any genetic susceptibility, however, environmental factors are thought to play a role in determining how the condition manifests in a given

Table 1 Examples of teratogens and times of maximal gestational sensitivity

Maximal sensitivity	Example agents
First trimester	Androgens (virilization)
	Carbamazepine (malformations)
	Ethanol (fetal alcohol syndrome)
	Thalidomide (days 34–50, limb reduction)
Second and/or third trimester	Coumadin (embryopathy)
	Iodides (fetal hypothyroidism)
	Tetracyclines (tooth enamel changes)

Summarized from http://www.uspharmacist.com/ (accessed on April 2007).

Table 2 FDA use-in-pregnancy ratings

A	Adequate, well-controlled studies in pregnant women have not shown an increased risk of fetal abnormalities to the fetus in any trimester of pregnancy.
B	Animal studies have revealed no evidence of harm to the fetus, however, there are no adequate and well-controlled studies in pregnant women. Or animal studies have shown an adverse effect, but adequate and well-controlled studies in pregnant women have failed to demonstrate a risk to the fetus in any trimester.
C	Animal studies have shown an adverse effect and there are no adequate and well-controlled studies in pregnant women. Or no animal studies have been conducted and there are no adequate and well-controlled studies in pregnant women.
D	Adequate well-controlled or observational studies in pregnant women have demonstrated a risk to the fetus. However, the benefits of therapy may outweigh the potential risk. For example, the drug may be acceptable if needed in a life-threatening situation or serious disease for which safer drugs cannot be used or are ineffective.
X	Adequate well-controlled or observational studies in animals or pregnant women have demonstrated positive evidence of fetal abnormalities or risks. The use of the product is contraindicated in women who are or may become pregnant.

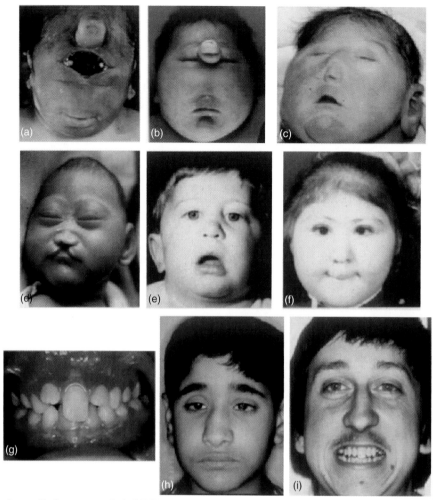

Figure 4 The spectrum of holoprosencephaly (HPE). Wide variation in holoprosencephaly ranges from malformations that are incompatible with life to suble changes in otherwise health individuals. The fetus in the upper part shows severe lack of midline development including a single, unseparated eye (cyclopia) and a severely underdeveloped nose (the fleshy tube, or proboscus, on the forehead). The mildly affected person in the lower right corner has HPE manifestations limited to a single central incisor.

individual. The most convincingly related environmental factor is maternal diabetes although other factors, including cholesterol metabolism and cholesterol-lowering drugs, are under investigation as possible contributors.

Screening and Detection

The current standard of care for pregnancies in the US, and some other industrialized countries, is to offer screening for common pregnancy complications. Many such complications have the potential to produce children with birth defects. Although the exact protocols for such screening are evolving, the following section outlines the general categories of techniques in use. At the time of birth, routine examination of the newborn may reveal malformations that are either diagnostic or suggestive of a birth defect. Often, further medical workup is required to make a definitive diagnosis. Even with the best available

resources, the root cause of a number of congenital malformations is never discovered. In such cases medical issues must be addressed as they arise.

Social/Psychological Impact

The diagnosis of a birth defect during pregnancy, or at the time of delivery, is an unwelcome event for any family. Even if a clear-cut diagnosis can be made expeditiously, the family often has to make difficult decisions about whether to continue an affected pregnancy, how to cope with a special needs child, and how to prepare for a different life than they were expecting. The family may have to simultaneously proceed with grieving and face the financial and time-related hardships of a prolonged hospitalization. Many parents find that a particularly difficult aspect of a birth defect diagnosis is that they have to communicate unexpected news to other family members.

It is said that parents of a newborn with birth defects must "grieve the loss of the child they were expecting before they can accept the child they have."

Birth Defect Mechanisms and Classification

Classical Nosology

A general framework exists for categorizing and naming birth defects. Historically, the framework was useful when little biochemical or molecular understanding of the origins of congenital malformations was available. It is currently useful both as an educational framework and as a set of diagnostic anchors that can help an experienced clinician to narrow the field of potential diagnoses.

Categories of dysmorphic mechanisms

Smith's Recognizable Patterns of Human Malformation by Dr. Kenneth Lyons Jones is a classic dysmorphology textbook. It defines several types of structural defects. **Figure 5** describes the relations between the defined types.

Malformations

Malformations encompass dysmorphic features that are the result of the abnormal formation of a particular tissue. Malformation sequences describe the situation when a malformation causes a cascade of subsequent events that affect other, related tissues. SHFM, described above, is an example of a malformation.

Pierre Robin sequence is an example of a malformation sequence. In the classic description of Pierre Robin, a small (micrognathic) or posteriorly positioned (retrognathia) jaw is the initiating malformation. The malpositioning of the jaw causes an elevated position of the tongue, which in turn interferes with the joining of the shelves of tissue that become the palate. At birth, the child has micrognathia or retrognathia, a malpositioned tongue that can interfere with breathing, and a cleft palate (**Figure 6**).

Deformation

A deformation results from external (to the fetus) forces acting on otherwise normal tissue. As with malformation, subsequent consequences can be grouped with the initial deformation to form a deformation sequence. A syndrome that is often used to illustrate the principle of a deformation is the Potter sequence. In the Potter sequence, an inadequate amount of amniotic fluid results in a restrictive intrauterine space. The resulting fetal compression results in deformation of the face, hands, and feet (**Figure 7**).

Disruption

A disruption birth defect occurs when normal tissue is destroyed as the result of external forces acting on it. For instance, amniotic banding is a type of birth defect in which a child is born with missing limbs, or other body parts. An otherwise normal looking limb will appear as if it were amputated at some point along its length. Sometimes, a small residual remnant (hand, foot) will be visible past the site of the apparent amputation. One theory to explain this phenomenon is that the limb gets tangled in folds of amniotic tissue at some point during fetal development. The fold then cuts off blood supply to the distal part of the developing limb, destroying it, and causing a disruption birth defect (**Figure 8**).

Dysplasia

Dysplasia is defined as the abnormal organization of a tissue. As an example, the large family of 'skeletal dysplasias' includes many different conditions. Osteogenesis imperfecta, for instance, is often called 'brittle bone disease' and features easily broken bones. The fragility of the bones in affected individuals is caused by the abnormal production of collagen, a protein that gives bending strength to bone. Achondroplasia, another type of skeletal dysplasia, is caused by a specific DNA change in the fibroblast growth receptor gene (*FGFR3*) that disrupts normal bone development. One characteristic feature of achondroplasia, and some other skeletal dysplasias, is a short upper arm bone, or humerus (**Figure 9**).

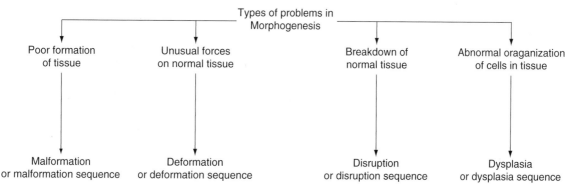

Figure 5 A scheme for the general classification of the causes of birth defects. See text for further explaination. Modified from Jones KL (1988) *Smith's Recognizable Patterns of Human Malformation*, 4th edn. Toronto: W.B. Saunders.

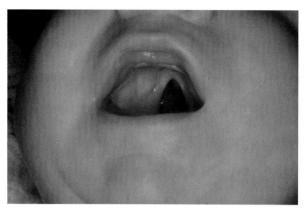

Figure 6   A cleft palate malformation in a newborn. One cause of cleft palate is the Pierre Robin sequence described in the text.

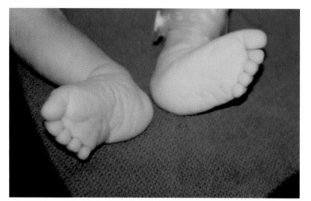

Figure 7   Feet in Potter sequence. The feet of this a newborn child are abnormally positioned secondary to the lack of adequate interuterine space during gestation.

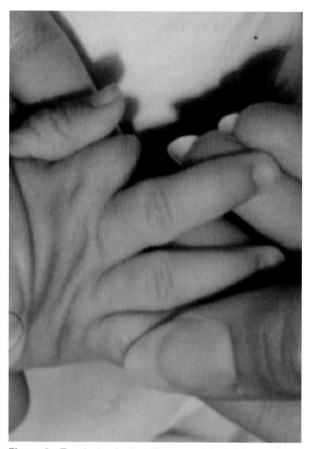

Figure 8   Terminal reduction. The second (index) finger of this child's hand is shortened and missing distal elements such as the fingernail and some joints.

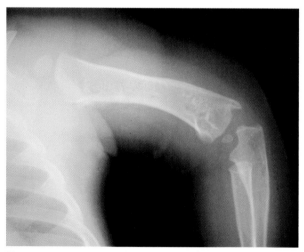

Figure 9   Short humerus. The X-ray photograph pictured shows the bone of the upper arm (the humerus) in an infant. In this case, the humerus is significantly shorter than expected relative to its width and the size of the other bones.

Categories of grouping

The fact that the study of birth defects arose from a descriptive science is reflected in the use of grouping terms that attempt to convey the frequency with which constellations of congenital variations are seen in the same individual. The word 'syndrome', for instance, conveys a greater degree of unifying causation than the word 'association'. The term 'developmental field defect' was used to describe birth defects that occur in a group of tissues that are near one another during fetal development. Our growing molecular understanding of the causes of birth defects is having a dramatic and ongoing affect on how birth defects are grouped.

An example of the evolution of birth defect naming is a group of conditions related to the chromodomain helicase DNA-binding protein (*CHD7*) gene. What is now known as the CHARGE syndrome was originally called an association. CHARGE is an acronym for 'coloboma' (an eye defect), 'heart defects', 'choanal atresia' (a defect of the skull and upper airway), 'retarded growth and development', 'genital abnormalities', and 'ear anomalies'. Component members of the association were grouped together because they often appeared together in individuals. Although the etiology for CHARGE was not known initially, the use of a grouping ('association') allowed for the accumulation of clinical experience with affected children. It is now known that many persons who were diagnosed with the CHARGE association (~2/3) have changes in the *CHD7* gene. The presence of a unifying molecular etiology prompted a change in the name from CHARGE association to CHARGE syndrome. The function of the *CHD7* gene is not well understood. However, it bears strong resemblance to other members of the *CHD* family. Genes in the *CHD* family are thought to be involved in the control of chromatin maintenance during embryogenesis. Specifically, *CHD* genes may have the ability to control the tightness with which DNA is wound. Unwound regions of DNA are readable; tightly round regions of DNA are generally not readable. Therefore, *CHD* genes may help to determine which regions of the DNA are available for reading. The general nature of the proposed *CHD* activity is consistent with a the finding that many systems are affected in CHARGE syndrome.

Grouping by embryonic age

Birth defects may occur during characteristic periods of embryogenesis. Teratogens, for instance, are often categorized according to when maximum sensitivity to the agent occurs during gestation (**Table 2**). By noting the timing of a drug exposure during pregnancy, the possible origins of a birth defect may be refined.

Birth Defects: Diagnosis and Treatment

Birth defects may be diagnosed during gestation, at birth, or at any time after birth. The timing of diagnosis depends on whether the diagnostic tools being used (e.g., prenatal ultrasound or physical examination) are capable of detecting the manifestations of the birth defect, and, the extent to which the severity of the consequences of the birth defect prompt further investigation and medical workup. Some conditions are routinely screened for both during pregnancy and in the neonatal period. Such screening provides increased sensitivity for birth defect detection and increased opportunities for utilizing any available therapeutic interventions.

Screening and Testing of Pregnancies

Routine pregnancy care includes a number of screening tests. Blood glucose screening, fetal ultrasound, and fetal growth monitoring can all provide evidence of increased risk for a birth defect. Genetic testing and/or screening are offered for pregnancies that are identified as being at elevated risk. Selected techniques are discussed in the following section.

Ultrasound

Fetal ultrasound is used for a variety of purposes including the verification of gestational age. The sensitivity of ultrasound to find signs of a birth defect generally increases with gestational age. At-risk pregnancies can be monitored by serial ultrasounds so as to monitor the growth of portions of the body associated with specific fetal anomalies. Femur length, for instance, can be used to screen for certain types of skeletal dysplasia. A deficient rate of head growth is suggestive of a deficiency in brain growth. Ultrasound is not 100% sensitive, however. In Down syndrome, a chromosomal abnormality associated with a number of minor and major malformations (see below), the sensitivity of prenatal ultrasound is only 50%.

Screening for chromosomal aneuploidy – Down syndrome

Women who are found to be at elevated risk for birth defects may be offered diagnostic testing if it is available. The distinction between a screening test and a diagnostic test is critical to understand. A screening test produces a revised estimate of the probability of a particular outcome. For instance, the risk for a birth defect might be 1/10 before a screening test and either 1/5 or 1/1000 afterward. A diagnostic test gives a definitive diagnosis, yes or no, within the limitations of the sensitivity and specificity of the test. The principles of pregnancy screening and diagnosis are illustrated by the chromosomal condition Down syndrome.

Some birth defects are the result of an abnormal number of chromosomes (aneuploidy). Many chromosomal changes, both aneuploidy and nonaneuploid, result in a spontaneous premature termination of gestation. The most common aneuploidy to survive to birth is Down syndrome, caused by an extra copy of chromosome 21. At birth, individuals with Down syndrome have variable degrees of characteristic facial features and other minor and major malformations. Many individuals with Down syndrome survive to adulthood, although all have some degree of cognitive impairment and other, usually manageable, medical issues.

The likelihood of a Down syndrome pregnancy increases with increasing maternal age, being approximately 0.05% at the age of 35 years and 20% in the late 40s. Screening for Down syndrome is routinely offered to women 35 years old or older. The technology for screening is evolving, but generally involves a combination of blood tests and specialized ultrasound-based fetal measurements. As noted above, the result of screening is a modification of a risk estimate based on the age of the woman carrying the pregnancy. Once the revised risk estimate becomes available, the pregnant woman might elect to undergo diagnostic testing. Diagnostic testing can be done using chorionic villous sampling (CVS) or amniocentesis. CVS is a technique whereby pregnancy-associated nonfetal tissue is sampled using either a needle passed through the abdominal wall or a thin tube inserted through the vagina and cervix. Amniocentesis samples amniocytes shed from the fetus into the amniotic fluid, also using a needle passed through the abdominal wall. In either case, the exact configuration of the fetal chromosomes can, within the limitations of tests, be ascertained and a diagnosis made.

Pregnancy screening and diagnostic procedures can also detect some other aneuploidies.

Molecular screening

Sometimes direct measurements of material or fetal DNA are the best option for detecting whether a fetus is at risk of, or affected by, a heritable illness. DNA testing for illness is termed 'molecular testing' by tradition. Most of the common molecular testing used during pregnancies does not relate to illnesses that present as birth defects. However, if a family has had a previous pregnancy affected by a birth defect, molecular testing may have a role. If the birth defect was attributable to a specific, known DNA change, subsequent pregnancies can be tested to see if a similar DNA change is present. The rapid rate at which specific illnesses are being linked to specific DNA changes suggests that scope of molecular screening and testing will continue to expand.

Birth Defects Newly Diagnosed at Birth

When a birth defect is first recognized at the time of birth, a larger range of diagnostic and therapeutic interventions are possible. The scope and number of such procedures is large; only selected topics are discussed here.

The first step in dealing with an unexpected birth defect is to address the immediate health implications for the affected child, which may range from incompatibility-with-life to nonexistent. Once medical stabilization is underway, an organized process of diagnostic thinking follows. Some birth defects can be diagnosed by characteristic sets of features visible during physical examination. Individually, the features may be relatively common in the general population. For instance, a single palmar crease can be used along with other physical features to make a diagnosis of Down syndrome. By itself, however, single palmar creases can be found in individuals with no other related medical issues (**Figure 10**).

If an exact diagnosis is not possible with examination alone, further testing can often narrow the diagnostic possibilities. It should be remembered that many physical anomalies present at birth are 'idiopathic' and a cause is never identified. Initial testing should be tailored to the findings at hand, but may have to be broad if the available evidence does not allow the field of possibilities to be successfully narrowed. For birth defects that involve

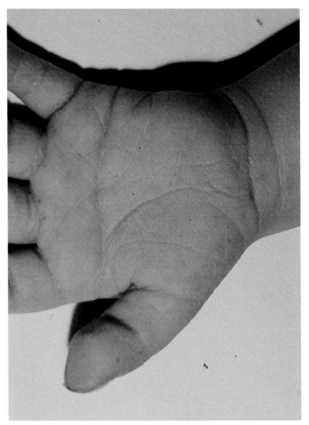

Figure 10   Single palmar crease. The prominent horizontal crease on this child's hand streches across then entire palm. Any cause of reduced fetal movement may cause this pattern and it is present among many, otherwise normal individuals.

multiple sites or organ systems, a genetic screening procedure such as karyotyping or microarray analysis is usually included in the workup. Karyotype analysis may show a microscopically visible difference from normal chromosome structure. Major organ systems such as the heart, brain, and kidneys are often evaluated to look for diagnostic clues that are not externally visible, and to rule out structural abnormalities with health implications. A careful examination of the eyes, face, and extremities can often provide valuable information.

Unexpected cleft lip and palate provides an example of how birth defects are evaluated. Assuming that no other physical anomalies have been discovered, and that the infant is stable (breathing and swallowing issues have been assessed and addressed), the diagnostician will consider the following information.

Cleft lip plus cleft palate (CL/P) is fairly common, occurring in approximately 1/700 births. The formation of the palate and upper lip occur by day 60 of fetal development, so the origins of the observed defect can be attributed to a particular span of time during pregnancy. The textbook *Principles and Practice of Medical Genetics* by Emery and Rimoin notes that approximately 70% of cases of CL/P are nonsyndromic meaning that there is no known causes and no other associated abnormalities. Of those that are syndromic, The Online Mendelian Inheritance in Man database lists more than 400 conditions in which CL/P is a feature. Alterations in many different genes can produce CL/P, some of them producing syndromes with subtle characteristic features that might not have been detected during an initial examination. Van der Woude syndrome, for instance, is associated with clefting and may produce lip pits. A repeat examination may be useful in detecting subtle clues.

A thorough pregnancy history is taken to exclude exposure to teratogens and pregnancy complications. Smoking, alcohol use, and some prescription drugs are under investigation or known to increase the risk of CL/P. A family history is taken to ascertain whether clefting or evidence associated with clefting-related syndromes has been seen in other family members.

Once all of the available information has been collected, a diagnostic plan will be implemented. If no diagnostic leads have been unearthed, common CL/P-associated conditions, including Stickler syndrome and DiGeorge syndrome, may be tested. Cardiac echocardiography may be considered due to the fact that some of the syndromes that produce clefting may also produce structural cardiac anomalies. A karyotype will likely be obtained to look for evidence of mosaic trisomy 13 (a chromosomal aneuploidy associated with clefting) or other microscopically visible chromosome anomaly. A counseling strategy will be defined and implemented to help the family to cope with the unexpected finding and to plan for future workup. Follow-up with a multidisciplinary team specializing in CL/P will be arranged. Although simplified, the presented CL/P example highlights some of the processes by which unexpected birth defects are evaluated and diagnosed.

Preventing Birth Defects

The prevention of disease is always better than the curing of disease. Most known birth defects are not preventable with current medical technology. However, particularly in industrialized countries, there have been some notable prevention successes. The prevention of iodine and folic acid deficiencies of pregnant women has had dramatic impacts on congenital hypothyroidism and neural tube defects (NTDs), respectively. Universal immunization has reduced the rate of birth defects due to some transmissible diseases. The careful treatment of women with phenylketonuria, while they are pregnant, reduces the likelihood of birth defects in their offspring. Further research in other areas, including diabetes, holds promise for additional improvements in prevention.

The story of folic acid supplementation is illustrative of a success in birth defect prevention. NTDs are a subclass of birth defects that involve the development of the spine during embryogenesis. Spina bifida, where the vertebrae fail to close over the spinal cord in one or more places, is an example of an NTD. NTDs are not rare, occurring as often as 1/1000 pregnancies. In the 1950s, it was observed that NTDs were more common among women of low socioeconomic status. That finding suggested that some environmental factor might have a role in NTDs. Further support for an environmental theory included the fact that most women who have children with NTDs do not have a family history of NTDs themselves, making a genetic cause less likely. In the 1960s, it was discovered that folate deficiency caused birth defects in animal models. In 1991, a landmark British study showed that folic acid supplementation reduced the risk that a woman with a history of an NTD-affected pregnancy would have a subsequent NTD-affected pregnancy. In 1992, the FDA recommended that folic acid be given to all women of childbearing age, and, that folic acid be added to certain grain-based foods including some types of flour, rice, and bread. In 1998, the Food and Drug Administration (FDA) mandated folic acid supplementation in these foods. The consensus among researchers and epidemiologists is that folic acid therapy has been a success. For instance, a 2001 Cochrane Library meta-analysis examined the data regarding the success of folic acid supplementation in preventing NTDs. The authors found the following: "Periconceptional folate supplementation reduced the incidence of neural tube defects (relative risk 0.28, 95% confidence interval 0.13 to 0.58)."

Birth Defects Research

Biomedical research has had a profound effect on our understanding of birth defects. What was once described by appearance alone can now be understood in terms of specific developmental mechanisms. In addition, advances in surgical technique are showing promise for interventions that may prevent or mitigate the effects of some birth defects. The following two examples demonstrate how an understanding of cellular and molecular mechanisms, and the use of pioneering therapies, are improving what medicine has to offer to families affected by birth defects.

Three-Dimensional Limb Patterning

The development of a limb from a few embryonic cells to a complex adult structure requires a highly coordinated set of cell interactions. One aspect of limb development that lends itself to a discussion of the relations between research and birth defects is limb patterning (**Figure 11**). The upper extremity (arm, forearm, and hand) is an asymmetric structure. Anatomical terms are used for orientation. The proximal portion of the upper extremity is

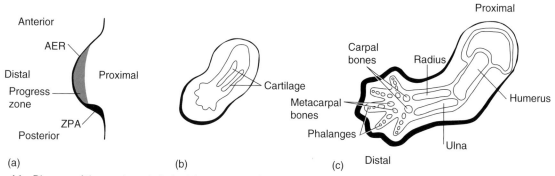

Figure 11 Diagram of the an ebryonic limb at three stages of development. Note the anterior/posterior and proximal/distal orientation of the limb as referred to in the text. The dorsal/ventral orientation (palm and back of hand) are not labeled. Specific cellular pathways, outlined in the text, control the patterning of the limb as is develops.

that closest to the shoulder; the distal most portions are the tips of the fingers. The dorsal part of the upper extremity includes the back of the hand, the back of the forearm, and, roughly, the outside of the arm; the ventral parts include the palm, the inside of the forearm, and the inside of the arm. Finally, the anterior portion is the thumb, the radial side of the forearm, and the 'front' of the arm; the posterior portion is the fifth finger side of the hand, the ulnar side of the forearm and the 'back' of the arm. The hand provides a ready example of the asymmetry of the upper extremity – the thumb is a significantly different structure than the fifth finger. Given that the upper extremity starts fetal life as a small lump of cells, specific cellular machinery must direct the patterning of the growing limb. Furthermore, it would be expected that defects in such developmental processes would have specific consequences for the fully developed extremity. A partial schematic of the structural and molecular signaling mechanisms in the developing limb will be used to highlight how a molecular understanding of limb development can be used to understand birth defects.

Anterior–posterior patterning

The earliest recognizable upper-extremity limb structure in a developing embryo is the approximately paddle-shaped limb bud. Early in embryogenesis, several recognizable structures are present. The edge of the limb bud that is furthest away from the fetal body forms an apical ectodermal ridge (AER). The edge of the limb bud that is closest to the tail contains a group of cells that form a zone of polarizing activity (ZPA). The ZPA cells generate a signal (comprising a molecule called sonic hedgehog or Shh) that defines the posterior aspect of the growing limb. On the other side of the limb bud, Shh signaling is repressed by a second molecule, Gli3, defining the anterior edge of the limb. These and other cellular signals are used to define the differential development of the anterior and posterior structures of the upper extremity.

Proximal–distal patterning

The AER remains at the tip, or distal end, of the limb as it grows. Experimental removal of the AER in model organisms will halt the lengthening of the limb. An important group of molecules involved in AER signaling are the fibroblast growth factors (FGFs). As the limb grows, cascades of signaling interactions form a pattern of gene expression that differs from the proximal to distal portions of the limb. One set of genes involved in forming this pattern is the homeobox, or *Hox* genes.

Dorsal–ventral patterning

As with the other developmental patterning axes, dorsal-ventral (DV) patterning is mediated by complex cascades of cell-signaling molecules. One such molecule that defines the dorsal portion of the limb is Wnt7a. Wnt7a acts on specific embryonic tissues to promote the expression of Lmx-1b, which in turn promotes a dorsal character to local tissues. On the ventral side of the limb, the molecule 'engrailed' En-1 inhibits Wnt7a, which in turn downregulates expression of Lmx1b. The result is a default, ventral limb surface.

Pallister–Hall syndrome

Dorsal–ventral, anterior–posterior, and proximal–distal limb patterning mechanisms interact during development to create a fully formed upper extremity. Genetic changes to the gene's coding for the involved signaling processes have the potential to disrupt normal limb formation. Pallister–Hall syndrome (PHS) results from a subset of mutations of the *GLI3* gene. PHS includes several congenital abnormalities including polydactyly (extra fingers), upper airway abnormalities, and specific brain tumors. Relevant to limb development, PHS can cause postaxial polydactyly, a defect in anterior–posterior limb development. PHS is an example of the general rule that developmental signaling molecules are involved in the development of numerous systems throughout the body. Polydactyly can be caused by a number of inherited conditions and is often not attributable to any known cause. Making a diagnosis of PHS in a child with polydactyly allows clinicians to use the historical experience with PHS to design strategies for counseling and therapy.

Nail Patella syndrome

Nail patella syndrome (NPS) classically includes abnormalities of the finger/toenails, knees, elbows, and pelvic bones. In addition, a substantial number of individuals will have kidney and eye involvement, often appearing later in life. NPS is caused by mutations in the *LMX1B* gene, and is therefore categorized as a defect of dorsal–ventral patterning. Several classic NPS features make intuitive sense as disruptions of dorsal–ventral patterning including the nail abnormalities and the presence of abnormal or missing kneecaps (both dorsal structures). Making a diagnosis based on congenital abnormalities allows for screening for later manifestations of the disease.

Retinoic acid embryopathy

Retinoic acids are chemically related to vitamin A. Some types of retinoic acid are used medicinally, including the severe acne treatment Isotretinoin. Isotretinoin use in pregnant women is contraindicated because of an association with a well-described set of birth defects. Retinoic acids normally formed in the body participate in the regulation of limb development and interact with other signaling systems such as the *Hox* genes. It is not surprising, therefore, that some cases of retinoic acid embryopathy include limb reduction defects – a disruption of proximal–distal patterning.

Three-dimensional limb patterning: Conclusions

Research into the molecular underpinnings of development continues. Further insight into the mechanisms by which normal development proceeds will continue to improve our ability to understand how and why specific birth defects occur.

Fetal Surgery

Another area of research focuses on the prenatal treatment of birth defects. Fetal surgery involves gaining surgical access to the fetus, usually through the mother's abdominal wall and the wall of the uterus. Once the fetus is exposed, surgery on the fetus is performed, followed by repair of the mother's uterine and abdominal incisions. Often a conservative surgical approach is taken, whereby a minimal set of interventions is designed to help the fetus survive to birth. After birth, additional medical and surgical treatments may be necessary. For instance, in 1981, a pioneering surgery performed by Dr. Michael Harrison involved a fetus that had a congenital blockage of urinary bladder outflow called a posterior urethral valve. A shunt was placed, allowing the passage of urine and decompressing the dangerously expanded bladder. After birth, the congenital anomaly was definitively repaired and the child developed normally. Similar, and more complicated, surgeries are being developed to treat a number of amenable conditions.

Summary

The term birth defect encompasses a wide variety of birth anomalies resulting from deviation in normal embryogenesis. There is a continuum of variation among newborns, including some variation that has detrimental health consequences for the child. Medical geneticists and other specialists evaluate children with birth defects in an attempt to classify them as an aid in counseling and treatment. Although the number of recognized mechanisms for birth defects is growing, the root cause of the majority of birth defects is unknown. Birth defects continue to pose a large medical and economic burden on individual families and on society. Biomedical research is expanding our ability to diagnose and treat birth defects.

See also: Birth Complications and Outcomes; Developmental Disabilities: Cognitive; Developmental Disabilities: Physical; Down Syndrome; Genetics and Inheritance; Intellectual Disabilities; Neuropsychological Assessment; Prenatal Care; Prenatal Development; Screening, Newborn and Maternal Well-being; Screening, Prenatal; Teratology.

Suggested Readings

Epstein CJ, Robert PE, and Wynshaw-Boris A (2004) *Inborn Errors of Development.* Oxford: Oxford University Press.

Jones KL (1988) *Smith's Recognizable Patterns of Human Malformation,* 4th edn. Toronto: W.B. Saunders.

Jones KL (2005) *Smith's Recognizable Patterns of Human Malformation,* 6th edn. Philadelphia, PA: W.B. Saunders.

Martin JA, Kochanek KD, Strobino DM, Guyer B, and MacDorman MF (2005) Annual summary of vital statistics – 2003. *Pediatrics* 115(3): 619–634.

Rimoin DL, Connor JM, Pyeritz RE, and Korf BR (2006) *Emery and Rimoin's The Principles and Practice of Medical Genetics.* Oxford: Elsevier.

Robbins JM, Bird TM, Tilford JM, *et al.* (2007) Hospital stays, hospital charges, and in-hospital deaths among infants with selected birth defects – United States, 2003. *Morbidity and Mortality Weekly Report* 56(2): 25–29.

Relevant Websites

http://www.cdc.gov – Centers for Disease Control and Prevention – Basic Facts About Birth Defects (accessed 18 September 2007); Birth Defects: Frequently Asked Questions (21 March 2006).

http://marchofdimes.com – March of Dimes Foundation.

Birth Order

D L Paulhus, University of British Columbia, Vancouver, BC, Canada

Glossary

Between-family research – Involves the comparison of individuals from different families. Therefore, when birth orders are compared, the large genetic differences between families are not controlled and add much extraneous variance.

Big five personality traits – Factor analyses of comprehensive personality questionnaires typically yield five super-factors. They are named extraversion, agreeableness, conscientiousness, neuroticism, and openness to experience. This simple taxonomy of personality has proved useful in describing birth-order effects.

Birth order – The numerical sequence of a child's arrival into a family. Environmental theories focus on the functional order (actual rearing order) whereas biological theories include all births.

Intellectual achievement – This term subsumes scholastic performance (e.g., GPA, number of years of attained education) and performance on scholastic-related tests (e.g., standardized intelligence tests, SATs).

Self-fulfilling prophecy – The dynamic process whereby an expectation about birth order difference becomes a reality.

Social development – In this article, the term social development is simplified to refer to age changes in personality and political attitudes.

Teaching effect – The long-term intellectual advantage conferred by the opportunity to help one's younger siblings.

Within-family research – Within-family research involves data collected on at least one sibling as well as the target individual. In many studies, one family member reports on the whole set of siblings including himself or herself. It does not seem to matter which family member does the reporting because family members tend to corroborate each other's judgments.

Introduction

The notion that birth order has an influence on child development has undergone several cycles of popularity and disrepute. This uneven history applies to birth-order effects on social development as well as intellectual development, although the two literatures have unfolded quite independently. In this article, a brief history of this cycling of popularity in each of these literatures is provided followed by elaboration on the major theories and research.

Most discussions of the topic focus on differences between firstborn and laterborn children. This simplification results in part from a reluctance on the part of researchers to differentiate among birth orders with small frequencies. While there are sufficient numbers of first- and secondborns in most samples, the frequencies are small for thirdborns and higher. Hence all laterborns are lumped together for analysis purposes. This article will follow suit in focusing almost entirely on the firstborn vs. laterborn differences – with the exception of some notable findings regarding middle-borns and lastborns.

A Brief History

Social Development

Alfred Adler, the second born of six children, was weak, sickly, and continually tormented by his older brother. In early childhood, Alfred envied his brother and felt that they were always in competition. But he worked hard to overcome his handicaps and became a popular member of the community. His success was such that his older brother grew to resent him.

Undoubtedly, these family dynamics played a role in Adler's seminal writings about the psychology of birth order. His ideas, published largely in the 1920s, anticipated many of the later perspectives on the subject. He suggested, for example, that firstborns are typically given more family responsibilities than laterborns and are expected to set an example. Consequently, they often become authoritarian and construe power as their natural right. This attitude can eventuate in an insecurity around the possibility of being 'de-throned' by laterborns.

Adler went further to write that achievement expectations are high for firstborns and they attempt to live up to them. Laterborns, in contrast, try to compensate for their inferiority in size and power by turning to alternative notions of achievement. For example, the laterborns might turn to more social or creative endeavors. Thus Adler's writings addressed both of the two primary domains, intellectual and personality development.

It was not until the 1960s that mainstream researchers raised the legitimacy of studying birth-order effects in personality development. Stanley Schachter, for example, conducted a series of archival and laboratory studies of birth-order differences. Other prominent psychologists (e.g., Robert Sears, Phillip Zimbardo, Edward Ziegler, and Mary Rothbart) all added to the body of research on birth order as well as its credibility. During this period, sociologists and economists also contributed both theory and data to the birth-order literature.

Although popular right up to the 1970s, the credibility of birth-order effects on personality faltered badly in 1983 with the publication of a comprehensive review by Ernst and Angst. The scope of their review was impressive: virtually every published study was included. After adding a variety of controls, including gender and family size, the authors concluded that associations between birth order and personality traits were minimal.

The reputation of birth-order effects remained in quiet disrepute until the 1996 publication of Frank Sulloway's book, *Born to Rebel*. In applying a bold new theoretical perspective, Sulloway revitalized the view that personality and social attitudes differ systematically across birth order.

Sulloway's treatment was persuasive, in part, because he offered two complementary forms of evidence. One was a catalog of captivating stories about the family life of historical figures. The second form of evidence was a meta-analysis of the large number of studies on personality and birth order. To great advantage, he organized the studies within the influential 'Big Five' or Five Factor Model of personality. That model is now generally accepted as the best organizational system (taxonomy) of personality traits.

Using that organizational system, Sulloway's meta-analysis of the apparently chaotic literature exposed a clear pattern. In particular, firstborns were more conscientious and socially conservative but less agreeable and open to experience than laterborns. These claims form the hub of debates that continue to swirl around birth-order effects on social development. Follow-up studies from other quarters have varied from highly supportive to highly critical of Sulloway's claims.

Intellectual Development

Scholarly interest in the relation between birth order and achievement can be traced to 1874 when Francis Galton published *English Men of Science: Their Nature and Nurture*. The book chronicled the lives of 180 eminent men from various scientific fields. Galton found that 48% of them were eldest sons, far higher than would be expected by chance. Anticipating later arguments, Galton provided three speculations on how the birth-order difference might come about. First was the impact of the primogeniture tradition: firstborn sons were given priority in the

inheritance of family wealth. Accordingly, they would be more likely have the financial resources to continue their education. Second, firstborns were more likely to be treated as companions by parents and be assigned more mature responsibilities than their younger siblings. Galton's third speculation was that, in families with limited financial resources, firstborns received more attention and better nourishment than other siblings.

The latter two notions remain central to current debates regarding birth-order effects. Although the primogeniture tradition has waned, recent surveys by anthropologists confirm that firstborns occupy special status in every human society. Other things being equal, they are awarded more respect and given priority in legal, religious, and social matters, even when all siblings are grown to maturity.

Almost a century of sporadic studies of intellectual development yielded inconsistent associations with birth order, partly because the sample sizes were insufficient. It was not until Robert Zajonc's research in the 1960s that massive data sets were given theoretical scrutiny in major psychology journals.

Zajonc's analyses provided persuasive evidence that the intellectual achievement of firstborns tends to surpass that of other birth orders. This advantage applies across a wide range of measures including school grades, intelligence quotient (IQ) scores, and SATs. Partly due to his credibility as a hard-nosed scientist, Zajonc's theoretical and empirical analyses were taken seriously: much of his research and follow-up studies were published in medical, economics, and hard-science journals. As detailed below, his work provoked an avid interest that continues to this day.

Theories of Birth-Order Differences

As noted above, the literatures on social and intellectual development have only minimal overlap. The various theories of birth order have developed primarily in the context of one field or the other. In reviewing the five most important theories, however, this article will attempt to draw out implications for both social and intellectual development.

Confluence Model

Proposed by Zajonc, this theory explains the firstborn advantage in terms of the intellectual environment evolving within the family. With only two propositions, the theory was able to explain birth-order effects as well as intellectual deficits deriving from five other family constellation effects: family size, close child spacing, multiple births, and being lastborn or an only child.

The first proposition of the model is simply that intellectual stimulation of children has enduring benefits for their later intellectual success. Only firstborns have a period of time where they receive 100% of their parents' attention. For secondborns, the maximum quality time involves sharing the parents' attention with the firstborn. With each successive child, the available parental attention gets watered down even further. In addition, the linguistic environment becomes increasingly less mature as more children enter the family. The second proposition of the confluence model was that lastborns miss out on the intellectual stimulation involved in teaching younger siblings. We consider that second proposition in the section below on lastborns.

Zajonc's first proposition does not seem radical or especially controversial: in retrospect, it seems more like commonsense. But he spelled out the various consequences and quantified them in a simple but persuasive arithmetic formula. To represent the quality of the intellectual atmosphere at any point in a child's development, one simply has to calculate the current mean mental age in the family. Integrated over the childhood years, this mean is higher for firstborns: they receive the most intellectual stimulation because they spend a larger portion of their time in a high-quality atmosphere. This stimulation stays with them in the form of superior cognitive abilities.

Intellectual deficits due to family size also follow from this watering-down mechanism. Increasing the spacing between children helps modulate this watering down effect by allowing the mental ages of the older children to increase before adding the new contributor of zero mental age. Finally, the extra deficit seen in children of multiple births follows from the extra drop in average mental age due to the addition of several zeros to the equation.

Although they are seldom spelled out, implications for social development can also be derived from the Confluence Model. Differential parental attention, even out of practical necessity, should affect the nature of the parent–child relationship across the birth order. Firstborns should be more attuned to their parents' aspirations for their children, more needy of their parents' approval, and expect to maintain the special status they enjoyed as children in future social settings. Together, these sequelae could eventuate in the different personality trait and value profiles typically found across the birth orders.

Resource Dilution Model

This theory, originally proposed by the economist Judith Blake and extended by the sociologist, Downey, goes beyond the Confluence Model to argue for a more comprehensive decrease in resources for each successive child. In particular, there is a progressive watering down of financial and educational sources such as books, travel, and tuition. Differences in such concrete parental resources across birth orders can culminate in different scores on IQ tests.

For example, parents with limited incomes may not be able to afford to send all their children to college. Any limitation in the opportunity for higher education will certainly diminish the likelihood of intellectual achievement. In combination with the decrease in parental attention, these other drawbacks handicap laterborns relative to firstborns. As noted for the Confluence Model, any special status, even if endowed arbitrarily by financial practicalities, may have implications for social development.

Writers adhering to the Resource Dilution Model seldom allude to differences in social development across the birth order. Nonetheless, it seems reasonable to speculate that the differential allotment of financial resources could influence personality. The model is consistent with a small number of studies suggesting that firstborns feel more entitled to special treatment and that laterborns experience more resentment and jealousy.

Parental Feedback Theory

This theory suggests that parents adjust their parenting style as they move from the firstborn to laterborns. This adjustment is not out of financial or attentional necessity, but out of increasing comfort and decreasing anxieties. The result is that parents are less demanding of laterborns, especially with regard to their school performance. Beyond the firstborn, parents may allocate their love and approval in a manner that is less contingent on the child's achievement.

In one of the few experimental studies examining the transmission of birth-order effects, Irma Hilton observed how mothers treated children in a laboratory setting. In the waiting-room, firstborn children were observed to remain physically closer to their mother, often holding on for security. After the children returned from a putative 'testing session', mothers were told that their child had performed extremely well or extremely poorly – based strictly on random assignment. Observation via a one-way mirror revealed that mothers of firstborns gave contingent feedback: if told their child performed well, mothers coddled and praised the child. If told their child performed poorly, mothers berated the child. Laterborns, however, received noncontingent treatment: mothers responded to the child as they had before the testing session – regardless of performance feedback.

It is easy to see how such differential treatment could set off rather different developmental trajectories for firstborns and laterborns. In firstborns, superior intellectual achievement should be accompanied by a number of personality traits: they should possess higher achievement motivation, a greater concern with approval from parents and subsequent authorities. In turn, such qualities may well diminish their popularity among peers. The need for approval from authorities should also engender more conservative political attitudes in firstborns.

Family Niche Theory

In Frank Sulloway's theory, parents play only indirect roles. Instead, birth-order effects unfold during the inevitable competition among siblings as they struggle for a family niche. Firstborns, having the first choice of niche, attempt to please their parents in traditional fashion, namely, by good performance at school and by generally responsible behavior. But, as other siblings arrive, firstborns must deal with threats to their natural priority in the sibling status hierarchy. The resulting adult character is conscientious and conservative.

Laterborns must contest the higher status of firstborns, while seeking alternative ways of distinguishing themselves in the eyes of the parents. Accordingly, they develop an adult character marked by an empathic interpersonal style, a striving for uniqueness, and political views that are both egalitarian and antiauthoritarian. In short, they are 'born to rebel'. This attempt to address birth-order differences in political orientation is unique to Family Niche Theory.

Although designed to explain birth-order differences in personality, the Family Niche Theory is not without implications for intellectual development. In fact, it makes predictions about two aspects of intellectual life – achievement and creativity. Firstborns strive to achieve via traditional academic means – conscientious striving, to be specific. This development begins with their attempt to please their parents via school success. Although traditionally distinguished as ability vs. motivation, the tight overlap between intelligence and conscientiousness has become more evident in recent work. Laterborns, in contrast, seek out creativity, even radical revolution, in their intellectual lives.

Prenatal Hypomasculinization Theory

Drawing on earlier work by Maccoby and others, Jeremy Beer and John Horn have developed a biologically based theory suggesting that the birth orders already differ at birth. The argument does not postulate an average genetic difference in the birth orders but a difference in their exposure to hormones. Previously called the 'tired mother' syndrome, the notion is that, with each succeeding male child birth, mothers expose their babies to lower levels of masculinizing hormones.

Beer and Horn derived their theory from recent findings indicating that the likelihood of male homosexuality increases with the number of older brothers. The common mechanism, they argue, is the progressive immunization of mothers to the hormones that masculinize the male fetus. Thus male children with older brothers are 'hypomasculinized' in both their sexual orientation and their personality characteristics.

According to Beer and Horn, this process eventuates in certain parallels between sex differences and birth-order effects. For example, males and firstborns should exhibit higher levels of competitive achievement whereas females and laterborns should exhibit more cooperation and flexibility. Firstborns should also be more disagreeable, and show more masculine interests. This hypothesized pattern of birth-order differences is consistent with the empirical evidence cited by Sulloway, Zajonc, and others.

To date, however, there is little direct evidence to support Beer and Horn's hypomasculinization theory. Yet the possibility that firstborns and laterborns already differ at birth is intriguing and should trigger further research on biological differences across birth orders.

Contrasting Mechanisms

Even among those writers who accept that children of different birth orders do differ in systematic ways, the disagreement over explanatory mechanisms is striking. According to the Prenatal Hypomasculinization Theory, the differences are already set at birth. For the Parental Feedback Theory, it is a change in parents' comfort level that is responsible for birth-order differences. For the Resource Dilution Theory, it is the diminishing availability of resources that aid education. For the Confluence Model, it is the devolving quantity and quality of intellectual stimulation. For the Family Niche Theory, birth-order effects are propagated by accompanying differences in age, size, knowledge, and status in the family: The oldest child will always be the oldest. Size, knowledge, and maturity differences will eventually even out but status differences can remain well into adulthood.

Of course, people spend most of their lives outside the purview of the family home and its unique interpersonal dynamics. Not surprisingly, then environmental theories typically suggest that birth-order forces on social and intellectual achievement should diminish with time. Even if accepting that the power of such differences eventually wanes, most psychologists – and lay observers, for that matter – believe that early environmental factors have a unique and enduring impact.

Modern Data: The Importance of Research Design

Each of the above theories has some intuitive appeal. But there remain serious questions about the data supporting the very existence of birth-order differences. As with many developmental debates, the key claims are not testable via laboratory-controlled experiments. Under contemporary mores, we cannot – or, rather, will not – randomly assign babies to different birth orders. Instead, social scientists can offer only correlational data and hope to clarify the developmental processes via statistical arguments.

The most persuasive birth-order studies entail a large sample of participants evaluated in an efficient experimental design that includes multiple control provisions to handle potentially contaminating variables. One critical design issue is whether the data are collected within families or between familes. Within-family studies involve a comparison of the siblings within each family. If firstborn Jason and secondborn Mark are raised entirely in the same family setting, then they are matched (in large part) on factors such as family socioeconomic status (SES), parents' child-rearing strategies, parents' personalities, family events, and many other environmental factors. Of special importance, the researcher need not be concerned with genetic differences because, on average, they do not differ among offspring of the same parents. All of these controls make for a fair comparison of Jason and Mark with respect to birth order.

In between-family studies, however, none of those controls are in place. If chosen randomly from a classroom, subject pool, street interview, or telephone survey, Jason and Mark are bound to differ on a host of environmental and genetic factors. Because those variables contribute their own (often larger) sources of variation, any birth-order differences will tend to be obscured.

Because birth-order effects are relatively small, large sample sizes are of special importance for teasing out the differences. In the case of within-family research, it is difficult to take seriously any study comprising fewer than several hundred families. For between-family studies, even larger samples are required. Because so many other factors add noise to the measurement, birth-order differences do not become apparent with fewer than 500 participants from a relatively homogeneous sample.

Debates over these methodological issues have created comparable levels of controversy in the research literatures on social and intellectual development. Yet the controversies have played out in rather different fashion in the two literatures.

Intellectual Development

In virtually every cross-sectional survey, a consistent advantage for firstborns continues to appear. Firstborns are over-represented among university students, among Nobel Prize winners, and on virtually any other concrete measure of intellectual achievement (e.g., IQ tests, SATs). Such birth-order differences, first communicated in a scientifically persuasive fashion by Zajonc and colleagues, continue to emerge in modern samples.

For the most part, however, such clearcut birth-order effects were observed in between-family (i.e., cross-sectional) data. A variety of confounds (e.g., SES, family size) make such results ambiguous. As Joseph Rodgers and others have demonstrated, when such confounds are

removed, birth-order effects on measures of intellectual achivement often disappear. Unfortunately, when such important variables are statistically confounded, it is difficult to distinguish which variables are genuine effects and which variables should be controlled. By removing the effects of variables that may have similar causal mechanisms to birth-order effects, such analyses may be 'throwing out the baby with the bathwater'.

As of the writing of this article, the empirical pendulum seems to have swung back to favor the claims made by Zajonc and others. Several Norwegian researchers have recently analyzed data from virtually the entire population of their country. In 650 000 families, firstborn children showed a clear advantage in IQ, educational attainment, and later adult income.

Apart from the largest sample size, this research has the most rigorous controls, including family size and SES. The fact that education is free in Norway helps mitigate the counter-argument that family finances play a determining role. So does the finding that the birth-order effects were actually stronger for children with highly educated mothers.

Social Development

In studies of personality and social attitudes, as well, the empirical debate about birth order is characterized by inconsistency. Sulloway and others have offered large data sets to support the idea that firstborns are more conforming and conscientious whereas laterborns are more agreeable, open to experience, and politically liberal. In response, other reputable scientists have disputed the size and importance of such birth-order differences.

Again the debate may turn on the choice of within-family vs. between-family designs. In this case, however, the advocate and skeptic views are reversed. Birth-order effects are evident in within-family designs whereas minimal results emerge from between-family designs. The within-family design is typified by the method used in a 1999 article by Paulhus and colleagues. They asked a variety of large samples to report on their own families. In one study, for example, participants were asked to rate themselves and their siblings on the Big Five personality traits and on political attitudes. Results firmly supported Sulloway's predictions.

Most recently, Healey and Ellis outlined the conditions that yield the clearest birth-order effects in personality: (1) when firstborns are compared with secondborns, (2) when the age difference is 2–4 years, and (3) when children reared apart are excluded. Again, these within-family patterns confirmed predictions from Sulloway's Family Niche Theory.

With respect to between-family studies, a prototypical example is the study conducted in 1998 by Tyrone

Jefferson and colleagues. Their data came from large archival samples that included both personality and birth-order data. On self-report measures, they found no significant birth-order effects on personality. On peer-ratings, the only significant finding was the usual conscientiousness advantage for firstborns. Few other studies can boast the feature of peer-raters: they provide a more objective perspective from outside of the family.

These conflicting results may have a simple resolution. Studies with weak or null birth-order effects always involve a comparison of individuals from different families. But, as noted above, families differ on a wealth of influential variables and a full range of appropriate controls is seldom available. Within-family data provide a natural control procedure for all between-family differences, including their largest contributor – genetics.

Most readers will be aware of the recent confirmation of substantial genetic effects on both intellect and personality. This consensus is helpful in understanding the conflicting conclusions drawn from within- vs. between-family studies: within-family designs remove a large component from the equation, namely, mean genetic differences between families. Accordingly, birth-order differences emerge more clearly in within-family studies.

The burgeoning behavioral genetics literature also supports birth-order claims in another way. Second to genetics, the primary source of variance is personality, values, and, even political orientation is within-family environmental variance. In other words, there are family dynamics at work making siblings more different than expected by random genetic effects. Sibling rivalry and differential parental treatment of different birth orders are likely to be part of these within-family dynamics.

Stereotype Effects

To repeat, within-family studies of personality inevitably show clear birth-order effects. Whether one asks the firstborn or laterborns, there is agreement on who is more conscientious. Moreover, this agreement seems to last a lifetime.

Yet some critics dismiss the importance of that within-family consensus, arguing instead that putative birth-order effects derive entirely from within-family stereotypes. As children grow up with siblings of different ages, real differences in size, power, maturity, and knowledge govern the intersibling dynamics. When asked later to compare their siblings, say at the age of 30 years, all family members tend to concur on the traditional family story about how the children differ. Beyond that, these critics argue, the stereotypes have no impact on people's lives.

But research from the social and developmental psychology literature indicates that self- and other-stereotypes run deeper than that. In fact, adult samples show the same

pattern and size of birth-order effects as much younger samples, even when the adults have been living apart from their siblings for many years. The stability of these perceptions across the lifespan undermines the accusation that they are artifactual and makes a stereotype perspective difficult to distinguish from standard conceptions of personality.

Alternatively, is it possible that birth-order differences in that perceptions of one's siblings are a fiction inculcated by stereotypes acquired from other sources? It is hard to believe that, throughout their lives, siblings systematically ignore bona fide evidence of their brothers' and sisters' actual traits in favor of erroneous stereotypes. It seems far more reasonable to believe that such stereotypes flourish because they have (at least) a kernel of truth. Critics would have to argue further that initially false stereotypes can endure a lifetime without having any impact on personality. According to social psychological research, however, one should expect some reification due to self-fulfilling prophecies. Can the stereotypes, the self-perceptions, and the peer-perceptions all be faulty?

Critics such as Judith Rich Harris hold an intermediate position in conceding the reality of within-family personality differences, but caution that the differences remain just that – within the family. In other words, birth-order differences have no effect on life outside of the family home: only on home visits do the old familiar patterns emerge. Many readers will relate to that experience. Nonetheless, that experience may not be an insignificant portion of adulthood. Many children do go on to spend a significant amount of their adult life involved in continuing interactions with the family of origin.

Harris's notion of circumscribed personality differences is also compromised by a number of recent studies reporting on more concrete differences outside of the family. For example, firstborns show more dismissive attachment styles in later life whereas laterborns disproportionately choose occupations that involve social interaction.

To summarize, the stereotype critique is an attempt to explain away the robustness of within-family personality differences as shared fiction. Such counter-arguments must always be taken seriously, but in this writer's opinion, there is simply too much evidence for the reality of birth-order differences.

Further Complexities

Firstborns, Only-Children, and MiddleBorns

Most of this article has purposely simplified birth-order issues by limiting the discussion to a comparison of firstborn children to all laterborns. The primary reason was that, in most respects, differences among laterborns are not as apparent as is the contrast with firstborns. Yet there are a few issues where middleborns and lastborns do stand out.

The unique findings for lastborn children include both good news and bad news. As noted earlier, Zajonc found an extra decline in the intellectual achievement of lastborns – above and beyond the gradual decline due to successive birth order. A thirdborn child, for example, fares more poorly if no younger children are added to the family. This finding was confirmed in the recent large sample and tightly controlled Norwegian data (described above).

Zajonc explained this anomaly in terms of the so-called 'teaching effect'. Firstborns (and older siblings in general) often have to answer questions posed by their younger siblings and assist with their homework. At the time, the older siblings may experience these tasks as onerous. Rather than a burden, Zajonc argued, this teaching opportunity actually benefits earlier-borns, perhaps by forcing them to engage more deeply with the material they are teaching. That claim is quite consistent with the tutoring research in educational psychology, which shows that teaching benefits the teacher at least as much as it benefits the student. A lack of such opportunities can thus explain why lastborn children show an extra deficit in intellectual achievement and why only, children do not achieve as highly as other firstborns.

The good news for lastborns lies in the personal popularity that ensues from their birth order. In surveys of comparative popularity, the lastborns are voted the 'favorite child' more often than any other birth order. This popularity may well reflect an inevitable tradeoff with personal achievement. Peers prefer others who are noncompetitive and more socially oriented than achievement oriented.

These arguments can also be applied to only-children. The fact that they are also lastborns, may explain why their academic achievement does not match up to that of other firstborns; the fact that they are also firstborns may explain why they are not as popular as other laterborns.

The tradeoff between the respect accorded to firstborns and the personal popularity of lastborns often leaves the middleborns feeling left out. Studies by Canadian researchers, Salmon, Daly, and Wilson has confirmed that, in various ways, the middleborns feel less attached to the parents. For example, they are less likely than either first- or lastborns to nominate their mothers as their favorite family member.

Gender

Compared to the prominent role that gender plays in many developmental issues, it has made surprisingly little difference in birth-order studies. The intellect and personality profiles that emerge for females are comparable to those emerging for males.

Certainly, over the long history of birth-order research, a number of statistical interactions have been reported where gender was involved. When found, the results of a particular combination – say, firstborn males with secondborn females and thirdborn males – were not difficult to explain with a 'just-so story'. But the fact that such interaction effects rarely replicate suggests that the original findings were due to chance. With larger families, the number of possible combinations escalates quickly. With increasing parity in male vs. female achievement, any such interactions with birth order may eventually vanish. For these reasons and others, recent research has paid little attention to possible gender differences in birth-order effects.

A couple of recent findings constitute exceptions to this rule. An interesting finding reported by Sulloway was that secondborn boys often develop with firstborn personalities if the firstborn is a girl. Perhaps boys do not see firstborn girls as competitors and react only to their brothers. Or parents may still place more value on the firstborn male child.

The lack of difference in the size of birth-order effects has played a role in evaluating the Prenatal Hypomasculinization Theory. The credibility of that theory is weakened by the fact that its predicted larger effect size in males has not materialized in recent (tightly controlled) research.

Summary

The impact of birth order on social and intellectual development seems at once self-evident and empirically elusive. When found, the pattern is consistent: firstborns are the most intelligent, achieving, and conscientious, whereas laterborns are the most rebellious, liberal, and agreeable. In competition to explain these profiles are such diverse theories as Differentiatal Parental Feedback, Resource Dilution, Family Niche, and Prenatal Hypomasculinization.

The difficulty in confirming these birth-order differences is disconcerting. Although intially evident in most large-sample studies, the differences often disappear when key variables are controlled. The fact that significant reverse effects (e.g., firstborns less conscientious than laterborns) are rarely found, suggests that birth-order effects are at work, but that they are masked by certain research designs. Even statistical experts cannot seem to agree on how to tease apart birth-order effects from those of family size and SES.

The fact that birth-order differences are small to begin with makes them especially difficult to confirm. Indeed, all the contending theories predict small differences. In the case of IQ differences, for example, the expected firstborn vs. secondborn difference is only two IQ points. The effect sizes of birth order pale in comparison with sex differences, and most important, with temperament differences instilled by genetic and congenital factors.

As Jerome Kagan has pointed out, stereotypes about birth order are widespread and have a powerful intuitive appeal. But surely this wide appeal derives, at least in part, from some real commonality in human experience. Those of us with siblings have spent considerable time evaluating our relationships with them. The consensus within our families emerged long before we learned about birth-order stereotypes. The fact that most adults are eventually made aware of these stereotypes does not undo their validity. Even stereotypes can have a self-fulfilling effect as family members strive to live up to their expected roles.

At this point in the history of birth-order research, the informed reader must live with the fact that experts disagree and the continuing empirical debates are abstruse. Nonetheless, in this writer's opinion, the current weight of evidence favors the view that birth order does matter for both intellectual and social development.

See also: Attachment; Family Influences; Social and Emotional Development Theories; Temperament.

Suggested Readings

Bjerkedal T, Kristensen P, Skjeret GA, and Brevik JI (2007) Intelligence test scores and birth order among young Norwegian men (conscripts) analyzed within and between families. *Intelligence* 35: 503–514.

Healey MD and Ellis BJ (2007) Birth order, conscientiousness, and openness to experience: Tests of the family-niche model of personality using a within-family methodology. *Human Evolution and Behavior* 28: 55–59.

Jefferson T, Herbst JH, and McCrae RR (1998) Associations between birth order and personality traits: Evidence from self-reports and observer ratings. *Journal of Research in Personality* 32: 498–509.

Paulhus DL, Trapnell PD, and Chen D (1999) Effects of birth order on achievement and personality within families. *Psychological Science* 10: 482–488.

Rodgers JL, Cleveland HH, van den Oord E, and Rowe DC (2000) Resolving the debate over birth order, family size and intelligence. *American Psychologist* 55: 599–612.

Salmon CA and Daly M (1998) The impact of sex and birth order on familial sentiment: Middleborns are different. *Evolution and Human Behavior* 19: 299–312.

Schachter S (1963) Birth order, eminence and higher education. *American Sociological Review* 28: 757–768.

Sulloway FJ (1996) *Born to Rebel: Birth Order, Family Dynamics, and Creative Lives.* New York: Pantheon.

Zajonc RB (1976) Family configuration and intelligence. *Science* 192: 227–281.

Brain Development

D Fair and B L Schlaggar, Washington University School of Medicine, St. Louis, MO, USA

Glossary

Afferent – A neural projection carrying ascending information from the periphery to the nervous system or providing input from one brain region to another (e.g., thalamocortical afferent).

Architectonics – The arrangement of cells (e.g., cytoarchtectonics) or other attributes such as molecular markers (e.g., chemoarchitectonic) particularly in the cerebral cortex. A feature that contributes to the characterization of a neocortical area.

Broca aphasia – A syndrome characterized by nonfluent verbal language expression coupled with the sparing of language comprehension classically, but not necessarily, due to a lesion in Broca's area within the left inferior frontal lobe.

Cortico-cortical – Neuronal projections connecting regions within the cerebral cortex.

Efferent – A neural projection carrying information away from the nervous system to the periphery or carrying the output from one brain region to another.

Epigenetic – A factor that can change the activity of genes without changing their structure. Often used to refer to a factor that interacts with genetic factors to influence phenotypic expression.

Genetic – A factor or mechanism, particularly in development, that is largely the consequence of genes.

Glia – Non-neuronal cells of the nervous system that perform a variety of functions. These cells include: astrocytes, oligodendrocytes, radial glia, Schwann cells, satellite cells, and microglia.

Perinatal – The period at or around the time of birth.

Prenatal – The period occurring prior to birth.

Retinotopic organization – A term describing the topographic representation of the retinal surface in the regions of the brain devoted to processing visual information.

Somatotopic organization – A term describing the topographic representation of the body surface in regions of the brain devoted to processing somatic motor and sensory processing.

Thalamocortical – Ascending neuronal projections from the thalamus to the neocortex. An example of an afferent projection.

Topography – Referring to the general property in the nervous system that adjacent points on a sensory surface (or within a brain area) are represented in adjacent points in brain regions to which the sensory surface (or brain area) projects.

Voxel – A three-dimensional unit of volume used in magnetic resonance imaging (MRI).

Wernicke aphasia – A syndrome characterized by fluent but nonmeaningful verbal language expression coupled with a lack of language comprehension classically, but not necessarily, due to a lesion in Wernicke's area within the posterior and superior aspect of the left temporal lobe.

Introduction

Tracing its roots back to 3000 BC, the Egyptian Edwin Smith Surgical Papyrus is the earliest written report of brain function. From this document and others, we recognize that the Egyptians deemed the heart, not the brain, as the seat of intelligence and cognition. By the fourth century BC, Hippocrates and one of his greatest enthusiasts, Galen (second century AD), ignited a sharp transition concerning the origins of thought. It became clear that the early Egyptian beliefs, although championed by greats such as Aristotle, were mistaken. At the 'heart' of thought is the brain.

Cognition is a consequence of brain function. The new acquaintance whose name you just learned, the time of your meeting tomorrow, or your spouse's birthday (presuming you have not forgotten) are all encoded in the brain. Throughout one's lifetime, this type of information/experience will continually modify brain structure.

The first two decades of life in particular represent a period of extraordinary developmental change in sensory, motor, and cognitive abilities. One of the goals of cognitive neuroscience is to link the complex behavioral milestones that occur throughout this period with the intricate changes of the neural substrate. In our view, developmental neurobiological mechanisms are complex and the linkage to the equally elaborate emergence or refinement of cognitive skills is not straightforward.

Overview

This article offers an overview on selected topics of human brain development with an emphasis on the neocortex to provide a context for relating brain development to cognitive development. The article starts with a brief description of prenatal cortical organization and characterizes the formation of the neocortex by the time of birth. This first section chronicles the genesis of neocortical layers and the initially coarse construction of neocortical areas. The second section on postnatal brain development describes lessons learned from the development of visual cortex. This section highlights the importance of both genetic and epigenetic influences on development and their relation to behavior. Also emphasized are the concepts of critical and sensitive periods which, although viewed by many in the context of behavior, are a reflection of changes in neural circuits. The third section continues with a characterization of the progressive and regressive events that manifest in the postnatal period. Such phenomena have been increasingly used in cognitive developmental models, but often metaphorically. The fourth section focuses on developmental neuroplasticity, emphasizing that neuroplasticity is mostly advantageous, sometimes undesirable, and often insufficient. The article's final section is devoted to neuroimaging. The discussion accentuates how noninvasive neuroimaging is a promising and provocative tool for the study of cognitive development, but cautions against the tendency for overinterpretation.

Prenatal Cortical Organization

Genesis of Cortical Layers

The neocortical sheet is a laminar structure consisting of six layers (**Figure 1(a)**). This laminar organization is a consistent feature across the neocortex and is relatively conserved among most mammalian species. The mature six-layer structure consists of neurons generated from precursor cells located in the ventricular proliferative zone. These neurons reach their final destination, for the most part, by traversing along radial glial guides toward the cortical surface and organize in an inside-out pattern. Neurons born early form the deepest layers of the neocortex and those born later form the most superficial layers (**Figure 1(b)**). Despite the radial migration of most neurons, inhibitory (γ-aminobutyric acid (GABA) ergic) interneurons, which play an important role in neuroplasticity and critical period determination (see the section titled 'The development of ocular dominance columns' later), are first born in the ventral (subcortical) portion of the proliferative zone and migrate tangentially into the developing cortex (**Figure 1(c)**).

In humans, mature layer formation begins with an initial accumulation of neurons outside the ventricular zone called the preplate (**Figure 1(d)**). By approximately 8 weeks, gestation, the continuous migration of newborn neurons splits the preplate, forming three temporary layers. These are the marginal zone, cortical plate, and subplate (going from pia to ventricle) (**Figure 1(d)**).

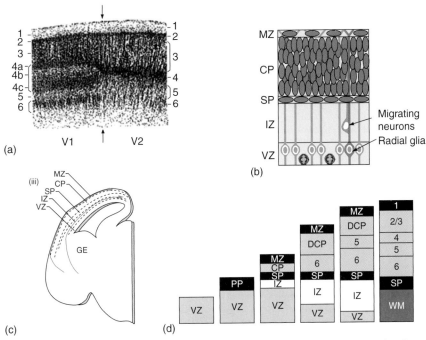

Figure 1 (a) The six layer organization of the neocortex. Provided is an image by Korbinian Brodmann of an 8-month-old human fetus. A sharp transition in the cytoarchitecture, most clearly identified in layer IV, is observed between V1 and V2 of the visual cortex. (b) The radial migration of neocortical neurons. Most neurons migrate radially along radial glial guides from the ventricular zone (VZ), through the intermediate zone (IZ), and aggregate in the cortical plate (CP). (c) The tangential migration of interneurons. Most neocortical interneurons (i.e., GABAergic interneurons) are generated in the ganglionic eminences (GE) and migrate tangentially through the IZ and marginal zone (MZ) to the neocortex. (d) Layer development in the neocortex. Neurons first generated in the VZ form the preplate (PP). The continuous migration of newborn neurons splits the preplate into the MZ, CP, and subplate (SP). In the developing cortical plate (DCP) the earliest born neurons form the deepest layers of the neocortex and those born later form the most superficial layers. (a) Adapted from Brodmann K (1909) *Vergleichende Lokalisationlehre Der Grosshirnrinde in Inren Prinzipien Dargestellt Auf Grund Des Zellenbaues*. Leipzig: J. A. Barth. (d) Adapted from O'Leary DD and Nakagawa Y (2002). Patterning centers, regulatory genes and extrinsic mechanisms controlling arealization of the neocortex. *Current Opinion in Neurobiology* 12(1): 14–25, with permission from Elsevier.

After subplate formation, there is a significant rise in the synaptic contacts within it. By approximately 22 weeks, gestation, the subplate is about 3–4 times thicker than the cortical plate. This growth of the subplate is accompanied by the initial (transient) projections to it from the thalamus, basal forebrain, and brainstem. As the cortical layers mature (continuing through the perinatal period), the transient synaptic contacts in the subplate are reorganized and form distinct contacts in the maturing lamina. Accompanying the loss of transient synapses is a progressive loss of subplate neurons.

The most drastic anomaly and subsequent cognitive deficits resulting from defects in neocortical neuronal migration is lissencephaly. Lissencephaly is a general term describing several disorders of migration that result in an abnormally thick, smooth, and agyric brain pattern. Affected individuals usually suffer from epilepsy and developmental delay/mental retardation.

Another anomaly, periventricular heterotopia (PH), refers to collections of neurons ectopically located (i.e., periventricularly) often under otherwise normal-appearing cortex. PH results in a less severe phenotype than lissencephaly, but is also thought to occur secondary to migration

defects. Affected patients usually present with seizures in late adolescence but with normal or near-normal levels of intelligence.

Developmental dyslexia is the most common and carefully studied learning disability affecting both children and adults. Highly heritable, dyslexia is characterized by reading difficulties in individuals with otherwise normal intelligence and access to education. Recently identified dyslexia susceptibility genes appear to have roles in neuronal migration suggesting that dyslexia could, at least in part, be due to a defect in this process.

Development and Differentiation of Functional Areas

The neocortex is composed of numerous morphologically and functionally distinct areas. The developmental differentiation of neocortical areas likely corresponds to certain aspects of maturing sensory, motor, and cognitive abilities. Therefore, understanding the developmental differentiation of neocortical areas is critical to understanding cognitive development. (In functional imaging, the term 'area' has often been used interchangeably with the term 'region', such that the actual meanings have often been lost. The term 'region' is a general term that describes a circumscribed portion of cortex (or subcortex) that may consist of an area, multiple areas, or portions of an area; e.g., 'the parietal region' or 'region of interest'.)

Defining a neocortical area

The first to champion neocortical localization of functions was Emanuel Swedenborg (1688–1772). He accurately reasoned that localization was the only way to explain how higher cognitive functions were not equally affected by diverse brain injuries. Franz Joseph Gall (1758–1828), the founder of phrenology, later claimed that the brain was divided into 27 separate 'organs', each of which corresponded to specific mental faculties. Although his work on phrenology was harshly criticized and proven incorrect, Gall's insistence that the brain was divided into specific areas with distinct functions influenced other prominent figures. Paul Broca's (1824–1880) landmark case of his aphasic patient Leborgne (famously referred to as 'Tan'), as well as work by other pioneers such as Carl Wernicke (1848–1905) and Joseph Jules Dejerine (1849–1917), spawned an era of investigation that was deeply rooted in cerebral functional localization.

After over a century of work, it has become evident that ascertaining a complete collection of neocortical areas is not straightforward. Historically, the greatest difficulty in identification stemmed from a failure to establish consensus criteria for defining an area.

Current consensus is that neocortical areas can be classified based on four properties: function, architectonics, connections, and topology (FACT) (**Figure 2(a)**). Each one of these properties has its strengths and weaknesses for revealing the underlying cortical organization and none in isolation can adequately depict it. (Obtaining all of these properties in humans represents an additional level of complexity.) For instance, single-unit recording or various neuroimaging strategies might potentially identify area borders based on function. Yet, the considerable similarity in the functional properties of many brain areas makes it difficult to differentiate them based solely on their function.

The oldest and most recognized method for identifying neocortical areas is architectonics. This point is highlighted by the distinguished rendition of areas introduced by Korbinian Brodmann (**Figure 2(b)**). Brodmann's version of the parcellation of the neocortex is but one of numerous versions that emerged in the last century albeit with the greatest staying power. Although differences in cytoarchitecture have played a prominent role in area identification, architectonic transitions are neither necessary nor sufficient for marking an area boundary. Results of architectonics can be ambiguous due to limitations in staining techniques, cortical folding patterns, or from internal area heterogeneity.

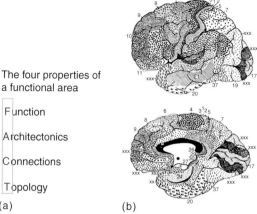

The four properties of a functional area

Function

Architectonics

Connections

Topology

(a)   (b)

Figure 2   (a) The four properties of a functional area. The mammalian cerebral cortex neocortex is composed of several morphologically and functionally distinct areas that are defined based on these four properties. (b) Cortical areas as defined by Korbinian Brodmann with architectonics. Further analyses have shown that some of these areas are also functionally and connectionally distinct, while other areas have been further differentiated with connections, function, and topography. Although Brodmann's areas are the most recognized, there are several other renditions of cortical areas using architectonics. Adapted from Brodmann K (1909) *Vergleichende Lokalisationlehre Der Grosshirnrinde in Inren Prinzipien Dargestellt Auf Grund Des Zellenbaues*. Leipzig: J. A. Barth.

Defining areas with patterned connections has also been used extensively. The deduction is that the connection boundaries of an area that receive input from another, likely define that area; however, as with architectonics, inter-area heterogeneity, as well as, individual subject connection variability, and 'noise' associated with tracer injections makes it difficult to mark area boundaries with cortical connections consistently.

Topography has also been a promising method for identifying area boundaries. Retinotopic mapping, for example, has further delineated some areas of the visual system originally defined by Brodmann with architectonics (**Figure 2(b)**). However, while somatotopic and retinotopic maps in early sensory and visual areas make topographic boundaries straightforward, topography degrades higher in the cortical hierarchy making area boundaries less discernible. As pointed out by Jonathan Horton, "What constitutes topography in regions concerned with language, motivation, or personality?"

When available, all or a combination of these sources of information that is, FACT should be used to provide a confident definition of a functional area.

Area differentiation: Protomap vs. protocortex

The extent to which genetic vs. epigenetic influences shape phenotype has been an intense discussion for over a 100 years. Throughout the 1990s, the controversy encompassed two predominant theories of neocortical area differentiation: the protomap and protocortex hypotheses. The protomap hypothesis suggested that cortical areas are prespecified and that the progenitors in the ventricular proliferative zone code area assignments as neurons are born. In contrast, the protocortex hypothesis suggested that the nascent cortical sheet is relatively homogeneous, and that neocortical areas are specified, in large part, by ascending afferent information. The controversy, now resolved (both theories are in large part correct), has led to a better understanding of how both epigenetic and genetic determinants drive the differentiation of area-specific features and areal boundaries.

Function

The specificity of areal function in the prenatal brain is not well known. This statement is not intended to suggest that neocortical areas are not functioning prior to birth. The neocortex of a fetus or premature infant is capable of receiving and responding to afferent activity. Prenatal behavior is not limited to subcortical functions. Nonetheless, using function to delineate neocortical areas in humans during the prenatal period is presently not feasible.

What can be said is that function of a particular area is malleable. In a remarkable series of experiments, Mriganka Sur and colleagues have shown that in ferrets, visual afferents can be surgically redirected to innervate presumptive auditory targets, such that they will produce physiological response to visual stimuli in auditory cortex. These rewired ferrets perceive visual cues as visual even though what is normally auditory cortex is being activated by the stimuli. These experiments and others show that, early in development, the specific function of a cortical area is malleable, can be driven by experience, and does arrive from a relatively multipotent phenotype.

Architectonics

In normal circumstances, however, by the time humans are born, an initial broad parcellation of the neocortical architecture has taken place. The primary sensory areas, including primary somatosensory, visual, and auditory cortex, can be identified by the presence of a prominent granular layer IV. In the frontal lobe, motor cortex can be differentiated by the presence of large pyramidal neurons in layer V called Betz cells and the absence of a prominent granular layer IV.

There are several lines of evidence that suggest that at least the early architectonic fingerprint is regulated by determinants inherent to the proliferative zone and independent of afferent (i.e., thalamocortical) or other epigenetic influences. However, this initial parcellation can be altered. Brad Schlaggar and Dennis O'Leary have shown that, embryonic visual cortex when transplanted to parietal somatosensory cortex in newborn rats can develop architectural and connectional features

typically unique to somatosensory cortex. Other experiments have demonstrated that thalamic input can modify gene expression such that the zone of expression will eventually match that of the thalamic innervation. Although the basic phenotype of an area can occur mostly under molecular control prior to thalamic influence, the degree and range of the phenotype is affected by afferent input.

Connections

The innervation of the neocortex by thalamic afferents constitutes the first step in the creation of processing circuits. Initial area-specific, thalamocortical projections appear to be primarily controlled by guidance molecules, but activity-dependent mechanisms that refine these projections are superimposed on this process.

Early cortical projections have widespread distributions, which are refined during maturation. The restricted adult projection patterns emerge through the elimination of functionally inappropriate axons and axon branches. This process is influenced by neural activity and culminates in the characteristic patterning of callosal, intracortical, and subcortical projections.

Topography

The general topographic organization of the cortex appears to be present very early in development. Adjacent points in the thalamic nuclei project to adjacent points in the corresponding area. This arrangement exists prior to initial thalamic contacts and is likely regulated initially by guidance molecules. As with the cortical connections described above, area specific topography will continue to be tuned over age.

Postnatal Brain Development

The most notable global developmental difference between humans and other primates is the protracted period of time between birth and maturity. This prolonged developmental span is believed to allow for an extended period of juvenile learning, but is also an extension for substantial brain growth, circuit organization, and myelin formation. Very little information is available detailing these events in the human or even nonhuman primate, which makes it quite difficult to adequately relate brain development with cognitive development. Despite this caveat, the following section summarizes some basic developmental events that occur in the postnatal period in hopes of providing the framework for making this link.

Particular emphasis is placed on experience-dependent and experience-independent developmental processes. As alluded to previously, distinguishing between processes dependent on experience and those independent of experience is rather complex. The terms 'experience-dependent' and 'experience-independent' are used very loosely in the literature despite the absence of a strict dichotomy. Even within each term, further semantic partitioning is often required to clarify discussions. William Greenough has proposed that 'experience-dependent' brain changes correspond to those arising from environmental influences specific to the individual. For example, synaptic changes corresponding to the events learned in a history class would be considered 'experience-dependent'. Others have used the term 'learning' to describe such changes.

In contrast, Greenough suggests that the effect of environmental influences on brain structure common to all members of a species should be considered 'experience-expectant'. The suprachiasmatic nucleus (SCN), our internal clock, provides a helpful example of an 'experience-expectant' neural system. The SCN 'free runs' at a greater than 24 h period, such that without the sun, our sleep–wake cycle (or day) would be longer than 24 h. The SCN is, however, 'experience-expectant' in that it 'expects' the patterned light of the sun to rise and fall on a 24 h period. The SCN matches its internal cycle to the 24 h time period such that our sleep–wake cycle matches the cycles of a day. Other terms such as 'primal' or 'maturation' have been used to describe this type of change.

In the context of cognitive development, human language acquisition *per se* can be considered experience-expectant, which of the multitude of languages a child learns is experience dependent.

Of note, that the term 'experience' is not synonymous with 'activity'. Processes independent of experience can also be driven by neural activity. For instance, spontaneous waves of activity that exist in the retina prior to eye opening may influence the formation of ocular dominance columns (see section titled 'The development of ocular dominance columns' below). Thus, in the following discussions, processes related to experience and those which are strictly 'activity-dependent' are distinct.

The term molecular in the discussion below is used to signify developmental changes that occur independent of neural activity. They occur in the context of the internal environment and are independent of outside stimuli or spontaneous activity. Despite the pragmatic definition used here, it should be appreciated that molecular cues and genes can be affected by activity-driven processes. It should also be noted that drawing strict borders between these types of processes is very difficult, and as our knowledge of brain development accumulates, certain brain properties might be reassigned to different developmental categories.

Lessons from Visual Cortex

Studies of the primary visual cortex have been fundamental to understanding the mechanisms that regulate the construction of brain circuitry. The next section focuses on the development of ocular dominance columns. The formation of these circuits, guided by both a variety of molecular cues and patterns of neural activity, inform us about principal developmental concepts.

The development of ocular dominance columns

The initial description of ocular dominance columns (ODCs) and their formation was Nobel Prize winning work done by David Hubel and Torsten Wiesel in the 1960s. By using electrophysiologic recordings in the cat primary visual cortex, they were able to show that information originating from either the left or the right eye is differentially represented into columns of cortical layer IV. Since these initial accounts, ocular representation in the primary visual cortex has been one of the most thoroughly studied brain systems (**Figure 3**).

Currently, two accounts of ODC development exist. One account proposes that early developmental projections from the two eyes initially overlap in the visual cortex. During the first postnatal months, visual experience promotes a preferential stabilization of some of the cortical connections and elimination of others resulting in a segregation of layer IV projections into well-delineated columns (**Figure 3(b)**). It has now been shown that retinal inputs are not required for column formation. Activity in the form of spontaneous correlated burst of activity from the thalamus can account for their formation. This process is likely to be the result of a Hebbian-type mechanism whereby connections are strengthened by correlated activity and weakened by uncorrelated activity.

The second account suggests that 'activity-independent' mechanisms are responsible for matching appropriate targets from the lateral geniculate nucleus (LGN) to layer IV of the visual cortex. According to account, molecular cues guide axons to appropriate targets with fairly precise branching. These initial projections result in highly segregated columns with limited production of exuberant connections.

Due to evidence on both sides, it is likely that both accounts influence column formation. Molecular cues underlie the initial coarse formation while activity dependent mechanisms are required to refine and maintain them. It is also likely that the relative influence of either model is species-dependent.

ODC formation has been the leading model for studying 'sensitive' and 'critical' periods. Historically these terms have been used interchangeably, while more recently, distinct definitions have emerged. A sensitive period is the time-window during which an experimental or natural manipulation of a system will strongly affect the system's development. A critical period is the point at which the effects of the manipulation are irreversible.

These concepts are often viewed from the perspective of behavior, but work in the visual system has been crucial to our understanding of these principles. Particularly

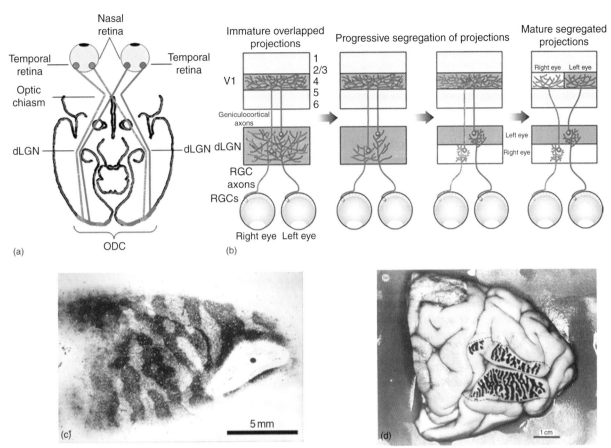

Figure 3 (a) In humans and in other mammals, retinal ganglion cells from the temporal retina of one eye and the nasal retina of the other eye encode and project the visual information from half the visual world to the same hemisphere. Ganglion cells from the temporal retina project to the ipsilateral hemisphere whereas those arising from the nasal retina cross the optic chiasm to project to the contralateral hemisphere. These projections to each hemisphere (and the lateral geniculate nucleus (LGN)) maintain their eye-specific segregation by terminating on discrete portion of the neocortex. This is the anatomical basis for ocular dominance columns (ODC). (b) There are now two primary accounts of how ODC develop. The account described, here asserts that the eye-specific patterned connections results from small-scale axon elimination. Immature exuberant projections from the two monocular inputs initially overlap in both the dorsal lateral geniculate nucleus (dLGN) and primary visual cortex (V1: layer 4c). A competitive process driven by correlated activity drives eye-specific stabilization of some of the cortical connections and elimination of others resulting in a segregation of projections in the dLGN and later layer IV projections in V1. (c) An example of ODC (identified with cytochrome oxidase) in a human section through layer 4c. (d) Projected pattern of ODC along the medial surface of the occipital lobe (outlined is the V2/V1 border). The example in (c) is located in the lower right portion of this illustration. (b) Adapted from Luo L and O'Leary DD (2005) Axon retraction and degeneration in development and disease. *Annual Review of Neuroscience* 28: 127–156 (c) Adapted from Horton JC and Hedley-Whyte ET (1984) Mapping of cytochrome oxidase patches and ocular dominance columns in human visual cortex. *Philosophical Transactions of The Royal Society of London, Series B, Biological Sciences* 304(1119): 255–272, figure 10b, with permission from The Royal Society of London.

relevant are experiments that elucidate the effect of monocular deprivation on ODC formation. Chronic closure of one eyelid, early in life, reveals plastic mechanisms in which the size of ODC representation corresponding to the sutured eye is significantly reduced. Conversely, the layer IV representation of the unaffected eye increases its size. In nonhuman primates, the sensitive period for this process begins at birth and slowly wanes until ∼10 weeks of age.

Interestingly, sensitive and critical periods seem to be linked to inhibitory inputs. In ferrets, the maturation of inhibitory GABA circuits lags behind excitatory

circuits. The timing of this maturation correlates with the critical period. Reducing GABA function in transgenic mice can keep the critical and sensitive periods open indefinitely. Increasing GABA-mediated inhibition can prematurely shorten the span of the sensitive period.

Although often overlooked, sensitive and critical periods are a reflection of the underlying neural circuits. Demonstrating the relationship between brain anatomy and complex behavior is difficult, yet there is nevertheless a direct link between them. For example, the eyes and consequently the retina of each eye are slightly displaced such that each views the visual world from a slightly different perspective. As a result, objects represented in the retina are offset in respect to the fovea. This disparity is encoded by cells in the visual cortex and helps with our very precise ability to perceive depth, based on binocular cues. The development of stereopsis occurs after birth, during the same period as ocular dominance column formation. The rapid development of an infant's ability to perform tasks associated with binocular vision, including stereoacuity, is time locked to the end of the development of ODC, and is blocked with manipulations that prevent normal ODC formation. Thus, ODC formation is integral to the development of stereopsis.

Research on critical and sensitive periods in ODC has informed the treatment of pediatric disorders such as strabismus and congenital cataracts. Furthermore, such research has also provided perspective when considering critical and sensitive periods in relation to social, cognitive, and perceptual abilities. But caution should be taken to not overinterpret critical period data to guide specific cognitive or educational interventions without appropriate evidence to support the interventions.

Progressive and Regressive Neural Events

The development of the mammalian nervous system encompasses a wide variety of both progressive and regressive neural phenomena. The following focus is on those events most commonly referenced in cognitive neuroscience and psychology, but by no means do these represent the entire collection of progressive and regressive observations throughout maturation. We emphasize in this section that although phenomena such as myelination and synaptic pruning fit in nicely to many current theories, several questions arise regarding specifics. We recognize that cognitive neuroscientists and psychologists are motivated to integrate such phenomena into developmental models; however, the details should not be overlooked.

Broad progressive events
General growth
Humans exhibit substantial brain growth between birth and adulthood. Adult brains are approximately four times larger than infant brains. This brain growth is not linear. Maximal growth rate occurs around birth and by 6 years of age, the brain is approximately 95% of the size of the adult brain. The bulk of this early growth comes from a variety of sources including increases in synapses and dendrites (neuropil), as well as myelination. Importantly, the production of new neurons does not contribute to general brain growth (see the following).

Neurogenesis
With the exception of a few brain regions, it has become fairly well accepted that most of the neurons we are ever going to possess have migrated and are in place by the time we are born. This idea was first proposed by the founder of the neuron doctrine, Santiago Ramon y Cajal, who stated that neurogenesis is an exclusively prenatal event. While primarily true, there are three systems that continue neuron production into the postnatal period. These are the olfactory bulb, the dentate gyrus of the hippocampus, and the cerebellum.

Despite 40 years since the initial discovery of adult neurogenesis, little progress has been made as to the function in mammals. Albeit controversial, some work suggests that adult neurogenesis affects behavior. For example, in rodents exercise-induced proliferation of neurogenesis in the hippocampus can lead to increased performance in spatial learning. Conversely, pharmacologically induced decreases in neurogenesis can negatively affect behavior. Similar types of findings have been observed in the olfactory stystem.

Some insights may be gained from pathologic conditions. Brain injury can lead to abrupt modifications. General injuries such as ischemic infarction will lead to increased neurogenesis. This phenomenon can occur not only in normally neurogenic regions (i.e., hippocampus), but also at dormant proliferative sites. Sometimes these newborn neurons deviate from their normal trajectory and migrate to the injured area implicating their role in recovery.

Myelination
Although glial cells are not highly referenced in this article, their importance cannot be understated. Glia, the most abundant cells in the nervous system, account for approximately 90% of cells in the brain. Historically, their function has been limited to a structural neuronal support role. Even the name 'glia', being Greek for 'glue', downplays the importance of these cells for brain development and overall function. Among the many functions of glia (e.g., synaptic formation, radial glia guides, neurotransmitter uptake, blood–brain barrier), one of the most critical is the development of myelin sheath.

Myelination is the most commonly referenced neuroanatomical additive event that occurs postnatally. Increased myelination proceeds from sensory to motor, and last in association areas, roughly following the hierarchical organization introduced by Felleman and Van Essen in 1991. The glial cells responsible for myelination in the central nervous system (CNS) are oligodendrocytes. Like the majority of cortical neurons, oligodendrocytes arise and migrate to position prior to birth from a region in the ventral telencephalon.

White matter maturation continues at least through the second decade of life. Because of this, myelination has often been used to explain the development of several cognitive abilities. However, it is unlikely that increases in myelination can account for all of cognitive development. As Joaquin Fuster has pointed out, myelination is not a prerequisite for functional axons. Unmyelinated connections are still capable of transmitting information. Nonetheless, there remains a clear and well-established link between myelination and the functional properties of an axon. Increased efficiency of signal propagation following the addition of the myelin sheath may be important for enhancing the integration between already functioning regions.

Although oligodendrocytes migrate into their positions by the time of birth, their slow, postnatal maturation makes them particularly vulnerable to injury. Periventricular leukomalacia (PVL) is an injury of the periventricular white matter that results in cerebral palsy. It is the leading cause of brain injury in the preterm infant. The most predominant feature of PVL is the disturbance in myelination likely secondary to ischemic and inflammatory damage to immature oligodentrocytes.

With the rapid advancement of magnetic resonance imaging (MRI), particularly diffusion tensor imaging (DTI), researchers are now uncovering several other developmental disorders that may partially be secondary to abnormal white matter integrity. These disorders include schizophrenia, autism, dyslexia, and developmental stuttering.

Synaptogenesis and intracortical connections

From approximately 30 weeks, gestation, through the first 2 postnatal years, there is substantial growth in synaptic contacts throughout the cortex. The relative region-wise timing of this synaptic exuberance continues to be debated. Pasko Rakic and colleagues suggest that synaptic numbers increase uniformly throughout the cortex during this period, whereas Peter Huttenlocher and colleagues argue that synaptic numbers follow the myelination pattern described above, such that primary sensory areas increase prior to association areas. Several factors may underlie the discrepancies. Foremost are the species differences examined. Peter Huttenlocher investigated human postmortem specimens, while Pasko Rakic examined nonhuman primates. Differences in sample size and experimental techniques may also have contributed to the discrepancies.

Synapse formation is always accompanied by synaptic loss. During the period of rapid synaptogenesis, more synapses are generated than eliminated, resulting in a net gain. At the end of this period, synapse numbers plateau. The duration of this plateau is not well known in humans, but what follows (see the section titled 'Broad regressive events') is a regressive phase with a net loss of synapses that continues through adolescence. The accumulative synapse remodeling during development results in peak numbers in childhood being approximately 140–150% of those in adulthood.

By approximately 9 months of age, the development of short- and long-range (axonal) connections between brain regions is thought to be complete. In the visual system, these initial connections reflect the hierarchical organization observed in the adult brain.

Considering the various types of synapses (excitatory, inhibitory) and connections (i.e., intracortical, thalamo-cortical, cortico-cortical – including feedforward and feedback, callosal), the gross measure of synapse number *per se* may be misleading if, for example, different types of connections have differing trajectories. This issue and others will be examined further below.

The most familiar disease linked to abnormal synapse formation is fragile X syndrome. Fragile X syndrome is the most common inherited cause of mental retardation. Although there are several general developmental defects in individuals with fragile X syndrome, the abnormal morphology of dendritic spines points to a strong relationship between synapse formation and the cognitive deficits seen in those afflicted with the mutation. The partial overlap in neurobehavioral symptoms and structural abnormalities has led some to suggest a similar pathology for those affected with autism.

Broad Regressive Events
Axon retraction/synapse elimination
Several complementary techniques have revealed that the elimination of axons and synapses involves at least two different mechanisms – micro or small-scale (local retraction of axon branches) and macro or large-scale (degeneration of axons over long distances) modifications. The segregation of visual input into eye-specific patterns in the visual cortex is a classic example of a micro-change (**Figure 3(b)**). In contrast, transient inter-hemispheric connections in primary visual cortex observed during infancy are an example of a macro-change.

Dennis O'Leary and colleagues have shown in rodents that cortical layer V neurons, after early extensive interstitial branching, acquire functional appropriate connections through selective collateral elimination dictated by the cortical area in which the neuron is located (macro-change; **Figure 4**). In newborns, axons projecting from both motor and visual areas share various brainstem and spinal cord targets. Over development, neurons in motor areas eventually lose their collateral branches to visually related targets, but retain the branches to those involved in motor control. In contrast, neurons in visual areas lose all collateral projections to the spinal cord and other targets involved in motor control, and retain projections to targets involved in vision. Interestingly, this area-specific, axon elimination and complementary maintenance is not perturbed by heterotopic transplantation of visual cortex into presumptive primary motor cortex or vice versa (**Figure 4**).

Certain aspects of this process should be considered with regard to the development of cognition. Is axon elimination uniform across all layers of the cortex? Are there different developmental trajectories for feedback and feedforward

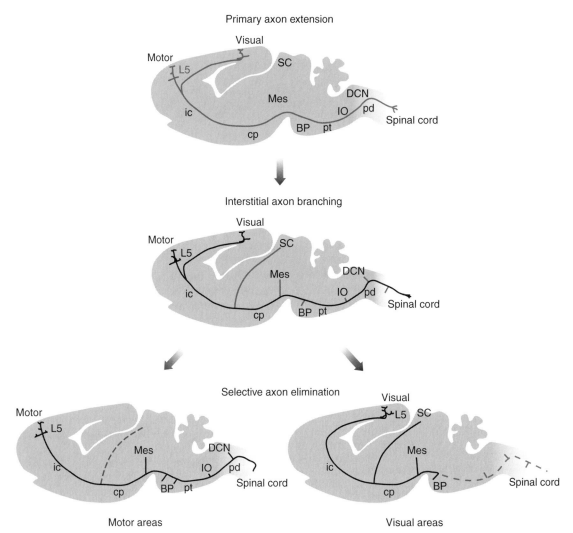

Figure 4 Provided is an example of large-scale axon elimination to develop area-specific subcortical projections of neocortical layer 5 neurons. Immature projections exhibit extensive subcortical interstitial branching, such that motor and visual neurons in newborns project to common targets in the brainstem and spinal cord. Through maturity functional appropriate connections are acquired through selective axon elimination, dictated by the cortical area in which the neuron is located. Motor area neurons eventually lose their collateral branches to visually related targets, but retain the branches to those involved in motor control. In contrast, the visual area neurons lose all collateral projections to the spinal cord and other motor-related targets and retain projections to those involved with vision. BP, basilar pons; cp, cerebral peduncle; DCN, dorsal column nuclei; ic, internal capsule; IO, inferior olive; Mes, mesencephalon; pd, pyramidal decussation; pt, pyramidal tract; SC, superior colliculus. Adapted from Luo L and O'Leary DD (2005) Axon retraction and degeneration in development and disease. *Annual Review of Neuroscience* 28: 127–156.

connections or inhibitory and excitatory synapses? How about for cortico-cortical, thalamo-cortical, or calossal connections? The limited amount of data available prevents a satisfactory characterization of these questions in animal models, let alone humans. However, existing evidence suggests that the developmental trajectory for both synapses and connections depends, in part, on their classification.

Some cortico-cortical connections have developmental trajectories that differ from those of the callosal and subcortical connections examined by O'Leary and colleagues. Although not a strict criterion, in the adult neocortex, feedforward projections generally arise from supragranullar layers (II, III), whereas feedback

projections generally arise from the infragranullar layers (V, VI). In primates, the hierarchical nature of these projections appears to be in place early in development, but the laminar organization differs from the adult. In developing primates, a large proportion of feedback connections in early visual areas arise from the supragranular layers that are absent in the adult. Over development, a prolonged remodeling of these projections occurs. Massive pruning of early formed supragranular connections ensues with parallel production of projections from infragranular layers. In contrast, feedforward pathways appear to be largely established during early development, and rely much less on large-scale elimination of inappropriate connections.

There is also evidence suggesting that inhibitory and excitatory connections are remodeled on different time-scales. Some investigators have shown a developmental lag in remodeling of inhibitory compared to excitatory connections. The remodeling of inhibitory connections becomes more focused from a more exuberant early pattern and is shaped by experience.

Cell death

Synaptic elimination and axonal retraction only partially account for the elimination of transient connections. Programmed cell death (or apoptosis) is also a prominent developmental mechanism that leads to a net loss of synapses. One example of apoptosis occurs in the developmentally transient subplate. Peak rates of apoptosis occur during the prenatal period but continue postnatally. Apoptosis is thought to be important for matching post-synaptic and presynatpic cell numbers.

Developmental Neuroplasticity

The term neuroplasticity is used in a variety of contexts but in general, refers to the brain's ability to organize itself in a novel way in response to perturbation. Perturbations can come in the form of natural experience (e.g., extended piano training), sensory loss (e.g., blindness), or intrinsic conditions (e.g., infarction).

In previous sections we briefly discussed the brain's ability to respond to significant perturbations in the developing visual cortex. The rewiring and transplantation experiments outlined above are classic examples of the brain's plastic capabilities. These types of plastic responses to extreme perturbations are available only in the earliest stages of development.

Significantly altering sensory experience can also lead to striking brain alterations, even in the adult. Michael Merzenich and colleagues showed in the adult owl monkey that correlating the inputs of two adjacent fingers by surgically connecting the skin surfaces leads to a significant change in the hand topography of the somatosensory cortex. The correlated inputs arising from the joined fingers abolished the normal discontinuities in digit representation.

Naturally occurring perturbations also provide insights into the remarkable capacity of the brain to organize itself in various ways. For instance, several groups have shown that the visual cortex of congenitally blind patients responds to tactile sensations associated with Braille reading.

Perinatal Stroke

Perinatal stroke is a particularly salient model which highlights the brain's capacity for novel organization after injury. Children with perinatal stroke are capable of acquiring relatively normal cognitive function, such as language, after experiencing a cortical insult that in adults would often lead to devastating lifetime disabilities. The fact that persistent language deficits frequently do not follow perinatal stroke even when extensive left hemisphere injury has occurred, underscores the brain's robust adaptive capabilities.

Although this developmental plasticity has long been recognized (e.g., Sigmund Freud described this clinical phenomenon in an excellent monograph written early in his career), its underlying neurobiological mechanisms are not well characterized. The most commonly referenced mechanism supporting normal language development after perinatal stroke is a functional reorganization to the contralateral hemisphere. However, other reports suggest, at times, intra-hemispheric reorganization supports the development of normal language after perinatal stroke.

A third perspective which encompasses both of these findings has recently surfaced. This model revolves around a developmental theory of typical development advanced by Mark Johnson called interactive specialization. Interactive specialization takes into account two aspects of typical development: (1) that regressive processes are not random, but selective, reduce connectivity between specific regions, span a protracted period of time, and are in large part activity-dependent, and (2) that the response properties of any given brain region are dependent on the brain's hierarchical organization. The functional specialization of regions is mediated by functional integration among feedforward, feedback, and lateral signals. There is ample anatomical support for the proposal that the brain is organized such that 'higher' cortical processes influence 'lower' cortical areas through reciprocal hierarchical networks. For example, both single-unit recordings and functional MRI (fMRI) studies have shown that attention to specific visual attributes can alter receptive fields and stimulus-driven activations. It is likely that for many regions of the brain, bottom-up, top-down, and lateral processes interact dynamically, continuously recalibrating neuronal responses to behaviorally relevant stimuli.

Johnson and colleagues suggest that these types of interactions between brain regions over development lead to the specialization of the adult brain. At birth, cortical regions and pathways are biased (due to intrinsic connectivity and architectural features) in their information processing properties, yet they are highly connected, and much less selective than in the adult (i.e., 'broadly tuned'). Johnson predicts that shortly after birth, multiple regions and pathways within the brain will be partially active during specific task conditions. But as these pathways interact and compete with each other throughout development, changes in the regional response properties will likely be observed and particular pathways will eventually dominate for specific task demands.

The interaction-driven theory of normal development described here would anticipate, as pointed out by Pamela Moses and Joan Stiles, that normal developmental mechanisms will interact with early onset pathology to yield an atypical adult functional neuroanatomy. If the balance of regional competition is altered by removal of one or more of the competing regions, as in the case of perinatal stroke, an alternate organization and developmental time course would ensue, that would largely depend on the timing, location, and the size of the stroke.

Several reports have bolstered this view. At the forefront, is work conducted by Maree Webster, Leslie Ungerleider, and Jocelyne Bachevalier. They have shown that in normally behaving adult macaques, visual recognition memory requires the interaction of inferior temporal lobe (area TE) and the medial temporal lobe (entorhinal and perirhinal areas). While inferior temporal cortex connectivity overlaps considerably between adults and infants, substantial elimination and refinement of initially widespread projections occur during development. For example, while area TE projects to the medial temporal lobe in infants and adults, an adjacent region, area TEO, projects to the medial lobe structures only in infant monkeys – suggesting elimination of transient projections. As opposed to lesions of TE in the adult animal, TE lesions in infant monkeys results in the sparing of visual recognition memory. This ability, it appears, is partially subserved by the maintenance of the normally transient TEO projections.

Adverse Effects of Neuroplasticity?

Although mostly advantageous to the developing organism, plasticity is not always beneficial and often insufficient. Residual motor function following a perinatal stroke often remains impaired into adulthood. Deficits of spatial cognition following early brain injury, albeit, more mild in children, also persist through childhood. More pronounced and multimodal deficits are associated with a diverse group of neurological disorders including fragile X syndrome. The plasticity mechanisms discussed here are not potent enough to overcome these types of perturbations.

In other conditions, extensive neuroplasticity can negatively affect function. Plasticity associated with repeated movements (such as in musicians) can lead to focal hand dystonia characterized by disabling co-contraction of agonist and antagonist muscles. The pain associated with phantom limb syndrome is also believed to be associated with cortical reorganization following amputation. It has also been suggested that in some instances altered brain activity in adults experiencing stroke can actually impair performance.

Neuroimaging Techniques for the Study of Human Development

Imaging Caveats

While most of our knowledge of postnatal human brain development comes from animal models with an extrapolation to human ontogeny, the development of noninvasive neuroimaging techniques now allows for detailed assessments of human brain maturation. MRI, in particular, has provided intriguing insights into human developmental processes both structurally and functionally.

The difficulties of relating cognition to changes in brain anatomy and physiology cannot be overstated. Although captivating, the current use of information from MRI in educational, judicial, and other social domains is often a grand misrepresentation of our current understanding of how the underlying neurobiology relates to behavior. New imaging techniques are undoubtedly valuable, but it needs to be realized that the growth of these techniques is still in the nascent stages.

One particular challenge with structural MRI involves resolution. Every MRI image is a collection of voxels (usually on the millimeter scale), any one of which consists of a mixture of neurons (axons, dendrites, cell bodies), glia (inlcuding myelin), and blood vessels. When studying gray matter development, the inability to identify all the components of a voxel makes it difficult to determine how these properties relate to maturation. For example, in some instances, increased myelination could potentially be misinterpreted as gray matter loss.

Studying the development of task induced activations with fMRI has its own conceptual and methodological concerns that require consideration. These issues include: (1) making direct statistical comparisons, (2) choosing adequate comparison tasks, (3) using a common stereotactic space, (4) assessing and accounting for performance between children and adults, and (5) using appropriate statistical measures to arrive at developmental differences and trajectories throughout development.

Irrespective of these caveats, noninvasive neuroimaging techniques performed on humans have already allowed us to explore uncharted territory in development and disease, and will continue to be a key instrument for the study of cognition.

Structural Magnetic Resonance Imaging

The current descriptions of white and gray matter development with MRI mostly agree with results from earlier histological work. The most consistent finding in white matter maturity is the linear protracted development which advances into young adulthood. In contrast, gray matter development consists of mostly nonlinear changes.

Studies differ on the details, but in general, there appears to be a differential peak in gray mature volume (or density) between childhood and early adolescence that begins to decline first in sensorimotor areas and later in dorsolateral, prefrontal, parietal, and temporal regions. This general description of white and gray matter development is only a partial account of a markedly complex process.

Several research groups have successfully linked these types of structural changes over age with behavior by correlating the MRI findings with performance on specific tasks. For instance, Elizabeth Sowell and colleagues have shown that cortical thinning of the dorsal frontal and parietal lobes negatively correlates with the verbal portion of the Weschler's intelligence scale and that left hemispheric thinning negatively correlates with verbal intelligence quodient (IQ).

Functional Magnetic Resonance Imaging

fMRI allows one to display changes in task-driven activation patterns over development. It can be used in conjunction with structural MRI to study the relation of brain anatomy to brain physiology. Recent evidence suggests that the general rules of structural maturation (sensory/motor maturation followed by association areas) do not necessarily hold true for 'functional' maturation.

Association areas are not idle during specific tasks. They are actively involved in information processing despite their structural immaturity. For example, Tim Brown, Steve Petersen, and Brad Schlaggar have documented multiple developmental changes in lexical processing regions. Lower-order processing regions, particularly in the right occipital and temporal cortex, began in children as positively activated and progressed to either nonresponsive or minimally responsive in adults. Regions in the inferior frontal gyrus and fusiform gyrus, which have been historically implicated in several aspects of language processing, were equally activated in children and adults. Similar types of findings have been noted elsewhere, suggesting that many regions, despite being structurally immature, can exhibit adult-like activation patterns for specific tasks.

This observation prompts several questions. Are 'functionally' mature but anatomically immature (or regions with mature structure while functionally immature) regions exercising the same operations? How do such regions relate to behavior?

The interactive-driven view of regional specialization discussed earlier suggests that the function of any given region is mediated by the activity-dependent interactions with other regions. Perhaps then (in some cases), the role of brain regions that are functionally mature, but structurally immature (i.e., myelination, pruning, etc.), are distinct between children and adults because the regions differ in their relations to other brain regions.

Work by Donna Thal and Elizabeth Bates may address this issue. They have shown that lesions in the cerebral hemispheres in children manifest quite differently than comparable lesions in the adult brain. Thal and Bates found that the first stages of language development in children after perinatal stroke are affected by lesion location. However, the types of developmental delays are not as expected based on the typical sequela of the same focal lesions in adults. For example, early delays with word comprehension were more common with right temporal lesions, rather than left temporal (Wernicke's area) lesions as commonly observed in adult populations. Wernicke-type lesions in children resulted in expressive delays, a symptom more common in a left inferior frontal gyrus (Broca's area) lesion in adult patients.

These results suggest that the role of a given region is not necessarily matched to that of the adult. Perhaps the increased efficiency of signal propagation following the addition of the myelin sheath is important for enhancing the functional integration between regions in the cortex. This integration may result in a different network structure and hence different roles for any particular region.

Review

This article reviewed selected topics of human brain development with a focus on the neocortex. The purpose of this article was to provide a context for relating brain development to cognitive development. The prolonged developmental span in humans is important for a variety of developmental processes including brain growth, circuit organization, synaptic pruning, and myelin formation. While cognitive neuroscientists and psychologists are obliged to integrate these phenomena into developmental models and theories, there should always be an appreciation of the details. Along the same lines, the promising tools of neuroimaging have the capacity to fill many of the information gaps that currently exist in human development, but caution should be practised not to overextend interpretations.

In this article we also highlight the importance of both genetic and epigenetic phenomena in the development of neural systems. The behavioral phenomena we observe throughout development (such as critical and sensitive periods) are a reflection of changes in the underlying neural circuits. These differential aspects of brain development and the caveats that accompany them should provide a solid basis for building associations between brain and cognitive development.

Acknowledgments

The authors would like to thank our financial supporters. These include: The Washington University Chancellor's Fellowship (D.A.F.), UNCF*Merck Graduate Science Research Dissertation Fellowship (D.A.F), NIH NSADA (B.L.S.), The McDonnell Center for Higher Brain Function (B.L.S.), and The Charles A. Dana Foundation (B.L.S.).

See also: Brain Function; Cerebral Palsy; Cognitive Development; Critical Periods; Hippocampus; Milestones: Cognitive; Neurological Development.

Suggested Readings

Allendoerfer KL and Shatz CJ (1994) The sub-plate, a transient neocortical structure: Its role in the development of connections between thalamus and cortex. *Annual Review of Neuroscience* 17: 185–218.

Bates E (1999) Plasticity, localization, and language development. In: Broman SH and Fletcher JM (eds.) *The Changing Nervous System*, pp. 214–253. New York: Oxford University Press.

Belmonte MK and Bourgeron T (2006) Fragile X syndrome and autism at the intersection of genetic and neural networks. *Nature Neuroscience* 9(10): 1221–1225.

Brodmann K (1909) *Vergleichende Lokalisationlehre Der Grosshirnrinde in Inren Prinzipien Dargestellt Auf Grund Des Zellenbaues.* Leipzig: J. A. Barth.

Brown TT, Lugar HM, Coalson RS, *et al.* (2005) Developmental changes in human cerebral functional organization for word generation. *Cerebral Cortex* 15: 275–290.

Casey BJ, Tottenham N, Liston C, and Durston S (2005) Imaging the developing brain: What have we learned about cognitive development? *Trends in Cognitive Science* 9(3): 104–110.

De Graaf-Peters VB and Hadders-Algra M (2006) Ontogeny of the human central nervous system: What is happening when? *Early Human Development* 82(4): 257–266.

Elman JL, Bates EA, Johnson MH, *et al.* (1996) *Rethinking Innateness: A connectionist perspective on development.* Cambridge, MA: The MIT Press.

Fair DA, Brown TT, Petersen SE, and Schlaggar BL (2006) A comparison of anova and correlation methods for investigating cognitive development with MRI. *Developmental Neuropsychology* 30(1): 531–546.

Fair DA, Brown TT, Petersen SE, and Schlaggar BL (2006) FMRI reveals novel functional neuroanatomy in a child with perinatal stroke. *Neurology* 67: 2246–2249.

Finger S (2000) *Minds Behind the Brain.* New York: Oxford University Press.

Friston KJ and Price CJ (2001) Dynamic representations and generative models of brain function. *Brain Research Bulletin* 54(3): 275–285.

Fuster JM (2003) *Cortex and Mind: Unifying Cognition.* New York: Oxford University Press.

Gould E and Gross CG (2002) Neurogenesis in adult mammals: Some progress and problems. *Journal of Neuroscience* 22(3): 619–623.

Greenough WT, Black JE, and Wallace CS (2002) Experience and brain development. In: Johnson MH, Munakata Y, and Gilmore R (eds.) *Brain Development and Cognition: A Reader,* 2nd edn., pp. 186–216. Oxford: Blackwell.

Gross CG (2000) Neurogenesis in the adult brain: Death of a dogma. *Nature Reviews Neuroscience* 1(1): 67–73.

Guillery RW (2005) Is postnatal neocortical maturation hierarchical? *Trends in Neuroscience* 28(10): 512–517.

Hensch TK (2005) Critical period plasticity in local cortical circuits. *Nature Reviews Neuroscience* 6(11): 877–888.

Horton JC (2000) Boundary disputes. *Nature* 406(6796): 565.

Horton JC and Hedley-Whyte ET (1984) Mapping of cytochrome oxidase patches and ocular dominance columns in human visual cortex. *Philosophical Transactions of the Royal Society of London. Series B, Biological Sciences* 304(1119): 255–272.

Horton JC and Hocking DR (1997) Timing of the critical period for plasticity of ocular dominance columns in macaque striate cortex. *Journal of Neuroscience* 17(10): 3684–3709.

Inder TE and Volpe JJ (2000) Mechanisms of perinatal brain injury. *Seminars in Neonatology* 5(1): 3–16.

Innocenti GM and Price DJ (2005) Exuberance in the development of cortical networks. *Nature Reviews Neuroscience* 6(12): 955–965.

Johnson MH (2001) Functional brain development in humans. *Nature Reviews Neuroscience* 2(7): 475–483.

Johnson MH (2005) *Developmental Cognitive Neuroscience,* 2nd edn. Malden: Blackwell.

Katz LC and Crowley JC (2002) Development of cortical circuits: Lessons from ocular dominance columns. *Nature Reviews Neuroscience* 3(1): 34–42.

Katz LC and Shatz CJ (1996) Synaptic activity and the construction of cortical circuits. *Science* 274(5290): 1133–1138.

Knudsen EI (2004) Sensitive periods in the development of the brain and behavior. *Journal of Cognitive Neuroscience* 16(8): 1412–1425.

Kostovic I, Judas M, Petanjek Z, and Simic G (1995) Ontogenesis of goal-directed behavior: Anatomo-functional considerations. *International Journal of Psychophysiology* 19(2): 85–102.

Lenroot RK and Giedd JN (2006) Brain development in children and adolescents: Insights from anatomical magnetic resonance imaging. *Neuroscience and Biobehavioral Reviews* 30(6): 718–729.

Leuner B, Gould E, and Shors TJ (2006) Is there a link between adult neurogenesis and learning? *Hippocampus* 16(3): 216–224.

Levitt P (2003) Structural and functional maturation of the developing primate brain. *Journal of Pediatrics* 143(Supplement 4): S35–S45.

Lopez-Bendito G and Molnar Z (2003) Thalamocortical development: How are we going to get there? *Nature Reviews Neuroscience* 4(4): 276–289.

Luna B and Sweeney JA (2004) The emergence of collaborative brain function: MRI studies of the development of response inhibition. *Annals of the New York Academy of Sciences* 1021: 296–309.

Luo L and O'Leary DD (2005) Axon retraction and degeneration in development and disease. *Annual Review of Neuroscience* 28: 127–156.

Marin O and Rubenstein JL (2001) A long, remarkable journey: Tangential migration in the telencephalon. *Nature Reviews Neuroscience* 2(11): 780–790.

Moses P and Stiles J (2002) The lesion methodology: Contrasting views from adult and child studies. *Developmental Psychobiology* 40(3): 266–277.

Mountcastle VB (1997) The columnar organization of the neocortex. *Brain* 120(Pt 4): 701–722.

Mukherjee P and McKinstry RC (2006) Diffusion tensor imaging and tractography of human brain development. *Neuroimaging Clincs of North America* 16(1): 19–43.

O'Leary DD and Nakagawa Y (2002) Patterning centers, regulatory genes, and extrinsic mechanisms controlling arealization of the neocortex. *Current Opinion in Neurobiology* 12(1): 14–25.

O'Leary DDM, Schlaggar BL, and Tuttle R (1994) Specification of neocortical areas and thalamocortical connections. *Annual Review of Neuroscience* 17: 419–439.

Palmer ED, Brown TT, Petersen SE, and Schlaggar BL (2004) Investigation of the functional neuroanatomy of single word reading and its development. In: Sandak R, Poldrack RA, and Manis FR (eds.) *Scientific Studies of Reading: The Cognitive Neuroscience of Reading*, pp. 203–223. Mahwah, NJ: Lawrence Erlbaum.

Paus T (2005) Mapping brain maturation and cognitive development during adolescence. *Trends in Cognitive Science* 9(2): 60–68.

Price DJ, Kennedy H, Dehay C, *et al.* (2006) The development of cortical connections. *European Journal of Neuroscience* 23(4): 910–920.

Qi Y, Stapp D, and Qiu M (2002) Origin and molecular specification of oligodendrocytes in the telencephalon. *Trends in Neurosciences* 25(5): 223–225.

Rakic P (2002) Neurogenesis in adult primate neocortex: An evaluation of the evidence. *Nature Reviews Neuroscience* 3(1): 65–71.

Schlaggar BL, Brown TT, Lugar HM, *et al.* (2002) Functional neuroanatomical differences between adults and school-age children in the processing of single words. *Science* 296: 1476–1479.

Schlaggar BL and O'Leary DD (1991) Potential of visual cortex to develop an array of functional units unique to somatosensory cortex. *Science* 252(5012): 1556–1560.

Schummers J, Sharma J, and Sur M (2005) Bottom-up and top-down dynamics in visual cortex. *Progress in Brain Research* 149: 65–81.

Shaywitz SE (1998) Dyslexia. *The New England Journal of Medicine* 338 (5): 307–312.

Sur M and Leamey CA (2001) Development and plasticity of cortical areas and networks. *Nature Reviews Neuroscience* 2(4): 251–262.

Sur M and Rubenstein JL (2005) Patterning and plasticity of the cerebral cortex. *Science* 310(5749): 805–810.

Toga AW, Thompson PM, and Sowell ER (2006) Mapping brain maturation. *Trends in Neurosciences* 29(3): 148–159.

VanEssen DC (1985) Functional organization of primate visual cortex. In: Peters A and Jones EG (eds.) *Cerebral Cortex*, pp. 259–329. New York: Plenum.

Webster MJ, Ungerleider LG, and Bachevalier J (1995) Development and plasticity of the neural circuitry underlying visual recognition memory. *Canadian Journalof Physiology and Pharmacology* 73(9): 1364–1371.

Brain Function

M de Haan and M Martinos, University College London Institute of Child Health, London, UK

Glossary

Cerebral cortex – The outer layer of the brain sometimes called the 'gray matter' in reference to its color. In humans, it plays a central role in many complex brain functions including memory, attention, perceptual awareness, 'thinking', language, and consciousness.

Developmental cognitive neuroscience – An evolving field that investigates the relations between neural and cognitive development.

Episodic memory – A type of explicit memory that refers to the memory of specific events within a spatio-temporal context.

Explicit memory – The conscious, intentional recollection of previous experiences and information.

Implicit memory – Is characterized by a lack of conscious awareness in the act of recollection. Typical instances of implicit memory include priming and procedural memory.

Long-term memory – Information that is in permanent store, but that need not be consciously accessible at all times. This includes memory for things that happened between a few hours to many years ago. Long-term memory includes memory for facts and events (explicit memory) as well as memory for skills and rules (implicit memory).

Plasticity – The adaptive capacity of the nervous system. Plasticity can occur during the normal course of learning or development, in response to brain injury or in response to alterations of sensory input.

Semantic memory – A type of explicit memory that refers to the memory for general knowledge and facts that are not tied to a specific spatial and temporal context.

Subcortex – This is the part of the brain that is underneath the cortex and that is considered older than the cortex in evolutionary terms. Its functions include basic ones such as the control of breathing and heart rate.

Working memory – A system for the temporary storage and manipulation of information. For example, working memory allows one to keep in mind a telephone number long enough to dial it once reaching the telephone.

Introduction

The human brain develops rapidly in the first years of life. At birth it is only 25% the size of an adult's brain, but by 5 years of age it is already 90% of the adult size. In parallel with this rapid brain development, dramatic changes occur in the child's ability to perceive, think about, and act in the world. However, until recently researchers studying psychological development and those studying neural development were working independently, with the former taking little notice of changes in the brain and the latter failing to consider the functional implications of such changes. Fortunately, the past decade has seen the emergence of a new field of study, called developmental cognitive neuroscience, which is devoted to studying the relationship between brain and cognitive–behavioural development. This field of study tackles complicated questions such as how the functional organization observed in the adult brain comes about, and how infants' and children's experiences influence the emergence of brain function. These questions are important to ask during early development for several reasons. First, any comprehensive model of brain or behavioral development must take both brain and behavioral factors into account, as there is little doubt that immaturity of the brain plays a major limiting factor in behavioral functioning in children or that children's interactions in the world in turn influence the brain. Second, a primary concern for those studying development is to understand how atypical development occurs and to develop intervention methods to optimize development in such cases. To accomplish this goal, it is very informative to know things such as how the brain has been injured, how this injury might influence further development of other brain regions, and the degree to which further development of this brain area is influenced by experience.

These questions are important to ask during early development for several reasons. First, any comprehensive model of brain or behavioral development must take both brain and behavioral factors into account, as there is little doubt that immaturity of the brain plays a major limiting factor in behavioral functioning in children or that children's interactions in the world in turn influence the brain. Second, a primary concern for those studying development is to understand how atypical development occurs and to develop intervention methods to optimize development in such cases. To accomplish this goal, it is very informative to know things such as how the brain has been injured, how this injury might influence further development of other brain regions, and the degree to which further development of this brain area is influenced by experience.

These lines of questioning have led to three views regarding how brain development and the emergence of behavior are interrelated. In this articles we first outline the three viewpoints of functional development: the maturation view, the interactive specialization view, and the skills learning view. Studying the different predictions of these viewpoints benefits from the use of neuroimaging tools that allow measurement of changes in brain structure during development and/or changes in the pattern of activation. Different tools are available, that differ in their strengths and weaknesses in measuring the brain and in their suitability for use with children. An overview of these different tools is provided in **Table 1.** We will describe studies using such tools to illustrate the theoretical merits and shortfalls of the three different views of functional brain development by using examples from the development of four specific functions: face processing, long-term memory, working memory, and language.

Theories of the Functional Development of the Human Brain

The adult cerebral cortex shows the characteristic of functional specialization. That is, certain areas of the cortex are particularly important for certain functions. For example, as will be described in more detail below, the fusiform face area is important for processing of faces and Broca's area is important for producing language. A key issue for those studying the functional development of the brain is to understand how this functional specialization comes about: is it present from the beginning, or does it emerge as a result of the developmental process? A related question asks, is this organization specified in the genes, or is it a consequence of input from the environment? Grappling with these questions has led researchers to identify three different processes by which cortical brain development can relate to the emergence of new functions: maturation, interactive specialization, and skill learning (see **Figure 1**). These options are not mutually exclusive, and it is possible that all these processes contribute to development. However, the three processes do make somewhat different predictions about how brain activity is expected to change when functions emerge and how the brain responds to injury and a typical experiences early on in life.

Maturation

In this view, the functional specialization observed in the adult brain is predetermined, and new functions emerge as the brain regions responsible for those functions mature (see **Figure 1(a)**). In other words, new skills come about supported by the maturation of previously immature brain areas necessary to perform these skills. Under these circumstances, we would expect to see an increase in brain activation with emergence of a skill, as a previously immature, silent brain region matures and becomes active. The maturation viewpoint predicts that damage to a brain region would result in a failure of the function it normally subserves to develop and that atypical experience would lead to a different rate of maturation of the skill in question.

Interactive Specialization

In this view, the cortex of the brain initially does not show the pattern of functional specialization observed in adults, although it may have some regional computational biases (see **Figure 1(b)**). Functional specialization comes about as a result of brain activity, and the function of a particular region is partly determined by its pattern of connection to other brain regions and how these regions are activated.

Table 1 Some of the methods used in the study of brain development

Tools	Description	Suitability for infants and very young children
Electroencephalography (EEG)	EEG is the measurement of the ongoing, spontaneous electrical activity of the brain by recordings from electrodes placed on the scalp.	EEGs are frequently used in experimentation with infants and children because the process is noninvasive.
Event-related potentials (ERPs)	Similar to EEG, but records brain's electrical response to a given stimulus or action. In contrast to an EEG, ERPs are time-locked to stimulus presentation or response execution. ERPs have a good temporal resolution but more limited spatial resolution.	ERPs are frequently used in experimentation with infants and children because the process is noninvasive.
Magnetic resonance imaging (MRI)	Structural MRI is a noninvasive method for imaging the morphology of brain structures with good spatial resolution.	MRI is very sensitive to motion artifact, but structural information can be successfully obtained while infants or children are sleeping.
Functional magnetic resonance imaging (fMRI)	fMRI computes brain activity by measuring levels of blood oxygenation in the brain. This is a good tool for studying the localization of function in adults.	Since fMRI is very sensitive to motion artifact and it usually requires participants to be awake, it is typically not used with children younger than about 7 years of age. However, there have been a few studies involving younger ages.
Magnetic encephalography (MEG)	MEG is a noninvasive technique that measures the magnetic activity produced by neural electrical activity. In contrast to ERPs and EEG, MEG has good spatial resolution and, in contrast to fMRI, has a good temporal resolution.	MEG is a fairly new technique that is suitable for use with infants and children. Has been used to study sensory development in fetuses.
Near-infrared spectroscopy (NIRS)	NIRS is a form of optical imaging that computes changes in blood oxygenation and indirectly measures levels of activity in different brain regions.	It is increasingly being used with infants and young children as it is noninvasive and is not greatly affected by motion artifacts.
Positron emission tomography (PET)	PET is a nuclear medicine medical-imaging technique which produces a three-dimensional image or map of functional processes in the body.	Not suitable for use with typically developing children because a radioactive isotope needs to be injected in the subject and, therefore, the technique is invasive.

Different cortical areas are thought to compete with one another to take on a functional role. This process of competition serves to 'sharpen' the set of functions a particular region carries out. As a result, the emergence of function is associated not just with increased activity of the particular brain region that will ultimately serve a function. Rather, it involves changes in brain activation over several regions that can include decreases in activity in regions that lose the competition and are no longer involved in that function. In the interactive specialization view there is still scope for skills to develop if the brain is damaged. This is because the initially broad functioning of the cortex would allow the possibility of flexibility in the region that would take on a particular function. That is, a new area could 'win' the competition for carrying out a particular function if the area that normally would do so was removed from the competition through injury. This would allow functional compensation, but with an atypical pattern of brain activation linked to that function as the interactive specialization view allocates an important role to experience, it predicts that atypical experience could lead to qualitative changes in the pattern of regional specialization (the brain areas that subserve a function are

different from normal) rather than just quantitative changes (a simple change in the rate of development as in the maturation view).

Skill Learning

A third idea regarding functional brain development is that the acquisition of new abilities during infant and early childhood development involves similar or identical processes that occur when adults acquire new complex skills (see **Figure 1(c)**). In other words, in some circumstances it might be that humans require similar neural substrates or draw upon the same processes to acquire new functions regardless of whether they are infants or adults. This view emphasizes continuity in process across the lifespan. When adults are learning new skills, unskilled performance is often associated with activation of different brain regions than is skilled performance. If infants acquire new functions in the same way, a similar change in the areas of brain activation would be expected as new skills emerge in development. Damage to areas important for the acquisition of skills would be expected to result in widespread and long-lasting deficits

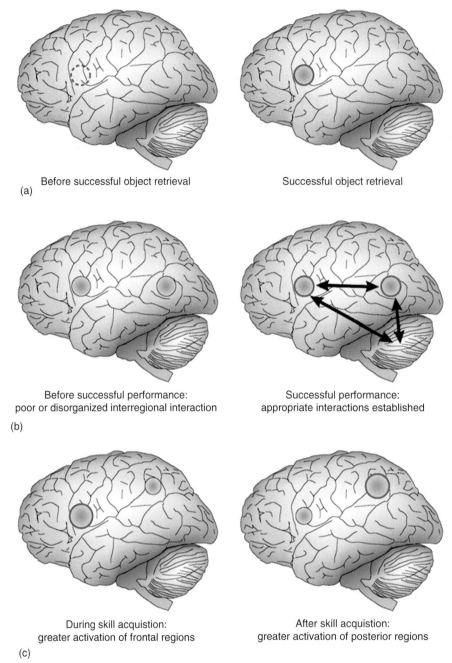

Before successful object retrieval   Successful object retrieval
(a)

Before successful performance: Successful performance:
poor or disorganized interregional interaction   appropriate interactions established
(b)

During skill acquistion: After skill acquistion:
greater activation of frontal regions   greater activation of posterior regions
(c)

Figure 1   Theories of functional development. (a) Maturational, (b) interactive specialization, and (c) skill learning.

in emergence of new future skills. In this view, early deprivation of normal experiences could be easily compensated in the future so long as the mechanisms of skill acquisition are still in place and could be utilized to acquire the skill in question.

A third idea regarding functional brain development is that the acquisition of new abilities during infant and early childhood development involves similar or identical processes that occur when adults acquire new complex skills (see Figure 1(c). In other words, in some circumstances it might be that humans require similar neural substrates or draw upon the same processes to acquire new functions regardless of whether they are infants or adults. This view emphasises[2] continuity in process across the life span. When adults are learning new skills, unskilled performance is often associated with activation of different brain regions than is skilled performance. If infants acquire new functions in the same way, a similar change in the areas of brain activation would be expected as new skills emerge in development. Damage to areas important for the acquisition of skills would be expected to result in widespread and long-lasting deficits in emergence of new future skills. In this view early deprivation of normal experiences could be easily compensated in the future so long as the mechanisms of skill acquisition are still in place and could be utilized to acquire the skill in question.

The three views of functional development offer different theoretical merits to the study of development. Below we illustrate these merits as well as shortcomings through examples from four specific domains of perceptual-cognitive development.

Functional Development of the Human Brain: Examples

The maturation, interactive specialization, and skill-learning views each describe how brain development relates to emergence function. Below we describe the emergence of four functions, face processing, working memory, long-term memory, and language, and evaluate how well the different view of development of brain function can account for the data in each of these domains.

Face Processing

Adults are usually quick and accurate at noticing faces and perceiving the different types of social information they display such as whether a person is familiar, how he is feeling, and where he is directing his gaze. A large body of work suggests that this expertise is supported by brain regions that are specialized for the function of processing facial information. This idea was initially put forward based on studies of patients who had lost the ability to recognize faces following brain injury. Some of these patients were severely impaired in recognizing faces, but were still able to recognize other types of objects. This was not simply because faces are more difficult to recognize than other objects, because other patients show the opposite pattern, with poor object recognition but good face recognition. This 'double dissociation' of spared and-impaired abilities across the two groups of patients provides classic neuropsychological evidence that different brain regions are important for face and object processing. More recently, this idea has received further support from brain imaging studies of adults without brain injury. These studies have identified a region on the underside of the brain called the 'fusiform face area' because it shows greater activation for faces than for other types of objects or even other body parts. Together, the data from brain-injured patients and from brain-activation studies in intact adults are consistent with a 'content-specific' view of how the brain processes faces: certain brain areas are devoted to processing certain stimulus content, in this instance faces.

Other researchers have challenged this content-specific view, and argue that the data can better be explained in terms of a process-specific view. According to this idea, the fusiform face area is part of a brain network important for the process of acquiring expertise in perceptual discrimination, rather than a brain network specialized for processing the domain of faces. The fusiform gyrus (see **Figure 2**) just appears to be specialized for processing faces because this is the only category for which most humans acquire expertise. The content of other areas of expertise are much more variable among individuals depending on their particular lifestyles and interests. In support of this view, studies have shown that just as activation in the fusiform face area is greater for faces than objects, activation in the same area is greater when adults view objects from their individual areas of expertise than objects from other categories. For example, activation is greater for birds than cars in bird-spotting experts and for cars than birds in car experts. These types of results show that activation in the fusiform face area can be related to acquisition of perceptual expertise, although it is important to note that the fusiform face area is still more active for faces than even these expert categories.

Developmental studies of the emergence of face processing and its brain correlates can provide insight into this debate because they focus on the period when babies are first learning about faces. Those who have spent time with young infants likely know that they are very interested in faces. From just a few hours after birth, infants are generally interested in visual patterns and will shift their eye gaze and often their heads to keep a moving pattern in view. However, they do so for longer if the elements of the pattern are arranged like a face than if they are not. Within hours to days of birth, infants can also recognize the mother's face: they look longer at it than a stranger's face, even when cues from the mother's voice and smell are eliminated.

Newborns' interest in faces could be seen as evidence that the fusiform face area is functioning as a content-specific 'face' area from birth, since this interest is present from birth and before infants could have acquired expertise. This interpretation would be consistent with a

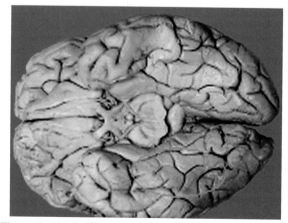

Figure 2 The fusiform gyrus in the adult brain.

maturation view: a function (orienting to faces) is possible because a brain area (fusiform gyrus) is mature. However, current theories and evidence suggest that the story may not be quite so simple. According to one influential theory put forth by John Morton and Mark Johnson, newborns' preference for orienting to faces is not mediated by the fusiform face area, but instead is supported by a subcortical system called 'Conspec'. In this view, Conspec is a reflex the newborn is born with, that works to ensure that infants frequently orient to faces in the first weeks of life. This 'face-biased' visual input then provides input to a more slowly developing cortical system, the 'Conlern', which would include the fusiform face area and other regions. Initially, Conlern functions as a general-purpose visual object-processing system, but with the help of input provided by Conspec, it begins to develop into a content-specific cortical face-processing area. In this interactive specialization explanation, the development of the cortical face-processing system can be seen in two ways: (1) the brain areas involved in face processing gradually come to respond more specifically to faces (respond more to faces and less or not at all to other stimuli), and (2) the cortical area that responds to faces becomes more focal and less widespread.

It is difficult to test the maturation vs. the interactive specialization accounts directly, because the functional magnetic resonance imaging (fMRI) techniques commonly used in adults are not well suited for studying typically developing infants (see **Table 1**). A rare exception is one study that used positron emission tomography (PET) (see **Table 1**) to examine activation of the fusiform gyrus in 2–3-month-olds, and found greater activation to faces than to pairs of lights matched in luminance to the faces. This result could be seen as consistent with the maturation viewpoint as it demonstrates that the fusiform face area is functioning by 2 months of age. However, it does not indicate whether the fusiform is specifically activated by faces (more than other objects) or more generally activated by patterned visual stimuli (compared to diodes). Studies of older children suggest that the second interpretation could be correct, because even up to 10 years of age children do not show the adult-like pattern of greater activation of the fusiform face area for faces than objects (e.g., houses).

Results from event-related potential (ERP) studies support the idea that Conlern is less 'tuned in' to faces in infants than it is in adults. These studies have focused on the development of a brain electrical response called the N170. The N170 probably reflects activation of the fusiform face area and related face-responsive brain regions. In adults, the N170 is larger and often occurs more quickly for faces than other objects. Infants as young as 3 months of age (the earliest age tested), show an N170-like response that is greater to faces than to control patterns created to have similar low-level perceptual

properties but are not recognizable as a face. However, this response is not as face specific as it is in adults. For example, adults show a specific N170 response to the upright, human face, which is different even from the response to similar patterns like an upside-down face or a monkey face. Three-month-old infants do not show this specialized response, and it is not until 12 months of age that infants begin to show a more adult-like response specific to upright human faces. This pattern of findings is consistent with an interactive specialization view, where regions of cortex sharpen their functions and become more specialized with development.

Further studies suggest that the process by which infants tune in to human faces might also involve a loss of abilities. For example, in an interesting study, De Haan and colleagues found that 6-month-old infants are better than 9-month-old infants or adults at recognizing individual monkey faces. These results can be explained by the view that the development of processing faces involves a change from a more general-purpose visual-processing system to one focused on the particular cues that are especially useful for discriminating the many different faces we encounter in a lifetime. For example, a general-purpose processing system might give equal importance to eye color and the spacing between the eyes. However, eye color is likely to be of limited use for uniquely identifying many different human faces, whereas spacing between the eyes is actually a very useful cue. Thus, a specialized system would tune in to these types of useful features and tune out the less useful or unnecessary ones to establish a fast and reliable face perception system. However, this specialization might be a hindrance in some circumstances, such as recognizing monkey faces. This is because the cues that are most useful for recognizing human faces are not the same ones that are best for recognizing monkey faces. Thus, if we automatically apply our human-face-processing strategy, tuning in to cues useful for human faces and tuning out others, we will not perform very well. In this case, a general-purpose system would do better as it would process a broader range of information. Thus, 6-month-old infants are able to recognize the monkey faces because they have processed a broader range of facial features, whereas 9-month-old infants and adults apply a more specialized strategy that is less optimal. Interestingly, a follow-up study provided evidence that this shift between 6 and 9 months is dependent on experience: when 6-month-old infants were given experience in discriminating monkey faces in the form of picture books, they retained their monkey-face discrimination abilities at 9 months.

Does this evidence for the importance of experience mean that the development of face processing can be equated with the process of perceptual skill learning in adults? There remains debate on this issue. However, one line of evidence suggests that the learning that occurs in infancy may not be the same as that which occurs in adulthood. In particular, visual experience in the first months of life is critical for the normal development of face processing and, if this experience is missing, it cannot be compensated for by perceptual learning later in life. This conclusion comes from studies conducted by Rick Legrand, and Daphne Maurer and colleagues in children born with cataracts. Cataracts are a clouding of the lens that prevents patterned light from reaching the retina of the eye. These cataracts can be removed, after which the child receives contact lenses to help them see normally. Research has shown that such visual deprivation in the first months of life results in permanent deficits in perceiving certain types of facial information. This is true even though the children are tested years after their sight has been restored. These results suggest that the developing cortical face-processing system must receive visual inputs during the first months of life in order to develop normally.

Working Memory

The prefrontal cortex is a large region, covering almost one-third of the entire cortex in adults (see **Figure 3**). It is believed to be involved in high-level information processing that allows us to do things such as formulate and carry out plans and behave appropriately in response to novel or challenging situations. Neuroanatomical studies suggest that the prefrontal cortex is quite slow to mature. This is consistent with the observation that many of the skills attributed to the prefrontal cortex appear to be absent in infants and to emerge slowly in childhood and adolescence. However, an important body of research documents that the prefrontal cortex is not dormant in the first years of life but functions from at least the second half of the first year of postnatal life.

The best evidence that the prefrontal cortex functions in infancy comes from studies of spatial memory. A well-known psychologist, Jean Piaget, had documented a series of developments in spatial and object memory over the

first 2 years of life. In the first 4 months of life, infants have a limited object concept, and when an object disappears from view they behave as if it no longer exists and do not search for it ('out of sight, out of mind'). From this time, babies' abilities improve, progressing from searches for partially hidden objects to full searches even when there are multiple hiding locations. In the middle of this time, at about 8–12 months, infants typically make a search error called the 'A not B' error. This can be seen in the classic search task, where there are two hiding locations. An object is first hidden in one location (A) and the infant is allowed to find it successfully. After a few successful hiding and searches at location A, the object is hidden in location B. The 'A not B' error occurs when infants continue to search in the location where they have previously successfully retrieved the object (A) rather than the location where they most recently saw it hidden (B). This occurs even though the delay between hiding and search is only a few seconds. By 12 months of age, infants' performance improves and they tend to search correctly even up to delays of 10 s. This is believed to be due to improvements in infants' abilities to guide their actions (in this case, searching) based on information that is not perceptually available but that must be held in memory (the correct location of the hidden object). Some researchers noticed the similarity between the 'A not B' task and a task of spatial memory that was known to rely on the prefrontal cortex in monkeys. Indeed, when they tested monkeys on the 'A not B' task, they found that damage to the dorsolateral region of the prefrontal cortex caused animals to make the 'A not B' error. Damage to the parietal cortex, a region involved in spatial processing, or the hippocampus, a region involved in other aspects of memory, did not affect performance on the 'A not B' task. These results support a maturational view that the improvement in human infants' performance between 8 and 12 months of age relies on development of the dorsolateral prefrontal cortex.

Further evidence in support of this view comes from brain imaging studies linking frontal activation to performance on object permanence and 'A not B' tasks. In one study, the relation between frontal brain activation as measured by near-infrared spectroscopy (NIRS; see **Table 1**) and emergence of object permanence was studied in infants longitudinally between 5 and 12 months of age. As the maturational view would predict, the researchers found that the ability to search for a hidden object was related to an increase in the NIRS measures of frontal activation. It is important to note that the researchers did not measure activity in other brain regions, so could not test the prediction of the interactive specialization view that the emergence of the skill would be accompanied by more widespread changes in brain activity. In a different line of studies, researchers used electroencephalography (EEG; see **Table 1**) and found that infants who were successful in performing the 'A not B' task

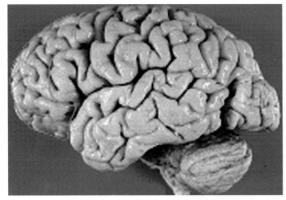

Figure 3 Lateral view of the prefrontal cortex in the adult brain.

with increasingly longer delays across the second half of the first year of life showed changes in EEG over the frontal regions. In both the NIRS and the EEG studies, the links between frontal activity and object permanence or 'A not B' performance were specific. For example, there were no links between frontal activity and general playing with toys or the ability to inhibit reaching to a novel toy.

These studies using reaching tasks to investigate object permanence and A-not-B performance have suggested that these abilities emerge in the second half of the first year of life. However, studies using different methods have found evidence that these skills may be present even earlier, raising the possibility that prefrontal regions are also functioning even earlier in life. These studies use looking time, rather than reaching behavior, to assess what infants know about hidden objects. Infants are shown a display such as a ball rolling behind a box. They are then shown two outcomes when the box is lifted: either the ball appears behind the box (possible outcome) or it is gone (impossible outcome). Infants as young as 3.5 months tend to look longer at the impossible outcomes in these types of tasks. Infants' surprise when the object is missing, as shown by their longer looking, suggests that they do understand that objects continue to exist when not visible. If so, then why do infants fail search tasks at this age? One reason might be that the strength of mental representations of hidden objects is still weak at this age, and not yet sufficient to allow infants to guide their searching behavior. Another possibility is that infants do remember the hidden object, but find the performance demands of a search task (which requires the child to carry out an action e.g., lift a cover in order to obtain the goal) more difficult than a simple looking task.

Within the prefrontal cortex, there is evidence that the projections it receives involving the neurotransmitter dopamine are particularly important for cognitive function. For example, the levels of dopamine within the dorsolateral prefrontal cortex increase over the period that infant monkeys improve on cognitive tasks such as the 'A not B' and object retrieval tasks, which are known to require these regions. Evidence that the same might be true in humans comes from studies of children with the rare genetic disorder phenykeltonuria (PKU). This is a disorder which disrupts the chemical pathway for producing dopamine and so results in lowered levels of this neurotransmitter. PKU is treatable with a special diet, which is very effective in preventing the global and severe developmental delay that occurs if it is untreated. However, even with this special diet the levels of dopamine are not completely normal. The functions of the prefrontal cortex can still be affected, especially due to this region being particularly sensitive to fluctuations in dopamine levels. Thus, evidence that children with PKU show poorer performance than comparison infants would provide further evidence in support of a link between the prefrontal cortex and working memory. Studies performed by Adele Diamond have provided just such evidence: infants who followed the special diet, but who still had a moderately atypical chemical profile, performed worse on the 'A not B' task than did comparison infants or infants with PKU who had only a mildly atypical chemical profile. Do infants grow out of these problems? At older ages, the 'A not B' task is less useful because it is too easy: any differences between the PKU and comparison groups might be obscured by 'ceiling effects' where both groups do well. Instead, older children must be tested with age-appropriate tasks that still tap the key abilities to hold information in mind and resist a dominant response. With this in mind, Gerstadt and colleagues tested pre-schoolers on a more advanced task that is, the 'day–night' task. In this task children are required to say 'night' when presented with a picture of the sun and 'day' when presented with the picture of the moon. This requires them to keep information in mind (the rule of which word to associate with which picture) and inhibit a dominant response (to say 'day' to the sun). The results of such testing showed that the pattern of deficits observed in infants persisted at the older ages: children with PKU and moderately high phenylalanine levels showed impairments. These impairments do not reflect general developmental delay, as the same children did well on tests designed to tap the functions of the parietal cortex and the hippocampus. These findings also point to a specific impairment related to the malfunctioning of the prefrontal cortex and, thus, seem to support the maturation approach to development.

Long-Term Memory

Memory is utilized almost in every single human activity: recognizing a face, riding a bicycle, and remembering a past holiday are all typical instances of ways in which memory serves our everyday life. In the past 50 years, a lot of progress has been made in the field of memory research. For one, the importance of the medial temporal lobes (MTLs) for memory has been established. This is partly due to the description of a patient, HM, who became densely amnesic following the bilateral resection of his MTL. After his surgery, HM became unable to form new memories, yet, he was still able to learn new skills. In contrast, patients with damage to other regions of the cortex such as the basal ganglia circuit show the reverse pattern of performance, that is, they can form new memories, yet, they have difficulty in learning a new skill. These seminal findings have led researchers to advance a theoretical distinction between explicit and implicit memory.

Implicit memory refers to a set of abilities that influence overt behavior without requiring the conscious recollection of doing so (e.g., the ability to play a piano; which a pianist seems able to just 'do', without being able to verbalize how), and explicit memory refers to the types of memories that can be brought to mind as an image or proposition without the need for perceptual support. Explicit memory includes both the memory for events known as episodic memory (e.g., remembering what color dress the piano teacher wore at the last lesson), and, the memory for facts known as semantic memory (e.g., remembering how many keys a piano has).

For some time it was thought that infants relied solely on implicit memory, with explicit memory developing only later. This was partly due to the fact that many of the tasks used to assess infant memory involve indirect measures (e.g., motor performance) similar to those often used to assess adult implicit memory, and partly due to the phenomenon of infantile amnesia, wherein most adults can recall very little from the first 3–4 years of their lives.

Explicit memory

More recently, evidence has begun to accumulate to suggest that MTL-based memory is present from very early in life. During infancy and early childhood, recognition memory has often been studied using the visual paired comparison (VPC) task. In the VPC task, subjects are familiarized with two identical items followed by the presentation of the familiar item coupled with a novel one. Attending to the novel stimulus for a longer duration is taken as a sign of recognition. Using this task, Olivier Pascalis and colleagues found that babies aged 3–4 days old can detect novel stimuli following a delay of 2 min. As children grow older they are able to withstand longer and longer delays, with babies aged 3 months old, withstanding delays of 24 h on the VPC. Performance on the VPC seems to rely on MTL regions, and more specifically, on the hippocampus (see **Figure 4**), because lesions to this structure lead to impairments in both infant monkeys and human or monkey adults. This suggests that MTL memory systems are operating, at least to some extent, early in life. This interpretation is consistent with the maturation view: the maturity of the hippocampus allows the memory function to operate.

As infants grow, their recognition memory becomes less constrained by the perceptual similarity between the original encoding context and the testing context. In one experiment, 6-, 12-, 18-, and 24-month-olds were tested on a variation of the VPC. In this variation, objects were presented against a different background during testing from the one presented during encoding. Only the 18- and the 24-month-old babies were unaffected by this change. This increasing flexibility in being able to recognize familiar items in different contexts probably reflects the progressive development of the hippocampus since adult humans and monkeys with bilateral lesions to this structure fail this task. Specifically, success on this task may rely on the development of the dentate gyrus,

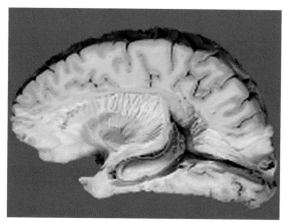

Figure 4 The hippocampal formation in the adult brain.

that is, the major source of input from extrahippocampal regions to the hippocampus.

Another task believed to rely on the MTL is the delayed nonmatch to sample task (DNMS). In particular, lesions to the MTL lead to impairments in the DNMS task if such lesions are suffered during adulthood. In this task the subject is presented with a sample object. The subject must then remove the object to reveal a reward. Following a delay, the sample object is presented along with a novel object. Opting for the novel object is rewarded. Human infants do not succeed on this version of the task until approximately the age of 15–21 months even with brief delays of 5–10 s. This is puzzling given their early proficiency on tasks such as the VPC. Interestingly, changing the demands of the task from receiving a reward for displacing the novel object to allowing the child to play with the novel object (i.e., stimulus = reward) lowers the success threshold to the age of 6 months, with infants at this age withstanding delays of 10 min. This has led investigators to believe that, while that success on the DNMS task requires recognition, successful performance also requires additional skills such as inhibition processes and an understanding of stimulus–reward relationships. According to the maturational view, the slower development of these additional skills set the pace for developmental improvements in performance on this task.

Further evidence for the importance of the MTL in infant memory comes from studies using the deferred imitation paradigm. In the deferred imitation task, the examiner models a sequence of events for the child (e.g., placing a marble into a cup and then shaking the cup to make a rattle), and, following a delay, the child needs to reproduce this sequence of events without the benefit of prior practice. Research on the neural underpinnings of the deferred imitation task has pointed to the hippocampus as the most likely candidate. Namely, individuals who have sustained bilateral hippocampal damage during childhood or around the time of birth show impairments on this task. The same holds true for individuals that have suffered damage to the MTL during adulthood. Performance on this task shows a gradual improvement with age with dramatic improvements from the age of 6–24 months consisting of the ability to remember longer sequences and withstand longer delays. Specifically, infants at the age of 6 months show delayed memory for single actions after a 24-h delay. By the age of 9–10 months, infants begin to show memory for the sequence of actions. Moreover, by the age of 18 months infants can perform normally even when the props used in the testing phase are different from the ones used in the presentation phase. Once more this finding points to the increased flexibility of memory-retrieval processes in older infants. However, similar to the DNMS task, given the taxing nature of the deferred imitation task, it is very likely that other regions such as the prefrontal cortex contribute to the development of the requisite processes for successful performance on this task.

Further insights about the relative contributions of MTL structures to memory come from the study of children who have suffered damage to the hippocampus during infancy. These children exhibit disproportionate episodic memory deficits in the face of relatively intact semantic memory processes. One interpretation is that semantic memory processes are supported by parahippocampal regions during development and, are, thus, spared in these children. This interpretation is consistent with the maturational view, whereby perturbations in the maturation of the hippocampus disable the emergence of its corresponding function, that is, episodic memory. However, these findings may also point to the relative degrees of plasticity inherent in different memory processes. Namely, it is plausible that semantic memory is relatively spared because it is better able to reorganize in the face of injury, whereas episodic memory processes are more affected because they are less able to do so. Functional MRI studies of the pattern of activation for semantic memory would help to decide between these possibilities: the maturational view would predict a typical pattern of activation, whereas the second view would predict an atypical (reorganized) pattern.

Implicit memory

Not many studies have examined the neural correlates of implicit memory during development. One area that has received some interest is the development of conditioning. Conditioning refers to the formation of a contingency between two previously unrelated stimuli through either an association with a cue (i.e., classical conditioning) or a reward (instrumental conditioning). A common classical conditioning paradigm used both in the infancy and animal literature is the eye-blink paradigm. In this paradigm, a puff of air is blown on to the eye causing the eye to blink. This is always preceded by a tone. With time the subject responds by blinking their eye upon hearing the tone. Infants between the ages of 10 and 30 days start to show the eye-blink response. It has been shown that lesions to the cerebellum in rabbits and human adults disrupt both the acquisition and the retention of a conditioned association. Interestingly, lesions to the hippocampus do not. Seeing as the cerebellum is known to mature early in life, the reliance of classical conditioning on the cerebellum seems plausible.

Another conditioning paradigm that was originally developed by Carolyn Rovee-Collier and colleagues and has been used extensively with infants is the mobile conjugate reinforcement paradigm. In this task, a ribbon is attached to the infant's leg on one end and to a suspended mobile on the other end. In order to animate the mobile the infants need to learn to kick the relevant leg. Infants as young as 3 months old are able to learn the relationship between leg kicking and mobile movement, and, they raise their kicking rate well above baseline levels to achieve this. Infants can then be tested following a given delay to see if they have retained the learned behavior. Significant improvements in the retention of this association have been recorded across the first year of life with 12-month-old infants retaining the behavior for up to 8 weeks. However, there has been very little research on the neural bases mediating performance on this task to verify if regions such as the cerebellum are involved, or how changes in activation of this or other brain areas are related to developmental improvements in memory.

Language

Understanding the mechanisms behind language acquisition has fascinated scientists for many years. Recently, the introduction of new techniques such as neuroimaging has enabled neuroscientists to probe deeper into the mysteries of human language acquisition. Infants start babbling at around 6–8 months, move on to the one-word stage by 10–12 months, and speak in full sentences by the age of 3 years. This developmental path is followed by both deaf and hearing subjects and is irrespective of cultural background. Given our unique position in the animal kingdom regarding language and the universal nature of language acquisition among humans, one natural question that has been raised is whether humans possess innate mechanisms dedicated to language acquisition.

One way to answer this question has been to study the emergence of the hemispheric specialization for language. Research with adults has shown consistently that language processes are disproportionately subserved by the left hemisphere with lesions to this hemisphere often leading to speech apraxia or aphasia. This phenomenon has given rise to a large corpus of research in search of a reason for this hemispheric predilection. In other words, is the left hemisphere a more suitable habitat for language processes? And if so, is this tendency evident from the very beginning of language acquisition? Three lines of inquiry have been pursued to attempt to answer these questions: (1) anatomical studies, (2) functional studies, and (3) neuropsychological studies.

Whereas the left and right hemispheres might appear to be the mirror image of one another, there are some structural differences that differentiate the two. Relevant to language, the left planum temporale and the left Sylvian fissure have been found to be larger than their right counterparts. Do these structural differences reflect the impact of experience on the immature brain or do they constitute an inherent predisposition? Overall, structural studies with infants have revealed that, during prenatal development, the right hemisphere has a head start over the left, with structures such as the superior temporal gyrus appearing 1 or 2 weeks earlier on the right side. However, the planum temporale and the Sylvian fissure have been found to be larger on the left side even during prenatal life. Importantly, these areas exhibit such structural asymmetries irrespective of auditory stimulation as has been shown by studies of twins and congenitally deaf individuals. Moreover, the brains of individuals with dyslexia, a language disorder that affects reading and writing abilities, have been found to lack these asymmetries.

Are these structural asymmetries mirrored in functional asymmetries in language processing? In other words, do these structural differences contribute or lead to the left hemisphere's dominant role in language processing? Converging evidence from fMRI, ERP, and NIRS studies seem to suggest that functional asymmetries in response to language processing are evident in newborns and young infants. For example, Mehler and colleagues, used NIRS to try and detect hemispheric differences in response to normal speech, backward speech, and silence in neonates. Backward speech shares a lot of acoustic properties with normal speech making it a good baseline for the study of speech perception. Left temporal regions were found to be significantly more active in response to normal speech than to silence or backward speech; pointing to an early left hemispheric predisposition for speech perception.

In a different paradigm, 3-month-old infants were exposed to trials of four identical syllables (e.g., da), which were either followed by a fifth identical syllable (e.g., da) or a fifth deviant one (e.g., ba). ERPs, time locked to the presentation of the fifth syllable, were recorded. Recordings from a left posterior temporal site were found to be sensitive to phonetic information as evidenced by a different neural response to the presentations of the deviant and the identical syllable. Again, these findings point to the early presence of neural proclivities in speech perception.

If language processes have so consistently shown a preference for the left hemisphere, what happens to

children who suffer damage to this hemisphere early on in life? It has been found that most children who suffer an early injury to their left hemisphere go on to acquire language abilities within the normal range, though, sometimes at the lower end of the spectrum. This finding is in direct contrast with the adult picture where damage to the left hemisphere often leads to aphasia. Interestingly, when direct comparisons between children with lesions on the left hemisphere and children with lesions on the right hemisphere are made, a very similar picture emerges for the two groups. A delay in language acquisition is observed in both groups relative to controls pointing to the setbacks of early injuries to overall functional development. These results support the interactive specialization theory that predicts functional compensation given an early injury. In other words, whereas some evidence does suggest that there is an early left hemispheric bias for language perception, it seems that this early bias can be overridden in the face of early injury. This conclusion is consistent with the results of PET studies of children who had neurosurgery to treat drug-resistant epilepsy. These studies suggest that, especially for language perception tasks, early left hemisphere injury is associated with enhanced activation of right-hemisphere areas. Thus, a more diffuse, widespread distribution of language networks in children compared to adults might be the explanation for children's greater recovery from left-hemisphere injury.

Summary and Conclusions

Although much of the basic architecture of the brain is laid down before birth, important aspects of neural development continue to occur after birth. The field of developmental cognitive neuroscience is aimed at understanding how these developments in the brain relate to the emergence and development of cognitive–behavioral skills. Researchers in the field have used a variety of techniques, including those described in **Table 1**, to document the changes in brain activity that occur during typical development and to understand how these processes are altered by brain injury or disease.

Investigations in the fields of face processing, working memory, long-term memory, and language have all examined how the brain areas known to mediate these skills in adults are related to advances in their development. In some cases, a maturational view, in which developments in skills are linked to the maturation of a particular brain region, appears to provide a good explanation of development. For example, there is good evidence that maturation of the dorsolateral prefrontal cortex is linked to development of spatial working memory. In other instances, the interactive specialization or skill-learning views appear to provide a better explanation. For example, development of face processing seems to involve a change from a more widespread, general-purpose visual-processing system to a more focal, face-specific one which would be consistent with both views. Additional evidence that the timing of visual experience is critical for normal development of face processing is in favor of the interactive specialization view, as the skill learning view would not expect this result. However, further studies are needed to provide a more complete account of functional brain development. For example, testing the differing hypotheses of the maturational and interactive specialization accounts requires consideration of changes in activation over the whole brain. Studies that consider activation in only limited brain regions (e.g., only dorsolateral prefrontal cortex and not other regions) might provide results consistent with the maturation view but do not truly rule out other views.

Studies of the plasticity of the brain in response to injury or alterations in sensory input illustrate that there is both remarkable flexibility as well as some constraints on functional brain development. For example, investigations of memory development following early, bilateral hippocampal injury suggest, consistent with the maturational or skill-learning views, that such early injury can result in permanent deficits. In contrast, studies of the development of speech and language following early unilateral cortical injury indicate that there is remarkable flexibility as speech and language outcomes are much better than when similar injury occurs during adulthood. These types of results are more consistent with an interactive specialization view.

The field of developmental cognitive neuroscience has clearly benefited greatly from technological advances that have allowed study of brain structure and function even in human infants. In the future, further improvements in technology and research methods will likely also prove critical, as will the integration of diverse approaches such as behavioral, neuroimaging, genetics, and pharmacology. These types of studies will provide a fuller picture of the mechanisms involved in the functional development of the human brain.

See also: Brain Development; Cognitive Neuroscience; Face Processing; Habituation and Novelty; Hippocampus; Memory; Object Concept; Speech Perception.

Suggested Readings

Diamond A (1996) Evidence for the importance of dopamine for prefrontal functions early in life. *Philosophical Transactions of the Royal Society of London B. Biological Sciences* 351: 1483–1493.

Dehaene-Lambertz G, Hertz-Pannier L, and Dubois J (2006) Nature and nurture in language acquisition: Anatomical and functional brain-imaging studies in infants. *Trends in Neurosciences* 29: 367–373.

De Haan M, Mishkin M, Baldeweg T, and Vargha-Khadem F (2006) Human memory development and its dysfunction after early hippocampal injury. *Trends in Neurosciences* 29: 374–381.

Johnson MH (2001) Functional brain development in humans. *Nature Neuroscience Reviews* 2: 475–483.

Johnson MH (2004) *Developmental Cognitive Neuroscience.* Oxford: Blackwell Publishing.

Johnson MH (2005) Subcortical face processing. *Nature Neuroscience Reviews* 6: 766–774.

Nelson CA, de Haan M, and Thomas KM (2006) *Neuroscience of Cognitive Development: Experience and the Developing Brain.* Hoboken, NJ: Wiley.

Nelson CA and Luciana M (eds.) (2001) *The Handbook of Developmental Cognitive Neuroscience.* Cambridge, MA: MIT Press.

Relevant Websites

http://www.pbs.org – PBS.
http://en.wikipedia.org – Wikipedia, The Free Encyclopedia.
http://www.zerotothree.org – ZERO TO THREE.

Breastfeeding

R A Lawrence, University of Rochester School of Medicine and Dentistry, Rochester, NY, USA

Glossary

Alveolus – A glandular acinus or terminal portion of the alveolar gland where milk is secreted and stored, 0.12 mm in diameter. From 10 to 100 alveoli, or secretory units, make up a lobulus.

Colostrum – The first milk. This yellow, sticky fluid is secreted during the first few days postpartum and provides nutrition and protection against infectious disease. It contains more protein, less sugar, and much less fat than mature breast milk.

Induced lactation – Process by which a non-pregnant female (or male) is stimulated to lactate.
Lacteal cells – Cells that produce the milk.
Lactiferous ducts – The main ducts of the mammary gland, which number from nine to 15 and open onto the nipple. They carry milk to the nipple from the alveoli where it is made.
Lactogenesis – Initiation of milk secretion.
Lobulus – A subunit of the parenchymal structure of the breast made up of 10–100 alveoli, or secretory units. From 20 to 40 lobuli make up a lobus.
Mammogenesis – Growth of the mammary gland.
Oxytocin – A chemical hormone synthesized in the cell bodies of neurons located mainly in the hypothalamus. Oxytocin stimulates the ejection reflex by stimulation of the myoepithelial cells in the mammary gland.
Prolactin – A hormone present in both males and females and at all ages. During pregnancy it stimulates and prepares the mammary alveolar epithelium for secretory activity. During lactation it stimulates synthesis and secretion of milk. At other ages and in the male it interacts with other steroids.

Introduction

The human infant is born to breastfeed. It is not just a simple matter of nutrition but survival and optimal growth itself. Mammals nurse their young. At birth, the offspring finds the teat and latches on. The length of feeds, the duration of feeding at the breast, and the time for weaning are unique to each species and vary with the offspring's level of development at birth and rate of growth and development. The human infant is the most immature of the mammalian species at birth with the exception of marsupials who are suckled in a pouch until they can forage for themselves. The very survival of the mammalian offspring depends upon successful nourishment from the mother. Those who are too weak or those whose mother has insufficient milk do not survive, which is the basic principle of survival of the fittest in all other species until modern times.

The human infant was equally challenged and often succumbed because of the lack of wet nurses or adequate nourishment if the mother could not, or would not, breastfeed. Mortality in infancy was high especially among nobility who farmed their infants out to wet nurses.

Because artificial feeding utilizing bovine milk is readily available, breastfeeding has become less common. The tremendous benefits of being breastfed are continually being confirmed with strong scientific studies and public health initiatives are encouraging women to breastfeed. The true understanding of the significance of breastfeeding to human development begins with understanding the benefits of being breastfed or more dramatically the risks of not being breastfed.

The Benefits of Being Breastfed for the Human Infant

Nutritionally, human milk is designed for the human infant. The constituents are specific for the needs of the human species. No other mammal drinks another species' milk. Although emphasis in nutrition has been on physical growth, the compelling nutritional needs of the human infant are for brain growth. The brain will double in size in the first year of life. The nutrients in human milk are ideal for this growth when in the proper ratios and accompanied by enzymes and ligands that facilitate the digestion and absorption of these nutrients including the microminerals.

Unique features of human milk include the presence of cholesterol, which is not found in infant formulas. No matter how the maternal diet is manipulated with high or low cholesterol and high or low fat, the cholesterol levels in her milk are consistent until weaning is completed. Cholesterol is a constituent of nerve myelination and the biochemical basis of many essential enzymes. Docosahexaenoic acid (DHA) and arachidonic acid (AA), two essential fatty acids that have been associated with brain and neural tissue growth, are found in high levels in human milk and minimally in bovine milk. Controlled studies of both term and preterm infants have demonstrated clearly that those infants who are fed mother's milk (or donor human milk) are more advanced on developmental scales, have greater visual acuity, and auditory acuity compared to those fed on cow's milk, infant formula, and even formula supplemented with DHA. Human milk promotes brain growth so that infants reach their full potential.

Protein is another important dietary constituent and each mammalian species' milk has a milk protein important to the offspring. Human milk has less protein than cow's milk, but it is predominantly whey protein which is easily digested and absorbed. The amino acid profile is equally important. Human milk is low in tyrosine and phenylalanine, which are essential but in excessive amounts can be detrimental to the brain. This is seen in individuals who lack sufficient enzymes to digest phenylalanine, which results in elevated levels and mental retardation as seen in untreated phenyl ketonuria (PKU) disease. An amino acid, taurine, is also necessary for brain growth and present in human milk but not in cow's milk. Although the adult can produce taurine so it is not considered an essential amino acid for adults, infants cannot produce it well and require it in the diet for proper brain development.

Even carbohydrate plays a significant role as the energy for the brain. Human milk has the most lactose of all mammalian milk. When the size of the brain (weight of brain per size of adult of species) is compared, the more lactose in the milk the larger the brain of the species (**Figure 1**). The human has the largest brain per pound of body weight and is the most advanced intellectually.

Clearly, human milk has been carefully engineered biochemically to support and promote the growth of the human brain.

Health Benefits of Being Breastfed

It has long been observed that breastfed infants have fewer infections. Data are dramatic in illustrating the rise in infant death from infection in developing countries as the rate of breastfeeding falls. Fifty per cent of infants in developing countries die in the first year of life if not breastfed. In industrialized countries, the rates of diarrhea, otitis media, and respiratory infections are considerably higher in bottle-fed infants. This is confirmed by increased use of antibiotics and the number of hospitalizations occurring in bottle-fed infants. The infection

protection constituents in human milk have been well studied and are numerous. The value of colostrum, the first milk, is recognized for all species and dairymen and other specialists in animal husbandry know that the offspring cannot be deprived of the colostrum or it will die of infection. Human milk contains human blood cells including T-lymphocytes and macrophages. Other immunologic components include lactoferrin, lysozyme, and many antibodies as well as cytokines and interleukins; all are important host defense mechanisms against bacteria, viruses, and other agents of infection.

The immunologic significance of human milk has been demonstrated in a number of epidemiologic studies reporting that breastfeeding for 4 months or longer can provide some immunologic protection against some childhood-onset diseases. It was noted that a rise in childhood-onset diabetes in Scandinavia was associated with a decline in breastfeeding. In 1991, a prospective long-term study among children in Finland showed a significantly lower incidence of childhood-onset diabetes in children at high risk if they had been breastfed for at least 4 months. Other epidemiologic studies have shown a decreased incidence of insulin-dependent diabetes in breastfed children. A decreased incidence of childhood cancers, especially

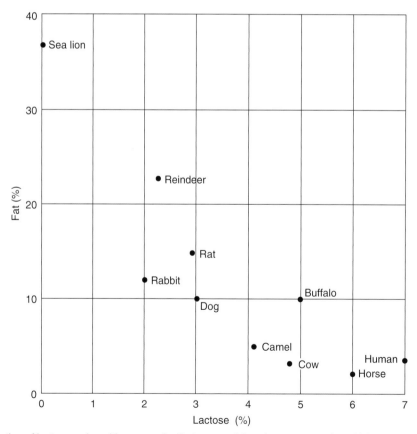

Figure 1   Concentration of lactose varies with source of milk. In general, less lactose, more fat, which can also be used by newborn animal as energy source. Reproduced from Lawrence RA and Lawrence RM (2005) *Breastfeeding: A Guide for the Medical Profession*, 6th edn, figures 4–11, p. 137. Philadelphia, PA: Elsevier, with permission.

leukemia, in children breastfed for at least 4 months has been demonstrated by other investigators. Reduction in the incidence of inflammatory bowel disease and celiac disease was similarly associated with breastfeeding. A large number of bioactive factors have been identified and measured in human milk over the course of lactation. Protection against allergy or at least postponement of the onset of allergic symptoms (eczema, rhinitis, and asthma) in children at high risk for developing allergies (parent or sibling with symptoms) has been associated with exclusive breastfeeding. That bovine protein causes allergies in humans is central to this issue and thus the absence of exposure to bovine protein while breastfeeding prevents allergies. Cross-species intolerance of other milk is demonstrated by the fatal effects of bovine protein on rabbits.

The only contraindications to being breastfed are galactosemia, a genetic ezyme disorder, where the enzyme that digests lactose is missing. HIV or AIDS is the only infection in the mother that contraindicates breastfeeding in developed countries. In developing countries where alternatives to breastfeeding are of poor quality and at risk for contamination, it may well be better to exclusively breastfeed if the mother is HIV positive.

Benefits for the Mother Who is Breastfeeding

Breastfeeding is also the physiologic completion of the reproductive cycle. The breast prepares for lactation during pregnancy as it responds to the milieu of hormones from the placenta and ovaries. The alveolar system develops and arborizes developing a lining of lacteal cells which prepare to make milk. When the placenta is delivered, the breast is ready to respond to the hormones prolactin and oxytocin to produce and release milk. The oxytocin also causes the uterus to contract. As a result, uterine involution is initiated promptly, there is less total blood loss, and there is earlier return to prepartum state in lactating women.

Women who breastfeed also return to their normal weight utilizing the stores deposited during pregnancy to produce milk. In the long range, breastfeeding women are also at less risk for long-term obesity if they breastfeed. Although it would seem counter-intuitive that a woman who breastfeeds would be at lower risk for long-term osteoporosis given the amount of calcium and phosphorus provided to the infant in the milk, the body protectively absorbs more minerals during lactation and during the weaning process. Thus, women who have children and breastfeed them are at lowest risk for osteoporosis, those who have children but do not breastfeed them are at modest risk, and those women who never have children (or breastfeed) are at the greatest risk for osteoporosis.

Epidemiologic studies of breastfeeding women have also demonstrated a lower risk of breast cancer in women who breastfeed. Ovarian cancer has also been shown to be reduced with increasing numbers of pregnancies and increasing duration of time spent breastfeeding, possibly related to the decrease in estrogen during pregnancy and lactation.

Disadvantages of Breastfeeding

Disadvantages to breastfeeding are those factors perceived by the mother as an inconvenience to her because there are no disadvantages for the normal infant. A common concern for some mothers is the fact the mother has to feed the baby every feeding unless she pumps her milk and someone else feeds the baby. The responsibility of caring for a newborn and the time and energy involved is a major cause of maternal fatigue. The infant's father should be encouraged to participate. If the infant is going to receive a bottle of pumped milk or formula, providing it during the night allows the mother some uninterrupted sleep. This may be necessary for the employed mother, unless there is on-site daycare. In traditional societies, the infant is carried with the mother at all times whether she is working in the fields or working in the home and feeding on-demand is traditional.

In addition, certain maternal medications may be a contraindication such as antimetabolites, street drugs, and a few of the antipsychotic medications. There are excellent references available to consult so that breastfeeding is not unnecessarily stopped.

Physical Growth of Breastfed Infants

It has been determined by an international committee of experts that the ideal mechanism for measuring childhood growth is the growth of the exclusively breastfed baby who is otherwise healthy and well-cared for. The important concept is that growth standards should be how children should grow, not how an average number of children do grow. The growth charts that have been in use in the US and elsewhere for a number of years were prepared by the Center for Disease Control by taking multiple small studies of children, sick and well, fat and thin, mostly formula-fed and occasionally breastfed. These data were averaged, and the curves were smoothed and these are the growth charts that have been used for a decade or more in the US. The World Health Organization (WHO) supported an international study to learn how children should grow. Eight thousand six hundred children were followed very closely in six countries (Brazil, Ghana, India, Norway, Oman, and the US) of the world for the first 2 years of life and a little less closely until all of the children had reached 5 years of age. These children were exclusively breastfed for at least 4 months and partially for 1 year. The mothers were healthy. The infants were healthy, and they received sufficient medical care and adequate nutrition. These growth charts have been released by the WHO and make an international standard possible. The experts observe that when nutrition was adequate and health was cared for, the children all over the world grow with a similar pattern.

Psychological Impact of Breastfeeding

The breast prepares to nurse the young during pregnancy. For the many decades when women embraced artificial feeding and wished to raise their children by the book, they sought formulas to relieve them of the total responsibility of nourishing their young. The psychologic benefit of breastfeeding was described during the era of attachment research and 'bonding' to one's infant shortly after birth. The work of Righard and others showed that an infant when placed on its mother's abdomen right after birth would find its way to the breast unassisted in 20 min, crawling, squinching, and wiggling toward the breast. The baby would also latch on and begin to suckle unassisted. Other adaptations of this research showed that when one breast was washed before birth, the infant chose the unwashed breast. Other studies described challenging infants to identify their own mother's nursing pads by presenting the infant with several nursing pads. The infant always chose his mother's. The interpretation of these studies was that infants recognize their mothers by many sensory means including smell.

Animal studies of parenting behavior in various species have shown a strong urge for the female of the species to groom, coddle, and feed her young after delivery. The male of species when given hormones, prolactin, oxytocin, the hormones of lactation, were also stimulated to groom and nurture the offspring. When other females of the species who were not pregnant were given these hormones, they too would groom and nurture a newborn of the species.

In the human, the two hormones that are in abundance during lactation are prolactin which stimulates the production of milk and oxytocin which stimulates its release from the breast. These hormones increase the nurturing tendencies of the human mother as well. The observation has been made that when a woman cradles her baby to the breast, the distance between the mother's eyes and the infant's eyes are equal to the estimated visual range of a young infant. In addition, when the baby suckles, it stimulates the release of oxytocin and prolactin so the mother reflexively reacts to this stimulus at her breast.

The psychologic benefit to the infant of being breastfed is not easily measured, but it can be observed that when the mother breastfeeds, she has to feed the baby herself. Without extreme effort of pumping and bottling she cannot pass the baby off to someone else to feed. If she's breastfeeding, she has to be there for the infant. So the infant has the advantage of being held closely, abdomen-to-abdomen, of being able to look into the mother's eyes, to hear the same heart beat that it had heard *in utero* and smell the same smells. Breastfeeding is an act of comfort and support for the infant in addition to the nutritional advantages so that the psychologic benefits of being breastfed have also been the benefits that have been unchallenged by artificial feeding.

Social Structure

In traditional societies, childbirth is an important event and involves the women of the family, the clan, and the community. In traditional societies, one generation lives with the older generation. Families are big and families stay together. Young children grow up knowing that breastfeeding is the norm. A young girl has the opportunity to observe the young women in the family nourishing their newborns by breastfeeding. It is a natural act. When the young woman comes of age to have her own children, there are generations of women who are experienced and supportive of breastfeeding. The culture mothers the mother, providing her with special foods, teas, and personal care. She is relieved of household obligations and her work is done for her by other women in the clan. The culture supports childbirth and supports breastfeeding so that infants can be breastfed exclusively for months and continued for years. In industrial countries, in our modern times, the community is not supportive of childbirth or breastfeeding. There is no family structure with knowledgeable, experienced nursing mothers available to help the new mother. That is the source of one of the greatest problems with breastfeeding today. Women do not have the social experience of how breastfeeding happens, and there is not a circle of women in their surroundings ready to help them nurse their infant.

It is clear that significant changes in the number of women who initiate breastfeeding and successfully and exclusively breastfeed for at least 6 months will not change markedly until children grow up knowing that breastfeeding is a norm. To that end, a curriculum has been developed by the department of education in the state of New York in cooperation with the department of health. This curriculum directed at children from kindergarten through high school was developed and trialed in the 1990s and is available on the website of the New York State Health Department. A national advertising campaign was initiated in 2004 by the office of Women's Health at the National Institutes of Health (NIH). This campaign used up-to-date advertising tactics to sell a social concept. The campaign was outspoken, assertive, and 'in-your-face'. It had been well established that the general public, and women in particular, knew that breast-feeding was best but that was not a strong enough message to persuade many young women to initiate breastfeeding. The campaign was based on the tagline, "Babies were born to breastfeed" but the message was that there was risk of otitis media, allergy, and other illnesses if one fails to breastfeed one's infant. Whether this approach will significantly change the behavior of mothers is debatable.

Guilt, Anger, Failure, and an Informed Decision

Considerable emotion is precipitated when the subject of breastfeeding and its great benefits is presented to some audiences. Many practitioners indicate that they do not discuss the decision about infant feeding in the prenatal period because they do not wish to 'cause guilt'. Some women, who are well educated and very knowledgeable about the benefits of breastfeeding but have failed to succeed at breastfeeding, express tremendous anger. In some cases, the failure is due to the system and is due to a lack of support and a lack of experienced healthcare providers to assist the mother. Sometimes, however, the cause is a biologic failure to produce milk. There are no formal studies to indicate how many women are unable to actually make an adequate amount of milk. Women with very unusual breast anatomies have been observed to have difficulty breastfeeding. These are women with asymmetric breasts, extremely small breasts, tubular breasts, and extremely inverted nipples. In some occasions, women with extremely large breasts are unable to produce sufficient milk, but the statistical probability of this event has not been calculated. Family history of several generations of women in a given family unable to produce sufficient milk for their offspring would suggest in some cases a genetic origin. The risk of possible failure or insufficiency is very low and does not justify the failure to help a mother make an informed decision about how she will feed her infant based on her circumstances and wishes.

The Anatomy and Physiology of Breastfeeding

Gross Anatomy

The mammary gland as the breast is medically termed comes from the Latin word for breast, 'mamma'. The human mammary gland is the only organ that is not fully developed at birth. It experiences dramatic changes in size, shape, and function from menarche to pregnancy, lactation, and ultimately involution (**Figures 2** and **3**). The gland undergoes three major phases of growth and

development before pregnancy and lactation: phase I – *in utero*; phase II – during the first 2 years of life; and phase III – at puberty.

Embryonic development of the breast begins with the milk streak in the embryo which appears at the fourth week when the embryo is only 2.5 mm long. It becomes a milk line or ridge during the fifth week and continues to develop from thereon. By 16 weeks, gestation, the branching stage of breast development has produced 15–25 epithelial strips that represent future secretory alveoli. The mammary glands of both male and female fetuses from 13 to 40 weeks, gestation have been carefully studied and the development outlined by ultrasound. At birth, the mammary gland is an important anatomic sign-post with which clinicians determine the gestational age of the infant. At the time of birth, in a full-term infant, the nipple is visible, there is a rudimentary duct system and the body of the breast is palpable at 3–4 mm. The breast is relatively quiescent until puberty in the female. Organo-genesis begins in most females between 10 and 12 years of age during which the ductal system begins to sprout resulting in the growth of breast parenchyma with its surrounding fat pad. The ductal tree extends and gener-ates its branching pattern, lengthening the existing ducts and growing lateral buds at the side of the ducts that will ultimately form alveolar buds that will develop into the milk-making part of the gland. The breast continues to mature stimulated by the hormones of the menstrual cycle. This microscopic growth continues monthly until about 28 years of age unless interrupted by pregnancy and lactation. Once menses are established and ovulation has occurred, there has been a sufficient hormonal stimulus to the breast so that the breast can continue to mature and develop lacteal cells capable of making milk. Adoles-cents who become pregnant are anatomically capable of nursing their infants. The ductal system in the breast develops in a very organized manner. The alveoli develop a lining of lacteal cells that are capable of producing the milk that passes into the duct system. The duct system carries the milk toward the nipple. There are about 16–18 ducts that empty into the nipple, each duct draining mul-tiple lobuli and each lobuli containing several alveoli.

The breast is prepared for full lactation from 16 weeks, gestation without any active intervention from the mother. It is kept inactive by a balance of inhibiting hormones that suppress the target cell response during pregnancy. In the first few hours of the first day postpar-tum, the breast responds to the change in hormonal milieu and the stimulus of the newborn sucking to produce and release milk. When the placenta is delivered, the inhibit-ing hormones that have blocked the breast from produc-ing a full amount of milk are gone. The energy expenditure during lactation has suggested an efficiency of human milk production that is greater than 95%. It is estimated that the energy cost of human lactation

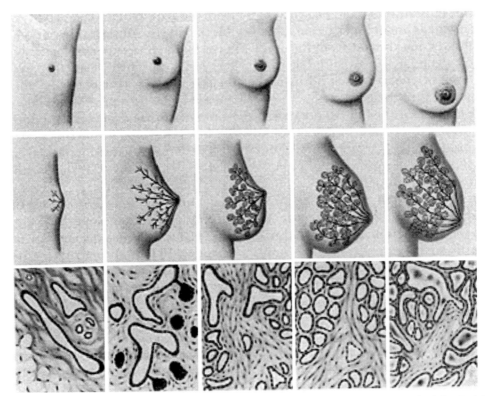

Figure 2   Female breast from infancy to lactation with corresponding cross-section and duct structure. (a–c) Gradual development of well-differentiated ductular and peripheral lobular-alveolar system. (d) Ductular sprouting and intensified peripheral lobular–alveolar development in pregnancy. Glandular luminal cells begin actively synthesizing milk fat and proteins near term; only small amounts are released into the lumen. (e) With postpartum withdrawal of luteal and placental sex steroids and placental lactogen, prolactin is able to induce full secretory activity of alveolar cells and release of milk into alveoli and smaller ducts. Reproduced from Lawrence RA and Lawrence RM (2005). *Breastfeeding: A Guide for the Medical Profession*, 6th edn, figure 2–3, p. 43. Philadelphia, PA: Elsevier, with permission.

is minimal. In contrast to most organs that are fully developed at birth, the mammary gland undergoes most of its morphogenesis postnatally in adolescence and in adulthood. Lactation is an integral part of the reproductive cycle of all mammals including humans.

The hormonal control of lactation can be described in relation to five major stages in the development of the mammary gland. (1) embryogenesis; (2) mammogenesis or mammary growth; (3) Lactogenesis or initiation of milk secretion; (4) lactation or full milk secretion; and (5) involution when the infant is weaned. Some women can express colostrum after about 16 weeks, gestation. The ability to produce milk can be measured by measuring the level of lactose and alpha lactalbumin. Stage I of lactogenesis actually starts during parturition and is heralded by significant increases in lactose, total protein, amino globulins, and by decreases in sodium and chloride and the gathering of substrate from milk production in the mother's system. The composition of prepartum secretion is fairly constant until delivery. Lactogenesis is initiated in the postpartum period by a fall in the plasma progesterone while prolactin levels remain high. The initiation

of the process does not depend on suckling by the infant until the third or fourth day when the secretion declines if milk is not removed from the breast.

Stage II lactogenesis includes the increase in blood flow and oxygen and glucose uptake as well as the sharp increase in citrate concentration considered a reliable marker for lactogenesis stage II. Lactogenesis stage II at 2–3 days postpartum begins clinically when the secretion of milk is copious, and biochemically when the plasma alpha lactalbumin levels peak. The major changes in milk composition continue for 10 days changing from colostrum to mature milk.

The establishment of a mature milk supply once called galactopoiesis is now referred to as stage III of lactogenesis. The composition of the milk over these first 10 days has been carefully studied, and the pattern is consistent among mothers.

The Role of Prolactin

Human prolactin is a significant hormone in pregnancy and lactation. Prolactin also has a range of actions in

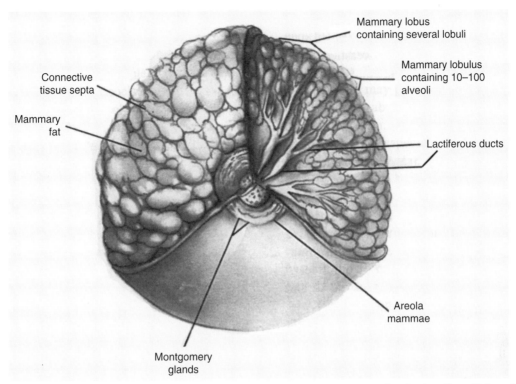

Figure 3 Schematic drawing of the breast to illustrate its orderly construction where the aveoli make the milk and it flows through the duct system to the nipple and is ejected when the breast is stimulated. Reproduced from Lawrence RA and Lawrence RM (2005). *Breastfeeding: A Guide for the Medical Profession*, 6th edn, figures 2–7, p. 48. Philadelphia, PA: Elsevier, with permission.

various species that is greater than any other known hormone. Prolactin has been identified in many animal species whether they nurse their young or not. Prolactin, however, has been shown to control nonlactating responses in other species and has been identified with more than 80 different physiologic processes. Prolactin exists in both the male and the female of the species and normal prolactin levels vary physiologically throughout life. Baseline levels of prolactin are essentially the same in normal males and females. Both experience a rise in prolactin levels during sleep. There is normal diurnal variation in levels as well. Many normal activities change the prolactin levels including stress, anesthesia, surgery, and exercise. During lactation, prolactin stimulates the lacteal cells to produce milk. The high prolactin levels that existed during pregnancy gradually drop in the first 2 weeks postpartum if the breast is not stimulated to produce milk. When prolactin levels are measured during lactation, it is noted that lactating women maintain a high baseline of prolactin and when the breast is stimulated, the prolactin levels surge to almost double the baseline. It is this surge that has been associated with successful milk production. It has also been observed that the prolactin levels do not rise unless the breast is stimulated either by the infant's suckling or by pumping even though the mother may be in contact with her infant (**Figure 4**).

The Role of Oxytocin

Oxytocin was the first hormone studied in relation to breastfeeding and to the let-down reflex. It was studied in detail because it was measurable and could be isolated in the laboratory. Oxytocin, however, is not just a female hormone. It is produced by both males and females and not just during reproduction in the female. It is credited with producing increased responsiveness to closeness, openness to relationships, and nurturing. The oxytocin circulating during breastfeeding has been credited with producing calm, lack of stress, and an enhanced ability to interact with the infant. The calm and connectedness system is part of a system of nerves and hormones that together trigger these effects. Oxytocin is a polypeptide found in all mammalian species and works through a mechanism whereby it activates receptors on the outer surface of the cell membrane. Oxytocin is produced in the supraoptic and periventricular nuclei of the hypothalamus. Receptors have been identified for oxytocin in the uterus and in the breast. It is well known that oxytocin causes the uterus to contract and in the early days of lactation, the mother will be aware that the uterus is cramping down when the baby goes to breast. It is this effect that allows the mother to recover postpartum more rapidly if she breastfeeds. Oxytocin has been called the hormone of calm, love, and healing. In relation to lactation, however, it is the hormone that

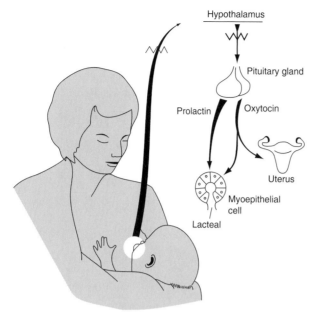

Figure 4 Diagram of ejection reflex arc. When infant suckles breast, mechanoreceptors in nipple and areolar are stimulated, which sends a stimulus along nerve pathways to the hypothalamus, which stimulates the posterior pituitary gland to release oxytocin. Oxytocin is carried via the bloodstream to the breast and uterus. Oxytocin stimulates the myoepithelial cells in the breast to contract and eject milk from the alveolus. Prolactin is responsible for milk production in the alveolus. Prolactin is secreted by the anterior pituitary gland in response to suckling. Stress such as pain and anxiety can inhibit let-down reflex. Sight or cry of infant can stimulate release of oxytocin but not prolactin. Reproduced from Lawrence RA and Lawrence RM (2005). *Breastfeeding: A Guide for the Medical Profession*, 6th edn, figures 8–18, p. 290. Philadelphia, PA: Elsevier, with permission.

stimulates the let-down of the milk in all species. The neuroendocrine control of milk ejection is illustrated in **Figure 5**. It is noted that when the nipple is stimulated, it sends a message via the afferent nerves to the brain and to the hypothalamus and stimulates the release of both oxytocin and prolactin through neuroendocrine control.

The amount of milk produced by a fully lactating woman averages between 750 and 1000 ml a day if she is nursing a full-term, singleton infant. Milk production, however, can exceed that depending on the demand as the breast produces the amount of milk that is demanded by the suckling infant or pumping. Therefore, women with twins and triplets can produce considerably larger volumes. Women have also briefly nursed quadruplets and quintuplets. The problem is not the ability to make milk but the time to feed the infant. Nursing twins is usually done simultaneously with the two infants at the breast initially. Studies done to try to force breastfed babies to take more milk by having their mothers pump after every feeding so they increase their production have shown that breastfed babies cannot be overfed. The mechanism of this is not clear, although it is assumed that there is a constituent of human milk that turns off the infant's appetite at an appropriate time.

In the human, lactogenesis occurs slowly over the first few days postpartum as progesterone levels drop. Women experience milk come in as the feeling of fullness between 40–72 h usually corresponding to the degree of parity with multiparas sensing this more quickly than primiparas. The volume of milk increases over time for the first 2 weeks starting at less than 100 ml a day and increasing to about 600 ml a day at 96 h.

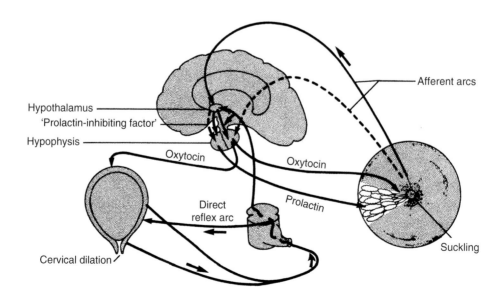

Figure 5 Neuroendocrine control of milk ejection. Reproduced from Lawrence RA and Lawrence RM (2005) *Breastfeeding: A Guide for the Medical Profession*, 6th edn, figures 3–9, p. 80. Philadelphia, PA: Elsevier, with permission.

The Composition of Human Milk

The biochemistry of human milk encompasses a mammoth supply of scientific data and information, most of which has been generated since the 1970s. Each report or study adds a tiny piece of the complex puzzle of the nutrients that make up human milk. The answers to some questions, however, still elude us. Even the simple question of the volume of milk consumed at a given feeding is a scientific challenge. The methodology must be accurate, reproducible, noninvasive, and suitable for home-use night and day, and must not interrupt breastfeeding. Advances in analytic methods bring greater sensitivity and speed to the analysis of milk composition. Milk brings both nutrients and non-nutrient signals to the neonate. All milk contain the nutrients for physical growth and development. When the offspring develops rapidly as in the case of rats, mice, and rabbits, the milk is nutrient-dense. When it develops slowly as in the human and other primates, the milk is more dilute. All mammalian milk contains fat, carbohydrate, and proteins, as well as minerals, vitamins, and other nutrients. The organization of milk composition includes lipids and emulsified globules coated with a membrane, colloidal dispersions of proteins as micelles, and the remainder as a true solution. At no other time in life is a single food adequate as the sole source of nutrition and wellbeing. Human milk is not a uniform body fluid but a secretion of the mammary gland of changing composition. The fore milk which is the first milk secreted at a feeding differs from hind milk. The fat content increases as suckling continues during a feeding. Colostrum differs from transitional and mature milk. Milk changes over time of day and as time goes by. As concentrations of protein, fat, carbohydrate, minerals, and cells differ, physical properties such as osmolarity and pH change. The impact of changing composition of the physiology of the infant gastrointestinal tract is beginning to be appreciated. Many constituents have dual roles, not only nutrition but infection protection, immunity, or a host of other effects.

The 200 or more constituents of human milk include a tremendous array of molecules whose descriptions continue to be refined as the qualitative and quantitative laboratory techniques are perfected.

Not only does human milk contain all of the nutrients necessary for the human infant to grow, i.e., protein, fat, carbohydrate, minerals and vitamins, but it also contains enzymes that facilitate the digestion and absorption of these nutrients. The enzymes that digest starch, sugars, fats, and proteins have been identified. There are also hormones in human milk, especially prolactin and oxytocin, but also steroid hormones such as gestagens, estrogens, corticoid, androgens, and opiate, like peptides that contribute to the infant's growth. Enzymes and hormones lose their activity when the milk is pasteurized. The presence of thyroid hormones is very important for the newborn infant, especially because it protects the hypothyroid infant in the first few weeks of life. There are many other compounds whose role has not been completely identified.

The host-resistant factors and immunologic significance of human milk is very important in the protection of the human infant against infection. Some of the most dramatic and far-reaching advances in the understanding of the immunologic benefits of human milk have been made using newer techniques to demonstrate the specific contributions of the numerous 'bioactive factors' contained in human milk. The multifunctional capabilities of the individual factors, the interactive coordinated functioning of these factors, and the longitudinal changes in the relative concentrations of them over the duration of lactation make human milk unique. The immunologically active components of breast milk make up an important aspect of the host defenses of the mammary gland in the mother. At the same time, they complement, supplement, and stimulate the ongoing development of the infant's immune system. There are cellular components of human milk in greater number in the colostrum but persisting throughout lactation. The cells are b-lymphocytes and phagocytes, and are active protectors against infection. The cells also include cells from the T-cell system and the B-cell system as well as polymorphonucolecytes. There are hormonal factors such as immunoglobulins which include secretory IgA, IgM, and IgG. Other protective factors include resistance factor, lysozyme, lactoferon, and interferon. Interleukins and cytokines have also been identified in human milk. Their levels are persistent and they have been identified as protective against various infections. Nucleotides have also been isolated and studied. It is well established that human milk not only protects against infection within the gastrointestinal tract but outside the gastrointestinal tract as well.

The Initiation of Successful Breastfeeding

Successful nursing depends on the successful interaction of a mother and an infant with appropriate support from the father, the family, and available healthcare resources. Both mothers and infants vary. No simple set of rules in hospitals can be outlined to guarantee success. In fact, one of the difficulties has been a rigid system established for initiating lactation in the hospitals that did not fit all mother–infant couples. Nowhere in medicine do one's personal interests or prejudices become more evident than in the area of counseling about childbirth and breastfeeding. Having a child does not make one an expert on the subject. Conversely, not having a child does not preclude the development of exceptional skills. Historically, rigid dogmas have directed the management of lactation. In an effort to replace these with more rational management, new dogmas have arisen. The key to the management of a mother–infant nursing couple is establishing a sense of confidence in the mother and supporting her with simple answers to her questions when they arise. Good counseling also depends on understanding the science of lactation. Then, when a problem arises, a mechanism is already in place for the mother to receive help from her physician's office before the problem creates a serious medical complication.

The ability to lactate is characteristic of all mammals from the most primitive to the most advanced. The divergence of suckling patterns, however, makes it urgent that human patterns be studied specifically. Some aquatic animals such as whales nurse under water. Others such as the seal and sea lion nurse on land. A variety of postures are assumed by different terrestrial animals. Nursing may be continuous as in the joey attached to a marsupial teat or widely different intervals characteristic of the species and parallel to the nutrient concentrations of the milk. The interval may be a half an hour in the dolphin, an hour in the pig, a day in the rabbit, 2 days in the tree shrew, or almost a week in the whale and hippopotamus. Although many anatomic distinctions exist as well between species, the principal mechanism of milk removal common to all mammals is the contractile response of the mammary myoepithelial under hormonal influence of oxytocin released from the neurohypothesis. The key function in all species is effective control of milk delivery to the young in the right amount at the appropriate intervals. This requires a reproduction system, exit channels, prehensile appendage, an expulsion mechanism, and a retention mechanism. The primary, secondary, and tertiary ducts form an uninterrupted channel for the passage of milk from the milk producing alveoli to the mammary duct system. The principal object of the suction produced by the facial musculature of the young is to draw the nipple into the mouth and retain it there. Positive pressure is used to expel the milk from the gland. While many species have been studied in the past, study of the human infant has until recently used artificial feeding with artificial nipple and bottle. The mechanism of suckling at the breast is quite different than sucking on an artificial nipple.

When the human infant is put to breast for the purpose of feeding, there are simple guidelines to facilitate the process. The baby should be held so that mother and infant are facing each other and the infant is looking squarely at the breast without turning his head. All normal infants are known to have a rooting reflex which is the response to the stimulus around the mouth toward the direction of the stimulus. Therefore, in order to have the baby latch on to the human nipple, the center of the lower lip of the infant should be stimulated preferably by the maternal nipple. This will cause the infant to extend the tongue and draw the nipple and areolar into the mouth and compress it against the hard palate. The grasp is sealed with the lips. The infant tongue remains in place compressing the elongated areolar and nipple against the hard palate with peristaltic motion. The peristaltic motion begins at the tip of the tongue, goes to the posterior pharynx and down the esophagus, initiating peristaltic motion all the way through the intestinal tract. With this peristaltic motion, the milk is ejected and carried along to the posterior pharynx and swallowed as one action. Therefore, while breastfeeding, it is not necessary for the infant to suck and then swallow as it is with an artificial nipple and bottle where a bollus is formed and the infant has to swallow it. This difference is critical to the smooth ejection of the milk and swallowing it. It is also significant that infants as immature as 28 weeks, gestation can suckle at the breast when they cannot suck and swallow from a bottle. Normal full-term infants are able to go to the breast shortly after delivery and will latch onto the nipple and areolar and begin to suckle. Infants have been sucking and swallowing amniotic fluid *in utero* since about 16 weeks, gestation and this is the continuation of that physiologic process.

After the initial feeding shortly after birth, the infant usually goes into a deep sleep and recovery phase. The infant usually does not waken to feed again for 4 to 6 h. They maintain their blood glucose very well with this short feeding. It is estimated that the infant receives between 20 and 30 ml of colostrum in this first feeding. Colostrum contains many immunoglobulins, antibodies, and some nutrition. In the first few days, it is appropriate to feed the infant on demand, that is, when the infant begins to rouse and shows signs of hunger. Signs of hunger include moving around, bringing the fist to the mouth, sucking on the fist, and becoming aware of one's surroundings. Crying is a late sign of hunger and may interfere with the peaceful latching onto the breast. In the first 96 h of life, the infant will probably nurse every 2–3 h and at least 8–12 times a day. In the first month of life, infants continue to feed 8–12 times in 24 h. The emptying time of the stomach of human milk is 90 min. This is in comparison to the emptying time of formula which is 3 h and the emptying time of homogenized milk which is about 6 h. The rapid digestion and absorption of human milk requires more frequent feedings. Frequent feeding is not a sign of inadequate milk; it is a sign of the digestibility of the feeding. As the infant takes a greater volume at the time of feeding, the frequency of eating in 24 h gradually reduces to eight times and then six times a day. It is recommended by the American Academy of Pediatrics and the WHO that infants be exclusively breastfed, that is, receive no other liquid or food except breast milk (by breast or bottle) until about 6 months of age. It is not necessary to begin solid foods until after this time when the infant has begun to not only hold his head up but to sit up and is able to take food into the mouth and swallow it.

Weaning

There is no set time for weaning the human infant. After 6 months of exclusive breastfeeding, the infant should be introduced to appropriate solid foods and gradually over the next 6 months develop a feeding schedule of three meals a day and several snacks. Breastfeeding is continued with the feedings and usually a large breastfeeding first thing in the morning and the last at night. By this time, the infant should be sleeping through the night. Different species wean their offspring at different times. Various calculations for the appropriate weaning time of the infant have been based on other anthropologic determinants. The American Academy of Pediatrics has said that the infant should be exclusively breastfed for 6 months, continue to breastfeed until a year, and then for as long thereafter as mother and child wish. In developing countries, infants are breastfed an average of 4.2 years. Because of the social stigma in some countries, mothers do what is referred to as 'closet nursing', that is, they continue to nurse their infants beyond a year or two in private. It has been shown that the composition of human milk does continue to provide infection protection and other nutrients to the infant. It is well documented that many children beyond 2 years are nursed usually first thing in the morning, last thing at night and at periods of stress while on a full diet. This is referred to as comfort nursing.

The technique for weaning an infant involves preferably a slow process where one feeding a day is discontinued and then another allowing the breast to adjust for lower production and allowing the child to accommodate. The older child who is feeding only once or twice a day may finally be distracted during a period of feeding or substitute activity provided so that the final feeding can be discontinued. Abrupt weaning is to be avoided except in an emergency because of its stress on the infant and its distress to the lactating breast which continues to make milk and may become painfully engorged.

Nursing the Adopted Child and Induced Lactation

There are women who have been unable to conceive and adopt a child in order to parent. Some of these women wish to at least provide nourishment for this child they could not. It is possible to induce lactation. The process requires commitment and dedication to the process. It involves stimulating the breast mechanically with a good electric pump on a regular basis increasing the length of time and the frequency. It should in a week or two involve pumping every 3–4 h for at least 20 min around the clock. The double pump system stimulating both breasts simultaneously saves time and increases the stimulus to the breast. It is unusual to be able to exclusively breastfeed the adopted child, but it is not the volume of milk but the relationship that is important.

A woman who has previously lactated and may wish to relactate following early weaning or to feed an adopted child can do so. The process is similar in terms of pumping as described above. In this case, however, the breast responds more quickly and effectively. Relactation can succeed in producing enough milk to exclusively breastfeed the infant.

See also: AIDS and HIV; Allergies; Asthma; Attachment; Endocrine System; Failure to Thrive; Feeding Development and Disorders; Healthcare; Immune System and Immunodeficiency; Newborn Behavior; Nutrition and Diet; Social Interaction; Suckling.

Suggested Readings

Baumslag N and Michels DL (1995) *Milk, Money and Madness: The Culture and Politics of Breastfeeding.* Westport, CT: Bergin and Garvey.

Huggins K (2005) *The Nursing Mother's Companion.* Boston, MA: Harvard Common Press.

La LL (2004) *The Womanly Art of Breastfeeding,* 7th edn. Schaumburg, IL: La Leche League.

Lawrence RA and Lawrence RM (2005) *Breastfeeding: A Guide for the Medical Profession,* 6th edn. Philadelphia, PA: Elsevier.

Meek JY (ed.) (2002) *AAP. New Mother's Guide to Breastfeeding.* New York: Bantam Books.

Schanler R (ed.) (2006) *AAP, ACOG. Breastfeeding Handbook for Physicians.* Elk Grove Village, IL: American Academy of Pediatrics.

Stuart-Macadam P and Dettwyler KA (1995) *Breastfeeding: Biocultural Perspectives.* New York: Aldine De Gruyter.

Relevant Websites

http://www.ars-grin.gov – Agricultural Research Service, United States Department of Agriculture.

http://vm.cfsan.fda.gov – Center for Food Safety and Applied Nutrition, Food and Drug Administration.

http://www.consumerlab.com – Consumer Lab.

http://www.herbmed.org – HerbMed.

http://nccam.nih.gov – National Center for Complementary and Alternative Medicine, National Institutes of Health.

http://www.ncahf.org – National Council Against Health Fraud.

http://www.quackwatch.com – Quackwatch.

Categorization Skills and Concepts

L M Oakes, University of California, Davis, Davis, CA, USA

Glossary

Categories – Collections of items that have common features; often these groups are in the world.

Categorization – Forming groups of objects that share some commonalities.

Concepts – Mental representations of groups of objects that share some set of commonalities.

Correlated attributes – The relation between multiple features that help to define some categories (e.g., birds have wings, feathers, and fly).

Dishabituation – A significant increase in looking time that infants often exhibit when shown a novel stimulus, following habituation to one or more stimuli.

Habituation – A method for assessing infants' perceptual and conceptual abilities. Habituation is a process by which infants' looking time decreases over repeated presentations with a familiar stimulus.

Multiple habituation – A method in which multiple related stimuli are presented during the habituation phase of an experiment. If infants detect the relation among those stimuli, their looking time will habituate, or decrease, over trials.

Prototype – A summary representation of a category that includes the average feature values of the encountered instances.

Representation – The mental storage of information.

Sequential-touching – A method for assessing infants' concept and categorization skills that depends on their spontaneous touching of items. When infants are presented with multiple items from two categories simultaneously, they often will touch in sequence items from one category.

Introduction

Categories and concepts are essential to understanding and organizing the rich environment around us. At every moment, we encounter new and familiar objects and people. It would be overwhelming to respond to each object, person, and event as completely new and unique. Creating a separate memory for each encounter with an item or event would result in an enormous amount of stored information, making cognitive processes cumbersome at a minimum, and quite likely some aspects of cognition would be impossible. Grouping objects into categories, therefore, helps us to more efficiently and effectively learn and remember the objects, people, and events we encounter. Moreover, understanding the referents of labels such as dog, cup, shoe, and tree requires categorizing different instances as dogs, cups, shoes, and trees, and having formed a representation that relates those instances together. Finally, inferring that a new dog will bark, that a new cup can contain liquid, that a new shoe can be put on one's foot, and that a new tree will have leaves, also depends on categorizing and forming concepts. Thus, many aspects of cognition (including memory, language, and problem solving) critically depend on concepts and categorization skills.

It should be obvious that these skills are important even in infancy. Young infants are faced with an enormous number of new objects, people, and events everyday. Making sense of this incredible amount of information is facilitated not only by remembering individual instances, but also by remembering what those instances have in common, perhaps even remembering similar items in clusters or groups. Recalling previously encountered collections of objects or general categories of actions will certainly help infants learn new words – although it is possible, and even likely, that infants learn commonalities among objects and events by hearing labels. The point is that concepts and categorization skills are an important part of cognition early in life.

Defining Categorization, Categories, and Concepts

Before discussing what we know about concepts and categorization skills in infancy, we must define our terms.

Often, the terms concepts and categories are used interchangeably, and rarely are they explicitly defined. However, understanding how infants categorize and form concepts depends on being clear about what we mean by these terms. Thus, as a first step we define these terms.

Categorization

Categorization refers to the process of forming groups of objects that share some common features. For example, dogs are all generally dog-shaped, eat dog food, wag their tails, and have puppies. Of course, this definition leaves open what a feature is – are features characteristics that you can see (such as shape, color, texture)? Can behaviors (e.g., barks, tail-wagging) and nonobvious properties (e.g., DNA) be features? The lack of a clear definition of feature contributes to a debate in the literature about whether all instances of categorization involve the same process or set of processes. For example, would we form a group of animals with the same general body and head shape in the same way as we would form a group of animals that move independently, are alive, and have similar digestive systems? Researchers have asked do we have a single process or set of processes for forming any group of items with common features, or do we have different processes or set of processes for forming groups depending on the kind of features being considered? For example, we may have one domain-general mechanism, or set of mechanisms, that allow us to detect commonalities among any kind of entities (pictures of colored shapes, speech sounds, abstract ideas) and form groups of common items. By this view, although there may be different kinds of features, the differences between those features are not important for understanding how we form categories. Alternatively, we may possess domain-specific mechanisms for dealing with different kinds of features – we use one set of mechanisms for forming groups of visually similar entities, another set for forming groups of similar speech sounds, another for forming groups of similar abstract ideas, and so on. By this view, the kind of feature being considered determines the particular categorization process invoked.

Another issue is whether categorization is itself a single process or whether categories result from several processes (such as memory, perception, and attention) working together. For example, we may possess a mental process, categorization, which is used to form groups of similar items. Like attention, perception, and memory, categorization may be a characteristic of the mind. The input to this process would be information from those other mental processes, and the output would be representations, or stored information, of a group. Alternatively, categorization itself may not be a process, but rather representations of groups of similar objects may emerge as a result of attention to, perception of, memory for, and comparison of instances. As these processes worked together, representations in memory would become linked or new representations of groups of objects would be formed. Regardless of which side one takes in such arguments, categorization refers to forming groups of objects with common features.

Categories

Categories are groups of objects that share a common feature or set of features. Usually, categories are thought to be out in the world, and often, but not always, such groups have labels (e.g., cat, flower, and car). When researchers talk about categories, it can sometimes seem like objects belong to only a single category – a particular Basset Hound, for example, is a dog. In reality, any individual item belongs to several categories – that Basset Hound is not only a dog, she is a pet, she is one of Marty's belongings, she is an animate creature, and she is a Californian. So, despite the fact that it is clear that categories are groups of similar objects, it is not always clear what real categories are (e.g., are whales only categorized with mammals, or can they also be categorized with sea creatures such as fish or sharks?).

Central to this debate is what one considers important features of categories. Features that are easily perceived – such as color and shape – are often dismissed as not important for real categories. Whales are not fish, for example, because superficial similarities in shape do not define real categories. Deep, conceptual features – such as warm-blooded and gives birth to live young – are more central. Such features often cannot be directly observed, but must be acquired through scientific methods or from an expert. For some researchers, both easily perceptible and less-obvious features are equally good types of commonalities. For others, deep, conceptual commonalities are key to true categories.

Categories are often organized hierarchically – dogs are types of animals, poodles are types of dogs. The point is that some categories encompass other categories – as you move up the hierarchy, categories become broader and more abstract (e.g., a dog and a worm are only similar in the most abstract of ways, and the features they share in common are extremely general). As you move down the hierarchy, categories become narrower and characterized by high levels of perceptual similarity (e.g., all pugs look alike, and it may be difficult to tell two pugs apart). Understanding the relation between different categories requires understanding that the same object can be categorized in many different ways. There has been significant debate in the literature about whether infants first form very general categories and only later become sensitive to narrower, exclusive categories lower in the hierarchy, or whether infants first learn the narrow, perceptually similar categories and gradually combine them to form the higher-level categories. Although there is general consensus that infants are sensitive to broad categories first, there remains significant debate about whether infants can flexibly categorize items at different hierarchical levels depending on the context.

Concept

The term concept generally refers to one's mental representations of categorical groups. Representation simply refers to how the information is stored mentally, or in the head. Not surprisingly, there is a significant debate regarding what is necessary for something to be considered a concept. Is any stored information about a collection of objects a concept? Is an infant's representation of what a group of objects looks like, without any additional information (e.g., is alive, has a brain), a concept? Or, must children store deep, nonobvious features for a representation to be considered a concept? For some researchers, concepts include both kinds of information and it is impossible to make a distinction between representations that include only perceptual information and those that include both perceptual and other kinds of information. For others, representations that include only information such as what an object looks like are not concepts, but rather are perceptual representations. For these researchers, concepts must include information about nonobvious features (e.g., can move independently, drinks), and the distinction between perceptual representations and concepts is extremely important.

There is also controversy about how to obtain evidence of concepts in infants. We can ask older children questions, probing their knowledge about categorical groups (do they know that dogs have puppies, are alive, etc.). Determining that infants have such knowledge is more difficult. Some have argued that the actions that infants perform on objects (e.g., making a dog drink from a cup) provide insight into infants' conceptual knowledge. However, even this evidence is quite controversial – how do we know that when infants make a small toy dog drink from a cup that it reflects their understanding that real, live dogs drink? This issue is yet to be resolved. Regardless of which position a researcher takes, a generally agreed upon definition of concept is that it is the mental representation of a category.

How Do We Measure Infants' Categorization, Categories, and Concepts?

How do we assess infants' concepts and categorization skills? Two methodological innovations in the late 1970s and early 1980s led to an explosion of studies examining

early categorization and concepts. The following paragraphs briefly describe these methodological innovations and the procedures that emerged from them. The final two sections discuss what we have learned about infants' categorization, as well as a new framework for understanding what we have learned. The results described in these final two sections all came from the procedures described in the following paragraphs.

The Multiple Habituation Procedure

The first methodological innovation was the development of the multiple habituation procedure. In the standard habituation procedure, which has been used since the 1960s to study memory and perception, infants are shown a single stimulus on a series of trials (see **Figure 1**, top). Initially, they exhibit a high level of looking, indicating high levels of interest or processing. But as the same stimulus is shown repeatedly, looking duration decreases, indicating a waning of interest or processing. Typically, once infants' looking reaches 50% of the level that it was at the beginning, they are said to have habituated. Following this attenuation of looking, infants will show an increase in looking when a novel stimulus is shown – for example, a circle following a series of presentations with a square.

In the multiple habituation method, several different items are presented during the habituation phase (see **Figure 1**, bottom). Categorization is tested in this procedure by presenting several items from the same category – for example, several different dogs, faces, or vehicles. Researchers make inferences about how infants categorized those dogs, faces, or vehicles from how they respond to new items from the category. If infants are sensitive to the adult-defined category presented during familiarization, they should fail to dishabituate to the new items from the same category.

In 1979, Leslie B. Cohen and Mark Strauss published one of the first studies using this method to examine infants' categorization of faces. In this study 18-, 24-, and 30-week-old infants were habituated to one or more pictures of faces. Some infants received habituation trials with a single picture, some infants received habituation trials with different pictures of the same face, and some infants received habituation trials with pictures of multiple faces. Infants then were tested with a new female face. The logic was this: if infants remember the particular face(s) or picture(s) presented during the habituation phase, they should dishabituate, or increase their looking to, the new female face. But, if they have recognized the commonalities among the multiple faces or pictures seen during familiarization and the novel female face, they should fail to dishabituate (i.e., not show a significant increase in looking) to the novel female face. Only 30-week-old infants appeared to categorize the

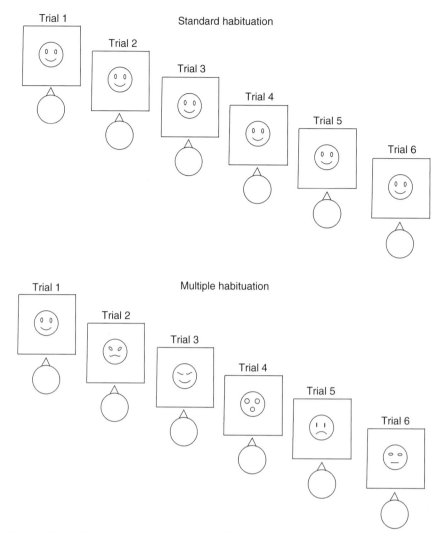

Figure 1 Schematic depiction of the sequence of events in visual habituation tasks. In the standard habituation task (top), infants are shown a single stimulus on several successive trials. In the multiple habituation task (bottom), infants are shown several different stimuli on a series of successive trials. In each version of the task, infants should look a long time on the first few trials, but their looking duration should decrease on subsequent trials as they learn about and remember the stimulus or type of stimuli.

faces. They dishabituated when familiarized with a single picture, but not when familiarized with multiple pictures of the same face or with multiple faces. The very youngest infants seemed to learn a specific picture, apparently not even categorizing different pictures of the same face as instances of a single person.

The multiple habituation procedure became the primary tool for studying infants' categorization, and many studies have shown infants respond to categories in this procedure. There are many different variations of this procedure – some researchers habituate infants as described above, others use a fixed familiarization phase; some studies present infants with pictures of objects, other present infants with real objects; in some studies items are presented one at a time and in other studies items are presented two at a time. One question that has not yet been resolved is how these methodological differences contribute to how infants categorize in these studies. Two variations of this procedure that have become widely used to study infants' categorization will be described next.

A variation: Familiarization and visual paired-comparison

Most modern studies of categorization provide infants with a relatively short familiarization period (e.g., six exposures to items from a category). Importantly, in this variation, two pictures are presented at the same time, rather than one picture as was more typical in early studies (see examples of each procedure in **Figure 2**). Thus, although infants only receive six exposures or trials, on each exposure they see two different category exemplars.

Following such familiarization, infants are then shown a pair of new stimuli, one from the nowfamiliar category and one from a different category. For example, after familiarization with horses, an infant might see a new horse paired with a dog. The logic is that if familiarization with horses causes infants to recognize, remember, or learn a narrow category that includes horses but not dogs, they will prefer the picture of the new dog to the picture of the new horse. But, if familiarization with horses causes infants to recognize, remember, or learn a broad category that includes both horses and dogs (e.g., four legged animals), then they will look equally at the new dog and the new horse. Several studies have shown that infants as young as 4 months respond to adult categories such as dog or horse when tested in the visual familiarization task.

Another variation: Object examining

Another variation that is widely used is the object examining task. The main innovation here is that rather than showing infant pictures of objects, the experimenter actually hands infants small plastic replicas of category items – for example, birds, horses, or cars (see **Figure 3**). In all other respects the task is the same as a familiarization procedure: infants are first familiarized with a series of items from within a category, and then their responding (measured as looking) to new items from the now familiar category and from a different category is assessed. The reasoning is that following familiarization with an adult-defined category such as dog, infants who are sensitive to this narrow category will be more interested in (i.e., look longer at) a novel item from a different category, such as a bird or horse, than in a novel item from the now-familiar category. Studies using this task have revealed that infants between 6 and 14 months of age are sensitive to a categorical distinction such as dog vs. horse, that perceptual similarity becomes less important in infants' attention to categories such as animal, and that infants seem to recognize global or broad categories

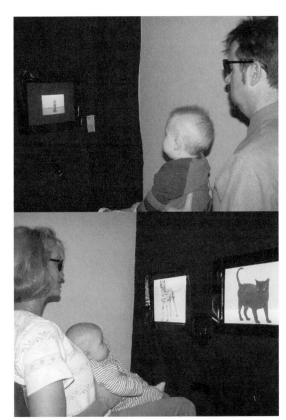

Figure 2 Examples of habituation (top) and visual familiarization (bottom) procedures used to test infants' categorization. In each procedure, infants are seated on a parent's lap, facing one or two computer monitors. The parent is provided with glasses that occlude their view of the stimuli to eliminate bias from parental response to the stimuli. An observer, seated out of sight, records infants' looking time via videorecording (in each photograph a camera is below the computer monitors) or peep holes in the display. In habituation (top) infants are presented with a single stimulus on each trial until their looking time decreases. In visual familiarization (bottom) infants are presented with two stimuli on each trial for a fixed number of trials.

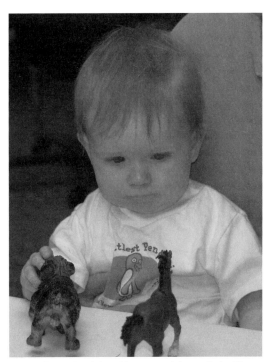

Figure 3 Example of the object-examining task. In this task, infants are presented with real, three-dimensional replicas of category items and they are allowed to play with them in any way they choose – touch, mouth, or bang them. Infants are presented with several stimuli from one category on successive trials, and then they are tested with items from the familiar category and from a novel category (as depicted here). Infants' duration of looking on each trial is recorded.

(e.g., animal, vehicle) at an earlier age than they notice exclusive or narrow categories (e.g., dog, airplane).

The Sequential-Touching Method

The second methodological innovation stemmed from studies published in the early 1980s on very young children's spatial classification or sorting. Infants and young children cannot be instructed to place similar objects into piles. Investigators observed, however, that when presented with collections that involved two groups of objects (e.g., four yellow pillboxes and four blue balls), infants in the first and second years of life engage in sequential touching of the items, or touching of several similar items in sequence. Jean Mandler adapted this task to study infants' responses to categories such as kitchen things vs. bathroom things or dogs vs. horses. In the typical version used to study categorization, infants are presented with a tray of objects from two adult-defined categories for a fixed period of time (e.g., 2 min), and their touching is observed (see **Figure 4**). If infants engage in more sequential touching of objects from one or both categories than would be expected by chance (i.e., if they selectively touch in sequence items from one category, effectively ignoring the items from the other category), then they are labeled categorizers. The proportion of children who categorize given the particular categorical contrast and/or stimulus set is used to draw conclusions about the development of categorization between 12 months and 3 years of age. Although it is difficult to uncover categorization in infants younger than 1 year with this task, a number of studies have shown that children between 12 months and 3 years of age will respond to a wide variety of categorical contrasts (e.g., animal vs. vehicle, dog vs. fish) in this task.

What Do We Know about Infants' Categorization and Concepts?

Do Infants form Categories and Concepts?

The earliest questions about infants' categorization and concept development addressed using these methods were simply do infants have concepts, or can they form categories? As described above, Cohen and Strauss's study suggested that young infants might not be able to categorize – it was not until infants were 7 months old that they responded to a category of female face – a remarkable finding given the extensive experience infants that must have had with female faces in the first 30 weeks of postnatal life.

However, subsequent studies have shown that in fact infants can and do form categories much earlier than this first study suggested. Indeed, in their original article Cohen and Strauss pointed out that infants younger than 30 weeks might be able to respond to different kinds of categories. Work in the more than 25 years since this article was published has revealed that the younger infants in this seminal study may have failed to categorize because aspects of the procedure may have made it difficult for them to detect the commonalites and differences among the items. For example, young infants may have difficulty categorizing when items are presented one at a time, as they were in this study. When items are presented in pairs, infants as young as 3–4 months can categorize male faces as different from female faces, dogs as different from cats, and triangles as different from squares. Some evidence suggests that sensitivity to these categories may emerge by 2 months, and that even newborns are sensitive to categories such as crosses vs. circles and speech sounds in some contexts.

It may be tempting to conclude that categorization is an extremely early emerging ability, and that infants form

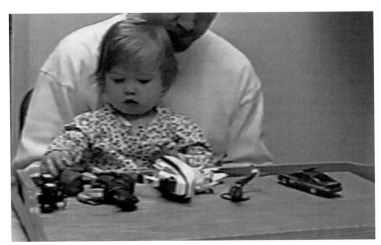

Figure 4 Example of the sequential-touching task. In this task, infants are presented with a collection of items – typically from two different categories (animals and vehicles depicted here) for a fixed period of time. Infants are allowed to manipulate and explore the objects in any way they choose. The objects they touch, and the order they touch them, are coded.

adult-like categories in the first weeks after birth. However, studies have revealed qualitative changes in infants' categorization. For example, younger infants tend to respond to broader categories than do older infants (e.g., treating dogs and horses as part of the same category, perhaps because they have four legs and tails). Different researchers have observed this pattern using different stimuli, methods, and categories. Therefore, one of the few areas of consensus in the field is that categorization in infancy develops from broad to narrow – early on, infants more easily form relatively broad categories, and later they become more sensitive to features that define narrow categories.

However, this developmental difference is not a simple broad-to-narrow developmental trajectory. For example, whether or not infants of a given age respond to a relatively broad or relatively narrow category depends on how similar the items are presented during habituation or familiarization. Infants (3–4-month-old) familiarized with a set of dogs that are very similar to one another respond to the difference between dogs and cats; infants at the same age who are familiarized with a set of dogs that are more different from one another, in contrast, respond instead to a more general category that apparently includes both dogs and cats. In addition, young infants respond to a narrow category if items are presented two at a time, side-by-side, but respond to a broader category if those same items are presented one-at-a-time. Although we do not yet know why young infants' categorization is influenced by such factors, it appears that when infants have to remember items from one trial to the next, as they do in the standard multiple habituation task, they have difficulty noticing the subtle commonalities among items that they can notice when two items are presented side-by-side.

Some problems of interpretation

So, what do we know about infants' categorization and concepts? Clearly, infants can categorize. Moreover, because in many of the studies of infants' categorization, adult-defined categories were used, these results also show us that infants are sensitive to adult-defined categories. These studies asked whether infants detect boundaries between categories important and salient for adults – dogs vs. cats, animals vs. vehicles, kitchen things vs. bathroom things. The fact that infants responded to those categories in experimental tasks provides evidence that they can be sensitive to those commonalities and differences that define such adult-defined categories. Importantly, such results do not tell us that these categories are the most salient or important from an infant's perspective – or even that the infant came to the laboratory with the knowledge of those categories. But, they do show that in particular experimental procedures with particular experimental stimuli, infants are sensitive to those categories.

Although these studies reveal infants' sensitivity to particular commonalities, it is not clear what they tell us about their 'concepts'. Recall that concepts are the mental representations infants (or adults) possess about categorical groups. Representations refer to the information stored in the head. How do we examine infants' representations? When we talk about concepts we typically are referring to enduring, long-term representations that have been built up over time. We do not usually mean a representation formed a moment ago, on the basis of limited exposure to a set of stimuli (though in theory such a representation could be a concept). One's concept of dog, for example, is based past encounters with dogs – what they look like, how they behave, and how to treat them. Habituation or familiarization procedures do not allow us to easily disentangle what infants knew before coming into the laboratory, or what categories are salient in their day-to-day lives, from what they learn over the course of the habituation period of the session. Researchers are just beginning to compare infants' responding to highly familiar stimuli (e.g., dogs for infants with dogs at home) to their responding to less familiar stimuli (e.g., dogs for infants who do not have a dog at home) to determine how learning outside the laboratory influences infants' behavior in tasks like visual familiarization. Moreover, concepts generally are thought to include more than perceptual similarities among items, and it is difficult to determine whether infants' responses in visual habituation reflect recognition of perceptual similarities vs. deeper similarities

Most researchers do assume that infants have concepts of one sort or another – although there is wide disagreement about the nature of those concepts and how to assess them. As soon as infants are old enough to act on objects, they demonstrate that they know more than just what objects look like. Infants can learn that some objects rattle when shaken, and they will attempt to shake objects that look like other rattles but not objects that do not look like those other rattles. Late in the first year, infants put toy telephones to their ears, shoes to their feet, and pretend to drink out of empty cups. These actions suggest that infants have represented more than what objects look like, and that they have concepts of those objects. Clearly, once infants are between 12 and 14 months of age and they begin to label objects, their use of labels seems to reflect concepts.

The question is whether the procedures described earlier tap such concepts? When infants are familiarized with a series of pictures of dogs and then they dishabituate to a picture of a cat, does that reflect their dog concept? When infants are given a collection of animals and vehicles and they touch the animals in succession, does this reflect an animal concept? This difficult question has been at the center of many disagreements in the study of infant categorization. There are two sources of this disagreement:

first, researchers have debated whether there are separate processes for forming concepts and perceptual categories, and second, researchers have argued whether some tasks are better able to tap concepts than others. For example, some argue that tasks in which infants touch and pick up real objects (such as small plastic replicas of objects) tap concepts better than do tasks in which infants simply look at pictures. The reasoning is that these tasks induce a kind of processing that goes beyond simply what objects look like, whereas tasks in which infants only look at pictures emphasize what objects look like. However, the same factors seem to influence infants' categorization in both types of tasks (e.g., variability among the stimulus items), casting doubt that one task is better at tapping concepts than the other.

In addition, studies using habituation of looking time have revealed more than simply infants' perceptual categories. For example, several studies have revealed a curious developmental trend that may reflect changes in infants' conceptual understanding. Specifically, younger infants actually notice more kinds of features than do older infants. This trend may reflect younger infants being open to more possibilities about how features in the world are related (e.g., the shape of a vehicle's headlights might predict whether or not the wheels go around, causal agents can have static or dynamic features), and older infants, who have learned more about the world, are constrained in the kinds of information they attend to and learn (e.g., only the appearance of the wheels predicts whether or not they go around, causal agents are more likely to be animates and thus have dynamic, moving parts). Thus, at least for some kinds of stimuli, visual habituation tasks can reveal more than infants' attention to perceptual similarities.

It is likely that this particular debate will never be fully resolved; it is extremely difficult to determine precisely what underlying representations are assessed with a particular procedure. What is clear is that although we can draw strong conclusions about infants' categorization and the categories to which they are sensitive, the conclusions we can draw about infants' concepts are more limited in large part because it is not clear how infants' behaviors in the tasks available to us reflect their underlying concepts.

What Are the Processes of Categorization in Infancy?

What we can know with confidence is something about the processes of categorization in infancy. That is, we can know not only that infants form categories, but how they form them. Understanding how infants form categories was the main focus of many early studies. Researchers examined infants' categorization of line drawings of made-up animals, schematics of faces, or geometric shapes to uncover how infants form categories. Three main characteristics of infants' categorization reveal similarities to adult categorization, suggesting continuity across development.

Summary representations

First, infants form summary representations for categories. When faced with a collection of similar items, one can either remember each individual item or form a single representation that summarizes what those items have in common. A summary representation might take one of several forms. It might include a list of features that is characteristic of all the items in a category – for example, chairs have four legs, a seat, and a back. This type of summary representation does not seem to accurately reflect how people think about categories, however. People are sensitive to the fact that categories do not have a set of necessary and sufficient features (i.e., a set of features that every member has, and that if an instance has those features it must be a member of that category), but rather category boundaries are fuzzy. For example, bean-bag chairs have none of the features in the list above and yet they are categorized with other chairs, and couches have all of those features and yet they are not categorized as chairs. It is possible, of course, that we have just not identified the right set of features, and that in fact all chairs do have some features in common that are unique to chairs.

Moreover, infants may not be aware that category boundaries are fuzzy; they may think of categories in terms of a set of necessary and sufficient features. For example, infants may have categories for objects with faces (e.g., animates), wheels (e.g., vehicles), or legs (e.g., a group that includes furniture, animals, and insects). Of course, such summary representations would cause infants to make categorization errors, at least from an adult perspective (consider the objects with legs example). An item may be miscategorized and incorrectly included in an adult-defined category if it has a highly salient feature in common with many items in that category. Indeed, in one famous example, a child used the word ball to refer to the moon and a round candle. Errors like these may indicate that the child has formed a summary representation for ball in which the feature round is necessary and sufficient. Carolyn Mervis suggested that these mistakes reflect child-basic categories – categories that are similar in some ways to adults, but that depend on slightly different features, or depend more heavily on some features, than do the categories of adults. Examples like these are not definitive, however. It is also possible that a child who calls the moon and a round candle a ball is simply noting the similarity between those different categories, or struggling to find the word in his or her limited vocabulary that comes closest to describing the object at hand. That is, errors in labeling may not reflect errors in categorization.

One summary representation that does take into account the fuzziness of category boundaries is a prototype. A prototype is the average of the category. Your prototype of a dog, for example, is not a real dog, but rather is the average of all the dogs you have seen in your life. When learning a series of animals with different length necks, therefore, the prototype formed will have the average of those neck lengths. Infants as young as 3–4 months of age have been shown to form prototypes, and by 10 months, infants clearly average features in the way just described. The formation of prototypes is a particularly important aspect of categorization. It helps to explain how items can be grouped together in a category even if they do not have all the same features, and how come some items seem to be better members of a category than others (e.g., a Labrador is a better dog than a Chihuahua). This propensity to form prototypes may be responsible for infants' preferences for attractive faces, and their ability to learn about some kinds of categories more quickly and efficiently than others.

Exemplar representation

A second finding that demonstrates that infants' categorization is like adults, is that infants not only form summary representations, but they also learn the individual exemplars. A debate in the adult literature is whether categories are represented as summary representations or as collections of instances. Theorists have pointed out that the exact same behavior is predicted if people represent individual instances or if they represent a summary representation such as a prototype. For example, a Labrador may seem like a better example of a dog either because it is closer to one's stored prototype of dogs, or because it is similar to more representations of individual dogs (e.g., German Shepherds, Golden Retrievers). Recently, however, it has become clear that adults learn the individual instances in some cases and form summary representations in others. That is, when building a representation of a new category, adults both encode individual items and they create summary representations such as prototypes of those items.

Like adults, infants also learn both individual items and form summary representations when familiarized with a series of items from within a category, depending on the particular task. When learning a collection of new female faces, for example, 3–4-month-old infants whose primary caregiver is a woman learn those individual faces; they do not learn to respond to a category of female face that is distinct from the category male face. When those same infants are familiarized with a collection of male faces, in contrast, they fail to learn the individual faces and instead respond to the category of male face as distinct from the category female face. Thus, infants can represent both the individual items and form summary representations, and which they do initially or more easily depends on factors such as the familiarity or the difficulty of the category to-be-learned.

Correlated attributes

Finally, infants' categorization is like adults in that they are sensitive to the correlations between features. For adults, categories are not defined by lists of unrelated features, but rather are defined by clusters of features (e.g., animals with feathers have wings and tend to fly). These correlated attributes are better predictors of category membership than the presence of particular features – for example, bats have wings but not feathers, and they are not birds. Thus, the presence of wings alone is not sufficient to identify an object as a bird.

Infants too are sensitive to correlated attributes. When familiarized with a completely novel category and unfamiliar items, infants in the first year of life can learn that individual features covary (e.g., that animals with long necks have wings; that purple objects squeak). By the end of the first year, infants can use those correlations to categorize objects. If, for example, they are familiarized with several long-necked animals with wings, they will recognize that a new, never before seen long-necked animals with wings is a member of the category.

Moreover, like adults, the particular correlations that infants learn are constrained by their knowledge and experience. Adults have an easier time learning correlations that make sense given their existing knowledge (e.g., a vehicle that travels in the jungle has wheels with large treads) than correlations that are arbitrary given their existing knowledge (e.g., a vehicle that travels in the jungle has numbers on the license plate). Similarly, the correlations among features that infants learn can be constrained in this way. As described earlier, infants in their second year restrict their attention to correlations that are consistent with how general principles of correlations exist in the world – for example, noticing that the appearance of wheels can predict whether or not they go round, but not that the appearance of different part can predict whether or not the wheels go round.

Summary

By focusing on the process of categorization, therefore, we have a picture of infants' categorization as similar to adults', and as a result we conclude that there is relative continuity in development. However, it is important to remember the discontinuities described earlier. Despite the fact that infants form categories in much the same way as do adults, the resulting categories they form are not identical to those of adults. Infants form broader categories, based on a larger set of correlations among features, and that include many items that have commonalities not salient to adults. The source of these discontinuities, and how to reconcile them with the clear continuity in categorization, is under debate. One possibility is that with development infants discover or become sensitive to different kinds of features. Or, what constitutes a feature might change. That is, the process itself might not develop, but the input to the process may change. Another possibility is that there is continuity in some kinds of categorization (e.g., perceptual categorization), but that other kinds of categorization (e.g., conceptual categorization) emerge later or undergo developmental change. We will gain insight into the answers to these questions by further understanding how infants form categories.

A New Framework for Categorization

A new framework of concepts and categorization is emerging, and it has the potential to change dramatically our understanding of how infants categorize, and how that categorization is related to categorization later in life. Concepts often appear to be conceived of as static representations. Researchers ask question such as: Do infants know particular categorical distinctions? Are infants' categories like those possessed by adults? When do particular concepts emerge? When examining adults' concepts, researchers have sought to identify the set of necessary and sufficient features that define those concepts, to understand how concepts are related to one another, and to examine how categories are represented.

Categories are not static, however. Rather, they are dynamic and flexibly constructed given the context and items to be categorized. Although categories may be groups of objects out in the world, how those objects can be grouped is constantly changing with the introduction of new information, salience of particular features, and encounters with new examples of existing categories. Consider, for example, an individual whose encounters with dogs included only German Shepherds, Labrador Retrievers, and Golden Retrievers. When first encountering a Chihuahua, this individual likely will not recognize it as a member of the category dog. But, as additional information is gathered (e.g., the Chihuahua is labeled dog) and subtle details are attended to (e.g., like other dogs, the Chihuahua growls and wags its tail), this individual's dog category will change to include Chihuahuas, and other similar dogs. This will change the summary representation as well as the instances stored as members of the category.

This process likely happens frequently to infants as they encounter many new objects, events, and people everyday. Consider the infant described earlier who called both the moon and the round candle a ball. If this labeling error reflects a categorization error, this child's category

ball must be undergoing significant change. At the time of the labeling error, his category ball includes many round things, and does not seem to be restricted to round things that are used in games. At this point his category ball likely does not include nonspherical balls such as footballs. As this child learns about other aspects of balls – that balls are often thrown, hit, or kicked, and that balls are typically toys – his category will continue to change and his view of round objects in the world will reorganize. Indeed, this is the central idea behind the notion of child-basic categories – they are categories that map onto adult-defined categories (such as ball), but imperfectly so because the child either ignores or is unaware of some features or places too much weight on others. Such child-basic categories become adult categories as children learn more about how adults typically weight different features in making category judgments. Clearly, therefore, overdevelopmental time categories are not static. Infants do not simply learn or acquire adult-defined categories and those representations do not go unchanged over time.

Categories are dynamic on even more fine-grained timescales. Our representations for categories change - from moment-to-moment. As they learn new categories in the laboratory, adults' representations change from trial-to-trial. Similarly, infants' representations for a collection of category items change across trials. Infants who are given fewer vs. more familiarization trials with exactly the same stimulus items show evidence of having represented the items differently, suggesting that representations changed with more encounters with items. Similarly, infants' sequential-touching behavior changes over time – they shift their touching patterns from comparing individual items to examining items all from within one category, from focusing on only a single category to touching items from both categories, and from relying on one dimension (e.g., shape) to relying on a completely different dimension (e.g., material) to determine which objects to touch in sequence. Thus, regardless of the categories shown or the procedure used, infants' behavior in experiments does not reflect static, unchanging categorical representations. Not only do representations change with increased development and knowledge, but also from moment-to-moment in the context of an experimental session.

The view of categorization in infancy as a dynamic, online process has implications for how we think about infants' behavior in experimental tasks assessing categorization – particularly tasks in which infants' behavior evolves over time. When encountering a series of items in a familiarization procedure, for example, infants are actively categorizing. Infants may categorize the first stimulus in relation to their existing knowledge – though this is less likely when they encounter completely novel stimuli. If they do categorize novel stimuli, they probably begin this process once they realize that there are several different stimuli that share commonalities – perhaps as early as the first trial when presented with multiple different items at one time, and by the time they have encountered two or three different stimuli on different trials. But, this early categorization is based on encountering few stimuli. If those stimuli are extremely similar, then infants may form an overly narrow category (such as one that only includes Golden Retrievers and Labradors) that must be broadened as new items are encountered. If the first stimuli are extremely dissimilar, infants may initially form a very broad category (such as a ball category that includes any round object) that gets narrower with experience. As infants are exposed to additional items, their category changes – the new items are included in the category, and the boundaries of the category change as a consequence. This process occurs throughout the session.

This can also happen in sequential-touching tasks, in which infants freely explore collections of objects. In this case, infants likely learn new features of the objects as they explore them, and their categories change. Consider an infant faced with a collection of balls and blocks. As she first explores those objects, the differences in shape may drive her behavior. But, as she picks up and touches the objects, she may discover that some are squishy. At this point, she may form narrow categories (e.g., squishy blocks and hard blocks) or she may shift her categorization to this new dimension (e.g., squishy things and hard things, regardless of shape). The point is that even when the infant is investigating objects on her own, she can constantly discover new commonalities and revise her categorization of objects online.

Summary: The Development of Categorization

The preceding sections have detailed changes in infants' categorization behavior. We have seen continuity in how infants form categories and changes in the kinds of categories infants respond to. But what exactly is developing? Categorization itself may not develop – from shortly after birth infants show evidence of forming some categories. But, as described earlier, infants' categories do change. One possibility is that as infants' perceptual and motor abilities develop, they become able to use a wider range of features to form categories. Thus, there may not be discontinuity in the process of categorization, but rather there may be discontinuity in the input to that process. Changes in perceptual and motor abilities may also allow infants to consider more information at one time than can younger infants. That is, quantitative changes in capacity limitations may allow the infant to consider more simultaneously information when categorizing. Indeed, there is a large literature revealing that as they develop, infants remember more information and for longer periods of time. Such changes may contribute to changes in categorization.

Not only do the categories that infants attend to change over time, but they attend to those categories in increasingly more situations with development. Infants' limited memory and other cognitive skills may make it difficult to carry out the comparisons necessary to detect commonalities and differences among items in some demanding contexts. The point is that developmental changes in categorization may in fact reflect changes in infants' other abilities.

In summary, infants' concepts and categorization skills are critically important for their developing understanding of the world. Infants can detect commonalities among items from an extremely early age. Although we do not yet know exactly why or how infants' concepts and categorization skills develop, there is a large body of evidence providing rich descriptions of those changes. This work will provide the foundation for our future understanding of the processes of categorization in infancy.

See also: Cognitive Development; Cognitive Developmental Theories; Habituation and Novelty; Language Development: Overview; Learning; Memory; Language Development: Overview; Perceptual Development; Reasoning in Early Development.

Suggested Readings

Cohen LB and Strauss MS (1979) Concept acquisition in the human infant. *Child Development* 50: 419–424.

Mandler JM (2004) *The Foundations of Mind: Origins of Conceptual Thought.* New York: Oxford University Press.

Mareschal D and Quinn PC (2001) Categorization in infancy. *Trends in Cognitive Sciences* 5: 443–450.

Oakes LM and Madole KL (2003) Principles of developmental change in infants' category formation. In: Rakison DH and Oakes LM (eds.) *Early Category and Concept Development: Making Sense of the Blooming, Buzzing Confusion*, pp. 132–158. New York: Oxford University Press.

Quinn PC (2002) Beyond prototypes: Asymmetries in infant categorization and what they teach us about the mechanisms guiding early knowledge acquisition. *Advances in Child Development and Behavior* 29: 161–193.

Rakison DH (2003) Parts, motion, and the development of the animate-inanimate distinction in infancy. In: Rakison DH and Oakes LM (eds.) *Early Category and Concept Development: Making Sense of the Blooming, Buzzing Confusion*, pp. 159–192. New York: Oxford University Press.

Rakison DH and Oakes LM (eds.) *Early Category and Concept Development: Making Sense of the Blooming, Buzzing Confusion*. New York: Oxford University Press.

Reznick JS and Kagan J (1983) Category detection in infancy. In: Rovee-Collier C and Lipsitt LP (eds.) *Advances in Infancy Research*, vol. 2, pp. 78–111. Norwood, NJ: Ablex.

Cerebral Palsy

M L Campbell, A H Hoon, and M V Johnston, Kennedy Krieger Institute, Baltimore, MD, USA

Glossary

Associated disorders – Other medical conditions that are seen in individuals with cerebral palsy.

Birth asphyxia – Asphyxia is a condition of severely deficient supply of oxygen to the brain, which, if prolonged, will result in metabolic acidosis and shutdown of body functions. The diagnosis of birth asphyxia is based on several criteria, including signs of fetal distress, serious brain dysfunction affecting breathing, movement, swallowing, as well as seizures in newborn infants.

Brain malformation – An abnormality in brain formation occurring early in fetal development, which is permanent, and which may be caused either by a genetic factor or by prenatal event(s) that are not genetic. It is a common cause of cerebral palsy.

Cerebral palsy (CP) – CP describes a group of disorders of the development of movement and posture, causing activity limitation, that are attributed to nonprogressive disturbances that occurred in the developing fetal or infant brain. CP may be categorized into spastic, extrapyramidal, and mixed forms. It is the result of underlying medical disorders or risk factor(s), which result in abnormal brain formation or injury.

Dyskinetic – Impairment in the ability to control movements; an alternative term to extrapyramidal CP.

Dystonia – A neurological movement disorder characterized by involuntary muscle contractions, which force certain parts of the body into abnormal, sometimes painful, movements or postures. Dystonic CP is a type of extrapyramidal cerebral palsy.

Etiology/risk factors – Underlying contributing or causal factors leading to the development of a medical condition such as CP.

Extrapyramidal – Refers to central nervous system structures (i.e., outside the cerebrospinal pyramidal tracts) that play a role in controlling functions.

Intellectual disability (mental retardation) – Mental retardation is a disability characterized by significant limitations both in intellectual functioning and in adaptive behavior as expressed in conceptual, social, and practical adaptive skills. This disability originates before age 18 years. The term mental retardation is being replaced by 'intellectual disability.'

Low birth weight – Birth weight of less than 2500 g.

Magnetic resonance imaging – an imaging tool used to identify structural abnormalities. It is very helpful in determining the underlying medical diagnosis in children with CP.

Management – A comprehensive approach to care including medical, rehabilitative, and alternative therapies that reflects the needs of the person. It is lifelong and changes over time with the person.

Mental retardation – See Intellectual disability above.

Periventricular leukomalacia – White matter injury in the developing brain, often occurring between 24–34 weeks of gestation in children with CP.

Prematurity – The current World Health Organization definition of prematurity is a baby born before 37 weeks of gestation, counting from the first day of the last menstrual period.

Spasticity – An involuntary increase in muscle tone (tension) that occurs following injury to the brain or spinal cord, causing the muscles to resist being moved.

Very low birth weight – Birth weight of less than 1500 g.

Introduction

Cerebral palsy (CP) describes a group of lifelong neurological disorders that affects muscle tone, posture, mobility, and hand use. CP is the most common cause of motor disability in childhood. It affects approximately 2/1000 children, with about 8000 young children diagnosed yearly in the US. Functional limitations vary in severity from isolated gait disturbance to an inability to move, and may change over time. Many children have associated problems with speech, cognitive processing, seizures, eye movements, and swallowing, as well as orthopedic deformities. The majority grow into adulthood, and can lead successful lives with appropriate supports.

CP is a clinical diagnosis that serves to identify the child as needing specific rehabilitative services, as well as a commonly understood term for families to use with others to describe their children, that is, "My child has cerebral palsy." From a medical perspective, CP results from a wide range of genetic or acquired risk factors and disorders that disrupt developing areas of the brain that control movement (**Table 1**). While muscles are secondarily affected, it is not a primary disorder of either muscle or nerve. Clinicians work to establish the underlying medical diagnosis that results in CP. This knowledge is useful in determining treatment, prognosis, and recurrence risk, as well as in allaying feelings of parental guilt or responsibility.

Knowledge of normal brain development as well as of the effects of deleterious genetic and environmental factors is beneficial in understanding developmental disabilities such as CP, as well as mental retardation (MR) and autism. When the disruption affects developing motor pathways, the result is CP; and when it affects cognitive parts of the brain, the result may be intellectual disability (MR), autism, or learning disabilities.

Effective management requires a thorough understanding of the causes, manifestations, and management options for children with CP. Maintaining open lines of

Table 1 Risk factors for cerebral palsy

Maternal factors
- Maternal age (<20 years or >35 years)
- History of infertility
- Previous pregnancy loss or neonatal death
- Prior-born child (elder sibling) with CP
- Thyroid disease
- Diabetes

Pregnancy complications
- Maternal infection (e.g., chorioamnionitis)
- Preclampsia
- Prolonged rupture of membranes
- Placental abruption
- Placental insufficiency

Infant attributes
- Prematurity/low birth weight
- Multiple gestation
- Male gender
- Growth retardation
- *In vitro* fertilization

Neonatal morbidities and interventions
- Genetic disorders
- Congenital brain malformations
- Hyperbilirubinemia (kernicterus)
- Perinatal stroke
- Chorioamnionitis (placental infection)
- Birth asphyxia
- Postnatal steroids
- Pneumothorax
- Prolonged exposure to mechanical ventilation
- Prolonged hypocarbia

In childhood
- Brain infections (i.e. bacterial meningitis, viral encephalitis)
- Vascular episodes (stroke)
- Brain injury-accidental, nonaccidental

communication between a wide variety of medical, rehabilitative, and social practitioners who treat the myriad of presentations and variable patterns of clinical expression is also required (**Table 2**). Early identification of CP as well as associated disorders facilitates the establishment of effective management plans. Inclusion of children and their families in decision making is critical to optimizing function.

Progress in brain imaging and epidemiology has provided new insights into underlying causes, while advances in medical care such as therapeutic botulinum toxin (Botox) and intrathecal baclofen (ITB) offer improved treatment options. In the past, interventions such as rubella immunization and Rhogam to Rh-negative women have eliminated CP from these causes. Important areas of current research focus on the prevention of prematurity/low birth weight as well as strategies to treat brain injury in high-risk newborns.

History

The cause(s) of CP has been vigorously debated for 150 years. In 1862, Sir William Little attributed CP to problems with childbirth, which has had ongoing effects in the legal arena to the present. In 1897, Sigmund Freud offered an alternative view point, suggesting that the cause of CP was of prenatal origin. ("Difficult birth, in certain cases...is merely a symptom of deeper effects that influence the development of the fetus.") Sir William Osler added the broader overview of the combination of brain anatomy, etiology, and extremity involvement in the diagnosis and classification of CP.

The modern view of CP is concordant with the views of Freud and Osler, emphasizing prenatal antecedents in the majority of affected children. This understanding has been supported by the work of Dr. Karin Nelson from the National Institutes of Health (NIH). In the 1980s Dr. Nelson and colleagues, in a study examining mothers and newborns to determine the cause of CP, found that the majority of cases were due to prenatal factors, which were often unknown. Current understanding has built on Nelson's research, to shift the focus from single causes to the recognition that in many affected children, there is a cascade of contributory factors acting along causal pathways, with the challenge being the identification of these pathways.

Classification

Movement requires the coordinated passage of simultaneous messages through different tracts in the brain, and then onto the spinal cord, peripheral nerves, and finally to the muscles themselves. CP type is related to the location(s) in the brain of white or gray matter injury or abnormal

Table 2 Associated disorders

	Spastic diplegia	Spastic quadriplegia	Hemiplegia	Extrapyramidal	Hyptonia/ataxia
Tone	Spasticity	Spasticity	Spasticity	Rigidity	Hyptonia
Extremity involvement	UE = LE	LE > UE	LE = UE	Unilateral	UE > LE
Movement disorders	Clonus, spasms, toe walking	Clonus, spasms	Clonus, spasms	Dystonia, chorea, athetosis	Ataxia
Speech/ swallowing	Mild impairment	Impaired	Intact	Impaired or absent speech	Variable
Cognitive impairment	Mild–moderate, learning disorders	Moderate–severe	Intact to mild	Intact to moderate	Variable
Associated problems	Strabismus, orthopedic deformities	Orthopedic deformities, epilepsy	Epilepsy	Orthopedic deformities, genetic–metabolic disorders	Undiagnosed genetic–metabolic disorders

LE – lower extremities; UE – upper extremities; Clonus – movements characterized by alternate contractions and relaxations of a muscle, occuring in rapid succession. Clonus is frequently observed in conditions such as spasticity (from website WE MOVE); Athetosis – involuntary, relatively slow, writhing movements that essentially flow into one another. Athetosis is often associated with chorea, a related condition characterized by involuntary, rapid, irregular, jerky movements (from website WE MOVE).
From Hoon AH and Johnston MV (2002) Cerebral palsy. Asbury AK, McKhann GM, McDonald WI, Goadsby PJ, and McArthur JC (eds.) *Diseases of the Nervous System*, 3rd edn., pp. 568–580. Cambridge: University Press. Associated disorders refers to other medical conditions a child with CP may have.

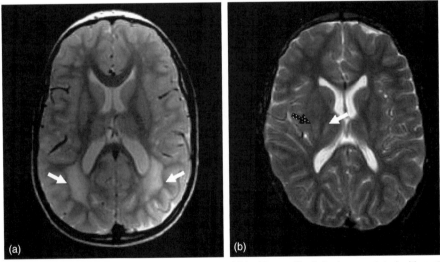

Figure 1 Brain MRI in two children with CP. (a) The MRI is from a child with spastic diplegia associated with preterm birth. The arrows point to areas of injury in white matter ('brain wiring'). (b) The MRI is from a child with dystonic CP associated with hypoxic–ischemic injury (low oxygen) in an infant born at term. The black speckled arrow shows injury in the putamen and the white arrow in the thalamus.

formation (**Figure 1**). One of the tracts, the cortical spinal tract (CST) or pyramidal tract, sends messages for motor movement to the spinal cord. If this tract is injured during brain development, a person will often develop spastic CP. Recent information indicates that pathways in the brain involved in processing sensory information may also lead to spastic CP.

A second motor control pathway tract in the brain provides messages for the control of movement. This pathway runs in a loop from the cortex of the brain through the basal ganglia/thalamus and back to the cortex, thereby regulating movement. If this tract is injured during brain development, a child may develop extrapyramidal CP, often with dystonia. Injury to this tract may be caused by problems with oxygen, increased jaundice at birth, or genetic diseases.

CP may be classified on the basis of neurologic examination, limb involvement, or degree of functional impairment. Each classification system has strengths and limitations that should be considered in selection and use.

A commonly employed classification with clinical implications for treatment is that based on examination into spastic, extrapyramidal, and mixed forms (**Figure 2**). Spastic CP is characterized by increased muscle tone (resistance similar to that felt in opening a clasp knife), greater leg than arm involvement, and increased risk for orthopedic deformities including hip dislocation and scoliosis. Seventy per cent of affected children have primarily spastic or spastic/mixed forms of CP, which can be sub-categorized into diplegia, quadriplegia, and hemiplegia. Diplegia refers to primary lower extremity involvement; quadriplegia to four limb involvement; and hemiplegia when one side of the body is affected. It should be noted that the causes of hemiplegia differ significantly from other types of CP in which both sides of the body are affected.

Extrapyramidal forms constitute the remaining 30%, and include dystonic, rigid, choreic, ataxic, and hypotonic subtypes. Another term used is dyskinetic CP, which refers to an impairment in the ability to control movements. Dystonia refers to sustained muscle contractions associated with involuntary twisting or repetitive movements, and is seen rather than felt on examination. Rigidity refers to increased tone similar to the sense of pulling taffy, and is

related to muscle co-contraction. Chorea can be thought of as the opposite of dystonia, and is characterized by jerky, irregular, and rapid involuntary movements. Ataxia refers to abnormalities in balance and coordination. Hypotonia refers to decreased muscle tone. Children with extrapyramidal CP often have underlying genetic disorders.

Mixed forms manifest both spasticity and extrapyramidal signs. With careful examination, many children have mixed signs. In these cases, medical treatment is often determined by the primary neurological finding(s).

As highlighted in the recent WHO publication, *The World Health Organization International Classification of Functioning, Disability and Health*, it is important to characterize the functional consequences of disorders such as CP. To classify ambulation, the gross motor function classification system (GMFCS) is often used, with five levels based on functional mobility or activity limitation. The manual ability classification system (MACS) is a recently described instrument with good reliability to classify upper extremity function. These instruments serve as a basis for discussion of clinical characteristics among healthcare professionals as well as in research studies of treatment interventions.

Causes

Prenatal genetic and environmental risk factors are linked in 70% of children with CP, while birth asphyxia plays a role in 10–20%. Risk factors which have been linked to CP include prematurity, low birth weight, multiple

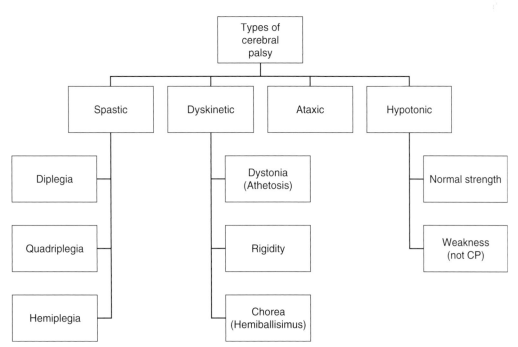

Figure 2 Neurological classification of cerebral palsy. Please see Section 'Classification' for description of terms.

gestation, maternal thyroid abnormalities, intrauterine viral infection, male gender, hereditary disorders of clotting, placental abnormalities, hypoxia–ischemia, and signs of intrauterine infection/inflammation.

Using brain magnetic resonance imaging (MRI) as the primary diagnostic tool, children with CP can be categorized into four groups, based on the timing of the underlying etiological factors: (1) disorders of early brain formation, (2) injury associated with prematurity, termed periventricular white matter injury or periventricular leukomalacia (PVL), (3) disorders presenting in the term infant, and (4) a heterogeneous group of disorders occurring in early childhood. Although there is variability in the developing and developed world, approximately 30% are associated with brain malformations, 40% with prematurity, 20% in term infants, and 10% to postnatal causes.

Diagnosis

A careful history and neurological examination remain key components to diagnosis. Symptoms may appear soon after birth or during early childhood. Risk factors in infants include significantly increased or decreased muscle tone, difficulty with head control, delay in rolling, sitting, crawling, standing, or early preferential hand use.

A minority of children, including those with genetic disorders, may appear normal and then manifest motor delay later in infancy or early childhood. A well-recognized example of this is the genetic disorder glutaric aciduria, type 1. In this condition, infants or young children develop normally until metabolic stress such as chickenpox or dehydration leads to acute signs of neurological injury, which progresses to extrapyramidal forms of CP.

In addition to a thorough physical examination looking for findings suggesting a genetic cause, a careful head measurement of head circumference (i.e., too small (microcephaly) or too large (possible hydrocephalus)), and neurological examination of eye movements, tone, and reflexes is important. Infants have reflexes such as the Moro (a quick body movement leading to arm movements that resemble an embrace), which normally disappear by 3–6 months of age. Persistence of this and other infant reflexes are neurological signs of possible CP. Tone abnormalities as described above are also concerning findings. Additionally, if an infant favors 1 hand or side of his/her body strongly before 1 year of age, this may be of a sign of hemiplegic CP.

If the history suggests that the child is losing abilities, or the examination shows unusual findings such as muscle weakness, this may represent one of a wide variety of other neurological disorders of childhood that do not fall under the diagnosis of CP. Referral to a pediatric neurologist is strongly recommended in this situation. In some cases, it may take several examinations to determine whether the findings are consistent with CP or represent another type of neurological condition.

Brain scans using ultrasound, computed tomography (CT), and MRI provide images of the brain structure. In children with CP, MRI has revolutionized understanding of the underlying cause in many children, with abnormalities seen in 70–90% of children with CP. MRI findings can serve to guide additional diagnostic testing, including blood and urine studies.

Management

The overall management of children with CP is directed toward optimizing function and quality of life. This is accomplished by improving muscle strength and endurance, enhancing academic success, facilitating personal hygiene, treating associated disabilities, and preventing medical complications such as slow growth, joint problems, and reduced bone mass/fractures.

The human brain grows rapidly from conception through early childhood. During this time connections between the brain and white matter pathways that make up the central nervous system are forming. During childhood approximately half of these connections are maintained due to stimulation through activity, and half are eliminated. This concept is called plasticity. This is the underlying medical basis of early intervention.

Early identification of motor delay or CP is important in establishing effective treatment. In this regard, the federal law IDEA (Individuals with Disabilities Education Act) provides the framework for eligibility and service delivery. This law has two important components, the Infants and Toddlers Program for children 0–3 years of age, and the Child FIND program for those aged 3–21 years. There are three ways to qualify for the free, home-based services in the Infants and Toddlers Program: grater than 25% delay in developmental abilities, high probability for developmental delay (e.g., Down syndrome), and atypical development. To qualify for Child FIND, a diagnosis such as CP, learning disability, or MR must be established.

Therapy and treatment options are key in successfully managing CP. A multidisciplinary team approach provides the best management strategy, leading to optimal outcome. A team may variably include a developmental pediatrician, pediatric neurologist, physiciatrist, orthopedic surgeon, physical therapist (PT), occupational therapist (OT), speech pathologist, social worker, and a psychologist. Collectively, this team focuses on an individual treatment plan for each child. Prior to beginning an intervention, treatment goals should be clearly established between the family and careproviders. For example, these goals might include ease of care, improved function, decreased risk for orthopedic problems, or reduction in pain/spasms.

Children with CP develop to their full potential when treatment programs optimize motor capabilities, minimize orthopedic deformities, and address associated impairments. Management can be divided into rehabilitative, medical, and surgical components, and may include physical/occupational therapy, therapeutic botulinum toxin, oral medications, and orthopedic and neurological surgery (**Figure 3**). Physical and occupational therapists provide the backbone of management for the child, in addition to providing education and guidance for family members.

In children with CP, there is a small but growing number of genetic disorders, including dopa responsive dystonia, glutaric aciduria type 1, and methymalonic acidemia, which may significantly improve or be ameliorated with appropriate medical therapy. Clinicians involved in diagnosis should maintain an eye of vigilance for these conditions.

Physical and Occupational Therapy

PTs and OTs provide initial and ongoing treatment for children with CP. From early infancy and throughout life, PT and OT services are important in treatment. PTs focus on optimizing functional abilities and preventing orthopedic deformities, using a variety of individually tailored approaches, including stretching, strength training, bracing, gait training, and equipment, including wheelchair, walker, and prone stander. OTs evaluate and treat oromotor/swallowing dysfunction, as well as upper extremity use in the context of activities of daily living and also address occupational/vocational opportunities.

Therapeutic Botulinum Toxin

Therapeutic botulinum toxin is used to for treating focal spasticity and dystonia. For this therapy, affected muscle groups are isolated and injected with a small amount of botulinum toxin. The toxin binds to the nerve endings and weakens the isolated muscle group. Balance can then be re-established with other muscles across the affected joint(s). It often works best when combined with casting or splinting. This therapy lasts for about 3 months and may be repeated. Botulinum toxin injection is relatively safe and can be used in the upper extremities, lower extremities, and neck.

Oral Medications

Medications such as baclofen (Lioresal), diazepam (Valium), trihexyphenidyl (Artane), and carbidopa/levodopa (Sinemet) can be used to treat CP (**Table 3**). Oral medications are variably used to improve function; decrease spastic, rigid or dystonic hypertonicity; as well as to decrease muscle spasms in certain situations (e.g., diazepam use after orthopedic surgery).

When using oral medications, the functional benefits for the person must outweigh the side effects of the medication. Often these medications have unwanted side effects that need to be monitored to make sure they are not interfering with functional outcomes. Furthermore, abrupt withdrawal of these oral medications may also cause unwanted effects as children become accustomed to these medications and must be weaned off slowly. Finally, these medications should periodically be tapered off to see if any recognized, positive effects are medication-related.

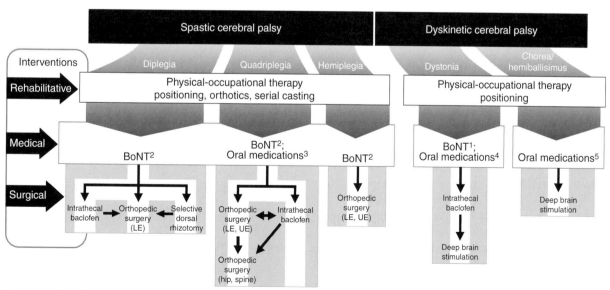

Figure 3 Algorithm of treatment. Puscavage A and Hoon AH (2005) Spasiticy/cerebral palsy. Singer HS, Kossoff EH, Crawford TO, Hartman AL (eds.) *Treatment of Pediatric Neurologic Disorders*, p.17. New York: Dekker. LE, lower extremities; UE, upper extremities.

Table 3 Medications that have been used for cerebral palsy and other motor disorders

Oral medication	Drug class	Spasticity	Rigidity	Athetosis	Dystonia	Chorea	Hemiballismus	Epilepsy
Baclofen	GABAergic	X	X	X	X			
Carbamazepine	Neuronal stabilizer					X	X	X
Clonazepam	Benzodiazepine	X				X		X
Clonidine	α2 agonist	X						
Dantrolene	Direct muscular	X	X		X			
Diazepam	Benzodiazepine	X	X	X	X			X
Levodopa	Dopaminergic	X			X			
Reserpine	Dopamine depleter					X		
Resperidone	Neuroleptic					X	X	
Tiagabine	GABA agonist	X						X
Tizanidine	α2 agonist	X						
Trihexyphenidyl	Anticholinergic			X	X			
Valproate	Neuronal stabilizer					X	X	X

Oral medications that are commonly prescribed in treating cerebral palsy and other childhood motor disorders.

Orthopedic Surgery

A baseline evaluation from an orthopedic surgeon experienced in the management of children with CP is of benefit both for recommendations on nonsurgical treatment as well as for planning potential, later surgical interventions. Despite all efforts, some children with spastic forms of CP will require orthopedic surgery. The orthopedic management of spasticity is directed toward reducing deformity and thereby facilitating function, utilizing tendon lengthenings/transfers, bony osteotomies, and joint fusion procedures. The current approach is often multilevel soft tissue procedures initially, with later bony procedures as required. Computerized gait analysis for preoperative planning may be of benefit.

Intrathecal Baclofen Pump

For people with severe spastic or dystonic CP, for whom other less-invasive approaches have not been beneficial, ITB may be a treatment option. Children, adolescents, and adults who are interested in this treatment should be initially evaluated in a clinic with expertise in ITB management. After a small test dose of baclofen given by spinal tap to assess effectiveness, a pump is placed in the abdomen and connected to a catheter threaded into the intrathecal cerebrospinal fluid-filled space around the spinal cord. The pump works by pumping tiny amounts of baclofen directly into the cerebrospinal fluid, which is then absorbed in the spinal cord reducing spasticity. Instead of using milligrams of baclofen orally, the ITB system uses micrograms of baclofen. As it is targeted more directly to the nervous system, it is often more effective than oral baclofen for people with severe CP. After the pump is installed, it is refilled every 2–6 months. Dosing can be increased or decreased at any time painlessly with a special programmer. A pump can last for 7–9 years before the battery runs down, requiring replacement.

Selective Dorsal Rhizotomy

Selective dorsal rhizotomy (SDR) is a permanent neurosurgical treatment option for children with spastic diplegic cerebral palsy, most commonly associated with prematurity and PVL. During surgery, 30–50% of sensory nerve fibers that run in the lower part of the spinal cord (lumbar/sacral area) are selectively cut. By cutting these nerve fibers, leg spasticity is reduced and function improves. As with ITB, initial evaluation should be made by a team with experience in SDR, including the necessary postoperative rehabilitation.

Complementary and Alternative Therapies

When conventional approaches do not lead to the anticipated improvements, families may explore other treatment options, including complementary and alternative medicine (CAM). CAM includes acupuncture, cranio-sacral therapy, myofascial release, therapeutic taping, diet and herbal remedies, electrical stimulation, constraint-induced training, chiropractic treatments, massage and hyperbaric oxygen therapy (HBOT). While there are individual reports and testimonials of dramatic improvements with various alternative therapies, some carry risk such as middle-ear dysfunction requiring PE tubes with HBOT. Furthermore, rigorous studies have not been conducted to assess efficacy. As with any treatment, families and clinicians should consider cost, efficacy, and potential side effects before embarking on one of these approaches.

Living Considerations

Accessibility

While the American Disabilities Act (ADA) passed in 1993 has significantly improved accessibility, it may still be a continuing problem for individuals with CP and other disabilities. Accessibility concerns may range from housing, to transportation, or using the bathroom. Parents of a child with CP need to be prepared for the ever-changing accessibility needs as their child grows. Working with local agencies can assist in addressing these issues. As children with CP grow, movement may begin to decrease and mobility options may change. These changes will affect accessibility.

People with CP are often assumed to have MR, recently termed intellectual disability. This is not always true. Having CP does not mean the person is mentally retarded or intellectually disabled. People with CP can have normal intelligence. Sometimes a person with CP cannot vocally communicate, but they are able to understand what is going on around them. Evaluations by both psychologists and speech language pathologists can be of great benefit in assessing cognitive abilities and recommending approaches to optimize communication.

As mandated by the federal law IDEA, children with CP and other disabilities should have educational opportunities in a least restrictive environment. There is an extensive list of rights that parents have in protecting their child's right to an appropriate education. An individualized education plan (IEP) is often used in education planning for children with CP. This plan is created with the assistance of a local school system. An IEP planning team must have a chairperson, special educator, and general educator. Parents should and are often invited to participate in the planning and must be given 10 days notice before the IEP meeting is to happen. Depending on the services provided in the IEP, input from a PT, OT, and/or speech language pathologist may be required. Parents have to approve the IEP which are reviewed minimally once a year. The IEP team can also suggest assistive technology to continue the child's access to education. Also, as the child with CP grows and is cognitively capable, he/she should be a part of the IEP planning process.

Another option in education is a 504 plan. A 504 plan is a legal document provided for children who have a physical and/or mental impairment that substantially interferes with one or more major life activities. Impairments can include attention deficit hyperactivity disorder, asthma, diabetes, and problems with speech, reading, and vision. Specialized accommodations are given to assist the child in the learning environment. Examples are scribe, proximal seating, extended time for daily classroom activities, and repetition of directions. A child may have either an IEP or a 504 plan.

Adult Outcome

As children with CP now live longer, as adults they face new challenges, including medical care, accessibility, and vocational opportunities. The risks for ischemic cardiac disease, cerebrovascular disease, cancer, and trauma are increased compared to the general population. While medical care is well established for children with CP, it is more fragmented for adults. Some adult medical practitioners are not familiar with management in CP and do not know how to integrate the care these persons have previously received into customary adult care.

Adults with CP should look for practitioners experienced in the care of chronic disorders, and who provide the additional time that may be required for evaluation. They also need to recognize the importance of preventive care, and report any changes in neurological function, as they may represent new impairments secondary to the underlying motor disorder.

Mobility as well as the ability to perform activities of daily living should be carefully monitored, as some adults will experience slow declines over time. Cognitively intact individuals should be provided with instruction in practical matters such as hiring quality personal aids and caregivers, as well as in self-advocacy and in seeking employment opportunities.

Conclusions

Individuals with CP often seek the same opportunities for employment, living, marriage, families, and recreation/leisure as others in their communities. Dramatic innovations in diagnosis using MRI and new treatment modalities offer improved quality of life, treatment options, and educational opportunities. The challenge is providing these opportunities in a changing medical environment, where a premium is not always placed on coordination of care.

See also: Birth Complications and Outcomes; Developmental Disabilities: Cognitive; Developmental Disabilities: Physical; Intellectual Disabilities; Premature Babies; Special Education.

Suggested Readings

Bax M, Murray G, Peter R, *et al.* (2005) Proposed definition and classification of cerebral palsy. *Developmental Medicine and Child Neurology* 47: 571–576.

Elaine G (1998) *Children With Cerebral Palsy: A Parents' Guide,* 2nd edn. Bethesda, MD: Woodbine House.

Eliasson AC, Krumlinde – Sundholm L, Rosblad B, *et al.* (2007) Using the MACS to facilitate communication about manual abilities of children with cerebral palsy. *Developmental Medicine and Child Neurology* 49: 156–157.

Freeman M and Steven B (1998) *Cerebral Palsy: A Complete Guide for Caregiving.* Baltimore, MD: The Johns Hopkins University Press.

Hoon AH and Johnston MV (2002) Cerebral palsy. Asbury AK, McKhann GM, McDonald WI, Goadsby PJ, and McArthur JC (eds.) *Diseases of the Nervous System*, 3rd edn., pp. 568–580. Cambridge: University Press.

Jane FL, Sherri C, and Margaret M (1997) *Keys to Parenting a Child With Cerebral Palsy (Barron's Parenting Keys).* New York: Barron's Educational Series.

Keogh JM and Badawi N (2006) The origins of cerebral palsy. *Current Opinions in Neurology* 19: 129–134.

Krigger KW (2006) Cerebral palsy: An overview. *American Family Physician* 73: 91–102.

Mark L and Batshaw MD (2002) *Children with Disabilities,* 5th edn. Baltimore, MD: Paul H Brookes.

Nelson KB and Ellenberg JH (1986) Antecedents of cerebral palsy. Multivariate analysis of risk. *New England Journal of Medicine* 315: 81–86.

Palisano R, Rosenbaum P, Walter S, *et al.* (1997) Development and reliability of a system to classify gross motor function in children with cerebral palsy. *Developmental Medicine and Child Neurology* 39: 214–223.

Puscavage A and Hoon AH (2005) Spasiticy/cerebral palsy. Singer HS, Kossoff EH, Crawford TO, Hartman AL (eds.) *Treatment of Pediatric Neurologic Disorders* p.17. New York: Dekker.

Relevant Websites

http://www.aamr.org – American Association on Mental Retardation (AAMR).

http://www.eparent.com – Exceptional Parent.

http://idea.ed.gov – IDEA (The Individuals with Disabilities Education Act).

http://www.ucp.org – United Cerebral Palsy (UCP).

http://www.wemove.org – WE MOVE (Worldwide Education and Awareness for Movement Disorders).

Child and Day Care, Effects of

A Clarke-Stewart and J L Miner, University of California, Irvine, Irvine, CA, USA

Glossary

Center-based care – Children in center-based care are cared for in an institutional setting where children of different ages are typically divided into different classes. Teachers in childcare centers often have received some training in child development and may offer an educationally enriching curriculum. Center-based care may have different sponsors, including churches, schools, colleges, universities, social service agencies, Head Start, independent owners and chains, and employers.

Child–adult ratio – A ratio of the number of children in a childcare setting to the number of adults available to care for them. This ratio is used as one index of childcare quality.

Childcare – Care of children in their home, in someone else's home, or in a center, where care and education are provided by a person other than the parent. Childcare settings vary in location, design, and quality.

Childcare quality – The quality of care children receive in various day-care settings is typically related to child–adult ratios, the sensitivity and educational training of the caregivers, the physical space available in the setting, the structure of the program, and the level of stimulation provided in the environment.

Home-based care – Care of children provided in a residential setting, typically by a woman who cares for several children of varying ages. These care providers rarely offer a structured 'curriculum' *per se*, although they may follow a rough schedule of activities during the day. This form of care may alternatively be known as family childcare.

NICHD SECCYD – One of the most comprehensive studies of early child development and childcare, the Study of Early Child Care and Youth Development (SECCYD) was initiated in 1991 with support from the National Institute of Child Health and Human Development (NICHD). The study has tracked over 1000 children from 10 sites across the US for over 15 years to determine how and in what ways childcare experiences influence children's development.

Introduction

Since the 1970s, a dramatic shift has taken place in who cares for infants and young children. In the US, in 1977, approximately 4.3 million children under the age of 5 years were cared for by someone other than their mothers for a significant portion of each week; by 1997 that figure had tripled to 12.4 million children who were regularly in some type of nonmaternal 'childcare' or 'day care' (see **Figure 1**). Although there is some indication that the increase in the number of children in childcare has begun to taper off, there is no indication that the trend

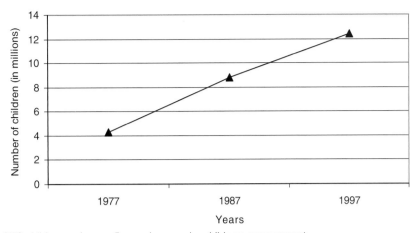

Figure 1   Number of US children under age 5 years in a regular childcare arrangement.

will be reversed any time soon. Of the 12.4 million children estimated to be in care in 1997, nearly 5 million were children under the age of 3 years, with more than half of these infants and toddlers (2.6 million) in childcare at least 35 h per week. In other industrialized countries, many children do not begin care until their second or third year because of the availability of paid parental leave programs. Regardless of when children enter care, however, there is much concern among parents and psychologists about how childcare affects children's social and emotional well-being and intellectual development. Empirical research suggests that childcare experiences can influence children's development in both positive and negative ways.

Types of Child Care

Working parents typically choose from the following options for childcare: care in their own home, care in someone else's home, or care in a center. The distribution of these types of care in the US is shown in **Figure 2**. Each option has advantages and disadvantages for children and families. Care in the family's own home is a popular choice for parents of infants. In this setting, care is typically provided by an adult relative; less commonly, by a babysitter or nanny. Educational programs and group activities with peers are uncommon in this type of care.

A family childcare home is a care arrangement in which an adult (almost always a woman) cares for a small group of children – usually of different ages – in her own home. These care providers rarely offer a 'curriculum' *per se*, although they may follow a rough schedule of activities; the typical day is relatively unstructured and may include free play, outside time, lunch and snacks, and nap time.

The third option for families seeking childcare is a childcare center. A center may provide care for fewer than 15 children or more than 300; on average, in the US, there are about 60 children in a center. About one-third

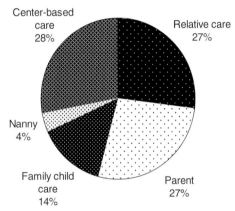

Figure 2   Primary care arrangements for children under 5 years with employed mothers.

of the children in centers attend full time (at least 35 h per week). They are usually divided into classes or groups according to their age. Most children in childcare centers are 3 or 4 years old, although, in the US, center care for infants and toddlers has risen in popularity. Teachers in childcare centers tend to be young women with training in child development. They are likely to offer children educational opportunities and the chance to play with other children of their age in a child-oriented, safe environment. In fact, center care is significantly safer than care in a private home in terms of protection against child fatalities from violence.

Although childcare centers generally offer educational opportunities and protections for children, there is wide variability in the regulations imposed on these facilities, making the quality of care provided to children vastly different from place to place. In the US, in states with the most stringent standards, specific educational and training requirements are in place for lead teachers and assistants; child–staff ratios are very low (e.g., one caregiver for every three or four infants or toddlers); and the total group size is restricted in an effort to maintain a calmer, quieter, less stressful, and chaotic environment. At the other extreme, in some states, there are no educational requirements for center-based teachers, child–staff ratios allow for as many as 10 or 12 toddlers to be supervised by one adult, and group sizes are unregulated.

Differences across states in childcare home regulation are even greater than those for center regulation, in part because there is such wide variability in what is considered a 'childcare home' and what kinds of arrangements should be subject to licensing. On average, states allow anywhere from five to 10 young children in a family childcare home to be cared for by a single-adult caregiver. Only 16 states require preservice training or orientation for family childcare providers. Thirty-five states require some ongoing annual training after a care provider is licensed, but the amount of training varies widely, from as little as 4 h to as much as 20 h.

Effects of Childcare Participation

Studies of childcare have offered compelling evidence that participating in childcare has implications for children's development in several areas, including physical health, language, cognition, social skills, and emotional well-being.

Physical Health and Development

To find out whether children in childcare differ from children who stay at home with their mothers in physical development and health, a number of researchers have included pediatricians', teachers', and mothers' reports of children's physical health and growth and standard tests of children's motor abilities (walking, jumping, throwing a ball, handling tools). The results of their investigations suggest that there is both good and bad news associated with being in childcare.

In terms of physical development, being in childcare advances children's motor development and activity, increases height and weight, and decreases the likelihood of pediatric problems – for infants and young children from poor families. However, if children already come from families who provide ample nutrition, adequate health services, and opportunities for exercise and activity, no benefit in physical development accrues from going to a childcare center; there is no advantage in growth or motor skills for middle-class children attending childcare programs.

There is, however, a difference in health. Children in childcare centers, whatever their family backgrounds, get more diarrhea, influenza, rashes, colds, coughs, and ear infections than children at home. On average, children in childcare have two to four gastrointestinal episodes per year, whereas children at home have only half that many.

The explanation most often offered for the higher rates of illness in childcare is that children there are exposed to more pathogens because they are with other children.

One important concern about childhood illnesses is whether children who experience more of them – because they are in childcare – suffer delayed language and cognitive development. In one prospective study of children in childcare, chronic ear infections were found to be related to lower verbal ability at age 7 years. However, in a much larger study, the National Institute of Child Health and Human Development Study of Early Child Care and Youth Development (NICHD SECCYD), with other potentially confounding factors controlled, there were no associations between the frequency of communicable illnesses and language competence at the age of 3 years.

The NICHD SECCYD is a multisite longitudinal study following 1300 children from birth through adolescence. Families from nine different states were selected, soon after their babies were born, in 1991. Children were then observed in whatever childcare arrangements their parents chose for them, allowing researchers to observe the full range of quality and variety of types of care available throughout the US. Observations of the children's experiences at home and, later, in school supplemented these observations and helped researchers understand the effects of childcare in the context of other influences on the children's development. Assessments of children's social, cognitive, language, and physical development were conducted every year from age 1 to 15 years. Future findings from the NICHD SECCYD will clarify whether there are longer-term consequences of early childcare-related illnesses.

Intellectual Development

Almost all studies of childcare in the preschool period – in Canada, England, Sweden, Czechoslovakia, Bermuda, and other countries, as well as the US – show that care in a decent childcare facility seems to have no detrimental effects on children's intellectual development. Only when care is of very poor quality have researchers found that scores on tests of perception, language, and intelligence were lower for children attending a childcare center than for children of comparable family backgrounds being cared for by parents at home.

In the NICHD SECCYD, participation in center care was related to intellectual development. After researchers adjusted for other influences, such as the parents' education, income, and psychological adjustment, children in centers performed better on cognitive tasks at 24 months, had better receptive language skills at 36 months, and had better memory skills at 54 months than children with no center care prior to 54 months of age.

A number of other studies have shown similar results. Children in childcare centers did better on tests of verbal fluency, memory, and comprehension. Their speech was more complex, and they were able to identify other people's feelings and points of view earlier. This apparent advantage of childcare attendance occurs most often for children who attend high-quality programs. In the US where care, on average, is mediocre in quality, an analysis of 59 studies revealed no reliable differences in intellectual skills between children who attended care and children who did not. In Europe, however, where childcare tends to be of higher quality, there have been more consistent results suggesting positive effects of care on children's intellectual and language skills. Participation in a childcare center is also more likely to benefit children's intellectual development if the children come from unstimulating homes. Although it is impossible to give a blanket statement about the effect of childcare on children's intellectual development, research suggests that care in a center, care of high quality, and care for disadvantaged children is more likely to have positive than negative effects.

Social Competence

Research on the effects of childcare participation on children's social competence has focused on children's relationships with their peers. Because children in childcare have daily experience with other children, researchers had anticipated that their social skills with their peers would be advanced. To see if this was true, researchers observed children in their childcare settings as they interacted with each other. They also brought pairs of children into play situations in the laboratory and watched their interactions, and they tested children on their willingness to cooperate and help each other. The results of their studies suggest that childcare can promote social skills.

Children with experience in childcare have been observed to have more complex and mature interactions with their peers than children with no childcare experience. They sustain their play longer and respond more appropriately and immediately to the other children's behavior, share materials, and behave empathically. They maintain coordinated action at younger ages than is customarily observed. In the NICHD SECCYD, children who had spent more time in childcare settings with other children were more positive and skilled in their peer play in childcare at age 3 years, even controlling for family background factors and other child characteristics. These effects may last. In elementary school, these children have been observed to have more friends and to be more popular. In one 14-year longitudinal study, the amount of time children spent in care prior to 3 years of age shaped their social skills with peers, and these individual differences in social competence remained stable through childhood and early adolescence. Like the differences in intellectual competence, differences in social competence with peers appear frequently, but not invariably.

Behavior Problems

Although some children appear to benefit from group childcare experiences in terms of social skills, there is also evidence that childcare can predict behavior problems in young children. Some studies show that children in childcare tend to be less polite, agreeable, and compliant with their mother's or caregiver's demands and requests, less respectful of others' rights, more irritable and rebellious, more likely to use profane language, louder and more boisterous, and more competitive and aggressive with their peers than children who are not or were not in day care. These differences in aggressive and noncompliant behavior appear in tests and natural observations, in the childcare center and on the playground, with adults and with other children, with strangers and with parents – although differences are smaller when the quality of childcare is good. The differences are also more marked for boys and children from lower-income families. Evidence from the NICHD SECCYD suggests that children who spend more time periods in center care before they enter school exhibit higher levels of these behavior problems throughout their elementary school years.

How can we integrate these negative differences with the differences in positive social behavior? Are children in childcare more socially competent or less? Are the same children both polite and difficult to manage? To answer this question, researchers in the NICHD SECCYD examined which children were displaying behavior problems and which were socially competent. Results indicated that the children who were high in social skills were not high in aggressiveness; positive social behavior and negative social behavior were unrelated to each other. Therefore, it seems likely that childcare promotes social advancement in some children and leads to behavior problems in others. More intelligent children may be advanced in the social realm, just as they are in the intellectual sphere, and childcare may contribute to their knowledgeability, self-sufficiency, and ability to cooperate. Less intelligent children may learn in childcare that antisocial behavior can get them what they want; they become determined to get their own way in the group and, without the social skills to achieve this smoothly, become aggressive, irritable, and rude as a consequence of being in childcare. In the NICHD SEC-CYD, children who exhibited more positive social behavior with peers scored higher on tests of cognitive ability; children who exhibited more negative social behavior with peers scored lower.

Childcare apparently can be a venue for learning social graces or a breeding ground for aggression. However, if there is an inherent risk of developing poor interaction patterns in childcare, it may be the result of spending a lot of time in the setting. A full day of care is stressful for young children, and the more hours children spend in care, the more likely they are to be aggressive (then and later on) – especially if the care is of low quality. Megan Gunnar and colleagues studied cortisol levels of children in childcare centers. Cortisol is a hormone produced by the hypothalamic–pituitary–adrenocortical system, which increases when demands exceed the individual's coping resources. They found that cortisol levels of children in centers were the same as or lower than cortisol levels in children at home in the morning, but by mid-afternoon had risen to a significantly higher level than when the children stayed home; children in half-day preschool programs did not show this increase in cortisol.

Relationship with Mother

Many psychologists and parents have worried that when children, especially infants, are separated from their mother for 8 or 10 or more hours a day, they will not enjoy as close and secure a relationship with her as children who are at home all day. Fortunately, research has shown that infants in childcare do form attachments to their mothers.

But the real question is whether the 'quality' of the relationship these children have with their mother is as emotionally secure as the relationship of infants who are raised exclusively at home. Putting together data from studies of infants in childcare that included the standard method of measuring children's attachment relationships, the Strange Situation, research reviewers at the end of the 1980s found that infants who were in childcare full time, compared with infants in childcare part time or not at all, were more likely to be classified as having insecure avoidant relationships with their mothers.

However, researchers later questioned these findings, wondering whether the Strange Situation was in fact the best method for measuring relationships between mothers and children who attended regular childcare. In the Strange Situation, an infant plays with toys in an unfamiliar room, is left by the mother alone in the room with an unfamiliar woman, plays with and is comforted by that woman in the mother's absence, and then the mother returns and picks up the infant. If the infant greets the mother with pleasure and proximity, that infant is classified as having a secure relationship; if the infant ignores or avoids the mother, the attachment is classified as insecure avoidant. However, children in childcare experience a separation from their mother on a daily basis; researchers wondered if the Strange Situation was the best way to assess the mother–child relationship for children who were more used to separations and reunions.

In a study in the Netherlands, researchers compared infants who attended a licensed childcare center for at least 20 h per week with infants reared exclusively at home. They, too, found that in the Strange Situation infants in childcare were more likely to be classified as having an avoidant attachment. But these researchers also

measured the infants' heart rates to determine how stressed they felt in the Strange Situation. When they were alone in the room, home-reared infants who were classified as avoidant showed major increases in heart rate; they indeed felt stressed. However, the childcare infants who were classified as avoidant showed only small heart rate increases. They apparently experienced only very mild stress. In a modified procedure, the researchers added strange sounds to the episodes preceding reunion with mother to increase the stressfulness for all infants, and, indeed, all of them showed increased heart rates. In this more stressful procedure, 70% of the childcare infants who had been classified as avoidant in the original Strange Situation were now classified as secure. In another evaluation of the validity of the Strange Situation for childcare children, researchers in California presented children with somewhat frightening novel events instead of separations from mother; these researchers, too, found that childcare children were as likely to be secure as home-care children in this more equitable assessment.

Meanwhile, studies conducted since 1990 have shown that the link between childcare participation and insecure avoidant attachment is less likely to be significant – even in the Strange Situation – when other child and family variables are statistically controlled. In the NICHD SEC-CYD, infants were assessed in the Strange Situation at 15 months of age and in a modified Strange Situation with longer separations at 36 months. No significant main effects of being in infant childcare were found at either assessment. The best predictor of attachment security was the mother's observed sensitivity with the child at 6 and 15 months and the mother's psychological health. If children were in care for more than 10 h a week or were in poor quality care and their mothers were rated as insensitive, their attachment security suffered; otherwise, childcare was not related to attachment quality.

Variations in Care

But childcare comes in many shapes and sizes, and, as we have already hinted, these variations make a difference in whether and how children's participation in care is related to their social, emotional, and intellectual development. Most critical for children's development is the overall quality of care.

Assessing Childcare Quality

Childcare quality is a complex construct and must be assessed with expert judgment and sensitivity. The best evaluations come from observing children's actual experiences in the care arrangement. This is time intensive and requires extensive training. Popular measures for assessing childcare quality include the Early Childhood

Environment Rating Scale (ECERS) and the Infant/ Toddler Environment Rating Scale (ITERS). These measures focus on the furnishings and room arrangement, the planning and scheduling of activities, the amount and accessibility of materials, and the caregiver's behavior with the children. Researchers in the NICHD SECCYD used a different instrument – the Observational Record of the Caregiving Environment (ORCE) – which involves a behavior checklist and qualitative ratings of the caregiver's interactions with a particular child. It, too, requires extensive training so that the observer can evaluate, for example, caregiving qualities such as sensitivity and intrusiveness. These measures of childcare quality have consistently been found to be linked to developmental outcomes in children.

Effects of Childcare Quality

In high-quality childcare, caregivers encourage the children to be actively engaged in a variety of activities, have frequent, positive interactions that include smiling, touching, holding, and speaking at the child's eye level, respond promptly to children's questions or requests, and encourage children to talk about their experiences, feelings, and ideas. Using global assessments of childcare quality, investigators have found that childcare quality is significantly associated with children's cognitive development. Associations have been observed in locations as far ranging as Bermuda, Sweden, Switzerland, and Singapore, as well as Canada, the UK, and the US, and in home childcare settings as well as centers, for children ranging in age from 1 to 4 years. The most compelling evidence that childcare quality affects children's cognitive development comes from experimental studies in which children from high-risk families placed in high-quality programs consistently make gains – at least as long as the high-quality care is maintained.

In the social realm as well, children attending high-quality childcare settings benefit. Such children are judged by observers and their caregivers to be more sociable, considerate, compliant, controlled, and prosocial. They are more interested, better adjusted, and have higher self-esteem, while children in low-quality care are angry, defiant, and cry more. Moreover, childcare quality is related to children's cortisol levels. Children who receive more attention and stimulation from their caregivers in childcare are less likely to have increased cortisol levels over the course of the day; their cortisol levels look just like those of children who stayed in the peaceful environment of their own homes.

In the NICHD SECCYD, after controlling for family variables that might be associated with childcare quality or the child's ability level, researchers found that children in higher-quality care, in both home-based and center childcare settings, received higher scores on tests of cognitive and language abilities, exhibited more positive and skilled interactions with peers in childcare, and were reported by their caregivers to have fewer behavior problems, at ages 2 and 3 years. At age 4 years, they performed better on a battery of standardized tests that assessed their abilities to read letters and words, do simple math problems, complete words, remember and recite back sentences, and understand and use language. They were less impulsive in a test of attention, and they were rated by their caregivers as being more socially skilled and as having fewer behavior problems. Childcare quality was not related to the children's attachment relationship with their mother or their interaction with a good friend, but consistently, across a variety of measures, it predicted social and intellectual competence. In fact, children in high-quality childcare performed better than children in full-time maternal care; children in low-quality care performed worse.

Physical Environments

Space is one aspect of the childcare environment that affects how caregivers and children behave, and in turn has implications for the quality of care provided to children. In particular, the amount of space per child has been a focus of research. In one study of children in classrooms ranging from 27 to 52 ft^2 per child, those in the more crowded classrooms scored lower on a test of cognitive style. When the space per child is very limited (less than 25 ft^2 per child), children have been observed to be more physical and aggressive with their peers and more destructive with their toys; they spend more time doing nothing and less time interacting socially. Fortunately, in the US, licensing standards for childcare centers and childcare homes ensure space that exceeds this lower limit (more than 35 ft^2 per child). Therefore, space may not be a major influence on children in most childcare settings, at least those that meet licensing requirements.

Another aspect of the physical environment related to childcare quality is the variety of play materials available to the children. If few materials are available and the equipment is fixed, inflexible, or limited, children spend their time watching, waiting, cruising, touching, imitating, chatting, quarreling, and horsing around with peers; their play is of low complexity and little intellectual value. However, when materials are varied, age-appropriate, and educational, children do better. In the NICHD SECCYD, children in settings with more stimulating, varied, and well-organized materials (including materials to stimulate math, movement, music, language, art, and play) received higher scores on tests of language comprehension and short-term memory.

Space of good physical quality can be found in both homes and centers, although materials that elicit high-level constructive play (puzzles, block, art) may be more common in centers, and opportunities for free play and tactile exploration (water, sand, dough, pillows) more likely in homes. In either kind of setting, the quality of the physical space affects the adults' behavior as well as the children's. In physical settings of good quality, caregivers can both allow children freedom to explore and spend their time demonstrating constructive activities with the materials available, rather than supervising and scolding all the time. In the NICHD SECCYD, across ages and different types of care – centers, childcare homes, in-home sitters, and grandparents – caregivers consistently provided more sensitive and responsive attention when the physical environment in the setting was safer and more stimulating.

Childcare Activities

Another dimension of childcare quality is the type of activities that make up the program. Childcare arrangements range from strictly care-oriented programs, where activities involve only daily routines and children are left to amuse themselves, to explicitly educational programs, in which children are offered a rich diet of intellectual and social stimulation. The first type of program is more likely to provide low-quality care; the second, high-quality care. In more educationally oriented programs, children spend more time in constructive and complex play with materials, engage in more cognitively challenging tasks, converse more with adults and other children, and score higher on tests of intelligence and achievement. These educational programs are particularly beneficial for older preschool children. When children spend their time in childcare just playing around with other children or watching television, they experience less 'rich' play and are less competent. In the NICHD SECCYD, children who watched more television while they were in the childcare setting received lower scores on math, vocabulary, and language comprehension. Thus, there is evidence that an educational program promotes more intellectually valuable experiences in the care setting and that these experiences have positive consequences for children.

Additionally, there is evidence that the amount of 'structure' in the childcare program is related to quality and to child outcomes. Some programs are highly structured and controlled – the children's activities are planned and directed by adults; others are unstructured and chaotic. Programs that blend prescribed educational activities with opportunities for free choice, that have some structure but also allow children to explore a rich environment of objects and peers on their own without teacher direction, seem to have the most benefit. In such programs, children have more opportunities for constructive activity, for learning academic skills, for problem solving, and for acquiring social skills. They also acquire positive motivation and persistence and these qualities aid them in later academic settings.

Regulable Features of Child Care

Features of the childcare environment that can be regulated by government agencies also contribute to the overall quality of childcare and are sometimes used as proxy measures of quality. These features include the number of children in the childcare group or class, the ratio of children to adults, and the level of the caregiver's education and training.

Group sizes and child–adult ratios

The size of the group of children in a childcare setting has been shown to relate to children's social competence and self-control. Although being with other children offers opportunities to develop and hone social skills, there are limits to how big the group should be. In large classes – classes of more than 20 children – researchers have observed that children are more likely to look apathetic, cry, and act hostile.

Having too many children for each caregiver to look after can be especially detrimental. Child–caregiver ratios range substantially from one childcare setting to another. In childcare homes, there may only be one or two children for the caregiver to attend to. In centers in the US, government-mandated ratios range from four to 13 toddlers per caregiver, from seven to 20 preschool children. With too many children for one caregiver, children suffer. Studies have shown quite consistently that overall quality of care suffers when child–adult ratios are high. Caregivers in classrooms with high child–staff ratios are less sensitive, responsive, and positive. Children have less contact with the caregiver, spend more of their time playing with other children, and less time in intellectual activities; fewer of their questions are answered; their conversations are shorter; and contact with the caregiver is more likely to involve prohibitions, commands, corrections, and routines. The children are less likely to develop secure attachment relationships with their caregivers, and, on tests, they receive lower scores for language and communication skills.

The benefits of having low child–caregiver ratios showed up clearly in the Cost, Quality and Child Outcomes Study – a study of 100 centers in each of four states, California, Connecticut, Colorado, and North Carolina: lower child–adult ratios were related to better overall quality of care, more caregiver sensitivity, and more effective teaching; children in classes with lower child–teacher ratios also had higher pre-reading scores. In the Florida Child-Care Quality Improvement Study, when child–staff ratios in preschool centers were reduced from 6:1 to 4:1 for infants and from 8:1 to 6:1 for toddlers, overall classroom quality improved, teachers became more sensitive, and children engaged in more cognitively complex play with other children and classroom materials, gained more in cognitive development, and were more securely attached to their teachers.

In childcare homes, where there are seldom more than six children for one caregiver, the link with child–caregiver ratio is not as strong – it may even be in the opposite direction. When caregivers who are taking care of more children provide higher-quality care, however, this may be because they are more professional than caregivers with few children – they have higher levels of education and training and are more likely to belong to a professional organization and to have their home licensed for childcare. In the NICHD SECCYD, it was found that when caregivers' training and education were controlled, the number of children in the childcare home was not related to the quality of care or child outcomes. In the California Licensing Study, when researchers added two more children to existing childcare homes, they found that the quality of caregiving declined. Thus, it does seem that, even in homes, children and caregivers benefit from relatively low child–caregiver ratios.

Caregiver education and training

Caregivers and children also benefit when the caregiver has a higher level of education and training. Research has shown that, all other things being equal, caregivers with higher education and training provide the highest quality of care to children. Educated caregivers are more involved and affectionate with the children in their care. They are less authoritarian – restricting children less and encouraging them more. They implement more developmentally appropriate practices in the classroom and provide richer literacy environments.

Furthermore, correlational studies suggest that when caregivers have higher levels of training and education, children do better both intellectually and socially. In a number of studies, children whose caregivers had higher levels of education and/or specialized training were more involved, cooperative, persistent, competent in play, and learned more. They also did better on standardized tests of cognitive and language development. In one study of low-income African American children in community centers, for example, girls whose teachers had 14 or more years of education did better on tests of cognitive and receptive language skills than children whose teachers had less education. In the Cost, Quality, and Child Outcomes Study, children whose teachers had associate or bachelor degrees in early childhood education or who reported that they had received training at community workshops had higher scores on a test of receptive language. In the NICHD SECCYD, caregiver training was related to higher caregiving quality and this in turn was related to advanced cognitive and social competence in children. In childcare homes, for example, children with a college-educated caregiver scored seven points higher on the Reynell Developmental Language Scales at age 3 years. Children with college-educated caregivers in centers also displayed advanced school readiness skills

and better language comprehension at age 3 years and greater cognitive and social competence at age 4 years. In studies comparing children's behavior before and after caregiver training programs, too, researchers have found that children make significant gains in complex social and cognitive play and become more securely attached to their caregivers after the training. In brief, then, the amount of training and education caregivers receive has demonstrable effects on the quality of care they provide and significant consequences for children's behavior and development. However, merely providing caregiver training or reducing child–adult ratios is not a guarantee of increased childcare quality – these aspects of the childcare environment must be translated into more sensitive patterns of caregiving to produce higher-quality care and more positive consequences for children.

Effects of Entering Childcare in Infancy

Much debate has surrounded the issue of the best age to first enroll children in childcare. Parents, pediatricians, and psychologists have worried that care in infancy, when children are particularly vulnerable and family relationships most fragile, is a risk factor for child development. The problem with studying the effects of being in care in the first year of life, however, is that the researchers have found it difficult to disentangle the effects of timing of care (the age at which the child enters care) and amount of care (the cumulative hours of care), because the two go together. Parents seldom have their children in care as infants, keep them home as toddlers, and send them back to care during the preschool years. Typically, children who are placed in care early on end up having experienced more care by the time they reach school age. We will probably not have an unambiguous answer to the question of whether there are long-term harmful effects of care in infancy until researchers study the issue experimentally.

Although we do not yet have a definitive answer to the question, it does appear that infant care is not likely to be harmful – at least in moderation. Researchers in a study in the UK headed by Kathy Sylva found that there were no detrimental effects on cognitive development if infants were placed in a childcare center while their mothers worked, rather than being with relatives or friends, and, in fact, these children did better in terms of intellectual abilities and social skills with other children than children who did not start childcare until they were 3 years or older. In a large, nationally representative study in the US, as well, entry into care in the first year was not related to lower cognitive scores, and, for boys from high-income families, there were even positive effects. A number of researchers have found that children who begin care in infancy are more likely to have externalizing problems – they are less compliant, act out more, have less tolerance of frustration, and are more aggressive. However, these effects are not found in all studies, and when researchers have controlled for the number of different care arrangements children have been in or how much care they attended in the preschool years, the apparent negative effects of childcare in the first year of life disappeared. In the NICHD SECCYD, the age of entering care was unrelated to children's behavior problems at 2 and 3 years of age, and it was impossible to determine whether the effect of care on externalizing problems at age 4 years was the result of care in infancy or cumulative hours overall. No significant effects of age of entry on attachment security were found when children were assessed in the Strange Situation at 1 or 3 years of age either. Thus, at the present time, there is no compelling evidence that beginning care in infancy has detrimental effects on children's development.

Moreover, the emotional issues involved in starting care after infancy are just as strong as earlier – perhaps stronger. When toddlers enter care in their second year, they are just as distressed as infants who enter care in the first year, and their cortisol levels rise over the course of a day in care as much as infants' do. Entry into care in the second year elevates the likelihood that children will develop an ambivalent relationship with their mother. It can also lead to behavior problems. In the NICHD SEC-CYD and other studies, children who first entered childcare around their second birthday exhibited more behavior problems in childcare than children who began care earlier. In brief, there is no reason to believe that children who enter care as toddlers do better than children who enter as infants. Research on children of this age shows that they are particularly negatively affected by being in large classes with many children and few caregivers.

Effects of Childcare for Different Children

Childcare researchers have also investigated whether there are particular kinds of children who are more at risk if they are enrolled in childcare. Children's gender, temperament, and health all have been examined to see if they create different susceptibilities and lead to different consequences in childcare.

Boys and Girls

Across the board, boys are more vulnerable to events in the environment and girls more resilient. Therefore, researchers have questioned whether the gender difference between males and females is reflected in different effects of childcare on boys and girls. Specifically, they have wondered whether boys are at greater risk in childcare than girls. Research findings in this area suggest the answer is a weak 'maybe'. There is some limited evidence that there is a slight disadvantage for boys in childcare.

First, in some studies, boys (but not girls) placed in childcare in infancy have been observed to develop less secure relationships with their mothers. Second, boys (but not girls) from high-income families have been found to have somewhat slower intellectual development if they are in full-time childcare as infants rather than with their mothers. Third, earlier entry into low-quality care has been associated with problem behavior and anger for boys (but not girls). However, most researchers using large samples, including the National Longitudinal Study of Youth; the Cost, Quality, and Child Outcomes Study; and the NICHD SECCYD, have failed to find significant interactions with child's gender. One significant interaction in the NICHD SECCYD indicating that boys (but not girls) with many hours of care were less likely to be securely attached to their mothers was not replicated at a later assessment or in another study, and an analysis of the effects of childcare on infant–mother relationships for a combined sample from more than 12 different studies found no significant differences overall for boys and girls. If there is a difference in the effect of childcare for boys and girls, therefore, it is small and unreliable.

Vulnerable Children

Researchers have further pondered whether children with certain vulnerabilities, such as emotional or physical health problems, are differentially affected by childcare. One emotional problem they have studied is having a difficult temperament in infancy. Infants with difficult temperaments cry intensely and often, are difficult to soothe, adapt to change and new situations slowly. Several researchers have studied whether this temperament style makes it harder for children to adjust to being in childcare. They have found that infants with difficult temperaments are indeed disadvantaged when they enter care. Compared with easy infants, difficult ones are less positive in their emotions, receive less attention from caregivers and the other children, and make less progress in their cognitive development during the first year in a childcare center. Their difficult temperament makes their initial adjustment to childcare difficult. Eventually, they adjust to being in the group, but even then, their problems may not be over. In one study, researchers found that infants with difficult temperaments were more susceptible to the effects of being in long hours of care. If, as infants, they were easily frustrated, they were more likely to develop externalizing behavior problems; if they were especially distressed by novelty, they were more likely to develop internalizing behavior problems (anxiety, depression). Children who are highly anxious may have persistent problems in childcare. Group care seems to be especially stressful for them, and they show more marked physiological reactions to childcare than other children. The length of the day, the need to sustain interactions with many other children, and the need to interact with multiple adults are challenges that tax their emotional resources and coping competencies.

This theme is echoed when children have physical health problems. Data from the National Longitudinal Study of Youth suggest that childcare provides enriching input that can enhance children's development only if the children are healthy. In fact, for infant boys with health problems, socioemotional and motor development were most helped by being cared for more extensively by their mothers. One health problem that has received researchers' attention is otitis media (ear infection). Children with frequent ear infections may suffer hearing loss and have difficulty attending to speech. In childcare environments where the child has little one-to-one interaction with adults, persistent background noise, and caregivers who are not sensitive to the child's hearing loss, the negative effects of otitis media on children's language and social development are exacerbated.

How Large are the Effects of Childcare?

As we have described, research indicates that there are reliable effects of childcare on children's development. But how large and how meaningful are these effects? In general, differences between children in childcare and at home, between children in high-quality care and low, and between children who attend care for many hours or few are relatively small. In examining the direct effects of childcare quality on children's intellectual development, for example, researchers in the NICHD SECCYD found that a 1 standard deviation (SD) increase in childcare quality between age 3 and 4.5 years was associated with an increase of 0.5 to 1.5 points, or about one-tenth of a SD increase, on standard cognitive tests.

One way that researchers have evaluated the size of childcare effects has been to compare them to the effects of parenting. Results of these comparisons in the NICHD SECCYD suggest that the effect sizes associated with type and quality of childcare are approximately half the size of parenting effects for cognitive skills (e.g., the effect size of childcare quality on school readiness was 0.39; whereas for parenting, the effect size was 0.83).

Another way researchers have evaluated the size of childcare effects is to compute the percentage of children in different childcare conditions who fall above a certain value. In the NICHD SECCYD, researchers used this method to evaluate the effect of spending a lot of time in care on children's externalizing behavior problems. Children who had spent a lot of time in care were not in the 'clinical' range of externalizing behavior problems. However, more of them received high scores on the externalizing scale – at least 1 SD above the mean.

When they were 4.5 years old, 24% of the children who were in care for more than 45 h per week received high scores, compared with 15% of those in care for 10 to 45 h and only 2% of those in care for less than 10 h. Differences were most marked at the time the children were still in care and became less marked in subsequent years.

The Family's Place

Childcare providers are neither the first nor the most important of the child's caregivers. Those distinctions go to parents. Even if children spend 40 h a week in childcare, this leaves 128 h for them to be at home with their parents – and that time counts. The observations and interviews researchers have used to study parents and children suggest quite clearly that, even though the parents of children in childcare are not intimately involved in their children's experiences at all times, they do continue to have an influence on their development – just as much as if the children were at home.

Researchers interested in the relative contributions of family and childcare to children's development have included both family and childcare variables in their analyses and compared the two. Such studies generally indicate that family factors predict children's development more strongly than childcare factors. Family variables such as mothers' education, sensitivity, and stimulation, and the quality of the home environment have been found to predict children's cognitive, language, and social development more strongly than whether or not the child attended childcare, the type of childcare attended, or the quality of the childcare program. In the NICHD SECCYD, children's security of attachment to mother, self-control, compliance and problem behavior, interactions with peers, and cognitive and social competence were all related to maternal sensitivity more strongly and consistently than they were related to childcare quality or the amount of time the child was in care. Furthermore, children who spent a lot of time in childcare (at least 30 h a week) were influenced by their families just as much as children who were in care for less than 10 h a week. Children had better cognitive and language development if their parents were more affluent and well educated and their mothers were more sensitive and warm – regardless of whether they were in childcare for many hours. In other words, family influence is not 'lost' when children are in care.

However, it has also been found that the best models for predicting children's development include both family and childcare factors. In such analyses, researchers have found that optimal development is supported when children receive high-quality care, stimulation, and encouragement in both home and childcare settings – or conversely, children's well-being is most likely to be impaired when children are in poor-quality childcare as well as unstimulating home environments. Childcare does not operate alone; it is part of a complex network of influences on child development. Moreover, within this network, influences are related and overlapping – for example, better-quality childcare is more likely for children from more stimulating homes, because more highly educated parents provide both environments – making it more difficult for researchers to isolate the effects of childcare.

Summary

The quality of research on childcare has improved substantially over the years, providing more insight into how childcare affects young children's social, emotional, and intellectual development. With large-scale studies such as the NICHD SECCYD, researchers have shed light on the development of children both in and out of childcare.

In terms of physical health, children in childcare do appear to experience more illnesses than children not in care, most likely because they are exposed to more germs from other children. However, the increased number of illnesses does not appear to have long-lasting consequences for children's development. In fact, in terms of intellectual development, children who attend childcare centers may benefit compared with children who have no center experience. Children in high-quality childcare centers perform better on tests of language, memory, and cognitive tasks than children at home or in low-quality centers, and this is especially true for children from impoverished backgrounds. Although center care appears to benefit children's intellectual performance, it is impossible to state categorically that center care is optimal for all children. The quality of care children receive – in any setting – in combination with the needs of the particular child, is what is related to positive outcomes.

Socially, children in childcare arrangements with other children appear to benefit from the increased opportunities to interact with peers. In most, but not all studies, results suggest that children in childcare have advanced peer interaction skills and are more independent and assertive than children not in childcare. These increased opportunities for interaction may be particularly beneficial for children without siblings. However, there is also evidence that children in childcare, especially those who spend large amounts of time in childcare of poor quality, display more behavior problems than children who remain at home. These problem behaviors may be related to spending time in a stressful environment with a number of other children, but the precise reasons are not yet fully understood.

The quality of the childcare setting has significant implications for how childcare affects children's development. High-quality care can promote desirable qualities such as prosocial behaviors and advanced cognitive abilities. When the care setting is of poor quality, however, such as when there is a high child–caregiver ratio, inadequate physical space, or an insensitive and unresponsive caregiver, children are observed to be more hostile and aggressive and to perform poorly on achievement tests.

The timing of entry into childcare appears to have few major effects. Children who enter care as infants do not seem to suffer in terms of relationships with their mothers, social competence, or intellectual development – unless they spent a lot of time in care. Boys and children with difficult or shy temperaments may react poorly to childcare, as may children with chronic health problems.

Thirty years after research on childcare began, the topic continues to be important. As the number of children beginning care early in life remains high, it is vital to use the best research tools to more fully understand how childcare experiences affect children. Thus, research into the effects of childcare will likely continue to be a focus of attention, and it is hoped that as the research improves, so too will our understanding of the mechanisms underlying childcare effects and our ability to improve the quality of care for all children.

See also: Friends and Peers; Maternal and Paternal Employment, Effects of; Imitation and Modeling; Preschool and Nursery School; School Readiness; Social Interaction.

Suggested Readings

Clarke-Stewart A and Allhusen VD (2005) *What We Know about Childcare.* Cambridge, MA: Harvard University Press.
Howes C and Sanders K (2006) Child care for young children. In: Spodek B and Saracho ON (eds.) *Handbook of Research on the Education of Young Children,* 2nd edn., pp. 375–391. Mahwah, NJ: Lawrence Erlbaum.
Lamb ME and Ahnert L (2006) Nonparental Childcare. In: Damon W, Lerner RM, Renninger KA, and Sigel IE (eds.) *Handbook of Child Psychology, Vol. IV, Child Psychology and Practice,* 6th edn., pp. 950–1016. New York: Wiley.
NICHD Early Child Care Research Network (ed.) (2005) *Child Care and Child Development: Results from the NICHD Study of Early Child Care and Youth Development.* New York: Guilford.
NICHD Early Child Care Research Network (2006) Child-care effects for the NICHD Study of Early Child Care and Youth Development. *American Psychologist* 61: 99–116.
Vandell DL (2004) Early Child care: The known and the unknown. *Merrill-Palmer Quarterly* 50: 387–414.

Relevant Websites

http://www.nccic.org – National Child Care Information Center.
http://www.naeyc.org – The National Association for the Education of Young Children.
http://www.nichd.nih.gov – The NICHD Study of Early Child Care and Youth Development.

Circumcision

A Taddio, The Hospital for Sick Children, Toronto, ON, Canada and University of Toronto, ON, Canada

Glossary

Circumcision – The surgical removal of the foreskin from the penis.
Phimosis – Inability to retract the foreskin due to a narrow preputial ring.
Preputial – Space between the foreskin and penis.

Introduction

Circumcision is the most common surgical procedure performed in male infants in the newborn period. It originated over 5000 years ago and has become an important ritual in several cultures worldwide. Circumcision has been promoted for its health benefits including protection against urinary tract infection (UTI) in infancy, penile cancer, and HIV infection. However, it may cause numerous complications, including pain, bleeding, infection, phimosis, meatitis, and other adverse effects. Nonritual circumcisions are not routinely recommended by medical associations except in very select situations. For infants undergoing circumcision, established analgesic interventions should be used to minimize pain. These include injectable and topical local anesthesia, sucrose solutions, and acetaminophen.

History of Circumcision

Male newborn infant circumcision is the most common planned surgical procedure, estimated to be practiced by 15% of the world's male population. It is the subject, however, of continual controversy. It has been debated as a matter of hygiene, religious ritual, and infant mutilation. Although the origin of the procedure is unknown, one of the first records of circumcision dates back to Egyptian times, over 5000 years ago. The reason for circumcision may have involved a hygienic measure to combat either dry, dusty, and hot environments or disease. However, it may also have been performed for punitive reasons (attenuation of castration, as castration was often a mortal injury), as a pubertal or premarital rite, as an absolution against vaginal blood, or as a mark of slavery. For certain religious and cultural groups, circumcision is an important ritual. These groups include members of the Jewish religion as well as Muslims, black Africans, Australian aborigines, and other groups around the world. There is a sociological component of circumcision as well. Circumcision provides a rite of passage for socialization and kinship.

Written documentation of circumcision appears in the Bible, where the ritual was elevated to a religious act by the Jews. In the biblical Covenant of God with Abraham, the 'father of the Hebrew nation', circumcision is described as a sacrifice of the foreskin to be performed in male newborn infants on the eighth day of life. The act of circumcision, however, has been continuously debated. In the early Christian era, Christians debated the need to be circumcised in order to be 'saved' and circumcision was cast as a mutilation, which led to a revolt of Jews against Rome.

The health benefits of circumcision became prominent in the nineteenth century. One of the theories promoted at that time was that circumcision prevented masturbation, which was the source of a variety of illnesses. Circumcision was believed to cure or prevent alcoholism, seizure disorders, asthma, gout, rheumatism, and many other conditions. In addition, circumcision was believed to act as a moderator of excessive sexual activity, making the practice consistent with mid-Victorian attitudes that sex was sinful. Circumcision became widespread in English-speaking countries around the turn of the century and was reinforced in the US in World War II due to hygiene-related urogenital disorders in noncircumcised soldiers. It began to be questioned by the medical community shortly thereafter, however, due to unproven benefits, and by the 1970s, national medical organizations declared that the procedure was not medically justified. The practice became virtually abandoned in the UK and uncommonly performed in other English-speaking countries except for in the US. In fact, the practice of routinely performing nonritual circumcisions in male newborn infants is unique to the US. In the US, the potential health benefits of circumcision were increasingly evaluated to justify prophylactic circumcision. In contrast, in Europe, studies of the natural course of foreskin development were undertaken that supported the conclusion that the procedure was unnecessary. There was also increasing opinion in Europe that experiences in the newborn period influenced emotional and psychological development in infants. Accordingly, gentle birth and gentle newborn experiences were promoted rather than violent birth and painful newborn circumcision.

In 1985, in response to consumer demand for accurate information regarding circumcision, the National

Organization of Circumcision Information Resource Centres (NOCIRC) was founded. It held its first international symposium in 1989 and declared that all human beings have a right to an intact body, including foreskin. Moreover, it stated that parents do not have the right to surgically remove the normal genitalia of their children and that physicians have the responsibility to refuse such procedures. It proclaimed that physicians practicing circumcision are violating the first maxim of medical practice, "First Do No Harm," as well as the United Nations Declaration of Human Rights "No one shall be subjected to torture or to cruel, inhuman, or degrading treatment." There is an ethical dilemma due to the fact that circumcision is not a medically necessary procedure but acceptable based on religio-cultural beliefs. From a human rights perspective, the right of parents to authorize the procedure on their child's body for nontherapeutic reasons has been questioned. The best interests of the child are of paramount importance in such debates.

What is Circumcision?

The male penis consists of a shaft, glans (cone-shaped end), urethral opening, foreskin or prepuce (redundant fold of penile skin which overlaps the glans penis), and the frenulum (fold of skin connecting the inner foreskin to the glans penis). The skin on the penile shaft and the outer part of the foreskin are similar; both are keratinized, stratified squamous epithelium, and barriers against infectious microorganisms. The inner part of the foreskin, however, is a mucosal membrane, and rich in blood supply and nerves, making it very sensitive to touch. At the time of birth, the foreskin is fused to the glans penis, and may not be retracted. If left unmanipulated, the foreskin separates from the glans penis over the first few years of life through physiologic processes including growth of the penis, accumulation of skin debris, and intermittent penile erections.

Circumcision is a surgical procedure that involves removal of the foreskin. Removal of the foreskin during circumcision removes a natural protective barrier for the glans penis against irritation and infection from urine and feces. There is a compensatory effect, however, in that there is development of a thicker, tougher skin over the glans which may be more resistant to disease.

The Debate over the Health Benefits of Circumcision

There is currently widespread debate over whether newborn infants should be routinely circumcised for health reasons. There are studies demonstrating both health benefits and complications of the procedure. Circumcision has benefits as a preventive health measure against conditions such as UTI, penile inflammatory conditions, penile cancer, and sexually transmitted diseases, including HIV. However, the benefits are counter-balanced by risks, including surgical complications and infant pain.

Benefits of Circumcision

There are several potential benefits of circumcision on future health (**Table 1**).

Urinary tract infection

Circumcision is associated with a significant reduction in the risk of UTI in boys. It has been postulated that the foreskin of uncircumcised males, which fosters a moist and warm environment, is conducive to the growth of infectious organisms. Circumcision may therefore prevent periurethral colonization of the foreskin by bacteria. Assuming that the incidence of UTI is 1% in the first year of life, there is a 10-fold reduction in risk to 0.1%. Most UTIs are uncomplicated, that is, they are diagnosed and treated with antibiotics, and do not lead to significant morbidity and mortality. A reduced incidence of UTI, therefore, does not lead to medical justification of routine circumcision of normal boys.

Male infants with a history of recurrent UTI or underlying renal tract anomalies, however, may benefit from circumcision. In these infants, the risk of UTI ranges from 10% to 30%, respectively and circumcision may result in a substantial reduction to 1% to 4% respectively, in the occurrence of UTI.

Penile cancer

Circumcision may protect men against penile cancer. However, penile cancer is a rare disease; the incidence of it is extremely low (approximately 1 case per 100 000 males in the US). Furthermore, there is a strong relation between penile hygiene and penile cancer, such that good hygiene practices significantly reduce the risk of cancer. It has been estimated by Learman that more than 322 000 neonatal circumcisions would be required to prevent one

Table 1 Potential benefits of circumcision

Prevention of
Urinary tract infection
Penile cancer
Sexually transmitted disease, including HIV
Penile inflammatory disorders

case of penile cancer per year and estimates suggest that there are more deaths from circumcision than penile cancer each year.

Sexually transmitted diseases, cervical cancer, and HIV infection

There is some evidence that circumcision reduces the risk of sexually transmitted diseases such as herpes, gonorrhea, Chlamydia, and syphilis. In a recent report by Fergusson, it was estimated that the overall rate of sexually transmitted diseases could be reduced from 10% to 5% if all males were circumcised.

In addition, circumcision has been suggested to be protective against cancer of the cervix in women. Uncircumcised men have been reported to be three times more likely to be carriers of human papilloma virus, the infectious agent that is involved in development of cervical and penile cancer, compared to circumcised men (absolute risk, 20% vs. 6%, respectively).

There are many factors that are believed to be important to overall infection transmission rates. For example, geographic location, access to healthcare, lifestyle, race, socioeconomic status, and sexual habits, all affect the pathogenesis of venereal diseases. The higher risk of sexually transmitted diseases in noncircumcised males is postulated to be due to the combined effects of overgrowth of pathologic organisms and susceptibility of the uncircumcised penis to minor trauma and ulcerative disease during intercourse. This, in turn, increases the risk of infection in males and their sexual partners.

In addition, it is believed that the inner mucosal layer of the foreskin contains cells that are specific HIV-target cells and therefore facilitate HIV transmission. The role of circumcision in the epidemiology of disease appears to be particularly important for HIV infection, a condition that is associated with significant worldwide morbidity and mortality. Circumcision has been investigated as a prophylactic measure against HIV infection in regions of high disease prevalence and poor hygiene, such as sub-Saharan Africa. In a recent report by Auvert, there was a 60% reduction in the chance of becoming infected with HIV in sub-Saharan African men who were circumcised at the beginning of a 14-month observation period when compared to a group of men who were circumcised after the observation period (absolute risk, 1.8% vs. 4.4%, respectively). It is believed that, in the future, male circumcision will become a valuable tool for reducing men's (as well as women's) risk of acquiring HIV infection in communities where HIV is common.

Penile inflammatory disorders

There are some medical complications that may arise in the development of the foreskin in uncircumcised boys that eventually leads to them being circumcised at an older age. These include phimosis (intractable foreskin), paraphimosis (retraction and constriction of the foreskin), balanitis (inflammation of the glans penis), posthitis (inflammation of the foreskin), and balanoposthitis (inflammation of the glans penis and foreskin). Being circumcised at a later age is associated with increased costs due to a more complicated surgical procedure that may involve general anesthesia as well as profoundly negative psychological effects in the child from fear of mutilation and pain. Circumcision has been advocated by some as a preventive measure against a future need for circumcision in boys by preventing these conditions from arising. The percentage of uncircumcised male children who require circumcision later due to such complications is not well documented; in the US, it is estimated to be in the order of 5–10%.

It is important to make a distinction between 'physiological' and 'pathological' phimosis. Pathologic phimosis is due to recurrent infection of the foreskin causing scarring and narrowing of the preputial (between foreskin and penis) ring. Physiologic phimosis is an asymptomatic non-retractible foreskin.

The underlying reason for many cases of postinfancy circumcision has been suggested to be from forceful retraction of the foreskin of infants by parents who are trying to 'clean' the penis, leading to complications. The proper care of the foreskin in uncircumcised male infants involves avoiding any manipulation. During infancy, when the foreskin and glans are attached to one another, external washing with water are sufficient. When the foreskin and glans have separated, then retraction of the foreskin and washing with water can be done.

Phimosis is a diagnosis that is often made prematurely, but does not usually require treatment. It has been suggested that the diagnosis of phimosis should not be made before the age of 18 years in uncircumcised males, and that some men are never able to retract their foreskins. In these men, the preputial space cleans itself by secretions of the prostate, seminal secretions', and mucin secretions. The sloughed epithelial cells (smegma), which lubricate and protect the glans, are broken down by these secretions as well. If treatment is needed, then nonsurgical techniques such as gentle stretching and topical steroids (such as 1% hydrocortisone or 0.05% betamethasone 2–4 times daily for 4–12 weeks) are tried first before circumcision is performed.

Complications of Circumcision

Circumcision is a surgical procedure and like all surgical procedures, is associated with a risk of complications (**Table 2**). The overall complication rate is estimated to be between 2% and 10%.

Table 2   Potential risks of circumcision

Acute	Subacute
Pain	Phimosis
Bleeding	Skin bridges
Infection	Inclusion cyst
Amputation	Meatitis
	Urethrocutaneous fistula
	Penile sensation
	Inadequate foreskin removal

Pain

Circumcision causes intense pain in newborn infants, which is evident by striking changes in physiologic and behavioral parameters. The physiological manifestations of pain include activation of the sympathetic autonomic nervous system and the 'flight or fight' response. Behavioral responses to pain including facial grimacing expressions and intense crying. In the immediate postoperative period, continued pain from the wound causes frequent crying and fewer infant interactions with mothers. Circumcision pain may cause long-term changes in pain responsiveness as well, due to 'imprinting' of pain in their memory. In two separate studies, the pain response during routine immunization 4 to 6 months after circumcision was more intense among circumcised male infants compared with uncircumcised male infants. Administration of anesthesia for circumcision partially attenuated the hypersensitivity to future pain.

Until recently, most circumcisions were performed without the benefits of analgesia. Analgesics were not administered routinely because of the beliefs that newborn infants could not feel pain or that it was unimportant to manage, and because of concerns regarding the adverse effects of drugs. It is now recognized that the pain from circumcision is clinically important and should be prevented. Effective pain-relieving medications have also been evaluated and demonstrated to be safe. In its 1999 policy statement on circumcision, the American Academy of Pediatrics (AAP) recommended that if circumcision is to be performed on an infant, adequate analgesia should be provided.

Bleeding

Bleeding occurs in approximately 1% of circumcision cases. Most episodes are minor and only require administration of pressure on the wound for treatment. Rarely, bleeding can be sufficiently severe so as to require medical intervention (such as silver nitrate, dilute lidocaine with epinephrine, fibrin, sutures, blood transfusion). Circumcision is contraindicated in infants with clotting abnormalities or a family history of bleeding disorders due to the risk of hemorrhage.

Infection

Infection occurs slightly less frequently than bleeding. Usually, infection is limited to the wound and can be treated with topical antibiotics. Rarely, however, it spreads and can become life threatening.

Phimosis

Phimosis is a relatively common complication of inadequate circumcision. It results from the insufficient removal of foreskin and resulting contraction of scar tissue during healing. It may lead to obstruction of urinary flow and infection as well. When present, this complication leads to a repeat circumcision in approximately 10% of cases.

Skin bridges

Skin bridges are connections between the penile shaft and glans penis. They are thought to form when the glans sustains a minor injury during circumcision and fuses with the circumcision wound. Skin bridges can cause curvature of the penis and pain during erections. They are treated by surgical division under local anesthesia.

Inclusion cysts

Inclusion cysts are due to excess skin that folds inward or by smegma in the circumcision wound. They can become large and infected, and may require surgical excision.

Urethrocutaneous fistula

A fistula (abnormal opening) may be caused by injury to the urethra, usually by entrapment of the urethra in the clamp during circumcision or necrosis during aggressive treatment for circumcision bleeding. The complication requires surgical intervention in the first year of life.

Meatitis

The foreskin protects infants from irritation due to incontinence. Meatitis, an inflammatory condition of the urethral opening (meatus), is a common complication of circumcision, occurring in between one-tenth to one-third of infants. It is caused by injury to the penis from ammonia in wet diapers. Meatal stenosis (narrowing of the urinary opening), ulcers, and infections may also occur. These complications virtually never occur in noncircumcised penises.

Penile injury

Traumatic penile amputation is a complication of circumcision. The risk of this complication is not well documented but believed to be rare. It is immediately treated with microreplantation with variable success. The risk of amputation appears to be greater for the Mogen clamp compared to other instruments.

Inadequate foreskin removal

The amount of foreskin removed may be either insufficient, excessive, or the pattern of removal may be asymmetrical, leading to poor cosmetic results. In addition, penile curvature from scarring skin may occur.

Penile sensation

The clinical significance of removal of the foreskin on genital sensations has been a topic of much debate. Studies do not demonstrate a clear pattern of effects of circumcision on either penile sensation or sexual satisfaction. It has been suggested, however, that these studies are flawed because the differences are not in sensation of the glans, but rather, in alterations in sensation due to removal of the foreskin. The foreskin is rich in nerves and provides a gliding mechanism for the penis during intercourse. Moreover, the foreskin contains smegma, the exudates of dead skin cells and oils that may prevent the loss of vaginal secretions. The presence of the foreskin may therefore enhance sexual satisfaction in noncircumcised males. In support of this hypothesis are testimonials from men who were circumcised as adults claiming that sensitivity was lost after the foreskin was removed and that sexual satisfaction was harder to achieve.

A number of circumcised men have subsequently undergone foreskin restoration, a practice that dates back to the first century, in order to regain their lost foreskin.

The Circumcision Decision

Deciding whether or not to circumcise an infant is a decision that is currently made by the parents. In Western society, the overall rate of complications from circumcision is similar, if not higher, than the rate of benefits, and circumcision has not been accepted as a universal public health measure by national organizations such as the AAP and the Canadian Paediatric Society (CPS). Despite these position statements, however, there is no clear effort to try to eliminate routine circumcision by the medical community. There are other important determinants of circumcision practices, such as infant-specific benefit–risk assessments, cultural, religious, and economic considerations.

Parents' decisions about whether or not to circumcise their male infants are usually based on religious, cultural, or personal beliefs rather than on the health benefits and risks of circumcision, and, as such, circumcision will continue to be practiced unless attitudes and traditions are altered. The most prevailing reasons for nonritual circumcision are that the boy's penis must be identical to the father's penis or that the appearance of an uncircumcised penis is unappealing. It must be recognized that these attitudes are culturally based, and the reverse may be said for circumcised penises by the majority of the world's population who does not circumcise.

The Circumcision Procedure

It is currently estimated that 60% of male infants in the US are circumcised. For these infants, every possible measure should be undertaken to ensure that the procedure is successful and that the infant's pain is minimized.

Careful inspection of infant anatomy is important before performing circumcision as the procedure is contraindicated in the presence of certain anatomic abnormalities, including; hypospadias and epispadias (birth defects of the penis and urethra in which the urethra opening does not appear at the head of the penis but on the ventral or dorsal aspect of the shaft), displacement of urethral meatus, and an abnormally short penile shaft. This is because foreskin tissue may be needed for their future repair. In addition, circumcision is contraindicated in infants with bleeding disorders or a family history of bleeding disorders due to the potential for postcircumcision hemorrhage. It is contraindicated in preterm and ill infants as well, until they are able to safely undergo the procedure.

Circumcision can be performed using a variety of surgical techniques. Clinicians are trained and experienced in the use of a particular technique, and each technique offers advantages and disadvantages over the other methods. In all cases, circumcision is accomplished by causing a crush injury to the foreskin with an instrument while shielding other parts of the anatomy. The physician must estimate the amount of foreskin to excise, detach it from the glans penis and leave the surgical instrument in place sufficiently long enough to ensure that there is a stoppage of bleeding (hemostasis) before amputation of the foreskin. The most common instruments used to perform circumcision are the Gomco clamp (67%), Mogen clamp (10%), and Plastibell device (19%). With the Gomco and Mogen clamps, the foreskin is crushed and surgically excised. In the Gomco technique, a bell-shaped apparatus is placed over the glans penis and under the foreskin to protect the glans. Then a ring is placed over the foreskin and tightened to crush the foreskin. The foreskin is excised and the clamp removed. With the Mogen clamp, the foreskin is lifted in an upward and outward direction and the clamp is secured around it. While the clamp is secured the foreskin is cut distal to the clamp. With the Plastibell device, a plastic shield is placed between the glans and the prepuce and secured in place with a ligature. The majority of the prepuce is removed, but the plastic ring is left in place and falls off after approximately 10 days with the remaining foreskin after it has become necrotic and sloughed off. The Mogen clamp is commonly used in religious circumcisions. It involves a less complicated procedure, resulting in a shorter surgery time (1 or 2 min vs. 5–10 min for the Gomco clamp) and less pain. The Gomco clamp is considered safer and is used more commonly in medical settings. It is associated, however, with an increased risk for excessive foreskin removal, leaving an abnormally denuded penis. The Plastibell seems to be associated with a higher risk of complications, including infection, urinary retention, strangulation of the tissue, and necrosis of the glans.

Postoperative management of the circumcised penis involves dressing changes and topical administration of petroleum jelly on the healing glans until it is fully epithelialized, which usually takes about 1 week to 10 days. The penile skin is retracted regularly while it is healing to ensure that there are no adhesions forming onto the glans. The glans will appear 'raw' and have a yellowish exudate crust until healed.

Analgesia for Circumcision

There are well-established analgesic techniques for the management of circumcision pain (**Table 3**).

Local anesthetic infiltration techniques

Infiltration of local anesthesia is considered the most effective single analgesic method for attenuating circumcision pain. The injection technique most commonly used to administer the anesthetic is the dorsal penile nerve block (DPNB). First described in 1978, DPNB involves injecting local anesthetic at the base of the penis at Buck's fascia. More recently, penile block has been performed by injecting local anesthetics in the foreskin or the subpubic space; these techniques, however, are less common than the DPNB.

The anesthetic usually used is lidocaine solution because of its proven tolerability and efficacy in newborns. Vasoconstrictive drugs such as epinephrine (adrenaline) are never added because they can reduce blood supply to the penis and lead to necrosis. The total lidocaine dose used usually ranges from 0.4 to 1 ml (0.2 to 0.5 ml per injection) of a 1% solution, maximum dose 7 mg kg^{-1}. To allow time for the anesthetic to work, it is administered 3–5 min prior to the procedure.

DPNB is estimated to reduce crying time by about 50% in newborn infants undergoing circumcision. In addition, it reduces heart rate changes due to pain by about 35 beats per minute. It also maintains oxygen saturation in the blood at a 3% higher percentage, where normal values are greater than 95%.

Unfortunately, lidocaine infiltration techniques cause pain due to the additional needle puncture required for their administration as well as the burning sensation from the local anesthetic itself. Studies in adults demonstrate that pain from lidocaine can be attenuated by warming the solution to body temperature prior to injection and by injecting it slowly. In addition, the solution can be neutralized by the addition of sodium bicarbonate in a ratio of 1 part sodium bicarbonate to 9 or 10 parts lidocaine. Topical anesthesia with lidocaine–prilocaine cream prior to DPNB can decrease the pain of needle puncture. DPNB is associated with injection-related complications in 7% of cases; most commonly bruising and/or hematoma.

Although very effective, anesthetic infiltration techniques have not been demonstrated to prevent pain in all infants. Treatment failures have been at least partially attributed to technical failures and/or waiting an insufficient period of time for the anesthetic to work before performing the procedure.

There is a potential for serious adverse effects from systemic lidocaine toxicity (e.g., seizures, dysrhythmias) that might occur after inadvertent intravascular injection of the drug. Intravascular administration is easily avoided if negative pressure is used to check for the absence of blood during injection of the local anesthetic. DPNB does not appear to affect the time to first urination postoperatively and future erections in infancy.

Topical local anesthesia

Lidocaine–prilocaine 5% cream (EMLA) is an oil in-water emulsion made up of equal parts of two local anesthetics, lidocaine and prilocaine. Lidocaine–prilocaine has demonstrated efficacy in reducing pain during circumcision. The usual dose is between 0.5 and 2 g applied to the penis for 60–90 min under an occlusive dressing.

Table 3 Analgesics for circumcision pain

Route of administration	Agent
Injection	Lidocaine
	Chloroprocaine
	Bupivacaine
Topical	Lidocaine–prilocaine (EMLA)
Oral	Sucrose
	Acetaminophen (postoperative)

Lidocaine–prilocaine is less effective than local anesthetic infiltration methods, reducing crying time by 15% and heart rate by 15 beats per minute.

Lidocaine–prilocaine is frequently associated with transient minor skin reactions such as pallor or redness. Allergic reactions are rare. Methemoglobinemia is the main systemic adverse effect that can occur following its use, and is a condition characterized by the inability of hemoglobin in red blood cells to carry oxygen to the tissues. To date, methemoglobinemia has only been reported when overdoses of lidocaine–prilocaine cream were administered to infants. Fortunately, if it occurs, methemoglobinemia is easily diagnosed by changes in skin color (cyanosis) due to poor tissue oxygenation. It is also completely reversible by giving methylene blue, the antidote.

Sucrose

Sucrose (sugar) solutions have analgesic and calming effects when given to young infants undergoing painful procedures. It is not clear how sucrose works to reduce pain but it may involve stimulation of the body's own natural painkillers (opioids) and distraction. Sucrose can be administered by mouth approximately 2 min prior to circumcision using either a syringe or a nipple/pacifier that is dipped in a sucrose solution. Sucrose solutions can be prepared by mixing one packet of sugar with 10 ml of water. Sucrose is not as effective as local anesthetics, however, and is optimally used in combination with local anesthetics. In addition, specially designed restraint chairs can be used that minimize infant stress associated with being restrained.

Acetaminophen

Postoperative circumcision pain management has been a neglected area of study despite general recognition that there continues to be pain from the wound after the procedure and that urination, defecation, and diaper and dressing changes will further intensify postoperative pain. Although lidocaine used during circumcision may provide some postoperative pain relief, the duration of action is only a few hours. The duration of postoperative pain relief can be increased with the use of longer-acting local anesthetics during circumcision, such as bupivacaine. However, acetaminophen should be added to reduce postoperative pain. Acetaminophen is safe in newborn infants and has been administered in doses of $15\,mg\,kg^{-1}$ by mouth prior to circumcision and 6-hourly postoperatively for the first postoperative day. It is unclear whether acetaminophen or other pain relievers are useful or safe beyond the first postoperative day because they have not been evaluated. However, given that it takes up to 10 days for the wound to heal, it is likely that infants continue to feel pain beyond the first post-operative day and that they should receive analgesics.

Summary

This article summarizes the most common planned surgical procedure in male infants, circumcision. It reviews the history of circumcision, the benefits and risks associated with the procedure, and the analgesic interventions used to minimize pain. Current evidence suggests that routine circumcision is not advocated based on health benefits. However, the benefits of circumcision must be weighed against the risks for individual infants and their families.

See also: AIDS and HIV.

Suggested Readings

Alanis M and Lucidi R (2004) Neonatal circumcision: A review of the world's oldest and most controversial operation. *Obstetrical & Gynecological Survey* 59: 379–395.
American Academy of Pediatrics (1999) Circumcision policy statement. American Academy of Pediatrics: Task force on circumcision. *Pediatrics* 103: 686–693.
Auvert B, Taljaard D, Lagarde E, Sobngwi-Tambeisoce J, Sitta R, and Puren A (2005) Randomized, controlled intervention trial of male circumcision for reduction of HIV infection risk: The ANRS 1265 Trial. *PLoS Medicine* 2(11): e298.
Brady-Fryer B, Wiebe N, and Lander JA (2004) Pain relief for neonatal circumcision. The Cochrane Database of Systematic Reviews. *The Cochrane Library, 4*.
Castellsague X, Bosch FX, Munoz N *et al.* (2002) Male circumcision, penile human papillomavirus infection, and cervical cancer in female partners. *New England Journal of Medicine* 346(15): 1105–1112.
Declaration of the First International Symposium on Circumcision (1989) *The Truth Seeker* 1(3): 52.
Fergusson DM, Boden JM, and Horwood J (2006) Circumcision status and risk of sexually transmitted infection in young adult males: An analysis of a longitudinal birth cohort. *Pediatrics* 118: 1971–1977.
Holman JR, Lewis EL, and Ringler RL (1995) Neonatal circumcision techniques. *American Family Physician* 52: 511–518.
Learman LA (1999) Neonatal circumcision: A dispassionate analysis. *Clinical Obstetrics and Gynecology* 42: 849–859.
Lerman SE and Liao JC (2001) Neonatal circumcision. *Pediatric Urology* 48: 1539–1557.
Singh-Grewal D, Macdessi J, and Craig J (2005) Circumcision for the prevention of urinary tract infection in boys: A systematic review of randomised trials and observational studies. *Archives of Diseases in Childhood* 90: 853–858.
Taddio A (2001) Pain management for neonatal circumcision. *Paediatric Drugs* 3: 101–111.

Relevant Websites

http://www.aap.org – American Academy of Pediatrics, Dedicated to the Health of all Children.
http://www.cps.ca – Canadian Paediatric Society.

Cognitive Development

C D Vallotton and K W Fischer, Harvard Graduate School of Education, Cambridge, MA, USA

Glossary

Cardinality – In development of numeracy, the principle that the last number counted in a set represents the size of the set.

Conservation – The awareness that quantity remains the same despite change in appearance.

Décalage – Unevenness in development characterized by level of performance varying across tasks or situations.

Joint attention – The sharing of attentional focus between two people, either on one another (dyadic, e.g., mutual eye contact) or on a third entity such as an object, event, or idea (triadic, e.g., mutually looking at an object or talking about an idea).

Looking time paradigm – A methodology for studying infant cognition prior to speech by measuring differences in the time infants look at presented stimuli. Longer looking times are richly interpreted by some as indicating mental states or attitudes including 'surprise' and 'preference'.

Number line – A one-dimensional representation of numbers as sequential integers along a continuum (e.g., –3, –2, –1, 0, 1, 2, 3).

Numeracy – Contracted form of 'numerical literacy'; a proficiency with numbers and measures which requires an understanding of the number system, a set of skills for manipulating mathematical information, and ability to reason about quantitative and spatial problems.

Object permanence – A term coined by Jean Piaget to describe the knowledge that objects remain in existence even when they are perceptually obscured (e.g., knowing that a ball hidden from view by a cloth still exists under the cloth).

Ordinality – In development of numeracy, the principle that numbers follow a constant sequential order.

Perspective taking – The ability to take and coordinate different perspectives on the same thing at one time, either multiple perspectives by a single perceiver or different perspectives by multiple perceivers.

Reflex – Species-specific instinctive action elements dependent on stimulation or body position (e.g., sucking an object placed in the mouth).

Representation – Mental manipulation of concrete aspects of persons, objects, or events (e.g., "I like cookies." or "Five is bigger than two.").

Sensorimotor actions – Flexibly controlled actions and perceptions of objects, people, and events (e.g., reaching for a rattle that is seen or heard).

Skills – Organized elements of behavior – including motor actions, thinking, and feeling – that an individual can control in order to meet the demands of a given physical or social context. Skills develop in a complexity hierarchy.

Social cognition – The processing of social information, including perception, encoding, storing, retrieving, and applying social information.

Social referencing – Using another's perception of a situation in order to determine one's own response to it. In infancy, usually referencing a caregiver for information in a novel situation.

Theory of Mind (ToM) – The understanding that others have mental processes – including desires, beliefs, and perceptions – that differ from one's own.

Introduction

In this article we describe early cognitive development from the Neo-Piatetian perspective on the construction of cognitive skills. We first describe the general framework of 'skill theory' including basic tenets and useful metaphors. Then we describe cognitive development through each of three domains – including physical causality, numeracy, and social perspective taking – from infancy through approximately 5 years of age. Finally, we describe sources of variation in cognitive skills both within and across children.

The Shape of Early Cognitive Development

The current state of knowledge of early cognitive development reveals both innate knowledge in the newborn infant and consistent development through eight levels (grouped into three tiers) of cognitive skills through the first 5 years. At first, infants relate to their world by acting on it physically – reacting to stimuli and developing expectations about things and people, then acting with more and more control and forethought. In the middle of their second year, children begin to relate to the world through mental representations, holding information in mind and manipulating it, often accompanied by physical action. Though this consistent sequence of development is underpinned by spurts in brain growth, new levels do not arise simply from maturational processes, but through co-occurrence and coordination of increasingly complex skills. Despite the consistencies across children from which the knowledge of levels is derived, there is both intra- and inter-individual variation in development, showing development to be a web of interconnected skills, rather than a unidirectional ladder.

Cognitive Development as a Web

Developmental ladders and staircases

Development is often conceptualized as a ladder, a metaphor in which development moves along consistent progression toward higher stages. The metaphor of the ladder has three characteristics: (1) development follows a straight line; (2) it progresses along fixed steps in a single sequence; and (3) it is conceptualized as a forward or upward progression, without deviation to the side or movement backwards.

Developmental webs

A better metaphor than the ladder is the web of individual development, in which an individual child moves concurrently along multiple strands in different skill domains. Cognitive development encompasses many different skills developing at different rates along various trajectories toward unique developmental endpoints, and interacting and integrating with one another to produce complex behavior. The web metaphor portrays cognitive development as the complex constructive process that it is, moving along independent strands that can be linked. Indeed the web only begins to capture the complexity of development, which also involves movement up and down along strands and influences between strands. In general, (1) skills vary within a range along a strand, not just at a single level or step; (2) links between domains of development (i.e., social and cognitive development, or neurological and cognitive development) exert bidirectional influence on one another and help to explain the dynamic nature of development; and (3) development involves not only forward progression along a strand but also moving backward along a strand in order to solidify the strand or reshape earlier skills to create a new skill. This article describes several strands of cognitive development, including infants' developing understanding of physical principles, numeracy, and perspective taking. The focus on the strands for these domains grounds understanding of the developmental process, which is always based in children's specific actions in particular situations (**Figure 1**).

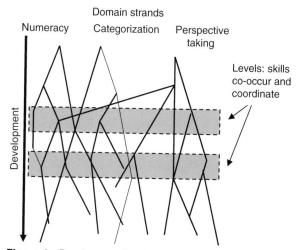

Figure 1   Developmental web showing two levels emerging in three domains. Intra-individual development of domains occurs along independent but intersecting strands, while emergence of new levels corresponds across domains. Adapted from Fischer KW and Bidell TR (1998) Dynamic development of psychological structures in action and thought. In: Damon W and Lerner RM (eds.) *Handbook of Child Psychology. Vol. 1: Theoretical Models of Human Development*, 5th edn., pp. 467–561. New York: Wiley.

The Process of Developing Cognitive Skills

Skills: The Child in Context

A skill is an ability under one's control within a specific context or task, from visually and motorically exploring a new object to maintaining multiple roles during pretend play. The concept of cognitive skill looks to performances in contexts that elicit them, rather than treating knowledge as an object that is obtained and housed in one's mind. Thus, skills are task specific, but potentially transferable to new contexts through a process of building connections. For example, a 2-year-old child may develop the skill of using a spoon to feed herself cereal from a bowl. She knows what is required in this situation, and is able to control her actions in order to do it. She may be able to transfer this skill to using a spoon to feed herself apple sauce from a bowl, but eating noodles from a bowl or eating beans with a fork are different enough that she needs to work to adapt her skill to this new situation. A skill is not a characteristic of the child herself, nor of the situation, but of the child in context – both social and physical.

Individual skills developed in later childhood and adulthood vary more in timing and sequence than those developed in early childhood (through age 5 years), where the first eight skill levels show moderate consistency within predictable age ranges. **Table 1** shows the age ranges based on research on a number of skill domains, but more research will be needed to specify the amount of variation in skill levels across children and domains.

Table 1   Levels of cognitive development in early childhood

Skill level age of emergence	Description and example of skills	Skill structure
Single reflex 3–4 weeks	Simple responses to stimuli: infant looks at object moving through visual field; grasps object placed in hand.	
Reflex mapping 7–8 weeks	Using two reflexes together in relation: infant extends arm toward object being looked at when posture allows.	
Reflex system 10–11 weeks	Coordinated system of simple responses: infant opens hand while extending arm to object being looked at when posture allows.	
System of reflex systems, which creates a single sensorimotor action 15–17 weeks	Coordination of two reflex systems: infant extends arm to and grasps object, sometimes adjusting reach as needed and relatively independent of particular postural configuration – which creates a single sensorimotor action, the flexible reach for an object.	
Sensorimotor mapping 7–8 months	Coordination of two sensorimotor actions, such as visually guided reach for an object: infant looks at an object and uses the visual information to grasp it.	
Sensorimotor system 11–13 months	Complex relations of sensorimotor action relations into a flexible system, such as visual/manual exploration: Infant uses visual information to manipulate an object and moves it to see different aspects of it. Infant also relates sound and vocalization, using single word to label object.	
System of sensorimotor systems/single representation 20–24 months	Integration of sensorimotor action systems into a concrete representation: child manipulates an object to make it carry out actions in pretend play, making a doll walk or talk, talking into a toy telephone. Child also describes attributes or actions of objects and people.	

Continued

Table 1 Continued

Skill level Age of emergence	Description and example of skills	Skill structure
Representaitonal mapping 3.5–4.5 years	Coordination of two representations: child uses two objects in pretend play, relating their actions to one another, e.g., pretending that a mother doll is taking care of a child doll. Child also uses language to coordinate attributes and actions appropriate to ascribed roles, such as mother and child.	

Reference: Fischer KW and Hogan AE (1989) The big picture for infant development: Levels and variations. In Lockman J and Hazen N (eds.) *Action in Social Context: Perspectives on Early Development* pp. 275–305. New York: Plenum.

The Process of Development: Getting from One Level to Another

Convergent findings in infant research demonstrate several periods of rapid change in cognition, producing discontinuities – or shifts in level – in cognitive skills. These periods of rapid change are 2–4 months, 7–8 months, 12–13 months, 18–21 months, and 4 years. Evidence indicates that changes in growth patterns of brain activity and anatomy underpin these cognitive changes. While this association of cognition and neurology shows maturational underpinnings for shifts in skill levels, it does not alone explain development from one level to the next; available evidence does not indicate one direction of causality between cognitive skills and brain growth.

Skill levels form a hierarchy that develops in consistent sequences within domains through the process of co-occurrence and coordination. Levels are grouped into tiers, three of which fit within early childhood: the tiers of reflexes, sensorimotor actions, and representations. At each tier, children differentiate and combine the skills under their control so as to move through a series of levels that eventually create the elements of the next tier. Reflexes, which are simple elements of voluntary action by the infant that depend on inborn patterns of posture and action, differentiate and combine to produce sensorimotor actions. Sensorimotor actions, which are flexible and coordinated actions on objects, are in turn differentiated and combined to develop a third type of cognitive unit – representations. Representations are mental manipulations of concrete aspects of persons, objects, and events. This process of skill elaboration goes on through development at later years into the 20s until abstract systems are combined to create principles in adults. However, the scope of this article covers cognitive development only in early childhood – to the level of representational mappings.

Children develop more complex cognitive skills through the processes of co-occurrence and coordination. The co-occurrence of two skills, both elicited by a given task, induces the child to notice them in contrast to one another, differentiating and coordinating them by using them together in subsequent tasks. Through these processes, increasingly complex skills are built in a sequence for each domain, like cognitive building blocks. **Table 1** provides a list of the skill levels, the typical ages at which they emerge, a description and example of each level considered in this article, and a pictorial representation of the skill level to facilitate understanding of the process of building complex cognitive skills through co-occurrence and coordination of simpler skills.

Like the type of skill, developmental level is not a characteristic of the child, but of the child in context. The level of a skill, as well as its type, depends on the task itself – including the demands of the task and the child's familiarity with it. Other important sources of variation in skill level come from within the child (such as internal state) and from the supports provided to the child within the environment. Because the type and level of skill vary within their context, the structure of skill development is different from that proposed by Jean Piaget, who contended that children develop through generalized stages of skills; that is, a child at a particular stage would exhibit behaviors consistent with that stage in all tasks. Instead, children operate at different skill levels for different tasks, depending on the sources of variation described above. This is why *décalage* – a term used by Piaget to describe the variation in cognitive abilities across tasks and situations in a single child – is the rule rather than the exception in cognitive development. Even highly educated adults will use very basic, low-level skills when they encounter an unfamiliar task. Variation across children shows that the progression from one level to the next is not as simple as expectable maturation (i.e., it is not merely a result of brain development). Instead, progressing from one level to another is a result of a combination of factors, including children's maturation and their experiences performing skills.

Typical Development of Early Cognitive Skills

This section describes the typical development of early cognitive skills, from newborn through age 5 years, from the earliest levels measured through representational mappings. The focus is on the development of skills for physical causality, numeracy, and perspective taking. (To date, much less is known about the 'reflex levels' – when children are typically 4 weeks to 4 months – than the other levels. We present only what is known currently, thus descriptions of cognitive development through these levels are limited.)

Physical Causality

From basic reactions to physical objects, to intentional actions on and with objects, followed by representations of object functions and relationships to one another, young children build their knowledge of physical causality. The ability to categorize objects by their functions (or effects) contributes to children's use of objects in problem solving, as well as understanding of causality.

Newborn

Within their first few days and weeks of life, infants can demonstrate that they already know a number of things about the physical world. Newborn infants will follow a moving object with their eyes, and infants demonstrate a perception of the world as three-dimensional – for example, by showing an expectation that objects remain the same size across situations. Throughout the early months, infants' actions are seriously limited by the particular postures or positions of their body. For example, they can look at an object on the side that their head is turned, but have great difficulty following the movements of that object past the midline, where they would have to shift their body to position their head for looking at it.

Reflex levels

Single reflexes. By 3–4 weeks of age infants have a number of reflexes under their control for responding to stimuli. Some of those most widely studied are looking toward or away from an object presented in their visual field, grasping an object placed in their hand, and kicking their legs, all of which depend on their being in a specific bodily position supporting the reflex.

Reflex mappings. Beginning between 7 and 8 weeks, infants can coordinate two reflexes into a reflex mapping. For example, they can extend their arm toward an object held in their view, provided that their posture places the arm in a position to reach in that direction.

Reflex systems. Around 10–11 weeks, infants can coordinate two reflex mappings to create a reflex system. For example, they will open their hand while extending their arm toward an object held within view, so long as their body is in a position that supports this movement. Further, 3-month-old infants seem to know that they can cause events, though because their actions are primarily reactions they do not know how they carry out a means to cause the event. By tying one end of a ribbon to a mobile, and the other to an infants' foot, infants will learn that when they kick that foot, the mobile moves, and will even remember to kick that foot if shown the same mobile a week later. But if the ribbon is loose, they do not figure out how to vary their action to take account of the loose ribbon. Also, when the ribbon is removed, the infants still kick their foot, as if expecting the mobile to move. They seem to have learned 'when I act, interesting things happen'.

Sensorimotor levels

Sensorimotor actions. At 15–17 weeks, infants can combine reflex systems into single sensorimotor actions. At this point, they begin to guide and shape their reaching, extending an arm toward an object to grasp it in different positions or to adjust to it as they open their hand, and sometimes adjusting their hand if the object changes position. Through this level, children come to use a variety of different actions on objects, including systematically dropping them from wherever they sit; importantly, they do not yet look to see where objects land when dropped.

Sensorimotor mappings. At 7–8 months, infants combine sensorimotor actions into simple sensorimotor mappings. For example, an infant can now reach and grasp an object (a sensorimotor action) in order to pull it closer and look at it. Infants at this level demonstrate an ability to distinguish between objects as either animate or inanimate. By 10 months, infants show awareness of basic 'billiard ball' causality, the idea that action is caused by something. Infants who have habituated to a toy car moving after being bumped by another moving toy will look longer when the conditions of causality are not met – for example, when the car begins to move before the other toy has bumped into it to cause the movement. Infants also demonstrate expectations consistent with an understanding of forces besides personal agents, such as gravity: They will drop an object to the floor and look accurately to see where it landed (**Figure 2**).

Sensorimotor systems. Around their first birthday, 11–13 months of age, infants begin to combine sensorimotor mappings into systems of sensorimotor actions. An infant at this level can actively explore objects by moving them around to see different sides. As a result of this new coordination, infants begin to understand more about objects' relations to and effects on one another, and can use this knowledge to achieve their own goals. For example, a 12-month-old child can pull a cloth to obtain a toy that is sitting on it, and will not pull the cloth when the toy is not resting on it (indicating an understanding of the relationship of support), whereas an infant at the previous level will pull the cloth hoping to obtain the object, even if the object he wants is held above it.

Controversy in interpretation:
Rich vs. conservative explanations of infant behavior
Rich interpretation of infants' behavior is the explanation of infant behavior in terms of adult-like mental activity. Some of the methods we describe in this article– such as habituation and looking time – lead to results which could be interpreted in either a rich or conservative way. An ongoing controversy affecting both the theory and methods in the field of cognitive development asks, is a rich interpretation of infant behavior or a conservative one more accurate and more useful for understanding the results found in studies of infant cognition?
Marshall Haith put forth a heuristically challenging argument against rich interpretation, asserting that such explanations of infant behavior provoke a number of problems in the field of infant cognition. These problems include claims that very young infants have knowledge they could not yet have reasonably acquired, use of concept, minimally supported claims that certain knowledge is innate, and undermining the study of development of cognitive skills over time.
On the other hand, many developmental theorists who are proponents of the 'competent infant' idea believe that conservative interpretations of infant behavior too often underestimate infants' abilities, failing to see the thoughtfulness and intentionality behind the behaviors of preverbal children.
From the perspective of Skill Theory, infant behaviors such as longer looking times or preferential head-turning likely indicate the beginning building blocks of cognitive skills which will be elaborated and generalized into their adult forms later in development.

Figure 2   Controversy in interpretation: rich versus conservative explanations of infant behavior.

Infants begin to recognize rudimentary object categories based on appearances. For example, they will show surprise when a duck emerges from an occluder instead of the toy car that went behind it. Prior to this, infants rely on trajectory as the primary way to identify an object. Infants at this level will group toys by object type (e.g., plastic farm animals vs. toy cars), whereas at 9 months they show less interest in or understanding of object categories.

Representations

Single representations. Around 20–24 months, young children coordinate systems of actions to produce concrete representations. It is at this point that they begin to engage in symbolic play with objects (e.g., representing toys as having particular uses or roles), rather than functional play (simple actions on and with objects). Related to their growing understanding of causality, children's skills for manipulating objects have also become far more complex. A child will use a tool to get a toy that is out of reach, showing emerging planning and problem-solving skills.

At this level children begin to categorize objects spontaneously, carefully organizing objects by category (e.g., horses vs. pencils), stacking them or lining them up, a skill that displays their capacity to represent concrete properties of objects. Combining their skills in categorization and causality, 2-year-olds will begin to categorize things by function rather than form. Alison Gopnik and colleagues showed children a series of novel objects that looked similar to one another. Then children were shown that two of the objects shared a function. The experimenter told the children that one of the objects was a blicket, and asked which of the other objects was also a blicket. Two-year-old children chose the object that had performed the same function as the first blicket rather than one more similar in appearance.

Understanding why an object does what it does is another matter, however. Toddlers often still act as though they believe that they (or another person-agent) must act in order to make interesting events happen. When interacting with a wind-up toy, they will physically manipulate the object to perform its action rather than using the mechanism. Even when the physical demands of the task are far easier than locating and manipulating a winding mechanism, toddlers will still use more of their own force than necessary to cause an event. Besides winding up a toy unnecessarily, they will push a toy car down a ramp rather than releasing it to roll.

Representational mappings. At this level, emerging around 3.5–4.5 years of age, children begin to relate concrete representations to one another. Categorization skills are more sophisticated, and children will categorize objects by their underlying natures, and use categories to appropriately ascribe characteristics. With this more sophisticated understanding of categories, pretend play becomes much more common and complex; for example, in play children will ascribe roles to the dolls and maintain them throughout the dolls' interactions with other objects or people, using the doll as an actor or cause.

Children at this level display their knowledge of causality through their use of language, beginning to give reasonable causal explanations, as well as to make simple if–then causal predictions about simple mechanisms and interactions in the physical world, based in representations of how events occur in relation to each other. Children's explanations of the causes of events often reveal, however, the same kind of teleologic reasoning that their earlier actions displayed; young children explain events as happening 'so that' rather than 'because of'. For example, Piaget discovered that when children explain events in nature, they indicate that rain is occurring 'so that we will have water', the sun goes down at night 'so that we can sleep'. At this level, causes are still thought of as forces coming from active agents for a purpose under the agent's will, such as that the wind makes the clouds move.

Throughout early development, children's understanding of the causes that can produce effects grows from a sense of self as agent, to others as agents, and then to interactions among agents and objects. Around the age of 5 years, children understand that forces other than people cause events, but these forces are still personified. Indeed, listening to adults' explanations of events may reveal that though they may know better, they do not lose the tendency to explain events in terms of personal forces and teleological purposes.

Numeracy

Research with infants and others has revealed two number systems that people develop to mentally track numbers – an exact system used for small numbers and an approximate system used for estimating larger magnitudes. Within their first year of life, infants' abilities to differentiate varying magnitudes improve to detect smaller ratios, while the specific numbers an infant can track using the precise system increase during the course of early childhood, from just 1 and 2, to 3, then 4, then to having a sense of the number line. After children understand the number line, and sets or groups of objects represented by a single number, they begin combining components of their knowledge of number in increasingly complex ways.

Newborn

Based on habituation with looking time, infants have been shown to recognize the difference between one, two, and three objects under a variety of different testing conditions. This has been found as young as 2 and 3 days after birth. Newborn infants' attraction to visual stimuli with high contrast, and to the edges or boundaries of objects within their view, gives an indication of infants' attention to identifying singular objects, or ones. Even in their first few days, infants appear to notice the constancy (and change) of object number, though this is limited to smaller numbers.

Reflex levels

During the reflex levels, infants are learning to parse visual stimuli in appropriate places and identify objects as unitary, counting as one thing. Along with attention to high-contrast edges, infants rely on motion to determine the boundaries of an object. For most objects, what moves together stays together as one object; and infants have been shown to expect this basic principle. A lamp and coffee table both made of the same color wood may appear to be one oddly shaped object because there are no stark contrasts; however, when someone lifts the lamp from the table, anyone – including the infant – can see that they are two distinct objects. It is no surprise, then, that from birth infants will follow a moving object with their eyes. Research has also shown that infants are more sensitive to the trajectories of an object than to its surface characteristics in determining the identity of an object. An infant will show more surprise and curiosity (looking longer or looking again) toward an object emerging from an occluder when the object has changed speed or direction than when it has changed shape and color. Research with congenitally blind children and adults who have gained sight confirm the necessity of motion for identifying singular objects; a newly sighted individual tends to parse visual stimuli in inaccurate places until they see an object move.

Sensorimotor levels

Sensorimotor actions. Can infants add and subtract? At this level, studies by Karen Wynn with a series of physically impossible events show that infants as young as 4.5 months of age can add $1+1$. Infants saw one Mickey doll on a puppet stage, then saw the Mickey doll occluded. Next they saw a hand come from the side of the stage to place another Mickey doll behind the occluding screen; when the screen came down, they saw either one or two Mickey dolls. They had not seen the dolls together before, so that the scene of two dolls on the stage was technically novel, not seen before. However, they looked longer when there was only one Mickey on the stage, showing that they expected to see two, because one had been added to the other; they were surprised at seeing only one. The same experiment works for subtraction; when infants are presented with two Mickey dolls, and one Mickey is taken away (from behind an occluder), infants expect to see only one and show surprise when there are still two. Further, the infants showed surprise with three objects as well, leading to the conclusion that 5-month-old babies do indeed know that $1+1=2$, not 1, and not 3.

Sensorimotor mappings. Using habituation of sucking rhythm (rather than looking time), experimenters have shown that young infants' sense of numeracy extends to sounds as well, for tones and for syllables; after being habituated to words with three syllables (or to three tones), infants respond with more interest (more vigorous sucking) when they hear words with two syllables (or two tones). Testing infants' numeracy across sensory modalities, Elizabeth Spelke and colleagues used looking time to show that infants between 6 and 8 months of age prefer to look at slides of two objects while hearing two drum beats, and three objects while hearing three drum beats. It appears that infants can identify the number of sounds they hear and compare it to the number of objects they see, leading to the conclusion that infants' perception of small numbers is both general and cross-modal.

In addition to advances in the system for precise number, infants at this level are learning to discriminate smaller ratios of magnitude in both the visual and auditory modality. Using looking time as an indicator of novelty recognition, Spelke and colleagues have shown that 6-month-old infants can estimate magnitude differences of 2.0 (e.g., 8 vs. 16 objects, 5 vs. 10 objects), while 9-month-olds can detect ratios of 1.5 (e.g., 8 vs. 12 objects) as well as 2.0. Using a method in which infants' head turning toward sounds is used to indicate recognition of novelty, Spelke and colleagues have more recently shown that this same timing of magnitude recognition holds in the auditory modality as well.

The statistically sensitive infant. Another way of detecting where one thing ends and another begins is by determining the likelihood that a particular thing will follow another. In the phrase 'happy day', how does a child learn that the syllables 'hap' and 'py' are one word and 'day' another, rather than 'hap pyday'? A series of studies has shown that infants as young as 8 months old are incredibly sensitive to the statistical likelihood of word syllables being heard together, thus defining the boundaries of words within a continuous stream of sound. Other studies have shown that the same learning mechanism works with visual as well as auditory stimuli – that infants expect to see a certain sequence of pictures when they have seen the same sequence a number of times, and that they look longer when the sequence is violated.

Sensorimotor systems. Around 12 months of age, infants begin to distinguish small numbers more consistently. Conservation of number in the early months of life rarely goes beyond the number three; in only a few tasks in a few different laboratories have infants been seen to correctly conserve (or add) 4, as opposed to three, and no infants under 12 months have been observed to distinguish 4 from 5 or 6. However, getting to three is important because it shows that babies are not just distinguishing 'one' vs. 'more than one'. Around 15 months of age, after learning basic addition and subtraction of small numbers, infants combine these skills to develop a sense of basic ordinality of quantity; that is, that three is larger than two. When offered a choice, they will select the larger group of toys.

Representational levels

Single representations. At this level, young children are beginning to use verbal language more fluidly; they have a representation of the concept of more and can respond to verbal instructions to choose a line with more things in it. Piaget made the observation that for large numbers, young children will choose the longer line of objects when asked to choose which line has more. However, subsequent studies have shown that for very small numbers children around age 2 years will choose the correct line of objects (i.e., marbles) when asked which line has more, even when the one with less is made to look longer.

Representational mappings. Coordinating the concept of the ordinality of the number line with the concept that a group of objects can be represented by a single number, children between 3 and 4 years old learn the rule of cardinality; that is, for counting the objects in a set, the last number counted is the number of objects in the set. However, overgeneralization of this rule can lead to miscounting until children coordinate it with more sophisticated understandings of object identity. For example, when asked to count how many pencils are lying on the table, if one pencil is broken in two, a child of 3 or 4 will count it as two pencils, concluding that there is one more pencil in the set than there really is. Children at this level understand that you count each object in order to derive the number in the set, but have not yet coordinated what they know about object identity to take into account the unity of the broken object.

There is a spurt in the development of numeracy between ages 4 and 5 years, as many children construct the number line. That is, with the knowledge that numbers are ordinal, children at this level extend their representations of individual numbers – 1, 2, 3, and 4 – into a number line, for numbers in general to 5 and beyond.

Perspective Taking

Perspective taking is one of the foundational skills for social interaction, including the recent line of research on 'theory of mind'. Understanding of self and other is reconstructed at every level of cognitive development, and perspective taking begins at the reflex levels with actions in response to others, and in response to self. This is followed by the coordination of perceptions of and actions toward self and other at the sensorimotor levels. Then at the representational levels, children begin to coordinate increasingly complex combinations of both similarities and differences between self and other, including the very challenging matter of differences in perspectives, thoughts, and beliefs. Though Piaget described children's ability to take and coordinate multiple perspectives on a single object or concept as developing around the age of 8 or 9 years, extensive research shows how precursory skills for perspective taking develop in the first 5 years.

Newborn

From the time they are 1 h old, infants can respond to others by imitating a few simple actions they see on another's face; there is the most evidence for newborns' ability to imitate sticking out the tongue.

Reflex levels

Single reflexes. At this level, infants respond to social stimuli, such as turning toward a familiar voice or staring at a likeness of a face, so long as their bodily position supports this reflex action.

Reflex mappings. Beginning between 7 and 8 weeks of age, infants can bring together two reflexes into a coordinated reflex mapping. For example, while looking into their mother's face, they will look at her eyes in response to her voice. Infants also show some sensitivity to contingencies during interaction and begin to show upset when interactions are noncontingent, that is, when mother's responses to changes in her infant's looks and vocalizations are delayed or nonexistent.

Reflex systems. At 2–3 months of age, infants can coordinate reflex mappings into a smooth system of social responses, smiling and cooing in response to facial and vocal cues from adults, so long as their bodily position facilitates these actions. They begin to coordinate their own facial expressions, gestures, and vocal productions with those of others for a contingent turn-taking game with rhythms similar to conversation.

Sensorimotor levels

Philosophers and scientists have debated whether self-knowledge is derived from knowledge of other, or the reverse. Scientists Sandra Pipp, Kurt Fischer, and Sybillyn Jennings examined the development of knowledge of self and other in children aged from 6 months to 3 years by posing a series of sequentially ordered tasks requiring children to demonstrate knowledge of self and mother in two different domains: agency (action on self and other), and features (recognition of attributes of self and other). Their findings show that self- and other-knowledge develop concurrently, and the primacy of each depends on the particular domain (agency or feature). This is another example of intersecting strands of the developmental web.

Sensorimotor actions. At this level, infants begin to act toward others – typically caregivers – in order to elicit desired responses. They no longer merely respond, but regularly initiate interactions and manipulate others' actions. Relatedly, they begin to recognize and respond to distinct internal states in others, differentiating (as tested by looking time) facial expressions consistent with the emotions of anger, fear, and surprise in female adults. At 6 months of age, infants can interpret the goals of human actions and show surprise (longer looking time) when the human hand reaches for the 'wrong' thing (a different goal object). The effects are not seen when the task is done with a nonhuman, mechanical hand. Now that infants guide their own actions to facilitate their own rudimentary goals, they interpret others as capable of having goals or interests that likewise guide action.

Sensorimotor mappings. At this level, infants begin to coordinate a few social behaviors in order to engage, re-engage, and direct the attention of others. Around 9 months of age, infants begin to consistently engage in joint attention with another person, coordinating their actions to engage in and direct others' attention. As early as 7 months of age, infants will attempt to re-engage partners in dyadic joint attention, coordinating sensorimotor actions (such as gaze following, and some early pointing) to engage in dyadic joint attention. By the age of 10 months, infants use these behaviors consistently in social interaction, as when they engage a partner during exploration of a toy.

In the study by Pipp and colleagues, infants successfully performed agency and feature tasks directed at oneself and mother. In the agency tasks, they followed the experimenter's request to feed themselves or their mother a Cheerio. Agency toward oneself developed slightly earlier than agency toward mother. In contrast, the order was reversed for features: Infants at this level successfully identified mothers' features in the mirror, but not their own.

Identification of features in others at 10 months of age includes attribution of social qualities and rudimentary psychological attributes to individuals. In studies by Karen Wynn and colleagues, infants saw one object (a geometric shape with eyes) apparently trying to climb a hill, and another object either apparently helping or hindering the first in reaching its goal. In a subsequent scene, the climbing object moved to be near either the helper or the hinderer. Infants showed surprise (longer looking time) when the climber moved next to the hinderer rather than the helper. This surprise indicates that infants attribute some quality to the helper and hinderer, and expect a certain reaction or disposition toward these two individuals from the climber. Given the opportunity to play with one of these objects after viewing the climbing scene, infants chose the helper object rather than the hinderer.

Sensorimotor systems. Around 12 months of age, infants' social cognition becomes noticeably more complex. They begin coordinating their responses with those of others, looking to others to guide their own actions – using social referencing in ambiguous situations. There is a rapid increase in the use of pointing to direct others' attention at this level, and more consistent following of adults' pointing. Infants will also imitate adults' actions on objects, even when there is a delay between when they see the action and when they get an opportunity to imitate it. At this level infants seem to be learning that they are like others, which informs their responses to their environment. They are aligning their attitudes and actions to those of others.

In the Pipp study of self and other, infants around 12 or 13 months of age passed a simple visual recognition task by identifying in a mirror a sticker placed on theirs or their mothers' hands. On average, infants passed this visual identification task for mother at earlier ages than for self and similarly identified rouge on their mother's face in a mirror around 15 months, 3 months earlier than they identified rouge on their own face.

At 14 months of age (but not 12 months), infants coordinate more complex social information to make distinct attributions of psychological dispositions. Studies by Wynn and colleagues used an elaboration of the climber task previously described in which there were two climbers; one climber was helped and one was hindered by the same third object. Children showed the expectation that the climber who was helped would move toward the third object while the climber who was hindered would move away. That is, they did not perceive the third object as simply a helper or hinderer, but differentiated how each climber would react to it based on their experiences.

Representations

Single representations. As children move into the representational levels, they hold aspects of self and other in mind beyond their immediate experience. Around 18 months of age, children understand that other people's desires or preferences can differ from their own. Alison Gopnik and colleagues have conducted a series of experiments in which an experimenter demonstrates by facial expression and voice that she prefers broccoli to goldfish crackers, then places her hand right between the two bowls of food and asks the infant to give her more. Whereas a 14-month-old will give the experimenter more crackers (the food that the child prefers), an 18 month old will give the experimenter more broccoli. At 18 months of age children understand that the experimenter's desires are discrepant from their own. Though children as young as 18 months of age have learned that desires can differ, they do not yet understand that perspectives can differ (the apparent discrepancy here is discussed below). The inability to coordinate multiple perspectives results in humorous behavior by toddlers, such as a child covering his own eyes as he walks past his parents with a stolen cookie, believing that if he could not see them, then they could not see him.

Showing insight into others as agents with intentions, 18-month-olds will re-enact the intended, rather than observed, act of a person. Children can infer the adult's intention by watching the failed attempts, and will later do the intended action, rather than the one they saw actually done. The fact that they do not show this same intention reading when they observe a machine 'fail' at a task reveals that they are attributing internal goals and intentions to human beings.

Self and other representations show differences similar to those earlier in infancy. For the feature tasks in the Pipp study, infants recognized rouge on their own nose at around 18 months, after they had noticed it on their mother's nose, thus developing skills for other before self. For the corresponding agency tasks, infants' pretended to feed themselves and their mothers or to drink. They succeeded at this task earlier for self than for mother, thus developing skills for other before self.

Between 18 and 36 months of age, infants coordinate more and more complex representations of self and other as agent and self and other features. In the Pipp agency tasks, infants acted on another person within a prescribed interaction, interacted with another within a prescribed role, and eventually represented two distinct social roles (mother and baby) in pretend play tasks. In the features tasks, children identified increasingly abstract aspects of self and other by correctly responding to a series of questions involving self and other in spatial location ("Where is (Mommy/Child's name)?"), as actors ("Who did that?"), as owners ("Whose is that?"), in familial relationships ("Who do you belong to?"), and in gender categories ("Are you a boy/girl?"). There was a small advantage for self-knowledge in the agency tasks and for other knowledge in the features tasks. At 3 years of age, children have built complex representations of others' characteristics, including concrete visual perspectives, so that they can hide a toy so that they can see it but another person cannot.

Representational mappings. At 3 years of age, children have already built complex representations of self and other, representing basic roles and identifying features. Around 3.5–4.5 years, they begin to coordinate these representations in more and more complex ways, coming to understand many differences between self and other, as well as differences between current and past self – in visual perspectives, thoughts, and beliefs. Children use language to explicitly contrast representations. Conversations between young children and their parents recorded in the CHILDES database reveal children between 3 and 4 years of age pondering contrasts between characteristics of people such as what they like and do not like.

Between 3.5 and 4 years of age children begin to represent different, contradictory perspectives of their own and others. Extensive research on 'theory of mind' uses false-belief tasks, which require coordination of representations of beliefs and perceptions in self and other – for example, differentiating between one's own current and past beliefs, or contrasting one's current knowledge with another's ignorance of a fact or circumstance. In order to pass these tests, it is necessary to understand that people have thoughts and beliefs about things, that people can have different thoughts and beliefs from one another, that those thoughts and beliefs can change, and that those thoughts and beliefs can be wrong, as well as right.

By the age of 5 years (typically), children understand that other people have minds in a way that has much in common with adults' understanding. They know that a number of different internal states exist and have specific characteristics: thoughts, beliefs, emotions, desires, and perceptions. But they still have difficulty coordinating all of these together until later skill levels at older ages.

Variation within Children

For any given cognitive skill, the child has a developmental range of performance with a lower bound called the functional level – the child's best performance with low support – and an upper bound called the optimal level – the child's best performance with high support. The concept of developmental range is related to, but more specific than, 'Vygotsky's zone of proximal development', which includes 'scaffolding', in which a more skilled partner such as the parent actually performs part of the task jointly with the child. Developmental range encompasses what children can do on their own with and without support, or in optimal or functional (ordinary) conditions. Thus, both individual characteristics (interest, emotion) and environmental contexts shape within-individual variation in cognitive performance.

Tasks

Different tasks often produce variation in performance, even if they appear to elicit similar cognitive processes. An example of the effect of task context on children's performance in numeracy is found in studies of children's ordinality concepts. Children aged between 2 and 4 years will choose the line with more M&Ms when asked which they would rather eat, but when asked about two rows of marbles, "Which one has more?" the 3- and 4-year-olds choose the longer line with fewer marbles. When the choice in the task is based on the child's own motivation (with the M&Ms), children answer correctly, but when they must interpret the experimenter's meaning, they often answer incorrectly. More generally, different tasks elicit different performances in children, even when the tasks seem to have similar cognitive demands. Counting M&Ms is not the same as counting marbles, and counting people or buildings are different in other ways. Children perform differently in different tasks, even when adults see the tasks as equivalent.

Environmental Support

There are countless ways in which the child's environment can facilitate or hinder performance of cognitive skills. Support can take many forms, from physical support of an infant to facilitate a reach, to priming with information to facilitate success on a task. The most consistent example of varying performance in early childhood is the difference between what infants can do when they are or are not in an optimal physical position, usually supported. Eye–hand coordination illustrates this difference dramatically: A 2-month-old infant seeing an object may be able to reach out and grasp it successfully when he or she is positioned at just the right angle to both see the object and easily reach it. In other postures, however, this behavior disappears immediately; and the infant cannot perform it consistently in many different postures until several months later.

Emotions and Internal States

In order to perform at an optimal level, a child's internal state must facilitate performance. State limitations are especially severe for young infants, who must be alert but not overly aroused in order to produce organized voluntary reflex actions. Internal states continue to play a role in cognitive performance throughout life, but as children gain more self-regulatory skills, they are better able to manage internal states in order to attend to cognitive tasks.

Emotional states continue to shape performance and development even when children are alert and focused on a task. Emotions evoke particular patterns of activity and bias behavior in certain directions. In representing self and other, for example, most children from a young age naturally represent the self as good and nice, and they reserve representations of bad and mean for other people. In examining children's stories about positive and negative social interactions, Kurt Fischer, Catherine Ayoub, and their colleagues found this bias of 'me nice but you mean' in most young children. Of course, children sometimes have great fun pretending to be the bad guy as well, but in most situations children growing up in stable, supportive environments show biases toward representing self as positive.

Variation Across Children: Webs and Pathways

Given the consistent sequence and typical timing of development of early skills, what explains the variations that occur across children? Variation in development involves both the speed and timing of change and distinct pathways arising from divergent experiences, abilities, and cultures.

Timing

Children typically develop at different rates across domains, while keeping the same general order of development. They progress through the same levels of skill development, but move more quickly than their peers in some domains and less quickly in others. These variations stem from both different experiences in particular domains and diversity in ability and motivation across children. Some cognitive tests – the Bayley Mental Index, for example – test cognitive skills as if they are a singular set. In such tests, items are ordered in a single line and children are expected to pass nearly all items subsequent to their own highest item. In contrast to this, others – such as the Uzgiris-Hunt test – are composed to examine children's cognitive skills in particular domains of knowledge. In such tests, items are grouped by domain, and children may have a different score, and different ranking amongst their peers, on each scale. Either of these tests can identify differences in timing of skill development, but use of a unitary intelligence test leads to interpretation of children as generally lagging behind or speeding ahead of their peers, whereas use of a domains test leads to identification of differences across distinct contexts and tasks.

Adaptations and Perturbations

Within a traditional framework of development as a ladder going in a single direction, any variation or discrepancy is interpreted as either delay or pathology. However, within a framework viewing development as a web of interconnected skills that arise in specific contexts, variations in development are adaptive responses to differing circumstances. For example, children's representations of self and others as nice and mean in social interactions develop along distinct webs for abused children compared to nonabused and for shy compared to outgoing children.

In stories about positive and negative interactions children who have been maltreated develop along distinct but equally complex pathways: Most commonly, they build complex stories about mean interactions, and simpler, less developmentally mature stories about nice interactions. If assessments require them to focus on positive interactions, their development appears delayed or retarded, but assessments that include their natural focus on negative interactions demonstrate that they function at normal developmental levels in that domain.

Emotional differences in temperament shape children's pathways in another way: Shy (temperamentally inhibited) children show typical development of the representation of self and other as positive (nice, good), but even in pretend play, they often avoid and resist representations of negative interactions, especially when they themselves are represented negatively (as mean or bad). Uninhibited, outgoing children, in contrast, often relish pretending to be mean and thus develop richer, more complex negative representations of self and other.

Cultural Variation

People organize their mental tools and the structure of their cognition through the symbols of their cultures, as Lev Vygotsky emphasized. Cultures shape people in the developmental pathways for building the tools through participation in everyday social interactions and cultural rituals, thus producing important differences in meaning systems. In an example of cultural differences in dominant symbols, people in China and the US give different prominence and meaning to concepts and tools related to shame. Chinese culture makes shame prominent from an early age and uses it to direct behavior constructively toward prescribed cultural norms. It includes tools for coping with shame and related transgressions and teaches these tools and concepts from an early age through storytelling routines by parents and other adults. European-American culture uses shame differently, minimizing its explicit use with young children and providing few culturally supported tools for dealing with it. As a result, Chinese and European-American children differ greatly in their understanding of shame, Chinese children showing basic use of the concept by 2 years of age, whereas European-American children do not use it until around age 7 years. In China, shame is a cultural tool that shapes individual relations to social groups. In the US it is something that is unhealthy and should be avoided. People from each culture have difficulty understanding what shame means in the other.

Cultures shape important differences in many other domains as well, reflecting differential emphases in language and cultural practices. For physical properties of objects, Korean children develop an early focus on actions, while English-speaking children focus more on objects. Korean mothers use more verbs in their speech to children, talking about actions; whereas English-speaking mothers use more nouns, doing more object labeling. Korean children go on to solve certain action puzzles earlier than English-speaking children, such as obtaining an out-of-reach object using a rake-like tool. English-speaking children, in contrast, categorize objects more frequently at earlier ages than the Korean children. In general, cultures create variation in timing and developmental pathways of early skills in particular domains.

Summary

Early cognition proceeds from reflex (reaction) to intentional action and then to representation, building increasing complexity through a series of levels of development. Children build distinct, individual skills, and developmental pathways vary across domains within children, with each child forming a developmental web with many strands. At the same time cognitive skills develop in a predictable, standard hierarchical sequence in each strand, and the timing of major advances seems to correspond to spurts in brain growth. Each individual child varies her or his level of functioning dramatically as a function of support, context, and interest. Different children vary in the specific strands that they build and their connections among strands. These variations result from differences in experience – including both benign cultural experiences and extreme experiences such as abuse – as well as in interests and temperament. In developing skills and understanding, children adapt to the specific environments where they live and to their individual abilities, emotions, and other characteristics.

See also: Amnesia, Infantile; Bayley Scales of Infant Development; Categorization Skills and Concepts; Cognitive Developmental Theories; Developmental Disabilities: Cognitive; Mathematical Reasoning; Milestones: Cognitive; Object Concept; Piaget's Cognitive-Developmental Theory; Taste and Smell; Theory of Mind; Reasoning in Early Development; Vygotsky's Sociocultural Theory.

Suggested Readings

Case R (1992) *The Mind's Staircase: Exploring the Causal Underpinnings of Children's Thought and Knowledge.* Hillsdale, NJ: Erlbaum.

Fischer KW, Ayoub CC, Noam GG, Singh I, Maraganore A, and Raya P (1997) Psychopathology as adaptive development along distinctive pathways. *Development and Psychopathology* 9: 749–779.

Fischer KW and Bidell TR (1998) Dynamic development of psychological structures in action and thought. In: Damon W and Lerner RM (eds.) *Handbook of Child Psychology. Vol. 1: Theoretical Models of Human Development,* 5th edn., pp. 467–561. New York: Wiley.

Fischer KW and Bidell TR (2006) Dynamic development of action, thought, and emotion. In: Damon W and Lerner RM (eds.) *Theoretical Models of Human Development. Handbook of Child Psychology,* 6th edn., vol. 1, pp. 313–399. New York: Wiley.

Fischer KW and Hogan AE (1989) The big picture for infant development: Levels and variations. In: Lockman J and Hazen N (eds.) *Action in Social Context: Perspectives on Early Development,* pp. 275–305. New York: Plenum.

Gardner H (1983) *Frames of Mind: The Theory of Multiple Intelligences.* New York: Basic Books.

Gopnik A, Meltzoff A, and Kuhn PK (1999) *The Scientist in the Crib: What Early Learning Tells Us about the Mind.* New York: William Morrow & Company.

Gruber HE and Vonèche JJ (eds.) (1977) *The Essential Piaget: An Interpretive Reference and Guide.* New York: Basic Books.

Mascolo MJ and Fischer KW (2005) The new constructivism: The dynamic development of psychological structures. In: Hopkins B, Barre RG, Michel GF, and Rochat P (eds.) *Cambridge Encyclopedia of Child Development,* pp. 49–63. Cambridge, UK: Cambridge University Press.

Stern DN (1991) *Diary of a Baby.* London: Fontana.

Vygotsky LS (1978) *Mind in Society: The Development of Higher Psychological Processes.* Cambridge, MA: Harvard University Press.

Wellman H (1992) *The Child's Theory of Mind.* Cambridge, MA: MIT Press.

Wellman H, Cross D, and Watson J (2001) Meta-analysis of theory-of-mind development: The truth about false belief. *Child Development* 72: 655–684.

Relevant Websites

http://www.gse.harvard.edu – The Dynamic Development Laboratory.

http://www.lectica.info – The Lectical Assessment System for Skill Complexity.

http://naeyc.org – National Association for the Education of Young Children.

http://www.unige.ch – The Jean Piaget Archives.

http://www.piaget.org – The Jean Piaget Society.

http://www.marxists.org – The Lev Vygotsky Archive.

Cognitive Developmental Theories

G S Halford, Griffith University, Brisbane, QLD, Australia

Glossary

Cognition – Includes thinking, language, learning, memory, and perception.

Epistemology – Theory of knowledge.

Natural kinds – Things that occur in nature, such as plant and animal categories. As children mature their understanding of the world and their ability to reason about it both increase to a remarkable extent. Cognitive developmental theories are designed to account for this process. There were important pioneering theories by Piaget and Vygotsky, whose ideas are still influential, but their ideas have been incorporated into a number of new theories, which we will outline.

Prototypic category – One based on the most typical example of the category (e.g., a prototype of the dog category would be the most typical dog). Prototypes are acquired automatically by exposure to examples of the category and are possibly the earliest categories to develop.

Transitive inference – If there is a relation
R between a and b, and between a and c, and if R is a
transitive relation, then the relation R will exist
between b and c (i.e., a R b and b R c implies a R c).
An example would be a > b and b > c implies a > c.
Unary relation – A relation with one argument.
An example would be category membership – Rover
is a dog, meaning that Rover is a member of the set
of dogs.

Introduction

Theories of cognitive development are reviewed, beginning with pioneering theories by Piaget and Vygotsky. Neo-Piagetian theories which integrated Piagetian theory with other conceptions of cognition were developed by McLaughlin, Pascual-Leone, Case, Fischer, and Chapman. Complexity theories propose that children become capable of dealing with more complex relations as they develop. Information processing theories, neural net theories, dynamic systems theories, and theories of reasoning processes all provide models of the reasoning processes employed by children at different ages. Micro-genetic analysis methods are used to study the processes of transition from one level of thinking to the next.

Piaget and Vygotsky

Some of the core ideas in Piaget's theory will be outlined first, because it is the most comprehensive and elaborate of the early theories of cognitive development.

A central idea in Piaget's theory was genetic epistemology, meaning that we can investigate how we understand the world by studying the way that understanding develops in children. An important aspect of our knowledge of the world is that objects are real and permanent, existing independently of our perception of them. How we know this has been an issue for philosophers for centuries. Piaget investigated how knowledge of object permanence develops in children by studying their reactions to vanished objects. The idea is that if an infant knows that an object still exists when it disappears from view the infant should look toward the point of disappearance, and perhaps even try to reach for the no-longer-visible object. Piaget concluded that infants had only a very rudimentary understanding of object permanence in the first few months, and that it developed over the first 2 years in a succession of substages. His conclusions have been modified by subsequent researchers, some of whom have claimed that infants' object permanence knowledge is greater than Piaget recognized. Nevertheless, the idea of

investigating children's object knowledge by assessing their reactions to vanished objects persists.

Another of Piaget's key ideas was that 'logic is the mirror of thought', meaning that logic reflects properties that are inherent in thought. He tried to define the nature of thought by specifying the logico-mathematical concepts to which it was equivalent. These included notions such as function, operation, group, and lattice. However, he did not claim that human thought conformed to logic as understood by logicians, but was based on quasi-logical ideas that he termed 'psycho-logics'. Research in the latter half of the twentieth century led to reduced emphasis on logic as a basis for human reasoning. We now judge human reasoning by our ability to adapt to the environment rather than by conformity to logic. However, some of Piaget's important observations of children's reasoning remain.

Piaget carried out very extensive empirical investigations of the development of infants' and children's cognitions, and he attempted to define the way children's reasoning developed by a succession of distinct psychologics, that have come to be known as 'stages' of cognitive development. The first was the sensorimotor stage, lasting from birth to about 1.5–2 years, characterized by structured, organized activity, but not thought. During this stage actions become integrated into a self-regulating system of actions. Piaget believed that the concept of objects as real and permanent emerged as this structure was elaborated. The preoperational stage lasted from approximately 2–7 years, and during this time symbolic functions were developed, including play, drawing, imagery, and language. Thought at this stage was conceptualized in terms of what Piaget called 'function logic', the essential idea of which is representation of a link between two variables. At the concrete operational stage, lasting from 8 to about 14 years, thought was conceptualized in terms of what Piaget called 'groupings' which included a logical operation. The essential idea here is the ability to compose classes, sets, relations, or functions, into integrated systems.

Concepts such as conservation (invariance of quantity, number, weight, and volume), seriation or ordering of objects, transitive inference, classification, and spatial perspectives emerge as a result of the more elaborate thought structures that develop during this time. At the formal operational stage, beginning in adolescence, the ability to compose concrete operations into higher level structures emerges, with the result that thought has greater autonomy and flexibility.

Cognitive development depended, according to Piaget, on 'assimilation' of experience to cognitive structures with 'accommodation' of the structure to the new information. The combination of assimilation and accommodation amounts to a process of self-regulation which Piaget termed 'equilibration'. He rejected theories based on association, arguing that these were inadequate to account for cognitive development, an idea that was often thought almost heretical at the time, but would be widely accepted now. In many ways his conceptions anticipated modern conceptions of information processing and dynamic systems, to be discussed later.

A core topic in Piaget's cognitive development research was conservation, which entails recognizing that quantity remains constant when transformed without adding or removing anything. Suppose a child is shown two equal quantities of liquid in identical glasses, then one quantity is transformed by pouring it to a taller and narrower vessel. The classical Piagetian finding is that young children tend to say the quantities are now unequal because the taller and narrower vessel makes the quantity appear more. The finding that young children give nonconservation answers to these tests has been replicated many times, but there has been little consensus about the correct explanation.

Children who give nonconservation answers typically recognize that the quantity will be the same again if poured back; they also recognize that it is still the same material, that nothing has been added or subtracted and that there have been changes in two variables, that is, that the quantity increased in height but decreased in breadth. There has therefore been some mystification as to why they give nonconservation answers. Proponents of the Piagetian position held that nonconservation answers represented genuine lack of understanding of the invariance of quantity, but others have proposed that the child was misled by the transformation which made it appear that quantity or number had changed, that the child misinterpreted the words used in the test believing, for example, that 'more' referred to height of the liquid, rather than to quantity, or that there was a conflict between knowledge of conservation and appearance of the display. We will consider how some more recent theories deal with this issue later in this article.

The work of the Piagetian school has been controversial, but his empirical findings have been widely replicated. That is, children have been found to perform as Piaget reported 'on the tests he used'. For example, children undoubtedly respond as Piaget observed on the conservation tests discussed above. The major challenges to his findings have been based on different methods of assessment, the claim being that his methods underestimated the cognitive capabilities of young children. However, these claims have also been subject to controversy, and the assessments that were proposed as improvements on Piaget's techniques have not always been validated. There were also some hundreds of studies designed to train children in the concepts that they did not understand, thereby demonstrating that cognitive development could be accelerated, and depended more on experience than on development of thought structures. However, these studies did not completely eliminate limitations to what children understand at particular ages. The stage concept has also been criticized on the grounds that development is gradual and experience-based rather than sudden or 'stage-like', and the concurrence between acquisitions at the same stage has often not been as close as Piagetian theory might be taken to imply.

The work of Vygotsky was the other major influence on research into the development of thinking, and his contribution is becoming increasingly influential even today. Three of Vygotsky's most important contributions were his ideas on the relation between thought and language, his emphasis on the role of culture in the development of thinking, and the zone of proximal development. Early in the history of cognitive development research there was considerable debate as to whether thought depends on language development, or the reverse. Vygotsky proposed that thought and language have different origins both in evolution and in development. Language was initially social in character, while problem solving was initiated in motor processes. Language and thought develop independently for some time after infancy; then the young child develops egocentric speech, which is the beginning of the representational function. Finally, children develop 'inner speech' which serves the symbolic function of thought. Vygotsky emphasized the interaction between biological maturation and social experience. As the child matured, language became an increasingly important influence on the development of thought, and was the chief means by which culture was absorbed by the child.

Vygotsky proposed that the cultural input was essential to the cognitive development of the child. One of its effects is to help the child to relate concepts to larger systems, while another is to increase conscious awareness of what the child has learned. However Vygotsky emphasized that culture was not simply absorbed, but there is an interaction between spontaneous development and cultural influences, each contributing to the other. Instruction is founded on development, but instruction also influences development. This is consistent with contemporary ideas on neural plasticity, which mean that environmental factors, including experience and learning, can influence the development of the structure and function of the brain. An important consequence of this theory is Vygotsky's concept of the 'zone of proximal development', which means that new developments occur close to existing cognitive abilities. New concepts are not simply absorbed but are integrated with the mental processes of the developing child. Consequently instruction will be effective to the extent that it introduces concepts that relate to, but are just slightly more advanced than, the child's current level of understanding.

In some respects this is consistent with Piaget's notion that new knowledge is assimilated to existing structure. This is part of a larger picture in which both Piaget and Vygotsky saw cognitive development as an active organizing process that tends toward an equilibrium with its own internal processes and with the external environment. However Vygotsky differs from Piaget in placing more emphasis on cultural input, including formal instruction, in the cognitive development of the child. Piaget's work had greater early influence, but the impact of Vygotsky's work is increasing at what appears to be an accelerating rate. Both continue to have an influence on education theory.

Neo-Piagetian Theories

There were several theorists who proposed alternative explanations for Piaget's observations of children's cognitive development. G. Harry McLaughlin proposed that a child's reasoning was determined by the number of concepts that could be considered simultaneously. He proposed that the progression from infancy, to early childhood, middle childhood, and adolescence, which Piaget characterized as sensorimotor, preoperational, concrete operational, and formal operational stages, respectively, required $2^0 = 1$, $2^1 = 2$, $2^2 = 4$, and $2^3 = 8$ concepts to be considered simultaneously. This lead was taken up by a number of theorists who proposed that children's cognitive development was driven primarily by increased ability to process information. Juan Pascual-Leone proposed that cognitive development depended on increases in central-computing space, or M-space, that corresponded to the number of separate schemes that they could coordinate. The value of M was $a + 1$ at age 3 years age and increased to $a + 7$ at age 15 years. He showed empirically that older children can understand more complex concepts than younger children, where complexity is defined by the M-space required to represent the concept.

Robbie Case proposed that cognitive development depends on children learning to make better use of the available capacity. Total processing space is constant over age, but older children use their capacity more efficiently, leaving more of it available for other tasks. Short-term memory span increases with age because rehearsal becomes more efficient, using less capacity, and leaving more capacity for storage. In a number of ingenious experiments, Case and his collaborators showed that if adults' rehearsal efficiency was reduced to that of 5-year-olds' by using unfamiliar materials, their short-term memory spans for the same materials were correspondingly reduced to those of 5-year-olds'. Case also acknowledged subsequently that processing capacity increases with maturation of the nervous system, particularly the frontal lobes. Case's work remains important because it was the first to demonstrate that cognitive development depends on the efficiency with which available information processing capacity was utilized.

Later Case developed the concept of central conceptual structures, a network of semantic nodes and relations that passes through a sequence of four major neo-Piagetian stages, that Case labels sensorimotor, interrelational, dimensional, and vectorial (or abstract dimensional). Each major stage is divided into substages known as unifocal, bifocal, elaborated coordination, and a preliminary substage that represents the transition from one major stage to another. Progression through the substages was due to working memory growth, under the influence of both maturation and experience. The transition to higher stages is achieved by coordinating two existing structures into a higher order structure. Case's theory has been applied to many domains, including science, mathematics, space, music, understanding narrative, social roles, and motor development.

Kurt Fischer's theory was based on development of cognitive skill in controlling sources of variation in a person's behavior, and was influenced by the work of Jerome Bruner as well as Piaget. There are three major stages or tiers: the sensorimotor, representational, and iconic symbolic. As with Case's theory, there is a recurring cycle of four levels within each of the major stages, and the highest level of one stage is shared with the lowest level of the next.

There is much common ground among the neo-Piagetian theories, and Michael Chapman attempted to capture this in an integrated theory. He argued that the capacity required for a given form of reasoning depends on the number variables to which values have to be assigned. The class inclusion concept has been found difficult for young children and is usually mastered in middle childhood. Suppose a child is shown some red beads (A), and some blue beads (A′), all of which are wooden (B), and there are more red than blue beads. Therefore, A and A′ are included in B (A and A′ = B), so there must be more of B than of A. However, young children have difficulty recognizing this and commonly say there are more red beads than wooden beads. Solving the problem depends on recognizing A = red beads, A′ = blue beads, and B = wooden beads. That is, it entails assigning values to three variables (A, A′, and B) that represent the classification hierarchy. Therefore, Chapman seems to have been the first to realize explicitly that the best way to analyze complexity of cognitive tasks is to determine the number of variables that have to be instantiated in parallel.

Complexity Theories

Two complexity theories have been developed. One is 'cognitive complexity and control theory', by Douglas Frye and Philip Zelazo and the other is 'relational complexity theory' by Graeme Halford, William Wilson, and Steven Phillips. Relational complexity refers to the number of entities that are related in a single cognitive representation. It corresponds to number of slots or 'arity' of relations.

A binary relation has two slots: for example, larger-than(—,—) has a slot for a larger entity and one for a smaller entity. Each slot can be filled in a variety of ways, such as larger-than(elephant, mouse), larger-than(mountain, molehill), etc. Complexity of relations can be defined by the number of slots:

- Unary relations have one slot: for example, class membership, as in 'dog' (Fido).
- Binary relations have two slots: for example, 'larger' (elephant, mouse).
- Ternary relations have three slots: for example, 'addition' (2, 3, 5).
- Quaternary relations: for example, proportion (2/3 = 6/9).

Because each slot can be filled in a variety of ways, a slot corresponds to a variable or dimension. Thus, a unary relation is a set of points on one dimension, a binary relation is a set of points in two-dimensional space, and so on. In general, an *N*-ary relation is a set of points in *N*-dimensional space.

The relational complexity metric has been applied to cognitive development, to adult cognition, to higher animals such as chimpanzees, and to industrial contexts, including air traffic control. There is a broad correspondence between levels of relational complexity and the phenomena that Piaget attributed to stages. Unary, binary, ternary, and quaternary relations correspond to preconceptual, intuitive, concrete operational, and formal operational stages, respectively. Processing capacity is an enabling factor, and concept acquisition is a function of experience, given that the relevant processing capacity is available.

The median ages at which each level of relational complexity is attained are: unary relations at 1 year, binary relations at 1.5–2 years, ternary relations at 5 years, and quaternary relations at 11 years. Concepts at a given level of complexity do not develop synchronously but are acquired by a biological growth function. Thus, ternary relations are acquired by approximately 20% of 4-year-olds, 50% of 5-year-olds, and 80% of 8–9-year-olds.

Empirical evidence suggests that a quaternary relation is the most complex that adults can process in parallel, though a minority of people can probably process quinary relations under optimal conditions. This is a 'soft' limit, meaning that increased complexity produces increased errors and decision times, rather than sudden failure. In order to handle more complex concepts, mechanisms for reducing processing loads are required. Relational complexity theory includes two such mechanisms, conceptual chunking and segmentation.

Conceptual chunking involves recoding concepts into less complex relations. However, there is a temporary loss of access to chunked relations. For example,

'speed = distance/time', is a ternary relation, but speed can be recoded into a unary relation, 'speed' (60 kmph) as when speed is indicated by the position of a pointer on a dial. However, the chunked representation does not permit us to answer questions such as "How does speed change if we cover the same distance in half the time?" To answer that we have to return to a representation of the ternary relation.

Segmentation entails breaking tasks into less complex steps, which can be processed serially. Strategies and algorithms are common ways of doing this: for example, adding one column at a time in multidigit addition.

Chunking and segmentation skills are important components of expertise. They are two of the processes that increase with age, and therefore they are important factors accounting for the way cognitive development occurs.

We will illustrate how relational complexity theory can be used to analyze class inclusion, using the example of fruit, including apples and nonapples. Fruit is assigned to the superordinate slot because it includes apples and nonapples fruit. Thus, the assignment of classes to slots in the hierarchy depends on processing the relations between the three classes, and is ternary relational.

Transitive inference is also ternary relational, according to relational complexity analyses. Consider the problem: Tom is taller than Mike, Peter is taller than Tom, who is the tallest? Process analyses have shown that the transitive inference can be made by integrating the premises into the ternary relation, Peter 'taller than' Tom 'taller than' Mike. It has been shown that integrating the premises into this ternary relation is the main information processing demand of transitive inference, and the main cause of cognitive effort, in adults and in children.

Transitivity and class inclusion are superficially different, yet they are structurally similar, and both entail ternary relations, as **Figure 1** shows. This is an example of how tasks can have equivalent relational complexity despite different domains and different test procedures. Transitivity and class inclusion were both originally Piagetian tasks, and the

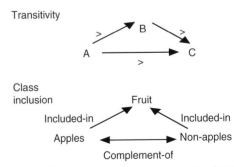

Figure 1 Similar structures in two ternary relational tasks, transitivity and class inclusion. Adapted from figure 4 in Halford GS and Andrews G (2004) The development of deductive reasoning: How important is complexity? *Thinking and Reasoning* 10: 123–145.

causes of children's failures have been controversial, but it appears that complexity is at least part of the explanation for the difficulties that young children encounter in these tasks.

Relational complexity theory proposes that understanding of conservation depends on learning relations between quantity and other dimensions like height and width. Piaget proposed that children have to recognize that when (for example) liquid is poured into a taller and narrower vessel, any increase in height is compensated by a decrease in breadth. Paradoxically however, children may recognize that width of a liquid had decreased as its height increased, yet they still fail to conserve. Relational complexity theory proposes that the reason is that compensation is a binary relation, between length and density, whereas understanding conservation requires a ternary relation between three variables, height, width, and quantity. This means conservation is a ternary relational concept, which should be attained at a median age of 5 years, like other ternary relational concepts. There is empirical evidence that supports this.

Cognitive complexity and control theory assesses complexity by the number of hierarchical levels of rule required for the task. A simple task consists of rules that link two variables, such as an antecedent and a consequent, whereas complex tasks have higher-order rules that modify the lower level rules. This adds another level to the hierarchy. The theory has been applied to the dimensional change card sort task, which has been an effective predictor of a number of important cognitive attainments in children, including theory of mind, to be considered later in this article. In a simple sorting task, cards are sorted by shape or by color, but not both. For example, a green circle might be assigned to the green category (indicated by a template comprising a green triangle) and a red triangle to the red category (indicated by a template comprising a red circle). In a complex task, sorting depends on whether the higher-order rule specifies sorting by color, as above, or by shape. If sorting is by shape, the green circle is sorted with the red circle and the red triangle is sorted with the green triangle. Children typically process a single rule by 2 years of age, a pair of rules by 3 years of age, and a pair of rules embedded under a higher order rule by 4 years of age. The transition to higher levels is achieved by reflective abstraction.

Information Processing Theory

Some theories have been based on the idea that thinking is basically a computational process that can be simulated on a computer. One example of this approach is the Q-SOAR theory of Tony Simon and David Klahr, designed to simulate children's acquisition of number conservation in a study by Rochel Gelman. Children are asked to say whether two equal rows are the same or different based on

counting, then are asked to say whether they are still the same (or different) after a transformation, such as lengthening one of the rows. Inability to answer this question is represented in Q-SOAR as an impasse. The model resolves this by quantifying the sets before and after the transformation, noting that they are the same, and develops the knowledge that the transformation did not change the numerosities, and therefore is conserving.

Implicit vs. Explicit Cognition

Human reasoning, whether in children or adults, is not always conscious and intentional, but sometimes occurs at a more automatic, unconscious level. This distinction has been captured in a theory developed by Andy Clark and Annette Karmiloff-Smith. Implicit knowledge is 'knowledge in the system but not knowledge to the system'. Implicit knowledge is an effective basis for performance, but is not consciously accessible, and cannot be modified intentionally or strategically. At the next level knowledge is accessible and modifiable, but not available to consciousness and cannot be reported verbally. At the highest level, knowledge is explicit, conscious, and can be described verbally. The transition from implicit to explicit knowledge is produced by representational redescription, which means that we take our implicit knowledge and represent it verbally, in a conscious form. An example would be when we first learn a skill we can perform the relevant actions, but it is only later that we can describe what we do, and then we become capable of consciously modifying our performance.

Knowledge of Mental States

One of the most important acquisitions children make is theory of mind, or understanding of other people's mental states. One test for theory of mind is the appearance-reality task. Children are shown a cutout of a white bird, which is then covered by a blue filter, and they are asked what color the bird is really (white), and what color does it appear when viewed through the filter (blue). Children below about 4–5 years of age have difficulty recognizing that the way the bird appears is different from the way it really is, and they tend to answer that the bird is white and looks white, or that it is blue and looks blue.

Both cognitive complexity and control theory and relational complexity theory propose that inferences about other people's mental states are difficult for young children because of their complexity. Understanding the difference between the way things appear to people and the way they really are depends on understanding the relations between three variables. First there is the attribute of an object (its color in the example above). Then there is the way an object appears to people (the percept), such as whether they see it as white or blue. The relation between these two is influenced by a third variable, the viewing condition, such as whether the object is viewed directly or through a colored filter. Ability to understand how objects appear to people depends on ability to relate these three variables.

Dynamic Systems Theories

Dynamic systems theories are complex and sophisticated, but we can present the essential ideas. Technically, a dynamic system is a formal system the state of which depends on its state at a previous point in time. Dynamic systems are self-regulating, meaning that they are the result of the interaction of variables, and processes, which combine spontaneously to achieve a stable state or equilibrium. One reason why dynamic systems are important to cognitive development is that they can account for different types of cognitive growth that have been observed. That is, development is sometimes slow and steady, while at other times sudden jumps occur, resulting in new levels of functioning that appear quite different from anything that was there before. This is what led theorists like Piaget to propose that cognitive development occurs in stages, such that entirely new cognitive processes emerge when the transition is made to a new stage. Dynamic systems can show how a complex, self-regulating system can emerge from the interaction of a few variables, and offer natural interpretations of concepts such as equilibration and self-regulation which are at the core of the theories of both Piaget and Vygotsky. Links have also been made between dynamic systems models and neural net models, to be considered below.

An example of a dynamic system would be children's acquisition of the concept of conservation, considered earlier. Children of 3–4 years of age typically think that, when liquid is poured from a short and wide to a tall and narrow vessel the quantity increases, because they see the large increase in height. Understanding that the quantity remains constant typically develops spontaneously, and often appears quite suddenly, so that in a short time the child might switch from being sure that the quantity increases to being sure that it is constant. What appears to happen is that the children start to relate the three variables of height, width, and quantity, as mentioned before. That is, they realize that when you take both height and width into account, quantity remains constant, or is 'conserved'. Here the three variables are brought into a new integration, which creates a new form of stability. These variables are also related to observations that nothing was added when the liquid was poured, and that it would be same again if poured back. Quantities that formerly seemed to increase or decrease as liquids were poured from vessel to vessel are now seen as invariant over those transformations, and a lot of additional information is integrated with this conception. This development is spontaneous and is not usually taught. Indeed, attempts to teach it might be ineffective until the child is ready to make the new integration. This illustrates the self-regulating nature of cognitive development.

Microgenetic Analysis

Microgenetic methods entail detailed analysis of strategies that children use in reasoning about a concept. This enables researchers to obtain information about the processes of cognitive development, the factors that influence it, about individual differences, and how strategies for performing tasks are formed. When they reason children typically have more than one strategy available at any one time, so strategies progress in overlapping waves, and development consists of selectively strengthening some strategies. An example would be investigation of how young children learn to reach for an object using one of a number of tools supplied. The task was set up so that success depended on choosing a tool that was long enough and that had an appropriate fitting on the end. For some children modeling was used to demonstrate the appropriate action. Later a hint, suggesting the right tool to use, was provided. Children's actions progressed over the course of three problems. They began by reaching without using the tool, or by asking the parent to obtain the toy. Later they learned to use the tool. Once the children learned a tool-using strategy, it was transferred to new problems by analogical reasoning. Detailed individual differences in strategies were observed, and proficiency in the immediately preceding component of a strategy was the best predictor of progression to the succeeding component. This can be seen as illustrating dynamic development. That is, children spontaneously make the transition to a new strategy when they have mastered the previous one, consistent with Vygotsky's theory. Spontaneous switching between different strategies is also sometimes observed before the transition.

Microgenetic analysis has also been applied to conservation acquisition. Children often progressed from explanations of conservation based on one dimension (e.g., height) to explanations based on the transformation, such as pouring from one vessel to another. There are many different patterns, and the number of different explanations given was a good predictor of conservation acquisition, consistent with dynamic systems theories. It also suggests that cognitive development can progress by different pathways, but also that conflict between height and quantity might be a stimulus to conservation acquisition. Resolution of this conflict by recognizing that compensating changes in height and width are consistent with constant quantity would be a strong motivation for conservation. This would integrate the three relevant variables, as proposed earlier in this article.

Theories of Reasoning Processes

Some theories have been devoted to understanding processes that are employed in reasoning. One of the most fundamental processes in human reasoning is analogy. Indeed, natural, everyday reasoning, as distinct from reasoning by a specialist, might be considered more analogical than logical. Analogies are used in mathematics, science, art, politics, and many other areas of life. Analogy is also important in knowledge acquisition, so new concepts can often be explained by analogy with a concept that is already understood. For example, electricity might be explained by analogy with water running down pipes, so diameter of a pipe corresponds to diameter of a conductor, and so on. Concrete teaching aids, such as those used in school mathematics, are essentially analogs.

Analogy is a mapping from a base or source to a target, where both source and target are defined as sets of relations. Typically, relations are mapped but attributes are not, and the relations that are mapped are those that enter into a coherent structure. The mapping is validated by structural correspondence between relations in source and relations in target. Proportional analogies have the form A : B : : C : D (e.g., horse : foal : : cat : kitten). Horse is mapped to cat and foal to kitten. The relation, parenthood, between horse and foal corresponds to the relation between cat and kitten. Performance on these analogies depends on children having the requisite knowledge of relations. Thus young children could understand the analogy between melting chocolate and melting snowmen (i.e., solid chocolate:melted chocolate::solid snowmen:melted snowmen) because the relations between solid and melted chocolate or between solid and melted snowmen, were familiar to them. Relational complexity theory predicts that if the requisite knowledge is available, analogies based on unary relations should be possible at 1 year, those based on binary relations should be possible from 2 years, and those based on ternary relations at 3 years.

Much human reasoning has been found to be performed by mental models, in which the premises are represented by concrete analogs. We will consider conditional reasoning, using a major premise, p implies q, represented symbolically as $p \rightarrow q$. This premise would be represented initially as a mental model with this form:

$$p \qquad q$$
$$\cdots$$

The first line represents a state of affairs in which p and q are both true, and there is a link between them. The dots on the next line represent implicit recognition that other possibilities exist. The representations will be 'fleshed out' to give explicit representation of other possibilities as follows (where $\neg p$ represents 'not p'):

$$p \qquad q$$
$$\neg p \qquad \neg q$$

Then the representation is further fleshed out as follows:

$$p \qquad q$$
$$\neg p \qquad \neg q$$
$$\neg p \qquad q$$

This corresponds to the standard (canonical) interpretation of a conditional. The fleshing out is governed by availability of examples in memory. Consider, for example, the premise:

If X is a dog, then X has legs.

We might represent this mentally as:

Dog legs.

Now it is easy to retrieve from memory cases of nondogs that have legs (e.g., tables). Now we elaborate our mental model as follows:

Dog legs

Table legs.

Now we will not commit the fallacy of inferring that if something has legs it is a dog (known as affirmation of the consequent) because our mental model includes cases of nondogs with legs. Inference is also influenced by the complexity of the resulting representation, and children of 5–7 years of age will only be able to add one relation to the simplest model, which permits some correct inferences but also makes them susceptible to fallacies.

Theories of Infant Cognitive Development

A number of theories have been devoted to understanding the precursors of later reasoning in the cognition of infants. Image schemas in the first year of life are seen as building blocks of later reasoning. They include self-motion, animate-motion, agency, path, support, and containment. Image schemas are really implicit concepts, and they comprise linked elements, but the components of an image schema are 'fused' and are not accessible to analysis. Infants are also known to recognize cause, possibly based on some kind of innate predisposition.

Infants also have an ability to represent structure, independent of content. Seven-month-old infants who listened to 2 min segments of utterances such as 'ga ti ga', of 'li ti li' could distinguish between further utterances that that were either consistent (e.g., 'wo fe wo') or inconsistent (e.g., 'wo fe fe') with the original sequences. This was indicated because infants paid more attention to utterances with the novel structure, even though both consistent and inconsistent sentences had novel content. Thus they appeared to have represented the structure independent of content. Although this is a long way from the reasoning of adults or even older children, it is an important step toward symbolic processes.

Infant's tendency to attend to novel or surprising events has also been used to assess their quantitative knowledge. First, infants aged 6–12 months discriminate between displays with different numbers of elements, indicating that they have some conception of number. From around 5 months of age they recognize when something has been added or removed from a small set of objects. Infants also demonstrate awareness of ordinal relations such as ascending or descending sequences.

Some theories attempt to explain infant's early quantitative knowledge. The 'accumulator model' proposes that the nervous system has a pulse generator that generates activity at a constant rate, and there is a gate that opens to allow energy through to an accumulator. The gate opens for a set amount of time for each item. The total energy accumulated is an analog representation of number. For example '–' represents one, '——' represents two, and '———' represents three. According to the 'object file model' infants construct an imagistic representation of the experimental scene, creating one object-file for each object in the array. These representations of numerosity are implicit in the sense that there is no distinct symbol for the numerosity of the set and there is no counting process.

Infants can form prototypic categories. This has been demonstrated by familiarizing infants with a number of different exemplars of the same category (e.g., dogs, horses), then testing them with a novel stimulus from the same category (different dog and different horse) and a novel stimulus from a different category (a car). Children as young as 4 months of age showed a preference for the car, indicating that they had formed a category of dogs and horses.

Categories develop rapidly in early childhood and even young children can make inductive inferences about categories that go beyond observable properties. Children as young as 2–3 years of age can infer properties on the basis of category membership, even when the relevant properties are unobservable to the child. For example, if they are shown a picture of a dog and told that dog has a spleen inside, they will attribute this property to other dogs, rather than to similar animals that are not dogs.

One theorist, Susan Gelman, proposed that children's categories are based on 'essences', which are unobservable properties that cause things to be the way they are.

Children's categories are based on early recognition of the causal basis of the properties of natural kinds. Another theorist, Frank Keil, showed that categories remain constant, despite changes in appearance or attempts at transformation. By about 8 years of age children recognize that an animal cannot be converted into a different species by changing their appearance or the way they behave. By contrast, children recognize that artifacts can be transformed. Thus, they appear to recognize that natural kinds have certain essential properties that are inherent in their composition, whereas artifacts can be transformed by human intervention.

Understanding the Physical World

Young children's understanding of physical phenomena is influenced by their ability to deal with complexity. If children are shown an apparatus in which a marble inserted at the top either always exits below the insertion point, or always crosses to the other side, 3-year-olds could predict on which side the marble would exit. Thus, they succeed if the exit point is always predictable by the insertion point. However if exiting below the insertion point, or crossing to the other side is indicated by whether a light was on or off, they do not succeed. Thus they succeed when the exit point was influenced by only one variable, point of insertion, but they could not take account of both side of insertion and the light signal. Four-year-olds could handle both. Thus, ability to handle the extra variable increases with age.

Children's concept of the Earth has been investigated by asking them to draw the Earth with people living on it, and to indicate the position of the sun, moon, and stars. Their conceptions reflected their attempts to reconcile what they had been taught about the Earth being a sphere with their everyday observations that it appears flat. Thus, children would draw a flattened sphere with people standing on top, or a hollow sphere with a horizontal platform inside for people to stand on, or even dual earths, one round and one flat. Their mental models showed some consistency, so if they thought that the Earth was spherical they were less likely to think it was possible to fall off the edge. To integrate everyday observation with what they are taught about the Earth being spherical children would need to know that, for example, the huge diameter of the earth makes it appear flat at any point on it surface.

The balance scale was widely used in Piagetian investigations of cognitive development. It comprises a beam balanced on a fulcrum with equally spaced pegs on each side on which weights can be placed. The beam balances when the product of weight and distance on the left equals the product of weight and distance on the right. Children's understanding of the balance scale can be defined by rules that develop progressively: with rule I children consider only weights, with rule II they also consider distance, but only if weights on the two sides are equal. With rule III they consider weight first, then distance but had difficulty if the greater weight occurred on one side and greater distance on the other. With rule IV they apply the correct principle, according to which the side with the greater product of weight and distance, goes down, but if the products on the two sides are equal the beam balances. Children progressed from rule I at 5 years of age to rule III in adolescence. Rule IV tended to be rare even in adults.

Children as young as 2–3 years have some understanding of the balance scale, and they can discriminate weights with distance constant, or distances with weight constant. These entail representing the binary relation between two weights, or two distances, according to relational complexity theory.

Neural Net Model of Balance Scale Understanding

Neural net models are designed to simulate cognitive processes by units that are connected together by variable weights. The units represent collections of neurons, and they have theoretical activation values that are intended to correspond to activations of neurons. The connection weights simulate associations between different sets of neurons. Activation is transmitted from one set of neurons to another with a strength that is determined by the weights (**Figure 2**).

Neural net models have been widely used to account for cognitive processes in children and adults and we will illustrate these developments with a model by J. McClelland of children's understanding of the balance scale, shown in **Figure 2**. There are four sets of five input units. One set, on the left side, represents number of weights, from one to five, and another set represents the number of weights on the right. The remaining sets represent number of steps from the fulcrum, on both left and right sides. There are four 'hidden' units, which are connected to the input units by connections with variable connections. Two of the hidden units compare weights and two compare distances. The hidden units are connected to the output units by another set of variable connections. Activation spreads from input to hidden units and then to output units. In **Figure 2**, the weights and distances from the fulcrum are indicated by black input units. Activations of the hidden and output units are also represented by units filled in with black. The weights are adjusted by a learning process so that the output units predict the balance state. The model's performance shows a good correspondence to the development of children's knowledge of the balance scale. Neural net models are often important for the properties that emerge as they learn. In this case, training of the neural net results in the units representing larger weights, or larger distances, having greater connections to the hidden units. Thus, metrics

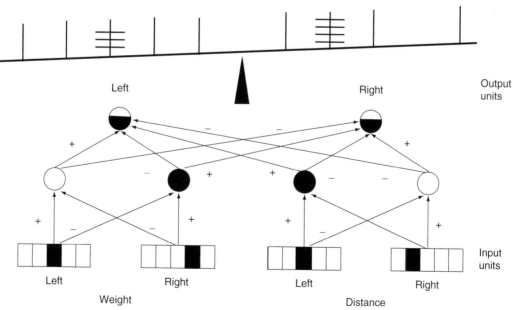

Figure 2 Neural net model of the balance scale by J. McClelland. Shading represents activation of neural units. The input units represent the weights and distances from the fulcrum and correspond to the weights and distances shown on the balance. Adapted from figure 22.1 from Halford GS (2005) Development of Thinking. In: Holyoak KJ and Morrison RG (eds.) *Cambridge Handbook of Thinking and Reasoning*, p. 533, by permission of Cambridge University Press.

for weight and distance emerge as a result of training, and are not predefined in the net. This is a good example of the way structure can emerge from experience with a set of phenomena.

Neuroscience Approaches

Some of the theories considered so far propose that children become able to process more complex concepts as they grow older. A probable reason for this is that the amount of information they can process at any one time increases with age. That is, their information processing capacity increases. Evidence has also been provided by Robert Kail that speed in making decisions increases with development, and this might be linked to an increase with capacity.

A theory by Quartz and Sejnowski suggests a neural basis for this capacity increase with development. They propose that, with age, increase in the number of synapses and the number of connections between neurons is responsible for this increase in capacity, and that this process is influenced by interaction with the environment. This enables more complex representations to be constructed as the child matures. It also has the important implication that brain capacity is influenced by what the child experiences during development. Spurts in brain growth, especially in the frontal regions of the brain have also been linked to transitions to higher levels of reasoning.

Summary

Cognitive developmental theories are attempts to define and explain the changes in children's concepts, their thinking and their understanding of the world, over the course of development. The pioneering theories of Piaget and Vygotsky were taken up by a number of other theorists, including the neo-Piagetian theorists, but also by others who emphasized the acquisition and organization of knowledge, and the increasing complexity of children's cognition as they developed. Some theories emphasize the nature of reasoning processes, including analogy and mental models, while others, such as dynamic systems theories and neural net models, define the processes of development. Still others, such as microgenetic analyses, are primarily concerned with ways of analyzing the development of children's cognition. Theories of the origin of cognition in infancy are also reviewed.

Although there are many theories, they tend to be complementary rather than contradictory. Cognitive development is influenced by increases in capacity to process complex information, as emphasized by neo-Piagetian theories, but acquisition and organization of knowledge are also important. Theories of reasoning processes do not contradict any of this, but relate it to what we know about how reasoning is carried out. Earlier theories, including those of Piaget and Vygotsky, have not been so much discredited as absorbed into larger bodies of knowledge. This has given us a much richer understanding of how thinking develops in children.

See also: Categorization Skills and Concepts; Cognitive Development; Humor; Milestones: Cognitive; Neonativism; Object Concept; Piaget's Cognitive-Developmental Theory; Reasoning in Early Development; Vygotsky's Sociocultural Theory.

Suggested Readings

Case R and Okamoto Y (1996) The role of central conceptual structures in the development of children's thought. *Monographs of the Society for Research in Child Development* 61: v-265.

Fischer KW and Bidell TR (2006) Dynamic development of action and thought. In: Lerner RM (ed.) *Handbook of Child Psychology: Volume 1, Theoretical Models of Human Development,* 6th edn. Hoboken, NJ: Wiley.

Halford GS (2005) Development of thinking. In: Holyoak KJ and Morrison RG (eds.) *Cambridge Handbook of Thinking and Reasoning.* Cambridge: Cambridge University Press.

Halford GS and Andrews G (2004) The development of deductive reasoning: How important is complexity? *Thinking and Reasoning* 10: 123–145.

Halford GS and Andrews G (2006) Reasoning and problem solving. In: Kuhn D and Siegler R (eds.) *Handbook of Child Psychology: Volume 2, Cognitive, Language and Perceptual Development,* 6th edn. Hoboken, NJ: Wiley.

Holyoak KJ and Thagard P (1995) *Mental Leaps.* Cambridge, MA: MIT Press.

Siegler RS (2006) Microgenetic analyses of learning. In: Kuhn D and Siegler R (eds.) *Handbook of Child Psychology: Volume 2, Cognitive, Language and Perceptual Development,* 6th edn. Hoboken, NJ: Wiley.

Cognitive Neuroscience

M H Johnson, University of London, London, UK

Glossary

Autism or autism spectrum disorder (ASD) – A neurodevelopmental disorder that affects development and subsequent behavior including markedly abnormal social interaction, patterns of interests, and communication. Estimates of ASD within a population range from 1 in 200 to 1 in 1000.

Cerebral cortex – A brain structure, composed of neuron cell bodies, found in vertebrates. In humans it is a highly developed structure, responsible for many higher-order functions like language and information processing.

Event related potentials (ERPs) – A set of voltage changes contained within a period of electroencephalogram (EEG) that are time-locked to an event, for example, presentation of an object. This is a noninvasive technique with excellent temporal resolution.

Extinction – A milder form of spatial neglect in which stimuli in one half of the visual field are neglected only when there is a competing stimulus in the opposite field, and not when presented in isolation.

Functional magnetic resonance imaging (fMRI) – A form of neuroimaging that uses magnetic resonance to measure hemodynamic responses in relation to neural activity. This is a noninvasive technique, which allows fine spatial localization.

Near infrared spectroscopy (NIRS) – A neuroimaging technique that uses infrared resonance to measure changes in blood and tissue oxygenation in a noninvasive way, and is a relatively new form of neuroimaging. This is often used in research with infants as an alternative to magnetic resonance imaging, because it is less sensitive to movement.

Plasticity – The ability of the brain, especially in our younger ages to compensate for change. Also refers to the ability of brain regions to take on a variety of different function during early development. Plasticity decreases as brain structures become more specialized during development.

Spatial neglect – A brain-damage syndrome in which patients appear to detect visual events contralateral to their damaged hemisphere. For example, right hemisphere damage can cause a patient to fail to read the left-hand side of a book.

Synapses – Found in the nervous system, these are specialized junctions of cells that facilitate communication between cells and allow neurons to form interconnected neural circuits.

Williams syndrome – A rare genetic disorder occurring in fewer than 1 in 20 000 people. It is characterized by a distinct difference in facial features, sociable demeanor, and developmental delay in certain areas.

Introduction

Cognitive neuroscience has emerged over the past decades as one of the most significant research directions in all of neuroscience and psychology. More recently, the scientific interface between cognitive neuroscience and human development, developmental cognitive neuroscience, has become a hot topic. Part of the reason for the renewed interest in relating brain development to cognitive, social, and emotional change comes from advances in methodology that allow hypotheses to be generated and tested more readily than previously. One set of tools relates to brain imaging – the generation of 'functional' maps of brain activity based on either changes in cerebral metabolism, blood flow, or electrical activity. The three brain-imaging techniques most commonly applied to development in normal children are event-related potentials (ERPs), functional magnetic resonance imaging (fMRI), and near infrared spectroscopy (NIRS). Another methodological advance is related to the emergence of techniques for formal computational modeling of neural networks and cognitive processes. Such models allow us to begin to bridge data on developmental neuroanatomy to data on behavioral changes associated with development. A third methodological innovation is the increasing trend for studying groups of developmental disorders (such as autism and Williams syndrome) together alongside typical development. Thus, rather than each syndrome being studied in isolation, comparisons between different typical and atypical trajectories of development are helping to reveal the extent and limits on plasticity.

Brain Development

Brain development may be divided into that which occurs prior to birth (prenatal) and that which takes place after birth (postnatal). While some of the same developmental processes can be traced from pre- to postnatal life, in postnatal development there is obviously more scope for influence from the world outside the infant. A striking feature of human brain development is the comparatively long phase of postnatal development, and therefore the increased extent to which the later stages of brain development can be influenced by the environment of the child. Some degree of plasticity is retained into adulthood, but this may decline with age.

By around the time of birth the vast majority of cells are in their appropriate adult locations in the human brain, and all of the major landmarks of the brain, such as the most distinctive patterns of folding of the cerebral cortex, are in place. A number of lines of evidence indicate that substantive changes take place during postnatal development of the human brain. At the most gross level of analysis, the volume of the brain quadruples between birth and adulthood. This increase comes from a number of sources such as more extensive fiber bundles, and nerve fibers becoming covered in a fatty myelin sheath that helps conduct electrical signals (myelination). But perhaps the most obvious manifestation of postnatal neural development as viewed through a standard microscope is the increase in size and complexity of the dendritic tree of many neurons. Less apparent through standard microscopes, but more evident with electron microscopy, is a corresponding increase in density of functional contacts between neurons, synapses.

Peter Huttenlocher and colleagues have reported a steady increase in the density of synapses in several regions of the human cerebral cortex. For example, in parts of the visual cortex, the generation of synapses (synaptogenesis) begins around the time of birth and reaches a peak around 150% of adult levels toward the end of the first year. In the frontal cortex (the anterior portion of cortex, considered by most investigators to be critical for many higher cognitive abilities), the peak of synaptic density occurs later, at around 24 months of age. Although there may be variation in the timetable, in all regions of cortex studied so far, synaptogenesis begins around the time of birth and increases to a peak level well above that observed in adults.

Somewhat surprisingly, regressive events are commonly observed during the development of nerve cells and their connections in the brain. For example, in the primary visual cortex the mean density of synapses per neuron starts to decrease at the end of the first year of life. In humans, most cortical regions and pathways appear to undergo this 'rise-and-fall' in synaptic density, with the density stabilizing to adult levels during later childhood. The postnatal rise-and-fall developmental sequence can also be seen in other measures of brain physiology and anatomy. For example, Harry Chugani and colleagues have observed an adult-like distribution of resting brain activity within and across brain regions by the end of the first year. However, the overall level of activity (as measured by glucose uptake) reaches a peak during early childhood that is much higher than that observed in adults. These rates returned to adult levels after about 9 years of age for some cortical regions. Recently, magnetic resonance imaging (MRI) has been used to study the postnatal development of brain structure. Using this method, the consensus is that brain structures have the overall appearance of those in the adult by 2 years of age, and that all the major fiber tracts can be observed by 3 years of age. Some reports suggest that after a rapid increase in gray matter up to 4 years of age, there is then a prolonged period of slight decline that extends into the adult years. Whether this decline is due to the dendritic and synaptic pruning described above remains unknown. Changes in the extent of white matter are of interest because they reflect interregional communication in the developing brain. Although increases in white matter continue through adolescence into adulthood, particularly in frontal brain regions, the most rapid changes occur during the first 2 years. For example, at around 8–12 months of age the white matter associated with the frontal, parietal, and occipital lobes becomes apparent.

Three different perspectives on human postnatal functional brain development are currently being explored. The first of these approaches, the maturational perspective, has the goal to relate the maturation of particular regions of the brain, usually regions of the cerebral cortex, to newly emerging sensory, motor, and cognitive functions. Evidence concerning the differential neuroanatomical development of cortical regions is used to determine an age when a particular region is likely to become functional. Success in a new behavioral task at this same age is then attributed to the maturation of a newly functional brain region, with maturation often assumed to be an 'all or none' phenomenon, or at least to have a sudden onset. Typically, comparisons are then made between the behavioral performance of adults with acquired lesions and behaviors during infancy. One example of this approach comes from the neurodevelopment of visual orienting and attention, where several researchers have argued that control over visually guided behavior is initially by subcortical structures, but with age and development, posterior cortical regions, and finally anterior regions, come to influence behavior.

In contrast to the maturational approach in which behavioral developments are attributed to the onset of functioning in one region or system, an alternative viewpoint assumes that postnatal functional brain development, at least within the cerebral cortex, involves a process of organizing interregional interactions: 'interactive specialization'.

Recent trends in the analysis of adult brain-imaging data have proceeded on the assumption that the response properties of a specific region may be determined by its patterns of connectivity to other regions and their current activity states. Extending this notion to development means that we should observe changes in the response properties of cortical regions during ontogeny as regions interact and compete with each other to acquire their role in new computational abilities. The onset of new behavioral competencies during infancy will be associated with changes in activity over several regions, and not just by the onset of activity in one or more additional region(s). In further contrast to the maturational approach, this view predicts that during infancy patterns of cortical activation during behavioral tasks may be more extensive than those observed in adults, and involve different patterns of activation. Within broad constraints, even apparently the same behavior in infants and adults could involve different patterns of cortical activation.

A third perspective on human postnatal functional brain development has been termed the skill-learning hypothesis. Recent neuroimaging evidence from adults has highlighted changes in the neural basis of behavior that result as a consequence of acquiring perceptual or motor expertise. One hypothesis is that the regions active in infants during the onset of new perceptual or behavioral abilities are the same as those involved in skill acquisition in adults. This hypothesis predicts that some of the changes in the neural basis of behavior during infancy will mirror those observed during more complex skill acquisition in adults.

Domains of Cognitive and Behavioral Change

Developmental cognitive neuroscience has, to date, only been applied to some aspects of perceptual and cognitive development. While several exciting new areas of development are beginning to be explored, we have selected three domains in which perhaps the most progress has been made: the developing 'social brain', speech and language acquisition, and the development of frontal cortex functions.

Developing a Social Brain

One of the major characteristics of the human brain is its social nature. As adults, we have areas of the brain specialized for processing and integrating sensory information about the appearance, behavior, and intentions of other humans. A variety of cortical areas have been implicated in the 'social brain' including the superior temporal sulcus (STS), the fusiform 'face area' (FFA),

and orbitofrontal cortex. One of the major debates in cognitive neuroscience concerns the origins of the 'social brain' in humans, and theoretical arguments abound about the extent to which this is acquired through experience.

The ability to detect and recognize faces is commonly considered to be a good example of human perceptual abilities, as well as being the basis of our adaptation as social animals. There is a long history of research on the development of face recognition in young infants extending back to the studies of Robert Fantz more than 40 years ago. Over the past decade numerous papers have addressed cortical basis of face processing in adults, including identifying areas that may be specifically dedicated to this purpose. Despite these bodies of data, surprisingly little remains known about the developmental cognitive neuroscience of face processing.

Some authors have speculated that the preferential responding to faces observed in newborn infants may be largely mediated by a subcortical visuomotor pathway, whereas later developing abilities to recognize individual faces (on the basis of internal features) are mediated by the ventral stream of visual cortical processing. Much research over the past decade has focused on this 'two systems' view. With regard to a newborn's responses to faces, the majority of behavioral studies to date have found some evidence for sensitivity to face-like patterns. Although views still vary as to the specificity of this newborn bias, John Morton and the present author have speculated that the newborn bias is mediated largely by subcortical visuomotor pathways. This proposal was made for several reasons: (1) that the newborn preference declined at the same age as other newborn reflexes assumed to be under subcortical control, (2) evidence from the maturation of the visual system indicating later development of cortical visual pathways, and (3) evidence from another species (the domestic chick). Due to the continuing difficulty in successfully using functional imaging with healthy awake newborns, this hypothesis has, as yet, only been indirectly addressed. One line of indirect evidence comes from adult neuropsychological and functional imaging studies, and specifically evidence from adult patients with spatial neglect and extinction. Spatial neglect is a brain-damage syndrome in which patients do appear to detect visual events contralateral to their damaged hemisphere. For example, right hemisphere damage can cause a patient to fail to read the left-hand side of a book. Extinction is a related deficit of attention in which patients can orient to stimuli in the contralesional field, but only if they are presented in isolation and are not presented simultaneously with a stimulus in the opposite visual field. Patrick Vuilleumier and colleagues report that these patients extinguish a face much less often than other stimuli. While a variety of explanations of this phenomenon are possible, one view is that damage to cortical circuits in adults releases inhibition of a subcortical face bias. Finally, functional imaging studies in adults that include subcortical structures in their analysis, have observed activation of a range of subcortical structures in addition to the well-known areas of cortical specialization.

Turning to the neurodevelopment of face processing during infancy and childhood, it can be seen that several laboratories have examined changes in ERPs as adults view faces. In particular, interest has been focused on an ERP component termed the 'N170' (because it is a negative-going deflection that occurs after around 170 ms) that has been strongly associated with face processing in a number of studies on adults. Specifically, the amplitude and latency of this component vary according to whether or not faces are present in the visual field of the adult volunteer under study (see **Figure 1**). An important aspect of the N170 in adults is that its response is highly selective. For example, the N170 shows a different response to human upright faces than to very closely related stimuli such as inverted human faces and upright monkey faces. While the exact underlying neural generators of the N170 are currently still debated, the specificity of response of the N170 can be taken as an index of the degree of specialization of cortical processing for human upright faces. For this reason Michelle de Haan and colleagues undertook a series of studies on the development of the N170 over the first weeks and months of postnatal life.

The first issue to be addressed in these developmental ERP studies is when does the face-sensitive N170 emerge?

In a series of experiments de Haan and colleagues have identified a component in the infant ERP that has many of the properties associated with the adult N170, but that is of a slightly longer latency (240–290 ms) (see **Figure 2**). In studying the response properties of this potential at 3, 6, and 12 months of age they discovered that (1) the component is present from at least 3 months of age (although its development continues into middle childhood) and (2) the component becomes more specifically tuned to respond to human upright faces with increasing age. To expand on the second point, it was found that while 12-month-olds and adults showed different ERP responses to upright and inverted faces, 3- and 6-month-olds do not. Thus, the study of this face-sensitive ERP component is consistent with the idea of increased specialization of cortical processing with age, a result also consistent with some behavioral results (see below).

While definitive functional imaging studies on face processing in infants and young children are still awaited, evidence for increased localization of cortical processing of faces comes from a recent fMRI study of the neural basis of face processing in children compared to adults by Alessandra Passarotti, Joan Stiles, and colleagues. In this study, even when children and adults were matched for behavioral ability (in a face-matching task), children activated a larger extent of cortex around face-sensitive areas than did adults.

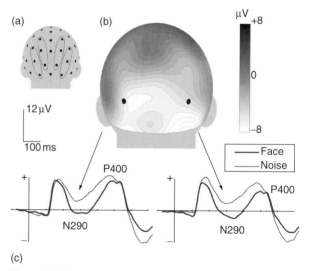

Figure 1 Scalp-recorded event-related potentials in response to faces and matched visual noise presented to adults. (a) Location on the scalp of the electrode sites selected for study (back of the head). (b) A scalp surface voltage map illustrating the distribution of voltage associated with the N170. (c) ERP waveforms from left and right posterior temporal recording sites (black spots on panel b). Note the increased amplitude of the N170 waveform to faces. Reproduced from Halit H, Csibra G, Volein A, and Johnson MH (2004) Face-sensitive cortical processing in early infancy. *Journal of child psychology and Psychiatry* 45:1228–1234, with permission from Blackwell Publishing.

Figure 2 Scalp-recorded event-related potentials in response to faces and matched visual noise presented to 3-month-old infants. (a) Location on the scalp of the electrode sites selected for study (back of the head). (b) A scalp surface voltage map illustrating the distribution of voltage associated with the infant equivalent of the N170 (N290). (c) ERP waveforms from left and right occipital temporal recording sites (black spots on panel b). Note the increased amplitude of the N290 waveform to faces. Reproduced from Halit H, Csibra G, Volein A, and Johnson MH (2004) Face-sensitive cortical processing in early infancy. *Journal of child psychology and Psychiatry* 45:1228–1234, with permission from Blackwell Publishing.

Converging evidence about the increasing specialization of face processing during development comes from a behavioral study that set out to test the intriguing idea that, as processing 'narrows' to human faces, then infants will lose their ability to discriminate nonhuman faces. Olivier Pascalis and colleagues demonstrated that while 6-month-olds could discriminate between individual monkey faces as well as human faces, 9-month-olds and adults could only discriminate the human faces. These results are particularly compelling since they demonstrate a predicted competence in young infants that is not evident in adults.

Moving beyond the relatively simple perception of faces, a more complex attribute of the adult social brain is processing information about the eyes of other humans. There are two important aspects of processing information about the eyes. The first of these is being able to detect the direction of another's gaze in order to direct your own attention to the same object or spatial location (eye-gaze cueing). Perception of averted gaze can elicit an automatic shift of attention in the same direction in adults, allowing the establishment of 'joint attention'. Joint attention to objects is thought to be crucial for a number of aspects of cognitive and social development, including word learning. The second critical aspect of gaze perception is the detection of direct gaze, enabling mutual gaze with the viewer. Mutual gaze (eye contact) provides the main mode of establishing a communicative context between humans, and is believed to be important for normal social development. It is commonly agreed that eye-gaze perception is important for mother–infant interaction, and that it provides a vital foundation for social development.

In a series of experiments with 4-month-old infants using a simple eye-gaze cueing paradigm (in which a stimulus face looked left or right), Teresa Farroni and colleagues have established that it is only following a period of mutual gaze with an upright face that cueing effects are observed. In other words, mutual gaze with an upright face may engage mechanisms of attention such that the viewing infant is more likely to be cued by subsequent motion. In summary, the critical features for eye-gaze cueing in young infants are (1) lateral motion of elements and (2) a brief preceding period of eye contact with an upright face.

Following the surprising observation that a period of direct gaze is required before cueing can be effective in infants, the authors investigated the earliest developmental roots of eye contact detection. It is already known that human newborns have a bias to orient toward face-like stimuli (see earlier), prefer faces with eyes opened, and tend to imitate certain facial gestures. Preferential attention to faces with direct gaze would provide the most compelling evidence to date that human newborns are born prepared to detect socially relevant information. For this reason Farroni and colleagues tested healthy human newborn infants by presenting them with a pair of stimuli,

one a face with eye gaze directed straight at the newborns, and the other with averted gaze. Results showed that the fixation times were significantly longer for the face with the direct gaze.

In a further experiment, converging evidence for the differential processing of direct gaze in infants was obtained by recording ERPs from the scalp as infants viewed faces. Babies, 4 months of age, were tested with the same stimuli as those used in the previous experiment with newborns, and a difference was found between the two gaze directions at the time and scalp location at the previously identified face-sensitive component of the infant ERP discussed earlier. The conclusion from these studies is that direct eye contact enhances the perceptual processing of faces in infants during the first months.

Beyond face processing and eye-gaze detection there are many more complex aspects of the social brain such as the coherent perception of human action and the appropriate attribution of intentions and goals to conspecifics. Investigating the cognitive neuroscience of these abilities in infants and children will be a challenge for the next decade.

Language Acquisition and Speech Recognition

Is language biologically special? This motivating question refers to the extent to which the human infant is predisposed to process and learn about language, and the extent to which the underlying neural circuits are 'pre-wired' to process language input. Two cognitive neuroscience approaches to this question have been taken. The first approach addresses the issue of whether there are particular parts of the cortex critical for primary language acquisition, or whether a variety of cortical areas can support this function. The second strategy has been to attempt to identify neural correlates of speech-processing abilities present from very early in life, before experience is thought to have shaped cortical specialization.

Language acquisition has become a focal point for studies designed to investigate the extent to which particular cortical areas, such as Broca's and Wernicke's areas, are 'pre-wired' to support specific functions. Two main lines of research have been pursued, with one set of studies examining the extent to which language functions can be subserved by other regions of the cortex, and another line of research concerned with whether other functions can 'occupy' regions that normally support language. The first of these approaches has been pursued through investigations of whether children suffering from perinatal lesions to the classical 'language areas' of cortex can still acquire language. The second approach has involved the testing of congenitally deaf children to see what, if any, functions are present in regions of cortex that are normally (spoken) language areas.

If particular cortical regions are uniquely pre-wired to support language, then it is reasonable to assume that damage to such regions will impair the acquisition of language regardless of when the insult occurs. This implicit hypothesis has motivated a good deal of research, the conclusions of which still remain somewhat controversial. Eric Lenneberg argued that, if localized left hemisphere damage occurred early in life, it had little effect on subsequent language acquisition. This view contrasted with the effect of similar lesions in adults or older children, and many congenital abnormalities in which language is delayed or never emerges. This view lost adherents in the 1970s, as evidence accumulated from studies of children with hemispherectomies suggesting that left hemisphere removal commonly leads to selective subtle deficits in language, especially for syntactic and phonological tasks. Similar results have also been reported for children with early focal brain injury due to strokes. These findings were compatible with studies of normal infants showing a left hemisphere bias at birth in processing speech and other complex sounds, and led some researchers to the conclusion that functional asymmetries for language in the human brain are established at birth, and cannot be reversed. This view was reinforced by a number of neuroanatomical studies that have shown differences between parts of left and right cerebral cortex in adults. For example, Norman Geschwind and Walter Levitsky reported that the left planum temporale (an area associated with language processing) was larger than the right in 65% of adult brains studied. A number of groups have looked for similar differences in infant brains as evidence for prespecified language abilities. As early as the 29th week of gestation, the left planum temporale is usually larger on the left than on the right in human infants. It is important to remember, however, that (1) this asymmetry is probably not specific to humans and (2) gyral and sucal measures only tell us about the quantity of cortical tissue within a region, and cannot therefore be used to argue for the detailed specific pre-wiring assumed by some to be necessary for language-specific processing. In addition to these reservations about the neuroanatomical evidence, many of the secondary sources that summarized the work on hemispherectomies and/or early focal injury failed to note that the deficits shown by these children are very subtle – far more subtle, in fact, than the aphasic syndrome displayed by adults with homologous forms of brain damage. Significantly, most of the children with left hemisphere injury who have been studied to date fall within the normal range, attend public schools, and certainly do better than adults with equivalent damage.

The other approach to studying the extent to which the cortical areas supporting language-related functions are prespecified is to see whether other functions can occupy such regions. This issue has been investigated in a recent fMRI study in which hearing and deaf participants were scanned while reading sentences in either English or American Sign Language (ASL). When hearing adults read English, there was robust activation within some of classical left hemisphere language areas, such as Broca's area. No such activation was observed in the right hemisphere. When deaf people viewed sentences in their native ASL they showed activation of most of the left hemisphere regions identified for the hearing participants. Because ASL is not sound-based, but does have all of the other characteristics of language including a complex grammar, these data suggest that some of the neural systems that mediate language can do so regardless of the modality and structure of the language acquired. Having said this, there were also some clear differences between the hearing and deaf activations, with the deaf group activating some similar regions in the right hemisphere. One interpretation of the right hemisphere activation is that it is evoked by the biological motion inherent in sign, but not spoken, language. A third condition addressed the issue of whether there is a sensitive period for the establishment of left hemisphere language. In this condition, deaf people read English (their second language, learned late) and did not show activation of the classical left hemisphere language regions, suggesting that, if a language is not acquired within the appropriate developmental time window, the typical pattern of adult activation does not occur.

The other general approach to investigating the extent to which language is biologically special involves attempting to identify language-relevant processes in the brains of very young infants. One example of this concerns the ability to discriminate speech-relevant sounds such as phonemes. Behavioral experiments have demonstrated that young infants show enhanced (categorical) discrimination at phonetic boundaries used in speech such as /ba/-/pa/. That is, like adults, a graded phonetic transition from /ba/ to /pa/ is perceived as a sudden categorical shift by infants. This observation was initially taken as evidence for a language-specific detection mechanism present from birth. However, over the past decade it has become clear that other species, such as chinchillas, show similar acoustical discrimination abilities, indicating that this ability may merely reflect general characteristics of the mammalian auditory processing system, and not a an initial spoken language-specific mechanism.

In a further line of behavioral experiments, Janet Werker and colleagues reported that, although young infants discriminate a wide range of phonetic contrasts including those not found in the native language (e.g., Japanese infants, but not Japanese adults, can discriminate between 'r' and 'l' sounds), this ability becomes restricted to the phonetic constructs of the native language around 12 months of age. If brain correlates of this process could be identified, it may be possible to study the mechanisms underlying this language-specific selective loss of sensitivity. Christaine Dehaene-Lambertz and colleagues presented their infants with trials in which a series of four identical syllables (the standard) was followed by fifth that was either identical or phonetically different (deviant). They recorded high-density ERPs time-locked to the onset of the syllable and observed two voltage peaks with different scalp locations. The first peak occurred around 220 ms after stimulus onset and did not habituate to repeated presentations (except after the first presentation) or dishabituate to the novel syllable. Thus, the generators of this peak, probably primary and secondary auditory areas in the temporal lobe, did not appear to be sensitive to the subtle acoustical differences that encoded phonetic information. The second peak reached its maximum around 390 ms after stimulus onset and again did not habituate to repetitions of the same syllable, except after the first presentation. However, when the deviant syllable was introduced the peak recovered to at least its original level. Thus, the neural generators of the second peak, also in the temporal lobe but in a distinct and more posterior location, are sensitive to phonetic information.

Researchers have used functional imaging methods with greater spatial resolution to investigate early correlates of speech perception. Dehaene-Lambertz and colleagues measured brain activation with fMRI in awake and sleeping healthy 3-month-olds while they listened to forwards and backwards speech in their native tongue (French). The authors assumed that forward speech would elicit stronger activation than backward speech in areas related to the segmental and suprasegmental processing of language, while both stimuli will activate mechanisms for processing fast temporal auditory transitions. Compared to silence, both forwards and backwards speech activated widespread areas of the left temporal lobe, which was greater than the equivalent activation on the right for some areas (planum temporale). These results provide converging evidence for the ERP data discussed earlier. Forwards speech activated some areas that backward speech did not, including the angular gyrus and mesial parietal lobe (precuneus) in the left hemisphere. The authors suggest that these findings demonstrate an early functional asymmetry between the two hemispheres. However, they acknowledge that their results cannot discriminate between an early bias for speech perception or a greater responsivity of the left temporal lobe, for processing auditory stimuli with rapid temporal changes.

In summary, therefore, a reasonable working hypothesis is that regions of the left temporal lobe are most suitable for supporting speech recognition. This suitability likely comes from a combination of spatial and temporal factors that may predispose this region to the processing of rapid temporal stimuli. Other regions of cortex are probably also important for the acquisition of language, and can substitute for the left temporal region if required. Language is only 'biologically special' in the broadest sense in which the human species' typical environment interacts with the architecture of cortex and its developmental dynamics to generate representations appropriate for the domain.

Frontal Cortex Development, Object Permanence, and Planning

The region of the frontal lobe anterior to the primary motor and premotor cortex, the prefrontal cortex (PFC), accounts for almost one-third of the total cortical surface in humans and is considered by most investigators to be critical for many higher cognitive abilities. In adults, types of cognitive processing that have been associated with frontal cortex concern the planning and execution of sequences of action, the maintenance of information 'on-line' during short temporal delays, and the ability to inhibit a set of responses that are appropriate in one context but not another. The frontal cortex shows the most prolonged period of postnatal development of any region of the human brain, with changes in synaptic density detectable even into the teenage years, and for this reason it has been the region most frequently associated with developments in cognitive abilities.

Two alternative approaches to the relation between frontal cortex structural development and advances in cognitive ability in childhood have been taken. One of these is the attempt to relate structural developments in the frontal cortex at a particular age to changes in certain cognitive abilities. A refinement of this approach is that the frontal lobes are composed of a number of regions that subserve different functions and show a different timetable of maturation. The alternative approach is based on the assumption that the frontal cortex is involved in acquisition of new skills and knowledge from very early in life, and that it may also play a key role in organizing other parts of cortex. According to this latter view, regions of frontal cortex are important in many cognitive transitions primarily because of their involvement in the acquisition of any new skill or knowledge. A corollary of this is that frontal cortex involvement in a particular task or situation may decrease with increased experience or skill in the domain. There is currently evidence consistent with both of these approaches.

One of the most comprehensive attempts to relate a cognitive change to underlying brain developments has concerned marked behavioral changes around 8–10 months of age. In particular, Adele Diamond and colleagues argued that the maturation of PFC during the last half of the human infant's first year of life accounts for a number of transitions observed in the behavior of infants in object permanence and object retrieval tasks. One of the behavioral tasks they have used to support this argument comes from Jean Piaget who observed that infants younger than 8 months often fail to retrieve accurately a hidden object after a short delay period if the object's location is changed from one where it was previously and successfully retrieved. Infants often made a particular perseverative error in which they reach out to the hiding location where the object was found on the immediately preceding trial. This characteristic pattern of error was cited as evidence for the failure to understand that objects retain their existence or permanence when moved from view. By around 9 months of age, infants begin to succeed in the task at successively longer delays of 1–5 s, although their performance remains unreliable up to about 12 months of age if the delay between hiding and retrieval is incremented as the infants age.

Diamond and colleagues tested monkeys in a modification of the above object permanence task. Consistent with the observations on human infants, infant monkeys failed to retrieve the hidden object. Further, adult monkeys with lesions to the dorsolateral region of the PFC (DLPC) were also impaired in this task. Lesions to some other parts of the brain (parietal cortex, or the hippocampal formation) did not significantly impair performance, suggesting that the DLPC plays a central role in tasks that require the maintenance of spatial or object information over temporal delays.

Further evidence linking success in the object-permanence task to frontal cortex maturation in the human infant comes from two sources. The first of these is a series of electroencephalogram (EEG) studies with normal human infants, in which increases in frontal EEG responses correlate with the ability to respond successfully over longer delays in delayed response tasks. The second source is work on cognitive deficits in children with a neurochemical deficit in the PFC resulting from phenylketonuria (PKU). Even when treated, this inborn error of metabolism can have the specific consequence of reducing the levels of a neurotransmitter, dopamine, in the DLPC. These reductions result in these infants and children being impaired on tasks thought to involve parts of the PFC such as the object-permanence and object-retrieval tasks, and being relatively normal in tasks thought to depend on other regions of cortex.

Having established a link between PFC maturation and behavioral change in a number of tasks, Diamond has speculated on the computational consequence of this aspect of postnatal brain development. Specifically, she suggested that the DLPC is critical for performance when (1) information has to be retained or related over time or space and (2) a prepotent response has to be inhibited. Only tasks that require both of these aspects of neural computation are likely to engage the DLPC. In the case of the object-permanence task, a spatial location has to be retained over time and the prepotent previously rewarded response inhibited. One experiment suggests that the PFC maturation hypothesis is not the whole story, however, and that some modification or elaboration of the original account will be required. Rick Gilmore observed that infants succeed on a task that requires temporal spatial integration over a delay at a much younger age than is indicated by the object-permanence tasks. In addition, studies by Renee Baillargeon and others entailing infants viewing 'possible' and 'impossible' events involving occluded objects have found that infants as young as 3.5 months look longer at impossible events indicating that they have an internal representation of the occluded object. In order to account for the apparent discrepancy between these results and those with the reaching measures, some have provided 'means-ends' explanations, arguing that infants are unable to coordinate the necessary sequence of motor behaviors to retrieve a hidden object. To test this hypothesis, Yuko Munakata and colleagues trained 7-month-olds to retrieve objects placed at a distance from them by means of pulling on a towel or pressing a button. Infants retrieved the objects when a transparent screen was interposed between them and the toy, but not if the screen was sufficiently opaque to make the object invisible. Since the same means-ends planning is required whether the screen is transparent or opaque, it was concluded that 'means-ends' explanations cannot account for the discrepancy between the looking and the reaching tasks. Munakata and colleagues proposed an alternative 'graded' view of the discrepancy implemented as a connectionist model.

The maturational approach to PFC development has also been extended to later childhood and adolescence. The results from a variety of behavioral tasks designed to tap into advanced PFC functions have demonstrated that adult levels of performance are not reached until adolescence or later. For example, participants from 3- to 25-years-old were tested on the CANTAB (Cambridge Neuropsychological Testing Automated Battery), a well-established and validated battery of tests previously used on adult human and animal lesion populations. This battery assesses several measures including working memory skills, self-guided visual search, and planning. Importantly for developmental studies, the battery is administered with touch-screen computer technology, and does not require any verbal or complex manual responses. Using the CANTAB, Maria Luciana and Charles Nelson found that while measures that depend on posterior brain regions (such as recognition memory) were stable by 8 years of age, measure of planning and working memory had not yet reached adult levels by age 12 years of age.

While such behavioral measures are useful as marker tasks, it is even better to use functional imaging while children perform tasks likely to engage PFC regions. This strategy has been adopted by Betty Jo Casey and colleagues who use fMRI to compare children and adults in working memory and inhibition tasks. In one experiment they used event-related fMRI while children and adults were engaged in a 'go-no-go' task. In this task, participants had to suppress their response when presented with a particular visual item within an ongoing sequence of stimulus presentations. The difficulty of the task was increased by increasing the number of 'go' items that preceded the 'no go' character. Successful response inhibition was associated with stronger activation of prefrontal regions for children than for adults.

Also, while in adults the activation of some prefrontal regions increased with increasing numbers of preceding 'go' trials (consistent with increasing need for inhibition), in children the circuit appeared to be maximally active for all trial types. Along with the poorer behavioral performance of children in this and other inhibitory tasks, these findings suggest that the functional development of some PFC regions is important for the mature ability to inhibit prepotent tendencies. The greater activation seen in children will be discussed further.

An alternative approach to understanding the role of the PFC in cognitive development has been advanced by several authors who have suggested that the region plays a critical role in the acquisition of new information and tasks. By this account PFC involvement in the object retrieval tasks is only one of many manifestations of PFC involvement in cognitive change. From this perspective, the challenge to the infant brain in, for example, learning to reach for an object, is equivalent in some respects to that of the adult brain when facing complex motor skills like learning to drive a car. A concomitant of this general view is that the cortical regions crucial for a particular task will change with the stage of acquisition. Three lines of evidence indicating the importance of PFC activation early in infancy have given further credence to this view: (1) fMRI and positron emission tomography (PET) studies, (2) psychophysiological evidence, and (3) the long-term effects of perinatal damage to PFC.

The limited number of fMRI and PET studies that have been done with infants have often surprisingly revealed functional activation in PFC, even when this would not be predicted from adult studies. For example, in an fMRI study of speech perception in 3-month-olds, Christaine Dehaene-Lambertz and colleagues observed a right dorsolateral DLPC activation that discriminated (forward) speech in awake, but not sleeping, infants. Similar activation of DLPC was found in response to faces at the same age. While this is evidence for activation of at least some of the PFC in the first few months, it remains possible that this activation is passive as it does not play any role in directing the behavior of the infant. Two other recent lines of evidence, however, suggest that this is not the case.

While developmental ERP studies have often recorded activity changes over frontal leads in infants, some recent experiments suggest that this activity has important consequences for behavioral output. These experiments involve examining patterns of activation that precede the onset of a saccade. In one example, Gergely Csibra and colleagues observed that pre-saccadic potentials that are usually recorded over more posterior scalp sites in adults, are observed in frontal channels in 6-month-old infants. Since these potentials are time-locked to the onset of an action, it is reasonable to infer that they are the consequence of computations necessary for the planning or execution of the action.

Further evidence for the developmental importance of the PFC from early infancy comes from studies of the long-term and widespread effects of perinatal damage to the PFC. In contrast to some other regions of cortex, perinatal damage to frontal cortex and PFC regions often results in both immediate- and long-term difficulties. For example, Mark Johnson and colleagues studied infants with perinatal focal lesions to parts of the cortex in a visual attention task. Damage to parietal cortical regions would be expected to produce deficits in this task in adults, but only infants with perinatal lesions to the anterior (frontal) regions of cortex were impaired suggesting that these regions were involved to a greater extent in the task in infants than in adults.

In conclusion, the prolonged anatomical development of the frontal cortex has led some to characterize the functional development of this region in terms of the differential maturation of different areas. However, increasing evidence indicates that at least parts of the PFC are functional in the first few months, and that these regions may be important for the acquisition of new skills and structuring of others parts of cortex.

Conclusions

While, even in the domains selected for review, much research remains to be done, the results that have been obtained to date suggest a simple maturational account of the development of human brain function is unlikely to accommodate the wide ranging evidence for experience-dependent effects. Similarly, while there may be some fruitful parallels to be drawn between neural changes associated with complex perceptual and motor skill acquisition in adults, and changes in brain functionality during early development, several of the primary features of early functional brain development (such as the existence of sensitive periods and initial biases) do not sit well with this approach. According to the interactive specialization view, cortical regions become specialized for particular functions partly through an activity-dependent process extending from prenatal into postnatal life.

The cerebral neocortex appears to be on a slower developmental pathway than other regions of the brain. Further, this relative delay may be exacerbated in species such as our own that have a long gestation period. Subcortical regions such as the cerebellum, hippocampus, and thalamus clearly undergo some postnatal changes, and these may be, at least partially, a response to changes in their interconnectivity with the cortex. As a whole, the human cerebral cortex has not reached adult levels of specificity at birth, but it appears that some regions of cortex may be relatively delayed compared to others. This leads us to the further question of whether all domains of cognition follow the same timetable of cortical specialization.

Some domains of cognition, such as language, appear plastic in the sense that regions of cortex are not exclusively dedicated to them from birth, but other domains, such as face processing, may have fewer options. Less extensive plasticity does not necessarily imply strict genetic determinism, however, because functions more closely tied to sensory input or motor output are likely to be more restricted to the cortical regions that have the appropriate information in their input. For example, face recognition is necessarily restricted to structures on the visual 'what' (ventral) pathway because it requires both visual analysis and encoding of particular items within a category. Language may be less constrained in the sense that it is less restricted to particular information-processing routes within the cortex. Thus, a key point about the emergence of localization of functions within the cortex is that the restrictions on localization may be more related to which cortical routes of information processing are viable for supporting the functions, rather than being due to pre-wired intrinsic circuitry within regions of cortex.

During prenatal development, spontaneous activity in sensory systems appears to play an important role in contributing to the differentiation of cortical regions. In early postnatal life infants contribute further to the specialization of their brain by preferentially orienting and attending to certain types of stimuli, such as faces. Later, social experience and interaction with caregivers may contribute further to the specialization of late developing parts of the cerebral cortex. Much of later postnatal brain development, therefore, can be viewed as an active process to which both children and their caregivers contribute. Thus, studying the postnatal emergence of cortical specialization for different cognitive functions offers the possibility of new perspectives not only on the study of perceptual and cognitive development in healthy human infants, but also for social development, education, and atypical developmental pathways. The new theoretical and methodological advances of developmental neuroscience will allow these advances.

Acknowledgments

Sections of text in this article are adapted from Johnson (2005), and the author is grateful to his various colleagues and collaborators who commented on those works for their indirect contribution to the present article. The writing of this section was primarily funded by the UK Medical Research Council (PG 9715587) and Birkbeck, University of London.

See also: Brain Development; Brain Function; Face Processing; Hippocampus; Object Concept; Speech Perception.

Suggested Readings

Casey BJ and de Haan M (2002) Imaging techniques and their application in developmental science. *Developmental Science (Special Issue)* 5: 265–396.
de Haan M (2001) The neuropsychology of face processing during infancy and childhood. In: Nelson CA and Luciana M (eds.) *Handbook of Developmental Cognitive Neuroscience*, pp. 381–398. Cambridge, MA: MIT Press.
de Haan M and Johnson MH (2003) *The Cognitive Neuroscience of Development.* Hove, UK: Psychology Press.
Halit H, Csibra G, Volein A, and Johnson MH (2004) Face-sensitive cortical processing in early infancy. *Journal of Child Psychology and Psychiatry* 45:1228–1234.
Johnson MH (2005) *Developmental Cognitive Neuroscience: An Introduction,* 2nd edn. Oxford: Blackwell.
Kingsbury MA and Finlay BL (2001) The cortex in multidimensional space: Where do cortical areas come from? *Developmental Science* 4: 125–142.
Marcus GF and Fisher SE (2003) FOXP2 in focus: What can genes tell us about speech and language? *Trends in Cognitive Sciences* 7(6): 257–262.
Mareschal D, Johnson MH, Sirois S, Spratling M, Thomas M, and Westermann G (2007) *Neuroconstructivism: How the Brain Constructs Cognition.* Oxford: Open University Press.
Nelson CA (1995) The ontogeny of human memory: A cognitive neuroscience perspective. *Developmental Psychology* 31(5): 723–738.

Colic

D R Fleisher, University of Missouri School of Medicine, Columbia, MO, USA
R Barr, University of British Columbia, Vancouver, BC, Canada

Introduction

The term 'colic' implies abdominal pain of intestinal origin. However, it has never been proved that colicky crying is caused by pain in the abdomen or anywhere else. Although infant colic is not considered to be a functional gastrointestinal disorder, the abdominal pain attribution persists and pediatric gastroenterologists receive referrals of babies with refractory colic or infants who cry excessively due to unsuspected colic. Therefore, familiarity with the 'colic syndrome' is necessary for the avoidance of diagnostic and therapeutic misadventures.

Definition

Colic has been described as a behavioral syndrome of early infancy involving large amounts of crying, long crying bouts, and hard-to-sooth behavior. Although colic-like crying may occur in infants who are sensitive to cow's milk proteins, by definition, infant colic is not caused by organic disease. Infant colic was defined heuristically by Wessel as "paroxysms of irritability, fussing or crying lasting for a total of more than three hours per day and occurring on more than three days in any one week." Crying bouts start and stop suddenly without obvious cause and are more likely to occur late in the day. Colicky crying tends to resolve spontaneously by 3–4 months of age or, in the case of babies born prematurely, 3–4 months after term.

Normal infants cry more during the early months of life than at any age thereafter. T. Berry Brazelton studied crying in normal infants and found that, on average, crying peaks at about 6 weeks and then steadily diminishes by 12 weeks of age. Ronald Barr, another researcher in this field, confirmed Brazelton's data and concluded that the normal 'crying curve' of healthy infants is not the result of pain. Colic "is something infants do, rather than something they have," according to Dr. Barr.

Epidemiology

About 20% of infants are perceived by their mothers to be colicky by Wessel's criteria. However, the prevalence of infant colic is influenced by parents' perceptions of the intensity and duration of crying bouts, the method by which data on crying are collected, the psychosocial well-being of the parenting couple and culturally determined infant care practices. Barr found, in his study of caregiving practiced by Kung San hunter-gathers of the Kalahari Desert, that the frequency of onsets of crying conform to the Brazelton–Barr 'crying curve', but the amount of crying was much less than in Western cultures. This may be due to the almost continuous contact between mother and infant and the consistently prompt comforting responses provided to the infant within the family group. Apparently, babies around the world start to cry with similar frequency, suggesting that an inherent trigger for crying is 'built-into' all babies, regardless of their cultural origin. What varies is the length of crying after of onset of fussiness and the comforting response patterns from culture to culture and between parents within a cultural group.

Clinical Evaluation

Many disorders cause irritability and crying that can mimic colic, including cow's milk protein intolerance, fructose intolerance, maternal drug ingestion during pregnancy causing withdrawal irritability in the infant, infantile migraine, gastroesophageal reflux disease (GERD), and anomalous origin of the left coronary artery with meal-induced angina. The colicky crying pattern results from organic disease in 10% or less of colicky babies. Behaviors associated with colicky crying, for example, prolonged bouts, unsoothable crying, crying after feedings, facial expressions of pain, abdominal distention, increased gas, flushing, and legs over the abdomen, are not diagnostic clues indicative of pain or organic disease but they do explain and justify parents' concerns.

A presumptive diagnosis of colic can be made in any infant under 4–5 months of age whose crying has the temporal features of infant colic, who has no signs of central nervous system (CNS) or intrinsic developmental difficulties, is normal on physical examination, and has normal growth patterns. If necessary, it is reasonable to apply a time-limited therapeutic trials appropriate for each of the causes most frequently suspected. Switching to a predigested (hypoallergenic) formula or omitting milk and milk products from the diet of the mother who breastfeeds should result in a rapid and sustained remission of colic-like crying due to cow's milk protein sensitivity. A similarly time-limited trial of medication that suppresses the production of stomach acid should relieve crying caused by acid reflux. If these diagnostic possibilities are correct, an unequivocal improvement should occur within 48 h.

The satiated infant's response to nonanalgesic, nonnutritive soothing maneuvers (such as rhythmic rocking and patting 2–3 times per second in a quiet, nonalerting environment) may quieten the baby. The soothing may work for as long as it goes on, but the infant may resume crying as soon as it is stopped. Repeatedly demonstrating such an 'off–on–off–on' sequence caused by stimulation that could not eliminate pain but does quieten colicky crying has great diagnostic and therapeutic value. Similarly, some colicky infants are quieted during rides in a car. Parents report what has been called the 'red light–green light' phenomenon in such babies: the car rolls and the baby is quiet; the car stops for a red light and the baby starts to fuss; the light turns green, the car moves, and the baby quiets down again. Parents have reported that they approach distant red lights with deliberate slowness when their colicky baby is in the car. 'Off–on' and 'red light–green light' phenomena would not be expected to affect crying caused by bodily pain or hunger.

Physiologic Features

Significant differences have been found in comparisons of colicky infants and infants who did not cry excessively, such as increased muscle tone, heart rates during feedings, ease of falling asleep and soundness of sleep, stool patterns, postprandial gallbladder contraction, and other features. However, none of these findings have been shown to be more than epiphenomena or have provided a basis for successful treatment in 90% or more of babies with colic syndrome. In addition, no differences between colicky and noncolicky infants were found with respect to the activity of the gastrointestinal tract, indicators of intestinal disease, amounts of gas within the intestines, or the frequency of flatulence.

Current evidence suggests that colicky crying is behavior originating in the brain rather than the gastrointestinal tract. Colicky babies have been shown to have different temperament characteristics. Another hypothesis for the genesis of colic is based on differences in infants' reactivity (i.e., the excitability and/or arousability of behavioral and physiologic responses to stimuli) and infants' inherent ability to self-regulate responses to stimuli and benefit from externally applied soothing procedures. Colicky infants seem to have more difficulty with state transitions. For example, a noncolicky baby may feel sleepy, be put down, perhaps fuss briefly, but easily makes the transition from being sleepy to being asleep. A noncolicky infant may feel hungry, fuss briefly, be offered a bottle, feed well, and enter a state of calm satiety. A colicky infant, by comparison, may feel hungry, fuss, and be offered a bottle; he begins to suck, but soon stops, stiffens, arches, and begins crying. He is still hungry and likes his food, but he cannot make the transition to comfortable feeding and relief of hunger. Colicky infants may be difficult to feed, especially if feeding is attempted in an environment that has distracting sights and sounds or by a feeder who is exhausted and tense. Infants with 'active temperaments' are keenly receptive and intensely reactive to stimuli in their environment and this characteristic makes transitions to states of greater comfort more difficult for them as indicated by 'crying for no good reason'.

Psychologic Features

Understanding infant colic requires an appreciation of the subject experience and development of the infant, the mother, their dyadic relationship, and the family and social milieu in which they exist.

At about 2–3 months of age, normal infants become more attentive, socially responsive, and aware of the distinction between 'self' and 'other'. They become better able to sooth themselves and interact and give pleasure to their caregivers. This developmental shift occurs at about the age that colic subsides. These developmental advances are smoother if the infant's temperament is easy, the mother is caring, intuitive, and self-confident, and if the dyadic relationship between them proceeds with smooth reciprocity.

Parents usually have conscious and unconscious ambivalence toward their infant. If the infant is not fussy or difficult to regulate, and if the circumstances of their lives are pleasant, positive feelings predominate and family life is happy. However, if the infant is colicky, resentful feelings may rise to the surface of the mother's awareness. Recognition of angry feelings toward her own infant triggers anxiety and guilt which may prompt her to intensify her efforts at being 'a good mother'. If she is unsuccessful at controlling her baby's crying, her guilty anxiety and her reaction to it may develop into a vicious cycle causing profound physical and emotional exhaustion. This is made more likely when the mother's relationship with her partner is unsupportive. This stressful state impairs her ability to sooth her infant and causes her to doubt her competence as a mother. The emergence of adversarial or alienated feelings toward the unsoothable infant lowers the threshold for abuse. Infant colic may then present as a clinical emergency. Even in noncritical cases, excessive crying may be associated with transient developmental delay in the infant and family dysfunction 1–3 years after the infant's birth.

Management

Any measure that parents perceive as definitely helpful is worth continuing, provided it is harmless. If there is a question of milk intolerance or acid reflux as causes of crying, a time-limited therapeutic trial of a hypoallergenic formula or medication to suppress stomach acid production is warranted. Relief in such cases should become apparent within 48 h. However, in more than 90% of cases, management consists not of 'curing the colic', but of helping parents get through this challenging period in their infant's development. There are at least 12 elements to consider in helping parents of a colicky infant, as shown in **Table 1**.

A problem that has been given too little attention is 'persistent colic', for example, the 9-month-old who has not grown out of colicky crying and is still difficult to feed and keep asleep. The following hypothesis is offered as an explanation. An overwrought infant during the first 3 months needs help in developing self-regulation. Responsive soothing prevents the onsets of fussing from progressing to hours-long periods of crying. Parents expect colic to subside by about 3 months and when it does, expectations are fulfilled and the care of their infant becomes a lot less arduous. In the case of the 9-month-old infant who still does colicky crying, the infant is mature enough to be capable of self-regulation. However, his parents are exhausted, somewhat desperate, and fearful that something potentially tragic affects their baby because he is still crying months beyond the age it should have ceased. By this time, the parents' interaction with their infant is dyssynergic, that is, the reciprocal quality of their interaction has been lost and caregiving tends to increase, rather than relieve tension. The author has hospitalized such infants in a quiet private room devoid of colorful or noisy or otherwise

Table 1 Elements to consider in helping parents of a colicky infant

- A painstaking history that elicits a detailed picture of the baby's symptoms is important; superficial assessment and casual reassurance are not helpful to worried parents. A clinical interview can explore conditions of family life, past, and present, that may impair coping.
- Although colicky crying is seldom caused by disease, it is valuable to acknowledge the severity and importance of the stress that parents experience and how disruptive the crying is to family life.
- It is helpful to schedule a consultation during a time when the infant is likely to be fussy so that the clinician can experience first hand what the parents have been going through. This is also an opportunity to observe parents' attempts at soothing as well as the infant's soothability as he/she attempts to quieten the baby.
- A thorough, gentle physical examination by a diligent, open-minded physician is necessary to search for organic disease (the parent's chief concern).
- Consider 'an alternative to the pain hypothesis' for colicky crying, namely, that crying is a manifestation of normal development in infants with active temperaments who have more difficulty with 'state transitions' during the early months of their lives.
- A colicky baby taxes even the most experienced, devoted parents who fail to sooth because of their own state of exhaustion. Infants sense parents' tension and react to it with more crying. Parents need to know this, especially if they are attempting to double and re-double their efforts while ignoring their own needs for sleep and respite.
- If the infant is healthy, confirm it unequivocally to the parents, along with the realistically optimistic outlook for subsidence of colic by 3–5 months of age.
- It is crucially important to identify methods for calming the infant, at least temporarily. Review and demonstrate the list of common techniques, such as rocking and patting, secure swaddling, rhythmic rolling back and forth in a pram, car rides, pacifiers, or monotonous noise. Because crying bouts gain momentum rapidly, they are easier to stop if soothing maneuvers are applied promptly.
- Management must be individualized. Find out what has worked in the past and what is easiest for each family. Then, support them in doing it their easy way. Avoid stock recommendations regarding feeding, burping, or holding techniques, especially if they may increase the infant's or the mother's stress. For example, burping after every ounce is a recommendation based on the unsubstantiated notion that swallowed air causes colic. Actually, such repeated interruptions make feedings frustrating for both infant and mother.
- Parents of colicky babies often experience feelings of hostility and rejection toward the baby they love and want, whether they are aware of them or not. The more conscientious the parent, the more prone he or she is to self-reproach. The clinician should reflect upon 'irrational parental guilt', how 'normal', if not unavoidable, it is, and (in the words of Dane Prugh) "incise and drain it!"
- Parents' needs must be recognized. Many minimize or deny their anguish and fatigue. They need 'time off', that is, scheduled times when they can withdraw from caring for their infant, leave the house, indulge in rest or recreation, and return to their baby refreshed. Such free time is helpful provided it is scheduled in advance and a competent surrogate caregiver is available. In addition, mothers need a 'rescue' arrangement, a pre-arranged contingency plan whereby a trustworthy relative or friend can promptly take over, should the mother suddenly feel overwhelmed. The more confident the parents are that help is accessible, the less vulnerable they feel and the less likely they will need a rescue. Parents of nocturnal criers need sleep. They might divide the night into two 4 h shifts. The parent who is 'off' can continue sleeping and the parent who is 'on' knows that, when his or her shift is over, sleep is guaranteed. Four hours of guaranteed sleep is likely to be more restful than 8 h of apprehensive dozing in anticipation of the next crying bout.
- The physician's promise to remain available enables parents to continue to cope with their colicky infant without turning to unnecessary diagnostic medical procedures or false 'cures'. They need to know that some intercurrent ear infection has not supervened, or if it has, that it will be treated without delay.

distracting stimuli. The mother is relieved of having to care for her baby by herself so that she can rest or recreate herself away from her baby for at least some of the time. Feeding and other infant care procedures are done by a sensitive, comfortable, nonjudgmental nurse. Although the infant is responded to promptly, he is allowed to sleep and is protected from procedures that cause pain of frustration. Usually, a surprising change becomes apparent within 24–48 h in the form of much less crying and easier feeding. Diagnostic interviews with the parents are carried out to elucidate sources of stress in their family life and to dismantle the tendency to self-blame. If this hypothesis is correct, it implies that young colicky babies need very attentive holding and other comforting maneuvers, while persistently colicky older babies benefit from being shielded from overly attentive holding during times they need the peace and quiet that permits them to regulate themselves and relax.

See also: Feeding Development and Disorders.

Suggested Readings

Barr RG (2001) 'Colic' is something infants do, rather than a condition they 'have': A developmental approach to crying phenomena, patterns, pacification and (patho)genesis. In: Barr RG, St. James-Robert I, and Keefe MR (eds.) *New Evidence on Unexplained Early Infant Crying: Its Origins, Nature and Management*, pp. 87–104. Johnson & Johnson Pediatric Institute.

Barr RG, Konner M, Bakeman R, and Adamson L (1991) Crying in !Kung San infants: A test of the cultural specificity hypothesis. *Developmental Medicine and Child Neurology* 33: 601–610.

Brazelton TB (1962) Crying in infancy. *Pediatrics* 29: 579–588.

Fleisher DR (1998) Coping with colic. *Contemporary Pediatrics* 15(6): 144–156.

Liebman W (1981) Infant colic: Association with lactose and milk intolerance. *JAMA* 245: 732–733.

Murray L and Cooper P (2001) The impact of irritable infant behavior on maternal mental state: A longitudinal study and a treatment trial. In: Barr RG, St. James-Roberts I, and Keefe MR (eds.) *New Evidence on Unexplained Early Infant Crying*, pp. 149–164. Johnson & Johnson Pediatric Institute.

St. James-Roberts I (1997) Distinguishing between infant pussing, crying, and colic: How many phenomena? In: Sauls HS and Redfern DE (eds.) *Colic and Excessive Crying – Report of the 105th*

Ross Conference on Pediatric Research, pp. 3–14. Columbus, OH: Ross.

Stifter CA and Bono MA (1998) The effect of infant colic on maternal self-perceptions and mother-infant attachment. *Child: Care, Health, and Development* 24: 339–351.

Treem WR (1994) Infant colic, a pediatric gastroenterologist's perspective. *Pediatric Clinics of North America* 41: 1121–1138.

Critical Periods

D B Bailey, RTI International, Research Triangle Park, NC, USA
J-L Gariépy, The University of North Carolina at Chapel Hill, Chapel Hill, NC, USA

Glossary

Canalization – Process by which the initially great range of potential narrows as a result of differentiation and the acquisition of structure and function.

Critical moment – A term usually used by cell biologists to refer to a precise moment in cell division and differentiation.

Critical period – A clearly defined period of time in the development of an organism in which an event or experience must occur in order for to it to have its greatest impact.

Experience-dependent – Process taking place throughout life by which neural nets are formed in response to idiosyncratic stimulation to facilitate individual adaptation to new environmental challenges.

Experience-expectant – Process taking place during a critical period when neural nets are formed in the presence of stimulation that is ubiquitous in the environment and common for the species (see also 'induction').

Facilitation – The growth or consolidation of an already developed structure, function, or behavior accelerated by stimulation or experience.

Induction – A developmental event (appearance of a new structure, function, or behavior) whose occurrence during a critical period is entirely dependent upon the presence of a specific stimulation.

Maintenance – Continuous stimulation at a certain level is necessary to maintain a fully developed capacity at an optimal level of function.

Plasticity – The ability of an organism to change its phenotype in response to changes in its environment.

Sensitive period – A more loosely defined period of time in which an event or experience is more likely to have a strong effect than another time.

Introduction

A critical period generally is considered to be a point in the life of an organism in which a specific type of environmental experience is likely to exert its greatest influence. The existence and nature of critical periods has been a topic of considerable discussion among both biologists and social scientists. Research on critical periods has come primarily from the fields of embryology, neurobiology, and ethology, relying on both naturalistic and experimental studies with animals. Considerable progress has been made, but much remains to be learned both about the existence of critical periods and the basic underlying mechanisms by which they occur.

Critical periods in human development have been difficult to study scientifically, as it is virtually impossible to perform experiments on humans to prove or disprove the presence of critical periods or the effects of variations in timing of basic sensory experiences. Recently, however, researchers have used naturally occurring instances of neglect or abuse to document their effects on the developing child. As we discuss some of these cases, it will be useful to keep in mind that unlike animal studies where presumed operative factors can be brought under strict experimental control, these studies only provide correlative evidence, albeit quite compelling in some cases. In general, however, strong bases for practical application of critical periods to justify social policy initiatives (e.g., early intervention programs for infants and toddlers) or specific educational practices or experiences (e.g., early exposure to certain types of music or to a second language, based on the assumption of critical periods for musical development or second language acquisition) are still largely lacking. Typically, this debate pits the notion of a permanent deleterious effect of missed opportunities against concepts of plasticity (the possibilities for learning and brain reorganization to accommodate to new situations, even when a critical period has been passed). Despite the paucity of human research, some important conclusions can be drawn about the timing of experiences in promoting optimal infant and early childhood development.

Basic Concepts of Critical Periods

History and Terminology

In general, a critical period might best be considered a hypothetical construct to help us think about the relative power of an experience at one time vs. another. In its most classic and restrictive sense, a critical period is a point in development where a particular experience is absolutely necessary. If the experience is provided, then it allows for development to proceed normally. If the experience does not occur within the defined critical period, the organism is irrevocably damaged, or at least limited in possibilities for future growth, and it is virtually impossible to recover, even if comparable experiences are provided later. In a similar vein, exposure to environmental toxins or negative life experiences (e.g., trauma or abuse) have also been studied from a critical periods perspective. Here the question is whether the toxin or experience has an especially damaging effect during a particular age or period of development.

The concept of critical period was first articulated in the 1920s by Charles Stockard, who showed that birth defects in fish embryos, resulting from extreme temperatures or toxic chemicals, were more likely to occur or to be more serious during a period of rapid cell growth. Since the defects were less serious or in some cases did not appear if the embryos experienced toxic exposure before or after this period of rapid cell growth, he described this phenomenon as a critical moment. A decade later, Hans Spemann extended this concept to what we now refer to as stem cells. Stem cells are cells derived from embryos that are unspecialized. That is, they are not associated with a particular part of the body and therefore cannot perform any specialized functions. However, they are capable of dividing and thus replicating themselves for an extended period of time. At some point in development, likely as

a result of complicated interactions among genetic, biochemical, and environmental influences, cells begin to become differentiated, a process whereby a stem cell acquires a specialized function, for example, as a heart, liver, or bone cell. Spemann used the term 'critical moment' to describe the point of differentiation. Before the critical moment, the cell has an open future, with the potential of becoming a variety of cell types. After the critical moment, the cell becomes differentiated, acquiring a specialized function. Once differentiation occurs, the cell has a constrained future, as differentiation is irreversible. In other words, once a stem cell becomes a heart or liver cell, it loses its potential for any other function.

Moving the concept of critical moments from the cellular level to the behavioral level was advanced by studies conducted by Konrad Lorenz, beginning in the 1930s. Lorenz observed that baby ducklings learned to recognize their mothers through visual and auditory exposure. Once this occurred (a process referred to as imprinting), the ducklings demonstrated that they had learned this distinction by following their mothers but not other adults. Lorenz systematically exposed ducklings to their parents, other animals, or even to humans or moving mechanical objects, varying the timing and duration of these exposures. He found that there was a very short window of time, between 9 and 18 h of age, during which imprinting was most likely to occur. Ducklings would imprint on, and subsequently follow, whatever moving object they saw during this period of time. This led to amusing situations where baby ducks were seen to prefer following humans, other animals, or even mechanical objects if they were exposed to them during what he called a critical period.

Perhaps the most well-known example of a clear critical period is the classic research conducted by David Hubel and Torsten Wiesel in the early 1970s. This research was based on the fundamental assumption that the development of the visual system depends on visual input from the environment. The discovery that normal biological development is dependent upon experience was important, but the key finding for critical periods was that the experience has to be provided during a particular time in development. Hubel and Wiesel deprived kittens of sight (by surgically closing one eye) at various periods of time during early development, systematically varying the timing and duration of visual deprivation. When the kittens were deprived of sight for 65 days, starting at 10 days of age, there was a massive reduction (from 98% to 16%) in the number of brain cells that responded to visual stimulation once the eye was opened. Essentially the kittens were blind in one eye and no amount of subsequent visual stimulation was able to allow the kittens to recover sight. But if the visual deprivation occurred before or after this period of time, the effects were less severe and, by a certain age, minimally damaging. The lack of an essential experience during a critical period of development prevented normal brain development from occurring and had a permanent damaging effect. Hubel and Wiesel preferred the term 'sensitive period' to 'critical period', however, implying a longer period of time with less well-defined boundaries.

Since these early experiments scientists have continued to look for critical periods in a number of different animals and for a range of behaviors. One example in which much research has occurred is the development of birdsong, which has been studied in a number of different species. This research shows the importance of young birds, being exposed to adult birdsong during the first year of life. Young birds who do not get this feedback produce abnormal songs as adults, even if exposed later to birdsong that is normal for their species.

Thus the history of critical periods is deeply rooted in embryology, ethology, and neurobiology, based on experimental manipulations of animals during key periods in development. A variety of terms have been used to describe the phenomenon whereby an event or stimulus has more effect at one time during development or another. Three terms have been used in the scientific and popular literature, terms that range in specificity and subsequent implications:

1. *Critical moment.* A term typically used by cell biologists to refer to a precise moment in cell division and differentiation.
2. *Critical period.* A clearly defined period of time in which an event or experience must occur in order for it to have an effect.
3. *Sensitive period.* A more loosely defined period of time in which an event or experience is more likely to have an effect than another time.

How Might Critical or Sensitive Periods Work?

If one accepts the assumption of the existence of critical or sensitive periods, the next set of questions regards their nature. What are the mechanisms by which a critical or sensitive period might operate? There has been an explosion of research in neuroscience and genetics in the past decade, and much has been learned about basic neurocognitive processes. However, we are far from a complete understanding of the mechanisms by which critical or sensitive periods operate.

While a definition of critical or sensitive periods as points in time when a specific stimulation is absolutely necessary for normal growth to occur has descriptive validity, it does not tell us why that stimulation should be necessary in the first place. In the 1983 edition of the *Handbook of Child Psychology*, Gilbert Gottlieb offered the following explanation: critical periods are points in time when novel structures are differentiating from an

undifferentiated mass of cell. For that process to lead to species-specific outcomes, species-specific forms of stimulation must be encountered that are capable of bringing about the expression of the right configuration of genes in a timely manner. As Mae Won Ho and Jablonka observed, critical periods exist because natural selection does not select only for genes but genes along with a species-specific developmental context that includes endogenous and exogenous stimulus events. Accordingly, normal genetic expression depends upon the predictable and timely recurrence, generation after generation, of these stimulus events. On this view, there is as much information for development in the species-specific developmental context as there is in species-specific genes. Failure to encounter this information in a timely manner can potentially derail development from its normal course.

A compelling example is Gottlieb's own research on the origin of the capacity of the young Mallard duck to discriminate and to respond to the 'maternal assembly call' of its own species right after birth. A developmental analysis of this so-called instinct led to the discovery that depriving the developing embryo of hearing its own voice 3 days before hatching (by applying a biodegradable glue on its vocal chords) completely prevented the establishment of the postnatal capacity to selectively respond to the maternal assembly call of its own species. In order to determine which specific acoustic features of the self-produced prenatal vocalizations may have this inductive effect, Gottlieb conducted experiments by which he was able to determine that it was the rate of repetition of the prenatal vocalizations, not their pitch or absolute frequency, that had this inductive effect. More specifically, he was able to show that it was prenatal exposure to the natural variability ($2-6$ notes s^{-1}) around the average rate of the maternal call (4 notes s^{-1}) that was crucially important for developmental induction. Indeed, devocalized embryos exposed to an invariant embryonic call of 4 notes s^{-1} showed no postnatal preference for their species-specific maternal assembly call.

For the past few decades, explanations of critical periods have relied heavily on synaptic elimination and learning. A well-accepted theory supported by research is that during the early years of life there is a proliferation of these connections, more than is needed for successful human function. Naturally occurring events reinforce certain connections which are retained. Those not used are pruned, and new connections are less likely to develop after a certain period. If true, this theory provides a partial explanation for how a critical period might work and why it might be irreversible or at least difficult to reverse. If we start out with many connections, some of which are selectively kept as a result of experience and some eliminated, and if new connections are rarely made, then once the possibility of a connection is lost due to pruning (as is likely the case in the visual development of kittens whose eyelids were surgically closed for defined periods of time) it would be virtually impossible to regain it, unless some form of reorganization or compensation is possible.

William Greenough introduced in 1987 the concept of 'experience-expectant' processes. By this term he meant two things. First, during the period of 'brain growth spurt' that begins a few months before birth and lasts through the first 2 years of development, an overabundance of dendrites and synaptic connections are produced, 'in expectation' of the stimulation to come. Second, consistent with the concept of critical periods, for neural development to proceed normally, certain forms of stimulation must be encountered. This expected stimulation can be of a very general form, including stimulation that is often ubiquitous (e.g., patterned light) or inevitable (e.g., like exposure to gravity), but nonetheless absolutely necessary for normal development to occur. Stimulation that plays such a role is described by Gottlieb as having an inductive effect, that is, its absence always causes development to derail from its normal course. A striking example of induction is the development in the frog of neural connections between the muscles of the hind legs and the motor regions of the brain. At birth, about 200 neurons are present that 'compete' for establishing this connection. Through exercise, only one of these neurons survives in the adult animal. Depriving the young of exercise leads to a situation where many more neurons survive but none is capable of supporting the long leaps that frogs perform.

According to Gottlieb, next to induction, there are two additional roles that experience may play in development. One is that experience may have a facilitative effect. This effect is not as strong as the first because in this case normal development takes place anyway in the absence of stimulation, but its course may be accelerated when such stimulation is present. Research by Arnold Gesell illustrates this role well, showing that practice, while not necessary for the acquisition of walking, is speeded up when parents provide early and repeated opportunities for practice. The second role that experience may play is that of maintenance. Here a given function, such as vision, is fully acquired and mature in its organization but still depends upon a constant stream of stimulation to maintain its full functionality. This role is illustrated in deprivation experiments where subjects kept in the dark for extended periods report having to 'readjust' to natural light before they can see normally again. Although different in their effects, induction, facilitation, and maintenance jointly illustrate the general principle of bidirectionality between structure and function. As explained by Gottlieb, this principle implies that organs do not mature and acquire structure in a vacuum but do so guided by the exercise of their function.

The strengthening and pruning of synaptic connections can only be part of what is almost certainly a much more complicated story. First, in humans there is a relatively long period of time for synaptic development (likely continuing for at least the first 6 years of life) and synaptic pruning (likely a lifelong process). Lifespan perspectives on development show that learning occurs at all ages and thus our ability to define the onset and ending of critical periods in human development precisely is quite limited. Second, mechanisms that stimulate the onset or conclusion of a sensitive period are not well understood, and likely comprise a combination of genetic, developmental, biochemical, and environmental conditions. Third, the formation and pruning of synapses most likely occurs in the context of complicated hierarchies of neural circuits operating at different levels of complexity and integration. This is probably true even for simple behaviors, but the development of complex systems such as language or social development suggests the likelihood of multiple sensitive periods, one building on another at increasing levels of complexity and interdependence. In general, however, there is under natural conditions a 'trade-off' between the plasticity of the early stages and the eventual acquisition of functional organization. This process in which the acquisition of function entails the loss of initial plasticity has been called 'canalization' by Gottlieb.

Alternative Conceptions of Critical Periods

So far, our discussion of the concept of critical periods has been guided exclusively by a biological conception of the term. This was a good place to start because, as mentioned earlier in this article, it is in the context of developmental biological studies that this phenomenon was first observed and described. But concepts that originated in one discipline are often altered when other scientific disciplines borrow them. For a while psychologists have used this concept in ways that departed little from its original definition. But the range of phenomena of interest to them, belonging as they do to the behavioral and cognitive domains, has motivated other uses of the term, some of which are quite different from the original meaning. These alternative conceptions are seen especially in research on social development and educational psychology.

Consider, for example, the way in which the term 'critical period' was recently used by Doris Entwisle in an article where she reports on the effects of transition to formal schooling among poor African-American children. She described this event as a critical period because, for these children especially, the transition to school represents a time when factors outside the child become as important as factors inside the child. In other words, a shift is taking place during this transition in the locus of the parameters that control behavior and development. To illustrate this, Entwisle explains that African-American families often promote a conception of the self that is based on 'who you are' as opposed to 'how you perform'. By contrast, European-American families tend to create a context in which how you perform is an important factor in the acquisition of a sense of self. Thus, for European-American children, the same sense of self that promoted adjustment to the family environment now promotes integration to the school system. For African-American children, by contrast, school entry entails a discontinuity in how the self is being experienced and used to organize behavior. Here, a critical period is created by the fact that the transition to formal schooling entails a reorganization of the alignment between internal and external systems.

School entry is not necessarily a critical period for every child. The criticality of this period is not obtained by age alone or the onset of a specific maturational event, but through the fact that a particular person-situation context, often highly idiosyncratic, is being created. We illustrate this with another example from the psychology of education. Jackie Eccles showed that in the standard school system, the transition to middle school is often the beginning of a downward spiral for some adolescents that is characterized by low academic performance, low self-esteem, and school drop-out. By comparison, among the adolescents who experience the less traditional K-to-12 system, the observed proportion of these problems is much lower. She explains the difference between the two groups of students by the fact that in the first case, and to a lesser extent in the second, there is a mismatch between the needs of adolescence for belongingness, autonomy, and competence, and the large, impersonal, bureaucratic environments that inner city schools often create for their students. Another interpretation of her finding is that adolescence is a natural period of developmental vulnerability because the internal changes taking place within the individual demand a fresh alignment between biological, cognitive, behavioral, and socioecological systems. In this context, a K-to-12 system would reduce disequilibrium and personal distress by maintaining at a time of internal reorganization a familiar environment that is more likely to satisfy those basic needs of adolescence. Accordingly, how 'critical' the transition to adolescence (or to middle school) will be for each individual is not determined solely by entering adolescence *per se*, but depends upon the social circumstances in which this developmental event takes place for different individuals.

Other terms such as 'windows of opportunity' also appear in the popular literature. And the educational literature uses terms such as 'teachable moment' to refer to times when the power of a learning experience is maximized. By the time of school entry, for example, the young child is generally eager to learn what the school has to teach. For many children, the opportunity to learn the things that the adults know, like reading and counting, promises greater freedom and independence. Our best teachers know this

and take full advantage of the high motivation to learn. Again, this period is not defined maturationally but rather by an event whose timing is determined by society, and subject to variation across cultures. Other researchers have recommended the term 'optimal period' to characterize a preferred time in which interventions or experiences might have the greatest impact. All of these terms share the common assumption that the timing of experience is important but that its impact may be highly variable and more or less specific in its effects. Inherent in the concepts of critical moments and critical periods are the necessary nature of the experience during precise periods of time and the irreversibility of their developmental impact. While there is nothing of absolute biological necessity about the acquisition around 6 years of age of literacy and numeracy in our American culture, failing to do so at this age has developmental consequences.

Most scientists now prefer the term 'sensitive period' over 'critical period', since precise critical periods, with the possible exception of biological maturation, are difficult to prove and may be rare. Furthermore, new revelations about the plasticity of human development and the possibilities for dramatic new treatments that could repair damage previously thought to be irreversible have forced scientists to rethink the malleability of organisms in new and important ways. Less dramatic but equally important is the fairly recent realization that the brain retains throughout life the capacity to form new connections and to reorganize itself in response to new adaptive challenges. While the evidence comes mostly from research with monkeys and rodents, there is no reason to believe that the findings cannot be generalized to our species. William Greenough also coined the term 'experience-dependent' processes. With this term he meant to highlight the fact that while there may be critical periods for the acquisition of the neural substrates that support species-specific capacities, there seems to be no time limit for the organization of neural substrates supporting individual adaptations. According to Greenough, the essential difference between the two processes is that, in the first case, dendritic proliferation is guided by genetic activity and endogenous growth factors, while in the second, dendritic proliferation is triggered by external (or self-imposed) demands for the acquisition of new capacities, new adaptations, or new skills, hence the term 'experience-dependent'. In this second case, the process by which a functional neural substrate is eventually sculpted is essentially the same: overproduction, competition, pruning, and functional validation (through exercise), of the surviving connections.

There are reasons to believe that the likelihood of finding critical periods in development diminishes as an inverse function of the degree of complexity achieved in various species through developmental differentiation. According to this principle, the more development gives rise to a hierarchy of complex systems that provide for highly sophisticated levels of behavioral control, the less likely would be the existence of tightly specified critical periods. In this regard it is informative to know that in the wake of John Bowlby's attachment theory in the early 1970s – a theory that is based largely on the ethological framework and findings obtained in avian species and apes – many suspected the existence of critical periods for attachment in human development. Lorenz, after all, had demonstrated that chicks that do not encounter a moving object within a narrow period of time shortly after birth, or became imprinted to the wrong object, suffered irreversible consequences observable well into adult life. Klaus and Kennel were among the first to take upon themselves the task of documenting the existence of such periods for attachment to the mother in our species. To do this they took advantage of situations in which mothers had been hospitalized for varying lengths of time following the birth of their babies. If a strict critical period for attachment had been operating, babies of mothers with longer stays in the hospital should have shown increasing difficulties to bond to their mother. This is not what they found. To be sure, these babies needed a longer period of habituation to their mother to warm up to her. Once this initial period passed, however, they were perfectly capable of forming an affectional bond to her. Note also that the flexibility of the attachment process in our species is further demonstrated by the capacity of the human baby to form multiple attachments.

Implications of Critical and Sensitive Periods for Early Childhood Development

What are the general implications of research on critical and sensitive periods for infant and early childhood development? Three broad conclusions are important to recognize: (1) the existence of critical periods in humans is difficult to prove scientifically; (2) without proof of critical periods, timing of experiences is still a relevant concept; and (3) emerging research on human plasticity and future treatment opportunities may provide important insights into previously held assumptions about unalterable courses of development.

Discerning Critical Periods in Humans

We can conclude with confidence that critical or sensitive periods in their original biological sense do exist, but this conclusion is largely based on research showing that animals who have been deprived of naturally occurring experiences or exposed to a toxic stimulus show varying degrees of severity and permanence of impairment depending on the timing of those experiences in development. Researchers cannot manipulate human experiences

in the same way that they are able to manipulate animal experiences, and thus conclusions about critical periods in human development are more difficult to prove. It is possible, however, to approach this question by using natural occurrences of neglect, maltreatment, or exposure to noxious stimuli. In this section we examine what can be learned from studies of such occurrences. As a note of caution, let us mention at the outset that these are quasiexperimental studies, that is, the subjects in this kind of research are taken as they come, and without the possibility of controlling (as we do under laboratory conditions) for extraneous variables that may leave in place alternative explanations. To make the matter more complex, we would like, ideally, to determine whether critical periods exist not just in the formation of biological structures, but also in the developmental organization of adaptively significant behaviors, including as examples the formation of an affective bond with the caregiver, the acquisition of language, or the capacity to think abstractly. The added difficulty in these cases is that multiple neurobiological systems develop and build on each other over time to support these complex behaviors. In turn, this organization generates buffering or compensatory mechanisms not present in simpler systems or, for that matter, in species where development does not advance at that level of sophistication.

It is well known that the ingestion of lead is associated with impaired intellectual development in children. Does it matter when the exposure to lead occurs? An experiment to answer this question definitively for humans would require systematic exposure to varying amounts of lead for varying periods of time at multiple points in development. For obvious reasons, such a study would be unethical. The only way this question can be answered is by finding a large group of children who are diverse with respect to whether they were ever exposed to lead, when that exposure occurred, how much lead was ingested, and for how long. Precise estimates of all these variables are difficult to determine, since most of these data would have to be retrospective. Nonetheless, some research suggests that late in the prenatal period may be the most dangerous time for lead exposure, with effects that are more severe and more permanent than exposure at other times. Although not definitive and not a real test of a true critical period, findings such as these can help policy makers decide when and where to invest limited resources to maximize potential benefit or minimize potential harm. Of course, there is no period in development when lead exposure is good, but evidence showing that exposure during a particular window in development has much more severe or less reversible consequences than other windows would be quite useful.

Perhaps the best-known and least controversial example of a critical period in human development is the period during prenatal development when the embryo differentiates into a male if it possesses a Y chromosome or remains a female otherwise. This happens around the ninth week of gestation in our species when the Y chromosome becomes involved in the synthesis of testosterone and its release in a brain that is still undergoing differentiation and functional organization at a rapid pace. This inductive event (see Gottlieb's research mentioned earlier) is accompanied by the masculinization of the genital ridge and the brain as a whole. Not only does this transformation take place within a well-specified window of time, but the effects of testosterone release in the brain during this period are also irreversible. Once a male, always a male! We have here a very good example of a critical period at the level of biological structures, whose mechanism of action has been documented in other species, and which we know takes place via the same mechanism in our own because errors of nature have confirmed it.

Now let us consider possible critical periods at a level of organization that may be of more immediate interest to the parent, the teacher, or the psychologist. A number of recent inquiries into naturally occurring instances of early deprivation have begun to provide some answers. One of the most informative studies in this regard is the Bucharest Early Intervention Project conducted by Nathan Fox, Charles Nelson, Charles Zeanah, and Dana Johnson at the Universities of Harvard, Minnesota, and Tulane. In this project, children previously institutionalized in Romanian orphanages were placed (at 9, 18, 30, or 42 months of age) into high-quality foster care in the US and Canada and observed over a number of years for placement effects on their intellectual, social–emotional, and brain development. Compared to a noninstitutionalized group of Romanian children, those in their study were characterized by a variety of risk factors including alcohol and prenatal drug exposure, as well as abandonment, and social and material deprivation. Prior to their placement in foster care, these children were behind on virtually every measure of interest to a child developmentalist. The Bucharest Early Intervention group hypothesized that the effects of foster care placement on the various domains of developmental interest to them would fall into two broad categories. They expected developmental domains that are guided by an experience-expectant process to be quite resistant to improvements in the quality of care, while other domains, less restricted to a narrowly defined critical period, would follow an experience-dependent process and be more open to the therapeutic effects of foster care.

In general they found that the infants who were institutionalized for the shortest periods of time and who were placed early in foster care were those who showed the most rapid and complete recovery from exposure to early adverse conditions. Moreover, with the exception of physical growth, they observed improvements in virtually every developmental domain measured, including IQ, emotional expression, attention, and social development. It is perhaps through the patterns of emotional

attachment these institutionalized infants displayed that the effects of early deprivation were the most striking. Zeanah and his colleagues noted among these children the prevalence of a syndrome called 'reactive attachment disorder'. This syndrome consists of two opposing but equally maladaptive responses to social contact characterized by emotional withdrawal and inhibition in the first case, and by indiscriminate, disinhibited approach of strangers in the second case. Remarkably, the research team found that over the course of 2 or 3 years of foster care placement and the experience of sensitive, child-centered interactions, these atypical behaviors were progressively replaced by more organized attachment patterns (e.g., using the caregiver as a secure base for exploration). Equally striking was the fact that the physiological measures indexing cognitive and behavioral competence (e.g., magnitude and amplitude of brain activity) also showed the same pattern of improvement over time. In every case, the effects of intervention were a joint function of the age at which placement in foster care took place and how much time had elapsed since placement. Clearly, these effects show that the cognitive, behavioral, and social–emotional domains (including their neuronal support), quite unlike basic cellular processes of differentiation, all develop following an experience-dependent process. In other words, they are not fixed by stimulation encountered within a narrow window of time. In fact, the only developmental domain that appeared so constrained by early experience in this quasiexperimental research was physical growth, a biological process more intimately tied to cell differentiation and growth.

Another domain that may be amenable to a similar type of analysis is language development – when is it important for children to hear language during the early years of life? Again, for ethical reasons it would be impossible to systematically vary the time at which a child first heard language. However, it is possible to study the rare cases of children who were deprived of language under conditions of abuse or neglect. A well-known case study is that of Genie, a child who was discovered after 13 years of being placed by her father in complete isolation in the backroom of a house where she had been strapped to a potty chair, beaten, abandoned, and prevented from any contact with other humans. By the time she was discovered, Genie could not speak, presumably because she had never been spoken to. The fact that sustained attempts to teach her language failed (she had considerable language, but could not master the most elementary rules of grammar) was taken by the linguist Eric Lennenberg as evidence that there is a critical period for the acquisition of language, somewhere between 2 and 13, he claimed, after which the biological window closes, making it impossible to acquire language later. As popular as this case has been, it remains what it is: a single case. Moreover, as pointed out by Kevin McDonald at the California State University, Genie is also an account of trauma, abuse in the extreme, and attachment failure – all factors that may, just the same, explain her failure to learn language. Under any circumstance, it would be extremely difficult to determine whether or not there is a critical period for the acquisition of language. Exposure to language in our species is virtually inevitable. Consider, for example, what a normal mother who just gave birth does when the nurse places the newborn in her arms for her to hold: she places her child in a face-to-face position and talks to her. Isolating exposure to language while keeping other factors approximately constant would be very difficult indeed.

It is also possible to study children with hearing impairments who have received a cochlear implant, an electronic device surgically implanted under the skin of individuals with congenital deafness to stimulate the auditory nerve and provide some sounds that otherwise would be impossible to detect. Studies of children who have received cochlear implants at varying ages suggest that children who receive implants before 4–6 years of age have superior speech and language development than those who receive implants after 8 years of age. These data do not provide positive proof of the existence of a critical period, but do provide useful information for clinicians and parents about the optimal timing of exposure to sound and the best time to have this surgery performed. There is, however, more than exposure to sound that is at play in the acquisition of language. The symbolic (gestural) and interpersonal aspects must also be factored in.

Time and Timing in Development

Although full proof of critical periods in humans remains difficult to obtain, timing, without a doubt, is an important consideration in infant and early childhood development. A number of researchers have pointed out that most animal research on critical periods has focused on basic sensory processes such as vision, whereas most questions of timing in human development address much more complex behaviors. These complex behaviors are likely to have less well-defined, and probably less constrained, windows of time; in fact once a window has opened for a set of skills to be learned, it is likely to remain open for a very long time. However, there is considerable evidence that although a window for learning may be open for a long time, competent development of a skill builds on earlier experiences and learning may be more difficult as one gets older.

The development of literacy is a clear example of these points. Many nonreading adults can be taught to read, and thus there is not likely to be a critical period for learning to read, at least in the classic way that developmental biologists have defined critical period. However, it is much more difficult for an adult to learn to read and the reading fluency of adult learners is likely to always be at a lower level than those who learn to read earlier. Likewise, most first graders

learn to read. However, the ease of learning to read is enhanced by exposure to critical experiences during the early childhood years, such as rhyming games to build awareness of similarities and differences in sounds and early exposure to books and print materials. Thus, while the infant and preschool years may not be considered a critical period for literacy development, they pose a window of opportunity where exposure to certain types of experiences provides an important foundation that enables more complex learning to occur earlier and with greater ease.

In practice, the use of the term 'critical period' may not be accurate or useful in infant and early childhood development. And statements arguing that the first 3 years of life constitute a critical period are misleading and likely to evoke strong reactions from commentators who have a more specific definition of critical period. Nonetheless, we can convincingly argue that the first 3 years are foundational, and that appropriate experiences during this period of life make later learning easier and more efficient.

New Horizons for Human Plasticity

Some have criticized the concept of critical periods as painting an unnecessarily pessimistic view of the possibilities for change. Metaphors such as 'as the twig is bent, so shall the tree grow', suggest that certain bad experiences or the lack of key learning opportunities could permanently alter development or limit the possibilities for future growth. In reality, opportunities for learning and growth occur throughout our lives. Research on brain injury has shown how neural reorganization can occur, even in injuries sustained later in life. Neurobiological research is leading to greater understanding of learning processes, and it is possible that in the future targeted medications or gene therapy may reverse or enhance processes previously considered unalterable.

Research on stem cells could force further rethinking of the concept of critical periods, since it could lead to cell-based therapies that could repair or replace damaged tissue. For years scientists thought that only embryonic stem cells (those obtained from human embryos) had therapeutic potential because they were undifferentiated and had the potential to develop into any human tissue. Recent research suggests that adult stem cells, that previously were considered to be limited in potential therapeutic use to the tissue from which it came (e.g., bone marrow), might actually be able to lead to the creation of cells in a completely different tissue.

Three Examples of the Importance of Timing

Although critical periods may not strictly apply to early human development, timing is clearly important, and concepts such as foundational skills, windows of opportunity, and optimal times for learning are quite relevant. Three examples are presented to show the importance of timing of critical periods in early development.

Newborn Screening

Newborn screening is a public health program that rests heavily on the urgency of early treatments to prevent morbidity or mortality caused by a range of endocrine, metabolic, or genetic disorders. Newborn screening can be traced back to the 1960s when Robert Guthrie developed a screening test for phenylketonuria (PKU) using blood spots. PKU is an inborn error of metabolism in which the body is unable to process phenylalanine, a critical amino acid. If untreated, PKU results in mental retardation, small head circumference, and behavioral problems, signs of which begin to emerge by 6 months of age. However, PKU is easily treatable by a diet that dramatically lowers phenylalanine in food. To be maximally effective, the diet must be started early. Changing the diet at a later point in development does not undo the damage done if PKU is not treated early.

Today every state operates a newborn screening program for all newborns. All newborn screening is based on a fundamental assumption that early identification of children with selected conditions allows for treatments that must be provided early if they are to be effective. Interestingly, research on critical periods has not typically been used to justify newborn screening; rather the justification is generally on a condition-by-condition basis, showing the devastating consequences of the condition and the benefits of timely treatment. For example, children with galactosemia, a recessive genetic trait that emerges when two carriers have children, lack a liver enzyme required to digest galactose, most commonly found in milk products. A build-up of galactose in cells is toxic and can quickly result in liver disease, cataracts, mental retardation, or death, which can occur as early as 1–2 weeks of age. An early and immediate change in diet to a soy-based formula can prevent these consequences, although individuals with the condition must eliminate galactose and lactose from their diets throughout life. Thus timing of the identification and treatment of this condition is essential, a critical period of identification that must occur within the first few days of life. However, this is an example of a critical period in which the window for the period never closes.

Rapid changes in technology, new treatment options, and advocacy efforts have led to significant expansions in newborn screening. Technology will soon allow the possibility of identifying conditions for which there is no current treatment or for which treatment during the earliest weeks or months of life may not be critical. Debates about the desirability of expanded screening have already begun. Some have argued that no condition should be identified unless it meets a standard closely associated with critical periods, with proof that early identification and early treatment is necessary and would be less effective or even ineffective if provided later; others argue that a critical-period type standard is too restrictive, and that information about a condition has other potential benefits to families and society, even if no treatment currently exists. Important discussions are needed about the ethical, legal, and social issues that accompany rapid expansion of genetic knowledge and the benefits of newborn screening for conditions where the urgency of timing early treatments does not yet exist.

Early Education

In the 1960s, many psychologists, educators, and policy makers became concerned about the alarming rate of childhood poverty in the US and evidence showing the devastating effects of poverty on school achievement, psychosocial adjustment, and adaptation to adult life. Prominent psychologists such as Benjamin Bloom argued that since the early years were the time of greatest developmental change, this period provided the greatest opportunity for lifelong impact on development. Arguments such as these led to the establishment of Head Start (a national program of preschool education for young children living in poverty), early intervention programs (a national program of services for infants and toddlers with disabilities and their families), and a new generation of longitudinal research on the benefits of early education.

Recent follow-up studies of children participating in early intervention and preschool programs, conducted after these children had entered the early adult years, have provided strong evidence of the long-lasting benefits of early education. Children from low-income environments who participated in high-quality early intervention programs have better academic, social, and economic outcomes than children who did not participate in such programs. Data from studies such as these have provided strong support for those who advocate for initiatives such as expanding Head Start to include all eligible children, lowering the age of Head Start entry, improving quality of early childhood programs, or providing universal prekindergarten education.

But this research has not been able to answer the more fundamental question of whether the early childhood years constitute a critical period. Since some advocates have invoked the critical-periods argument in support of improving or expanding early childhood education programs, this predictably has elicited reactions from those who do not agree. Opponents argue that since no proof exists to support the general notion that the early childhood years constitute a critical period in development, increased expenditures for early childhood programs are not warranted.

This debate will never be answered conclusively. The question of whether the infant and preschool years constitute a critical period is too broad and too diffuse, and no amount of research could be done that would provide a 'gold-standard' scientific answer. But most scientists and advocates agree on two general points. First, windows for learning open at birth and remain open for a very long time, probably all of our lives. A task force on the science of early development sponsored by the National Academy of Sciences concluded:

> Early experiences clearly affect the development of the brain. Yet the recent focus on "zero to three" as a critical or particularly sensitive period is highly problematic, not because this isn't an important period for the developing brain, but simply because the disproportionate attention to the period from birth to 3 years begins too late and ends too soon.
>
> Shonkoff and Phillips (2000, p. 7)

It is difficult to argue that the early childhood years constitute a critical period, since much learning happens after this period. However, almost everyone agrees that the early childhood years are foundational. Recent publications have used the metaphor of 'brain architecture', suggesting that early childhood experiences have a uniquely powerful effect on the brain and on how brain development manifests itself in terms of later cognitive development and social processes. While claims of a critical period may be exaggerated, the early childhood period is the only time in human life that provides the foundation for all future learning.

Second, substantial data exist showing that variations in the quality of early childhood environments and experiences are associated with variations in later outcomes in school and postschool success. On the side of negative factors, clearly many children experience environmental circumstances (e.g., poverty, maternal depression, abuse or neglect, dangerous neighborhoods) that have adverse effects on academic and social development. As a result many children enter school not ready for the demands and expectations of the school environment. On the positive side, early intervention and preschool programs have been shown to improve outcomes, and high-quality programs result in better outcomes than poor-quality ones. Collectively these data show that what we do with children during the early years does make a long-term difference, irrespective of the proof of critical periods that provides sufficient evidence to support special attention to the early years.

Thus far our discussion of early education has focused on children at risk for compromised development, either as a result of poverty or as a result of a disability. A related dimension of early education has a different goal, namely the enhancement of accelerated or optimal performance by providing enriched opportunities for advanced skills

early in life. For example, a number of commercial products can be purchased to help parents maximize early development through exposure to a wide range of experiences. Two questions could be asked of these products or activities: (1) do they work? and (2) must they be provided early in life to have the greatest impact? Two specific areas in which the critical-periods question has been asked are early exposure to music and early exposure to a second language. The assumption underlying both of these is that the early years constitute, if not a critical period, an important window of opportunity to maximize the likelihood of mastering a musical instrument or learning a second language.

As with all areas of human development, it is virtually impossible to answer these questions from a critical-periods perspective. Consider the case of learning a second language. Research clearly shows that early exposure to a second language can have important short-term and long-term benefits, and many of the findings are, indeed, age-related. Of course many adults would be able to learn a second language. But research clearly shows that learning a second language is more difficult for an adult than for a child, it takes longer for the adult to learn the second language, and the adult is less likely to perfect the accents and the phonetic features associated with native speakers of the second language. Thus early exposure to a second language (and probably to music or lots of other experiences) is not a necessary requirement for success, but it can make it easier to attain success or mastery. Of course, many other factors also contribute to success, among which include genetic features, motivation, enjoyment, and the context in which the experiences occur.

Critical Life Events and Teachable Moments

The notion of critical or sensitive periods is based on the fundamental assumption of a time at which an organism is particularly sensitive to environmental input. But usually this is based on a developmental or age-based perspective – the critical or sensitive period begins when the organism has reached a biological stage in development where the organism is biologically ready and environmental input is most needed, thus grounding critical periods in a maturational perspective.

An alternative view builds on notions such as critical life events or teachable moments as critical times for proper environmental input. Rather than relying on maturation reaching a certain point before input becomes important, this perspective assumes that when certain events happen, those events provide a window of opportunity for maximal environmental influence, irrespective of maturational level. The organism's maturational level might determine the nature and complexity of the environmental input likely to be most beneficial, but has relatively little bearing on the need for this input.

Some educators or psychologists refer to these times as teachable moments, a time at which the individual needs or wants to learn something or receive some type of input to help cope with a situation. Many examples of these moments are evident in the daily life of any individual. A toddler wants a toy she cannot reach or a peer has just grabbed a toy from her. A preschool child picks up a picture book and, after turning a few pages, stops and looks intently at one particular page that has piqued his interest. A preschool child just learns that her grandmother has passed away. A 5-year-old walks into a kindergarten class for his first group educational experience. A middle-school child gets stumped in the middle of a complicated experiment. A high-school student loses a friend in a tragic accident or a parent walks out of the home. An adult needs to know how to insert artwork into an electronic slide presentation.

These events vary enormously in nature and magnitude, but they share a common feature in that they all represent times during which environmental input is needed at that moment in time. Input provided before or after the event might be helpful, but the time immediately following the event is the moment in which environmental input might be most effective. When two toddlers are arguing over a toy, it provides a brief but important window of time for an adult to help model and support appropriate ways to take turns, share, and communicate. The preschool child who shows interest in a book is much more ready for appropriate and supportive adult input to build early literacy skills than one who is busy working with clay. The first day of kindergarten is a time of special vulnerability; how adults and peers interact with that child and the child's feelings of competence and belongingness on that particular day or week will have a powerful impact on feelings about school for many days to come. In all of these events, the timing of experience is likely to be a key feature in learning or adaptation.

This might be one of the most useful ways to think about critical moments – experiences that create periods of vulnerability and openness to information, feedback, or support. Hopefully research in the coming decade can be more focused on the match between the needs of the organism and the timing of environmental input.

See also: Bilingualism; Brain Development; Head Start; Language Acquisition Theories; Teratology.

Suggested Readings

Bailey DB (2002) Are critical periods critical for early childhood education? The role of timing in early childhood pedagogy. *Early Childhood Research Quarterly* 17: 281–294.

Bailey DB, Bruer JT, Symons FJ, and Lichtman JW (eds.) (2001) *Critical Thinking About Critical Periods*. Baltimore: Paul H. Brookes Publishing.

Bruer JT (1999) *The Myth of the First Three Years: A New Understanding of Early Brain Development and Lifelong Learning*. New York: The Free Press.

Gottlieb G (1983) The psychobiological approach to developmental issues. In: Mussen PH (ed.) *Handbook of Child Psychology*, 4th edn. vol 2, pp. 1–26. New York: Wiley.

Greenough WT, Black JE, and Wallace CS (1987) Experience and child development. *Child Development* 58: 539–559.

Hensch TK (2004) Critical period regulation. *Annual Review of Neuroscience* 27: 549–579.

Knudsen EI (2004) Sensitive periods in the development of the brain and behavior. *Journal of Cognitive Neuroscience* 16: 1412–1425.

Knudsen EI, Heckman JJ, Cameron JL, and Shonkoff JP (2006) Economic, neurobiological, and behavioral perspectives on building America's future workforce. *Proceedings of the National Academy of Sciences USA* 103: 10155–10162.

Michel GF and Tyler AN (2005) Critical period: A history of the transition from questions of when, to what, to how. *Developmental Psychobiology* 46: 156–162.

Shonkoff JP and Phillips DA (2000) *From Neurons to Neighborhoods: The Science of Early Childhood Development*. Washington, DC: National Academy Press.

Crying

B M Lester and L L LaGasse, Warren Alpert Medical School of Brown University, Providence, RI, USA

Glossary

Acoustics – The science or study of sound or sound waves including the quality and characteristics of sound.

Formants – High-frequency regions in a sound spectrum that determine the characteristic quality of a sound.

Innervation – The amount or degree of stimulation by nerves.

Phonetics – The science or study of speech sounds including production, transmission, and reception.

Prosody – The intonation and melody patterns of an utterance.

Introduction

Within the first few minutes of an infant's birth, the physician examines the newborn and completes the Apgar score which includes a strong, lusty cry (respiratory effort) as a sign of the infant's health. Crying requires an infant to perform a complicated and sophisticated set of physiological activities that involve the brain and the respiratory, motor, and vocal systems. Crying helps physiology by increasing pulmonary (lung) capacity. But there is more to crying than meets the ear. Crying is a critical survival mechanism. Infants' cries attract responsible adults who provide basic needs for the otherwise helpless infant.

Cry as Language

Crying is an infant's first language. The word infancy comes from the Latin 'infans' which means speechless and, while infants do not have speech, cry is their language and how they communicate. Infant crying contains linguistically salient aspects of human speech that are physiologically based and adapted for communication. Human speech is divided into linguistic and paralinguistic components. The linguistic component refers to the elements that become syllables and words to be organized into phrases and sentences by rules of syntax.

Qualitative aspects of speech, the intonation patterns, inflection, stress, intensity or loudness, and general melody constitute the paralinguistic component. These so-called prosodic features of speech have their acoustic correlates in the timing (duration), intensity (loudness), and fundamental frequency (pitch) of phonation. It is these features that convey attitudes, and emotional states such as happiness or anger that enable us to recognize familiar voices, in short, what makes human speech sound human. Communication relies heavily on these prosodic features of speech which are basically independent of linguistic elements.

Prosodic features come first in the vocal repertoire of the human infant, the cry. Cry is all prosody. Infants communicate their feelings, their needs, and their wants through the prosodic features of pitch, loudness, melody, and intonation. It is the language the infant uses before words.

Crying and Evolutionary Biology

Crying is attention getting. Humans are particularly responsive to higher pitched sounds – the maximum acoustic response of the ear is above 800 Hz (cycles s^{-1}). Ambulance sirens get our attention because they are high pitched and because they change, they are dynamic. Likewise with infant cries. From an evolutionary point of view, we would say that infant cries are programmed to be at certain pitch levels as an infant's survival mechanism. Crying is an acoustical umbilical cord that keeps the infant close to the mother.

Distress Calls Among Mammals

Ethologists, who study animal behavior, call crying a proximity-promoting and -maintaining signal, that is, it encourages the mother to stay with and sooth her infant. There are many species that can immediately locomote or in some way follow their mothers at birth. But relative to other species, human infants have a long period of dependency. Many species including cats, bats, elephants, seals, and reindeer, and of course chimps, use a distress vocalization or cry to signal the mother when the infant is isolated, hungry, or cold. The prosodic features of speech described above are the major component of the vocalization of most mammals. Researchers have analyzed acoustic patterns and shown that nonhuman distress vocalizations show similar patterns to human infant cries. Infants depend on vocal communication to signal distress or otherwise summon the mother close. Crying turns out to be an exceptionally effective survival mechanism. It is an information transmission system that sends affective messages, for example, hunger, pain, and need for attention. Another example of how this remarkable system works comes from opera.

Attention Getting Potential for Sound Similar to Distress Cries

Opera singers project their voices so that they carry throughout a theater and can be heard without a microphone and above the orchestra. They are able to do this by using a technique called singer formant in which they shift their voice into a higher vocal register to which human hearing is more sensitive. This is the same vocal register as sirens and infant cries.

Crying as a Biosocial Phenomenon

Crying is a biological siren, alerting the caregiving environment about the needs and wants of the infant and motivating the listener to respond. Infant crying is a biosocial phenomenon. Crying provides information about the biological integrity of the infant and is a social signal that affects parenting. Thus, there are two aspects to the cry. There is the cry itself, which is innervated by the cranial nerves modulating the autonomic nervous system and signals emergency status. There is the salience of the cry to any potential caregivers in the environment, producing a visceral reaction that compels action. Infant cries have both infant and caregiver in a state of strong sympathetic nervous system activation. This state is commonly described as the fight or flight response, except in this case only the caretaker can take action. Bystanders may take flight or distance themselves from the crying infant (e.g., avoidance of crying infant in supermarket or airplane) while responsible potential caregivers fight to stop the crisis by alleviating infant's distress which turns off the siren.

Thus, cry is not just an infant behavior, but rather it is a part of a behavioral system in the human species that assures survival of the helpless neonate by eliciting others to meet basic needs. The process underlying the behavioral system can be disrupted in two general ways. First, the cry signal may be poor or atypical. Second, the caregiver may have atypical reactions to the cry including both under- and over-responsiveness. Either situation can compromise the effectiveness of the behavioral system. The most extreme case is shaken baby syndrome in which aversive cries trigger aggression toward the infant rather than toward the reason the infant is crying.

The Science of Infant Crying

History

Before computers, measuring acoustic cry features was difficult and imprecise. The first attempt at ascribing meaning to the acoustic structure of cry patterns was in 1838 by William Gardiner, who used musical scores to illustrate the cry of a spoiled child, as well as other human cries and nonhuman vocalizations. In 1855, Charles Darwin used photographs and line drawings to describe crying and other emotional expressions. The science of infant crying began in 1906 after the invention of the Edison phonograph when T. S. Flateau and H. Gutzmann recorded the vocalizations of 30 infants on wax cylinders and used a phonetic alphabet and musical notation to describe features of the cry. In 1942, G. Fairbanks applied acoustic methods to analysis of the pitch of infant hunger cries.

The invention of the telephone by Thomas Bell introduced the concept of separating the speech signal into different frequency components. In the late 1940s, the sound spectrograph was invented by R. K. Potter at Bell Laboratories, which is still considered to be the most important instrument in phonetics or the study of speech sounds. The sonograph is a mechanical device which produces pictures called sonograms, which are a visual tracing of various frequencies of speech. Soon after in 1951, A. Lynip reported the first spectrograms of infant vocalizations.

Sonograms

Sonograms were made by having pens make marks on pressure-sensitive paper. The marks represented regions of energy where sound was concentrated. The pressure-sensitive paper is calibrated in Hz (cycles s^{-1}) so when you play a cry, each dark strip means a cry frequency and the length of each mark is how long the cry sound lasted. The lowest strip is the fundamental frequency or pitch. These sonograms were somewhat like inkblots: a mix of fact and fantasy and open to interpretation. At best they provided descriptive information of visual patterns. Despite limitations, the sonograms could distinguish infants with various medical complications as shown in **Figure 1**.

The advent of high-speed computer technology has improved the science of cry analysis with enhanced quantitative methods. In the digital age, acoustical analysis is done by computer and measurement is very precise. Years ago, we developed a special computer acoustical analysis system just for infant cries. This is important because the vocal tract of an infant is shaped differently to that of an adult. In fact, the infant's vocal tract is similar to the chimpanzee. If computer algorithms are based on the adult vocal tract, there will be errors in the calculation of the acoustical data. The acoustical analysis we conduct is computed every 250 ms to 0.25 s. So, four times a second we get a complete acoustical profile of the infant: Four times a second was chosen because that is how fast the vocal cords can change. The acoustical data are summarized and we build an overall cry profile of the infant, called a criogram (**Figure 2**). A description of many of the measures shown in the criogram including amplitude or loudness (called energy in the criogram), phonation or voicing, dysphonation or noisy cries, hyperphonation or very high pitch, duration of cry sounds between breaths is found in **Table 1**.

Crying and Acoustics

Crying is, of course, sound, so ultimately cry analysis amounts to analyzing sound or acoustics. Sounds are composed of sound waves that vibrate at different frequencies. The science of acoustics breaks apart sound waves into their component frequencies. For example, when we play a C note on a piano, the piano key forces the hammer to hit a string that vibrates at a specific frequency producing the C sound. The infant's vocal cords are somewhat like piano strings; they vibrate and produce the sound that we hear as cry. But it is not just the vibration of the vocal cords that determine the sound of the cry. A C note on a piano sounds different than a C note on a violin even though the C sound vibration is at the same frequency. These notes sound different because they, like most sounds, are not composed of single, or sine, wave but of many waves vibrating at different frequencies called complex waves.

Acoustics does for sound what a prism does for light. In the same way that a prism shows us that the color we see is

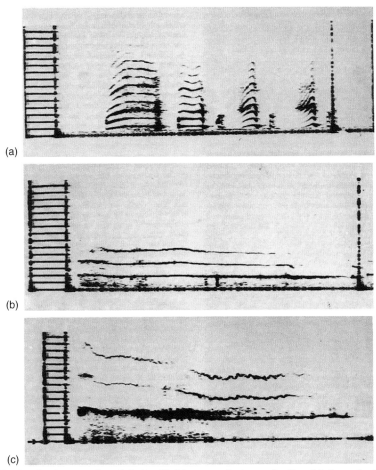

Figure 1 Sonograms of pain cries from newborn infants with (a) one medical complication; (b) six medical complications; and (c) the syndrome cri du chat (cry of the cat). From Zeskind PS and Lester BM (1978) Acoustic features and auditory perceptions of the cries of newborns with prenatal and perinatal complitions. *Child Development* 49: 580–589.

really composed of many colors at different frequencies of the color spectrum, acoustic analysis takes sound and shows its component frequencies. Complex sound waves are generated because the initial sound, the fundamental frequency, is affected by its surroundings. The C on the piano is produced in a different acoustical environment or resonating chamber than the C on the violin producing a complex waveform that makes these sounds unique. The resonating chamber is what changes the sound and gives it its unique qualities or richness. These are called resonance frequencies or formants and they refer to how the sound is modified, changed, and filtered by the resonating chamber.

The infant vocal tract is a resonating chamber. Sound generated by vibration of the vocal cords in the larynx or voice box vibrates at a particular frequency. The sound is modified as it travels through the throat, upper airway, and mouth producing a distinctive cry. Each infant's cry is acoustically unique. The fundamental frequency (f_0) is the basic pitch of the cry: the number of cycles s^{-1} (Hz) that the vocal cords are vibrating. Most infants have an f_0 around 250–450 Hz. Their vocal cords are vibrating

300–400 times s^{-1}. The sound waves become more complex as the sound travels up the airway producing concentrations of sound energy or clusters of frequencies at higher levels called formants. There are formant frequencies around 1200–1400 Hz (first formant, F1) and 3200 Hz (second formant, F2) produced in the upper airway. A description of cry characteristics is found in **Table 1**.

Anatomy and Physiology of Cry

This section has more detail on how the vocal tract and the brain produce a cry. Infant crying comprises a rhythmic alternation of cry sounds (utterances) and inspiration. Crying is part of the expiratory phase of respiration with sound or phonation produced by the larynx, which is identified in **Figure 3(a)**, and contains the vocal cords or folds and the glottis (opening between vocal folds). The larynx has three functions: swallowing (a flap called the epiglottis covers the opening of your larynx to keep food out of the wind pipe), breathing (glottis is fully open), and

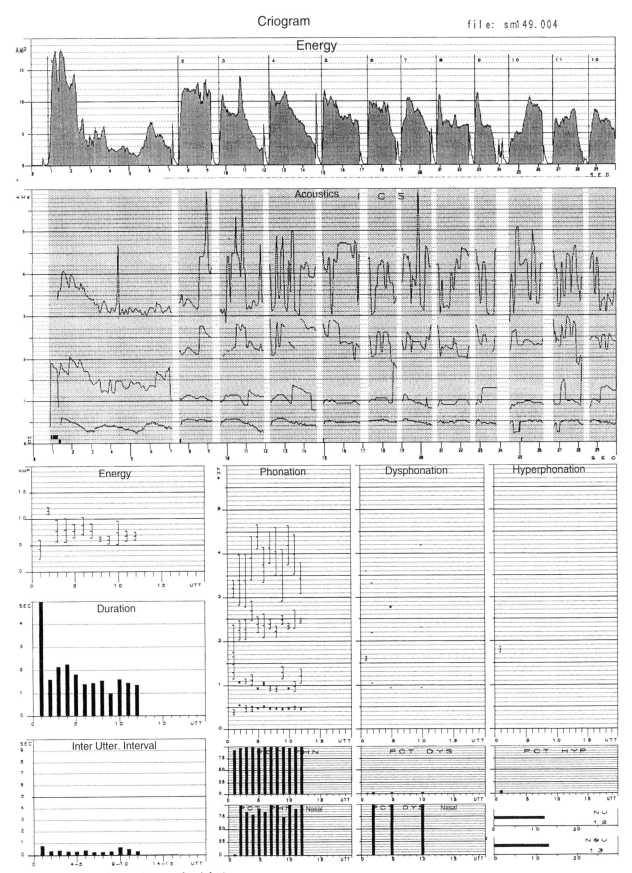

Figure 2   Cry profile or criogram of an infant cry.

Table 1   Cry characteristics and biological mechanisms

Characteristic	Definition	Biological mechanism
Cry latency	Time from known stimulus to onset of the first utterance (cry sound)	Arousal from limbic–hypothalamic system
Threshold	Number of applications of stimulation to elicit a cry	Arousal from limbic–hypothalamic system
Utterances	Number of cry sounds across cry	Neural control of respiratory system
Short utterances	Number of unvoiced sounds across cry	Unstable respiratory control
Phonation	Cry mode resulting from vocal fold vibration between 350 and 750 vibrations or cycles s^{-1} (Hz)	Neural control of muscular tension in vocal folds and air flow through the glottis
Hyperphonation	Cry mode caused by a sudden upward shift in f_0 to >1000 Hz	Neural constriction of the vocal tract
Dysphonation	Cry mode caused by noisy or inharmonic vibration of the vocal folds	Unstable respiratory control
Cry mode changes	Number of times cry modes change during an utterance	Instability in neural control of the vocal tract
Fundamental frequency (f_0)	Base frequency during vocal fold vibration, heard as the pitch of the cry	Vagal input to larynx and lower vocal tract
First formant (F1)	Frequencies centered at first resonance of f_0, approximately 1100 Hz	Neural control of size and shape of upper vocal tract
Second formant (F2)	Frequencies centered at second resonance of f_0, approximately 3300 Hz	Neural control of size and shape of upper vocal tract
Duration	Time (ms) from onset to offset of cry utterance	Neural control of respiratory system
Duration of inspiration	Time (ms) between cry utterances or interutterance interval. Breath holding is evaluated by the second inspiratory period	Neural control of respiratory system
Amplitude	Intensity or amplitude of the cry (dB). Heard as loudness	Neural control of respiratory system and capacity
Variability in f_0	Changes in f_0	Instability in neural control of the larynx and lower vocal tract
Variability in F1, F2	Changes in formant	Instability in neural control of upper vocal tract
Variability in amplitude	Changes in intensity or loudness within an utterance or averaged across utterances	Instability in neural control of the respiratory system

From LaGasse LL, Neal AR, and Lester BM (2005) Assessment of infant cry: Acoustic cry analysis and parental perception. *Mental Retardation and Developmental Disabilities* 11: 83–93.

voice production (glottis is closed). When air is forced through adducted (closed) vocal cords, the increased air speed due to passage through a constricted tube (Venturi tube effect) results in a drop in air pressure (the Bernouilli principle), causing the vocal cords to open and close rapidly (approximately 250–450 Hz in normal, healthy newborns). This vibration is the f_0 and is heard as the pitch of the cry.

The lower vocal tract (below the larynx and glottis), which includes the lungs and trachea (windpipe), is closely associated with the autonomic nervous system. In addition to f_0, the lower vocal track produces sound characteristics such as amplitude or loudness, and the rhythm of expiratory cry sounds and inspiration (inhalation as well as breath holding). The sound from the larynx and lower vocal tract is modified by the size and contour of the upper vocal track, which consists of the nasal cavity, the mouth, and the pharynx or throat also shown in **Figure 3(a)**. The upper vocal tract shapes the sound to produce resonant frequencies or formants (F1 and F2), which are frequency bands above f_0. These formants are similar to a sound chamber of a musical instrument, which makes it easy to distinguish, say, a C note from a piano vs. a guitar.

In addition to the infant vocal tract, **Figure 3** also shows the vocal tracts of adults (b) and chimpanzees (c). As seen in the figure, the infant vocal tract is smaller than the adult vocal track. Further, the shape of the newborn's vocal tract is more like a chimpanzee than a human adult with the larynx positioned higher in the vocal tract. Subsequently the larynx drops to the adult position in the vocal track beginning at 6 months and achieving adult positioning by 2 years of age.

The newborn infant's vocal tract, like the chimpanzee, has another capacity that a human adult does not. The chance of choking on food is much greater in adults than for infants who can easily breathe and feed at the same time. Infants can do this because of the high position of the larynx as well as a biological lock that forms between the soft palate at the back of the infant's mouth and the epiglottis or the flap that covers the trachea or windpipe during swallowing. This lock makes the infant's respiratory pathway (nose to lungs) completely separate from the alimentary pathway (mouth to stomach via the esophagus). The unique capacity for speech in humans is related to the lower position of the larynx. The cost for this capacity is greater vulnerability for inhaling and choking on food.

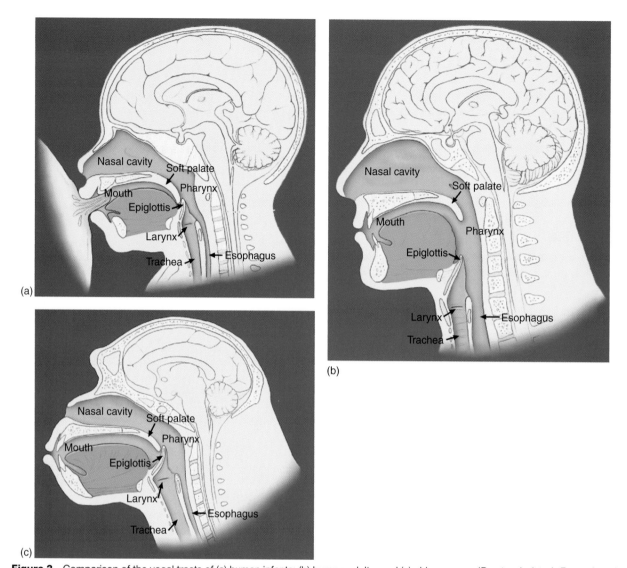

Figure 3 Comparison of the vocal tracts of (a) human infants; (b) human adults; and (c) chimpanzees (*Pan troglodytes*). Reproduced from Davidson TM (2003) The great leap forward: The anatomic basis for the acquisition of speech and obstructive sleep apnea. *Sleep Medicine* 4: 185–194, with permission from Elsevier.

Neurological Basis of Cry

Neonatal cry arises from aversive internal or external stimulation and is produced by coordination among several brain regions including the brainstem, midbrain, and limbic system. The lower brainstem controls the muscles of the larynx, pharynx, chest, and upper neck through the vagal complex (cranial nerves 9–12) and the phrenic and thoracic nerves. Variation in tension on the larynx muscles, cricothyroid and vocalis, and the abdominal respiratory muscle are thought to be responsible for f_0. There are three identifiable cry modes of vocal fold vibration: basic cry or phonation or f_0, high-pitch cry or hyperphonation (1000–2000 Hz), noisy or turbulent cry or dysphonation. Damage in the vagal cranial nerve complex

is related to atypical patterns of f_0. Rapid shifts or variability in f_0 suggests instability in the neural control system. The brainstem also controls the contour and cross-sectional airway of the upper vocal tract, which shapes the formant frequencies. Only the first two formants are usually measures: F1 occurs at approximately 1100 Hz and F2 at approximately 3300 Hz.

Other regions of the brain participate in cry initiation (limbic system, hypothalamus), configuration of the cry pattern (midbrain), and motor coordination of respiration, larynx, and articulation (reticular activating system). The extensive vagal innervation and central nervous system (CNS) coordination underlying infant cry result from modulation by the autonomic nervous system. Cry initiation, including latency to cry or threshold (amount of stimulation required), has been associated with sympathetic arousal. Respiratory modulation and temporal patterning of the cry (long or short duration of cry utterance, number of utterances, duration of inspiration, number of unvoiced sounds or short utterances) reflect parasympathetic mechanisms of homeostasis. Modulation of the overall contour of f_0 as well as the amplitude of the cry reflects both faciliatory and inhibitory autonomic mechanisms. **Table 1** provides a quick reference for cry characteristics currently used in infant assessment. The first four characteristics evaluate the entire cry (cry latency to short utterances); the remaining characteristics are calculated for each cry utterance and may be averaged across utterances.

In addition to measuring the acoustical characteristics of the cry the amount of crying per day can also be measured. This is typically done either by placing a recorder near or attached to the infant or by asking the mother to complete cry diaries. Cry diaries divide the 24-h day into boxes that represent blocks of time such as 30 min. The mother places a check mark in each time block. The blocks are added as a measure of how much the infant cried that day.

Crying and Culture

Crying is universal, all infants cry, but crying has different meanings in different cultures. Most of the studies about infant crying have been done in the western societies, such as the US, Canada, England, and Denmark. Anthropological studies show that in some cultures, infant crying is considered to be an unusual event and an ominous sign. Among the indigenous tribes of Tierra del Fuego infants rarely cry and then only when they are sick or in pain. In fact, infant crying is so rare that adults say that hearing it is the equivalent of having a terrible earache. In the Celebes, the Torjada believe that an infant's cry will bring a curse upon its parents. In other cultures, for example, India, colic refers to the kind of crying that indicates there is something wrong with the infant's digestion. The word colic comes from colon.

In contrast, the Kurds of Turkey and Iran say that crying is necessary to develop the voice, and leave their infants to cry. Taiwanese mothers see crying as a form of infant exercise, and even have a proverb that says, a child must cry to grow.

A survey of over 180 societies showed that infants cried less when they were carried. Mothers were able to tell from the infants' movement when their infant was starting to get upset and intervene before the infant cried. This suggests that, to some extent, crying is influenced by cultural practices. We should not be surprised that infants cry more in western societies. When infants are in a different room the culture is teaching them to cry to get attention.

Normal Development of Crying

Infants have two types of cry when they are born, a basic cry and a pain cry. These cries are clearly, audibly different to the human ear and reliably distinguished by acoustical analysis. The pain cry as implied means the infant is in pain and signals an emergency. The pain cry is usually high-pitched, loud, has a sudden onset, and includes long periods of breath holding. Often the very first 'waaaaah' or cry burst is prolonged and followed by a long period of breath holding. The basic cry is used for everything else. It is most typically heard when the infant is hungry, and is often called the hunger cry. The basic cry has a more gradual buildup and a lower pitch. It does not have the long periods of breath holding or the frantic, emergency quality of the pain cry.

The basic and pain cry of the newborn to 4–6-week-old infant has a reflexive or automatic quality to it. At about this time the infant's nervous system matures to a point where the infant gains voluntary control of the vocal cords. This is a prerequisite for the development of speech and language. The infant also learns that he or she has this control and also learns to use the cry to summon the mother and this cry sounds less reflexive like and more intentional. The amount of crying also reaches its peak at 4–6 weeks of age when approximately 20% of infants cry as much as 3 h per day. This is also when colic is most frequently diagnosed.

Crying and Colic

Colic is a special case when crying becomes problematic for the family. Colic used to be equated with excessive crying that only focuses on the amount of crying. However, this definition fails to consider other characteristics of the cry, impact of the crying on the infant's development and on the family. Other characteristics associated with colic include high f_0, high amplitude and increased duration of inspiration or breath holding suggesting pain, as well as physical signs in the infant (e.g., tight muscles, hard stomach, legs pulling up), the sudden onset, and episodic quality and inconsolability. When viewed from a psychosocial perspective, colic is diagnosed when crying at this age is problematic because it interferes with the infant's development (e.g., causes sleep or feeding problems), the developing parent–child relationship or the family.

There are probably multiple causes of colic. Gastroesophageal reflux (GER) occurs when acids from the stomach leaks into the esophagus, causes pain, and produces a pain cry. Colic can also be due to sleeping or feeding problems, or to hypersensitivity that reflects an immature nervous system. Although colic typically occurs in infants from 3 weeks to 3 months, the acoustic characteristics like high f_0, high amplitude, and longer duration of inspiration may precede the onset of colic itself, suggesting greater vulnerability to stress in these infants, which may be a precipitating condition in the development of colic. Colic is not caused by bad parenting. Colic does often, however, disrupt the infant–parent relationship and psychosocial treatment addresses the parents' issues as well as the infant's issues.

Crying and Temperament

In infancy, crying is part of the definition of temperament. Infants who cry a lot are said to have difficult temperament and there can be overlap between difficult temperaments and colic. Crying fits into the temperament constructs of reactivity and regulation that are thought to be linked to underlying autonomic processes. Infants with a short latency to cry onset are more reactive while infants who fail to cry or take a long time to cry despite aversive conditions are under-reactive. Infants who are difficult to sooth as well as have acoustic characteristics such as dysphonation, frequent cry mode changes between voicing and noise, and rapid shifts of variability in f_0 have poorer regulation.

The advantage of cry analysis to the measurement of temperament is that cry can be measured earlier than standard measures of temperament, which are typically evaluated no sooner than 3 or 4 months of age. Also, temperament measures are often based on parent report, whereas cry provides a more objective measure. In keeping with the definition of a temperamental trait as a biological disposition that is relatively stable and enduring across context and time, highly reactive or under reactive infants who are also poor regulators (as identified from neonatal cry) may be more likely to have difficulty with behavioral regulation or inhibitory control during later infancy and childhood.

Crying in Older Infants

Older infants cry for additional reasons to those mentioned above. Fear, stranger anxiety, and temper tantrums are common causes of crying in older infants. The reasons why these and events like them trigger crying are varied. Crying occurs when higher (brain)-level coping strategies are no longer effective in regulating the infant's behavior and more primitive (e.g., limbic) structures take over. As frontal areas of the brain develop, higher-level executive function abilities develop, which enable the infant to develop better coping strategies and inhibit the build up in emotion that leads to crying. This breakdown in coping can occur because the infant is frustrated. For example, receptive language develops before expressive language

and infants can become frustrated because they know what they want to say, but they cannot say it. Learning new things often presents the infant with conflicts so that gains in one area are at the expense of gains in another area. Learning to walk is exhilarating and brings feelings of autonomy and mastery. But it also brings the fear of rupturing the attachment relationship to the mother. In these cases crying is often a form of coping in which there is a temporary regression, or step backward, that enables the infant to regroup and then master the new challenge.

Crying and Medical Problems

Research relating cry features to medical problems are divided into three general categories: (1) significant medical problems that may be identified by standard techniques, (2) medical problems that are currently undetectable until it is too late for treatment, and (3) medical conditions that place the infant at risk for poor outcome but the prognosis is not clear.

Severe Medical Problems Already Identified

Much of the early interest in infant crying was based on the use of cry acoustics in the diagnosis of medical syndromes or damage to the CNS listed in **Table 2**. The genetic syndrome cri du chat (cry of the cat) is caused by deletion of the short arm of chromosome 5 and is virtually diagnosed by the distinctive high-pitched cry. In addition to other trisomy conditions (trisomys 13, 18, and 21), other conditions related to acoustic cry characteristics include asphyxia, undefined brain damage, Down syndrome, hydrocephalus, hypothyroidism, Krabbe's disease, encephalitis,

Table 2   Cry characteristics associated with severe medical conditions

Medical condition	Cry characteristic
Asphyxia	↑ f_0, ↑ f_0 variability (biphonation), ↑ subharmonic break, ↑ or ↓ duration
Brain damage	↑ f_0, ↑ f_0 variability (biphonation), ↓ duration, ↑ threshold, ↑ latency, ↑ short utterances
Cri du chat	↑ f_0
Down syndrome	↓ f_0, ↑ f_0 variability, ↓ amplitude
Hydrocephalus	↑ f_0, ↑ f_0 variaiblity, ↑ latency
Hypothyroidism	↓ f_0
Krabbe's disease	↑ f_0
Meningitis (bacterial)	↑ f_0, ↑ f_0 variability (biphonation), ↓ duration
Trisomy 13, 18, and 21	↓ f_0

and bacterial meningitis. As seen in **Table 2**, the most common changes in cry characteristics associated with severe medical conditions are higher f_0 and more variability in f_0. In most cases with these known medical conditions there are other clinical signs. However, cry characteristics may have diagnostic value even for infants already diagnosed with CNS damage. For example, infants with severe asphyxia or bacterial meningitis with the most abnormal cries had the poorest prognosis.

Severe Medical Problems Not Yet Detectable

The fact that cry changes are related to known significant medical conditions opens the door to determining if specific cry characteristics are associated with medical problems for which currently available methods of detection are inadequate. Sudden infant death syndrome (SIDS) is a common but unexplained cause of death in infants between 1 month and 1 year of age. Limited research has shown that infants with alterations in the cry that indicate vocal constriction (high F1) and poor control over the vocal tract (increased cry mode changes) are more likely to die of SIDS. However, the sensitivity of this cry test does not meet criteria for clinical application. Cry analysis has also been applied to the identification of infants with less obvious CNS damage.

Medical Conditions Increasing Risk for Poor Development

The category of the at-risk infant is a broad category of infants who may have poor neurodevelopmental outcome based on potential neurological insult. Factors thought to compromise the fetal brain include prematurity, intrauterine growth retardation, prenatal exposure to environmental toxins such as lead or to illegal drugs (cocaine, heroin, marijuana, methamphetamine) or legal drugs (tobacco and alcohol), and prenatal exposure to psychotropic drugs that mothers take for conditions such as depression (the class of drugs commonly refereed to as selective serotonin reuptake inhibitors (SSRIs) or serotonin-noradrenaline reuptake inhibitors (SNRIs), such as fluoxetine sold as Prozac). As shown in **Table 3**, studies have consistently found that these at-risk factors do alter the acoustical characteristics of the cry. Although high f_0 and increased cry mode changes are found in at-risk infants, many other characteristics have been observed. There is also some, more limited research suggesting that alterations in cry acoustics in these infants predict later neurobehavioral status. These studies are important because at-risk infants may have undetected CNS damage and cry analysis may be able to identify these infants when no other symptoms are present. Treatment programs can then be developed for affected infants.

Cry Perception

Crying is a biosocial phenomenon. It tells us about the infant but it also impacts the parenting environment. As a form of communication, crying is a dyadic event, much of the early infant/parent relationship is negotiated around crying. The cry is the infant's contribution, the parent's contribution is twofold. First, the parent has to interpret the infant's cry (e.g., hunger or pain). Second, the parent has to act on that interpretation and provide appropriate parenting.

Table 3 Cry characteristics associated with potential neurological insults

Medical condition	Cry characteristics
Low birth weight (<2500 g), small for gestation	↑ duration, ↓ f_0, ↑ f_0, ↑ f_0 variability (biphonation)
Preterm infants	↑ f_0, ↑ f_0 variability, ↓F1 variability, ↓ amplitude associated with ↓ BSID (18 months), ↓ duration, ↑F1, ↓ amplitude with ↓ cognitive scores (McCarthy. 60 months), ↑ short cry utterances with ↓ developmental outcome (30 months)
Hyperbilirubinemia	↑ f_0, ↓ duration, ↓ latency, ↑f_0 variability, Unstable glottic function [mode changes], ↑ F1 variability, ↑ phonation
Lead exposure	Low % nasalization, ↓ number of cries, ↑ f_0
Prenatal opiate exposure	↑ hyperphonation, ↑ short utterances, ↑ f_0, ↑ duration of 1st cry utterance associated with withdrawal symptoms, increased likelihood of abnormal cries
Prenatal cocaine exposure	Direct effects (excitation) ↑ duration, ↑ f_0, ↑ F1, ↑ F1 variability; Indirect effects via growth retardation (depression) ↑ latency, ↓ amplitude, ↑ dysphonation, ↓ cry utterances, ↑ short cry utterances, ↓ hyperphonation, ↓ F2, 2nd utterance
Prenatal marijuana exposure	Shorter cries, ↑ dysphonation, ↑ f_0, ↑ f_0 variability, ↓ F1, ↑ mode changes, ↑ F2
Prenatal alcohol exposure	↑ dysphonation, ↑ F1, ↓ threshold, ↑ hyperphonation, ↓ F1
Prenatal tobacco exposure	↑ f_0, ↑ F2, ↑ F2 variability
Prenatal methamphetamine exposure	↓ threshold, ↑ variability in f_0, ↑ variability in amplitude, ↑ mode changes, ↑ dysphonation, ↑ short utterances, ↑ variability in dysphonation

BSID, Bayley Scale of Infant Development.

Adult Reactivity to Infant Cries

Adults and even older children are similar in how they react to infant cries. In research, adults rate cries along dimensions of how the infant sounded (e.g., sick, distressed) and how the listener felt (e.g., aroused, angry, sad). Adults can clearly distinguish the cries of normal infants from the cries of infants with the array of medical problems discussed earlier. In other studies, physiological responses such as changes in heart rate or electrodermal properties of the skin change when adults listen to abnormal cries. **Table 4** shows how adults respond to specific changes in cry characteristics. Overall, increases in f_0 are most consistently associated with negative cry perception. Specifically, cries with higher f_0 were rated as more aversive, sick, urgent, angry/sad, distressing, and arousing. Other cry characteristics that yield negative cry ratings include increased variability of f_0, increased dysphonation, increased hyperphonation, increased duration, and decreased amplitude.

Table 4   Cry characteristics that affect cry perception

Type of cry	Perception/response
↑ f_0	Rated as more aversive, sick, urgent, angry/sad, distressing, and arousing.
↑ f_0 variability	Rated as more urgent, sick, angry/sad, distressing, and arousing.
↑ dysphonation	Rated as more intense, distressed, urgent, sick, aversive, grating, piercing, distressing, arousing, and discomforting. Reported shorter latency to caregiving.
↑ hyperphonation	Rated as more aversive and sick.
↑ duration	Rated as more urgent, sick, piercing, grating, aversive, distressing, arousing, and discomforting.
↑ utterances	Rated as more distressed. Reported shorter latency to caregiving.
↓ amplitude	Rated as more urgent, sick, aversive, grating, piercing, arousing, and discomforting.

Listener Characteristics Affect Cry Perception

Cry perception is also affected by characteristics of the listener. For example, crying has been implicated in child abuse and studies have found that mothers who previously abused a child had stronger reactions to even normal cries than mothers with no history of child abuse. Other characteristics that affect how normal cries are perceived and caregiver responses include gender, culture of the listener, parental personality characteristics, parental age, parity or number of children, parents vs. nonparents, and maternal learned helplessness, depression, and cocaine use. **Table 5** describes how listeners with these characteristics perceive normal cries and what they would do to alleviate the infant's distress compared to a control group of listeners. Also included are measures of physiological response in the listener, which provide further evidence for the biological basis of the cry behavioral system operating in both infant and caregiver. Characteristics

Table 5   Listener characteristics that impact cry perception

Factors	Perception/response
Gender	Men rated cries as more aversive, as eliciting more irritation and anger, and rated infants as more spoiled than women. Mothers rated cries as more likely to evoke sympathy and evoke caregiving than fathers.
Culture	Cuban-American and African-American mothers rated cries of at-risk infants as less aversive than Caucasian mothers. Caucasian mothers were more likely to pick up and cuddle, Cuban-American mothers were more likely to give a pacifier, and African-American mothers were more likely to 'wait and see'.
Parental personality/ characteristics	Parents rated as more empathic, higher neuroticism, higher extraversion, and lower conscientious had more sensitive responses to infant distress.
Parental age	Younger parents were more likely to rate cry as aversive; however, teenage mothers of at-risk infants rated ↑ f_0 as less aversive and a 'better' cry, ↑ range of f_0 as less piercing, and ↑ variability in and wider range of F1 as less piercing and less irritating. Older parents reported longer latency to caregiving.
Parity	First-time mothers were more likely to rate cry as a problem. Primiparous parents had higher skin potential levels and found cries to be the most arousing.
Parent vs. nonparents	Mothers more accurately identified pain cries than nonmothers. Fathers more accurately identified hunger and pleasure cries than nonfathers. Parents rated cry as less aversive and distressed than nonparents. Mothers were more likely to provide caregiving in response to cry.
Maternal learned helplessness	Mothers in the learned helplessness condition showed dampened physiological and behavioral responses to subsequent crying.
Depression	Depressed mothers showed ↓ sensitivity to changes in f_0, rated ↑ f_0 as less arousing and less salient, and were less likely to provide caregiving.
Cocaine user	Cocaine-using mothers rated cries as less arousing, aversive, urgent, and sick; they reported that they would be less likely to pick up and feed the infant and more likely to give a pacifier or 'wait and see'.

such as depression, drug addiction, and teen motherhood may be risk factors not only for misperception, but also for subsequent inappropriate provision of care.

Goodness of Fit

Characteristics of the infant and characteristics of parent can combine to determine the child's developmental outcome. Goodness of fit is a construct that comes out of the temperament literature and has been applied to cry. A good fit is when there is a match between the characteristics of the child and characteristics of the parent, with positive developmental outcome as the expected result. When there is a mismatch between child and parent characteristics the prognosis is negative. In cry research with preterm infants, the known aversive quality of the cry (high pitch) was determined by acoustical analysis. Infants of mothers who correctly identified their infant's cry as aversive or as not aversive when the infant was 1 month old, had higher mental and language scores at 18 months than infants whose mothers misperceived their infant's cry signal. Mothers who were better able to read their infants cry signal seem to provide the kind of parenting environment that optimizes development.

Conclusion

Infant crying is a rich source of information. Crying provides information about the biological integrity of the infant. Crying is also a social signal that affects the parent's response to the infant and the developing parent–infant relationship. Crying is also an example of the biological basis of human behavior and the multifaceted ways in which our species adapts to environmental demands.

See also: Attachment; Birth Defects; Colic; Emotion Regulation; Newborn Behavior; Pragmatic Development; Self-Regulatory Processes; SIDS; Temperament.

Suggested Readings

Davidson TM (2003) The great leap forward: The anatomic basis for the acquisition of speech and obstructive sleep apnea. *Sleep Medicine* 4: 185–194.

Golub HL and Corwin MJ (1982) Infant cry: A clue to diagnosis. *Pediatrics* 69(2): 197–201.

LaGasse LL, Neal AR, and Lester BM (2005) Assessment of infant cry: Acoustic cry analysis and parental perception. *Mental Retardation and Developmental Disabilities* 11: 83–93.

Lester BM (1984) A biosocial model of infant crying. In: Lipsitt L and Rovee-Collier C (eds.) *Advances in Infant Research*, pp. 167–212. Norwood, NY: Ablex.

Lester BM (1987) Prediction of developmental outcome from acoustic cry analysis in term and preterm infants. *Pediatrics* 80: 529–534.

Lester BM (2005) *Why Is My Baby Crying?* New York: Harper Collins.

Lester BM and Boukydis CF (1990) No language but a cry. In: Papousek H, Jurgens J, and Papousek M (eds.) *Nonverbal Vocal Communication: Comparative and Developmental Approaches*, pp. 41–69. New York: Cambridge University Press.

Zeskind PS and Lester BM (1978) Acoustic features and auditory perceptions of the cries of newborns with prenatal and perinatal complications. *Child Development* 49: 580–589.

Zeskind PS and Lester BM (2001) Analysis of infant crying. In: Singer LT and Zeskind PS (eds.) *Biobehavioral Assessment of the Infant*, pp. 149–166. New York: Guilford Publications.

Demographic Factors

L M Casper and P Kitchen, University of Southern California, Los Angeles, CA, USA

Glossary

Child well-being indicators – Measures used to track the macro- and microlevel health and socioeconomic demographic contexts in which child development occurs. They are also used to compare the status of children in different contexts over time and to ascertain whether contexts are improving or worsening over time. Child well-being indicators include measures of mortality, vaccinations, nutrition, and education.
Demographic factors – Macro- and microlevel health and socioeconomic demographic contexts that affect child development.
Less-developed countries (LDC) – LDCs are defined as countries with technologically underdeveloped, pre-industrial economies, or states that are structurally handicapped.
Macrolevel contexts – Health and socioeconomic demographic factors at the country, state, county, and city level affecting everyone in a given location. Education and healthcare systems and the economy are examples of macro contexts.
More-developed countries (MDC) – MDCs are defined as countries with technologically advanced, industrial economies, or states that are structurally sound.
Microlevel contexts – Health and socioeconomic demographic characteristics of individuals. Examples of micro contexts are family income and educational attainment.

Introduction

Demographic factors are extremely important for understanding the context in which child development occurs.

Some children are raised in affluent countries with booming economies and highly developed social infrastructures and supporting services such as education and healthcare. Other children are raised in poor countries that find it difficult to provide enough food, water, and adequate shelter for their populations. Even within countries, children are raised in vastly different contexts: some grow up in urban settings where crime and violence are commonplace, whereas others grow up in suburbs replete with shiny new community recreation centers. These macrolevel contexts can be defined broadly as demographic factors and provide the backdrop for infants' and children's development.

The physical, social, emotional, cognitive, and behavioral development of children depends, in large part, on the environment in which they are raised. Children require the basic necessities of nutritious food, clean water, adequate shelter, and basic healthcare to develop normally. Once these requirements are met, development can be enhanced through things such as the education of children and parents, and high levels of family income. Parents with high levels of education and high incomes are better able to provide and to draw on resources to enhance the development of their children than are parents with little education and low incomes. The demographic, social, and economic attributes of individuals provide another context for child development; these attributes are known as microlevel demographic characteristics. The economic, social, and political environment – the macrolevel demographic context of a country, city, or community – also affects the amount of resources parents can provide to their children, regardless of the parents' own characteristics. Thus, both micro- and macrodemographic characteristics must be considered when examining children's development.

Demography, also called population studies, is the study of human populations. The word demography is derived from the two Greek roots 'demos' – people and 'graphy' – written knowledge relating to a specific science.

In 1855, Achille Guillard was the first to use this term in his book *Éléments de Statistique Humaine ou Démographie Comparée*. Demographers study the causes and consequences of population size; population growth or decline; the social and economic characteristics of countries, cities, communities, and populations; and levels of and trends in mortality, fertility (childbearing), migration, marriage, divorce, and cohabitation. These are the demographic characteristics and processes that shape the macro- and microlevel contexts in which infants, children, adolescents, and indeed adults, age and develop.

Demographers and social scientists use indicators of demographic factors to understand whether the contexts in which children are developing are getting better or worse over time. Indicators are also used to track the change in children's health and social and economic well-being over time, and to compare children across different social, economic, and demographic contexts. Child well-being indicators include measures of mortality, vaccinations, nutrition, and education. Researchers in some fields, as well as some policy makers and nongovernmental organizations such as United Nations Children's Fund (UNICEF), even refer to these types of measures as indicators of children's development. Health and socioeconomic demographic indicators of context can be macro (as in the case of the number of doctors in a country or a country's gross national product) or they can be micro, such as the educational level of parents and children. Demographers use these types of indicators to examine the broad context of child well-being.

In sum, vast global inequality exists today. The abundance of resources found in more-developed countries (MDCs) contrasted with the lack of resources in the impoverished countries of the 'third world' set the context for child development. Access to healthcare, education, clean water, transportation, food, and shelter all affect the ability of a child to live, grow, develop, and thrive. Child development, in this sense, is part of the social and economic environment in which each child exists. Countries that have economic and political stability, an abundance of healthcare workers, and a functioning public educational system are ones in which children are the least likely to fall victim to disease, poverty, exploitation, and malnutrition. Children whose parents have a lot of personal resources, including high levels of education and income, are also more likely to have accelerated development. Indicators of child well-being can be used to track children's health and well-being directly as well as the types of environments in which they develop.

Less-Developed Countries and More-Developed Counties

Researchers at universities, the United Nations, UNICEF, the World Health Organization (WHO), and other organizations generally split countries into two categories: less developed countries (LDCs) and MDCs. LDCs are generally defined as countries with technologically underdeveloped, pre-industrial economies, or states that are structurally handicapped. By contrast, MDCs are defined as countries with technologically advanced, industrial economies, or states that are structurally sound. The classification system the United Nations uses to categorize countries is widely accepted by demographers. Recently, the United Nations has used three categories – least developed, developing, and industrialized – based on three criteria. To be identified as 'least developed', a country must have (1) low income (generally under US$ 900 per capita gross domestic income (GDI) averaged over 3 years); (2) a human resource weakness as measured by a human assets index based on indicators of nutrition, health, education, and adult literacy; and (3) economic vulnerability as measured by an economic vulnerability index based on indicators such as the instability of agricultural production and of exports of goods and services. In 2003, least-developed countries under this definition included 50 countries.

Examples of least-developed countries include Afghanistan, Cambodia, Nepal, and Senegal. Australia, Denmark, Greece, Japan, and the US are examples of industrialized countries. The middle category, developing countries, consists of all of the least-developed countries in addition to countries that are slightly better off, but are not yet classified as industrialized countries, such as Argentina, Bolivia, Kenya, and Vietnam.

When discussing child well-being indicators, it is important to consider the demographic, social, and economic conditions experienced by children all around the world. In MDCs, the majority of children have access to, at minimum, the essentials of survival: food, water, shelter, and medical care. Thus, child well-being in MDCs tends to be measured with indicators of educational achievement, school expenditures, and behavioral and cognitive development. In LDCs, children face hunger, homelessness, and a lack of water (or sanitized water). These children have little or no access to medical care or immunizations, live near environmental hazards, and some are even exposed daily to armed conflict. In these countries, child well-being is more likely to be measured with indicators of disease, malnutrition, and exposure to war. Furthermore, MDCs also tend to have more developed data infrastructures and statistical systems from which to derive measures of children's well-being. Thus, many more measures are available for children in MDCs than for those in LDCs.

Given the vast gulf between these health and socioeconomic demographic contexts, the different types of indicators of child well-being they engender, and the relative lack of data in LDCs, this article focuses on those indicators that exist in both MDCs and LDCs.

Health Demographic Factors

Child Mortality

Among the most important indictors demographers use to assess child well-being are the infant and child mortality. Infant mortality is a good indicator of societal development because as the standard of living in a country goes up, the health of babies improves earlier and faster than the health of older people. The highest probability of death for children occurs within the first year of life. Infant mortality is measured by the number of deaths of children under the age of 1 year for every 1005 live births that occurred during the year. The simple number of infant deaths in a country does not provide a good comparison across countries because larger countries will have more deaths just because they have a greater number of babies at risk of dying. The denominator – per 1000 live births – takes into account the size of the population, making the figure easier to compare across countries.

In the least developed countries, 97 out of every 1000 children born in 2005 died before they reached the age of 1 year (**Table 1**). The corresponding figures were 57 deaths per 1000 live births in all LDCs and 5 deaths per 1000 births in the MDCs. Infant mortality is the highest in Western and Central Africa where, on average, 108 infants die for every 1000 babies born. In Eastern and Southern Africa, the rate is a little lower at 93 deaths per 1000 births (**Figure 1**). South Asia has the next lowest rate at 63 deaths per 1000 births, followed by 43 per 1000 in the Middle East and North Africa, 29 per 1000 in Central and Eastern Europe, rates are as low as 6, 5, 4, and 3 per 1000 for the US, UK, France, and Sweden, respectively.

In the LDCs the probability of dying remains high throughout childhood due to the lack of food, water, medicine, and sanitation. The under-5 mortality is an indicator of the level of death in early childhood and is measured by the number of deaths of children under the age of 5 year for every 1000 live births in that year. In 2005 under-5 mortality in the least-developed countries was 153 deaths per 1000 live births (**Table 1**). In all LDCs, the comparable figure is about half that at 83 deaths per 1000 live births. In MDCs, it is one-fifteenth as high, at 6 deaths per 1000 live births.

Western and Central Africa are home to many of the least-developed countries in the world. The macrolevel context for young children growing up in these countries is one of poverty, pestilence, disease, and often violence. The countries have fragile healthcare infrastructures that do not provide preventive care or treatment of illness, disease, or injury. These countries also experience socioeconomic stagnation due to conflicts, wars, political instability, and AIDS. It is thus not surprising that this region of the world has the highest under-5 mortality at 190 deaths per 1000 live births (**Figure 1**). Thus, nearly one in five African children will die before reaching their fifth birthday. Almost half of the 10.1 million deaths that occur in the world to children under 5 years of age occur in sub-Saharan Africa. Sub-Saharan children not only have the highest chance of dying before they are 5 years of age – they are losing ground to children in other parts of the world. Under-5 mortality is now seven times higher in the African region than in the European region; the rate was 'only' 4.3 times higher in 1980 and 5.4 times higher in 1990. Deaths in early childhood are increasingly concentrated in the African region (48% of the global total in 2005, up from 30% in 1990). Almost one-third of under-5 deaths occur in South Asia. Thus, Africa and South Asia account for nearly four out of five early childhood deaths. One-half of child deaths are concentrated in just six countries: China, the Democratic Republic of the Congo, Ethiopia, India, Nigeria, and Pakistan.

While the main causes of under-5 mortality can be traced to only a handful of diseases and ailments, the macrocontexts that allow such diseases to flourish must be taken into account. Lack of economic resources and geographic isolation affect the availability of preventive vaccines, healthcare workers, emergency care, transportation to health facilities, and proper nutrition. Lack of sanitation and freshwater help diseases to spread. In developed countries, everyone has access to clean water and adequate sanitation facilities. By contrast, in the least-developed countries only 59% of the population are using improved drinking water sources and only 36% are using adequate sanitation facilities.

The macropolitical context must also be taken into account; levels of infant mortality are staggering in areas besieged by war and political conflict. In fact, most of the countries where under-5 mortality exceeds 200 deaths per 1000 live births have experienced major armed conflict since 1999. It is no surprise that countries with a great number of resources have the lowest under-5 mortality rates – again, child well-being is part and parcel of the economic and social environment in which each child lives.

The main causes of death for children under 5 years of age around the world are pneumonia, diarrhea, malaria, measles, AIDS, and perinatal diseases. Of those deaths occurring to children aged 1 month to 4 years, 29% are due to pneumonia, 27% are due to diarrhea, 12% are due to malaria, 6% are due to measles, and 5% are due to AIDS (**Figure 2**). Combined, these diseases account for four in five deaths to children aged 1 month to 4 years. Apart from AIDS, all are considered highly curable or preventable in industrialized countries. Diarrhea is remedied with oral rehydration therapy, malaria can be prevented with a wide range of antimalarials and medical treatment or the use of protective netting, measles can be prevented with immunization, and pneumonia can be cured with antibiotics.

Table 1   Indicators of child wellbeing in least-developed countries, less-developed countries, and more-developed countries

Indicators	Least developed countries	Less developed countries	More developed countries
Mortality			
Infant mortality in 2005 (number of deaths of infants under 1 year per 1000 live births)	97	57	5
Under-5 mortality in 2005 (number of deaths of children aged 0–4 years per 1000 live births	153	83	6
Maternal mortality in 2000 (adjusted number of pregnancy-related deaths of women per 100 000 live births)	890	440	13
Immunizations (in 2005)			
Percentage of infants immunized against diphtheria, pertussis, and tetanus	76	75	96
Percentage of infants immunized against measles	72	75	92
Percentage of infants immunized against hepatitis B	41	54	64
Percentage of newborns protected against tetanus	64	69	NA
Nutrition			
Percent of infants with low birth weight	19	16	7
Percent moderately and severely underweight (aged 0–4 years)	35	27	NA
Percent with moderate and severe wasting (aged 0–4 years)	10	10	NA
Percent with moderate and severe stunting (aged 0–4 years)	42	31	NA
HIV/AIDS			
Number of children aged (0–14 years) living with HIV, 2005	1 100 000	2 300 000	13 000
Healthcare			
Percentage of pregnant women aged 15–49 years who received prenatal care	59	71	NA
Percentage of births attended by doctors, nurses, or midwives	35	60	99
Economy			
Per capita gross national income (US$) in 2005	383	1 801	35 410
Percent average annual rate of inflation (1990–2005)	59	18	2
Percentage of the population living on less than US$ 1.00 a day	41	22	NA
Education[a]			
Percentage of girls enrolled in primary school	72	85	95
Percentage of boys enrolled in primary school	77	89	97
Percentage of girls enrolled in secondary school	29	51	93
Percentage of boys enrolled in secondary school	33	52	91
Percentage of girls attending secondary school	20	43	NA
Percentage of boys attending secondary school	22	46	NA
Literacy rate of women (percentage aged 15 years and over who can read and write)	50	72	NA
Literacy rate of men (percentage aged 15 years and over who can read and write)	70	85	NA
Fertility			
Total fertility rate (the average number of children women have by the end of their reproductive years)	4.9	2.8	1.6

[a]The percent of children enrolled in school is actually a net enrollment ratio, which is the number of children enrolled in school who are of the official age for that school level (primary or secondary) expressed as a percentage of the total number of children of official age for that school level. The percent of children attending secondary school is a net attendance ratio, which is the number of children attending secondary or tertiary school who are of the official secondary school age expressed as a percentage of the total number of children of official secondary school age. Percentages for secondary school for LDCs exclude China.
NA, data not available.
Adapted from various tables in UNICEF (2007) *The State of the World 's Children: Women and Children – The Double Dividend of Gender Inequality.* New York: UNICEF, with permission from UNICEF. The UNICEF report includes data from several sources: UNICEF, World Health Organization, United Nations Population Division, United Nations Statistics Division, World Bank, Demographic and Health Surveys (DHS), Multiple Indicator Cluster Surveys (MICS), USAID, UNAIDS, UNESCO.

Of those deaths occurring to children under 5 years of age, over one-third occur to neonates – infants dying in the first 28 days of life. The main causes of neonatal death are complications associated with premature birth, sepsis/pneumonia, and asphyxia.

Immunizations

Immunizations are an extremely effective way of preventing disease and death among children and adults around the world. Vaccines are commonplace in MDCs

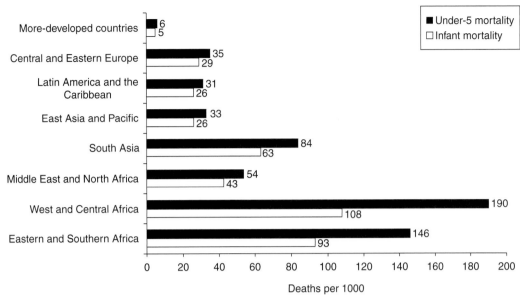

Figure 1   Infant and under-5 child mortality, by region of the world: 2005. Infant mortality is the number of deaths per 1000 live births in 2005. The under-5 child mortality is the number of deaths of children aged 0–4 years per 1000 live births of children under 1 year of age. The Central and Eastern Europe category includes the Commonwealth of Independent States. Adapted from **Table 1** in UNICEF (2007) *The State of the World's Children: Women and Children – The Double Dividend of Gender Inequality*. New York: UNICEF, with permission from UNICEF. The data in the table are from UNICEF, World Health Organization, United Nations Population Division, and United Nations Statistics Division.

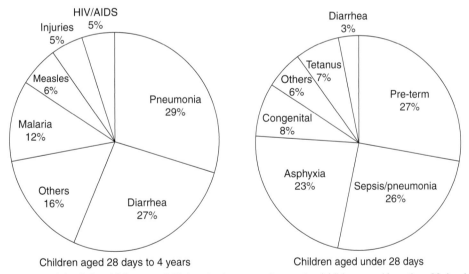

Figure 2   Major causes of death for children aged 28 days to 4 years and neonates (children aged less than 28 days): 2000–2003. Undernutrition is an underlying cause of 53% of deaths among children aged 0–4 years. Source: World Health Organization. Adapted from data accessed at http://www.who.int/child-adolescent-health/OVERVIEW/CHILD_HEALTH/child_epidemiology.htm on 23 April 2007, with permission.

and are becoming more widespread in LDCs, in part due to the WHO's expanded Programme on Immunization. However, many children in LDCs still have not been immunized. In 2005, 96% of 1-year-old children in MDCs were immunized against diphtheria, whooping cough, and tetanus, compared with only 75% in all LDCs and 76% in the least-developed countries (**Table 1**). The percentages are similar for measles immunizations. Immunizations against hepatitis B are lower all over the world: 64% of children under 1 years of age in industrialized countries have received HepB3, compared with 54% in LDCs and only 41% in the least-developed countries.

WHO estimates that 1.4 million children under the age of 5 years died in 2002 of diseases preventable by widely used vaccines. Over 500 000 of these deaths were caused by measles; nearly 400 000 by Hib (or *Haemophilus influenzae* type b); nearly 300 000 by pertussis (or whooping cough); and 180 000 by neonatal tetanus.

Immunizations are extremely cost-effective compared with the cost of combating the disease once contracted. Even though the price of vaccination is considered low in MDCs, in LDCs, the costs can be a significant burden. In the mid-1990s, vaccines to provide 'basic' coverage for tuberculosis, polio, diphtheria, tetanus, whooping cough, and measles cost about US$1 per child. Inclusion of vaccines for hepatitis B and Hib raises the vaccine cost alone to US$7–13 per child (not including administration and injection equipment) in the developing world. When vaccine administration is included, the costs amount to between US$20 and 40 per child. This is quite a financial burden considering that the average per capita gross national income (GNI) in least-developed countries was only US$383 in 2005 and 41% of the population survived on less than US$1 per day. However, with the help of various global initiatives, including the Expanded Programme on Immunization and the Global Alliance for Vaccines and Immunization, as well as funding from charitable organizations and governments, immunization rates are rising around the world.

Nutrition

In MDCs, children typically have access to an abundance of food, including meat and fresh fruits and vegetables. But, to the dismay of health professionals, they also have access to highly processed fatty foods that tend to be high in calories and low in nutritional value. By contrast, in many LDCs, children are staving to death due to famine and food scarcity. They lack well-balanced diets and the vitamins and minerals they need to develop normally. In these countries, even when food is available, many families cannot afford to pay for it. WHO estimates that one out of three people in LDCs are affected by vitamin and mineral deficiencies. Undernutrition has been implicated either directly or indirectly in 53% of deaths to children under 5 years of age worldwide.

Even when undernutrition does not result in death, it can have grave effects on the well-being of children; undernutrition causes or is linked to several serious consequences, including, impaired physical development, learning disabilities, mental retardation, poor health, infections, and blindness. Undernutrition among pregnant women in developing countries contributes to infants being born with low birth weight. Low-birth-weight infants are more likely to die prematurely, be born with birth defects, and develop at a slower pace. In the least developed nations, 19% of babies are low birth weight

(under 2500 g) (**Table 1**). By contrast this figure stands at 7% for MDCs. In the least developed countries, more than one-third of children aged 0–4 years are moderately or severely underweight, 10% are experiencing moderate or severe wasting, and 42% are experiencing moderate or severe stunting. In MDCs, comparable data are not available, suggesting that these conditions may not be prevalent enough to measure in these countries.

While undernutrition continues to have disastrous consequences for children in LDCs, child obesity is posing a serious health risk for children and adults in MDCs and somewhat surprisingly, is also beginning to concern LDCs. Obesity is not an issue to be taken lightly; it poses a major risk for serious diet-related noncommunicable diseases, including diabetes mellitus, cardiovascular disease, hypertension and stroke, musculoskeletal disorders such as osteoarthritis, and certain forms of cancer. The effects can range from diminishing quality of life to premature death. In 1995, 18 million under-5 children were overweight. By 2005, this figure was estimated to be 20 million.

HIV/AIDS

The HIV/AIDS epidemic is of great significance to the study of child well-being. In 2005, 2.3 million children in the world were living with HIV (**Table 1**). Nearly half of the world's children with HIV reside in the 50 least-developed countries. In fact, almost all of them reside in the LDCs – only 13 000 live in MDCs. Children are not only born with HIV, but in many countries significantly large numbers are infected every year. According to the United Nations, an estimated 6000 youths under the age of 15 year around the world become infected with HIV each day – one every 14 s.

Of great concern for child development is the vast number of children who have been orphaned by AIDS. In 2005, UNICEF estimated that 15.2 million children aged 0–17 years were orphaned by AIDS. By 2010, the number of children orphaned by this disease is projected to reach 25 million. Four out of five of the world's AIDS orphans live in sub-Saharan Africa, including over 1 million orphans each in Kenya, South Africa, Tanzania, Uganda, and Zimbabwe. Effectively, this epidemic is creating a generation of children who are growing up without the benefits of parents to help them attend and succeed in school, and receive proper nutrition, medical attention, and guidance. Children without parents are more likely to be used in forced labor and be exploited.

Even when children are not orphaned by AIDS, many have sick parents and must assume the role of household head because their parents are incapable of earning a living and caring for their children. Because the vast majority of children in these situations are in the least-developed countries, the social, economic, and healthcare infrastructures are not equipped to assist these adolescents when they are forced prematurely to assume adult roles. Given the lack of support, many adolescents are obliged to sacrifice educational and other opportunities to build human capital to support their families and survive.

Access to Healthcare

Access to healthcare is directly related to the economic resources and political climate of the country – the macrocontext. The poorest countries have the least amount of healthcare and the fewest skilled doctors. The availability of heathcare in different regions of the world is also inversely related to the burden of disease; countries with the lowest relative need for healthcare workers (the healthiest populations) tend to have the largest numbers of them. For example, the Americas, including Canada and the US, suffer only 10% of the global burden of disease, yet are home to 37% of the world's healthcare workers. By contrast, Africa endures about 15% of the global burden of disease, but has access to only 3% of its medical personnel.

Another indicator of access to healthcare is the amount of money a country spends on it. The Americas as 50% of the money spent for healthcare worldwide, whereas Africa is the beneficiary of only 1%, even including grants and loans from other countries (**Figure 3**).

Maternal Mortality

Access to proper maternal and child healthcare is vital to ensure the health, safety, and proper development of the newborn and the mother before, during, and after birth. The absence of adequate medical care in LDCs directly affects children and pregnant mothers. According to the WHO, each year over 300 million women in LDCs suffer from short-term or long-term illness brought about by pregnancy and childbirth. More than half a million die each year due to complications of pregnancy and child birth (including 68 000 as a result of unsafe abortions), leaving behind motherless children who are more likely to suffer developmental delays and premature death. Maternal mortality is an indicator of the annual number of deaths from pregnancy-related causes and is measured by the annual number of maternal deaths per 100 000 live births. The ratio is 890 per 100 000 live births in the least-developed countries, 440 per 100 000 in all LDCs, and only 13 per 100 000 in the MDCs. It is the highest in sub-Saharan Africa, followed by South Asia (**Figure 4**).

The countries with the highest infant and maternal mortality and the highest incidence of low birth weight babies are also the ones with the worst healthcare infrastructures. The poor outcomes in these countries can be linked in part to relatively low levels of prenatal care and the higher probability of giving birth without a skilled attendant present. In the least-developed countries, only 59% of women received prenatal care. This figure compares with 71% in all LDCs. Prenatal care of some sort is nearly universal in MDCs (**Table 1**).

A similar pattern is evident regarding giving birth with a skilled attendant present. In the MDCs, 99% of births occurred with a doctor, nurse, or midwife present. For all LDCs, this figure falls to 60% and for the least-developed nations it stands at 35% (**Table 1**).

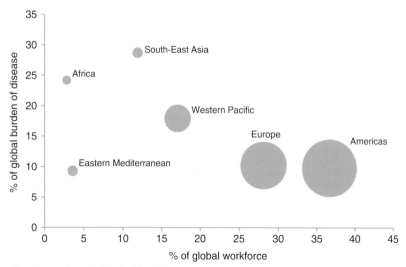

Figure 3 Distribution of health workers by level of health expenditure and burden of disease, by World Health Organization region. The vertical axis shows burden of disease, the horizontal axis, the number of health workers, and the size of the dots represents total health expenditure. Reproduced from World Health Organization (2006) *The World Health Report: Working Together for Health.* Geneva: World Health Organization, with permission.

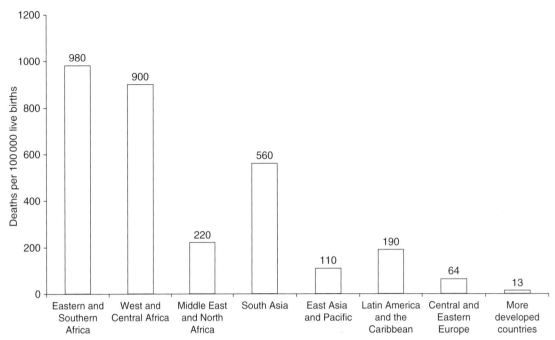

Figure 4   Adjusted maternal mortality, by region of the world: 2000. The maternal mortality is the number of deaths of women from pregnancy-related causes per 100 000 live births in 2005. The Central and Eastern Europe category includes the Commonwealth of Independent States. Adapted from table 8 in UNICEF (2007) *The State of the World 's Children: Women and Children – The Double Dividend of Gender Inequality*. New York: UNICEF, with permission from UNICEF. The data in the UNICEF table are from UNICEF and the World Health Organization.

Thus, in the 50 least-developed countries, two-thirds of births occur without the assistance of a healthcare professional.

Socioeconomic Demographic Factors

Economy

Among the most important macrolevel demographic factors for child development is the economy of the country in which a child is raised. On the one hand, countries with robust, strong economies have the resources they need to build and support strong social infrastructures. They typically have better, more developed healthcare systems, sewerage and sanitation services, good schools with mandatory attendance, social services to support people in need, reliable infrastructure that supports commerce (e.g., roads, ports, and airports), good public transportation systems, effective law enforcement, strong defenses, and strict human rights regulations. Children who grow up in these environments have many resources upon which they and their families can draw. Parents also have access to information and services to help their children reach their greatest potential. On the other hand, countries with instable, weak economies do not have money to build or maintain strong social infrastructures. In addition, they are often crippled by war, famine, disease, and pestilence.

Children growing up in poor economies thus experience a double affect: they do not have the social infrastructures that provide resources, information, and services, yet they must endure the harmful, unhealthy environment brought about by poverty, war, famine, disease, and pestilence.

In the least-developed countries, in 2005 the per capita GNI was US$383 and 41% of the people lived on less than US $1 a day on average (**Table 1**). This contrasts sharply with the MDCs where the GNI was US$35 410 and, on average, no one lived on less than US $1 a day. The GNI for all LDCs was US$ 1801 and 22% of the people in these countries lived on less than US $1. The GNI tends to be highest in Northern European countries, such as Norway with a GNI of US$ 59 590, and lowest in African countries such as the Democratic Republic of the Congo with a GNI of US$ 120. Thus, on average each Norwegian (including children) lives on US $163 per day, whereas each Congolese lives on 27 cents per day. Inflation also tends to be higher in LDCs than MDCs. The average annual rate of inflation was 59% in the least-developed countries, 18% in all LDCs, and only 2% in MDCs (**Table 1**).

The US spent an average of US$6103 on healthcare per person in 2006, the most of any country in the world. The WHO estimates that the minimum amount needed per person, per year to provide 'basic, life-saving services' is only US$35–50. Sixty-four of the member states spend under $50 per person, per year for healthcare, a figure that includes private spending, government spending, and private donations.

Education

Education is central to child development both directly and indirectly. Children's own education helps them to develop cognitively, behaviorally, socially, and emotionally. A parent's education also affects a child's development; more highly educated parents have access to more information on proper parenting techniques, nutritional guidelines, healthy hygiene practices, and proper healthcare behavior. They are also more likely to know how to navigate social institutions such as schools and healthcare facilities effectively and how to interact successfully with their staff. The macrolevel education of a country can also affect child development. The more people in a country who are educated, the less likely disease is to spread, and the more likely people are to earn better incomes.

In the least-developed countries, 72% of girls and 77% of boys of primary school age are enrolled in primary school. These figures compare with 85% of girls and 89% of boys in all LDCs. Nearly all children of primary school age in MDCs who are able to do so either attend school, have private tutors, or are home-schooled (**Table 1**).

Enrollment in secondary schools is lower. In the MDCs 93% of girls and 91% of boys of secondary school age are enrolled in school. Comparable figures for all LDCs are 51% for girls and 52% for boys and in the least-developed countries are 29% for girls and 33% for boys. Even though students are enrolled, they do not always attend classes. In the least-developed countries, 20% of girls and 22% of boys of secondary school age attend secondary schools compared with 43% of girls and 46% of boys in all LDCs. Comparable figures are not available for MDCs.

A good indicator of the level of education of parents and of a country as a whole is the adult literacy rate because it can be compared across countries even when they have different educational systems. In the least-developed countries, 70% of men, but only 50% of women aged 15 years and over, are able to read and write. This compares with 85% of men and 72% of women in all LDCs. Adult literacy is near universal in MDCs. One of the best ways to increase children's well-being is to raise the literacy rate in the LDCs, particularly among women, since they are the ones who become pregnant and are more likely to care for children.

Age Composition

One of the more important macrolevel demographic contexts for child development is the age distribution of a population. Demographers use two indicators to examine the age distribution of the population: the dependency

ratio and the population pyramid. The age–sex composition of a country is a good indicator of the challenges that country faces in providing healthcare for its children and seniors, education for its youth, employment opportunities for its young adults, and food for everyone. MDCs typically have lower fertility, mortality, and population growth rates, whereas LDCs have higher levels of fertility, mortality, and population growth. The difference in these population processes generally results in younger populations for LDCs, with a higher proportion of young dependents to support, and older populations in MDCs which tend to have populations that are more evenly distributed across all age groups.

The dependency ratio is the ratio of children and the elderly to working-aged adults and is a measure of how many dependents each working-aged adult must support. Intuitively, the fewer dependents each working adult has on average, the more resources can be allocated to that dependent. The total dependency ratio in the US was 51 in 2000 – there were 51 people younger than 15 years or older than 64 years for every 100 people aged 15–64 years. Thus, in the US, each working-age person would have to support about 0.5 of a person. The average dependency ratio for other developed countries was similar at 48 per 100. By contrast, LDCs have a much higher dependency ratio of 61 per 100.

The relative youth of the LDCs can be seen more clearly by examining the youth dependency ratio. The youth dependency ratio is the number of children under age 15 years per 100 people of working age in the population. The average child dependency ratio for the world in 2000 was 48 children under 15 years for every 100 people aged 15–64 years. For all MDCs (with the exception of the US), the average was 26 per 100. For all LDCs, the average is about double at 53 per 100. The youth dependency ratio is particularly high in sub-Saharan Africa at 84 per 100. In LDCs, where youth dependency ratios are extremely high, the need for adults who can provide medical care, education, and other services is urgent. Youth dependency is an especially serious problem in countries ravaged by HIV/AIDS because many adults are ill or dead, leaving relatively few people of working age to support the social and economic infrastructure and care for children.

Examination of the elderly dependency ratio tells just the opposite story and reflects the relatively low life expectancy of people in developing countries. The elderly dependency ratio is 22 people aged 65 years and older for every 100 working-aged people in the MDCs (with the exception of the US) and only 8 per 100 in the LDCs.

Population pyramids are pictorial representations demographers use to describe the age and sex composition of populations. On the one hand, LDCs have pyramidal shapes with large proportions of the population at young ages and small proportions at older ages. On the other hand, MDCs have barrel shapes with a more even distribution across age groups. The largest portion of the population in MDCs tends to be in the working ages, with smaller portions of the young and old alike (**Figure 5**).

Total Fertility Rate

Central to the discussion of child development is the amount of resources available to children. In this sense, the number of children per family is an important indicator of child well-being – simply put, the more children in each family, the fewer resources each child is able to receive.

These resources include food, clothing, medical care, and even time spent with parents. On average, women have many more children in LDCs than in MDCs. The high level of childbearing in LDCs is the single largest contributor to the youth dependency problem discussed in the previous section. In 2005, the United Nations Population Division estimated the total fertility rate (TFR) in the least-developed countries at 4.9 (**Table1**). This means that the average woman in least-developed countries will give birth to five babies by the end of her reproductive years. The

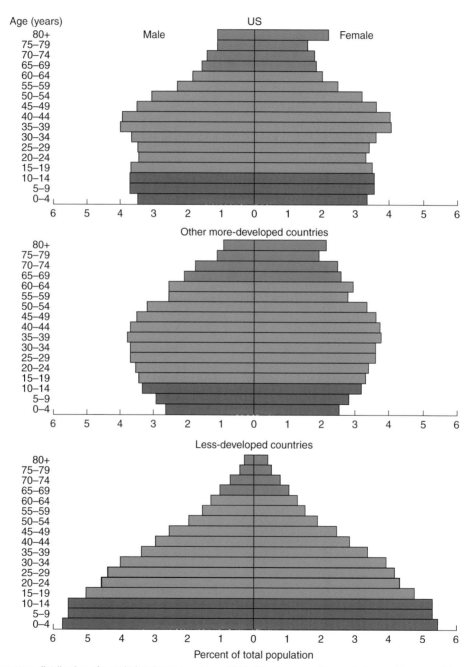

Figure 5 Percentage distribution of population, by age and sex: 2000. Blue bars, youth population; red bars, working-age population; and green bars, elderly population. Reproduced from figure 5 in McDevitt TM and Rowe PM (2002). *The United States in International Context: 2000.* C2KBR/01–1. Washington, DC: The U.S. Census Bureau.

TFR was 2.8 for all LDCs and only 1.6 in MDCs, well below the level needed for the MDC population to replace itself. Women in West and Central Africa have the highest TFR, averaging 5.7 live births in their lifetime. In 2005, the average number of children born to women exceeded seven in four countries: Afghanistan, Niger, Timor-Leste, and Uganda.

Orphans

In addition to AIDS, harsh social and economic conditions and armed conflict can cause many parents to die. Overall, 133 million children are currently orphaned by all causes around the world. The numbers are highest in sub-Saharan Africa (47 million) and South Asia (38 million). Without parents to protect them, children are exposed to any number of physical and psychological hazards. Child well-being is affected not only by the social, economic, and political conditions of a country, but whether a child can grow and thrive, depends on the strength, durability, and availability of parental figures and relationships.

Child Labor

Among the many health and developmental concerns surrounding child well-being, child labor is often overlooked in part because it is illegal in industrialized countries. In the *State of the World's Children 2007*, published by UNICEF, data show that in the least-developed countries, 29% of children aged 5–14 years are involved in child labor. In LDCs (excluding China), this number drops to 17%. Child labor is even normative in some countries, exceeding 50% in many sub-Saharan nations. Even though child labor is normative in some countries, it can have a devastating impact on children's lives.

It can place them in an unsafe environment, open the door to exploitation, and expose them to conditions that are detrimental to their health at crucial times during their development. No less serious for the development of children is that child labor robs children of their childhood – a critical time for educational, behavioral, cognitive, social, and emotional development.

The International Labour Organization estimates that 8.4 million children have been forced against their will into debt bondage or other forms of slavery, prostitution and pornography, trafficking, armed conflict, or other illicit activities (**Figure 6**). The 8.4 million total excludes the category of trafficked children found in the figure because of the risk of double-counting. Boys tend to be trafficked for forced labor in commercial farming, petty crimes, and the drug trade, whereas girls mainly appear to be trafficked for commercial sexual exploitation and domestic service.

Additional legislation, economic sanctions and aid programs, intervention, international cooperation, and education are needed to ensure that children are not subjected to such horrific conditions and are able to live, grow, and experience childhood free from violence, exploitation, and bondage.

Conclusion

Demographic factors are essential to the study of child development. Children grow up in countries with very different health and socioeconomic contexts that vary on both the macro- and microlevel. Macrolevel contexts are defined by demographic factors at the country, state, county, and city level, and they affect everyone in a given location. Education and healthcare systems and

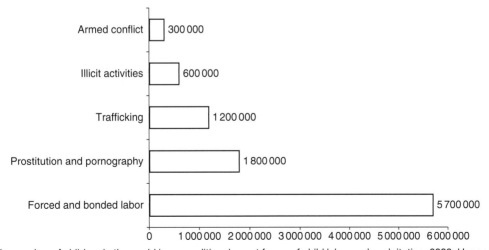

Figure 6 The number of children in the world in unconditional worst forms of child labor and exploitation: 2000. Unconditional worst forms of labor correspond to those outlined in article 3 of the International Labour Organization Convention No. 182. Adapted from table 9 in International Labour Organization (2002) *Every Child Counts: New Global Estimates on Child Labour*. Geneva: International Labour Organization.

the economy are examples of macrolevel contexts. Micro-level contexts are defined by the demographic character-istics of individuals. Examples of microlevel contexts are family income and educational attainment. Child well-being indicators are measures used to track macro and micro health and socioeconomic demographic contexts over time and across populations.

The health and socioeconomic demographic contexts of children throughout the world vary vastly and portray a picture of 'haves' in the MDCs and 'have nots' in the LDCs. Children in MDCs grow up in the context of developed social and economic infrastructures that pro-vide good healthcare, schooling, and an abundance of food. By contrast, children in LDCs often do not even have access to healthcare, schools, or food. Thus, children in LDCs face significant challenges to normal development. Children in the sub-Saharan region of Africa and those in South Asia grow up in the least hospitable health and socioeconomic demographic contexts. One cannot help but think of the remarkable resiliency of the Asian and African children who do survive childhood, given the overwhelming demographic factors they overcome.

See also: AIDS and HIV; Birth Complications and Outcomes; Family Influences; Family Support, Interna-tional Trends; Healthcare; Literacy; Mortality, Infant; Nutrition and Diet; Obesity.

Suggested Readings

International Labour Organization (2002) *Every Child Counts: New Global Estimates on Child Labour.* Geneva: International Labour Organization.

McDevitt TM and Rowe PM (2002) *The United States in International Context: 2000. C2KBR/01–1.* Washington, DC: The U.S. Census Bureau.

UNICEF (2006) *The State of the World's Children: Excluded and Invisible.* New York: UNICEF.

UNICEF (2007) *The State of the World's Children: Women and Children – The Double Dividend of Gender Inequality.* New York: UNICEF.

World Health Organization (2005) *The World Health Report: Make Every Mother and Child Count.* Geneva: World Health Organization.

World Health Organization (2006) *The World Health Report: Working Together for Health.* Geneva: World Health Organization.

http://www.who.int/child-adolescent-health/OVERVIEW/ CHILD-HEALTH/child_epidemiology.htm.

Relevant Websites

http://www.oecd.org – Organization for Economic Co-operation and Development.

http://www.childinfo.org – Statistical information from the United Nations Children's Fund.

http://www.worldbank.org – The World Bank.

http://www.who.int – The World Health Organization.

http://www.unicef.org – United Nations Children's Fund.

http://www.unfpa.org – United Nations Population Fund.

http://www.census.gov – United States Census Bureau.

Dentition and Dental Care

S Schwartz, J Kapala, and J M Retrouvey, McGill University Health Centre, Montréal, QC, Canada

Glossary

Class I occlusion – Normal occlusion or normocclusion.

Class II occlusion – The upper jaw and/or the upper teeth protrude excessively over the lower jaw and/or over the lower teeth.

Class III occlusion – The lower jaw and/or the lower teeth protrude excessively over the upper jaw and/or over the upper teeth.

Cross bite – Some or all of the upper teeth close on the lingual side of the lower teeth.

Deep bite – The overlap of the maxillary incisors over the mandibular incisors is excessive when the mouth is closed. Sometimes the mandibular incisors impinge on the palatal gingiva.

Dentition – Primary dentition, or baby teeth, followed by mixed dentition and then permanent dentition.

Distal – Toward the sides or back, away from the center or midline.

Gingiva – The gums (soft tissue).

Labial – Toward the outside, cheeks and lips.

Lingual – Toward the inside, tongue and palate.

Mandibule – The lower jaw or lower arch.

Maxilla – The upper jaw or upper arch.

Mesial – Toward the center.

Midline – The imaginary vertical line running down the center of the face from the forehead to the chin. At the jaw it divides each jaw into two, forming four quadrants, by convention named: 1st quadrant: the upper right; 2nd quadrant: the upper left; 3rd quadrant: the lower left; and 4th quadrant: the lower right.

Names of the teeth in the primary dentition – Beginning at the midline and progressing distally: central incisor; lateral incisor; canine; first molar; second molar. There are no bicuspids or premolars in the primary dentition.

Occlusion – The meeting of the maxilla and mandible in a closed position.

Open bite – One or several teeth in both arches fail to meet when the mouth is closed.

Palate – The roof of the mouth.

Introduction

Primary teeth, commonly called 'baby teeth', are important for the child, not only during his baby years, but also because they will have a lasting effect on the child's permanent teeth and occlusion and on his general welfare in later life. The eruption of the primary teeth contributes to the harmonious development of the child – they allow him to chew his first solid food, help him make his first sounds and pronounce his first words, allow him to smile and communicate, and are an intrinsic part of his personality. They keep the space for a more proper alignment of the permanent teeth. They give early notice of future anomalies, and give the parent or practitioner the opportunity to intervene at a more appropriate time to help change oral habits and help the prognosis.

The Primary Dentition

Growth and Development

At birth the face of the child is wide and short. The oral cavity is just starting to develop, and the upper jaw, or maxilla, and particularly the lower jaw, or mandible, are underdeveloped in comparison to the upper facial structures such as the orbits. Teeth are not present and the tongue occupies the whole oral cavity. The mandible is very small and the chin point is recessed, but will develop during the first years of life. During these first years, the maxilla and the mandible will grow in a downward and forward direction.

Craniofacial growth is a highly complex phenomenon, and different theories have tried to explain facial growth. None has been totally successful.

Development of the Primary Dentition

The primary dentition consists of 20 teeth (**Figure 1**). Each dental arch contains 10 teeth, namely four incisors, two canines, and four molars. The primary teeth are smaller and have a shallower anatomy than their successors. Their enamel is thin, with an average of 0.5 mm thickness compared to 1.0 mm or more in the permanent dentition. Primary teeth are also whiter due to increased water content.

The development of the human dentition evolves in a predictable pattern that can be divided into six different stages, starting from birth and ending with the complete adult dentition. The first stage is the one of interest here, namely from birth to the eruption of the 20 primary teeth that occurs just before the child reaches 3 years of age.

At birth, the primary teeth are at different stages of 'budding', but all of them are present even though they cannot be seen in the oral cavity. The buds are stacked up inside the jaws that are still diminutive and have not yet expanded enough to let all the primary teeth align. The jaw bones will elongate in an anteroposterior direction when the baby is 6–8 months old.

The development of dental hard tissue begins *in utero* between 4 and 6 months. At birth the molars may have traces of enamel covering their cusps, whereas the enamel of the incisors is almost entirely formed. In the case of the permanent dentition, there are no or just traces of enamel formation.

The first teeth to erupt are the mandibular central incisors (6–10 months). They are followed by the maxillary central incisors (8–12 months), the maxillary lateral incisors (9–13 months), and the mandibular lateral incisors (10–16 months). The mandibular and maxillary first molars emerge between 16 and 19 months, and the baby has a smile with gaping holes in the four corners of his mouth, imitating an older adult who would be partially edentulous. The canines appear shortly after, between 19 and 23 months. Finally, the second molars make their entrance, starting at 27 months with completion at 33 months (**Figure 2**).

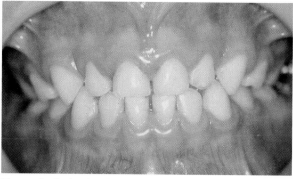

Figure 1 A classic primary dentition: the 20 teeth are straight, well aligned, and have a pleasing appearance.

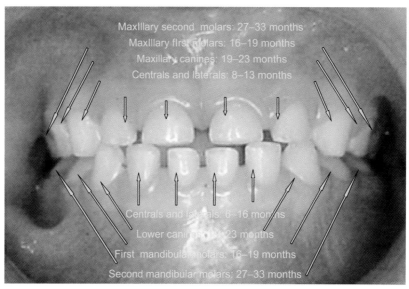

Figure 2 Sequence of eruption of the primary dentition.

A great variation in eruption time may exist. It is not rare to witness a 1-year-old child who has not yet 'grown a tooth'. This delay is still within normal limits, especially when the child has been born prematurely, and delay may be expected for his entire dentition. A few rare syndromes are associated with premature eruption of the primary dentition. Delayed eruptions are more frequent, Down syndrome being the most representative group. What is most important is the sequence of the eruption. A baby whose molar erupts ahead of the incisor presents a highly abnormal situation, either of syndromic or systemic cause (e.g., one of the first symptoms of histiocytosis X disease is the premature eruption of a posterior tooth). Another consideration pertains to the symmetry of the dentition. Each tooth should erupt within 3 months of its sister tooth on the other side. If it does not, one should determine the cause of the delay, such as absence of the bud, lack of space, or extra tooth, and treat it accordingly.

Natal and Neonatal Teeth

Natal and neonatal teeth are teeth present at birth (natal) or present within 30 days postpartum (neonatal). They have an incidence of 1 in 2000–3000 births. Eighty-five per cent of them erupt in the mandibular incisor area. Infants 'born with a tooth' are in good company: Churchill, Napoleon, Gandhi, to name a few, had one tooth at birth. Natal and neonatal teeth rarely are supernumerary teeth. They are almost always genuine primary lower incisors. Since they are very immature and have a minimal length of root development, they appear grayish and are hypermobile. They can irritate the baby's tongue during suckling and/or irritate the mother's nipple if the baby is breastfed. Pediatricians and parents may worry that the child might aspirate these highly unstable teeth. Even though there are no reports of such accidents, these teeth are usually extracted. It is true that, if retained, they are a constant irritant, and even cause occasional bleeding of the neighboring mucosa because of their great mobility. If they are kept they stabilize with the maturation of their roots, but they will always look small, hypoplastic, and unusual. As a rule, they have a shorter lifespan. If they have the advantage of keeping the space for their succeeding partners, it is for a limited time only.

Natal and neonatal teeth frequently occur in children born with a cleft lip and palate. In these children, the teeth are situated in the maxilla, on each side of the alveolar cleft(s). They serve no purpose, and should be extracted as soon as possible. Their extractions are not difficult, as they are attached to the mucosa only, and a minimum of local anesthetic is required. The teeth are not really extracted, but are carefully separated from the gum, and then removed by the use of gauze rather than by the classic pliers to prevent the baby from aspirating or swallowing them.

Teething

Teething has been blamed for fever, rashes, loss of appetite, infections, diarrhea, bronchitis, and even convulsions. Thorough investigations have failed to prove these allegations. Surely, some babies have more difficulties with teething than others. Parents may notice a change in color of the mucosa around the erupting tooth of a complaining or 'cranky' child, which may explain the behavior. These are temporary disturbances. Sometimes, a slight soft elevation of the mucosa appears where the tooth is supposed to erupt. The small mass is soft, darker, and shows different shades of blue. It is an 'eruption hematoma' or 'eruption cyst'. The dental sack gets filled with blood, probably due to a recent local irritation. It is not painful and will break down spontaneously when the tooth erupts. Parents sometimes get quite worried when they see these lesions and need to be reassured. A surgical opening of the cyst is not necessary, and is rarely done.

The Occlusion

The primary dentition has a different and simpler occlusion than the permanent dentition.

The primary teeth are perpendicular to the occlusal plane, which is an imaginary plane that separates the maxillary and the mandibular molars, and passes through the contact between the most anterior maxillary and mandibular incisors.

The midlines for the upper and lower arches are almost always harmonious and properly centered. The dental arches are oval in shape, and these shapes vary with the facial morphology.

The primary dentition has fewer and less apparent occlusal variations than the permanent dentition because a large amount of growth is still to come with the eruption of the permanent dentition.

Prevalence of Potential Malocclusion

Of interest to the dentist is the prediction of a potential malocclusion. The occlusion in the primary dentition is a good predictor of future occlusal relationships in the permanent dentition, because in the vast majority of cases, the occlusion in the permanent dentition of the child will be similar to that of his primary dentition.

It is possible to assess the anteroposterior type of occlusion that will probably be present in the adult by observing the position of the primary canines.

Malocclusions

A class I relationship, or normocclusion (**Figure 3**), is considered the desirable dental occlusion. In the primary dentition normocclusion occurs when the upper canine occludes in the embrasure between the lower canine and the first primary molar, and exists 70% of the time. When the upper primary canine shows a position anterior to the mandibular canine, which takes place approximately 25% of the time, it is assumed that the permanent dentition will probably end up as a class II malocclusion (protrusive upper incisors and/or receded mandible). In less than 5% of the cases, the upper canine shows a disproportionate distal positioning of the upper canine

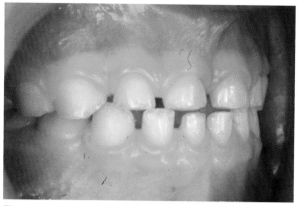

Figure 3 Normocclusion in the primary dentition, right side. The maxillary canine tip is positioned in the embrasure between the mandibular canine and the mandibular first molar.

that will probably end up as a class III malocclusion (retrusive upper incisors and/or protrusive mandible).

Using the primary canines as the guide to occlusion, the canines occlude symmetrically on the left and on the right side of the children 80% of the time, while the remaining 20% shows an asymmetric arrangement.

Cross Bites

It is another pediatric anomaly, occuring in about 12% of the population, and exists when some or all of the upper teeth close on the inside (lingual side) of the lower teeth. When untreated, this condition may lead to serious problems in later life. This condition is generally not genetic, but is often caused by environmental conditions, such as thumb-sucking.

Malocclusions often do not appear very severe in the primary dentition, but they tend to worsen with growth and dentoalveolar development.

Interdental Spaces

Also known as Baume's spaces, interdental spaces frequently occur, especially between the incisors and the canines. These spaces do not increase in size once the teeth are erupted. The assessment of spacing is made exclusively on the basis of whether or not a space exists. In approximately 24% of children there is spacing between all the teeth in the maxilla, and in approximately 15% of children there is spacing between all the teeth in the mandible.

Crowding

Crowding in the primary dentition is defined as close contact between teeth, and exists in about 4% of children in the maxilla and 5% in the mandible.

Contrary to popular belief, the arch width of the anterior part of the maxilla and the mandible will not increase dramatically in size past the age of 3 years. The position of primary dentition is therefore a fairly valid predictor of potential problems regarding arch size and tooth size disharmonies.

If 6 mm of interdental spacing is present in the primary dentition, there will probably be no crowding in the permanent dentition. If there is no interdental space between the primary teeth, there is a 60% probability of crowding in the permanent dentition. Finally, if crowding is present in the primary dentition, severe crowding will be present in the adult dentition.

Problems Associated with the Primary Dentition

Enamel Defects

The formation of dental enamel (amelogenesis) takes place in two phases. In the first phase, the matrix is formed. When the matrix formation is disturbed, whether on its entire surface or in spots or bands only, the tooth is declared hypoplastic. In a second phase of amelogenesis, the matrix becomes calcified. If this calcification is defective in whole or in part, the tooth is declared hypocalcified.

Enamel hypoplasia is less frequent in the primary dentition than in the permanent dentition. This variation is due to the fact that much of the enamel has been formed while in the safety of the uterus. Thus, children who are born prematurely will miss most of this protected process of amelogenesis, which takes place mostly in the third trimester of pregnancy. Premature infants will show more enamel defects than their counterparts who were born at term. Studies show that, even if the teeth of these children look normal clinically, enamel hypoplasia will often be visible under the scanning microscope. It is believed, though not proven, that difficulties encountered by neonates at birth and shortly afterwards can affect the cells responsible for amelogenesis, the ameloblasts. The teeth most often affected are the second primary molars, because enamel has not yet formed on the tips of the cusps as it normally would have in babies at term.

Enamel constitutes the outer defense of the tooth. An enamel deficiency affecting the whole dentition is usually referred as amelogenesis imperfecta (AI). It is a developmental defect that affects primary and secondary dentitions, and occurs in 1 in 14 000–16 000. The enamel that is present is always defective. The defects take different forms, as there are many types of AI. The teeth may appear small and yellowish (the enamel is too thin and the darker color of the dentin shows through) with sharp edges and sharp cusps throughout the dentition. The enamel may be pitted with irregular shapes, and some teeth may be more involved than others. The problem with AI is that the teeth can be very sensitive to cold, to sweets, and to touch. The children so affected tend to avoid certain foods and resist brushing their teeth. Dental caries occur early and frequently, and the treatment of these teeth is quite challenging. Regardless, the condition should be dealt with as soon as it is diagnosed and before the carious lesions become severe. There are many new dental materials that help in the treatment, such as the new composite materials and new bonding agents.

Dentin Defects

The dentinal tissue lies between the dental pulp and the outer enamel. It represents the bulk of the tooth. It contains 86–90% collagen, mostly type I. A collagen I disturbance will result in defects of the dentin with variable severity. These defects are known as dentinogenesis imperfecta (DI). The teeth are typically grayish with a component of yellow. They have normal clinical shapes and sizes initially, but tend to fall apart easily because of their brittleness. The attachment between the normal enamel and the defective dentin is poor.

There are two main types of DI: one type is associated with osteogenesis imperfecta and exists when defective type I collagen is responsible for the bone disease. A second type affects the teeth only, without bone involvement. This second type is an autosomal-dominant disorder. The teeth exhibit the same general characteristics of unsightly color and brittleness in both types of DI. The defect is consistently severe, affects primary and secondary dentitions, and has 100% penetrance. This is not the case with the first type, where the primary dentition is usually more affected than the permanent dentition, where different teeth in the same arch might show different involvement, and where the penetrance varies from one patient to another. DI teeth also display bulky crowns and slender roots.

The DI defects have multiple dental implications. When the enamel chips away, the tooth is no longer protected and the poor-quality dentin wears off quickly. Decay ensues and pain follows. The bulky crowns lead to crowding, and the slender roots make the teeth more fragile.

Development Disturbances Affecting the Oral Structures

Unless the reader functions in a hospital setting, he is unlikely to encounter many of these rare syndromes. However, a few of these conditions should be mentioned, because they have important dental implications.

Down Syndrome

Down syndrome or trisomy 21 is one of the best-known chromosomal syndromes. As mentioned earlier, the dentition erupts and develops very slowly. The teeth are often smaller, and are not well aligned, because a large tongue and a hypotonic muscle function inhibit a harmonious alignment. The primary dentition is usually complete, but not the permanent dentition. Typically, these children exhibit chronic periodontal disease despite acceptable home care, so much so that chronic periodontal disease is considered part of the syndrome. A defect in neutrophil chemotaxis may be at least partially responsible for the alveolar bone destruction. About 40% of these children have associated cardiac defects, and antibiotic prophylaxis for subacute bacterial endocarditis is a necessary adjunct to the dental treatment.

Cerebral Palsy

Cerebral palsy is not a specific disease, and does not have a consistent presentation, but it is frequent enough (1 in 200 live births) to merit consideration. The oral implications reflect the severity of the symptoms. The children have poor motor-control movements. They may display constant and erratic grinding of their teeth (bruxism), and if so, their molars flatten down and gradually loose their cusps. They have a high degree of periodontal break down, possibly because of the unusual forces they apply on their dentition, possibly because of poor oral hygiene, or because of a combination of both. Malocclusions are frequent, which is easily understandable. Normocclusion occurs more often when there is a proper balance of pressure exerted by the tongue, lips, and cheek muscles, and this balance is compromised in this condition. Malocclusions may develop during primary dentition, and progress into the mixed dentition. These children may drool more than the other children of their age, but treatment of this condition is reserved for older patients.

Cleft Lip and/or Cleft Palate

Cleft lip and/or cleft palate is a birth defect that occurs in 1 in 750–1000 births in North America. Care for these babies is usually provided in health centers where multidisciplinary teams are available. The defect has a direct effect on the oral structure.

The cleft may involve the lip only, in which case the dental arch of the maxilla remains intact, but the pressure exerted by the scar tissue of the lip after its surgical closure (the lip is more rigid and somewhat shortened) prevents the front of the maxilla from developing normally, and causes the upper incisors to tilt toward the palate.

When the cleft involves the alveolus (dental arch or bone hosting the teeth), it usually continues the full length of the palate, separating the maxilla into two sections if the cleft is unilateral, or into three sections if the cleft is bilateral. The cleft interrupts or disturbs the budding of the teeth, which are usually deformed or missing in this area altogether. A bilateral complete cleft is the most severe form of the syndrome. The bilateral discontinuation of the maxillary arch means that the posterior segments are not necessarily collapsed at birth but start to move toward the palate midline shortly afterwards, and that the premaxilla (the anterior part of the arch that hosts the four incisors) is positioned forward and even looks as though it is attached to the nose.

When the posterior segments collapse, they place themselves behind the anterior segment (premaxilla) which is maintained in a forward position; the gap between the nasal and the buccal components of the lip becomes wider and more difficult for the surgeon to close. Several methods may be used in order to approximate the separated portions. The most popular is the placement of a prosthesis ('bulb' prosthesis) on the detached portion of the anterior alveolar ridge, and to bring this segment back toward the palate with the help of elastic material attached to a cap. Other clinicians may choose to use an elastic band directly placed on the segment of the upper lip and attached to the cap or to the cheeks. The continuous pressure applied to the premaxilla allows flattening of the philtrum area, and the surgeon can suture the cleft lip with less undue tension. The suturing of the lip takes place at approximately the 10th week of life for otherwise healthy children who have good weight (10 lb). Some clinicians advocate the placement of a prosthesis (called a plate or obturator) to cover the newborn palatal cleft. It is believed that this appliance facilitates sucking because it closes the oronasal opening, and also has the advantage of keeping the posterior alveolar segments separated, thus maintaining a more normal rounded arch. After the lip closure, another obturator is fabricated to accommodate the growth of the infant's mouth. Other clinicians prefer to treat the baby only with the lip closure, advocating, for example, that sucking is not improved with an obturator, and that an elongated nipple for the milk bottle will deposit the milk sufficiently far enough down the throat, thus avoiding the palatal opening and preventing the passage of milk into the nasal area. They believe that the maxilla can be rapidly expanded when orthodontic treatment begins at an older age, and that there is no need to subject the very young children to years of wearing the prosthesis. They have enough to endure as it is, such as orthodontics, alveolar bone grafts, other maxillofacial surgeries, rhinoplasty, and scar revisions. General consensus has not been achieved.

Most of these children do not possess a full complement of permanent teeth. Their primary dentition is more stable, except when the cleft includes the alveolar bone, in which case some maxillary primary incisors may be missing. Often, teeth bordering the cleft areas are hypoplastic and deformed. In all cases, maxillofacial growth has been disturbed, and malocclusions are inevitable.

Robin Sequence

The Robin sequence combines a cleft palate, a very small chin, and a tongue that has a tendency to fall backward. Its incidence varies from 1 in 2000 to 1 in 8500 births. This large variation is explained by the differences in defining the sequence. Sometimes a diagnosis of a micrognathia (a small chin) is given too hastily and is incorrect.

Besides a very retrusive chin, children with the Robin sequence generally fare better than their cleft lip/palate counterparts. They still lack some permanent teeth, but rarely experience missing primary teeth. At the neonate stage, because their chin is so short, their relatively large tongue is pushed into a retroposition, which may cause an airway obstruction. Again, several treatment approaches are possible. One of these is to fabricate an obturator that will direct the tongue into a more anterior position, causing the chin to develop anteriorly and 'catch up'.

Early Childhood Caries

Early childhood caries (ECC) has been defined by the American Association of Pediatric Dentistry (AAPD) as any carious lesion present in the primary dentition of a child under the age of 6 years. Severe early childhood caries (S-ECC) has been defined as any sign of smooth-surface caries present in the dentition of a child under the age of 3 years.

ECC affects between 5% and 7% of the general population, and reaches higher rates in certain groups, such as in Inuit communities, where it can reach an incidence as high as 75%. Caries are the most prevalent chronic infectious disease of the young, and are responsible for considerable pain, frustration, and expense. They have a definite impact on the child's quality of life and they are a public health problem. This is an important subject and merits a more thorough discussion.

Dental caries is a chronic infection caused by the microorganism *Streptococcus mutans,* which when develops without interference, leads to carries. *S. mutans* is hardly present in the infant's mouth at birth in part because it requires the presence of teeth in order to continue to exist in the oral cavity. It is believed that the mother transmits her own bacteria to her child through her saliva (kissing, tasting, sharing food, etc.). Mothers who have high levels of dental caries and *S. mutans* have children with higher numbers of dental caries than children whose mothers have healthy teeth.

Streptococcus mutans alone cannot cause the disease. The interaction of three variables is necessary to produce a carious lesion: *S. mutans,* fermentable carbohydrates, and teeth. One must also include a time factor, as the microorganisms must adhere to a sufficient length of time on the enamel surface in order to damage it. The carious process starts with the decalcification of the enamel, but it does not happen continuously. Enamel does remineralize if the attack is not continuous and not too aggressive. Saliva is the main natural agent in remineralization, and acts as a buffer. This explains in part why the lower incisors, closest to the sublingual salivary glands and therefore more protected by the saliva, are less vulnerable to decay.

Unfortunately, in very young children with newly erupted teeth (that are more susceptible to demineralization than older mature teeth), the speed and the extent of the damage is not as controllable as for older children.

The literature points to the same causes for S-ECC: most consistently blamed is the frequent and prolonged consumption of liquids containing fermentable carbohydrates, such as juices, sodas, and even milk. Some authors blame prolonged breastfeeding, when the baby gets fed on demand, including at night, but this claim is not endorsed by the whole profession. Precise data do not exist.

Dental plaque is a complex transparent pellicle that adheres to the dental enamel. It harbors the remnants of carbohydrates, which in turn interact with the *S. mutans* that have colonized the teeth. Other bacteria join in. The microorganisms use the carbohydrates and produce acid. The acid starts the decalcification. The rate of the decalcification will depend upon the number of *S. mutans* colonies, the frequency and duration of the carbohydrate intake, and the length of time the dental plaque is allowed to remain on the teeth. Thus, it becomes evident that good oral hygiene (tooth brushing, removal of the dental plaque) and healthy diet (avoiding frequent consumption of fermentable carbohydrates) are essential weapons against ECC.

Unfortunately, this common-sense knowledge does not seem to have reached those communities where the children have a high rate of this disease. If statistics show a general decline in tooth decay in some industrialized countries, they also confirm an alarming increase in the rate of S-ECC in certain social groups. This discrepancy can be explained as follows: ECC is predominant in certain ethnic subpopulations in North America and in low-income families. It is well known that these groups are the most difficult to reach, for they are well entrenched in their social habits, or they have a limited concept of prevention. The problem is further complicated by the fact that the children may be very young when the carious lesions begin (as early as 8 months of age). By then the babies have gotten used to the 'sweet stuff', and this will make it very difficult for parents to stop this deleterious habit. It may take years before a dentist has a chance to help, and because the S-ECC carious lesions progress rapidly, they remain undetected until it is too late for prevention.

Prevention of ECC

In the past, pediatric dental care started when the child started school, around the age of 5 years. Now this is deemed much too late. The AAPD believes that the dental profession now has enough tools to prevent and control caries in young children. It makes the following recommendations: tooth cleaning should start, and *ad libitum* nocturnal breastfeeding should be stopped, as soon as the first tooth erupts. Infants should not be put to sleep with a bottle containing fluid given other than water (the care should feed the baby in her arms and put the child to bed when the milk bottle is finished). The consumption of milk or carbohydrate-fermenting liquid should be supervised. The first dental appointment should be at age 1 year, or 6 months following the eruption of the first tooth. This would allow the detection of a precocious lesion at a stage when the process could be reversed, given appropriate care. It would give the clinician the opportunity to find out if the child is at risk for ECC, and provide the parents with appropriate counseling. Finally, the mother or the primary caregiver who has a high rate of untreated caries should make an attempt to decrease her level of *S. mutans* in order to lower the risk of infecting the baby.

Fluoride

Fluoride is known to reduce the risk of caries and promote enamel remineralization. When present in the community water supply (systemic form) at an optimal concentration (0.7 ppm), it has been proven to reduce caries by 40–60% and the cost of dental healthcare by 50%. Systemic fluoride is incorporated into the enamel and dentin of developing teeth and makes them more resistant to demineralization provoked by acids. Furthermore, systemic ingested fluoride becomes incorporated in the saliva, that, as a buffering agent, offers added protection to the teeth. When combined with other measures, such as the use of fluoridated toothpaste (topical form), it further reduces this risk. Since most of the enamel of the incisors and the first permanent molars is formed between the ages of 3 months and 3 years, the intake of fluoride during this period is of the utmost importance.

Most of the larger cities in North America have incorporated fluoride in their community water. When they have not, the level of natural-occurring fluoride is usually below 0.3 ppm. In such cases, fluoride supplements are advised. These vary according to the age of the child, the amount of natural fluoride already present in the water, and the risk the child has to develop dental caries. The risk is calculated according to the following factors: does the child get his teeth brushed regularly (a child under the age of 3 years cannot do the brushing by himself), does he have a medical condition that prevents good oral hygiene or weakens his general defense such as a reduced salivary function, does his diet predispose him to the disease, and does his mother and/or do his siblings have a history of caries?

Supplements are not given before the age of 3 months, and thereafter the amount given is calculated according to the age of the child and the amount of naturally occurring fluoride in the water. Tables for the amount of required dosage are readily available. Care must be taken to prevent the child from swallowing too much toothpaste (no more than a small pea-size quantity of paste or gel per brushing).

If the combination of city water fluoride, toothpaste fluoride, and other sources such as baby formula raise the intake of fluoride to 2.0 ppm, the risk of fluorosis appears. Fluorosis is mild, moderate, or severe. Mild fluorosis consists of minute white spots covering the tooth that in time fade and blend, giving the tooth a very white appearance that in fact is not esthetically unappealing. Moderate fluorosis, characterized by brown spots, and severe fluorosis, characterized by a brown or molted tooth, is very rare in North America.

Certain groups have expressed concern about the addition of fluoride to domestic water, citing, among other reasons, the danger of fluorosis. But fluorosis may begin to occur only when the level of fluoride reaches more than twice the optimal level, which today is established at 0.7 ppm, having been reduced from 1.0 ppm in 2004 largely because of these concerns. In most cases, fluorosis is a minor cosmetic defect that should not be cause for alarm.

The World Health Organization has declared that the addition of fluoride to the community water has been one of the most successful public health measures of the past decades.

Gingivitis and Periodontal Disease

Gingivitis

Gingivitis is an inflammation of the gingiva. In young children it is usually limited to the marginal gingiva, which is the band of unattached soft tissue that borders the teeth, and is normally about 1 mm wide. The inflammation may be either localized or generalized. Since the inflammatory process is limited to the marginal gingiva, there is neither formation of gingival pockets nor resorption of the underlying alveolar bone. It is not rare in children and is reversible. It is called dental plaque-induced gingivitis. The plaque forms originally on the tooth surface, and contains a limited number of bacterial species. These are streptococci and Gram-positive bacteria, and are largely responsible for dental caries. The plaque matures when it is not removed mechanically, and other bacteria join in, including the Gram-negative ones that are predominantly responsible for gingivitis. A complex aggregate of microorganisms is formed and it has the ability to adhere to the tooth and provoke gingival inflammation around the sulcus. A good regimen of plaque removal and brushing should restore health to the gingiva, and flossing can be initiated as soon as there is contact between the teeth, first by the caregives and then by the child. There are flossing devices readily available at most pharmacies that allow most children to begin flossing on their own as early as 4 years of age.

A generalized gingivitis that does not respond to simple oral health attention within 2 weeks should alert the clinician. Differential diagnoses include a viral infection, a change in the neutrophil count (such as in leukemia), or a systemic disease (such as in Crohn's disease). These are rare situations in the young pediatric group, except for primary herpetic infection. This infection, which is common among children aged 1–3 years, is a stomatitis affecting the mucosa of the entire oral cavity (herpetic viral infection). A severe gingivitis precedes the stomatitis. The entire gingiva is red, slightly swollen, and may bleed spontaneously. The condition lasts 2–4 days. Shortly afterwards small white vesicles appear. They are very painful, last 12–15 days, after which the tissue heals without scarring.

Periodontitis

Periodontitis is the progression of the inflammatory process to the underlying tissue and to the alveolar bone. It is rare in infants and very young children. It is associated with systemic diseases or syndromes, such as the Papillon–Lefèvre syndrome, hypophosphatasia, Langerhans' cell histiocytosis, leukemia, cyclic neutropenia, and agranulocytosis. In some cases, the teeth become mobile and may even fall out spontaneously. In Down syndrome, neutrophil and immune cell functions are deficient, and this may play a role in the pathology.

Ranula

Ranula is a benign lesion caused by a partial obstruction of the duct of a major sublingual salivary gland. This obstruction causes the gland to fill up with mucus and to enlarge slowly below the tongue. It is easily treated. Other minor salivary glands can also obstruct causing smaller lesions (mucoceles), and are rare in babies.

Traumatic Injuries and Treatment Options

The greatest incidence of trauma to the primary dentition occurs between the ages of 2 and 3 years. Some of these accidents can be avoided (safe surroundings, good car seats, proper supervision), some cannot (simple falls, collisions with other children). The injuries can involve the teeth alone, the supporting tissue, or both. The maxillary and mandibular anterior areas, as well as the chin, are the sites most commonly injured. When the chin is hit, the blow is transmitted to the site of the articulation of the lower jaw (condyle) of the child and to his posterior teeth. This type of injury can lead to condyle fractures and/or to fractures of the molars. The goal of the clinician who assumes treatment of traumatized primary dentition is to ensure the safety of the permanent dentition.

The loss of primary maxillary incisors does not lead to a shortening of the space for the permanent dentition because the arch space for the incisors is maintained by the canines. These canines emerge before the age of 2 years, that is, before the high incidence of trauma. Since the incisors are almost always the injured teeth, the damage remains limited to a temporary diminution of the height of the alveolar bone. However, the loss of a canine or a molar will automatically lead to disturbances in the permanent dentition.

Injuries to the teeth include infarctions, crown fractures, and root fractures. Infarctions (cracks through the enamel) are unimportant and rarely lead to complications. Equally unimportant are minor crown fractures (enamel only or enamel with a small surface of dentin). When the fractures involve a deeper part of the dentin, the chances for sensitivity of the tooth and necrosis of the pulp become higher. In such cases the damaged surfaces should be coated with a protective material. Necrosis of the tooth is unavoidable when the fracture is across the pulp and when treatment is not immediate. Root fractures are frequently undetected, especially when they take place at half the apical length of the root.

Injuries to the supportive tissue occur when the tooth is concussed or loosened (subluxation) or displaced (luxation) or expelled from its alveolar socket (avulsion). Concussed teeth have inconsequential repercussions, whereas injuries where the periodontal ligament and the alveolar bone have been damaged have harmful consequences. Young children have a very elastic alveolar bone that will easily yield to blows. This explains why, when there is a blow to the dentition, permanent teeth are mainly broken in older children, whereas in younger children primary teeth are more often displaced or intruded. It is typical of the incisors in very young children to be driven into the alveolar bone and disappear completely from sight (total intrusion). Intruded primary teeth will eventually re-erupt spontaneously, but their prognosis is poor because the alveolar bone has been crushed and the apical area has been irreversibly damaged.

An intrusion or a severe luxation of a primary tooth can have a harmful effect on its permanent successor if the latter is still in the process of forming. Immediate extraction of the traumatized tooth does not necessarily avoid injury to the underlying tooth, because the damage, if any, to the underlying tooth, happened at the time of the blow.

A discolored primary incisor, otherwise intact, strongly suggests a previous trauma. The incisor appears yellow when the pulp space is calcified (which by itself is not severe pathology), and shows different shades of gray when it has become necrotic. The clinician should be alerted in such a case, because such a 'dead' tooth does not cause any pain or discomfort, but may indicate a chronic infection with the presence of a larger zone of inflammation around its apex, thus creating a serious risk of damaging the permanent tooth or hindering its path of

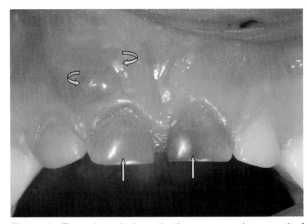

Figure 4 Two primary incisors that have necrosed as a result of a trauma (ascending arrow). The curved arrows show the presence of an abscess that is ready to burst.

eruption (**Figure 4**). Sometimes the inflammation or chronic infection around the root of the necrosed tooth will develop into an obvious abscess.

Avulsed primary teeth should not be reimplanted. The chances of failure are high and the health of the permanent successor may be jeopardized.

The Impact of Oral Habits and Other Environmental Factors on the Oral Structure in the Primary Dentition and the Future Dentition

Equilibrium of Environmental Forces

The portion of the jaw bones that hosts or will host the teeth (dentoaveolar process) develops during the primary tooth eruption, and the dentition will position itself in equilibrium between the pressure exerted by the tongue and that exerted by the muscles of the lips and cheeks. This pressure contributes to the position of the teeth on the arches. Intermittent short-duration pressure such as masticatory force does not have a significant impact on the position of the teeth. However, light and continuous force can move teeth very effectively. This is a unique characteristic of the periodontal ligament. It results in bone resorption on the pressure side, and bone apposition on the tension side of the dental root. Certain oral habits or abnormal tongue positions have the potential of upsetting the equilibrium. If the equilibrium is broken, for example by thumb-sucking, undesirable tooth movement will result, and a malocclusion may develop.

Non-Nutritive Sucking

In the newborn, the sucking reflex is normal. A newborn child will suck if an object is placed in his mouth. This reflex disappears once the baby is placed on a coarser diet. If the child continues to suck on an object or his thumb,

the habit becomes a non-nutritive sucking habit. Almost 50% of infants have a non-nutritive sucking habit that continues until 24–28 months of age. The most common objects used for a non-nutritive sucking habit are the thumb and the pacifier.

Pacifier and Thumb-Sucking

Thumb-sucking and pacifiers alter the occlusion in different ways. Many infants who suck their thumbs or use pacifiers develop occlusal interference between the upper and lower primary canines. This usually self-corrects if the use of the pacifier is discontinued. The anterior open bite is normally more obvious and visible earlier in pacifier-suckers than in thumb-suckers. Both are associated with tongue-thrust during swallowing. With pacifier-suckers, the open bite corrects itself spontaneously when the habit stops, despite the tongue-thrust. Sucking a pacifier is more clearly related to posterior cross bites in the primary dentition than is thumb-sucking.

Usually, thumb-sucking lasts longer than using the pacifier, and therefore has a greater potential to create a malocclusion. The severity of the malocclusion will depend on the duration of the habit, on the amount of force placed on the dentition, as well as the position of the thumb in the oral cavity.

Parents should not worry unless this habit continues beyond age 3 years. It normally stops by the time the permanent incisors erupt.

Normal Swallowing Pattern and Tongue Thrusting

During mature swallowing, the lips relax, the tip of the tongue places itself behind the maxillary incisors, the bolus or saliva is pushed toward the throat, the posterior teeth contact, and the bolus descends into the pharynx. The oral preparation phase of the swallowing pattern is voluntary, while the esophageal one is involuntary. This mature swallowing pattern does not fully take place until the child is 4–5 years old.

Tongue thrusting, known alternately as 'infantile swallow', 'deviate swallow', or 'visceral swallow', is normal in the newborn and will eventually evolve into mature swallowing. This evolution is coincident with the eruption of the primary dentition and the dentoaloveolar development. A tongue-thrusting 'habit' is a combination of an abnormal tongue position, where the tongue is positioned too anteriorly, and an abnormal movement of the tongue during swallowing.

Tongue thrusting has been associated with anterior open bites, protrusion of the maxillary incisors, narrowing of the upper dental arch, and speech problems such as lisping. Treatment is usually limited to observation, as the vast majority of the infants will eventually develop the mature swallowing pattern. Once this mature pattern is adopted, the protrusive incisors and the open bite will tend to self-correct.

Mouth Breathing

Nasal breathing is believed to be the 'normal' breathing pattern of the newborn. Nose breathing, mature swallowing patterns, and correct chewing patterns are all more frequently associated with a normal occlusion and with harmonious facial features. However, mouth breathing occurs in about 30% of the newborn without harmful effects. Chronic mouth breathing is frequently associated with orthodontic problems. The fact that the mouth is constantly open results in a downward movement of the mandible and the appearance of a long face. The upper molars in the permanent dentition may erupt beyond their normal length, contributing to an altered pattern of the growing mandible. Chin point is displaced distally and inferiorly, and contributes to the convex profile and the formation of an anterior open bite.

Orthodontic Intervention in the Young Patient

Orthodontic treatment is generally not initiated in the very young patient because of the lack of psychological maturity of the patient who may be reluctant to wear fairly complicated appliances, and because the patient is in primary dentition, and will show a large amount of growth in the years to come. Corrections in the very young patient, even if successful, may not be permanent, and may require a second phase of treatment in the mixed dentition.

Simple treatment, such as the correction of occlusal interference between the upper and lower canines may correct a posterior cross bite, and simple appliances may be used to correct functional anterior cross bites where the upper incisors are in occlusion behind the lower incisors.

The role of the orthodontist in very young patients is mainly to observe and intervene only in a very limited fashion.

See also: Birth Complications and Outcomes; Birth Defects; Breastfeeding; Cerebral Palsy; Crying; Down Syndrome; Feeding Development and Disorders; Nature vs. Nurture; Nutrition and Diet; Premature Babies; Safety and Childproofing.

Suggested Readings

American Academy of Pediatric Dentistry (2005–2006) Reference Manual, Vol 27, No 7.
Andreasen JO, Andreasen FM, Bakland LK, and Flores MT (2003) *Traumatic Dental Injuries, A Manual,* 2nd edn. Oxford: Blackwell Munksgaard.
Enlow D, Hans M, and McGrew L (1996) *Essentials of Facial Growth.* Oxford: Elsevier.
Kawashima S, Peltomäki TH, Sakata H, Mori K, Happonen RP, and Ronning D (2002) Craniofacial morphology in preschool children with sleep-related breathing disorder and hypertrophy of tonsils. *Acta Paediatrica* 91(1): 71–77.
Larsson E (1994) Artificial sucking habits: Etiology, prevalence and effect on occlusion. *International Journal of Orofacial Myology* 20: 10–21.
McDonald RE and Avery DR (2004) *Dentistry for the Child and Adolescent,* 8th edn. Oxford: Mosby.
Miller MJ, Martin RJ, Carlo WA, *et al.* (1985) Oral breathing in newborn infants. *Journal of Pediatrics* 107(3): 465–469.
Orenstein MD and Susan R (2006) Oral, pharyngeal, and esophageal motor disorders in infants and children: Part 1 Oral cavity, pharynx and esophagus, GI motility online. Available at http://www.nature.com/gimo/contents/pt1/full/gimo38.html (accessed 14 July 2007).
Profitt WR and Fields HW (1999) *Contemporary Orthodontics.* Oxford: Elsevier.

Depression

J Luby, M M Stalets, and A Belden, Washington University School of Medicine, St. Louis, MO, USA

Glossary

Affect – The observable manifestations of a subjectively experienced emotion.

Anaclitic – A term coined in 1946 by psychiatrist Rene Spitz to refer to children who became depressed after being separated from their mothers during the second 6 months of life.

Anhedonia – The experience of having an inability to derive pleasure in normally pleasurable acts.

Dysphoria – A mood of general dissatisfaction, restlessness, depression, and anxiety; a feeling of unpleasantness or discomfort.

High-risk state – Being more likely than the general population to experience a disorder, on the basis of certain characteristics or conditions.

Mood – One's predominant emotion or state of mind.

Introduction

Depression is a clinical mental disorder characterized by a constellation of dysphoric emotional and behavioral symptoms lasting for 2 weeks or longer. Sad or irritable mood and anhedonia, the inability to experience pleasure and joy, are the core symptoms of depression but physiological changes, known as 'vegetative signs' that include changes in sleep or appetite, are also known to occur. A clinical depressive episode is distinct from a sad mood or a transient grief reaction and is not a normative phenomenon. The prevalence rate of the disorder in adults varies with age and gender but depression stands as a major public health problem worldwide. However, the recognition and treatment of the disorder varies by culture and socioeconomic conditions. Women in the childbearing years are at the highest risk with an estimated 10% prevalence rate while there are lower (approximately 2%) prevalence rates in males and prepubertal females. Therefore, the majority of the population will never experience a clinical depressive state even when faced with significant losses and/or stresses. While the exact etiology of this disorder remains unclear, numerous research groups have established the finding that the onset of the disorder is based on both genetic and psychosocial risk factors. Along these lines, it appears that individuals with a genetic vulnerability to the disorder are at highest risk for an episode of depression when they experience stressful life events. However, it is also possible to have a depressive episode without a prior family history of the disorder although this is less common.

Historically, a distinction between 'reactive' or psychosocially based vs. 'endogenous' or biologically based depression was made in adults. However as more data about the etiology of the disorder has been ascertained, this distinction no longer appears to be valid or clinically useful. Despite findings from numerous studies demonstrating that there is a neurobiological basis to depression, there is currently no medical test that can be used to confirm or disconfirm the diagnosis. The diagnosis of depression is made based on clinical interview and mental status examination of the patient. For an adult, this involves an 'interview' in which thought content and predominant feeling states are assessed among other things. For young children, this includes interviewing parents or caregivers to obtain a detailed history about key emotions and behaviors and also requires direct observation of the child and the parent/child dyad to assess play skills and interests as well as psychosocial and relationship functioning. A 'mental status' examination of a young child should involve observations of play behavior and developmental skills and abilities within a dyadic context.

The Concept of Depression Arising in the Very Young

Depression has been recognized as a mental disorder in adults as early as the late nineteenth century or perhaps earlier. For decades after its discovery, developmental theorists asserted that it would be developmentally impossible for a child to experience a depressive episode. This was based on the notion that prepubertal children would be cognitively and emotionally too immature to experience the complex negative emotions, such as shame and guilt, known to be integral to a depressive state. In contrast, the clinical observations of Rene Spitz (described below) and the theoretical work of psychiatrist John Bowlby documented infantile depression and had an impact on some areas of mental health and public policy. However, despite these descriptions, mainstream child mental health clinicians suggested instead that children would manifest 'masked' symptoms of depression, such as somatization or aggression, in lieu of the typical depressive affects and this presumption was widely accepted in clinical practice despite the absence of empirical evidence.

Subsequently, in the early 1980s child psychiatrist investigators, Dennis Cantwell and Gabrielle Carlson, provided empirical data demonstrating that children as young as 6 years of age could display typical symptoms of depression similar to those that characterize the adult disorder. In their landmark paper entitled 'Unmaking masked depression in children and adolescents' they noted that while 'masked' nonspecific symptoms such as stomach aches, were also observed in depressed children, the 'typical' symptoms known in depressed adults and described in the standard diagnostic manual used to define diagnoses in psychiatry, the Diagnostic and Statistical Manual (DSM), were also the most common and specific markers of the disorder in children. These findings and the findings from many other studies that followed replicating and expanding these data revolutionized public health as it opened the doors to the recognition and treatment of depressive disorders in children.

One reason why the discovery of a depressive syndrome in children was so delayed is perhaps related to the underlying idea and wish that childhood should be a joyful time of life. It is unpleasant and difficult to imagine a depressed child. Based on this historically, and to a surprising degree still currently, there is a great deal of social resistance to the idea that depression can arise in childhood. Following this pattern of thought, the idea of depression arising even earlier in life, during the infancy and preschool period, is particularly difficult to imagine or accept. As will be reviewed in this article, there is now empirical evidence demonstrating that depression can arise in children as young as 3 years of age when developmentally adjusted symptom manifestations are assessed.

While empirical studies of clinical depression have not yet been conducted in infants and toddlers, clinicians have observed and described infant depression for some time. Numerous high-risk studies demonstrating that the infants of depressed mothers had a greater tendency for negative affect and depressed mood (described below) have also provided support for the idea that depressive affects could occur much earlier in life than previously recognized. These data on early alterations in emotion development set the stage for an exploratory study of clinical symptoms of depression in children between the ages of 3 and 6 years. However, to date there are no available controlled studies that have investigated the question of whether children under the age of 3 years can experience a clinical depressive syndrome.

Theory and Early Observations of Infant Depression

Psychoanalyst Melanie Klein was perhaps the first to elaborate on the idea that a form of depressed affect arose in infancy. She postulated that infants experienced a normative and transient 'depressive position' at approximately 8 months of age. Klein proposed that this very early affective experience represented feelings of guilt that emerged as a result of the infant's aggressive impulses toward the caregiver. This theory has had little clinical application or utility in the practice of mainstream mental health and was never empirically tested. Current developmental data demonstrating that the capacity for guilt does not arise until approximately 3 years of age would suggest that this theory does not represent a valid phenomenon. Therefore, Klein's 'depressive position', which postulates a nonclinical developmental form of depressive experience, is now of interest as a unique historical developmental theory.

The first published observations of depressed affect arising in infants date back to the mid-1940s when psychoanalyst Rene Spitz provided compelling reports of withdrawal, apathy, depressed mood, and failure to thrive in institutionalized infants. He suggested that these infants who had been separated from their primary caregivers and placed in institutional settings were displaying a syndrome he referred to as 'anaclitic depression'. This name was based on the idea that this was a depression that arose secondary to separation from the caregiver with whom the infant was developing a close relationship to and was dependent upon. Even more astonishing, these infants who were deprived of the opportunity for primary caregiving relationships displayed failure to thrive despite adequate nutrition and physical care. Remarkably, these delays were found to diminish significantly after the child and primary caregiver were reunited. Spitz also described 'hospitalism' in which infants institutionalized very early in life, prior to the development of a close relationship

with a caregiver, displayed more severe delays, which were thought to be largely irreversible. Despite how remarkable this finding of depressed affect and physical growth retardation arising from psychosocial deprivation was, Spitz's observations had little impact on the practice of mainstream child psychiatry for decades (despite impact in other areas of public health).

High-Risk Studies of Infants of Depressed Mothers

Numerous empirical studies since the mid-1980s have shown that maternal depression is a risk factor for a range of poor developmental outcomes in children. Maternal depression that extends beyond the transient experience of 'baby blues' and crosses the threshold into a clinical postpartum depression, or a more chronic major depressive syndrome, has been shown to be associated with impairments in caregiving capacities. Findings indicate that overall depressed mothers are less responsive, display less positive affect, and gaze at their infants less than nondepressed mothers. Other findings suggest that mothers experiencing chronic depression had difficulty providing an adequate level of social stimulation for their babies. Depressed mothers touched their babies less, engaged in fewer games, and participated in fewer activities with their infants compared to nondepressed mothers. Depressed mothers have also been found to talk to their infants less than nondepressed mothers do. Further, studies show that instead of reciprocating their infants' smiles, depressed mothers are more likely to look sad and anxious while interacting with their infants. Importantly, these alterations in parenting have been associated with a range of negative emotional developmental outcomes in infancy and early childhood.

These 'high-risk' studies focusing on the mood and affective responses of the infant offspring of depressed mothers have produced converging evidence demonstrating that maternal depression occurring early in a child's development may have adverse effects on the emotional development of infants. Observational paradigms in which infants' facial expressions and motor activity in response to evocative events are observed and systematically rated have been designed so that inferences about the emotional states of infants can be made. Such indirect measures are necessary due to the infant's inability to make his/her feeling states clear. Jeffrey Cohn and Edward Tronick are two developmental psychologists who were among the first to design such a paradigm to test the emotional states of 6-month-old infants in response to their mother's varying emotional expressions. These investigators designed a paradigm known as the 'still face' during which the infant is seated directly facing their mother. Mothers are instructed to respond to their infants and then to behave in an artificial emotionally unresponsive fashion in an effort to simulate a depressed and withdrawn affect, similar to what might be seen during a depressed state.

The response of the infant to their mother's lack of reciprocal emotional response has been shown to vary significantly depending upon the infant's past experience of parenting. Studies utilizing this paradigm in healthy vs. depressed mothers and their infants have indicated that infants of depressed mothers are less active, more withdrawn, and display less positive affect than infants of nondepressed mothers at baseline. Of key importance was that in response to mother's expression of a 'still face' these infants displayed less protest than the infants of nondepressed mothers. These behaviors suggested that infants of depressed mothers are accustomed to maternal nonresponse and do not protest or experience this as an unusual event. These findings were among the first to demonstrate the sensitivity of very young infants to the emotional states of their caregivers. The implications of this are very important as they confirm that early interpersonal and environmental factors may have a material impact on key aspects of development in the infant and very young child.

Maternal Depression and Its Influence on Infants' Psychobiological Processes

In addition to inferring differences in infants' emotional responses based on their facial expression or bodily movements, investigations that look at other physiological markers of reactivity, such as brain activity and heart rate variability, are important for understanding influential factors of emotion development during the infancy period. Specifically, developmental researchers use electroencephalogram recordings to trace connections between electrical activity in the brain and ongoing thoughts as well as emotions being experienced by infants. Records from an EEG are obtained from a series of electrodes that rest on the scalp allowing participants to move freely, making this technique especially useful for infants. Richard Davidson is a psychologist who was among the first researchers to show that the left side of the frontal area is associated with approach-type emotion reactions, which are positive emotions, such as joy in adults. These same researchers also demonstrated decreases in left frontal brain activity in depressed adults. Withdrawal and inhibitory emotion reactivity, such as fear, have been traced to the right side of the brain. Numerous studies using both adults and very young infants have found that asymmetries in frontal lobe activation and function are related to discrete emotions. That is, results indicate that right frontal lobe activations are more likely to occur during crying and sadness, whereas relatively stronger left frontal lobe activation occurs during happiness.

Psychologist Geraldine Dawson proposed that differences in individual children's frontal lobe activation might be a result of life experience as opposed to innate biological factors. The significant role that parents play in infants' emotion development capacities (e.g., regulation, expression, and understanding), undoubtedly accounts for some of the difference in children's frontal brain activity. Empirical findings indicate that during pleasurable and playful interactions with their primary caregivers, infants typically show greater activation in the left frontal area. Dawson and colleagues found that infants of depressed mothers show no difference in left and right activation of the frontal area, indicating that infants of depressed mothers may not find mother–child interactions highly pleasurable. These results support the notion that caregiver socialization can influence frontal asymmetries. One of the critical questions to be answered by future longitudinal studies is whether measures of frontal lobe activity and asymmetry can predict vulnerability for emotional disorders both concurrently as well as later in life.

Overall, these developmental studies provided key evidence for the previously unrecognized relative emotional sophistication of infants and very young children. That is, these data demonstrated that infants were aware of and sensitive to emotional factors and events in their environments at very early stages in development. Such findings opened the door to the idea that early alterations in mood and affect could occur and might be early markers of risk and/or signs of clinical depressive syndromes. There now is an established body of evidence pointing to the emotional sensitivity of young children. This literature was reviewed by a group of scholars in 2000 resulting in the conclusion that "all children are born wired for feeling" and that the quality of early environments and relationships sets the stage for later emotional health.

Empirical Studies of Preschool Depression

Several clinical reports of depression occurring in preschool children were published in the 1970s, but little systematic study of the disorder was conducted until the 1980s when Javid Kashani and colleagues published a series of papers examining the existence of depression in preschoolers using standard diagnostic criteria. These studies identified several children from community and clinical samples who met the widely accepted adult Diagnostic and Statistical Manual (DSM)-III criteria for major depressive disorder (MDD), providing further evidence for the existence of preschool depression. However, the disorder defined by the DSM criteria for adults was found to occur less frequently in preschool children compared to the prevalence rates in older children, adolescents, and adults. In addition to children who met full diagnostic DSM-III MDD criteria, larger numbers of children who displayed concerning depressive symptoms, but failed to meet diagnostic criteria were also identified. These findings led to the suggestion that due to their immature development, preschool age children may manifest symptoms differently than how they were described in the DSM system. These authors therefore suggested that modifications to DSM criteria to identify depressed preschoolers should be explored.

Using age-appropriate diagnostic measures that assessed for age-adjusted symptoms of depression, data from two independent samples of preschoolers in the St. Louis metropolitan area have demonstrated that preschool children can manifest a clinical depressive syndrome. An example of a developmental adjustment would be a focus on negative or sad themes in play, instead of focusing on the emergence of this in thought as would be done in adolescents or adults. Based on the preschool child's more limited verbal abilities, the content of play themes is the best representation of the young child's mental preoccupations. Developmental modifications to the standard DSM criteria, which were developed largely for application to adults, are outlined in **Table 1**.

Symptoms of preschool depression have been found to cluster together and to differentiate depressed preschoolers from those with other psychiatric disorders.

Similar to Cantwell and Carlson's findings with older children mentioned earlier, depressed preschooler children exhibited higher rates of 'typical' depressive symptoms rather than 'masked' symptoms. Thus, age-adjusted typical symptoms serve as the best clinical markers even in this very young age group. One distinction between depressive episodes arising in children compared to adults is that the symptom of irritability may present instead of the symptom of sadness. For this reason the DSM system has one adjustment in the formal criteria for depression for children and it is that irritability may present instead of sadness. The symptom of anhedonia, or the inability to experience pleasure, emerged as a key and highly specific symptom of depression in young children. This means that the presence of this symptom serves as a specific marker of depression and can distinguish this from other psychiatric disorders. In the young child, anhedonia may manifest most typically as an inability to enjoy activities and play. Because joyfulness and the experience of pleasure in play is a key focus of the young child's life, it stands to reason that the absence of this is a marker of an aberrant developmental process. Depressed preschooler children who demonstrate anhedonia appear to have a more severe and biologically based subtype of depression. While the presence of this symptom is cause for concern, it alone does not mean that the child is depressed. If this symptom arises and endures over a period of several days to weeks, referral to a mental health clinician is warranted.

One issue that arises when considering psychiatric diagnoses for very young children is the prevailing attitude that most children 'grow out' of their difficulties and therefore they are of little clinical consequence. However, available evidence indicates that this is not true for many preschool-onset mental disorders. Along these lines, there is evidence that preschool depression is stable over a 6-month period, which is a considerable span of time considering the rapid development that occurs in the preschool period. Further, depressed preschoolers appeared impaired both developmentally and functionally when compared to normally developing peers. These findings indicate that preschool depression is of significant consequence to the young child as it affects their general functioning and development and therefore warrants intervention. Further, evidence of impairment is especially important in establishing the validity of preschool depression, as impairment is a key marker of a clinically significant mental disorder according to the DSM system.

In addition to a specific and stable symptom constellation, other key markers of diagnostic validity have been found in depressed preschoolers. Similar to findings of studies of older depressed children and adults, depressed preschoolers have a greater family history of depression and similar disorders than comparison groups. Biological markers, specifically changes in stress hormone reactivity, have been found in depressed preschoolers lending additional support for the validity of the disorder.

As recognition of the existence and clinical characteristics of depression arising during the preschool period is

Table 1   Diagnostic criteria for MDD adapted for preschool age

1. Depressed, sad, or irritable mood for a portion of the day for several days, as observed (or reported) in behavior.
2. Markedly diminished interest or pleasure in all, or almost all, activities or play for a portion of the day for several days (as indicated by either subjective account or observations made by others).
3. Significant weight loss or weight gain (not explained by normal growth) or decrease or increase in appetite nearly every day.
4. Insomnia or hypersomnia nearly every day.
5. Psychomotor agitation or retardation nearly every day (change that is observable by others).
6. Fatigue or loss of energy nearly every day.
7. Feelings of worthlessness or excessive or inappropriate guilt (that may be only evident as persistent themes in play).
8. Diminished ability to concentrate on a task, or indecisiveness, for several days (either by subjective account or as observed by others).
9. Recurrent thoughts of death (not just fear of dying), recurrent suicidal ideation without a specific plan, or a suicide attempt or a specific plan for committing suicide. Suicidal or self-destructive themes may be evident only as persistent themes in play.

relatively new, little information is currently available regarding the prevalence of the disorder. However, the most recent and most diagnostically and developmentally specific data indicate that 1.4% of preschoolers exhibit depressive symptoms consistent with DSM-IV criteria. As there are over 6 000 000 children between 3 and 6 years of age in the US, as many as 84 000 preschoolers in the US may be experiencing clinically significant depressive symptoms and it is reasonable to speculate that the vast majority of these children are not identified as depressed or offered treatment to target depressive symptoms.

Given findings supporting the validity and clinical importance of preschool depression, continued investigations of the characteristics and antecedents of this early-onset disorder are warranted. Longitudinal studies following depressed preschoolers through school age would provide important information about the longer-term outcome of these children. Biological correlates of depression, including differences in the size and shape of specific brain regions have been found in depressed compared to healthy adults. A study in which depressed preschoolers and healthy control children undergo structural brain imaging is currently underway and may provide additional support for the validity of preschool depression and clues to the neurobiological mechanisms that underlie the disorder. An important area of investigation is the determination of factors that place preschoolers at risk for depression. Information relating to family history of psychiatric disorders, history of stressful events, and the quality of the parent–child relationship is being investigated and may ultimately allow for early identification of those preschoolers at greatest risk.

Depression in Infants and Toddlers

Although empirical investigations of clinical depression in infants and toddlers are not yet available, an alternative developmentally sensitive diagnostic system entitled the 'Diagnostic classification of mental health and developmental disorders in infancy and early childhood' has outlined diagnostic criteria and symptom descriptions designed for application to this younger group. This diagnostic system is based on the experience and clinical observations of a multidisciplinary group of mental health clinicians and has also been informed by the available empirical database; a revised edition was published in the year 2005. A section on depression of infancy and early childhood outlines proposed developmental translations of depressive symptoms that encompass two diagnostic categories: major depression and depressive disorder not otherwise specified (NOS). These categories are designed to apply to infants and toddlers and may provide a useful framework for future empirical investigations. In addition, the DC:0–3R includes a unique

category entitled 'Prolonged bereavement/grief reaction' that addresses the more transient depressed affect that may arise after the loss of a primary caregiver.

Numerous compelling clinical observations of depressed infants and toddlers, as well as the findings of alterations of affect in infants at high risk for mood disorder previously described, suggest that a clinical depression can arise at this earlier point in development. However, at this point in time controlled investigations of clinically significant depression among children younger than 3 years of age have not yet been conducted. Therefore, this is an area in which empirical studies are needed to inform diagnostic classification and clinical identification.

Treatment

As preschool depression has only recently been recognized and efforts to validate the disorder are still in progress, treatment studies have not yet been done. Therefore, there are currently no empirical treatment studies to guide clinical practice for depressed preschool children at this time. However, in general, psychotherapeutic treatments for preschool-aged children are conducted in a dyadic context. That is, the preschooler and their primary caregiver are seen together as a 'couple' and the relationship becomes the focus of the treatment. This is based on the idea that the young child cannot be viewed as an independent entity and is inextricably dependent upon the caregiver for emotional and adaptive functioning. Therefore, a primary focus of treatment is on helping the caregiver to meet the emotional needs of the young child better and to facilitate positive development. It should be noted that the dyadic focus does not imply that the etiology or cause of mental problems in young children is based on relationship problems. Rather, because the primary caregiver plays such a key role in the young child's life they are in a unique position to facilitate healthy development.

In keeping with the importance of the dyadic relationship in the treatment of early onset psychopathology, the DC:0–3 system includes a unique relationship axis. This represents an addition to the multiaxial diagnostic system as it is outlined in the DSM system that focuses on the quality of the caregiver–child relationship and considers that disorders may be specific to relationships in young children. In addition to consideration of this issue, the objective is for the clinician to assess the caregiver–child relationship and consider this not only as a potential source but also as the context in which the child's symptoms arise and therefore may be ameliorated.

When considering specific treatments for preschool depression, lessons learned from treatment studies of older school-age depressed children might be highly applicable and informative. Early treatment studies of depressed children using antidepressant medications, known as tricyclic antidepressants, demonstrated that these medications, known to be effective in adults and adolescents, did not show the same efficacy in depressed children. Specifically, outcomes of children treated with tricyclic antidepressants were no better than those taking a sugar pill or placebo. This finding had a profound impact on our understanding of the treatment of childhood mental disorders overall, suggesting that children could not be viewed as miniature adults and that the treatments and pathophysiology of childhood mental disorders may be unique and certainly worthy of independent study.

More recently, new medications for the treatment of depression have been developed. This class of medicines, so-called selective serotonin reuptake inhibitors or SSRIs, have proved to be very useful in the treatment of adult depression as they have proved efficacy and a more favorable (well-tolerated) side-effect profile. One double-blind, placebo-controlled study of an SSRI medication in children demonstrated that it was efficacious for the treatment of childhood depression. Concerns about this class of medication and the potential for increasing suicidality has resulted in the Food and Drug Administration (FDA) issuing stronger warning labels and therefore have mitigated clinical enthusiasm for their use in child populations. Whether increases in suicidal ideation or behavior arise as a result of the medication or the underlying illness remains unclear and is the subject of investigation.

In addition to medication, psychotherapies have also proved effective in the treatment of depression. In particular, two therapeutic modalities have been the subject of empirical investigation. Cognitive-behavioral therapy (CBT) is one modality that has been well tested and proved effective in adults. CBT focuses on identifying and correcting the negative cognitive distortions known to occur in depression. This means that depressed individuals are known to perceive events in a more negative way than how those without depression might perceive the same events. Several studies have demonstrated efficacy of CBT in adolescent as well as school-age children. Its application to younger preschool-age children may be limited by immature cognitive development. Another form of psychotherapy called interpersonal psychotherapy (IPT) has also demonstrated efficacy in the treatment of depression in adults and adolescents. Studies of its use and adaptation in school-age children are now underway. This modality focuses on the impairments in interpersonal functioning that arise as a result of depressed mood. The application of these psychotherapeutic treatments to even younger preschool-age children has not yet been explored but may be promising.

Due to the lack of necessary empirical data exploring the safety and efficacy of antidepressant medications in young children, the first line of treatment for preschool depression should be psychotherapeutic. Parent–child

relational therapies have been developed for the treatment of other preschool disorders and efforts are underway to adapt and test some of these techniques for application to depression. While it seems clear that parenting and other key aspects of the psychosocial environment are important mediators of outcomes for young children, studies that address the effects of interventions in these domains are needed.

Clinical Vignette

Identifying depression in a preschool-aged child is difficult and not immediately obvious even to the well-trained child mental health clinician. Depressed preschoolers often do not have an obvious sad or withdrawn mood as is often true of severely depressed adults. Therefore, detailed questions about the child's pattern of behavior and play are essential and may be highly informative. A line of questioning that focuses on negative mood, affect, and internalizing symptoms is important as caregivers tend to pay less attention to these behaviors and instead focus on behaviors that are disruptive. Because of this, it is not uncommon for these symptoms to be present but not spontaneously reported by the parent/caregiver. The following is a case example that highlights some of the features that are typical markers of depression arising at the preschool age.

RK is a 4.2-year-old Caucasian male who was referred by his mother due to concerns about frequent episodes of extreme irritability. Mother reported that RK often seemed to "wake up on the wrong side of the bed." That is, he often would appear angry, irritable, and withdrawn for no apparent reason. Further questioning revealed that during these times RK also did not seem interested in his favorite play activities. One fall, this symptom was so severe that he did not want to go trick-or-treating for Halloween, a holiday he typically enjoyed greatly and looked forward to participating in. Mother also reported upon specific questioning that RK's play involved very negative themes in which dangerous events were taking place and/or harm was befalling the play character. This seemed to be a recurrent play preoccupation and not just a transient interest. In addition, he had a restless sleep pattern characterized by multiple night awakenings. He also did not seem to enjoy his favorite foods such as pizza during these periods. RK did not express feeling sad directly but did endorse this symptom when it was approached as "not feeling as happy as your siblings or other kids seem to feel."

This case underscores the importance of irritability as a marker of depression in early childhood. It also highlights the central feature of the symptom of anhedonia as was evidenced in the child's lack of interest in Halloween festivities. The absence of clear and overtly expressed sadness is also an important issue as depressed preschoolers tended to report themselves as 'less happy' rather than 'sad' on an age-appropriate puppet interview. In addition, the presence of vegetative signs such as disturbances in sleep and changes in appetite were also evident.

Conclusion

Depression is a serious clinical disorder that has long been recognized in adults, has been recognized in school-age children for more than 20 years but has only more recently been recognized as occurring in early childhood during the preschool period. Early case reports suggested that some preschool children meet standard diagnostic criteria for MDD, despite the fact that these criteria were developed primarily for application to adults. Two independent studies utilizing symptom criteria adjusted to reflect the developmental level of the preschool child have identified groups of depressed preschoolers. Among these depressed preschoolers, the symptom of anhedonia, or the inability to experience pleasure or enjoyment of activities and play, emerged as a clinically significant and specific symptom. These studies also found important markers supporting the validity of the disorder similar to those found in studies of depressed older children and depressed adults, including a stable symptom constellation, greater family history of affective disorders, and physiological correlates. Although systematic studies of depression in infants and toddlers are not yet available, a compelling body of case reports and clinical experience suggest the disorder can arise at this even earlier stage of development. Given that preschool depression has only recently been the subject of systematic investigation, little information is available regarding effective treatment of the disorder when it onsets during the preschool period. Information available from the treatment of other preschool-age disorders indicates that treatment should target the parent–child relationship. Studies of early intervention may be particularly important as early intervention in mental disorders may represent a window of opportunity for more effective treatment. Additional study of the risk factors and treatment of this disorder are therefore warranted both for the benefit of preschool children as well as for their implications on the lifelong trajectory of the disorder.

See also: Abuse, Neglect, and Maltreatment of Infants; Adoption and Foster Placement; Attachment; Emotion Regulation; Mental Health, Infant; Parental Chronic Mental Illnesses; Postpartum Depression, Effects on Infant; Screening, Newborn and Maternal Well-being; Social Interaction.

Suggested Readings

Downey G and Coyne JC (1990) Children of depressed parents: An integrative review. *Psychological Bulletin* 108: 50–76.

Kashani JH and Carlson GA (1987) Seriously depressed preschoolers. *The American Journal of Psychiatry* 144(3): 348–350.

Luby JL (2000) Depression. In: Zeanah ChL (ed.) *Handbook of Infant Mental Health,* 2nd edn., pp. 382–396. New York: The Guilford Press.

Luby JL, Heffelfinger AK, Mrakotsky C, et al. (2002) Preschool major depressive disorder: Preliminary validation for developmentally modified DSM-IV criteria. *Journal of the American Academy of Child and Adolescent Psychiatry* 41(8): 928–937.

Luby JL, Heffelfinger AK, Mrakotsky C, et al. (2003) The clinical picture of depression in preschool children. *Journal of the American Academy of Child and Adolescent Psychiatry* 42(3): 340–348.

Luby JL, Mrakotsky C, Heffelfinger AK, et al. (2003) Modifications of DSM-IV criteria for depressed preschool children. *American Journal of Psychiatry* 160(6): 1169–1172.

Developmental Disabilities: Cognitive

S L Pillsbury, Richmond, VA, USA
R B David, St. Mary's Hospital, Richmond, VA, USA

Glossary

Echolalia – The repetition of that which is said.
Etiology – The origin or cause of a medical disease or condition.

Language pragmatics – the set of rules governing the use of language in context. This includes factors such as intention; sensorimotor actions preceding, accompanying, and following the utterance; knowledge shared in the communicative dyad; and

the elements in the environment surrounding the message.
Prosody – The element of language which concerns intonation, rhythm, and inflection.
Semantics – The study of meaning in language, including the relations between language, thought, and behavior.
Syntax – The way in which words are put together in a sentence to convey meaning.
Verbal auditory agnosia – The inability to understand spoken words; pure word deafness.
Verbal dyspraxia – Also referred to as childhood apraxia of speech; a nonlinguistic sensorimotor disorder of articulation characterized by the impaired capacity to program the speech musculature and the sequencing of muscle movements for the volitional production of phonemes (sounds).

Introduction

Developmental delay is the failure to achieve developmental skills at are appropriate for the age of the child. This article will concern itself with developmental disabilities in the area of higher-order cognition, as seen in preschoolers.

It is universally agreed that early recognition of developmental disability is key in optimizing functioning of the child. The age at which identification is possible varies with the nature of the developmental disability. For example, some disabilities at occur relatively infrequently, but which carry high morbidity (such as cerebral palsy, severe degrees of mental retardation, sensory impairments (blindness, deafness), lower-functioning autism spectrum disorders (ASD), and severe communication disorders), are more likely to be diagnosed in the preschool years. In contrast, disabilities at occur much more frequently but which carry lower morbidity (such as learning disabilities, mild-to-moderate mental retardation, attention deficit hyperactivity disorder (ADHD), higher-functioning ASD, Asperger's syndrome, and higher-order language disorders) often will not be diagnosed until school age. Early identification, with appropriate referrals and interventions, permits counseling for families and planning for the child's future. The child's progress is monitored over time, and interventions are modified as needed. In the very young child, there is obviously a great deal of uncertainty, both in diagnosis and prognosis. The passage of time allows greater diagnostic precision and therefore better targeted therapies.

Not all children subsequently diagnosed as having developmental disabilities will have any identifiable risk factors at birth, and many causes of developmental disability are unknown. When risk factors are present from birth or early infancy, they may be isolated or multiple, and they may interact in complex ways. Multiple risk factors can have an additive effect. However, the absence of risk factors does not guarantee typical development. Risk factors present from birth or early infancy may include:

- genetic syndromes (patterns of malformation), chromosomal abnormalities, malformations of the central nervous system (CNS);
- prematurity (although prematurity alone is a weak risk factor; its effect is probably mediated by those complications that are seen commonly in low birth weight infants such as metabolic disturbances like hypoglycemia, severe chronic lung disease, CNS hemorrhage, sepsis, and infections of the CNS);
- conditions in pregnancy that interfere with uteroplacental circulation, oxygenation, or nutrition, which may lead to premature delivery or intrauterine growth retardation (a baby who is significantly smaller than would be predicted for his gestational age). Among these are pregnancy-induced hypertension, chronic maternal disease (such as hypertension, renal disease, and cyanotic congenital heart disease), maternal smoking, and maternal drug abuse;
- congenital infections (including cytomegalovirus, rubella, HIV, herpes, congenital syphilis, congenital toxoplasmosis);
- adverse prenatal and perinatal events such as hemorrhage from placenta previa or placental abruption, severe maternal disease (e.g., infections, seizures, trauma), anoxia (especially when the newborn is symptomatic in the immediate neonatal period with seizures, hypotonicity, and feeding problems); and
- sociocultural factors including poverty, poor access, to or underutilization of, medical care, lower maternal educational levels, physical or mental illness in the mother or other caregivers, abuse, and neglect.

Imagine the complex interplay of factors when an infant is delivered prematurely to a young, single mother who has a history of mental illness and substance abuse, and infant neorate intensive care unit (NICU) with multiple medical needs.

The pediatrician is the professional in the best position to identify the infant at increased risk for developmental disability. A careful developmental history should be obtained for milestones in the entire range of development (gross motor, fine motor, language, social, and adaptive). Parent histories may be influenced by the level of their knowledge of typical child development, and also by their readiness to acknowledge delays or atypicalities when present. The physician must, therefore, supplement the history by observation of the child's acquired skills in the office. When indicated, formal standardized screening instruments may be administered. The

Denver Developmental Screening Test has been popular for many years because of its ease of administration. Another such tool, the Cognitive Adaptive Test/Clinical Linguistic and Auditory Milestone Scale (CAT/CLAMS), yields a developmental quotient (DQ) which correlates well with the 'mental development index' of the more labor-intensive Bayley Scales of Infant Development, at least in healthy children without risk factors for developmental delay. Developmental progress should be tracked over time, as the outcomes will be markedly different depending on whether the infant has suffered the consequences of a prenatal or perinatal event without ongoing insult, vs. an ongoing risk factor such as abuse or neglect, vs. a neuro-degenerative disease, for example.

Earliest indicators of developmental disability may include abnormalities of tone, feeding problems, and poor response to stimuli. As the child's neurological development normally progresses from primitive reflexes and postural responses to volitional movement patterns, the persistence of primitive reflexes may be an early sign of developmental atypicality, as are asymmetries and tone abnormalities (both hypertonicity and hypotonicity). Motor delays later in the first year, such as delays in sitting and crawling, may be noted. Language and behavioral abnormalities are commonly noted in the second and third years of life. Indicators of ADHD and learning disability may be seen by the time the child is ready to enter kindergarten.

Significant progress has been made in the area of early identification and intervention for infants with developmental disabilities and those at high risk over the years since 1975, when the Education for All Handicapped Children Act (EHA) was passed. Public Law 94-142 mandated "free and appropriate education in the least restrictive environment" for children with disabilities. In 1977, this was extended to include ages 3–21 years (although coverage from ages 3 to 5 years was optional). PL 99-457 in 1986 added coverage for infants and toddlers below age 2 years with disabilities. It provided for Individual Family Service Plans (IFSPs) for the delivery of individualized services to the families of these infants and toddlers. The EHA was reauthorized in 1991; PL 101-476 gave the new title of Individuals with Disabilities Act (IDEA). Its key components were: (1) identification of children with learning-related problems; (2) evaluation of the health and developmental status of the child with special needs, determining present and future requirements for intervention, with the formulation of a plan to address each area of need with appropriate services; (3) provision of those services, both educational and related services; and (4) guaranteed due process. Under PL 101-476, it was now assured that children with disabilities and their parents were as entitled to a free and appropriate education as were those without disabilities. Children from birth to age 3 years continued to have a written plan of service known as the Individual Family Service Plan (IFSP). From age 3 to 21 years, this written plan is known as an IEP or Individual Education Plan. Autism and traumatic brain injury (TBI) were included for special education coverage by PL 101-476.

As reauthorized in 1997 in PL 105-17, so-called 'Part C' called for the provision of early intervention services for all infants and toddlers with disabilities through the creation of statewide, coordinated, multidisciplinary, interagency programs. This law did not mandate these services, but did provide partial reimbursement for their cost. Currently, all 50 states have established early intervention services for children from birth to age 3 years, addressing developmental issues in the physical, communicative, cognitive, and psychological realms, as well as self-help skills, with the goals of minimizing disability and enhancing the ability of families to meet their children's special needs. It was hoped that early intervention would also decrease the costs for special education services once they reached school age. It was left to the states to define developmental delay for the purposes of establishing eligibility for services. Services are provided both to children with demonstrated delays and to those with biological conditions that place them at high risk for delays. The states may provide services to children at risk due to environmental factors, at their discretion. At the heart of 'Part C' of PL 105-17 is its focus on family involvement and family support. Evaluation, assessment, and planning are all subject to the approval and participation of the family. Likewise, intervention services are optional. When provided, these services are rendered in the most natural setting possible, such as in the home or the day care center. Follow-up surveys of families who have been involved in early intervention services under 'Part C' have shown a generally high level of parent satisfaction with their access to services, and also their perception of optimism for the future and their own competence in caring for and advocating for their children (although less positive outcomes were seen in minorities, in families with children with complex medical needs, and in single-parent families). Another important component of PL 105-17 was the extension of coverage to include ADHD.

Any child who presents to the physician with developmental delays should receive a thorough general and neurological examination. Particular attention should be paid to growth abnormalities, congenital anomalies which may suggest a genetic syndrome or chromosomal abnormality, a congenital CNS disorder, or intrauterine infection. Abnormal skin markings may be seen in genetic disorders as well as congenital infections. Abnormalities of the eyes, heart, limbs, and abdominal organs should also be noted. As the number of physical anomalies increases, the likelihood of a genetic disorder also increases. Neurological examination should especially include measurement of head circumference, examination of the cranial nerves, and assessment of tone. Persistent primitive reflexes and any asymmetries should be noted. Testing of hearing and vision should next be undertaken, using tools that are appropriate for the age and developmental level of the child. Other tests may be obtained, as clinically indicated, and may include genetic studies, imaging of the brain using magnetic resonance imaging (MRI), electroencephalogram, and others. Specialists from pediatric neurology, cardiology, orthopedics, genetics, psychology, or psychiatry may be consulted. However, referral to the early intervention program need not be delayed pending the results of these evaluations. The early intervention interdisciplinary team includes evaluation by occupational therapists, physical therapists, speech and language pathologists, and social workers. With the establishment of the IFSP (or IEP), services may then be initiated with the cooperation of the families, with systematic monitoring of the progress of the child and his family.

Cognitive disorders in infancy and early childhood fall into three fundamental domains: disorders of communication, disorders of socialization, and visual–spatial disabilities. Within each domain, there is a continuum of severity and complexity. There is also considerable overlap among the identified domains.

Individual variations in cognitive style and temperament include activity level, rhythmicity, approach to new stimuli, adaptability, intensity, mood, perseverance, distractibility, and threshold to stimulation. Areas of concern may include withdrawal from novel stimuli, slowness to adapt, intensity of response, a predominantly negative mood, shyness, and withdrawal. While these are not considered disorders in the infant and preschool children, they may predispose the child to problems later in life.

Disorders of Communication: Developmental Language Disorders

Normal Language Development

There is a large degree of variability in the rate at which language is acquired (first words anywhere from 6 to 30 months), as well as variability in the rate of acquisition of different linguistic components, such as phonology, lexical retrieval, and syntax. For example, many normal children concentrate their early acquisition of new words to nouns, while others add verbs and adjectives at a similar pace. There is no explanation at present for these differing styles of learning, and both groups of children are normal. The wide variability in so-called normal children can make it difficult to distinguish children with developmental language disability from those normal children with an idiosyncratic initial delay who will eventually catch up. Typically, developing children have good receptive language by 2 years of age, with an expressive vocabulary of 50 words or more and some two-word phrases. A general rule of thumb suggests that, after children develop a 50-word expressive language repertoire, other individual words, as well as phrases and sentences, will generally follow quickly. Expressive language is usually well-developed by 3 years of age.

There is clear evidence that children with developmental language disorders (DLDs) are at risk for a variety of social–emotional problems in older childhood and in adult life. There are certain 'red flags' for DLDs. Children with DLDs may demonstrate early problems relative to other oral functions such as sucking, swallowing, and chewing. Infants who fail to vocalize to social cues or to vocalize two syllables by age 8 months are suspect. Slightly older children are at risk if they acquire new words only slowly and with great difficulty, if they rely too much on contextual cues for understanding of language, if their social interactions are limited to getting their needs met, if they produce few or no creative utterances of three words or more by age 3 years, and if they show little attention and interest for language-related activities such as book reading, talking, or communicating with peers. By age 3 years, typically developing children have developed symbolic, imaginative play.

Currently available tools for assessment of early language development can result in both underdiagnosis and overdiagnosis of DLDs. As many as 40% of children identified as having DLDs in the first 2 years of life may no longer retain that diagnosis at age 3 or 4 years. Ten per cent of these children are 'normal' at school age, while others have had their diagnoses refined to mental retardation, ASD, and others. It is preferable in the very young child to overidentify DLDs, as delay in diagnosis and treatment may have long-term social, behavioral, and educational implications.

Children at Risk

Risk factors for DLDs include parental mental retardation or a family history of DLDs. Premature and small-for-gestational-age infants are also at greater risk. There is a higher concordance rate in monozygotic vs. dizygotic twins, suggesting that environmental influences alone are insufficient to explain the occurrence of DLDs. A number of gene loci have been identified, implicating 13q, 16q, and 19q as candidate genes for further study. An autosomal dominant mode of inheritance is frequently seen, but there is variability in penetrance as well as in expressivity. For the purposes of this discussion, children with significant hearing loss and those who have identifiable brain lesions have been excluded.

Subtypes of Developmental Language Disorders

DLDs are described based on the linguistic area which is most significantly disturbed.

Articulation and expressive dysfluency disorders

Phonologic (pure articulation) disorders. Most children will speak intelligibly by 2 years of age. By age 3 years, fewer than 15% of children have unintelligible speech. Minor articulation defects, such as a distortion of the "th" and "r" sounds may persist with little consequence. Phonologic awareness, however, is critically important in the acquisition of normal reading skills, and children with delayed phonologic acquisition are at greater risk for developmental reading disorders at school age.

Dysfluency (stuttering and cluttering). Some degree of dysfluency is common as language skills develop, particularly as the mean length of utterance reaches six to eight words between 3 and 4 years of age. Some children with dysfluency may be relatively fluent for days or weeks at a time, then experience a protracted interval of relative dysfluency. Both stuttering and developmental dysfluency may be influenced by factors such as the complexity of the thought to be expressed and by being rushed or when excited, happy, or angry. Between-word dysfluencies include interjecting 'um' in a sentence, repeating a phrase, or revising the sentence structure in midstream. Within-word dysfluencies include repetitions of individual sounds or syllables, prolongations of sounds, and blocks. Stuttering is a disorder in the rhythms of speech, in which an individual produces a disproportionately large frequency of within-word dysfluencies compared to normally fluent peers, particularly at grammatically important points in the sentence. It often is a genetic trait, and occurs more frequently in children with other DLDs as well as with mental retardation. It is equally common in boys and girls at its onset, but is three times more likely to persist in males. Associated behaviors such as head, torso, or limb movement, audible exhalation or inhalations immediately prior to the dysfluency, and visible muscle tension in the orofacial region are signs that the child is becoming aware that talking is difficult. In younger children, the earliest and most frequently observed associated behaviors involve the eyes (such as blinking, squeezing the eyes shut, side-to-side movements of the eyes, and consistent loss of eye contact with the listener). These behaviors are seen in stutterers and usually not in children who are simply developmentally dysfluent. Most children who begin to stutter at preschooler age will recover without specific therapy, especially those with onset prior to age 3 years, if family history is either negative or characterized by spontaneous resolution, and if there are no coexisting speech and language or learning problems. Cluttering, by contrast, is characterized by echolalia, palilalia (compulsive repetition in increasing rapidity and decreasing volume), incomplete sentences, perseveration, poor articulation, and stuttering, seen in children with fragile X syndrome.

Verbal dyspraxia. This condition, in which children are extremely dysfluent, as often been called 'dilapidated speech'. Language is produced only with great effort. Phonology is impaired, including omissions, distortions, and substitutions. Language comprehension is preserved, and intelligence is normal.

Disorders of receptive and expressive language

Each of these disorders has a receptive component. While receptive language is heavily dependent upon attentional factors, reception may be impaired independently. Reception is dependent upon spoken rate, register, and dialect. It is a mistake to assume that the child who appears to be paying attention also understands what is said to him. In an emotionally charged context, reception breaks down further.

Phonologic syntactic syndrome. This condition, which is very common, is characterized expressively by disturbances in phonology, particularly in consonant sounds and consonant clusters in all word positions. The child is extremely difficult to understand, and grammatical forms are atypical. Semantics, pragmatics, and prosody are normal. Associated neurological problems are particularly common in this DLD, as are problems with feeding (sucking, swallowing, and chewing).

Verbal auditory agnosia (VAA). Children with VAA do not understand meaningful language, despite normal hearing. VAA may be seen as a DLD, or may be acquired in association with a form of epilepsy in which the epileptogenic portion of the brain involves the receptive language areas of the temporal lobe. The prognosis for this disorder is generally poor, although better in children with the acquired variety.

Higher-order language syndromes

Lexical syntactic syndrome. This common disorder is characterized by dysfluent speech, the consequence of word-finding difficulties, and a deficiency in syntactic skills. In the absence of finding the appropriate word, the child may 'talk around' the word in what are referred to as paraphasias. Speech is intelligible because phonology is normal. The child's language production is better for repetition than for spontaneous speech. Comprehension is normal except when the child is required to process very complex utterances.

Semantic pragmatic syndrome. Children with this condition are fluent, even verbose, but their large vocabularies belie the difficulty they have with meaningful conversation and informative exchange of ideas. Their chatter and often formal style may give the impression of a high intelligence quotient (IQ). Speech in this disorder has been described as stilted, pedantic, or professorial; the speech quality may be mechanical, monotonous, or 'sing-song'. Children with this disorder are often unable to respond to 'who', 'what', 'where', 'when', or 'why' questions, but may appear to exhibit great eloquence in subject matters of their interest or fixation. Comprehension is impaired. This syndrome is often seen in high-functioning autistic children.

Outcome for Children with Developmental Language Disorders

Preschool language skills predict later reading ability. DLDs are associated with problems such as ADHD, behavioral and emotional problems, and academic underachievement. While articulation and fluency problems may be the most obvious, more subtle disorders of comprehension may be misdiagnosed as conduct disorders or oppositional-defiant disorder, owing to the emotional 'meltdowns' in the child who is chronically unable to understand the intentions and expectations of others, both peers and adults. These communication problems may persist into adulthood in more than half of these children, impairing both social interactions and career success. The effect of speech and language therapy on outcome, particularly in the more significantly affected children, is still a matter of debate. This therapy is labor intensive, and may need to continue for prolonged periods of time. Continued association with more normally conversant children in the daycare or preschool setting is helpful.

Autism Spectrum Disorders

The clinical presentation (phenotype) of autism is highly variable, as is its natural history. There are several hypotheses regarding the essential cognitive deficit in ASD. The theory of mind blindness is related to the theory of mind; it implies that individuals on the autism spectrum lack the capacity for understanding or sensing another individual's state of mind. In other children with ASDs, the ability to solve problem, to shift sets, and to plan to reach a goal are deficient. A third theory suggests that children on the autism spectrum fail to integrate information, and are deficient in Gestalt ('big picture') formulations.

ASDs are characterized by the triad of impaired socialization skills, impaired verbal and nonverbal (body language) communication skills, and restricted areas of activity and interests.

Classic autism, as originally described by Leo Kanner in 1943, is estimated to occur in approximately 1 in 1000 individuals. ASD occurs in a wide degree of severity in an estimated 1 in 150 individuals (2007). Fifty to seventy per cent of autistic individuals can be determined to have demonstrated impairments since birth (e.g., using, scoring of the infant's interpersonal interactions on home videos). There is a subtype of ASD children, 30–50%, who evidently were typically developing children until language/autistic regression occurred at a mean age of 21 months (range 12–36 months) under the influence of unknown triggers (including potentially infectious, immunologic, or psychosocial stressors).

Etiology

There is strong evidence for a genetic (probably multigenic) basis for ASD, including a recurrence risk of 4.0–9.8% in subsequent children in a family with one autistic child. In fact, the recurrence rate would certainly be higher but for the stoppage rule, that is, parents with one severely affected child often do not have more children. Lower-functioning autistic adults commonly lack the social skills that lead to successful interpersonal relationships, decreasing their likelihood of becoming parents. Males predominate at a rate of approximately four to one. There is a high concordance rate in monozygotic twins, approximately 90%. Several candidate genes have been identified. In addition, children with certain identified genetic disorders (including fragile X syndrome, phenylketonuria, tuberous sclerosis, Angelman's syndrome, and Cornelia de Lange syndrome) may demonstrate autism symptoms.

Future genetic studies may demonstrate that, for the autism spectrum, there are indeed many specific genotypes, as opposed to a single defective gene. In addition, confusion may arise, since there are many other conditions overlap with ASDs, including obsessive–compulsive disorder, Tourette's syndrome, and ADHD. Elements of these disorders are often found within ASDs. Future research depends upon standardization of diagnostic criteria to a research level of certainty.

The recent 'epidemic' of individuals identified to be on the autism spectrum is believed by most in the scientific community as being a manifestation of increased awareness and better identification, although research is also ongoing in the areas of possible environmental triggers, especially for those children who have appeared to undergo autistic and/or language regression in the second year of life.

Diagnosis

The diagnosis of autism is based upon the presence of specific criteria. The meeting of these criteria represents an analysis of phenotype, as opposed to genotype. It is currently possible to make a reliable diagnosis of ASD at age 24–36 months in many cases, with stability of the diagnosis up to 9 years later. The DSM-IV criteria are less useful in younger toddlers, below 24 months of age, and it is not yet well understood how early symptoms map onto later symptoms. Research is ongoing to develop reliable markers as young as 6 months of age, using the lack of joint attention as the operational definition of infantile autism. (Joint attention is a platform for language development, closely linked to abstract rule-learning, which is measured by three-point gaze shifts and following the look and point gestures of others.) Children subsequently diagnosed with 'congenital' ASD typically demonstrate language delays by 14 months of age, slower than normal language development from 6 to 36 months, and decreased initiation of communication for social or instrumental purposes. Because most children who present early are identified due to delays in acquisition of language skills, the first specialist consulted is commonly the audiologist. Sixty per cent of 14-month-olds later diagnosed with ASD, and 90% of 24–36-month-olds with ASD, were seen to have stereotyped patterns and interests. In endeavoring to make the diagnosis early, the clinician is challenged to differentiate the child with early signs of autism from the range of normal variability in development.

Tools to aid in early diagnosis of autism include detailed questionnaires, specific interview techniques, and blinded reviews of home videotapes. Using such tools, it can be seen that a subset of autistic children underwent global regression prior to 24 months of age. Most studies have shown that the majority of these children with autistic regression had minor impairments prior to the onset of the regression, however. This phenomenon is well described but poorly understood. It is crucial to assess the hearing of all children with language impairment, using, when necessary, brainstem auditory evoked responses (BAERs, otherwise known as auditory evoked potentials). This will distinguish those children who are language impaired or on the autism spectrum from those who are severely hearing impaired. (Although there are children who are severely hearing impaired who also meet criteria for ASD, in general, children who are hearing impaired alone do not manifest the degree of social impairment seen in autism. However, It can be very difficult to diagnose ASD in the deaf child with significant behavior problems such as severe ADHD.) The skilled evaluation of a speech and language pathologist is also mandatory. The gamut of language disorders described earlier in this article can be seen in ASD individuals, and severely language impaired children may also present diagnostic confusion with autism, especially, again, when behavior is abnormal. Seventy to eighty-five per cent of children with ASD are mentally retarded on standard testing; this group has a generally poorer prognosis. In contrast, many other children on the autism spectrum are above average, or even superior in intellectual functioning. It should be borne in mind that measurable mental retardation does not always translate to functional retardation. Many ASD adults may take advantage of their relatively well-preserved or even enhanced abilities in the visual–spatial domain, performing such repetitive functions as data entry. Preschool children with ASD may demonstrate excellent skills in puzzle construction, for example.

Core Deficits

Social competence

Social incompetence is the hallmark of ASD. It represents a lack of intuitive social skills. Children on the autism spectrum are unable to sense the emotional state of others. The 'theory of mind' refers to the concept that autistic individuals demonstrate a lack of awareness of the internal state of others. Their play is often parallel rather than interactive, with little symbolic play. When they do engage in interactive play, they generally take a passive role. At the extremes, they may be socially unavailable, aloof, or with an intense stranger anxiety, but there are also those autistic children who are socially impaired by virtue of being 'too social', with a dramatic lack of apprehension of strangers, and a willingness to go off with anyone, even those who are totally unknown to them. Most autistic children are withdrawn and exclude themselves (and their idiosyncrasies lead their peers to exclude them as well, thus reducing their opportunities to pattern social behaviors and language on typically developing peers). Autistic individuals may exhibit difficulties with personal space, which may be represented as a reluctance to have their own individual space invaded, with no corresponding reluctance to invade the personal space of others. Lack of eye contact was thought at one time to be the hallmark of ASD, but this is not necessarily the case. Children on the autism spectrum may look at, through, or beside others. There is, for the most part, an impairment in the quantity or quality of eye engagement. Children with ASD demonstrate a lack of interest in the human face, and are more likely to concentrate on the mouth or other parts of the correspondent, rather than his eyes. (Brain imaging in adults has shown differences from normals in the cortical areas involved in the perception of facial emotion.) Sharing and turn-taking are almost always impaired. Verbal autistic children give the impression of talking 'at' others (see the discussion of language impairments in ASD, below). Socially impaired children who are also nonverbal are often hyperactive, inattentive, and aggressive. Tantruming and uncontrollable screaming – what parents often refer to as 'meltdowns' – are common, especially in the children under age 3 years. These children may be inconsolable. Sleep is often disturbed. Bowel and bladder training are difficult and often significantly delayed. Self-injurious behavior may be present, particularly in lower-functioning children. Many consider the difficulties in the domain of social cognition to be the most essential feature of ASDs.

Language impairment

Verbal and nonverbal communication deficits are an essential part of the autism triad. Language generally parallels intelligence. Echolalia, while occasionally seen as a brief developmental interlude in normal children, and infrequently seen in persistent fashion in pure DLDs, is common in children on the autism spectrum. (Echolalic speech often portends the development of more fluent speech, and therefore it is not necessarily a bad sign.) As previously stated, a thorough assessment of hearing and the evaluation of a skilled speech and language pathologist are essential. In low-functioning children, verbal auditory agnosia, phonologic-syntactic, and lexical–syntactic language disorders are seen. In higher-functioning children, pragmatic and semantic deficits are characteristic. This includes deficits in who/what/where/when/how questions and in language turn-taking. In addition, prosody is frequently impaired, such that these children speak in monotone rather than in well-modulated speech. Hyperactivity and inattention relate inversely to language competence in autistic children under the age of 3 years. It is the consensus that language competence at 5 or 6 years of age quite accurately predicts long-term prognosis, since language, as suggested earlier, determines intelligence, which then relates to functionality. Chances for a child who remains nonverbal at the age of 8 or 9 years becoming linguistically competent are very poor.

Restricted range of behaviors, interests, and activities

ASDs are characterized by behaviors that are, from the viewpoint of the observer, odd or idiosyncratic. Activities such as toe-walking, twirling, licking, flapping, rocking, opening and closing doors, and manipulating light switches are seen. Again, there is variability, and these stereotyped behaviors may appear only for a brief period of time, only to be replaced by another oddity. Motor stereotypies are repetitive actions which are complex, involuntary, and purposeless, which are carried out with predictable form, amplification and location, in the autistic individual. They usually have their onset prior to the age of 2 years. Although tics may also be seen not uncommonly in ASD individuals, these brief, uncomplicated movements usually have onset after 6 or 7 years of age. Inattention and hyperactivity may be seen in ASD, thus overlapping with ADHD. Children with ASD may have great difficulty with transitions, and may become overfocused on certain activities, especially in the visual–spatial realm, such as assembling puzzles. As mentioned earlier, preschool children with ASD behavioral idiosyncrasies serve to accentuate their social isolation.

Treatment

There have been no long-term studies comparing those individuals who received early intervention vs. those who did not. Nevertheless, it is a good presumption that early intervention can improve eventual outcomes. Of the treatment modalities currently available, those that use an operant conditioning approach appear to be the most efficacious. While all use an applied behavior analytic approach (operant conditioning), individual protocols may vary widely. There is currently no way to compare protocols.

Pharmacologic treatment is presently limited to helping with related problems (such as stimulants for hyperactivity, medications to help with sleep, anticonvulsants for coexisting seizures, etc.). Speech and language therapies and social skills training may also be of benefit.

Visual–Spatial Disabilities

Visual–spatial disabilities (VSDs) involve perceptual organization, memory, and imagery. The literature on visual–motor and spatial disabilities in young children is very limited, but impairments in this domain will significantly impact future academic success. For example, the preschooler child's ability to copy geometric shapes has predictive value for reading and math in elementary school. Motor execution is the fundamental medium for expression of function or dysfunction in the visual–spatial areas. While there are motor-free tests for perceptual dysfunction, more commonly VSDs are grouped with related disorders of motor execution as perceptual–motor disabilities, and it can be difficult to consider one without the other.

Traditional IQ tests demonstrate VSDs by the discrepancy between verbal and performance IQ. Subtests include those of design copy and memory, picture memory, and mental rotation. Other easily administered tests include requesting a child to draw or copy shapes, or to draw a human figure. There are age-related norms for both these tasks.

The etiology of visual–spatial and motor deficits in the preschooler is in the right hemisphere of the brain, as evidenced in studies comparing the copying and drawing skills and the ability to create spatial arrays of toys in children with known injury (e.g., stroke) involving the right vs. the left hemisphere.

A long-term follow-up study of premature babies demonstrated that the inability to copy a circle and a low score of sorting blocks (by shape, color, and size) at age 4 years correlated with hyperactivity and an abnormal neurological examination (so-called 'soft signs') at age 7 years. The 30% of children in the same study who showed poor execution of copying of a cross had a higher rate of diagnosis of learning disabilities and also of neurological soft signs. Poor performance on a maze task correlated with learning disabilities, ADHD, and abnormal neurological examination. Conversely, those children who demonstrated proficiency in the copying of a square at age 4 years, and who had high scores on block-sorting tasks, were actually at decreased risk for learning disabilities and hyperactivity.

Referral to occupational therapy is important for the preschooler with suspected visual–spatial and motor disability. Treatment using perceptual training programs has been reported to be helpful. Disability in this realm may seriously impair the child's perception of the world. Understanding the relationship between visual–spatial and motor disability and subsequent learning disabilities may aid in earlier recognition and special education for learning disabilities and nonverbal learning issues.

A related issue which will be discussed elsewhere in this encyclopedia merits mention here. Disorders of motor planning and execution are difficult to separate from disorders in the visual–spatial and motor realm. Disorders of motor execution can accompany paralysis, spasticity (most forms of cerebral palsy), and movement disorders, but there are also a variety of disorders of motor execution that do not result in apparent alterations in strength, tone, or posture, but rather manifest themselves by clumsiness and inadequate performance of motor acts. The true incidence of higher-order motor deficits in first graders in regular schools is estimated at somewhere between 2% and 12%. These higher-order motor abnormalities are best detected if age-appropriate sequences of individual motions are performed under an examiner's observation. These disorders of motor execution frequently occur concurrently in children diagnosed with ADHD or learning disabilities. It is a mistake, however, to consider disorders of motor execution only in the context of other conditions, because impairments in this realm are disabling in themselves.

Historically, these children often will not have met gross motor milestones such as independent walking (usually met by 10–15 months of age), climbing stairs by themselves (normally 14–24 months), riding a tricycle (2–3 years), and riding a bicycle (4–6 years). They may also have failed to meet fine motor milestones, including holding a cup (10–14 months), executing buttons and snaps (3–4.5 years), printing their own name (4.5–6 years), and tying shoe laces (4.5–6 years). On examination of gait, their walking, running, skipping, tandem gait, hopping on one foot, and climbing stairs are clearly impaired. Upper extremity functions, including finger-tapping, wrist-turning, button-pressing, finger-nose-finger, copying, drawing, and writing are also impaired. The ability to imitate nonsense gestures (dyspraxia), to pantomime to command, and to use actual objects are similarly impaired. There are age-standard normative values for performance on the Purdue Pegboard and the subtests of Kaufman that relate to hand movements and spatial memory.

In the vernacular, a clumsy child is the classic 'klutz'. The purely clumsy child exhibits slow and inaccurate fine or gross motor performance deficits. The abnormality does not pertain to impairment of strength or tone, but rather to speed and dexterity. Clumsiness is a primary cause of school failure in early grades where the demands for motor task performance are great.

Synkinesis is an involuntary movement of voluntary musculature that occurs during the course of a voluntary action. One test that can elicit this, the Fog test, requires that a child walk on the sides of his feet, either on the insides or the outsides of the sole. When the child performs this maneuver, especially when a narrow base is demanded, the arms and hands frequently enter into distorted postures. Of synkinetic movements, mirror movements are the most commonly appreciated form of synkinesis. In a finger-tapping test, mirror movements commonly occur. The demonstation of synkinetic movements can be a part of normal developmental variation, but it is abnormal in children older than 6 years. Clumsy children often exhibit synkinetic movements as well.

Dyspraxia represents a characteristic failure in a complex, voluntary act, which is more easily recognizable when the child attempts to learn more complex motor sequences. Because dyspraxia is generally not evaluated apart from clumsiness or synkinesis, it is often not described, therefore, in preschool children. The failure to appreciate the presence of dyspraxia may result in a child being regarded as lazy, oppositional, or unintelligent, and this can result in poor self-esteem, poor motivation, and poor conduct. These children may demonstrate delays in self-care skills such as dressing and grooming themselves. Their specific problems may relate to buttoning, snapping, zipping, dressing, tying shoe lace, manipulating combs, toothbrushes, and particularly scissors. They may manifest as refusing to attempt tasks such as writing and coloring, at the same time that gross motor skills may be age appropriate. Most children with dyspraxias have no identifiable brain abnormality. Dyspraxias can be elicited by asking younger children to pantomime, for instance, blowing a kiss, or waving goodbye. A child can also be asked to fold a piece of paper, fit it into an envelope, or roll up a paper to use as a pretend telescope. Practice has a significant influence in children with dyspraxia as well as in clumsy children. Unlike the clumsy child, successful performance of the task by the dyspraxic child does not improve when extra time is allotted. It is particularly important in preschoolers to note that clumsy children and children with dyspraxia are more likely to injure themselves, and parents may be wrongly accused of abuse.

Other disorders of cognitive functioning that will be seen in the older child are not well studied in preschoolers. These include deficits in memory and in executive function. Executive function refers to the ability to maintain an appropriate set of procedures for problem solving, to attain a future goal. It involves the intention to inhibit a response to defer it, to formulate a sequential, strategic plan of action, and to encode relevant material in memory for future use. Preschoolers developmentally do not demonstrate significant skills in such future-oriented behavior, but some difficulties may be noted with self-regulation, selective attention, and vigilance.

Summary

Cognitive disorders in infancy and early childhood may have long-term consequences, for learning, for social success, and for employment. Cognitive limitations and the child's frustrations related to them may lead to significant behavior problems as well. The goal of early identification and early intervention is to recognize those children at risk, as well as those children demonstrating signs of cognitive disabilities. Appropriate consultations and therapeutic regimens may then be utilized to optimize the outcomes for the child and his family. The challenge for future research is to develop the tools for practitioners to aid in early recognition. Greater precision in diagnosis will also aid in the understanding of the natural history of cognitive disabilities, and also their causes. Longitudinal studies will be helpful in comparing the outcomes of varying educational and other treatment modalities.

See also: ADHD: Genetic Influences; Autism Spectrum Disorders; Birth Defects; Genetics and Inheritance; Grammar; Language Development: Overview; Learning Disabilities; Milestones: Cognitive; Neuropsychological Assessment; Pragmatic Development; Risk and Resilience; Vision Disorders and Visual Impairment.

Suggested Readings

Nass R and Ross G (2005) Disorders of cognitive function in the preschooler. In: David RB (ed.) *Child and Adolescent Neurology,* 2nd edn., pp. 486–510. Oxford: Blackwell.
Rapin I (ed.) (1996) *Preschool Children with Inadequate Communication: Developmental Language Disorder, Autism, Low IQ.* Cambridge: Cambridge University Press.
Tuchman R and Rapin I (eds.) (2006) *Autism: A Neurological Disorder of Early Brain Development.* Cambridge: Cambridge University Press.

Developmental Disabilities: Physical

R R Espinal and M E Msall, University of Chicago, Chicago, IL, USA

Glossary

Acute lymphoblastic leukemia (ALL) – The most common type of childhood leukemia and the most common childhood cancer.

Antifolate – A substance that blocks the activity of folic acid. Antifolates are used in chemotherapy for cancer, since folate is necessary for the production of new cells.

Antineoplastic – Acting to prevent, inhibit, or halt the growth of tumors.

Arabinoside – A chemotherapeutic agent.

Ataxia – Unsteady and uncoordinated movements of limbs or torso due to cerebellar dysfunction.

Cerebral palsy (CP) – A disorder of movement and posture due to central nervous system impairments with associated impairments often occurring in vision, audition, communication, learning, manipulation, perception, and neurobehavioral control.

Clean intermittent catheterization – Technique to manage neurogenic bladder.

Congenital heart disease – Structured malformations of the heart that impact on cardiac structure or function.

Conotruncal cardiac abnormalities – Abnormalities of the outflow of the heart and great vessels. The entities include tetralogy of fallot, transposition of the great arteries, truncus arterisois, and higher interrupted aortic arch.

Cranial radiation therapy (CRT) – Treatment targeted to destroy leukemic or tumor cells in the brain.

Creatine phosphokinase (CPK) – A chemical found in muscle fibers that is released into the bloodstream when the muscles undergo damage and breakdown.

Deep hypothermic circulatory arrest (DHCA) – Specialized procedure during cardiac surgery to stop the heart so as to carry out safe surgical repair.

Dexamethasone – A steroid medication given to cancer patients undergoing chemotherapy to counteract side effects.

Dysphagia – Difficulty swallowing.

Dystrophin – A rod-shaped cytoplasmic protein that connects the cytoskeleton of a muscle fiber to the surrounding extracellular matrix through the cell membrane. Its gene's locus is Xp21.

Encephalopathy – Disorders caused by the impairment or damage of the brain.

Endothelium – Inner lining of blood vessels.

Excitotoxic neuronal death – Mechanism of cell death through predominance of excitatory neurotransmitter amino acids and small molecules.

Extrapyramidal – Organization of the basal ganglia, thalamus, and cerebellum for motor control of fine, balance, and precision movements.

Extremely low birthweight (ELBW) – <1000 g at birth.

Factor 5 leiden mutations – One of the family of proteins responsible for hemostasis (when bleeding is stopped).

Fludarabine – A chemotherapy drug commonly used to treat a type of leukemia known as chronic lymphocytic leukemia.

Focal calcifications – Abnormalities on comarter tomography in specific brain regions.

Glucocorticoid – A steroid with anti-flammatory properties.

Hematopoiesis – Development of blood cells.

Hemiplegia – A condition where paralysis is present in the vertical half of the patient's body (i.e., right arm and right leg, or left arm and left leg).

Homocysteine – A type of amino acid.

Hydrocephalus – Abnormal accumulation of cerebrospinal fluid (CSF) in the brain.

Hypoplastic heart syndrome – A cardiac malformation resulting in an underdeveloped left heart.

Hypoxemic–ischemic–reperfusion injury – Inability of impaired brain regions to tolerate the metabolic demands of restored blood flow.

Intensification (or consolidation) therapy – A type of high-dose chemotherapy often given as the second phase (after induction therapy) of a cancer treatment regimen for leukemia.

Interrupted aortic arch – Abnormalities in the ascending aorta.

Intrapatrum asphyxia – Interference with blood–gas exchange during labor and delivery that occurs with a major sentinel event and meets the criteria specified by the American College of Obstetricians and Gynecologists, which can lead to cerebral palsy.

Intrathecal chemotherapy – Receiving chemotherapy through the spinal canal.

Leukoencephalopathy – A rare side effect caused by methotrexate and/or radiation on the central nervous system of leukemia patients. Chemotherapy and/or radiation cause the destruction of the myelin sheaths covering nerve fibers, which results in motor and cognitive impairment.

Magnetic resonance imaging (MRI) – A procedure that allows imaging of internal organs. Cranial MRI allows visualization of the central nervous system.

Meninges – Membrane that covers the central nervous system.

Merosin – Biochemical that is part of stabilizing muscle cells.

Microangiopathy – Abnormal disease process affecting small blood vessels.

Necrosis – Accidental death of living cells and tissues brought on by injury, infection, cancer, and inflammation.

Necrotizing enterocolitis – Infection and gangrene of the neonatal intestine.

Nelarabine – A chemotherapeutic agent.

Nucleoside analogs – An artificially made nucleoside that interferes with the replication of DNA and RNA of viruses.

Oligodendroglia – A type of central nervous system cell.

Pentostatin – A chemotherapy drug commonly used to treat types of leukemia.

Perioperative hypoxemia – Abnormally low blood oxygen level during surgery which meets the criteria specified by the American College of Obstetricians and Gynecologists (ACOG) standard.

Periventricular leukomalacia – White matter injury in brain tracts near the ventricles.

Quadriplegia – A condition where decreased voluntary movement is present in all four limbs.

Remission-induction therapy – Initial treatment with anticancer drugs to decrease the signs or symptoms of cancer or make them disappear.

Secondary ventricular dilation – Increased ventricular size after intraventricular hemorrhage (IVH).

Sepsis – Bacterial infection of the blood.

Synkinesia – Symmetric overflow movements.

Systemic lupus erthematosis – An autoimmune disease.

Tetralogy of fallot – A cardiac malformation.

Truncus arterisois – Abnormalities of the aortic and pulmonary blood vessels resulting from a common trunk.
Velo-cardio-facial syndrome (VCFS) – A cleft palate, heart defect, and facial malformation syndrome.
Ventriculoperitoneal (VP) shunt – A catheter that redirects cerebrospinal fluid from the inner space of the brain (ventricles) to the abdominal cavity (peritoneum).
Very low birthweight (VLBW) – <1500 g at birth.

Introduction

Over the past 20 years, major advances in maternal–fetal medicine, neonatology, and translational developmental biology have resulted in unprecedented survival rates for very preterm (29–32 weeks' gestation) or extremely preterm (≤28 weeks' gestation) infants. These very low birthweight (VLBW; 1001–1500 g) and extremely low birthweight infants (ELBW; ≤1000 g) who received neonatal intensive care currently have survival rates of 90% and 80%, respectively, and include over 50 000 US children per year. In addition, with advances in pediatric cardiology and oncology, over 90% of children with congenital heart disease and over 90% of children with leukemia are surviving. These children number over 30 000 in the US per year. Although increasing numbers of these children with life-threatening disorders are surviving, they face neurodevelopmental disabilities as a result of their treatments affecting systems that contribute directly or indirectly to brain structure and function. We focus on prematurity as a model disorder for understanding several important pathways of the cerebral palsy (CP) syndromes of diplegia, hemiplegia, and quadriplegia. These neurodevelopmental motor impairments remain a substantial sequelae in 30% of infants weighing <500 g, 15–20% of infants weighing <750 g, and 10% of infants weighing 750–1499 g. These rates of functional motor disability contrast with 3–5% rates among infants weighing 1501–2499 g and 0.2–0.5% in infants weighing >2500 g. In addition, the developmental consequences of CP impact on children's vision, hearing, perception, communication, learning, and neurodevelopmental process are high. We also highlight advances in leukemia as this malignancy has significant lessons for how chemotherapy or radiation therapy in early childhood impact the developing brain. Major advances in congenital heart disease will also be examined due to the increased attention to neuroprotection during cardiac surgery. The management of muscular dystrophy (MD) and spina bifida (two of the most common causes of physical disability in children)

is also discussed. These five groups of pediatric physical disability can help us to understand the critical importance of a biopsychosocial, functional, and family ecological approaches to pediatric physical disability.

The International Classification of Functioning Model Applied to Early Childhood Disability: Cerebral Palsy after Prematurity, Leukemia, Congenital Heart Disease, Muscular Dystrophies, and Spina Bifida

One key framework for understanding child disability is the International Classification of functioning (ICF) model, proposed by the World Health Organisation (WHO) in 2001. This model describes a child's health and wellbeing in terms of four components: (1) body structures, (2) body functions, (3) activities, and (4) participation. Body structures are anatomical parts of the body such as organs and limbs, as well as structures of the nervous, visual, auditory, and musculoskeletal systems. Body functions are the physiological functions of body systems, including psychological functions such as being attentive, remembering, and thinking. Activities are tasks done by children and include walking, climbing, feeding, dressing, toileting, bathing, grooming, communicating, and socially interacting. Participation means involvement in community life, such as playing with peers, preschool education experiences and attending family activities such as visiting relatives, attending religious services, or going shopping. The ICF model also accounts for contextual factors in a child's life, including environmental facilitators and environmental barriers as well as personal factors. Environmental facilitators include family leave policies, daycare and early education accessibility, and comprehensive health insurance. Environmental barriers include negative attitudes of others, lack of legal protections, and discriminatory practices. Personal factors include age, gender, interests, and sense of self-efficacy, and these can be facilitators or barriers. **Table 1** illustrates application of the ICF model to a 3-year-old girl with diplegic CP after extreme prematurity, a 4-year-old with hearing loss and communicative delays after leukemia, a 5-year-old with hyploplastic left heart repair who developed hemiplegia and attention deficit hyperactivity disorder (ADHD) after an embolic cerebral vascular accident (CVA), a 5-year-old boy with Duchenne's muscular dystrophy (DMD), and a 6-year-old girl with spina bifida.

Understanding Cerebral Palsy

Of the 4 000 000 children born each year in the US, 2 per 1000 will go on to have one of the CP syndromes. These syndromes include spastic diplegia, hemiplegia,

Table 1 ICF model scenarios in preschool children with physical disability

Dimension	Girl, 3 years	Boy, 5 years	Girl, 4 years	Boy, 5 years	Girl, 6 years
Pathophysiology Molecular/ cellular mechanisms	800 g, 27 weeks' gestation, periventricular leukomalacia	Left CVA after staged hypoplastic heart repair	Hearing loss and communicative delays, leukemia in remission	DMD; absence of dystrophin	Meningomyelocele with Ventriculoperitoneal shunt
Body structures and body functions Organ structure/ function	Asthma Spastic diplegia Speech delays	Hemiplegia; neurobehavioral and adaptive delays	Speech and adaptive delays, 50 db sensory hearing loss bilaterally	Speech delay	Perceptual and learning delays, neurogenetic bladder, decreased sensation of feet
Activity (functional) strengths Ability to perform essential activities: feed, dress, toilet, walk, talk	Indoor walking with AFOs; drinks with straw; likes to pretend play with dolls	Climbs slide and goes down easily; loves talking books	With hearing aides, can carry on conversation. Speech understood by mother 50% of time	Knows colors; counts to 5	Able to decode words; independent in dressing
Activity (functional) limitations Difficulty in performing essential activities	Unable to climb steps; unclear speech unless repeats	Difficulty with fasteners, difficulty waiting and transitioning; impulsive	Inattentive in large groups; speech not understood by peers	Cannot run; speech understood by his mother 50% of time	Cannot run or peddle 2 wheel bicycle; wears pull-up diapers
Participation Involvement in community roles typical of peers	Plays in parallel with peers	On T-ball baseball team	Loves to dance and skate	Loves to swim	Enjoys singing
Participation restrictions Difficulty in assuming roles typical of peers	Misses day care due to asthma; uses supplemental nutrition products	YMCA will not let him use play-ground because he falls too much	No audiological consultation available to rural school	Cannot play ice hockey with brother	Excluded from YMCA swimming because of diapers
Contextual factors: environmental facilitators Attitudinal, legal, policy, and architectural facilitators	Has asthma care plan; participates in Hanen speech therapy group	Mother unable to attend pediatric stoke support groups because of her work hours	Community speech therapist works closely with teacher and school therapist	Community pool does not require climbing stairs	Active wheel chair sports leagues at YMCA
Contextual factors: environmental barriers Attitudinal, legal, policy, and architectural	Loves to watch brother age 5 use computers; on waiting list for speech therapy	Denied life insurance policy	Kindergarten has signing as only education option	Kindergarten class on 2nd floor	No nurse at school to supervise CIC

CVA, cerebral vascular accident; AFO, ankle-foot orthosis; DMD, Duchenne's muscular dystrophy; VP, Ventriculoperitoneal; CIC, Clean intermittent catheterization.

and quadriplegia as well as dyskinetic (extrapyramidal) disorders. Advances in epidemiology, neuroimaging, and habilitiative interventions have included a revised consensus definition of CP, application of noninvasive neuro-imaging modalities that allow examination of central nervous system (CNS) structure, and a gross motor and manual ability functional classification systems. The epidemiological advances have reinforced what was initially discovered the 1960s in the US Collaborative Perinatal Project. In this prospective study, approximately 50 000 pregnant women were recruited between 1959 and 1966. Their children were prospectively followed from birth to age 7 years. Fewer than 10% of the children who went on to have a confirmed diagnosis of CP had intrapatum asphyxia. Major etiological contributions to CP included CNS dysgenesis, (e.g., microcephaly and holoproencephaly), non-CNS malformations (e.g., congenital heart disease), and congenital infections (e.g., rubella, Cytomegalovirus, and toxoplasmosis).

The value of neuroimaging in understanding the timing of CP was recently highlighted in the European Cerebral Palsy cohort of children born between 1996 and 1999. This population-based sample involved children from London, Edinburgh, Dublin, Lisbon, Stockholm, Tubingen, and Helsinki. Almost one in three children had the diplegic pattern of CP, one in four had hemiplegia, and one in five had quadriplegia. In addition, one in five had extrapyramidal CP with either ataxia or dyskinesia. White-matter abnormalities were present in 43% overall. However, this abnormality of white matter was present in 71% of children with diplegia, 33% of children with hemiplegia, and 33% of children with quadriplegia. These white-matter lesions highlight a critical window of timing in that the period of vulnerability is maximal between 24 and 34 weeks' gestation. Currently, extensive research efforts are occurring to understand the complex causal pathways of inflammation and infection and their role in oligodendroglia vulnerability.

In the European cohort, malformations were as common as cortical–subcortical damage and occurred at the frequency of 9%. The latter abnormality is the most common lesion associated with term infants with neonatal encephalopthy. This selective vulnerability of cortical and subcortical areas reflects the complex metabolic demands of these regions in term infants. Of children with basal ganglia and thalamic injury, 76% had dystonia manifested by involuntary movements, difficulty with fine motor skills, and difficulty with oral-motor control. Of children with hemiplegic CP, 27% had focal infarcts, reflecting a neonatal stroke. It is in this group that maternal disorders of coagulation need to be investigated comprehensively. These disorders include maternal systemic lupus erythematosis that can cause clotting to a critical brain region and biochemical genetic factors (factor 5 Liden mutations and protein C abnormalities).

Advances in neuroimaging have helped in the understanding of timing and extent of lesions in children with CP. However, almost one in eight children with CP have a normal magnetic resonance imaging (MRI) scan. This group awaits the promise of neurogenetic technologies. Among the term and near-term children with CP, approximately 20% have the MRI abnormality of cortical–subcortical or basal ganglia injury. Since a large number of this population had emergency cesarean section deliveries, fewer than 10% of infants have had CP because of delayed obstetrical interventions. These 1996–1999 estimates are similar to the Collaborative Perinatal Project outcome data as well as data from the Western Australian CP registry for children born between 1975 and 1980.

In summary, ongoing population data, coupled with neuroimaging and functional measures, have helped advance our knowledge about the complexity of CP syndrome. Several lessons emerge from this.

1. More than half of the children with CP continue to be term gestation.
2. Multiple pregnancy increases the risk of CP.
3. Poor intrauterine growth increases the risk of CP.
4. Children born at 28–36 weeks' gestation contribute the largest numbers of CP preterms. Fewer than 10% of children with CP are less than 28 weeks of gestation. However, these children have high rates of communication, cognitive, and adaptive delays.
5. Approximately 10% of the children born in Europe with CP were born at less than 28 weeks of gestation.
6. Children with CP have high rates of recurrent seizures and sensory impairment. More than one in four children experience epilepsy, and more than one in three experience strabismus, restricted visual fields, and refractive errors. Seven per cent experience hearing impairment.
7. The majority of CP is not severe. Approximately three in five children have hemiplegia or diplegia. These children with two limb involvement can be considered as having less challenges in performing motor and functional skills compared with children with quadriplegia and four limb involvement. Overall, children with hemiplegia and diplegia have an excellent prognosis for ambulation. One hundred per cent of children with hemiplegia and 90% of children with diplegia are able to walk.
8. Quadriplegia is the least common, but most severe form of CP. The majority of children with quadriplegia have sitting challenges, manipulative challenges, and communicative difficulties as well as comorbidities of dysphagia, seizures, and recurrent pneumonia. Only 9% of children with quadriplegia can walk. In addition, there are multiple severe neurodevelopmental functional challenges, medical frailty, and high rates of caregiver stress.

From a population standpoint, reducing the prevalence of CP requires understanding pathways of the CP syndromes in term infants, improved understanding and management of moderately preterm infants, and reducing CP in survivors of multiple births. More than 750 000 children and adults in the US have a CP syndrome. Their lifetime medical cost is estimated at $1 million per individual. In this respect, disproportionate attention to perinatal hypoxemic–ischemic encephalopathy in term infants and neurodevelopmental complications in extreme prematurity lead to the erroneous perception that these two risk groups of children account for the majority of cases of children with CP.

Current advances in developmental medicine highlight pathways of risk and protection for specific groups of children with one of the CP syndromes. These pathways include twins and higher-order multiples, children with intrauterine growth restriction, children with malformations, children undergoing congenital heart surgery, and children with neonatal seizures and encephalopathy. Major neuroscience research has focused on mechanisms of white-matter injury in preterm infants and neuroprotection of term infants with severe encephalopathy. It is critically important that all health, education, and rehabilitation professionals understand that a substantial number of term and near-term children do not have a simplistic cause for their motor disability. In this way, families can be helped and the general public can understand the need for addressing shortcomings in current knowledge. Most importantly, health and educational professionals can work together to optimize functioning, learning, and participation of children with CP.

Understanding Leukemia

Childhood cancers result in neurodevelopmental disabilities related to tumors, complications of surgical resection, or exposure to chemotherapy or cranial irradiation. Leukemias and brain tumors account for over one-half of all new childhood cancers. About one-third of childhood cancers are leukemias. Of the leukemias, acute lymphoblastic leukemia (ALL) is the most common, affecting 2400 children annually in the US. ALL accounts for 75% of all pediatric leukemia cases. In the 1990s the 5-year event-free survival rates for childhood ALL ranged from 70% to 83% and the overall survival exceed 90%. The current understanding of ALL has led to intensive treatment based on phenotype, genotype, and risk. Specific treatment approaches differ but consistently emphasize: (1) remission–induction therapy to eliminate greater than 99% of the initial tumor burden, (2) intensification (or consolidation) therapy, and (3) continuation treatment to eliminate residual leukemia. Primary treatment modalities include systemic or intrathecal chemotherapy, cranial irradiation, and bone marrow transplantation. Therapy directed at the central nervous system (CNS), which starts early in the clinical course, is given for varying lengths of time. These interventions depend on the child's risk of relapse, the intensity of systemic treatment, and whether cranial irradiation is used. Research over the last 20 years has demonstrated that CNS treatment with radiation and/or chemotherapy frequently results in neurocognitive dysfunction. Difficulties in memory, attention, information processing speed, visuospatial skills, and executive functioning result from these side effects. Challenges to both preschool developmental and school-age academic output can also occur. There is evidence that the

incidence of severe cognitive impairment is decreasing with the diminished use of craniospinal irradiation. Intrathecal chemotherapy for CNS prophylaxis is the current preferred intervention that is used to prevent CNS relapse. The disease and treatment appear to have little impact on the brain structures and processes that are in place prior to onset of disease and treatment. The younger the child, therefore, the more global and severe the delayed effects will be. Factors predictive of increased risk for late effects are younger age at the time of CNS prophylaxis, female gender, high radiation dose, treatment with dexamethasone, and a history of CNS relapse.

Mechanisms of Central Nervous System Injury in Childhood Cancer

The principal mechanisms responsible for delayed neurotoxic effects are white-matter damage, disruption of the blood–brain barrier, vascular insults, and calcifications. Regions with greater myelin density, such as the right frontal lobe, are particularly vulnerable. The distribution of white-matter density may explain why some neurocognitive functions are affected and others relatively spared. The developing brain may be more susceptible to damage, because newly synthesized myelin has higher metabolic activity that makes it more susceptible to the toxic effects of therapy. Leukoencephalopathy may be detected by MRI in over 20% of patients after therapy for ALL. Correlation between these neuroradiographic findings and a neuropsychological injury has been inconsistent. However, a recent study demonstrated that smaller white-matter volumes were directly associated with challenges in attention, intelligence, and academic performance.

Cranial Radiation Therapy

Cranial radiation therapy (CRT) prophylactically is given to all patients to prevent leukemic cells from infiltrating the CNS and consists of CNS chemotherapy alone or in combination with cranial irradiation. Multiple studies demonstrate that of anticancer treatments, CRT appears to have the greatest impact on demyelination and therefore on delayed neurocognitive effects. CRT is also linked to necrosis with secondary ventricular dilation, vascular compromise, and focal calcifications in the basal ganglia and periventricular region. There is ongoing controversy between dose of CRT and risk for neurotoxic effects. Initial studies noted significant declines in intellectual functioning, especially in younger children. A meta-analysis of 30 studies of cognitive function conducted before 1988 concluded that those completing CRT experience on average a 10-point decline in intelligence quotient (IQ). Decreased rates of neurotoxic effects were noted when

CRT was reduced from 2400 to 1800 cGy. There is no consensus on a safe dose. Cranial irradiation has largely been replaced by intrathecal or systemic chemotherapy. In most clinical trials, cranial irradiation is still recommended for those with a high risk for relapse such as initial CNS leukemia, T-cell ALL, or high tumor burden. The combination of CRT and chemotherapy results in the most severe neurotoxic effects and is attributed to injury to the blood–brain barrier.

Chemotherapy

Three classes of chemotherapeutics are most responsible for late neurotoxic effects: nucleoside analogs, glucocorticoids, and antifolates.

Nucleoside Analogs

The antineoplastic effect of nucleoside analogs depends on their inhibition of DNA and RNA synthesis. Agents in this class include cytosine, arabinoside, pentostatin, fludarabine, and nelarabine. Irreversible neurotoxicity and leukoencephalopathy may be evident weeks to months after exposure to these agents. The mechanism of action for the neurotoxic effects is unclear but appears to be related to duration rather than dose or dose intensity.

Glucocorticoids

Glucocorticoids may induce transient mental status, changes, confusion, or disturbances of sleep, affect, or memory. Their mechanism of action is through their action on the hippocampus, which is the brain structure with the highest concentration of glucocorticoid receptors. Glucocorticoids inhibit glucose utilization by neurons and glia, increase the concentration of glutamate in the hippocampal synapse, and lead to excessive stimulation of postsynaptic receptors and excitotoxic neuronal death. Dexamethasone has a longer half-life and better penetration into the cerebrospinal fluid than prednisone or prednisolone. As a result, dexamethasone is used more frequently in induction and continuation therapies.

Antifolates

The primary antifolate is methotrexate (MTX). As their name implies, antifolates function as antineoplastic agents by interfering with folate-dependent biochemical systems. Steady-state cerebral spinal fluid (CSF) folate concentrations are two to three times those of serum values. The high concentration of folate in the CNS is essential for cell replication and to establish and maintain axonal myelination. MTX given intravenously at high doses decreases hematologic, testicular, and CNS relapses. All patients treated with MTX are at risk for delayed neurocognitive effects. The risk of late effect is related to individual dose and cumulative exposure to high-dose ($>500\,\mathrm{mg\,m^{-2}}$) intravenous therapy. Chronic MTX exposure is thought to deplete folate in the brain and thereby increase homocysteine levels. Homocysteine is toxic to the endothelium promoting occlusive vascular disease and microangiopathy, which results in focal neurological deficits. Homocysteine and its metabolites are excitotoxic amino acids and may contribute to the development of seizures. The disruption of folate-dependent systems may lead to demyelination and subsequent neurological manifestations such as developmental delay, intellectual disability, depression, seizures, or leukoencephalopathy.

Leukemia: Neurocognitive and Academic Outcomes

Declines in cognitive functions and academic performance are extensively described in the literature on long-term outcomes of childhood leukemia. However, the long-term educational and social outcomes of childhood cancer survivors have not received the same attention. Initially, the Canadian Late Effects Study examined parent-reported educational and social outcomes among child and adolescent survivors of childhood cancer and compared these survivors to a population-based control group with no cancer history. Significantly, more survivors than controls experienced poorer educational outcomes as measured by attending a learning-disabled program (19% vs. 7%), a special-education program (20% vs. 8%), or failing a grade (21% vs. 9%). Similarly, data from the Childhood Cancer Survivor Study (CCSS) described the self-reported utilization and factors associated with the use of special education services. The CCSS identified 12 430 English- or Spanish-speaking patients under the age of 21 years at time of cancer diagnosis from 25 sites across the US and Canada between January 1970 and December 1986. All children in this study had survived 5 years from the date of diagnosis. Younger ages at initial diagnosis were associated significantly with higher needs for special education services. Among children with leukemia, the odd ratios (ORs) for special education services were 4.40 (3.75–5.16) in preschool, 3.30 (2.66–4.00) at 6–10 years, and 1.70 (1.26–2.24) at 11–15 years. Although all academic subjects were affected, mathematics, English, and science were the most impaired. In a study of 593 survivors of ALL and 409 sibling controls from the Children's Cancer Group, there was a 3.6-fold increase in risk for special education services. Importantly, survivors had the same likelihood as their siblings of completing high school and entering college (93.6% vs. 92.4%). The likelihood of entering college (51.9% vs. 57.2%) and earning a bachelors degree (56.6% vs. 54.2%) were also similar among survivors and their siblings. However, a more recent study revealed that survivors of childhood leukemia were 60% more likely to complete high school and 30% more likely not to complete college than their siblings regardless of their treatment modality. In addition, the survivors enrolled in special education services were as likely to complete high school as those enrolled in special education without a history of cancer. The implication of these data is that with appropriate intervention, many survivors of childhood leukemia can complete high school. However, many are at continued risk for school failure and educational underachievement. Treatment modality may offer insight into the differences in educational outcomes. The OR of not finishing high school was 2.50 (1.76–3.67) for those who received only CRT; 1.0 (1.33–2.42) for those who received both intrathecal MTX and CRT; 1.60 (1.11–2.18) for those who received intrathecal MTX only.

Social Outcomes

There is increasing interest in the social and emotional outcomes of long-term cancer survivors. As the number of survivors have increased, it is increasingly important to understand the impact of cancer survival on personal sociobehavioral and emotional adjustment. Young adult survivors of childhood cancer are more likely to be underemployed and have lower income. Parents of survivors were more likely to report that their child did not have close friends or use friends as confidants. Children with leukemia often miss many days of school that may contribute to difficulties with social integration. The emotional adjustment of childhood cancer survivors may be measured in terms of overall functioning, self-esteem, anxiety, and depression. In 2001, Stam and colleagues reviewed several studies that addressed emotional adjustment of cancer survivors and reported that overall emotional adjustment of cancer survivors as a group was within normal limits. A closer look at the components of emotional adjustment revealed that on measures of self-esteem, cancer survivors felt better about their intellectual and academic status, behavior, and overall happiness than controls. A limitation of these studies is their reliance on self-report measures of general psychological functioning such as anxiety and depression rather than on measurement of the stress of a chronic illness. However, one in five of young adults with a history of cancer meet criteria for post-traumatic stress disorder (PTSD) as compared to one of the 10 abused adolescents and one in 20 typical adolescents.

Developmentally adolescents strive to become independent and define an identity. The adolescent cancer patient, however, is faced with a life-threatening illness that results in pain, physical changes, and forces them to depend on others. The degree to which late effects of cancer affect the quality of life of the increased number of long-term survivors of ALL needs to be better

addressed. Strategies that augment supportive care services, increase patient's network size and reliance on formal and informal social ties will enhance long-term survivor's quality of life.

Understanding Congenital Heart Disease

Congenital heart disease occurs with a frequency of 7 per 1000 livebirths and are the leading cause of death between ages 1 month and 1 year. Early postoperative CNS sequelae such as stroke and seizures occurs in a small percentage of children with congenital heart disease. Since cardiovascular and neurologic systems develop in tandem during the first trimester, a variety of genetic etiologies often accompany congenital heart disease. These include well-known disorders of chromosome number (Trisomy 21,18,13) contiguous gene disorders (e.g., Williams syndrome and DiGeorge syndrome), and syndromes with multiple congenital malformations (e.g., CHARGE and VACTERL). Microdeletions of the 22nd chromesome (22q11) are associated with conotruncal cardiac abnormalities, including tetralogy of fallot, interuped aortic arch, or truncus arterisois. This contiguous gene disorder includes branchial arch abnormalities that impact facial structure, endocrine, immunological, and auditory systems including the DiGeorge phenotype. Other individuals with complex genetic abnormalities in this chrosomal region have the velo-cardio-facial syndrome with an associated range of communicative, learning, and behavioral impairment.

Early Childhood Outcomes

The study of preschool outcomes of children with congenital heart disease receiving open heart surgery in infancy took place in Montréal, Boston, and Philadelphia. Children with hypoplastic heart syndrome were excluded. In a cohort of 118 survivors, 83% were followed at 18 months and adaptive behavior was measured by using either the WeeFIM or the Vineland Scales. Children demonstrated functional limitations in self-care, mobility, and social cognition. Only 21% of the cohort was functioning in basic skills similar to peers. Moderate functional disability was noted in 37%, and severe functional disability in 6%. On the Vineland Scales, functional difficulties in daily living skills were documented in 40%, with more than half of the children having poor socialization skills. Factors enhancing the risk for functional disabilities included perioperative neurodevelopmental status, microcephaly, length of deep hypothermic circulatory arrest (DHCA), length of stay in the intensive care unit, age at surgery, and maternal education. This study demonstrates that there were high prevalence rates of functional limitations that significantly impacted development and community care. In addition, there are opportunities for prospectively examining supports that enhance parent management skills involving some of these functional challenges.

The mechanisms of CNS injury in children with congenital heart disease are complex. Up to 25% of neonates with congenital heart disease have periventricular leukomalacia before surgery. Other mechanisms for CNS vulnerability include perioperative hypoxemia and hypotension, cardiac arrest, hypoglycemis or hyperglyncemiia, hyperventilation, and hyperthermia. Bellinger and colleagues described the developmental and neurological status of children at 4 years of age after DHCA or low-flow cardiopulmonary bypass (CPB). With circulatory arrest the risk to the CNS is through hypoxemic–ischemic–reperfusion injury. With low-flow CPB, embolic complications are associated with increase time of extracorporeal circulation as happened in our illustrated third case. At the age of 4 years, the mean IQ was 93 with no difference in treatment groups. At the age of 8 years, more than one in three required special education services and one in 10 had repeated a grade. Compared to children without cardiac disorders, there were higher rates of speech, developmental, learning, and attention problems. In particular, there were significant challenges in visual spatial, visual motor, executive functioning, working memory, sustained attention, and higher-order language skills. In addition, there was difficulty with hypothesis generation and coordination skills. The DHCA group did worse on motor and speech functioning while the low-flow CPB had worse impulsivity and behavioral disorders.

Model Multicenter Studies for Understanding Mechanisms Impacting on Central Nervous System Structure and Function after very low birthweight and extremely low birthweight

Several recent studies have examined postneonatal processes that might be amenable to interventions that decrease risks for long-term CNS dysfunction. One study in the US Neonatal Network examined neurodevelopmental impairments among 6093 survivors who weighed between 401 and 1000 g and were born between 1993 and 2001. Among those without infection, 29% had neurodevelopmental impairments. Approximately 50% of infants with sepsis, sepsis and necrotizing enterocolitis, or meningitis had neurodevelopmental impairments. One of the disquieting outcomes of this study was in the high rate of early cognitive developmental disability. This intellectual disability occurred in one in five of infants without infection and in as many as two in five of those with infection. One mediator of this effect was brain growth at 36 weeks postmenstrual age. In children with infection, 40–60% had acquired microcephaly (e.g., head circumference of less than 10% on standardized premature infant charts). In contrast, only 25% of children without infection had acquired microcephaly.

A different approach was undertaken in a randomized clinical trial of inhaled nitric oxide (INO) to reduce death and chronic lung disease in very preterm infants in respiratory failure. In this study the children receiving INO had lower rates of death, chronic lung disease, severe intraventricular hemorrhage (IVH), and perviventricular leukomalacia. In addition, the risk of neurodevelopmental disability in the survivors at the age of 2 years who had received INO was 53% less than controls. This occurred predominantly by decreasing cognitive developmental disability. Thus, current studies are now underway to determine neuroprotective properties of INO in preterm infants.

Kindergarten Functional Status, School Function and 10-Year Health-Related Quality of Life

Children with severe neonatal retinopathy of prematurity (ROP) are the sickest, tiniest, and most medically fragile of VLBW and ELBW infants. Functional assessment of daily living skills in mobility, self-care, and social cognition was used at age 5.5 years to over 1000 children less than 1250 g enrolled in the National Institutes of Health (NIH) sponsored multicenter randomized trial of cryosurgery for ROP. Overall, 88% of these children were followed across 23 centers. As severity of retinopathy increased, functional status declined, and severe disability in social roles increased. The complexity and severity of disability was worse in the children with severe ROP and unfavorable visual acuity compared to children with severe ROP and favorable visual acuity. The former group had a rate of motor limitations of 43% compared to 5% for the latter group; a rate of self-care limitations of 78% compared to 25%; a rate of continency limitations of 51% versus 4.5%; and a rate of communicative/cognitive limitations of 67% vs. 22%. Functional limitations in children without ROP were low: motor – 4%, self-care – 7%, continency – 4%, and communicative/cognitive limitations – 8%. Multiple logistic regression analysis revealed that favorable visual status, and favorable 2-year neurologic score predicted functional status at age 5.5 years. The favorable 2 year neurological score included the absence of microcephaly, seizures, and/or hydrocephalus. Access to health insurance and African-American race also contribute to better functional status at age 5.5 years. At the age of 8 years, the group that had the most severe retinopathy of prematurity was examined with respect to developmental and educational outcomes. Favorable visual and functional status at kindergarten entry and higher socioeconomic status were associated with significantly lower rates of special education services and below grade level educational achievement. Factors that were significantly associated with an increased risk for special education services included minority status, poverty, lack of access to a car, and supplemental social security income because of disability and poverty. In multivariate regression analysis the key predictors of special educational services at the age of 8 years were unfavorable visual status and unfavorable functional status at 5.5 years.

Additional studies at the age of 10 years involved parental assessments with the Health Utilities Index (HUI) for Health Related Quality of Life (HRQOL). The proportion of sighted children with limitations in four or more HUI attributes of mobility, speech, dexterity, cognition, emotion, pain, or hearing was 6.4% compared to 47% in children with blindness or low vision. The median HUI score was 0.87 for sighted children compared to 0.27 for blind/low-vision children ($p < 0.001$) with scores of 1 indicating perfect health and scores of 0 reflecting death. Thus, this series of outcomes at ages 5.5 years, 8 years, and 10 years revealed the value of both functional assessment and health-related quality of life for children receiving new technologies. Both the pathways that lead to more severe ROP and the processes that allow for preservation of some visual functioning after threshold ROP are involved in the severity of neuromotor, adaptive, and communicative disability at kindergarten entry. In addition, understanding mechanisms involved in extreme prematurity that result in the absence of retinopathy are critically important for preserving CNS neuromotor, higher cortical integrity, and participating with peers.

Understanding Muscular Dystrophy and Neuromuscular Coordination

MD is not one single disorder, but a collection of over 30 genetic/hereditary disorders with progressive degeneration of skeletal muscles. Many of these conditions have onset in childhood (including Duchenne's (DMD), Becker's (BMD), and congenital muscular dystrophy (CMD)) while others do not begin showing symptoms until adulthood or middle age (distal MD, myotonic MD, and oculopharyngeal dystrophy). Some, such as DMD, eventually lead to paralysis and death of the young adult in teenage years due to respiratory and/or cardiac muscle failure. However, for several MDs (such as myotonic MD), the progress of the disorder can be very slow and can span several decades before the adult becomes severely disabled. It is no surprise, therefore, that treatment and prognosis varies depending on the type of the disorder. To date, no effective treatment or cure has been found for MD. In this section, we examine the four most common types of MD and their impact on child health and development.

Duchenne's Muscular Dystrophy

DMD is an X-linked recessive disorder due to a mutaton on the *Xp21* gene. This gene is one of the largest known genes and has greater chance for mutation or deletion compared to other smaller genes. DMD occurs in 1 in 3500 live male births and is the most common form of MD. Onset of symptoms occurs in the preschool years. Early symptoms include delay in learning to walk, waddling gait, difficulty running, climbing stairs, or rising from the floor. By the ages 3–4 years, the child is not able to keep up with his peers. By age 12 years, the vast majority of children with DMD requires wheelchairs for mobility and can be community independent with the use of electric motor chairs. On average, boys with DMD tend to have a lower IQ with increased rates of, slow learning (IQ 70–85), and mild intellectual disability (IQ 55–69) than their healthy counterparts. The child's muscles, including the heart, becomes progressively weaker, and life expectancy is shortened to late teens or early 20s.

Apart from clinical examinations during early childhood (from characteristic pattern of weakness in limbs such as lordotic gait), DMD can be identified through series of tests such as measuring the creatine phosphokinase (CPK) level, ultrasound of the muscles, and muscle biopsy. Patients with high level of CPK in their blood, and a deficiency in dystrophin in their muscle can be confirmed as having DMD. Prenatal tests (such as chorionic villus sampling, amniocentesis, placental biopsy, and fetal blood sampling) can also be used to determine if the child has MD prenatally. DNA analysis can provide definite diagnosis, carrier detection, and prenatal diagnosis.

Although support and treatment are available through physical therapy, surgery (to treat scoliosis), and rehabilitation engineering, deterioration of both respiratory and cardiac muscles lead to death in young adulthood.

Congenital Muscular Dystrophy

The term CMD refers to a group of inherited disorders. Unlike DMD, muscle weakness for CMD patients is present at birth, and children of both genders are affected. Six forms of CMD have been identified to date: three without structural brain abnormalities, and three with structural brain change. In the latter case, significant intellectual disability can accompany severe muscle weakness.

A significant number of children with CMD without intellectual disability have either reduced or absence of merosin in their muscle tissue. Infants with merosin-deficient CMD often have severe weakness, decreased muscle tone, floppiness, congenital contractures, as well as breathing and swallowing problems. Although some children learn to stand or walk with supportive devices, the majority of children with merosin-deficient CMD achieve sitting unsupported as their best motor skill. Elective surgeries should be performed with great caution as there is high risk for pneumonia or ventilator dependence.

Spinal Muscle Atrophy

Spinal muscle atrophy (SMA) is a motor neuron disease where the degeneration of anterior horn cells of the spinal cord impact on voluntary muscle strength and control. SMA affects proximal muscles (muscles closest to the trunk – i.e., shoulders, hips, and back) than the distal muscles. SMA is an autosomal-recessive genetic disorder, and for a child to be affected by SMA, both parents must be carriers. There are four types of SMA: type 1 (severe SMA, also known as Werdnig–Hoffman disease), type 2 (intermediate SMA), type 3 (mild SMA, also known as Kugelberg–Welander disease), and type 4 (adult onset).

The age of onset for type 1 SMA is often prenatal, with mothers reporting decreased movements in the last trimester. Infants with type 1 SMA show symptoms of severe hypotonia and weakness, difficulty in sucking and swallowing, as well as respiratory problems immediately after birth. Infants adopt a frog-like posture, have paucity of movement, poor head control, bell-shaped chest, diaphragmatic breathing (appearing to breathe with their stomach muscles because of weak intercoastal muscles), weak cry, absent of tendon reflex, and internal rotation of arms. These infants struggle to lift their heads against gravity or accomplish motor milestones. They are unable to roll or sit unsupported. Infants with type 1 SMA are prone to respiratory infections, and the majority die of pneumonia and respiratory failure in the first year of life.

The age of onset for type 2 SMA is usually between 6 and 12 months with diagnosis before the age of 2 years. Children with type 2 SMA exhibit symptoms of weakness of legs and inability to stand or walk. Although some may sit unsupported when placed in a seated position, many are not able to sit upright by themselves without assistance. Typical clinical signs of type 2 SMA are inability to take full weight on legs, symmetrical weakness of legs and hypotonia, fasciculation of tongue, tremor of hands, absence of deep tendon reflexes, and various respiratory problems. Children with type 2 SMA have normal facial movements, as well as normal-to-above-average intelligence. Their muscle weakness is not progressive, and some children show signs of functional improvement, although many have long-term disability.

Type 3 SMA has more variable onset and age of presentation, from the second year of life to adolescence. Most children are diagnosed in the preschool years. Children are able to stand and walk without assistance, but generally have difficulty in running, climbing steps, and jumping. They also fall more frequently and have difficulty in getting up from sitting on the floor or a bent over position. Hand tremors and tongue fasciculation can be present with varying degrees of severity. Children with type 3 SMA have a very good chance at long-term survival, given preventive pulmonary care, nutrition supports, and rehabilitation intervention.

Myotonic Dystrophy

Myotonic dystrophy (also known as Steinert's disease) affects about 1 in 8000 people worldwide. The age of onset can be anywhere from birth (congenital myotonic dystrophy) to 80 years, and muscle problems can range from none to severe. Two main muscle symptoms of myotonic dystrophy are weakness and inability to relax a muscle grip. Patterns of muscle weakness include weakness of face and jaw muscles, droopiness of the eyelids (ptosis), weakness of neck, hands, and lower leg muscles, as well as thighs, shoulders, and trunk. This leads to loss of facial expression, indistinct speech, difficulty in writing, lifting, and fine movements, as well as clumsiness and unsteadiness. One of the common characteristics of myotonic dystrophy is the difficultly in relaxing hand muscles, resulting in the inability to quickly let go after a hand shake.

Although myotonic dystrophy can potentially be debilitating, deterioration is not rapid, and the disorder rarely changes its rate of progression if the onset was in adulthood. Most children and adults with myotonic dystrophy will not require a wheelchair during their lifetime. This is because large muscles needed for weight bearing and walking are only moderately affected. However, myotonic dystrophy can cause distressing symtoms including irregular heartbeat, frequent respiratory infections, choking, difficulty in swallowing, constipation, poor vision due to cataracts, and daytime sleepiness.

Congenital myotonic dystrophy, in contrast, has serious consequences. Infants born to mothers with myotonic dystrophy have severe hypotonia at birth, facial weakness, and a wide range of breathing and swallowing difficulties. It is believed that the infant is affected in the uterus, with mothers noting poor movements during the second and third trimester. From birth, infants display muscle weakness with floppiness, poor sucking, difficulty swallowing, respiration insufficiency, severe hypotonia, facial weakness, and weakness of limbs. Although early childhood respiratory insufficiency may be life threatening, once the child survives infancy, respiratory problems improve, as do swallowing difficulties and hypotonia. However, children with congenital myotonic dystrophy often have delay in communicative and cognitive milestones.

While many physicians rely on family history to diagnose myotonic dystrophy, there are several tests that can be performed to confirm the diagnosis. The myotonic dystrophy gene and the level of muscle protein creatine kinase level can be detected through blood tests. Muscle tests include biopsy and electromyography (EMG).

Early Childhood Outcomes of Muscular Dystrophy

MD is a collection of a wide range of inherited muscular disorders with varying degrees of symptoms, management and, treatment. However, few children with MD lead an independent life or hold a job without assistance. Management includes physical therapy, orthopedic surgery to manage deformity and scoliosis, occupational therapy to enhance the use of electronic switches, special education supports to optimize academic and social participation, nutrition supports, cardiac care for cardiomyopathy, and respiratory care (especially noninvasive ventilators).

Understanding Spina Bifida

Spina bifida (also known as meningomyelocele) is a birth defect that occurs during the first month of pregnancy. One or more vertebral laminae fail to fuse in the midline. The defect usually occurs in the midback (thoracic), lower back (lumbar), or at the base of the spine (sacral). There are three types of spina bifida, depending on the severity of the condition: spina bifida occulta, meningocele, and meningomyelocele.

Spina bifida occulta is the mildest and the most common form of spinal bifida. Although the bones of the spine are incomplete and the spine therefore is incomplete, the malformation (i.e., opening of the spine) is covered by a layer of skin. Children with spina bifida occulta rarely have neurological symptoms or have any developmental or medical complications. This condition does not require neonatal surgery.

Meningocele is the bulging of meninges of the spinal cord through an opening in the vertebrae; thus, the baby is born with a sac protruding from its back. However, no spinal cord is present in the sac and few nerves are affected. Surgery is needed to repair the affected area. Some minor neurological sequelae can be present.

Meningomyelocele is the most serious type of spina bifida where the sac protruding from the infant's back contains spinal cord and nerve tissues. There is no skin covering the malformation, which means that the spinal cord is exposed to amniotic fluid that has open communication with the external environment; therefore, meningitis can occur. The infants require surgery at 24–48 h of birth in order to enclose the CSF and the damaged nerves surrounding the malformation. Children with meningomyelocele have neurological symptoms including muscle weakness, sensory loss, and neurogenic bowel and bladder.

Early Childhood Outcomes of Spina Bifida

As with any congenital malformations, childhood outcomes of spina bifida depend on various factors, namely, the level of the lesion, how early treatment is administered, how soon or often rehabilitation services are received, constant monitoring from caregivers and teachers, and experience healthcare professionals.

One of the most serious consequences of spina bifida is hydrocephalus, secondary to the Arnold Chiari malformation. In order to prevent additional neurological damage from increased intracranial pressure, ventricular–peritoneal shunting is performed. Although the effects of shunt treatments on cognitive and motor abilities are still under debate, several studies suggest that children with spina bifida who did not require the aid of shunts performed better in intelligence tests than patients with spina bifida who underwent shunt treatments. This reflects a decreased risk from increased intracranial pressure. However, without shunts, children with hydrocephalus can have increased impaired mobility, seizures, and cognitive decline. Children with meningomyelocele and hydrocephalus, are more likely to experience difficulty in fine motor and visual perceptual skills. For this group, the severity of the motor impairment and/or cognitive performance depends largely on the location of the lesion: the higher the lesion, the more severe the impairment.

Upper extremity incoordination is common in spina bifida. It includes fine motors difficulties such as use of hands and fingers for grasping and manipulation of objects, as well as speed and strength during bilateral hand use and motor planning ability. These impairments can lead to the child having challenges with grooming, dressing, bathing, and toileting. However, it is important to note that lack of dexterity of movement in children with meningomyelocele can be the result of limited opportunity to use their hands for manipulation, play, or everyday activities.

See also: Cerebral Palsy; Learning Disabilities; Intellectual Disabilities; Mental Health, Infant; Milestones: Physical.

Suggested Readings

Bellinger DC, Wypij D, Kuban KC, *et al.* (1999) Developmental and neurological status of children at 4 years of age after heart surgery with hypothermic circulatory arrest or low-flow cardiopulmonary bypass. *Circulation* 100(5): 526–532.

Biggar WD (2006) Duchenne muscular dystrophy. *Pediatrics in Review* 27(3): 83–88.

Dickerman JD (2007) The late effects of childhood cancer therapy. *Pediatrics* 119(3): 554–568.

Doherty D and Shurtleff DB (2006) Pediatric perspective on prenatal counseling for myelomeningocele. *Birth Defects Research Part A, Clinical Molecular Teratology* 76(9): 645–653.

Krageloh-Mann I and Horber V (2007) The role of magnetic resonance imaging in elucidating the pathogenesis of cerebral palsy: A systematic review. *Developmental Medicine and Child Neurology* 49(2): 144–151.

Msall ME and Tremont MR (2002) Measuring functional outcomes after prematurity: Developmental impact of very low birth weight and extremely low birth weight status on childhood disability. *Mental Retardation and Developmental Disabilities Research Reviews* 8: 258–272.

Msall ME, Phelps DL, Hardy RJ, *et al.* (2004) Educational and social competencies at 8 years in children with threshold retinopathy of prematurity (ROP) in the CRYO-ROP multicenter study. *Pediatrics* 113: 790–799.

Skinner R, Wallace WH, and Levitt G (2007) Long-term follow-up of children treated for cancer: Why is it necessary, by whom, where and how? *Archives of Disease in Childhood* 92(3): 257–260.

Wernovsky G, Shillingford AJ, and Gaynor JW (2005) Central nervous system outcomes in children with complex congenital heart disease. *Current Opinion in Cardiology* 20(2): 94–99.

World Health Organization (2001) *International Classification of Functioning Disability and Health.* Geneva: WHO.

Diarrhea

J A Rudolph, Children's Hospital Medical Center, Cincinnati, OH, USA
P A Rufo, Children's Hospital Boston, Boston, MA, USA

This article is reproduced from the *Encyclopedia of Gastroenterology*, volume 1, pp 585–593, 2004; © Elsevier Inc.

Glossary

Acute diarrhea – A diarrheal illness of less than 14 days duration. Acute diarrheal disease in children is most often the result of self-limited viral infections. Management includes prompt assessment and repletion of hydration status. Evaluation for an etiologic process is generally not warranted unless there is an associated finding such as blood in the stool or systemic symptoms.

Chronic diarrhea – A diarrheal illness of greater than 14 days duration. Chronic diarrhea in children can be due to either infectious or noninfectious processes. Evaluation for a specific etiology is indicated. Management of comorbid conditions such as poor growth or malnutrition is essential.

Colitis – Any inflammatory process affecting the colon. Colitis usually presents clinically as bloody diarrhea, abdominal cramping, and tenesmus.

Congenital diarrhea – A group of diarrheal illnesses that are present from birth. Congenital diarrhea can be the result of either a specific genetic defect in a secretory or absorptive pathway or abnormal intestinal development.

Gastroenteritis – A diarrheal process that affects the upper gastrointestinal tract and presents most typically as an acute watery diarrhea. Gastroenteritis usually denotes an acute diarrhea that is infectious and self-limiting.

Hemolytic uremic syndrome – A sequela of *Escherichia coli* O157:H7 colitis. This toxin-mediated microangiopathy results in a triad of hemolytic anemia, thrombocytopenia, and renal failure. The occurrence of the syndrome is generally limited to children under 10 years of age.

Inflammatory diarrhea – A diarrheal illness in which the predominant pathologic finding is an invasion of the intestinal epithelium by immunocytes. This type of diarrhea can be the result of either a normal immune response to an abnormal environment, as in infection, or an abnormal immune response to a normal environment, as in inflammatory bowel disease.

IPEX syndrome (Immunodysregulation, polyendocrinopathy, and enteropathy: X-linked) – An inherited X-linked syndrome that results from a mutation in the *FOXP3* gene in humans. It is characterized by autoimmune enteropathy and multiple endocrinological abnormalities including diabetes mellitus, hypothyroidism, and hemolytic anemia.

Osmotic diarrhea – A diarrheal illness that is driven by osmotic forces that promote a net flux of water out of the interstitium and into the intestinal lumen. A stool sodium level of mEq l^{-1} and an osmotic gap of greater than 100 mosm l^{-1} suggest an osmotic diarrhea.

Secretory diarrhea – A diarrheal illness that is driven by the active secretion of salt and water by intestinal epithelial cells. A stool sodium level of greater than 70 mEq l^{-1}, an osmotic gap of less than 100 mosm l^{-1}, and a failure of the diarrhea to respond to a controlled fast suggest a secretory diarrhea.

The frequency and consistency of stool can vary considerably from individual to individual, as well as in the same individual over time. There has therefore remained a lack of a consensus as to how diarrheal illness should be defined. Investigators have employed a number of qualitative and quantitative dimensions of stool output to address this issue in the past. For the most part, children pass between one and three stools, or approximately 5–10 ml of stool per kilogram of body weight per day. As such, investigators have begun to use these benchmarks as the upper limits of normal in their identification of subjects in studies addressing acute or chronic diarrheal disease.

Regulation of Intestinal Fluid Secretion and Absorption

The mucosa lining the gastrointestinal tract must reconcile daily a seemingly contradictory array of physiologic tasks. These conflicting responsibilities include the maintenance of a tight barrier against potentially virulent bacterial and viral pathogens in the intestinal lumen, while at the same time presenting a selectively permeable interface through which to carry out immune surveillance and nutrient absorption. In this context, intestinal fluid secretion can serve both defensive (flushing away pathogens and toxins) and homeostatic (maintenance of mucosal hydration necessary to facilitate enzymatic digestion) purposes.

Stool output in humans is a composite of ingested, secreted, and absorbed fluid intermixed with residual dietary matter and cellular debris. Adults typically ingest approximately 2 l of fluid per day and produce an additional 9 l in the form of salivary, gastric, small intestinal, and pancreato-biliary secretions, to complete the process of digestion. The small intestine and colon have evolved highly efficient intercellular and transcellular pathways for the reabsorption of the vast majority (approximately 99%) of this intestinal fluid, and the average adult will pass only approximately 200 g of stool per day. This balance between fluid secretion and absorption is therefore quite tight. Any microbiologic, dietary, pharmacologic, or hormonal input that affects cell membrane transporters and/or the intercellular tight junctions responsible for fluid absorption can tip this net fluid balance in favor of secretion (or reduced absorption) and thereby trigger the increased stool output observed in patients with diarrheal illnesses.

The cellular basis for salt and water secretion in the intestine, as well as in other hydrated mucosal surfaces in the body, depends upon a vectorial transport of Cl^- ions by specialized epithelial cells. Intestinal crypt epithelial cells use basolateral membrane Na^+/K^+-ATPase pumps as well as the Na^+- and K^+-coupled cotransporter NKCCl to accumulate Cl^- ions above their electrochemical

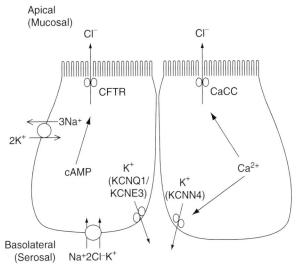

Figure 1 Intestinal crypt epithelial cells use basolateral membrane Na^+/K^+-ATPase pumps as well as the Na^+- and K^+-coupled cotransporter NKCCl to accumulate Cl^- ions above their electrochemical gradient. The subsequent opening of Cl^- channels located in the apical membrane of enterocytes permits sequestered Cl^- ions to move down their electrochemical gradient and into the intestinal lumen. The parallel activation of plasma membrane K^+ channels conduct K^+ outside, thereby sustaining the inside-negative cell membrane potential that is necessary to initiate and maintain a Cl^- secretory response.

gradient (**Figure 1**). The subsequent opening of Cl^- channels located in the apical membrane of enterocytes permits sequestered Cl^- ions to move down their electrochemical gradient and into the intestinal lumen. The parallel activation of plasma membrane K^+ channels conducts K^+ outside, thereby sustaining the inside-negative cell membrane potential that is necessary to initiate and maintain a Cl^- secretory response.

Fluid secretion in the intestine is tightly regulated by endocrine as well as neuroenteric mechanisms that utilize either cyclic nucleotides ($3',5'$-monophosphate (cAMP) or cyclic GMP (cGMP)) or Ca^{2+} as second messengers. Cyclic nucleotide-dependent agonists initiate Cl^- secretion through the parallel activation of the apical membrane Cl^- channel CFTR (the cystic fibrosis transmembrane receptor) as well as the basolateral membrane K^+ channel KCNQ1/KCNE3. In contrast, agonists utilizing Ca^{2+} as a second messenger activate the apical membrane Cl^- conductance CaCC in concert with the basolateral membrane K^+ channel IKl (KCNN4). The net movement of Cl^- ions into the intestinal lumen imparts a transiently negative charge to this extracellular compartment and positively charged Na^+ ions move via paracellular pathways in response. The osmotic force generated by transported Cl^- and Na^+ ions pulls water molecules along to effect net fluid secretion. The activity of CFTR is regulated primarily by cAMP- and cGMP-dependent protein kinases. In contrast, Ca^{2+}-dependent Cl^- secretion in the intestine conducted by CLCA appears to be limited by the generation of the intracellular down-regulatory intermediates inositol-3, 4,5,6-tetrakisphosphate, and phosphorylated extracellular signal-regulated kinase.

Whereas Cl^- secretion drives intestinal fluid secretion, fluid absorption is mediated primarily by the vectorial transport of Na^+ ions out of the intestinal lumen and into the interstitium. Na^+ transport can be electrogenic (as in the case of apical Na^+ channels), Na^+-coupled, or electroneutral. The accumulation of absorbed Na^+ ions in the tissue interstitium favors the subsequent movement of Cl^- ions and water molecules out of the intestinal lumen via transcellular and paracellular pathways, thereby effecting salt and water uptake. Na^+ channels have been identified in the apical membrane of the epithelium of the gastrointestinal tract. By acting in a coupled fashion with basolateral membrane Na^+/K^+- ATPase pumps, these channels permit lumenal Na^+ ions to move down their electrochemical gradient and into the cell. The favorable Na^+ gradient established by Na^+/K^+ pumps has also been exploited by the small intestine to promote nutrient absorption. SGLT1 is the Na^+-coupled glucose transporter expressed along the apical membrane of enterocytes. Similarly, Na^+ uptake in the small intestine is effected through Na^+-coupled amino acid transporters that are present along the enterocyte brush border. Finally, the Na^+/H^+ exchanger NHE-3, expressed in the apical membrane of enterocytes, appears to mediate electroneutral Na^+ transport in the intestine.

The tasks of intestinal fluid secretion and absorption are separated spatially along the length of the crypt–villus axis through a segregation of relevant plasma membrane channels and transporters. Cells newly differentiated at the crypt base display a primarily secretory phenotype and express high levels of CFTR. As these cells mature and migrate up the axis to take up more villous positions, they express increasing numbers of absorptive proteins including NHE-3 and Na^+-coupled glucose and amino acid transporters. Stool output is therefore the net product of intestinal fluid secretion originating in crypt cells (which occupy approximately one-third of the crypt–villus axis) and fluid absorption from villus cells (which take up the remaining two-thirds of the crypt–villus axis). Any disorder damaging surface villi, and thereby decreasing the villus/crypt ratio, will selectively decrease mucosal absorptive potential and cause increased stool output. This explains the increased stool output observed in patients with celiac disease, postviral syndromes, and giardiasis.

Approach to the Child with Diarrhea

Diarrhea can be classified on the basis of several descriptive factors (acute vs. chronic, inflammatory vs. noninflammatory, infectious vs. noninfectious, secretory vs. osmotic) that aid in the diagnostic approach. These include the duration of the illness, the existence of a secretory or osmotically driven mechanism, the presence or absence of a pathogen, and the degree of mucosal inflammation. Although the pathogenesis of diarrheal disease can be explained by a discrete process in some patients, increased stool output is more often the result of a combination of factors. As such, patients with inflammatory diarrhea can present with a secretory component due to the local release of endogenous secretagogues. Clinical diagnosis rests on an understanding of the close interplay between environmental and host factors in these patients.

Central to the diagnosis of a diarrheal illness is the clinical context in which it presents. Characteristics of the individual, such as age, are often the first clue in determining an etiology. This is most apparent in the case of congenital diarrheas, which present exclusively within the first few days of life. Components of the child's overall health, such as atopy or immunodeficiency, can also suggest a particular etiology. Environmental factors, including diet, must also be taken into consideration in the diagnostic approach to the pediatric patient with diarrhea. In the setting of infectious diarrhea, an exposure history such as an ill contact at home or in daycare, a recent travel history, or contact with a pet or animal, can sometimes provide useful epidemiologic information when attempting to understand how a pathogen may have been acquired.

The character of the stool itself is often helpful when arriving at a specific diagnosis. Stool that is both watery and voluminous in nature suggests an abnormality in the absorptive or secretory function of the small intestine. In contrast, crampy abdominal pain, tenesmus, and the presence of frank blood in the stool suggest colitis or large bowel disease.

Several aspects of diarrheal disease in children merit special consideration. Children, and most especially infants, are more susceptible to dehydration than their adult counterparts. This is due both to their greater overall body surface area relative to their weight and to a dependence on caregivers, who may be less likely to offer fluids to or feed a child who is vomiting or appears ill. Poor growth and malnutrition can also become a factor in children when diarrhea is chronic in nature. During infancy and early childhood, a large proportion of caloric intake is devoted to growth. Diarrheal disease, resulting in inadequate intake or poor nutrient absorption during this critical developmental period, can alter weight gain and, in severe cases, result in stunted linear growth.

The scope of the remaining article will discuss the causes, evaluation, and treatment of diarrheal disease in infants and children. By convention, the discussion will be segregated into infectious causes and noninfectious causes with a special reference to age of onset where appropriate.

Infectious Diarrhea in Children

Infectious diarrhea is usually of acute onset in a previously healthy child. Fortunately, most causes of infectious diarrhea are self-limited and require only symptomatic care. However, if left untreated, acute diarrheal illness can progress to chronic diarrhea in some patients. Fever is a common associated symptom of infectious diarrhea and vomiting is not unusual, especially if the infection occurs in the upper gastrointestinal tract (i.e., gastroenteritis). In general, infectious diarrheas are secretory or mixed secretory/osmotic in character. Toxin production, pathogen adherence, or frank tissue invasion all can contribute to increased Cl^- secretion by affected epithelial cells. When pathogenic invasion of the epithelium occurs, there is usually an inflammatory component to the diarrhea as well. Pathogens that cause diarrhea can be viral, bacterial, or parasitic.

Viruses are the most common cause of acute infectious gastroenteritis in children (**Table 1**, part A). There are several reasons for the preponderance of cases of viral diarrheas. The naive immune system of an infant has not been exposed to many of the viral pathogens present in the environment. In addition, daycare provides group settings that facilitate the transmission of enteric and respiratory viral diseases.

Rotavirus is the most common viral pathogen. All children exposed to rotavirus, regardless of whether or not they manifest symptomatic diarrhea, will develop circulating antibodies to this pathogen. The decreasing incidence of rotavirus in adults is thought to be due to the protective effect of these antibodies. Rotaviruses are small, wheel-shaped viruses approximately 70 nm in diameter. Of the four major groups (A, B, C, and D), type A viruses are the most important in children. The virus invades the epithelium and promotes an inflammatory response that ultimately contributes to the destruction of the villous surface. However, the frequency and severity of stool output in these patients does not correlate closely with the degree of intestinal damage observed endoscopically or histologically. This has led to the speculation that there are other pathogenic mechanisms that contribute to the malabsorption and net fluid losses observed in these patients. Although villous destruction can be severe in rotaviral disease, recovery is rapid in most patients and symptoms typically resolve in 2–7 days.

Caliciviruses, including the Norwalk and Norwalk-type agents, are the second leading cause of pediatric viral diarrheas. This group of viruses presents in a

Table 1 Etiology of pediatric diarrhea

Infectious diarrhea	Noninfectious diarrhea
A. Viral pathogens	D. Inflammatory
Rotavirus	Inflammatory bowel disease
Adenovirus	Celiac disease
Norwalk agent	Allergic enteropathy
Calicivirus	Autoimmune enteropathy
Astrovirus	Graft-vs.-host disease
Coronavirus	E. Noninflammatory
B. Bacterial pathogens	Congenital diarrheas
Campylobacter spp.	Congenital chloride
Salmonella spp.	diarrhea
Shigella spp.	Congenital sodium
Escherichia coli	diarrhea
Enterotoxigenic	Microvillus inclusion
Enteropathogenic	disease
Enterohemorrhagic	Tufting enteropathy
(shigatoxin producing)	Carbohydrate transporter
Enteroadherent	defects
Enteroinvasive	Dissacharidase deficiency
Yersinia spp.	Amino acid transporter
Vibrio spp.	defects
Aeromonas spp.	Pancreatic insufficiency
Plesiomonas spp.	Bile acid transport defects
Clostridium difficile	Abetalipoproteinemia
C. Parasitic pathogens	Acquired diarrheas
Giardia lamblia	Toddler's diarrhea
Cryptosporidia	Short bowel syndrome
Cyclosporidia	Small bowel overgrowth
Entamoeba histolytica	Antibiotic-associated
Nematodes	diarrhea
Cestodes	Münchausen's syndrome
(tapeworms)	Secondary lactase
Trematodes	deficiency

similar fashion to rotavirus, with the exception that the diarrhea is usually milder. Astroviruses are similar to calciviruses and are a common cause of diarrheal illness. Adenovirus (serotypes 40 and 41) is a well-established cause of viral diarrhea and has a slightly longer incubation period and a longer course than rotaviral disease. More recently, Torovirus has been implicated as a potential cause of diarrhea in children. However, more definitive epidemiologic data concerning this pathogen are currently lacking.

Bacterial infections can also cause diarrheal disease in infants and children (**Table 1**, part B). As in the case of viral diarrhea, the onset of bacterial illness is usually acute and presents with fever and sometimes vomiting. Because the most common forms of bacterial diarrhea are invasive, bloody diarrhea is often reported in these patients. Specific types of bacterial illness have been reported to occur more commonly in specific age groups. *Campylobacter jejuni*, for instance, has a bimodal distribution of onset with the first peak occurring in children from 1 to 5 years old and a second peak in adolescents. Nontyphoid

Salmonella enteritidis can cause a bacteremia in infants and in immunocompromised hosts. *Shigella* species can be found in the toddler age group, but is not a commonly isolated pathogen in the US. *Clostridium difficile*, an important cause of antibiotic-associated diarrhea in adults, is not usually a pathogen in infants. *C. difficile* toxin can be found in up to 10% of healthy newborns and is even more prevalent in neonatal intensive care units. The reason for the inability of this organism to cause diarrhea in infants remains unclear. Based on animal studies, it is thought that the receptor for this toxin is developmentally regulated and absent in early infancy. *Vibrio cholerae* causes a prototypical bacterial secretory diarrhea. It produces a toxin composed of two subunits. The B, or binding, subunit displays a pentameric form that binds selectively to the ganglioside GM_1. The A, or active, subunit is internalized by intestinal epithelia, alters signal transduction, and leads to increased production of cAMP and Cl^- secretion. Other forms of toxin-producing organisms include enterotoxigenic *Escherichia coli*, the pathogen responsible for traveler's diarrhea, and organisms responsible for acute food poisoning such as *Staphylococcus aureus* and *Bacillus cereus*. *E. coli* O157:H7 is an important pathogen in children. This enteropathic *E. coli* adheres to the intestinal lumen and produces a toxin that is absorbed and causes the hemolytic–uremic syndrome.

Parasitic disease causing diarrhea is far less common in industrialized countries (**Table 1**, part C). One notable exception is *Giardia lamblia*, which is especially prevalent in the daycare setting. *Giardia* can present as an acute diarrheal illness or as a more chronic process. The mechanism by which this organism causes diarrhea is not fully understood. There is no gross alteration in intestinal architecture or evidence of a significant immunologic response. There are multiple other parasites that can cause diarrheal disease in children. However, these occur much less commonly and will not be discussed further.

Noninfectious Diarrhea in Children

Occasionally, a child will present with a diarrheal illness that is not self-limiting. Fever may or may not be present and other comorbidities, such as growth failure and malnutrition may be prominent. Stool cultures are negative. The etiology of diarrheal disease in these patients can be broadly classified as being inflammatory or noninflammatory in nature, based on clinical history, physical examination, and biochemical workup. Similar to patients with infectious diarrhea, the increased stool output observed in these patients is typically the result of a combination of pathogenic mechanisms.

Inflammatory Diarrhea

The intestine displays a tremendous capacity to generate an immune response based on the presence of numerous effector immunocytes that lie within the intestinal mucosa and submucosa. More recent data have demonstrated that intestinal epithelial cells themselves also possess the ability to process lumenal antigens and present them to the underlying immune cells. The intestinal epithelium is in constant contact with the external environment. It is subsequently in a constant state of low-grade inflammation (often referred to as 'physiologic inflammation') that is the result of the epithelium playing its role in the surveillance of and response to the broad array of dietary, microbiologic, and toxigenic stimuli present within the intestinal lumen. When the degree of mucosal inflammation is severe enough to affect the absorptive and secretory function of the intestine, diarrhea ensues.

A number of immune defects or imbalances can affect the intestine (**Table 1**, part D). Inflammatory bowel disease is example of an inflammatory diarrhea that is likely the result of a genetically driven dysregulated immune response to the lumenal environment. It is also likely that genetic predisposition may leave some individuals vulnerable to an exaggerated immune response to dietary antigens that are usually not perceived to be a threat to intestinal function. This may explain the incidence of allergic enteropathies in some children. In patients with celiac disease, or gluten-sensitive enteropathy, there is an immune-mediated response to a protein present in wheat and related grain products. Although these patients can show marked diarrhea, they more commonly present with a failure to thrive precipitated by the introduction of wheat-containing solid foods between 6 and 9 months of age.

Autoimmune disease can target the intestinal epithelium itself and antibodies directed against enterocytes contribute to the severe inflammation and tissue destruction observed histologically in these patients. The IPEX syndrome is an X-linked autoimmune enteropathy that is associated with polyendocrinopathy and results in high morbidity and mortality. The gene defect is thought to lie within the *FOXP3* gene and it has been shown to encode the protein scurfin, a regulator of T-cell function in mice. The important role played by lymphocytes in maintaining intestinal barrier function can be appreciated in the context of bone-marrow transplant recipients. Diarrhea is a major feature of graft-vs.-host disease, a clinical condition in which donor lymphocytes recognize host intestinal epithelial cells as being foreign. Activated immunocytes subsequently initiate a destructive process that is manifest histologically as increased epithelial cell apoptosis and clinically as a secretory or inflammatory diarrhea.

Noninflammatory Diarrheas

Children can also suffer from diarrhea that is neither infectious nor inflammatory in nature. These diarrheal illnesses can be broadly categorized into congenital or acquired forms (**Table 1**, part E). Congenital diarrheas are most often the result of abnormal gene expression, resulting in a clinical presentation within the first week of life. Congenital chloride diarrhea is caused by a mutation in the down-regulated in adenoma (*DRA*) gene, thought to be a colonic chloride transporter. This disease presents uniformly *in utero* with polyhydramnios. Severe diarrhea and abdominal distension appear shortly after birth and profound electrolyte disturbances can occur in these patients if not resuscitated promptly. In contrast, the cause of congenital sodium diarrhea is not known but is thought to be due to a functional uncoupling of sodium and hydrogen exchange in the intestine. No mutations have been described in the known Na^+/H^+ exchangers in the intestine to date. The clinical presentation of congenital sodium diarrhea is similar to congenital chloride diarrhea with the exception that stool chloride levels in these patients are typically lower and the stool pH tends to be more alkaline. In addition to defects in ion transporters, there have been a number of diseases that have been described with altered transport of glucose, galactose, and amino acids. Gastrointestinal symptoms vary from defect to defect. Amino acid transport defects often have extraintestinal manifestations whose consequences far outweigh changes in bowel patterns.

Congenital diarrheas can also be caused by genetic defects that result in the malabsorption of the products of digestion such as carbohydrates and fat. Congenital disaccharidase deficiencies are rare and result in an osmotically driven diarrhea. Much more common are the transient and secondary deficiencies in mucosal disaccharidase levels that result from small intestinal injury or inflammation. Fat malabsorption can also present with diarrhea of variable severity. Congenital fat malabsorption can be the result of pancreatic insufficiency, seen in patients with cystic fibrosis, or due to specific genetic defects such as abetalipoproteinemia. Fat malabsorption is characterized by varying degrees of greasy and malodorous stools. Finally, congenital disorders of the intestinal architecture can lead to diarrhea. Microvillus inclusion disease is a rare autosomal recessive disease that is characterized by severe watery diarrhea at birth. Diagnosis is based on a histologic demonstration of marked or complete villous atrophy and electron microscopic evidence of intracellular microvillus inclusions and absent or rare microvilli.

There are multiple acquired forms of pediatric diarrhea that can be characterized as being noninfectious and noninflammatory in nature. Often, these diarrheas result from a predisposing insult that diminishes the ability of the intestinal mucosa to absorb nutrients, thereby contributing to an osmotic diarrhea. The most common example of this is toddler's diarrhea or chronic nonspecific diarrhea of childhood. There is no underlying inflammatory or biochemical abnormality that drives the increased stool output seen in these young children. In many cases, these patients will respond to a reduced dietary intake of fruit juices. Because many of these juices contain large amounts of sorbitol, an indigestible carbohydrate, they can induce an osmotic diarrhea. As such, the diarrhea will resolve in most patients within a few days after removal of the offending juice. Other examples of acquired and primarily noninflammatory diarrheas that fall into this category include antibiotic-associated diarrhea, short bowel syndrome, and small bowel bacterial overgrowth. Additionally, Münchausen's syndrome-by-proxy must always be considered in children with diarrhea and no predisposing factors.

Laboratory Evaluation of Diarrhea

Laboratory evaluation of the pediatric patient with diarrhea varies with the suspected cause and is dictated by the clinical picture. Any suspicions about potential inflammation or bacterial infection should be addressed immediately. Evaluation of acute diarrhea is usually limited to cases in which a given patient is presenting with systemic symptoms or comorbidities. Chronic diarrhea must always be evaluated, especially in the context of poor growth or malnutrition. An evaluation that proceeds in a logical and stepwise manner generally results in the most expedient and cost-effective diagnosis.

The first step in the evaluation process is to determine whether or not the presenting patient's symptoms are most consistent with an inflammatory or noninflammatory process. This can be done by an examination of the stool for gross or occult blood or the presence of fecal leukocytes. Previous studies have also demonstrated the sensitivity and specificity of biochemical assays for fecal lactoferrin, a constituent of neutrophil granules. Patients with infectious or inflammatory diarrhea will typically present with rectal bleeding or overt (positive fecal leukocyte smear) or biochemical evidence (lactoferrin) of fecal white blood cells. In contrast, these studies should be negative in patients with noninflammatory (viral, osmotic, or secretory) diarrheal disease. Nonetheless, although these markers may increase the yield of sending stool cultures, they do not exclude intestinal inflammation and any final decision about pursuing an infectious workup must be made on clinical grounds.

If there is clinical or biochemical evidence of an inflammatory process, then routine stool cultures remain the gold standard in the search for a bacterial cause of diarrhea. Most hospital-based laboratories have a standard panel of cultures associated with common pathogens including *Campylobacter*, *Shigella*, *Salmonella*, and *Yersinia enterocolitica*. Many hospitals also routinely screen for *E. coli* O157:H7. The identification of some pathogens relies on the detection of a particular toxin that is produced by the bacteria and released into the stool. *C. difficile* is perhaps the best recognized pathogen in this class.

The diagnosis of parasitic disease is most often made by a close microscopic evaluation of the stool for ova and parasites. The identification of *Giardia* and Cryptosporidia has been further facilitated by the development of enzyme-linked immunosorbent assay (ELISA)-based stool tests. It is imperative to know a specific laboratory's capabilities and limitations prior to interpreting the results of any stool, toxin, or parasitic studies.

Most 'noninflammatory' diarrheal disease is viral in nature. However, routine evaluation of stool for viral pathogens is not often useful because of the self-limiting nature of the disease process in the vast majority of patients, the specialized nature of obtaining viral cultures, and the expense of detecting specific viral pathogens. One notable exception is the rotavirus stool antigen test. This commercially available ELISA-based test provides relatively rapid results that can assist both in patient care and in making decisions about the need for isolation of hospitalized patients. Other viral stool tests include polymerase chain reaction-based screening for viral DNA in the case of adenovirus. However, these more costly and specialized tests are typically reserved for the evaluation of immunocompromised patients, in whom targeted supportive or antiviral therapy is much more critical.

Characterization of the stool can be helpful for determining the nature of noninflammatory diarrheal illness. Stool evaluation for fat, pH, and reducing substances is important in determining whether or not there is an underlying malabsorptive process. The presence of 'neutral' fat in the stool suggests some deficiency in the production or delivery of pancreatic (lipase) or hepatic (bile acid) secretions into the intestinal lumen. An increase in 'split' fat in the stool indicates a primary inability of enterocytes to perform fat absorption. Reducing substances are the result of undigested carbohydrates making their way into the large intestine. The presence of these fecal sugars can be readily assessed with commercially available colorimetric strips or test solutions. It must be remembered that sucrose is a nonreducing sugar. As such, stool must first be pretreated with an acid solution to make this nonreducing sugar detectable. Undigested carbohydrates, as well as dietary fiber, are consumed by bacteria in the large bowel and generate short-chain fatty acids. Carbohydrate malabsorption can therefore also be assessed by a fall in stool pH.

Stool electrolytes can help to determine whether or not a diarrheal process is secretory in nature. In general, a stool Na^+ concentration of greater than $70 \, mEq \, l^{-1}$ is indicative of a secretory process. The stool osmotic gap, calculated by:

$$([Na^+] + [K^+]) \times 2 - stool \; osmolarity$$

where $[Na^+]$ = concentration of Na and $[K^+]$ = concentration of K are useful in distinguishing between osmotic and secretory diarrheal disease. An osmotic gap greater than $100 \, mosm \, l^{-1}$ are suggests an underlying osmotic process. Similarly, whereas osmotic diarrhea will typically respond to a dietary fast, secretory diarrheal diseases are driven by processes that are independent of exogenous (dietary or pharmaceutical) factors.

The ability to study the large and small intestine of patients using videoscopic endoscopy has greatly advanced the ability to diagnose and treat diarrheal disease in pediatric and adult patients. Clinicians are now able to assess the gross appearance of the lining of the small and large intestine, obtain biopsy samples for histologic examination, measure directly mucosal disacchari-dase levels, collect pancreatic and biliary secretions, and sample fluid from the small intestine for quantitative culture.

Blood tests can often prove to be useful adjuncts to stool studies. Peripheral eosinophilia may point to an underlying allergic disease. Decreased serum albumin levels can suggest malnutrition or a protein-losing enteropathy. Specialized serum tests such as the detection of antibodies directed against tissue transglutaminase are highly predictive of celiac disease. However, for most patients, blood work plays a supportive role in the workup of diarrheal disease. Results from serologic studies most often suggest an etiology that will need to be confirmed by more definite stool or endoscopic studies.

Treatment of Diarrheal Disease

The treatment of pediatric diarrheal disease can be divided into symptomatic and curative therapies. First and foremost in the treatment of any child with diarrhea is a prompt assessment of hydration status. For most cases of mild to moderate diarrhea, oral rehydration solutions are the first line of therapy. When oral intake is limited secondary to an altered mental status or when severe dehydration or shock is present, intravenous replacement of fluid and electrolytes can be lifesaving. Once the patient is adequately hydrated, the diet may be readily advanced. The provision of adequate calories is critical to maintain an anabolic state that will provide the metabolic fuel necessary to promote epithelial restitution. The advantages of enteral supplementation should not be overlooked as lumenal contents have been shown to be trophic to the intestinal epithelium. A transient lactose intolerance may occur in either acute or chronic diarrhea. This can be addressed using soy, rice-based, or lactose-free milk products. High-fructose and sorbitol-containing drinks are palatable, but should be avoided due to the increased osmotic load they place on an already compromised epithelial lining. Other supportive measures that have been used include antisecretory agents, antimotility agents, and resin binders. These agents decrease overall stool output by slowing intestinal transit. Although clinically beneficial in most cases, clinicians must be wary of the possibility that these agents can contribute to third-spacing of body fluid in distended and pharmacologically atonic intestinal loops.

Specific therapies that are designed to treat the underlying cause of diarrhea can be employed. This includes antibiotic use in certain forms of infectious diarrheas. In general, however, antibiotics should be avoided in patients with diarrheal disease unless there are systemic consequences of the diarrhea, such as that observed with *Salmonella* infections in infants and the elderly. Inappropriate antibiotic use can lead to resistant organisms or prolong the carrier state. Notable exceptions include infectious diarrheas that may become chronic if left untreated, such as diarrhea caused by *C. difficile* and *G. lamblia*.

Other specific therapies for diarrheal disease in pediatric patients include the following: immunosuppression in the immunologically mediated diarrheas such as inflammatory bowel disease or autoimmune enteropathy; specific replacement of electrolytes in the case of the congenital chloride and sodium diarrheas; or enzyme replacement therapy in patients with pancreatic insufficiency or lactose intolerance. Removal of an offending agent, such as gluten-containing foods in celiac disease, lactose in lactase deficiency, or specific dietary antigens in congenital or acquired protein intolerances, can be critical in certain diarrheal illnesses.

Summary

The intestine is a site of competing physiologic processes including salt and water secretion, nutrient absorption, and immune surveillance. Stool output is subsequently the net product of opposing secretory and absorptive capacities that are separated geographically along the length of the intestine as well as along the length of the crypt–villus axis. Any disruption of these tightly regulated homeostatic processes can lead to altered stool formation and the development of pathologic diarrhea. In most cases, these illnesses are selflimited in nature and respond favorably to supportive measures. Nonetheless, pediatric diarrheal diseases remain a significant cause of morbidity and mortality worldwide.

The diagnostic approach to diarrheal disease in children differs substantially from that pursued in other age groups. Consideration must be given to congenital or developmental etiologies not seen in adult populations. Moreover, because children are still growing, the impact of chronic diarrheal processes on linear growth and physical development must also be addressed. Evaluation of diarrheal disease in pediatric patients should proceed in a stepwise fashion that begins with an indepth clinical history and includes a limited number of microbiologic and biochemical tests. Physicians with a firm grasp of the epidemiology and pathogenesis of diarrheal illness in children will be better positioned to pursue a rational approach to the diagnosis and management of their pediatric patients with these common and potentially debilitating illnesses.

See also: Demographic Factors; Immune System and Immunodeficiency; Mortality, Infant; Nutrition and Diet.

Suggested Readings

American Academy of Pediatrics. Provisional Committee on Quality Improvement, Sub-committee on Acute Gastroenteritis (1996) Practice parameter: The management of acute gastroenteritis in young children. *Pediatrics* 97: 424–435.

Corrigan JJ and Boineau FG (2001) Hemolytic–uremic syndrome. *Pediatrics in Review* 22: 365–369.

Fuller CM, Ji HL, Tousson A, Elble RC, Pauli BU, and Benos DJ (2001) Ca(2+)-activated Cl(−) channels: A newly emerging anion transport family. *Pfluger's Archive* 443: S107–S110.

Guandalini S (2000) Acute diarrhea. In: Walker WA, Drurie PR, Hamilton JR, and Watkins JB (eds.) *Pediatric Gastro-Intestinal Disease*, pp. 28–38. Lewiston, NY: B. C. Decker.

Jensen BS, Strobaek D, Olesen SP, and Christophersen P (2001) The Ca^{2+}-activated K$^+$ channel of intermediate conductance: A molecular target for novel treatments? *Current Drug Targets* 2: 401–422.

Keely SJ and Barrett KE (2000) Regulation of chloride secretion: Novel pathways and messengers. *Annals of New York Academy of Sciences* 915: 67–76.

Ramaswamy K and Jacobson K (2001) Infectious diarrhea in children. *Gastroenterol. Clinics of North America* 30: 611–624.

Rudolph JA and Cohen MB (1999) New causes and treatments for infectious diarrhea in children. *Current Gastroenterology Reports* 1: 238–244.

Sandhu BK (2001) Practical guidelines for the management of gastroenteritis in children. *Journal of Pediatric Gastroenterology and Nutrition* 33: S36–S39.

Schroeder BC, Waldegger S, Fehr S, et al. (2000) A constitutively open potassium channel formed by KCNQ1 and KCNE3. *Nature* 403: 196–199.

Sellin JH (1993) Intestinal electrolyte absorption and secretion. In: Feldman M, Sharshmidt BF, and Sleisenger MH (eds.) *Gastrointestinal and Liver Disease.* pp. 1451–1471. Philadelphia, PA: W. B. Saunders.

Sicherer SH (2002) Food allergy. *Lancet* 360: 701–710.

Vanderhoof JA (1998) Chronic diarrhea. *Pediatrics in Review* 19: 418–422.

Velázquez FR, Matson DO, Guerrero ML, et al. (2000) Serum antibody as a marker of protection against natural rotavirus infection and disease. *Journal of Infections Diseases* 182: 1602–1609.

Wildin RS, Smyk-Pearson S, and Filipovich AH (2002) Clinical and molecular features of the immunodysregulation, polyendocrinopathy, enteropathy, X linked (IPEX) syndrome. *Journal of Medical Genetics* 39: 537–545.

Discipline and Compliance

J E Grusec, A Almas, and K Willoughby, University of Toronto, Toronto, ON, Canada

Glossary

Autonomy – Being able to make independent choices, to freely choose one's own actions, and make one's own decisions. It is considered by many to be a universal feature of human functioning, even in cultures that place high value on respect for authority and group cooperation.

Collectivist and individualist cultures – General terms used to differentiate cultural contexts where greater emphasis is put on individualist values such as autonomy and individuation as opposed to collectivist values such as group harmony and interdependence.

Identification – The process whereby an individual takes on characteristics of a model and incorporates them into that individual's self-image.

Minimal sufficiency principle – From attribution theory, the idea that pressure to conform or comply should be just sufficient enough, and no more, to produce compliance.

Moral development – Changes reflecting interactions between biology and culture that pertain to actions perceived to be binding on individuals. These are actions that 'should' or 'ought' to be performed by members of the social group, as opposed to those that involve choice.

Power assertion – Discipline techniques such as punishment, threat of punishment, withdrawal of privileges and material rewards, and social isolation, whereby the greater power of the socializing agent is used to control the behavior of the individual with less power.

Reasoning – A discipline technique involving statements about social norms and explanations for why a given action is inappropriate and/or unacceptable.

Socialization – The process whereby new members of the social group are assisted by older members in the acquisition of behaviors, values, attitudes, and beliefs of their group. It involves the transmission of cultural information from one generation of members to the next, although the new members seek out information as well as contribute to changes in the culture.

Withdrawal of love – A discipline technique involving psychological isolation and expression of parental disappointment.

Introduction

How to get children to do what they are asked to do is a significant problem for many, if not most, parents. Certainly most children do not comply with every parental request and there are variations in how much noncompliance different parents are willing to tolerate. Nevertheless, excessive early failures to comply are harbingers of later problems with delinquency and antisocial behavior. Accordingly, researchers have found that young children who show high rates of noncompliance with the directives of parents and teachers have difficulty in regulating their own behavior and, as they grow older, display increasing amounts of aggression toward their parents and their peers as well as perform poorly in academic settings. Often these children are referred for therapeutic intervention, with these interventions focusing to a considerable extent on training parents so that they learn how to encourage greater levels of compliance. This is not to suggest that noncompliance is the only source of problematic development. Modeling, for example, is a powerful socialization force, with exposure to aggressive, impulsive behavior promoting these same actions in observers. This article, however, focuses on children's refusals to follow suggestions, commands, and directives, and it deals with reactions of agents of socialization to those refusals.

Noncompliance, it should be noted, is not totally undesirable. Its initial appearance in young children is considered a developmental milestone – a manifestation of their growing physical and cognitive abilities and of their desire to behave in an autonomous or independent fashion. Just as we are uncomfortable with children who show great amounts of noncompliance, so too are we uncomfortable with those who are too compliant. In fact, certainly in a Western industrialized or so-called individualistic cultural context, one of the major aims of raising children is to encourage them to make independent decisions and to use their own judgment (although part of parenting, of course, involves helping them make wise judgements). When noncompliance or independence of thought is encouraged, however, the form it takes is important. Angry, spiteful, defiant noncompliance is more indicative of future difficulty than is noncompliance accompanied by calm exchange and explanation. Even reasoned noncompliance, however, is not universally valued. It is less acceptable, for example, in cultures that place greater emphasis on family harmony and group interdependence, so-called 'collectivist' cultures. Even in these cultural groups, however, evidence suggests that feelings of autonomy and freely chosen action, even if that freely chosen action involves deferring to authority and complying with the requests of the authority figure, are important for positive psychological functioning.

Although recognizing that some degree of and some forms of noncompliance are positive features of children's socioemotional development, developmental psychologists have devoted considerably more time to trying to understand the conditions under which children are most likely to go along with the wishes of their parents. Research has indicated that there are a number of such conditions. Parents who are responsive to their children's needs for security and protection, for example, have been shown to have more compliant children: at least part of the reason for their compliance may be that these children trust their parents to make requests that are in their best interests. Researchers have also shown that parents who are responsive to their children's requests will have children who are, in turn, more likely to comply with their requests. This tendency to reciprocate is a basic part of human nature and a feature of children that can be used to good advantage in fostering a disposition to compliance and cooperativeness: it will be discussed briefly at the end of this article. However, the pathway to compliance that has received the greatest attention from developmental psychologists involves discipline strategies that are employed by parents in response to their children's noncompliance. The role of discipline in the development of children's compliance will, therefore, be the main focus of this article.

Throughout the article the emphasis is on parents as disciplinarians. Certainly, other individuals who are responsible for socializing children also use discipline – that is clearly evident in the school system. However, at least in Western nations, most discipline is left to parents. Schools are limited in the kinds of discipline they can impose and individuals who are not relatives or teachers rarely impose discipline, as such action would be negatively received both by parents and by the children whom they disciplined. Discipline to a considerable extent, then, is a significant part of the parent–child relationship alone.

Internalization of Values as a Goal of Discipline

Discipline frequently is considered part of the socialization process during which parents attempt to gain compliance from their children and, ultimately, have them internalize or take over the values, norms, and standards of society as their own (or, more precisely, the norms and standards of behavior of society as presented by the parents). The second part of the process – internalization of norms, standards, and values – requires that children learn to self-regulate or govern their own behavior so

that parental requests and external consequences for compliance or failure to comply are no longer necessary. The distinction is really one of the child's motivation for compliance. Initially, the child may obey parental requests because of fear of punishment or hope of reward. Ultimately, however, the motivation for compliance comes from internal sources, such as a belief in the inherent correctness of an action, a desire to see the self as a good person, or guilt about harming others. Appropriate discipline leads to the latter sort of motivation, whereas inappropriate discipline discourages internalization and leaves the parent dependent on external pressures to achieve compliance. Moreover, external pressure may work when the parent is with the child, but it obviously is not effective in the absence of the parent's surveillance. Thus, in a setting where constant surveillance is not possible, it is highly desirable that people go along or comply with social norms even when there is a very low probability that their noncompliance will be discovered.

Historical Views of Discipline

Interest in how children learn to comply with the dictates of society has a long history. Recent history, however, begins with the work of Freud and psychoanalytic theorists whose ideas reduce to the fact that children are inevitably frustrated by the control their parents exert over them and that this frustration produces feelings of hostility toward the parents. The hostility is repressed, however, because children fear they will be punished for expressing it – in particular that they will be abandoned or rejected. As a result they adopt or take over, in a relatively unmodified form, parental rules and prohibitions and, by so doing, are helped to maintain the repression as well as make it more likely they will receive parental approval. Psychoanalytic theorists also suggested that children acquire a general motive to copy parental behavior, including their actions as disciplinarians. In this way, hostility to the parent is turned inward and takes on the form of guilt. The only way to avoid feelings of guilt is to comply with parental commands and wishes and to block from conscious awareness wishes to deviate from parental norms.

Although Freud and his followers presented a richly detailed picture of children's development, their approach, based almost entirely on clinical observation, was lacking in scientific rigor. That rigor was supplied by psychologists working in the 1940s and 1950s who wanted to develop a theory of human behavior that combined the clinical approach of the psychoanalysts with the scientific approach of behavioral psychologists who were developing a theory of learning based on reinforcement principles. Robert Sears was an important figure in this movement, and his research and theory moved the understanding of compliance and discipline forward to a significant degree. According to

Sears, children find their parents' behavior reinforcing because the parents and their actions are associated with the reduction of drives such as hunger and thirst. (Later, psychologists focused on other drives such as physical contact, comfort, and need for protection.) Becoming like the parent, including internalizing the parent's standards of behavior and values, was thus motivated by the pleasure that came from reproducing behavior associated with the parent. In a landmark investigation of 379 families who lived in the area of Boston, Massachusetts, Sears and colleagues showed that children who displayed the highest levels of conscience – an indication that they had internalized or taken over the norms and standards of behavior of their parents – were those whose parents relied on withdrawal of love (i.e., social isolation, expressions of disappointment) as a technique of discipline, particularly if those parents showed high levels of warmth to their children. The idea was that warm parents who withdrew love when their children failed to comply compelled the children to reproduce their parents' behavior in order to provide themselves with the reinforcement that had been withdrawn. Parents who relied on withdrawal of material rewards or physical punishment, contrast, were not setting conditions that led to incorporation of or identification with parental values. The superiority of psychological approaches to discipline over those that were materially or physically oriented (e.g., power-assertive techniques) is a notion that continues to guide the thinking of researchers, although it has become a somewhat more nuanced one in the last decade or so.

The Three Faces of Discipline

A move forward in thinking about compliance and discipline came with the work of Martin Hoffman who expanded the concept of psychological discipline to distinguish between power assertion, withdrawal of love, and reasoning and explanation. Hoffman suggested that, over time, children change their views of rules as being arbitrary and imposed from external sources to feeling that those rules are rational and objective. This change is facilitated by parents and others who minimize the obviousness of their surveillance of the child and who accompany their demands for compliance with reasoning. For Hoffman, one form of reasoning was especially important, that which emphasized the negative impact that the child's noncompliance or deviant behavior had on others (i.e., 'other-oriented' reasoning). For example, parents who tell children to be kind to others and point out that meanness hurts a person's feelings will be more effective because they draw on the child's empathic capacity to experience and appreciate the discomfort of others. In this way they ensure that the child will try to make amends and/or show greater consideration in the future. Children who are punished in a power-assertive fashion,

that is, who experience physical punishment or withdrawal of privileges, can escape the unpleasant consequences of their antisocial actions when the parent or other agent of socialization is no longer around to threaten or impose sanctions; in contrast, feelings of guilt are self-produced events that cannot be escaped. Withdrawal of love, in Hoffman's analysis, was more effective than power assertion, again because the anxiety it produced over fear of abandonment and loss of love was self-induced. In contrast to existential guilt (the kind of guilt aroused by focusing on harm to others), however, there was little the child could do to reduce the unpleasant feelings it aroused. Only in the case of existential guilt could relief from unpleasant feelings be achieved by making amends to the harmed party.

In sum, Hoffman proposed that there were three major categories of discipline that parents could impose when children failed to comply. These were (1) power assertion, including the administration of, or the threat of administration of, physical punishment, deprivation of privileges and material goods, and verbal disapproval; (2) love withdrawal, including isolation and refusal to communicate; and (3) reasoning, with particular emphasis on pointing out the consequences of the child's actions for others, although including other explanations for why compliance was desirable as well as appealing to the child's pride. Moreover, the scientific literature that had accumulated by the time Hoffman was exploring these ideas – the early 1970s – supported his arguments about the relative effectiveness of these various approaches.

Why, in addition to the reasons given above, should these approaches to discipline have different outcomes with respect to encouraging the child to go along with parental wishes? Power assertion may make children angry because it threatens their feelings of autonomy – people do not like to be forced to do things they do not wish to do. Power assertion may well lead to reactance, a desire to do just the opposite of what is requested. In addition, adults who use power-assertive discipline provide a model of aggression and teach that the best way to achieve one's aims is through the use of force and imposition of greater power. None of these are issues in the case of withdrawal of love or of reasoning, with the latter (as noted above) having the additional advantage of fostering children's capacities to empathize with the plight of others.

Other explanations for the superiority of reasoning over power assertion have invoked a variety of cognitive mechanisms. Social psychologists have long noted that people frequently search for explanations for their own behavior and that of others. These explanations then guide their subsequent actions. In the case of power-assertive discipline and compliance, the linkage between behavior and explanations for behavior is rather clear. Suppose a child complies and searches for an explanation for that compliance. If the discipline has been power assertive and/or threatening, the explanation is obvious – compliance occurred as a way of avoiding punishment from an external agent. Suppose, however, the discipline is more subtle or mild, or involves reasoning about how the well-being of others might be threatened. Now the explanation for compliance is less clear. It cannot be that compliance occurred in order to avoid punishment. It must be that compliance occurred for some internally motivated reason. The latter, of course, is akin to internalization of standards which is seen to be a major goal of gaining compliance.

Considerable research has indicated that modest pressure to comply is ultimately more successful than strong pressure. Relative to children who are harshly punished, for example, children punished mildly for playing with a particular set of toys are less likely to play with these same toys at a later point in time when they are alone and believe that no one is watching them. The 'minimal sufficiency principle' has been used to explain the fact that levels of discipline that are greater than the minimal amount necessary to produce compliance will be associated with noncompliance in the future, once children understand that their noncompliance can no longer be detected. Contrast, discipline or motivation that is just at the level needed to gain compliance will more certainly be associated with future compliance in the absence of surveillance because external attributions are difficult to make and, therefore, internal attributions take their place. Here, then, is a compelling explanation for why power-assertive techniques are less effective than more psychologically oriented techniques, particularly reasoning. It should be noted, of course, that some pressure is necessary to achieve initial compliance so that there is a need to find a reason for compliant behavior. That pressure might include mild punishment or threat of punishment, including social disapproval.

Other cognitive explanations have been offered for the superiority of reasoning and explanation (including other-oriented reasoning), beyond that of attribution theory. It has been argued, quite reasonably, that reasoning makes the contingencies between action and outcome more evident to the child. Often, in the highly charged emotional atmosphere generated by angry and punitive socialization agents, the specific disciplinary message or concern can be vague or even misunderstood. In addition, children who are angry and aroused have less cognitive capacity available to pay attention to cues in the surrounding environment or to understand why their behavior might be unacceptable. Reasoning provides material on which the child can build, organizing concepts and principles in a way that strengthens underlying cognitive structures pertaining to acceptable action. During the course of reasoning, parents may realize that children have constructed an event in a different way than the parent has, and thus the occasion for clarification and

exchange of views is offered. No matter what the explanation, however, developmental psychologists have usually adhered to the position that reasoning is a superior approach to power assertion or, at least, reasoning accompanied by just sufficient power assertion to promote compliance.

Limitations of the Reasoning/Power Assertion Dichotomy

Simple answers to complex questions rarely work. And this is certainly true of discipline and compliance. Although, on average, reasoning accompanied by modest amounts of power-assertive behavior is a useful recommendation for child rearing, the situation is rather more complicated.

To begin with, the evidence, although favoring the use of reasoning, is not always consistent. Moreover, not all analyses of the discipline process focus squarely on the centrality of reasoning. Second, reasoning and power assertion take on many forms, and these are not always similar in their impact. For instance, reasoning can be other-oriented, as already noted, or it can refer to norms, statements of appropriate behavior, elaborated arguments, or meaningless talk. Power assertion includes withdrawal of privileges and social isolation, as well as the use of corporal punishment, and it is not clear that these are comparable in their impact. Third, research evidence indicates that the context in which discipline is administered has a significant effect on its impact. Thus, a single discipline prescription that works across all situations does not appear to exist. Fourth, there is considerable evidence that parents do not always use the same discipline strategy but change it as a function of a number of variables, including a variety of features of their children, as well as the nature of the noncompliance under question. And, finally, an additional concern has to do with causal connections between discipline and compliance. Although many researchers infer that different forms of discipline produce differences in the ease with which children comply, others have argued that noncompliance elicits power-assertive forms of discipline. Thus, rather than power assertion making children noncompliant, it can be argued that noncompliant children produce power-assertive parents who are driven by frustration or the need to become more harsh and directive in their parenting. In the next sections we talk about each of these limitations.

There Is Inconsistent Evidence and Other Approaches Have Been Successful

Power-assertive discipline has been shown to work effectively when used by fathers, when used in a lower socioeconomic context, and when used in cultures outside that of Western industrialized nations. Thus, the evidence for the centrality of reasoning is inconsistent. In addition to inconsistent evidence, it has been shown that power-assertive practices, when used appropriately, can have positive outcomes for compliance. Some of the most compelling work in the area of discipline and compliance has been conducted by Gerald Patterson and his associates and, for them, the central feature of discipline is the effective management of reinforcement contingencies. More specifically, they are less concerned with ideas of internalization than with the importance of control of the child's actions by environmental contingencies.

Adopting a social interactional perspective on the process of child compliance, Patterson has studied family interactions where aggression and conflict have reached problematic levels. He has described how parents, children, and siblings change each other's behavior over time, and how these patterns of change differ from those in families where aggression is not a problem. Patterson has employed two approaches to this description. The first involves a molar approach in which predictors of child compliance are identified, and the second, a molecular approach in which specific mechanisms that lead to non-compliance are described.

In the molar analysis, Patterson has identified discipline and monitoring as two important features of parenting that make direct contributions to compliance. Positive features of discipline include consistency and follow-through whereas negative features include: not insisting that rules be obeyed, talking/'reasoning' rather than requiring compliance, and threats of or actual physical punishment. In order for families to function well, Patterson has demonstrated that rules need to be clear and spelled out and consistently applied to all children in the family; moreover, these rules need to change to be appropriate to the developmental level of the child. With repeated reinforcement for following rules or engaging in appropriate behavior over hundreds of trials, children's behavior becomes overlearned and automatically exhibited. Because of this overlearning, then, concepts such as internalization or internal attributional processes are unnecessary to explain why behavior would continue in the absence of surveillance. The second essential feature of effective parenting is monitoring. In his research, Patterson noted that boys who spent considerable time with deviant peers were more likely to be noncompliant, and that parents of these antisocial boys had only vague notions of where their children spent their free time. He demonstrated that parents who were trained to find out what their children were doing and with whom they were doing it were in a much better position to apply appropriate reinforcement than were those who were not so trained.

Although the identification of effective discipline as consistency of reinforcement and monitoring provides a useful analysis of compliance, it does not explain how these features bring about changes in the child. It is in the careful observation of family interactions that Patterson and colleagues were able to establish dynamic features of these interactions. They viewed the child as actively seeking out certain members of the family on the basis of the reinforcement those members provided as well, in turn, as shaping the behavior of those family members. In problem families, the events that happen during conflict are of central importance because the patterns of reinforcement are different from those in better-functioning families. Thus, in problem families, prosocial behavior tends not to be rewarded, whereas antisocial or coercive behavior is rewarded. Specifically, in families where excessive noncompliance exists, a situation is set up whereby children are negatively reinforced for coercive or noncompliant actions. For example, a child who is playing games on his computer rather than cleaning his room may receive a command from his parent to stop what he is doing immediately, to which he responds with whining because he wants to continue playing. The parent could respond to this irritating behavior with a more forceful intervention that would escalate the interaction or, alternatively, give up and leave the child to play. If the parent gives up, the child will stop whining, and the parent has achieved the goal of removing the current source of irritation (whining) and the aversiveness of conflict escalation, although not the goal of child compliance. The child, in turn, has achieved his goal of playing on the computer, with his whining strategy negatively reinforced because it has led to the cessation of or escape from an aversive event – the parents' demand for compliance. It is in this kind of parent–child interaction, then, that noncompliance flourishes.

Not only does the kind of mismanagement described here lead to problems in the home, but it also can lead to serious antisocial behavior outside the home. Noncompliant children often face rejection from their parents. They are also rejected by their peers because they have generalized their coercive behavior from their parents to the peer group and because they have failed to learn a more positive interactional style. Low self-esteem, therefore, is a frequent accompaniment of their antisocial behavior. These children are more likely to associate with deviant peer groups who are more likely to accept them and to engage in antisocial activities such as substance abuse. Ultimately, they have an increased probability of unemployment, disrupted partner relationships, and incarceration in adulthood.

Different Forms of Power Assertion and Reasoning Exist

Power assertion includes a variety of strategies including withdrawal of privileges, social isolation or time-out, coercion, threats, verbal punishment, and the use of corporal (or physical) punishment. Also included is parental intrusiveness, sarcasm, derogation, and guilt manipulation, with these latter subsumed under the label of psychological control. Psychologists have paid considerable attention to psychological control, contrasting it with behavioral control, which involves reasonable setting of rules for children's behavior and their reasonable (i.e., nonautonomy-threatening) enforcement, as well as monitoring of children's activities. Low levels of behavioral control are more often associated with noncompliance whereas high levels of psychological control tend to be linked to so-called internalizing problems, such as low self-esteem, anxiety, depression, low self-reliance, and self-derogation.

The form of power assertion that has generated the most attention and debate over the past few decades is corporal punishment. More specifically, the question of whether or not corporal punishment produces more negative outcomes than other disciplinary techniques has been the center of much concern. Indeed, the debate has spread to legislative bodies, and corporal punishment has been outlawed in a number of countries. In contrast, some surveys indicate that it has been used in North America at one time or another by 97% of American parents and 48% of Canadian parents and it appears unlikely that these countries will outlaw its use in the very near future. Arguments against the use of corporal punishment include the plausible suggestion that it provides a model for aggressive behavior and teaches its recipients that hitting and brute force are acceptable and successful ways to solve problems. Critics fear that corporal punishment can quickly escalate into physical abuse, although there is some evidence that the antecedents of the two are rather different. Surveys of the very large research literature on the topic do suggest that corporal punishment is associated with a number of negative outcomes including increased aggression, lower levels of moral internalization, and mental health problems; however, few studies have been designed so that it is possible to determine that corporal punishment has actually played a causal role in these associations. In other words, it may be that children's problematic behavior causes parents to resort to harsh forms of power assertion, or it may be that some shared genetic predisposition makes it more likely that both parents and their children display physical aggression. The evidence does suggest, however, that corporal punishment is effective for obtaining short-term compliance.

Although the conclusion of many opponents of corporal punishment is that other forms of power assertion are less harmful, there is not a great deal of carefully controlled research that speaks to the issue. Nevertheless, there is some suggestion that corporal punishment is more effective in reducing noncompliance when compared to withdrawal of privileges or verbal disapproval and threats. There are important caveats, however, and these are that the corporal punishment is used to back up milder tactics such as reasoning and time-out and that it is used in a controlled and moderate manner that does not involve parental anger accompanied by verbal abuse.

As with power assertion, there are also different forms of reasoning. Already discussed is the efficacy of reasoning that refers to the negative impact of a child's noncompliant behavior on others. Parents can also reason in a way that draws children's attention to their own emotions, intentions, and desire to make amends, and they can also use reasoning to communicate rules, social norms, and standards of appropriate behavior. Few studies have actually compared the effectiveness of different forms of reasoning, although there is some evidence that other-oriented reasoning about the needs and welfare of others can be more useful in gaining compliance than is reasoning that refers simply to norms and rules. In addition, other-oriented reasoning is associated with children's empathy and prosocial behavior, perhaps because it encourages children to take the perspective of others and to therefore act in a way that is not harmful to other individuals.

The Impact of Power Assertion and Reasoning Depends on the Context in Which They Are Administered

In their original work described above, Sears and colleagues underlined the importance of warmth in making withdrawal of love effective: withdrawn love from an unloving parent is obviously not going to be very punitive. Hoffman also argued that warmth was an important aspect of the parent–child relationship and that children whose parents are approving and loving are more likely to want to please them by complying. Warmth increases children's self-esteem, and children who are high in self-esteem also tend to be more compliant and willing to go along with the wishes of others. Another important aspect of family relationships has to do with parental protection and comfort. Thus, parents who respond sensitively to their children's distress set the conditions for their children's secure attachment, with securely attached children confident that they will be kept safe by their caregiver as well as helped to cope with distressing situations. In this context as well, children are more likely to respond more positively to parental attempts at control or obtaining compliance, in part at least because the control takes on a different meaning when it is delivered in an atmosphere of trust than when it is delivered in one of anger.

The importance of context and the meaning it imparts to a particular discipline strategy is evident from research that addresses the use of power-assertive techniques in working-class settings as well as in different cultural contexts. In both these settings, strong power-assertive parenting is less likely to undermine children's socialization than it is in the middle-class, Western industrialized settings where most of the research on discipline and compliance has been conducted. In part, the difference is a reflection of different values that parents have in these different contexts, as well as other features of parenting with which the discipline is associated and which alter its meaning to the child.

With respect to different values, some researchers have argued that the requirements of success in socialization differ depending on social class. Different skills are needed by working-class breadwinners who more frequently find themselves in jobs where they must please supervisors and accept the authority of others. In contrast, middle-class breadwinners are somewhat more likely to find themselves in jobs where they can work independently and are less reliant on the wishes of others. Thus, autonomy and independence and the internalization of standards of behavior may be valued more in middle-class contexts as a socialization outcome, whereas obedience and acceptance of authority may be more important in lower SES contexts. The two settings differ as well in the level of protection that children may need, with lower-class neighborhoods, on average, being more physically dangerous as well as more likely to present temptations for antisocial conduct. Where danger is great, immediate compliance becomes even more important. For both reasons, then, discipline strategies that serve different values may be differently employed.

Similarly, there are cultural differences in the standards parents try to instill in their children. Those favoring an individualist approach value independence and autonomy where the aim of parenting is to move the child from a position of dependence on the group to one of self-sufficiency (albeit in the context of positive relationships with others). Parents in a collectivist culture (e.g., Chinese) value group harmony and interdependence to a greater degree, and the aim of parenting is to move the child from early separateness from the group to closer affiliation with the group. This distinction between individualist and collectivist values has frequently been criticized and it is certainly not without its problems. For example, there is variability within any given cultural group in the extent to which autonomy or interdependence is valued. Nevertheless, the distinction has been of considerable use in trying to understand cultural differences in the exercise of control and of the impact of that control on children. And, again, the research evidence points to power assertion as less detrimental, on average, in collectivist societies than it is in individualist societies, in part at least because it encourages the learning of cultural values in one case and not in the other.

In keeping with the argument about socialization values, researchers have noted that those characteristics of parents that accompany power assertion differ depending on the values of the parent and the cultural or social group with which the parent identifies. In middle-class, industrialized societies, parents who resort to harsh power assertion have been found to be lower in warmth and more rejecting of their children than those who do not resort to harsh power assertion. In other words, in a culture where the use of power assertion is less approved, it is more likely to be associated with frustration and anger when it does occur. Moreover, it appears that it may be the low levels of warmth that are more important in determining child problems. In other cultural contexts, however, warmth and rejection are not associated with the use of power assertion and so there are fewer problems as a result of its use. 'No-nonsense' parenting, for example, which involves some force in the control of children but is also part of a caring relationship, seems to have positive effects when used by African-American mothers. Similarly, Chinese mothers are more power assertive or authoritarian in their parenting, but the authoritarianism is considered part of a loving teaching relationship and, when this is the case, the impact of authoritarianism is less likely to be negative. Finally, when power assertion is used more frequently and is considered to be an important part of responsible child rearing, it becomes normative, and the normativeness affects the meaning its use imparts to the child. Thus, children appear to be less negatively affected by physical punishment when they receive it in a culture where it is a more accepted disciplinary practice.

There Are Variations in the Use and Effectiveness of Discipline Strategies

Advice to rely predominantly on reasoning as a discipline strategy would suggest that successful parents use reasoning in all situations. And yet the evidence suggests that this is not the case, and that parents do not always behave in the same way in every disciplinary situation. There are a number of reasons for this variability. One set may lie in characteristics of parents who, when tired or angry, may be less patient or less accommodating to their children than when they are in a more benign state.

Also, parents have different goals when they are disciplining their children, and these goals are achieved by different discipline strategies. Most researchers have focused on the goal of socializing children or teaching them values that they internalize and carry within themselves (even when the value involves deference to authority). Even Patterson and colleagues assume that if acceptable behaviors are carried out frequently enough they will become strong habits that persist through different situations and with different people. But, sometimes agents of socialization have other goals in mind. They may, for example, simply wish to obtain immediate obedience or compliance, without an interest in long-term outcomes. Taking out the garbage when asked becomes more paramount than internalizing a norm of proper contribution to family functioning. And sometimes parents are more focused on maintaining a good relationship with their children and reducing the level of conflict, even if the probability of compliance is lessened. In each of these cases different discipline strategies may be appropriate.

Internalization of values, at least in a Western context, seems best accomplished through the use of reasoning, associated with a modest level of power assertion. Immediate compliance is more likely to be obtained through punishment or threat of punishment. And the maintenance of a positive relationship is facilitated through negotiation and compromise.

Aside from varying states and goals of parents, however, there is also evidence that strategies interact with characteristics of the child, the parent, and the situation to produce different responses or outcomes. These interactions are central to a proper understanding of discipline and compliance and so we describe some of them here. In many cases, the evidence lies in the observation that effective parents make distinctions depending on the characteristics of the child and the situation and the inference is that they must have grounds for their action which have to do with the need for a match between discipline and event.

The child's developmental status

It seems obvious that children's level of cognitive sophistication will determine their ability to understand different forms of reasoning and that this fact will affect their compliance. Abstract reasoning, for example, works better with an older than a younger child. Younger children are less able to deal with or understand humor or sarcasm and so may be differently affected by its presence. Sometimes parent discipline is confusing in that the content does not always match the tone of the parent's voice and some research suggests that content carries more weight in this situation than it does for older children. Younger children also evaluate physical punishment as being more fair and acceptable than do older children and, as a result, may be less likely to react against it. They also accept that their parents have the right to make demands over a large range of activities, whereas older children increasingly view a number of situations as ones over which their parents have no right to make decisions or to exercise control. These various differences mean that the form of reasoning will need to be modified as children age and that the range of misdeeds for which discipline is seen as appropriate and acceptable will also change.

The child's sex

Sex differences in aggressive and rough and tumble play appear to be mediated by the presence of hormones. When females are exposed prenatally to elevated levels of androgens (male sex hormones), for example, they act more like males in these areas. Similarly, during adolescence, increases in testosterone (another male sex hormone) are associated with increases in antisocial behavior, even after accounting for visible signs of puberty. These differences in aggression, which appear to be hormonally regulated, mean that males may be more likely to be punished and, therefore, more likely to react more negatively to the stress caused by higher levels of parental discipline.

The child's temperament

Temperament is a constitutionally situated feature of an individual's characteristic attentional and motor reactivity, as well as ability to self-regulate. Differences in temperament appear early in childhood and are a good predictor of various features of the child's development. Early inhibition, for example, is linked to later internalizing problems such as depression and anxiety. Impulsivity or unmanageability and irritability are linked to later externalizing problems such as aggression and noncompliance. Parenting, however, also makes its own contribution to these outcomes. Moreover, it is clear that temperament and parenting frequently interact in their effects. Thus, difficult temperaments can be modified by careful parenting, and children who are temperamentally easy can remain that way even when they are exposed to harsh discipline.

Investigators interested in the interactions between parenting and temperament have considered three general classes of temperament: novelty distress, fearfulness, or unadaptability; negative emotional reactivity or difficult temperament; and dysregulation or unmanageability and resistance to control. For each of these characteristics there is considerable evidence that parenting (and parent discipline) has a different impact, depending on the extent to which the child displays these temperament characteristics. For example, fearful children, who withdraw from or are slow to adapt to novel or risky situations, are especially likely to respond well to reasoning as a discipline strategy. Fearless children, in contrast, are not so influenced by this approach. Nor do they become more compliant when exposed to more power-assertive techniques but, rather, seem to become more compliant when they have a cooperative and responsive relationship with their mothers. Children who are high in negative emotional reactivity are quintessentially 'difficult', reacting adversely to frustration and overstimulation. They respond particularly poorly to intrusive and emotionally negative parenting, as well as show higher levels of noncompliance when cared for by mothers who are depressed or nonresponsive rather than by those who are more sensitive to their needs and wants. Inconsistent parenting or discipline has also been linked more strongly to noncompliance in these difficult children. Finally, children who are temperamentally unmanageable or resistant to control may actually respond better or become more compliant as a result of higher levels of control on the part of their parents than might otherwise be considered appropriate.

The nature of the child's misdeed

Sometimes parents vary their discipline as a function of the nature or form of noncompliance. Reasoning and power assertion are used together, for example, when children fail to comply with directives having to do with physical or psychological harm to others, whereas power assertion is rarely used in response to the failure to be helpful to others (at least in a Western industrialized cultural context, although not necessarily in a collectivist context). The content of reasoning is also tailored to be appropriate to the particular content of noncompliance. For example, parents are more likely to talk about the rights and needs of others in the case of acts that are harmful to others and the importance of custom and social order when reacting to bad manners. Another feature of noncompliance that affects parent discipline is its severity and frequency: more serious noncompliance and repeated acts of defiance are likely to elicit parents' use of alternative and harsher forms of discipline.

Just because parents react differently to different events does not mean, of course, that they are utilizing the most effective strategies. Nevertheless, there is some evidence that this may well be the case. Children do evaluate the fairness of reasoning more highly if they view it as appropriate. For example, if parents focus on harm done to others in the case of behaviors that are physically or psychologically harmful and on rules in the case of failure to follow arbitrary social conventions. Rigid parents and those who are abusive are more likely to use power-assertive interventions regardless of the situation at hand, whereas parents who are more successful in their approach to child rearing tend to vary their responses.

Socialization is a Bidirectional Process

To this point we have argued as though the relation between compliance and discipline is one-way, with the parent's discipline determining the nature of the child's compliance. It has long been recognized, however, that the process of socialization is bidirectional and that not only do parents influence children, but children also influence parents. Thus, as noted above, the relation between discipline and compliance could just as well occur because noncompliant children elicit strong and power-assertive discipline from their parents.

There are several ways in which the direction of the effect can be ascertained. These include the conduct of experiments and the deliberate manipulation of different discipline strategies in order to observe the nature of their effects. To do this in a truly realistic manner, however, is difficult. Intervention studies are another way in which researchers attempt to test hypotheses about the direction of effect, with parents whose children's noncompliance is at a clinical level instructed in better management practices: if compliance in the intervention group increases relative to compliance in an appropriate comparison group then the direction of causality is clear, although that does not mean that the form of intervention

is necessarily a reflection of what actually happens in the natural course of events. Another way in which researchers make inferences about direction of effect is to measure children's compliance and parent discipline at one point in time and then measure compliance at a later point in time. If parent discipline is related to compliance at the later point while controlling for initial levels of compliance, then the inference that the discipline was at least partially responsible for changes in compliance is a reasonable one to make. All these approaches have been used to attempt to answer the question of whether discipline is really causally related to compliance, with the simplest answer being that the evidence supports the conclusion that discipline affects compliance and that children's compliance also affects the discipline they receive. In other words, there is a bidirectional relation between discipline and compliance.

Another possibility with respect to the direction of relations between discipline and compliance is that they are each determined by a third variable such as the genetic make-up of both parent and child. Thus, genetically based temperamental difficulty in children would manifest itself in noncompliance, whereas genetically based difficulty in parents would manifest itself in harsh and power-assertive parenting. Studies in which the impact of genes on children's and parent's behavior is controlled for indicate that parenting does indeed have an environmentally mediated effect on children's aggression. One interesting exception is the relation between parents' use of corporal punishment and children's aggression which does not appear to be environmentally mediated. In other words, children's bad conduct appears to provoke their parents into using corporal punishment, rather than the corporal punishment being responsible for bad conduct. This is a striking conclusion and it will be important to see if the finding can be replicated.

A New Look

Given all the caveats associated with the relation between discipline and compliance, Grusec and Goodnow offered a somewhat different way of viewing the nature of effective socialization. They argued that rather than effective discipline being thought of in terms of specific strategies, it should be thought of as an intervention that makes it easy for children to (1) perceive accurately the message or value that the parent is trying to convey, and (2) to accept that message, that is, to behave in accord with it in a willing fashion. A number of features of discipline affect how easy or difficult it is for the child to understand correctly what is required. Messages need to be clearly expressed (e.g., "Do not hit your brother" yields a clearer message than "I am so angry with you"). Children's attention needs to be captured – this suggests that some level of negative arousal is necessary to focus that attention but not too much or the attention will be focused on the negative arousal rather than the message. Messages need to fit with the child's existing cognitive schemas; that is, they should be comprehensible and make sense. Finally, an indication of the importance of compliance to the parent needs to be included in the message.

Grusec and Goodnow also examined three different categories of children's acceptance of the disciplinary message. The first involves the appropriateness of the discipline to the particular misdeed – discussion of other people's feelings is probably less convincing in the case of bad table manners than in the case of insulting someone or calling them names. Due process needs to be seen to be served. The parents' actions need to be perceived as well-intentioned and the nature of their intervention a common one in the cultural context in which the family lives (e.g., spanking may be viewed as more acceptable by children whose siblings and friends are also spanked as opposed to children who see themselves singled out for this particular form of punishment). The discipline must be appropriate to the child's temperament, age, and sex. The second category of acceptance involves motivational processes designed to facilitate internalization, including the arousal of empathy, the setting of conditions that encourage identification with the parent, and knowledge of the importance of compliance to the parent (e.g., consider the child who complies only after an explosion of anger from a formerly calm parent and who comes to the sudden realization that the parent is really serious about the request). Finally, actions need to feel self-generated, with threats to autonomy minimized.

Ultimately, of course, certain strategies will work better than others. Reasoning and explanation accompanied by moderate amounts of power assertion may work best to gain children's compliance. The importance of a positive relationship between parent and child is also evident. Even behavior management, a power-assertive approach, fares better in a context where there are high levels of warmth and acceptance. But what the research evidence does indicate is that there is considerable variability in what constitutes power assertion or excessive control or a good relationship depending on the context in which it takes place. Ultimately, it is how the child interprets the meaning of an intervention that determines its effectiveness.

Another Approach to Compliance

Discipline, which involves regulation, control, and correction is not the only way in which children's compliance is achieved. In this final section we note that there are other ways in which compliance can be gained. For instance, Eleanor Maccoby has proposed that the relationship between parent and child can exist in such a way as to lessen the parent's need for coercive strategies or discipline techniques. The key to this type of relationship is mutual reciprocity in the form of responsiveness to each other's needs and wishes. Responsiveness forms the foundation upon which all future interactions will be based and therefore, in part at least, determines the success of socialization. If the parent is responsive to the child early on and is willing to meet the child's reasonable needs and wishes in a reliable way, the child will internalize a similar sense of obligation to the parent and, in turn, become more willing to go along with parental requests. It is then less necessary for the parent to use pressure or power assertion to gain compliance from the child. The usefulness of this analysis is demonstrated in a study in which mothers were instructed during a play session to comply with their children's requests. Compared to a group of children who played with their mothers as they normally did, these children were more likely to comply with their mothers' later requests for compliance. The link between this mutually compliant orientation accompanied by shared enjoyment has been shown to contribute to children's subsequent moral development, presumably because the shared cooperation between mother and child makes the child more likely to comply with maternal wishes and the shared enjoyment makes the child more willing to internalize the values of the mother.

Conclusion

The study of compliance and discipline has a long history in developmental psychology, beginning with Freud and continuing to the present. The more we learn, the more complex the topic becomes. Nevertheless, certain key points have emerged in research to this point in time. One is that the identification of some sort of absolute directive with respect to discipline is bound to be misleading. It is the meaning of discipline to the child who receives it that is the primary determinant of its efficacy. And the same event can have a very different meaning depending on a whole host of variables, many of which have been described in this article. When discipline is perceived to be the result of love and caring on the part of the parent it is much more likely to be successful than when it is seen to be a result of arbitrariness and hostility. A second point is that effective discipline depends on the clarity of the message provided by the disciplinarian as well as the consistency with which the content of that message is applied. The clearer and the more consistent the parent's actions, the greater is the level of compliance that will be achieved. A third feature of our analysis is that both the goals and the values of parents determine the nature of the discipline they administer. Not all parents share the same values nor do they all have the same goals. Even the same parent may have different goals at different points in time, or be willing to alter his or her value system. This feature, in turn, determines the form of discipline that is administered and its outcome. Compliance, as we have noted, is an important achievement in the developmental process, and the conditions that facilitate it remain a primary concern for researchers who are focused on the optimal conditions for children's growth.

See also: Independence/Dependence; Imitation and Modeling; Parenting Styles and their Effects; Routines; Self-Regulatory Processes; Social and Emotional Development Theories; Social Interaction; Socialization in Infancy and Childhood; Temperament.

Suggested Readings

Grusec JE and Hastings PD (eds.) (2006) *Handbook of Socialization.* New York: Guilford.
Killen M and Smetana J (eds.) (2005) *Handbook of Moral Development.* Hillsdale, NJ: Erlbaum.
Kuczynski L (ed.) (2003) *Handbook of Dynamics in Parent–Child Relationships.* Thousand Oaks, CA: Sage.

Divorce

J S Wallerstein, The Judith Wallerstein Center for the Family in Transition, Corte Madera, CA, USA

Glossary

Attachment – Defined as an enduring bond between the child and a specific caregiver. Once attached the child relies on that figure as a source of security when distressed and a secure base for exploration of a wider world.

Disorganized attachment – Refers to a type of insecure attachment that is identified on the basis of contradictory and disturbed patterns of attachment. It is associated with major disturbances in care giving and is believed to reflect the effects of disrupted caregiver signals and frightened or frightening caregiver behavior.

Joint custody – Refers to shared parental responsibility for the child after divorce. Joint physical custody refers to the child's residing in both parental homes after divorce with substantial amounts of time in each home.

Introduction

The few existing studies and clinical reports of the large numbers of children, age 3 years and younger in divorcing families report high anxiety, distress, and a range of troubling symptoms, especially among boys, which appear to reflect diminished parenting, family strife, and the departure of one parent from the home. Remedies proposed including overnight stays with father and joint custody are shown in three studies to depend largely on a cooperative relationship between parents including good communication and restored parenting. These are difficult to achieve after divorce by court order or to maintain voluntarily over time.

Population in the United States

Children 6 years of age and younger represent one-half of the million children each year whose parents divorce in the US. An estimated three-quarters of these children are age 3 years or younger. Despite these numbers there have been very few studies of the impact of divorce on infants or very young children, either at the breakup or during their long years in the divorced or remarried family. In actuality, the number of young children who experience family rupture exceeds reports in official records because of the thousands of children who are born to the large and growing numbers of cohabiting adults whose relationships, as we have learned, breakup more frequently than those who are legally married. According to US census reports in 2000 the number of unrelated adults of the opposite sex living together rose from 1.6 million in 1980 to 4.9 million in 2000. This trend affects individuals in all ethnic and racial groups. The decline in legal marriage has led inevitably to fewer legal divorces. These statistics of decline may or may not reflect decrease in divorce in the general population which is estimated at 35% or 40% of first marriages.

Divorce Research on Very Young Children

Assessing the impact of divorce on infants and very young children has been daunting. Few family researchers or mental health practitioners who advise the court have the requisite knowledge of infant and early childhood development to observe youngsters in this group properly, or to design appropriate criteria for recording or assessing change. There are fewer norms and far fewer tests appropriate for measuring infant and early childhood development than there are for older youngsters. Individual development during these early stages is likely to run along a much more idiosyncratic track, characterized by widely varying tempos and highly individualized patterns of regression and progression, as the child readies herself, on her own timetable, to ascend each step on the developmental ladder.

Research on divorce has led to increased concern about the suffering of children, and the long-term detrimental effects of divorce on children of all ages, as studies throughout the 1980s and 1990s revealed significant disparities in psychological adjustment between children in divorced and married families. The few studies that have included the youngest preschool children, including toddlers, have employed different methodologies and yielded uneven, sometimes puzzling results. Thus, a large study in 2000 by Clarke-Stewart and colleagues, reporting data from the Child Health and Human Development Study of Early Child Care, which assessed over 1000 children from ages 0 to 3 years from intact, never married, separated, and divorced families, found that when maternal education and family income were controlled by statistical analysis, the only significant differences that children from separated and divorced families showed was in cognitive achievement. These findings are in marked contrast to the work of others. Findings reported by Hetherington and colleagues, from a large, longitudinal study of children who were of 4 years of age at the divorce, and probably 1 or 2 years younger at the decisive separation, found a wide range of symptomatic behaviors during and after the marital breakup, and that 20–25% suffered with psychiatric symptoms in young adulthood. Another 25-year study by Wallerstein and colleagues, which followed children age 2–18 years at the time of the divorce into their adulthood, reported difficulties during childhood, troubled adolescences and problems in establishing enduring adult relationships with the opposite sex, including fewer marriages and a higher incidence of divorce. The youngest children were the most vulnerable.

Research addressed specifically about children 3 years old and younger has included only one major investigation which used attachment methodology, which is specifically designed to observe and assess parent–child relationships at this young age. Another study that is ongoing in collaboration with the courts has focused on the adjustment of very young children who spend one or two nights a week in the father's home. A third small study has clinically assessed very young children who spent half their time in each parent's home over several years. Additionally, case reports from clinicians who work with toddlers in play and clinical settings provide insight into their feelings, thoughts, and perceptions. None of this work reports outcome more than 3 years after the divorce. The absence of long-term data seriously limits our understanding of the impact and its sequellae.

Diminished Parenting during and after Divorce

A large body of pediatric and psychological science speaks to the special needs and vulnerabilities of infants and young children, and their sensitivity to the quality and consistency of their parental care. The human child is exquisitely attuned to her caregivers, and requires similar attunement from parental figures that come to know her well. The perceptions and interactions of the child, during her very early years, are critical to her rapidly developing view of herself, the world around her, and the people she needs in order to survive. Recent psychological advances have called attention to the importance, for the parenting of infants and very young children, of being able to acknowledge the very young child as a separate and distinct individual from the parent and to respond to the child accordingly. Divorcing parents, who are often overwhelmed by their own emotions, may have difficulty in distinguishing their own feelings and wishes from those of their children. This shows up especially in the parents' failure to recognize that the child may have a different relationship with the other parent and may enjoy and benefit from contact that is unacceptable or even reprehensible to the adult.

Tragically, despite the love of parents for their children, and their wish to shield them from the stress of the breakup, the period of divorce and its aftermath is one of diminished parenting. This diminished parenting, which includes less time with the child, less patience, greater irritability, and less empathy with the child's feelings, has been widely observed and reported in divorcing families with children of all ages The divorcing parent typically struggles with many contradictory, conscious, and unconscious, feelings and impulses in regard to the children. These include guilt over the family disruption, increased worry about the child, greater dependence on the child's presence to assuage the pain of loneliness, and greater need for the child's love and approval, along with the urge to escape the responsibilities of parenting, and find solace in new relationships. Additionally, the stress on many parents of young children is grounded realistically, in the greater likelihood of sinking into poverty, which many families face as a result of the divorce. This hits parent and child hardest in the long working hours of parents who carry several low-paying jobs and the sub-standard childcare for many of the youngest children. All of these psychological and economic factors converge in the acute stress that the parent experiences, which, in turn, impacts heavily on their parenting. The steep decline from what may have been nurturing, devoted parenting in mothers and in fathers during the marriage may be of short duration, or it may be lasting.

Special Vulnerability of Very Young Children

Since infants and young children depend on consistent, sensitive parenting and round the clock attention and care, it is reasonable to assume that the disruption of household routines and parental nurturance that divorce brings in its wake is likely to be more total, and more overwhelming, for little children than for their older siblings. Their experience is different and may be more difficult at the outset, and during the years, that follow than the experience of older youngsters. The youngest children have hardly any capacity for delay, their needs are all consuming, and always in the present. Unlike older children who are familiar with teachers, and often a range of caregivers, the youngest children are more dependent on familiar caregivers. Strangers can evoke panic. Adults can use language to prepare the school-age child, even the kindergarten-age child, for the breakup. They can explain to older children that the disorganiza-tion they are experiencing is likely to be short lived, and they can provide older children with pleasurable experi-ences to offset their worry. But this support for older children has no counterpart in the younger child's experi-ence for whom explanations are incomprehensible, and the parent's lap is the only comfort. There is no trip to the cinema or ball game that can assuage the infant's anxiety or calm his restlessness. Moreover, the school-age child or adolescent at the breakup is likely to have interests and relationships outside the immediate family that may help to mute the trauma. Relationships in school and commu-nity can, in many instances, even sustain their develop-ment over the years that follow the divorce. Little children lack these or other means for helping or comforting them-selves. They depend entirely on the recovery and avail-ability of the parental figure or figures. These timelines are not predictable. One researcher has suggested that house-hold stability is reestablished on average 2 years after the

breakup. This can encompass or exceed the life of the young child.

Additionally, the infant's or young child's relationship with the nonresidential parent may well be different from that of the older child. Although both parents are more likely to care for infants and young children today than in earlier times, the shared loving care of the new infant occurs far less in the troubled marriage than in the happy marriage, where the baby represents the love that binds the couple together. The relationship of many fathers to their young children depends on the mother's active encouragement. The very young child may have had far less opportunity than older children to form relationships with both parents that can survive the tensions, frequent transitions, and household differences inherent in co-parenting in two separate homes, especially when parents lack trust or empathy for each other. Among those couples who separated in the third trimester of the pregnancy in one study, some fathers did not attend the delivery. One longitudinal study of older preschool children found that some fathers withdrew their interest in their children, especially in the boys, years before the actual breakup.

Another issue of particular relevance to infants and young children as compared with older children is the much higher probability of remarriage for both divorced parents who are likely to be relatively young at the breakup. The remarriage may include children from a prior marriage, as well as new half siblings, who, studies tell us, are often preferred by the step-parents over the child from the first marriage. Since the incidence of divorce is much higher in remarriage than in first mar-riage, the chance of the very young child experiencing a second or third divorce is higher than that of the child who is older at the first divorce. Parent and step-parent may be cordial or rivals. Contrast, the very young child, and especially the infant, who has hardly known her biological family, may experience the happy remarriage as her original intact family and share none of the issues of concern to the child of divorce. These are among the many questions for which we lack answers. In the absence of long-term studies we can also wonder whether a young person who has hardly experienced the married family would come to adulthood with greater anxiety about enduring relationships in the family, and surely less knowledge about married life, than the child who has experienced living within a two-parent family. Many chil-dren from divorced families remember periods when the parents got along well together. Many adolescents are able to trace the downward course of their parent's marriage, and to draw conclusions for their own guidance, from what they regard as their parents' mistakes. This is not the experience of the younger children who have not had the opportunity to observe or internalize an image of the couple together.

Children's Responses During the Breakup

One of the few sources of knowledge about the affective and behavioral responses of the very young child during and immediately after the divorce are reports by psychologist and attachment researcher Alicia Lieberman whose book *The Emotional Life of the Toddler* includes several chapters on children of divorce based on her clinical observation of 120 such children during the years 1978–1992 at the infant–parent program, established in the Department of Psychiatry of the University of California at San Francisco School of Medicine. Referrals to this program came from concerned parents and from the courts requesting evaluation for custody decisions.

Lieberman's observations coincide with clinical observations of 50 children in this age group who were brought by the divorcing parents to the Judith Wallerstein Center for the Family in Transition for advice in explaining the divorce to the child and in planning postdivorce arrangements. The children were not brought for treatment. They were observed by well-trained clinicians for three play interviews which were recorded in detail. The families seen at the Judith Wallerstein Center represented a more affluent, better-educated, and ethnically and racially more homogeneous group than those seen at the medical school which drew an ethnically and racially diverse city population. Despite class, ethnic, and racial differences in the two groups, the similarities in the young children's responses are striking.

While divorce may ultimately be the right course for the parents, it is rarely a mutual or calm decision in a family with children. Unlike the common perception, overt conflict is not part of daily life in most families who divorce, but typically occurs during the last act of the dying marriage. Inevitably there are unhappy scenes. What this means for the young child is not only disruption in household routines, and disorder, but witnessing or overhearing parents in tears, shouting recriminations, and sometimes wild behavior, including throwing objects and sometimes, personal violence in adults whose behavior has previously been restrained. Children are frightened and disoriented by a parent's loss of control. They have no way of understanding the events, or of recognizing that the parent's anger is unlikely to continue indefinitely. Moreover, the aggression that they witness is harmful to their development when, as toddlers, they are busily engaged in learning to control their own aggressive impulses in their earliest social interactions with peers.

The departure of one parent from the home after fighting is very upsetting. Children worry about the parent who leaves and are often haunted by terrifying fantasies of disasters that had befallen that parent. Some observations of 2- to 3-year-olds may capture their terror and their panic responses. One frightened 3-year-old clung to his father refusing to separate for any cause. We discovered from his play in our office that he thought that his mother had been burned up in a fire that had occurred in his neighborhood a few weeks before she left. In his anxiety the child had conflated the two occurrences. A toddler who had been speaking clearly suddenly began to stutter, and soon would hardly speak, shortly after his dad left. Another toddler stood silently all day, like a little sentry, in the middle of the living room which was the last place that he had seen his departing mother, and he screamed uncontrollably when his frantic dad tried to put the exhausted child to bed. He was clearly awaiting her return, and perhaps was convinced that if he left, she would disappear forever.

Typically young children are worried about being forgotten when divorced parents are distracted and preoccupied, or distressed. Children who had been well cared for since their birth are no longer sure that their care will continue as before, whether they will be fed, whether they will be comforted and, at bottom, whether they will survive as the two giants who tower above them appear to be lost in their hurt, their anger, and their disappointment with each other. In playrooms equipped with toy animals these youngsters frequently play out stories of dangerous monsters who grab food from little lambs, and other helpless small animals. These behaviors occurred in households where there was relatively little overt conflict, as well as in those where they were openly fighting. Young children also show a wide range of symptoms especially at bedtime, and other moments of separation from the parent. Their tantrums and other expressions of being upset at these times reflect the child's anxiety and fear of abandonment. Going to sleep calmly is almost impossible for children who are fearful of awakening in an empty house. Several 2-year-olds refused to lie down in their cribs and clung to the rails crying bitterly at bedtime. Going to group settings, or to babysitters, is fraught with anxiety that they are being sent away forever. Understandably they may scream and refuse to go. Symptomatic behaviors may include regressive soiling, transitory stuttering or facial tics, distractibility, clinging, whining and often a high rise in aggression including tantrums, hitting the family pet, or the parent, and most often playmates and siblings.

Response to High Parental Conflict

It is generally agreed that witnessing or overhearing conflict between the parents is a major factor in the maladjustment of all children. What is less known is that toddlers and infants are deeply affected by domestic strife. They react with intense anxiety and disorganization.

Infants cried or behaved with increasing irritability and restlessness when the parents fought or spoke angrily to each other when they met to deliver the baby to the other parent for the visit. All symptomatic responses in the children are greatly magnified when the parents argue, or taunt each other. Tragically one of the effects of conflict between the parents is that it reduces the parenting of each parent, especially the capacity of each parent to comfort the frightened child. Parents are sometimes immobilized by their shame about what they have done, or what they allowed to happen. Mothers confessed that they felt unable to pick up or comfort the fretful infant following an angry encounter with the father.

Not all toddlers are able to express their fears. Although many toddlers and 3-year-olds show their feelings dramatically, others run and hide, or remain silent. Their secret worries can reappear then or later in nightmares, in renewed bedwetting, or soiling, in unexplained sudden terrors, or phobias. One 3-year-old who appeared to be unusually quiet in her behavior during the very angry breakup developed acute fear of her mother dying several years later. She would panic whenever she heard an ambulance siren.

Violence by parents, whether threatened or enacted, is always terrifying and always traumatic, typically requiring sustained psychological treatment for parent and child. The children fear that they, or a sibling, or a beloved pet, will be hurt or destroyed. They worry about the survival of one or both parents. The terror that they experience often haunts them as an intrusive unbidden memory and as a repetitive nightmare for many years, sometimes well into adulthood. Tragically, most of such instances in the child's life are not treated professionally until they resurface to create serious problems during the youngster's adolescence.

Responses to Parental Depression

Toddlers are sensitive to a parent's withdrawal or depression. Clinicians report toddlers sitting on the lap of a depressed mother, reaching up to stroke her cheek and comfort her. It is amazing to watch toddlers, hardly out of their infancy, taking an almost parental role. One 3-year-old comforted his distressed mother whose lover had left her. Patting her shoulder he said "It's not fair. He shouldn't quit in the middle." This behavior, which psychologists have labeled role reversal, appears typically after divorce when the parent is severely depressed and hardly able to function, and the child prematurely assumes the role of comforting parent.

The Wish to Restore the Intact Family

Toddlers and 3-year-olds remember the intact family fondly, and are eager to restore it. One little girl refused to listen to her mother sing a lullaby. "That is Daddy's song," she insisted. Another child, whose mother introduced her to her new boyfriend several months after the separation, refused to greet him "You're not my Daddy," she protested loudly. All of the toddlers and 3-year-olds who entered our playroom placed mother and father dolls carefully in the same playhouse bed. One 3-year-old placed the entire family together in the playhouse bathtub. She could tolerate no separation at all. Household scenes which the children constructed with toy figures typically included both parents and all of the children seated around the kitchen table.

Self-Blame

Young children are often convinced that their misbehavior led to the divorce. They are rarely offered simple explanations that they can comprehend, and they make no connection between the parent's anger, or even overt fighting, and their decision to separate. Their self-blame is a heavy burden that rarely comes to light, and can persist for many years. When parents or others assure the child that the parents did not divorce because of anything that the child did, as mental health experts advise parents routinely to do, this often fails to reassure the little children. Years later, children confide in a parent or a trusted teacher that they had made so much noise that one parent got mad and left the home. Only as the child's cognitive capacity matures is the adult able to help relieve the child's self-blame and consequent fear of punishment.

Gender Differences

There is evidence from several studies that little girls have an easier time than the boys at the breakup and its immediate aftermath. This may also be true of toddlers but the data are insufficient. The parental separation, and the departure of the father, caused anxiety as well as sadness and longing among the little girls, but did not routinely trigger the aggressive hitting or tantrums that the little boys show. Girls seem more able to adjust to multiple caregivers as well. The girls may feel protected by their close bond and identification with their mother. They seem able to deal with the father's absence by inventing comforting fantasies of their father's presence. One 3-year-old reported proudly about her father whom she had not seen for many months "My daddy sleeps in my bed every night." By contrast, the boys showed significantly more aggression, including wild temper tantrums, and angry defiance of the mother, after the father left the home. Their emerging identification with the father may lead them to blame the mother for the father's leaving, and to fight with her for his return, and to protect themselves against a similar fate.

Improvement and Follow-up

Some children improve dramatically after the breakup, especially if they have been forced during the marriage into undue closeness with a disturbed parent, or have been frightened by the violence. One previously sleepless 2-year-old boy, who would repeatedly bang his head against the wall in agony when he overheard his mother being beaten by his father in the adjoining room, recovered his capacity to sleep through the night as soon as they separated. Another troubled 3-year-old child, whose alcoholic father would climb into the bed nightly during the marriage, showed a striking developmental advance in his social and cognitive development within a year after the marital breakup.

Although we lack sufficient data to generalize about how long these early symptomatic responses to the breakup endure in toddlers and 3-year-olds, we may assume, in accord with reports of the behavior of 4- and 5-year-olds, that most severe symptoms are muted or disappear as family life and household are stabilized and parenting restored. Sometimes new symptoms appear several years later in 3-year-olds who showed few changes during the breakup. We may assume that the duration of symptoms that appeared during the breakup is considerably longer in those children who were functioning at less than optimal levels during the troubled marriage, or whose parents remained troubled, or depressed, or were psychiatrically ill. Additionally, symptoms in the children may continue or worsen or become chronic in families where substance abuse, depression, or parental conflict continue or rise during the postdivorce years. When parents stabilize their lives, however, well-functioning children can improve and resume their usual routines. We have no comprehensive reports on the incidence of full recovery in these very young children or their parents.

Providing Help at the Breakup and Immediate Aftermath

The support and encouragement that the child receives before and during the divorce crisis can make a very important difference in the child's suffering and symptoms. Parents can be very helpful in soothing young children, in explaining the divorce in simple words, in assuring them that they are loved by both parents, and that they will continue to be cared for. Spending extra time with these young children, especially at bedtime, is helpful when the gathering dusk, and the separation imposed by sleep, loom ahead, giving rise to anxieties about being alone with the night shadows. Calming bedtime rituals, including lullabies and simple stories about children or little animals that were lost or frightened, and were happily rescued by loving parents, can usually diminish the child's night-time anxiety and sleeplessness. Parents who are able to establish friendly relations with the other parent can help the child. Transitions between the two homes can be eased by pleasant goodbyes and assurances of a warm welcome on the child's return. Short distances between the two homes, which appear negligible to the adult, may be as challenging to the little child as a foreign journey. In some families, older siblings protect the youngest child. In many families, grandparents, or other family members as well as caregivers, are actively supportive. Teachers and caregivers describe young children climbing into their laps or clinging to them during the entire day during and after the divorce. Their attention and affection can be very reassuring to the child.

Postdivorce Dilemmas

The shape of the optimal postdivorce family for the very young child is an unresolved issue for parents, mental health practitioners, and the courts. Traditionally, the central nurturing unit for infant and toddler has been that of mother and child. Psychologists and pediatricians agree on the importance of providing a calm, stable environment as each mother learns to understand and respond sensitively to her child, and the child learns to recognize and trust the mother, becomes attached to her, and through her consistent nurturance achieves a view of a trustworthy and safe world. Protecting this developing relationship between mother and child has been considered central to the child's psychological and physical wellbeing, and to her future growth. Fathers have also been able to carry the role of sensitive caregiver for the child and have done so successfully.

Recently, some mental-health practitioners have attributed the psychological problems that many children of divorce show over the postdivorce years to the diminished role of the father. Basing their views on the significant contribution of the involved father to the children's development in the functioning intact marriage, some psychologists and judges have favored a postdivorce family comprising two separate households and two parents, who would cooperate fully in co-parenting their children, beginning in infancy. Their proposal includes overnight stays for infants and toddlers in the father's home. In those situations where the mother was reluctant to allow overnights at the father's home, some psychologists and judges have favored court orders which would override the mother 's hesitancy, including her concerns that the child would not have safe or adequate care in the father's home.

In actuality, the demands of the workplace have led to increased numbers of such households following divorce. Reports from several states including California and Wisconsin indicate that over one-third of children of age 2 years and younger, spend overnights in the father's home in separated and divorced families. Parents have many motives for choosing this arrangement. The most common reason for shared care of the infant and toddler is to accommodate the work schedule of two parents, both of whom work full time. Other parents believe that the child benefits from shared time in each parent's home because it enables the child to have a relationship with both parents that comes closer to approximating the intact family. Still other parents want time and freedom for themselves, and decide to share the care of the child in order to limit each individual's responsibility and commitment of time.

Despite the spreading practice of divorced parents sharing the care and housing of very young children, the psychological community, the courts, and many parents, remain undecided regarding the initial and long-term effects of overnight stays at the father's home for very young children. Their questions abound: Does going back and forth frequently between the two parental homes benefit the very young child or does it destabilize his life and imperil his development? When parents disagree about such arrangements, and seek the court's judgment, because one parent, usually the mother, fears for the baby's safety and comfort in the father's home, how can the court resolve their conflict in ways that protect the child? Does the court have fail-proof, or even adequate, ways for assessing the parent's concerns about proper care of the child? Are overnight stays at this young vulnerable age really critical in order to maintain the father–child relationship? Is the mother–child relationship hobbled, and her self-image as a competent parent, impaired as a result of court orders that override her judgment and intuition? The problem is complex because disputes that bring the parents to an adversarial court may be based on reality concerns regarding the health and well-being of the child, or they may reflect the hurtful sequellae of a humiliating, painful divorce. How to balance the interests of the very young child with those of each parent, recognizing that the decision has consequences for the parents' immediate and long-term relationship with their children, and with each other? Only three studies have addressed these issues. Their findings are dissimilar.

Very Young Children in Overnight Visits with Father

Attachment Research

Psychologists and leading attachment scholars, Judith Solomon and Carol George, undertook a longitudinal study of attachment patterns of 145 children who were 12–18 months old at baseline, seeking to find out under what conditions infants should spend overnights with their divorced fathers, with what benefit or detriment to the baby or toddler. The study, launched in the 1990s, included three family groups: (1) separated families in which the child stayed overnight for one or two nights weekly with the father. Most of these overnight stays had been court-ordered over the mother's objection; (2) families where father visited regularly but the child did not spend overnight in his home; and (3) a comparison group of intact families with children of the same age. Children were assessed separately with each parent at baseline, and tested 1 year later, in accord with the standard attachment protocol (the Strange Situation) which is designed to assess the child's reaction to brief separation and reunion with the parent and to classify the security of the child's relationship with the parent in accord with well-established clinical and experimental categories.

Results showed that two-thirds of the infants who had overnight stays with the father were classified as having disorganized or unclassifiable attachments to their mother. Those children who were not court-ordered to leave the mother's home were assessed as significantly more secure in their relationships. Additionally, the child's visits with the father, including overnights, did not lead to secure attachments with him. Parental conflict and low communication between the parents about the infant were associated with disorganized attachment to both parents. Results from assessment of the children 1 year later showed that children who exhibited disorganized attachment earlier continued to have difficulties in their relationships. Children who had mostly court-ordered overnights with father performed as well as others, during a challenging problem-solving session, but separation from the mother led more frequently to angry provocative behavior than was seen in the other two research groups. Paradoxically, children who spent overnights with the father were more troubled by separation from the mother 1 year later, than those who had spent all their nights in the mother's home.

The finding that the majority of the young children who were court-ordered to spend overnights in their father's home showed disorganized attachment is serious, because this classification, in accord with attachment theory, reflects high anxiety in the child along with distrust of the parent's interest or capacity to protect him. Everett Walters and colleagues and others have shown that the majority of attachment patterns formed in early childhood are likely to endure into adulthood. Disorganized attachment in infancy may portend serious relationship problems in the years that lie ahead.

In examining the context of the infant's experience it appeared that the families of children who showed disorganized attachment were characterized by hostility between the parents and poor communication regarding the child. Their failure to communicate translated into inability to plan together to ease the transitions between homes, or to coordinate caregiving routines. Additionally, mothers in these homes experienced their parenting as disabled by the court orders. They felt that their fundamental responsibility to protect their child was undermined by the court orders and that they had been rendered helpless to parent. The researchers concluded that when divorced families are able to create a supportive caring environment for the child, when the parents get along with each other, and can talk together to plan for the child, and when the plan does not disrupt the mother's sense of competence in caring for her child, overnight stays for children aged 12 and 18 months can be beneficial to the child. Lacking these three critical conditions, overnight visiting with the father was judged detrimental.

Children's Adjustment in Overnights with Father

Another still ongoing program, which also addresses the benefits and drawbacks for the very young child of spending one or two nights weekly at the father's home after the breakup, was guided by a different psychological approach and reached different conclusions. The child's symptoms and behavior, as observed by the parents, were reported to psychologist Marsha Kline Pruett and colleagues, who in cooperation with the Connecticut Courts launched a combined intervention and research study, which included 161 children, aged 0–6 years, of whom 32 were aged 2 years or younger at baseline. Families were recruited by interviewing court-bound divorcing families, inviting them to participate in a court-related intervention whose aim was to ease the impact of divorce on their young children. Twenty-five per cent of the children had no overnights at the father's home, 31% spent one overnight weekly, and 44% spent two or more overnights at the father's home. Follow-up data were obtained from parents about 132 children, 15–18 months later.

The investigation relied entirely on both parents' observations and some caregiver reports, and responses to specific questions about the children, about the level of conflict between the parents, details of the parenting plans and their implementation, and the number of caregivers for the child. Researchers examined frequency of overnight stays, schedule consistency, number of caregivers, and correlated these with the children's adjustment as reported by both parents.

Parental reports about the children often differed, as expected, but both agreed that when parent–child

relationships were troubled, children showed a wide array of symptoms and difficult behaviors. Parental conflict was significantly linked to difficulties among the older preschool children, but was less relevant to the adjustment of the youngest children, who may have been less aware or less exposed to the hostility. Parental conflict increased parental distress significantly, leading to sleeplessness and somatic symptoms in mothers, and unhappiness, along with aggressive, and destructive, behavior in fathers. In accord with former reports, there were observable gender differences. Boys seemed more troubled and more vulnerable to change in adult care. They suffered more when the number of caregivers increased. Girls were less symptomatic and seemed to benefit, or enjoy, the higher number of caregivers and the overnights. All of the children suffered when schedules were inconsistent, but this was most distressing to the youngest children.

Follow-up findings months later confirmed the centrality of the parent–child relationships, and the relationship between the parents, to the psychological functioning of the child. They also showed the close connection between parental conflict and diminished parenting, reduced parental involvement with the child, and a wide range of symptomatic behavior in the children, especially among the older preschool youngsters. It was encouraging to find that several of the older preschool children looked especially well at follow-up and had established good relationships with their father. We may assume that parental interaction and both parent–child relationships had improved in their families.

One important finding was that the father's involvement with his children diminished in the presence of parental conflict. There may, in fact, be a special vulnerability of the father to the mother's rejection of his active participation as a parent. By contrast with the father, the difficulties that were reported in the mother–child relationship reflected her worry and depression, consequent to the breakup, or chronic symptoms, but did not appear related to the parental strife. Other findings of this study are that the overnights were of far less moment than the family relationships in which they occurred. Apparently, the overnights by themselves did not contribute to the children's difficulties, neither did they protect children against the stressful impact of poor parenting, or the effect of hostility between parents, or the irregularities of schedules.

One drawback of this study is its reliance on reports furnished by parents who are neither trained nor disinterested observers, and the absence of direct assessment of the children by experts in early childhood. As a result, the study lacks information about the child's mood and feelings, her relationships with each parent, her understanding of the parental separation, or of her repeated transitions between the two households. One key finding, which may be helpful in planning psychological interventions, is the reported discouragement among many fathers about parenting their children, in the face of the mother's anger or rejection of their participation. The history of the marital failure may be critical, as well, in explaining the lasting resentments of parents, and their inability, or refusal, to forgive the other parent.

Children in Joint Physical Custody

Joint physical custody, where the child resides alternately in both homes, has been proposed as a way to mute the effects of divorce by continuing the child's close and frequent contact with both parents. Accordingly, during the early 1980s Rosemary McKinnon and Judith Wallerstein conducted a longitudinal study of 26 young children in joint physical custody ranging in age from 14 months to 5 years. Six of the 26 were under 3 years old, the remaining 20 were aged 3–5 years, with the preponderance close to the 3-year mark. The goal was to evaluate the child's experience and psychological adjustment in this new family structure. The custody arrangements that were arranged by the parents without court intervention provided for half-time in each parent's home. Most children spend half of each week with each parent. Others changed homes weekly. Children and parents were evaluated comprehensively, by professional personnel, at baseline, and at 6-month intervals, over the 2–3 years from baseline. Conflict between the parents was low at the outset of the study.

Over the years of the study the majority of parents became increasingly dubious about dividing the child's time in half between them. Clinical assessment of the children at 6-month intervals showed increased difficulties in the social and psychological adjustment of the majority of the children. Many of the difficulties that the children experienced reflected the nonacrimonious lack of agreement between the parents on schedule and everyday routines for the children. Diet, discipline, and sleeping arrangements for the young child were often strikingly dissimilar in the two homes. Thus, in several homes where the child slept with the parent in one home and was expected to sleep in her own room in the other, the child was soon sleepless in one of the homes. Different food habits in each home were also a problem. Food labeled junk food and forbidden in one home was fully acceptable in the other. Not surprisingly, children became cranky eaters or refused the food in one of the homes. Parents were often resistant to changing their habits or compromising their values to ease the child's adjustment to divergent rules. They also resisted changes in schedules to suit the other partner when work schedules changed, or

business trips intervened. Living in two homes was not reassuring to a few children who felt homeless. One 3-year-old refused to enter the car to travel to the other parent. He protested "Mommy has a home. Daddy has a home. Jimmy has no home." The youngest children seemed better able to negotiate the transitions between the two homes than the 3- to 5-year olds, who were more troubled as the study progressed over several years. Children who deteriorated badly, over the life of the study, suffered with inadequate parenting from both of their parents.

The investigators concluded that the quality of the parent–child relationship was critical and that the custody arrangement by themselves did not significantly hurt or protect the child. Parents in this study were not in divorce-related conflict. Nevertheless, they were reluctant, over the years of the study, to modify their priorities and parenting values in order to maintain the joint custody arrangement. The study showed that successful joint custody depended on parental accord and willingness of both parents to make sacrifices in order to maintain consistency in routine care of the child and to accommodate to the changing needs of the other parent. Differences between parents became increasingly complicated to resolve as new adults, and sometimes new babies, and older children, entered the family.

Conclusions

Overall this article, which covers the available research and clinical observations about the infant and very young child during divorce and its immediate aftermath, shows the anxiety and symptomatic responses of very young children and the bewilderment of their parents. The limits of current knowledge about how to bring about change on the child's behalf, to formulate social policy, or to make firm recommendations to parents or the courts are evident as well. The time duration of the few studies that we possess is much too short to allow even limited predictions about outcome. We are left with many concerns for these vulnerable youngsters and with unanswered questions about how to protect their childhood in the postdivorce family. The major scientific work lies ahead.

Clinical reports show encouragingly that early acute symptoms are likely to subside or disappear as the child's care is restored, as the household order is re-established, and as parents regain their emotional equilibrium and reorganize their lives. There is less known about the persistence of acute symptoms, which may become chronic, or attachment disorders, which may be built into the child's expectation of future relationships in divorced homes where the parents emotional difficulties or angers persist or worsen. Although the widespread use of mediation programs has achieved success in reducing litigation, it has not succeeded in bringing about friendly cooperation for the majority of divorced parents in conflict. Moreover, mediation does not address the heartbreak, or loneliness, or fear of the future which many adults experience which can surely rock the parent's self-confidence as a parent or his or her empathy, or even devotion, to parenting the child.

It is apparent that there is an urgent need for a wide range of studies by researchers with specialized skills for observing and understanding infant and young child development. The research would have many priorities and, at best, would lead to informed social policy and knowledge-based guidance for parents. One priority would be to study individual differences and identify subgroups among the very young children and their families along a range of parameters including their vulnerabilities and resilience at the divorce crisis and during the years that follow. A second priority would be investigations which would follow the young child for many years in order to distinguish the relationships, responses, and subsequent adjustment of the very young child from the experiences of older children. Are these youngsters at greater risk than their older siblings? If so, should parents be advised to delay their filing for divorce until the youngest child has reached a less vulnerable age? Are the very young likely to show no long-term effects considering their limited awareness of the family events swirling around them at the breakup. Or, to the contrary, will the influences of being raised in the divorced and remarried family in the total absence of memory of the intact family, impact powerfully on their view of themselves and their expectations in adolescence and adulthood?

Long-term study of parent–child and step-parent–child relationships are also needed. Do these relationships run a different course in families when the child is very young at the breakup? Is the bonding between mother and child interrupted or strengthened by a divorce that may leave the mother alone with a new baby and an uncertain future? Under what circumstances can we assume that fathers of infants are likely to maintain their love and commitment as parents until the child reaches adulthood? And what role do step-parents play in these complex equations considering the likelihood that most young parents will remarry or cohabit.

The major policy that has been proposed, of divided care of very young children, is neither supported nor negated by what we know thus far. There is no evidence in these studies that overnights by themselves are helpful in protecting children, or that they are uniformly hurtful. Nor is there evidence that father–child relationships, or young children, flourish in divided custody or in overnight stays. What matters is the total family context in which shared or divided care of the child occurs.

What matters, as well, is the history of the marriage and why it failed, and whether one or both parents are able or willing to forgive the trespass or disappointment that led to the marital rupture, or the hurt that followed.

It seems reasonable to posit that if the parenting in both homes is reasonably restored, and if the overnights and the divided custody occur within a trusting relationship between parents who are devoted to their children, overnights can promote two parent–child relationships which child and parent can enjoy with pride and mutual profit. Loving relationships with two moral, committed, mentally healthy parents, who trust each other, are invaluable to every child. Whether two such relationships can be created, or catalyzed, out of the ruins of a failed marriage in divided households, with the help of mediation, psychotherapy, or education remains to be demonstrated but is surely possible with enhanced understanding.

See also: Attachment; Family Influences; Marital Relationship; Risk and Resilience.

Suggested Readings

Amato PR, Booth A, Johnson DR, and Rogers SJ (2007) *Alone Together How Marriage in America is Changing.* Cambridge, MA: Harvard University Press.

Clarke-Stewart KA, Vandell DL, McCartney K, Owen MT, and Booth C (2000) Effects of parental separation and divorce on very young children. *Journal of Family Psychology* 14: 304–326.

Kline MP, Williams TY, Isabella G, and Little TD (2003) Family and legal indicators of child adjustment to divorce among families with young children. *Journal of Family Psychology* 7(2): 169–180.

Lieberman AF (1993) *The Emotional Life of the Toddler.* New York: The Free Press.

McKinnon R and Wallerstein JS (1986) Joint custody and the preschool child. *Behavioral Science and the Law* 4(2): 169–183.

Pett PA, Wampold BE, Turner CW, and Vaughan-Cole B (1999) Paths of influence on preschool children's psychosocial adjustment. *Journal of Family Psychology* 13(2): 145–164.

Pruett MK, Ebling R, and Isabella GM (2004) Critical aspects of parenting plans for young children: Interjecting data into the debate about overnights. *Family Court Review* 2(1): 39–59.

Solomon J and George C (1999) The development of attachment in separated and divorced families: Effects of overnight visitation, parent and couple variables. *Attachment and Human Development* 1: 2–33.

Solomon J and George C (1999) The Effects of Overnight Visitation in Divorced and Separated Families: A Longitudinal Follow Up. In: Solomon J and George C (eds.) *Attachment Disorganization*, pp. 243–264. New York: Guilford Press.

US Census Bureau (2000) *Statistical Abstract of the United States 2000.* Washington DC: Government Printing Office.

Wallerstein JS and Blakeslee S (2003) *What About the Kids? Raising Your Children Before, During, and After Divorce.* New York: Hyperion.

Wallerstein JS, Lewis JM, and Blakeslee S (2000) *The Unexpected Legacy of Divorce: A 25 Year Landmark Study.* New York: Hyperion.

Waters E, Merrick S, Treboux D, Crowell J, and Albersheim L (2000) Attachment security in infancy and early adulthood: A twenty-year longitudinal study. *Child Development* 71(3): 684–689.

Down Syndrome

D J Fidler, Colorado State University, Fort Collins, CO, USA

Glossary

Attachment – An enduring bond between a child and his or her caregiver.

Autism – A neurodevelopmental disorder that is manifested in terms of deficits in core social relatedness, language and communication deficits, and narrow/repetitive interests.

Behavioral phenotype – The observable expression of behavioral traits; in this case a profile of behaviors associated with a specific genetic disorder.

Genotype – The genetic make-up, manifested in the form of genes on chromosomes, of an organism.

Joint attention – The use of verbal or nonverbal communicative forms (eye contact, gesture, vocalization) to direct and focus a partner's attention on an object or an event with the purpose of social sharing.

Nondisjunction – When paired homologs fail to migrate to different cells during cell meiotic division.

Nonverbal requesting – The use of verbal or nonverbal communicative forms (eye contact, gesture, vocalization) to regulate another's behavior in order to obtain object, initiate action.

Phenotype – The observable expression of traits, based on genetic make-up and environmental influences.

Introduction

Down syndrome is a genetic syndrome, occurring in from 1 in 650 to 1 in 1000 live births. In 95% of cases, Down syndrome is caused by nondisjunction during cell division, resulting in an extra chromosome 21 (trisomy 21). Most cases of Down syndrome involve a nondisjunction during the first meiotic cell division, with mothers contributing the extra chromosome in 85% of cases. When nondisjunction occurs after fertilization, this leads to mosaic Down syndrome, where one line of cells in the developing fetus contains the extra copy of chromosome 21 and a second line of cells in the developing fetus does not. In a small percentage of cases, Down syndrome is caused by a translocation of genetic material on chromosome 21. Risk for Down syndrome is associated with maternal age. The pathways from genotype to phenotype in Down syndrome are currently not well characterized. However, current studies aim to identify how the additional chromosomal material on chromosome 21 impacts upon the developmental process.

Down syndrome was first described in the 1860s by John Langdon Down, who observed the clustering of specific physical and psychological features in a subgroup of individuals with cognitive impairments in medical settings. At that time, an unfortunate association was made between the craniofacial appearance of individuals with this clustering of symptoms and the physical features of specific ethnic groups. Modern genetic research has completely dispelled any a link between ethnic origin and Down syndrome. The discovery of the chromosomal cause of Down syndrome (trisomy 21) was made in 1959 by Jerome LeJeune. Since then, many notable advances have been made in this population, including increases in the life expectancy of individuals with Down syndrome (average life expectancy in the late 50s), as well as improvements in developmental outcomes and quality of life.

Though Down syndrome can be diagnosed clinically, a chromosome analysis is still considered necessary in order to confirm the clinical impression and to identify the underlying type of chromosome disorder. Common physical features associated with Down syndrome include a distinctive craniofacial structure, brachycephaly (abnormally wide head shape), short neck, congenital heart defects, anomalies of the extremities, muscular hypotonia, and musculoskeletal hyperflexibility. Most individuals with Down syndrome are born with a unique craniofacial appearance that includes palpebral fissures, epicanthal folds, Brushfield spots, flat nasal bridge, dysplastic ear, and a high arched palate.

Medical Issues in Early Development

Thorough physical examinations are recommended throughout the neonatal period, with careful monitoring of the systems that are most frequently impaired in this population. Approximately 50% of individuals with Down syndrome are born with congenital heart disease, with atrioventricular defects most commonly observed. Because of these vulnerabilities, examination by a pediatric cardiologist is recommended shortly after birth and monitoring is recommended throughout childhood. Cardiac anomalies are the main cause of death in children with Down syndrome, especially within the first 2 years of life. Beyond heart-related issues, routine screening for errors of metabolism and compromised thyroid function are recommended though the first few years of life. Children with Down syndrome are also vulnerable to congenital abnormalities of the gastrointestinal tract. A small number (3–4%) of infants with Down syndrome are born with congenital cataracts, which should be corrected with glasses or contact lenses.

Behavioral Features in Young Children with Down Syndrome

Beyond these health issues, there is evidence that Down syndrome predisposes children to a distinct behavioral phenotype (profile of behavioral outcomes). Research findings since the late 1960s suggest that individuals with Down syndrome are predisposed to relative strengths in visual processing, receptive language, and some aspects of social relatedness. Relative deficits have been reported in the areas of verbal processing, expressive language, and some aspects of motor functioning (see **Table 1**). More recent studies have also explored how Down syndrome impacts the development of problem-solving skills and personality motivation as well. It is important to note that not every individual with Down syndrome will show all aspects of the behavioral profile. Rather, the likelihood that individuals with Down syndrome will show this pattern of outcomes is elevated relative to other children with developmental disabilities.

This article focuses on the manifestations of these phenotypic outcomes in early childhood. While a great deal of research has been conducted on older children, adolescents, and adults, existing research on early development in Down syndrome has shed light on how these phenotypic outcomes emerge and develop over time. Rather than considering these areas of strength and weakness as static outcomes in middle childhood and adulthood, the developmental approach can offer a picture of the early manifestations of later, more-pronounced outcomes. Identifying these early pathways gives researchers and practitioners a critical opportunity to target these early pathways before they develop and become more pronounced in later ages.

Cognition

Overall Intelligence Quotient

The majority of children with Down syndrome score in the mild (55–70) to moderate (40–55) range of cognitive impairment, though the range of outcomes includes mild-to-profound impairments. Over the first few years of life, most children with Down syndrome make steady gains in mental age, but they do not make these gains at the same rate as other children without disabilities. Thus, their Intelligence Quotient (IQ) scores tend to become gradually lower throughout childhood. Longitudinal studies report mean IQs in the 60s or 70s in children under the age of 3 years and mean IQs in the 40s and lower 50s in children aged 5–7 years. By the time children are between 9 and 11 years, studies report average IQs in the upper 30s and lower 40s. Two points should be noted. First, children with idiopathic mental retardation do not show this pattern of decline. Thus, this developmental trajectory appears to be somewhat specific to children with Down syndrome. Second, these IQ declines do not reflect a loss of skills or developmental regressions in Down syndrome. Rather, children with Down syndrome appear to be making gains at increasingly slower pace; thus, the differential between their performance and those performed by typically developing children becomes more pronounced over time. There is some controversy regarding whether children with Down syndrome are following a delayed or deviant pathway of development. Nonetheless, in most cases, children with Down syndrome will continue to show a differential between their chronological and developmental ages throughout their lives.

Instrumental thinking and problem solving

Older children with Down syndrome show atypical performance on goal-oriented problem-solving tasks in laboratory settings, like puzzle completion. But in young children with Down syndrome, more subtle evidence of an emerging deficit in problem solving can be observed with close examination. Two early components of problem solving will be considered: the ability to represent goals, and the ability to chain behaviors together strategically in order to achieve those goals. While there is relatively little evidence that

Table 1   Phenotypic outcomes associated with Down syndrome

Developmental domain	Phenotypic outcome
Cognition	Visual > verbal processing; deficits in instrumental thinking; cognitive slowing
Speech, language, and communication	Receptive > expressive language; strengths in joint attention; deficits in nonverbal requesting
Social functioning	Strengths in core social relatedness; higher rates of insecure attachment
Motor development	Hypotonia; hyperflexbility; deficits in motor planning
Maladaptive behavior	Increased risk for comorbid autism spectrum disorder
Personality, motivation	Decreased persistence; inconsistent performances
Families/parenting	Lower levels of family stress than other disabilities; maternal directiveness

children with Down syndrome show difficulty with the ability to represent goals, young children with Down syndrome may show delays in the foundational skills related to strategizing.

In studies of sensorimotor development, infants with Down syndrome have been shown to develop competently in some areas, with sequences and structures observed as similar to development in typically developing infants, albeit in a delayed fashion. However, in the area of means–end thinking, infants with Down syndrome have been shown to develop more slowly, and there is evidence that they demonstrate unusual developmental structures in this area as well. For example, unlike typically developing children, milestones in the area of means–end performance early in infancy in Down syndrome are statistically unrelated to milestones achieved later in infancy. In addition, infants with Down syndrome take longer to transition from one stage to the next in the area of means–end thinking than they do in other areas of sensorimotor development. These difficulties seem to persist beyond infancy, as toddlers with Down syndrome show shorter goal-directed chains of behavior and poorer-quality strategies on problem-solving tasks. Instrumental thinking deficits may also be observed in the early communicative profile of young children with Down syndrome, who tend to show difficulties with nonverbal requesting, an instrumental communicative behavior that is associated with means–end thinking.

Information Processing

Research on information processing in older individuals with Down syndrome suggests that visual processing is an area of relative strength and verbal processing is an area of distinct challenge. While there is only limited information on functioning in these domains during early development, infants with Down syndrome may show subtle evidence of this profile in some ways as well. Competence in the area of visual imitation and some components of visual memory have been reported in some studies, while other areas, such as visual acuity, visual exploration, and the development of eye contact, seem to be relatively delayed. Early auditory/verbal processing appears to already be impaired, as atypical auditory brainstem responses have been reported in infants with Down syndrome, and poor vocal imitation (not visual; see below) has been reported in this population as well.

It is important to note that these findings are modest evidence for early manifestation of later phenotypic outcomes, and it may be that the true early forms of this profile only become evident later in development. Furthermore, while these findings have been reported in studies of specific areas of functioning in isolation, studies that attempt to compare early performances in auditory and visual processing in infancy and toddlerhood do not report pronounced dissociations at these early stages of development. This may be because the split is truly not evident at this early stage of development, or these findings may be related to the measures selected, which may confound the measurement of visual performances with other receptive skills.

Speech, Language, and Communication

Speech and language skills are delayed relative to nonverbal abilities in young children with Down syndrome. In terms of speech and expressive language, atypical vocalizing is already evident in infants with Down syndrome from 2 to 12 months, who produce atypical prelinguistic phrases compared to those produced by typically developing infants. In contrast with the relatively strong visual imitative competence in young children with Down syndrome, vocal imitation is attenuated. Decreased vocal imitation in Down syndrome has been shown to be associated with lower expressive and receptive language skills.

In the first 6 months of life, infants with Down syndrome also produce more nonspeech-like sounds than speech-like sounds, which may negatively impact the later development of normal vocal behavior. Additionally, delays in age of onset of canonical babbling have been found in some studies of development in infants with Down syndrome. The early development of canonical babbling is linked to later communication milestones, as the age of onset of canonical babbling is correlated with later performances on measures of early social communication in toddlerhood. Several factors may contribute to delayed babbling in Down syndrome, including maturational delay, hypotonia, and conductive hearing loss in infancy.

Nevertheless, other aspects of prelinguistic vocal development seem to be on par with typically developing infants, including the amount of vocalization produced, developmental timetable of vocalizations, and characteristics of consonants and vowels produced during babbling. The rhythmic organization of babbling in infancy in Down syndrome is similar to patterns observed in typically developing infants; however, infants with Down syndrome generally take longer to finish a prelinguistic phrase. These slower, longer vocal turns may contribute to the increased rates of conversational 'clashes' with a parent, a finding often observed in parent–infant dyads in this population.

Children with Down syndrome generally show delays in the transition from prelinguistic communication to meaningful speech. Studies report an average productive (signed or spoken) vocabulary of 28 words at 24 months, 116 words at 36 months, 248 words at 48 months, and 330 words at 72 months. Other studies that include only spoken (not signed) vocabulary words report that only 12% of 1-year-olds with Down syndrome have produced their first words, and only 53% of 4-year-old children have vocabularies larger than 50 words, a level that would be on par with a typically developing 16-month-old child. This profile appears to be delayed even when compared to children with other developmental disabilities at the same overall developmental level.

There is some debate regarding whether children with Down syndrome have the same 'vocabulary spurt' observed in typically developing children. Some researchers have reported that vocabulary growth in Down syndrome does follow an exponential trajectory, but it takes place later in development than would be expected based on overall mental age. Others argue that only some children with Down syndrome show such a vocabulary spurt, but vocabulary acquisition does not follow the exponential trajectory of a vocabulary spurt in many other individuals with Down syndrome.

Despite showing pronounced delays in expressive language, receptive language seems to develop with relative competence in Down syndrome, with the majority 0–5-year-old children showing mental-age-appropriate receptive language skills. While the majority, but not all, children with Down syndrome show a profile of receptive over expressive language, there may be different pathways leading to this outcome. One group of children with Down syndrome may show expressive language impairments from the onset of first words, while other children may only begin to show pronounced lags when the morphosyntactic (grammatical) demands of more complex language become a factor.

Most children with Down syndrome begin to generate two and three word utterances by the time they are 3 or 4 years old. In general, these utterances are similar to the early word combinations observed in typically developing children. Early grammar develops slowly and is related to vocabulary size, while more complex grammar is specifically delayed relative to vocabulary. Some have noted the paradox that while children with Down syndrome do not have difficulties with word order in simple utterances, many have difficulties with more complex morphosyntactic forms (third person, past tense) later in development.

It has also been noted that children with Down syndrome often opt to use short, telegraphic sentences rather than more complex syntactic forms, a finding that has been linked to both atypical language development and motivational/persistence issues in this population. Young children with Down syndrome produce fewer grammatical words (e.g., prepositions) than would be expected for their developmental level, a finding that persists throughout development. Case reports in the literature describe slow and steady syntactic development until roughly 5 years of age, after which children may hit a plateau. Yet, though children with Down syndrome may show less-complex morphosyntactic forms as they develop, it has been noted that they use language in pragmatically appropriate ways for their mental age. Thus, early in development, children with Down syndrome appear to be using language to achieve the same types of interpersonal goals as other children without Down syndrome at the same developmental level.

In terms of early communicative competence, some areas seem to be intact while others are impaired. Referential pointing in young children with Down syndrome emerges prior to the production of referential language, a sequence observed in typically developing children. Young children with Down syndrome show mental age-appropriate levels of nonverbal joint attention according to most studies. In addition, despite deficits in expressive language development, the early use of gestures in children with Down syndrome seems to be intact, with some reports of relatively stronger gestures in young children with Down syndrome compared with controls matched for word comprehension. Children with Down syndrome have been shown to use more advanced gestures than expected based on their general language development levels, and a wider repertoire of functional, symbolic, and pretending gestures than typically developing children at the same language level.

Yet, even in the context of these communicative strengths, other aspects of early communicative competence seem to be impaired. In particular, young children with Down syndrome show deficits in nonverbal requesting behaviors, or the use of eye contact, gesture, or vocalization for instrumental purposes. Rates of nonverbal requesting across a structured social communication assessment are associated with the quality of strategies performed by children with Down syndrome on a nonverbal problem-solving task. This suggests that deficits in nonverbal requesting may stem primarily from difficulties with early instrumental thinking, and not difficulties with early pragmatics or nonverbal communication.

Social–Emotional Development

There is evidence that infants with Down syndrome follow a typical pathway in the development of visual imitation in infancy. Some 1- and 3-month-old infants with Down syndrome have been found to imitate facial displays (tongue protrusions and mouth openings), as typically developing infants do. It is argued that these skills enable young infants with Down syndrome to engage dyadically with a caregiver in competent ways that establish the origins of social relatedness.

Later into infancy, increased looking directed at parents has been observed in studies of infants at 4, 6, and 9 months of age. Increased looking in infants with Down syndrome has also been shown in ambiguous situations, but not in social referencing contexts. Infants with Down syndrome also produce more vocalizations (emotional, melodic) when interacting with people rather than objects. Toddlers and preschoolers with Down syndrome display relative strengths in certain types of nonverbal social interaction including more reciprocal turn taking and other social initiations (object shows, invitations) when compared with typically developing children.

While very young children with Down syndrome show less exploratory behavior with toys, their trajectory of play development is similar to that observed in typically developing children. As they develop, children with Down syndrome spend less time engaging in functional play with objects and increase the amount of time they spend engaged in sociodramatic and symbolic play. When they do reach the symbolic play stage, children with Down syndrome show pattern of themes and language usage in their play similar to those observed in typically developing children.

In terms of attachment, conflicting accounts of outcomes have been reported. Some studies suggest that children with Down syndrome and MA-matched children without Down syndrome show similar attachment behaviors during Ainsworth's strange situation. In addition, a typical positive association between distress intensity and contact maintenance, and a negative association between distress intensity and distance interaction is observed in this population, suggesting that the organization of attachment behavior in Down syndrome is structured in similar ways to the organization of attachment in typically developing children.

Other reports suggest that attachment in Down syndrome is unlike attachment in typically developing children. Children with Down syndrome show less intense and shorter duration separation distress than children at similar mental and chronological ages. They also show less proximity seeking, contact maintenance, and resistance than other children. In terms of attachment quality, the frequency of insecure attachments in Down syndrome may be elevated compared to other children with disabilities at similar developmental levels. According to some reports, roughly 45% of children with Down syndrome are rated as type B (secure) in their attachment. Another large percentage of children are classified as type D (insecure unclassifiable), a finding that has been attributed to diminished negative reactivity in young children with Down syndrome. As a result, some suggest that the strange situation may not be an appropriate measure of attachment in this population. Others argue that increased numbers of category D classifications is attributed to hypotonia, ataxia, abnormal motoric responses in children with Down syndrome, and as a result, it is suggested that motor atypicalities be considered in the larger picture of attachment classification in this population.

The emotional displays of young children with Down syndrome were first described as muted relative to typically developing infants. Subsequent studies conducted with more objective coding systems reported that frequent low-intensity smiling in young children with Down syndrome occurred in the context of typical rates of other emotional displays, including high-intensity smiling.

In addition, young children with Down syndrome are also more likely to send ambiguous nonverbal signals that are more difficult for adults to interpret than the signals of other children at similar developmental levels.

Two important issues should be considered in the exploration of early social development in Down syndrome. While early social skills seem to be developing with some measure of competence in young children with Down syndrome, it is important to acknowledge that such strengths may not persist later in development, as the demands of social interactions become more sophisticated and cognitively mediated. In addition, emerging relative strength in social functioning may interact with other areas of development that are lagging behind. There is evidence that from 12 to 30 months, young children with Down syndrome make greater gains in the area of social development than in the areas of motor development and cognitive development. It may be that young children with Down syndrome come to rely on their strengths in social functioning and adopt social strategies at times when other instrumental strategies may be more effective or appropriate. A discussion of this profile is presented in the following section.

Temperament, Personality, and Motivation

Studies of temperament in young children report no temperament differences between infants with Down syndrome and typical infants in early infancy (at 2 months), and later at 12–36 months. Other studies, however, report that toddlers with Down syndrome are rated as of more positive mood, more rhythmic, and less intense than CA-matched children. These findings echo the findings of increased predictability, increased positive mood, and decreased persistence in older children with Down syndrome. However, nearly one-third of young children with Down syndrome show signs of difficult temperament as well, a possible precursor to stubbornness and other behavior problems.

There may also be a connection between problem solving and motivational orientation in Down syndrome, where children with poor problem-solving skills may abandon challenging tasks more readily than children who can strategize more effectively. When faced with cognitive challenges, many children with Down syndrome avoid the tasks with both positive and negative behaviors. In young children, these behaviors can include refusing to look at a task, struggling out of a chair, or sudden crying behavior. Older children may use their relative strengths in social functioning to engage the experimenter and distract them from the task at hand. These behaviors may contribute to inconsistent performances observed over time in children with Down syndrome, who may select appropriate or inappropriate behavioral strategies as a function of motivation related to a given activity.

Poorer mastery motivation continues to be evident throughout early development in Down syndrome. Preschooler children show lower levels of task persistence and higher levels of off-task behavior during simple play tasks than MA-matched comparison group children. Toddlers with Down syndrome show increased rates of toy rejection, indicating that 'quitting out behaviors' still impact performance at this point in development. Children with Down syndrome also show lower levels of causality pleasure, exhibiting fewer positive facial displays during goal-directed mastery behaviors than developmentally matched typically developing children.

Yet, other findings suggest that by the time they reach 3–4 years of age, most children with Down syndrome place great value on success during goal-directed tasks. Compared with developmentally matched children, preschool-aged children with Down syndrome show higher rates of pride-related displays during laboratory tasks. Children with Down syndrome show more positive affect, more frequent pauses to regard their finished product, and increased referencing of their caregiver upon successfully completing their task than other children at similar developmental levels. This suggests that though children with Down syndrome may have greater difficulty reaching desired goals, they are no less rewarded by the experience of having achieved their goals than other children.

Motor Development

Motor development is an area of pronounced delay in many children with Down syndrome. Atypical development of reflexes, low muscle tone, and hyperflexibility are often observed in infancy. As a result of hypotonia, infants with Down syndrome often show unusual postures and leg positions. The achievement of important motor milestones is also delayed during early development in Down syndrome: the average age for rolling over in Down syndrome is 8 months (typical = 5 months); sitting without support happens on average at 9 months (typical = 7 months); walking with support happens on average at 16 months (typical = 10 months); standing alone happens on average at 18 months (typical = 11 months); and walking alone happens on average at 19 months (typical = 12 months).

Later in early childhood, children with Down syndrome show difficulty with the development of motor planning, in the form of reaching and other essential skills. Children with Down syndrome have been shown to exhibit poorer-quality reaching strategies on an object-retrieval task, where children must reach for a desired object through the one side opening of a clear box. In order to reach efficiently, rather than coordinating eye gaze and reach, many children with Down syndrome use less optimal strategies such as looking through the open side of the box, straightening up, and then reaching for the desired object. This type of motor planning is essential for the development of efficient day-to-day adaptive skills, and thus, targeting these skills in early development may have downstream effects on important skills for later independent functioning.

Maladaptive Behavior

While children with Down syndrome show lower levels of maladaptive behavior than children with other developmental disabilities, rates of behavior problems and comorbid psychiatric diagnoses are still elevated relative to typically developing children. Disruptive disorders is the most common category of psychiatric diagnosis in individuals with Down syndrome under age 20 years (6% attention deficit disorder, 5% conduct/oppositional disorder, 6% aggressive behavior). In contrast, the most common diagnoses in adults with Down syndrome over the age of 20 years are major-depressive disorders (6%) and aggressive behavior (6%).

While temperament at 12 months is a predictor of maladaptive behavior outcomes at 45 months in children with other developmental disabilities, no such association has been observed in children with Down syndrome. The lack of association between infant temperament and later behavior problems suggests that young children with Down syndrome may make a more pronounced shift in the development of behavior problems than do typically developing children and children with other developmental disabilities. This may impact the parenting experience, as parents of children with Down syndrome may observe changes in their child's behavior that are more unexpected during early childhood, which may necessitate different parental supports to adjust to the changing needs of the child.

There is an increasing awareness of the subgroup of individuals with Down syndrome who also meet criteria for autism spectrum disorders (ASDs). Current estimates of comorbidity prevalence range between 3% and 10% of children with Down syndrome meeting criteria for autism or ASD. However, diagnosing these two disorders, especially in early childhood, remains a complex task. To meet criteria for comorbid autism or ASD and Down syndrome, problems in core social relating must be evident, not attributable to alternative explanations (such as very low cognition, very poor motor skills, or depression and mood problems) and persistent (across a variety of contexts and functioning levels). It may be that existing standardized assessments of autism symptoms may overidentify children with Down syndrome as having autism or ASDs if they function below 18 months cognitively, exhibit signs of depression, irritability, and extreme mood lability, and/or show significant impairment with regard to initiation of motor movements. Thus, obtaining an accurate diagnosis of the presence or absence of an ASD in children with Down syndrome must be performed by a clinician with both an expertise in autism and an understanding of the complex manifestation of autism in individuals with Down syndrome.

Families of Children with Down Syndrome

Parenting Behavior

Atypical patterns are observed from the earliest measures of parent–infant interaction in Down syndrome. The establishment of eye contact is delayed by several weeks and the amount of eye contact increases at a slower pace over the first few months of life. However, once infants with Down syndrome have reached a peak level of eye contact with their caregiver, they engage in higher levels of mutual eye contact than typically developing infants. Most typically developing infants experience a decline in the amount they gaze toward their caregiver by 5 or 6 months. In contrast, some studies report infants with Down syndrome at these ages and beyond who are focused overwhelmingly on their caregiver's face, and in particular, their eyes. This higher level of eye contact is maintained throughout much of early development in Down syndrome. Mothers of infants with Down syndrome have been shown to use more tactile and kinesthetic stimulation than mothers of typically developing infants on laboratory interaction tasks. Infants with Down syndrome have also been shown to be less distressed and responsive to the still-faced paradigm than typically developing infants in laboratory settings.

During toddlerhood, parent–child relations in Down syndrome also have some unique features. Mothers of toddlers with Down syndrome make significantly more frequent unsuccessful invitations to their child for interaction. Mothers of toddlers with Down syndrome are also found to be more directive and less child-dependent during their interactions with their children. Along these lines, mothers of children with Down syndrome have been found to take longer and more frequent turns during their parent–child interactions than other mothers. They also tend to change the topic of conversation more frequently than other mothers when interacting with their child, and they show increased rates of verbal clashing, where both mother and child speak at the same time.

However, maternal directiveness does not appear to be associated with a lack of maternal sensitivity, and maternal stimulation behaviors are associated with child's developmental level, suggesting that these strategies may be adaptive given the child's disability. In addition, maternal sensitivity has been linked to other factors, including education level and family income. In intervention studies where mothers have been taught techniques to improve their responsivity to their child, increased responsive behavior and a less intrusive style can be successfully fostered. In addition, certain child factors may impede a parent's ability to be responsive, including hypotonia, ambiguous emotional signaling, temperament, and distractibility.

Parents of preschool-aged children with Down syndrome give more praise to their children during goal-directed tasks. During laboratory tasks, parents of preschool-aged children with Down syndrome have been shown to use higher rates of praise than parents of developmentally matched typically developing children, perhaps in order to encourage their child to stay on task. In contrast, there are some reports that parents of children with Down syndrome use less positive verbalizations during naturalistic settings. Instead, in these naturalistic environments, mothers of children with Down syndrome have been found to do more explaining, modeling and demonstrating, labeling, and questioning for the purpose of teaching some concept, as well as more frequent attempts to assist their child complete tasks of daily living.

Yet other features of parenting behavior, such as parent input language, are similar across groups. For example, parents of children with Down syndrome give similar levels of language input (e.g., grammatical complexity; type of information) as mother of children at similar developmental levels without disabilities. In addition, some parenting behaviors that are considered more beneficial for typically developing children are also more beneficial for children with Down syndrome. For example, as observed in typically developing children, higher rates of joint attention in parent–child dyads are associated with greater gains in child receptive language in children with Down syndrome, too.

In general, mothers of children with Down syndrome report different goals in their interactions with their children. More often than in typically developing children, mothers of children with Down syndrome report that they feel a need to use interactions as a chance to teach their child a new skill or practice an existing one. Rather than enjoying interaction as a time to play with their time, mothers are more likely to request that their child produce higher-level behaviors than those they are producing spontaneously. This may be an adaptive parenting response given the child's developmental delays, but it may also lead to a different parenting style than observed in the general population.

Family Stress

Many parents experience life disruption and elevated stress upon receiving a diagnosis of Down syndrome for their child. This may be related to the quality of the interaction they experience with the diagnosing clinician, who may not have the counseling skills or the specific knowledge based to provide information to new parents of young children with Down syndrome. In one study, 60% of parents reported dissatisfaction with the experience of receiving their child's Down syndrome diagnosis, citing issues related to a lack of sympathy on the part of the healthcare professional, the inability of the professional to share information that would reduce anxiety, and delays in receiving information.

However, after an initial period of adjustment, families of children with Down syndrome report lower levels of stress and higher levels of rewardingness than families of children with other developmental disabilities. Some studies show similar patterns of cohesiveness and adaptation in families of children with Down syndrome and families of typically developing children. Lower levels of stress have been attributed to several possible factors: lower levels of behavior problems observed in children with Down syndrome relative to other children with disabilities, perceived maturity in Down syndrome, social relatedness in Down syndrome, age and higher socioeconomic status of parents of children with Down syndrome, and the support networks available for families of children with Down syndrome in their communities.

While cross-sectional studies of outcomes in families of children with Down syndrome have shed much light on the family experience, it may also be important to explore how the family experience changes over time. There is evidence that families of young children with Down syndrome may experience a different stress trajectory than families of other children with disabilities, starting out with lower levels of maternal stress at 1 year of age, but showing equivalent stress levels when the child reaches preschool age. Since family responses are closely associated with specific child characteristics, it may be the case that as the child's behavioral profile changes during development, and phenotypic characteristics become more pronounced, family stress responses shift.

Conclusion

Children with Down syndrome are predisposed to a specific profile of strengths and weaknesses in development, which include strengths in visual processing and social relatedness, and deficits in verbal processing, motor planning, and expressive language. Parents of children with Down syndrome are also likely to show certain parenting characteristics, often as a result of the behavioral profile associated with Down syndrome, though often

families show lower levels of parenting stress than families of children with other types of developmental disabilities. Future challenges include characterizing the emergence of the Down syndrome behavioral phenotype and identifying developmental trajectories of educationally relevant areas of functioning. With a greater understanding of the nature and course of development in Down syndrome, it may be possible to develop increasingly more sophisticated interventions for children and their families.

See also: Attachment; Autism Spectrum Disorders; Developmental Disabilities: Cognitive; Developmental Disabilities: Physical; Genetics and Inheritance; Intellectual Disabilities; Language Development: Overview; Learning Disabilities; Maternal Age and Pregnancy; Preverbal Development and Speech Perception; Temperament.

Suggested Readings

Block ME (1991) Motor development in children with Down syndrome: A review of the literature. *Adapted Physical Activity Quarterly* 8: 179–209.

Carr J (1995) *Down's Syndrome: Children Growing Up.* Cambridge: Cambridge University Press.

Dmitriev V (2001) *Early Intervention for Children with Down Syndrome: Time to Begin.* Austin: Pro-Ed.

Hodapp RM and Fidler DJ (1999) Special education and genetics: Connections for the 21st century. *Journal of Special Education* 33: 130–137.

Jarrold C, Baddeley AD, and Hewes AK (1999) Genetically dissociated components of working memory: Evidence from Down's and Williams syndrome. *Neuropsychologia* 37: 637–651.

Miller JF (1999) Profiles of language development in children with Down syndrome. In: Miller J, Leddy M, and Leavitt LA (eds.) *Improving the Communication of People with Down Syndrome*, pp. 11–40. Baltimore: Brookes Publishing Inc.

Mundy P, Sigman M, Kasari C, and Yirmiya N (1988) Nonverbal communication skills in Down syndrome children. *Child Development* 59: 235–249.

Pueschel SR and Pueschel JK (1992) *Biomedical Concerns in Persons with Down Syndrome.* Baltimore, MD: Paul H. Brookes Publishing Co.

Sigman M and Ruskin E (1999) *Monographs of the Society for Research in Child Development, 64: Continuity and Change in the Social Competence of Children with Autism, Down Syndrome, and Developmental Delays,* 114pp. Oxford: Blackwell Publishing.

Spiker D (1990) Early intervention from a developmental perspective. In: Ciccehtti D and Beeghly M (eds.) *Children with Down Syndrome: A Developmental Perspective*, pp. 424–448. New York: Cambridge University Press.

Emotion Regulation

R A Thompson, S Meyer, and R Jochem, University of California, Davis, Davis, CA, USA

Glossary

Effortful control – Refers to the capacity to inhibit a dominant response (such as getting angry or frustrated) and initiate a subdominant response (such as turning away or constructive problem-solving), and is a temperamental quality associated with the development of individual differences in emotion regulation.

Emotion dynamics – Concern the intensive and temporal qualities of an emotional response, and often reflect emotion-regulatory processes.

Emotion regulation – Consists of the internal and external processes responsible for monitoring, evaluating, and modifying emotional reactions (especially their intensity and timing) to accomplish one's goals.

Emotional competence – The ability to feel as people want to feel, to enlist their emotions adaptively to accomplish their goals, and to respond emotionally in social situations both genuinely and tactically.

Goodness of fit – The concept of goodness of fit between a child's temperamental qualities and caregiving practices describes how certain practices have different consequences for children who differ in temperament. A 'good fit' between the child's temperament and parenting practices means that temperament and caregiving are well coordinated. A 'poor fit' means that caregiving practices conflict with the child's temperamental qualities, and this can undermine emotion regulation.

Introduction

The first 5 years witness significant advances in developing skills for emotion regulation. The newborn infant's uncontrollable crying evolves into the 5-year-old's capacities to seek comfort, use words to communicate feelings, and deliberately move away from or avoid emotionally distressing situations. Researchers are interested in these developing skills because the growth of emotion regulation is multifaceted, building on emerging capacities for self-control arising from brain maturation, conceptual growth, socialization, and other sources. Parents and practitioners are interested in these developing skills because individual differences in emotion regulation are important, underlying differences in psychological wellbeing, social competence, and for some, emergent psychopathology.

Our goal is to profile current understanding of the origins of emotion regulation in the early years and its significance for psychological development. In the following section, emotion regulation is defined and some of the ways by which children's capacities for managing their feelings unfold with increasing age are outlined. Methodological challenges of studying emotion regulation, especially in infants and young children are considered. In the sections that follow, (1) processes of brain development relevant to emotion regulation, (2) temperamental influences, (3) how capacities for managing emotion are influenced by the child's conceptual growth, such as understanding of emotion, and (4) social influences on emotion regulation, especially those from parents and peers are considered. Throughout, it is proposed that although emotion regulation is a developmentally extended process, infancy and early childhood establish the foundation for lifelong capacities and individual differences in emotional self-control.

What Is Emotion Regulation?

Emotion regulation concerns how individuals manage emotional experience for personal and social purposes. More specifically, emotion regulation consists of the internal and external processes responsible for monitoring, evaluating, and modifying emotional reactions (especially their intensity and timing) to accomplish one's goals.

Viewed in this manner, the development of emotion regulation is a more complex, multifaceted processes than we commonly think. First, emotion regulation can target positive as well as negative emotions, and can include diminishing, heightening, or simply maintaining current levels of emotional arousal. Emotional self-control is apparent, for example, in preschooler children's abilities to constrain exuberance in formal social situations, enhance feelings of sadness to elicit a parent's nurturance, or control anger when provoked by a peer. Second, emotion regulation involves emotional self-monitoring and personal evaluations of emotional experience as well as strategies for modifying one's feelings. Included in this developmental process, therefore, are children's developing concepts of emotion and understanding of social expectations for emotional self-control, as well as knowledge of how to feel better. Third, emotion regulation involves both external and internal influences. Early in life, infants rely significantly on others for soothing their distress, but these social influences are increasingly supplemented by the child's own self-initiated efforts to manage feelings, such as through internal distraction or leaving a distressing situation. Over time, of course, children become increasingly self-reliant in their emotion management, but individuals are always influenced by others in their emotion-regulatory efforts (such as when another provides a sympathetic ear).

Finally, emotion regulation should be viewed in terms of the person's goals for managing feelings. It is easy to perceive a fussing child or dour adolescent as emotionally dysregulated when, instead, their goals for managing emotions might be very different from those of the adult observer. A toddler who cries petulantly for candy (but ceases after the parent accedes) and a moody adolescent may be emotional tacticians in ways that reflect capable, not deficient, capacities for emotion management, even though their behavior is undesirable in the eyes of most adults. In a similar manner, children with reserved temperamental profiles, or who have experienced difficult family challenges, may be managing their feelings using strategies (such as social withdrawal) that would not necessarily reflect competent emotion regulation in others. This means that evaluating individual differences in emotion regulation and the adequacy of a child's self-control strategies must take into consideration the child's characteristics, background, and goals for emotion regulation. We must always ask what are the child's goals for managing feelings in the circumstances in which they are doing so.

Taken together, therefore, children who are emotionally well regulated are capable of altering how long or how deeply they feel as they try to accomplish emotional self-efficacy – that is, feeling as they want to feel in specific situations to accomplish their goals. Developing the skills to do so involves growth in psychological understanding, language, and emotional self-awareness, and is deeply influenced by social experience, brain development, and culture. Several developmental progressions characterize the growth of emotion regulation throughout infancy and childhood, including:

- growth from emotional management by other people to increasing self-regulation of emotions;
- increasing breadth, sophistication, and flexibility in children's use of emotion-regulatory strategies, and their growing ability to substitute effective for ineffective strategies;
- enlisting other developing capacities into emotion-regulatory efforts, such as emerging language, attentional control, and strategic thinking;
- developing sensitivity to context in emotion regulation: understanding the social requirements for emotion management and that the strategies that work best in some situations might not be appropriate in others;
- increasing use of emotion-specific regulatory strategies (such as managing fear but not anger through encouraging self-talk);
- emerging complexity in the social and personal goals underlying emotion regulation as young children increasingly learn to regulate their feelings to manage social interaction, support self-esteem, and accomplish other psychologically sophisticated purposes; and
- the emergence and growing stability of individual differences in emotion-regulation goals, strategies, and general style.

The development of emotion regulation is important because individual differences in emotion regulation are associated with social competence. Young children with greater skill in emotion regulation are more socially competent in their interactions with peers and more cooperative with caregivers. Emotion regulation is also relevant to psychological well-being and developing risk for affective psychopathology. Even in the preschool years, children exhibiting poorer capacities for emotional self-control are at risk of developing later internalizing (such as depression) and externalizing (such as aggression) problems. Of course, these problems are also associated with other risk factors, such as temperamental vulnerability and aversive parent–child interactions, that also contribute to a child's self-regulatory difficulties. Emotion-regulation problems are thus an important avenue by which other difficulties in a child's life may give rise to psychological problems.

Research on the development of emotion regulation in the early years is challenging. Infants and young children cannot report on their emotional experiences or self-control strategies as adults can. Instead, developmental scientists must infer emotion regulation in the behaviors that young children show in emotionally arousing situations, such as gaze-aversion, self-comforting, or seeking proximity to a caregiver. They must be interpretively cautious, of course, since each of these behaviors indexing emotion regulation is multidetermined: sometimes gaze-aversion reflects efforts to manage exposure to an emotional elicitor, but sometimes it does not. The particular circumstances in which young children are observed (such as whether the situation is familiar or unfamiliar, or a caregiver is present) also have profound effects on how capably infants and young children can manage their feelings. This helps to explain why individual differences in emotion regulation are not very strong when measured across situations. Young children may be good at managing their feelings in some circumstances, but relatively poor in others, and this makes the search for stable individual differences in self-regulation more difficult.

These are challenging considerations, and they have led to the development of innovative and complex research strategies that promise new insight into the growth of emotion regulation in the early years. These include using multiple convergent measures of emotion regulation that simultaneously index behavior and physiological arousal, examining differences in emotion management in contrasting situations with different expected emotional demands, and using microanalytic sequential analyses to examine the influence of self-regulatory strategies on emotional expressions over short periods of time. In addition, some researchers examine emotion-regulatory processes as they are revealed in the 'emotion dynamics' of a young child's behavior. These dynamics concern the intensive and temporal qualities of an emotional response, and include: (1) the latency from an emotion-eliciting event until the response begins, (2) the rise-time until peak emotional intensity is achieved, (3) the peak intensity of the emotional response, (4) the recovery time until emotional behavior has reached baseline, (5) the range of emotional responding over a period of time, and (6) the lability of emotional responding over that period. These emotion dynamics can be studied to reveal developmental changes in emotion-regulatory capacities, individual differences in a child's emotional self-regulatory style, and, in some cases, the emotional features of affective psychopathology (such as the diminished emotional range of a depressed child or the fluctuating emotional intensity of bipolar disorder).

Taken together, these and other research strategies contribute to contemporary enthusiasm for studying emotion regulation in young children, and an expanding research literature on this topic.

Neurobiological Development and Early Emotion Regulation

Emotion regulation is multifaceted, so it is no surprise that the neurobiological foundations of emotion regulation are comparably complex and multifaceted. Multiple brain regions and neurohormonal processes are associated with emotion and its management. Significant advances in emotion regulation occur during the early years as components of these systems begin to mature. From the beginning, however, emotion-regulation capacities unfold through the interaction of neurobiological maturation and caregiver support – the familiar nature–nurture dynamic.

At its core, emotion regulation arises through the interplay of biological systems that excite and inhibit arousal. These systems are immature at birth, but they are active. Subcortical structures such as the amygdala and hypothalamus function in concert with the hypothalamic–pituitary–adrenocortical (HPA) axis to activate sympathetic nervous system activity and arouse the newborn. Parasympathetic nervous system activity, which has an inhibitory effect, is still quite immature early in infancy, and this accounts for the immediate, 'all or none' quality of early arousal manifested in raucous crying that the infant cannot regulate.

Although even newborns are equipped with rudimentary behaviors that function to manage arousal, such as self-sucking and motor activity, young infants are highly dependent on their caregivers for emotion management. In an adult's provision of rocking, warmth, and a soothing vocal tone (as well as practical ministrations, such as feeding and diaper changing, that may alleviate the source of distress), young infants experience emotional relief in the context of close human contact. More broadly, the caregiver's contingent and sensitive responding helps to maintain the baby's emotional arousal within manageable limits in the context of social interaction. These early experiences of emotion regulation are important catalysts for the development of secure parent–child relationships, and they contribute to the growth of social expectations by the baby that the adult's arrival will provide relief, which is also a source of relational security and emotional management. Some studies have shown, for example, that fussing infants begin soothing when they can hear the caregiver's approaching footsteps. Unfortunately, in some families insensitive or inappropriate parental responding is more typical, together with the emergence of patterns of emotional responding in young infants that reflect more limited parental support for emotion regulation, such as lack of soothability or emotional withdrawal. These early experiences of social interaction also contribute to the development of insecure parent–child relationships that undermine the child's emotional regulatory capabilities.

By 3–4 months, parents begin to notice a decrease in the baby's fussiness and an increase in soothability, along with greater alertness and social responsiveness. These reflect progressive maturation of inhibitory systems, particularly in the parasympathetic nervous system and areas of the prefrontal cortex, that permit greater control of attention and arousal. By 6 months, as a consequence, infants can begin looking away from an upsetting stimulus and refocus on objects or activities that distract them from what is distressing. They also become more capable of becoming distracted by the efforts of caregivers. By 10–12 months, there are also some indications of emotional response inhibition, such as when a baby tries to fight back tears when watching mother leaving.

During the second year, the baby's emotional arousal becomes more graded and less labile. These changes are attributable, in part, to the progressive maturation of several areas located toward the front of the brain that contribute to self-control: the anterior cingulate cortex, the dorsolateral prefrontal cortex, the ventromedial prefrontal cortex, and the orbitoprefrontal cortex. The maturation of these areas gradually enables young children to acquire greater capacities for attentional control and redirection, the inhibition of impulsive responses and the substitution of more reasoned responses, the enlistment of working memory into planning sequential tasks, and other abilities. As these areas mature, they exercise inhibitory control over lower brain areas (including the hypothalamus and amygdala) and thus permit more careful and reflective problem-solving, behavior, and emotional responding. Although temper tantrums remain characteristic of toddlers, these slowly maturing neurobiological capacities gradually enable young children to momentarily inhibit emotional outbursts to respond more constructively, reorient attention toward or away from events with anticipated emotional consequences, and enact simple strategies (such as seeking a security object or a caregiver) when upset. Parents exploit these developing capacities by helping their young children to enlist distraction or problem-solving strategies for managing everyday emotional demands.

These frontal inhibitory areas have a very extended developmental timetable, with maturation complete not until early adulthood for some areas. This helps to explain why the growth of emotion regulation (and more generally of self-control) unfolds in sophistication and scope throughout childhood and adolescence. Beginning in early childhood, however, the emerging capacity to inhibit initial emotional impulses and respond more reflectively has many consequences. It not only allows young children to begin to cope constructively with their feelings of distress or anger, but it may also contribute to the emergence late in the second year and early in the third year of self-conscious emotions such as pride, guilt, shame, and embarrassment. These emotions require self-awareness and the recognition of how others are evaluating oneself, which is built on the capacities for reflective thought that are facilitated by the neurobiological foundations of self-control. These neurobiological foundations may also contribute to the developing ability to consider the effects of one's emotional expressions on others and to modify them accordingly. It is thus not surprising that beginning in the fourth year, preschooler children begin applying social display rules to manage their outward displays of emotion, such as looking pleased when opening a disappointing gift in the presence of the gift-giver (although they can alter their emotional expressions for personal benefit at earlier ages).

Taken together, therefore, the neurobiological foundations of emotion regulation begin to emerge in infancy and early childhood, although considerably more remains to mature in years that follow. Nevertheless, these developing neurobiological capacities not only equip the child with growing ability to exercise emotional self-control, but also enable caregivers to assist in the child's emotion management more effectively. A 4-year-old who is awaiting the mother's arrival at the end of the day in childcare with growing distress, but can say to herself the words that provide self-comfort (e.g., "Mommy's coming" "we go bye-bye") is enlisting a simple strategy for emotion management that is built on the foundation provided by early brain maturation

Temperamental Influences

Another biological foundation to developing emotion regulation is temperament. Temperament consists of the early emerging, stable behavioral characteristics that make people unique. These characteristics include dominant mood (either positive or negative), soothability, irritability, proneness to fear or anger, inhibition, and other qualities related to emotion, its expression, and its regulation. Temperamental qualities are apparent at birth and are typically viewed by parents and practitioners as harbingers of mature personality. However, it is important to recognize that as infants mature, the psychobiological systems on which temperament is based also change and temperamental qualities evolve. Temperament is therefore not unchanging and immutable, and there can be considerable change in temperamental qualities, especially in the early years. Indeed, growth in emotion regulation contributes to changes in temperament and its manifestations over time: temperamentally fussy or withdrawn infants gradually mature into preschooler who can manage their difficult moods or shyness better. But temperament also influences the growth of emotion regulation.

One way that temperament influences emotion regulation is how certain temperamental qualities contribute to the intensity and persistence of emotionality that requires management. Some children – such as those who are temperamentally more prone to fearfulness or anger, or less soothable – face different challenges in regulating their emotions than do others who are temperamentally more placid. Heightened proneness to negative emotion is likely to make emotional self-regulation more difficult. This is important to remember in evaluating how emotionally well regulated are different children. One child may appear to be somewhat more prone to distress than others, but if this child is temperamentally negatively reactive this behavior may actually reflect considerable effort at emotion regulation rather than deficits in self-regulation. The effectiveness of a child's emotion-regulatory efforts must be interpreted in light of the child's temperamental qualities.

Another way that temperament influences the growth of emotion regulation is that young children with certain temperamental profiles tend to use distinct emotion-regulation strategies. Developmental studies indicate, for example, that temperamentally fearful or wary children tend to use avoidance, self-soothing, and seeking proximity to a caregiver to manage their feelings. Other temperamental qualities are even more directly associated with emotion regulation. Young children who are temperamentally more soothable, for example, are more easily calmed by caregivers when distressed. Likewise, a temperamental quality called 'effortful control', which refers to the capacity to inhibit a dominant response (such as getting angry or frustrated) and initiate a subdominant response (such as turning away or constructive problem-solving), facilitates emotion regulation. Young children who are temperamentally inhibited also appear to be more emotionally self-regulated. It is important to remember, however, that effective emotion regulation does not always involve emotional suppression, and that some circumstances require vigorous and forceful (and internally managed) emotional expressions, such as when children must stand up to a bully. Children who are always emotionally subdued may be exhibiting problems with emotion regulation as well as children who are emotionally undercontrolled.

Finally, temperament may interact with caregiving influences to shape the growth of emotion regulation through the interaction of the child's emotional qualities with characteristics of the caregiving climate. Researchers use the concept of 'goodness of fit' between a child's temperamental qualities and caregiving practices to describe how certain practices have different consequences for children who differ in temperament. In a family with parents who are outgoing and a child who is temperamentally inhibited, for example, parental and child styles do not mesh well, and the poor fit (e.g., parents who enjoy encounters with new people and experiences; a child who is threatened by these) can leave offspring unhappy and emotionally dysregulated. By contrast, a good fit between child temperament and caregiving qualities (such as when all family members are reserved) can create manageable emotional demands for the child and foster emotion regulation. A good fit between temperament and caregiving can also be created when parents create an emotional climate for the home that complements rather than conflicts with the child's qualities (e.g., parents give a shy child much time to accommodate to new situations and people). Creating a 'good fit' between a child's temperament and the caregiving climate can be more challenging when the temperamental profile is more extreme (such as negatively reactive) because in these circumstances children require greater support in their efforts to manage their feelings. Thus caregiving practices mediate between the child's temperament and the quality of emotion regulation that develops.

In this respect, an analogy may be useful. Temperament can be viewed as the canvas of a person's life. As all painters know, each canvas has its own special properties – its tautness, texture, how well it accepts paint – that affect how the painter works. If the artist strives to paint the same way on each canvas, the result will sometimes be outstanding art and sometimes disappointment. Likewise, with respect to parenting, one approach does not fit all children. Parenting must be adapted to children's unique temperamental characteristics, and how parents do so affects the growth of emotion regulation.

Viewed in this light, how parents perceive the temperamental qualities of offspring is crucial. Parental expectations for the child's emotionality are likely to color how parents introduce new or challenging experiences to the child, their organization of everyday routines, and how they respond to the child's emotional expressions. Such perceptions, when accurate, can contribute to a good fit between children's temperamental qualities and caregiving support. In contrast, children may have, as a result, fewer opportunities to acquire new strategies for emotion regulation in contexts, such as peer interactions, where parental assistance is not immediately available. But a foundation of parental support for competent emotion regulation at home is an asset to children's capacities to manage their feelings in other circumstances, and thus sensitive and accurate understanding of children's emotional style and temperament is an important contributor to the growth of emotion regulation. Unfortunately, it is sometimes true that young children's feelings receive insensitive reactions – such as when stressed parents ignore or dismiss the child's expressions – which undermines caregiving support for competent emotion regulation.

Taken together, temperamental individuality is an important part of the developmental context in which early emotion-regulatory capacities grow. Temperament individualizes the emotional experiences that children must learn to manage, shapes the strategies that may be most constructive, and defines the kinds of caregiving practices that are likely to facilitate emotion regulation.

Understanding Emotion, Self, and Emotion Regulation

Emotion regulation incorporates how people monitor and evaluate their emotional experiences. It is the appraisal that one is feeling differently from how one wants to feel that often motivates efforts to manage feelings. In addition, managing feelings involves knowing what you must do to feel differently in particular circumstances. Consequently, advances in young children's understanding of emotion, the causes and consequences of emotional experiences, strategies of emotion management, social rules for emotional displays, and the self have significant influences on the growth of emotion regulation in early childhood.

Early childhood witnesses, in fact, significant advances in young children's understanding of emotion. Toddlers' earliest verbal references to their internal states include naming their emotions, often accompanied by comments about the causes of their feelings, and sometimes including self-regulatory references (e.g., "Scared. Close my eyes."). In early childhood, children are constructing an understanding of the prototypical situational causes of basic emotions like distress, fear, and anger (e.g., falling down, being hit by another). They are also beginning to comprehend the internal origins of emotional experience, such as how feelings arise from one's perception of emotionally arousing events, sadness arises from unfulfilled desires, anger from blocked goals, surprise from unrealized expectations, etc. Consistent with this developing understanding, older preschooler children are also becoming aware of the privacy of emotional experience: that one's feelings do not have to be disclosed to others and, in fact, other people can be fooled about what one is feeling. This contributes not only to deliberate deception (e.g., feigning ignorance about a broken dish) but also to emotional displays that intentionally mask underlying feelings. Although preschooler children cannot yet clearly explain why they do so, by age 4 or 5 years they are likely to display happiness when opening a disappointing gift in the presence of the gift-giver (especially if they are girls) or minimizing their fear in the presence of a bully.

In addition, preschooler children are capable of enlisting several constructive strategies of emotion regulation and, on occasion, talking about them, such as leaving an emotionally arousing situation, removing or restricting one's perception of emotionally arousing events, seeking comfort from a caregiver, and other behavioral strategies. Their dawning understanding of emotion-regulation strategies is based on their developing understanding of emotion: that one's perception of emotionally evocative events leads to feelings (which can be managed by looking or moving away), that anger derives from impeded goals (which can be managed by doing something else), that surprise derives from unfulfilled expectations (which can be managed by seeking information), and similar insights. (Sometimes their emotional understanding undermines emotion regulation, such as when toddlers' awareness of the association between sadness and unfulfilled desires causes them to become fixed on getting what they want in order to feel better!) It is not until middle childhood, however, that children acquire a more fully psychological conception of emotion regulation involving the mental events by which feelings can be managed, such as through internal distraction, redirection of thoughts, cognitively reframing the situation, evoking conflicting emotions (such as thinking of happy things in scary situations),

and similar strategies. Even so, it is apparent that the growth of emotion understanding and of the causes and contexts of emotional expressions facilitate awareness of strategies of emotion regulation that enable young children to develop into more competent emotional tacticians as they mature conceptually.

The development of emotion regulation is also influenced by the growth of self-awareness. Early childhood witnesses significant advances in self-understanding, manifested especially in the appearance of many forms of self-reference after 2 years of age: self-descriptions (e.g., "me big!"), assertions of competence and insisting on "do it myself," assertions of ownership, verbal labeling of internal experiences such as emotions, categorizing the self by gender and in other ways, and growing sensitivity to evaluative standards applied to the self. This period is also when self-referential emotions emerge as young children increasingly exhibit pride in their accomplishments, guilt about their misbehavior, shame when behavior reflects negatively on the self, and embarrassment when effusively praised. The emergence of self-referential emotions introduces greater complexity into the young child's emotional experience, and also creates new challenges for emotion self-regulation as young children become increasingly concerned with how they appear in the eyes of others. Consequently, efforts to manage feelings now become enlisted to manage emotional expressions (such as sadness and fear) that might evoke disapproval from parents and peers, as well as to manage feelings of embarrassment, guilt, shame, and other compelling self-referential feelings when they occur. Indeed, young children's efforts to avoid situations that evoke guilt or shame is an important emotional resource to early socialization.

Because emotion regulation is oriented toward the self – it is, after all, motivated primarily by the desire to feel how one wants to feel – it is not surprising that further advances in self-understanding are associated with later growth in emotion regulation. In particular, as children further mature, emotion-regulatory efforts become increasingly devoted to maintaining self-image and self-esteem in complex social situations involving peers, and children develop self-regulatory styles that fit well with emergent personality and self-awareness.

Family Socialization of Emotion Regulation

Emotion regulation in young children develops in concert with the influences of parents, siblings, and peers who teach children how to think about, express, and respond to emotions in culturally and contextually appropriate ways. Throughout the early years, in particular, emotionally salient interactions with parents contribute to children's emerging capacities to manage their feelings. In this section, we consider two kinds of parental influences on the socialization of emotion regulation: the influence of direct parental interventions and strategies (such as reactions to the child's emotions, specific instruction about emotion regulation, and conversations about emotion) and the influence of the broader emotional climate of the home.

Beginning at birth and continuing throughout life, emotions are calmed, enhanced, and managed by other people. Parents efforts to soothe a distressed infant, or maintain the child's joyful demeanor during play, contribute to regulating the baby's arousal and to emerging social expectations that certain people are reliable sources of pleasure and relief. In the years that follow, parents intervene in many ways to manage the feelings of offspring. Parents may distract young children who have become distressed or angry to calm them, look reassuring in potentially upsetting circumstances, assist in solving problems that children find frustrating, or try to alter the child's interpretation of negative events (e.g., "It is just a game"). They also seek to manage the feelings of young offspring by proactively structuring children's experiences to make emotional demands predictable and manageable; for example, many parents schedule naps and meals to accord with their knowledge of the child's temperamental qualities and tolerances, and choose activities that are congenial to young children's needs and capabilities. This is part of what is involved in enhancing the 'goodness of fit' between the child's emotional and temperamental qualities and the caregiving climate. In this sense, parents manage the emotions of their children through direct interventions and by the organization of daily experience.

Parents are also tutors in emotion regulation. They suggest constructive approaches to managing heightened feelings that are within the young child's capacities, such as encouraging toddlers to 'use words' instead of hitting and urging problem-solving resolutions to disputes between siblings. When anticipating challenging situations, parents may also rehearse strategies for emotional coping. In our own research, we have observed parents coaching their shy children before beginning tasks that require young children to interact with unfamiliar research assistants in the parent's absence.

Research studies indicate that caregivers who offer warm, sensitive support during emotional challenges have offspring who are more capable of constructively managing their emotions, such as through problem-solving or distraction. By contrast, when parents are critical or punitive, young children are less likely to regulate their emotions constructively but instead vent their frustration or distress. Supportive assistance from parents can influence young children's immediate emotional experience so children can practice skills of emotional self-control. Parental warmth and support also contributes to young childrens' developing beliefs about the manageability of their feelings and how to use social support in doing so. Young children

learn, in other words, that others can assist in controlling strong feelings, and that they need not be overwhelmed by heightened emotional arousal. This knowledge provides a foundation for the growth of self-initiated strategies of emotion regulation, and over time, there is a gradual transition from the child's reliance on external sources of emotion regulation to greater self-regulation as parents intervene less frequently.

Parental interventions are guided by parents' evaluations of the child's emotions, as noted earlier. Parents respond to the emotional expressions of offspring in many ways: they can do so sympathetically, dismissively, critically, denigratingly, punitively, or may simply ignore what the child is feeling. These evaluative responses, conveyed through verbal comments, facial expressions, and other behavior, convey broader messages about the appropriateness of the child's feelings, the extent to which those feelings are valued by the adult, and the adult's availability as a source of support for emotion regulation. Although parents respond to children's emotional outbursts in various ways in different circumstances (even the most sympathetic parent will have difficulty being patient with petulent crying at a family gathering), the general tenor of the parent's responsiveness can influence the growth of emotion regulation in offspring. Supportive and constructive parental responses contribute to the development of more adaptive and constructive strategies of emotion regulation in young children.

By contrast, reactions that are punitive, critical, or dismissive impair children's self-regulatory development in several ways. In the immediate situation, these responses are likely to exacerbate the child's negative reactions, making it more difficult for the child (or the parent) to manage the child's feelings. More broadly, recurrent parental criticism or dismissiveness contributes to developing beliefs that the child's feelings are unjustified, unimportant, or should not be expressed, or that the child is emotionally incompetent. They also contribute to insecurity in the parent–child relationship that can make it difficult for children to seek assistance or talk about their feelings with the parent on future occasions. A toddler who is always told that "big boys (girls) do not cry" may struggle alone to manage feelings of sadness with this emotion judgment as a continuing influence.

Parent–child conversations about emotion are another forum for growth in emotion understanding and emotion regulation. Even before young children are good conversationalists, they are attentive to, and participate in, conversations about everyday events in which emotions (including the child's own feelings) and their causes and consequences are discussed. These conversations may concern a sibling's temper tantrum, the child's previous misbehavior, the reason for an adult's joyful exclamation, an event that provoked fear, or other common experiences involving feelings. These conversations are important to the growth of emotion understanding because they provide clarity and insight into emotional experiences that interest young children but may be confusing because of the invisible psychological processes that are involved (e.g., desires, goals, intentions, and expectations that influence emotion). The content and structure of parent–child conversations about everyday events can provide a window into the psychological influences on emotion and emotion self-regulation. Two studies have found that the frequency, complexity, and causal orientation of emotion-related conversations between mothers and their 3-year-olds predicted the child's emotion understanding at age 6 years, and other research has similarly documented the influence of such conversations on the growth of early emotion understanding.

Conversations about emotion are important to the development of emotion regulation because parents frequently comment on expectations for emotion management and how feelings should be regulated. Indeed, young children are frequently coached about how to use distraction to reduce frustration ("why not do something else?"), how to manage the physiological accompaniments of emotional arousal ("take a deep breath and calm down"), how to use adaptive modes of expression that might have more emotionally satisfying consequences ("use words rather than hitting"), and other ways of regulating their feelings. One ethnographic observational study of families in an urban, low-income community documented mothers' diverse strategies for teaching their 2–3-year-old daughters how to manage their emotions in ways that were consistent with the values of the community. Mothers encouraged their daughters, for example, to defend themselves in hostile situations by coaching their children through evocative role-playing scenarios, modeling appropriate behaviors, and rehearsing specific strategies of anger expression and self-control. As a consequence, their daughters developed a rich repertoire of expressive modes for conveying anger, but were also capable of regulating its arousal and expression consistently with the rules of the community. Furthermore, these mother–daughter conversations revealed some of the internal processes of emotion regulation to offspring, such as which emotions should be regulated, under what circumstances, and how to do so.

As this study illustrates, culture and subcultural values are important to the family socialization processes involved in emotion regulation. Families in Western and non-Western cultures differ, for example, in their beliefs concerning the emotions that are most appropriate to feel and express in social situations (such as whether shame or anger is the most appropriate emotional response to interpersonal difficulty) and appropriate modes of emotional expression. Diversity in subcultural beliefs within societies reveal that local values concerning, for example, the expression of anger or fear, are important for young children to acquire, as illustrated by the ethnographic study

previously described. There are even differences in parental socialization of emotion regulation by gender, with one study concluding that parents have a greater relational focus in their conversations with daughters compared with sons to help them cope with distress produced by common frustrating events. Family socialization of emotion regulation is, therefore, relative to the cultural values of the broader society concerning emotion and its management.

These elements of family socialization – parental interventions, evaluations of the child's feelings, parental warmth and support, coaching in emotion regulation, and conversations about emotion – collectively contribute to the broader emotional climate that characterizes family life. In addition, the frequency and intensity of the parents' emotional expressions contributes to the family emotional climate. Young children's ability to cope adaptively with their own intense feelings is undermined when they are faced with frequent and intense expressions of negative emotion in the home, especially when the emotions are directed at them. The emotional climate also contributes to children's developing representations of emotion as family members are models of how and when to express emotion and manage their feelings, and contribute to children's expectations for emotion in the world at large (e.g., are emotions threatening? empowering? irrational? uncontrollable?).

A large research literature indicates that an emotionally positive family climate is associated with self-soothing behaviors in infants and children's enhanced self-regulatory capacities, and an emotionally negative family climate is associated with more negative and mixed outcomes. These associations are stronger for infants and toddlers compared with older children, suggesting their particular sensitivity to the family emotional climate. The effects of a negative family climate vary depending on whether the negative emotions are submissive (e.g., sadness, embarrassment) or dominant (e.g., anger). Negative dominant emotions expressed within the family create a rejecting and fearful environment that undermines the development of emotion self-regulation in offspring. Young children from these families exhibit less constructive coping strategies and more venting of emotions. By contrast, children in negative submissive families have greater opportunities to acquire constructive strategies for emotion regulation because the family environment is less hostile. In these settings, for example, young children can observe and participate in comforting a distressed family member or rectifying misbehavior.

The intensity and duration of negative emotion expressed within the family is also important. Family members who exhibit sadness, anger, or fear at moderate intensity can also provide children with constructive models of coping. By contrast, heightened expressions of negative emotion may overwhelm the child and undermine effective coping. This is illustrated by studies of children living in maritally conflicted families who become acutely sensitive to parental disagreements and have difficulty managing the distress these arguments create in them. In a similar manner, young children of chronically depressed caregivers also experience emotion-regulatory difficulties because of their emotional involvement in the parent's affective distress and the parent's unavailability as a coping resource for the child.

Taken together, young children develop emotion-regulatory capacities in family environments that shape their understanding of emotion, strategies of emotion management, and capacities for coping with heightened emotion in themselves and family members. Central to these family environments are the relationships that young children share with their caregivers.

Parent–Child Relationships and Emotion Regulation

In infancy and early childhood, emotional wellbeing depends on the quality of children's relationships with their caregivers. The security, mutual cooperation, and warmth of these relationships enables children to seek parental support when upset, with the trust that the parent will be a reliable source of assistance. For the same reason, the security of parent–child relationships is also a foundation for the growth of emotion regulation. The sensitivity of parental care and the trust inherent in secure relationships enables children to become more emotionally self-aware and to develop a more flexible capacity to manage their emotions appropriate to circumstances. By contrast, young children in insecure relationships are more prone to emotional dysregulation, especially in stressful circumstances, that may be manifested in heightened levels of negative emotions or, alternatively, in suppressing the expression of their feelings.

A variety of research studies show that by early childhood, children in secure parent–child relationships exhibit more competent and constructive emotion-regulation capacities compared to children in insecure relationships. One reason is that their parents are likely to be more sensitive to the child's emotions and needs, and thus provide more helpful assistance in the child's emotional coping. Another reason is that a secure parent–child relationship enables young children to talk more readily about their feelings with their caregivers, especially negative feelings that may be confusing or disturbing, and to expect a helpful response. Several research studies have shown that securely attached preschooler children talk more about emotion in everyday conversations with their mothers, and their mothers provide offspring with richer insight into emotion during these conversations. This may help to explain why as early as age 3 years, securely attached children are more advanced in emotion understanding than are insecure

children. The mutually cooperative orientation of a secure relationship also makes young children more receptive to parental messages about emotion or coaching about emotion regulation. Taken together, there are multiple reasons why young children in secure relationships develop greater competencies in emotion regulation: the enhanced sensitivity of parental assistance in managing emotions, the insight derived from parent–child conversations about emotion (especially emotions that children may find more difficult to comprehend), and the greater cooperation that enhances children's responsiveness to parental coaching about emotion regulation.

These conclusions indicate that the relational context in which emotion regulation develops in the early years is important. Although the warmth and security of parent–child relationships overlaps with (and, to some extent, derives from) the family socialization influences described earlier, children's security in close relationships is 'more than the sum of the parts' of parental influences. Because relational security derives from the quality of the emotional connection between partners, children derive from secure relationships the trust and confidence needed to manage strong feelings. In doing so, they also develop confidence in themselves as emotional beings.

Peer and Sibling Influences

Although most research on the socialization of emotion regulation has focused on the robust influence of parents, many studies highlight the importance of siblings and peers. To be sure, similar socialization influences occur in each kind of relationship. As with parents, young children derive from their conversations with siblings and peers knowledge about emotion and expectations for emotional self-control. Indeed, young children talk about their feelings more frequently with their friends and siblings than they do with their mothers, and these conversations also contribute to children's developing emotional understanding. Siblings and peers are each potent models of emotional expression and self-control, and interactions with other children enable young children to learn about managing their feelings in social contexts that are similar to but also distinctly different than those of parent–child interaction.

Relationships with siblings and peers also offer unique opportunities for the development of emotion regulation. One reason is that the activities children share with siblings and peers present unique demands for managing emotions. Other children are less capable social partners than are parents, and are less likely to accommodate to the child's feelings and needs readily. Developing capacities to coordinate one's behavior with that of another child is important to social competence with peers and siblings, and this requires that children manage their own feelings of exuberance, frustration, distress, and anger, and sensitively attend to the other child's feelings, in order to get along with another.

Peer interactions often entail disagreements over conflicting intentions and desires, and socializing with other children requires developing skills for managing one's own feelings and coping with the emotional expressions of others. Indeed, emotion understanding and emotional self-control are cornerstones of social competence with peers because they contribute to the development of friendships, curb aggressive behavior, and enhance opportunities for prosocial initiatives. Thus young children who are emotionally perceptive and capable of managing their own feelings in social interaction are likely to be more successful in their peer encounters. Moreover, pretend play presents preschooler children with further opportunities to develop emotion-regulation skills as imaginative activity offers a safe arena for exploring negative emotions and to rehearse and reflect on self-regulatory strategies from a distance. Peer relationships are also important for helping children learn the rules for emotional expression and self-control for the peer culture that may be different from those of the family. Understanding how to be appropriately assertive but not aggressive is important to successful peer relationships, for example, but parents respond much differently to a child's assertiveness at home. Peer relationships thus enable children to learn the 'feeling rules' for the peer environment that they will use outside the home for emotion self-regulation, and which children coordinate with the feeling rules of the home.

Many of the emotional characteristics of peer interaction also characterize young children's interactions with younger or older siblings. Siblings are both more challenging and more understanding than are parents or peers. Sibling relationships are typically characterized not only by a high rate of conflict but also a broad range of emotion, from sympathetic comfort to angry confrontation. These relationships thus present children with a daunting variety of emotional demands requiring emotion self-control. Moreover, in contrast to peers, sibling relationships are family relationships that endure over time, forcing children to learn how to negotiate, bargain, and accommodate to someone who will always be with you. Viewed in this light, some of the greatest challenges to emotional self-regulation are encountered in young children's interactions with their siblings.

Taken together, it is clear that a full account of the social influences on the development of emotion regulation must include influences from other children. In the emotional demands they impose, examples they provide, and the conversations they share, peers and siblings are important catalysts for the growth of emotion self-control and for children's understanding of how to regulate their feelings successfully in the unique cultures that children share with each other.

Conclusion

The development of emotion regulation in the early years is important because a child's capacity for emotional self-control is deeply connected to the growth of emotional wellbeing, social competence, and risk for emotional disorders. Indeed, emotion regulation underlies emotional competence: the ability of children to feel as they want to feel, to enlist their emotions adaptively to accomplish their goals, and to respond emotionally in social situations both genuinely and tactically. The development of emotion regulation is one of the most salient differences in the emotional behavior of newborns and adults as children progressively acquire capacities to manage their feelings in ways that are self-initiated, situationally appropriate, culturally guided, flexible, effective, and strategic. It is also a core foundation of personality development as children's preferred styles for managing their feelings become incorporated into the broader network of temperamental qualities, self-referential beliefs, and dispositions that are at the core of personality.

As this review of research indicates, however, emotion regulation is not a unitary developmental phenomenon but rather an integrated network of loosely allied developmental processes arising from within and outside the child. Many aspects of neurobiological development, conceptual growth, temperamental individuality, and social influence are enfolded into developing capacities to manage emotions. To the developmental scientist, this makes the development of emotion regulation a uniquely integrative field of study because understanding how young children learn to manage their feelings requires analysis at multiple levels of development. To practitioners, this means that emergent problems in emotional self-control can have many sources, including troubled family environments, temperamental difficulty, neurobiological problems, or a combination of these and other causes. The complexity of these developmental processes underscores how the integration of scientific and practical concerns is likely to yield further understanding of the growth of emotional self-control and the application of this knowledge to assisting young children who need assistance.

Young children have many reasons for trying to manage their feelings, of course. They do so to feel better when distressed, manage fear, enhance positive well-being, strengthen relationships, comply with social rules, promote constructive coping, and for many other reasons. The manner in which this occurs as it is colored by temperament, guided by close relationships, enlivened by an emerging sense of self, prepared by brain maturation, and structured by emerging concepts of emotion is a fascinating developmental story worthy of further study.

See also: Abuse, Neglect, and Maltreatment of Infants; Adoption and Foster Placement; Anger and Aggression; Attachment; Brain Development; Crying; Depression; Empathy and Prosocial Behavior; Endocrine System; Family Influences; Fear and Wariness; Friends and Peers; Humor; Independence/Dependence; Lead Poisoning; Marital Relationship; Mental Health, Infant; Parental Chronic Mental Illnesses; Parenting Styles and their Effects; Postpartum Depression, Effects on Infant; School Readiness; Self Knowledge; Self-Regulatory Processes; Separation and Stranger Anxiety; Shyness; Smiling; Social and Emotional Development Theories; Social Interaction; Socialization in Infancy and Childhood; Temperament; Theory of Mind.

Suggested Readings

Denham S (1998) *Emotional Development in Young Children.* New York: Guilford.

Eisenberg N, Cumberland A, and Spinrad TL (1998) Parental socialization of emotion. *Psychological Inquiry* 9: 241–273.
Eisenberg N and Morris AS (2002) Children's emotion-related regulation. In: Kail R (ed.) *Advances in Child Development and Behavior,* vol. 30, pp. 190–229. San Diego: Academic.
Harris PL (1989) *Children and Emotion: The Development of Psychological Understanding.* Oxford: Blackwell.
Kopp CB (1989) Regulation of distress and negative emotions: A developmental review. *Developmental Psychology* 25: 343–354.
Miller PJ and Sperry L (1987) The socialization of anger and aggression. *Merrill-Palmer Quarterly* 33: 1–31.
Saarni C (1999) *The Development of Emotional Competence.* New York: Guilford.
Thompson RA (1994) Emotion regulation: A theme in search of definition. In: Fox NA (ed.) *Monographs of the Society for Research in Child Development, Vol. 59 (2–3), Serial no. 240: The Development of Emotion Regulation and Dysregulation: Biological and Behavioral Aspects,* pp. 25–52. Ann Arbor: Society for Research in Child Development.
Thompson RA and Meyer S (2007) The socialization of emotion regulation in the family. In: Gross J (ed.) *Handbook of emotion regulation,* pp. 249–268. New York: Guilford.

Empathy and Prosocial Behavior

J Robinson, University of Connecticut–Storrs, Storrs, CT, USA

Glossary

Altruism – A motivation to act for the benefit of others without apparent self-gain.
Emotions and basic emotions – Feeling states reflected in facial, gestural, and verbal expressions; basic emotions appear early in development and appear to have universal or species-typical expressions (e.g., joy, fear, anger).
Empathy – A higher-order emotion that reflects a connection to another's experience; the expression of caring and concern for another.
Perspective taking – Understanding the intentions and feelings of another.
Prosocial – Behaviors intended to benefit or help others.
Socialization – Responses by parents and caregivers that encourage desired behaviors and limit undesirable behaviors.
Sympathy – Vicarious emotional response of caring concern for another arising from the experience of empathy.
Theory of mind – The child's understanding of the apparent feelings, thoughts, and intentions of others.

Introduction

This article defines empathy and prosocial behavior and presents the major theories of empathy development and evidence supporting it in the empirical literature. It describes the principal methods of study of empathy and prosocial behavior. It considers the current knowledge of heritable and environmental influences on early empathy and prosocial behaviors. Disorders of empathy include autism spectrum and conduct disorders as well as early maltreatment within the family. Future directions for research include a greater emphasis on cross-cultural parental values and socialization strategies that promote early empathy development.

Empathy and Related Terms

Empathy is one of the 'higher-order' emotions that typically emerge as the child comes to a greater awareness of the experience of others, during the second and third years of life. (In contrast, 'basic emotions' such as distress, joy, anger, or sadness emerge in the first months of life; they appear before infants understand social intentions.) Empathy is an emotion that specifically arises in the context of someone else's emotional experience, and reflects a resonance or connection to the other's experience. Empathy is thus a personal, emotional response to the emotional state of another. When researchers study empathy, they are most often interested in whether the individual experiences an emotional shift toward or closer to the other person's emotion; sometimes an emotional response is considered empathic only if it mirrors the emotion of the other (i.e., if it is a close match). In this sense, empathy might be considered a form of emotional contagion. However, most researchers agree that empathy is an 'other-directed' emotional experience and is not purely contagion because 'true' empathy reflects awareness of the distinctness of the self and another and its focus is the other's emotional experience.

Although it is possible to observe an empathic response in a variety of emotional situations (e.g., being happy about another's success), empathy customarily describes a caring or sympathetic response to the distress or suffering of another. Investigators of empathy in very early childhood have most often studied children's emotional and behavioral changes in response to someone who becomes sad or distressed. The related term 'sympathy' was used by Nancy Eisenberg and colleagues to describe the vicarious emotional response of caring concern for another. In her nomenclature, sympathy arises from the experience of empathy. Others, however, have defined empathy as synonymous with sympathy. We will use the term empathy throughout this article to mean the expression of caring concern to the distress of another with the goal of relieving their distress. 'Compassion' is yet another term that connotes empathy and refers to the experience of shared suffering and concern for another, including the motivation to aid another, but it is used less commonly in psychological research.

Feelings of empathy can be expressed through the display of basic emotions such as sadness or concerned interest. Empathy can also motivate behaviors that express care or concern for the other such as approaching the other, seeking information about what caused the distress, and prosocial and helping behaviors (such as hugging or offering a toy or band-aid). 'Prosocial' behaviors (i.e., behaviors intended to benefit or help others) are also of great interest to researchers studying the development of morality and peer relations and many studies of preschool- and school-age children have been conducted on this topic, in addition to empathy. The repertoire of responses of very young children can include all of the above-described behaviors, including prosocial behavior, as well as more 'self-focused' or empathic distress responses that likely reflect contagion of emotion rather than empathy.

Development of Empathy

In the 1970s, Martin Hoffman provided seminal ideas that have continued to influence thinking about early empathy development when he proposed that the development of altruistic motivations arise from the synthesis of emotional responses and cognitive abilities that typically develop in the first years of life. The child's emotional response to another in distress, or empathic distress, is the core feeling state of empathy. Hoffman argues that empathic distress is an involuntary, evolutionarily adaptive response to the suffering of another wherein the individual experiences the other's painful emotional state. He and others have argued that such automatic distress responses to the distress of another reflect behaviors that evolved in mammals to support caring for the young of the species. Observing infants in the newborn nursery become vigorously distressed shortly after another infant began to cry suggested that there might be a reflexive component to empathic distress. Mimicry also contributes to the emotional response tendency to experience the distress of another. Human infants begin to mimic or imitate facial, vocal, and postural muscle movements of caregivers in the first weeks and months of life and the 'afferent feedback' (i.e., feedback to the central nervous system from the peripheral nervous system) resulting from imitating a distressed caregiver may contribute to the infant's own feelings of distress. In the first year of life, the infant may feel distress globally in the presence of a distressed other (either reflexively, imitatively, or both) and the source of the distress is unclear to the infant. In Hoffman's theory, this global empathic distress stage reflects a self-oriented phase in empathy development that is a precursor to more mature empathic distress.

The development of a cognitive sense of others, indicated through person permanence (i.e., that another exists as a separate physical entity), perspective taking, and role taking, are crucial to the development of more mature empathy feelings. Current theories about this, called 'theory of mind', seems to capture well Hoffman's idea that young children begin to understand that others have inner states, such as beliefs, intentions, and feelings that are different from one's own. The young child's theory of mind develops gradually over the first 3–5 years of life, but toward the end of the first year of life we can infer from the infant's tendency to look to others for emotional reassurance when in uncertain circumstances that he/she is aware of the emotions of others. The cognitive awareness that others have feelings that are different from one's own thus sets the stage for other-oriented empathy to emerge.

During the second year of life, children's responses to another's distress become more complex as they begin to develop more awareness of the experience of others or other-oriented understanding. Between the first and second birthdays, when witnessing another's distress children typically pause and may spend several seconds looking intently at the victim or even babbling with an inquiring tone. This is illustrated in **Figure 1**, where we see a

16-month-old girl reacting to her mother as she simulates injuring finger. Mother has been sewing and when prompted, she exclaims as she pretends to have stabbed her finger. For several seconds the little girl watches her intently as mother rubs her finger and moans in pain. Early in the second year, this inquisitiveness may be accompanied by blended emotional expressions of distress and concerned interest as seen in **Figure 1**. Clearer emotional expressions of empathic concern are more commonly observed toward the end of the second year.

At this stage, which Hoffman terms 'quasiegocentric empathic distress', children's sober attentiveness appears to signal that they have identified that the source of distress is outside of themselves. Children's emerging cognitive understanding of others as separate also transforms the egoistic, global empathic distress response to a sympathetic distress response. At this age, most children attend to the distressed other for a substantial period (e.g., 5–10 s) and become only moderately aroused rather than highly aroused or distressed themselves. In **Figure 2** the little girl takes a few moments' break from attending to mother; she has stopped looking at her and for 3 or 4 s looks for a bit of snack that she had dropped a few moments before. The strong impression is that she is regulating her own emotions by breaking her attention away from mother. As suggested by Mary Rothbart, a temperament researcher, the ability to sustain a focused attention on something outside of themselves may serve an important role in helping children to regulate their emotional responses in this situation. In **Figure 3**, she moves closer to mother and expresses a great deal of concern on her face.

Between the second and third year of life there is a gradual increase in the frequency of children's prosocial attempts to assist the person in distress. Prosocial behaviors of young children include trying to distract, offering a toy, patting or hugging, and by age 3 years, expressing verbal sympathy ('Are you okay? You will be okay!'). In **Figure 4**, the young girl looks inquiringly to mother as she gestures a sign for 'all done' and her mother responds that the finger is feeling better.

Figure 1   Sixteen-month-old responding to maternal distress simulation.

Figure 3   Sixteen-month-old approaches and expresses great concern.

Figure 2   Sixteen-month-old re-grouping attention during maternal distress simulation.

Figure 4   Sixteen-month-old attempts distraction during maternal distress simulation.

Hoffman calls this stage 'more veridical empathy,' and he argues that one difference between prosocial acts observed during the previous stage and this one is the fittedness or accuracy of the child's attempts to assist someone in distress. Earlier, prosocial attempts have a strong egocentric quality, for example, the child may offer the distressed person a comfort object, but no studies have actually tested this idea. In **Figure 4**, when the 16-month-old signs 'all done' to her mother, we might also consider this an egocentric attempt at distraction.

By 36 months, the typical child is attentive to, and makes relatively complex verbal inquiries about, the distress ('What happened?' 'Does it hurt?'), he/she expresses moderate-to-strong sympathetic concern toward the person in distress, and about one-third of the time he/she attempts to relieve the mother's distress through a prosocial action and/or sympathetic statements. By age 3 years, the majority of children are able to respond empathically when in the presence of a distressed individual. Among the children whose language abilities allow more sophisticated expressions of interest and concern, it is quite common for children to communicate verbally about their concern for whether the pain is subsiding. However, in Hoffman's theory, the most mature form of empathy allows one to also express sympathy for those who are not immediately present (e.g., to feel empathy for a child with chronic illness who is in the hospital) and this does not typically appear until middle childhood when children have more advanced cognitive insights about self-identity. Thus, empathic responses are most likely when young children are in the presence of someone who is distressed.

Methods of Study

Some of the earliest systematic observations of children younger than age 3 years responding to someone in distress were conducted in the 1970s and involved a crying infant as a stimulus. These and later infant-cry studies used an audiotape of a crying infant in order to standardize the duration and level of distress of the stimulus event. When the responses of newborn infants were studied, live observers in the newborn nursery recorded only the occurrence of a cry response and changes in the state of infant (i.e., whether the infant was awake or drowsy). The stimulus tape was quite long (6 min), allowing time for the newborn to arouse and respond to it. However, when investigators began to study the responses of older infants and toddlers to infant cry, the cry stimulus tape was much shorter (1 min or less) and videotapes were made so that raters could examine behavioral responses and emotional changes in detail.

Carolyn Zahn-Waxler, at the National Institute for Mental Health, and colleagues have been the primary investigators of the emotional, behavioral, and biological empathy responses of toddlers and preschooler children. The earliest study of toddlers involved three research methods in addition to the infant-cry stimulus. First, mothers were trained to identify situations at home in which the child caused or witnessed distress in another. Mothers were trained to record both the event and the child's response to the event in a diary according to specific instructions to ensure that certain details about the situation were included. Second, researchers simulated distress events (e.g., an injury or a coughing fit) and videotaped children's responses. Third, mothers were trained (by observing examiners simulate distress and through oral and written instruction) to simulate similar events at home and to record responses using a similar diary method as for the naturalistic events.

Raters looked for the following types of child responses to distress events (both written and videotaped): (1) spontaneous prosocial behaviors, (2) emotional responses (concern, distress, and positive emotions), (3) hypothesis testing or attempts to understand the situation, (4) self-referential behaviors such as imitating the behavior of the person in distress, and (5) callous/aggressive behavior. These features of children's responses are now commonly studied in empathy research in early childhood and the rating scales developed by Zahn-Waxler in this early study are the basis for the standard assessment tool of empathy in children under 3 years of age. The simulation of distress, primarily through feigning injury, has also become the most commonly studied stimulus used in research of children between 1 and 3 years of age.

Researchers have conducted injury simulations in both home and laboratory settings. When performed either by mother or by an examiner, each simulation is standardized and typically lasts for 90 s. The simulation is typically preceded by settling the toddler into play with an interesting toy or objects so that there is a common 'baseline activity' from which the simulation will potentially draw the child's attention. Once settled, the simulation begins when the 'victim' pretends to hurt his/her finger on a sharp object or by snapping a clipboard or pretends to hurt his/her knee when trying to get up from a chair or off the floor. During the next 30 s, the victim verbalizes about the pain ('Ow, I hurt my finger! It hurts! Ouch!') and holds and rubs the injury. The victim refrains from looking or talking directly to the child during this time. In the next 30 s, the victim verbalizes that the injury is beginning to feel better and that rubbing has helped. At the end of this phase (i.e., at 1 min), the victim says, 'It's all better now'. The videotape continues to record the child's behavior for an additional 30 s to capture any delayed responses while the examiner and mother talk about the distress event and transition to another activity. Examiners use a stopwatch during their own simulations to time the 30 s epochs and they give mothers covert signals (e.g., knocking on a laboratory observation window) during their simulations. Mothers learn how to simulate distress by hearing verbal instructions and watching the examiner simulations; the examiner also offers mother a few quick reminders just before her simulation. An important part of the standardization of the procedure is the level of emotional distress that is simulated, and both examiners and mothers are asked to display a moderate level of pain in their voice and on their face as they hold and rub the injury. Mothers are typically very responsive to these instructions and ratings of the credibility of all simulations indicate that over 90% of mothers are able to create a credible distress stimulus on their first try.

Research assistants, who have had considerable training and practice, typically 40–60 h, rate the videotapes of the children's responses. They rate responses for the five behaviors described above on 3–5-point scales of intensity. In addition, they rate several behaviors that add to the description of the toddlers' responses, such as approach/avoidance of the victim, arousal level, and ignoring or maintaining active play. A global rating of empathy allows the rater to provide an overall or summary ranking of the child's response on a 7-point scale. This scale ranges from 1 (no concern, uninterested, or callous response) to 7 (moderate-to-strong concerned responses that include prosocial actions in addition to approach and inquiry; absence of any callous response). Most children in the 1–3 year age range score near the midpoints of this scale, a rating of 3 (sobered, sustained attention, mild or brief concern and no prosocial actions), 4 (sustained attention with mild-to-moderate concern, may show a fleeting prosocial act), or 5 (concerned responses including at least one clear help-oriented or victim-oriented behavior).

Maternal report methods involving the use of a diary of naturalistic distress event responses have not been as widely used since Zahn-Waxler and colleagues' early study. It is a time-intensive approach that might make it difficult for many of today's working mothers to participate. However, such a diary method has the advantage of providing rich information under varied, naturalistic circumstances that are typically not repeatable in a research laboratory. Future studies of early empathy might include more information about mother's attempts to socialize their young children toward greater empathy and prosocial responses; relatively little is known about that topic although a great many studies have investigated how parents socialize their children away from aggressive and antisocial behavior.

Influences on Empathy

Although the development of empathy has been described here in prototypical terms, young children have very diverse empathy responses that are influenced by many factors. These factors can be broadly classified as heritable (i.e., due to genetics) and environmental. In very early childhood, family socialization practices that encourage empathy and prosocial responses are very important influences. These can include caregiving behaviors in which parents model how members of their family respond when someone is in distress as well as direct instruction about what to do. However, as we will see, aspects of the family that are not specifically about empathy training can also exert an influence on children's empathy, for example, parent sensitivity and warmth in their relationship with the child, are important as are sibling relationships. We will also see that children's capabilities, including temperament, specific and general cognitive skills, and the ability to regulate emotion influence their empathy responses, and these capabilities are themselves influenced by genetics and environmental experiences. We will begin this section by briefly discussing gender differences in early empathy and prosocial behavior.

Gender differences. Across ages and methods of study, girls have generally been observed to show greater empathic concern, hypothesis testing, and prosocial behavior than boys. This has been described by some as part of a more emotionally expressive interactive style for females. However, the gender differences that have been observed in response to simulated distress are generally small in magnitude. Although most studies of early empathy development have involved children from socioeconomically advantaged homes, in a large study of first-born, ethnically diverse, low-income children by Robinson and colleagues, somewhat smaller but still significant gender differences were also found.

Another way to think about gender differences is that the sources of individual differences differ between boys and girls; for example, boys might respond differently than girls to similar socialization demands or that heritability might play a stronger role for females (perhaps because of evolutionary pressures) than males. Little support has been found for gender differences in how heredity and environment influence empathy at this age, although studies of emotion processes and empathy among older children show much more gender differentiation of both types (i.e., both mean levels and differences in the processes of development). In sum, although gender differences in average levels of empathy are observed in early development, gender similarity in the processes that influence empathy may be more the rule in the first 3 years of life.

Heritable influences. One of the largest studies of children's empathy, the MacArthur Longitudinal Twin Study (MALTS), was designed to address the question of whether early-observed empathy responses to the distress of another were influenced by genetics or heritability as well as by the environment. One other study, of adults, had examined the question of whether prosocial behaviors, specifically altruism, empathy, and nurturance, were heritable. In that large study of adult twins, participants responded to questionnaires about their own behavior and the results supported the hypothesis that altruistic behaviors had a substantial genetic influence. However, genes can influence behavior differently over time because they turn on and off during development. Thus, finding that prosocial behavior is heritable in adulthood might not necessarily mean that genes would also influence empathy in the first 3 years of life.

In the MALTS, over 400 families with infant twins were recruited to participate when the children were 14 months old. The study intended to follow children over the next 2 years of their lives, but has been extended to follow them into adolescence. Using the simulation of injury methods, a maternal interview about child empathy behaviors and her responses to children in distress contexts, as well as the infant-cry stimuli described earlier. The study investigated empathy at 14, 20, 24, and 36 months of age (and later, at ages 5 and 7 years). Children's responses to mother and examiner simulations of distress were videotaped during home and laboratory visits as were their responses to the infant cry played in the laboratory for a total of five responses at each age. The study found that heritability was a moderate influence on both prosocial behaviors and hypothesis testing from 14 months of age onward, and that concern was weakly influenced by heritability. Callous/indifferent responses to the victim's distress showed a pattern of weak but significant genetic influence at 14 and 36 months of age but not at 20 and 24 months. At the latter ages, shared environmental influences such as those described below were significant. The investigators created a composite, overall empathy from the individual behaviors and found that genetic influence began at 14 months and contributed to continuity in children's empathy responses over time; additional genetic influences were added at 20 months but not later. The role of the common environment, that is, family practices that are shared or in common to both twins, however, was strongest at 20 months and contributed to continuity in empathy behavior through age 36 months. This may mean that in early childhood, all the genetic contributions appear during the early developmental stages (a sort of genetic launching pad) and that family practices common to both twins, such as observing dad respond to mom when she is in distress, are strongest in the middle of the second year of life. This is when children are in between the quasiegocentric empathy and more veridical empathy stages described by Hoffman. Such differences in the timing of genetic and environmental influences provide some support for empathy developing in a stage-like way rather than continuously. However, our knowledge base about the timing of genetic influences during infancy and toddlerhood is limited to this one study and future studies may certainly alter these ideas about heritable influences on development in this age period.

Environmental influences. Socialization within the family is the starting point for thinking about influences on empathy and prosocial behaviors. Socializing a child is the process of inculcating in the child the values and behaviors of the family; the socialization of empathy, in part, involves the sharing of a value system. Thus, parents' beliefs and attitudes about the importance of responding empathically are potential influences to consider. Recent research by Robinson and colleagues has found that maternal preconceptions about the importance of empathic caregiving behaviors, that is, her attitudes and beliefs 'before' the birth of her children, exert an import influence on her own caregiving behavior as well as the child's later observed empathy responses when she simulates distress. Thus, how we think and behave as teenagers or young adults (which in turn are genetically as well as environmentally influenced) sets the stage for how we may model and instruct our own children to respond when others are distressed.

Several studies have shown that maternal warmth and sensitivity are associated with more empathic responses in their toddlers. In addition, children who are securely attached have been observed to have stronger empathy responses than insecurely attached children. Recent findings showed that qualities of the relationship, including the child's responsiveness and eagerness to involve mother in play, contributed to the child's response to the simulated distress, not only by mother but also by the unfamiliar examiner. The presumed mechanisms of how sensitive caregiving affects children's empathy are through social learning mechanisms, that is, observing and imitating a sensitive caregiver and through internalizing qualities of the caregiving relationship. To the extent that the child internalizes such qualities as the felt wellbeing from being cared for and the responsiveness that is offered by the empathic caregiver, beyond what the caregiver specifically does, the child will also show greater empathy in response to the distress of others.

Direct instruction on how to behave when family members or friends are distressed (e.g., telling the child to hug an injured sibling or to give back a toy that was taken without asking) is also an important, although less well-studied influence in very early childhood. Zahn-Waxler's early study of naturalistic empathy events documented by mothers also included mothers' responses to distress events. When mothers directly instructed toddlers about prosocial action, children were found to show more prosocial behavior during the simulation events. In addition, mothers' explanations of the harmful consequences of the toddlers' aggressive behavior and the internal distress state of the injured person ('Look, you made him sad!'), coupled with firm discipline and high expectations for mature interpersonal behavior together contributed to children showing greater empathy at a later age. Thus, reactions during aggressive events that promote understanding of others' feelings, and encourage prosocial behavior were effective in promoting empathy development.

Several other researchers have found similar positive associations between parental induction of prosocial and controlled emotional responding with observed and self-reported empathy among older children. These studies of older children also found that when parents used threats of punishment and other power assertion techniques, children were less likely to respond prosocially. All of the existing studies, however, are based on parent report of their practices, not observation. Given the dearth of information about socializing empathic responses and prosocial behavior in toddlers, future studies of direct instruction and other socialization techniques among families of diverse socioeconomic and cultural backgrounds would greatly enrich our knowledge on this topic.

Finally, sibling relationships within the family are a potential influence on children's empathy development during the first 3 years of life. Following an in-depth, descriptive study of a small number of siblings, Judy Dunn and colleagues have conducted numerous studies of how sibling relationships influence development. Their research provided rich insights into how the sibling relationship provides opportunities for modeling prosocial action and cooperation. Both being an older sibling to a younger child as well as being the younger sibling, provide benefits for learning about prosocial behavior. In one study, they recruited British families who had siblings where one was a toddler; the siblings were observed at home 6 months apart. They found that the larger the age gap between the siblings, the more likely it was that older siblings would act prosocially. For younger siblings, the age gap did not matter but when their older sibling was more prosocial toward them, they were more likely to be prosocial 6 months later. A cycle of positive reinforcement of their relationship was also found because when the younger sibling was earlier more prosocial toward the older sibling, the older sibling was later observed to be more prosocial toward the younger. Interestingly, higher observed conflict between siblings did not predict lower prosocial behavior at the later observation. Although not frequently studied together, empathy and agonistic/aggressive behavior in the early years are not always inversely related. That is, toddlers who are more aggressive are not necessarily less empathic. In some studies, there is a positive correlation between these two behaviors and in others there is no correlation. However, among older children the inverse relationship between empathy and aggressive behavior is consistently observed.

Children's capabilities. We have already alluded to one temperament ability, being able to sustain attention toward outer events, as a capability that supports the child's emotion regulation when others become distressed. Three other aspects of early temperament that describe different aspects of the child's emotion regulation have also been investigated: behavioral inhibition/fearfulness, sociability and positive engagement, and difficult temperament style (i.e., negatively reactive and difficult to soothe). Children who are more inhibited, shy, or fearful show less empathic concern toward strangers than less inhibited children; differences in their responses toward mother have not been found. Studies have also found that more sociable and positively engaging children show greater empathy across a range of ages, including the toddler years. These associations underscore the interpersonal or social nature of empathy responses. That is, if children are temperamentally reserved or shy we are less likely to observe the kind of interpersonal displays to someone unfamiliar that convey concern or help when they are in distress. More commonly, these responses are observed among children who are generally more sociable and outgoing. However, temperament does not affect children's responses to mother. These reserved responses are quite different from uncaring responses that are also observed; the reserved child is attending and quite aroused during the distress event while the uncaring child responds with little arousal or attention. Studies that include measures of heart rate or respiration rate may be help to distinguish between a reserved–uncaring response (where minimal heart or respiration changes might be observed) from a reserved–highly aroused response (where increased heart rate and respiration might be seen).

Two groups of children appear prone to uncaring/indifferent responses to others' distress; infants who have higher difficult-temperament scores and infants who are underaroused during an emotion challenge. Difficult-temperament infants were less likely to display empathy or act prosocially as toddlers. This may be because children become labeled 'difficult' in part because of their general difficulty in regulating emotion during challenging or emotionally charged situations. We do not know if they are more aroused by the distress of another than more 'easy-going' temperaments but this seems likely to be the case. The underaroused infants, on the other hand, may represent quite a different temperamental type with a long-term inclination toward low response to emotional expressions of others.

Cognitive and language abilities may also influence early empathy development and there is some evidence from Robinson and colleagues that in early development more advanced cognitive and language abilities are associated with more empathic responses toward mother when she simulates distress. It will be interesting to see whether cognitive abilities continue to influence empathy development beyond the early years; it may be that once the core abilities of person permanence and perspective-taking emerge that variation in cognitive ability does not have a significant role. As suggested earlier, cognitive and language skills are in part heritable and in part influenced by the early caregiving environment.

Research from several groups support Hoffman's ideas about the influence of self-recognition and perspective-taking abilities. Between 18 and 30 months, toddlers who were able to recognize themselves in a mirror were more likely to have higher empathy responses.

Imitation has also been implicated as one source of influence in the earliest phases of empathy development because the feeling of empathic distress may arise as a result of the infant's general inclination to imitate behaviors in caregivers. However, as we will see in the next section on disorders of empathy, not all children are able to imitate emotional expressions, and so it may be that children who show little empathic distress have low motivation to imitate others. An investigation by Kochanska and colleagues that examined typically developing toddlers' inclination to imitate mothers during a teaching task also studied children's empathic response to mother's simulated distress but no association was found between these two behaviors at age 22 months. This study did find that empathy was greater among children who had been more responsive to mother in infancy, although imitation *per se* was not studied in infancy. This may be a fruitful line of future inquiry on child capabilities that support early empathy development.

Disorders of Empathy

There are two principal disorders that involve deficits of empathy: the autism spectrum disorders (ASDs) and conduct disorder. Although children are not diagnosed with a conduct disorder in the first 3 years of life, in the past 10 years, children with ASDs are more frequently being diagnosed between 18 months and 3 years of age. This has largely been a result of the field of developmental disorders more specifically delineating the core deficits of autism and the development of reliable and valid diagnostic tools.

Social communication is considered to be the overarching core deficit among children diagnosed with ASDs. ASD is a developmental disability where verbal and nonverbal communication and social interaction are significantly impaired. The impairments are generally evident before age 3 years. It is called a spectrum disorder because children can range in how broadly or significantly they are functionally impaired. However, in the milder forms of ASD, Asperger's syndrome or high functioning autism, the child's social communication deficits are still substantial and they differ from autism disorder primarily in the degree of general cognitive and language impairment. Indeed, children with Asperger's syndrome have normal or above-normal intelligence and frequently have special talents that early interventionists capitalize

on in their work with children with these diagnoses. Although children who are less functionally impaired by ASD may have normal vocabulary and sentence construction, their social use of language (called praxis) is noticeably different from typically developing children. They usually do not have the musicality (called prosody) in their speech that communicates the feeling aspect of our intentions. ASD is highly heritable, and it is common for other members of the same family to show the milder autism spectrum symptoms.

All children diagnosed with ASD have early appearing deficits in social interaction and communication, including deficits in the ability to differentiate emotional facial expressions, basic imitation abilities, and empathy. Simon Baron-Cohen and colleagues have authored several papers on the symptoms and early diagnosis of ASD. In a study of toddlers diagnosed with autism disorder, these investigators matched them with two comparison groups. One comparison was with children with cognitive delays, but no ASD, and the second group was a typically developing group. All children were 20 months old and the groups of children with autism disorder and cognitive delays had a similar mental age (i.e., their general cognitive abilities were about the same). An examiner simulated distress (using shortened distress periods than the standard described above) and invited the children to imitate four different motor actions. All of the typically developing children and developmentally delayed children looked at the examiner's face while only 40% of the children with autism disorder did so. And, while 68% of the typically developing children and 44% of the developmentally delayed children showed empathic concern, none of the children with autism disorder did so. In another study of children under 3 years of age with ASD, facial/vocal imitation tasks as well as object/action imitation tasks were impaired among children with ASD compared to similarly created comparison groups. In addition, the degree of imitation impairment correlated significantly with the degree of severity of autism symptoms.

There are two commonly used tools for identifying children at risk for ASD: the autism diagnostic observation schedule (ADOS) and the autism diagnostic interview-revised (ADI-R). A trained examiner working directly with the young child (e.g., by presenting standardized imitation and communication opportunities) gathers information for the ADOS, while the ADI-R is an interview conducted with the child's caregiver when the child is at least 18 months of age. Both tools allow a clinician to determine whether the child's disabilities meet criteria for diagnosis on the autism spectrum.

Children who are diagnosed (usually after age 3 years), with conduct disorder show a persistent pattern of aggressive and nonaggressive acts that threaten to harm others, theft and deceitfulness, and serious social rule violations. Children may begin to show some of these behaviors between ages 2 and 3 years, and usually have impairments of empathy and guilt. Although older children diagnosed with conduct disorder do not typically show the same type of social communication difficulties as children diagnosed with ASD, they do express great difficulty in feeling empathy and often misread or misunderstand the emotion cues and intentions of others. In one study of preschool-age children who were at risk for conduct disorder, empathy responses did not differ as a function of level of problem behavior. However, severity of conduct symptoms did correlate with their empathy responses by the time the children were 7 years old.

Conduct disorder is also a heritable disorder, although environmental circumstances such as maltreatment, parental alcoholism, or severe marital discord may trigger it in individuals who are genetically vulnerable. These adverse life events may particularly contribute to a family climate of low empathy and parenting practices that are more insensitive to the young child, eliminating one potential protective influence for children who are genetically vulnerable, a warm and predictable family environment. Conduct disorder is much more commonly diagnosed among males than females. Children with early-appearing symptoms of conduct disorder are at risk for lifetime persistent conduct problems, juvenile delinquency, and adult criminality. Effective treatments for children diagnosed with conduct disorder include functional family therapy, multisystemic therapy, and cognitive behavioral approaches that focus on socioemotional skills and anger management.

It is also worth underscoring that children from maltreating environments have been found in several studies to show lower empathy toward their peers. These studies suggest that 'disordered environments' such as those where caregivers physically and emotionally harm children can also mute the developing empathy responses of the children. However, not all children who have been maltreated are uncaring and being placed in the care of empathic, sensitive caregivers is likely to contribute to empathy and prosocial behavior developing in the children over time.

One final note on disorders associated with empathy is that Zahn-Waxler and colleagues have hypothesized that the experience of too much empathy along with excessive feelings of responsibility and guilt in early childhood can predispose sensitive females, in particular, to later depression. In this theory, precociously appearing empathy in early childhood, while not maladaptive, may signal that the individual in the longer term may be vulnerable to feelings of over-responsibility and depression during childhood. More prospective research is needed to address this topic as well as the more general topic of how genetic predispositions and environmental circumstances affect specific children's empathy development adversely.

Conclusions

The roots of empathy development are in the young child's family of origin and the child's constitution. Our species has been adapted to a prolonged period of immaturity that has also necessitated an extended period of parental care. Both biological preparedness (through the tendency to automatically feel distressed in response to the distress of others) and an extended training period in other-oriented empathy responding within the family and among peers contribute to the successful continuation of this cycle. We must appreciate that both of these adaptations are necessary for broad segments of the population to develop mature empathy. Ensuring young children's parents are caring and available is one of the important roles of societies. Some industrial societies do this by encouraging work and family responsibilities shared between mothers and fathers within the culture and by offering supportive policies that provide for paid family leave from work for new parents and the development of high-quality, affordable childcare. High-quality childcare, in particular, helps families to feel that their values of responsive and empathic care for their children are not disrupted when both parents work.

An important goal of future research in early empathy development is investigating how diverse families support their young children to express empathy and prosocial behavior. We must better understand the goals of families in diverse cultures and the socialization strategies that they value and utilize during the first 3 years of life to promote prosocial development. Providing parents and other caregivers to notice the early developing empathic responses of their toddlers is an important step toward supporting their child's prosocial development.

See also: Anger and Aggression; Attachment; Autism Spectrum Disorders; Emotion Regulation; Siblings and Sibling Rivalry; Social and Emotional Development Theories; Social Interaction; Socialization in Infancy and Childhood; Temperament; Theory of Mind.

Suggested Readings

Charmon T, Swettenham J, and Baron-Cohen S (1997) Infants with autism: An investigation of empathy, pretend play, joint attention, and imitation. *Developmental Psychology* 33: 781–789.

Dunn J and Munn P (1986) Siblings and the development of prosocial behavior. *International Journal of Behavioral Development* 9: 265–284.

Emde RN and Hewitt J (2001) *Infancy to Early Childhood: Genetic and Environmental Influences on Developmental Change.* New York: Oxford University Press.

Kiang L, Moreno AJ, and Robinson JL (2004) Maternal preconceptions about parenting predict child temperament, maternal sensitivity, and children's empathy. *Developmental Psychology* 40: 1081–1092.

Kochanska G, Forman DR, and Coy KC (1999) Implications of the mother–child relationship in infancy for socialization in the second year of life. *Infant Behavior and Development* 22(2): 249–265.
Robinson JL, Zahn-Waxler C, and Emde RN (1994) Patterns of development in early empathic behavior: Environmental and child constitutional influences. *Social Development* 3: 125–145.

Zahn-Waxler C, Radke-Yarrow M, and King RA (1979) Child-rearing and children's prosocial initiations toward victims of distress. *Child Development* 50: 319–330.
Zahn-Waxler C, Radke-Yarrow M, Wagner E, and Chapman M (1990) The development of concern for others. *Child Development* 63: 126–136.

Endocrine System

S E Watamura, University of Denver, Denver, CO, USA

Glossary

Catecholamines – Any of a group of organic compounds (amines derived from catechol) that have important physiological effects as neurotransmitters and hormones and include epinephrine, norepinephrine, and dopamine.
Endogenous – Originating or produced within an organism, tissue, or cell.
Exogenous – Derived or developed from outside the body; originating externally.
Glucocorticoids – Any of a group of steroid hormones, such as cortisol, that are produced by the adrenal cortex; are involved in carbohydrate, protein, and fat metabolism; and have anti-inflammatory properties.
Hormone – Chemical signals secreted into the bloodstream that have effects on distant tissues.
Precursor – A biochemical substance, such as an intermediate compound in a chain of enzymatic reactions, from which a more stable or definitive product is formed.

Introduction

The endocrine system is one of two main systems in the body responsible for communication and regulation of body functions, coordinating functions such as growth, metabolism, and reproduction. The other main communication and regulation system is the nervous system. The endocrine system utilizes hormones for communication and regulation, and these are typically released into the blood. In contrast, the nervous system utilizes primarily cell-to-cell synaptic communication. While divisions are helpful for classification and specification, it is important to bear in mind the multitude of bidirectional effects between all the major systems of the body, including those between the endocrine system and the nervous system.

Hormones are chemical messengers derived from the major classes of compounds used by the body more generally, such as proteins and lipids. Endocrine coordination via the release of hormones into the blood stream presents unique problems of production, distribution, and mechanisms for action. Receptors on cell surfaces and in the nuclei of cells play a critical role in the endocrine system. Without the active participation of receptors, hormones would be incapable of executing their wide-ranging effects. Another critical aspect of the endocrine system to be discussed is regulation of hormone action via feedback loops.

Over- or under-production of hormones, autoimmune diseases, and genetic mutations can lead to a variety of endocrine disorders, some of which are life-threatening if left untreated. Diseases to be discussed below are those that are most likely to influence infant and child development, including diabetes, growth disorders, thyroid disease, adrenal insufficiency, and the rare disorder, Cushing's syndrome.

Endocrine functions also change across the lifetime, so that unique endocrine processes and profiles are found during pregnancy, in the fetus, during puberty, and in aging. Those changes that accompany pregnancy and those that are important during the fetal period will be discussed.

Endocrine System Components

Glands

A gland is a group of cells that produces and secretes chemicals. Using precursors in the blood or cell, glands create completed chemical products and release them in the body. Some glands release chemicals that act locally, for example two exocrine glands are the sweat and salivary glands that release secretions in the skin or inside the mouth. The primary glands of the endocrine system are the pituitary gland, the pineal gland, the adrenal gland, the thyroid and parathyroid glands, the pancreas, and the reproductive glands (ovaries and testes) (see **Figure** 1). Other nonendocrine organs, including the brain, heart, lungs, kidneys, liver, thymus, skin, and placenta, are also capable of producing hormones.

The hypothalamus is a region of the brain connected to the pituitary gland with short capillaries. It translates signals from the nervous system into secreted hormones that activate the pituitary to produce or inhibit specific hormones in response to the hypothalamic signal. Signals that are translated by the hypothalamus into hormones that stimulate or inhibit the pituitary are the result of both external conditions like temperature or light, and internal conditions, like fear, that have been processed by other areas of the brain and communicated to the hypothalamus by neuronal signaling. Thus, the hypothalamus serves as a critical link between the nervous system and the endocrine system.

The pituitary (also called the hypophysis) is sometimes referred to as the master gland, because it regulates the hormonal production of many other glands. It is located at the base of the brain just beneath the hypothalamus. It has two main sections, called the anterior pituitary (or adenohypophysis) and the posterior pituitary (or neurohypophysis). The anterior pituitary produces hormones that affect the thyroid, adrenal, and reproductive glands. The posterior pituitary produces antidiuretic hormone, which helps control water balance and oxytocin. The pituitary also produces endorphins, which act on the nervous system to reduce sensitivity to pain.

The two parts of the pituitary arise from very different kinds of embryonic tissue, and receive inputs from the hypothalamus in very different ways. The anterior pituitary is a classic gland in the sense that it is composed primarily of cells that secrete hormones. The anterior pituitary receives hormonal signals from the hypothalamus via a direct blood supply. The hypophyseal artery branches into a capillary bed in the lower hypothalamus, where hypothalamic hormones are secreted. This capillary bed drains into the hypothalamic-hypophyseal portal veins, which further branch into capillaries within the

anterior pituitary. Hypothalamic hormones travel the short distance from the hypothalamus to the anterior pituitary via this unique vasculature. This vasculature allows very small amounts of hypothalamic hormones to quickly and effectively signal the pituitary. In contrast, the posterior pituitary is more like an extension of the hypothalamus in that it is composed largely of the axons of hypothalamic neurons.

The pineal gland is a tiny gland located in the middle of the brain. The pineal gland produces melatonin during the dark phase of the day/night cycle. In mammals, the pineal receives signals from the suprachiasmatic nucleus (SCN) which maintains the biologic clock of the organism. The clock follows a roughly 24 h rhythm even in a continuously dark environment, however, via signals from the eye, the clock is entrained to the external day/night cycle. The pineal begins producing melatonin around 3 months after birth, and the pineal is larger in children, decreasing in size and in melatonin production after puberty.

The thyroid gland is a large gland located in the front part of the lower neck and is shaped like a bowtie or butterfly. The primary function of the thyroid is to regulate metabolism through the production of thyroid hormone. Attached to the thyroid are four glands which function together and are called the parathyroid glands. The parathyroid glands produce parathyroid hormone which, in conjunction with calcitonin released by the thyroid, regulates the level of calcium in the blood.

The adrenal glands are triangular glands, one located on top of each kidney. The adrenal glands have two anatomically and functionally distinct parts, the outer part called the adrenal cortex and the inner part called the adrenal medulla. The adrenal cortex produces glucocorticoids and mineralcorticoids. These hormones help the body regulate salt and water balance, respond to stress, participate in metabolism, regulate the immune system, and are involved in sexual development and function. The adrenal medulla produces the catecholamines epinephrine (adrenalin) and norepinephrine (noradrenalin) involved in the first line response to stress.

The pancreas produces the hormones insulin and glucagon which act together to maintain homeostasis in blood glucose levels. If blood glucose levels rise above the set-point, insulin production increases to drive glucose into cells, which decreases blood glucose, and if blood glucose levels fall below the set-point, glucagon production increases to increase blood glucose.

The reproductive glands, or gonads, are the ovaries in females and the testes in males. The ovaries are located in the pelvis and produce ova as well as release the hormones estrogen and progesterone. Both hormones are involved in healthy female sexual development, menstruation and pregnancy. The testes are located in the scrotum and produce androgens critical for healthy male sexual development, sexual behavior, and sperm production.

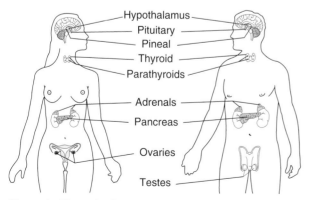

Figure 1 The endocrine system.

It is beyond the scope of this article to discuss the intricacies of each gland and the vast array of hormones produced. The focus here has been on the main glands and the best-characterized and most critical hormones. However, this necessitates leaving out the less well-characterized and more auxiliary hormones. For example, the glucocorticoid cortisol and the mineralcorticoids are mentioned; however, the adrenal cortex alone produces over 50 known steroid hormones.

Primary Endocrine Hormones

Hormones are chemical signals secreted into the blood stream that have effects on distant tissues. Hormones do not share a chemical structure and are not defined by their molecular components; rather they are defined by their actions. Hormones are produced primarily by the endocrine glands. Many hormones are produced in cascades, beginning with a signal from the hypothalamus that stimulates the pituitary to produce a hormone that then stimulates another gland (e.g., the thyroid gland, adrenal gland, or the reproductive glands) to produce a third hormone (see **Table 1** for an overview of some of the main hormones and their representative effects). Common hypothalamic hormones include corticotropin-releasing factor (CRF), gonadotropin-releasing hormone (GnRH), prolactin-releasing factor, prolactin-release inhibiting factor, growth hormone-releasing factor, somatostatin

(growth hormone release inhibiting factor), and thyrotropin-releasing factor. As can be seen from their names, these hormones typically stimulate or inhibit the release of hormones by other endocrine glands.

Common pituitary hormones include oxytocin, vasopressin, adrenocorticotropin (ACTH), lipotropin (LPH), thyroid-stimulating hormone (TSH), growth hormone (GH), prolactin (PRL), luteinizing hormone (LH), and follicle-stimulating hormone (FSH). Some hormones produced by the pituitary activate tertiary hormones (e.g., ACTH activates the adrenal glands to produce cortisol), and others have primarily direct effects on target tissues, for example, oxytocin. The pineal gland produces melatonin, which is critically involved in biologic rhythms like sleep/wake behavior. The adrenal gland produces glucocorticoids and mineralcorticoids in the adrenal cortex, and epinephrine (adrenalin) and norepinephrine (noradrenalin) in the adrenal medulla. Many of these adrenal hormones are involved in the body's response to physical or psychological stress. The thyroid gland produces thyroxine (T4), which is critically involved in metabolic processes, and the parathyroid glands produce parathyroid hormone which raises blood calcium levels. The pancreas produces insulin, which works as described above to lower blood glucose levels and glucagons, which raise blood glucose levels. The reproductive glands produce estrogens, progestins, and androgens, all of which support sexual development and reproductive function.

Table 1   The major hormones of the endocrine system

Hormone	Released by	Example effects
Corticotropin-releasing factor (CRF)	Hypothalamus	Stimulates the pituitary to produce ACTH
Gonadotropin-releasing hormone (GnRH)	Hypothalamus	Stimulates the pituitary to produce FSH and LH
Growth hormone-releasing factor	Hypothalamus	Stimulates the pituitary to produce GH
Somatostatin	Hypothalamus	Stimulates the pituitary to inhibit GH and TSH
Thyrotropin-releasing factor	Hypothalamus	Stimulates the pituitary to produce TSH
Adrenocorticotropic hormone (ACTH)	Pituitary gland	Stimulates adrenal cortex to secrete glucocorticoids
Follicle-stimulating hormone (FSH)	Pituitary gland	Stimulates sex cell development (ova and sperm)
Growth hormone (GH)	Pituitary gland	Stimulates growth and metabolism
Luteinizing hormone (LH)	Pituitary gland	Stimulates ovaries and testes
Oxytocin	Pituitary gland	Stimulates uterine and mammary gland cell contractions
Prolactin (PRL)	Pituitary gland	Stimulates milk production and secretion
Thyroid-stimulating hormone (TSH)	Pituitary gland	Stimulates thyroid gland to produce thyroid hormones (T3 and T4)
Triiodothyronine (T_3) and thyroxine (T_4)	Thyroid gland	Stimulate and maintain metabolism
Parathyroid hormone (PTH)	Parathyroid glands	Raises blood calcium
Glucagon	Pancreas	Raises blood glucose
Insulin	Pancreas	Drives glucose into cells
Epinephrine and norepinephrine	Adrenal glands (medulla)	Raise blood glucose; constrict some blood vessels; increase metabolism
Glucocorticoids	Adrenal glands (cortex)	Raise blood glucose; inhibit long-term growth and restorative processes
Androgens	Testes	Support sperm formation and male secondary sex characteristics
Estrogens	Ovaries	Stimulate uterine lining growth and female secondary sex characteristics
Progesterone	Gonads	Promotes uterine lining growth
Melatonin	Pineal gland	Regulates biological rhythms

As mentioned earlier, many other organs not classified as endocrine glands also produce hormones. For example, gastrin, secretin, motilin, vasoactive intestinal peptide (VIP), somatostatin, and substance P are produced in the gastrointestinal tract; the placenta produces estrogens, progestins, and relaxin; the liver produces angiotensin II; the kidney produces calcitriol; and the heart produces atrial natriuretic peptide (ANP). While hormones are critical for health, and are capable of inducing wide-ranging effects, they are completely ineffective without appropriate receptors on target cells.

Hormone Secretion, Receptors, and Regulation

Circulating hormone levels are controlled both by the secretion pattern of glands and by binding and clearance rates. Some glands secrete hormones in short distinct pulses in response to discreet signals, for example, the secretion of insulin in response to blood glucose levels. Other hormones are secreted in rhythms coordinated by external cycles, for example, melatonin in response to day/night cycles and estrogens in response to 28-day lunar cycles. Still other hormones appear to be secreted continuously, for example, prolactin. Many hormones also have both baseline and stimulated secretion patterns, for example, cortisol is produced in basal levels following day/night cycles, and is also produced in response to stress or challenge. In the nervous system, neurons accumulate inhibitory and excitatory signals from other cells that together determine whether the particular neuron will 'fire', thus producing its own signals. Similarly, endocrine cells accumulate signals from hypothalamic hormones, peripheral hormones, nutrients, and many other mechanisms, and these accumulated signals determine the pattern and quantity of hormone secretion. This allows for subtle changes in hormone secretion in response to dynamic changes in the individual's external and internal environment.

Receptors for hormones are typically classified as either cell-surface receptors or nuclear receptors, with nuclear receptors found in the nuclei of cells. Because hormones are released into the bloodstream and are circulated all through the body, receptors are a critical mechanism of control for when, where, and how hormones exert their influence. Receptors determine which cells are affected by which hormones, the timing and degree of effect, and the nature of the effect.

Regulation occurs through at least three mechanisms: the release and binding of hormones to regulate circulating levels; the location, number, and nature of receptors; and the feedback loops elicited by hormones and their receptors that control subsequent release. Feedback loops occur when the end-stage hormonal product of a particular organ acts as an inhibitory or excitatory signal for the hypothalamic and/or pituitary hormone. For example, in the hypothalamic–pituitary–adrenal axis (HPA-axis), the hypothalamus produces CRF, which stimulates the pituitary gland to produce ACTH, which stimulates the adrenals to produce glucocorticoids (cortisol in humans). Receptors for cortisol in the hypothalamus allow cortisol levels to signal the hypothalamus to inhibit production of CRF, thus ending the cycle.

Endocrinology of Pregnancy

Preconception

Both female and male reproductive capacity is initiated and regulated by hormones. In fact, while the sex of the fetus is genetically determined by the X or Y chromosome inherited from the father (XX is female, XY is male), without the appropriate release of hormones during fetal development fetal sexual development is impaired (see the section titled The endocrinology of fetal development). In adult males, GnRH from the hypothalamus signals the pituitary to produce LH and FSH. LH stimulates the testes to produce androgens, primarily testosterone, and FSH promotes the development of sperm. Testosterone has an inhibiting effect on secretion of GnRH by the hypothalamus, completing the negative feedback loop.

In females, GnRH from the hypothalamus also stimulates the pituitary to produce LH and FSH. In females these hormones stimulate the ovaries to produce estrogens, and stimulate the follicle to mature a single oocyte (egg cell). Low levels of LH and FSH during this phase (called the follicular phase of the ovarian cycle) inhibit the pituitary in a negative feedback loop similar to that for males. However, as the follicle matures, estrogen production increases sharply. These high levels of estrogens stimulate the pituitary in a positive feedback loop, and the resulting high levels of LH supports the final maturation of the follicle and ovulation (the rupturing of the follicle and release of the oocyte). As estrogen production increases, the lining of the uterus (called the endometrium) thickens with blood and nutrients that could support an embryo. After ovulation (the luteal phase), LH stimulates the remaining follicular tissue to develop into a glandular structure called the corpus luteum. The corpus luteum responds to LH by producing progesterone and estrogen, keeping the pituitary inhibited with the negative feedback of these hormones. These hormones continue to foster development and maintenance of the endometrium, which begins secreting a nutrient fluid that could support an embryo prior to implantation. At the end of the luteal phase, if there is no implantation of a fertilized ovum, the corpus luteum disintegrates, the levels of progesterone and estrogens drop, and the pituitary hormones rise again, promoting the growth of new follicles, beginning the cycle anew. This same drop in progesterone and estrogens causes spasms of the arteries in the endometrium depriving it of blood, and the endometrium is shed in menstruation.

Conception and Gestation

After fertilization of the egg by the sperm (conception), the resulting zygote begins dividing. This process continues, leading to a ball of cells that travels to the uterus. This ball of cells becomes a fluid-filled sphere, known as a blastocyst, which will implant into the prepared endometrium. After implantation (approximately postconception day 10), the pregnancy is in the embryonic stage. The embryo begins secreting embryonic hormones that alert the mother's body to its presence and that exert control over the mother's reproductive system. For example, the embryo releases human chorionic gonadotropin (HCG), which maintains the secretion of progesterone and estrogens by the corpus luteum through the first few months of fetal development. This prevents disintegration of the corpus luteum, and thus the resulting drop in progesterone and estrogen that would initiate menstruation. HCG can be measured in the mother's urine and is used in many pregnancy tests.

The outer layer of the blastocyst and the endometrium give rise to the placenta, a unique organ that is composed of both fetal and maternal tissue. This organ allows for the exchange of nutrients and waste via the umbilical cord. The high levels of progesterone initiate changes in the mother's body, including increased mucus in the cervix which forms a protective plug, enlarged breasts, and as a result of the negative feedback loop, inhibition of the ovarian and menstrual cycles.

In the second trimester, HCG declines, the corpus luteum deteriorates, and the placenta produces the progesterone necessary to maintain the pregnancy. Interestingly, however, the placenta lacks or has reduced levels of several enzymes important for generating estrogens and progesterone and therefore requires precursors from both the mother and the fetus. This dependence has led to conceptualizing the placenta as part of a maternal–fetal–placental unit.

Estrogens have several actions during pregnancy. They help with the uptake of low-density lipoprotein (LDL) cholesterol that is important for placental steroid production; they increase utero-placental blood flow; they increase endometrial prostaglandin synthesis; and they prepare the breasts for lactation. However, conditions resulting in drastically decreased estrogen production do not result in miscarriage, thus suggesting that estrogens are not essential for maintaining pregnancy.

Parturition

Parturition (or birth) is initiated by a combination of local regulators (prostaglandins) and the hormones estrogen and oxytocin. Estrogen reaches its highest levels in the last weeks of pregnancy and promotes the development of oxytocin receptors on the uterus. Oxytocin is produced both by the fetus and by the mother, and stimulates uterine contractions and the production of prostaglandins that enhance contractions. These hormones are part of a positive feedback loop that encourages the progression of labor.

Endocrinology of Fetal Development

In addition to the role of hormones in maintaining pregnancy and initiating birth, hormones are critically involved in the development of the fetus. Hormones govern sexual development, are critical for growth, and help to prepare the fetus for birth and life outside of the womb.

Nutrients and wastes are exchanged via the placenta to support the fetus; however, the fetal hormonal environment is largely independent of the mother's. This separation is maintained in part by the impermeability of the placenta to many hormones. Hormones that cross the placenta are typically those that are lipid-soluble or smaller molecules and include cortisol, estradiol, thyroid hormones, and catecholamines. The placenta also produces many hormones, including hypothalamic, pituitary, adrenal, and gonadal hormones as well as growth factors and endorphins. The placenta has its own regulatory mechanisms. Placental hormones may modulate fetal and maternal processes, and in turn maternal and fetal hormones may regulate placental hormone production.

While many of the hormones in the fetal environment are hormones that are active later in life, some operate primarily in the fetal period. Hormones are also often present in different levels in the fetus and may be converted to inactive forms more readily than in the mother. Clearance rates also may differ. For example, in the fetus cortisol, estradiol, and thyroid hormone are quickly converted to inactive forms (cortisone, estrone, and rT3, respectively) that do not have the same effects that they would have later in life. In addition, the complexity of hypothalamic–pituitary control of hormone production requires the cortex, midbrain, hypothalamus, vasculature, and peripheral systems to have matured. While these structures and systems are maturing, circulating hormone levels may differ due to inadequate maturation. Increased conversion to inactive forms may protect the fetus from high circulating hormone levels while these regulation and clearance mechanisms mature.

As the fetus matures, changes in hormone levels support growth and development. For example, TSH and T4 increase over the fetal period and support bone and central nervous system (CNS) maturation, including neurogenesis, migration, differentiation, dendritic and axonal growth, synaptogenesis, and myelination. The fetal endocrine system is also responsive to stress, for example, in response to insufficient oxygen. Fetal catecholamines are the primary stress hormones in the fetus. Catecholamines are critical for fetal cardiovascular function and survival.

Some hormones, though present, are relatively inactive. For example, insulin and glucagon levels are both high, but secretion, clearance, and responsivity to glucose are all impaired. This may be due to the relative stability of glucose levels maintained by the placenta, because both pre-term and full-term infants quickly utilize glucagon and insulin to regulate blood glucose levels after birth. Insulin and insulin-like growth factors (IGFs) are, however, critical for normal growth in the fetal period.

In week 7 of gestation, the gonads of both male and female fetuses begin to differentiate. In males, androgens initiate the development of male primary sex organs, and later in the fetal period androgen release in the male fetus inhibits the cyclicity of the pituitary gland. In female fetuses, or in male fetuses without sufficient androgens, the female primary sex organs will develop, and the cyclicity of the pituitary will be preserved to later give rise to the menstrual cycle.

The endocrine system is also critically involved in preparing the fetus for birth and life outside of the protected uterine environment. The fetus must suddenly initiate breathing, defend against hypothermia, and prevent hypoglycemia and hypocalcemia. A cortisol surge near term supports this transition, and is achieved both by increased fetal cortisol production and decreased conversion of cortisol to cortisone. The surge supports surfactant synthesis in the lung which is critical for breathing, increases liquid reabsorption in the lung, supports duct closure, and stimulates small intestine and liver processes. Mice deficient in CRF or glucocorticoid (GC) receptors die within the first 12 h of life. Catecholamines also surge dramatically and support increased blood pressure, increased glucagon, decreased insulin, lung adaptation, and heat generation in fat tissue.

In humans and other mammals, the decreasing levels of progesterone in the mother following birth release the inhibition of the pituitary and allow the production of prolactin. Prolactin then stimulates milk production 2–3 days after birth, and oxytocin controls the release of milk from the mammary glands. Oxytocin release is regulated at least in part by the hypothalamic hormone α-melanocyte-stimulating hormone.

Endocrine System and Normal Growth

The endocrine system is critically involved in normal growth. GH, IGFs, and receptors are all involved in normal growth, and disruptions can result in abnormal growth (see the section titled Endocrine disorders relevant for infants and children).

Mean levels of GH decrease across childhood from high values in the neonatal period to lower values at early puberty. Levels then increase during mid-to-late puberty, and decrease again from late puberty across the

lifespan. Maximum GH secretion occurs at night, especially during the onset of the first slow-wave sleep. Sleep, nutrition, fasting, exercise, stress, obesity, estrogen, and testosterone all affect GH secretion.

IGFs are a family of peptides that are partially GH dependent, and that mediate many of the actions of GH. IGF-I is important for both fetal and postnatal growth and is involved in normal fertility. Levels of IGF-I are low in newborns, rise during childhood, and attain adult levels at the onset of puberty. Levels then rise to two to three times the adult levels. After 20–30 years of age, levels begin an age-associated decline. IGF-II is a major fetal growth factor and is involved in placental growth. IGF-II levels are also lower in the newborn, but adult levels are reached by 1 year and there is little-to-no decline across the lifespan. IGF-II may be involved in inhibiting excessive growth.

Androgens and estrogens do not significantly contribute to normal growth before puberty; however, during puberty they enhance GH secretion and stimulate IGF-I production. Growth during puberty is both growth in height and in bone mineral density. Thyroid hormone is also very important for postnatal growth, and untreated hypothyroidism in the infant can lead to profound growth failure and a lack of skeletal maturation (see section titled Thyroid disease). Thyroid hormone appears to influence growth by having a permissive affect on GH.

Endocrine System and Stress

Physical and psychological stress is managed in part through the efforts of the endocrine system. When a stressor is perceived by the brain, two stress systems are activated. The first is the norepinephrine sympathetic adrenal medullary system (NE-SAM), and the second is the HPA-axis. The NE-SAM system has immediate results, known as the 'flight or fight' response. This includes increased heart rate, increased pupil dilation, and increased respiration. These effects are mediated by epinephrine and norepinephrine, both produced by the adrenal glands. Epinephrine and norepinephrine are also neurotransmitters produced by nonendocrine cells in the brain where they have a number of important effects.

The first response to stress by the NE-SAM system is metabolically costly, burning through reserve glucose stores. Thus, one of the main goals of the stress systems is to generate energy. The second stress system, the HPA-axis helps to generate energy. Once activated by a stressor, the HPA-axis works to replenish the glucose stores that were quickly depleted by the NE-SAM system in preparation for a response to a second stressor or to a prolonged response to the initial stressor. All three components of the HPA-axis are part of the endocrine system; however, it is the glucocorticoid hormone released by the adrenal gland which has effects throughout the body. This hormone is cortisol in humans and corticosterone in non-human animals.

Endocrine Disorders Relevant for Infants and Children

Endocrine disorders can be the result of a variety of genetic mutations affecting the endocrine glands, hormones, and receptors. Endocrine disorders can also be the result of tumors, both malignant and benign, and can result from autoimmune destruction of endocrine glands or tissues. An important endocrine disorder is congenital hypothyroidism (also known as cretinism), resulting from iodine deficiency in the fetal, infant, or early childhood periods. Many endocrine disorders can be treated by administration of exogenous hormones, and hormones are also used to treat nonendocrine disorders, for example, steroid hormones are widely used to treat inflammation or to suppress immune response.

Diabetes

Diabetes is characterized by insufficient insulin production, insulin resistance, or both. Treatment may involve insulin replacement through intramuscular injection or oral medication to support the effective utilization of insulin by the body. Insulin-dependent diabetes (sometimes also called type 1 or juvenile diabetes) results from a genetic predisposition to the destruction of insulin-producing cells of the pancreas by the immune system. Noninsulin-dependent diabetes (also known as type 2 diabetes) results from a change in the ability of the body to utilize insulin and is often associated with obesity

Type 1
Type 1 diabetes typically does not affect infants and young children; however, it usually emerges prior to adulthood. While the genetic predisposition is present from conception, the autoimmune destruction of pancreatic insulin-producing cells occurs later in childhood. In children who develop diabetes prior to adolescence, it can be particularly difficult to maintain proper treatment as lifestyle changes may make regular meals and insulin injections difficult. However, failing to keep the glucose/insulin balance that the body intricately maintains in the absence of diabetes puts the individual at risk for a wide range of life-threatening illnesses such as cardiovascular disease and kidney failure.

Gestational
During pregnancy, the high glucose demands of the fetus are supported in part by increased insulin resistance. In gestational diabetes, the mother's body becomes unable to maintain a normal blood glucose level during pregnancy. Gestational diabetes is usually not followed by type 1 or type 2 diabetes after birth in either the infant or the mother. However, it must be treated during pregnancy and even when treated it does lead to effects on the fetus such as significantly increased birth weight.

Growth Disorders

Growth problems can be classified into four types: primary growth abnormalities as a result of genetic defects; growth problems secondary to chronic diseases or endocrine disorders; growth problems that result from prenatal or postnatal environmental factors; and those of undetermined origin. Primary growth disorders include the osteochondrodysplasias, for example, short-limbed dwarfism, which are genetically transmitted disorders of cartilage, bone, or both, and chromosomal abnormalities such as Down syndrome or Turner's syndrome. In some primary growth disorders, GH treatment can be beneficial, for example, in Turner's syndrome. Chronic diseases that delay growth include malabsorption and gastrointestinal diseases, chronic liver disease, chronic renal disease, chronic anemias, severe pulmonary disease, chronic inflammatory diseases, and HIV. In general, treatment of the primary disease is necessary to restore growth. For diseases such as severe pulmonary disease and chronic inflammation, glucocorticoids are widely used; however, minimizing exposure to glucocorticoids to the minimal dose and frequency necessary to control the underlying disease is important, as glucocorticoids used clinically can delay growth.

Endocrine disorders that involve GH as well as those that do not directly involve GH can result in delayed growth. Those that directly involve GH include central hypothalamic–pituitary dysfunction, failed or reduced pituitary GH production, and GH insensitivity. These symptoms may reflect underlying genetic abnormalities, congenital malformations, head trauma, brain inflammation due to infection, tumors, radiation therapy, or psychosocial maltreatment. Birth size may be normal or near normal, although severe early-onset GH disorder infants may be very small. Delayed growth may begin in the first few postnatal months, and by 6–12 months growth may deviate from the normal growth curve, with proportions remaining relatively normal. Weight-to-height ratios are often increased, musculature is poor, fontanel (skull bone) closure is often slow and facial bone growth may be particularly delayed. Infants with genetic GH disorder are below the 5th percentile in height, and also have documented abnormal growth velocity. As with chronic diseases, the underlying cause of the GH disorder should be treated, however, GH supplements can also be helpful and are generally regarded as safe. Prior to 1985 and the creation of recombinant DNA GH, however, GH was derived from the pituitary of cadavers, was in short supply, and has since been found to be a transmission route for Creutzfeldt–Jakob disease.

Hypothyroidism, which occurs in newborns and can develop in childhood (see section titled Thyroid disease), can retard growth. However, when treated promptly, growth and adult height are normal. Cushing's syndrome, though rare, has more severe effects, resulting in impaired skeletal growth. This impaired growth can only be partially ameliorated with exogenous GH and children may not attain target adult height. Vitamin D deficiency, which can cause rickets in infancy, can also result in delayed growth.

A number of environmental factors are very important for normal growth, the most common being malnutrition. There are two types of malnutrition, marasmus, which results from an overall deficiency in calories, and kwashiorkor, which results from mostly protein deficiency. Children with marasmus have a generally wasted appearance, while those with kwashiorkor typically have a protruding belly and very thin limbs. Stunting of growth due to malnutrition can have life-long consequences. Furthermore, protein and calories are essential for normal brain growth, thus depending on the duration and severity of malnutrition, children may have developmental delays or permanent cognitive deficits as well as an impaired immune system.

During the prenatal period, environmental agents and maternal factors can also affect normal growth. Maternal malnutrition and maternal drug use may be particularly detrimental to prenatal growth. In addition, the most common cause of infants born large for their gestational age is maternal diabetes.

Inorganic failure to thrive or psychosocial dwarfism

Infants or young children who are otherwise normal but are experiencing the extreme stress of severe maltreatment sometimes fail to grow despite a normal diet and no apparent problems with their endocrine system. When these children are relocated to a more positive environment, growth is restored, indicating that the psychosocial stress they experienced temporarily inhibited growth. Both GH and HPA-axis disruptions have been documented in this population, however GH treatment is not usually beneficial until the psychosocial situation is improved.

Precocious Puberty

The average age of puberty has decreased steadily over the last 100 years in industrialized nations and is influenced by genetic and environmental factors. Precocious puberty is premature development of body characteristics that normally occur during puberty. The declining age of the onset of puberty complicates the diagnosis of precocious puberty. Currently, in girls, precocious puberty is when any of the following develop before 8 years of age: breasts, armpit or pubic hair, mature external genitalia, or first menstruation. In boys, precocious puberty is when any of the following develop before 9 years of age: enlarged testes and penis, armpit or pubic hair, or facial hair.

The main causes of precocious puberty are structural abnormalities in the brain and hormone-secreting tumors. Medications can temporarily surpress sexual hormone secretion, however, some tumors require surgical removal. Children (of both sexes) with early sexual development are also more likely to have psychosocial problems. Their early sexual development can result in self-esteem problems, depression, acting out at school and home, and alcohol and illegal substance abuse.

Thyroid Disease

Thyroid disease is relatively common, and includes hyperthyroidism, hypothyroidism, thyroid cancer, and enlargement of the thyroid, or thyroid goiters. Hyperthyroidism results from overproduction of thyroid hormone and includes symptoms of upregulation of metabolism such as feeling hot, losing weight or increased appetite, heart palpitations, trembling hands, nervousness, insomnia, increased bowel movements, and fatigue by the end of the day. Treatment includes both temporary and long-term procedures. Short-term solutions include β-blockers, which temporarily inhibit the effects of increased thyroid hormone while not inhibiting its actual production, and antithyroid medications that temporarily block the ability of the thyroid to make thyroid hormone. More permanent treatments include oral doses of radioactive iodine and surgery. Radioactive iodine treatment is the recommended treatment as iodine is selectively absorbed by the thyroid allowing selective destruction of thyroid cells. However, with both surgical removal of the thyroid and radioactive iodine treatment, the patient must subsequently be treated for hypothyroidism as their thyroid will no longer produce sufficient quantities of thyroid hormone. 'Hypothyroidism' may affect as many as 10% of women and can result from iodine deficiency, treatment of hyperthyroidism as described previously, from temporary inflammation of the thyroid, or as the result of autoimmune destruction of thyroid hormone-producing cells. Common symptoms include weight gain, fatigue, cold intolerance, constipation, and depression. Hypothyroidism can be treated effectively with oral hormone replacement therapy.

Iodine deficiency is a serious problem, with approximately 130 countries and 29% of the world's population living in areas with insufficient iodine in the soil. This occurs most often in mountainous regions, including the Andes, Alps, and Himalayas, and in lowland regions far from oceans such as central Africa and eastern Europe. In some countries, iodine is supplemented in table salt or via the water supply. Where iodine is deficient and not supplemented, thyroid hormone synthesis is impaired. If this occurs in adults, symptoms are as described above for hypothyroidism. If this occurs in the fetus, infant, or young child, it results in the serious condition known as congenital hypothyroidism or cretinism. This condition affects CNS development and maturation and results in permanent mental retardation, neurological defects, and growth abnormalities.

Incomplete thyroid development is the most common defect, occurring in approximately 1 of every 3000 births. It is twice as common in girls as in boys. Typically few symptoms are present, as the deficiency is mild. However, infants with severe hypothyroidism have a distinctive appearance, including a puffy face, dull look, and a large and protruding tongue. Symptoms include poor muscle tone, poor feeding, choking, constipation, prolonged jaundice, and growth delays. If untreated, cognitive and growth delays are severe. However, with early diagnosis and replacement hormone therapy beginning in the first month of life normal intelligence and growth are expected. Most states routinely screen newborns for hypothyroidism to allow early diagnosis.

Thyroid disease during pregnancy

Both hypo- and hyperthyroidism can influence the regularity of a women's menstrual cycle and thus the chance that she may become pregnant. During pregnancy, women with previously undiagnosed hypo- or hyperthyroidism may assume their symptoms are related to the pregnancy. In the case of hypothyroidism, symptoms such as fatigue and weight gain are common effects of both hypothyroidism and pregnancy. Doses of exogenous thyroid hormone typically need to be increased during pregnancy to maintain appropriate blood levels. As thyroid hormone is critical for normal CNS maturation, it is important that hypothyroidism be treated. Maternal hypothyroidism has been associated with a 5–10 point IQ deficit in the child. Hyperthyroidism during pregnancy is the more serious condition, because it is associated with increased risk of birth defects and miscarriage, and this condition is difficult to treat during pregnancy without effects on the thyroid of the fetus. Women of childbearing age with hyperthyroidism may be encouraged to choose a permanent treatment of hyperthyroidism (described previously) prior to becoming pregnant.

Addison's Disease or Adrenal Insufficiency

This condition results from insufficient production of glucocorticoids by the adrenal cortex. Symptoms may include weight lost, weakness, fatigue, low blood pressure,

abdominal pain, nausea, dehydration, and sometimes darkening of the skin.

Primary adrenal insufficiency, which can be acute or chronic, is most often caused by destruction of the adrenal gland as the result of an autoimmune disorder. However, destruction of the gland can also occur in other ways, including tuberculosis (TB) infection. Usually, adrenal insufficiency results when over 90% of the adrenal cortex is destroyed. Primary adrenocortical insufficiency is rare, it can occur at any age, and it is equally common in women and men.

Primary adrenal insufficiency is sometimes also one part of a more widespread endocrine disease called polyendocrine deficiency syndrome. Adrenal insufficiency may be accompanied by underactive parathyroid glands, slow sexual development, diabetes, anemia, chronic candida infections, and chronic active hepatitis. Polyendocrine deficiency syndrome is likely inherited because frequently more than one family member tends to have one or more endocrine deficiencies.

Secondary adrenal insufficiency may be caused by hypothalamic–pituitary disease, or it may result from suppression of the hypothalamic–pituitary axis by exogenous steroids or endogenous steroids (i.e., those produced by tumors). Extensive therapeutic use of steroids has greatly contributed to increased incidence.

Symptoms of adrenal insufficiency usually begin gradually. Adrenal insufficiency is most clinically problematic during times of physiologic stress. In the absence of corticosteroids, stress results in hypotension, shock, and is sometimes fatal. Adrenal insufficiency is treated with oral replacement hormones. Cortisol is replaced orally with hydrocortisone derivatives. If aldosterone is also deficient, it is replaced with oral doses of a mineralocorticoid such as fludrocortisone acetate. Patients receiving aldosterone replacement therapy are usually also advised to increase their salt intake. Because patients with secondary adrenal insufficiency normally maintain aldosterone production, they do not require aldosterone replacement therapy. Individuals with adrenal insufficiency who need surgery or who are nearing childbirth will be given additional injections of hydrocortisone and saline.

Cushing's Syndrome and Cushing's Disease

Cushing's syndrome and Cushing's disease are hormonal disorders caused by prolonged exposure of the body's tissues to high levels of the glucocorticoids. Sometimes called 'hypercortisolism', it is relatively rare and most commonly affects adults aged 20–50 years. Effects of excessive glucocorticoids can occur either as a result of endogenous glucocorticoid administration used to treat autoimmune diseases or as the result of a pituitary tumor. As with most illnesses, symptoms may vary, but most people have upper body obesity, rounded face, increased fat around the neck, and thinning arms and legs. Children tend to be obese with slowed growth rates. Other symptoms include fragile and thin skin which bruises easily and heals poorly. Purplish pink stretch marks may appear on the abdomen, thighs, buttocks, arms, and breasts. The bones are also weakened. Most people have severe fatigue, weak muscles, high blood pressure, and high blood sugar. Irritability, anxiety, and depression are common. Women usually have excess hair growth, and menstrual periods may become irregular or stop. Men have decreased fertility with diminished or absent desire for sex.

Symptoms may result from benign tumors of the pituitary, benign or malignant tumors outside the pituitary, or adrenal tumors or abnormalities of the adrenal glands. Rarely, children or young adults develop small cortisol-producing tumors of the adrenal glands.

Treatment depends on the specific reason for cortisol excess and may include surgery, radiation, or chemotherapy. If the cause is long-term use of glucocorticoid hormones to treat another disorder, the dosage will be gradually decreased to the lowest dose adequate for control of that disorder. Once control is established, the daily dose of glucocorticoid hormones may be modified to avoid symptoms.

A number of investigators are working to understand this disease better, including improving strategies for prevention and treatment. Type 1 diabetes is one of the diseases that may benefit from stem cell therapies, as it results from selective autoimmune destruction of insulin-producing cells in the pancreas.

Exogenous glucocorticoids are widely used in clinical practice. Particularly relevant for infants and young children are studies examining the short- and long-term effects of exogenous glucocorticoid administration to pre-term infants. This treatment is highly effective in promoting lung development to allow independent breathing. However, there may be short- and long-term effects of this treatment.

See also: Abuse, Neglect, and Maltreatment of Infants; Attachment; Birth Complications and Outcomes; Brain Development; Breastfeeding; Depression; Emotion Regulation; Failure to Thrive; Fear and Wariness; Genetic Disorders: Sex Linked; Genetic Disorders: Single Gene; Illnesses: Autoimmune Rheumatological Diseases; Immune System and Immunodeficiency; Nutrition and Diet; Obesity; Physical Growth; Prenatal Development; Sleep; Stress and Coping.

Future Directions

Active areas of research involving the endocrine system in development include studies examining the stress systems, including their involvement in initiating labor before full term, hormonal involvement in attachment and bonding, treatment and prevention of diabetes, and the effects of treatment with exogenous glucocorticoids.

A number of investigators, including the author, are examining the development of the HPA-axis in the first few years of life as influenced by factors such as maternal stress during the prenatal period, early caregiving experiences, and temperament. Laboratories exploring HPA-axis development typically also look at factors related to individual differences in cortisol reactivity or basal cortisol levels, including attention, inhibition, and health.

Growing interest in the endocrine system's role in bonding and attachment is reflected in a number of studies of pair-bonding in monogamous species, particularly voles. Oxytocin and vasopressin appear to be critical in effective pair-bonding in this species. It may also be the case that these hormones are critically involved in bonding in humans, both in the bonding of adult mates and the bond between parents and infants.

Type 2 diabetes is becoming increasingly common in the US, at least in part because of the rising rates of obesity.

Suggested Readings

Campbell NA and Reece JB (2005) *Biology,* 7th edn. San Fransisco: Benjamin Cummings.

Greenspan FS and Gardner DG (2001) *Basic & Clinical Endocrinology,* 6th edn. New York: McGraw-Hill.

Larsen PR, Kronenberg HM, Melmed S, and Polonsky KS (2003) *Williams Textbook of Endocrinology,* 10th edn. Philadelphia: Elsevier.

National Institutes of Health. The Endocrine and Metabolic Diseases Information Service – Addison's disease. http://www.endocrine.niddk.nih.gov/pubs/addison/addison.htm (accessed on 16 July 2007).

National Institutes of Health. The Endocrine and Metabolic Diseases Information Service – Cushing's syndrome. http://endocrine.niddk.nih.gov/pubs/cushings/cushings.htm (accessed on 16 July 2007).

National Library of Medicine. National Institutes of Health – Addison's disease. http://www.nlm.nih.gov/medlineplus/endocrinesystemhormones.html (accessed on 16 July 2007).

National Library of Medicine: National Institutes of Health. Health Information – MedlinePlus, Fetal Development. http://www.nlm.nih.gov/medlineplus/ency/article/002398.htm (accessed on 16 July 2007).

Relevant Websites

http://www.diabetes.org – American Diabetes Association, Gestational Diabetes.

http://www.hgfound.org – Human Growth Foundation.

http://www.jdrf.org – Juvenile Diabetes Research Foundation International.

http://www.hormone.org – The Hormone Foundation.

Exploration and Curiosity

A Baxter, University of South Alabama, Mobile, AL, USA
H N Switzky, Northern Illinois University, DeKalb, IL, USA

Glossary

Competence and intrinsic motivation theories of curiosity – Theories that believe that curiosity is the result of humans' motivation to master our own environments.

Constructivist – The belief that children build their knowledge through their experiences.

Curiosity – The desire to learn or know about anything; inquisitiveness.

Drive theories of curiosity – Theories about the arousal of curiosity that propose that curiosity leads to the unpleasant sensation of arousal that is reduced by exploration.

Exploration – Relatively stereotyped perceptual-motor examination of an object, situation, or event the function of which is to acquire information.

Incongruity theories of curiosity – Theories of curiosity arousal that posit that as humans try to make sense of the world their violated expectations about the world lead to curiosity.

Play – Intrinsically motivated behaviors and behavioral sequences, accompanied by relaxation and positive affect, that are engaged in for 'their own sake'.

Introduction

Exploration and curiosity are two forces that shape the development of infants and young children. It is difficult to tease the two apart and distinguish where one ends and the other begins. There is almost a symbiotic relation between the two. For example, when infants are curious about an object, person, or event, they explore the source of the curiosity. This exploration either satisfies the curiosity or arouses more curiosity. Thus, children either decrease their exploratory behaviors or explore more, respectively. A decrease in exploratory behaviors may be accompanied by the initiation of play with the object or person. Similarly, exploration may lead to curiosity as children discover something that was previously unknown. Both curiosity and exploration stem from similar factors: uncertainty, discrepancy/incongruity, and novelty.

While both curiosity and exploration are important to child development, they have not received equal attention from researchers. Exploration has been studied more intensively than curiosity. This is probably due to the nature of the two phenomena. Exploration is readily observable but curiosity is a mental process that is inferred from affect and behavior. In fact, many studies of exploration have just assumed that children are exploring because they are curious. Researchers have investigated situations in which curiosity precedes exploration but they have ignored those in which exploration has led to the arousal of curiosity.

This article describes both of these concepts. It defines each one, charts its developmental progression in the first 3 years of life, looks at its relation to other developmental skills, and places it within the larger context of the developing infant and toddler.

Curiosity

Most theories of child development posit that curiosity plays an important role in fostering optimal development. It is a behavior believed to be present across the lifespan although the nature of its expression changes qualitatively with development. Curiosity is identified as a factor that led to human adaptation, survival, and progress throughout history. Developmentally, curiosity plays a large role in facilitating infants' and children's cognitive, social, emotional, and physical development. Developmental theories as diverse as Piaget's and Freud's viewed curiosity as a central component in development. Curiosity is the motivational component behind children's exploration. Curiosity has both positive and negative influences on people's behaviors. However, curiosity has not been well researched. This is probably due to the fact that it is mental state inferred from children's actions which makes its measurement difficult. As a result, we may not have a full understanding of the construct of curiosity.

Despite almost universal acceptance of the importance of curiosity in facilitating development, there is no consensus about what curiosity 'is'. Defining it is difficult. There is no agreement about what causes curiosity or whether it is a unitary construct or separate constructs. Some researchers view curiosity as a drive, although they disagree whether it was a primary or secondary drive. They debate whether it is internally or externally motivated. To some it is a passion for knowledge that individuals have while others believe it is a response to specific stimulus characteristics. People's 'appetite for knowledge' is thought to be curiosity.

The research on curiosity reflects the ambiguity of the concept. Multiple researchers and theorists proposed multiple multidimensional models of curiosity. Separate cognitive, physical thrill-seeking, and social thrill-seeking curiosities are posited. Some wrote about an information-seeking or cognitive or epistemic curiosity. Others described a sensory curiosity related to sensation-seeking activities. Still others thought of curiosity as a 'drive to know'. There were discussions as to whether curiosity was a state or trait variable. Research studies designed to define and understand curiosity better resulted in different conceptualizations of curiosity as well as different patterns of relations among skills related to curiosity.

Most models of curiosity view it as something (i.e., drive, emotion, cognitive state) aroused by discrepancy, uncertainty, and/or other stimulus properties leading to stimulus exploration to resolve discrepancy or uncertainty. Exploration, in turn, provides the individual with information (e.g., cognitive, perceptual, and/or sensory information) about the stimulus and additional experience with the stimulus. This either decreases the need to explore the stimulus or leads to more engagement with the stimulus. In addition, many models realize that curiosity is not aroused in all individuals by the same stimuli. Curiosity can be stimulated by objects, people, or events in the environment; whether an object, person, or event acts as a stimulus for curiosity is a function of the individual's previous experiences. Children also differ in whether they are curious about objects or people. Thus, individual differences in curiosity and its arousal are expected.

These contrasting views of curiosity are, at least partially, the result of measurement and conceptual variations among the researchers. Some studies found a great deal of overlap among the different types of curiosity they measured. Because of the methodological difficulties inherent in truly understanding what curiosity is, many researchers chose to study what causes curiosity. The major theories are reviewed below.

Theories of Curiosity

Psychologists studied the underlying causes of curiosity since its inception. Early psychologists viewed curiosity as an emotion. They believed that stimuli either elicited curiosity or fear in children. Curiosity led children to explore the environment; fear inhibited exploration. At one time curiosity was considered a basic instinct. Some theorists viewed curiosity of having a positive affective valence. They believed that the information and experiences gained through curiosity led to feelings of personal growth and competence. Other theorists differentiated curiosity into state and trait curiosity. State curiosity

is situation-specific whereas trait curiosity describes individuals who generally approached most situations with curiosity. Research with adults indicated that measures of trait curiosity (e.g., descriptions of a propensity to learn new things and experience interest in many things) and state curiosity (e.g., curiosity at a specific moment in time), while positively and significantly correlated with each other, measured separate and distinct phenomena.

More recent theories viewed curiosity as derived from a drive resulting from either incongruities or a competence motivation. Thus, they viewed curiosity as a motivational construct that was central to initiating children's exploration of their worlds. They saw curiosity as an intrinsically motivated 'passion' that individuals have to acquire information. Others likened curiosity to an 'appetite' for acquiring knowledge. Data indicating that highly curious children explored environments equally in the presence or absence of their mothers but those less curious children explored more in the presence of their mothers than with an unfamiliar experimenter supported this conceptualization.

The major theories about the underlying causes of curiosity are discussed below. They differ in whether they view curiosity as arising from a drive, stimulus, or competence and intrinsic motivations.

Drive theories

Drive theorists viewed curiosity as a drive that caused arousal, which is an unpleasant state for children. As a result, children explored the object or situation leading to the arousal. The exploration led to a reduction of the arousal. Support for curiosity as a drive came from animal studies where even in the absence of physiological needs, exploratory behavior still occurred. Curiosity was motivating in and of itself and, if left unsatisfied for a period of time, the curiosity drive grew. If children had the opportunity to interact with the stimulus that aroused curiosity, the curiosity decreased with time as the drive became satisfied. This satiety was very object specific however, and not generalized as seen with other drives. Thus, curiosity about a specific stimulus can be satisfied but another item can lead to more curiosity almost instantaneously.

Among drive theories, however, there were points of disagreement. Some saw curiosity as a primary drive and others as a secondary drive (i.e., the result of a more basic drive). Some believed that curiosity arose from a lack of homeostasis within the individual. Others thought it was triggered by the stimulus.

Freud, for example, believed that curiosity was a personality characteristic. As with most of Freud's theory, the personality trait of curiosity emerged from children's sex drives. He posited that biological urges and ego mechanisms worked together to guide exploratory behaviors. Exploration was a way for children to cope with mastering the social behaviors and conflicts inherent in life. He believed that children sublimated their sex drives in one of three ways. One way was through inhibition of thought processes which was contrary to curiosity. Another was through compulsive brooding about events and objects. The final way was through the development of a type of generalized curiosity.

Perhaps the most well-known curiosity theorist was Berlyne. To him, curiosity had its basis in the stimulus and the incongruity it aroused in the individual. The stimulus' incongruity resulted from its novelty, complexity, or uncertainty. To him, curiosity had two dimensions that he labeled perceptual-knowledge curiosity and specific-diversive curiosity. Perceptual curiosity was a drive response to novel stimuli that was decreased by exploration of, and continued exposure to, the stimuli. Knowledge or epistemic curiosity reflected humans' quest for knowledge and information. Specific curiosity reflected our seeking of particular bits of information about stimuli of moderate complexity, novelty, and incongruity. Diversive curiosity was our tendency to seek out stimulation. He posited that it was related to boredom. Thus, when curiosity is strong and fear or anxiety is low, diversive exploration occurs. When curiosity is relatively weak and anxiety is high there will be stimulus avoidance. Others viewed diversive curiosity as sensation or thrill seeking not as curiosity at all.

Whether curiosity is a drive or not is unresolved; it may never be resolved. In terms of our understanding of curiosity it may not be an important issue. What is important to understanding curiosity is that it is an aversive state that arises from the environment and leads to specific responses (i.e., exploration or attaining information) to be satisfied. If it is not satisfied it will intensify. While both internal (e.g., previous experience) and external (e.g., objects, people, events) factors influence curiosity their relative impacts are not the same; the nature of the stimulus and children's previous experiences jointly determine whether curiosity is aroused.

Incongruity theories

Incongruity theorists believed that curiosity is our reaction to incongruous events or information. Piaget viewed incongruity leading to curiosity in young children. To him curiosity was the impetus for children's constructivist activities (i.e., infants' and toddlers' building their understanding of the world from their experiences with the world). To him, children were innately curious. He saw curiosity as responsible for children's striving for new information and novel forms of stimulation. He saw the importance of both 'cognitive' and 'sensory' curiosity. He believed that curiosity sprung from children's interactions with, and attempts to make sense of, the world around them. He believed that curiosity arose from environmental events that violated children's expectations of how the world worked. To Piaget curiosity resulted from the disequilibrium produced by children trying to assimilate discrepant, new information into existing cognitive structures. Piaget believed that the relation between curiosity and incongruity varied as a function of the degree of incongruity. To him, information that was not very discrepant was readily assimilated into existing knowledge structures. Information that was too discrepant was ignored. Information that was moderately discrepant led to curiosity and the accommodation of existing cognitive structures based upon the information gained after the arousal of curiosity.

Kagan, in his description of the four basic human motives, acknowledged curiosity through his 'motive to resolve uncertainty'. He expanded on traditional views of incongruity, which had focused on the violation of expectations, by including the uncertainty arising three different sources. He added the uncertainty that arises from incongruities between incompatible ideas and that arises from incompatible behavior and ideas. Finally, he added the uncertainty raised by our inability to predict the future. This broader-based view of the dimensions of uncertainty related to curiosity has given us a fuller understanding of the contexts from which curiosity evolves. While many theorists included uncertainty in their models of curiosity, Kagan pointed out that uncertainty can result from inconsistencies in ideas and behaviors as well as thinking about the future.

Competence and intrinsic motivation theories

Another cadre of theorists believed that curiosity came from our motivation to master our environment. This 'competence' or 'effectance' motivation led to curiosity. We are motivated to be in control of the world around us. Curiosity helps us figure out how things work so we can understand the world and, as a result, feel competent. For example, Gibson attributed much of infants' perceptual learning and development to exploratory and curiosity behaviors. Curiosity has been linked to the presence of uncertainty and unexpected events in infants. Studies indicated that uncertainty may be a very important setting event for curiosity. It appears to be more salient than novelty in the arousal of curiosity. Situations that do not confirm expectancies maintain curiosity as measured by infants looking toward and attending to events. This suggests that infants use curiosity to understand and feel competent in their environment.

Other theorists viewed curiosity as a mechanism that evolved to resolve conflict. They suggested that unusual events led to conflict because they did not fit children's understanding of the world. Curiosity drives children to understand the stimuli lead to knowledge and reduces the conflict. As a result, the examination of unusual events becomes an approach to future situations because the decrease in conflict reinforces the future use of that strategy. Such a use of curiosity helps infants feel competent.

Types of Curiosity

Despite these disparate views on the origins of curiosity, there is some general agreement that there are at least two types of curiosity. One is focused on the acquisition of knowledge and information and is termed 'epistemic curiosity'. The other is focused on perceptual or sensory experiences and is called 'perceptual curiosity'. Perceptual curiosity can involve physical or social sensory experiences.

Curiosity, as a psychological phenomenon, is unusual because of its highly transient nature. Once resolved, curiosity quickly disappears. Often there is disappointment when the curiosity is resolved. When activated, curiosity tends to be very intense. Individuals expend a great deal of effort toward resolving the curiosity. These factors all make studying curiosity very difficult. Researchers have struggled with their definitions of curiosity and deciding how to measure it. Debates raged about whether curiosity can be thought of as a 'state' or 'trait' variable. Is it a primary or secondary drive? Does curiosity arise from the situation, or object, or is it an internal phenomenon?

While curiosity emerges from incongruity/uncertainty, stimulus complexity, and/or novelty, the relations among these variables are complex. Data indicate that too little uncertainty, complexity, or novelty leads to boredom not curiosity. We rarely see curiosity arising during interactions with simple, familiar, and understood objects or situations.

In contrast, too much uncertainty, complexity, or novelty distresses children and results in avoidance behaviors. This can lead to fear; a protective response to encounters with dangerous objects or situations. Thus, mild-to-moderate levels of uncertainty, stimulus complexity, or novelty are associated with curiosity in young children. In such situations, children act on their curiosity to learn more about the object and the object's complexity or novelty. This helps them to understand the object and decrease their uncertainty. It is important to remember, however, that children's individual experiences influence their perceptions of the stimulus. For example, a novel stimulus may be novel to some children and not novel to others. Similarly, the objects associated with curiosity in young children do not lead to curiosity in older children or adults.

Curiosity manifests itself in many different ways. One interesting manifestation of curiosity is through impulsive behavior. When curious, individuals often immediately act to obtain the information that will dissipate the curiosity. This resulted in great discoveries and progress. However, curiosity also leads children, and adults, to act in ways that are not always in our best interests. With development we are better able to inhibit some of this impulsiveness, but some of it remains with us.

Social curiosity motivates us to interact with others. For example, the presence of a caregiver or a peer may lead to curiosity in children. Children then begin to interact with the partner which continues the social interaction and in many cases leads to the desire to engage in more such interactions in the future. In addition, the interactive partner often feels positive about the interaction because the children's curiosity has kept them involved and engaged in the interaction. Many of the social games that caregivers play with young children are based on children's social curiosity.

Developmental Progression

The nature of children's expression of curiosity changes with age and development. Young infants' curiosity is inferred through their use of prolonged visual attention to a stimulus. With age and developing perceptual and motor skills, curiosity becomes the instigator of reaching for, touching, and manipulating the object in question. With further development and the advent of speech, verbal questions about the stimulus become how curiosity is expressed. With still further development, cognitive activities become the manifestation of curiosity. Reasoning is used to fill in the missing information or resources are accessed (e.g., books, the internet) to fill in the gap. There are indications that children's curiosity increases with their mental age, such that within a group of children those with higher mental ages are also more curious.

Infants use their perceptual, physical, emotional, communication, and social skills in service of curiosity. However, there are qualitative differences in curiosity during the first 3 years of life. These differences are described below.

During the first 6 weeks of life, infants' curiosity is difficult to assess. They look around at their surroundings but little sustains their attention for too long. At about 6 weeks of age, infants' curiosity is channeled toward their own hands. They begin to show some interest in their hands and they look at them from every angle as they make new and interesting movements. Next, they begin to use their hands to engage the environment. They bat at objects, explore surfaces for different textures, and grab anything within reach. Once they have the item in hand, they will manipulate it. These manipulations are usually accompanied with intense visual inspection.

At around 6 months of age, infants have become very accomplished with using their hands to explore their environments. This puts them in contact with many new objects about which they are very curious. They are beginning to understand cause and effect relations and they curiously look for these relations during their interactions with the physical and social environments. Much of their curiosity is in response to perceptual information.

Some theorists believe that at about 8 months of age infants enter a peak period of curiosity. As they develop self-locomotion through crawling and walking, the objects that were so far away now become objects of curiosity for them. They are able to get to those objects and satisfy their curiosity about them. Objects that do not lead to curiosity in adults are fascinating to infants.

During the second year of life, infants continue to try and understand cause and effect relations. Their curiosity is shifted somewhat. They are less interested in objects *per se* but they are more curious about what objects can do. They begin to enjoy mechanical toys and they spend a great deal of time observing the consequences of different actions on objects. Their exploratory behaviors decrease and the infant becomes more interested in investigating the effects of their motor behaviors. Thus, their curiosity becomes more focused on what they can do with the stimuli rather than just what the stimuli are or look like. It is also at this time that curiosity related to discrepancies between ideas and knowledge develops. This change reflects children's changing cognitive abilities; curiosity does not only lead to exploration but to additional thoughts.

As the second year of life comes to a close and the third begins, young children shift some of their curiosity to the development of their communication and social interaction skills. Their expression of curiosity begins to rely on their verbal skills. They begin asking the name for everything and their vocabulary is growing exponentially. After their vocabulary becomes fairly well established they begin posing questions. 'What' questions are the first to appear; they relate to children's development of specific cognitive concepts. These 'what' questions are followed by 'who is what' questions and then the 'why' questions. Children who ask more questions also explore toys more than children who ask fewer questions. Again we see the link between curiosity and exploration.

As children continue to develop language, they begin to spend more time in reflection. Curiosity again changes qualitatively. It becomes more internal and exists in children's thoughts. Curiosity becomes thinking about things and then acting on them through further thought, language, or physical actions. At this time there is also an increased focus on peer relationships. Curiosity is used to try and understand others' behaviors and reactions to other children's different social behaviors. It is through these mechanisms that children begin to learn appropriate and inappropriate social behaviors.

Facilitation of the Development of Curiosity

The environments in which children are raised can either support or interfere with the development of curiosity. There have been few empirical studies of the environmental variables that influence the development of curiosity specifically; however, there are many studies investigating the facilitation of cognitive development in young children. Studies supporting a preference for novelty suggest that a caregiving and physical environment that is

structured, yet varied and changeable, will support the development of curiosity and the practice of exploration skills. Too much variability however, within the environment will make it too difficult for children to understand their environments. Children's individual differences in experiences and processing capabilities are important. Too familiar an environment is as detrimental as an environment that has too much variability or too complex stimuli.

Additional research indicates that social components of the caregiving environment are also important in fostering curiosity. Although most studies in this area have been correlational and involved preschooler children and elementary-aged students, they provided evidence that social environments characterized as being positive, accepting, supportive of autonomy, and communication enhancing are associated with more curiosity in children. It also seems reasonable that a caregiving environment that fosters positive social emotional development might also support the development of curiosity. Characteristics of infants with secure attachments suggest that a caregiving environment that allows for the development of secure attachments will also lead to the development of exploration skills and a preference for novelty. A caregiving environment that is aware of and respects temperamental differences among young children should also encourage curiosity.

An environment that encourages physical development should support the development of the sensory, motor, and perceptual skills that are essential to curiosity. If infants or children have cognitive and/or physical disabilities or challenges, it is crucial that accommodations are used so that they can engage in curiosity and exploratory behaviors.

Other correlational studies reported that children who were raised in environments that encouraged autonomy had well-developed senses of curiosity. Similarly, the setting of limits encouraged curiosity, as did attending to children's excitement about new things. Directing children's attention to the sensory aspects of a new toy or situation also encouraged the development of curiosity. Finally a trusting and loving caregiver–child relationship fostered curiosity.

Relation to Other Developmental Skills

Curiosity is associated with cognitive development and intelligence. It is well established that a preference for novelty is associated with higher levels of intelligence, even in infancy. In adults, a preference for novelty is associated with creativity suggesting that there is a relation between curiosity, novelty, intelligence, and creativity.

Clearly curiosity, in its deployment in search of knowledge, influences what is learned as well as how much is learned. Curiosity provides a context for children to practice the many skills believed to be important to intellectual development. Skills such as accessing information to reduce uncertainty, detecting incongruous, complex, and novel stimuli, resolving conflicting solutions or explanations to problems, as well as hypothesis formation and testing are involved with curiosity. They are also highly valued cognitive skills. Finally, curiosity is a phenomenon that involves all developmental domains, not just cognition. Sensory skills, motor skills, communication skills, and social skills are all involved in curiosity and its expression.

Exploration

Exploration is an individuals' sensory, perceptual, and motor behaviors that are used to investigate an object, event, or person. Two different types of exploration have been described. The first is best typified by the way that young children wander around an environment, either familiar or novel, until something in the environment attracts their attention. The second type of exploration occurs when a stimulus catches children's attention, curiosity is aroused and they explore the stimulus to learn more about it. In the first example, exploration leads to curiosity and in the second curiosity leads to exploration. Psychologists have studied the first type of exploration only minimally while they have studied the second type extensively. In this article, we too will focus primarily on this second type of exploration: the exploration that results from curiosity.

Exploration is a behavior focused on the here and now. It does not have to be driven by curiosity. Children explore the objects they find readily at hand or the environment in which they find themselves. While they may use memory and schemas about previous situations, their exploration efforts are focused on where they currently are. Exploration is not a constant event, however. If children are initially very eager to explore an object their exploration of that object decreases with time, as they gather information about it. Exploratory behaviors decrease as children's familiarity with the object or event increases. Conversely, when children are hesitant to explore the object; there will be an increase in exploratory behavior across time and then a decrease as they come to 'understand' the object.

Exploration involves concentration and inspection of the object. In exploration, children use their perceptual, motor, and cognitive abilities. The purpose of exploration is to reduce uncertainty about the object, event, or person. Exploratory behaviors follow a pattern of paying attention to the object, event, or person, visual inspection of it, motor and perceptual examination, and finally physical interaction with the object or person.

The exploratory behaviors of infants and young children are well documented. Exploratory behaviors play important roles in the processes of learning about and adapting to the many environments in which children find themselves. It is through exploration that infants and young children learn about the characteristics of different objects and environments. Exploratory behaviors are related to characteristics of stimuli. Young children's exploratory behaviors result from several factors. One is an almost biological drive to explore the environment. Past experiences with objects and different environments support exploration. The state of the current environmental context drives exploration. Finally, exploration is often accompanied by neutral or mildly negative affect, suggesting that exploration is not necessarily pleasurable for the child.

Just like curiosity, exploration has been associated with stimulus incongruity/uncertainty, discrepancy, novelty, and complexity. However, exploration has different patterns of relations with these stimulus characteristics than curiosity.

Exploration is most commonly the response to moderate uncertainty. In young children the relation between stimuli uncertainty and exploration varies with the degree of uncertainty. If there is no uncertainty, children will not explore the stimulus. Too much uncertainty will lead to children inhibiting any interactions with the stimulus. A moderate degree of uncertainty, however, will lead to exploration. Thus, in reality, in uncertain situations, exploration often takes precedence over other behavioral alternatives.

There is a similar relation between discrepancy, the degree to which the stimulus is similar to known stimuli, and exploration. Stimuli that are too discrepant are not likely to lead to exploration. Stimuli that are not discrepant are more likely to encourage the activation of play or ignoring the stimulus than exploration.

The relation between novelty and exploration is different however. Research has indicated that, given the choice, infants and children prefer to explore novel toys and objects over familiar toys. Novel toys lead to higher rates of exploratory behaviors than more common toys. All novelty is not equal, however. It appears that toys with novelty along perceptual dimensions are more supportive of exploration than novelty focused on problem solving. This may be because the problem-solving toys supported more instrumental behaviors whereas the novel perceptual toys were more ambiguous and, therefore, encouraged more exploratory behaviors. No novelty induces boredom and little exploration in infants. A moderate degree of novelty seems to lead to exploration, whereas too much novelty tends to overwhelm the child and inhibit exploration, sometimes even leading to withdrawal or avoidance.

Stimulus complexity is also important to exploration. The relation between stimulus complexity and exploration is not consistent across the developmental period. Studies indicated that very young children do not explore stimuli that are too complex. For 2-year-olds, the relation between stimulus complexity and exploration is curvilinear; yet for older children it is linear. Two-year-olds explore stimuli with moderate amounts of complexity more than they do stimuli with relatively less or more complexity. Preschooler and young school-age children explore the most complex objects much more than less and moderately complex objects. Thus, with younger children the relation between stimulus complexity and exploration is the same as the relation discussed for novelty, uncertainty, and discrepancy. However, for older children more complex stimuli are associated with increased exploration and less complex stimuli with little exploration.

Modes of Exploration

Infants' and young children's exploratory behaviors are limited or facilitated by their skills in many developmental domains. In infants and toddlers, exploration often involves all of the senses. It is visual and tactile. It involves active manipulation, mouthing, and banging of stimuli, language, and thinking. We see qualitative differences in children's exploratory behaviors based upon other developmental skills.

Soon after birth, infants begin exploring their environment. Although they have a limited focal range, they will look to areas of high contrast. The caregiver's face soon becomes a favorite subject. With development, infants gain control of the movement of their eyes and learn to intentionally shift their gaze between people and objects of interest.

With further development comes visually directed grasping and reaching. This allows infants to initiate physical interactions with people and objects. Infants learn to change their grasp, to transfer objects from one hand to the other, and to put objects in their mouths to explore them more fully. Most studies of young children's exploratory behaviors have focused on their visual and manipulative behaviors. Manipulative behaviors that have been studied often include behaviors such as holding, banging, grasping, scratching, shaking, and hitting. Usually children engage in these behaviors while simultaneously looking at the object. Children learn how to vary their manipulation of objects based upon characteristics of the item. For example, children use both hands for larger objects.

Infants next learn to sit independently. This allows them to use both hands to explore and manipulate objects. This new position also allows them to visually explore more of their environment. They are now also able to manipulate objects and visually explore other parts of their environments. These changes prepare them for the next stage of exploration, being able to independently get to objects of interest.

The next development in exploration is self-locomotion. This new development allows infants to explore every nook

and cranny of their environments. They can move and explore their environments at will.

As language develops, children shift from physical manipulations of objects to using questions as a way to explore their environments. As with curiosity, with age exploration is transformed into more of a cognitive activity whereby children explore more through their thoughts than through the physical manipulations of items.

Developmental Patterns

Infants' limitations in perceptual, motor, and cognitive development impact their exploratory behaviors. Initially, infants' only mode of exploring objects or environments is through their use of vision. Newborns will look around their environments and fixate on objects or people of interest. Their limited visual acuity forces them to look to objects within their focal distance or places where there is high contrast. Often, they focus on caregivers' faces. This continued visual exploration leads to the recognition of familiar caregivers and the fear of strangers that emerges in response to new interactive partners.

During the first year of life children rely on one sensory modality (e.g., vision, tactile) at a time to explore. As development proceeds they develop control over their fingers and begin to develop visually guided reaching abilities. Once infants have learned to coordinate their hand and arm movements with visual information and have prehension skills, they begin to actively manipulate objects of interest. Simply looking at the stimulus no longer meets all of their exploratory needs. This typically occurs at about 5 months of age. At this age, infants not only explore through looking at an object. Most exploratory behaviors at this age involve reaching, grasping, and mouthing of objects. Infants use both hands as they shake, hit, and finger the object. They feel it, put it in their mouths, mouth it, and gum/chew it. They also listen to it. From this point forward, most exploration involves children manipulating objects at the same time as they visually inspect them. Five-month-old infants are typically able to manipulate many objects and have a very advanced exploratory repertoire. Their sense of touch becomes well refined.

With the development of crawling, and later walking, a larger environment is opened up for infants to explore. They venture out into their environments and explore areas they could not access before. They do not need any incentives to explore because the environment itself provides a very large incentive. Both the mundane and exotic are explored. It does not matter if it is a crumb on the floor or water in the toilet; it will need to be explored thoroughly. Two- and 3-year-olds use all sensory domains to explore.

Changes in infants' cognitive development drive them toward exploring their ever-growing collection of concepts and relations in both the physical and the social realms. Exploration helps infants to 'know' their world so that novelty can be detected more easily through its contrast with the familiar. Up until about 14 months of age, infants spend a great deal of time exploring their environments. After this time however, there is a decrease in exploratory behaviors and a corresponding increase in what have been called mastery behaviors. While exploration involves examination of an object by one or several methods, mastery behaviors involve the practice of specific skills using a variety of objects. This change relates to a concomitant change in working memory that allows for the inhibition of some behaviors and facilitates goal-oriented behaviors. We see infants spend less time exploring objects but more time dropping and throwing them. They tend to practice different motor skills more and explore less. By the time that they are 2 years old they typically engage in more mastery than exploratory behaviors. The decrease in exploration also occurs because thought, rather than exploration, emerges as children's response to stimuli that are inconsistent with their knowledge base.

Developmental differences exist in infants' examining, mouthing, and banging of objects. Examining behaviors (e.g., inspection and visual attention with or without manipulation) are believed to represent an exploratory behavior. Age, related differences were found in children's latency it to examine an object. Younger infants (7-month-olds) had a longer latency to examine the object than older infants (12-month-olds) even though the two age groups did not differ in the total amount of time they spent examining the object. This difference may reflect the younger infants needing more time to process the stimulus and organize their exploratory responses. In both age groups, examination of the object preceded mouthing and banging behaviors. As exposure to the object increased, examination decreased and mouthing and banging increased. These results suggest that examination is a primary exploratory behavior and mouthing and banging are secondary exploratory behaviors. These findings suggest that the latency to examine objects and the total amount of time infants spend examining objects represent different behavioral systems. Thus, what appears to change with development is not the amount of time that infants spend exploring objects but how quickly they begin examining objects. Therefore we should evaluate several dimensions of infant exploratory behaviors because they each tell a different story about the infant's abilities.

Influences upon the Development of Exploration

There are great individual differences in young children's exploration. These differences relate to individual differences in motivation, prior experiences, and the social environments in which children develop. The motivational differences are often attributed to individual differences in curiosity which are thought to have their roots in both the genetic and environmental factors.

Studies exploring the contributions made by parents and the environment to individual differences in exploratory behaviors found interesting results. Some studies indicate that the parents of children who readily explore their environments tend to supply their children with a great deal of information about the environment. Even when separated from the parent, these children spend a relatively large amount of time exploring the environment (e.g., questioning, toy manipulation, exploring the room). When interacting with a woman experimenter, children who did not explore environments as readily were more inhibited in their exploration and questioning than when observed with their mothers. These children, however, did explore environments when guided in the exploration by an adult. Thus, children have the requisite skills for exploration; but they do not regularly use these strategies to understand their environments. Other studies of the young children of mothers with depression, indicated that they explore less than children whose mothers are not depressed. The depressed mothers' withdrawn and less-stimulating interactive styles did not encourage their children to explore. Thus, it appears that the social environment can lead to individual differences in rate of exploratory behaviors and facilitate exploration in those who do not readily explore.

Individual differences among children influence their exploration. For example, children seem to differ in terms of how much ambiguity they can tolerate. Children who cannot tolerate a great deal of ambiguity tend to explore very little. Likewise, children of mothers with depression do not explore much. Children with visual impairments and/or motor impairments explore less than children without these impairments because of the limitations that their disabilities put upon the major systems used in exploration. Finally, individual differences in experiences influence the ambiguity, uncertainty, novelty, and complexity of objects. These differences, in turn, influence whether children explore or not.

Gender differences in exploratory behaviors are documented. Infant girls explored less and remained closer to their mothers in new settings. Boys explored objects more actively than girls. Research indicated that these gender differences in exploratory behaviors linger into childhood and beyond. These gender differences have been observed in terms of children looking at objects and their manipulation of objects although the reasons for the differences are not clear. Some have suggested that they reflect differences in reactions to new or unusual objects, while others suggest that these findings might be an artifact of boys' more active mode of play.

Facilitation of the Development of Exploration

The attachment relationship between caregivers and infants is related to infants' exploratory behaviors. The different attachment classifications are based upon infants' patterns of maintaining proximity to their caregiver and exploring the environment. The phrase often heard about infants who are 'securely attached' is that they use their caregivers as a 'secure base' for exploration of the environment. This is also seen as a measure of young children's competence. A caregiver–child relationship that encourages children's exploration and object mastery also supports the development of competence. Likewise, exploratory behaviors are less common in children with ambivalent (type 'C') attachments. These children explore the environment very little. These behaviors make sense in that if parents provide information about exploration and the environment infants are likely to explore and use the parental information. If the parent does not provide such support, children may be hesitant to engage the environment because the parent does not provide them with information about safety. Type C infants have parents who typically do not provide contingent information about the environment, so they are hesitant to explore.

To encourage exploratory behaviors it is important to have environments that are appropriately stimulating to children. The environment should contain a balance of new and old activities and objects. It should allow for exploration to occur through hands-on activities. Caregiver questions will maintain exploration; asking children what an object is, and what it does. Finally, the social environment should encourage exploration rather than discouraging it by saying things such as 'Stop making a mess' or 'Do not get into things that are not yours'. In addition, children should have time to explore and be allowed to explore until their curiosity is exhausted.

Relation to Other Developmental Skills

Exploratory behaviors have been linked to the development of cognitive, communication, social, perceptual, and motor skills. It is difficult to determine if developments in exploratory behaviors lead to changes in the other domains or if changes in these domains lead to changes in exploratory behaviors. Attention is also related to children's exploratory behaviors.

Exploration and play are treated as different phenomena in the research literature. Exploration is a process of collecting sensory information to reduce uncertainty or gather information. It is accompanied by neutral or mildly negative affect. It occurs prior to, but often leads to, play. Play, defined as pleasurable activities that are engaged in for their own sake, is often described as 'children's work'. While exploration is very similar across situations, play varies from child to child, object to object, and situation to situation. Exploration and play are children's responses to different questions. Exploration is the response to the question, 'What is this stimulus and what can it do?' Play is the response to the question, 'What can I do to this stimulus?' Thus, exploration is children's responses to stimuli, while play is centered on children acting upon the object. Exploration typically precedes play in children's interactions with unfamiliar stimuli. It satisfies their curiosity and then children either ignore the stimulus or integrate it into one of their play schemas. While uncertainty is linked to exploration, play does not occur when children are uncertain or fearful. However, repeated exposure to a stimulus decreases the novelty and thus exploratory behaviors while play behaviors increase with familiarity. Play and exploration also have different affective components. Children who are playing also smile and laugh; exploration is accompanied by neutral or slightly negative affect. Play and exploration lead to the increase in independence and decrease in dependence on the caregiver that emerges in most children during development. It is in the context of exploration and play with objects, people, and situations that children come to construct meaning of the physical and social worlds around them.

Children's exploratory behaviors have always been an important component of our construct of attachment. Bowlby believed that infants would be very hesitant to explore their environments if they did not feel secure about the mother's proximity and availability. Unless the infants were sure that the parent would be there if their exploration led them to a dangerous situation, they would not explore the environment. Members of the four different attachment classifications differ in terms of their exploratory behaviors during the strange situation's three different interactional contexts: interacting with the caregiver, interacting with the stranger, and being left alone in the room. The attachment classification differences in exploration reflect temperamental and parent–child relationship differences.

The relation between novelty and exploration is an interesting one. Infants strive to explore their environments to make what seems novel, at a particular point in time, become familiar. Once familiarity has been established, children no longer use exploratory behaviors but rather interact with the toy through play. It is against this new mural of familiarity that infants again search for novelty. As a result of this never-ending process, children amass a great deal of information about their environment. This preference for novelty, which seems to drive exploration, at least in part, has been found to be positively related to the subsequent assessments of children's intelligence. Other studies have indicated that attention to novelty is related to information processing abilities throughout life.

Exploration and Curiosity

It is thought that a lack of knowledge or understanding leads to curiosity, which in turn is associated with exploration in order to acquire the missing information. Studies comparing children with individual differences in curiosity found that children who were more curious explored novel toys much more than children who were less curious. However, when more conventional preschool toys were used the children did not differ in their manipulations of the toys but the high-curiosity children played with the toys longer than the low-curiosity children did.

Interrelations of Curiosity and Exploration

Early theorists posited that curiosity and fear were common reactions to novel stimuli. The fact that novel objects elicited both curiosity and fear suggested an antagonistic relation, although both emotions were considered to be adaptive. Curiosity was thought to support one's exploration of the environment, while fear typically led to withdrawal and avoidance. Thus, relatively novel stimuli were associated with curiosity, while more unusual stimuli typically led to avoidance or withdrawal. Thus, exploration is thought to arise from curiosity and its concomitant anxiety. Exploration is one strategy children can use to decrease this anxiety.

Some researchers have taken a social learning approach to understanding the development of curiosity and exploration in young children. They suggested that parents develop curiosity in their children through modeling and reinforcement. Mothers who explored toys with their children, encouraged children's continued exploration, reinforced children's exploratory behaviors, and answered children's questions had children who asked more questions and spent more time exploring new toys. Similarly, parents who were critical of children's exploratory behaviors had children who exhibited very few exploratory behaviors. Studies comparing children relatively high and low in curiosity interacting with researchers with different interactive styles have found that regardless of the adult's interaction style highly curious children had high levels of exploratory behaviors. In addition, low-curiosity children did not explore the environment differentially when interacting with an adult who demonstrated exploration and reinforced it, an adult who was responsive but not encouraging of exploration, or an adult who was unresponsive and inattentive. Their exploratory behaviors were at low levels in all three interactive conditions.

Many quality early childhood care environments focus on the facilitation of curiosity and exploration in addition to play and creativity. These four factors, when seen in young children are thought to be evidence of participation in high-quality caregiving environments. The lack of curiosity and/or exploration in young children is often seen as evidence of psychological problems.

Parents and caregivers need to be careful about facilitating exploration and curiosity. The encouragement of these two skills in young children can be generalized in ways that may be unsafe. For example the same curiosity that leads infants to play with a new toy, can also lead children to climb up the bookcases in the living room. Thus, attention must be paid to the environment and stimulation that are available to infants and toddlers and efforts must be made to keep children safe while they are exploring based on their curiosity.

Importance to Development

Studies of the cognitive development of infants and young children point to curiosity and exploration as important to cognitive development. These effects are not limited to the infancy and early childhood years, but appear to hold for the lifespan. In addition, many of the characteristics that lead to curiosity play a role in the development of information-processing skills in human beings. Important components of curiosity: a preference for novelty, a strong desire to reduce uncertainty, and selective attention to incongruity are also important to information processing development. Curiosity and exploration also influence development in other domains as well. Children's exploration of their environments and curiosity about what they encounter foster language development. What is an object's name? What does it do? Why does it do that? Exploration and curiosity also advance social development. Children learn about how to initiate and maintain interactions with people who arouse their curiosity and who support their exploration. Social environments that do not support curiosity and exploration lead to children who do not use these skills regularly to interact with their environments. Curiosity and exploration are also linked to motor development. From children learning to control the movement of their eyes and use their binocular vision, to reaching, grasping, and manipulating objects, to self-locomotion and getting into everything, curiosity and exploration drive the refinement of children's motor skills.

See also: Attachment; Attention; Cognitive Development; Family Influences; Fear and Wariness; Habituation and Novelty; Imagination and Fantasy; Motor and Physical Development: Locomotion; Motor and Physical Development: Manual; Perception and Action; Perceptual Development; Play; Reasoning in Early Development; Safety and Childproofing; Self-Regulatory Processes; Separation and Stranger Anxiety; Shyness; Temperament.

Suggested Readings

Berlyne DE (1954) A theory of human curiosity. *British Journal of Psychology* 45: 180–191.

Fowler H (1965) *Curiosity and Exploratory Behavior.* New York: Macmillan.

Görlitz D and Wohlwill JF (eds.) (1987) *Curiosity, Imagination, and Play: On the Development of Spontaneous Cognitive and Motivational Processes.* Hillsdale, NJ: Erlbaum.

Henderson BB (1984) Parents and exploration: The effect of context on individual differences in exploratory behavior. *Child Development* 55: 1237–1245.

Lowenstein G (1994) The psychology of curiosity: A review and reinterpretation. *Psychological Bulletin* 116: 75–98.

Reio TG, Petrosko JM, Wiswell AK, and Thongsukmag J (2006) The measurement and conceptualization of curiosity. *Journal of Genetic Psychology* 167: 117–135.

Spielberger CD and Starr LM (1994) Curiosity and exploratory behavior. In: O'Neil HF and Drillings M (eds.) *Motivation: Theory and Research*, pp. 221–243. Hillsdale, NJ: Erlbaum.

Switzky HN, Haywood HC, and Isett R (1974) Exploration, curiosity, and play in young children: Effects of stimulus complexity. *Developmental Psychology* 10: 321–329.

Weisler A and McCall RB (1976) Exploration and play: Résumé and redirection. *American Psychologist* 31: 492–508.

Wentworth N and Witryol SL (2003) Curiosity, exploration and novelty-seeking. In: Bornstein MH, Davidson L, Keyes CLM, Moore KA, and Kristin A (eds.) *Well-being: Positive Development across the Life Course*, pp. 281–294. Mahwah, NJ: Erlbaum.

Relevant Websites

http://scholastic.com – Curiosity: The fuel of development; Encouraging new triumphs. When and how to challenge your baby to try new tasks.

http://www.naeyc.org – Early years are learning years. ''I can do it myself'': Encouraging independence in young children.

http://www.evenflo.com – Month 5: Exploration and Experimentation.

http://www.sesameworkshop.org – Parenting essentials for moms and dads who grew up on Sesame Street.

http://www.gfi.org – Parenting with Gary and Annie Marie: Toddlers; Toddlers and curiosity.

http://childcare.about.com – Ways to spark creativity in a child.

http://www.zerotothree.org – ZERO TO THREE.

Face Processing

O Pascalis and D J Kelly, The University of Sheffield, Sheffield, UK

Glossary

Categorization – A mental ability to classify visually distinct stimuli as members of the same class at the superordinate level.

Conspecific – An organism belonging to the same species.

Event-related potentials (ERPs) – A measure of the brain's response to the presentation of a stimulus recorded via scalp electrodes. They provide information about the timing of the neurocognitive process occurring during an event.

Exemplar – A single item from a single category.

Forced choice task – Participants are required to identify a target stimulus (to which they have previously been exposed) presented with one or more distracter stimuli.

Preferential looking task – Indexes the level of interest of one stimulus from a pair by comparing the time spent looking at each of the stimuli. Preference is inferred from a greater looking time to one stimulus as compared to the other.

Prototype – A mentally stored averaged representation of all individual faces seen throughout a person's lifetime.

Visual paired-comparison task (VPC) – Indexes the level of interest for one stimulus in a pair after one of these stimuli has been learned during a prior familiarization period. Recognition is inferred from the participant's tendency to fixate toward the novel stimulus.

Introduction

From birth, infants keenly explore the visual world around them. Initially, poor acuity and sensitivity to contrast restricts the newborn to seeing little more than the person holding them. During the first weeks of life we can expect much of the newborns' visual attention to be directed toward the faces of their caregivers. This early, constant exposure to faces will influence the way we, as adults, process faces. Furthermore, we will never be exposed to any other category of complex visual stimuli so frequently during life. In brief, faces are omnipresent in our environment and are crucial for our social life.

There is a general consensus that we gradually become human-face-processing experts during ontogeny. A rudimentary face-processing system is present at birth, but it is far from mature and requires extensive visual experience with faces to develop into the sophisticated system seen in adults. Which components are present at birth, which develop first, and at what stage the face-processing system becomes adult-like are still hotly debated topics. The aim of this article is to review our current knowledge and understanding of the development of the face processing system from birth to adulthood, discussing changes through infancy, childhood, and adolescence.

We shall first summarize the major characteristics of the adult-face-processing system and review when they emerge during infancy and how they develop during childhood.

Face Processing in Adults

Face categorization

Guillaume Rousselet and colleagues have shown that faces can be visually detected even in very rapid presentation as short as 30 ms! This shows that faces are a salient stimulus from our environment that we can detect with ease.

Configural processing

Faces are a category of stimuli that, unlike most other objects, are homogenous in terms of the gross position of their elements (two eyes above the nose, nose above the mouth, etc.) and require discrimination on the basis of relational information (**Figure 1**), such as the distance between the eyes, or between lips and chin. The ability to process relational information, called configural processing, is posited to be the consequence of our vast experience with the human face. Configural processing consists of two types of relational information: First-order relational information refers to the configuration of facial features (i.e., two eyes situated above the nose etc.), whereas second-order relational information concerns the spacing between these facial features. One of the most important indicators of configural processing is the inversion effect; the fact that faces are recognized more accurately and faster when presented in their canonical, upright orientation than when presented upside-down. Importantly, the inversion effect for faces is disproportionate when compared to recognition accuracy for inverted nonface objects. It is believed that the second-order relational information required to accurately identify

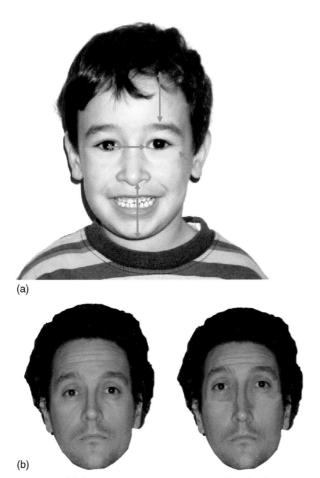

(a)

(b)

Figure 1   (a) Configural processing: is processing the distances (red arrows) between major face features such as the eyes, nose, and mouth, and between these features and face contours that specify differences among individuals. (b) Example of a face and its distortion of second order, the eyes are moved up and are further apart.

individual faces is disrupted by inversion, forcing a less accurate featural processing strategy.

Face Recognition

Human adults are able to recognize hundreds of distinct faces such as family, friends, and famous people whom they have never even met. Remarkably, we are even capable of recognizing people we have not encountered for many years despite changes of age, hairstyle, and paraphernalia (e.g., spectacles).

Face-processing specificity

The adult face-processing system has been found to be species-specific and not flexible enough to process faces of other species at an individual level. Whereas adults can recognize a human face seen for only a few milliseconds, they are unable to extend this ability to nonhuman primate faces. Pascalis and colleagues investigated the abilities of humans and monkeys to recognize faces from their own- and other-species using the Visual Paired Comparison (VPC) task. They showed that human participants were more accurate at recognizing human faces than monkey faces, while the opposite result was found in monkeys. These results have since been replicated in humans and extended to other nonhuman primate species.

Adults are typically more proficient at recognizing faces from their own-racial group, as opposed to faces from other-racial groups. This is commonly known as the other-race effect (ORE). One popular account of the ORE is the 'contact hypothesis', which asserts that the effect is a consequence of primary exposure to faces from a single racial category. An influential model proposed by Tim Valentine in the early 1990s suggests that faces are encoded in a face-space framework in terms of vectors in a multidimensional perceptual space. The origin of the space represents the average of all faces experienced by an individual with more 'typical' faces close to the origin and more 'distinctive' faces situated further out. All new faces will be encoded in terms of their deviation from the prototypical (average) norm. This model provides a possible explanation for the ORE. Every person's face space will be tuned toward the category of face most commonly seen within the visual environment. Subsequently, the face system will become most efficient at processing exemplars from that particular face category while becoming poorer at processing faces from groups encountered less often.

Inner and outer face features

Faces can be divided in two parts, the inner part of the face, which contains features that form the core of a configural representation of a face (i.e., eyes, nose, and mouth) and the outer part (i.e., hairline, chin, and ears) (see **Figure 2**).

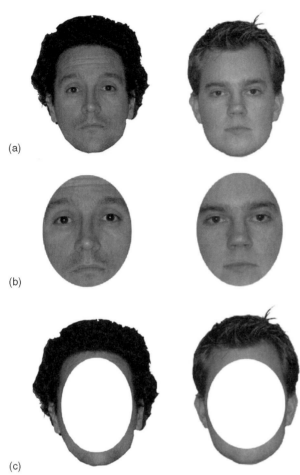

Figure 2 Inner/outer effect. Examples of the three types of stimuli: (a) full-face condition, (b) inner-face condition, and (c) outer-face condition.

The ease with which people are recognized from their inner and outer features changes across development and is mediated by familiarity. When a person is unfamiliar, adults find it easier to recognize that person from a photograph of just their outer features than from a photograph of just their inner features. As adults gain familiarity with a person, they become better able to recognize the inner parts of that person's face, to the extent that recognition on the basis of inner parts becomes superior to recognition of the outer parts of the face.

Development of the Face-Processing System

Many researchers agree that face processing involves an innate system that guides attention to faces, which is crucial for developing an expert face system. However, the system continues to undergo changes for many years before developing into the adult system. Face processing requires multiple levels that follow different development

trajectories, such as: face preference, face categorization, and individual recognition.

Face Processing During Infancy

Face preference

Faces are ubiquitous in the infants' environment, and almost certainly the most frequently encountered visual stimulus during the first few days of life. It is reasonable to assume that faces may consequently be processed preferentially compared to other categories of visual stimuli, and that they will be looked at for a longer period than other stimuli. In a series of experiments conducted during the 1960s, Robert Fantz found that infants as young as 1-month-old showed a consistent, spontaneous preference for face-like stimuli over nonface-like patterns. Some researchers hypothesized that movement is an important component of the face preference shown during the first week of life. This hypothesis has been supported by a number of studies, which have found that moving face-like patterns yield greater tracking behavior than nonface-like patterns in newborns tested just a few minutes after birth. Furthermore, many studies have observed a preference for static schematic face-like stimuli over equally complex visual stimuli such as scrambled schematic faces. One longitudinal study observed preferential tracking of faces over the first 5 months of life. They showed that 1-month-old infants track a schematic face-like pattern further than stimuli that possess facial features in an atypical configuration or nonfacial features (three squares) in a facial configuration. They also demonstrated that this preferential tracking appears to decline between 4 and 6 weeks after birth; preference for faces is not documented for older age groups. Recently, a preference has been demonstrated using photographs of real faces when paired with inverted faces. The inverted stimuli used were photographs of faces depicted from the crown of the head to the neck, which were featurally inverted (see **Figure 3**, left panel). This type of inversion contrasts with classical inversion, where the entire face is simply rotated 180° (see **Figure 3**, right panel). Therefore, it can be argued that the preference for the upright face may be due to the discrepancy between the neck and the positioning of the internal features.

Eyes and eye gaze also appear to be important factors in the elicitation of face preferences in the first few days of life. For example, newborns prefer to look at a female adult face with open eyes when paired with the same face with closed eyes and will look longer at a face when the eye gaze is directed straight at them.

Face categorization

Studies with newborns have shown that faces are preferentially oriented to when paired with other stimuli,

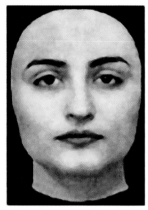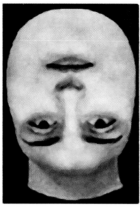

Figure 3 Example of stimuli: stimuli used by Macchi-Cassia *et al*. (2002).

demonstrating that faces are categorized rapidly after birth. It is unequivocal that the face-processing system develops very quickly from birth, with infants receiving more and more experience with faces from their visual environment. Initially, infants will gain experience with only a handful of face exemplars, most notably those of their parents. It has been demonstrated that infants begin to show evidence of face prototype formation at 3 months of age. Before this age, face recognition seems to be exemplar based; each face is encoded independently of all other faces that have been viewed. Paul C. Quinn and colleagues hypothesized that the face representation of 3-month-olds may be biased toward female faces as the mother is usually the primary caregiver during the first few months of life. They found support for their hypothesis, demonstrating that 3-month-old infants raised by their mothers prefer to look at female faces paired over male faces, even if the hair was removed. The authors also identified and tested a small population of 3-month-olds who had been raised by their father. In this instance, a preference for male faces was observed.

The role of experience early in infancy has been highlighted further by the finding that 3-month-old Caucasian infants prefer to look at faces from their own-racial group when paired with faces from other-racial groups as measured by the preferential looking task. Caucasian newborns tested in an identical manner demonstrated no preference for faces from either their own- or other-racial groups. These results have now been extended to a range of other racial groups. A preference for own-race faces has been demonstrated with 3-month-old Chinese, Ethiopian, and Israeli infants tested in their respective, native countries. Also, a population of Israeli-born, Ethiopian infants who had received exposure to both African and Caucasian faces were tested. These infants showed no preference for faces from either ethnic group. Collectively, these results clearly demonstrate how the early face-processing system is influenced by the faces observed within the infants' visual environment.

Face recognition

A third aspect of the face-processing system is face recognition; recognizing that a specific face has been encountered before. The ability to learn and recognize individual faces is paramount for the development of attachments and social interaction. Recognition of the mother is particularly important for the development of attachment and emotional bonds between mother and child.

Mother face recognition

Researchers have found that newborns show a visual preference for their mother's face when paired with a stranger's face under a range of different experimental conditions. Newborns are able to visually discriminate between their mother's and a stranger's face even when they have been matched for hair color and length. It has also been established that this preference is not dependent on the availability at auditory or olfactory cues during testing.

The preference is still observed after a delay of 15 min between the last exposure to the mother's face and testing. Newborns will also produce significantly more sucking responses to see a static image of their mother's face on a screen as opposed to a stranger's face. In addition, recognition of the mother's face has also been found at 1 and 3 months of age.

A recent study by Fatma Sai elegantly demonstrated that newborns only recognize their mother's face if a postnatal exposure to the mother's voice–face combination has been available. If the infant is denied the auditory input of the mother's voice shortly after birth, recognition of the mother's face is not demonstrated 'at this stage'. It appears that the mother's face is learned in conjunction with the mother's voice, which has been heard during gestation.

Unfamiliar face recognition

Newborns can also learn and discriminate face stimuli that differ only by their inner elements. Four-day-old infants habituated to a photograph of a stranger's face will show recognition both immediately after familiarization and after a delay interval of 2 min. It is important to note that the pictures of the faces used in these experiments are identical in the familiarization phase and during the recognition test. This raises an important question: is the newborn demonstrating simply picture recognition or are they recognizing a 'face'? While this is a valid criticism, the faces used in experiments such as the aforementioned are paired with care and subsequently recognition should not be achievable on the basis of gross differences such as hair style. Other studies have shown that infants at 3 months of age are able to learn a face on a certain point of view (e.g., full face) and recognize it on a new point of view (e.g., three-fourth profile). Therefore, true face recognition abilities are unequivocally present at 3 months of age, but are likely to be in place earlier.

Inner and outer face features

There are limited data available to help us understand whether newborns' face recognition relies on inner or outer faces features. It has been demonstrated that the newborns' preference for their mother's face is abolished when the outer contour of the head and the hairline is removed, suggesting that 3–4-day-old infants need both the outer features and inner part of the face for successful recognition. These findings were extended in a separate study that tested infants aged between 19 and 155 days. A preference based solely on the internal features was observed around 35 days. The external features presented alone elicited a preference for the mother's face from 4 months of age.

The importance of inner and outer information for unfamiliar faces has been investigated recently by Chiara Turati and colleagues with the VPC task. In their first experiment, newborns participated in one of three different experimental conditions: full-face, inner features, or outer features. In this instance, recognition was shown in all three conditions demonstrating that newborns are able to process both the inner or outer facial features. However, in the second experiment newborns were unable to recognize the familiar face when important changes occurred between the habituation and the test phase; namely the inner and outer face stimuli. These two stimuli sets looked profoundly different to each other and while newborns in experiment 1 showed recognition within each of these sets, the newborns in experiment 2 were unable to show recognition between the sets. These results are important as they demonstrate that recognition is not dependent on the availability of full-face information, but recognition of a full-face cannot be achieved if the changes between habituation and test are too great (**Figure 4**).

Configural processing

In the 1970s, Susan Carey and Rhea Diamond proposed that early in life, recognition of individual faces is based largely on the basis of individual facial features, such as distinctive eyes or noses. This is called 'featural processing'. They asserted further that it is only late in childhood, after sufficient exposure to faces that children begin to recognize individuals in an adult way, using configural information. This claim was originally based on evidence that children's recognition of faces is less affected by inversion (see below) than adults. This view remained and influenced the field of development of face processing for many years and is still widely accepted outside our field. A number of studies conducted with infants and children during the past decade have produced results which strongly challenge this account.

The inversion effect

One of the first of a series of more recent studies reported signs of the inversion effect from 7 months of age. After being habituated to two adult female faces, infants were

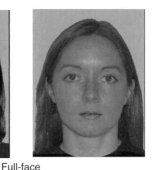

Full-face
condition

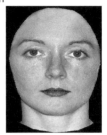

Inner features
condition

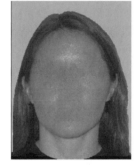
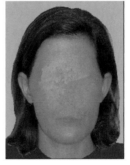

Outer features
condition

Figure 4   Example of stimuli used by Chiara Turati and colleagues.

tested with a composite face constructed from the internal features of one face pasted onto the outer features of the other face. The results showed that in the upright condition infants looked longer at the composite 'switched' face than the familiar face. This indicates that infants must have processed at least some configural properties of the face. However, infants could have combined all or only a number of internal and external features. In the inverted condition, infants did not look longer at the composite face than the familiar face, showing no evidence of being able to process the configuration of inverted faces. The 7-month-olds, showed an adult-like pattern of response; they processed the upright face configurally and the inverted sections featurally. Other researchers have extended this work using schematically drawn faces. It was found that 4-month-olds process the stimuli using just the facial features, whereas 6- and 8-month-olds used a more configural approach. In a follow-up study, real faces were used to determine which part of the face creates a switch effect. They compared small changes in either the eye or the mouth with bigger changes in the eye and mouth. They found that 10-month-old infants processed faces configurally for both small and big changes, whereas 6-month-olds processed eyes as features, but the mouth as part of the whole face. Four-month-olds processed both the eyes and mouth as features. These mixed results show that limited configural processing can be observed in young infants but is still very different from the configural system observed in adults.

Finally, other studies have indirectly provided evidence of configural processing in even younger infants. Certain infant face preferences have been found to be abolished by inverting face stimuli. These include newborns' preference for attractive over unattractive faces and 3-month-olds' preference for female over male faces. If these very young infants are not using configural information to make judgments of attractiveness and gender, then why are these preferences lost when the stimuli are inverted?

Face-processing specificity

Charles A Nelson has drawn a parallel between language development and face recognition development. Nelson proposed that the face-processing system develops from a broad nonspecific system at birth to a human-tuned face processor by the end of the first year of life. Furthermore, he suggests that the faces observed within the infants' visual environment shape and influence the developing face system. By adulthood, extensive experience with human faces has produced a prototype-based system which is tuned to the most frequently observed stimuli (i.e., human faces), with individual faces encoded according to their deviation from a prototypical norm. It has demonstrated that infants begin to show evidence of face-prototype formation at 3 months of age, but prior to this, face recognition is exemplar based; the face of each recognized individual is separately encoded. Infants recognize individuals by comparing the face seen in the visual environment with an internal representation. Later, individual faces will be encoded (and recognized) in relation to their deviation from the prototype.

It has been theorized that face recognition should start to become specific to human faces soon after 3 months as these are the only category of faces to which infants are regularly exposed. The ability to discriminate human and monkey faces has been tested in 6- and 9-month-old infants and adult participants using the VPC task. The youngest participants discriminated between individuals of both species whereas older infants and adults only discriminated between individuals of their own species. This result supports the notion that visual input during early infancy influences the development of the internal face prototype and face-processing abilities. A follow-up study investigated whether the more general processing

properties of the face-processing system could be maintained by exposing infants to faces from an unfamiliar species between the ages of 6–9 months. Six-month-olds were exposed regularly to Barbary Macaque monkey faces (MF) during a 3-month period and their ability to discriminate MF was then assessed at 9 months. Their discrimination performance was compared with a control group of 9-month-olds who received no training. A group of adults were trained and tested in the same way as the 6-month-olds. Furthermore, to determine the impact of extensive training, a group of primatologists were also tested. If the ontogeny of the face-processing system mirrors the development of language, exposure to the MF should broaden or extend the perceptual window through which faces are processed. The 9-month-old infants who had received 3 months of training were able to reliably discriminate the MF. In contrast, the 9-month-old control infants, the adults, and the primatologists could not. These findings demonstrate that passive exposure is sufficient to extend the infants' ability to discriminate between novel MF. Similar exposure time, however, is not sufficient for adults to learn to discriminate between novel MF.

Face Processing in Childhood

The vast majority of the face-processing literature from the 1970s reports that children younger than 10 years of age are poor at face recognition. A close inspection of more recent literature shows that children are performing far better than previously thought. The changes in our understanding of face processing in childhood have been brought about by newly designed tasks that are more appropriate for such age groups.

Face recognition

Children from 4 to 5 years of age can achieve 80% accuracy when recognizing faces in a standard forced choice task. Another study employed the use of picture books to assist face learning in children aged between 2 and 6 years. The 5–6-year-olds performed almost at ceiling, achieving approximately 93% accuracy, while 2–4-year-olds achieved roughly 73% accuracy. Despite these highly competent abilities, the face-recognition system does not mature fully until late adolescence. In fact, some studies have noted a developmental dip around 12 years of age, disrupting the linear trend observed during childhood. The dip may reflect either a change in the encoding strategy or biological changes occurring during puberty. It is clear that recognition *per se* is very efficient from an early age, but the way in which faces are recognized might differ across development.

Inner and outer face features

During childhood, the adult pattern of recognition for 'familiar' faces appears relatively late in development.

Internal face parts do not appear to be more important than external face parts for familiar face recognition until at least 7 years of age. At 5 years of age, recognition of familiar classmates' faces is equal or better when using external face parts as opposed to internal parts. There is little agreement over the precise timing of the switch from reliance on external to internal face part information for familiar faces. However, the transition appears to occur sometime between 7 and 14 years. Children show the same reliance on internal and external face parts as adults for the recognition of 'unfamiliar' faces, demonstrating an external face part preference.

Configural processing
The inversion effect
Susan Carey Rhea Diamond demonstrated a strengthening of the inversion effect between 6 and 10 years of age. Since then, many studies with children have demonstrated the existence of an inversion effect at a young age.

Face-Processing Specificity

Although there are limited studies to have looked at the specificity of face processing during childhood, it would appear that the system is becoming ever more refined. This inference is drawn from studies showing that children from 3 years of age demonstrate a strong ORE.

Electrophysiological studies
Face-selective electrophysiological activity has been observed in event-related potential (ERP) studies with adults. Most studies have reported a negative deflection elicited around 170 ms after stimulus onset (N170), which is believed to be occipital in origin. Although the N170 response is produced by all objects, the potential tends to be of larger amplitude and shorter latency for human faces. The N170 also shows sensitivity to stimulus inversion. The response produced for inverted human faces is of greater amplitude and longer latency compared to upright human faces. This effect is specific to human face stimuli and has not been observed for animal faces. In infants, a negative deflection is observed around 290 ms after stimulus onset with latency decreasing between 3 and 12 months of age. This is called the infant N290 and is a possible developmental precursor of the adult N170. A positive component is also observed around 400 ms (P400) after stimulus onset with latency again decreasing between 3 and 12 months of age. Despite the difference in polarity between the N170 and the P400, some authors have argued that the latter is a more probable candidate for the N170 precursor. Like the adult N170, the infant P400 peaks faster for faces than for objects and unlike the infant N290, it is sensitive to stimulus inversion. Michelle de Haan and colleagues have suggested that

both components are precursors of the adult N170 and will become integrated after the first year of life when face processing becomes more automated. Indeed, the most recent evidence suggests that this is already occurring at 12 months of age, although the N170 elicited is certainly not truly adult-like until at least 15 years of age.

Conclusions

It is clear that during the first days of life, newborns have a face representation which develops with subsequent visual experience. Exactly what is represented during the first week of life and how it is represented cortically remain controversial issues. John Morton and Mark H Johnson have suggested that there is a subcortical mechanism (CONSPEC), which contains a crude specification of the arrangement of the main facial features (eyes and mouth). It is thought that the role of CONSPEC is to trigger newborns' attention to face-like patterns, ensuring the development of the face system.

Other authors have suggested that the face representation in neonates could be innately provided by evolutionary adaptation or that proprioceptive feedback *in utero* (e.g., yawning, swallowing, etc.) could provide the newborn with a crude facial prototype. This would result in the infant entering the world with a representation of the human face that will develop rapidly through visual experience into a more detailed representation. This early representation may be integral in enabling the infant to imitate facial gestures, which they can do within the first days of life. A series of recent findings have found support for this view. One study found that a simple object (e.g., a prism) handled for just a few seconds by a newborn could subsequently be recognized visually. It had previously been thought that newborns were unable to produce cross-modal recognition.

Although the origins of the face representation are yet to be resolved, it is clear that rudimentary face-processing abilities are present at birth and subsequent facial input from the visual environment plays a crucial role in the systems' development. Daphne Maurer and colleagues have shown that infants with bilateral congenital cataracts fail to develop normal face-processing abilities in later life when the cataracts are not removed within the first few months of life. The face system undergoes many changes throughout infancy, childhood, and adolescence before finally reaching maturity. Even when the system is fully developed, considerable plasticity can still be demonstrated, although some abilities (e.g., other-species face processing) are lost in adults.

See also: Attention; Brain Development; Categorization Skills and Concepts; Habituation and Novelty; Intermodal Perception; Perceptual Development; Visual Perception.

Suggested Readings

Le Grand R, Mondltoch C, Maurer D, and Brent H (2003) Expert face processing requires visual input to the right hemisphere during infancy. *Nature Neuroscience* 6: 1108–1112.

Morton J and Johnson MH (1991) CONSPEC and CONLERN: A two-process theory of infant face recognition. *Psychological Review* 98(2): 164–181.

Pascalis O and Slater A (eds.) (2003) *Face Perception in Infancy and Early Childhood.* New York: NOVA Science Publishers.

Schwarzer G and Leder H (2003) *The Development of Face Processing.* Cambridge, MA: Hogrefe and Huber.

Failure to Thrive

D R Fleisher, University of Missouri School of Medicine, Columbia, MO, USA

Glossary

Constitutional growth delay – A temporary lag in growth affecting some healthy infants and toddlers having no discernable environmental cause.

Failure to grow (FTG) – Growth deficiency, indicated by a down-shift of growth curves to lower percentile channels over time. Its causes include disease, malnutrition, psychosocial adversity and all other factors having a negative impact on a child's fulfillment of his or her genetic potential for physical growth.

Failure to thrive (FTT) – A special form of FTG caused by aberrant nurturing involving dysfunction in the parent–child relationship that impairs nutritional and developmental wellbeing.

Fundoplication – A surgical procedure that narrows the opening between the esophagus and the stomach thereby impeding gastroesophageal reflux and its resultant vomiting.

Gastroesophageal reflux (GER) – Backwash of liquid contents from the stomach up into the esophagus. It occurs normally, especially after meals, and, as such, is not pathologic.

Gastroesophageal reflux disease (GERD) – Excessive GER that causes irritation of the linings of the esophagus or airway.

Gastrojejunostomy – A surgical procedure that creates an additional opening between the stomach and the small intestine (jejunum) to facilitate emptying of the stomach.

Genetic small stature – Inherited small stature in an individual who nevertheless grows (or has grown) at normal rates, that is, near, but parallel to the lower percentile channels for height. This kind of smallness should not be mistaken for growth failure.

Hyperphagia – Overeating.

Hyperphagic psychosocial dwarfism ('deprivation dwarfism' 'pseudo-hypopituitary dwarfism') – A form of FTT associated with failure of emotional attachment between child and principal caregiver, characterized by cessation of growth resembling organic hypopituitarism along with compulsive consumption of extraordinary amounts and kinds of food. Growth resumes and eating normalizes when the child experiences nurturing.

Hypertrophic pyloric stenosis – An organic disease acquired in the first month or two of life that presents with increasing vomiting and weight loss. It is caused by thickening of the walls of the outflow tract of the stomach resulting in progressive occlusion of the channel through which stomach contents normally pass into the intestine (duodenum). It usually requires surgical intervention.

Infant anorexia syndrome – Inadequate food intake resulting from conflict between infant and caregiver around feeding and eating.

Infant rumination syndrome (IRS) – An acquired self-stimulatory habit characterized by repetitive, nearly effortless regurgitation of recently ingested food into the mouth followed by re-swallowing and/or loss to the outside. It usually has its onset between 2 and 8 months of age as a result of insufficient reciprocal interaction between infant and caregiver.

Nurturing – The process by which children's physical and emotional needs are fulfilled sufficient for normal development.

Pyloroplasty – A surgical procedure that widens the normal passageway (the pylorus) from the stomach to the intestine (duodenum) to facilitate emptying of the stomach.

Small for gestational age – A term applied to newborns whose growth *in utero* had been less than normal.
Thriving – A global concept that includes normal physical, emotional, cognitive, and motor development.
Urine-specific gravity – A measure that reflects the amount of water in urine.

Introduction

Failure to thrive (FTT) is usually defined as growth failure due to any cause. In this article, however, FTT is defined as a special kind of growth failure, namely, growth failure caused by aberrant nurturing. Although organic diseases, such as kidney failure, cystic fibrosis, or congenital heart disease, may impair growth, a large proportion of infants and children who present with growth failure have no underlying organic disease. Instead, they fail to grow because of dysfunction within their parent–child relationships, and it is this kind of growth failure to which the term FTT is specifically applied.

Philosophic Considerations

Compassionate, efficient management is based on two concepts worthy of consideration before proceeding with a discussion of FTT: (1) the distinctions between the conventional biomedical model of practice and the biopsychosocial model of practice and (2) the definitions of 'functional', as contrasted with 'organic' disorders.

The conventional, biomedical model of practice dichotomizes illness as either organic or psychogenic, the former being materialistic and the province of medicine, the latter being insubstantial or 'mental' and the province of the mental health professions. However, advances in the neurosciences have increasingly shown the unity of mind–brain–body. Sustained emotional stress, for example, has been shown to cause alterations in the anatomy and function of the brain and other organs, many of which affect behavior and influence growth and development.

The biomedical approach to diagnosis is dichotomized and biased in favor of organic disease. The process is sequential: first rule out all plausible organic causes of growth failure and only then consider psychosocial factors. Even when psychosocial pathology is evident on first encounter, the clinician adhering to the biomedical model may avoid this aspect of growth failure out of concern that an organic disease might be missed. Unfortunately, medical diagnostic procedures are not necessarily benign, especially in infants and children. They often involve pain, anguish, exposure to ionizing radiation, and other adversities. The task of the clinician is to identify the causes of growth failure in a manner that not only diagnoses them correctly, but does so in a manner that is least stressful or damaging to the patient.

The biopsychosocial approach, as described below, avoids both organic and psychogenic biases and, instead, approaches both aspects of failure to grow simultaneously, pursuing a course that is flexible, influenced by the most likely elements as they emerge during the evaluation. This lessens the likelihood of what John Apley called 'wantonness of inquiry' during which "the doctor has the notion that if he excavates long enough and deep enough the answer [i.e., the organic diagnosis] will come up."

What are the characteristics that define and distinguish organic disorders and functional disorders? Primary organic disease involves objectively demonstrable tissue damage and resultant organ malfunction. By contrast, primary functional disorders present with symptoms produced by organs that are free of organic disease. The phenomena that produce symptoms occur as part of the repertoire of responses inherent in nondiseased organs. A runner's leg cramp, for example, causes intense pain and is at least temporarily disabling, but there is no disease, no organic pathology. It is in the repertoire of healthy muscle to go into spasm when sufficiently fatigued. Recovery from this very simple example of a functional disorder is effected by a period of rest, and the expected recovery tends to confirm the diagnosis of 'normal muscle cramp' as opposed to, say, a pathologic fracture.

The central nervous system modulates the activity of the stomach. Emotional stress has long been known to cause disordered gastrointestinal motility (as exemplified by the abnormal X-ray findings of narrowing of the outflow passage of the stomach in some infants with 'nervous vomiting'). The functional nature of such vomiting, and any other presentation of FTT, can be positively differentiated from organic disease by the infant's response to a procedure referred to as a 'therapeutic trial of comfort' (analogous to the rest period that relieves a muscle cramp). This therapeutic/diagnostic procedure is described below.

This is not to say that functional and organic disorders are mutually exclusive or that functional disorders cannot have serious or even fatal organic complications. However, it would be a mistake to focus on the disease caused by a functional disorder without attending to the underlying causes of that disorder. This error is exemplified by a recommendation for surgery to prevent vomiting in an infant with FTT due to unrecognized infant rumination syndrome (IRS).

Some Misconceptions About Failure to Thrive

FTT in infants and children is characterized by lags in somatic growth often accompanied by delays in cognitive, motor, and emotional development. The complexity of FTT has lead to conceptual over-simplifications.

One over-simplification is the notion that FTT is simply equivalent to undernutrition and that the problem can be cured by getting the child to eat more. Although poor nutrition may be the most salient feature in children with FTT, nutritional deficits are merely symptomatic of their causes. Unless poor nutrition is the result of organic disease or famine, the causes of inadequate food intake in FTT almost always involve problems in the relationship between child and principle caregiver. If the nature and causes of the dysfunctional relationship are not brought to light and addressed, clinical management of FTT is not likely to achieve optimal success.

Another misconception is that FTT signifies neglect. Although neglect may be the immediate cause of FTT in some cases, FTT can also occur in children of attentive, devoted parents.

Many published series of children with FTT were reported from large city hospitals serving medically indigent populations. As a result, the impression may be gained that FTT is a manifestation of poverty. However, FTT also occurs in middle- and upper-class families.

Defining the Presence and Causes of Growth Failure

Diagnosing growth failure on the basis of one-time measurements can be very misleading. It is helpful to plot a patient's growth before eliciting the history and then, during history taking, marking on the abscissa the ages of the child at which significant events occurred. The morphology of the curve may make cause-and-effect relationships stand out. A growth curve is far superior to a single set of measurements for evaluation of growth.

Among the many growth problems that may be mistaken for FTT, three warrant special mention: children who were born small for gestational age, children with constitutional growth delay, and children with genetic short stature. Infants born small for gestational age (i.e., intrauterine growth retardation) may be proportionate in length, weight, and head circumference, or their length and head circumference may be closer to normal than their weight. Those whose measurements are abnormally small but in-proportion are likely to have sustained an insult early in gestation, such as infection, chromosomal abnormalities, or exposure to teratogens; they generally have a poorer prognosis for catch-up growth. Those whose weights are disproportionately small for their lengths and head sizes are likely to have sustained an insult later in gestation (e.g., placental insufficiency, maternal malnutrition, or hypertension) and have a generally better prognosis for catch-up growth.

Constitutional growth delay is a normal variant seen in some healthy infants. Their growth progresses normally until sometime between 4 and 12 months when height and weight gains slow causing downshifts of their curves to lower channels. Slow or arrested growth persists for several months or more with spontaneous resumption of normal growth rates by 2–3 years of age, usually along channels lower than those followed prior to the period of delayed growth.

Children with genetic short stature are healthy and typically grow at normal rates, that is, parallel to, but at or below the 5th or 3rd percentile channels. There is usually a family history of small stature.

What Constitutes Nurturing and How Can Aberrant Nurturing Cause Growth Failure?

Emotional Aspects

Nurturing is the process by which infants survive, grow, and develop. It has emotional, cognitive, and nutritional aspects. It takes place within a dyadic relationship made up of the infant and its principal caregiver. In a well-functioning dyad, mother and infant interact in a reciprocal manner so that the infant's behavior and the caregiver's behavior are mutually regulatory. Actions, responses, and reactions occur during which the infant's cues are appreciated by the mother who then reacts in a timely, sensitive, contingent manner. These interactions, over time, foster the infant's sense of existing as an individual, physically and cognitively separate from his or her mother, and able to express a need with the expectation that it will be correctly interpreted and responded to. In order for a dyadic relationship to function, the mother must have first 'fallen in love' with her infant and 'claim' her baby as her own, physically and emotionally. 'Claiming' may occur immediately in the delivery room the moment the baby is put to her breast, it may occur some time later, or may not sufficiently take place at all. Without it, however, there is insufficient motivation for devoted caring and the capacity for feeling pleasure and pride in her baby and in her achievements as a mother, all of which foster effective nurturing.

The infant's capacity to engage in dyadic interaction depends upon its ability to manage transitions of state, including those required for feeding and eating. The infant *in utero* has little experience with changes in state. Feeding is continuous; there is no experience of hunger or thirst. The fetus experiences periods of sleep and wakefulness, but, unlike sleep and wakefulness in postnatal life, it does not occur in a changing environment, neither does it have much effect upon its environment. The intrauterine environment presents comparatively few stimuli for arousal and quiescence or changes in state.

Once out of the uterus, however, the infant is inundated by an environment that changes continuously.

The ability of an infant to make transitions from one state to another (e.g., hunger/thirst, followed by feeding, followed by satiety) depends upon, for example, the infant's temperament as well as the commitment, sensitivity, and skill of the mother.

Nutritional Aspects

Nutritional adequacy is comprised of a sequence of phenomena: (1) Adequate food must be available in the environment. (2) The child must take in adequate amounts of food. If food is not offered, older children can forage, but infants and toddlers depend entirely on their caregivers' recognition of, and responses to, their hunger signals. The dyadic relationship, of which feeding and eating is a major part, must work well enough to avoid feeding disorders that impair parents' ability to feed and their infants' ability to eat in a comfortable, satisfying manner. (3) Ingested food must be retained and not regurgitated or vomited before it can be digested. (4) The food must be digested before it can be absorbed. A child with cystic fibrosis, for example, may grow poorly, even though he or she may eat more than normal amounts of food, because the pancreas fails to secrete enough enzymes to digest the food that enters her intestine. (5) The lining of the small intestine is the surface through which the products of digestion enter the bloodstream. If this absorptive surface is reduced in area, as in celiac disease, weight gain and linear growth may be severely impaired. (6) As soon as the products of digestion have been absorbed, they may be utilized for growth. The presence of sufficient nutrients in the bloodstream promotes normal growth. However, severe physical and/or emotional stress can interfere with tissue-building and promote tissue-wasting so that the nutrients available for incorporation into growing tissues are, instead, washed out in the urine along with the waste products of normal metabolism.

This sequence of phenomena essential for growth is depicted in **Table 1**.

Defective nurturing can impair growth and development at steps 1, 2, 3, and 6. Each defect has a distinctive clinical pattern.

Irene Chatoor described six feeding disorders of infancy and early childhood, three of which are often associated with FTT: 'feeding disorder of state regulation' (onset during the first 2 months); 'feeding disorder of reciprocity' (onset between 2 and 6 months); and 'infantile anorexia' (onset during the transition from bottle or breast to spoon and self-feeding). During the first 2 months, successful feeding requires the infant to achieve the state of calm alertness. Infants who have difficulty with state regulation may be difficult to calm. Even though they may feel hungry, they may be unable to reach or maintain the necessary state of calm alertness or they may

Table 1   The sequence of nutrition and its impediments

	The sequence	The impediments
Step 1	The availability of food	Famine or starvation
Step 2	Ingestion of food	Infant feeding disorders: • Feeding disorder of state regulation • Feeding disorder of reciprocity • Infantile anorexia
Step 3	Retention of ingested food	Nervous vomiting syndrome Infant rumination syndrome
Step 4	Digestion of food	Pancreatic insufficiency (e.g., cystic fibrosis)
Step 5	Absorption of the products of digestion	Celiac disease
Step 6	Assimilation of absorbed food (anabolism and growth)	Stress

be too sleepy, too agitated, or too easily distracted by stimuli in the environment to complete or even begin a feeding. The infant may fail to gain weight. As the mother becomes more anxious, her ability to soothe her infant may deteriorate. A vicious cycle may become established, similar to the one that perpetuates 'persistent colic'.

Feeding disorder of reciprocity has its onset beyond 2 months of age. An infant older than 2–3 months has the ability to respond to, and initiate, social interactions with its caregiver. The dyadic relationship includes communicative behaviors such as mutual smiling, cuddling with holding, and mutual vocalizations. Feeding is a major focus of mutually pleasurably engagement during which cognitive and emotional development accompanies physical satisfaction and growth. If, during this stage of development, the caregiver is unresponsive to the infant's presence or signals, the infant may withdraw, become depressed, and fail to learn behaviors that foster nurturing. Instead of the social smile, he or she may avoid eye contact. Instead of cuddling and molding into a comfortable feeding position, he or she may be either stiff or limp. A caregiver's failure to attach emotionally to her infant impairs the infant's ability to become attached to the caregiver. It may become passive, lethargic, and disinterested in food. In the absence of environmental stimulation, such infants tend to engage in self-stimulatory behaviors, such as head-rolling or rumination. Growth and development lag; malnutrition and decreased resistance to infection may result in death of the infant or, if it survives, a damaged capacity to form attachments in later life.

Infant anorexia is another feeding disorder that may impair growth. Its onset occurs between 6 months and 3 years, a period during which the infant becomes increasingly able to function more independently and willfully. The offer of food by spoon normally becomes an interaction during which any discordant desires of toddler and mother need to be negotiated. This may be especially difficult in toddlers who, by temperament, are easily distracted. Feedings take too long. The parent tries harder to get food into their child's mouth. Attempts to trick or cajole the child into taking food are met with refusal to open the mouth, refusal to swallow, turning away, throwing food or utensils, or climbing out of the highchair. Parents become increasingly worried as the child seems to never be hungry, is increasing difficult to feed, and begins to lose weight. Chatoor found that "... difficult infant temperament was associated with higher mother–infant conflict during feeding ... the most difficult infants demonstrated the highest levels of conflict and growth failure."

Case Vignette: Infant Anorexia Syndrome

A 13-month-old boy, the product of an unplanned pregnancy, was born to a 36-year-old mother of two teenage children. He was described as an 'active baby' who had colic for the first 6 weeks. At about 4 months of age, the baby refused his formula. Different formulas were tried, but taken poorly. Spoon feedings became difficult and he needed to be coaxed or forced to eat. He chewed food put into his mouth, but refused to swallow it. Weight gain continued along the 10th to 25th percentile channels until 1 month prior to being hospitalized at 13 months of age.

At the time he was hospitalized, his weight was below the 3rd percentile. He refused almost every food except apple juice. After gaining the mother's acquiescence, we tested the premise that he would not feel hunger and that he would starve if left to feed himself without being fed by a caregiver. Accordingly, the infant was placed in a crib along with several bottles of milk and an abundance of finger-foods, but no apple juice. Foods were freely available for the taking, but no feeding was attempted. His mother and other observers mostly remained out of the room. After drinking and eating nothing for almost 18 h, he reached for a grape, put it into his mouth, chewed, and swallowed it. After a minute or two, he began drinking milk and avidly eating the cookies, grapes, and other finger-foods in his crib. During the remaining 4 days of hospitalization, meals consisted of finger-foods placed on the tray of his highchair. The dramatic improvement in his intake coupled with the unequivocal appearance of a weight gain trend made it easier for the mother to comply with our recommendations that she completely abstain from feeding him and, instead, provide food for him to feed himself. During an individual diagnostic interview, the father described his wife as "overprotective...even

the older ones, if they miss lunch, she's afraid they're going to die."

The baby was brought for a follow-up visit 12 days after his release from the hospital; his weight had risen from below the 3rd percentile to just below the 10th percentile. He was feeding himself 24 oz of milk a day plus meat, pasta, rice, and other finger-foods. He was still active and distractible.

FTT can occur in infants who eat more or less adequate amounts of food, but fail to retain what they have taken in. This is exemplified by two functional vomiting disorders of infancy: 'nervous vomiting' and IRS.

Before considering these two vomiting syndromes that impair growth, it is important to recognize that about 50% or more of healthy infants vomit or regurgitate during the first 6–18 months of life. This kind of vomiting occurs in normal infants and has been termed 'innocent vomiting'. It may range from effortless regurgitation to projectile vomiting. Its main characteristics are that there is no associated pain, nausea, loss of appetite, or underlying organic disease. It does not respond to dietary, positional, or pharmacologic measures. It does not impair weight gain, presumably because, if feedings are given liberally, to satiety, the infant compensates for what has been lost by taking in more. Innocent vomiting resolves spontaneously by or before 12–18 months of age. It is an important consideration in infants who fail to thrive because innocent vomiting plus growth failure may be mistaken for organic disease (e.g., gastroesophageal reflux disease or GERD). This may result in failure to diagnose and treat existing FTT and, instead, impose stressful diagnostic tests in pursuit of nonexistent diseases.

Nervous vomiting was described by the British physician, H. C. Cameron, in 1925. It is often associated with FTT. It may mimic hypertrophic pyloric stenosis radiologically. Nervous vomiting is a visceral reaction to stress and excitement causing the stomach's motility to be functionally altered, delaying passage of food into the intestine. Food is retained in the stomach longer than normally, keeping it as a reservoir for vomiting. Infant–mother interaction becomes increasingly distressed as vomiting increases and weight lags. A vicious cycle becomes established as a result of a breakdown in the nurturing relationship which, in turn, causes irritability, vomiting, and feeding difficulties leading to FTT (see **Figure 1**).

Once suspected, the diagnosis of FTT associated with nervous vomiting can be made by a positive response to a 'therapeutic trial of comfort'. This includes simultaneous efforts to relieve both the infant's and the parent's distress. Cameron wrote that, "treatment, if it is to be successful, must aim not so much at controlling the vomiting as at allaying the nervous unrest." This begins a therapeutic process that heals the parent–infant relationship and is necessary for continued improvement beyond the period of clinical management.

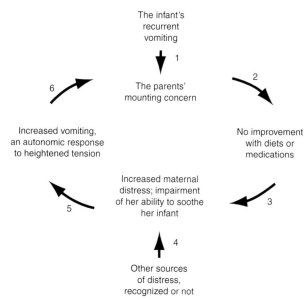

Figure 1 The vicious cycle of nervous vomiting. Modified from Fleisher D (1994) Functional vomiting disorders in infancy: Innocent vomiting, nervous vomiting, and infant rumination syndrome. *Journal of Pediatrics* 125(6) part 2: S84–S94.

Healing may be summarized schematically in terms of interrupting the vicious cycle shown in **Figure 1**. Vomiting must cease to worry the parents (step 1). To effect this attitudinal change, the parents must be shown that the infant's condition does not deteriorate when antireflux measures and dietary restrictions are dispensed with (step 2). This tests the premise that the infant is vulnerable to becoming ill when he or she is nurtured comfortably and normally, as though it were a healthy baby. This critical test may initially frighten the parents who fear that it will harm their baby. Therefore, it is best done in hospital on an infant's ward. The parents must first be assured that their child will be closely observed by a clinician who will supplement the parents' keen but unavoidably subjective observations with the objectivity of clinical practice. The infant is kept in a calm environment, comforted when fussy and fed liberally, as much as it wants whenever it wants it, vomiting notwithstanding. Burping is done during spontaneous pauses in feeding, not arbitrarily imposed in a way that interrupts a satisfying feed. Frustration and tension-producing procedures are avoided as much as possible. The baby's physical state is closely monitored by being weighed at least twice daily and by measurement of the specific gravity of spontaneously voided urine every shift. These data are charted graphically to aid in early detection of dehydration so that it can be treated without delay, in the unlikely event that it occurs.

Hospitalization also provides a supportive environment in which it might be possible for the mother to be relieved of burdensome outside responsibilities so that she can focus on the care and protection of her infant with the support of the nursing staff. Hospital personnel must not exacerbate the parent's irrational guilt for having 'caused' their infant's illness. Diagnostic tests to screen for anatomic and metabolic causes of vomiting and/or weight lag are necessary because the infant will experience illness, like any other normal child, after the hospitalization is over. The parent's previous experience of illness may have created a sense of their child's vulnerability, predisposing them to new worries that 'something may have been missed' or that the doctor, having based his or her diagnosis on clinical impression, may have been wrong. Therefore, a prudent series of diagnostic tests based on the differential diagnosis of the infant's presenting signs and symptoms is needed. Two caveats apply, however: (1) Stressful tests (e.g., over-night fasts in preparation for X-ray studies, blood drawings) should be minimized, spread out over several days, and not done in rapid succession so as not to cause stress-induced FTT in the hospital. This also allows sufficient time for a valid therapeutic trial of comfort. (2) 'Wantonness of inquiry' should be avoided. It is counter-diagnostic and counter-therapeutic.

Time is a powerful and humane diagnostic tool provided, (1) the physician and the family are reasonably confident that no hidden, potentially serious disease has been missed and (2) they can rely on their physician's open-mindedness, accessibility, and responsiveness should they have new concerns. When these two caveats are implemented, parents, who may have initially wanted 'everything done' to find the cause of their baby's growth failure, instead become deeply appreciative of the doctor's omission of unnecessary, distressing tests.

Confirmation of the diagnosis of FTT is only possible by the resumption of thriving. This can only be ascertained during continuity of care over time. Honest mistakes are always possible. The use of time in confirming the diagnosis is safer (and more satisfying) when the relationship between the doctor and the parents is one of mutual respect, loyalty, and accessibility.

Interrupting the cycle at step 4 involves diagnostic/therapeutic interviews of the parents, together and individually, to gain insight into the conditions of their lives and the sources of stress that interfere with nurturing.

Interruption of the cycle at steps 5 and 6 occurs when the parents' optimism returns as they see their baby begin to gain weight and fuss less, and as diagnostic tests confirm the absence of organic disease.

Case Vignette: Nervous Vomiting

A 6-month-old first-born girl was hospitalized for weight loss and recurrent vomiting since 6 weeks of age. She was weaned at 3 months because her mother felt her milk was 'drying up'. Thereafter, she continued to be fussy, vomited frequently, and passed loose stools. Diagnostic studies carried out during a 10-day hospitalization, which included a therapeutic trial of comfort, showed no disease.

She rarely vomited in the hospital and passed three to six stools of varying consistency each day. She was fed the same formula as before admission, as much as she wanted, whenever she wanted it, and she gained weight. Diagnostic interviews with each parent revealed severe marital discord that worsened after the mother discovered she was pregnant. The father objected to the pregnancy and became increasingly unavailable. The mother handled her baby hesitantly during the early days of hospitalization, but became more comfortable in caring for her as the infant improved. The baby was released from the hospital with no medications and the same formula that she'd been fed before admission. During the follow-up visit 2 months later, the mother reported that her infant's fussiness had greatly improved. Although she still 'spat-up' small amounts daily, this resolved by 15 months. The mother said, "I feel a lot better now knowing that nothing is really wrong with her. She's getting to be more fun." The mother was in the process of divorcing her husband. (**Figure 2** depicts the patient's weight course.)

IRS is the second functional vomiting disorder that causes FTT and can potentially lead to an infant's death from inanition. It typically emerges between 2 and 8 months of age, during the period in which infants normally develop an increasingly strong attachment to their caregivers. The development of IRS is predisposed by failure of reciprocal interactions and, in this respect, is similar to feeding disorder of reciprocity. The caregiver is emotionally distant, unable to sense the baby's needs, and poorly responsive to his/her signals. The baby learns to regurgitate

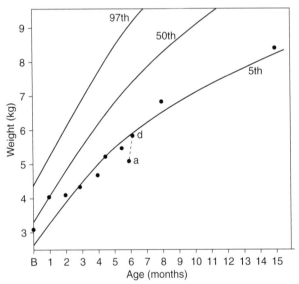

Figure 2 Weight of an infant with nervous vomiting at time of admission, a, and discharge, d, from hospital. The broken line indicates the weight gain during hospitalization (97th, 50th, and 5th percentile channels). Modified from Fleisher D (1994) Functional vomiting disorders in infancy: Innocent vomiting, nervous vomiting, and infant rumination syndrome. *Journal of Pediatrics* 125(6) part 2: S84–S94.

gastric content into its mouth for the purpose of self-stimulation and satisfaction of oral needs that would normally be supplied by the caregiver. Whereas older children and adults who habitually ruminate taste and re-swallow their regurgitated gastric content, infant ruminators are unable to contain all of it in their mouths and lose enough of it to the outside to cause progressive weight loss. Rumination is an acquired skill and is done voluntarily, without apparent nausea or distress. It ceases during sleep. It only occurs when the infant is awake, alone, quiet, and self-absorbed. It stops immediately when the baby senses the presence of another person, which might indicate the baby's intense need for social contact. The FTT does not improve with antireflux management, formula changes, arm restraints, or tube feedings. The parents' inability to nurture is one manifestation of more pervasive difficulties they have with interpersonal relationships.

The diagnosis of IRS can be confirmed by a therapeutic trial of comfort which, in this case, involves employment of a mother-substitute to hold, socially interact with, and feed the infant when it is awake. The mother-substitute should enjoy nurturing, be affectively attuned, empathic, and sufficiently observant to sense whenever the baby enters the self-absorb state of withdrawal that fosters rumination. He or she should respond to the baby with prompt engagement and interaction. A positive response consists of reversal of the weight loss trend followed by subsidence of rumination over time.

It is crucially important to foresee the likely reaction of the mother when the mother-substitute takes over the care of her baby for 1 week or more. Her feelings of guilt and resentment should be anticipated. Therefore, ask the parents' permission for this nursing procedure, the purpose of which is to help with data collection and childcare. The nurses who provide surrogate mothering should be perceived as helpers to the physician and parents, not didactic teachers of superior mothering technique. Diagnostic/therapeutic interviews with each parent alone and together strengthen rapport, provide opportunities for them to recognize their emotional pain, and explore their willingness to accept more formal psychotherapeutic help.

Case Vignette: Infant Rumination Syndrome

An 8-month-old boy began vomiting during an upper respiratory infection at 6 months of age. Respiratory symptoms cleared within a week, but frequent regurgitation persisted followed by progressive weight loss. He had been hospitalized twice for intravenous hydration and diagnostic testing. Dietary, positional, and pharmacologic measures failed to help. Finally, surgical fundoplication or gastrojejunostomy and pyloroplasty were offered, but declined by the parents. During his fourth and final hospitalization, it was noticed that his 'vomiting' ceased immediately each time he was engaged in social eye contact. The mother and maternal grandmother were enlisted in an effort to hold the patient while he was awake, watch for movements that preceded regurgitation, and immediately engage the patient. His weight stabilized, but the parents became exhausted within 2 days. The family agreed to employment of a special nurse to hold and nurture their baby. His weight increased during the next 6 days, accompanied by a decrease in urine-specific gravity indicative of improved hydration. Rumination became less frequent. He was discharged on the 20th hospital day and continued on an unlimited diet and no medication. The parents agreed to continue the special nurse at home for 10 days after his release from the hospital. On follow-up examination at 2.5 years, his weight was near the 5th percentile. He appeared well and had no further vomiting. **Figure 3** shows his weight course during hospitalization.

Table 2 compares the features of the two functional vomiting syndromes in infancy that cause FTT.

The failure of anabolism (step 6, **Table 1**) is a matter of controversy. Many accept as axiomatic that adequate nourishment automatically restores growth. The professor of social work, Dorota Iwaniec has written, "Aetiologic factors of inadequate intake of food are complex and varied, but the fact remains that all children who fail to thrive (for whatever reason) do not get sufficient calories into their systems." Nevertheless, the question remains: can emotional stress impair growth in the presence of adequate intake and retention of food? Can unhappiness stunt growth?

There are theoretical and experimental findings that would support this contention. Corticotropin-releasing factor (CRF) is a neuropeptide secreted by the hypothalamus. In nonstressed states, it regulates the cyclic secretion of adrenocorticotropic hormone (ACTH) by the pituitary gland which, in turn, controls the secretion of cortisol by the adrenal cortex. Stress increases CRF secretion, which results in activation, arousal, and anxiety. CRF is the physiologic mediator of unpleasant states associated with anger, sadness, or fear. Heightened CRF secretion has been shown in animal models to inhibit food intake and increase energy expenditure. Cortisol has catabolic effects when its secretion is increased during stress. A study of the sociologic, psychologic, and metabolic aspects of patients in the community of a metabolic ward revealed increased excretion of the products of tissue catabolism during stress, especially interpersonal difficulties between individuals having strong emotional

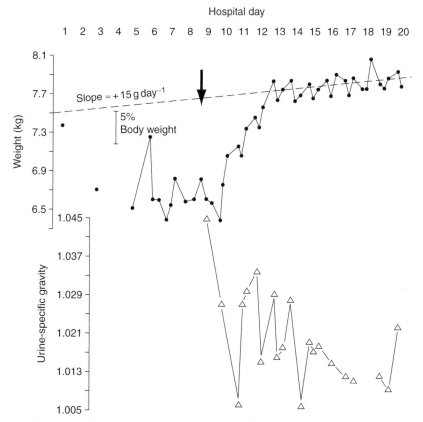

Figure 3 Response of an 8-month-old boy with IRS to 'holding therapy'. The arrow indicates the start of special nursing by a mother-substitute. There was subsequent catch-up weight gain and decreased urine specific gravity indicative of the infant's improved nutrition and hydration. Source: Fleisher DR (1979) Infant rumination syndrome: Report of a case and review of literature. *American Journal of Diseases of Children* 133(3): 266–269. © 1979 American Medical Association.

Table 2   Functional vomiting syndromes associated with FTT in infants

Features	Nervous vomiting	Infant rumination
The nature of vomiting	Involuntary Visceral Purposeless	Voluntary Behavioral Self-stimulatory
Age at onset	As early as newborn	After 2–3 months
Mothering	Attentive, but dys-synchronous; increases instead of relieves tension	Emotionally distant; little reciprocal mother–infant interaction
Typical circumstances	During the baby's response to environmental stimuli	In the absence of environmental stimuli
Prognosis if untreated	Failure to thrive	Failure to thrive and possible death
Treatment	Lessening excessive stimulation; alleviating the tension-producing quality of mother–infant interaction	Increasing environmental stimulation; satisfying the infant's needs by mothering

Modified from Fleisher DR (1994) Functional vomiting disorders in infancy: Innocent vomiting, nervous vomiting, and infant rumination syndrome. *Jounal of Pediatrics* 125(6) part 2: S84–S94.

ties. Widdowson's study of 100 children 4–14 years of age in two German orphanages following World War II showed that unhappiness may be associated with impaired weight gain and that the return of happiness was associated with catch-up weight gain in children given equal amounts of food.

Whitten *et al.* disagreed with Widdowson's conclusions. In 1969, they published a study in which infants 3–24 months of age with FTT were confined for 2 weeks in a severely deprived environment consisting of a windowless room in which they were fed by personnel who were instructed to not talk to, smile at, or hold them during feedings. They were handled only for basic physical care, except during infrequent, brief visits by parents. However, the infants were fed adequate amounts of food. They did not change their abnormal behaviors (auto-eroticism, watchfulness, apathy, or developmental lags), but they did gain weight (except for two who had feeding difficulties suggestive of Chatoor's infant anorexia syndrome). The authors concluded that children with FTT, but free of organic disease, never fail to grow provided intake of food is normal.

Uncritical acceptance of this concept raises two concerns. One has to do with the effect it might have on the clinical management of FTT; the other has to do with its failure to explain a type of FTT in which supra-normal amounts of food are consumed, but growth ceases in children who failed to develop emotional attachment to their mothers.

If FTT is viewed principally as a disorder of undernutrition that can be cured by food, the focus of management is shifted toward the immediate cause of growth failure (inadequate food intake) and away from its underlying cause (aberrant nurturing). However, the ultimate goal of treatment should be restoration of the mother's ability to love, attach, empathize, and find pleasure in her child, or if that's not possible, placement of the child with a surrogate caregiver who will provide adequate nurturing. If the child receives nutritionally and emotionally normal

nurturing, the child's nutritional state and growth are likely to normalize.

The type of FTT in which anabolism and growth fail despite supra-normal intake of food has been called hyperphagic psychosocial dwarfism. It typically presents in preschool and school-aged children and is characterized by functional growth arrest resembling organic hypopituitarism. When these children are tested to determine whether their pituitary glands can secrete growth hormone or their adrenal glands can secrete cortisol, secretion fails to occur. Growth hormone injections fail to restore growth. Typical behavioral features include bizarre overeating, consumption of unusual foods and liquids, a preoccupation with food and food hoarding, social withdrawal, offensive behaviors (e.g., fecal and urinary soiling, excessive attention-seeking), and attentional deficits. Invariably, there is a failure of emotional attachment between mother and child. Surprisingly, growth resumes, aberrations of eating and drinking abate, and function of the hypothalamic–pituitary–adrenal axis returns to normal when these children are admitted to a hospital or other nurturing environment. Moreover, the abnormal behaviors and growth failure may recur when the child is returned into his or her previous environment. Although this form of FTT has been recognized for more than 60 years, it is rare and the mechanisms by which it occurs have yet to be elucidated fully. Nevertheless, this form of FTT has theoretical as well as clinical importance.

Case Vignette: Hyperphagic Psychosocial Dwarfism, That Is, 'Deprivation Dwarfism'

An 8-year-old boy, the third of four siblings, was born to a 30-year-old mother. Birth weight was 8 lb, 5 oz. He thrived during his first year. His father deserted the family when he was 18- months- old. By 2 years of age, his weight lagged and bowel movements became more frequent. He was hospitalized in a community hospital at 2 years

9 months of age with complaints of diarrhea, abdominal bloating, and listlessness after meals. Celiac disease (gluten enteropathy) was suspected, a gluten-free, milk-free diet was prescribed and he was discharged to home. One month later he was allowed to eat Christmas dinner without restrictions. His family were 'frightened at the gigantic amounts of food' he consumed. His abdomen became acutely bloated to the extent that 'he couldn't get his breath'. At 4 years of age, the patient and his siblings were abducted by their estranged father who kept them for 5 weeks. Upon his return, he had a more ravenous appetite and passed loose stools three to fives times a day. Growth continued to lag. At 6 years, he was admitted to a university hospital for evaluation. Physical examination revealed an immature boy with a 'celiac habitus' (thin limbs, distended abdomen, and flat buttocks). However, no evidence of malabsorption, cystic fibrosis, or other disease was found, except that his bones were demineralized and at a 2–3 year level of maturation.

A gluten-free diet (devoid of wheat, rye, and barley) was ordered. Nevertheless, on the first hospital day, he was found in the ward kitchen eating a sandwich. Hoarded candy was found in his bed. He surreptitiously ate cookies and saved parts of his meal for later. Painful abdominal bloating and flatulence recurred almost daily. Bowel movements were passed an average of 1.7 times a day and all but two were formed. He wet his bed every night. He gained 9.25 lb in 25 days. He was discharged after 3.5 weeks with recommendations to continue his glutin-free diet, notwithstanding how impossible it had been to impose such a restrictive diet during his hospital stay.

Six months later, worsening behavior and difficulty in keeping him from eating glutin-containing foods prompted his enrollment in a school for handicapped children. He was notorious for eating his lunch en route to school. He stole food from other children's lunches, hid food in places where he could and eat it in secret, and was often found with bread crusts and leftovers in his pockets. "We have tried all methods, short of tying him up, to avoid his getting at food that makes him ill," his teacher wrote. At home, "it was guard duty 24 h a day . . . he couldn't stand to let a crumb of food on anyone's plate. When he goes to the store, he just looks at the food counter, not at the toys. Food is his way of life!" his mother said. Unusual appetites appeared, for example, he might eat five packages of gum, an entire 12-oz jar of peanut butter, or a bowl of pet food at one sitting. The patient was hospitalized for the third time for 1 month at 7 year 8 months of age. He had lost weight and had hardly grown during the 18 months since his previous hospitalization. His appearance, X-ray, and laboratory findings were essentially unchanged. Growth hormone responsiveness to an insulin challenge tested on the 13th hospital day was at the lower limit of normal (Re-testing done 7.5 months later showed a more robust normal response). Psychometric testing showed intelligence quotients (IQs) between 67 and 83, but he seemed depressed and poorly motivated during testing. Receptive and expressive language was 1.5–2.5 years below expected levels for his age. The patient was offered a regular diet and free access to food. His appetite decreased somewhat and gorging and bloating lessened. He played parallel to, rather than with, his peers. He showed a striking lack of loneliness for his mother who lived 3 h away and she visited only once during his hospital stay. One day, after not seeing or speaking with her for 2 weeks, he was offered the use of a telephone and help in making a call to his mother. He declined the offer without the slightest show of emotion. Similarly, his mother showed little evidence of missing her son. The patient's mother was 38 years old, mildly obese, depressed, and emotionally continent, other than for expressions of resentment toward her son. She described herself as "a change of life baby," her next older sibling being 15 years her senior. Her father died of cancer when she was a toddler and she grew up without a step-father. She described her mother as emotionally distant. "She never went out of her way or took an interest in how I did at school". "I raised myself on my own." Her pregnancy with the patient was unwanted. She felt depressed after his delivery and attributed that to not having anyone to help her at home with the new baby. Her husband, a construction worker, was 4 years younger than she and had not been heard from in 2.5 years. The patient's three siblings were 6–14 years of age, healthy and growing normally. After 27 days, the patient was transferred to a small pediatric convalescent facility. Gorging recurred during the first week there, then subsided. He gradually began to engage in play and defend his rights with his peers. He developed a friendship with the cook and began to show spontaneous affection. He returned home after 7 months. Binging recurred the first week at home and then subsided. When seen 3.5 months after his return home, he seemed well adjusted to the second grade. Although he was still quite interested in food, gorging, bloating, stealing food, and bizarre cravings were no longer evident. (**Figure 4** depicts his growth and physical appearance before, during, and after his stays in hospital.)

Clinical Presentations of Failure to Thrive

Knowing the features of each defined syndrome does not guarantee that the clinician who encounters a child with undifferentiated growth problems will easily recognize FTT. Giulio Barbero and Eleanor Shaheen identified four presentations of FTT: (1) growth lag with no other symptoms or signs; (2) growth lag with symptoms that mimic disease, but are not pathologic and not the cause of growth failure; (3) growth lag in the presence of child abuse; and (4) children with organic diseases who have

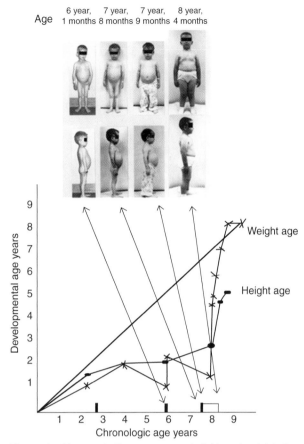

Age   6 year,   7 year,   7 year,   8 year,
      1 months 8 months 9 months 4 months

Figure 4   Changes in the appearance, height, and weight of a boy with 'hyperphagic Psychosocial dwarfism'. The ordinate indicates developmental age; the abscissa indicates chronologic age. The heavy diagonal line indicates the points at which developmental and chronological ages are equal. The solid rectangles on the abscissa indicate brief hospitalizations. The open rectangle indicates his 7-month stay at a pediatric rehabilitation facility. (All images conform to the same scale.)

super-imposed FTT caused by distortions in the parent–child relationship resulting from the parents' knowledge or perception that their child has a life-threatening or life-altering condition. The following case vignettes illustrate each of the four patterns of presentation.

Case Vignette: Type I – Failure to Thrive Without Other Signs or Symptoms

A 6-month-old girl presented with progressive weight lag since birth. She had no illness. Her appetite was described as 'excellent'. She had no diarrhea or frequent regurgitation and she was 'always in a good mood'. She was hospitalized for 10 days. Diagnostic evaluation showed no organic disease and she remained symptom-free. Her calorie count averaged $125 \, \text{cal kg}^{-1} \, 24 \, \text{h}^{-1}$ which was normal for her size. Her mother was 26-years-old who grew up as a rather indulged only child, the center of an adult world. She saw herself as a 'lousy housekeeper'

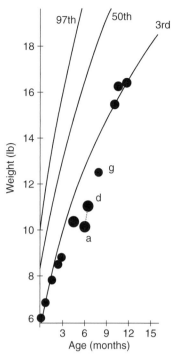

Figure 5   Type I FTT. a, weight on admission to hospital. d, weight at time of discharge. g, weight at the end of her 1-week stay in the care of her grandmother (97th, 50th, and 3rd percentile channels).

and her husband, a 30-year-old computer engineer, as a 'very perfectionistic, angry man'. The parents' interaction tended to perpetuate the mother's low self-esteem, inadequate performance, and depression, which, in turn, exasperated her husband. The family home had been severely damaged by a fire the previous year. The father's job was in jeopardy, and they were financially insecure. Although they denied any effects of these stresses during interviews, they were finally able to acknowledge their emotional distress during the pre-discharge conference. The baby was seen for follow-up 3.5 weeks after discharge. She had been cranky upon their arrival home from the hospital. The parents then went on a 1-week vacation leaving her in the care of the maternal grandmother who increased the frequency of feedings. The baby gained 11 oz during the 7 days she was in her care. Catch-up weight gain continued thereafter (**Figure 5**).

Type II – Failure to Thrive Presenting with Symptoms Suggestive of Organic Disease

A 7-month-old male infant was hospitalized for vomiting and weight loss during the preceding 3 months. Two upper gastrointestinal barium studies showed narrowing of the outflow tract of the stomach, delayed gastric emptying, and gastroesophageal reflux. Although endoscopy was unrevealing, biopsy of the esophageal mucosa showed mild inflammation. Interviews with each parent revealed that the family had moved three times during the preceding 2 years, most recently a few weeks before the patient's birth. The father, a young business executive, had been transferred from another state and assigned the task of making a large, but failing business profitable. He worked long hours. The mother, a successful career woman in her own right, spent most of her time in their new home mothering her 7-year-old son and new baby. She now lived far from friends and family and had not had time to make new friends. She was self-critical, perfectionistic, and felt at fault when anything went wrong at home. The relationships in this nuclear family had become critically stressed, but the crisis had not been apparent to them, partly because of their preoccupation with their increasingly sick infant and their frustrated attempts to find the elusive food intolerance they believed caused his vomiting. The infant was hospitalized for 12 days during which he was fed on demand, shielded from commotion, and comforted whenever he fussed. Vomiting decreased. The parents became aware of their state of emotional exhaustion during reflective discussions with the attending pediatrician. Their stated fear of loosing their child subsided as a weight gain trend began to emerge on the 4th hospital day. The patient was released from the hospital on no medications or special diet. Ten weeks later, the mother wrote, "he has gone for 6 days straight without even spitting up, much less vomiting!" (**Figure 6** depicts the weight course before, during, and after his hospitalization.)

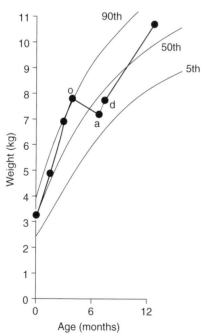

Figure 6 Type II FTT. o, onset of vomiting and weight loss. a, weight on admission to hospital. d, weight at discharge (90th, 50th, and 5th percentile channels). Adapted from Fleisher DR (1994) Functional vomiting disorders in infancy: Innocent vomiting, nervous vomiting, and infant rumination syndrome. *Journal of Pediatrics* 125(6) part 2: S84–S94.

Comment. This infant had gastroesophageal reflux and weight loss which became worse as treatments for food allergy and GERD failed and parental anxiety intensified. The possibility of nervous vomiting with FTT was considered and the approach to management was changed. The premise of organic disease was tested by omitting antireflux measures and elimination diets; this could be done provided the parents were assured that any signs of deterioration would be responded to immediately while he was closely observed in hospital. The crisis began to resolve when the parents' distress was discovered and acknowledged and when they experienced their infant's response to food and comfort as an unexpectedly welcome indication of his health rather than disease.

Type III: Failure to Thrive Accompanied by Physical Abuse

It would be wrong to assume that all children who fail to thrive are therefore abused. Such an assumption promotes judgmental attitudes that can wreck the collaborative relationship between clinicians and parents needed for successful outcomes. FTT commonly results not from abuse, but from dysfunction in the parent–child relationship, often despite the parents' sincere efforts to restore their child's health. Nevertheless, when abuse is part of the harm suffered by the child, its peril constitutes a pediatric emergency that mandates immediate hospitalization or other protective measures.

There are at least three kinds of abuse. The first kind is exemplified by the angry parent who loses control of hostile impulses and batters his or her child during a moment of frustration. The attack is not premeditated and the parents may be capable of genuine remorse afterwards. The second kind of abuse is exemplified by premeditated torture for sadistic pleasure. The third kind of abuse is also premeditated and is exemplified by the parent who creates factitious disease or suspicion of one. In the first kind of child abuse, the child must be protected from the parent, but the possibility may exist that, after treatment for anger management and/or psychotherapy, highly motivated parents may recover to the extent that it becomes safe for the child to return to their care. The second and third kinds of child abuse entail little or no possibility of safe restoration of the child to the care of its abusive parents.

Case Vignette: Type III Failure to Thrive – Factitious Disorder by Proxy (Also Known as Munchausen Syndrome by Proxy)

A 5-week-old girl born near term weighing 7 lb was said to have vomited her initial feedings and was switched to a soy-based formula. Alleged apnea spells in the newborn nursery prompted her discharge with an apnea monitor at 2 days of age. She was re-hospitalized at 4 days of age for unverified apnea and discharged 2 days later on a hypoallergenic formula. It was reported that vomiting occurred about 12 times a day and was often projectile. At 5 weeks, she had gained only 7 oz above her birth weight and she had a severe diaper rash.

At 8 weeks, she weighed only 1 oz more, her rash was more severe, and she was scrawny, irritable, and ravenously hungry. She was then hospitalized for 29 days. The patient consumed large amounts of formula with a calculated daily caloric intake that was twice that expected of a well baby. Nevertheless, she failed to gain. On the 11th day, it was noticed that her urines were invariably dilute. A sample of leftover formula and a sample of formula from an unopened bottle were taken from her room for chemical analysis. The sodium concentration in the small amount of leftover formula was found to be about half that of the formula from the unopened bottle. Despite the mother's assurances to the contrary, the formula had been diluted. The patient occupied a private room and was cared for exclusively by the mother and maternal grandmother. They always declined help from the nursing staff. Observations of feeding revealed that the mother related to her 10-week-old infant as though the baby were able to obey or disobey and respond to disciplinary measures. The mother and maternal grandmother fed the infant in an adversarial, teasing manner. In an individual interview with the father, he described his wife's medical history in terms that were typical of ongoing factitious disorder since childhood. On the 14th hospital day, the infant's malnutrition seemed life-threatening. She was transferred to the pediatric intensive care unit (ICU). The next morning, the attending, resident physicians and social worker held a conference with the family about the possibility that the patient's failure to gain weight might be caused by consumption of diluted or tainted formula. In order to test this possibility, we requested a 5-day trial during which the patient would remain in the ICU and be handled and fed only by the nursing staff. There was to be absolute physical separation of all friends and family members from the infant. They were told that, painful as the prospect of separation was, it was the only way we could explore the possibility that someone was doing something that was making their baby waste away. The parents reluctantly agreed to not approach the baby closer than 10 ft and a red line was taped to the floor encircling the crib which was placed directly in front of the nurses desk. The patient promptly gained weight. The diagnosis of Munchausen syndrome by proxy was confirmed and the baby was placed into high-quality foster care. When the foster-mother brought her for a follow-up visit 3.5 weeks after discharge, the patient was alert, no longer irritable, and had continued the catch-up pattern of weight gain that began in the pediatric intensive care unit (**Figure 7**).

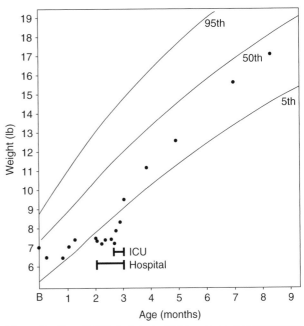

Figure 7   Type III FTT. An infant with Munchausen syndrome by proxy. The longer horizontal bar indicates the total period of hospitalization. The shorter bar indicates the part of her hospital stay during which she was closely observed in the intensive care unit, separated from her mother and maternal grandmother.

Type IV Failure to Thrive – Primary Organic Disease Complicated by Failure to Thrive

This form of FTT typically occurs in a child with an actual or perceived organic disease that may impair growth. Superimposed FTT occurs because the child's illness interferes with emotional attachment. For example, the parents may prevent themselves from attaching fully with their child because he/she might die and the anticipated grief might be overwhelming. Therefore, without being aware of it, they avoid 'getting too close'. Or, the demands of caring for their invalid child may seem daunting and even more difficult were their child unable to respond in ways that gratify and encourage his parents. The superimposed growth lag caused by aberrant nurturing of this type may not be appreciated by the clinician who attributes the growth problem entirely to the child's organic disease.

Case Vignette – Type IV Failure to Thrive

Mark was born to middle-class parents weighing 4 lb, 10 oz after a pregnancy complicated by premature rupture of membranes 1 week prior to the onset of labor. Delivery was by breech extraction which caused brachial palsy and multiple cranial nerve palsies, including an absent gag reflex. A feeding gastrostomy was constructed at 2 weeks, after which he received fully adequate amounts of nourishment by tube. He was admitted to a children's hospital at 18 months for evaluation of severe lags in growth and development. Developmental evaluation showed gross motor function at the 10-month level, communication at the 9-month level and a bone age of 6–12 months. Laboratory evaluation revealed no metabolic, absorptive, or infectious diseases. There had been no vomiting, diarrhea, fever, or unusual respiratory symptoms. He passed most of his time during the 22-day hospitalization lying in a crib, receiving almost continuous infusions of food via his gastrostomy tube. Feedings consisted of 60 oz of milk plus pureed solids and vitamins in excess of twice normal requirements. Nevertheless, he gained only 4 oz in 23 days. His growth failure and retarded development were attributed to brain damage. The patient was transferred to a pediatric convalescent hospital several hours drive from his parents' home. Mark's mother remained with him for the first week, after which she returned to her husband and three children and visited on weekends. Mark remained in the convalescent hospital for 4 months during which he developed remarkable catch-up growth and accelerated development. Oral feedings were begun during the second month and he fed entirely by mouth by the end of his stay, despite an improved, but still poor gag reflex. The gastrostomy was closed several months later because his swallowing was good enough to prevent aspiration pneumonia. Follow-up at 3.25 years of age showed a happy, sociable boy with a 20–30 word vocabulary. He rode his tricycle well. He had occasional 'mild trouble chewing bulky foods', but he had not experienced aspiration pneumonia. His growth spurt in the convalescent hospital had continued at home (**Figure 8**).

In an attempt to understand what had caused the patient's remarkable improvement, the mother was asked what she had experienced the day her son entered the convalescent hospital. "I was surprised the way they handled him; they just put him in the walker, in with the other children! I felt sad and depressed when I had to feed him by the tube and the others ate at the table." When asked how she felt at the end of the first week when she decided to return home, she said, "I felt torn by having to leave Mark, but I felt a little better because he looked as though he was improving. He walked in the walker, he was happier and he had some social interest. I didn't feel so afraid for him." She indicated that the experience allowed her to overcome her fears of and for her son, as well as the feeling that he was hopelessly damaged and more than she could manage. The experience seemed to have tipped the balance in her mixed feelings and enabled her to give and receive pleasure in her relationship with her child. She felt that his outlook for normal or near-normal mental and physical function was excellent – an expectation that was confirmed when the patient was last seen at 5 years of age.

Coda

In 1980, Donald Berwick wrote an article of enduring value on FTT. In it, he summarized principles for evaluation and management of FTT in terms of eight precepts which are re-stated in **Table 3**.

The multidisciplinary tasks required for management of FTT necessitate a team led by a biopsychosocial clinician. Otherwise, the imposition of the bias and time

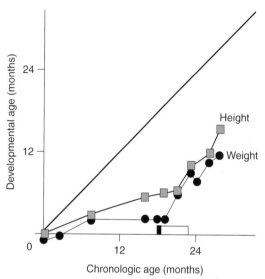

Figure 8 Type IV FTT. Growth of a prematurely born infant who sustained neurological damage during a traumatic delivery. Severe growth lag occurred during his first 18 months at home and during 22 days in hospital, despite adequate input of nutrients via feeding tube. Accelerated growth occurred during his 4-month stay in a pediatric convalescent hospital. (The ordinate indicates developmental age. The abscissa indicates chronologic age. The heavy diagonal line indicates the points at which developmental and chronologic ages are equal. The solid rectangle on the abscissa indicates the 22-day stay in hospital. The open rectangle indicates his convalescent hospitalization.)

Table 3 Principles of evaluation and management of FTT

- Involve the parents in the investigation and treatment of the child
- If the child is hospitalized, mold the physical climate of medical care to facilitate normal behavior and natural parent–child interactions
- Use laboratory investigations frugally and mainly to pursue clues from a careful history and physical examination
- Begin both evaluations and treatment with an interdisciplinary team
- Attend to interactional issues, even in the presence of organic disease
- Assure long-term follow-up
- Avoid moral judgments and threats of foster care
- Foster parental self-esteem

Adapted from Berwick (1980) Nonorganic failure to thrive. *Pediatrics in Review* 1(9): 265–270.

constraints characteristic of the biomedical model are likely to undermine the team's efforts and clinical success.

The naturalistic observations of clinicians and the data gathered by clinical investigators have created clear descriptions of FTT in its many types and presentations. However, much remains to be learned about FTT. Fortunately, contemporary neuroscience is beginning to elucidate relationships among experience, emotions, feelings, and health. The artificial designation of clinical problems as either mental or physical has outlived its usefulness. The development of a deeper, more integrated understanding of FTT shall bring us closer to its prevention and cure.

See also: Abuse, Neglect, and Maltreatment of Infants; Attachment; Colic; Endocrine System; Feeding Development and Disorders; Mortality, Infant; Nutrition and Diet; Physical Growth; Suckling.

Suggested Readings

Barbero GJ and Shaheen E (1967) Environmental failure to thrive: A clinical view. *Journal of Pediatrics* 71(5): 639–644.
Berwick DM (1980) Nonorganic failure to thrive. *Pediatrics in Review* 1(9): 265–270.
Chatoor I (2002) Feeding disorders in infants and toddlers: Diagnosis and treatment. *Child and Adolescent Psychiatric Clinics of North America* 11: 163–183.
Fleisher DR (1979) Infant rumination syndrome: Report of a case and review of literature. *American Journal of Diseases of Children* 133(3): 266–269.
Fleisher DR (1994) Functional vomiting disorders in infancy: Innocent vomiting, nervous vomiting, and infant rumination syndrome. *Journal of Pediatrics* 125(6) part 2: S84–S94.
Fleisher DR (1994) Integration of biomedical and psychosocial management. In: Hyman PE and DiLorenzo C (eds.) *Pediatric Gastrointestinal Motility Disorders*, pp. 13–31. New York: Academy Professional Information Services.
Heinrichs SC (2005) Behavioral consequences of altered corticotrophin-releasing factor activation in brain: A functionalist view of affective neuroscience. In: Steckler T, Kalin NH, and Reul JMHM (eds.) *Handbook of Stress and the Brain*, part 1. pp. 155–177. Amsterdam: Elsevier.
Iwaniec D (2004) *Children Who Fail to Thrive.* West Sussex: Wiley.
Powell GF, Brazel JA, and Blizzard RM (1967) Emotional deprivation and growth retardation simulating idiopathic hypopituitarism. *New England Journal of Medicine* 276(23): 1271–1278; 1279–1283.
Whitten CF, Pettit MG, and Fischoff J (1969) Evidence that growth failure from maternal deprivation is secondary to undereating. *Journal of the American Medical Association* 209(11): 1675–1682.
Widdowson EM (1951) Mental contentment and physical growth. *The Lancet.* June 16: 1316–1318.
Zeanah CH, Mammen OK, and Lieberman AF (1993) Disorders of attachment. In: Zeanah CH (ed.) *Handbook of Infant Mental Health*, 1st edn., pp. 332–349. New York: The Guilford Press.

Family Influences

B H Fiese and M A Winter, Syracuse University, Syracuse, NY, USA

Glossary

Confounded (variables) – When the effects of two or more explanatory variables or constructs on a response/outcome variable cannot be distinguished from each other.

Empirically based study – A research study designed specifically to test theoretical hypotheses using data.

Externalizing behaviors – Problem behaviors that are directed outward toward the social environment, such as aggression, disruptive and noncompliant behaviors, and attention or hyperactivity problems.

Family conflict – Controversy or quarrel among two or more family members.

Family rituals – Meaning and symbolic communication associated with family routines and activities that convey what it means to be part of a particular family and provide continuity in meaning across generations.

Family routines – Predictable, repetitive patterns of family activities (e.g., meals, bedtime, weekend, vacation) that involve instrumental communication (e.g., pertaining to what needs to be done and who will do it).

Heritable characteristics – Those characteristics of an individual that can be attributed to genetic factors.

Internalizing behaviors – Problem behaviors that are directed inward toward the self, such as depression, anxiety, withdrawal, and inhibition.

Interparental conflict/discord – Controversy or quarrel among two parents.

Marital conflict/discord – Controversy or quarrel among two married individuals.

Personal efficacy – A belief about the extent to which one's own efforts can influence or control events that affect his/her life.

Poverty line – The minimum level of income that an entity (typically a government) considers necessary for a family to achieve a standard of living that provides essential resources to all family members.

Socialization – A learning process by which one develops the skills and knowledge (e.g., the language, social skills, roles, and behavior patterns) to function in the culture and community in which they live.

Socioemotional adjustment – The relative social and emotional health of a person.

Temperament – The in-born, relatively stable, and enduring style or pattern of behavior that organizes and influences a person's personality and approach to the world. Overall temperament style is often classified along dimensions such as easy/flexible, difficult/feisty, and cautious.

Introduction

Families are organized systems that influence children within the larger world in which they reside. This entry considers how the child makes a contribution to family functioning and how families are organized systems embedded in their larger physical and social surroundings. Family routines, communication, and conflict are considered as sources of influences on infant and child development. Cultural variations, intervention strategies, and implications for public policy are introduced.

Family Influences on Infant and Early Childhood Development

Children's development is a family affair in many respects. First, to the extent that the environment (e.g., the community, the parents' educational backgrounds, family economic circumstances, and so on) allows, families provide food, shelter, and stability for children. Second, families are charged with helping children to develop emotionally. Children first learn about social and emotional interaction within the home; thus, their capability of coping with interpersonal situations hinges on this early experience. Third, family life builds upon past experiences, resulting in predictable routines and imparting of values across generations. The extent to which one generation is able to adjust successfully therefore depends in part on the extent to which the previous generation did so. However, the child is not a passive player in his or her unfolding story. Individual characteristics of the child such as gender,

temperament, and fit within the family as a group also contribute to overall health and well-being.

Because it is beyond the scope of this article to offer an exhaustive examination of family influences on child development, we endeavor to increase understanding and interest in how families are organized systems that influence children within the larger world in which they reside. Toward that end, we consider first how the child makes a contribution to family functioning and how families are organized systems embedded in their larger physical and social surroundings. Next, we examine multiple aspects of the family environment that shape infant and child development. We then place observations of researchers and clinicians in a cultural context that illustrate how families are embedded in cultures that in turn regulate child development. Finally, we conclude with notes pertaining to intervention and public policy.

Directions of Influence

Researchers' conceptualizations of how families influence child development have changed markedly over the past 50 years. At one time there was the notion that children were mainly influenced by their mothers, with the reason being that mothers were the primary caregivers and had the most contact time with children. This way of thinking led researchers to focus primarily on the dyad of mother and child, and important advances were made to suggest that responsive and secure forms of relationships were associated with more optimal outcomes for children. However, there is more to family life than how mothers and children get along. Researchers began to include fathers in their observations, although still not to the extent that they relied on direct observation of mother and child. The researchers concluded that not only do fathers play an active role in shaping children's lives but that the triad (mother, father, child) has an influence on development beyond that of either the mother–child or father–child relationship. Further, once fathers were brought into the picture it was possible to consider the role that marital relationships played in affecting parenting. For example, if couples are dissatisfied with their marriage there is a greater likelihood that they will experience difficulties in parenting which in turn affects their children. Thus, the first point to consider is that family influences on child development are not limited to mother–child, father–child, or even sibling–sibling effects but include how multiple relationships within the family are negotiated and influence each other. This is often referred to as a general systems model and is different from its predecessors that focused on individual functioning – either that of the parent or the child.

Associated with the general systems approach is the consideration of how children are influenced by, as well as how they influence members of the family. Children are not passive in family life. Indeed, such characteristics as gender and temperament will influence how the child fits in with the family as a whole. Take, for example, a male child born into a family of four older sisters. If the four older sisters vary considerably in their temperament and style there is a good chance the young boy may fit in quite well with few adjustments to the daily routine of the family. If, however, the son has a relatively active temperament and the daughters were all relatively calm babies there may be a disruption to the household that had not been anticipated. In this case, the child has a definite influence on the family. Over time, transactions occur between child and parents such that each affects the other. For instance, a fussy baby can be calmed by an experienced parent and soothed to sleep and in turn the parent can develop feelings of confidence. Alternatively, a baby can present challenges to an inexperienced parent and be difficult to calm, which reinforces feelings of ineffective parenting and in turn the child is more difficult to put to sleep.

An emerging body of literature suggests that family genetics as well as environmental characteristics influence child development. Studies of identical and fraternal twins who vary in shared genetic make-up suggest that there are aspects of personality development and reactions to stress that can be attributed, in part, to heritable characteristics. Most researchers in the field of behavioral genetics agree that biological contributions can also be understood in the larger context of family dynamics, with a focus on how biology and family processes interact in complex ways to influence child outcomes.

A fourth notion about family influences to consider is the extent to which their properties are universal. For quite some time, many researchers and theorists implicitly assumed that most well-functioning families were pretty much alike. These assumptions have been called into question as more attention has been paid to variations across cultures and ethnic backgrounds. Closer scrutiny of contextual influences on family functioning and its effect on child development has convinced researchers that a 'one size fits all' approach to describing healthy family processes is unlikely to be satisfactory. It is for this reason that we take a socioecological perspective when describing family influences on child development.

Families as Organized Systems within a Social Ecology

Urie Bronfenbrenner proposed that child development is regulated by a variety of contexts that are embedded in each other. As depicted in **Figure 1**, the child is at the center of an environment that includes many nested areas of influence that progress from very close and particularly

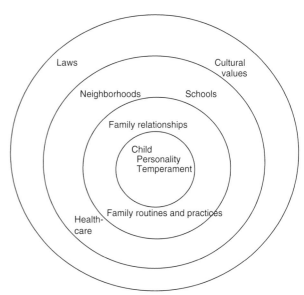

Figure 1 Ecological model of family influences on child development.

influential to more distant and indirectly influential. A child is most closely affected by the closest, or most proximal, level of organization – the family. At the next level, the child and family together are nested in their neighborhood and community, including daycare centers and schools, parents' workplaces, churches, and other area resources. At a third level, the child, family, and community are nested in the political and cultural context of the world in which they live. Each level is influenced by, and influences, other levels.

Consider the following scenario as an example of how this socioecological model is pertinent to family influences on child development. A child born prematurely may have a biological vulnerability that places him at risk for developmental delays. The responsiveness of his family in terms of providing language stimulation may affect the likelihood of whether he develops speech problems. If he shows signs of speech delays, the likelihood that he receives adequate early intervention will depend, in part, on the resources available in the neighborhood school system. The likelihood that the parents notice signs of speech delay or follow the advice of their referring pediatrician may be affected by cultural beliefs that they hold. In sum, a single risk of premature birth is likely to be influenced by multiple factors outside of the family home even though the child will be most directly exposed to his parents and siblings.

In order to understand these more proximal effects of the family it is important to understand families as organized systems. Imagine that a family consisting of two parents and three children is gathered in the kitchen preparing for the day. The father makes breakfast while the mom packs lunch for the children. Around the breakfast table, there is a quick check-in about everyone's day:

The mom reminds the older son that he has a field trip to the museum and also basketball that day; the youngest child has a music lesson after school and the middle child is volunteering at the neighborhood recreation center. That evening during a quick dinner, the Mom asks the children how the day went, and the oldest son tells everyone that he had an argument with a friend at school because the friend was picking on another boy. Everyone talks about how he might be sad, but why it is good that he refused to bully the other boy, and then Dad tells a story about how he once solved a similar argument with his best friend. The middle child chimes in about a time they had to break up a fight at the recreation center. Later, the older son helps his younger sister study for her spelling test. At bedtime, the father reminds the children to brush their teeth, the mom reads a story to the youngest child, and the father comes in to say goodnight to everyone before going over carpool plans with the mom. This scenario illustrates several elements of healthy family functioning. Families are charged with a host of tasks to insure the health and well-being of their children, and this family provided structure, protection, and guidance in six important areas:

1. Physical development and health, by providing shelter, meals, and a safe environment, and by encouraging healthy activities (e.g., basketball, tooth brushing).
2. Emotional development and well-being, by listening and caring about their children's feelings about arguments with peers and also by providing a warm and loving environment (e.g., kissing them goodnight and siblings helping with homework).
3. Social development, by supporting team activities (basketball) and volunteering in the community, and guiding children in peer relations.
4. Cognitive development, by valuing and helping with homework.
5. Moral and spiritual development, by helping them understand why bullying is wrong and encouraging community involvement.
6. Cultural and esthetic development, through supporting the trip to the museum and involvement in the community.

These six areas are at the core of what families must manage on a day-to-day basis in promoting the health and well-being of their children. One of the ways in which families organize their busy lives is through the creation of routines.

Family Routines and Child Development

Families are faced with multiple challenges in keeping the group together; they must balance the needs of individuals that differ in age and personality, connect the family to institutions outside the home, and provide some regularity and predictability to daily life. One way to consider a family as an organized system is to examine how, as a group, it promotes family well-being by dealing with these challenges through daily practices. These practices are often part of the family's daily routines.

While family routines tend to include a time commitment, some form of communication to get a task done, and are repeated over time, there is no standard definition of a family routine. What might be a routine practice for one family may not be for another family. However, routines should be differentiated from habits, because unlike habits routines are deliberate and organized.

The family scenario described previously demonstrates two common family routines: family mealtime (breakfast and dinner) and bedtime. In addition, families tend to also have routines such as weekend activities and annual celebrations (e.g., birthday parties, holiday meals). These routines tend to become more organized and predictable after the early stages of parenting an infant and into preschool and elementary school years.

The establishment of family routines speaks to how families organize behaviors that promote child well-being. Families are responsible for providing an environment where individuals can feel safe and secure (physically and emotionally) and where they can gain a sense of personal efficacy and feel that they belong to a group that cares for and nurtures them. The repetition of routines over time and the creation of family rituals may help establish such connections. The regularity of family routines and the sense of group belonging associated with routine events have been found to be associated with increased child well-being and mental health.

Family routines may also be important in promoting parenting practices that can promote healthy child adjustment. For example, there is some evidence to suggest that parents who have experience with childcare routines feel more competent when they have their first baby. However, it is important to understand that the relation between parent competence and family routines is likely the result of a series of transactions between the parent and children. For example, for infants, stable routines may result in an infant that is easier to soothe, more willing to take naps, and less likely to wake in the night. This predictability, in turn, could reduce parents' burden and concern, and increase their feelings of competence. Then, as parents engage in more caregiving activities they become more confident in their abilities and the routines themselves become more familiar and easier to carry out.

In addition to the regularity of family routines, the degree to which they are healthy for a family also depends on the emotional investment in the routine. As a family practices its routines over time, family members come to expect certain events to happen on a regular basis and they form memories about how family gatherings go. For some families, family gatherings are enjoyed and looked forward to because they are known to be a time of family warmth and support. Feelings of positive group belonging created during positive gatherings are associated with the increased well-being of children and families.

For other families, routine gatherings are not associated with positive feelings. Gatherings may be avoided because they might be a time of conflict and lack of support or because group gatherings feel like a burden. Alternatively, there may be few routines and instead the environment is very chaotic and therefore characterized by unpredictability, overcrowding, and noise. Feelings of burden or chaos may be particularly likely in families facing other types of stress such as poverty, dangerous living conditions, or family violence. Children raised in more chaotic environments have been shown to have more problems adjusting; for example, they have more difficulty reading social cues. Thus, in more disorganized and chaotic family environments children are more likely to experience a sense of unpredictability that gives them few guideposts for behavior or little sense of belonging to a supportive group. The energy of the group is often directed toward managing conflict and derision rather than establishing means to bring members closer to the fold. Over time, these repetitive interactions are linked to poorer behavioral and emotional outcomes for children. Let us now consider which aspects of family interaction are most closely associated with child development.

Family Interaction and Child Development

There are many aspects of family interaction that should contribute to healthy family functioning, including direct and clear forms of communication, effective problem solving, responding to the emotional needs of others, showing genuine concern about the activities and interests of others, and supporting autonomy. We focus on two family interaction patterns and how they impact child adjustment, namely family communication and family conflict. We discuss these interactions in terms of family systems theory; in other words, instead of considering each individual or even each pair of individuals, we focus on the family as a group that attempts to maintain some type of balance and integrity over time.

Family Communication and Child Adjustment

Communication has been defined in very different ways, but it appears that a central tenet of communication is the inclusion of symbols that can be shared with others. These symbols can unite and define a group. Surely, you can think of certain statements or ways of communicating that are central to your own family. Some are broader, group identifying communications; for example, 'We

called ourselves the Fabulous Five of First Street!' Other communication patterns are more specific, as in, 'My grandmother always patted me gently on my back when she wanted me to quiet down.' Of course, groups can also be separated with communication: 'We stopped talking to Uncle Bob after he said he disliked us', or 'Tell your father I said I am not talking to him'. Thus, family communication can be an indication of group identity.

Families have patterns of communication that convey relationship warmth and support. Variations in communication patterns during family routine events, for example, point to markers of family-level emotion regulation and are associated with child mental health. For example, during routine mealtime conversations families who indicate a genuine interest in their child's activities or are able to actively problem solve about emotional tribulations promote healthier behavioral adjustment in their children. Researchers speculate that over time, open and direct forms of communication support the child's ability to solve problems effectively on their own as well as serve as good role models for emotional regulation.

Family communication is also an arena in which child socialization takes place, and it is important to children's learning, cognition, and socioemotional adjustment. For instance, research regarding emotion socialization by Susan Denham, Judy Dunn, and their respective colleagues has demonstrated that greater discussion of emotions with preschool and young children is associated with greater emotional understanding and judgment skills, affective perspective-taking skills, and fewer problem behaviors. For example, families who use more elaborative and explanatory styles of communication about personal dilemmas with their preschool-age children provide a more solid base for their children in facing emotional disappointments, a normative event.

Similarly, how effective parent discipline is may depend on whether and how parents explain the transgression and disciplinary tactics to the child. Parents who clearly explain the reasons and consequences for discipline are less likely to use harsh forms of behavioral control and their children are more likely to respond immediately to discipline practices rather than develop coercive and manipulative forms of interaction. In other words, communication can turn everyday events into a learning experience.

For example, imagine that in the family scenario previously described, if the older son had bullied the other child and got in trouble at school. His parents might punish him with no television for 1 week and require that he apologize to the boy. At the minimum, the son learns that bullying will get him in trouble. His parents also have the opportunity to use this as time to elaborate on his feelings and the feelings of the boy who was being bullied. If they communicate to their son that his actions may have made the other boy feel sad and scared, just as the son had felt when he was once teased, and that the family has rules against causing emotional hurt to others, the communication has expanded to include the role that empathy and emotion can play when interacting with others. Research shows that the second scenario is most likely to result in the child really understanding and accepting the punishment, which is likely to be easier for his parents, too. In addition, the parents have also fostered emotional understanding and affective perspective taking by putting the son in the bullied boy's shoes; the likely outcome of this is empathy and an increased understanding of moral values. Empathy and emotional understanding is in turn associated with better social outcomes for children.

It is also important to take into account characteristics of the child when considering family communication. For infants, communication is often nonverbal but is still extremely important. For example, infants communicate through crying, so how the family responds to cries is a form of communication with that child. Consistently coming to the aide of the crying child and soothing him/her (e.g., by holding, patting, speaking in a soothing tone) can result in the child feeling more trusting that his/her needs will be met and therefore more physically and emotionally secure. Infants who have formed a bond with caregivers that is characterized by a sense of trust and security go on to develop more independence and their emotional and social development is likely to be healthier than children who learned early on that no one would be there for them.

Young, preschool, and school-age children are in a sensitive period of rapidly developing cognitive and language abilities that result in greater understanding of parental explanations; thus, communication is likely to be quite important to them as they begin to take part in discussions yet still look to parents for guidance. These children face the task of interacting with new adults and children at school, and so continued communication of emotional support as well as new discussion of social skills and problem solving is likely to be important to children of school age. As children age and their understanding increases, it is important for the family to guide the child while also allowing him/her to develop his/her own opinions and style.

Family Conflict and Child Adjustment

All families experience conflict. Everyday examples include mom and dad squabbling over whether to have chicken or pasta for dinner, parent and child disagreeing about pairing pink polka dots with orange tights for a school outfit, or siblings quarreling over who gets to ride in the front seat of the car. Families vary considerably in the overall amount of conflict they express on a given day, how disagreements typically progress, and how children are directly exposed to family conflict. The degree to which family conflict effects child development depends on several factors. For example, research by E. Mark Cummings and Patrick Davies, and John Grych indicates that if disagreements are resolved, mild or moderate conflict may not necessarily result in poorer child outcomes. However, sustained and/or unresolved conflict in the home can increase child adjustment problems. For example, physically or verbally aggressive disagreements can lead to feelings of fear and insecurity, which in turn can lead to poorer child adjustment. Similarly, if disagreements pertain to the child, it could lead to feelings of guilt, shame, etc., and result in poorer outcomes for the child.

Also, typical family conflict may be easier for children to cope with if the family is also very loving and warm to one another, but if conflict occurs within an environment that is chaotic or not supportive it may be more difficult for children to cope. Similarly, if the family is faced with greater levels of other stress – for example, financial problems, physical or mental illness, employment/work problems – conflict may add to that burden and result in poorer outcomes. Over time, conflictual family interactions can also reduce opportunities for effective problem solving and lead to greater negativity that can threaten the ability of the family to cope with stress.

Along with impacting children directly, conflict also impacts children by changing other aspects of the family environment. One primary example of this is when higher conflict leads to less warmth and less responsive parent–child relationships, which in turn puts children at risk for many different adverse outcomes. This can happen in many different ways, but consider the following examples: A husband and wife are arguing and the young daughter comes in to the room and asks for some homework help; dad snaps, 'Not now! Can't you see we're busy?!' Later the family is at the dinner table and the daughter and mom are disagreeing about table manners. These are both examples of 'spillover' effects, where a conflict in one relationship spills over into (impacts) another relationship. Such spillover effects indicate how discord in one relationship can soon impact the entire family.

In addition to depending on family characteristics, the impact of conflict on children can depend on characteristics of the children themselves. Children may react differently to conflict depending on factors such as gender, age, or general temperament. For example, compared to young girls, young boys seem to be more reactive to disagreements between their parents. However, when they are adolescents, the opposite tends to be true: Girls tend to be more vulnerable to interparental conflict than boys. Some theories support the notion that boys and girls are both vulnerable to family discord, but that boys tend to show it through externalizing behaviors (acting out, noncompliance, etc.) whereas girls tend to express their distress through internalizing problems (anxiety, withdrawal, etc.). Thus, a child's reactions to discord in the family are likely an expression of the discord itself as well as the gender and developmental stage of the child.

However, it is important to recognize that no matter how they show it, children are impacted by conflict in their environments even if it seems like they are not seeing it or do not understand it. Even infants recognize and are impacted by discord in the family. For example, infants in households with greater interparental discord respond less positively to parents than infants with parents who exhibit less discord. Moreover, discord between family members can impact parenting, which in turn impacts infant development (e.g., by making a parent less responsive to the baby's cries or less nurturing when interacting with the baby).

Family Stress and Child Development

What might be some of the reasons that families engage in conflict or interact in ways that are nonsupportive of their child's development? Researchers have attempted to tease apart the multiple influences on family life that can derail the positive effects of the collective group in nurturing young children. Two areas have received considerable attention: marital status and poverty.

Marital status. Families come in all shapes and sizes. Young children are raised in two-parent, one-parent, grandparent led, and foster homes and they may be raised in several types of these homes in their early years. During the last decade of the twentieth century, over 1 million children experienced divorce every year. In a summary of studies conducted over the last decade of the twentieth century, Paul Amato and colleagues found that children of divorced parents scored significantly lower than children with married parents on measures related to academic achievement, conduct, psychological adjustment, self-concept, and social relations. However, these differences in child functioning in divorced and maritally intact families are not evidenced in all children. One explanation is that marital discord plays an important role in how children are affected by parental divorce. It is not just the presence or absence of discord that is essential, but how conflict unfolds during the dissolution of the marital relationship. When children have not been exposed to discord prior to the divorce they experience more long-term difficulties in adjustment, perhaps due to an increase in conflict and stress following the divorce. In contrast, when there are relatively high levels of conflict prior to the divorce, dissolution of the marriage may actually be a relief for the child and there are fewer long-term effects on child adjustment. Thus, it is not divorce *per se* that results in poor adjustment in children but the context it occurs in.

For infants and young children, the effects of parental divorce appear to reside not so much with the break up of the marriage but with the nature of family relationships and the consequences of the divorce. When relationships are characterized by unresolved conflict or insecure and inconsistent parenting then children will fare less well in the long run. Likewise, when parents' negativity surrounding the divorce spills over onto the child, for example, through disparaging remarks made about the other parent or by putting the child in the middle of interparental disputes, child outcomes are likely to be less positive. Moreover, if disruption in the family is compounded with a change in economic circumstances, there tends to be more serious consequences to children. Indeed, economic hardship and poverty can affect family functioning in many ways.

Multiple risk and poverty. Children growing up in poverty are disproportionately affected by physical and mental health problems. At the beginning of the twenty-first century, the poverty rate was highest for younger children with 20% of children between the ages of birth and 5 years being raised in households below the poverty line. There are concerns that children exposed to poverty over long periods of time may be at increased risk for poor physical and socioemotional outcomes. Limited economic resources can have crushing effects on family life not only through its effects on the provision of basic needs but by its effects on relationships and parenting. For example, in studies of rural farm families in Iowa led by researcher Rand Conger, it was found that the downward turn of economic circumstances preceded marital distress and led to increases in hostile and coercive interactions between parents and their children. Economic strain has also been noted to be associated with maternal depression, a risk factor for compromised child outcomes. Extensions of these studies have found that in urban and ethnically diverse families, economic strain, maternal and paternal depression, marital conflict, and ineffective parenting style influence child outcomes in negative ways.

Typically, studies that examine the effects of poverty on child development consider multiple sources of stress on the family. Poverty is considered only a marker of a host of environmental risks that families are exposed to over time; in other words, it is not having less money that causes vulnerability, but it is all that comes along with it that matters. For example, one study considered the physical and mental health of children raised in poor rural communities and the multiple environmental risks they were exposed to including crowding, noise, housing problems, family separation, family turmoil, violence, single parent status, and lower parent education level. Consistent with the previous reports on multiple risks in less economically disadvantaged families, increasing numbers of risk factors were associated with more child psychological distress and feelings of less self worth. Other research has found that multiple risk factors including parent psychiatric illness, single parent status, parent education level, infant birth status, and neighborhood conditions predict preschool intelligence quotient (IQ) and early adjustment to school. What these studies have in common are findings that it is not a particular risk factor (e.g., marital status, education level) but the cumulative effect of stressors in the environment and on the family that predict child outcomes.

Risk conditions are compounded in nature and difficult to unravel. For example, the effects of family poverty of children's development depends on how long the poverty lasts and the child's age when the family is economically disadvantaged. Children raised in chronically poor households fare less well than children who experience poverty for brief periods of time as young infants. Similarly, single parent status also cannot be viewed as an isolated risk, as number of adults in the household has been identified as a marker of social capital known to be associated with many different child outcomes.

Perhaps one of the most difficult multiple risk contexts to disentangle is that of the overlapping effects of economic conditions and ethnic background. In many empirically based studies of family effects on child development, poverty is confounded with minority status. A recent exception is a study employing the National Longitudinal Survey of Youth. Robert Bradley and colleagues examined nearly 30 000 home observations of young children diverse in economic and ethnic backgrounds. Because of the relatively large sample size, the researchers were able to distinguish between poor and nonpoor European American, African American, and Hispanic American families. In general, they found that poverty accounted for most, but not all, of the differences between the racial and ethnic groups when it came to less-stimulating home environments (availability of books, having parents read to the child, parent responsiveness). There were some differences, however, that were attributed to ethnic background when controlling for poverty status. For example, European American mothers were more likely to display overt physical affection during the home observation than African American mothers. There were no ethnic group differences in the likelihood that mothers would talk to their infants or answer questions prompted by their elementary school-age children. Thus, the distinguishing characteristics most often associated with poor outcomes for children such as enriching home environments were more closely associated with low-income status than with ethnic background.

We provide these examples of multiple-risk to highlight the multifaceted context of family influences on child outcomes. It is not sufficient to note that children from poor families are at greater risk for developing certain physical and mental health problems than their more economically advantaged peers. Nor is it sufficient to

assert that children raised in warm and supportive households are less likely to develop mental health problems than children raised in harsh and rejecting environments. The consequences for children's development are too far-reaching to expect that family influence would be simple and uniform. Therefore, a consideration of the family must be sensitive to multiple avenues of effect while also taking into account that there is diversity in the ways in which families go about the tasks of raising children. Let us examine some of the cultural variations that come into focus when we consider family influences on child development.

Cultural Context of Family Influences

Families are embedded in culture and at the same time responsible for imparting cultural values to their children. While there are instances where elaborate and highly ritualized family events clearly mark belonging to a specific culture (such as christenings or naming ceremonies), many of the more mundane day-to-day practices are more subtly influenced by culture. For example, feeding practices of infants and the use of highchairs is integrally linked to cultural values. In western Anglo cultures, there are implicit values for independence and autonomy beginning at a relatively young age. In contrast, in Latino cultures and some Puerto Rican families, greater value is placed on good manners and staying close to the mother during the toddler years. In studies of family mealtime practices, Anglo mothers are more likely to be observed using highchairs and letting their infants and toddlers feed themselves. Robin Harwood and colleagues speculated that this type of feeding practice encourages independent behavior at a young age. Conversely, Puerto Rican mother were observed to hold their infants on their laps and spoon feed the child. The researchers proposed that this type of practice is based on a cultural value of good manners and young children being close to caregivers. This is not to say that either type of feeding pattern is 'better' than the other. Rather, each pattern is consistent with cultural-specific notions of what it means to be a family member.

Another example of cultural variations in family practices concerns discipline and personal transgressions. Children around the world get into trouble. However, the type of misdeeds that get them into trouble varies by culture. In western cultures, researchers have noted that parents will talk proudly of their young child's independence and boldness such as the 4 year old backing the car down the driveway. The same parent, however, will discipline the child for hurting his sister's feelings. In contrast, parents in Eastern cultures are more likely to punish their child for embarrassing them in public or for an act that causes shame to the family as a group. Of course, these are generalizations and there will always be subtle contrasts within each family. The point to be made here is that something as apparently simple as setting family rules or feeding a child in a high-chair is regulated, in part, by cultural context.

Altering Family Behavior

During infancy and early childhood, different types of problems may arise that call for intervention efforts aimed at changing family behavior. The intent of changing family behavior in these instances is to place the child back on course so that development can unfold and risk is minimized. There are at least four ways that professionals may engage with families to improve the health and well-being of infants and young children. The first approach requires making adjustments to the child's condition so that the family can maintain its daily routines and organized features to support the child's development. This type of intervention is often seen in cases where the child's condition is remedied through surgical interventions such as repairing a cleft palate so that the child can develop fully functional eating and speech practices. Another example would be instances where simple alterations in the family diet need to be made to accommodate a child's food allergy.

Most circumstances that call for professional advice, however, require fuller involvement by the family. A second type of intervention is family education. For families who are relatively well organized and experiencing relatively little environmental stress, educational programs aimed at specific problems can often be quite effective. For example, brief educational programs for toilet training or that target creating bedtime routines to address sleep problems have been helpful for worried parents of toddlers. Parenting education programs are also beneficial in teaching parents skills associated with discipline, setting routines, and effective listening.

The third type of family-based intervention falls under the category of redefining relationships. These types of interventions can be aimed at either adult relationships, parent–child relationships, or entire family relationships. As discussed previously, conflict can have a destructive and toxic effect on child development. In these circumstances it is important to work with parents and children to reduce conflictual patterns of interactions and promote effective conflict resolution and problem solving. Relationships sometimes need to be redefined if there has been a history of abuse and neglect. Parents who, as children, experienced harmful child raising experiences of their own often carry with them images of past hurt that can influence how they interact with their infants and young children. Sometimes referred as the 'ghosts in the nursery', it is important for parents to be able to distinguish their history of rejection and hurt in past relationships from their current relationships so they can be responsive to their young infant.

A fourth type of family-based intervention arises when there is a divorce in the family and relationships and routines need to be re-defined. In these instances, the break-up the marital relationship calls for a negotiation between the parents as to who will be responsible for different aspects of the child's care. This extends well beyond custody arrangements or who 'has the kid on the weekends'. Such issues as bedtime, diet, and peer and social activity choices need to be agreed upon for the child to have a consistent set of rules and routines.

Family-based interventions may be used at different points of the lifecycle. It is not unusual for transition points such as birth of a new sibling, move to a new neighborhood, or divorce to upset the family system. It is during these transitions that families often need to reorganize their routines, reassess their rules, and redefine their relationships. At times, professional advice can ease these transitions.

Family Influences and Public Policy

It is beyond the scope of this article to address the many ways that family life is a priority for public policy. However, we close with a few comments about the intersection between family systems, influences on child development and future public policy. Most policy makers would support the contention that the overarching goals of a family are to promote healthy child development: shelter, a safe environment, emotional support, and opportunities for educational, moral and esthetic enrichment. What is the subject of disagreement is how to set policy and invest public funds to meet the multiple needs of families in such a way that the benefits outweigh the costs to society. We have seen that families are not simple systems and that their influence on child development is complex. However, complexity is not an excuse for ignoring the very real needs of children and their families. There are several examples where changes in public policy have benefited children. Changes in child labor laws, advertising restrictions on children's television programs, and seat belt laws are just a few of the public laws and policies that have been put in place in the best interest of children. The best interest of families must also be preserved by providing for flexibility in the workplace that can accommodate the need for regular routines in the home, ensuring accessibility to low-cost transportation to assist families with multiple healthcare needs, and recognizing that multiple adults play a significant roles in children's lives; these are but a few of the arenas to which policy makers can direct their attention. Family life is not likely to become less hectic or less complicated over the next 50 years. However, the importance of family influences to child development will assuredly remain firm.

See also: Behavior Genetics; Birth Order; Demographic Factors; Divorce; Emotion Regulation; Marital Relationship; Mental Health, Infant; Mental Health, Intervention and Prevention; Parenting Styles and their Effects; Routines; Social Interaction; Socialization in Infancy and Childhood; Temperament.

Suggested Readings

Bronfenbrenner U (1979) *The Ecology of Human Development.* Cambridge, MA: Harvard University Press.

Clarke-Stewart A and Dunn J (2006) *Families Count: Effects on Child and Adolescent Development.* New York: Cambridge University Press.

Cummings EM, Davies PT, and Campbell SB (2000) *Developmental Psychopathology and Family Process.* New York: Guilford.

Denham SA (1998) *Emotional Development in Young Children.* New York: Guilford Press.

Fiese BH (2006) *Family Routines and Rituals.* New Haven, CT: Yale University Press.

Hetherington EM and Kelly J (2003) *For Better or For Worse: Divorce Reconsidered.* New York: W. W. Norton and Company.

Parke RD (2004) Development in the family. *Annual Review of Psychology* 55: 365–399.

Plomin R, DeFries JC, McClearn GE, and McGuffin P (2000) *Behavioral Genetics.* New York: Worth Publishers.

Family Support, International Trends

N Tolani and J Brooks-Gunn, Columbia University, New York, NY, USA

Glossary

Childrearing leave – Leaves from employment developed primarily as a supplement to maternity leaves or as a variation to parental leaves. These leaves are typically longer than maternity leaves and are paid at a lower level; as such, these leaves are often described as a 'mother's wage' because it may not be limited to parents who are employed.

Family leave – A social policy allowing employees to take a specified amount of time off work in order to deal with the needs of their families. Maternity, paternity, parental, childrearing leaves are subsumed within this category. Family leaves may be paid (taxable income) or unpaid.

Maternity leave – Job-protected leaves from employment for women prior to giving birth and following childbirth or adoption in some countries. Some countries provide compulsory 6–10 week leave period following childbirth. Policies in most countries allow women to combine pre- with postbirth leave periods.

Parental leave – Gender-neutral, job-protected leaves from employment that typically follow maternity leaves and permit either fathers or mothers to choose or share who will take advantage of this benefit.

Introduction

Infants and young children require a high level of investment of parental resources: financial, emotional, physical,

and mental. Reconciling the inherent conflicts between work obligations and families' needs are a source of stress for most employed parents. While the responsibilities associated with raising infants and toddlers may be universally felt, governmental supports provided to working parents vary widely between countries. Programs and policies designed to support the working parent can benefit parents and children, mitigating the negative impacts of poverty and stress and promoting school readiness and socialization. Most advanced industrialized nations, as well as a growing number of developing countries, have placed high priority on implementing child and family policies that support working parents and improve the health and well-being of children, such as family leave and early childhood education and care (ECEC). Understanding the linkages between social welfare policies and child well-being should be a priority for researchers, practitioners- and policymakers alike.

In Canada and in many countries belonging to the European Union (EU), the dominant model is to provide family-leave options for employed parents and ECEC for infants and children because they are considered beneficial for them regardless of their parents' employment status. (There are 25 member countries within the EU, including: Austria, Belgium, Cyprus, Czech Republic, Denmark, Estonia, Finland, France, Germany, Greece, Hungary, Ireland, Italy, Latvia, Lithuania, Luxembourg, Malta, Poland, Portugal, Slovakia, Slovenia, Spain, Sweden, the Netherlands, and the UK.) These programs are largely universal, voluntary, and available to families with children irrespective of their income. This reflects the prevailing attitudes of the Organization for Economic Cooperation and Development (OECD), a group of countries who believe that early childhood care and education are essential and inseparable components of services for children of working parents. (There are 30 member countries of OECD, including: Australia, Austria, Belgium, Canada, Czech Republic, Denmark, Finland, France, Germany, Greece, Hungary, Iceland, Ireland, Italy, Japan, Korea, Luxembourg, Mexico, the Netherlands, New Zealand, Norway, Poland, Portugal, Slovak Republic, Spain, Sweden, Switzerland, Turkey, the UK, and the US.) These policies also foster gender equality by strengthening links between mothers and employment, and encouraging fathers to participate more in caregiving at home. These attitudes are starkly contrasted with the current modality in the US, an industrialized nation with a robust economy and high labor force participation rates for mothers and father. Yet parents in the US are often required to create their own childcare solutions, which have exacted a significant toll on the health and well-being of working parents and their children. While many of the challenges related to reconciling work and family are not unique to the US, cross-national comparisons do suggest that the US is faced with a more acute set of circumstances than other industrialized nations. The conditions and outcomes of the disadvantaged working families in the US are particularly formidable when compared cross-nationally. Despite vast differences in geography, political climate, and racial diversity, many lessons may be learned from observing the benefits and services offered to parents in both English-speaking countries and industrialized European countries.

This article will review trends in policy approaches of advanced industrialized economies, comparing the US with other member nations of the OECD within three domains: prenatal and postnatal period policies (e.g., family leave), policies for the first and second year of life (e.g., early childcare programs), and policies for the third and fourth year of life (e.g. early childhood education programs). Policies within each time period are different. For example, prenatal and postnatal programs include maternity, paternity, and family-leave programs, as well as nutritional programs such as the Special Supplemental Nutrition Program for Women, Infants and Children (WIC) in the US. Programs designed to support working families with children of 1–2 years of age include family-leave policies and childcare programs. However, because so few countries provide working parents with leave benefits greater than 52 weeks (e.g., Sweden), the primary focus of this section will be childcare programs. Finally, programs for working families with children 3–4 years of age include childcare and early childhood education. This article will conclude with a discussion of the effects of these policies on the working parents and children they are designed to serve. It is important to note that pre- and postnatal policies are often embedded within a variety of health and social service programs (e.g., health insurance, immunization services) focusing on the medical care of infants and toddlers. The present article will not discuss these programs; interested readers are referred to sources within the 'Suggested readings' section provided at the end of this article for links to further information on these topics.

The Working Family: Sociodemographic Shifts and the Need for Support Programs

The implementation of support programs for families with young children, such as ECCE; maternity, paternity, and parental leaves; and tax benefits in industrialized countries, developed in partnership with demographic shifts occurring worldwide, such as rising female participation rates in the labor market and changes in family structures (e.g., single-headed households, female-headed households in particular). For example, according to US census data from 1995 to 2001, a labor force transformation occurred during the latter half of the twentieth century. Whereas 28% of women in the US were in the labor force in 1940, 38% were in 1960, and 60% in 2000. Because the most dramatic increase appeared in married mothers, participation in the labor market also had significant implications for the traditional nuclear family in the US. In 1960, approximately 25% of married mothers in the US were employed; however, by 2000, almost 75% of married mothers in the US were employed. Put another way, in 1930, approximately 55% of children in the US lived in homes where only one parent worked outside the home; by 2000, about 70% of children lived in homes where either one or both parents were employed.

In 1996, labor market shifts within the US were primarily driven by the enactment of the Personal Responsibility and Work Opportunities Reconciliation Act (PRWORA) established by the Temporary Assistance to Needy Families (TANF) which expects maternal welfare recipients to find employment when their babies are as young as 3 months of age. In 2002, more than half of all US mothers were employed when they gave birth and more than half returned to work within 3 months of childbirth. Sixty-three per cent of mothers returned to work before their children's first birthday. Earlier patterns of employment differed greatly; for example, in 1970, only 27% of mothers with infants were employed. With the exception of the Nordic countries, US maternal employment rates for mothers of young children exceed rates found in Continental European countries. Interestingly, cross-national data from the Luxembourg Income Study (LIS) suggest that parents in the US work more hours per week, on average, than dual-earning couples in Continental European and Nordic nations, like the UK and Sweden, respectively. This again highlights the unique set of challenges faced by working families in the US.

However, parents in the US are not alone in their struggles to reconcile demands between work and families. Employment rates of married and cohabiting mothers and fathers in the US match those in other wealthy, European nations. For example, in the late 1990s, Continental European countries reported married and cohabiting maternal employment rates ranged from 40% to 65%, while over 92% of married and cohabiting fathers were employed. While paternal employment rates within the Nordic countries matches those found within the Continental European nations, maternal employment was higher, exceeding 76%. Research conducted by the OECD has found that in both regions, however, over 60% of mothers with children aged 6 years or younger were employed. Labor market researchers have also reported that developing nations will witness similar trends in female employment rates; such increases will naturally correspond with increased demands for policies reconciling work and family issues. Upward trends in female employment are expected to continue and the policies that governments create to reconcile the conflicts between employment and parenthood will be critical. In sum, while increases in female labor market participation were seen in many countries, policy responses enacted by governments varied dramatically.

Pre- and Postnatal Policies

Family Leave: An Introduction

In nations that encourage employment, it is assumed all too often that mothers and fathers will be able to manage childrearing responsibilities with intensive work schedules. However, these responsibilities can prove overwhelming during the first few weeks after childbirth, and indeed the first few years of a child's life. Public family-leave policies are one of the most important benefits that governments can provide working families. Family-leave policies refer to a package of benefits, consisting of maternity leave, paternity leave, parental leave, and childrearing leave. Job-protected leaves following childbirth and adoption are the standard practice in industrialized nations, although the support packages provided by English-speaking countries are often less generous.

History of Family-Leave Policies in Europe

Paid maternity leaves originated as part of the first social insurance program enacted by Bismarck in Germany in the latter half of the nineteenth century. Similar social insurance programs, providing health insurance, paid sick leave, and paid maternity leaves, were subsequently developed in Sweden (1891) and France (1928). In 1919, the International Labor Organization (ILO) adopted the first convention on maternity protection, stating that

employed women are entitled to 6 weeks of paid maternity leave preceding childbirth and 6 weeks following childbirth (this was compulsory). These policies were revised and expanded further in 1952. However, significant trends occurred between the 1960s and the 1970s in the Nordic and Eastern and Central European countries when these countries implemented paid parental and childrearing leaves (respectively). During this time, as labor force participation rates of women also began to rise, the trend in OECD countries was that of increasing generosity: maternity leaves were extended and a greater sense of gender equity was created through the provision of parental leaves as a supplement to maternity leave policies. While such benefits were originally created for mothers, the 1970s witnessed an emerging concern for fathers in the form of paternity leaves. Around the same time, concern for children's well-being emerged with the implementation of childrearing leaves. In 1992, an EU directive mandating a paid 14-week maternity leave was adopted. In 2000, the ILO Convention on Maternity Protection was revised and adopted. In brief, this policy recommended a 14-week job-protected maternity leave (inclusive of 6 weeks prior to childbirth), and allowed for the extension of maternity leaves in the event of complications of illness due to the pregnancy or childbirth. It also stated that cash benefits should be publicly funded and 'at a level that ensures that the woman can maintain herself and her child in proper conditions of health and with a suitable standard of living'. Adopted as the new world standard, 128 nations provided at least some paid and job-protected maternity leave in 1999. Over the years, leave policies have gradually shifted from a narrower and inflexible approach, focused on traditional, gendered work and family issues to the broader realization that commitments conflicts between work and family obligations are also an issue for men given the increase in maternal employment rates. However, substantial variations exist between the benefits provided by these countries. Paid leaves average about 16 weeks, including 6–8 weeks of time off pre- and postbirth. In some countries (e.g., France) leave is mandatory, while in others it is voluntary. Most countries who do provide working parents with paid leaves stipulate the leaves to be maternity leaves (vs. paternity or parental leaves). In most OECD countries, the cash benefit provided while working parents are on leave ranges from 70% of the worker's prior wage to the full wage or maximum provided by the social insurance program funding the leave benefit. Benefits in the remaining OECD countries replace between 50% and 70% of the lost wages.

Family-Leave Policies in the US

The current family-leave policy began as a statewide phenomenon in 1867, wherein Wisconsin restricted the total number of hours women would work on a daily basis to 10, presumably because working beyond that point would have damaging consequences for their childbearing abilities. At the federal level, the Supreme Court enacted a similar ruling in 1908. Further policy shifts did not occur until 1978, when the Pregnancy Disability Act amendment to Title VII of the Civil Rights Act was passed. Whereas earlier legislation was based on protection of women's health, this amendment included gender and sex discrimination as a basis for future leave policies. Sex discrimination was legally expanded to include discrimination based on childbirth or related medical issues. In 1987, nine states had enacted unpaid maternity leave policies and by 1989 this figure had risen to 14. Despite its initial introduction in 1985, the Family and Medical Leave Act (FMLA) was not enacted until 1993 under President Clinton. This is a US labor law requiring employers with 50 or more employees to allow eligible workers to take up to 12 weeks of unpaid leave per 12 months for multiple reasons, including caring for a newborn infant; handling adoption or foster care placement issues; or caring for a sick child, spouse, or parent. That the leave is unpaid differentiates the US from all European countries which provide some type of income support for at least part of the leave period. The focus within the US-system is less on maternity protection and more on disability, which encompasses pregnancy and childbirth. Working mothers may return to the same position they held prior to childbirth, or to a position that is equal in pay, benefits, and responsibility. These employees are also entitled to a reinstatement of all benefits they received prior to taking the leave.

In 1996, President Bill Clinton signed into law the PRWORA, a comprehensive bi-partisan welfare reform plan that had dramatic effects on America's welfare system. Under the new law, welfare recipients are required to work after receiving 2 years of assistance, with only a few exceptions. Work was defined rather broadly and could include any of the following activities: unsubsidized or subsidized employment, on-the-job training, work experience, community service, 12 months of vocational training, or provide childcare services to individuals who are participating in community service. Single parents were required to participate in at least 20 h of employment each week during their first year and demonstrate an increase in working in hours in their second year, while two-parent families were required to work 35 h per week. At the time of enactment, PRWORA provided $14 billion toward childcare funding in order to facilitate transitions from welfare to work for parents receiving assistance. This law also maintained health and safety standards for daycare centers protecting the well-being of children. Finally, single parents with children under 6 years of age could not be penalized for failure to meet employment requirements if they were unable to find adequate childcare. In

fact, states were able to exempt single parents from the employment provision if they had children under the age of 1 year. In sum, while the US is undeniably less generous in its provisos for working parents than of countries, passage of both the FMLA and the PRWORA suggest that efforts are being made to acknowledge and make easier the struggles of employed parents, especially those whose lives are characterized by economic instability.

International Trends in Benefits

International variation exists in terms of eligibility criteria, duration of leave, generosity of benefits, and take-up of family-leave policies. In 2003, Janet C. Gornick and Marcia K. Meyers conducted a comprehensive review of work and family policies entitled *Families that Work*. Much of the information obtained on cross-national variation in family-leave policies has been obtained from this seminal review. Only two countries, among the OECD nations, do not legally provide some type of leave at the national level: South Korea and Switzerland. However, Switzerland does provide coverage at the state, or 'canton' level. In addition to paid leave, some countries provide a 'birth' or 'nursing allowance' or an in-kind childbirth package consisting of clothing or pharmaceutical items. These countries often also provide a supplementary leave in addition to the basic paid leave. For example, Germany provides an unpaid childrearing leave which allows parents to take off a third year after the birth of the child. This leave is in addition to 14 weeks of paid leave wherein mothers receive full wages and 2 years paid at lesser, flat rate. Half the countries providing a parental leave option match the salary normally earned by the worker, or the maximum covered by social insurance programs. Benefits provided by EU countries extend beyond the standard issued by the ILO as the average length of childbirth-related paid and unpaid leave (maternity, paternity, and parental) is 1.5 years. The duration of this paid leave averages 36 weeks, typically consisting of 14–16 weeks of paid maternity leave supplemented by either a paid parental or paid childrearing leave. In some cases, both pre- and postbirth maternity leaves are mandatory, while in other cases these components can be combined and used after childbirth.

An emerging trend is for parental leaves to be offered on a full- or part-time basis allowing the parent to remain employed part-time or work from home with job protection benefits. Among 29 industrialized countries, only eight do not offer extended leave policies: Australia, Ireland, Japan, South Korea, Mexico, the Netherlands, Turkey, and the US. Most researchers conclude that a primary weakness of the family-leave system within the US is that it does not allow for any paid leave for the majority of its workers. Eight countries now provide paid 'paternity' leaves, ranging in duration from 3 days to

6 months: Austria, Belgium, Denmark, Spain, France, Finland, and Sweden. In addition, 21 countries provide a supplementary parental leave, but in 13 of these countries, the leave is paid. Such benefits typically 'expire' when the child reaches 1.5 to 3 years of age. Finally, an EU directive has mandated the institution of family leaves such that parents can care for an ill child or family member. Enacted in 1992, this directive mandated a 14-week paid maternity leave as a health and safety measure for working mothers. In 1998, a 3-month parental leave was instituted, which led to the adoption of parental leave policies in several EU countries. In both the Nordic region and the EU, family-leave benefits often supplement maternity-leave policies, providing both mothers and fathers paid leave during their children's preschool years. Relatedly, leaves for family reasons are also often provided to parents throughout their children's lives. For example, in Sweden, parents are provided with paid time off to visit their children's schools.

According to the review conducted by Janet Gornick and Marcia Meyers, the Nordic countries provide the most generous package of benefits to working mothers. Family-leave policies in these countries offer mothers between 30 and 42 weeks of leave with full pay. However, benefits for the highest earning mothers are capped which simultaneously limits program expenditures for governments and provides working mothers with economic security and stability. Nordic governments spend the most on average for each employed mother (between $594 and $808 per month, in equivalent US dollars). Continental European countries offer a more modest package of benefits to mothers, offering between 12 and 16 weeks of full pay. While some of these countries also cap maximum covered earnings, these figures vary greatly. Governments in these countries spend slightly less on average per working woman ($67 to $465 per month, in equivalent US dollars) than Nordic countries. In comparison, English-speaking countries provide the minimum benefits to pregnant working women. While the duration of Canada's leave policy is generous (up to 50 weeks of paid leave), mothers are only provided with 55% of wage replacement, providing less economic security to these working mothers and their families when compared to the Nordic and Continental European countries. The disparities that exist between the duration of the benefit and the amount of the benefit (i.e., wage replacement) highlight the notion that a benefit is only successful if families can afford to take advantage of it.

Studies such as the review conducted by Janet Gornick and Marcia Meyers, and others independently conducted by Sheila Kamerman and Jane Waldfogel have consolidated the available take-up data on maternal paid leave. Results of their evaluative efforts suggest that most employed women take advantage of most if not all of their entitlement. However, data on unpaid leave are not available for most countries. Take-up of parental or paternity leaves is similarly high. Available data on the Nordic countries (e.g., Denmark, Finland, Iceland, Norway, and Sweden) suggest between 0.5% and 90% of eligible fathers take advantage of these benefits, with the exception of working fathers in the Netherlands. Outside of these countries, take-up rates among men are typically quite low – between 1% and 9%. Reasons for this vary, but research suggests that paternity leaves are shorter in duration and occur immediately after childbirth and are especially common in countries where gender equity in policies has been established and female labor force participation is higher. Paternal take-up rates parental leaves that are longer in duration are much lower, ranging from 1% in Germany to less than 5% in Denmark and Finland. Reasons for this are unclear.

In terms of working parents' access to family-leave benefits and eligibility determination, most countries restrict benefit eligibility to women employed for a minimum period of time prior to childbirth. Exceptions to this trend are Germany, wherein a childrearing leave is available to almost all parents, and several Nordic countries like Norway and Sweden, wherein a small minimum benefit is provided to women covered under health insurance and benefit eligibility is extended to housewives. In Denmark, even unemployed mothers are eligible for extended parental leave. The benefits provided by most countries are universally available, without regard for income level. Again, exceptions exist: for example, New Zealand provides income tested maternity leave to poor single mothers only, while Germany and France provide benefits that are income tested, such that only 80% of new parents qualify.

International Examples

Nordic countries

Family-leave policies enacted within the Nordic region (e.g., Denmark, Norway, and Sweden) provide an excellent example of gender equity in parental-leave policies, emphasizing the parental contributions that both men and women make. In these nations, extended paid parental leaves are offered mandating that at least 1 month of the extension is to be a 'use it or lose it' option for fathers. That is, if fathers do not take advantage of this extension, the benefit cannot be transferred and will be lost. One example often described in the literature is Sweden's Parental Leave and Parent Insurance policy, noted for its comprehensive, generous, and flexible stipulations. Parents are provided with an 18-month job-protected parental leave. Fathers are allowed an additional 1 month. In addition, parents of children 8 years of age and younger are legally protected to work 75% of their normal working time. These benefits can be shared between the parents and may be applied to full or part-time work.

Continental European countries

Within this geographical region, Spain provides a useful example because it is not a wealthy country and yet its family-leave policy may still be defined as generous and flexible. Working Spanish parents are provided with 16-week paid maternity leave at the time of childbirth, including 6 weeks prior to the birth paid at 100% of prior wages. This benefit is supplemented with an unpaid parental leave, available until the child's third birthday. In addition, parents with a disabled child or a child less than 6 years of age are legally permitted to work part time. Working parents in Austria are also guaranteed a generous and gender-equitable package of benefits around the time of childbirth. Working mothers are provided with a 16-week mandatory leave with a cash benefit that returns all lost wages. Two years of additional job-protected, paid leave is available to either parent as well. However, the last 6 months of this leave is paid only if the 'other' parent takes the leave (typically the father) prior to the child's third birthday.

English-speaking countries

The benefits provided by English-speaking countries are typically less generous than the benefits provided by EU countries. However, Canada is the only Anglo-American country to provide both paid maternity and paid parental leave. This system of benefits is funded through its unemployment insurance system. More information is available on the family-leave package available within the US. As mentioned earlier, in terms of generosity of benefits, the US is an outlier compared to other industrialized and developing nations. For example, 45% of workers in the US are not eligible under the provisos of the FMLA, in contrast to other OECD countries that provide coverage for almost all parents. Employees may be required to use accrued sick leave or vacation time to cover some or all of the leave. Employers also may deny leave to employees within the highest paid 10% of its workforce if permitting the leave would have negative consequences to the company. Political efforts are underway to expand parental and family-leave benefits within the US. Most state policy initiatives focus their efforts on two funding sources: the Unemployment Insurance (UI) system and the Temporary Disability Insurance (TDI) program. However, an extension of state UI programs would be costly and not fully cover all employed parents. In addition, the TDI benefit does not provide job protection so employees are not granted a right to return to work. Several states use state TDI monies to partially fund parental leave programs (e.g., California, Hawaii, New Jersey, New York, and Rhode Island), in conjunction with contributions from employers and employees. However, employers in the remaining 45 states are permitted to voluntarily participate in TDI plans for their employees.

Funding for Family-Leave Benefits

Most industrialized OECD countries fund family-leave benefits through the same system as sickness programs (e.g., statutory paid sick leave). In fact, 95% of these countries provide modest health and medical benefits (excluding US) to women at the time of childbirth. Comparative analyses such as those conducted by Sheila Kamerman, Janet Gornick, and Marcia Meyers have shown that even after combining the costs of maternity, paternity, and parental leave benefits, the total costs of administering these benefits totals to a small proportion of the overall gross domestic product (GDP). For example, these costs total 1% of Finland's GDP and 2% of Sweden's GDP. In sum, most OECD countries (and all EU countries) have supplemented existing maternity leave policies with parental leaves. Paid maternity leaves are indeed the international norm and the average duration of these leaves is often significantly longer than the 12-week, unpaid family leave provided by the US to employed parents of infants, as well as the provisions stipulated by the ILO. Substantial variation exists in the percentage of mothers and fathers who take up the benefits that are provided.

Policies for the First and Second Years of Life

Early Childcare: An Introduction

As the trend in most OECD member nations is to provide working mothers with some form of job-protected leave (i.e., maternity, family) during the child's first year of life, an issue of increasing importance for working parents as their children reach 1 year of age is finding adequate childcare. Childcare is a crucial form of support for working parents and their children. A recent study by the OECD in 2005 of several advanced industrial nations led to policy recommendation that governments provide a continuum of support for working families and their children in their early years, most notably public investment in day and evening childcare programs such that mothers are less hampered in their efforts to seek paid employment. This report also encouraged employers to provide more flexible working schedules for full- and part-time employees such that parents are better able to meet their children's needs and balance the obligations between work and family. In addition to providing an enriching environment for young children, high-quality group care for infants and toddlers can provide important support for families.

Coverage and Access

Governments may provide childcare directly or they may subsidize or reimburse some of the costs of care privately purchased by parents. Within the US, the existing childcare system faces problems with service provision and the quality of existing services. For example, the existing supply of childcare programs for infants and toddlers does not meet the needs of working parents – that is, there are not enough facilities to ensure that all working families with young children who need childcare are afforded those services. Moreover, it is estimated that half of existing home- and center-based infant care programs are rated poorly and are even negatively associated with children's long-term health and social outcomes. In addition, most services provided under the ECEC system in the US are private, although the government does play a role in financing a small proportion of programs through subsidies and tax policies. According to the US Department of Education, in 2002, 60% of working families in the US in which the youngest child is under age 5 years paid for licensed or informal care. Of these children, over 50% were cared for informally (e.g., by relatives, friends, or nonlicensed providers), while the remaining 48% were cared for at licensed childcare or family facilities. According to data published by the US government, only 3% of children under the age of 1 year, 4% of children between the ages of 1 and 2 years, and 13% of children between the ages of 3 and 5 years participated in publicly supported childcare programs, such as Head Start and Early Head Start (EHS). In addition, only 15% of eligible low-income families receive subsidized care. Similar patterns exist within Canada, Continental Europe, and the Nordic countries. For example, 2% of 0–3-year-olds in Germany and 23% of eligible 0–3-year-olds in France participate in the ECEC system. In contrast, participation rates of older children aged 3–6 years are universally high in European countries. For example, 78% of German children and 99% of French children in this age bracket participate in some type of early care or preschool program. Within the Nordic countries, the percentage of children under age 3 years enrolled in ECEC ranges from 20% (Norway) to 48% (Denmark). Percentages of children between 3 and 6 years of age enrolled in ECEC are higher as in the Continental European countries and range from 53% (Finland) to 82% (Denmark).

Descriptions of Leading Childcare Systems

In their cross-national comparison of supports in place for working families, Janet Gornick and Marcia Meyers identified two systems to be exemplars based on nearly universal access to publicly provided childcare: Nordic countries and the integrated systems found within France and Belgium. The public ECEC systems within Denmark, Finland, and Sweden are under the purview of the national social welfare program and uniquely offer a nearly universal entitlement for childcare prior to primary school. These childcare programs are often integrated with early educational experiences, sometimes called 'EduCare'. For example, since 1995 Swedish children are

entitled to public or private care from age 1 to 12 years. And while such entitlements were initially targeted toward unemployed parents, children in Denmark and Finland are now guaranteed childcare irrespective of their parents' employment status. Entitlement for childcare begins after the parental leave and is guaranteed through the beginning of primary school. Thus, between 25% and 75% of 1–2-year-old children are in publicly funded childcare programs in the Nordic countries. The percentage of children in publicly funded care increases as children age; for example, among older children (aged 3–5 years), 83–90% are in publicly funded childcare programs, and nearly all children are in public care in the last year before primary school. The primary mechanism for provision of these services is direct. National taxes account for 25–33% of the costs associated with providing these programs, while parental co-payments cover at most 20% of costs (varies across nations). Low-income families are often not required to pay fees; sliding scale applies to other families. Subsidies permitting parents to purchase childcare from private providers are not often utilized. For example, Finland and Norway recently began providing parents of children not enrolled in public childcare with a cash benefit to purchase private childcare – the near universality of childcare programs within the Nordic countries make the provision of subsidies less necessary. Finally, the Nordic countries also make the most substantial investments per child, ranging from $4050 (Denmark) to $4950 (Sweden).

In contrast, Continental European nations provide early childcare in conjunction with universal education programs. While still generous, these systems are less consistent than those programs found within Nordic countries. For example, Belgium and France provide universal public preschool programs for children as young as 2.5–3 years, but childcare is not an entitlement for students prior to entering public preschool programs ('ecole maternelle' in France and 'kleuterschool' in Belgium). By age 3 years, nearly all children are enrolled in preschool programs. However, publicly funded childcare programs for children aged less than 3 years are targeted toward needy families and not commonly available; thus, only 20–40% of children in this age group attend publicly funded childcare programs. In these countries, publicly funded childcare is also directly provided to parents of young children and is subsidized through national and local taxes and parental co-payments. Parental co-payments account for between 17% and 25% of the costs of care for children less than 3 years of age and are also set on a sliding scale dependent upon parental income. However, employers also contribute monies to cover costs associated with childcare services. As such, in both Belgium and France, care for children approximately 3 years of age is free for parents. As parents are responsible for purchasing care for children younger than 3 years of age, the French and Belgian governments permit deductions for these expenses from income taxes. The more modest system provided by these nations is also less costly for their governments – for example, France spends the equivalent of $3161 per child under school age (in equivalent US dollars) annually.

In stark contrast to these European systems, the US again ranks lowest with other English-speaking countries in their provision of early childcare, serving a mere 6% of children via means-tested programs. However, 53% of children aged 3–5 years are enrolled in publicly funded childcare. Most childcare costs are privately born. As reviewed earlier, the parental contribution to costs associated with childcare in the US is two to three times greater than in European nations. Within the US, costs are also inequitably distributed – low-income parents spend a greater proportion of their income (approximately 22% by recent estimates) on childcare than wealthier parents. Not surprisingly, the primarily private funding of the childcare system within the US does not present much of a financial burden to the government – data from the mid-1990s suggest that the US invested an average of $548 per month per child prior to beginning primary school. Welfare reform efforts led to the expansion of pre-kindergarten programs and modest increases in the average spending per child per month increased to $679 by 2000.

An oft-neglected component of childcare programs in policy discussions is the schedule of publicly funded childcare programs and schools which can help determine their utility and meaningfulness to working parents. For parents, the hours of operation for both preschool and primary school programs determine which schools are viable childcare options. As with other aspects of childcare programs, substantial cross-national variation exists. For example, the Nordic countries (e.g., Sweden and Denmark) provide public childcare during the preschool years on a full-day, full-year basis, often extending past 60 h per week. Before and after typical school hours, children in Sweden and Denmark may also attend supervised centers near the school. Hours of operation are much shorter and inconsistent in Continental European nations due to extended lunch breaks and afternoon closings (e.g., France and Belgium). Within the US, most pre-kindergarten programs (including Head Start) and public kindergarten programs are open for part of the day. When children begin primary school, they are offered continuous instruction throughout the school day. However, the length of the school week in US-based schools (33 h per week on average) is only half that of Swedish schools which average approximately 60 h per week of instruction. In sum, the US ECEC system is significantly less generous when compared to the system of early childcare and education provided in Nordic countries and Continental European countries. Janet Gornick and Marcia Meyers

concluded their cross-national comparisons by summarizing the lessons that the US should learn from these other systems of care, which include: (1) matching school schedules to parents' working hours in consideration of parents working nonstandard shifts to create alternative methods of service delivery; (2) increasing certification requirements and compensation for childcare professionals; and (3) because families with younger children may be less able to bear the burden of costs associated with higher-quality childcare, the federal government can assume the bulk of childcare costs, with parents assessed a sliding scale fee.

Funding of Early Childcare Programs

Within the US, states may use TANF funds to provide childcare to low-income families through vouchers for current and former recipients of welfare. Another funding stream for federal childcare is the Social Services Block Grant (SSBG). However, as states are able to choose the proportion of these funds they will devote to the provision of public childcare programs, usage of these funds is remarkably low. Recent estimates suggest that only 13% of all SSBG funds were earmarked for childcare services or vouchers. Further work is needed to better understand why states allocate a small proportion of SSBG funds to public childcare programs.

Cross-national variation also exists in the amount of costs parents are expected to bear themselves. Estimates vary, but it appears that parents within the US bear a larger burden than parents in European or Nordic countries. The US government covers 25–30% of the cost of childcare for children under age 3 years and for children aged 3–6 years. This amount includes tax credits to reimburse parents, subsidies to parents, or public care that is purchased for free or reduced cost. In stark contrast are the Nordic countries. Governments here fund the majority of child-care costs, ranging between 68% and 100%. Such variation is dependent upon a variety of factors, including the generosity of parental leave benefits. For example, in Germany, parental-leave benefits may be extended in place of childcare provision. A government's interpretation of their own role in providing ECEC for young children also contributes to the significant variation in childcare policies between countries. For example, within the US, infant and toddler care is viewed as a private decision and this perspective is reflected in the monetary investment made by the US government – $600 per year, per preschool-aged child (in equivalent US dollars). Annual investment in ECEC is significantly higher in the Continental European countries (e.g., France, $3000, in equivalent US dollars) and the Nordic countries (e.g., Sweden, $4500, in equivalent US dollars). That the US government views early childcare for young children as a private decision is also reflected in funding patterns.

Specifically, the US funds publicly provided childcare programs through tax benefits and subsidies, rather than creating high-quality programs and places for publicly funded care. This choice results in programs of widely varying quality that are difficult to monitor and evaluate despite local and state-level licensing efforts. In contrast, Nordic and several Continental European countries view childcare as a public responsibility and a public good and have instituted universal preschool care for children aged 3 years and older (Nordic countries recently guaranteed care for all children aged 1 year or older). Because a larger proportion of childcare services are offered through the social welfare or educational systems, quality can be regularly checked and providers held accountable.

Policies for the Third and Fourth Year of Life

Early Childhood Education: An Introduction

Early childhood education (ECE) is another type of family support policy provided by many nations. The success of these programs is best understood in the larger context of a nation's social and economic well-being. Throughout the past several decades, several factors have served as significant barriers to the implementation and expansion of ECE programs, including: a competition for resources to deal with other societal ills such as economic deprivation, an absence of highly qualified and trained ECE staff due to the lack of a national standardization or certification system, and an absence of governmental infrastructure to monitor and evaluate educational programming for young children. However, enrolment in ECE programs is rising throughout the world as public perceptions of the value of such programs are shifting.

History of Early Childhood Education Benefits

Recent years have witnessed a growing interest in ensuring access to these programs in urban as well as rural areas, and in both industrialized and developing nations. The provision of early childhood benefits to parents of young children began in Finland in 1985, although the idea of providing childrearing allowances to working parents was first observed in Hungary in 1967. Many countries offer childrearing allowances in addition to the family-leave benefits discussed earlier. The goals of these cash benefits are similar to those of family-leave policies as they allow mothers to stay at home with their young children. However, in contrast to the eligibility restrictions imposed upon recipients of childrearing allowances, early childhood benefits are provided to all parents of children aged 3 year sand younger regardless of income or employment status. These benefits also provide parents greater flexibility and choice in making their childcare decisions. For example, the provision of an early childhood benefit for

parents who are not using publicly funded childcare allows parents to choose between the available public care, private childcare, and parental care. Limited data exist on the take-up rates of this benefit, but in 2001, Jane Waldfogel's research suggested that the most popular option for parents of toddlers was publicly funded childcare.

While political and public interest has centered on the provision of educational programs for primarily for 3–5-year-olds, an understanding has emerged of the importance of programs for children under 3 years of age based on studies that illustrate the importance of the first years of life for social, cognitive, physical, and emotional development. For example, in 2003 the British government guaranteed positions in a half-day nursery school for 4-year-olds whose parents wanted to enroll their children. By 2004, the British government had also guaranteed positions for 3-year-olds in public nursery school programs. Other trends influencing this rising interest in the provision of ECE programs for children younger than 3 years are numerous and varied and include: an increase in rates of female labor force participation; the ratification of the United Nations Convention on the Rights of the Child (CRC), the World Summit for Children (1990), parents' need for alternative and affordable childcare, and the need to ensure young children's access to primary and secondary school. This last factor is also highlighted in the ratification of the United Nations' Millennium Development Goals (MDGs) by countries around the world. One of the eight MDGs identifies the achievement of universal primary education to ensure that all young boys and girls complete a full course of primary schooling as an objective by 2015. These MDGs were agreed upon by both developing and industrialized nations in an effort to reduce social inequalities and conduct a concerted attack on problems such as illiteracy, poverty, and gender discrimination. The identification of universal access to primary education clearly illustrates the growing acknowledgment cross-nationally of the importance of educational programming for young children.

Examples of Early Childhood Education Programs

As with family leave and early childhood care, substantial variation exists between industrialized nations in the type of educational programs offered to parents. The availability of such programs is highly dependent upon the particular demographic, social, economic, and political contexts of each nation. Due to space constraints, singular examples from each geographic region will be given.

Continental European countries

On the more generous end of the spectrum is France. Because children aged 3 years and above are legally guaranteed preschool education, the French government has instituted a highly developed universal preschool system for 2–6-year-old children. Most other Continental European countries, such as Belgium, Germany, and Italy, legally guarantee children a spot in preschool programs at 3 years of age. Almost 100% of 3-, 4-, and 5-year-olds and almost 20% of 2-year-olds attend the free 'ecolles maternelles', or 'maternal schools'. These preschools were formerly grouped with primary education services and governed by the Ministry of Education. The French system of ECE is managed by national and local level entities, including the Ministry of Social Affairs, Employment, and Solidarity and the Ministry of Health, Family, and Handicapped Persons. Each national body is responsible for different areas, such as defining the curriculum, recruitment, and training of ECE staff and developing licensing requirements for the different forms of nonschool ECE. 'Crèches collectives' are center-based services operated by the local municipal bodies and NGOs. Parent cooperatives, or 'crèches parentales', stimulate parental involvement in the daily activities of the crèche. Attendance at 'ecole maternelle' has had a positive impact on educational inequalities between advantaged and disadvantaged children in France. As illustrated before, equitable access to ECE programs supports the needs of both young children and their families. Studies have shown the length of time spent in 'ecole maternelle' is negatively associated with the chance of failing or repeating the first grade, especially for children from disadvantaged homes. These results inspired the French government to expand access to 'ecole maternelle' to include 2-year-old children.

Nordic countries

Another important example is Norway, which enacted its childrearing grant in 1998, officially titled the Cash Benefit for parents with Small Children Act ('kontantstotteloven'). The Norwegian system of 'barnehage', or ECE has in fact existed for hundreds of years but only recently improved access for young children. As with many European nations, the Norwegian system of ECE is integrated with early childcare services such that programs provide childcare for parents during working hours and provide educational services in alignment with the broader educational system. Funding for this program is obtained from three sources: the national government, the private owner of the center providing ECE programs, and parental subsidies. Approximately 2% of Norway's GDP was appropriated to family support programs in the late 1990s. Norway has over 6200 ECE centers; approximately, half of these are privately owned. Private institutions are smaller; not surprisingly, a greater percentage of young children are enrolled in public 'barnehager'. Access to barnehager also varies significantly across municipalities, although most provide ECE services to 50–60% of children. This policy provides a cash benefit to parents of children between 1 and 3 years of age and similarly to Finland, is provided to parents who are not using the publicly funded 'barnehage' or childcare

system. Families with a child in part-time 'barnehage' will be entitled to a partial cash benefit according to the number of hours in 'barnehage'. The cash benefit is financed by the state and will be at the same level as the subsidy provided to parents whose children attend 'barnehage'. This rationale was designed to increase options and choices for parents such that they would be free to choose whether they wish to stay at home with their young children, use some private form of childcare, or enroll their children in a 'barnehage' with state funding. However, because of this stipulation, Norway's system of private care has expanded significantly, concurrent with a decrease in growth of spaces in publicly funded childcare system. Controversy exists as to the merits and pitfalls of this policy given these unintended consequences. In 1996, a national curriculum was implemented by the Ministry of Children and Family Affairs called the Framework Plan, which is based on the Nordic tradition of 'EduCare', or combining education and care, and must to be used by all 'barnehager'. The Framework Plan provides the 'barnehage' with a centralized educational and pedagogical foundation. However, the Plan does not impose detailed guidelines or prevent variation at the local level. Despite a great deal of variation in curricula and activities for children across 'barnehager', systematic reviews of ECE programs in Norway have consistently pointed to a higher quality of care and education. Several goals remain unfulfilled for the Norwegian government, as attempts are made to: increase access for children aged 0–3 years, reduce disparities in quality between private and public 'barnehage', increase 'barnehage' funding in order to reduce financial burdens placed on disadvantaged parents and ensure that educational programming meets the needs of working parents.

English-speaking countries

The primary vehicles for publicly funded early educational programs within the US are Head Start and universal pre-kindergarten. Pre-kindergarten programs were developed more recently than Head Start and are designed to provide children with 1–2 years of education prior to kindergarten. These programs are typically funded at the state level, although local school districts are able to use federal funds to support such programming. Universal pre-kindergarten programs also often include a wide range of services such as health screenings, meals, and transportation. Most state programs target disadvantaged 3- and 4-year-old children. In 2002, estimates suggest that 14% of eligible 4-year-old children were enrolled in publicly funded, school-based pre-kindergarten programs. A second major early educational program in Head Start is a national, federally funded program that provides a variety of services ranging from health, nutrition, and education, to low-income children aged 3–4 years. This program was established in the 1960s to promote general well-being and school readiness in economically disadvantaged children. With this program, states are able to use monies to provide low-income parents with subsidies or contract with providers. As with other early educational interventions, short-term cognitive benefits have been found – however, these advantages diminish by the third grade for minority children. One potential explanation for these 'fading' effects is the quality of schools students attend during their primary years. Disadvantaged children are more likely to attend poor-quality schools that lack the necessary resources to provide students with adequate education. Another federally funded national program, EHS, was created in 1994 to enhance the development and health of infants and toddlers. This program combines a unique mix of services designed to enhance family and community partnerships, parent education, and service delivery for pregnant women. Results of a program evaluation conducted when children were 3 years old, pointed to modest impacts of program participation. Specifically, involvement with EHS did lead to improvements in achievement and interfamilial relationships, parenting practices, and education/employment of participating mothers.

Funding of Early Childhood Education Programs

Countries differ significantly in their levels of investment in early education. In general, wealthier OECD member nations tend to spend more while also stipulating that ECE programs take place in regulated centers with certified staff. As mentioned earlier, however, maintaining and evaluating such standards may not be the most cost-effective option for all countries and has in fact served as a barrier to implementation of any ECE programming. Countries around the world have not yet agreed upon universally applicable standards for investment in ECE services, but it is obvious that quality is an issue for most countries and there is a growing consensus on the most important dimensions of ECE services. A useful starting point in the literature is often the proportion of funds spent relative to a country's GDP. Such an indicator suggests the importance governments attach to these programs. Overall, OECD countries spend about 5.6% of their collective GDP on educational systems, including both pre-primary and university level schooling. In 1995, the European Commission on childcare recommended that countries should spend approximately 1% of their GDP on early childhood education and care. However, only the Nordic countries have reached this level of investment. For example, in 2004, Sweden spent over 2% of its GDP on integrated services for children. Another potentially useful indicator of governmental investment in ECE programs for young children is the amount of monies spent per child (limited to those who receive services). For example, the average annual amount spent per child in OECD nations is $4922 (in equivalent US dollars). However, both the US and the UK exceed

the OECD annual average, spending $8452 and $7881 per child (in equivalent US dollars), respectively. Such information can prove useful when disaggregating data to make comparisons across gender, race, and economic lines.

The question of who pays for funding for ECE programs is also an important one and the answer again varies across nations. Early education programs are rarely free as with primary education services. ECE programs may be primarily funded by government support in addition to parental subsidies, private donations or providers, and cash or in kind contributions from international agencies such as NGOs. In almost all countries, the governments pay the largest share of costs, with parents covering between 11% and 30% of costs (the exception is the US where parents bear 55–70% of costs associated with enrolling their children in ECE programs). The average level of expenditure from private sources in OECD countries is 17.9%. While such an arrangement places less financial burden on governments, greater responsibilities are assigned to parents who must provide fees for books, uniforms, and other materials. This naturally affects the accessibility of such services to low-income or moderately disadvantaged families and the quality of services accessible to disadvantaged families. In the EU, parents often pay an average of 25% of the costs associated with nonkindergarten early education and care, while in English-speaking countries in North America, parents are often required to absorb 50% of the costs. According to a 2004 policy brief issued by UNESCO on early educational financing, parents in North America are also required to pay these costs for a longer period of time as most states or provinces begin to provide publicly funded care and education at the age of 4–5 years.

Federal and state governments in the US have expanded funding recently for ECE programs. For example, in 2002, the federal government allotted $6.3 billion to local Head Start grantees, who served approximately 65% of eligible 3- and 4-year-olds. In 2005, Katherine Magnusson and Jane Walfdogel's research indicated that although state spending greatly varied, funding for separate, state-financed pre-kindergarten programs had increased by 250% to approximately $1.9 million in 2002. Thirty-nine states and the District of Columbia provide state-financed pre-kindergarten for at least some of their 3–5-year-olds, up from about 10 states in 1980. The largest source of federal funding for means-tested subsidies is provided to states through the Child-Care Development Fund (CCDF), a federal block grant. States are permitted to use CCDF funds to serve working families with incomes within 85% of the state median, although most states have set a lower threshold. States are also allowed to determine the level of parental co-payment, the level of provider reimbursement, and certification standards for childcare and education centers. The second funding stream within the US for means-tested assistance is the TANF block grant. States are allowed to transfer up to 30% of their TANF funds toward CCDF programs – approximately one half of states in the US do so. Federal support is also provided through tax credits to subsidize childcare costs for working parents – despite these funding sources, researchers agree that the current level of federal and state funding is not sufficient for parents who wish to enroll their children in childcare or educational program.

The patterns of governmental and parental investment also have implications for enrolment in ECE programs and services. Enrolment data are especially difficult to obtain given the proportion of ECE services that are privately managed. Most available enrolment data for pre-primary education combine information on 3–4-year-olds with available information on 5–6-year-olds. Such aggregate reporting may be misleading to researchers and policy-makers concerned with issues of access because higher rates of participation for older children (5–6 years) could skew overall enrolment rates and mask lower enrolment rates for younger children (3–4 years). These disparities are also thought to vary cross-nationally. The average enrolment of young children ages 4 years and younger in pre-primary education for OECD countries is approximately 70%. More than one-third of countries around the world have exceptionally low enrolment rates in pre-primary education, of fewer than 30%, such as those in sub-Saharan Africa. However, all countries in North America and Western Europe, as well as most countries within Central and Eastern Europe, report enrolments exceeding 50% of preschool-aged children. For example, in 2000, the US, Canada, and Finland reported gross enrolment rates between 50% and 70%, while Sweden, Denmark, Norway, Austria, and the UK reported gross enrolment rates between 70% and 90%. Few countries reported almost universal enrolment (e.g., Italy, Germany, Iceland, France, and Belgium). Coverage and enrolment rates differ for children less than 3 years of age. For example, only 30% of children under age 3 years are enrolled in pre-primary educational programs in France, while almost 60% of 1–2-year-old children are enrolled in Denmark.

Outcomes by Policy Type

Effects of Family-Leave Policies on Parents

Thus far, research on the effects of family-leave policies on parents has limited its scope to the effects of maternal leave on maternal outcomes, such as stability of wages and employment over time. Researchers such as S. B. Kamerman and C. J. Ruhm have found that while leaves shorter in duration do not have negative consequences for mothers, longer and more frequent leaves (e.g., 3 years) may negatively impact mothers in the form of reduced wages. Other researchers such as Janet Gornick

and Marcia Meyers have focused on the effects of taking unpaid maternity leaves that are shorter in length on the labor market. Their results were not so clear-cut. While some research suggests that taking such leaves has little impact on aggregate female employment rates, other studies have found that both paid and unpaid maternity leave packages increase the likelihood that women will return to their jobs within 3 months and have a substantial impact on female employment. Clearer evidence exists with regard to the positive effects of lengthier paid leaves (e.g., several months to 1 year in duration) on women's labor market attachment. For example, women with paid leave are more likely to work longer into their pregnancy and return to work within 1 year of childbirth to the same employer especially when compared to mothers who did not receive paid leave. These studies also found that lengthier maternity leaves (e.g., 2–3 years as found in some European countries) may in fact depress female employment rates given the high take-up rates, although it is ambiguous at which point the length of maternity leave becomes disadvantageous. Within the US specifically, studies on the effects of the FMLA on working women suggest that there is no impact on female participation in the labor force and that the costs to employers associated with voluntarily providing family-leave benefits are modest. While some women did take longer leaves after childbirth, most took only a brief leave or none at all. In addition, these studies have shown those women who could most benefit from extended leaves (poor, single mothers) do not have access to such benefits. Finally, some studies have suggested that parental leaves lead to less maternal stress, longer periods of breastfeeding, and provide a less expensive childcare alternative for working families.

Less is known about the effects of paternity leave policies on male employment patterns. Some studies have suggested that more generous parental leave policies have been positively associated with the time fathers are able to invest in their young children. However, these studies do suggest a higher level of employer resistance to paternal take-up of these benefits. Interestingly, paternal take-up has increased with the development of the 'use it or lose it' policy, wherein fathers are unable to transfer their benefits to female partners and will lose the benefits if they are not used within a certain period of time. For example, in Norway, paternal take-up increased from 5% to 70% after the 'use it or lose it' policy was implemented.

In sum, research has suggested several positive impacts of maternity leave policies on maternal outcomes, ranging from statistically significant decreases in women's unemployment and higher aggregate rates of female labor force participation to higher revenue for governments (vis-à-vis income tax payments). It is plausible that a final benefit may include reductions in stress related to work and family conflicts, but this outcome is more difficult to measure empirically.

Effects of Family-Leave Policies on Children

Limited evidence exists on the effects of family-leave policies on children. In 1998, C. J. Ruhm conducted a study which found that paid parental leave policies were negatively associated with low birth weight status and mortality rates of infants and children. It was hypothesized that parental leave policies had a favorable impact on pediatric health outcomes because parents are now able to devote more time to their children. Increasing generosity levels of parental leave policies were also associated with lower postneonatal mortality and mortality between the first and fifth years of life. Again, it was believed that because parents were able to spend more time with their children, they were better able to attend to their physical and psychosocial needs.

A greater amount of empirical evidence exists on the effects of maternal employment on child outcomes. Studies using US-based samples found modest, negative effects of maternal full-time employment during the child's first year of life on their cognitive development. These effects extended past preschool into primary school years (7 or 8 years). However, maternal employment begun 'after' the first year of life does not have a detrimental impact on young children and may even exert a positive influence on child outcomes. Longitudinal studies conducted within the UK also found that children whose mothers were employed during their preschool years experienced poor educational outcomes, and often completed less years of schooling, even after statistically controlling for family background characteristics such as income and levels of education obtained by the parent. This finding is less conclusive as other researchers have found such family background characteristics (e.g., low levels of household income and parent education), to have a significant negative influence on children during the preschool years. Despite the limited empirical evidence, the findings are suggestive of a need for programs and policies that promote positive outcomes and adequately support working parents and their children.

In sum, while the particulars of family-leave policies vary widely across nations, the goals of these policies are similar in that they seek to protect the health and emotional well-being of mothers, fathers, and their young children. Policies now seek to form a viable substitution for infant and toddler care outside of the home. Efforts to reconcile work and family conflicts include emphasis on gender equity – seen in paternity leaves. Consider these within a broader framework of employment and income support policies designed to reduce burden on employed parents.

Effects of Childcare Policies on Mothers

Policies designed to reduce the cost of childcare or increase the availability of care are positively associated with maternal employment rates. However, other studies on the effects of childcare costs in the US and maternal employment rates have suggested a consistently negative association. Specifically, estimates of maternal employment effect sizes range from 0.3 to 0.4, suggesting that if childcare costs were reduced by 10%, this would lead to a 3–4% increase in maternal employment. Studies examining Canadian data found similar results. Cross-national data on Nordic countries and Continental European are not readily available.

Effects of Childcare Policies on Children

Because childcare policies best suited for one family may not be the ideal solution for another family, research has found inconsistent associations between childcare policies and child outcomes. It is intuitively appealing that more flexible packages, such as those available to parents in France and several Nordic countries, wherein parents can choose the type of childcare best suited for their children during the first 3 years, would be beneficial to children. Extensive studies have been conducted on the contributions of childcare quality, childcare choices, and family characteristics (e.g., income, marital status) on indicators of child well-being. For example, in 1999, Margaret Burchinal found that children who are exposed to higher-quality care have fewer behavioral problems, higher cognitive and language development, better social skills, and better relationships with peers. These findings were again confirmed in 2000 by Deborah L. Vandell and Barbara Wolfe, who found significant associations between aspects of childcare settings, such as the environment, caregiver practices, child activities, staff–child ratios, class sizes, and certification of caregivers, and children's cognitive development, emotional adjustment, and school readiness. Each of these researchers has stressed that family background characteristics are still highly important – that is, childcare quality appears to moderate effects of disadvantage and parental education on child well-being (quality matters most for disadvantaged children).

Such patterns are visible cross-nationally, with similar results reported in the UK, France, and Sweden. For children aged 2–6 years, participation in higher quality ECEC programs positively impacted their social, cognitive, and emotional development, as well as levels of school readiness and achievement. This relationship between child-care quality and child development was moderated by financial disadvantage (i.e., these effects were more pronounced in poor children). Janet Gornick and Marcia Meyers note that longitudinal evaluations also found that children beginning childcare programs between 6 and 12 months of age scored higher on aptitude tests and received higher ratings on social behaviors and emotional regulation by teachers when compared to children who remained at home or entered care at a later age. Paid parental leave with full wage replacement and quality of caregiving staff were important factors operating in tandem with age of childcare entry. Preschool care for older children aged 3–4 years is widely accepted now in countries such as the UK. Instead of asking questions on whether maternal employment or childcare is to the detriment of the child, the focus has and should continue to move to the type of care that is most beneficial to infants and young children.

Effects of Early Childhood Education Programs and Policies on Parents

Less evidence is available as to the effects of ECE programs and policies on demographic indicators such as female labor force participation rates. However, some research has been conducted on the effects of children's participation in ECE programs and parental attitudes toward their children's education within the US. For example, several studies conducted in the 1980s found that parents' attitudes toward their children's schooling improved after their children successfully graduated form a preschool program. The educational aspirations of these parents toward their children's futures also increased as they gained a greater level of satisfaction from their children's educational accomplishments. Finally, parental involvement in their children's schools also increased – even though children who attended preschool had fewer behavioral problems, parents contacted their children's teachers more often. Studies on the effects of parental involvement on young children's cognitive and behavioral outcomes have shown such involvement to have a powerful impact on the benefits these young children receive from participation in ECE programs.

Effects of Early Childhood Education Programs and Policies on Children

Extant literature on the effects of early childhood educational policies is often intermixed with studies on the effects of early childcare policies, included in the previous section. While the most extensive and rigorous research on the outcomes and impacts of ECE programs has been conducted in the US, relevant and important research has also been conducted in other nations. Research in France, Sweden, the US, and the UK points to positive associations between participation in high-quality early education programs and the cognitive (school readiness and achievement), social, and emotional development of children. For example, attendance of French preschoolers at 'ecole maternelle' is associated with reductions in achievement inequalities. Students who were enrolled by 2 years of age performed better than students enrolled at 3 years on tests of oral and pre-reading skills, general knowledge tests, logic and premath skills, and classroom behaviors. Similarly, researchers in Italy found that children were better prepared for primary school if they had attended preschool and were also better prepared for preschool if they had even earlier educational experiences.

A longitudinal study conducted in Sweden found that children starting childcare between 6 and 12 months of age performed better on aptitude tests and were rated more positively by teachers on socioemotional functioning than children who entered care at a later age or children who were cared for at home. These positive effects remained when the children were 13 years old.

Positive effects of ECE programs are particularly strong for disadvantaged children, suggesting an interaction between poverty and the quality of these educational programs. In 1998, Lynn Karoly and colleagues conducted a cost–benefit analysis of early childhood programs, such as Head Start. Her results pointed to significant early gains in cognitive performance for poor children and later, more modest gains in achievement, grade completion, and high school graduation rates. Ensuing debates on the significance of 'fading' effects have centered on the quality of educational services disadvantaged children receive after attending programs like Head Start. A longitudinal evaluation was conducted in the UK on the Effective Provision of Pre-School Education Project (EPPE), which followed 3000 children from 141 preschool settings from ages 3 to 7 years as they transition from preschool to primary school. As Janet Gornick and Marcia Meyers reported, positive short-term effects on cognitive development and school readiness were found, more so for disadvantaged children than their wealthier counterparts. While preschool attendance can in fact narrow achievement gaps, the duration of these effects is still undetermined. Extant research has pointed to the benefits of preschool programs on socioecomotional, cognitive, and physical child outcomes for 2–5-year-olds (irrespective of parental employment). However, the debate on how to ensure equal access to high-quality childcare continues among policymakers and practitioners.

Conclusions

This article has considered the benefits and drawbacks to various policy options designed to support the working family during the time of childbirth. Such policies must provide both economic and employment security in order to be viewed as successful. In addition, these policies should be guided by principles of gender equity such that both parents are afforded the opportunity or legal right to take paid leave during their children's earliest years. Increasing the flexibility of policy programs would provide parents with choices; these choices are especially critical for low-income families as they could partition the leave into separate breaks. Comparatively, the US provides less support to families with children under the age of 3 than most advanced industrialized nations, and generally has weaker system of health and social services. These policy options have placed an added burden on parents of infants and toddlers in the US. Relatedly, little is known on the effects of childcare arrangements on child outcomes. Given that no single combination of childcare and education will work best for all children, it is important to advocate policy packages that are more flexible in nature, providing parents with the option to choose. Questions remain on how to best provide support to these families, but many options exist, including the expansion of parental leave, increasing funding support for early childcare and education benefits. In addition, most researchers conclude that steps should be taken to ensure higher quality of care for children of all ages such that the US is aligned with provisos of other advanced nations. In sum, family-leave policies and early care and education programs should be considered an essential component of every nation's social welfare system and provide crucial support for working parents who seek to effectively balance their familial responsibilities and improve the quality of life for their children.

See also: Family Influences; Maternal and Paternal Employment, Effects of; Parental Leave; Preschool and Nursery School.

Suggested Readings

Burchinal M (1999) Child care experiences and developmental outcomes. *Annals of the American Academy of Political and Social Science* 563: 73–97.

Gornick JC and Meyers MK (2003) *Families That Work: Policies for Reconciling Parenthood and Employment.* New York: Russell Sage Foundation.

Kamerman SB (2000) Parental leave policies: An essential ingredient in early childhood education and care policies. *Social Policy Report* 14(2): 3–15.

Kamerman SB (2003) Maternity, paternity, and parental leave policies: The potential impacts on children and their families. In: Tremblay RE, Barr RG, and Peters RD (eds.) *Encyclopedia on Early Childhood Development*, pp. 1–4. Montréal, QC: Centre of Excellence for Early Childhood Development.

Kamerman S and Gatenio S (2002) *Mother's Day: More than Candy and Flowers, Working Parents Need Paid Time-Off. The Clearinghouse on International Developments in Child, Youth, and Family Policies. Issue Brief, Spring 2002.* New York: Institute for Child and Family Policy.

Kamerman SB, Neuman M, Waldfogel J, and Brooks-Gunn J (2003) *Social Policies, Family Types, and Child Outcomes in Selected OECD Countries.* Working paper series (Social, employment and migration, No. 6). Paris: OECD.

Karoly LA, Greenwood PW, Everingham SS, *et al.* (1998) *Investing in Our Children: What We Do and Don't Know about the Costs and Benefits of Early Childhood Interventions.* Santa Monica, CA: RAND Corporation.

Magnusson KA and Waldfogel J (2005) Early childhood care and education: Effects on ethnic and racial gaps in school readiness. *Future of Children* 15(1): 169–196.

Organization for Economic Cooperation and Development (OECD) (2005) *Babies and Bosses: Reconciling Work and Family Life,* vol. 1–4. Paris: Author.

Ruhm CJ (1998) The economic consequences of parental leave mandates: Lessons from Europe. *Quarterly Journal of Economics* 113(1): 285–317.

US Department of Education (2002) *The Conditions of Education.* Washington, DC: Author.

Vandell DL and Wolfe B (2000) *Child Care Quality: Does It Matter and Does It Need to Be Improved?* Special report (November). Madison: University of Wisconsin, Institute for Research on Poverty.

Waldfogel J (2001) International policies toward parental leave and child care. *Future of Children* 11(1): 99–111.

Relevant Websites

http://www.childpolicyintl.org – The Clearinghouse on International Developments in Child, Youth, and Family Policies at Columbia University.

http://www.ecdgroup.com – The Consultative Group on Early Childhood Care and Development.

http://www.oecd.org – Organisation for Economic Co-operation and Development (OECD).

http://www.unesco.org – United Nations Educational, Scientific and Cultural Organization (UNESCO).

Fear and Wariness

J Kagan, Harvard University, Cambridge, MA, USA

Glossary

Anxiety – A state accompanying anticipation of a threat.

Discrepant – An event that cannot be understood immediately.

Fear – A state evoked by an imminent threat.

Neurotransmitter – A molecule secreted by the brain that influences the excitability of a neuron.

Temperament – A biologically based bias for a feeling state or behavior to an incentive.

Introduction

The components of human emotions like fear and wariness include a brain state produced by an event, a conscious feeling, an appraisal of the feeling, and, on some occasions, a behavioral response. There is serious controversy over the definition of an emotion because its components do not always occur together and each can occur without the others. For example, the sound of distant thunder will create a brain state, but there need not be any accompanying change in feeling, thought, or behavior. Similarly, a child watching snakes on a television monitor may think privately that he is afraid of snakes without experiencing any change in feeling or implementing any action. Because young infants in the first year are not consciously aware of their feelings, and do not have a rich set of thoughts, they necessarily have different emotional experiences than 10-year-olds. Some scientists, especially those who study the brain, define an emotion as a brain state, whereas psychologists rely primarily on a child's verbal statements and behaviors. Neuroscientists studying animals use brain profiles, or behaviors like body immobility, as signs of fear, whereas child psychologists and psychiatrists more often use children's descriptions of their feelings or their avoidant behavior as indicating a fear state. These two forms of evidence are different and do not refer to the same natural phenomenon. As long as this controversy remains unresolved, it is not useful to debate the correct definition of emotion. The more fruitful strategy is to discover the coherences among events, brain states, feelings, and behaviors.

Further, each of the above components of an emotion changes with development; hence, the meanings of 'fear' and 'wariness' should change, too. The meaning of fear applied to a 1-year-old who cries when a parent leaves the house is to be distinguished from the meaning of fear ascribed to a 13-year-old who reports that she is afraid she will harm herself. Thus, the concept of a single state of fear or wariness must be replaced with a family of emotions that, although related are, nonetheless, distinct and associated with particular causes, brain profiles, feelings, and behavioral reactions. Sentences that contain the words 'fear' or 'wariness' that fail to specify the origin of the child's feelings and the specific behaviors are not useful.

Thus, the word 'emotion' resembles the word 'weather'. Weather refers to the change in relations among wind velocity, humidity, temperature, barometric pressure, and form of precipitation. Occasionally, a unique combination of these qualities creates a blizzard, tornado, or hurricane. But wind, temperature, and humidity are always varying and usually do not produce the extreme phenomena we call blizzards and hurricanes. Meteorologists do not define weather, but determine the relations among the qualities they are able to measure. For the same reason, psychologists should study the relations among provocative events, brain states, feelings, thoughts, and actions and, when they discover coherence, they should give it a name. For example, scientists should come to some agreement on a name for the combination of seeing a potentially dangerous animal in a forest, anticipating physical harm, perceiving a rise in heart rate, and fleeing. But the name for that event should differ from the words selected to describe the state of a child who sees a frown on a parent's face, expects criticism for a misdemeanor, perceives a rise in heart rate, and apologizes. The causes and physiological and psychological features of the two events are different.

It is also necessary to distinguish between the changes in body and brain that are undetected and, therefore, unconscious, and those that are detected and become conscious feelings. The latter motivates children older than 3 or 4 years of age to interpret their change in feeling by privately selecting a word or phrase to name their state at that time. A thought experiment illustrates the usefulness of distinguishing conscious recognition of a feeling state from an undetected change in the body. Imagine three adolescents, each of whom is thinking about an arithmetic test to be given the following day. The first child detects increased muscle tension, thinks about those sensations, and decides they are due to worry over the examination. The second adolescent detects the same changes in muscle tension but decides it is because she has not had much exercise. The same changes in muscle tension in the third adolescent are undetected, but the parents recognize that the adolescent appears unusually tense. Although the imminent examination generated a bodily change in all three adolescents, psychologists should consider using three different terms to name the adolescents' emotions. It is an error to regard all three as fearful because their conscious states and behaviors are different. At the least, we should call these states fear1, fear2, and fear3.

Cultures differ in how they categorize human emotions. One example is found among those who live on the small, isolated atoll of Ifaluk in the western Caroline Islands of Micronesia. The residents use the word 'rus' to name the feeling that occurs when they are in a situation of imminent harm, such as a typhoon or a serious fight. Hence, the word 'rus' seems to resemble the meaning of the English word 'fear'. However, that is not a perfect translation because a mother will say that she feels 'rus' if her child has died. American parents are more likely to say they feel sad or depressed, not fearful. The Ifalukians use the word 'metagu' to name situations that have the potential to generate the emotion 'rus'. Examples include a future interaction with a stranger. Thus, the word 'metagu' comes close to what Americans would call 'social anxiety', but that translation is imperfect for 'metagu' is also used when a person anticipates encounter with a ghost or the anger of another.

English terms for emotions usually focus on the quality and intensity of the felt experience and less often differentiate emotions with respect to their origin or the context in which it occurs. The Ifalukians, by contrast, use different words to name emotions that have different origins. Americans would resemble Ifalukians if they used different words to describe how they felt when they lost a friendship, a valuable possession, or a loved one.

The Role of History

The state of fear assumed prominence in Medieval Christian Europe because of a preoccupation with God's wrath. St. Augustine, who thought fear was the fundamental emotion, regarded fear of divine punishment as a blessing because it helped humans behave morally. By contrast, contemporary Americans believe that fear and the related state of anxiety restrict the capacity for love and work. Neurotic symptoms, according to Freud, were learned behaviors whose purpose was to reduce the anxiety created by conflict over sexuality. By arguing that anxiety, but not realistic fears, could be eliminated by emptying the unconscious of its repressed wishes, Freud and his followers implied that anxiety was not a necessary emotion and it was possible to free everyone of this feeling.

Darwin's naturalization of the human emotions is another strand in the history of the concepts of fear and anxiety. Nineteenth-century scholars regarded an animal's state as natural and animals behave as if they are capable of fear. Hence, the emotion of fear was treated as an inherent quality of being human. But anxiety was not given the same treatment and was regarded as unnatural. If life's assignment is to control desires for pleasure, as was true in St. Augustine's century, fear is an ally and not an alien force. But if the day's assignment is to gain friends, seduce a lover, and take risks for status and material gain, fear and anxiety become the enemy. As history altered the daily scripts people were to follow, fear displaced desire as the emotion to subdue. If humans must restrain greed, lust, competitiveness, and aggression, self-control is a prerequisite. But a person's will is less potent when fear is the demon to be tamed because it is more difficult to rid oneself of fear than to control an action directed at gaining some desired state of affairs. As a result, history relegated will to the same ash heap of ideas where Newton's ether lies gathering dust. The belief that humans can and should be free of anxiety is one of the distinguishing illusions in Western thought. The assumption that anxiety is abnormal is incorrect. Feeling apprehensive before speaking to a large group of strangers is as naturally human as anger at being cut off by a motorist at an intersection. One reason why psychologists and psychiatrists regard anxiety as a sign of mental illness is that this emotion is a greater burden than anger in contemporary society where taking risks and meeting new people are required for adaptation and success.

Fear vs. Wariness

The traditional distinction between fear and wariness (this article uses the term 'anxious' as synonymous with wariness) is based on whether an unwanted or undesirable event is imminent, and, therefore, causes fear, or is anticipated in the future and, therefore, creates wariness. English speakers use words like 'afraid', 'frightened', and 'scared' to name the emotion produced by an imminent

threat, but more likely to use 'worried', 'concerned', 'troubled', or 'anxious' to describe an unwanted event that is anticipated. There are at least five different types of events that can evoke fear. They include the imminent possibility of: (1) pain or harm to the body (e.g., sight of a hypodermic needle); (2) the loss of a supportive affectionate relationship with another (e.g., learning that a parent is about to die); (3) failure on a task (e.g., either a test or an athletic performance); (4) criticism, punishment, or rejection by another; and (5) an unexpected or unfamiliar event that is not understood immediately (e.g., the sight of a turtle on one's pillow). There is a parallel set of five states of wariness or anxiety that results from the anticipation of each of the above events; for example, anticipating failure on a test to be given the following week. The state created by seeing a doctor with a needle about to inject a child with a influenza shot is not identical to the state created by anticipating the same influenza shot while at home.

Children from different cultures asked to state their fears (they phrased the question this way because many children do not understand the difference between fear and anxiety) nominated five different categories: physical harm, failure, criticism from others, the unknown, and animals with unusual physical features, like spiders, mice, and snakes. Some older children added a fear of rejection or unacceptability to others, and ~15% named a feeling of fear, which is called anxiety sensitivity. Children are not yet concerned with dying. The emotions of 1-month-olds and 10-year-olds following the sudden appearance of a large snake are different because older children know that snakes are potentially dangerous, are conscious of a change in feeling, and are able to flee. One-month-olds lack all three of these features. Therefore, the emotional states of infants and older children are distinctly different.

Fears of the Human Infant

Infants less than 6 or 7 months old have limited cognitive abilities and, therefore, are capable of a small number of fear states. One such state is provoked by an unexpected event (e.g., a sudden, loud sound or the sudden offset of a light). A second state is provoked by an event that has become a conditioned stimulus for a painful experience that occurred in the past (e.g., the sight of an older sibling who has been teasing the young infant). Because infants under 6 months of age do not yet have a reflective self-awareness, they do not interpret the bodily changes that accompany provocative events and, obviously, do not name them. The 5-month-old who cries to a sudden, loud sound can be likened to a mouse that suddenly becomes immobile to a bright light or to a sound that has become a conditioned stimulus for the delivery of mild electric shock. The attribution of fear to a young infant, or to a mouse, is based on the relation between an event and a behavior (i.e., between a loud sound and crying in the infant or between a bright light and immobility in the animal). There is no presumption of a conscious feeling or a symbolic interpretation. These feeling states are different from those experienced by older children and adolescents.

The maturation of the brain is accompanied by a transition at 8–12 months that enables infants to recall a past event and to compare it with the present in a process called working memory. As a result of these cognitive advances, events that are unfamiliar, or discrepant from the infant's knowledge, elicit a state called 'fear of novelty' if they are not understood immediately. A fear of novelty is less likely if infants can control the appearance of the unfamiliar event. For example, 9-month-olds show behavioral signs of fear if they are presented with a cymbal-clapping monkey unexpectedly, but display little sign of fear if they can control when the toy monkey makes its motions.

There are two universal fears that appear during the second half of the first year. One is a fearful reaction to unfamiliar people called 'stranger anxiety'. Infants who have acquired a relatively firm representation of the familiar adults who care for them experience fear if a stranger approaches because the face, posture, and gait of the unfamiliar person are unfamiliar and the infant tries, but fails, to relate the features of the stranger to their knowledge of the adults familiar to them. This failure evokes the fear of novelty and the infants may cry. A similar explanation applies to the second fear of this period, called 'separation fear', that occurs when a familiar caregiver leaves the infant alone, especially in an unfamiliar place. The child retrieves the representation of the caregiver's prior presence, but cannot relate the present situation with the caregiver absent to the representation of her former presence seconds earlier. As a result, infants cry. Even in the familiar home environment, a mother's departure produces more crying if the parent leaves by an unfamiliar exit, like the basement door, than if she leaves by the familiar front door. Children across many cultures first show fear following separation from a caregiver around 7–8 months, with peak occurrence at ~15 months, followed by a sharp decline in separation fear after the second birthday.

Infants also show a fear reaction when placed on a special apparatus called 'the visual cliff', which consists of a clear sheet of plexiglass that has one checkerboard cloth just beneath the surface of one side, and another cloth placed several feet below the transparent surface on the other side. This situation produces a perception of apparent depth. An infant is placed on a small platform between the safe and the apparently deep side. Although mothers call their infants to cross over the plexiglass on the deep side in order to approach them, most 8- and 9-month-olds appear fearful and do not do so, although a mother's smile can mute the intensity of the fear. One interpretation of their apparent fear is related to the fear of novelty for the infant cannot assimilate the inconsistent information coming from different sensory modalities on the deep side. The visual and vestibular information produced by the perception of depth implies a 'drop off', whereas the tactile information originating in the hands placed on the plexiglass implies a solid surface. If infants cannot assimilate the inconsistency between the two sources of information, they may experience a fear state and avoid crawling over the deep side. Animals, too, show avoidance of novelty, but, as in human infants, this reaction is not present at birth. Infant monkeys begin to display signs of fear to novelty ~2.5–3 months of age. Because monkeys mature at a rate three times that of human infants, this age corresponds to 8–9 months in the child. Although fear of the novel and the unfamiliar is different from the fear state to an unexpected, loud sound in 4-month-olds, both younger and older infants are not consciously aware of their feelings and do not interpret their psychological states.

The Second and Third Years

Continued maturation of the brain enables 2–3-year-old children to be aware of their feelings and to know that certain actions are punished by caregivers. As a result, a new emotion appears that might be called 'fear or wariness over adult disapproval or punishment' for violating a family standard on proper behavior. However, this new source of fear or wariness exploits the power of unexpected events. A mother who has just seen her 14-month-old spill milk on the tablecloth says, in a voice louder and with a face sterner than usual, 'Don't do that'. The unexpected parental action creates a state of uncertainty that resembles other hedonically unpleasant experiences, like pain, hunger, and cold. As a result, the child quickly learns that spilling food is followed by a similar chastisement, and the resulting feeling of wariness leads children to inhibit such acts. It is probably impossible for any parent to raise a child without interrupting some actions that are potentially harmful or violate a family standard. The child's representations of the actions parents prohibited, the parent's disciplinary response, and the child's feeling of uncertainty become linked to create a conditioned reaction of wariness whenever the child is in a situation associated with prior parental discipline.

Three-year-olds also show signs of wariness in situations where no punishments occurred in the past. For example, most 3-year-olds will hesitate, or not perform at all, if a parent asks them to act in a way that would violate a family norm; for example, pouring cranberry juice on a clean tablecloth, even though the child has never displayed the behavior and, therefore, had never been punished for it. The refusal implies that the child possesses a concept of prohibited actions that includes behaviors never punished in the past. This fear state, which psychologists call 'anxiety over disapproval', is the first stage of moral development and a conscience. During the second and third years, morality is supported by fear or wariness over the reactions of others. Several years later, the child's moral sense will be sustained by the emotion of guilt and the desire to regard the self as good. In addition, children are told, directly or indirectly, about objects and events that are potentially dangerous. Witches, goblins, ghosts, large animals, spiders, snakes, lightening storms, and kidnapping are examples. Many children say they are afraid of one or more of these events, even thought they have never encountered them and have not experienced any pain through those encounters. These sources of fear comprise a special set, different from the fears created by conditioning of a fear state to these or other objects.

The next transition, which occurs around 4 years of age, occurs when children are able to relate the distant past to the present, rather than the immediate past. Now children experience a state of fear if they are unable to understand why an event has occurred. This emotion is not a fear of novelty, but an emotion produced by a failure of understanding, or what some psychologists call 'cognitive dissonance'. That is, the state created by noting a frown on the parent's face and not knowing why is not equivalent to the state created by seeing a stranger.

A year or two later, children become aware of some of the social categories to which they belong and any behavior that is inconsistent with their understanding of the definition of the category elicits a state of wariness. All children learn a social category for their gender and at a later age acquire categories for their religion and ethnicity as well. Each social category is associated with the set of defining actions, beliefs, and qualities. Children want to maintain consistency between the features that define the categories to which they belong and their personal actions and qualities. If one of their behaviors or features is inconsistent with those of the category, a state of anxiety occurs. For example, 6-year-old boys know they should not wear lipstick. Although many boys have never been punished for doing so, they will feel anxious if they violate this feature of their gender category. Girls learn they are supposed to control extreme aggression and believe they ought to be loyal to that criterion. Children who belong to an ethnic or religious category feel an obligation to be loyal to its standards, and if not, will experience anxiety. Children are prepared to believe that words like 'boy', 'girl', 'Catholic' and 'Hispanic' refer to a set of fixed psychological characteristics and experience as much anxiety if they stray from those obligations as they would if someone called a bird a dog.

This emotional state is elaborated during adolescence when the detection of inconsistency among self's beliefs evokes a feeling of anxiety. For example, some adolescents experience uncertainty when they recognize the inconsistency between their belief that God loves humanity and their reflection on human catastrophe and cruelty.

The Amygdala

Scientists believe that the amygdala, a small structure located within the brain's temporal lobe, participates in many, but not all, states of fear and wariness. The amygdala, which is relatively mature before birth, consists of different collections of neurons, each with a distinct pattern of connectivity, neurochemistry, and functions. The three most important collections are called the basolateral, corticomedial, and central areas. The basolateral area receives information from vision, hearing, smell, taste, and touch and some information from the body. The basolateral area is reciprocally connected to parts of the brain that are involved in memory, autonomic activity, and parts of the motor system that mediate flight and attack. The corticomedial area primarily receives information from smell and taste. Although the central area receives some information from taste, vision, hearing, and the body, its most important input comes from the basolateral and corticomedial areas, and the central area is the origin of a large number of projections to diverse sites in brain and body. Some of the consequences of these projections include changes in heart rate, blood pressure, muscle tension, activity of gastrointestinal tract, and hormonal secretions associated with a conscious feeling of fear.

The qualities of the events that activate the amygdala, and the psychological states that can follow, are a source of controversy. The central disagreement is whether the amygdala reacts primarily to imminently dangerous events to produce a fear of harm, or to unexpected or unfamiliar events to produce a fear of novelty that resembles surprise. Dangerous events that can harm the person should create a state distinctly different from unfamiliar ones because not all unfamiliar events pose a threat and some threats – like a reprimand – are neither unfamiliar nor unexpected. The female Diana monkey displays a distinct vocal sound to the unexpected alarm call of a male leopard. However, the monkey does not vocalize to the same call a few minutes later, even though the leopard is still a threat, because the call is expected. The monkey vocalized when the potentially dangerous auditory event was unexpected, but not when the same sign of danger was anticipated.

Older children more often name snakes as a fear target rather than lions or tigers, even though they have encountered none of these animals, because of the snake's unfamiliar features. Snakes have an uncommon skin coloring, an atypical head-to-body ratio, and an unusual way of moving. The authors of the Tree of Knowledge allegory in the *Old Testament* probably chose the snake as tempter because of its discrepant features. Unfamiliar or discrepant events initially elicit a state of alert surprise; a state of fear occurs when these events cannot be understood or if the child has learned that they are dangerous. Snakes do not evoke behaviors indicative of fear in infants because they have not learned the usual shapes of animals and do not know that some snakes are dangerous.

One woman dated the origin of her fear of birds to an afternoon when, as a 7-year-old, she was watching Hitchcock's film, 'The Birds', in which large flocks of birds attacked humans. The woman remembered feeling very surprised by seeing birds, which she had regarded as benevolent, acting aggressively toward humans. The sharp disconfirmation of her childhood belief activated the amygdala and her thoughts became associated with amygdala activation and its bodily consequences. However, the phobia would not have developed if the woman had not been surprised by the bird's behavior. A 5-year-old boy developed a phobia of buttons for the same reason. The boy went to the front of the room to retrieve buttons from a bowl in order to finish a teacher-assigned task. He slipped as he reached for the bowl and all of the buttons spilled over him. This experience surprised and embarrassed him, and a phobia that began that day lasted for 4 years.

The amygdala is intimately connected to brain sites that mediate the feelings of fear, wariness, and anxiety. Two such sites are the insula and the orbitofrontal prefrontal cortex. In addition, some brain sites modulate the amygdala and mute its excitability. Two such sites are the anterior cingulate and the ventromedial prefrontal cortex. These latter sites mature later in development, and that is why older children are better able to control their feelings and behavioral signs of fear. Parents do not expect 2-year-olds to inhibit a scream to the sight of a spider, but do expect 10-year-olds to do so. Thus, by early adolescence, the brain structures involved in fear and anxiety are so interconnected it is difficult to know, or to predict, whether children are experiencing fear or wariness to danger, reprimand, discrepancy, or discrepancy. Further, there is considerable variation among children in their vulnerability to these emotions.

Temperamental Biases

Children differ in their susceptibility to the varied forms of fear, anxiety, or wariness. When the cause of the fear or anxiety is an unfamiliar or discrepant event that cannot be understood, children who frequently express fearful behaviors are called 'inhibited', whereas those who are less susceptible to this fear are called 'uninhibited'. These susceptibilities may be due to differences in inherited neurochemistry and are moderately stable over the first 15 years of life.

During the interval from the first to the tenth birthday, the most obvious behavioral sign of an inhibited temperamental bias is extremely shy behavior with unfamiliar peers or adults. Most young children are initially subdued for a few seconds or a few minutes when they meet a stranger, but most overcome their initial restraint quickly. A small proportion, ~15%, remain shy for a much longer period and are consistently subdued whenever they meet people they do not know. Of course, shyness with strangers can result from conditions other than a temperamental bias. Children who have been abused, severely punished, or isolated from others are usually shy, but their behavior is a result of these experiences. Thus, only some children who are excessively shy inherited a temperamental bias and, in addition, some who inherited a temperamental bias to be shy when young overcome this behavior and express their inhibited temperament in other ways. Therefore, shy behavior in children has many causes. Children do not inherit a tendency for shyness, but, rather, inherit a temperamentally based bias to react to discrepancy and unfamiliarity with initial caution and restraint.

Children who are consistently inhibited to unfamiliar people, places, and situations are different as early as 4 months of age. These children show high levels of motor activity and crying when presented with unfamiliar stimuli at 16 weeks of age. These infants thrash their limbs and cry when shown colorful mobiles or hear recordings of human speech. It is believed that these behaviors are a result of an excitable amygdala. By contrast, other 4-month-olds show minimal motor activity and no crying to the same unfamiliar events, and it is presumed that they possess a lower level of amygdala excitability. The former group of infants are called 'high-reactive' and the latter are called 'low-reactive'.

Unfortunately, it is not possible to describe the different brain chemistries that high- and low-reactive infants inherit. Some possibilities include variation in the concentration of, or the density of receptors for, gamma-aminobutyric acid (GABA), opioids, corticotropin-releasing hormone (CRH), norepinephrine, oxytocin, or vasopressin. For example, some infants are born with a compromise in the neurotransmitter GABA and, as a result, may have difficulty modulating extreme states of distress. Variation in norepinephrine and its varied receptors may affect the response to novelty, and it is easy to invent an argument that relates this variation to infant temperament. Opioids modulate the level of excitation in the brain and the autonomic nervous system. The brain sites that mediate pain and discomfort contain receptors for opioids, and possession of a high density of opioid receptors or a high concentration of opioids in the medulla should mute information coming from the body to the medulla. Therefore, children with greater opioid activity in the medulla might experience more frequent moments of relaxation, whereas those with compromised opioid activity might be vulnerable to more frequent bouts of distress. CRH is another molecule that has diverse influences as it influences many systems but, especially, the hypothalamic–pituitary–adrenal axis. One consequence of activity in this axis is secretion of the hormone, cortisol, by the adrenal cortex. Monkeys with high cortisol levels are more avoidant of novelty than animals with lower levels.

Finally, it is possible that the month of conception could influence temperamental qualities because the secretion of very levels of melatonin by the pregnant mother can affect the fetus. If conception occurs at the end of summer (late July through September) in the northern hemisphere, as the hours of daylight are decreasing, the embryo's brain will be maturing at the same time the mother is secreting greater amounts of melatonin. Adults conceived during the late summer months are at somewhat higher risk for extreme shyness during the preschool years.

However, the immaturity of our current knowledge relating the chemistry of the brain to varied temperaments makes it impossible to posit a lawful relation between a chemical profile and a particular temperament. It is also unlikely that any single gene for a neurotransmitter or receptor distribution is the basis for a temperamental type.

Developmental Consequences of Infant Temperament

When high- and low-reactive infants were exposed to a variety of unfamiliar events in the second year of life – strangers, clowns, unfamiliar rooms – more high-reactive infants showed consistent avoidance or crying compared with the 2-year-olds who had been low-reactive infants. At 4.5 years of age, high-reactive children were more likely to be shy and quiet when playing with unfamiliar children of the same sex and age, whereas low-reactive children were sociable and emotionally spontaneous. About 25% of 7-year-olds who had been high-reactive infants showed extreme levels of shyness, worried over storms and large animals, had frequent nightmares, and were occasionally reluctant to go to school. By contrast, 90% of low-reactive children did not show many of these symptoms.

At 11 years of age, more high- than low-reactive children showed four biological characteristics that are indirect signs of amygdalar excitability. First, they showed greater activation in the right, compared with the left, hemisphere as measured by electroencephalogran (EEG) recordings. Activity of heart, blood vessels, muscles, and gastrointestinal tract, which is increased during fear or anxiety, ascends to the brain and is more fully elaborated in the right hemisphere than in the left. As a result, children with greater bodily activity should show greater activation in the right hemisphere.

Second, high-reactive children also had a larger evoked potential from a structure in the brainstem that is part of the circuit for perceiving sound. Because the amygdala projects to this structure, children with larger evoked potentials probably possess a more excitable amygdala. Third, the 11-year-olds who had been high-reactive infants showed greater sympathetic tone in the cardiovascular system, and the amygdala enhances sympathetic tone on the heart and circulatory vessels. Finally, the high-reactives showed a larger event-related potential to discrepant visual scenes. Humans, as well as animals, show a distinct brain reaction to unexpected or discrepant events. The amygdala, which is also activated by discrepancy, sends projections to neurons in the cortex that are responsible for the event-related potential. Thus, high-reactives differed from low-reactives in four biological features that are under the influence of the amygdala.

However, the prediction that a high-reactive 4-month-old infant will not become a highly sociable, spontaneous child with low levels of biological arousal can be made with greater confidence than the prediction that a high-reactive infant is likely to become an extremely subdued adolescent with high biological arousal. Similarly, the prediction that a low-reactive infant will not be extremely shy and show signs of high biological arousal will be more often correct than the prediction that such infants will become extremely sociable and display low biological arousal.

Thus, high- and low-reactive temperaments constrain what each child will become as an adolescent, but these temperaments do not determine the adolescent profile. Further, by adolescence, social behaviors become dissociated from the individual's feelings and biology. Many high-reactives who were inhibited in the second year and showed biological signs of an excitable amygdala were not unusually shy as 15-year-olds. However, they described themselves as worriers who were often serious and tense. The biological characteristics of high- and low-reactive infants changed less over time than their public behaviors. This fact implies a dissociation between the biological processes that form the foundation of their temperaments and their social behavior. Life experiences can change a childhood profile of extreme wariness and shyness to a more normative pattern without eliminating the excitability of the brain structures that were the bases for the infant behavior.

A temperamental bias for high reactivity renders some children vulnerable to serious anxiety following a traumatic event. For example, only 10 of 40 California school children who were kidnapped and terrorized for 2 days developed serious post-traumatic stress disorder. During the winter of 1984, a sniper fired at a group of children on the playground of a Los Angeles elementary school, killing one child and injuring 13 children. One month later, professionals determined which children were experiencing extreme anxiety. Only about one-third were judged highly anxious, while an equal proportion was free of unusual levels of anxiety. Many of those who were judged to be very anxious had inhibited prior to the school violence. Thus, it is likely that many children who develop a serious anxiety disorder following a traumatic event were temperamentally predisposed to that reaction.

The consequences of a high-reactive temperament for later personality are influenced by each person's life history and cultural context. A high-reactive infant with overprotective parents will become more wary than one whose parents encouraged them to cope with challenge and unfamiliarity. Further, the implicit norm in contemporary America encourages children to be sociable and to seek, rather than to avoid, challenge. The opposite profile was dominant three centuries ago. Thus, the bases for a shy personality, compared with an ebullient one, in seventeenth century Colonial Massachusetts were different from the bases for the same profiles in contemporary Boston.

Adolescents who recognize that they are especially vulnerable to feeling anxious try to avoid situations that produce this feeling. These adolescents are likely to select vocations that allow them to control unpredictable events and frequent encounters with strangers (e.g., writer, computer programmer, or bench scientist). Adolescents who fail to attain that insight might select a career that contains more uncertainty than they are prepared to deal with effectively.

Acute Emotion vs. Mood

Fear and wariness are terms used for acute feelings that last a short time. However, some individuals have chronic feeling tones, called moods. Two adolescent girls who had been high-reactive infants and shy and fearful in the second year were relaxed and minimally defensive when interviewed by an unfamiliar woman. However, during the interview these girls reported that they feel wary in crowds, worry over the opinions of their friends, and feel uneasy when they violate one of their personal standards. These emotions are influenced by activity in the amygdala, the insula, and the orbitofrontal prefrontal cortex. When a subtle change in usual mood is mildly unpleasant, and ambiguous as to its cause, adolescents might decide that they feel shame, guilt, regret, illness, fatigue, or, in other cultures, possession by the devil. American children who experience this change in feeling are especially vulnerable to deciding that they probably violated a conventional standard. They might decide that they had been rude to a friend, told a lie, or harbored a cruel prejudice. In other cultures, these adolescents might interpret the same feeling of uneasiness as meaning that they had offended an ancestor or broken a taboo on eating. The adolescents who had been high-reactive infants reported that they felt bad whenever their parents chastised them for doing something wrong. Thus, children with a

high-reactive temperament are more highly motivated to examine their beliefs in order to ferret out the inconsistencies that create uncomfortable feelings of anxiety characteristic of introverted personalities.

Advice for Parents

Parents should appreciate that some children inherit a temperamental bias to be shy and anxious, especially in unfamiliar situations, and should not always assume that their parenting practices were the sole source of these traits in their children. This recognition should alleviate some of the guilt felt by parents who misinterpret their child's shyness as a sign that they must have done something wrong.

Parents of high-reactives usually adopt one of three strategies. Imagine three types of American families who have an easily aroused, distressed, high-reactive infant. One category of parents feels empathic for their apparently unhappy infant and become overly solicitous. These mothers pick their infants up as soon as they cry in order to sooth their distress. This routine strengthens the infant's tendency to cry at the first sign of novelty and, when older, to avoid unfamiliar events. The parent's reluctance to make their children unhappy motivates them to accept retreat from novelty, even though this acceptance increases the likelihood that, as the children age, they will become cautious in new situations. Continued acceptance of shy, inhibited behavior can lead, in time, to a timid 6-year-old.

A second group of equally affectionate parents holds a different philosophy. These families believe, often tacitly, that they should prepare their children for a competitive society in which retreat from challenge is not adaptive. These parents refrain from comforting their infants every time they cry and wait for the infant to regulate its own distress. Rather than accept retreat from challenge and unfamiliarity, these parents encourage their children to greet strangers and to approach unfamiliar children on the playground. They also praise their sons and daughters whenever they overcome their caution. Children raised this way are less likely to be avoidant when they begin school. This script is more common in sons than in daughters because of a sex-role bias in American culture that regards retreat from challenge as less acceptable in boys than in girls.

Finally, some high-reactives are born to parents who misinterpret their child's irritability as willful. These parents often become angry and punish what they interpret as misbehavior or disobedience. This regimen may exacerbate an already high level of limbic excitability. Because such children cannot always control their emotions, they become more irritable or, depending on the severity of the parental behavior, withdraw. Fortunately, this developmental course is less common than the first two.

Low-reactive infants are easier to care for because they smile frequently, cry less often, and sleep well. However, during the second and third years, some of these children resist parental demands for the control of aggression and tantrums because parental punishment does not generate strong fear. Now two developmental itineraries become possible. In the most common, which is a derivative of the American value on autonomy and freedom from anxiety, parents adopt a 'laissez-faire' approach and permit disobedience to all but the most serious infractions. These children are likely to become relatively exuberant and sociable, as long as their parents are consistent in the socialization of serious violations of norms on aggression, domination, and destruction. If parents are inconsistent in their punishment, and, in addition, there are peer temptations for asocial behavior during later childhood, these children, especially if boys, are at a slightly higher than normal risk for developing an asocial personality. If parents are unwilling to brook any opposition to their demands and punish all disobedience, their children are likely to react with angry tantrums. Continuation of this cycle can create a rebellious posture toward all authority.

Although many American and European parents assume that low-reactive infants and uninhibited children are the more desirable personality type, each temperament has both advantages and disadvantages in contemporary society. A technological economy requires a college education for a challenging, satisfying vocation. Students with higher grade point averages in high school are more likely to be accepted at better colleges and, therefore, have a higher probability of attaining a gratifying, economically productive career. High-reactive children raised in middle-class homes are more concerned with school failure and, therefore, more likely to have an academic record that will gain them admission to a good college. Following graduation, this temperamental type has many occupational options because American society needs adults who like to work in environments where they can titer the level of uncertainty. Adolescents who had been high-reactive infants often choose intellectual vocations because this category of work allows control of each day's events in settings where unanticipated interactions with strangers can be held to a minimum. Our society needs these vocational roles and those who fill them are rewarded with respect and financial security.

In addition, youth confront many temptations that promise pleasure, peer acceptance, and self-enhancement if they are willing to assume some risk. Driving at high speeds, experimenting with drugs, engaging in sex at an early age, and cheating on examinations are four temptations with undesirable consequences. Adolescents who experience anxiety as they think about engaging in any one of those behaviors will avoid these risks. Thus, there are advantages to a high-reactive temperament.

Low-reactive, uninhibited, children also enjoy advantages. A sociable personality and a willingness to take career and economic risks are adaptive in contemporary society. The 18-year-old willing to leave home in order to attend a better school or accept a more interesting job is likely to gain a more challenging position and greater economic return than one who stays close to home because of a reluctance to confront the uncertainties of an unfamiliar place. Thus, each of the temperaments has advantages and disadvantages.

The island of Cayo, Santiago, close to mainland Puerto Rico, contains a large number of macaque monkeys and no human residents. Observers visit the small island each day to make notes on each animal. The consistently timid, inhibited monkeys are more likely to die of starvation because, when food is put out each day, they wait for the other animals to feed first. If a timid monkey waits too long on too many days, it will starve to death. In contrast, the bold, uninhibited monkeys get sufficient food, but they are at higher risk of dying from wounds because of impulsive attacks on a stronger animal. Thus, the balance between the advantages and disadvantages of each temperamental bias depends on the challenges the child faces. The best advice for parents is contained in three suggestions:

1. Acknowledge your child's temperamental bias and do not assume that either your rearing practices or the child's willfulness are the only reason for their behavior.
2. Acknowledge your child's malleability and capacity for change. The infant's temperament does not determine what he or she will become 20 years later because temperament is not destiny.
3. Accommodate parental goals to the child's wishes. A regimen of rearing that takes into account both the parent's hopes and the child's desires can be found if parents are willing to search for it.

See also: Brain Function; Discipline and Compliance; Emotion Regulation; Endocrine System; Exploration and Curiosity; Future Orientation; Habituation and Novelty; Parenting Styles and their Effects; Self-Regulatory Processes; Separation and Stranger Anxiety; Shyness; Social and Emotional Development Theories; Temperament.

Suggested Readings

Kagan J and Fox N (2006) Biology culture, and temperamental biases, Vol. 3: Social, emotional, and personality development. In: *The Handbook of Child Psychology*, 6th edn., vol. 3. New York: Wiley.

Kagan J and Snidman N (2004) *The Long Shadow of Temperament*. Cambridge MA: Cambridge University Press.

Rubin KH and Asendorpf JB (1993) *Social Withdrawal Inhibition, and Shyness in Childhood*. Hillsdale, NJ: Erlbaum.

Feeding Development and Disorders

I Chatoor and M Macaoay, Children's National Medical Center, Washington, DC, USA

Glossary

Dyadic reciprocity – Effective communication patterns between infant and caregiver during feeding or play.

External regulation – When the caregiver regulates feeding using distraction, cajoling, or force-feeding the infant regardless of the infant's hunger or satiety cues.

Failure to thrive – Infants and children who fail to gain weight or lose weight.

Infants – Birth up to 12 months.

Internal regulation – The ability of the infant/toddler to recognize his/her own inner signals of hunger and satiety.

State regulation – The ability to maintain a homeostatic mood state (e.g., calm state of alertness for feeding).

Toddlers – 12–36 months.

Transactional model – This perspective emphasizes the reciprocal, bidirectoral influence of the communication environment, between the child and his/her caregivers.

Introduction

For most infants, feeding appears to be a natural process; however, approximately 25% of otherwise normally developing infants and up to 80% of those with developmental handicaps have been reported to have feeding problems. In addition, 1–2% of infants have been found to have serious feeding difficulties associated with poor weight gain. Feeding problems may derail an infant's early development and have been linked to later deficits in cognitive development, behavioral problems, and to eating disorders during childhood, adolescence, and early adulthood.

Regulation of feeding is established in the first few years of life when internal regulation of feeding by the child vs. external regulation of feeding by the parents may develop. Internal regulation is attained when the infant becomes aware of his own internal hunger and satiety cues and responds accordingly by eating or ceasing to eat. External regulation of feeding is characterized by the parent or caregiver assuming the role of regulation and feeding the infant regardless of whether the infant may be hungry or satiated.

A three-stage model of feeding development will provide a fundamental foundation of feeding behaviors, followed by a classification of feeding disorders (see **Table 1**).

Early Development and Regulation of Feeding

An important developmental process in the first years of life is the acquisition of autonomous internal regulation of feeding. The young child becomes increasingly aware of his internal hunger and satiety cues and responds accordingly by communicating his interest to eat when hungry and ceases to accept food when he recognizes fullness. Under ideal conditions, the infant is able to emit clear, unmistakable signals of hunger to his caregiver, who in turn, acknowledges these signals and responds by feeding the infant. The infant also signals when he is full by decreasing the frequency of sucking movements or by closing the mouth and not accepting any more food. When parents are attuned to their infant's hunger and satiety cues, they act as external scaffolds for the emerging abilities of the infant to self-regulate feeding. Positive reinforcement of the infant's cues during mealtimes allows the infant to trust that his basic nutritional needs will be satisfied. A nurturing and well-attuned caregiving environment establishes a solid foundation on which a secure transition from mutual regulation of feeding to the infant's own internal self-regulated feeding transpires.

The development of internally regulated autonomous feeding unfolds in three stages: (1) homeostasis, (2) dyadic reciprocity, and (3) transition of self-feeding.

Stage 1: Achieving Homeostasis

Whereas the fetus' nutritional demands were met unencumbered through the maternal umbilical cord *in utero*, upon birth, infants must actively and clearly signal hunger and satiety to caregivers. During the first few months of life, they must establish both rhythms of sleep and wakefulness, and feeding and elimination. Attaining and maintaining a calm state of wakefulness becomes a major developmental task during early feeding. Difficulties in achieving this state may result in inefficient attempts of feeding, as the infant may be too irritable or too sleepy to feed. Establishment of the infant–parent communication system is the key to establishing nutritional homeostasis – a state in which the infant's nutritional needs for growth

Table 1 A classification of feeding disorders by Irene Chatoor

1. Feeding disorder of state regulation
2. Feeding disorder of caregiver–infant reciprocity
3. Infantile anorexia
4. Sensory food aversions
5. Post-traumatic feeding disorder
6. Feeding disorder associated with a concurrent medical condition

and development are met. Appropriate perception of the infant's hunger and satiety signals by the parent and their contingent behavioral responses to the infant's signals solidifies the bonding and secure attachment between infant and caregiver.

Most infants establish a qualitatively distinct cry of hunger, in contradiction to other types of cries (e.g., pain, fear, tiredness). Ideally, these distinct cries become increasingly discernible for parents who become efficient at understanding the infant's cries during the first few weeks of life, thereby developing a communication system that allows the infant to express his needs. As the caregivers become more adept at differentiating among these cries and between infant hunger and satiety cues, they respond appropriately and thus facilitate the infant's internal regulation of feeding. However, if parents are challenged by confusing, indiscernible cues, they may respond erroneously by feeding or not feeding their infant, irrespective of their infant's state of hunger and satiety. Consequently, the infant's and the parents' distinction between hunger and satiety can become confused, which may lead to under- or overfeeding.

Stage 2: Achieving Dyadic Reciprocity

Most infants will attain state regulation (homeostatic mood state e.g., calm state of alertness for feeding) by 2–4 months of age. By that time they have become more social and infant–parent communication has become increasingly clearer. Parent and infant interactions are characterized by mutual eye contact and gazing, reciprocal vocalizations, and mutual physical closeness expressed through touch and cuddling. The adaptive infant mobilizes and engages caretakers and their interactions become increasingly reciprocal in nature. Reflexive hunger cues decrease, while intentional cues, such as vocalization for food, begin to emerge. In addition, body language that signals feeding cessation occurs with greater discern (e.g., refusing to open the mouth and turning the head). A more mature communication pattern evolves as infants begin to regulate caregivers actively and purposefully. Through receiving stronger and clearer hunger and satiety signals from the infant, caregivers are more apt to regulate the presentation and withdrawal of food accordingly. Feeding interactions become a mutually regulated process, satisfying both infant and caregiver.

The mutually regulated processes may become derailed when the infant's hunger signals are weak or inconsistent, and/or the parents lack attunement and are preoccupied with their own internal needs. These infants may be fed sporadically and inefficiently, resulting in inadequate nutrition. They may be at risk for developing a feeding disorder of failed reciprocity, as discussed later.

Stage 3: Self-Feeding Stage

The developmental stage of separation and individuation occurs between 6 months and 3 years of age. Motor and cognitive maturation processes enable the infant to become more physically and emotionally independent. Autonomy vs. dependency is negotiated daily during the feeding interactions. During each meal, mother and infant need to negotiate who is going to place the spoon in the infant's mouth. As the infant becomes more competent the parent needs to transition the infant to self-feeding. During this transition the infant not only needs to understand the difference between hunger and satiety, but also needs to differentiate the physical sensations of hunger and fullness from emotional experiences (e.g., eliciting comfort, affection, and feelings of anger or frustration).

In order for the infant to learn this differentiation, it is of paramount importance that caregivers are able to differentiate hunger and satiety cues from affective cues, and respond contingently. This includes offering food when the infant signals hunger, abstaining from offering food when the infant needs affection or calming through tactile sensations, terminating the meal when the infant appears satiated, and not insisting that the infant keeps eating until his plate is empty. This facilitative process by the caregivers ensures both the infant's awareness of hunger and satiety as well as the differentiation of physiological sensations of hunger and satiety from emotional needs.

Conversely, if the parent misinterprets the infant's emotional cues when he desires physical and emotional comfort by feeding the infant, the infant may confuse hunger with emotional experiences. He then may become conditioned to eat or refuse to eat when feeling sad, lonely, frustrated, or angry. External regulation of feeding may evolve based on the emotional experiences of the infant.

Ideally, the infant gives clear, discernible cues, and the parents interpret these signals correctly. If, however, the infant gives weak hunger cues, this may raise the parents' anxiety and cause confusion. As parents become concerned about how to deal with the nutritional needs for their infant's growth and development, they may try to override the infant's cues by feeding , even if the baby is not interested and may refuse to open its mouth. A once-envisioned happy and peaceful mealtime activity may evolve into a highly emotionally charged battleground, pitting the infant's food refusal against the parents' increasing concern about the low volume of food intake. Conversely, if an infant's weak signals of satiety or need for comfort are misinterpreted by overfeeding, parents may be unknowingly contributing to their infant's learning to eat when emotionally distressed or when seeking pleasure.

These early formative years are critical in the development of the child's internal vs. external regulation of eating. Maladaptive feeding patterns may emerge during these three developmental stages, as the infant and caregiver attempt to mutually regulate feeding behaviors. A feeding disorder may emerge if maladaptive feeding patterns become chronic, compromising the infant's growth and development. Three feeding disorders that emerge at these specific developmental stages will be discussed in the following section. In addition, other feeding disorders which can occur at various stages of development will be described.

Classification of Feeding Disorders

A Historical Perspective

Failure to thrive (FTT) was, and still is, a common term used in the medical diagnosis of feeding disorders. Clinicians initially distinguished between two forms of FTT: organic FTT, where a medical condition is considered to be the major reason for growth failure, and nonorganic FTT, where the growth failure was thought to reflect maternal deprivation or parental psychopathology. Mixed FTT and nonorganic FTT was a third category that was later added to describe growth failure that is related to a mixture of organic and environmental factors.

The diagnosis of FTT has been sharply criticized in the recent past. The main concerns are that many infants with feeding disorders do not demonstrate FTT and that it is a purely descriptive term for growth failure, rather than a diagnosis. Not until 1994, with the publication of the Diagnostic and Statistical Manual of Mental Disorders (DSM IV) by the American Psychiatric Association, was Feeding Disorder of Infancy and Early Childhood introduced as a diagnostic category to address a child's growth failure and specific feeding behaviors. Criteria include persisting failure to gain weight or significant loss of weight over at least 1 month; the disorder is not due to an associated gastrointestinal or other medical condition, another mental disorder, or by lack of available food; and onset is before 6 years of age. It is important to note that children who have feeding difficulties without weight problems or a feeding disorder associated with a medical condition are excluded from these limited criteria. Additionally, various subtypes of feeding disorders have been described by several authors that have not been addressed by this limited diagnostic definition. For example, subclassification systems of feeding disorders have been attributed

to various organic and nonorganic etiologies. Other authors have used different diagnostic labels to describe the same symptomatology, or the same label to describe different symptoms. Consequently, diagnostic dilemmas have arisen as to what constitutes a feeding disorder, and how one may differentiate one feeding disorder from another or from a subclinical feeding problem. These problems in the definition of feeding disorders have also led to confusion about how to treat specific feeding problems. MaryLou Kerwin, in her literature review of empirically based treatments for severe feeding problems, noted that treatments for feeding disorders exist but the question remains for whom they are appropriate, and when, and why.

We will present a classification system of feeding disorders, developed by Irene Chatoor. Dr. Chatoor was supported in her effort to develop diagnostic criteria for these six feeding disorders by a task force of national experts in infant psychiatry. These efforts by the task force in 2004 were supported by the American Academy of Child and Adolescent Psychiatry. The criteria were further revised with the help of a work group of experts in psychopathology of infants and young children for publication in a monograph by the American Psychiatric Press Incorporated. This classification differentiates six feeding disorders that have different etiologies and require different interventions. Each feeding disorder is characterized by specific symptoms that differentiate it from other feeding disorders and healthy eating. In addition, each feeding disorder has specific criteria of impairment relating to the child's nutritional state and/or threat to the child's health and development in order to differentiate clinical feeding disorders from subclinical forms.

Feeding Disorder of State Regulation

Diagnostic criteria

1. The infant's feeding difficulties start in the first few months of life and should be present for at least 2 weeks.
2. The infant has difficulty reaching and maintaining a state of calm alertness for feeding; he/she is either too sleepy or too agitated to feed.
3. The infant fails to gain age-appropriate weight or may show loss of weight.
4. The infant's feeding difficulties cannot be explained by a physical illness.

Clinical description

Infants who have difficulty with state regulation, who are either too sleepy or too distressed and agitated, and cannot reach a state of calm alertness during feeding, are unable to suckle effectively. Consequently, their intake of milk is inadequate, which causes concern and anxiety in the mothers. As the mother becomes increasingly anxious, she is unable to help her infant in state regulation, and mother and infant become trapped in a vicious cycle of mutually distressing interactions. Newborn infants with undeveloped central nervous systems, with cardiac, pulmonary, or gastrointestinal disorders, or other medical illnesses may be at greater risk for this disorder.

Review of the literature

There is a dearth of literature regarding this subgroup of infant feeding disorder. The only study by Chatoor and colleagues posited that infants diagnosed with a feeding disorder of state regulation (homeostasis), demonstrated less positive reciprocal interactions between infants and their mothers, compared to healthy eaters and their mothers. Future studies are needed to enhance an understanding of infant and parent characteristics associated with this feeding disorder.

Treatment

Understanding infant and parent characteristics is essential in treating this feeding disorder of state regulation. The treatment may focus on assisting the mother to modulate the amount of stimulation, especially at mealtime (e.g., feeding an excitable infant in a calm and low-lighted room or massaging an infant who has difficulty waking up to feed). The intervention may target maternal factors (e.g., treating anxiety or depression in the mother to help her deal more effectively with her challenging infant). In severe cases of poor weight gain or weight loss, temporary nasogastric tube feedings may be necessary to stabilize the nutritionally compromised infant and prevent further deterioration of the infant's condition. In other cases, a combination of interventions targeting the mother and the infant may be most helpful.

Feeding Disorder of Caregiver–Infant Reciprocity

Diagnostic criteria

1. This feeding disorder is usually observed in the first year of life, when the infant presents with some acute medical problem (commonly an infection) to the primary care physician or the emergency room, and the physician notices that the infant is malnourished.
2. The infant shows lack of developmentally appropriate signs of social responsivity (e.g., visual engagement, smiling, babbling) during feeding with the primary caregiver.
3. The infant shows significant growth deficiency (acute and/or chronic malnutrition, or the child's weight deviates across two major percentiles in a 2–6-month period).
4. The primary caregiver is often unaware of the feeding and growth problems of the infant.
5. The growth deficiency and lack of relatedness are not solely due to a physical illness or a pervasive developmental disorder

Clinical description

These infants usually look ill and malnourished. They may avoid eye contact, and appear withdrawn and listless. The mothers often do not seek help for their infants, until the infants become acutely ill with an infection, and their growth and development is severely compromised. Often, the infants show poor muscle tone and impaired coordination. When held they stiffen, scissor their legs, and hold their arms up in a surrender posture to be able to hold up their heads.

Impairment in caregiver–infant interactions and communication can be observed. The mothers appear unable to engage their infants in reciprocal interactions; they often appear detached and overwhelmed by their own emotional needs. They may have been unaware of any problems, as their infants' sleeping patterns lengthened and their feeding drive lessened. They may admit to propping bottles for feeding and spending minimal time with their infant. The mothers may be guarded and distrustful, and avoid contact with medical professionals.

Review of the literature

Maternal neglect has been thought to bring on this feeding disorder, and in the early literature this feeding disorder has been referred to as Maternal Deprivation or Environmental Deprivation. In accordance with these terms, the DSM-III referred to it as 'reactive detachment disorder of infancy' and highlighted 'FTT' as the central symptom. However, in 1994, DSM-IV modified reactive attachment disorder to a selected relatedness problem without including growth failure in the diagnostic criteria. Chatoor and colleagues posited that significant issues of attachment are at the root of this feeding disorder and they called it "feeding disorder of attachment." They reported that infants with a feeding disorder of attachment and their mothers demonstrated less positive engagement and dyadic reciprocity during feedings than healthy infants and their mothers. Additionally, mothers of infants with this feeding disorder received higher maternal noncontingency ratings than mothers of healthy infants. Several studies have reported high rates of insecure attachments in the infants, and described that the mothers often suffer from psychiatric disorders (e.g., substance or alcohol abuse, mood disorders, and personality disorders). Other studies have noted that the mothers are often from a lower socioeconomic background, and may have their own history of being victims of abuse and neglect.

Treatment

Treatments, ranging from home-based interventions to hospitalization, have been proposed. The early work by Selma Frailberg, proposed an in-home approach that focused on nurturing the mother, so she could begin to nurture her infant. A study that compared three types of outpatient treatment – short-term advocacy, family-centered intervention, and parent–infant intervention – found that no treatment was superior to the other. However, outpatient treatment should be an alternative solution only in cases of mild neglect with no evidence of depriving behaviors on account of the mother, when the child is more than 1 year old, the parents have an established support system, and have a history of seeking medical care for previous illnesses and for immunizations.

More commonly, the severely malnourished and developmentally delayed infant requires hospitalization for emergency treatment. During the course of the hospitalization, treatment should address the infant's nutritional state and developmental delays while the assessment of the mother and family environment occurs. It is important that a minimum of different nursing staff are assigned to the infant and that a primary care nurse assumes responsibility for continuity of care, and for a compassionate environment for the infant. In addition, the infants may need physical therapy to address generalized hypotonia secondary to tactile understimulation in their home environment. While the rehabilitation of the infant goes on, the degree of parental awareness and cooperation with the infant's treatment needs to be assessed, as the parent's involvement with treatment is predictive of the outcome for these infants. Unfortunately, there are cases where the primary caregiver may be severely impaired or live in a chaotic environment, and may be unable to become engaged with the infant. Alternative placements have to be considered in these cases, as attention to the infant's well-being and health cannot wait.

Infantile Anorexia

Diagnostic criteria

1. This feeding disorder is characterized by the infant's or toddler's refusal to eat adequate amounts of food for at least 1 month.
2. Onset of the food refusal often occurs during the transition to spoon and self-feeding, typically between 6 months and 3 years of age.
3. The infant or toddler rarely communicates hunger, lacks interest in food and eating, but shows strong interest in play, exploration, and/or interaction with caregivers.
4. The infant or toddler shows significant growth deficiency (acute and/or chronic malnutrition; or the child's weight deviates across two major percentiles in a 2–6 month period).
5. The food refusal did not follow a traumatic event to the oropharynx.
6. The food refusal is not due to an underlying medical illness.

Clinical description

These infants characteristically have a very poor hunger drive, display minimal to no interest in food or eating, are easily distracted by other stimuli, and prefer to explore and interact with others or their environment. Caregivers report that their major concerns are the infant's intense food refusal and growth failure. Despite a variety of parental interventions to help their infants eat (e.g., coaxing, distraction techniques, feeding while playing, etc.), these measures may temporize the situation in the short term, however, over the long term, prove unsuccessful.

Some mothers report that the infants were already easily distracted while breastfeeding and would stop feeding if somebody entered the room or the phone would ring. Most commonly, by the end of the first year, these infants take only a few bites of food and then stop, refusing to eat anymore, despite the parents' attempts to feed them. The parents become increasingly worried about the infants'/toddlers' poor food intake and use great efforts to get them to eat by distracting with toys or video movies; feeding while playing, coaxing, or cajoling; feeding while the infants are sleepy; or even force-feeding them. However, in spite of all these efforts by the parents, the infants/toddlers eat inadequate amounts of food to support growth. Initially, they fail to gain appropriate weight for age, and as their feeding problems continue, they grow at a slow rate. Children at 3 years of age may look like 2 year olds, and at age 10 years, they may have the height and bone age of a 7-year-old child. However, not all children with this feeding disorder become stunted in their growth; some grow at a normal rate and become extremely thin. Interestingly, their head size tends to progress at a normal rate, and their cognitive development is age-appropriate and sometimes superior.

Review of the literature

Several studies by Chatoor and colleagues and by Massimo Ammaniti and Loredana Lucarelli have examined child and parent characteristics associated with infantile anorexia. These studies have revealed that mother–infant interactional patterns are characterized by less dyadic reciprocity, high dyadic conflict, struggle for control, and increased talk and distractions during feedings. Additional studies have found that toddlers with infantile anorexia are rated by their parents as more negative, more irregular in their feeding and sleeping patterns, more willful on the one hand and more dependent on the other than healthy eaters. In addition, toddlers with infantile anorexia exhibited more anxiety/depression, more somatic complaints, and aggressive behaviors than healthy control children. These studies also revealed that more mothers of toddlers with infantile anorexia demonstrate insecure attachment patterns to their own parents and more dysfunctional eating attitudes, anxiety, and depression than mothers of healthy control children. In addition, these studies found that toddlers with infantile anorexia exhibited a higher rate of insecure attachment relationships to their mothers than healthy eaters, although the majority of anorexic toddlers (60%) showed secure attachment patterns. However, the significant correlation between the severity of malnutrition and the degree of attachment insecurity indicated that an insecure toddler–mother relationship is associated with a more severe expression of infantile anorexia.

It is of interest that studies at both sites by Ammaniti and colleagues and by Chatoor and colleagues showed that infantile anorexia occurs with the same frequency in boys and girls, in contrast to anorexia nervosa which is seen predominantly in females.

A recent study by Chatoor and colleagues demonstrated that on average, toddlers with infantile anorexia performed within the normal range of cognitive development, although the mental developmental index (MDI) scores of the healthy eaters were significantly higher than those of the infantile anorexia group. In contrast to some of the previous reports by other authors that cognitive development is negatively affected by FTT, correlations between MDI scores and the toddlers' percent ideal body weight did not reach statistical significance, whereas the toddlers' MDI scores showed a significant correlation with the quality of mother–child interactions during feeding and play, and the socioeconomic status of the family.

In addition, a recent pilot study by Chatoor and colleagues found increased physiological arousal and decreased ability to modulate physiological reactivity in toddlers with infantile anorexia compared to a control group of healthy eaters. This study raises the question whether the difficulty of recognizing hunger, which characterizes children with infantile anorexia, may be related to a heightened physiological arousal pattern.

All of these studies were cross-sectional, and consequently, they do not allow any firm conclusions about the causality of this feeding disorder. However, the developmental histories of these children reveal that from early infancy, these children display little interest in feeding and eat only small amounts, thereby triggering anxiety and worry in the parents who try to compensate for the infant's poor food intake by falling into maladaptive feeding patterns of distracting, coaxing, and even force-feeding. These parental behaviors further interfere with the toddler's awareness of hunger and fullness and lead to increasingly conflictual interactions between mother and child, both struggling for control. As a result of these struggles, the child does not learn to regulate eating internally, but eating becomes completely dependent on the interactions of the child with the environment.

Treatment

This transactional model described above has served as a basis for an intervention to facilitate internal regulation of eating in toddlers with infantile anorexia. A treatment was developed by Chatoor and colleagues that addresses the major components of the model: (1) the parents are helped

to understand the toddler's special temperament, namely the lack of awareness of hunger and fullness, and the toddler's intense interest in play and interactions with the caretakers; (2) the parents' anxiety and worry about the infant's poor growth, their difficulty in setting limits to the toddler's provocative behaviors around eating, and how these difficulties may relate to experiences with their own parents, are explored; and (3) the parents are provided with specific guidelines that detail how to structure mealtimes in order to facilitate the toddler's learning of hunger and fullness, and how to deal with the toddler's behaviors that interfere with feeding.

Sensory Food Aversions

Diagnostic criteria

1. This feeding disorder is characterized by the infant's/toddler's consistent refusal to eat specific foods with specific tastes, textures, smells, and/or appearances for at least 1 month.
2. The onset of the food refusal occurs during the introduction of a new type or taste of food (e.g., may drink one type of formula but refuse another; may eat carrots, but refuse green beans; may eat crunchy foods but refuse pureed food or baby food).
3. Eats well when offered preferred foods.
4. Does not show growth deficiency, but without supplementation, demonstrates specific dietary deficiencies (i.e., vitamins, iron, zinc, or protein), and/or displays oral motor and expressive speech delay, and/or starting during the preschool years, demonstrates anxiety and avoids social situations which involve eating.
5. The food refusal did not follow a traumatic event to the oropharynx.
6. Refusal to eat specific foods is not related to food allergies or any other medical illness.

Clinical description

This feeding disorder becomes apparent during the early years, when infants and toddlers are introduced to a variety of baby food and table food of different tastes and textures. Parents usually report that when specific foods were placed in the infants' mouths, the aversive reactions ranged from grimacing to gagging, vomiting, or spitting out the food. After an initial aversive reaction, the infants usually refuse to continue eating that particular food and become distressed if forced to do so. Some infants generalize their reluctance to eat one food to other foods that look or smell similar (e.g., an aversion to spinach may generalize to all green vegetables). Parents frequently report that these children just look at certain foods without ever trying them and are generally reluctant to try any new foods. Some children may even refuse to eat any food that has touched another food on the plate, while others will only eat food prepared by a specific restaurant or company (e.g., French fries from McDonalds).

If infants reject foods that require significant chewing (e.g., meats, hard vegetables, or fruits), they will fall behind in their oral motor development due to lack of experience with chewing, and they may show difficulty with articulation. If children refuse many foods or whole food groups (e.g., vegetables and fruits), their limited diet may lead to specific nutritional deficiencies (e.g., vitamins, zinc, and iron). In contrast, the restricted diet of some of these children may be very unhealthy and lead to early cholesterol problems. In addition, the children's refusal to eat a variety of foods often creates conflict within their families during mealtime, and may cause social anxiety as early as in preschool, when the children are faced with eating at school, or when they get older and want to participate in events that include eating (e.g., birthday parties, sleepovers, and summer camp).

Sensory food aversions are common and occur along a spectrum of severity with some children refusing to eat only a few foods, whereas others may be limited to eating a very restrictive diet. Therefore, the diagnosis of a feeding disorder should only be made if the food aversions result in dietary deficiencies, requiring that the child needs supplementation of certain nutrients (e.g., vitamins, zinc, iron, or protein) and/or are associated with oral motor delay and speech articulation problems. In older children, social anxiety (if the child fears or avoids social situations, which involve eating with others) should be considered as a sign of impairment.

In addition to their sensitivity to certain foods; many of these children experience problems in other sensory areas as well. Parents may report that some infants with sensory food aversions do not like to touch certain foods; struggle when they need a hair cut; become distressed when asked to walk on sand or grass; or do not like to wear socks, certain types of fabric, or labels in clothing; or that they may be hypersensitive to certain odors or loud sounds.

Review of the literature

Several authors have described children's difficulty to eat certain foods, but have used a variety of names (e.g., picky eaters, choosy eaters, selective eaters, food neophobia, and food aversion). These reports are usually based on clinical case studies, they do not delineate specific diagnostic criteria, and they do not differentiate between milder or more severe forms of food selectivity. Consequently, some authors report very high numbers (25–50%) of 'picky eating' in the general population. In our own diagnostic study of feeding disorders, which was conducted in a Multidisciplinary Feeding Disorders Clinic, 'sensory food aversions' was the most commonly diagnosed feeding disorder (31% of all referrals), and 13% of the children were comorbid for sensory food aversions and infantile anorexia.

Some studies have explored whether taste sensitivities are heritable. Various models of genetic transmission have been suggested, and in 2003 Un-kyung Kim and colleagues have described that specific polymorphism of gene *Tas2r* on chromosome 7q are related to taste sensitivity in adults. A relationship between genetic taste sensitivity to propylthiouracil (PROP) and food selectivity has also been reported in preschool children. However, Lean Birch and colleagues have demonstrated that certain aspects of the eating environment can also have a strong influence on the development of food preferences, and shape selective food refusal. In summary, previous studies indicate that genetic predisposition as well as the eating environment, affect toddler's food preferences. However, further studies are needed to understand the relation between genetic and environmental influences on the development of this feeding disorder.

Treatment

Once infants or toddlers experience a food as aversive, they tend to grimace, spit out the food, or they may even gag and vomit, when they try to swallow it. After a negative experience with a particular food, they tend to refuse to accept any more of that food. Parents frequently try to coax toddlers into eating foods that they reject by promising certain privileges, for example, watching television, or getting to play with a certain toy, or they withhold a favorite food to induce the toddler to eat more healthy foods. Unfortunately, these techniques have often a significant negative effect, they make the children more anxious and lead to oppositional food refusal in general. In contrast, if parents recognize and accept that the infant/toddler experiences specific foods as aversive, and they stop offering these foods, the children can relax and consequently, they are more willing to try new foods that they may be able to tolerate. Toddlers are particularly responsive to modeling by the parents, and they are more willing to try a new food if they can observe their parents eating it. However, these clinical experiences have not been tested in any systematic research studies.

Post-Traumatic Feeding Disorder

Diagnostic criteria

1. This feeding disorder is characterized by the acute onset of severe and consistent food refusal.
2. The onset of the food refusal can occur at any age of the child from infancy to adulthood.
3. The food refusal follows a traumatic event or repeated traumatic insults to the oropharynx or gastrointestinal tract (e.g., choking, severe vomiting, insertion of nasogastric or endotracheal tubes, suctioning) that trigger intense distress in the child.

4. Consistent refusal to eat manifests in one of the following ways, depending on the feeding experience of the child in association with the traumatic event:
   - refuses to drink from the bottle, but may accept food offered by spoon (although consistently refuses to drink from the bottle when awake, may drink from the bottle when sleepy or asleep);
   - refuses solid food, but may accept the bottle; and
   - refuses all oral feedings.
5. Reminders of the traumatic events cause distress, as manifested by one or more of the following:
   - may show anticipatory distress when positioned for feeding;
   - shows intense resistance when approached with bottle or food; and/or
   - shows intense resistance to swallow food placed in the mouth.
6. Without supplementation (e.g., specific fortified formula preparations, intravenous. fluids, nasogastric or gastrostomy tube feedings, or parenteral nutrition), the food refusal poses an acute and/or long-term threat to the child's health, nutrition, and growth, and threatens the progression of age-appropriate feeding development of the child.

Clinical description

The term post-traumatic eating disorder was first coined by Chatoor, Conley, and Dickson in a paper on food refusal in five latency-age children who experienced episodes of choking or severe gagging, and later refused to eat any solid food. These children were preoccupied with the fear of choking to death, if they were to eat any solid food. Infants and toddlers with a post-traumatic feeding disorder cannot verbally communicate their fear of eating, but display fear and intense resistance to eating in their behavior. Parents may report that their infants' refusal to eat any solid foods started after an incident of choking, or after one or more episodes of severe gagging. Other parents may report that the infant's refusal to drink from the bottle started after one or more severe episodes of vomiting. After the severe vomiting, the infant would cry at the sight of the bottle and would not drink from the bottle any more. Some parents may have observed that the food refusal followed intubation, the insertion of nasogastric feeding tubes, or major surgery requiring intubation or vigorous oropharyngeal suctioning.

Reminders of the traumatic event(s) (e.g., the bottle, the bib, or the high-chair), frequently cause intense distress, and the infants become fearful when they are positioned for feedings and presented with feeding utensils and food. They resist being fed by crying, arching, and refusing to open their mouths. If food is placed in their mouths, they intensely resist swallowing. They may gag or vomit, let the food drop out, actively spit out food, or store food in their cheeks and spit it out later to avoid swallowing it. The fear of eating seems to override any awareness of hunger, and infants who refuse all food, liquids and solids, require acute intervention to prevent dehydration and starvation.

Literature review

Similar symptomatology primarily in children has been described by other authors as fear of choking or choking phobia, and as dysphagia and food aversion. However, these studies are primarily clinically descriptive and do not provide any clear diagnostic criteria. Using the diagnostic criteria listed above, Chatoor and colleagues found that toddlers with a post-traumatic feeding disorder showed intense mother–infant conflict during feeding and demonstrated the most intense resistance to swallowing food.

However, more research is needed to understand this feeding disorder better.

Treatment

This feeding disorder can be life-threatening, and many infants with this disorder may require gastric tube-feedings. Although this intervention helps infants to survive, the tube-feedings interfere with their experience of hunger and further complicate the feeding disorder. The fear of swallowing and the lack of hunger make the children very resistant to oral feedings. However, one controlled, randomized treatment study by Diane Benoit and colleagues demonstrated the effectiveness of a behavioral intervention to overcome their resistance to oral feedings and to come off tube-feedings.

A more gradual desensitization approach has been used at The Children's National Medical Center. The infants' anticipatory anxiety to the feeding situation is assessed, and through gradual exposure to the bottle, the high-chair, and the feeding utensils, the infants or toddlers are helped to overcome their fear by looking at these objects and playing with them. Food is gradually introduced, and the children are encouraged to self-feed to give them more control. This gradual desensitization process is slow and only possible if adequate food intake of the children is guaranteed through bottle-feedings or tube-feedings.

Feeding Disorder Associated with a Concurrent Medical Condition

Diagnostic criteria
1. This feeding disorder is also characterized by food refusal and inadequate food intake for at least 2 weeks.
2. The onset of the food refusal can occur at any age and may wax and wane in intensity, depending on the underlying medical condition.

3. The infant or toddler readily initiates feeding, but over the course of the feeding, shows distress, and refuses to continue feeding.
4. Has a concurrent medical condition that is believed to cause the distress (e.g., gastroesophageal reflux, cardiac, or respiratory disease).
5. Fails to gain age-appropriate weight or may even lose weight.
6. Medical management improves but may not fully alleviate the feeding problems.

Clinical description

Some medical conditions are not readily diagnosed and food refusal may be the leading symptom. For example, food allergies can be difficult to diagnose in this young age group, and silent gastroesophageal reflux may be overlooked by pediatricians because the infant does not vomit, and vomiting is usually the leading symptom of reflux. However, with silent reflux, the infant may experience regurgitations of gastric content into the esophagus, causing irritation of the esophageal mucosa without any outward signs of reflux. In addition, there is considerable individual variability in the perception of pain. Consequently, only a small percentage of infants with gastroesophageal reflux exhibit distress during feeding or after vomiting. In order to understand the individual infant's experience with gastroesophageal reflux, it is important to observe the infant during feeding. Typically, infants drink one to two ounces of milk before reflux is activated, and the infants display vomiting or show signs of discomfort (e.g., coughing, wiggling, arching, and crying) and push the bottle away. Some infants can calm themselves and resume feeding until they experience a new episode of pain. However, some others cry in distress and become increasingly agitated while their caregivers try to continue feeding.

Some infants with respiratory distress may feed for a while and take a few ounces of milk or food until they tire out and stop feeding. In general, these infants consume inadequate amounts of food, fail to gain age-appropriate weight, or even lose weight. Good medical management improves the infants' feeding difficulties. Frequently, however, the feeding problems do not remit completely, and require further psychological interventions.

Review of the literature

In the early literature, a mixed category of FTT, which is caused by a combination of various organic and nonorganic problems, has been described, and it has been generally accepted that organic conditions can be associated with psychological difficulties and lead to severe feeding problems. Some authors have reported severe feeding problems such as food refusal, and taking more than 1 hour per feeding, in infants with gastroesophageal reflux. Others have described that the food refusal associated with gastroesophageal reflux was so severe that some infants had to be tube-fed, and that even after successful surgery, the severe feeding difficulties and growth failure continued. These studies are primarily descriptive and highlight the observation that the combination of medical and psychological factors can lead to complex feeding problems, requiring combined interventions.

Treatment

Because of the interaction of organic and psychological factors in producing feeding difficulties, treatment of these infants requires close collaboration between the pediatrician or pediatric specialist and the child psychiatrist or psychologist. Direct observation of infants with their primary caregivers during feeding is most important to monitor how much distress the infant experiences during feeding and whether the medical condition is being adequately treated or requires further intervention. The observation of the feeding also reveals how parents deal with the infant's distress, whether they become anxious and inadvertently heighten the infant's distress, thus escalating the feeding problems. These observations can be used to optimize the medical treatment, and assist the parents in developing different feeding strategies to help minimize the infant's discomfort during feeding. In situations in which the medical intervention cannot control the medical illness, when infants continue to experience distress during feedings and their caloric intake is limited, supplemental nutrition through nasogastric or gastrostomy tubes must be considered. The therapist must help the parents and the medical team deal with this difficult decision and work with them to maintain the infant's interest in feeding and develop age-appropriate feeding skills, while most of the nutrition is given via tube-feedings. In general, children with this feeding disorder require individualized interventions by an experienced team.

Comorbidity between various feeding disorder subtypes

In a diagnostic study by Chatoor and colleagues, 20% of the infants and young children examined showed comorbidity of two or more feeding disorders. The most frequent comorbidity observed was between infantile anorexia and sensory food aversions, although these feeding disorders occurred more frequently in the pure form. Characteristically, children with both feeding disorders are very selective about foods which they are willing to eat, but even when given their favorite foods, the children have a poor appetite and eat little. It is important to diagnose each feeding disorder because each has a different etiology and shows different responses to treatment. Toddlers with infantile anorexia respond to the regulation of mealtimes with an increase of appetite and improved food intake, whereas toddlers with sensory food aversions will eat less or nothing if offered aversive foods, regardless of how hungry they may be. Consequently, the treatment needs to address both feeding disorders.

A problem arises when the children exhibit symptoms that appear to fit the diagnoses of three or four different feeding disorders. We recommend that the symptoms should be differentiated as much as possible to understand what the underlying diagnoses may be, because, as point out earlier, each feeding disorder requires a different therapeutic approach.

See also: Abuse, Neglect, and Maltreatment of Infants; Colic; Failure to Thrive; Newborn Behavior; Nutrition and Diet; Self-Regulatory Processes; Social Interaction; Suckling.

Suggested Readings

Benoit D, Wang EE, and Zlotkin SH (2000) Discontinuation of enterostomy tube feeding by behavioral treatment in early childhood: A randomized control trial. *Journal of Pediatrics* 137: 498–503.

Birch LL (1999) Development of food preferences. *Annual Review of Nutrition* 19: 41–62.

Burklow KA, Phelps AN, Schultz JR, McConnell K, and Randolph C (1998) Classifying complex pediatric feeding disorders. *Journal of Pediatric Gastroenterology and Nutrition* 27(2): 143–147.

Chatoor I, Conley C, and Dickson L (1988) Food refusal after an incident of choking: A posttraumatic eating disorder. *Journal of American Academy of Child and Adolescent Psychiatry* 27: 105–110.

Chatoor I, Ganiban J, Surles J, and Doussard-Roosevelt J (2004) Physiological regulation and infantile anorexia: A pilot study. *Journal of American Academy of Child and Adolescent Psychiatry* 43: 1019–1025.

Chatoor I, Getson P, Menvielle E, O'Donnell, *et al.* (1997) A feeding scale for research and clinical practice to assess mother–infant interactions in the first three years of life. *Infant Mental Health Journal* 18: 76–91.

Chatoor I, Surles J, Ganiban J, Beker L, Paez LM, and Kerzner B (2004) Failure to thrive and cognitive development in toddlers with infantile anorexia. *Pediatrics* 113(5): e440–e447.

Frailberg S, Anderson E, and Shapiro U (1975) Ghosts in the nursery. *Journal of American Academy of Child and Adolescent Psychiatry* 14: 387–421.

Kerwin ME (1999) Empirically supported treatments in pediatric psychology: Severe feeding problems. *Journal of Pediatric Psychology* 24: 193–214.

Mahler MS, Pine F, and Berman A (1975) *The Psychological Birth of the Human Infant.* New York: Basic Books.

Fetal Alcohol Spectrum Disorders*

H Carmichael Olson, S King, and T Jirikowic, University of Washington, Seattle, WA, USA

Glossary

Active case ascertainment – 'Ascertainment' is the scheme by which individuals are selected, identified, and recruited for participation in a research study. 'Active case ascertainment' means that individuals are actively searched for so they can be identified.

Behavioral phenotype – A characteristic pattern of motor, cognitive, linguistic, and social observations that is consistently associated with a biological disorder.

Epigenetic – A factor that changes the phenotype (the observable characteristics of an individual) without changing the genotype (the entire genetic identity of an individual).

Neurobehavioral – Relating to the relation between the action of the nervous system and behavior.

Prevalence – The proportion of individuals in a population having a disease or disorder.

Teratogen – An agent capable of causing developmental malformations in the embryo or fetus when there is prenatal exposure.

*Portions of this article were previously published in: Carmichael Olson H, Jirikowic T, Kartin D, and Astley S (2007) Responding to the challenges of early intervention for fetal alcohol spectrum disorders. *Infants and Young Children* 20(2): 162–179. These sections, including Table 1, are used here with permission of © copyright owner Lippincott Williams and Wilkins, Inc.

Introduction

Prenatal alcohol exposure can lead to significant and lifelong developmental disabilities, now recognized under the umbrella term of 'fetal alcohol spectrum disorders' (FASDs). FASDs are a global public health problem, found worldwide and across all ethnicities. These lifelong birth defects are an especially important topic for those interested in infancy and early childhood. Not only are these birth defects preventable but, if they do occur, there is real hope that FASDs can be effectively treated. Basic animal research and human studies are building the case that diagnosis and intervention (especially early in life) can improve the developmental outcome of individuals affected by alcohol exposure before birth. Providers working in early intervention can be among the first to grasp the real hazards of prenatal alcohol exposure for children's development, referring for diagnosis, alerting caregivers to emerging learning and behavior problems, and connecting families with needed services.

Understanding the Problem of FASDs

A Spectrum Disorder

FASDs are increasingly understood as a spectrum or continuum of effects that result from prenatal alcohol exposure. Individuals with FASDs can show adverse physical signs from prenatal alcohol exposure (such as characteristic facial anomalies or various other bodily malformations), but do not always have these physical effects. Individuals with FASDs typically do show alcohol-related deficits in many different aspects of learning, behavior, and daily function, and these are the difficulties that are the most troubling (but also the most important) to identify, diagnose, and treat. These functional difficulties vary from individual to individual, who can have very different profiles of alcohol-related deficits.

Table 1 defines diagnoses on the fetal alcohol spectrum. The range of FASDs includes the full 'fetal alcohol syndrome' (FAS), defined by growth impairment, characteristic facial features, and evidence of significant central nervous system (CNS) damage and/or dysfunction. **Figure 1** shows the 'face' of FAS. But FAS is found in only a fairly small proportion of individuals affected by prenatal alcohol exposure and, unfortunately, the effects of alcohol are not readily recognized if they do not fit the classic FAS definition. There are children born with 'partial FAS' who manifest some but not all of the characteristic physical features, and so do not meet criteria for the full syndrome. But a larger number of children, adolescents, and adults have alcohol-induced impairments without the characteristic facial features, and may or may not have growth impairment. Their functional difficulties can be just as serious, or more so, than those seen in FAS. The term 'alcohol-related neurodevelopmental disorder' (ARND) has been applied to these conditions. There are also individuals born prenatally exposed who have other alcohol-related physical abnormalities of the skeleton and certain organ systems. These anomalies are referred to as

Table 1 Criteria for diagnosing fetal alcohol spectrum disorders

Diagnosis	Diagnostic features
Fetal alcohol syndrome (FAS)	• Growth deficiency: height or weight < 10th percentile • Cluster of characteristic minor facial anomalies: small palpebral fissures (eyeslits), thin upper lip, smooth philtrum (groove above the upper lip) • Central nervous system damage (evidence of structural and/or functional brain impairment) • Reliable evidence of confirmed prenatal alcohol exposure: not necessary if the cluster of characteristic facial features is present
Partial FAS (PFAS)	• Some of the characteristic minor facial anomalies • Growth deficiency: height or weight < 10th percentile[a] • Central nervous system damage (evidence of structural and/or functional brain impairment) • Reliable evidence of confirmed prenatal alcohol exposure
Alcohol-related neurodevelopmental disorder (ARND)	• Central nervous system damage (evidence of structural and/or functional brain impairment) • Reliable evidence of confirmed prenatal alcohol exposure

[a]*The 4-Digit Code provides more detail about PFAS, as there are some cases in which growth impairment is not a necessary criterion because the expression of facial features so closely resembles the classic FAS presentation.*
Reproduced from Carmichael Olson H, Jirikowic T, Kartin D, and Astley S (2007) Responding to the challenges of early intervention for fetal alcohol spectrum disorders. *Infants and Young Children* 20(2): 162–179, with permission of Lippincott, Williams & Wilkins, Inc. Adapted from Astley SJ (2004) *Diagnostic Guide for Fetal Alcohol Spectrum Disorders: The 4-Digit Diagnostic Code*, 3rd edn. Seattle, WA: University of Washington Publication Services. (*Manual for the 4-Digit Diagnostic Code* available from the University of Washington Fetal Alcohol Syndrome Diagnostic & Prevention Network). The '4-Digit Diagnostic Code' is a system used by clinicians and researchers to evaluate the effects of prenatal alcohol exposure. While the 4-Digit Code maps onto these widely known diagnoses, the system actually uses different descriptive terms to more precisely describe how children manifest PFAS and ARND, and provides more detail. There are other diagnostic guidelines that also include a category called 'alcohol-related birth defects' (ARBDs). ARBDs exist in the presence of confirmed prenatal alcohol exposure, but children who have ARBDs do not necessarily show the characteristic facial features. ARBDs are defined as) 'Any of a number of anomalies (such as heart or kidney defects) present at birth that are associated with maternal alcohol consumption during pregnancy'. National Institute on Alcohol Abuse and Alcoholism (2000) *Tenth Special Report to the U.S. Congress on Alcohol and Health: Highlights on Current Research*. Washington, DC: US Department of Health and Human Services, Public Health Service, National Institutes of Health. Available at http://www.niaaa.nih.gov/publications/10report/chap05.pdf (Accessed February 2005).

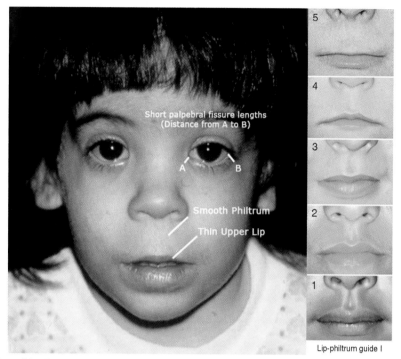

Lip-philtrum guide I

Figure 1   Based on the 4-Digit Diagnostic Code (Astley, 2004), the three diagnostic facial features of FAS are (1) palpebral fissure length (eyeslit length) of 2 or more standard deviations below the mean; (2) smooth philtrum (the vertical groove between the nose and upper lip) (Rank 4 or 5 on Lip-Philtrum Guide); (3) thin upper lip (Rank 4 or 5 on Lip-Philtrum Guide). Photo used by permission of Susan Astley, University of Washington.

'alcohol-related birth defects' (ARBD). Of course, a person cannot have FASDs unless there has been prenatal alcohol exposure, although documentation or valid reports of this is not always easy to obtain.

Prevalence of FASDs and Recognition of the Risk of Prenatal Alcohol Exposure

FASDs are now known to have a higher prevalence than has been understood in the past. Reported rates vary widely and depend on the population studied and the detection and diagnostic methods used. In 2002, the number of children in the US with the full FAS was estimated by the Centers for Disease Control and Prevention (CDC) at 0.2–1.5 per 1000 live births, with much higher rates in some communities. Prevalence rates of the full range of FASDs, including FAS, ARND, and ARBD, are believed to occur about three times as often as FAS. But the full extent of the problem is still being investigated, and over the past 10 years some have estimated the rates of the full range of FASDs as high as 9 or 10 per 1000 live births in the US, approaching the latest estimated prevalence of autism spectrum disorders. There are certainly selected communities in the US with higher prevalence rates.

Prevalence rates depend on the diagnostic and detection (case ascertainment) methods used. Diagnostic issues are discussed later in this article. With FASDs, active case ascertainment is the best method for identifying the full extent of this set of developmental disabilities. Passive ascertainment methods, such as counting birth defects when a child is born and listing them in a registry, can only identify a small number of alcohol-affected individuals.

FASDs have been found worldwide. While not all countries have reliable estimates of prevalence, there are countries that have uncovered higher prevalence rates in their population or in certain communities. In Italy, for example, researcher May and colleagues have used active case ascertainment in schools and documented recent population-based estimates at 35 cases of FASDs per 1000 children. In South Africa, May and coworkers have also used active case ascertainment in schools in a selected, highly impacted community, and have shown the highest reported prevalence in any community of 68.0–89.2 per 1000 children. This is uncommonly high and alarming. In response, researchers are working together internationally because the public health problem of FASDs has become such a grave concern in some countries and communities. Around the globe, communities and governmental agencies in many locations are working on a public policy response and on efforts to build service capacity. In Canada, for example, there have been grassroots and province-wide efforts at prevention, diagnosis, and/or intervention for individuals with FASDs and their families. In the US, as another example, there have been service system interagency coordinating efforts at the federal level, momentum to build state systems, and the emergence of local and national parent support organizations – all created in response to the public health problem of FASDs. Unfortunately, though, there are still many countries and communities, including many within the US, that have not recognized FASDs as a problem deserving a coordinated response.

While there remains great need for public and professional education worldwide, prenatal alcohol exposure is increasingly recognized as a risk factor for children's development. In the US, for example, screening for prenatal alcohol exposure is increasingly performed – especially for children in the child welfare, foster care, and adoption systems; for children born to women with chemical dependency; and for youth in the juvenile justice system. More frequently, young children with prenatal exposure to substances (including alcohol) are being recognized as at risk, referred for developmental assessment, and sent on for developmental services. Federal regulations in the US passed in 2003 and 2004 for the education, early intervention, and child welfare systems mandated that children involved in substantiated cases of child abuse and neglect be referred to early intervention systems. Educational regulations further required early intervention referral for children whose development is impacted by prenatal substance exposure. These regulations will certainly result in larger numbers of children in early intervention systems who are alcohol-exposed, who should be screened for the possibility of FASDs, and, when needed, referred for diagnosis. These children should be offered early intervention, when appropriate, either because of their at-risk status or for a diagnosed condition on the fetal alcohol spectrum.

Caregiving Burden and Costs

Families caring for individuals affected by prenatal alcohol exposure report many unmet needs for services and support, societal understanding, and modifications to existing service systems and their eligibility criteria. Parents report clinically elevated levels of stress when raising children with clearly diagnosed FASDs. Higher levels of parental stress have been clearly associated with raising a child who has complex neurobehavioral impairments, specifically executive-function deficits (that affect memory, planning, and organization), challenging problem behaviors, and poorer adaptive function.

The human costs of FASDs arise from high rates of debilitating problems in lifestyle and daily function found among affected individuals, such as mental health problems or disrupted school experiences, and the emotional and financial burdens on their families. In the US, there have been careful estimates of economic cost data for the full FAS – and these costs in care, rehabilitation, and lost productivity are striking. FAS is only one part of the larger fetal alcohol spectrum, but is actually one of the more costly

birth defects. In 2004, researchers found that the median of adjusted annual cost estimates to the US economy was $3.6 billion. Little is known about costs of the wider fetal alcohol spectrum beyond FAS, but additional affected individuals can only further raise already high societal costs.

The Teratogenic Effects of Alcohol and Prevention of FASDs

A growing body of animal literature, long-term human descriptive studies, clinical studies, and neuroimaging research all come together to make clear the negative impact that alcohol use during pregnancy can have on the developing child. These teratogenic effects of alcohol continue lifelong.

Alcohol as a Neurobehavioral Teratogen

As a neurobehavioral teratogen, alcohol interferes with normal fetal growth and CNS development through multiple actions at different sites. Alcohol is a very potent teratogen, altering the developing fetal brain and CNS largely through alcohol-induced disturbance in neurogenesis and synaptogenesis (how neurons and connections between neurons are formed). But the impact of alcohol on the CNS is found at the cellular, hormonal, neurochemical, structural, and functional levels – with a further complex interplay of genetic and epigenetic factors. Researchers are now carefully and intensively studying the biochemical mechanisms underlying alcohol's effects, in part because of hope that pharmacologic treatments might eventually be used to intervene with (or prevent) alcohol-related fetal injury. Researchers are also using technology such as ultrasounds, structural and functional neuroimaging, spectroscopy, and physical measures (including electroencephalograms (EEGs)) to understand more precisely how alcohol damages the brain.

Fetal effects of alcohol exposure differ depending on the amount, timing, and pattern of maternal drinking, so deficits are highly variable from one child to another. In general, the more a pregnant woman drinks, the greater the severity of persistent CNS deficits. Episodic (or binge) drinking that creates higher maternal peak blood alcohol concentrations is associated with greater fetal damage. On an individual basis, however, any amount of drinking during pregnancy can cause harm, and alcohol use at any time during gestation is associated with a higher risk of CNS dysfunction.

Prevention of FASDs

In 2005, the US Surgeon General released an updated advisory on drinking and pregnancy. Among other information, this advisory states that no amount of alcohol consumption during pregnancy can be considered safe, that cognitive and behavioral deficits resulting from prenatal alcohol exposure are lifelong, and that ARBDs are completely preventable. This advisory offers specific advice for women who are pregnant or considering becoming pregnant, and for their healthcare providers. But any early interventionist, educator, social service provider, or mental health provider working with families raising young children may be in a position to pass on information that will accomplish prevention of FASDs in current or future pregnancies.

Prevention of FASDs is necessary at a universal, community-wide level. Also vital are 'selective' prevention efforts targeted toward women who are pregnant or considering pregnancy and so at risk for gestational drinking, and 'indicated' prevention efforts directed toward women who are actually drinking during pregnancy and can be helped to cut down or stop.

Diagnosis of FASDs

Diagnosis in the Early Years

There have been systematic and successful research efforts aimed at how best to screen for risky drinking during pregnancy, and there is newer research focused on finding biomarkers of problematic gestational alcohol use. Research has also begun to focus on how to accomplish early detection of FASDs so that intervention can be applied. Different risk factors (such as the presence of prenatal alcohol use or small head size or a risk index based on parental characteristics) have been compared for their predictive value. Recently, ultrasound technology to image the neonatal brain has been explored as an early detection technique with suggestive but still preliminary findings. In part, this effort is prompted because of growing knowledge about the plasticity of the CNS, and because of intriguing findings in animal research on the positive developmental effects of enriched environments. Taken together, these findings strongly imply that early identification and intervention for children who are affected by prenatal alcohol exposure may be especially important, because CNS function might potentially be improved.

It is still difficult to identify children with FASDs in infancy and early childhood. This is because CNS damage or dysfunction can be hard to detect in the early years. Current FASD diagnostic guidelines require evidence of structural, neurological, and/or functional damage to the CNS. Evidence of structural/neurological damage may include such findings as microcephaly (small head size), abnormal neuroimaging, or seizure disorders. These can sometimes be found in infancy and early childhood. Evidence of CNS dysfunction includes standardized testing with scores that show significant global delays or, more commonly with FASDs, variability

revealed in significant gaps in skills, atypical patterns of development, or uneven profiles of learning strengths and weaknesses. It is often hard to gather clear evidence of CNS dysfunction when children are very young. The early developmental deficits of alcohol-exposed children may be subtle and yet important precursors to later problems. Tests used with young children often cannot detect variability in learning profiles or subtle problems, so young children will less often meet diagnostic criteria for FASDs. In addition, alcohol effects may often emerge most clearly as deficits in higher-level cognitive functions or complex information processing that typically develop at age 8 years and older. As a result, the problems of children with FASDs may not even become evident until well past the window for early intervention, around second to fourth grade.

Diagnostic Guidelines

While FASDs are usually treated as developmental disabilities or behavioral concerns, these diagnoses are actually considered to be medical conditions. Current practice guidelines suggest that diagnosing physicians work with an interdisciplinary or multidisciplinary team and follow well-defined FASD diagnostic guidelines. National diagnostic guidelines have been published in the US for FAS and in Canada for the broader fetal alcohol spectrum. There is a variety of more specific diagnostic systems and techniques that are continually evolving. Currently there exist FASD diagnostic criteria developed for clinical and research use that allow diagnosis of the full range of conditions making up the fetal alcohol spectrum (e.g., see **Table 1**). There is vivid research interest in ways to improve diagnosis of FASDs (such as three-dimensional facial imaging, or defining a 'behavioral phenotype' that can be discerned through psychometric testing). Neuroimaging and neurochemistry research is being carried out to learn more about how alcohol affects the brain, which may eventually assist in diagnosis.

Co-Occurring Conditions

Diagnosed FASDs commonly co-occur with developmental delays or deficits arising from other genetic or medical causes, which are sometimes recognizable in infancy. Beginning as early as the preschool years, FASDs quite often co-occur with psychiatric disorders (such as oppositional defiant disorder or attention deficit hyperactivity disorder (ADHD)). In the school years, children with FASDs are also often found to have academic problems or learning disabilities. For example, lowered achievement in pre-math and arithmetic skills are among the most persistent learning problems that have been identified in children across the fetal alcohol spectrum. A full understanding of an individual child's problems requires taking all co-occurring conditions into consideration.

Diagnoses on the fetal alcohol spectrum typically add an important dimension to the description of a child's problems as classified through use of medical or psychiatric diagnostic systems, or educational categories. For example, there is growing evidence that the attentional deficits caused by prenatal alcohol exposure may differ in type from those seen among children with ADHD from nonteratogenic causes. When children with prenatal alcohol exposure were directly compared with children with ADHD on measures of attention, they showed greater impairment in encoding information, while children with ADHD had more problems in focusing and sustaining attention. However, findings on the specific types of attention deficits among children affected by prenatal alcohol exposure vary depending on the sample being tested (clinical or nonclinical) and the assessment measures used. As another example, children prenatally exposed to alcohol may have associated medical complications and a complex profile of additional cognitive and learning difficulties. Therefore, knowing that a child has early-onset ADHD in the presence of FASDs can mean that commonly used interventions for ADHD (such as stimulant medication) may not have expected effects, and that a different set of treatment techniques may be needed. Learning how children affected by prenatal alcohol exposure compare to children diagnosed with other disabilities is another area of vivid research interest.

But there is a central practical benefit to understanding that a child has FASDs, which has been discussed by experienced clinicians such as Malbin. Knowing that an alcohol-exposed child's behavior problems may arise (at least in part) from that child's alcohol-induced neurological impairment can fundamentally change a parent's or provider's perception of the child's learning and behavior. This understanding can help the caregiver 'reframe' behavior problems, attributing the cause to underlying neurodevelopmental disorders rather than to willful misbehavior. For example, a 10-year-old boy with FASDs who does not cooperate when his mother asks him to 'get dressed to go to the ball game and clean your room' may actually be unable to do this because he (1) cannot mentally image what a 'clean' room looks like; (2) may not know how to organize and sequence the required behaviors; and/or (3) may not adequately process the higher-level language of his mother's multiple-step instructions. His mother will more successfully understand his behavior (and come up with effective ways to respond) if she 'reframes' her understanding of his lack of cooperation as actually arising from cognitive and language difficulties. Carmichael Olson and colleagues have made learning 'reframing' into a central treatment process in specialized behavioral consultation intervention to assist caregivers raising children with FASDs and challenging behavior problems.

Serving Children Across the Full Fetal Alcohol Spectrum

FAS is increasingly recognized as a diagnosed condition with a high probability of developmental delay, thus deserving early intervention. But there are many additional children, who do not have the full syndrome but do show diagnosable FASDs, who have a high likelihood of developmental delay and significant later psychopathology. Because of this, providing early intervention services for all children diagnosed with FASDs is vital. An even broader approach that is strongly encouraged is to classify developmentally vulnerable young children 'at-risk' because of prenatal alcohol exposure coupled with evidence of emerging learning problems and/or environmental risk. If an exposed, 'at-risk' child does not qualify for early intervention, or improves after intervention and then no longer qualifies, careful monitoring and re-evaluation at key developmental transitions is recommended. One important transition is before a child enters kindergarten; another is prior to entering second or third grade; a third is just before transitioning to middle school. One strategy that does not fit with current evidence is assuming that a young alcohol-exposed child will 'grow out of' apparently mild early delays, and therefore not providing early intervention services or developmental monitoring.

Young Children with Prenatal Alcohol Exposure and Their Families

What does current scientific evidence say about young children with prenatal alcohol exposure and their families? What are the implications for early intervention? Data come from larger clinical samples of young children referred for diagnosis, small samples of those diagnosed with FASDs, and large prospective longitudinal studies of the effects of prenatal alcohol exposure. Also referred to briefly in the following discussion are long-term studies of young children born polydrug-exposed, including alcohol, who typically have lives characterized by high levels of postnatal environmental risk.

Overview of Data from a Large Clinical Sample

There are no national data available describing FASDs, but large clinical samples are being gathered and examined. In 2007, data from a large clinical database of children with confirmed prenatal alcohol exposure referred for diagnosis of FASDs in Washington State were reviewed by Carmichael Olson, Jirikowic, Kartin, and Astley.

Highlights of their findings on nearly 800 children aged birth to 8 years are included here. The authors noted that conclusions from this database were somewhat limited by geographical context.

There was a very wide range of racial and ethnic backgrounds in this group of young children, with most no longer in the care of their birth parents and over 70% in foster/adoptive homes. The average age of identification was just over age 5 years, with only about one-third of children referred brought in during the early years before age 4 years. Surprisingly few families of these young children (<5%) were referred for diagnosis from school or early intervention settings. Instead, referrals came from medical, psychological, or social service providers. There was wide variation in prenatal alcohol exposure, ranging from reported maternal consumption during pregnancy of one single glass of wine to daily intoxication.

Of the nearly 800 children examined, only about 7% received a diagnosis of FAS or partial FAS. It is this smaller group who are most likely to be recognized and qualified for early intervention even though their CNS dysfunction is often no more severe than children with FASDs who lack the 'face' of FAS. There were 25% of this group of children aged birth to 8 years who were diagnosed with the equivalent of 'severe' ARND. This subgroup showed learning and behavior problems but were children for whom strong advocacy would likely be needed to obtain early intervention services. Importantly, there were 43% diagnosed with the equivalent of 'mild' ARND. This is a subgroup who may show more significant problems later in life but are far less likely to receive any services in infancy and early childhood, even if strong advocacy is exerted.

The authors discuss what might be early indicators suggesting a need for referral for diagnosis of FASDs. Interestingly, the presence of significant developmental delays does not seem to be a reliable early indicator, making early evaluation and identification difficult. More than 25% of these alcohol-exposed young children actually had early developmental profiles well within normal limits. Only just over half of the children referred for diagnosis showed testing evidence of marked developmental delay in the first 3 years of life – even among those with the full FAS or partial FAS.

The Developmental Impact of Prenatal Alcohol Exposure in Young Children

What are the effects of prenatal alcohol exposure on child development in the early years? In overview, it is well-established that heavy levels of prenatal alcohol exposure lead to neurobehavioral deficits, but the effects of lower levels of alcohol exposure are less clear. Neurobehavioral deficits from prenatal alcohol exposure are lifelong, but their impact on adaptive function emerges more clearly over time. These deficits can likely be improved or made worse by postnatal experiences. Gender differences in alcohol-related disabilities have rarely been examined in individuals with FASDs. Certainly, FASDs occur in both boys and girls, but the emergence of behavior problems across development may differ by gender. Limited data suggest that depressive symptoms may appear earlier for girls. It appears there are no published data available at this time about gender differences in cognition or learning. Problems in development resulting from prenatal alcohol exposure occur across multiple domains of function, as discussed below.

Problems in Early Cognitive Skills and Learning

In 2004, Kable and Coles found that infants born with higher-risk prenatal alcohol exposure had "difficulties with regulating the interactions between arousal level and the attentional system needed to provide optimal efficiency in processing environmental events" (p. 489). These infants were slow to initiate attention, and to encode visual and auditory information. They also demonstrated higher arousal levels. The authors speculated that the inability to modify arousal levels to encode information may result in slower responses to environmental events, which subsequently influences learning and later problem-solving and behavioral outcomes that depend on the interplay of these basic cognitive processes.

The enduring impact of prenatal alcohol effects on cognitive–behavioral processes and learning has been the subject of extensive research. In a 2007 literature review, Kodituwakku proposed that children with FASDs or prenatal alcohol exposure have a central, generalized deficit in complex information processing. Kodituwakku suggested that individuals with FASDs have difficulty integrating information from multiple regions of the brain and proposed this as part of the behavioral phenotype of FASDs. This overarching conclusion is drawn from synthesis of a large body of research evidence on FASDs and prenatal alcohol exposure. This evidence describes slow information processing, diminished nonverbal and verbal intellectual functioning, and difficulty with complex cognitive tasks.

In 2007, Carmichael Olson and colleagues reviewed the clinical research on infant, preschool, and older children with FASDs, which yielded consistent findings of deficits in attention, wide-ranging difficulties in higher-order cognitive processes called 'executive functions', visual–spatial processing deficits, other problems in information processing speed and efficiency, and difficulties with mathematical problem solving and achievement. There was a wide variety in type and extent of deficits across individuals. In infancy, adverse alcohol effects were found on basic conditioned learning and other cognitive functions that reveal an impact on specific brain regions (such as the cerebellum).

Long-term prospective studies of more moderate levels of prenatal alcohol exposure find associations with cognitive and learning difficulties through adolescence and beyond, though not all longitudinal studies find effects. Prenatal alcohol has been associated in group studies with mildly decreased performance on developmental scores in infancy and with mildly lowered intelligence quotient (IQ) scores in preschoolers. Newer longitudinal research with wider ethnic diversity and somewhat more highly exposed samples (using more sensitive measures) found an association between alcohol exposure and early, specific cognitive and achievement difficulties. For example, infants prenatally exposed to alcohol have shown deficits in information processing. In response to visual stimuli under laboratory conditions, infants with prenatal alcohol exposure reacted more slowly to changing visual stimuli and gazed longer at visual stimuli, suggesting slower and less efficient information processing. These information-processing deficits, which are considered indicators of intellectual abilities in both younger and older children, appear to persist at least into the early elementary years. Additional findings among older children include decreased processing efficiency under more complex cognitive conditions and a distinctive prenatal alcohol effect on number processing. In new research, substrates in brain structure and function of many of these functional difficulties are being explored.

Deficits in Speech, Language, and Communication

A variety of problems in speech, language, and communication have been described for clinical samples of school-aged children with heavy prenatal alcohol exposure, including delays in speech acquisition, impaired receptive and expressive language, and problems in speech production. Some school-aged children with FASDs show gaps between relatively better verbal abilities (such as vocabulary or basic grammar), and reduced capacity to use these skills effectively in social communication. Social communication is an important foundation for developing social relationships and exchanging information. So far, there has been limited clinical or longitudinal study of language and communication in infants and young children with prenatal alcohol exposure. Current clinical thinking by experts such as Coggins is that young children with FASDs can often acquire basic linguistic skills in a typical manner. However, they seem to have difficulty becoming socially competent communicators later on and, as they grow older, show deficits in more complex, higher-order language skills. For instance, children with FASDs have difficulty telling coherent and cohesive narratives, which is an important communication skill used in learning situations and interpersonal relationships.

Problems in Motor Development, Neurological Soft Signs, and Sensory Processing

Alcohol-exposed individuals consistently show impairments in the development of motor control. Clinical studies of infants reveal that motor delays occur more often and/or with increased severity in the presence of heavy alcohol exposure. Among younger children with diagnosed FASDs, delays in motor skills are generally mild-to-moderate in extent, and these preschoolers show poor movement quality. In clinical samples, significant deficits in visual-motor development (but not in motor-free visual perception), and in fine motor coordination, have been described.

Some longitudinal researchers examining population-based samples have found no measurable impact of more moderate levels of social drinking on motor development in infancy or early childhood. Other long-term studies have found clinically significant effects when examining psychomotor performance on standardized tests such as the Bayley Scales of Infant Development, but only among infants at high exposure levels. However, in infants with lower exposure levels, qualitative differences in motor behaviors, such as the ability to stand, walk, and imitate have been detected and described. In early childhood, longitudinal research has found increased levels of minor neurological soft signs among alcohol-exposed preschoolers, and a positive association between greater alcohol exposure and increased deficits in fine motor steadiness and balance.

In clinical samples, parents completing questionnaires on sensory-processing behaviors documented clinically significant difficulties modulating sensory information among their young children with FASDs. Specifically, Jirikowic described behaviors that suggest fluctuating responses (i.e., both over- and under-responsivity) to sensation that include tactile, auditory, and visual sensitivity, under-responsiveness to sensory information and sensation-seeking behaviors, and poor auditory filtering.

Difficulties in Adaptive Behavior and Social–Emotional Development

Difficult behaviors and social skill deficits that persist across time are a central concern among individuals of all ages with prenatal alcohol exposure in both clinical and longitudinal prospective samples. Mental health problems have been reported for a very large majority of individuals diagnosed with FASDs in natural history research, and there are elevated rates of psychiatric disorders in childhood and beyond. Clinical studies of children with FASDs use terms such as impulsive, distractible, and 'always on the go' to describe behavior in the preschool and elementary years. In addition, while younger children with FASDs have been described as engaging, verbal, apparently alert, and bright-eyed, they likely appear more functional than they actually are. Clinicians note they seem to lack social boundaries (e.g., by showing indiscriminate affection or seeking physical proximity to strangers). There is wide variability in the level of these deficits, ranging from subtle to severe.

For school-aged children with FASDs, studies often show clinically elevated attention, social, and sometimes internalizing behavior problems and (even more frequently) problems with externalizing behavior and aggression. Adaptive behavior and social skills among preschool and young school-aged children with FASDs are reported as lower than expected for age and intellectual level. Although decreased adaptive performance has been described across most adaptive domains including daily living, social, communicative and, to a lesser extent, motor function, the development of social and interpersonal relationships appears to be especially problematic for those affected by prenatal alcohol exposure. After the age of 8 years, these deficits in adaptive function and social performance in children exposed to alcohol appear relatively greater, even than clinic-referred peers with behavior or adjustment problems. This is likely the result of the lifelong neurological impairment of children born alcohol-affected.

Longitudinal studies begun in infancy report early problems in behavioral regulation associated with prenatal alcohol exposure, including mild-to-moderate irritability, poor habituation, sleep problems, and feeding difficulties. Among preschooler children, limited longitudinal data suggest that prenatal alcohol exposure has been associated with functional compromise that includes mild-to-moderate inattention and hyperactivity, and subtle-to-moderate impulsivity, and behavior problems.

Thinking in a more complex way about developmental systems reveals how the impact of prenatal alcohol exposure (coupled with maternal drinking) on children's social–emotional development reverberates across infancy and early childhood. For example, O'Connor and colleagues described an evolving process in which alcohol-exposed infants showed increased negative affect, and their mothers had difficulty responding to these babies (perhaps because of their own depression that may be associated with a tendency to drink, and/or because of difficulties presented by the child). In this evolving developmental process, the quality of the attachment between parent and child was then compromised, and these alcohol-exposed children grew into preschooler children who showed increased levels of depression at around age 4–5 years. This process has been demonstrated among middle-income families, and shown even more clearly among low-income families who have additional risk factors that augmented the severity of this negative cycle. Additional analyses of middle-income families through age 6 years revealed this process as primarily true for girls, but there were also separate trends or significant direct effects of prenatal alcohol exposure on developmental outcome in both boys and girls. All this suggests strongly that intervention could usefully begin in infancy, and could be targeted specifically at ameliorating negative child behaviors (and providing parental support).

The Importance of Cumulative Risk

The impact of prenatal alcohol exposure is often coupled with that of other prenatal drug exposures, poor prenatal care, and life with parents who struggle with chemical dependency. Children may live in environments with health and safety concerns, relationship problems, and chaotic lifestyles. Longitudinal studies have tracked the often negative developmental outcomes of children with prenatal substance exposures in these high-risk situations. Clinical studies have examined the difficult lives of children born substance-exposed. These studies are sharp reminders that children with FASDs may often spend at least part of their early years in an environment of cumulative risk. Thinking about FASDs from a developmental systems perspective emphasizes that cumulative risk could be especially harmful for a child made biologically vulnerable by prenatal alcohol exposure.

The Potential of Treatment for FASDs

A small body of research on animal models of FAS suggests that exercise training and appropriate enrichment of the learning environment have potential to improve behavioral and learning outcomes, and to reduce alcohol-induced injury to the brain and CNS – early in life but even during the equivalent of adolescence or early adulthood. Natural history study of a large clinical sample by Streissguth and colleagues highlights 'protective factors' that are related to reduced odds that 'secondary disabilities' (negative outcomes in lifestyle and daily function) will occur among individuals with FASDs. These protective factors, which include early diagnosis, good quality of caregiving during childhood, absence of parental substance abuse and trauma, and presence of appropriate social services, can be enhanced through intervention. There may be pivotal 'turning points' in development when intervention is most needed, with the idea of creating 'downstream effects' that push development in a positive direction. A first turning point is in early infancy, when preventive intervention may take advantage of neuronal plasticity and improve attachment quality. Secure attachment between parent and child is a powerful predictor of positive outcome in many developmental areas. A second turning point may be in preschool, a time when tailored educational techniques and focused efforts to help parents assist their children in learning behavior regulation may maintain developmental progress. Another set of turning points may occur during the time when a child grows from 5 to 11 years (and first enters kindergarten, then second or third grade, and then approaches middle school). Among children who are typically developing, these school years are a time of rapid cognitive and social development, when children master increasingly complex language and social skills, and when demands for self-reliance and self-regulation increase. For children with significant prenatal alcohol exposure or FASDs, it is during this period of time that alcohol-related deficits may start to become obvious (and troubling) to parents and teachers, and progress in the growth of adaptive behavior is likely to plateau.

A Developmental Systems Model Applied to Early Intervention for Children with FASDs and Their Families

Thinking about developmental systems can guide practice with young children with FASDs or who are at risk because of confirmed prenatal alcohol exposure. This type of thinking helps to make clear why early intervention for this population is especially important, and suggests that treatment models should be designed especially for the population served and subgroups within the population, and targeted to the individual needs of children being served. Developmental systems thinking also means that risk and protective factors specific to the population should be identified and considered.

Child and Family Characteristics in the Presence of FASDs: Risk and Protective Factors

A developmental systems model was created by Guralnick in 2001 to guide the design of early intervention with children who have special needs, and can usefully be applied to FASDs. Conceptualized from the standpoint of this model, prenatal alcohol exposure can be seen as placing a child at risk for (or creating) biological vulnerability and 'disabling child characteristics' or 'core deficits'.

Across all ages, possible disabling child characteristics for those with FASDs include sensory sensitivities and difficulty with behavioral regulation (especially in stressful or unstructured situations). In infancy, as mentioned earlier, children with FASDs may have deficits in basic cognitive processes, including the ability to regulate the interactions between arousal level and the attentional system needed to provide optimal efficiency in processing environmental events. Older children with FASDs may show a central, generalized deficit in complex information processing.

On an individual basis, the specific pattern of cognitive/behavioral deficits of a child with FASDs or significant prenatal alcohol exposure is individually variable and arises in multiple domains of development (perhaps in subtle ways). A particular child's pattern of deficits may emerge more clearly or become more debilitating over time. On an individual basis, then, a child's deficits and compensatory strengths must be comprehensively assessed.

Disabling child characteristics can disrupt existing family interaction patterns. This creates information and resource needs for parents, threats to parenting confidence, an impact on the parent–child relationship in the early years, and high levels of both caregiver and family stress. Comprehensive assessment of these 'environmental risk factors' (especially cumulative risk) is essential. One crucial family interaction risk factor identified through clinical experience centers around inappropriate caregiver reactions to deficits of the child with FASDs – deficits caregivers may not easily recognize as the result of alcohol-induced neurological impairment, but see as willful disobedience. This requires caregivers to learn 'reframing' and understanding that neurological impairment underlies problem behaviors.

Research supports the importance to developmental outcome of caregiver characteristics and quality of caregiving – in the early years and through middle childhood – for children born alcohol-exposed. In 2000, Coles and colleagues developed a cumulative risk index from data gathered during the neonatal period, based on maternal characteristics. They found this risk index was useful in predicting early growth and developmental compromise (at 6 and 12 months) in low-birthweight children exposed to alcohol before they were born. This risk index had more predictive power than when using other early indicators such as the presence of heavy alcohol exposure or small head size. Other pivotal family risk factors have been documented in natural history research on FASDs by Streissguth and workers. For this population, past or current parental substance use, exposure to violence, and poor quality of the childhood caregiving environment, are all significantly associated with the occurrence of poor outcomes later in life.

The developmental disabilities and early intervention literatures stress that protective factors, such as family strengths and coping skills, are also important and should be assessed. There are specific protective factors hypothesized by Carmichael Olson and colleagues as important for the population of children born with FASDs. In infancy, these include the quality of attachment security and treatment for parental alcohol abuse or dependency (a protective factor for a child of any age). Starting in the preschool years, protective factors include the parent's level of optimism and sense of parenting efficacy, use of specialized parenting practices, advocacy skill, knowledge of FASDs, use of respite and beneficial social support, and appropriate linkage to community resources and social services.

Developing Interventions Based on a Developmental Systems Model

Early intervention should aim to reduce both the impact of disabling child characteristics and cumulative environmental risk, while at the same time enhancing protective factors. For 'at-risk' young children with prenatal alcohol exposure who may or may not yet show clinically concerning behaviors, preventive interventions are needed. This could include careful and ongoing developmental monitoring, providing anticipatory guidance to parents, daycare providers, and school staff, general environmental enrichment, and very early substance abuse education.

For the diagnosed child, a comprehensive intervention program can first provide thorough and tailored assessment, so that all caregivers understand the child's 'disabling characteristics' and can 'reframe' their understanding of the child's behavior in light of neurological impairment. Then intervention can provide resource, social, and information supports, and direct treatment to enhance family interaction patterns, adjusted to the family's unique needs and targeted to the disability.

If the caregiving relationship is strongly affected, early in life it may be vital to provide ongoing infant mental health interventions that help caregivers behave in ways that foster attachment security, even in the face of negative affect or unresponsiveness from the infants born prenatally exposed. This may have to be adjusted depending on whether the caregiver is from a birth, foster, or adoptive home. Intervention will likely become 'multimodal' as a child with FASDs grows older. For example, if a preschooler with FASDs has sensory sensitivities and deficits in information processing arising from prenatal alcohol exposure, then environmental modification and supportive occupational therapy services, caregiver education about the child's cognitive processing, and advocacy for later special education, all might be necessary. Home visiting or clinic-based services to help caregivers understand alcohol-related brain damage, modify attitudes, learn specialized parenting skills, and learn effective advocacy to access existing community supports may be required when preschool-aged and school-aged children with FASDs have especially difficult behavior problems. Intervention models for such services have been developed by researcher Carmichael Olson and colleagues.

An Emerging Body of Evidence on Intervention for FASDs

General intervention approaches to FASDs have been developed based on expert opinion and research on related disorders, and are reviewed in a set of guidelines published by the National Center on Birth Defects and Developmental Disabilities and discussed by other authors. Carefully designed FASD intervention studies are beginning to show positive effects for children and families. The broader child treatment literature also suggests strong potential for positive change if early intervention is designed based on scientific data now available about child development and early psychopathology.

Examining the efficacy of relationship-based infant mental health treatments, which have been useful for families raising children with polydrug exposure, is an important next step in research on early intervention for children with FASDs and their families. New research by several investigators, funded by the CDC, including Carmichael Olson, Chasnoff, Coles, Gurwitch, and O'Connor, suggests that treatment for children with FASDs in the preschool and school years can (among other outcomes) improve social skills and/or reduce disruptive behavior, and improve self-reported parenting attitudes and knowledge. There are many field-initiated research projects underway examining a wide variety of interventions for FASDs, although mostly for older children and youth. Intervention for youth in juvenile justice, coordinated state service responding to FASDs, and intervention for high-risk chemically dependent women and their young children (who are likely drug- and alcohol-exposed) are only some of the interventions being actively explored. Successful intervention may prevent or reduce secondary disabilities and debilitating family strife – and intervention earlier in life has the potential for powerful positive change. Although only family and expert clinical experience support the idea – even as late as adolescence and adulthood, diagnosis and intervention for individuals with an FASD are deemed crucial (especially holistic treatment and management of associated mental disorders). Even later in life, intervention can likely reduce suffering and cost to affected individuals and those who care for them.

Acknowledgments

The authors gratefully acknowledge the Centers on Disease Control and Prevention (Grant No. U01-DD000038-02, awarded to Heather Carmichael Olson) and the National Institute on Drug Abuse (Grant No. 5 T32 DAO 7257-14) for postdoctoral support of Dr. Jirikowic for support and facilitation during preparation of this manuscript.

See also: ADHD: Genetic Influences; Attachment; Birth Defects; Cognitive Development; Depression; Developmental Disabilities: Cognitive; Intellectual Disabilities; Neuropsychological Assessment; Prenatal Development; Risk and Resilience; Sensory Processing Disorder; Teratology.

Suggested Readings

Carmichael Olson H, Jirikowic T, Kartin D, and Astley S (2007) Responding to the challenges of early intervention for fetal alcohol spectrum disorders. *Infants and Young Children* 20(2): 162–179.

Families Moving Forward Program (2007) Families moving forward. http://depts.washington.edu/fmffasd/index.html (accessed 13 July 2007).

Fryer SL, McGee CL, Matt GE, Riley EP, and Mattson SN (2007) Evaluation of psychopathological conditions in children with heavy prenatal alcohol exposure. *Pediatrics* 119(3): 733–741.

Kable JA and Coles CD (2004) The impact of prenatal alcohol exposure on neurophysiological encoding of environmental events at six months. *Alcoholism: Clinical and Experimental Research* 28(3): 489–496.

Kalberg WO and Buckley D (2006) FASD: What types of intervention and rehabilitation are useful? *Neuroscience and Biobehavioral Reviews* 31: 278–285.

Kodituwakku PW (2007) Defining the behavioral phenotype in children with fetal alcohol spectrum disorders: A review. *Neuroscience and Biobehavioral Reviews* 31: 192–201.

National Center on Birth Defects and Developmental Disabilities (NCBDD) (2004) *Fetal Alcohol Syndrome: Guidelines for referral and diagnosis.* Washington, DC: Centers for Disease Control and Prevention. Available in hard copy from CDC or at http://www.cdc.gov/ncbddd/fas/documents/FAS_guidelines_accessible.pdf.

O'Connor MJ, Frankel F, Paley B, *et al.* (2006) Controlled social skills training for children with fetal alcohol spectrum disorders. *Journal of Consulting and Clinical Psychology* 74(4): 639–648.

Spadoni AD, McGee C, Fryer SL, and Riley EP (2007) Neuroimaging and fetal alcohol spectrum disorders. *Neuroscience and Biobehavioral Reviews* 31: 239–245.

Spohr HL, Willms J, and Steinhausen HC (2007) Fetal alcohol spectrum disorders in young adulthood. *Journal of Pediatrics* 150(2): 175–179.

Streissguth A (1997) *Fetal Alcohol Syndrome: A Guide for Families and Communities.* Baltimore, MD: Paul H. Brookes.

Streissguth AP, Bookstein FL, Barr HM, *et al.* (2004) Risk factors for adverse life outcomes for fetal alcohol syndrome and fetal alcohol effects. *Journal of Developmental and Behavioral Pediatrics* 25(4): 228–238.

Substance Abuse and Mental Health Services Administration (SAMHSA) (2007) *FASD Center for Excellence.* Multiple fact sheets available at http://fascenter.samhsa.gov (accessed 13 July 2007).

Surgeon General's Advisory on Drinking and Pregnancy (2005, February). United States Department of Health & Human Services. Available at http://www.hhs.gov/surgeongeneral/pressreleases/sg02222005.html (accessed 13 July 2007).

Tenth Special Report to the US Congress on Alcohol and Health: Highlights on Current Research. Available at http://www.niaaa.nih.gov.

Fragile X Syndrome

M Y Ono, University of California, Davis, Medical Center, Sacramento, CA, USA
F Farzin, University of California, Davis, Davis, CA, USA
R J Hagerman, University of California, Davis, Medical Center, Sacramento, CA, USA

Glossary

Autism spectrum disorder – The category that includes PDDNOS and autism.

Cortisol – The glucocorticoid produced by the adrenal cortex upon stimulation by adrenocorticotropic hormone (ACTH) that mediates various metabolic processes, has anti-inflammatory and immunosupressive properties, and whose levels in the blood may become elevated in response to physical or psychological stress.

Dyspraxia – The impairment of the ability to perform coordinated movements.

Echolalia – The pathological repetition of what is said by other people, echoing.

Hyperarousal – The state of having excessive arousal.

Hypotonia – The state of having deficient tone or tension.

Macroorchidism – The condition of having large testicles.

Otitis media – The acute or chronic inflammation of the middle ear.

Strabismus – The inability of one eye to attain binocular vision with the other because of imbalance of the muscles of the eyeball, also known as lazy eye.

Introduction

Fragile X syndrome (FXS) is a heritable form of mental retardation that affects approximately 1 in 4000 individuals. Not only is it the most common cause of mental retardation, but it is also a common cause of developmental delay, that is, learning disabilities, in addition to psychological and behavioral problems among children.

Genotype

FXS is caused by a mutation on the fragile X mental retardation 1 (*FMR1*) gene, which was discovered in

1991 to be located on the lower end of the X chromosome of Xq27.3. This mutation involves an expansion of the CGG trinucleotide repeat within the promoter region of the *FMR1* gene. Depending on the size of the expansion, the gene can become silenced, resulting in a diminished or complete lack of transcription into messenger RNA (mRNA) and subsequent lack of translation into FMR1 protein (FMRP). The lack of FMRP leads to the classic phenotype of FXS.

In general, the degree of affectedness is determined by the size of the CGG repeat expansion. In normal individuals, the CGG repeat size is 5–44. The gray zone category is 45–54 repeats and the allele can become unstable when transmitted to the next generation. Individuals with 55–200 repeats are categorized as premutation carriers and they typically are more mildly affected by FXS compared to the individuals who have the full mutation. The full mutation is defined by having more than 200 repeats and usually involves the gene being completely methylated and thus silenced. Males with the full mutation and complete methylation do not produce FMRP and therefore display the most severe symptoms of FXS. In contrast, females with the full mutation are less affected because of their second X chromosome. Depending on their X activation ratio, that is, the proportion of cells that express the allele without the mutation, the amount and severity of symptoms will vary. There are also individuals who are mosaic, meaning they have some cells with the premutation and some cells with the full mutation or they have the full mutation but they are partially unmethylated, which allows them to produce some FMRP.

Physical and Medical Phenotype

During infancy, the physical features of a child with FXS commonly appear to be normal. Most individuals with FXS are not diagnosed until 3 years of age or older because there is a lack of awareness of this disorder among clinicians and the fact that the physical features are somewhat common among the general population. The classic features include prominent ears, long face, high-arched palate, hyperextensible finger joints, soft or velvet-like skin, and flat feet (**Figures 1** and **2**). However, approximately 30% of children and adults with FXS do not have obvious physical features.

Retrospective studies have found that many males with FXS also have various medical conditions. For example, approximately, 85% have otitis media (middle-ear infection), 36% have strabismus, 31% have emesis, 23% have a history of sinusitis, 22% have seizures, and 15% have failure to thrive in infancy. Loose connective tissue is thought to lead to some of these features, including hyperextensible finger joints, soft or velvet-like skin, flat feet, and otitis media.

Figure 1 Young boys with fragile X syndrome.

Figure 2 Young female with fragile X syndrome.

One study, conducted by Jean-Pierre Fryns in 1988, found an unexpectedly high rate of sudden infant death syndrome (SIDS) in babies with FXS. Eight per cent (17/219) of males and 4% (6/169) of females died of SIDS before the age of 18 months. Although SIDS had not been thoroughly studied in FXS, there were various hypotheses of why it occurred, that is, central nervous system (CNS) disturbances, hypotonia leading to obstructed airway, seizures, mitral valve prolapse (MVP), or cardiac arrhythmias. SIDS also occasionally occurs in older individuals with FXS and these cases are thought to relate to cardiac arrhythmias.

Neurobiology and Brain Development

FMRP is a regulator of translation that binds to approximately 4% of all mRNAs in the neuron. It typically suppresses translation such that the absence of FMRP leads to enhanced translation of many messages in the CNS. One pathway that is remarkably enhanced is the metabotropic glutamate receptor 5 (mGluR5), which leads to enhanced long-term depression (LTD). LTD is the weakening of a neuronal synapse thought to result from changes in postsynaptic receptor density, and subsequently results in weak and immature synaptic connections, particularly in the hippocampus and cerebellum and this is thought to be a significant cause of the mental retardation in FXS. Synaptic connections are weak and immature in the fragile X knockout (KO) mouse model and there is a lack of synaptic plasticity and pruning. The identification of enhanced mGluR5 activity in KO mice and humans with FXS suggests that mGluR5 antagonists might be a specific treatment for FXS. Studies in the KO mouse and Drosophila models of FXS have demonstrated benefits of the mGluR5 antagonist, 2-methyl-6-phenylethynyl-pyridine (MPEP). In addition, lithium, which downregulates the mGluR5 system, has been shown to be helpful for the KO mouse in decreasing seizures and improving cognition and for the Drosophila in enhancing cognition and lifespan. In humans with FXS, lithium improves behavior, but improvements in cognition have not yet been demonstrated, although studies are currently being conducted.

There is limited information regarding the brain development of infants and toddlers with FXS, primarily due to the late diagnoses. One study, by Hill Karrer *et al.* in 2000, used event-related potentials (ERPs) to measure electrical brain activity through the scalps of infants with FXS, Down syndrome (DS), and typical development when they viewed visual images. Results indicated exaggerated early visual processing, possibly related to the deficit in dendritic pruning, in infants with FXS.

Children with FXS typically have large heads and prominent foreheads in early childhood. Neuroimaging studies have shown overall larger brains with significant enlargement of the caudate and a smaller size of the cerebellar vermis. Those with FXS and autism also have larger heads than individuals with FXS without autism.

Learning, Cognition, and Perception

Learning and Cognition

Research to-date on infants and young children with FXS has been restricted by the ages included and methodologies used. In 1998, Don Bailey *et al.* published a study describing a prospective examination of the development of infants with FXS. This study administered three developmental screening tests (Denver-II, Battelle Developmental Inventory Screening Test, and Early Language Milestone Scale-2) and two comprehensive assessment measures (The Mullen Scales of Early Learning and the Receptive-Expressive Emergent Language Scales) to 18 infants and toddlers with FXS (13 boys and 5 girls). One of the main objectives of the study was to assess the predictive validity of individual screening measures that might be useful in picking up delay in children with FXS at early ages. It was found that all of the screening tests and both comprehensive assessment measures were successful at detecting developmental delays in most children with FXS.

Another set of findings by Don Bailey's group examined the early developmental trajectories of males with FXS by employing a longitudinal design. Children in this study varied in age from 24 to 72 months and were evaluated on overall development in the domains of cognition, communication, adaptive, motor and personal–social, using the Battelle Developmental Inventory. The application of hierarchical linear model analysis revealed that the overall rate of development in the boys with FXS was approximately half of that expected for typically developing children. Also, although stable patterns of development were observed within individual subjects, development was not uniform across the sample. Significant variability was reported among subjects in both the levels of performance at each age, and in the rate of developmental change over time. Communication and cognitive skills were typically lower than social, adaptive, and motor skills.

In early childhood, intelligence quotients (IQs) are in the borderline to mildly mentally retarded range in 80–90% of males with FXS. In contrast, 30–35% of females with FXS test in the mild mental retardation range and 20–30% are in the borderline normal range. IQs in the normal range are seen in 40–50% of females with FXS; however, approximately half of those testing in the normal range have learning disabilities. In males with FXS, the level of FMRP correlates with the overall IQ.

IQs typically decline in most males and in many females with age. By adulthood, most males with FXS have IQs in the 40s. However, approximately 15% of males have an IQ greater than 70 and are called high-functioning males. They usually have FMRP levels greater than 50% and their DNA pattern is unmethylated or mosaic. IQ decline is not seen in individuals with FMRP levels greater than 50%.

One of the common cognitive weaknesses in children with FXS is sequential learning. For example, processing of verbal directions that require keeping both the instructions and their sequence in mind is difficult for children with FXS. Also, learning how to read using the phonics system, and performing the steps involved in math functions can be a struggle. Executive function is another common weakness in individuals with fragile X since it is difficult for them to store information in their short-term working memory as they attempt to complete a task, such as putting pictures in a specific order. Lastly, children with FXS often think in concrete terms and have difficulty with concepts that they cannot see, hear, or touch. As a result, their abilities to solve problems and think abstractly are compromised.

Perceptual Development

Much of the evidence about perceptual development in FXS comes from the field of visual processing. A number of studies have suggested that FXS is not associated with a global deficit in visual processing, but rather, that deficits in this group are specific. In particular, a finding that emerges consistently across studies is that affected individuals perform worse on tasks that require coordinated visual–spatial activity. It has been shown that individuals with FXS are worse than those with DS on block construction and drawing completion tests. Similarly, Kim Cornish's group conducted a comprehensive study that compared males with FXS to chronological age (CA) and mental age (MA)-matched and (DS) controls on a series of visual–motor (e.g., the block design and object assembly subtests of the Wechsler Intelligence Scales for Children, the triangles subtest of the Kaufman-ABC test, and the Annett pegboard) vs. a visual–perceptual task (the gestalt closure task). Males affected with FXS performed worse than both the CA-matched and DS control groups on all visual–motor tasks, but performed relatively and significantly better than the DS group on the visual–perception task. Together, these findings converge on a picture of visual information processing in FXS in which performance deficits are consistently reported for tasks that require visual–motor but not visual–perceptual processing.

Most recently, two studies by Cary Kogan *et al.* provided both molecular and behavioral evidence for the claim that the visual–motor deficits evident in FXS are attributable to a selective M pathway and dorsal stream deficit. First, they conducted immunohistochemical staining of the lateral geniculate nucleus (LGN) of a normal human male, and showed high FMRP basal expression selectively within the magnocellular (M) layers, suggesting an increased susceptibility of these neurons to the lack of FMRP as occurs in FXS. They also performed staining of the LGN of a male FXS patient, which revealed, unlike the normal male, a population of small-sized neurons within the M layer, providing anatomical and morphological support for the idea that M pathway pathology exists in FXS. For the behavioral evidence, they tested male patients with FXS on tasks that probed either the M pathway or the P pathway and found that they had reduced contrast sensitivity only for those stimuli probing the M pathway. Lastly, they demonstrated that male patients with FXS performed poorly on a global motion task, but not on a form perception task, again suggesting a selective deficit of M pathway (dorsal stream) functioning in FXS.

Auditory processing is another domain that has been found to be impaired in FXS. One study by Maija Castrén assessed auditory processing using ERPs by comparing N1 responses (the occurrence of a negative peak in response to repeated presentations) to tones in school-aged children with FXS to the responses of CA-matched controls. N1 amplitude to standard tones was significantly larger in FXS than in controls. Perhaps most interestingly, children with FXS exhibited no habituation of N1, indicating increased sensory sensitivity for auditory stimuli in FXS. These results are in line with previously reported findings in adults with FXS.

Motor and Language Development

Motor

Clinically, young males with FXS often present with hypotonia, or low muscle tone, which can affect joint stability, fine and gross motor coordination, and sensory integration. For example, hypotonia can lead to a delay in sitting up on one's own and crawling or walking. Children with FXS also have difficulty with sensory integration, particularly integrating and processing touch, sound, sight, and movement. Children with FXS may also experience dyspraxia, trouble with planning a sequence of movements. Often they will toe walk. Approximately 60–90% of boys with FXS are tactile defensive, meaning they do not like people to touch them, the feeling of their clothing, and/or the texture of the food.

Language

While FXS is considered to be a common genetic disorder associated with a varying array of developmental delays, language impairment has been recognized as a characteristic hallmark associated with FXS. Still little is known about the very early developmental course of language in FXS, and its link to atypical sensory information processing that may affect subsequent language outcomes in social communication contexts. But, recent studies on the emergence of deviance in language in FXS have documented word-retrieval deficits, linguistic and cognitive profiles, and deficits in conversational skills. In addition, a study prospectively examined the developmental trajectories of receptive and expressive communication in 39 young males with FXS (aged 20–86 months) who were given a standardized language test. Results revealed marked delays in language development, but substantial individual variability. In general, it was found that expressive language skills were acquired more slowly over time than were receptive skills. Gain of receptive language progressed at about half the rate expected for typically developing children and expressive language progressed at about one-third the rate.

In general, boys with FXS will develop single words approximately at the age of 3 years and two-word phrases around 4 years. In contrast, girls with FXS usually have single words at the age of 2 years and full sentences at the age of 4 years. Once these children acquire the skill to speak, they tend to perseverate and have echolalia, characteristics often seen in autism. Eventually, the majority of children with FXS will develop functional speech; however, approximately 10–15% of males do not have useful speech, which may be related to dyspraxia.

Social and Emotional Development

An important aspect to understanding the early social development of infants and young children with FXS is to determine how these features are changed with the co-morbid diagnosis of autism spectrum disorders (ASD). Rates of ASD comorbidity range from 25% to 40% in males and 3% to 17% in females, and while a considerable amount is known about the overlap between FXS and ASD, much remains unanswered about the factors that predispose an infant with FXS to develop an ASD. Given that FXS can be diagnosed accurately in late infancy, the determination of ASD risk factors in FXS will also inform our understanding of the emergence of the symptoms of ASD in other more heterogeneous groups for which the etiology is unknown.

The behavioral phenotype of infants and young children with FXS includes a relatively pervasive pattern of anxiety symptoms and social problems that reach prominence and clinical attention by perhaps 3–5 years of age. It is critical to study this phenomenon early in development for several reasons. First, there is an overlap in anxiety-related social deficits in FXS (e.g., gaze avoidance and social withdrawal), raising questions about whether social anxiety, early in development, contributes to or predisposes an infant with FXS to more significant social problems or perhaps even autism in later childhood. Anxiety may be manifested by gaze aversion in novel social situations, withdrawn behavior and social isolation, distress with changes in routine and desire for sameness, obsessive compulsive behavior, and repetitive and tangential speech. Some stereotypic behaviors associated with autism, such as hand-flapping and hand-biting, seem to occur more often during periods of increased anxiety, stress, or excitement.

Eye gaze provides a link to social competency in infancy, in addition to being one of the earliest regulators of perceptual input, visual attention, visual processing, and integration. There are several published studies documenting abnormal eye gaze in children with FXS, and recently, it has also been suggested that deviant processing of gaze in FXS may be indicative of dysfunction of underlying neural systems serving these functions. In studies examining attention shifts in typically developing children vs. children with autism, the latter group demonstrated a lack of attention to social stimuli. This has obvious implications for the development of an infant's social system whereby dyadic interactions are not experienced the same. Studies on eye gaze in children with FXS found that the children with FXS are more avoidant of direct eye contact when someone looks at them, suggesting a reactive avoidance, as opposed to children with autism who avoid direct eye gaze whether people are looking at them or not.

Autism Spectrum Disorders

The characteristics of autism in FXS, including the hand-flapping, hand-biting, perseveration in speech, shyness, and poor eye contact, are often present, but some researchers believe the core social deficits typical of autism present less commonly. Usually, individuals with FXS tend to be interested in social interactions and they typically are aware of facial emotional cues from others. In contrast, a study by Sally Rogers *et al.* in 2001 involving 24 children with FXS, 27 children with autism, and 23 children with developmental delay (DD) ages 21–48 months, demonstrated that the FXS and autism group had a similar profile on the autism measures to the autism only group, but performed lower on the developmental measures, specifically the fine and gross motor domains. This study also found two distinct groups in the FXS sample, one group had autism and the other did not. The FXS group alone was identical to the DD group on the autism measures and the developmental instrument. The two distinct FXS groups may imply possible additional genetic contributions acting synergistically with the *FMR1* mutation, thus leading to the presence of autism.

There are not many studies that have evaluated the role of molecular outcomes, such as FMRP, in the presence of autism in FXS. In addition to the limited number of studies, the measures used to evaluate ASD or symptoms of ASD are not always the same, which also contributes to inconsistencies in the findings. For example, a study by David Hessl *et al.* in 2001 found that FMRP, after controlling for IQ, predicted symptoms of autism on the Autism Behavior Checklist (ABC) in girls with FXS; but not in boys. In contrast, Deborah Hatton *et al.* in 2006 found levels of FMRP were negatively correlated to scores on the Childhood Autism Rating Scale (CARS) in males and females. Specifically, low levels of FMRP were related to higher scores on the CARS, indicating more autistic behavior. A small study by Beth Goodlin-Jones *et al.* in 2004 found that individuals with the premutation and ASD had lower FMRP levels compared to individuals with the premutation alone using the CARS, autism diagnostic observation schedule (ADOS), and autism diagnostic interview-revised (ADI-R) or social communication questionnaire (SCQ).

In females with FXS, ASD is not as well studied, primarily due to the low incidence in females compared to males. The few studies on ASD prevalence in females with FXS have found rates of 3–17%. One study by Michele Mazzocco *et al.* in 1997 found that females with lower activation ration (percentage of cells expressing the normal allele) had more repetitive behaviors, a characteristic of autism. When autism is present in females with FXS, it indicates the most severe end of the spectrum of social anxiety and withdrawal.

Hyperarousal, Anxiety, and Attention Deficit Hyperactivity Disorder

Studies have looked at physiological indices of enhanced reactions to sensations. For example, the electrodermal responses (EDRs) in FXS, as compared to controls, to olfactory, visual, auditory, tactile, and vestibular sensations which were presented in a 'sensory challenge protocol'. They reported that the FXS group differed significantly from the controls demonstrating greater magnitude, more responses per stimulation, responses on a greater proportion of trials, and lower rates of habituation to these stimuli. In addition, within the group of FXS subjects, the EDR patterns were related to their FMRP expression. Because EDR activity indexes the sympathetic nervous system, and because of the link between FMRP and development in the limbic system, researchers have suggested that the overarousal to sensation often reported in FXS may be due to a sympathetic nervous system dysfunction, possibly caused by a deficiency or absence of FMRP.

Based on early studies documenting hypothalamic and endocrine abnormalities in individuals with FXS, a comprehensive in-home study of anxiety was conducted to examine salivary cortisol (a hormone related to levels of stress) levels in 109 children with FXS (ages 6–17 years) in comparison to their unaffected biological siblings. The results of the study documented that the children with FXS, especially males, had elevated baseline cortisol during the day and before bedtime, and they had a greater cortisol response to the diverse challenges of the home visit (meeting the examiners, undergoing neuropsychological testing, and engaging in tasks designed to elicit social anxiety), in comparison to their unaffected biological siblings. Salivary cortisol levels were positively associated with severity of behavior problems, predominantly withdrawn behavior, social problems, and attention problems. The association in FXS was present after accounting for several other factors shown to predict behavior in these children, including IQ, FMRP, parental psychopathology, and the quality of the home environment, indicating a unique association between anxiety and behavior in children with FXS. Approximately 70% of young males and 60% of young females with FXS have

anxiety. In some females with FXS, the anxiety is so severe that they can become selectively mute, where they are silent in some situations, usually at school, but are able to talk in other situations, usually at home.

Research by Maria Boccia and Jane Roberts in 2000 found autonomic dysregulation and hyperarousal in FXS by comparing heart rate variability in young males with FXS compared to CA-matched normal controls. Results showed those with FXS had a faster heart rate and lower parasympathetic activity, but similar sympathetic activity compared to the normal controls. The lowered levels of parasympathetic activity indicate autonomic dysfunction and another mechanism of hyperarousal. Hyperarousal and anxiety in children with FXS can often lead to aggression and tantrums. Studies have shown as high as 42% of young males and 28% of young females with FXS have aggression.

Research evaluating the prevalence of attention deficit hyperactivity disorder (ADHD) in young males with FXS has documented 70–80% meet criteria for ADHD, but studies have also indicated that with age, hyperactivity decreases. For example, 80% of young males with FXS compared to 54% of older school-aged males with FXS had ADHD. In females with FXS, ADHD is seen less frequently with studies showing approximately 35% meeting criteria, which is not significantly different from CA and IQ-matched female controls. The young females with FXS and ADHD typically have less hyperactivity compared to their male counterparts; however, other symptoms, that is, impulsivity and short attention span, can be a significant issue.

In summary, the socioemotional presentation of children with FXS involves symptoms of ADHD and autism with behaviors, such as hand-flapping, hand-biting, poor eye contact, hyperactivity, perseveration, anxiety, and tantrums. Two factors, low FMRP expression and a high degree of autistic behavior, are linked with poorer cognitive outcomes in young children with FXS. In particular, social and communication skills are typically lower for children who have both FXS and autism.

Involvement in Premutation Carriers

As stated earlier, individuals with the premutation have between 55 and 200 CGG repeats. Although carriers were initially thought to be completely unaffected, advances in our understanding of their clinical phenotype and the molecular abnormalities which occur in the premutation have changed this opinion.

Some children with the premutation present with clinical problems including cognitive deficits, mental retardation, autism, ADHD, or emotional difficulties. For children who are in the upper premutation range, a CGG repeat greater than 130, there is often a deficit of FMRP production which lead to features of FXS, including prominent ears and hand-flapping. The majority of individuals with the premutation do not exhibit the clinical features of FXS, although ADHD, shyness, and even social avoidance, can be common. A recent study by Farzin *et al.* compared boys with the premutation who presented clinically (probands) to boys with the premutation who were identified in families after cascade DNA testing (non-probands) and to brothers who do no have the fragile X mutation. The probands demonstrated ADHD in 93%, whereas the nonprobands had ADHD in 38% and the typical controls had ADHD in 13%. Most striking was the symptoms of ASD which occurred in 71% of the probands, 8% of the nonprobands, and 0% in the typical controls. The probands were significantly different in the number of autistic features from the typical controls; but the nonprobands also demonstrated a significant increase in autistic features compared to the typical controls. Shyness and social aloofness were common in the nonprobands even when they did not meet criteria for ASD. The importance of these findings is that a clinician should not ignore the finding of the premutation allele in a child who presents with developmental delay or behavioral problems.

As previously mentioned, some children with the premutation have lowered FMRP levels which lead to their fragile X involvement. However, we also see an increase of mRNA in premutation carriers. The *FMR1* mRNA levels are increased from two to 10 times the normal. Once the CGG repeat is higher than 200 and into the full mutation range, the level of mRNA drops to zero, particularly if the full mutation is completely methylated. Some individuals with the full mutation will have a lack of methylation or they may have mosaic status, which is a mixture of the premutation and full mutation alleles in the cells. In these cases, the level of mRNA may be elevated or below normal, but it is usually detectable. The elevated mRNA also includes the expanded CGG repeat and this forms a hairpin structure which is 'sticky' and will bind to other proteins in the neuron, leading to a dysregulation of other protein function. Some of the proteins bind to the elevated mRNA, myelin basic protein, αB-crystallin, and lamin A/C.

Research has shown that in individuals who are aging with the premutation, the elevated mRNA combined with the bound protein will form inclusions in the neurons and it is more likely for the neurons to die, leading to brain atrophy. These neuropathological changes lead to the development of the fragile X-associated tremor/ataxia syndrome (FXTAS) in approximately 40% of older adult male carriers and in a subgroup of older adult female carriers. The prevalence of FXTAS in aging male carriers includes 17% in individuals in their 50s, 38% in individuals in their 60s, 47% in individuals in their 70s, and 75% in individuals in their 80s. These numbers were seen in male carriers and the prevalence

of FXTAS is significantly decreased in female carriers, most likely related to the protective effects of the second X chromosome. The effect of the elevated mRNA in aging carriers is an RNA toxicity problem, which is similar to what occurs in myotonic dystrophy. We do not yet know if the problems that are seen in some children with the premutation relate to a developmental RNA toxicity effect. This will require further studies of children with the premutation.

The premutation is also the most common single gene associated with premature ovarian failure (POF). In population studies of POF, anywhere from 6% to 14% of women, will have the premutation with the higher rates related to studies of women who have familial POF. Overall, 20% of women with the premutation will have POF, menopause before age 40 years, although a higher percentage will have some degree of ovarian dysfunction including irregular periods or elevation of their FSH. Sometimes POF can occur in the 20s and so genetic counseling is important, particularly if a woman wants to have children.

The premutation can also be associated with psychiatric problems in adults and children. Anxiety and social phobia is a frequent problem, in addition to depression. Depression is more common in the adults, particularly when they have children affected by FXS. The emotional problems related to the premutation may also be related to an RNA toxicity effect in the CNS. The level of psychiatric problems in adult carriers is significantly different from controls and correlates with the degree of mRNA levels.

Newborn Screening

The technology now exists for screening blood spots in newborns for the *FMR1* mutation. Early identification will facilitate genetic counseling, which is critical for this disorder. Once a child is identified, the entire family tree should be reviewed. If the patient has a full mutation, then the mother is the carrier. If the mother's father is the carrier then all of the mother's sisters are carriers and are at risk of having children with FXS. Often extended family members will not have been identified and cascade testing in the family will reveal others who are involved with fragile X, either with premutation or full mutation involvement. Families should be referred to a genetic counselor for further counseling. Reproductive options are reviewed, including prenatal diagnosis, *in vitro* fertilization, egg donation, and adoption. Newborn screening also facilitates early and intensive interventions which can be molded to the needs of infants with FXS.

Treatment and Intervention

Early intervention and specialized services are known to improve outcomes for children with developmental disorders such as FXS, but no study has ever measured the progress of individual children before and after receiving early intervention. In light of this, and given the current discussions about the possible initiation of newborn screening procedures for FXS, there is a need to describe the early developmental profiles, which can be utilized to better understand the benefit of early interventions using behavioral, medical, and education-based treatments.

To reach full potential, a child with FXS may need speech and language therapy, occupational therapy, and physical therapy to help with the many physical, behavioral, and cognitive impacts of the disorder. For young children with FXS, or any type of developmental disability, the Individuals with Disabilities Education Act mandates the early intervention services, special education, speech and language, occupational, and physical therapies be provided for qualified individuals. To gain access to these services, the first step is to have the child assessed for specific strengths and needs. Pediatricians routinely refer children for developmental evaluations, although a developmental evaluation can also be scheduled by calling the local health department.

For very young children with FXS, early intervention services may include family counseling, home visits to help families with nursing, parent/infant programs to encourage language, play, and sensory development. During the preschool years (ages 3–5 years), the local public school system can provide information on the services that are available. Therapies are specialized interventions that can be accessed either through the public systems or privately through the healthcare system. Health insurance coverage of therapeutic interventions varies widely. Therapies may be offered as individual sessions where the child is pulled out of the classroom or integrated into the class routine. Special education programs should be created to fit a child's individual needs to modify classes and assignments. A special educator will often use a variety of tools to meet the needs of each child, such as special equipment, strategic room arrangement, behavior management system, visual communication techniques, and shortened assignments. It is important for parents to be involved in the various steps of their child's educational plan and to implement the recommended practices at home as well.

The association between FXS and autism makes it especially important to identify the children with FXS who also meet criteria for a diagnosis of autism. Several intervention strategies have been shown to be beneficial for individuals with autism, which may also benefit children with FXS and autism. For example, the applied behavior analysis (ABA) technique involves the application of basic behavioral practices (positive reinforcement, repetition, and prompting) to reach a desired outcome. Discrete trial teaching (DTT) is a primary methodology, but not the only instructional method used in ABA programs for individuals with autism. DTT involves breaking down skills into small sub-skills and teaching each subskill, intensely, one at a time. It involves repeated practices with prompting to insure the child's success. DTT also uses reinforcement to help shape and maintain positive behaviors and skills. Also, the Denver model, created by psychologist Sally Rogers, has been a successful treatment program for children with autism. The model is a developmental approach which has two foci, one on intensive teaching and the other on developing the social-communicative skills. The model advocates that social-communicative development originates from emotional relatedness, and so, emphasizes affective connection, relationship building, and understanding communication and emotional exchange between people. Overall, intensive behavioral interventions are helpful in young children with autism and may also benefit young children with FXS and autism combined.

Medication

If the social–emotional dysfunction is detectible in the young children with FXS, there is a great potential for early psychopharmacological intervention, in addition to behavioral therapies as discussed above, to address the antecedent difficulties before well-established maladaptive social, emotional, and behavioral patterns are established.

Early use of psychopharmacological agents, such as selective serotonin reuptake inhibitors (SSRIs), which increase serotonin levels at the synapse, can be helpful for autism and anecdotally in children with FXS who are 3 years or older. Folic acid therapy has also been used in infants and toddlers with FXS, but the results are not consistent. Benefits in behavior are seen in about 50%, but controlled studies are mixed regarding outcome. The advent of mGluR5 antagonists, that is, MPEP and lithium, may have a targeted effect in children with FXS, but studies on these medications have not yet been initiated.

Conclusion

FXS can affect the entire lifespan of an individual, often beginning with motor and language delays in infancy and toddlerhood; cognitive, socioemotional, and behavioral deficits in early childhood; and continued intellectual, behavioral, and emotional problems in adults. Thus far, there has been limited research on FXS in infancy and early childhood; however, research in this younger age range is growing as methods for earlier detection advance and a greater awareness develops in medical settings, schools, and communities overall. The earlier an individual is diagnosed, the sooner the symptoms can be treated, leading to better prognosis. This article reviewed the existing research on FXS in infancy and early childhood, in addition to involvement in older children and adults and intervention strategies.

Acknowledgments

This work was supported by NICHD (grants HD36071, HD02274), NINDS (grant NS044299), a collaborative agreement with the Center for Disease Control and Prevention (grant U10/CCU925123), and the M.I.N.D. Institute at the University of California at Davis. We also thank the families who have supported our research.

See also: ADHD: Genetic Influences; Autism Spectrum Disorders; Developmental Disabilities: Cognitive; Genetic Disorders: Sex Linked; Genetic Disorders: Single Gene; Genetics and Inheritance; Intellectual Disabilities; Learning Disabilities; Obesity; Screening, Newborn and Maternal Well-being; Screening, Prenatal; Sensory Processing Disorder; SIDS; Special Education; Theory of Mind.

Suggested Readings

Bailey DB, Jr. (2004) Newborn screening for fragile X syndrome. *Mental Retardation and Developmental Disabilities Research Reviews* 10: 3–10.

Bailey DB, Jr., Hatton DD, Tassone F, Skinner M, and Taylor AK (2001) Variability in FMRP and early development in males with fragile X syndrome. *American Journal on Mental Retardation* 106: 16–27.

Bailey DB, Warren SF, Hatton DD, and Brady N (2007) *Intervention Strategies for Young Children with Fragile X Syndrome*. Baltimore, MD: Brookes Publishing Company.

Berry-Kravis E and Potanos K (2004) Psychopharmacology in fragile X syndrome – present and future. *Mental Retardation and Developmental Disabilities Research Reviews* 10: 42–48.

Cornish KM, Sudhalter V, and Turk J (2004) Attention and language in fragile X. *Mental Retardation and Developmental Disabilities Research Reviews* 10: 11–16.

Hagerman RJ and Hagerman PJ (2002) *Fragile X Syndrome: Diagnosis, Treatment, and Research,* 3rd edn. Baltimore, MD: The Johns Hopkins University Press.

Mirrett PL, Bailey DB, Jr., Roberts JE, and Hatton DD (2004) Developmental screening and detection of developmental delays in infants and toddlers with fragile X syndrome. *Journal of Developmental and Behavioral Pediatrics* 25: 21–27.

Rogers SJ, Wehner EA, and Hagerman RJ (2001) The behavioral phenotype in fragile X: Symptoms of autism in very young children with fragile X syndrome, idiopathic autism, and other developmental disorders. *Journal of Developmental and Behavioral Pediatrics* 22: 409–417.

Scerif G, Cornish K, Wilding J, Driver J, and Karmiloff-Smith A (2004) Visual search in typically developing toddlers and toddlers with fragile X or Williams syndrome. *Developmental Science* 7: 116–130.

Relevant Websites

http://www.fpg.unc.edu – FPG Child Development Institute, The University of North Carolina at Chapel Hill.

http://www.fraxa.org – FRAXA Fragile X Research Foundation.

http://www.cdc.gov/ncbddd – National Center on Birth Defects and Developmental Disabilities, Centers for Disease Control and Prevention.

http://www.fragilex.org – The National Fragile X Foundation.

Friends and Peers

C Howes, University of California, Los Angeles, Los Angeles, CA, USA

Glossary

Cultural community – A group of people who participate in a shared set of practices and traditions. Children develop peer relations and peer friendships within particular cultural communities. With time, peer groups become cultural communities with their own practices and traditions.

Friendship – A friendship refers to a reciprocated relationship between two children. Friendships between young children can begin very early, before the first birthday, and include mutual preference and mutual affection. Friendships between young children provide a context for developing complex social interaction and for emotional support. Not all children in a peer group develop friendships with each other, although within peer groups of young children most children have at least one reciprocated friendship. Teachers and parents sometimes call all the children in the group 'friends', for example, "We are all friends in this classroom." This may confuse children and delay the development of reciprocal friendships.

Peer – Children are considered peers when they are similar in age. Peer interaction and peer friendships occur when children engage with and form relationships with age-mates. A close-in-age sibling or cousin can be considered a peer. When children are very young (infants and toddlers), children are considered peers if their age separation is 6 months or less. If children are preschoolers (3–4 years old), peers can have an age separation of 1 year. This is because social skills for interacting with peers develop more rapidly in earlier developmental periods.

Peer popularity and acceptance – Young preschool-age children who have experience in peer groups can reliably assess peer popularity and acceptance. They are able to discuss in a sociometric interview which of their peer group members they consider 'fun to play with' and which are 'not fun to play with'. Children rated by their peers as good playmates have more positive social skills.

Pretend – In 'pretend' children make believe that something that is not there is present symbolically. For example, a lion enters the play of two children. "Watch out, the lion is coming into the tent!" The lion can be totally imaginary or a stuffed toy imagined to be real.

Shared pretend play – Pretend play with a peer partner requires that the child manipulate symbolic transformations, imagine what is not there, and communicate the meaning of the imagined object or event to a partner.

Social interaction – Social interaction occurs when children direct social behaviors including verbal behaviors such as talking and nonverbal behaviors such as smiling, offering a toy to a peer while attending visually to the peer, and the peer responds to the child who directs the social behavior with a verbal or nonverbal behavior while attending visually to the peer.

Social status groups – Sociometric nominations or ratings result in social status groups (popular, rejected, neglected, and average). Popular children have high liking and low disliking profiles. Rejected children have low liking and high disliking profiles. Neglected children have low liking and low disliking profiles. Average children fit none of the other profiles.

Sociometric nominations and ratings – Sociometric nominations are a procedure that involves asking each child in a defined group to identify the children who he or she likes and dislikes as one means to determine peer acceptance. Children with reciprocated nominations of liking are usually considered friends. Children also can be assigned a sociometric rating based on the average of ratings provided by each of the children in the group.

Unilateral friendships – Friendships that are 'wished for' – not true friendships. Unilateral friendships occur when one child prefers another but the preference is not reciprocated. These friendships do not have the developmental advantages of reciprocated friendships.

Introduction

Very young children develop peer interactions and friendships very early, soon after their first birthdays. Children barely able to toddle play peek-a-boo, run-and-chase, or simple pretend games like "I am pulling you in my wagon. We are going to the store." By ages 3 or 4 years, play becomes sophisticated with elaborate scripts and costumes. Children can say and act out: "We are going camping, and pretend that there is a bear," and "You are the baby in your pajamas." "And I climb out of the tent and try to give the bear a cookie."

It is only within groups of peers that children develop both social interaction skills particular to peer interaction and construct social relationships particular to peers – friendships. The social interactions and relationships of the children within the group become the basis of that peer group's shared understandings and practices and a base for children's development individually. Through experiences within many different (and at times overlapping in membership) peer groups, children internalize representations of social relationships and of practices within peer groups that can influence their individual orientations to the social world as older children, adolescents, and adults.

This theoretical approach to examining peer social interactions and relationships is best described as relationship development nested within sociocultural contexts or cultural communities. In this framework a cultural community, as defined by Barbara Rogoff, is a group of people who participate in a shared set of practices and traditions. Peer groups are a kind of cultural community. As cultural communities peer groups construct shared understanding and meanings in forms that include shared scripts for pretend play, games, and conversations; knowledge of who hangs out with whom; who can and cannot be trusted to gossip without hurting other people's feelings; and, generally, ways to behave within the group. These shared understandings and meanings are the practices of the peer community. Peer communities have shared histories as well as practices. Children within a peer group that lasts over time can remind each other of events that have meaning only within the context of the peer group for example, "Remember when Sylvie was so mean to Nancy, and then they stopped being friends?"

An alternative, important theoretical approach to the study of friends and peers in young children is a social cognitive approach. This will not be covered in this article.

Much of the research on peer relations has been conducted as if all peer cultural communities were universally similar and thus developmental patterns were identical as well, regardless of the characteristics of peer groups. However, placing the development of peer interactions and friendships within cultural communities helps to resolve some of the contradictions that have persisted within the empirical literature on the topic. Take, for example, a basic question of whether infants and toddlers engage in peer interaction. As we will discuss below, research since the 1970s has documented that if 1-year-olds have regular playmates they engage in relatively sophisticated patterns of interaction. However, if such young children are observed with unfamiliar peers, they are fairly unskilled at interaction. Many parents and teachers believe that infants and toddlers receive no benefit from peer interaction. Thus, young children do not have regular playmates unless their families engage in practices that create baby peer groups such as enrolling the baby in childcare, getting together regularly with friends who have same-age children, regularly attending a play group, or having regular extended family gatherings with age-matched children. Therefore, in this example, the age of development of a particular social skill depends on whether a child belongs to a same-age peer group that meets regularly, and this, in turn, depends on parent practices and values.

As another example, consider that children are in peer groups as diverse as peer groups informally formed by living in a neighborhood or being part of a group of families who holidays together, or a formal one such as all the children in a childcare arrangement, or a church school. The practices or ways to do things within each of these peer groups are particular to the peer group. For example, cross-gender or cross-age friendships may flourish in informal neighborhood peer groups and be actively discouraged within classroom-based peer groups. Peer groups in some settings may facilitate cross-ethnic, cross-religious, or cross-class friendships forbidden within others. Since children simultaneously hold membership in overlapping peer groups they may have social practices that are competent in one peer group and incompetent in another. For example, ways to do things with peers in one group, such as children using their spoons to bang on the table, catching each others' eyes and laughing, may lead to being removed from the peer group in another group (such as banging on the table while smiling and laughing with a peer results in being put in 'time-out'.) These contradictions in practices for children who move between peer groups may actually be tied to their adaptive competence, a keen understanding, and awareness of the significance of social context.

Learning to Interact: Infants and Toddlers

So how do children become a group? How do they figure out how to engage with each other? How do they form friendships with some children and not others? When children only entered same-age peer groups as preschoolers, the answer was that children first learned social skills from their parents or perhaps from older siblings and cousins, and then applied them to peers at some later time. When children enter same-age peer groups as infants the answer is not as clear. From the 1920s and 1930s until the early 1970s, the study of peer relations in the US was dominated by early descriptive research about the ways in which individuals' social skills developed. Pioneer researchers in the 1930s based their conclusions on the cultural communities that were available to them to study. Because the earliest researchers studied naturally occurring peer groups of infants and toddlers (milk distribution centers in New York City's Central Park and orphanages), they concluded that infants and toddlers engage in games and other early forms of peer interaction. Mildred Partens, a researcher from the 1930s frequently cited in text books, studied children in nursery schools willing to enroll only those children who were 2 years 9 months of age and toilet-trained. She argued that all children progressed through a set of social participation categories from solitary play, through parallel play, to true social participation in the form of cooperative play. The 2-(almost 3)-year-olds were the solitary players, the 3-year-olds the parallel players, and the older children the truly social children.

In retrospect, the development captured in Partens' theory was closely linked to the context in which the observations typically took place: nursery schools. Nursery schools were part of a social movement that was based in a particular belief system. Within this belief system, it was assumed that the ideal child-rearing environment for preschoolers consisted of a 'stay-at-home mother' and, two or three mornings a week, a socialization experience with children of similar backgrounds. Children younger than preschool were not expected to play with peers or to develop friendships.

The first and most influential nursery schools were physically located in colleges and universities and served as laboratories for the study of child development and training professional teachers. Because research in peer relations and teacher training was connected with a particular social organization for 50 years, early childhood teachers in training and parents seeking advice were told that 2- and 3-year-olds were solitary or parallel players, while 5-year-olds were cooperative players.

Current research and theory on peer interaction supports an alternative perspective. The forms of play identified by Partens are not an invariant sequence based on age but, instead, are categories of play which tend to be used by children of all ages as the play they engage in with peers gradually shows more signs of structure, cognitive ability, and communicative ability. When research begins with a different set of assumptions and takes place in a different context, children as young as 10–12 months of age can be seen engaging at least one peer in cooperative play, which is more social than expected, based on early theories.

This alternative perspective was established because of two sociocultural changes: (1) research on peer relations moved out of laboratory schools, and (2) a social movement changed our beliefs about appropriate child-rearing environments. In the 1970s, the women's movement and changes in the structuring of the US economy led to an influx of mothers of very young children into the workforce. Part-day nursery schools could not accommodate the childcare needs of these women. There was a dramatic increase in full-day childcare centers that served infants and toddlers, instead of just preschoolers. These social changes changed the social context for constructing peer interactions, peer relationships, and peer social networks. In the social context of childcare, peer interactions and relationships developed within long periods of 'everyday life events' rather than relatively brief 'socialization experiences'. Children in childcare centers were with their peers every day and stayed there from breakfast time, through nap, until the end of the day. The nursery school experience might be compared to a date, while the childcare experience is more like living together.

Beyond structural changes in the social context of peer encounters, the women's movement changed the context in which the researchers were formulating their questions. They began to ask questions about the function of peers and peer relationships. Rather than simply being part of enrichment experiences, potentially peers could function to provide the child with experiences of social support, trust, and intimacy in the absence of the child's mother. Children who grew up together sharing the common resources of the childcare center might engage in close rather than conflictual interaction. Cross-sex or cross-ethnic peers who became friends in a different environment than a traditional nuclear family might form different kinds of relationships. In the context of these questions, research on developmental changes in the complexity of peer interaction structure began.

Casual observations of infants and toddlers within full-day childcare settings suggested that they were not behaving according to Partens' observations. Their play was more interesting and complex than that of an onlooker or parallel player. Faced with this discrepancy, beginning in the 1970s, researchers began to examine how infants and toddlers constructed their play encounters within peer groups in childcare centers. In 1972, Blurton-Jones published an influential collection of observational studies of peer behaviors using ethological methods, with careful attention to the description and functional meaning of behaviors. Ethology had its roots in biology and the study of animal behavior. It provided a method for close, detailed observations of behavior. Subsequently, a generation of researchers

including Eckerman, Davis, and Didow; Hay; Howes; Ross and Goldman; and Vandell applied ethological methods to the task of identifying and describing changing structures underlying peer interaction.

One descriptive system that emerged from this research was the Howes Peer Play Scale. Because each of the points of the scale represent increases in the complexity of play, it frames the discussion of how young children participate in constructing social structures within peer groups. There are several assumptions underlying the Peer Play Scale. One assumption is that a necessary condition for children to be considered friends is that adult observers can infer from their behavior that each child understood the other to be a social actor and that social actions between partners could be coordinated and communicated. Therefore, as a starting place, research had to establish that children behaved as if they had these understandings. A second assumption is that later development in social play could occur only as the child increasingly understood the role of the other, incorporated symbolic play, and communicated shared meaning. These assumptions are the bases for the behaviors that are captured in the Peer Play Scale. Initially, children are expected to show signs of each of these three components but, eventually, they are expected to use them fluidly and communicate negotiations with each other about their play.

There are two key aspects of early play that presuppose such social understanding: mutual social awareness and coordination of action. Together, these two markers represent the necessary components for what Howes called complementary and reciprocal play. Specifically, each play partner's actions reverse the actions of the other. A child chases his or her partner, then is chased. One child peeks at his or her partner, the partner says boo, and then peeks back. Research in cognitive and communicative development suggests that the representational underpinnings of these understandings are present in children as young as the toddler developmental period. Naturalistic observations established that toddler-age children constructing their peer interactions within full-time childcare centers were indeed engaging in complementary and reciprocal play.

Constructing Shared Meanings: Toddlers and Preschoolers

The next developmental step is to incorporate symbols into shared play. Children in peer-group settings first begin to use symbols or to play 'pretend' alone or with a competent adult player, but symbolic play soon enters the realm of peer play. Pretend play with a partner requires both that the child manipulate symbolic transformations and communicate the resulting symbolic meaning to a partner. The simplest form of social pretend play, called 'cooperative social pretend play' in Peer Play Scale terminology, occurs among toddlers in full-time childcare centers. Note that once more a developmental event is placed within a particular sociocultural context. Central to this level of structure of interaction is that children enact nonliteral role exchanges. Play partners integrate their 'pretend' actions by using a familiar 'pretend' theme or script such as a tea party. Similar to complementary and reciprocal play, cooperative social pretend play requires that children reverse the actions of the other but, in this form of play, the actions are nonliteral or symbolic. The actions of the children presuppose that each partner understands that each player may engage in the symbolic behaviors. The children are able to share understanding about the symbolic meaning of their play, but this is communicated through the implicit script of the play rather than explicit talk about the play. For example, when a toddler offers a cup to a partner who is holding a pitcher, the child is engaged in a very simple form of social pretend play compared to the preschool-age child who discusses the play script, sets the table, brings festively dressed toy bears to the table, and gives the bears a tea party. Nonetheless, toddler-age children are beginning to understand the role of the partner in constructing social sequences.

Despite the new skills incorporated into cooperative social pretend play, it remains a pale imitation of the well-developed fantasy play of older children. Toddlers have only just begun to transform symbols so their transformations are not fluid and may be only partially developed. By preschool age, children's symbolic, linguistic, and communicative development permits meta-communication about social pretend play. Children can plan and negotiate the sequences of symbolic actions with fluidity, modify the script as it progresses, and step out of the 'pretend' frame to correct the actions or script. These behaviors, such as those seen in a tea party for the toy bears, indicate the most structurally complex play captured in the Peer Play Scale. This play form is labeled 'complex social pretend play'.

The Peer Play Scale is based on a set of measurement assumptions: the play forms develop in the predicted sequence and children develop particular play forms before or during particular age intervals predicted by theories of cognitive and communicative development. Howes' longitudinal and cross-sectional studies which focused on validating the Peer Play Scale supported these two assumptions. Seventy-four per cent of the children followed the predicted sequence for emergence of play forms: complementary and reciprocal play, cooperative social pretend play, and complex social pretend play. Children who showed earlier emergence of complementary and reciprocal play also showed earlier emergence of cooperative and complex social pretend play.

As a measure of structural complexity, the Peer Play Scale makes no distinctions among positive, aggressive, or agonistic (instrumental aggression such as toy-taking) social bids. A structurally complex interaction could reflect prosocial behavior, a conflict, or any number of social styles. As is discussed further, research examining associations between the structure and content of peer interaction appeared later in the 1980s and 1990s.

Perhaps because it captures structural complexity rather than the content of peer play, the Peer Play Scale has been successfully used in cultures other than the US. Jo Ann Farver has used it to describe and examine peer interaction in such diverse cultures and ethnic groups as Mexican children, Latino and African-American Head Start children in Los Angeles, Indonesian children, and Korean-American children. Across studies, children's play was represented at each structural play level and play forms emerged at similar ages. When differences emerged, they were in the frequency of play forms rather than in different play forms or a different sequence of play. Farver suggested that the sociocultural context influences the style or frequencies of peer play. In particular the types of themes in pretend play appear rooted within children's particular cultural communities. Whether children play at wrapping the 'babies' in shawls and placing them on their backs or enacting the latest television superhero's antics is dependent on the practices of daily life within their cultural community.

Gender also appears to influence the style rather than the structure of peer play. There are well-established differences in the content of the play of boys and girls. However, consistent with the lack of cultural and ethnic differences in the structure of peer play, there appear to be few differences in the complexity of the structure of children's peer play. Girls and boys of the same age engage in structurally similar play when the content differs. For example, both a game of 'mother, sister, and baby' among girls and a game of 'the day the tigers ate the village' among boys are very likely to be rated as complex social pretend play.

However, consistent with a Vygotskian perspective, the complexity of the structure of play is influenced by the skill level of the play partner. If the sociocultural context of the peer group includes mixed-age children such as in family childcare homes, toddlers play more skillfully with their peers when their partner is somewhat older and presumably more skilled at play. In contrast, when the mother is present as in parent cooperative preschools the presence of the child's mother appears to reduce the complexity of peer interaction. Similarly, when adults engage with children in preschool settings, peer interaction is inhibited. Although adults are more skilled play partners than children, peer play in the presence of an adult may be less skillful because three-partner interaction is more difficult than dyadic interaction or because adult–child play content is different from child–child play content.

Forming Social Structures within Peer Groups: I'll Be Your Friend If ...

Young children's peer interaction is organized around networks of play partners. Within any peer group, there are children who prefer to play with each other, and children who prefer not to play with each other. Some children have no problem finding and keeping play partners, while others have difficulty entering peer groups and sustaining play. Children as young as 3 years old who have experienced full-time childcare in stable peer groups can describe these patterns of peer acceptance within their classroom. When shown pictures of all children in the classroom, they can identify by name each child in the classroom and reliably rate how much they would want that child as a friend. This picture of sociometric procedure provides a description of social status.

In the late 1970s, a series of longitudinal studies were published suggesting that children's peer acceptance in middle childhood was associated with positive mental health outcomes in adulthood. These studies precipitated a period of intense research into children's sociometric status and continues as an important area of research. The term 'sociometric' refers to ways of measuring peer acceptance and friendships. Sociometric nominations involve asking each child in a defined group to identify the children who he or she likes and the children who he or she dislikes as a means to determine peer acceptance. Children with reciprocated nominations of liking are usually considered friends. Children also can be assigned a sociometric rating based on the average of ratings provided by each of the children in the group. Social status groups (popular, rejected, neglected, and average) are formed using either nominations or ratings. Popular children have high liking and low disliking profiles. Rejected children have low liking and high disliking profiles. Neglected children have low liking and low disliking profiles. Average children fit none of the other profiles. Peer acceptance refers to high ratings or many positive and few negative nominations.

Landmark studies of sociometric status by Asher and others found that children who were classified as popular or more socially accepted were more friendly and prosocial, and less likely than children who were not accepted to engage in aggressive behaviors with peers. Furthermore, sociometric status tended to be stable over time. Although the original research was conducted in elementary school classrooms, similar findings have emerged in preschool settings. These findings based on sociometric measures raised two questions about the Peer Play Scale. First, if the Peer Play Scale represents socially competent behavior with peers, would children who engage in structurally complex play also have higher peer acceptance and engage in prosocial, nonaggressive interactions with peers? Second, is competent peer play stable over developmental periods? That is, does demonstrating marker behaviors of structurally complex peer interaction during an earlier developmental period predict engaging in marker behaviors of structurally complex peer interaction during a later developmental period?

Longitudinal studies were needed to answer these questions. Several conducted by Howes and colleagues found that within developmental periods, independent observers rated children who engaged in more complex play as more prosocial, sociable, and less aggressive, and children who engaged in more complex play at earlier developmental periods were observed and rated as more prosocial and sociable, and less aggressive and withdrawn, during subsequent periods.

Longitudinal studies in general suggest that social competency in peer relations may be quite stable from early childhood through adolescence. Some of this stability may occur because once children develop early social skills with peers they continue to use them and become the type of older children who are comfortable and easy interacting with peers. In contrast, the children who are shy or aggressive in early peer interactions have fewer opportunities to learn how to be comfortable and easy with peers as they become more isolated from positive peer interactions. This work supports the idea that peer groups become a form of cultural community with its own particular practices and traditions. When the peers in their peer group accept children and they indulge in prosocial behavior, it can be a wonderful context for individual development. The complement of this statement may also be true if the social status structure of a peer group comes to be part of its practices and traditions; children who are not well accepted have a difficult time.

The natural variation in familiarity and stability of peer partners provides an opportunity to examine sociocultural variations in children's experiences in peer groups. Several longitudinal studies suggest that it appears to be advantageous for children to enter peer groups as younger children and maintain a familiar and stable peer network. Some children who moved between childcare centers actually moved with a peer group, as opposed to moving to a childcare center with an entirely new peer group. Those children who were able to move with a peer group, and who therefore did not have to reestablish relationships, were more competent in peer interaction.

Of course some peer groups might promote more social competence than others. In general, children are more socially competent within a peer group if the peer group is within a childcare program, center, or family childcare that is more supportive of positive relationships. These findings point to the importance of adults structuring a childcare environment which supports the construction of positive and skillful peer interaction. In a study by Howes and colleagues in the 2000s, the social–emotional climate of the childcare environment as well as children's individual social competence and teacher–child relationships predicted social competence with peers 5 years after children were in childcare.

A report of the National Institute of Child Health and Human Development (NICHD) Early Child Care Research Study highlights these concerns. The NICHD study reported that at ages 2 and 3 years, children who had experienced positive and responsive caregivers and the opportunity to engage with other children in childcare were observed to be more positive and skillful in their peer play in childcare but not in play with a friend in a laboratory. When the children were 4.5 years old and in kindergarten, teachers, caregivers, and mothers rated children with more time in childcare as having more problematic relationships and behavior with others. While overall measures of childcare quality did not mediate this relationship between time in childcare and problematic behavior, this study did not include specific measures of adult structuring of peer relationships. As Maccoby's comment on this study suggests, perhaps it was not the amount of childcare but the amount of time in a kind of childcare that fosters individualistic rather than cooperative interactions with others.

How do adults structure a childcare environment that promotes positive peer relations? Not, it turns out, by coaching children in social skills. Adults working with naturally formed groups of children rarely engaged in such behaviors unless in an intervention program. A more promising line of research examined relations between children's attachment-relationship quality with the primary caregivers in the childcare setting and the development of children's social competence with peers. Adult caregiving behaviors directly influence children's attachment relationships that, in turn, influence peer relations. More specifically, more securely attached children have caregivers who are rated as more sensitive and responsive to the children in their care. In turn, children who are more securely attached to their childcare providers are more socially competent with their peers. When children's attachment relationships with their childcare provider are contrasted with their attachment relationships with their mothers, the childcare attachment relationship is more powerful in predicting social competence with peers. This may be because these adults are physically present with the children's earliest experiences within peer groups. The children can use the childcare provider as a secure base as they explore interactions and relationships within the peer group.

Childcare providers are not a stable presence in children's lives. In contrast, mothers usually remain constant. Changing childcare providers appears to have implications for children's social competence with peers. Children who had the most changes in primary childcare providers tend to be most aggressive with peers. Some of the caregiver changes can have a positive effect in the short term. If the child is able to construct a more secure relationship with the new provider than with the old, he or she may be able to, simultaneously, became more skilled with peers. However, it appears that the cumulative effect of instability in childcare caregivers is detrimental to the development of positive social competence with peers.

Broadening the study of peer relations to include a more careful examination of child–caregiver relationships has led to a particularly fruitful series of research studies. The evidence that children construct interactions and relationships with their peers from infancy onward highlights the importance of these early peer relationships. Likewise, the relationships that children form with the adults who supervise their encounters with peers influence social competence. It is noteworthy that both these types of relationships, with peers as friends and with childcare providers, are relationships outside of the family. While these results should in no way downplay the importance of family influences on child development, carefully examining alternative relationships underscores the value of social networks both inside and outside of the family.

Friendship as Affective Relationships

All dyadic relationships within the peer group are not interchangeable. Even the earliest studies of the construction of peer interaction among infants noted that babies seemed to form early preferences. Furthermore, sociometric inquiry rests on the premise of differential preferences within the peer group. But are early friendships affective relationships or merely preferences? Friendships are relationships based on mutual support, affection, and companionship. School-age children can articulate these qualities of friendship and tell an adult whether a friendship does or does not have these qualities. Infants and toddlers and even preschoolers do not have the verbal and cognitive capacities to articulate friendship characteristics. They simply say "she is my friend." So are infants and toddlers able to form real friendship relationships based on mutual support, affection, and companionship or are all young friends, including preschool friends, merely momentary playmates?

Again, both theoretical and sociocultural influences are important in shaping this emphasis on relationships. As intense research on social acceptance with peers and its correlates grew, so did a renewed interest in relationships as a context for development. In 1976, Hinde, an ethologist, introduced the theoretical notion that relationships are more than a bi-directional effect of the influences of both partners. According to Hinde, a relationship is a new formation understood by examining not only the behaviors of both partners and their contingencies, but the pattern of relating uniquely to the two individuals.

Furthermore, the processes of relationship formation embedded in Bowlby's attachment theory began to be understood to be also applicable to relationships other than the child–mother attachment. According to this reinterpretation of attachment theory, relationships (whether attachment or playmate relationships) develop through multiple and recursive interactive experiences. Recursive interactions are well-scripted social exchanges that are repeated many times with only slight variations. Examples include infant–caregiver interaction around bedtime or repeated toddler-age peer run-and-chase games. From these experiences, the infant or young child internalizes a set of fundamental social expectations about the behavioral dispositions of the partner. These expectations form the basis for an internal working model of relationships. Therefore, through repeated experiences of social and social pretend play with a particular peer, a child forms an internal representation of playmate relationships.

Thus, some playmate relationships become friendship relationships. It is important to note that both the structure and content of experiences interacting with a partner are part of the child's representation of the partner. Children who engage in more complex interactions are more likely to recognize the partner as a social other and construct a relationship. Furthermore, the content of the interaction is likely to influence the quality of the resulting relationship.

Sociocultural influences also played a role in the research interest in friendships between very young children. When infants and toddlers had daily experiences within the stable peer groups of childcare, parents and childcare teachers began to notice that the children had preferences among their peers. These preferential playmate relationships appeared to parents and teachers as friendships. Teachers and parents reported that these emerging friendships helped children separate from their parents at the beginning of the childcare day. The children would greet the preferred peer and join in play with him or her as part of the ritual for parents saying goodbye. The following sections discuss the process of friendship formation in very young children and the functions of these early friendships.

The Process of Friendship Formation When the Children Cannot (Yet) Talk About Friendships

An 'affective relationship' is one that includes feelings of affection or what would be called 'love' in adult–child relationships. Toddler-affective relationships have attributes of friendship common to the 'best friendships' which provide older children with emotional security and closeness. These early friendship relationships appear to be formed in a way similar to adult–child attachment relationships.

In a similar manner to the research on structural complexity of peer interaction, the friendship studies began with the collection of observational data. Since toddler-age children cannot report on their friendships, behaviors distinguish friendship relationships from playmate relationships in prelinguistic children. This results in some discontinuity in research about friendships because later research relies heavily on a child's ability to talk about friendships. For example, reciprocity of friendship is an important dimension of later friendship research, but cannot be explored in early childhood. Because early friendships can be defined only based on observed behavior, the earliest identified friendships must be reciprocal, with both partners engaging in defining behaviors.

Howes uses a three-part criterion for 'friendship': preference, recognition of the other as a social partner, and enjoyment. Within this definition, children must prefer the company of the friend over the company of others and must enjoy the time they spend together to be considered friends. To operationalize these constructs for observational research, the following criteria were developed: (1) proximity (being within 3 feet of each other at least 30% of the observational period), (2) at least one instance of complementary and reciprocal play to indicate recognition of the other as a social partner, and (3) shared positive affect (both children expressing positive affect while engaged in interactive social play). Shared positive affect to define friends is the stringent part of the definition and the hardest to observe. This supports the premise that early friendships are affective relationships based on mutual affection. Subsequent work found that these early friendships tended to be sustained if the children continued to be part of the same peer group.

Of particular interest, given the prevalence of same-gender friendships in preschool and older children, friendships formed in the early toddler period are as likely to be same-gender as cross-gender in composition. Further, cross- and same-gender friendships are equally likely to be maintained over time if formed in the toddler period. This is in sharp contrast to the friendships of preschool and school-age children who tend to form and maintain same-gender friendships.

Children in the late toddler period (24–36 months) have more friendship relationships than younger children, perhaps because they begin to differentiate between the different functions of friendship. Some peers become friends, while others remain playmates. To be considered a playmate, but not a friend, a pair meets the criteria for preference and the recognition of the partner portions of the friendship definition but not the affective-sharing component. In the late toddler period, friendships must still be identified by behaviors as children under 3 years of age cannot reliably complete a sociometric task. Again, the affective component of behaviorally defined friendship is critical in distinguishing friends from playmates.

By preschool age, children can play with children who they do not consider to be friends. Toddler-age children's social interactions are more fragile and more dependent on rituals and routines than the social interactions of preschoolers. Therefore, toddlers, more often than preschoolers, play only with their friends. We assume that this is because patterns of interaction between toddlers are highly ritualized.

Although behavior identification of friends is still a reliable measure in preschool, children can also reliably identify friends using sociometric ratings and nominations. Beginning at age 3 years and if they have a stable peer group, children select the same children in sociometric procedures as do observers. This ability to communicate friendship status to another is a new skill. Perhaps this skill develops because preschoolers are now able to communicate the meaning of the construct of friendship. They use the language of friendship to control access to play ("I'll be your friend if you let me play").

While preschool-age children are no longer dependent on rituals to sustain play, their play is still easily disrupted. For example, a pair of children may spend 10 min establishing the roles and scripts for a pretend play episode: "You be the lion and I'll be the little boy who finds you in the forest and then . . ." "No, I want to be a baby . . ." "OK, how about you be a baby lion and I'll be the little boy who finds you in the forest and . . ." "OK, and then when you take me home, you feed me with a bottle . . ." If a third child attempts to join, the play negotiations may have to start all over again and may not be successful. If the child attempting to enter the group is a friend, the other children appear more willing to undergo the negotiation process. Children who are rejected by peers (using sociometric measures) but who have reciprocated friendships are more likely to be able to enter play groups because of having friends within the group. Continuing the lion example, a child who is a friend might enter the play by saying, "Remember the time, I was the baby lion and then I got to be a great big lion and I roared but I didn't really hurt you. I'm going to be the daddy lion."

Recall, as discussed previously, just as there is stability in children's social interaction skills with peers (children who engage in complex play with peers in early developmental periods are the same children who are competent with peers in middle childhood), there is stability in children's friendship quality over developmental periods. This suggests an interrelation between these two components of social competence – social interaction skills and friendship.

The Functions of Friendships Between Very Young Children

Can peers provide other child experiences of social support, trust, and intimacy? Do children who grew up together sharing the common resources of the childcare center have a different kind of social interaction than acquaintances? Do cross-sex peers and cross-ethnic peers who became friends in the context of childcare form nontraditional relationships? Each of these questions describes a potential function of friendship: experiences of social support, trust, and intimacy; a context for mastering social interaction; and a context for engaging with children who are unlike the self. The first of these functions has received the most research attention; research on the third function is just emerging.

Friendships Provide Experiences of Social Support, Trust, and Intimacy

Children who are good friends as older children or adolescents derive feelings of social support, trust, and intimacy from these relationships. It is difficult to apply these constructs of social support, trust, and intimacy directly to the friendships of very young children. There are, however, several pieces of evidence that support the idea that children who form friendships as preverbal children in childcare do experience social support, trust, and intimacy within these relationships. The children who were used for the early case studies of friendship in the 1970s are now adults. Informal conversations with these children suggest that their toddler friend, although no longer a 'best friend' remains a person of importance in their lives. As discussed above, toddler friend pairs tend to remain stable friends. This suggests that toddler friendships function to provide affective support, rather than functioning merely as a context for play, when the child's life history allows for continuity of those friendships. Another support for the premise that toddler friendships have an affective component is that toddler-age children are more likely to respond to another child's crying if that child is a friend.

Children who sustain friendships from the toddler to the preschool period are able to use the context of play to explore issues of trust and intimacy. When pretend play was rated for self-disclosures, they were higher in a long-term friend group than in either a short-term friend or a nonfriend group of dyads.

Childcare Friendships as a Way to Learn How to Engage with Peers

In general, the development of social interaction skills has been treated as semi-independent of the development of friendship skills. However, more socially skilled children tend to have friends and children who have friends tend to be more socially skilled. In particular, children who engage in more complex play have less difficulty than less skilled children in entering play groups. One reason that these socially skilled children can easily enter play groups is that they are likely to be friends with children within the play group.

Friendships appear to be a particularly important context for the construction of complex peer interactions during early developmental periods. Infants and toddlers make the greatest increases in complexity of social play when they engage with stable friends, as opposed to acquaintances or playmates. Social pretend play, which involves the communication of symbolic meaning, also appears first within friendship dyads and then within playmate dyads. Likewise, preschoolers who had been sustained friends are better able to use communicative behaviors to extend and clarify pretend play than preschoolers who are more recent friends. Children who are friends do not have to simultaneously devise the game structure and integrate or communicate pretend meanings. Instead, they integrate new pretend meanings into well-developed and routine-like games. In a large landmark study within Head Start classrooms, Brian Vaughn and colleagues found similar relations between friendship formation and social competence.

The mastery of social skills within friendships is not limited to typical children. Friendships appear to facilitate conflict resolution and conflict avoidance in children enrolled in an intervention program for emotionally disturbed children. Toddler-age friends were more likely than acquaintances to avoid conflict. Similarly, preschool friends were less likely than acquaintances to misinterpret prosocial bids and more likely to avoid conflict by decreasing their agonistic bids.

Friendships as a Context to Engage with Children Who Are Unlike the Self

To the extent that children of different genders and ethnic backgrounds have different social styles, friendships appear to give children access to these diverse social styles. As discussed above, toddler-age children do not select their friends on the basis of gender. Instead, toddlers form and maintain cross-gender friendships into preschool. Likewise, in a study of young children in an ethnically diverse school, Howes found that young children were better able than older children to form and maintain cross-ethnic friendships. As we discussed in the introduction whether peer groups are diverse or homeogeneous depends on the cultural context of the peer group. As childcare institutions are segregated by income of parents the resulting peer groups do not cross social class and in many instances race lines, and thus children lose opportunities to form friendships with children unlike themselves. As pointed out in a review by Maccoby and Lewis, the cooperative vs. competitive tone set in the childcare arrangement may facilitate or inhibit social interaction and friendships among similar or dissimilar children.

Future Directions

The study of peer relations, what we have selected to study and, to some extent, what we have concluded as we studied children forming relations with peers, has been influenced by the sociocultural influences of the historical period of the research as well as the theoretical lens through which we view peers. We suspect that these joint forces will continue to influence the study of the processes by which children construct and maintain their relationships with peers. One emerging sociocultural influence that appears to be increasingly influencing research on peer relations is changing demographics in both urban and increasingly rural areas of the US. There is a large influx of families from societies that generally are considered more collectivist in their values than traditional families in the US. Within collectivist societies, there is a greater emphasis placed on the individual within the group than on the individual self. These collectivist values are not dissimilar to the values of the visionaries that opened childcare centers to infants and toddlers in the early 1970s. Those involved in the intersection of the women's movement and childcare in the 1970s also dreamed of a society that valued the collective and helped children to be prosocial, altruistic members of a group.

These values on creating a group based on cooperation are somewhat different than the premise expressed in traditional early childhood education that the development of the individual child is paramount. It is also different than another prevailing force that emphasizes academics rather than social relationships in preschool. As more children and families from subcultures based on collectivist ideas enter childcare settings, there may be renewed tension between constructing groups of children who support and help one another while simultaneously helping each child within the group to reach their individual potential. This tension emerges in the Maccoby and Lewis review and may influence the next set of studies of peer relations.

Attachment theory, once expanded beyond the study of early parent–child relationships, is also likely to continue be a powerful influence on the study of peer relationships. This work has suggested that attachments with alternative caregivers are influential to peer relations. Furthermore, descriptive studies have established that relationships between young children are stable affective bonds. These findings may lead researchers to move beyond the description of friendships toward the study of internal representations of friendships. The question of what internal representations are derived from early peer affective relationships and how these representations shape children's working models of relationships is far from answered. In this context, it is important that the earliest friendships appear to be based on some 'chemistry' that leads toddlers to prefer each other rather than on matches between children of similar gender and ethnicity. As classrooms become filled with children who come from very different cultural communities, the study of peer relations will need to address the question of whether providing children with opportunities to form important relationships with persons unlike themselves at young ages will predict respectful relationships with others unlike themselves as older children and adults.

See also: Attachment; Child and Day Care, Effects of; Gender: Awareness, Identity, and Stereotyping; Play; Preschool and Nursery School; Social Interaction; Socialization in Infancy and Childhood; Vygotsky's Sociocultural Theory.

Suggested Readings

Asher S, Singleton L, Tinsley B, and Hymel S (1979) A reliable sociometric measure for preschool children. *Developmental Psychology* 15: 443–444.

Blurton-Jones N (1972) Categories of child–child interaction. In: Blurton-Jones N (ed.) *Ethological Studies of Child Behavior.* Cambridge: Cambridge University Press.

Bowlby J (1969) *Attachment and Loss Vol. 1. Attachment.* London: Hogarth.

Brownell CA, Ramani GB, and Zerwas S (2006) Becoming a social partner with peers: Cooperation and social understanding in one and two year olds. *Child Development* 77: 803–821.

Dunn J (2005) *Children's Friendships: The Beginning of Intimacy.* Oxford: Blackwell.

Hinde R and Stevenson-Hinde (1976) Towards better understanding relationships. In: Bateson P and Hinde R (eds.) *Growing Points in Ethology.* New York: Cambridge University Press.

Howe N, Rinaldi CM, Jennings M, and Petrakos H (2002) ''No! The lambs can stay out because they got cozies'': Constructive and destructive sibling conflict, pretend play, and social understanding. *Child Development* 73: 1460–1473.

Howe N, Rinaldi CM, and Le Febvre R (2005) ''This is a bad dog, you know ...'' Constructing shared meanings during sibling pretend play. *Child Development* 76: 783–794.

Howes C (1988) Peer interaction in young children. *Monograph of the Society for Research in Child Development* 53(1), Serial No. 217.

Howes C and Lee L (2006) Peer relations in young children. In: Balter L and Tamis-LeMonda C (eds.) *Child Psychology: A Handbook of Contemporary Issues,* 2nd edn. vol. 2, pp. 135–152. New York: Taylor and Francis.

Howes C and Ritchie S (2002) *A Matter of Trust: Connecting Teachers and Learners in the Early Childhood Classroom.* New York: Teachers College Press.

Maccoby E and Lewis C (2003) Less Day care or different day care. *Child Development* 74: 1069–1075.

NICHD ECCRN (2001) Child care and children's peer interaction at 24 and 36 months: The NICHD study of early child care. *Child Development* 72(5): 1478–1500.

NICHD ECCRN (2003) Does the amount of time spent in child care predict socioemotional adjustment during the transition to kindergarten? *Child Development* 74: 976–507.

Rubin KH, Bukowski W, and Parker JG (1998) Peer interactions, relationships, and groups. In: Eisenberg N (ed.) *Social, Emotional and Personality Development,* 5th edn. vol. 3, pp. 619–700. New York: Wiley.

Future Orientation

N Wentworth, Lake Forest College, Lake Forest, IL, USA

Glossary

Anticipatory eye movements – Eye movements that are initiated before stimulus onset and that bring the fixation point to the location of an upcoming picture.

Expectation – Hypothesized mental construct that is inferred from anticipations or faster responses during predictable sequences compared to during irregular sequences.

S1–S2 paradigm – A procedure in which paired stimuli are presented over a number of trials with a fixed interval between the first stimulus (S1) and the second (S2); used to examine physiological and behavioral anticipation.

Saccadic eye movements – Small, rapid eye movements that quickly change the point of fixation; saccadic eye movements are present at birth.

Smooth pursuit eye movements – Slow, continuous changes in eye position that occur when the observer watches a smoothly moving target; smooth pursuit eye movements appear during the first few months after birth.

Visual expectation paradigm – A procedure in which infants watch a sequence of brief pictures in alternation with blank, no-picture, intervals; the spatiotemporal structure that governs the picture sequence can be regular, and thus predictable, or irregular; used to examine anticipatory and facilitated eye movements indicative of visual expectations.

Introduction

The ability to predict what lies ahead, and adjust our behavior accordingly, is one of the defining characteristics of intelligent adaptation to a complex and dynamic environment. This article considers the emergence of this ability and its early development across several action systems including vision, visually directed reaching, posture, self-produced locomotion, and social interaction. Skills that may facilitate acquisition of future-oriented capabilities early in life are discussed as are common physiological indicators of expectation and anticipation. Progress in the developmental study of future-oriented behavior will require creative solutions to difficult methodological and conceptual challenges.

Future Orientation

We spend much of our mental lives in the future – setting goals, for example, and planning means to accomplish them, anticipating future obstacles and developing strategies to bypass them, and envisioning the scenarios we hope for as well as others we fear. The future similarly governs many of our actions – we anticipate the trajectory of a ball so that our hands can arrive in time to catch it, we stiffen postural muscles in anticipation of picking up a bag of groceries, and we articulate speech syllables differently depending on what we intend to say next. This ability to predict what lies ahead, and to adjust our behavior accordingly, distinguishes expert from novice performance and is one of the defining characteristics of intelligent adaptation to a complex and dynamic environment. But how does this capability to envision the future develop? What are the earliest forms of its expression? Which environmental or experiential factors shape its developmental course? How does the child's awareness of time emerge and differentiate into a sense of the past, present, and future?

Although the future clearly plays an important role in shaping our thoughts and actions, investigation of the early development of future-oriented processes has lagged substantially behind research on how past and present circumstances influence the behavior of infants and young children. Thus, an impressive body of literature establishes that infants, from very early in life, attend to many dimensions of the sensory information that surrounds them and that they have preferences for certain types of stimulation over others. These preferences tend to highlight certain aspects of the environment, such as edges of objects or the speech of humans, giving infants differential exposure to, and practice with, processing those attended features. An equally impressive body of literature indicates that past experience can also shape infants' preferences and activities. For example, studies have shown that newborns will modify the components of their sucking in order to hear sounds to which they were exposed when they were *in utero*. While the knowledge base concerning the influence of the future on children's behavior is much less extensive, the research that has been

done suggests that infants look toward the future much earlier in life than was once thought possible, although the skills for doing so are quite limited at first and have a notably long time course.

Historical Roots

Developmental psychology's relative neglect of future-oriented processes may be traced, at least in part, to the mechanistic tradition of behaviorism. The behaviorist tradition rests on two classic discoveries. After Ivan Pavlov's discovery that salivation could be elicited by a bell simply by repeatedly pairing the bell with food, proponents of radical behaviorism attempted to explain all behavior in terms of the organism's basic biological predispositions in conjunction with the history of stimulus–response pairings to which the organism had been exposed. At about the same time as Pavlov discovered the essentials of classical conditioning, Edward Thorndike discovered the law of effect. Cats that were placed in puzzle boxes learned to escape by a process of trial and error; those responses that led to freedom were stamped in by the reinforcement of successful escape, while other responses dropped out. After Thorndike's discoveries, proponents attempted to explain all behavior in terms of the organism's reinforcement history. The discovery of the principles of classical and operant conditioning promised the early behaviorists mechanisms that might explain the emergence of intelligent behavior solely in terms of the past. Such mechanistic accounts were more parsimonious than other accounts available at the time that either invoked a nearly limitless list of instincts to explain intelligent behavior or that suffered from the logical flaw of teleology in which the cause of a behavior rests in the future and must work its way backwards in time to produce its effect. In behaviorism, in contrast, the source of explanations for the present rests in the past; the future is then free to unfold in a single linear direction that flows logically from out of the past.

It should be noted, however, that even the most ardent followers of Pavlov and Thorndike found it necessary to incorporate terms to represent the influence of the future in their behavioral equations. For example, Leonid Krushinskii, one of Pavlov's intellectual descendents, described studies of the extrapolation reflex of birds, such as pigeons, ducks, hens, crows, and rabbits. Animals were given the opportunity to eat from a moving food dish that eventually entered a tunnel. Krushinskii speculated that the animal's ability to extrapolate the movement of the food dish into the future became associated with the unconditioned response of eating in the vicinity of the dish. The capacity to make these associations explained the intelligent and future-oriented behaviors that Krushinskii observed in the animals such as running to the end of the tunnel to intercept the food that would soon emerge. Similarly, the fractional anticipatory goal response in Clark Hull's classic learning equation was Hull's clever way of bringing a representation of the goal backwards in space and time, to the start of the maze, giving the animal an incentive to run toward the goal. These examples illustrate a general tendency within the behaviorist tradition; in order to explain how animals were able to benefit from their past training, theorists had to credit their subjects with at least some capacity to look ahead into the future.

Research on Infants' and Young Children's Future-Orientation

Future-Oriented Processes in the Saccadic Eye Movement System

A number of research paradigms have yielded results that suggest that infants are capable of at least rudimentary future-oriented processing from a remarkably early age. In the 1980s, Marshall Haith and colleagues described the Visual Expectation Paradigm (VExP), a new technique for investigating young infants' visual expectations. In the VExP, infants watch a sequence in which very brief pictures alternate with blank intervals during which no picture is displayed (**Figure 1**). The pictures occur at locations according to either a random schedule, in which case successive locations are unpredictable, or according to a simple rule, such as left–right (L-R) alternation, in which case successive locations are 100% predictable. If infants detect the underlying structure that governs the predictable picture sequence, and if they use this information to form expectations about successive events, they can make anticipatory eye movements to the locations of upcoming pictures during the blank intervals before the pictures actually appear. Infants can also demonstrate expectations if they react more quickly to unanticipated pictures during predictable sequences than during irregular sequences. Thus, in this paradigm, an expectation is seen as a mental construct whose presence is inferred on the basis of two behaviors: anticipations and facilitated reactions.

Studies using the VExP have shown that by 2 months of age, infants can rapidly form expectations about simple L–R alternation sequences and that this ability quickly expands to encompass more complicated L–L–R and L–L–L–R sequences by 3 months. In addition, a number of studies have revealed that infants use information about the spatial, temporal, and event content properties of the picture sequence to form expectations about upcoming events. For example, in 2–1 and 3–1 sequences (e.g., L–L–R and L–L–L–R, respectively) infants differentiate the spatial location at which pictures predominate by returning to it more quickly and anticipating pictures there more frequently than at the location where pictures appear less often. When the temporal duration of the interval between pictures is shortened, for example from 1 s to 700 ms, infants' anticipations happen earlier; when the interval is lengthened to 1400 ms, infants' anticipations occur later. When stable picture content appears at one location, and varying content appears at the other, infants anticipate the former more readily than the latter. Finally, by 3 months of age, infants can use a contingent relationship between location and the content of the central picture to anticipate where the upcoming peripheral picture will appear, even when the peripheral pictures do not follow a regular L–R spatial pattern.

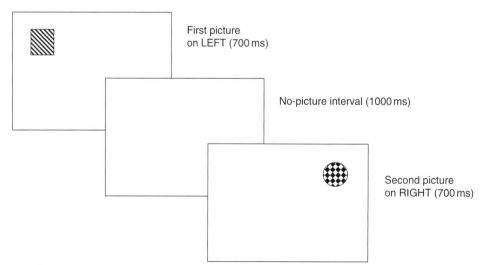

First picture on LEFT (700 ms)

No-picture interval (1000 ms)

Second picture on RIGHT (700 ms)

Figure 1 A schematic representation of events in the Visual Expectation Paradigm. A picture appears briefly (700 ms) on the left, followed by a blank no-picture interval (1000 ms), and then a new picture appears briefly on the right (700 ms). The sequence of 60–80 pictures is either regular (e.g., left–right alternation) or irregular. Eye movements that begin during the no-picture interval and go to the location of the upcoming picture are scored as anticipations.

Individual infants differ in the propensity to use spatial, temporal, and event content information to form and act upon expectations in the VExP, with the percentage of anticipations typically ranging from near zero to 30% at 3 months of age. Individual differences in the rate of anticipations appear to be fairly stable both within a single session of the VExP, with split-half correlation coefficients of approximately 0.5, and over the period of 1 week, with test–retest correlations being somewhat lower but still statistically significant. However, longitudinal analyses indicate that individual differences in anticipation rates may be somewhat less stable over longer periods, such as 1 month or more, until 6 months, when the median correlation between months averages nearly 0.70, excluding outliers. Individual differences in reaction times are similarly stable within a session, with correlations ranging from 0.45 to 0.69, and between sessions, with correlations between reaction time data collected at adjacent months typically more than 0.65, and also comparably high over periods longer than 1 month once infants reach 6 months of age. Moreover, individual differences in the VExP performance of infants at 3–4 months of age are related to other concurrent measures of cognitive processes, such as novelty preference and information processing speed; they predict later cognitive function measured at 3 years, and are correlated with parental intelligence quotient test scores.

Infants' performance in the VExP demonstrates an extraordinary ability to benefit from the regularity of the brief picture sequences. In most studies, the VExP sequence includes 60–80 picture presentations, and lasts approximately 2 min. Despite their short experience with this fairly rapid rate of information delivery, infants at 2 and 3 months quickly coordinate their eye movements with the structure of the task, anticipating some pictures and responding more rapidly than might otherwise be possible. Can infants retain information from such brief experiences and use it to anticipate more effectively at a later time? Two types of studies suggest that they can. First, infants typically perform better on their second VExP session than they do on their first, provided the two sessions occur within 1 week. Second, VExP performance in a L–R alternation sequence is generally better for infants who have seen a regular sequence on a prior occasion, even if the prior sequence followed a different regular pattern (e.g., up–down). Thus, the expectations that young infants form in the VExP appear to be retained for at least 1 week and generalize to other similar situations.

As impressive as infants' performance in the VExP is, it is vastly different from the behavior of adults in the same task who anticipate nearly every picture in a sequence of regular L–R alternation and who do so much earlier in the interpicture interval. The VExP is undoubtedly more difficult for infants than it is for adults. Infants have slower reactions, less working memory capacity, and cannot maintain a focused state of attention for nearly as long as adults can. In addition, infants may not have the same motivation to test and confirm their expectations as adults and older children do. Infants and adults no doubt bring different strategies to the task and it is possible that the infants' strategies are less task-appropriate or more difficult to inhibit. Studies have shown, for example, that, at 3 months of age, infants have a tendency to respond to picture offset in the VExP by looking further into the periphery, a tendency that must be reversed in a sequence of alternating pictures. Adults do not show this same tendency and, thus, do not have to overcome an incorrect initial response. Alternatively, infants' lower rates of anticipation may come from problems of detecting the underlying regularity in the sequence, of extrapolating from this regularity to expectations for the future, or of translating these expectations into acting before the pictures appear.

Although longitudinal analyses over the first year of life show that reaction times in the VExP speed up and become less variable, they do not show comparable changes in the rate of anticipation over the same period. If anything, the percentage of anticipation is approximately stable by 3 months, except for a decline between 9 and 12 months. What is the source of this unexpected decline in infants' tendency to anticipate? For one thing, the VExP is most likely a very different experience for a 3-month-old infant and a 12-month-old infant. Advances over the first 6 months of life in visual information processing, spatial memory, and the voluntary control of eye movements may make the VExP easier, and perhaps less captivating, for older infants. In addition, the appearance of reaching, grasping, crawling, sitting, and other developmental milestones during the first year means that many activities can compete for an older infant's attention compared to the number of tasks that can engage a 3-month-old. Thus, with a wider behavioral repertoire, and a potentially less engaging picture presentation sequence, it is possible that older infants are less motivated to anticipate upcoming pictures in the VExP compared to younger infants. Future studies will need to explore the cognitive, motivational, and behavioral factors that might limit infants' rates of anticipation at different ages in the VExP.

Future-Oriented Processes in the Smooth Pursuit Visual System

The saccadic eye movement system is functional from birth, and comes progressively under the infants' voluntary control during the first 6 months of life and beyond. The smooth pursuit system, in contrast, is not operational at birth but becomes functional some time around 2 months and improves thereafter. This onset of smooth pursuit tracking at approximately the same age as anticipatory saccades in the VExP may provide further evidence of future-oriented capabilities at this age.

The reasoning is as follows. Smooth visual tracking of a steadily moving target requires extrapolation of the spatial and temporal properties of the target's trajectory. Without such extrapolation, the target will have moved out of view by the time an eye movement has occurred, forcing a saccade to regain the target. Indeed, a number of investigators have shown that prior to 2 months of age, infants typically track a smoothly moving target in this fashion, with a sequence of saccades. Beginning at around 2 months of age, however, infants' tracking becomes smoother and, when a smoothly moving object stops suddenly, infants' eyes continue to move beyond the object, for a few hundred milliseconds, along the same trajectory they were tracking, suggesting that they were using predictive tracking prior to the object's sudden stopping.

Although VExP and smooth pursuit tracking studies provide consistent evidence of future-oriented processing in the visual system by 2 months of age, there is an important methodological difference between these two types of studies; in the VExP, infants' anticipatory eye movements are generated in the absence of visual stimulation whereas in smooth tracking of a moving object, the target is always available throughout its trajectory as the infant extrapolates to upcoming locations. Can young infants track an object's motion when the object temporarily disappears? Several studies in the 1970s and 1980s tested infants' abilities to track an object that moved smoothly across regions where the movement was hidden by a screen or tunnel. Although infants as young as 5 months reportedly tracked target movement across the occluding screens in these studies, it is unclear whether these infants actually forecasted the reappearance of the target on the opposite side of the occluding screen, or merely failed to stop tracking the target when it disappeared behind the barrier.

More recent studies, using improved measurement of eye position, suggest that although young infants may not extrapolate the motion of a target across an occluder on their first experience with it, they learn to do so quite quickly, at least when the target moves in a simple linear trajectory and is invisible for a relatively brief occlusion. By 6 months of age, infants can learn to associate specific target cues, such as shape or color, to predict the location at which a moving target will reappear from behind an occluder, even when the trajectory is a simple nonlinear one. Over the second half-year of life, infants learn to anticipate the reappearance of targets that move along yet more complicated trajectories, such as a circular path, although even at 12 months, this ability depends on target speed and the duration of the occlusion.

Future-Oriented Processes in the Visual–Manual System

For the adult, visually guided reaches are characterized by two phases: an approach phase, during which the hand is brought into the vicinity of the object; and a grasp phase, during which the fingers conform to the object's shape. The latter typically begins before the approach phase has ended. Like adults, infants as young as 5 months begin to close their hands in anticipation of contact with the target, at least on some of their reaches. This anticipatory adjustment becomes increasingly prevalent at 9 and 13 months of age, and happens progressively earlier in the approach phase. Moreover, infants preadapt their reaches and grasps in coordination with the target's direction, distance, shape, size, and orientation, with gross adjustments occurring early in development, especially when the infant's posture is stabilized, and finer adjustments appearing later, in the second half year of life.

Most remarkable, and indicative of future-oriented processing, are the findings presented by Claes von Hofsten and colleagues who report that infants from 4 to 5 months of age, at about the same age that they begin reaching for stationary objects, are able to catch objects that move in front of them in a radial path. Given the reaction time to initiate an arm movement, the inertial properties of the infant's arm, and the velocity of the object's motion, it would be impossible for infants to succeed in these catches if they aimed their reaches to the location where the target was when they launched the reach. Instead, to be successful, infants must have reached to a point ahead of the target when they initiated their reaches. This anticipatory aiming was confirmed on trials when the target motion was abruptly stopped while the reach was underway. What changed with development were the target speeds that infants could accommodate and the tendency to flexibly choose between ipsilateral and contralateral reaching strategies.

With a few notable exceptions, such as those described above, most research with infants and young children has focused on reaches for easy-to-grasp, stationary objects in an uncluttered area with the child receiving a fair amount of postural support. While it is clear from these studies that young infants use visual information to anticipate some required manual adjustments, naturalistic observations reveal many instances where infants' prehensile skills are far from smooth or effective. Consider a young child learning how to use a fork or trying to pick up a bar of soap. These tasks, and most daily activities, require the simultaneous coordination of many degrees of freedom to achieve true manual dexterity. In lifting an object, for example, adults jointly apply gripping and lifting forces whereas infants and young children apply these forces inconsistently, sometimes even applying the opposite of a needed force. How do infants and young children gain mastery over coordinating the multiple degrees of freedom required in their daily lives, and what role does predictive processing play in this development? Answers to these questions will require additional research that uses more dynamic and varied reaching tasks.

Future-Oriented Processes in the Postural and Locomotion Systems

In addition to using visual information to make anticipatory adjustments during reaching, infants also make anticipatory postural adjustments as well. In fact, some would argue that development of infants' skillful reaching must be preceded by development of postural control. Consider an infant sitting balanced on a parent's knee, receiving support only at the hips, while trying to grasp an object that is just beyond reach. Seated in this position, with limited support, the forces generated by arm and hand extensions would tend to perturb the infant's posture were they not counterbalanced by compensatory activity in the legs and trunk. Recordings of muscle activity in the trunk extensors of seated infants and the leg extensors of standing infants provide evidence of such early anticipatory postural control in infants who stabilize their posture before attempting to lean out for a reach or to pull open a drawer. Although such rudimentary anticipatory postural control has been found in infants who are just beginning to sit or stand, the timing of this control and the specificity of it improves with age and depends upon motor experience.

Anticipatory processes are also involved in the maintenance of equilibrium during self-produced locomotion. As infants learn to crawl, and then walk, they also learn to gather information through visual and haptic exploration, and they use this information to guide their movements around barriers, across gaps, down slopes, up stairs, through apertures, and across a variety of solid and deformable surfaces that support some types of locomotion better than others. For example, studies by Karen Adolph and associates show that 12-month-old infants who have just begun to walk, like younger infants who have accumulated a few weeks of crawling experience, will use the visual information of a drop-off to avoid the apparent deep side of a visual cliff. Once they have had a few weeks of walking experience, toddlers appear to recognize, in advance, the danger posed by challenges to their newly acquired walking skills, such as steep slopes or narrow bridges, and they often refuse to proceed unless there is some assistance, such as an adult's hand or a handrail. Before they continue, infants first test out the material properties of the surface they face and the type of assistance that is available.

Although infants and young children are quite skilled at picking up much of the visual and haptic information that is available to guide their movement, there are definite limits to what can be learned in advance of action. Adults have learned to interpret certain arbitrary visual stimuli, such as a mop, bucket, and caution sign, as cues to the types of surfaces that lie ahead, and they can use this information in an adaptive way. When they encounter a novel stimulus that can signal what lies ahead, adults very quickly learn to use the new cue to make anticipatory adjustments in their manner of walking, such as slowing down and taking smaller steps; as a consequence, adults are less likely to fall. Toddlers, however, require far more experience with the same novel cues before they learn to use them in an anticipatory fashion and, consequently, are much more likely to fall.

Future-Oriented Processes in Early Social Systems

The physical properties of objects, surfaces, and events provide the structure that infants can tap into to form expectations about the consequences of their own actions and to make predictions about what will happen next in the world that is beyond their control. Gravity consistently pulls objects in the same direction; one object cannot occupy a location where something else already exits; as an object approaches, its rate of expansion in our visual system indicates when it will be within reach, and so forth. Experience with repetitive sequences, as in the VExP, can provide another source of regularity that infants can use to anticipate upcoming events. It has been hypothesized that so too can the numerous repetitions of infant distress/caregiver response/infant relief cycles that occur throughout the first few months of life.

As new parents settle in to the routines of feeding, changing, burping, changing diapers, and rocking their infants, babies can begin to form expectations – both specific, as in what will happen next in the care sequence, and generalized, as in how likely is it that the social world will provide relief for the infant's distress. Perhaps these expectations explain why infants show less crying over time if parents are highly responsive to their needs in early infancy. Indeed, several studies have shown that during the first few months of life, infants will stop crying sooner when the caregiver intervenes, that infants will often stop crying even before they are picked up, and that this anticipatory quieting is specific to familiar caregiver. Of course, these results are far from definitive since it is not clear whether infants actually expect to be soothed, or whether some other explanation might account for their less persistent crying. For example, during the period from 2 to 6 months of age, infants may cry less because they develop more effective self-soothing strategies or they may stop crying when they see a familiar caregiver not because of an expectation for impending relief but because they have simply formed strong positive associations to the sight of their caregivers.

Although research has not definitively shown at what age infants begin to expect that their caregivers will be able to provide effective soothing, studies have shown that infants' early social interactions have an internal structure that can support expectation formation. For example, in early face-to-face interactions with their infants, parents often repeat sounds or gestures to which their infants have given a positive response. This combination of parental repetition of responses that are contingent on the infant's behavior is ideal for providing the basis for infants to form expectations about what their parents will do next as well as expectations about the consequences of their own behavior. Similarly, it is possible that the impressive degree of temporal coordination and turn-taking that develops between parents and infants in their early bouts of mutual gazing help infants form expectations about the general framework of social interaction; indeed, infants will protest a nonresponsive parent's face in the still-face lab procedure, an experimental paradigm in which parents adopt a neutral facial expression and stop responding contingently to the infant's behavior. Later, infants quickly learn the rules of common interactive games such as peek-a-boo and pat-a-cake and will play their role in alternation with their social partners. Thus, the daily routines that parents establish in caring for and playing with their infants and young children, and the family traditions that develop, for example, around celebrations such as birthdays and holidays, can provide the regularity that infants can use to predict what will happen next in their interactions with significant others.

Structured interviews with parents about what their infants and toddlers understand about the future suggest that parents find ample evidence in their daily interactions with their children for their infants' and toddlers' developing sense of what lies ahead. According to an interview study reported by Janette Benson and associates, parents of children from 12 to 42 months of age interpret their children's persistent attempts to accomplish a goal as evidence of a burgeoning orientation to the future. Also, parents rate their children as capable of participating in a familiar routine, and parents see this as further evidence of their young child's growing sense of the future. However, 12-, 18-, and 24-month-old children were rated as unlikely to be able to form an expectation about something unfamiliar based on being told that it will occur, while children older than 30 months were rated as showing some capacity to form expectations in this way. Parents rated other aspects of future-oriented understanding, such as wanting things done in the same order, doing things to prepare for the future, and pretending to be upset in order to get what is wanted, as increasingly characteristic of their children from 12 to 42 months of age. Parents believed that infants and young children learn about the future by being told in advance about what lies ahead. Thus, it is not surprising that parents reported talking to their children about what is going to happen in the future, even though they believed their children really did not understand what the future meant. Analyses of parents' speech to their children during brief lab tasks confirmed that parents talked about the future to their children more than they talked

to them about the past, and that this tendency to talk about the future generally increased from 14 to 36 months of age. However, parents talked twice as often about the present than either the past or the future for children in this age range.

Observations of parent–child conversations and the results of the parental questionnaire studies suggest a possible sequence through which infants pass on their way toward building a firm sense of the future. At first, parents credit their infants with the capacity of extending or repeating familiar patterns and in persisting in their efforts to achieve a goal. Later, infants learn to make preparations for those familiar patterns that they can anticipate. Finally, they can use information to make vicarious expectations about upcoming events that are not simple extrapolations of a familiar pattern and they also learn to manipulate someone else's expectations.

Early Tools for Learning about the Future

Learning Temporal Patterns and Ordered Sequences

Research suggests that infants and young children possess several perceptual and cognitive resources that can be applied to the task of detecting the structure that underlies predictable future events. Young infants are remarkably attuned to variations in visual and auditory information over time. For example, newborns detect the difference between two lights that differ only in the rate at which they flash and babies have preferences for some temporal rates over others; specific preferences depend on the infant's age and current level of arousal. Spatially patterned visual stimuli, such as checkerboards or stripes, can also be discriminated based on the rate of flashing and phase reversals (i.e., black switches to white and white to black). Similarly, infants can discriminate between two auditory streams that use the same syllable (e.g., /ba/ba/ba/) but present the syllable at different rates. Differences in the global temporal patterns of visual and auditory stimuli, such as the number of cycles per second, are detected as are differences in the local temporal properties, such as rhythm, and discrepancies in the synchrony of the auditory and visual information of a multimodal event. In addition to using basic mechanisms to track the rate, duration, simultaneity, and interval between the first and second auditory or visual event, infants, at least by 4 months, also detect and remember the order of specific elements within brief repetitive sequences. By 11–12 months of age, infants can also remember and reproduce short sequences of ordered steps that produce an interesting effect. In doing so, infants show that they know what comes next, at least within the short-term timeframe of the behavioral sequence they are imitating.

Learning Contingencies in Reinforcement Paradigms

In addition to having the requisite skills to begin parsing the temporal structure of events, young infants also possess abilities that let them learn important contingencies between their behavior and reinforcing events. For example, newborns have been classically conditioned by pairing the delivery of a sucrose solution with tactile stimulation of the infant's forehead. After a number of such pairings, infants make reliably more sucking and head-orienting responses following strokes to their foreheads than infants who have not received the forehead stroke plus sucrose pairings. By 2 months of age, infants can learn to increase their spontaneous kicking when doing so activates a mobile. By 3 months of age, infants appear to generalize beyond the specific stimulus–response contingencies on which they are trained. Consider two groups of infants; one group received the same mobile on two successive days of training while the other received two different mobiles, a new one on each of the two days. On the third day, infants either received a new mobile or one they had seen before. Infants who had received the same mobile on days 1 and 2 were more likely to kick if they received that same mobile again on day 3 than if they received a new mobile. In contrast, infants who were trained on two different mobiles were more likely to kick if they received a new mobile on day 3 than if they received one of the mobiles on which they had already been trained. Thus, infants in the first group had apparently learned expectancies for the specific mobile on which they had been consistently trained whereas infants in the second group had apparently learned expectancies for mobiles to vary from one day to the next.

Learning about the Future through Conversation

Language is a third tool that infants and young children have for forming expectations about upcoming events and sharing them with others. As noted earlier, parents admit that they speak to their young children about the future before they believe their children understand what is being said. Naturalistic studies of parent–child conversations confirm that parents review the day's events with their children and preview upcoming events as well. Through participation in conversations such as these, children presumably learn skills for envisioning themselves in time and for developing a sense of the past, present, and future.

The topics of children's conversion, and their mastery of the grammatical structures for denoting the spatial and temporal locations of events, have been examined to gain insight into the child's capacity to reflect on the past and to anticipate the future. Two cautions are necessary. First, valid inferences about young children's mental processes from analysis of their speech will most likely need to come from cross-linguistic studies. This is because languages differ in the way spatial and temporal relations are coded, and some coding systems may be easier for young children to master than others. In the same vein, different ways of coding particular spatial and temporal relations may exist within a single language and these codes may differ in the demands they place on the child's cognitive processes. Consider verb tense, for example. In English, past action is generally coded by adding '-ed' to the present tense while the future is coded by inserting an auxiliary verb (e.g., 'will') before the present tense. The fact that children who are acquiring English typically master the past tense before they master the future may indicate that they understand the past before they understand the future, or, alternatively, that they merely have more trouble mastering the copula grammatical form. Second, making inferences about young children's mental processes from their speech may not accurately reflect their conceptual competence. Research indicates that language production skills typically lag behind language comprehension and this difference in developmental rate may lead to underestimating what children know when we rely on analysis of what they talk about. However, it is also possible for young children to pick up on linguistic cues as they answer researchers' questions which may lead to overestimating children's knowledge. Thus, researchers need to be especially concerned with the reliability and validity of measures of young children's language behavior.

With these caveats in mind, cross-linguistic studies suggest that young children first speak about objects and events in their immediate experience. The child's earliest future-related comments involve short-term goals or desires, for example when a child requests "more" or "again" to express continuation of a pleasant experience. Children use the present tense to refer to things they remember as well as things they intend to do. For example, a child may say "I get it!" while running to another room to retrieve a toy. Eventually, children learn to use adverbs such as "yesterday" and "tomorrow" to denote an indefinite length of time before or after the present. Later, children refer to time in the future in terms of expected events or outcomes such as "When I go to school" or "When I grow up."

By 3 years of age, children can describe what typically happens during familiar activities such as going to the zoo or attending a birthday party. Thus, when asked to describe what will happen at an upcoming birthday party, young children might answer based on recollecting the scenes that are stored in a mental birthday party script or, alternatively, by imagining themselves in a future birthday party as it unfolds. To what extent can preschool-aged children actually envision themselves in a future setting? Research by Cristina Atance and colleagues suggests that 3-year-olds can envision themselves in a familiar scene, such as in the dark or in an igloo, imagine how being in the scene might affect them physiologically, such as make them scared or cold, and anticipate the types of items that might be useful in their imagined activity, such as a flashlight or a winter coat. However, this ability to envision oneself in a different situation and reason about what that might entail is fragile with young children being particularly vulnerable to choosing items that are semantically associated with the imagined scene rather than practically related to their most likely future needs.

Physiology of Future-Oriented Processes

Researchers who have sought physiological indicators of future-oriented processes in infants and young children have most often examined changes in heart rate and the electrical potentials of the brain. Heart rate is tightly coupled to the demands for information processing in infants and adults. When a novel stimulus is introduced, a coordinated orienting response ensues that combines responses that direct the sensory receptors toward the stimulus as well as integrates changes in respiration and heart rate, thought to maximize processing of the new stimulus. This heightened attention wanes with repeated exposures of the stimulus unless the stimulus signals that something of significance is about to occur. The S1–S2 paradigm has been used extensively to study adults' physiological responses in those cases when one stimulus signals that something important is about to occur. In this paradigm, paired stimuli are presented over a number of trials, with a fixed interstimulus interval (ISI) separating the signal stimulus, S1, and the second stimulus, S2, which is usually made distinctive or imperative in some way. For example, adults may be told to count the number of S2 stimuli or to press a response button when S2 is delivered. The adult's heart rate response varies with the specific details of the S1–S2 paradigm, such as the length of the ISI, and the nature of the response required to S2, but invariably includes a phase of deceleration that reaches its lowest point just before onset of S2. This deceleration phase is thought to reflect the buildup of anticipation for S2.

Studies of heart-rate changes in neonates who were undergoing classical conditioning have produced mixed results. Some have found the anticipatory heart-rate changes seen in adults, at least in some newborns, while others have found heart-rate deceleration only during early extinction trials, when the unconditioned stimulus (i.e., S2) should have occurred but did not. Studies with older infants have produced more reliable evidence of anticipatory heart-rate decelerations, during the intervals between S1 and S2, as well as heart-rate decelerations following the omitted instances of S2 on interspersed extinction trials. The literature in this area is consistent with the hypothesis that young infants may be able to notice that a pattern has been altered before they are able to anticipate what the pattern is. Thus, retrospective pattern detection may precede prospective pattern extrapolation.

One disadvantage of using heart-rate deceleration as an index of anticipation is that the time course of the heart-rate response is somewhat sluggish. A typical ISI in S1–S2 paradigms with infants is on the order of 5–10 s; when heart-rate deceleration has been found, it typically emerges 2s or more seconds after onset of S1 and takes 6s or more seconds to reach its maximum level. It is possible that young infants might be able to anticipate events over shorter intervals before they can extrapolate over intervals as long as 5–10 s. If so, the heart-rate response may not reflect the type of anticipation that young infants can most easily generate.

Recordings of the electrical potentials of the brain have also been collected during S1–S2 studies with adults. These studies have revealed a slow brain-wave response, the contingent negative variation (CNV) that reaches its most negative point just before onset of S2. A study of electrical cortical potentials preceding stimuli in the VExP suggests that a rudimentary form of the CNV may occur in 3-month-olds during the latter part of the 1000 ms intervals between picture presentations.

Electrical potentials of the brain that precede anticipatory saccadic eye movements have also been contrasted to the potentials that precede reactive eye movements in the VExP and related paradigms. In these studies, potentials during the 500 ms preceding 3–3.5-month-old infants' anticipatory saccades were more negative over frontal areas of the brain than were the comparable brain potentials before reactive saccades. In addition, a less robust positive potential was found approximately 100 ms before the onset of anticipatory saccades in some infants. These findings suggest that the programming of saccades in anticipation of upcoming stimuli involves different brain areas than the programming of reactive saccades.

The regular alternation of pictures between the two locations of the VExP, repeated over 60–80 occurrences, gives young infants many opportunities to observe the side-to-side transitions that govern this sequence. The highly repetitive nature of this sequence means that infants can form both local expectations, such as on this trial the next picture will appear on the left, as well as global expectations, such as left pictures will always follow pictures on the right. Other studies have used a spatial cuing paradigm to examine the effects of directing an infant's attention to one location by presentation of a brief cue at that location, and then presenting a subsequent stimulus either at the precued location or at the opposite location. In this paradigm, infants may expect a stimulus to occur at the cued location, at least locally, on some trials.

Cortical potentials that precede saccades to locations that have been cued have been compared to those that precede saccades to the other location in infants during the first 6 months of life. In this paradigm a robust positive cortical potential has been found in the 50 ms before the onset of saccades to the cued location, especially over the frontal areas of the brain; the amplitude of this positive potential before saccades to the cued location depends on age, with older infants more likely to show the effect. Again, these studies suggest that when attention is shifted to a peripheral location, and a target then occurs at that location, infants' programming of saccades to that location will have a different pattern of cortical activation than would occur to uncued, or unexpected, target locations.

Several studies have examined infants' cortical potentials during an oddball paradigm, a procedure in which participants watch two stimuli that appear in a semirandom sequence at a single location. One stimulus predominates, typically appearing on 80% of the picture presentations; the other stimulus, the oddball, typically appears on the remaining 20% of the presentations. In variants of this procedure, the frequent and rare pictures may occur on 80% and 10% of the trials, respectively, and a set of novel pictures may appear on the remaining trials, with a new picture chosen for each trial. Electrical potentials are typically averaged for each of the trial types from the interval just before picture onset through picture offset. In adults and children from 5 years of age through adolescence, two brain-wave responses are typically found: a positive potential with a latency of 300 ms after stimulus onset (P300) and a negative potential with a latency of 200 ms (N200). Both components typically vary as a function of the probability of the rare stimulus. The findings with infants from 2 months of age are different in form from those with children and adults. Although there are differences in the names that have been applied to the types of brain waves that have been observed in this procedure with infants, there is general agreement in finding a positive potential beginning at about 200 ms after stimulus onset, followed by a large, slow negative wave developing around 500–600 ms after stimulus onset, and a later positive slow wave peaking at approximately 1000 ms after stimulus onset. Most importantly, as with adults and children, the form of the averaged brain-wave response differs in infants for the frequent vs. rare pictures. The change in the form of the averaged brain potential is thought to reflect an expectancy that has been violated by introduction of the relatively rare picture into the sequence of repeated presentations of the frequent picture. Additional comparisons have been made of the averaged brain potentials for the frequent pictures that occur immediately before and after an oddball, respectively. Since the picture is the same in these two averages, differences in the form of the brain wave must reflect the effect of the intervening oddball stimulus.

Other Methods and Methodological Challenges

The oddball paradigm, described earlier, exemplifies a variety of techniques that have been used to gain insight into infants' and young children's expectations. These studies generally use two classes of events, rare and typical. Rare and typical events are either defined by differential familiarization during the course of the study, as in the oddball paradigm, or by the infants' experience prior to coming to the laboratory, as in the still-face procedure when the mother suddenly adopts a neutral facial expression and stops responding contingently to her infant. In either case, the experimental method involves a comparison of infants' responses following the rare event to their responses following the typical event. Presumably, when an infant gives a different response after the rare event, it indicates that a violation in the infant's expectations has occurred. In an effort to detect violations of infants' expectations, a wide variety of responses has been monitored including surprise reactions, crying or distress, suppressed motor activity, duration of looking, amplitudes and latencies of brain-wave components, heart-rate changes, and so forth. Unfortunately, these studies do not provide definitive information about infants' expectations. The rationale is as follows. It is possible that after seeing one, two, three, or more occurrences of a particular event, infants actively expect yet another repetition of this typical event to happen again and, when it does not, their expectations are violated leading to a surprise reaction. However, it is equally possible that the infant's surprise represents an after-the-fact reaction to something unusual that has just happened and it is also possible that the thing that has just happened is more difficult to process because it isn't typical. That is, it is possible that the infant was not actually expecting anything, in the sense of actively forecasting the future, at all. Thus, studies that rely on comparing infants' responses to typical and atypical events give ambiguous results: do infants' reactions reflect ad hoc responses to the rare event that just occurred, or do they reflect expectations that the typical event should have recurred but did not?

Several other research paradigms have produced results that seem relevant to questions about the early development of future-orientation. For example, studies of infants' and young children's object representation, inferential reasoning, visual and manual search strategies, and means-ends skills suggest that by 12 months, if not much sooner, infants have the knowledge sufficient to understand that solid objects will stop when they encounter a barrier, that infants can use conditional probabilities to infer causal structure, and that they can remember where to retrieve objects that have been hidden under covers in various locations. If infants do indeed possess this knowledge, it seems reasonable to expect that they could put it to use in a task that requires them to predict where a ball rolling down a slope will come to rest when it hits a solid wall that blocks its motion and, furthermore, that they should be able to use their predictions and memory to open a door at the ball's current location to retrieve the ball. However, it is not until 30 months or more that infants can perform reasonably well at this task, despite the suspicion that all of the requisite component skills are in place much earlier in development. Thus, acquisition of the knowledge or component skills needed to engage in future-oriented behavior does not necessarily imply that infants or young children will actually be able to use their skills to anticipate what is required or to act in preparation for what will occur. In contrast, infants and toddlers may be able to recognize after the fact that something odd has happened without having been able to predict what would happen in advance.

Finally, studies with adults have shown that location information may serve as an anchor for binding together the multiple properties of a single object. That is, the warmth of a campfire, its smell, the sight of the flame, and the crackle of the burning logs can all be bound together because they originate at the same location. Adults appear to use location information to help them recall object properties that were associated with that location. For example, in trying to recall a message that was once posted at a particular location, adults will look at that location, even when the message is no longer there. They look to this location, not in anticipation of seeing the message there, but as a way of retrieving the information that has been associated with the location. Studies suggest that infants may also use location information to bind together the multiple sensory inputs from multimodal events. For example, if young infants see and hear an event at a particular location, such as a talking face, when the auditory information is presented alone, infants look to the location where the face had previously appeared. But it is unclear whether infants expect to see the face at this location or, like adults, remember that the sound came from this location. Similarly, research paradigms that investigate infants' and toddlers' behavior during the interval between two events face the methodological challenge of showing that the child's behavior during the interval is truly future-oriented, predictive behavior of upcoming events rather than responses to the previous events or to an interruption in the flow of events. In the VExP paradigm described earlier, for example, infants typically made one response to picture offsets (i.e., a repetitive eye movement further into the periphery) but a different response to the upcoming picture (i.e., an eye movement in the opposite direction). Without this response difference, it would be difficult to determine when an infant was expecting upcoming pictures vs. remembering previous ones.

A Taxonomy for Classification of Future-Oriented Processes

Investigators who study learning and memory have differentiated many dimensions for thinking about past experience and for organizing the results of the numerous studies of how the past affects current behavior. There are temporal dimensions such as short-term, long-term, and the briefest of sensory registers; content dimensions such as episodic and semantic memories; and functional dimensions such as working memory or repressed memory. Similarly, investigators who study the impact of current stimuli on behavior have differentiated dimensions that are useful for classifying objects and events such as size, shape, color, number, complexity, and symmetry.

What are the relevant dimensions for thinking about future-oriented behavior? At present, no formal conceptual framework has been articulated, although a number of dimensions have been proposed (see **Table 1**). For example, one proposal has borrowed the episodic vs. semantic distinction from the study of memory and, in doing so, has speculated that infants and young children may have separate systems for thinking in general (semantic) terms about what will happen, such as Tuesday will come after Monday, and for thinking in specifically personal (episodic) terms about the future, such as "I will go to the park after lunch." Thus, according to this proposal, future-oriented thoughts can be classified on the basis of their content – events in the individual's own future vs. impersonal events in the world. Although the episodic-semantic distinction has stimulated a number of intriguing studies, it also raises some difficult questions, for example, where should we place a child's expectation that her mother will be sad when she discovers the broken vase? Do expectations about important social partners' future states belong in the personal-episodic category, the impersonal-semantic category, or somewhere in between? Thus, it is important to determine whether the episodic-semantic distinction defines a true dichotomy, a set of categories, or a self-other continuum.

As **Table 1** suggests, several additional dimensions have been proposed including the timeframe of expectations, ranging from very short intervals, measured in fractions of a second, such as those involved in making postural adjustments to accommodate intended actions, up to very long-term expectations that may extend beyond the person's own lifetime, for example in planning a legacy for one's grandchildren. In addition to unfolding according to different timescales, from milliseconds to decades, future-oriented behaviors differ in how time dependent they are. For example, in catching a falling object, certain hand and postural adjustments must happen at a particular time or in a particular sequence for the behavior to unfold smoothly whereas in other activities,

Table 1 Selected relevant dimensions for classifying future-oriented behavior

Dimension	Range
Subjectivity	Personal (episodic) future vs. general (semantic) future
Timeframe	Milliseconds to a lifetime and beyond
Space/time dependency	Anticipation that must occur at a particular location or by a particular time vs. anticipation that does not need to be coordinated with space or time
Behavioral system	Motor, perceptual, cognitive, social
Level of abstraction	Extrapolation of a repetitive pattern, prediction from a functional relationship, creative invention of a new solution to a problem
Purpose	Fantasy/imagination, preparation, planning, problem solving
Openness to modification	Ballistic prepared response vs. ongoing modification through feedback
Degree of conscious control	Automatic or habitual vs. effortful
Complexity	Anticipating a single event vs. anticipating multiple sequential or simultaneous events

such as getting ready for bed, the order and timing of events is less critical. Presumably, different timescales would place varying demands on a child's excitatory, inhibitory, coordination, and memorial resources but this remains to be established.

Future-oriented behavior occurs in different behavioral systems as well, including motor anticipation, for example, in reaching, grasping, and locomotion; perceptual and cognitive anticipation, for example, in cross-modal perceptual effects where the sight of an object can create an expectation for the object's texture, or where the context of a story primes expectations for theme-related words; or social anticipation, for example, when a child contemplates how a friend might respond to a particular birthday gift.

Future-oriented behaviors can also be differentiated by the degree of abstraction required ranging from simple extrapolation of a repetitive pattern into the future, as when we expect summer to follow spring; to prediction of a future event from knowledge of a functional relationship, as in expecting rain when dark clouds gather; to envisioning a new solution to a perplexing problem, as in creative invention.

Other dimensions include the purpose of the future-oriented behavior, for example for play or entertainment, as in daydreaming; for preparing for an expected event, as in putting things into the backpack to take to daycare; or for planning a strategy to achieve a goal, as in figuring out how to equitably divide the Halloween candy. Is the plan open to modification from feedback, as when a child shifts

from an "I want" request to an "I need" request in the face of parental resistance, or must the plan be executed without modification, as when a child plans to jump across a puddle? How much conscious control is required? For example, is the child getting ready for an event that has an established routine, such as getting ready to go to preschool, or is the child anticipating a novel event, for example traveling by a new mode of transportation? Anticipated events can also vary in complexity. For example, one child may anticipate only the next event in a sequence while another anticipates the next three or four steps. Similarly, some anticipated events are fairly simple, such as the light will go on when the switch is flipped, while others are quite complex, such as the toy will light up, make noise, and begin moving when the switch is flipped.

Although the dimensions listed in **Table 1** may help us think about the development of future-oriented behavior, they do not constitute a formal, systematic conceptual framework. There are many unknowns. Is the list exhaustive? Are the dimensions relevant to all classes of anticipatory behavior? How are the dimensions related to each other? What are the assumptions underlying each dimension? How can the dimensions best be operationalized into empirical behaviors? Further research is needed to answer these questions; doing so should help us define the trajectories along which future-oriented behaviors develop.

In conclusion, the study of future-orientation in infants and young children suggests that by 2 months of age, infants can use regularity of spatial, temporal, and content information to form expectations for upcoming events and then use these expectations to adapt their eye movements to the structure of the sequence of events they are watching. This ability is followed by similar achievements in other action systems as they come under the infant's voluntary control, from smooth tracking of visual motion to eye-hand coordination, postural control and self-produced locomotion, and social communication. Infants have several capabilities that facilitate their acquisition of a future-orientation including the ability to detect the structure of temporal patterns and ordered sequences; to learn about the contingent relationships that exist between actions and consequences or between contiguous or adjacent events; and to communicate using systems of symbols. Although evidence of future-orientation appears quite early in infancy, there is a protracted period of development during which infants and children learn to coordinate progressively more degrees of freedom and to go beyond the simple extrapolation of familiar patterns.

See also: Cognitive Development; Fear and Wariness; Habituation and Novelty; Imagination and Fantasy; Learning; Motor and Physical Development: Manual; Perception and Action; Routines.

Suggested Readings

Atance CM and Metzoff AN (2005) My future self: Young children's ability to anticipate and explain future states. *Cognitive Development* 20: 341–361.

Benson JB, Talmi A, and Haith MM (2003) The social and cultural context of the development of future orientation. In: Raeff C and Benson JB (eds.) *Social and Cognitive Development in the Context of Individual, Social and Cultural Processes*, pp. 168–190. New York: Routledge.

Haith MM, Benson JB, Roberts RJ, Jr., and Pennington BF (eds.) (1994) *The Development of Future-Oriented Processes.* Chicago: The University of Chicago Press.

Haith MM, Wentworth N, and Canfield RL (1993) The formation of expectations in early infancy. In: Rovee-Collier C and Lipsitt LP (eds.) *Advances in Infancy Research*, pp. 251–297. Norwood: Ablex.

Joh AS and Adolph KE (2006) Learning from falling. *Child Development* 77: 89–102.

Moore C and and Lemmon K (eds.) (2001) *The Self in Time: Developmental Perspectives.* Mahwah: Lawrence Erlbaum Associates Publishers.

von Hofsten C (2005) The development of prospective control in tracking a moving object. In: Riesern JJ, Lockman JJ, and Nelson CA (eds.) *Action as an Organizer of Learning and Development: Volume 33 in the Minnesota Symposia on Child Psychology*, pp. 51–89. Mahwah: Lawrence Erlbaum Associates Publishers.